T0180295

Lecture Notes in Computer Science 9921

More information about this series at http://www.springer.com/series/7407

Julia Handl · Emma Hart
Peter R. Lewis · Manuel López-Ibáñez
Gabriela Ochoa · Ben Paechter (Eds.)

Parallel Problem Solving from Nature – PPSN XIV

14th International Conference
Edinburgh, UK, September 17–21, 2016
Proceedings

Editors
Julia Handl
University of Manchester
Manchester
UK

Emma Hart
Edinburgh Napier University
Edinburgh
UK

Peter R. Lewis
Aston University
Birmingham
UK

Manuel López-Ibáñez
University of Manchester
Manchester
UK

Gabriela Ochoa
University of Stirling
Stirling
UK

Ben Paechter
Edinburgh Napier University
Edinburgh
UK

ISSN 0302-9743 ISSN 1611-3349 (electronic)
Lecture Notes in Computer Science
ISBN 978-3-319-45822-9 ISBN 978-3-319-45823-6 (eBook)
DOI 10.1007/978-3-319-45823-6

Library of Congress Control Number: 2016950392

LNCS Sublibrary: SL1 – Theoretical Computer Science and General Issues

Printed on acid-free paper

This Springer imprint is published by Springer Nature
The registered company is Springer International Publishing AG
The registered company address is: Gewerbestrasse 11, 6330 Cham, Switzerland

Preface

This LNCS volume contains the proceedings of the 14th International Conference on Parallel Problem Solving from Nature (PPSN XIV). This biennial event constitutes one of the most important and highly regarded international conferences in nature-inspired computation, ranging from evolutionary computation and robotics to artificial life and metaheuristics. Continuing with a tradition that started in Dortmund in 1990, PPSN XIV was held during September 17–21, 2016, in Edinburgh, Scotland, UK.

PPSN XIV received 224 submissions from 50 countries – an increase in both figures from the previous conference, demonstrating the continued and widening interest in the field. After an extensive peer-review process, where most papers were evaluated by at least four reviewers, the Program Committee Chairs examined all of the reports and ranked the papers. Where there was disagreement amongst the reviewers, the Chairs evaluated the papers themselves in order to ensure fair and accurate decisions. The top 93 manuscripts were finally selected for inclusion in this LNCS volume and for presentation at the conference. This represents an acceptance rate of 41.5 %, which guarantees that PPSN will continue to be one of the most respected conferences for researchers working in nature-inspired computation around the world.

PPSN XIV was enhanced by the inclusion of three distinguished keynote speakers representing facets of the field's future and the interfaces with other disciplines: Susan Stepney (University of York, UK), Josh Bongard (University of Vermont, USA), and Katie Bentley (Harvard Medical School, USA).

The meeting began with four workshops bringing together work in specialized areas: "Intelligent Transportation Workshop" (Neil Urquhart), "Landscape-Aware Heuristic Search" (Nadarajen Veerapen and Gabriela Ochoa), "Natural Computing in Scheduling and Timetabling (Ahmed Kheiri, Rhyd Lewis, and Ender Özcan), and "Advances in Multi-modal Optimization" (Mike Preuss, Michael G. Epitropakis, and Xiaodong Li). These workshops allowed researchers with similar interests to discuss and explore ideas in an informal and friendly setting.

PPSN XIV also included 16 free tutorials to give us all the opportunity to learn about new aspects of our field: "Gray Box Optimization in Theory" (Darrell Whitley), "Theory of evolutionary computation" (Benjamin Doerr), "Graph-Based and Cartesian Genetic Programming" (Julian Miller and Patricia Ryser-Welch), "Theory of Parallel Evolutionary Algorithms" (Dirk Sudholt), "Promoting Diversity in Evolutionary Optimization: Why and How" (Giovanni Squillero and Alberto Tonda), "Evolutionary Multiobjective Optimization" (Dimo Brockhoff), "Intelligent Systems for Smart Cities" (Enrique Alba), "Advances on Multi-modal optimization" (Mike Preuss and Michael G. Epitropakis), "Evolutionary Computation in Cryptography" (Stjepan Picek), "Evolutionary Robotics – A Practical Guide to Experimenting with Real Hardware" (Jacqueline Heinerman, Agoston E. Eiben, Evert Haasdijk, and Julien Hubert), "Evolutionary Algorithms and Hyper-heuristics"(Nelishia Pillay), "A Bridge between Optimization over Manifolds and Evolutionary Computation" (Luigi Malagò), "Implementing

Evolutionary Algorithms in the Cloud" (J.J. Merelo), "The Attainment Function Approach to Performance Evaluation in EMO" (Carlos Fonseca and Andreia Guerreiro), "Runtime Analysis of Evolutionary Algorithms: Basic Introduction" (Per Kristian Lehre and Pietro Oliveto), "Meta-model Assisted (Evolutionary) Optimization" (Boris Naujoks, Jörg Stork, Martin Zaefferer, and Thomas Bartz-Beielstein).

We wish to express our gratitude in particular to the Program Committee members and external reviewers who provided thorough evaluations of all 224 submissions. We would also express our profound thanks to all the members of the Organizing Committee and the local organizers for their outstanding efforts in preparing for and running the conference. Thanks to all the keynote, workshop, and tutorial speakers for their participation, which greatly enhanced the quality of the conference. Finally, we also express our gratitude to the sponsoring institutions, including Edinburgh Napier University, for their financial support, and the conference partners for participating in the organization of this event.

September 2016

Julia Handl
Emma Hart
Peter R. Lewis
Manuel López-Ibáñez
Gabriela Ochoa
Ben Paechter

Organization

PPSN XIV was organized by the School of Computing, Edinburgh Napier University, Scotland, UK.

Conference Committee

Conference Chairs

Emma Hart	Edinburgh Napier University, UK
Ben Paechter	Edinburgh Napier University, UK

Honorary Chair

Hans-Paul Schwefel	Dortmund University of Technology, Germany

Program Chairs

Julia Handl	University of Manchester, UK
Manuel López-Ibáñez	University of Manchester, UK
Gabriela Ochoa	University of Stirling, UK

Tutorial Chairs

Carola Doerr	Université Pierre et Marie Curie, France
Nicolas Bredeche	Université Pierre et Marie Curie, France

Workshop Chairs

Christian Blum	University of the Basque Country, Spain
Christine Zarges	University of Birmingham, UK

Publications Chair

Peter R. Lewis	Aston University, UK

Local Organization Chair

Neil Urquhart	Edinburgh Napier University, UK

Local Organizing Committee

Kevin Sim	Edinburgh Napier University, UK
Christopher Stone	Edinburgh Napier University, UK
Andreas Steyven	Edinburgh Napier University, UK

Steering Committee

Carlos Cotta	Universidad de Málaga, Spain
David W. Corne	Heriot-Watt University Edinburgh, UK
Kenneth De Jong	George Mason University, USA
Agoston E. Eiben	VU University Amsterdam, The Netherlands
Juan Julián Merelo Guervós	Universidad de Granada, Spain
Gunter Rudolph	Dortmund University of Technology, Germany
Thomas P. Runarsson	University of Iceland, Iceland
Robert Schaefer	University of Krakow, Poland
Marc Schoenauer	Inria, France
Xin Yao	University of Birmingham, UK

Keynote Speakers

Katie Bentley	Harvard Medical School, USA
Josh Bongard	University of Vermont, USA
Susan Stepney	University of York, UK

Tutorials

Gray Box Optimization in Theory

Darrell Whitley

Theory of Evolutionary Computation

Benjamin Doerr

Graph-Based and Cartesian Genetic Programming

Julian Miller and Patricia Ryser-Welch

Theory of Parallel Evolutionary Algorithms

Dirk Sudholt

Promoting Diversity in Evolutionary Optimization: Why and How

Giovanni Squillero and Alberto Tonda

Evolutionary Multiobjective Optimization

Dimo Brockhoff

Intelligent Systems for Smart Cities

Enrique Alba

Advances on Multi-modal Optimization

Mike Preuss and Michael G. Epitropakis

Evolutionary Computation in Cryptography

Stjepan Picek

Evolutionary Robotics: A Practical Guide to Experimenting with Real Hardware

Jacqueline Heinerman, Agoston E. Eiben, Evert Haasdijk, and Julien Hubert

Evolutionary Algorithms and Hyper-heuristics

Nelishia Pillay

A Bridge between Optimization over Manifolds and Evolutionary Computation

Luigi Malagò

Implementing Evolutionary Algorithms in the Cloud

J.J. Merelo

The Attainment Function Approach to Performance Evaluation in EMO

Carlos Fonseca and Andreia Guerreiro

Runtime Analysis of Evolutionary Algorithms: Basic Introduction

Per Kristian Lehre and Pietro Oliveto

Meta-model Assisted (Evolutionary) Optimization

Boris Naujoks, Jörg Stork, Martin Zaefferer, and Thomas Bartz-Beielstein

Workshops

Intelligent Transportation Workshop

Neil Urquhart

Landscape-Aware Heuristic Search

Nadarajen Veerapen, Gabriela Ochoa

Natural Computing in Scheduling and Timetabling

Ahmed Kheiri, Rhyd Lewis, Ender Özcan

Advances in Multi-modal Optimization

Mike Preuss, Michael G. Epitropakis, Xiaodong Li

Program Committee

Hernan Aguirre
Youhei Akimoto
Enrique Alba
Richard Allmendinger
Dirk Arnold
Anne Auger
Doğan Aydn
Jaume Bacardit
Helio Barbosa
Thomas Bartz-Beielstein
Roberto Battiti
Heder Bernardino
Hans-Georg Beyer
Mauro Birattari
Christian Blum
Peter Bosman
Pascal Bouvry
Anthony Brabazon
Jürgen Branke
Dimo Brockhoff
Will Browne
Alexander Brownlee
Larry Bull
Edmund Burke
Stefano Cagnoni
David Cairns
Ying-Ping Chen
Francisco Chicano García
Miroslav Chlebik
Sung-Bae Cho
Carlos Coello Coello
Ernesto Costa
Carlos Cotta
Kenneth De Jong
Antonio Della Cioppa
Luca Di Gaspero
Carola Doerr
Benjamin Doerr
Marco Dorigo
Johann Dreo
Rafal Drezewski
Aniko Ekart
Talbi El-Ghazali
Michael Emmerich
Andries Engelbrecht
Anton Eremeev
A. Sima Etaner-Uyar
Katti Faceli
Bogdan Filipič
Steffen Finck
Andreas Fischbach
Iztok Fister
Carlos M. Fonseca
Martina Friese
Marcus Gallagher
Jonathan M. Garibaldi
Mario Giacobini
Tobias Glasmachers
Brian Goldman
Roderich Gross
Walter Gutjahr
Jussi Hakanen
Hisashi Handa
Julia Handl
Jin-Kao Hao
Verena Heidrich-Meisner
Torsten Hildebrandt
Holger Hoos
Christian Igel
Hisao Ishibuchi
Christian Jacob
Thomas Jansen
Yaochu Jin
Laetitia Jourdan
Bryant Julstrom
George Karakostas
Joshua Knowles
Timo Kötzing
Krzysztof Krawiec
Halina Kwaśnicka
Joerg Laessig
Dario Landa-Silva
William Langdon
Frédéric Lardeux
Sanja Lazarova-Molnar
Per Kristian Lehre
Peter R. Lewis
Jingpeng Li
Xiaodong Li
Jiawei Li
Arnaud Liefooghe
Giosuè Lo Bosco
Fernando Lobo
Daniele Loiacono
Manuel López-Ibáñez
Ilya Loshchilov
Jose A. Lozano
Simon Lucas
Gabriel Luque
Thibaut Lust
Evelyne Lutton
Jacek Mańdziuk
Vittorio Maniezzo
Elena Marchiori
Carlos Martin-Vide
Giancarlo Mauri
James McDermott
Alexander Melkozerov
J.J. Merelo
Marjan Mernik
Silja Meyer-Nieberg
Martin Middendorf
Kaisa Miettinen
Edmondo Minisci
Sanaz Mostaghim
Boris Naujoks
Ferrante Neri
Frank Neumann
Giuseppe Nicosia
Michael O'Neill
Gabriela Ochoa
Pietro Oliveto
Yew-Soon Ong
Jose Ortiz-Bayliss
Gregor Papa
Gisele Pappa
Luis Paquete
Andrew J. Parkes
David Pelta

Justyna Petke
Silvia Poles
Petr Pošík
Mike Preuss
Robin Purshouse
William Rand
Khaled Rasheed
Tapabrata Ray
Eduardo Rodriguez-Tello
Andrea Roli
Thomas A. Runkler
Conor Ryan
Erol Sahin
Frédéric Saubion
Ivo Sbalzarini
Robert Schaefer
Andrea Schaerf
Marc Schoenauer
Oliver Schuetze
Michèle Sebag
Eduardo Segredo
Martin Serpell
Roberto Serra
Marc Sevaux
Jonathan Shapiro
Prad Kumar Shukla
Kevin Sim
Christopher Simons
Karthik Sindhya
Moshe Sipper
Jim Smith
Christine Solnon
Jorge Soria-Alcaraz
Cătălin Stoean
Jörg Stork
Thomas Stützle
Dirk Sudholt
Ponnuthurai Suganthan
Andrew Sutton
Jerry Swan
Daniel Tauritz
Jorge Tavares
Hugo Terashima
Germán Terrazas Angulo
Andrea Tettamanzi
Lothar Thiele
Dirk Thierens
Renato Tinós
Jerzy Tiuryn
Marco Tomassini
Alberto Tonda
Heike Trautmann
Vito Trianni
Elio Tuci
Tea Tušar
Nadarajen Veerapen
Sébastien Verel
Tobias Wagner
Markus Wagner
Lipo Wang
Elizabeth Wanner
Simon Wessing
Darrell Whitley
Man Leung Wong
John R. Woodward
Ning Xiong
Shengxiang Yang
Xin Yao
Gary G. Yen
Yang Yu
Martin Zaefferer
Aleš Zamuda
Christine Zarges
Qingfu Zhang

Contents

Genetic Programming

Multi-objective, Many-objective and Multi-level Optimisation

Parallel Algorithms and Hardware Issues

Real-World Applications and Modelling

Theory

Diversity and Landscape Analysis

Adaptation, Self-adaptation and Parameter Tuning

Online Model Selection for Restricted Covariance Matrix Adaptation

Youhei Akimoto[1(✉)] and Nikolaus Hansen[2]

[1] Faculty of Engineering, Shinshu University, Nagano, Japan
y_akimoto@shinshu-u.ac.jp
[2] Inria, Research Centre Saclay – Île-de-France, Palaiseau, France
nikolaus.hansen@lri.fr

Abstract. We focus on a variant of covariance matrix adaptation evolution strategy (CMA-ES) with a restricted covariance matrix model, namely VkD-CMA, which is aimed at reducing the internal time complexity and the adaptation time in terms of function evaluations. We tackle the shortage of the VkD-CMA—the model of the restricted covariance matrices needs to be selected beforehand. We propose a novel mechanism to adapt the model online in the VkD-CMA. It eliminates the need for advance model selection and leads to a performance competitive with or even better than the algorithm with a nearly optimal but fixed model.

1 Introduction

The covariance matrix adaptation evolution strategy (CMA-ES) [6] is a stochastic search algorithm for continuous optimization. It is considered a state-of-the-art algorithm for black-box scenarios. In the CMA-ES, candidate solutions are generated from a normal (Gaussian) distribution $\mathcal{N}(\mathbf{m}, \sigma^2\mathbf{C})$ with mean vector $\mathbf{m}$, step-size σ, and covariance matrix $\mathbf{C}$. Thanks to the adaptation of positive definite symmetric covariance matrix $\mathbf{C}$, the CMA-ES is known as an efficient optimizer for ill-conditioned and non-separable functions. On a quadratic function, it is empirically known [6] and theoretically supported [1] that the covariance matrix approximates the inverse Hessian, which turns the problem into a spherical function.

In the references [2,3,10,11], variants of CMA-ES with a restricted covariance matrix model are proposed. All of these approaches have common advantages and disadvantages over the standard CMA-ES. The advantages are mainly twofold. One is the internal complexity. As the covariance matrix is represented with a fewer number of parameters, its space complexity is improved. Moreover, computationally efficient update formulas for these restricted covariance matrices are employed, leading to an improvement in the internal time complexity. Therefore, they are promising when solving an optimization problem in a high dimension. The other advantage is the speedup in terms of the number of function evaluations required to adapt the covariance matrix. Since they have fewer parameters to be adapted, the update in one iteration is more reliable, allowing a

J. Handl et al. (Eds.): PPSN XIV 2016, LNCS 9921, pp. 3–13, 2016.
DOI: 10.1007/978-3-319-45823-6_1

greater learning rate. The disadvantage is that the restricted covariance matrix may not be rich enough to approximate the inverse Hessian of the objective function. If this is the case, the convergence rate will be very low. The mean vector will not approach to the optimum within a reasonable run-time.

The VkD-CMA [3] is a variant of the CMA-ES with a restricted covariance matrix mode. It parameterizes the covariance matrix with a diagonal matrix $\mathbf{D}$ and k vectors $\mathbf{V} = [\mathbf{v}_1, \dots, \mathbf{v}_k]$, i.e., $\mathbf{C} = \mathbf{D}(\mathbf{I} + \mathbf{V}\mathbf{V}^{\mathrm{T}})\mathbf{D}$. It is proved that the algorithm is equivalent to sep-CMA-ES [11] if $k = 0$ and is equivalent to CMA-ES if $k = d - 1$. Therefore, the VkD-CMA is considered as a generalization of these variants of the CMA-ES and allows us to control the model complexity by tuning the number of vectors, i.e., k, between diagonal and positive definite symmetric. However, k needs to be tuned in advance to exploit the structure of a function. Without a strong prior knowledge, it can be prohibitively expensive.

In this paper, we propose an online adaptation of the model complexity for the restricted covariance model used in the VkD-CMA, i.e., online k adaptation. The idea to increase k is to detect the condition that we observe when the covariance matrix model is not rich enough to approximate the inverse Hessian. The idea to decrease k is to check if the current covariance matrix is well approximated with a smaller k. We expect two advantages of the online k adaptation. First, it obviates the need for tuning of k, leading to a speedup in the preprocessing of optimization and turning the algorithm more user-friendly. Second, online adaptation of the model complexity may lead to a faster adaptation of the covariance matrix than the optimal but fixed k.

The rest of the paper is organized as follows. Section 2 is devoted to the introduction to the VkD-CMA. The proposed k adaptation mechanism is presented in Sect. 3. We conduct experiments in Sect. 4 to check how efficiently the proposed mechanism adapts k and to compare with variants of CMA-ES. In Sect. 5 we summarize our contributions and discuss a possible line of future work.

2 VkD-CMA

The VkD-CMA [3] is a variant of the covariance matrix adaptation evolution strategy (CMA-ES) [6–8] with a restricted covariance matrix model. As well as the other variants of CMA-ES, multiple candidate solutions are sampled from the multivariate normal distribution $\mathcal{N}(\mathbf{m}, \sigma^2\mathbf{C})$, they are evaluated on the objective function $f : \mathbb{R}^d \to \mathbb{R}$, and the distribution parameters, $\mathbf{m}$, σ, and $\mathbf{C}$, are updated using the candidate solutions and their fitness ranking. In the VkD-CMA, the covariance matrix $\mathbf{C}$ is parameterized with a d dimensional positive-definite diagonal matrix $\mathbf{D}$ and k orthogonal vectors $\mathbf{V} = [\mathbf{v}_1, \dots, \mathbf{v}_k]$, the latter of which is decomposed into a k dimensional nonnegative definite diagonal matrix $\mathbf{\Lambda}$ and a $d \times k$ dimensional matrix $\tilde{\mathbf{V}}$ with orthogonal columns of unit length. Then,

$$\mathbf{C} = \mathbf{D}(\mathbf{I} + \mathbf{V}\mathbf{V}^{\mathrm{T}})\mathbf{D}, \quad \text{or equivalently,} \quad \mathbf{C} = \mathbf{D}(\mathbf{I} + \tilde{\mathbf{V}}\mathbf{\Lambda}\tilde{\mathbf{V}}^{\mathrm{T}})\mathbf{D}. \tag{1}$$

The parameter, k, determines the richness of the covariance matrix mode. Let $\mathcal{M}_k$ be the set of matrices in the form (1). The set $\mathcal{M}_0$ with $k = 0$ is the set of diagonal matrices and $\mathcal{M}_{d-1}$ with $k = d-1$ is the set of arbitrary positive-definite symmetric matrices. The covariance matrix adaptation, i.e., update of $\mathbf{D}$, $\mathbf{\Lambda}$, and $\tilde{\mathbf{V}}$, is based on the projection of the covariance matrix from $\mathcal{M}_{d-1}$ onto its subset $\mathcal{M}_k$. The algorithm employing the two-point step-size adaptation (TPA) [5] is described below, followed by the description of the parameters appearing in the algorithm and their default values.

Algorithm. We initialize $\mathbf{m}^{(0)}$, $\sigma^{(0)}$ and $\mathbf{D}^{(0)}$ according to the initial search interval of a given problem. Let $\tilde{\mathbf{V}}^{(0)} = \mathbf{0}$, $\mathbf{\Lambda}^{(0)} = \mathrm{diag}(0,\dots,0)$, $\mathbf{p}_c^{(0)} = \mathbf{0}$, and $s^{(0)} = 0$. Let $t = 0$ and $r = k + \mu + 1$. Repeat the following steps until a termination criterion is satisfied.

1. If $t \geqslant 1$, generate a pair of symmetric points $y_\pm$ along the previous mean shift $\mathbf{dm}^{(t-1)}$ according to

$$y_\pm = \pm(\|\mathcal{N}(0,\mathbf{I})\| / \|\mathbf{dm}^{(t-1)}\|_{\mathbf{C}^{(t)}})\mathbf{dm}^{(t-1)}, \tag{2}$$

where the Mahalanobis norm $\|\mathbf{dm}^{(t-1)}\|^2_{\mathbf{C}^{(t)}} = (\mathbf{dm}^{(t-1)})^{\mathrm{T}}(\mathbf{C}^{(t)})^{-1}\mathbf{dm}^{(t-1)}$ is computed with the following formula: Let $u_1 = \mathbf{D}^{-1}\mathbf{dm}$ and $u_2 = \tilde{\mathbf{V}}^{\mathrm{T}}u_1$, then

$$(\mathbf{dm})^{\mathrm{T}}\mathbf{C}^{-1}\mathbf{dm} = \|u_1\|^2 + u_2^{\mathrm{T}}((\mathbf{I}+\mathbf{\Lambda})^{-1} - \mathbf{I})u_2. \tag{3}$$

Let $y_1 = y_+$, $y_2 = y_-$. If $t = 0$, generate y_1 and y_2 in the same way as the next step.

2. Sample $\lambda - 2$ independent random vectors $z_i \sim \mathcal{N}(\mathbf{0},\mathbf{I})$, for $i = 3,\dots,\lambda$, and compute y_i according to

$$y_i \leftarrow \tilde{\mathbf{V}}^{\mathrm{T}}z_i, \quad y_i \leftarrow ((\mathbf{\Lambda}+\mathbf{I})^{1/2} - \mathbf{I})y_i, \text{ and } y_i \leftarrow \mathbf{D}(z_i + \tilde{\mathbf{V}}y_i). \tag{4}$$

Let $x_i = \mathbf{m}^{(t)} + \sigma^{(t)}y_i$ for $i = 1,\dots,\lambda$.

3. Evaluate x_i on the given objective function f, and let the index of the ith best point among them be denoted by $i:\lambda$.

4. Compute the weighted average $\mathbf{dm}^{(t)}$ of the steps $y_{i:\lambda}$ and update the mean vector $\mathbf{m}^{(t+1)}$ according to

$$\mathbf{dm}^{(t)} = \textstyle\sum_{i=1}^{\mu} w_i y_{i:\lambda}\ , \quad \mathbf{m}^{(t+1)} = \mathbf{m}^{(t)} + c_m\sigma^{(t)}\mathbf{dm}^{(t)}. \tag{5}$$

5. If $t \geqslant 1$, we update $s^{(t+1)}$ and update the step-size according to

$$s^{(t+1)} = (1 - c_\sigma)s^{(t)} + c_\sigma(\mathrm{rank}(x_2) - \mathrm{rank}(x_1))/(\lambda - 1)\ , \tag{6}$$

$$\sigma^{(t+1)} = \sigma^{(t)}\exp(s^{(t+1)}/d_\sigma). \tag{7}$$

Let $h_\sigma = \mathbb{I}\{s^{(t+1)} < 0.5\}$. Otherwise, let $s^{(1)} = s^{(0)}$, $\sigma^{(1)} = \sigma^{(0)}$ and $h_\sigma = 1$.

6. Update the evolution path

$$\mathbf{p}_c^{(t+1)} = (1 - c_c)\mathbf{p}_c^{(t)} + h_\sigma(c_c(2 - c_c)\mu_{\mathrm{eff}})^{1/2}\mathbf{dm}^{(t)}. \tag{8}$$

7. Let $\mathbf{W} = [\alpha_c^{1/2}\tilde{\mathbf{V}}^{(t)}(\mathbf{\Lambda}^{(t)})^{1/2},\ (\mathbf{D}^{(t)})^{-1}\mathbf{Y}]$, where $\alpha_c = 1 - c_\mu - c_1 + (1 - h_\sigma)c_1c_c(2-c_c)$ and $\mathbf{Y}$ is a $d\times(\mu+1)$ dimensional matrix whose first μ columns are

given by $(c_\mu w_i)^{1/2} y_{i:\lambda}$ for $i = 1, \ldots, \mu$ and the last column is $c_1^{1/2} \mathbf{p}_c^{(t+1)}$. Let $r = \min(k + \mu + 1, d)$. Compute the thin singular value decomposition (SVD) of $\mathbf{W}$, denoted by $\mathbf{LSR}^\mathrm{T}$, where $\mathbf{S}$ is a $r \times r$ dimensional diagonal matrix whose diagonal elements are the singular values of $\mathbf{W}$ aligned in descending order, $\mathbf{L}$ and $\mathbf{R}$ are matrices of dimension $d \times r$ and $r \times r$, respectively, whose columns are the left and right singular vectors with unit length. Compute $\beta = \alpha_c + (d-k)^{-1} \sum_{i=k+1}^{r} [\mathbf{S}]_{i,i}^2$ and update $\tilde{\mathbf{V}}^{(t+1)}$, $\mathbf{\Lambda}^{(t+1)}$, and $\mathbf{D}^{(t+1)}$ as

$$\tilde{\mathbf{V}}^{(t+1)} = \mathbf{L}_{:,:k} \ , \quad \mathbf{\Lambda}^{(t+1)} = \frac{(\alpha_c - \beta)\mathbf{I} + \mathbf{S}_{:k,:k}^2}{\beta} \ , \quad [\mathbf{D}^{(t+1)}]_{i,i} = \frac{[\mathbf{D}^{(t)}]_{i,i} (\alpha_c + \sum_{j=1}^{r} [\mathbf{W}]_{i,j}^2)^{1/2}}{(1 + \sum_{j=1}^{k} [\mathbf{\Lambda}^{(t+1)}]_{j,j} [\tilde{\mathbf{V}}^{(t+1)}]_{i,j}^2)^{1/2}} . \tag{9}$$

Here, for any matrix $\mathbf{A}$, $[\mathbf{A}]_{i,j}$ denotes the (i,j)th element of $\mathbf{A}$ and $\mathbf{A}_{:i,:i}$ denotes the $i \times i$ upper left block of $\mathbf{A}$, $\mathbf{A}_{:,:i}$ denotes the first i columns of $\mathbf{A}$.

8. Compute the $2d$th root of the determinant of the new covariance matrix as $\gamma = \exp\left(\frac{1}{d} \sum_{i=1}^{d} \ln([\mathbf{D}^{(t+1)}]_{i,i}) + \frac{1}{2d} \sum_{j=1}^{k} \ln(1 + [\mathbf{\Lambda}^{(t+1)}]_{j,j})\right)$ and $\mathbf{D}^{(t+1)} \leftarrow \mathbf{D}^{(t+1)}/\gamma$ and $\mathbf{p}_c^{(t+1)} \leftarrow \mathbf{p}_c^{(t+1)}/\gamma$. Then, we have $\det(\mathbf{C}^{(t+1)}) = 1$.

Default Parameter Values. The default parameter values are summarized as follows. The population size $\lambda = \lfloor 4 + 3\ln(d) \rfloor$, the number of parents $\mu = \lfloor \lambda/2 \rfloor$, and the weights

$$w_i = \frac{\ln((\lambda+1)/2) - \ln(i)}{\sum_{i=1}^{\mu} (\ln((\lambda+1)/2) - \ln(i))} \quad (i = 1, \ldots, \mu) \ , \qquad w_i = 0 \quad (i > \mu) . \tag{10}$$

Let $\mu_{\mathrm{eff}} = 1/(\sum_{i=1}^{\mu} w_i^2)$. The learning rate for $\mathbf{m}$-update $c_m = 1$, the learning rate for s in TPA $c_\sigma = 0.3$, the damping parameter for TPA $d_\sigma = d^{1/2}$. The cumulation factor c_c, the learning rate for the rank-one update c_1 and the learning rate for the rank-μ update c_μ are as follows[1] (letting $a \wedge b = \min(a, b)$ for any real a and b)

$$c_c = \frac{4 + \mu_{\mathrm{eff}}/d}{(d + 2(k+1))/3 + 4 + 2\mu_{\mathrm{eff}}/d}, \ c_1 = \frac{2}{d(k+2) + 2(k+2) + \mu_{\mathrm{eff}}}, \ c_\mu = (1 - c_1) \wedge \frac{2(\mu_{\mathrm{eff}} - 2 + 1/\mu_{\mathrm{eff}})}{d(k+1) + 4(k+2) + \mu_{\mathrm{eff}}} . \tag{11}$$

Properties. With the restricted covariance matrix model, we achieve cheaper computational time and space complexity and faster adaptation of the covariance matrix than the CMA-ES. Its space complexity is $\mathcal{O}(dr)$, where $r = \min(d, k + \mu + 1)$, and its time complexity is $\mathcal{O}(dr^2 + dk\lambda)$ per iteration. If $r \ll d$, it is cheaper than the CMA-ES, which requires $\Theta(d^2 + d\mu)$ space and $\Theta(d^2\lambda)$ time complexity per iteration. Moreover, since there are fewer parameters to be adapted if k is smaller, it accepts relatively higher learning rates (11) than the default values used in the CMA-ES, resulting in faster adaptation of the covariance matrix. Therefore, we want to keep k as small as possible.

---

[1] The default c_1 is slightly different from the original setting in [3]. The value presented in the paper is slightly more stable for k close to zero.

On the other hand, if k is too small to approximate the inverse Hessian of the objective function, the VkD-CMA is not able to solve the problem efficiently. Empirically, we know that the convergence rate, defined as the slope of the step-size in log-scale, is proportional to $\mathrm{Cond}(\mathbf{AC})$ on a quadratic objective function $f(x) = x^{\mathrm{T}}\mathbf{A}x$. Therefore, we need to set k large enough to approximate the inverse Hessian of the objective function. However, since a prior knowledge on the problem is limited in the black-box scenario, it is hard to choose a reasonable k in advance. In the next section, we propose a mechanism to adapt k, i.e., the model richness of the covariance matrix, during the optimization process.

3 Adaptive Covariance Model Selection

Ideas. Let us consider solving a quadratic function with a positive definite symmetric Hessian $\mathbf{A}$. If $\mathbf{C} \propto \mathbf{A}^{-1}$, the quadratic function is identified with Sphere function $f(x) = \|x\|^2$. In this case, we can deduce that the optimal convergence rate (i.e., the slope of $\ln(\sigma)$) is approximately $0.5\lambda/d$ using the result given in [4][2]. We also empirically know that the convergence rate is approximately proportional to $1/\mathrm{Cond}(\mathbf{CA})$ if $\mathrm{Cond}(\mathbf{CA}) \gg 1$.

Consider the cases where the covariance matrix is richer than sufficient. It means that the inverse Hessian of the objective function can be approximated by the covariance matrix in $\mathcal{M}_k$ with a smaller k. Let us consider that we drop the ith vector $\mathbf{v}_i$, i.e., drop the ith column of $\tilde{\mathbf{V}}$ and ith column and row of $\mathbf{\Lambda}$, and let $\tilde{\mathbf{C}}$ be the resulting covariance matrix. Then, we have that $\mathrm{Cond}(\mathbf{A}\tilde{\mathbf{C}}) \leqslant \mathrm{Cond}(\tilde{\mathbf{C}}\mathbf{C}^{-1})\mathrm{Cond}(\mathbf{AC}) = (1 + [\mathbf{\Lambda}]_{i,i})\mathrm{Cond}(\mathbf{AC})$. It means that by dropping the ith component from the covariance matrix, we may increase the condition number at most by the factor of $1 + [\mathbf{\Lambda}]_{i,i}$. If $1 + [\mathbf{\Lambda}]_{i,i}$ is smaller than a given threshold β_{dec}, it is safe to remove $\mathbf{v}_i$. However, if the covariance matrix is in the middle of adaptation and $[\mathbf{\Lambda}]_{i,i}$ is still increasing, there is a chance that $1 + [\mathbf{\Lambda}]_{i,i}$ will grow up above the threshold. Therefore, we want to decrease k and drop some components from the covariance matrix only when $1 + [\mathbf{\Lambda}]_{i,i}$ is small enough and is regarded as not increasing.

Consider the cases where the covariance matrix is not rich enough. In this case we observe that $\mathbf{C}$ is kept roughly constant and σ converges very slowly. If we observe $\mathbf{C}$ not significantly changing (except the scaling factor), it implies that $\mathbf{C}$ is close to the optimal approximation of the inverse Hessian in the current covariance model. If the covariance model is rich enough, $\mathrm{Cond}(\mathbf{CA})$ will be close to 1 and the convergence rate will not be very small compared with the optimal convergence rate. If the covariance model is not rich enough, $\mathrm{Cond}(\mathbf{CA}) \gg 1$ and we will observe a slow convergence of σ with the convergence rate proportional to $\mathrm{Cond}(\mathbf{CA})$. Based on this observation, we detect insignificant change of $\mathbf{C}$ and slow convergence of σ. If all of $1 + [\mathbf{\Lambda}]_{i,i}$ are greater than a given threshold β_{inc} and both of the conditions are satisfied, we increase k.

[2] If we use only nonnegative weights as we do in this paper, the possible convergence rate halves.

Algorithm. Let $t_{\text{ada}} = 0$ be the number of iterations after the last k increase or the initialization. Initialize the exponential moving average M_* of the slopes of $\ln(\sigma)$, $\ln([\mathbf{C}]_{i,i})$ and $\ln(1 + [\mathbf{\Lambda}]_{i,i})$ by zero, where $*$ is either $\ln(\sigma)$, $\ln([\mathbf{C}]_{i,i})$ or $\ln(1 + [\mathbf{\Lambda}]_{i,i})$. The following steps are performed after Step 8 of the VkD-CMA algorithm.

1. Update the exponential moving averages according to

$$M_{\ln(\sigma)}^{(t+1)} = (1 - \alpha_{\text{exp}}^{(\sigma)}) M_{\ln(\sigma)}^{(t)} + \alpha_{\text{exp}}^{(\sigma)} \big(\ln(\sigma^{(t+1)}) - \ln(\sigma^{(t)}) \big). \tag{12}$$

The exponential moving averages for $\ln([\mathbf{C}]_{i,i})$ and $\ln(1 + [\mathbf{\Lambda}]_{i,i})$ are updated analogously with the decay factor $\alpha_{\text{exp}}^{(\mathbf{C})}$. Note that $\ln([\mathbf{C}]_{i,i}) = 2\ln([\mathbf{D}]_{i,i}) + \ln(1 + \sum_{j=1}^{k} [\mathbf{\Lambda}]_{j,j} [\tilde{\mathbf{V}}]_{i,j}^2)$.

2. If $t_{\text{ada}} > \mathrm{T}_{\text{exp}}$ and $1 + [\mathbf{\Lambda}]_{i,i} > \beta_{\text{inc}}$ for all $i \in [\![1, k]\!]$, we additionally check if (1) $|M_{\ln(\sigma)}^{(t+1)}| < \gamma_\sigma \min(0.5, 0.5\lambda/d)/\max(1, \beta_{\text{inc}}/10)$ and if (2) $\max_i |M_{\ln([\mathbf{C}]_{i,i})}^{(t+1)}| < \gamma_{\mathbf{C}}(c_1 + c_\mu)$. If all the above conditions are satisfied, update k as

$$k = \min(\max(\lfloor \kappa_{\text{inc}} k \rfloor, k + 1), k_{\max}) \tag{13}$$

and set $[\tilde{\mathbf{V}}]_{:,i} = (0, \dots, 0)$ and $[\mathbf{\Lambda}]_{i,i} = 0$ for all $i \in [\![k_{\text{old}} + 1, k]\!]$, where k_{old} is the value for k before updated. Let $t_{\text{ada}} = 0$.

3. If $t_{\text{ada}} > k\mathrm{T}_{\text{exp}}$ and there is at least one index $i \in [\![1, k]\!]$ such that $1 + [\mathbf{\Lambda}]_{i,i} < \beta_{\text{dec}}$, let J be the set of such indices. If there exist indices in J satisfying $M_{\ln(1+[\mathbf{\Lambda}]_{i,i})}^{(t+1)} < 0$, drop ith column from $\tilde{\mathbf{V}}$ and drop ith column and row from $\mathbf{\Lambda}$ for all such indices i and update k accordingly. Then, re-normalize $\mathbf{D}$ and $\mathbf{p}_c$ (perform Step 8 in the VkD-CMA algorithm again).

4. Let $t_{\text{ada}} = t_{\text{ada}} + 1$. If k is updated, update c_1, c_μ, c_c according to (11).

Default Parameter Values. The parameters appearing in the k-adaptation algorithm are described below, together with their default values.

- $\alpha_{\text{exp}}^{(\sigma)} = 0.5 \min(1, \lambda/d)/\max(1, \beta_{\text{inc}}/10)$, $\alpha_{\text{exp}}^{(\mathbf{C})} = 1/d$; The discount rate for the exponential moving average, $0 < \alpha_{\text{exp}} < 1$.
- $\gamma_\sigma = 0.1$, $\gamma_{\mathbf{C}} = 0.3$; The threshold to detect insignificant change of the step-size and the covariance matrix, $\gamma_\sigma > 0$ and $\gamma_{\mathbf{C}} > 0$.
- $\mathrm{T}_{\text{exp}} = \frac{2}{\min(\alpha_{\text{exp}}^{(\sigma)}, \alpha_{\text{exp}}^{(\mathbf{C})})} - 1$; The number of iterations to wait for k adaptation after the last k increase, $\mathrm{T}_{\text{exp}} \geqslant 0$. It is introduced to prevent k from oscillating. The sum of the weights for last $\mathrm{T}_{\text{exp}} + 1$ iterations, i.e., $\alpha_{\text{exp}}, \alpha_{\text{exp}}(1 - \alpha_{\text{exp}}), \dots, \alpha_{\text{exp}}(1 - \alpha_{\text{exp}})^{\mathrm{T}_{\text{exp}}}$, is $1 - (1 - \alpha_{\text{exp}})^{\mathrm{T}_{\text{exp}}+1} \approx 1 - \exp(-2)$. It implies that the last $\frac{2}{\alpha_{\text{exp}}} - 1$ iterations contain about 86 % of the information in the exponential moving average and the information in the exponential moving average is considered refreshed.
- $k_{\min} = 0$, $k_{\max} = d - 1$, $k_{\text{init}} = k_{\min}$; The minimum, maximum and initial number of vectors, $0 \leqslant k_{\min} \leqslant k_{\text{init}} \leqslant k_{\max} \leqslant d - 1$. The maximum value should be set smaller if the available memory and cpu time are limited.
- $\kappa_{\text{inc}} = 1.414$; The factor for increment of k.

- $\beta_{\text{inc}} = \beta_{\text{dec}} = 30$; This corresponds to the condition number $\text{Cond}(\mathbf{C}^{-1/2}\tilde{\mathbf{C}}\mathbf{C}^{-1/2})$ we accept, where $\mathbf{C}$ and $\tilde{\mathbf{C}}$ are the covariance matrix before and after k-decrease. The greater β_{inc} is, k tends to be smaller.

4 Experiments

We first test the performance of the proposed k-adaptation mechanism on the quadratic function $f(x) = \frac{1}{2}x^{\mathrm{T}}\mathbf{A}x$ with the inverse Hessian

$$\mathbf{A}^{-1} = (10^{-6}/2)\mathbf{D}_{\text{ell}}^{-1}(\mathbf{I} + (10^6 - 1)\mathbf{U}\mathbf{U}^{\mathrm{T}})\mathbf{D}_{\text{ell}}^{-1}, \tag{14}$$

where $\mathbf{D}_{\text{ell}}$ is a diagonal matrix whose ith diagonal component is $10^{3\frac{i-1}{d-1}}$, and $\mathbf{U}$ is a $d \times k_{\text{cig}}$ matrix whose columns are orthogonal to each other and of length one. Ten instances of $\mathbf{U}$ are generated by applying the same procedure to create $\mathbf{R}$ with $m = k_{\text{cig}}$ in Table 1. The inverse Hessian is then in $\mathcal{M}_{k_{\text{cig}}}$. In this experiment, the dimension is $d = 100$ and $\sigma^{(0)} = 2$, $\mathbf{D}^{(0)} = \mathbf{I}$, and $\mathbf{m}^{(0)}$ is generated from $\mathcal{N}(3 \cdot \mathbf{1}, 2^2\mathbf{I})$ for each run. All the other parameters for the VkD-CMA and for the k adaptation mechanism are set to the default values described in this paper.

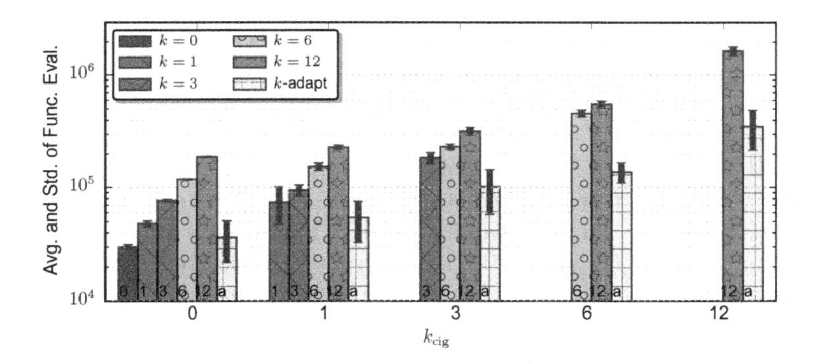

Fig. 1. Average and standard deviation of the number of function evaluations to reach f_{target}. VkD-CMA with fixed $k = 0, 1, 3, 6, 12$ and with the adaptive k are compared on the quadratic with the inverse Hessian (14) with $k_{\text{cig}} = 0, 1, 3, 6, 12$.

Figure 1 shows the average and standard deviation of the number of function evaluations till the target function value $f_{\text{target}} = 10^{-8}$ is reached over 10 independent runs. If $k < k_{\text{cig}}$, the target value was not reached within 10^7 function evaluations. If $k \geqslant k_{\text{cig}}$, the smaller the value of k is, the smaller the number of function evaluations to reach the target value is. Comparing to the fixed optimal $k = k_{\text{cig}}$, the adaptive strategy requires even fewer function evaluations except for the case $k_{\text{cig}} = 0$. Figure 2 reveals an advantage of the adaptive strategy. The VkD-CMA first adapts $\mathbf{D}$ on this quadratic function, then learns $\tilde{\mathbf{V}}$ and $\mathbf{\Lambda}$.

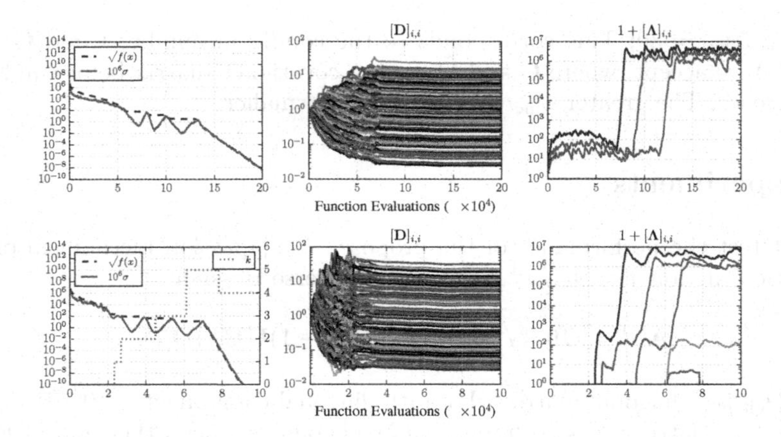

Fig. 2. A typical behavior of the VkD-CMA with fixed optimal $k = k_{\text{cig}}$ (top) and with adaptive k (bottom) on a 100-D quadratic with the inverse Hessian (14) with $k_{\text{cig}} = 3$.

At the beginning, the proposed k adaptation strategy keeps k to zero, resulting in the faster adaptation of $\mathbf{D}$ than the algorithm with the fixed k. Then, it increases k and finally learns the nearly optimal k.

Next, we compare the different variants of CMA-ES, namely, the VkD-CMA with the proposed k adaptation, the VkD-CMA with fixed $k = 1, \mu$, the sep-CMA-ES (SEP) [11], the CMA-ES with cumulative step-size adaptation (CMA-CSA) and the CMA-ES with TPA (CMA-TPA). The parameter values described in Sect. 2 are used for the VkD-CMA. The parameter values for the CMA-CSA and CMA-TPA are taken from the reference [3], and the parameter values for the SEP are taken from [11]. The comparison is done on the test functions summarized in Table 1. For all the functions, the target function value is $f_{\text{target}} = 10^{-8}$. Each run is considered successful if and only if the algorithm evaluates a candidate solution having a better function value than f_{target} before spending $10^5 d$ function evaluations. We conduct ten independent runs for each setting. The initial mean vector and step-size are $\mathbf{m}^{(0)} = 3 \cdot \mathbf{1} + \mathcal{N}(\mathbf{0}, 2^2\mathbf{I})$ and $\sigma^{(0)} = 2$ for all but f_{ros} and f_{rosrot}, where they are initialized as $\mathbf{m}^{(0)} = \mathbf{0} + \mathcal{N}(\mathbf{0}, 0.1^2\mathbf{I})$ and $\sigma^{(0)} = 0.1$.

Figure 3 shows the average number of f-calls. As reported in the references [2,3,11], the sep-CMA-ES and the VkD-CMA with $k = 1$ and $k = \mu$ can solve the functions with inverse Hessian in $\mathcal{M}_0$, $\mathcal{M}_1$, $\mathcal{M}_\mu$, respectively, and can not efficiently solve the functions with highly ill-conditioned inverse Hessian outside them. On f_{sph}, f_{cig}, f_{cigrot}, we do not observe a significant difference between variants compared to the other cases. See [2] for the detail. On the others, a variant with smaller covariance matrix model tends to solve the functions with inverse Hessian inside the model more efficiently.

The VkD-CMA with the proposed k adaptation mechanism succeeded to find the target value within the given budget for all scenarios. Moreover, its efficiency is competitive with or better than the other variants including

Table 1. Benchmark function suite. The d dimensional orthogonal matrix $\mathbf{Q}$ is constructed as follows. First all the elements are generated from the standard normal distribution and apply Gram-Schmidt procedure to orthonormalize its columns. The block diagonal orthogonal matrix $\mathbf{B} = \mathrm{diag}(\mathbf{Q}_1, \mathbf{Q}_2)$ is constructed from two orthogonal matrices $\mathbf{Q}_1$ and $\mathbf{Q}_2$ of dimension $d/2$ that are generated analogously to $\mathbf{Q}$. The $d \times m$ dimensional matrix $\mathbf{R}$, where $m = \lfloor 2 \ln(d) \rfloor$, is the first m columns of $\mathbf{Q}$. For $f_{\text{ellrotsub}}$, the inverse Hessian does not live in $\mathcal{M}_{\lfloor 2 \ln(d) \rfloor}$, but in its closure.

Name	Definition	(pseudo) inverse Hessian
Sphere	$f_{\text{sph}}(x) = \sum_{i=1}^{d} x_i^2$	$\mathcal{M}_0$
Cigar	$f_{\text{cig}}(x) = x_1^2 + 10^6 \sum_{i=2}^{d} x_i^2$	$\mathcal{M}_0$
Ellipsoid	$f_{\text{ell}}(x) = f_{\text{sph}}(\mathbf{D}_{\text{ell}} x)$	$\mathcal{M}_0$
Discus	$f_{\text{dis}}(x) = 10^6 x_1^2 + \sum_{i=2}^{d} x_i^2$	$\mathcal{M}_0$
TwoAxes	$f_{\text{twoax}}(x) = \sum_{i=1}^{\lfloor d/2 \rfloor} x_i^2 + 10^6 \sum_{i=\lfloor d/2 \rfloor}^{d} x_i^2$	$\mathcal{M}_0$
Ellipsoid-Cigar($k_{\text{cig}} = 1$)	$f_{\text{ellcig}}(x) = f_{\text{cigrot}}(\mathbf{D}_{\text{ell}} x)$	$\mathcal{M}_1$
rotated Cigar	$f_{\text{cigrot}}(x) = f_{\text{cig}}(\mathbf{Q}x)$	$\mathcal{M}_1$
Ellipsoid-Cigar($k_{\text{cig}} = \lfloor \ln(d) \rfloor$)	$f_{\text{ellciglog}}(x) = \frac{1}{2} x^{\mathrm{T}} \mathbf{A} x$ with $\mathbf{A}$ in (14)	$\mathcal{M}_{\lfloor \ln(d) \rfloor}$
subspace rotated Ellipsoid	$f_{\text{ellrotsub}}(x) = f_{\text{ell}}(\mathbf{R}^{\mathrm{T}} x)$	$\mathcal{M}_{\lfloor 2 \ln(d) \rfloor}$ (semi-positive)
rotated TwoAxes	$f_{\text{twoaxrot}}(x) = f_{\text{twoax}}(\mathbf{Q}x)$	$\mathcal{M}_{\lfloor d/2 \rfloor}$
2-blocks rotated Ellipsoid	$f_{\text{ellrot(2-blocks)}}(x) = f_{\text{ell}}(\mathbf{B}x)$	$\mathcal{M}_{d-1}$
rotated Ellipsoid	$f_{\text{ellrot}}(x) = f_{\text{ell}}(\mathbf{Q}x)$	$\mathcal{M}_{d-1}$
rotated Discus	$f_{\text{disrot}}(x) = f_{\text{disrot}}(\mathbf{Q}x)$	$\mathcal{M}_{d-1}$
Rosenbrock	$f_{\text{ros}}(x) = \sum_{i=1}^{d-1} 10^2 (x_i^2 - x_{i+1})^2 + (x_i - 1)^2$	$\mathcal{M}_{d-1}$ (non-quadratic)
rotated Rosenbrock	$f_{\text{rosrot}}(x) = f_{\text{ros}}(\mathbf{Q}x)$	$\mathcal{M}_{d-1}$ (non-quadratic)

CMA-ES on all but f_{ellrot}, f_{disrot}, and $f_{\text{ellrot(2-blocks)}}$ functions. To approximate the inverse Hessian of these functions, k needs to be increased nearly to $d - 1$. Then the proposed algorithm spends more function evaluations to adapt the covariance matrix than the CMA-ES does. The function $f_{\text{ellrot(2-blocks)}}$ has the inverse Hessian in $\mathcal{M}_{d-1} \setminus \mathcal{M}_{d-2}$ but we have $\min_{\mathbf{C} \in \mathcal{M}_0} \mathrm{Cond}(\mathbf{AC}) \leqslant 10^3$, which is relatively small. Even though the convergence speed of the sep-CMA-ES is slower than the CMA-ES due to this condition number, the adaptation time for the covariance matrix for the sep-CMA-ES is shorter and it requires fewer function evaluations to reach the finite target value of $f_{\text{target}} = 10^{-8}$. The proposed algorithm, however, tends to increase k as well as on the fully rotated Ellipsoid function.

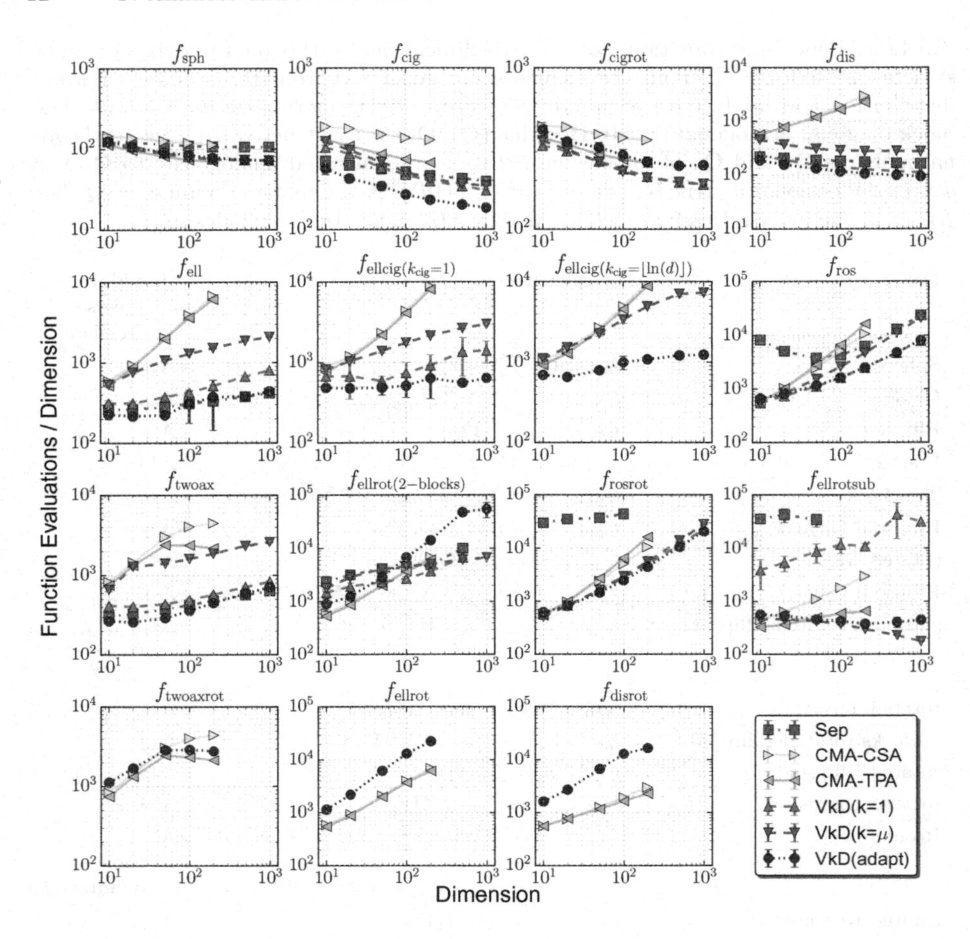

Fig. 3. Average and standard deviation of the number of function evaluations over ten independent runs v.s. dimension d. The data points displayed in the figures correspond to the setting for which the target function values are reached within the given budged ($10^5 d$) in all the ten runs. Note that the experiments have been done up to $d = 200$ for the CMA-CSA and CMA-TPA, whereas the maximum d is 1000 for the other algorithms except for f_{twoaxrot}, f_{ellrot}, f_{disrot} where the maximum d is 200.

5 Discussion

The proposed mechanism adapts the number k of vectors, i.e., the model complexity of the restricted covariance matrix, online for VkD-CMA. The proposed approach removes the need for pre-selection of the restricted covariance matrix model, which is the main shortage of VkD-CMA. The experimental results reveal that the proposed algorithm is competitive with a variant of the CMA-ES with a nearly optimal model of the restricted covariance matrices on most of the problems with limited variable dependencies, importantly without any tuning of the model in advance. On the fully non-separable functions we observe slowdown compared to the CMA-ES; it takes about 7.5 times more function evaluations at the worst case (f_{disrot}, $d = 200$). On f_{disrot}, the inverse Hessian has one smaller eigenvalue than the others, meaning that there is one direction in which the function value is sensitive. For such a function, the active covariance matrix update [9], i.e., assigning negative weights for unsuccessful candidates, is known to accelerate the covariance matrix adaptation in the standard CMA-ES. We expect that it improves the performance of the proposed algorithm. This is one of the main future work.

Acknowledgments. This work is partially supported by JSPS KAKENHI Grant Number 15K16063.

References

1. Akimoto, Y.: Analysis of a natural gradient algorithm on monotonic convex-quadratic-composite functions. In: Proceedings of Genetic and Evolutionary Computation Conference, pp. 1293–1300. ACM (2012)
2. Akimoto, Y., Auger, A., Hansen, N.: Comparison-based natural gradient optimization in high dimension. In: Proceedings of Genetic and Evolutionary Computation Conference, pp. 373–380. ACM (2014)
3. Akimoto, Y., Hansen, N.: Projection-based restricted covariance matrix adaptation for high dimension. In: Proceedings of Genetic and Evolutionary Computation Conference. ACM (2016, to appear)
4. Arnold, D.V.: Optimal weighted recombination. In: Wright, A.H., Vose, M.D., De Jong, K.A., Schmitt, L.M. (eds.) FOGA 2005. LNCS, vol. 3469, pp. 215–237. Springer, Heidelberg (2005)
5. Hansen, N., Atamna, A., Auger, A.: How to assess step-size adaptation mechanisms in randomised search. In: Bartz-Beielstein, T., Branke, J., Filipič, B., Smith, J. (eds.) PPSN 2014. LNCS, vol. 8672, pp. 60–69. Springer, Heidelberg (2014)
6. Hansen, N., Auger, A.: Principled design of continuous stochastic search: from theory to practice. In: Borenstein, Y., Moraglio, A. (eds.) Theory and Principled Methods for the Design of Metaheuristics. NCS. Springer, Berlin (2014)
7. Hansen, N., Muller, S.D., Koumoutsakos, P.: Reducing the time complexity of the derandomized evolution strategy with covariance matrix adaptation (CMA-ES). Evol. Comput. **11**(1), 1–18 (2003)
8. Hansen, N., Ostermeier, A.: Completely derandomized self-adaptation in evolution strategies. Evol. Comput. **9**(2), 159–195 (2001)
9. Jastrebski, G., Arnold, D.V.: Improving evolution strategies through active covariance matrix adaptation. In: 2006 IEEE Congress on Evolutionary Computation, pp. 9719–9726. IEEE (2006)
10. Loshchilov, I.: A computationally efficient limited memory CMA-ES for large scale optimization. In: Proceedings of Genetic and Evolutionary Computation Conference, pp. 397–404 (2014)
11. Ros, R., Hansen, N.: A simple modification in CMA-ES achieving linear time and space complexity. In: Rudolph, G., Jansen, T., Lucas, S., Poloni, C., Beume, N. (eds.) PPSN 2008. LNCS, vol. 5199, pp. 296–305. Springer, Heidelberg (2008)

Genotype Regulation by Self-modifying Instruction-Based Development on Cellular Automata

Stefano Nichele(✉), Tom Eivind Glover, and Gunnar Tufte

Norwegian University of Science and Technology, Trondheim, Norway
nichele@idi.ntnu.no

Abstract. A novel method for regulation of gene expression for artificial cellular systems is introduced. It is based on an instructon-based representation which allows self-modification of genotype programs, as to be able to control the expression of different genes at different stages of development, e.g., environmental adaptation. Coding and non-coding genome analogies can be drawn in our cellular system, where coding genes are in the form of developmental actions while non-coding genes are represented as modifying instructions that can change other genes. This technique was tested successfully on the morphogenesis of cellular structures from a seed, self-replication of structures, growth and replication combined, as well as reuse of an evolved genotype for development or replication of different structures than initially targeted by evolution.

1 Introduction

In biological systems, the process that produces phenotypes from genotypes, i.e., genotype-to-phenotype mapping, is a complex and intricate combination of interactions between the genotype and the environment. As a result, intermediate phenotypic stages emerge, which themselves influence the decoding/regulation of the genotype for the next phenotypic stage. It may be argued that genotypes possess the ability to self-modify [8]. As such, the development process is influenced not only by the instructions encoded in the genome, but also by the mutual interaction between genotype and phenotype, and with the environment. Biological genomes possess an intrinsic ability to evolve and adapt to novel environments, i.e., evolvability [2]. Modularity [5] is a key factor that contributes to evolvability. In fact, many biological networks are modular [1], e.g., brain networks, gene regulatory networks, metabolic networks. It turns out that it is easier to rewire a modular network with independent substructures than an unstructured network [9]. Kovitz [12] introduced the concept of "cascading design", a form of genotype coordination that allows to preserve the relationship between separate traits. He describes such coordination as *"a good house design, where the plumbing and electrical wiring connect water and electricity to the devices that need them. If you change the architectural plans for a house to move the bathroom from the northwest corner to the middle of the east wall, moving*

J. Handl et al. (Eds.): PPSN XIV 2016, LNCS 9921, pp. 14–25, 2016.
DOI: 10.1007/978-3-319-45823-6_2

Table 1. IBD instruction set. L, C, R, U, D represent Left, Center, Right, Up, and Down neighbors. n represents the number of CA cell states

Instr.	Operation	Description	Instr.	Operation	Description
AND	$N[i_1] = N[i_1] \wedge N[i_2]$	Bitwise AND	INC	$N[i_1] = N[i_1] + 1$	Increment
OR	$N[i_1] = N[i_1] \vee N[i_2]$	Bitwise OR	DEC	$N[i_1] = N[i_1] - 1$	Decrement
XOR	$N[i_1] = N[i_1] \oplus N[i_2]$	Bitwise XOR	SWAP	$N[i_1] \Leftrightarrow N[i_2]$	Swapping
NOT	$N[i_1] = \neg N[i_1]$	Bitwise NOT	ROR	$LCR \Rightarrow RLC$	Rotate right
INV	$N[i_1] = n - N[i_1]$	Inverse	ROL	$LCR \Rightarrow CRL$	Rotate left
MIN	$N[i_1] = min(N[i_1], N[i_2])$	Minimum	ROU	$UCD \Rightarrow CDU$	Rotate up
MAX	$N[i_1] = max(N[i_1], N[i_2])$	Maximum	ROD	$UCD \Rightarrow DUC$	Rotate down
SET	$N[i_1] = N[i_2]$	Replace	NOOP		No Operation

the walls and fixtures is not enough. You must also reroute all the pipes and electrical connections". If an evolutionary algorithm was to design such a house plan, *"then moving the bathroom across the house requires that many mutations occur simultaneously: one mutation for each segment of pipe that needs to move, one mutation for each doorway, etc. As coordination becomes more complex, the probability of making all the needed mutations simultaneously gets lower and lower"*. Biological evolution seems to possess some kind on intrinsic coordination. Lee Altenberg describes genes that affect the probability of mutation of other genes and subroutines in genetic programming [2].

In this paper, cellular automata (CA) are used as a test-bed model of development. Traditional CA transition tables are replaced by an algorithmic representation, i.e. program, which includes instructions that can modify the code itself. Such representation may allow the emergence of genotype coordination mechanisms that may encode different sub-processes. As such, different parts of the genome may be active at different stages of phenotypic development. The problems targeted include the morphogenesis of structures from a zygote, the replication of given shapes, both development and replication achieved by the same genotype, and the reuse of evolved genotypes for development and replication of different structures than those initially targeted by evolution.

2 Background

In nature, phenotypes are not determined only by their genotypes. Genes "behave" in relation to each other and in relation to the environment. Genes may even suppress other genes, i.e., methylation [18]. In fact, not all genes are active at all times and methylation is one of the factors that control gene expression. In the context of artificial evolutionary and developmental (evo-devo) systems, different models of gene regulation mechanism exist. Some models aim to be true to biology [15] whether other are more abstract models of development [10,11,13]. In this paper, the used model can be placed within abstract computational models of development based on CA [27]. Therefore, we do not aim at a truly biological model of development. The used two-dimensional CA evo-devo model includes the interplay between genotype, developing phenotype, and evolutionary developmental effects. Traditionally, CA rules are in the form of transition tables. One of the implications of such representation is that rules grow exponentially with the increase in cells neighborhood and number of cell states. As such, transition tables do not scale well.

2.1 Instruction Based Development

The idea of evolving instruction-based representations is a rather old approach [14]. Cartesian Genetic Programming (CGP) has been introduced by Miller and Thompson [17] for the evolution of programs as a directed graph. Sipper [22] proposed the evolution of non-uniform cellular automata with cellular programming. Bidlo and Skarvada [3] introduced the Instruction-Based Development (IBD) for the evolutionary design of digital circuits. Bidlo and Vasicek [4] exploited IBD for the development and replication of cellular automata structures. Even though their approach has been shown to improve the overall success rate, the number of available instructions was experimentally chosen. IBD allows representing CA transition functions by means of a sequence of instructions executed on local neighborhoods in parallel and deterministically update the state of each cell. Evolutionary growth of genomes has been used to evolve local transition functions starting from a single neighborhood configuration [19]. Using an instruction-based approach removes the problem of specifying all combinations in the transition table. The initial ruleset of IBD is presented in Table 1.

2.2 Scalability and Modularity

Biological organisms are the best example of scalability, ranging from simple unicellular organisms to multicellular organisms, where trillions of cells develop from a single cell holding the genome. The genomes of different species have variable lengths, as result of biological complexification mechanisms [16] through gene duplications [25] and continuous elaborations. In artificial evo-devo systems, complexification mechanisms have been used to allow variable length genomes. Stanley and Miikkulainen [23,24] introduced NeuroEvolution of Augmenting Topologies (NEAT), a method for the incremental evolution of neural networks. Nichele and Tufte [19] presented a framework for the evolutionary growth of genomes using indirect encodings. Trefzer et al. [26] investigated the advantages of variable length gene regulatory networks in artificial delelopmental systems.

One of the characteristics that allow variable length genomes to evolve and adapt to new environments is their modularity. Clune et al. [5] showed that a key driver for evolvability is the modularity of biological networks. Modular artificial evo-devo systems have been shown to have increased evolvability. Kovitz [12] has investigated the evolution of coordination mechanisms using a modular cascading design based on graph representation inspired by CGP [17]. Harding et al. introduced Self-modifying Cartesian Genetic Programming (SMCGP) [7], a form of CGP where genotype programs are allowed to modify themselves.

Taking inspiration from IBD and SMCGP, we propose a cellular automata developmental system based on self-modifying cellular programs.

2.3 Self-modifying Instrucion Based Developement

We propose Self-modifying Instruction Based Development (SMIBD) for cellular automata, where the instruction set in Table 1 is extended with the instructions in Table 2. As such, genomes have the ability to perform different functions during different developmental stages. Each cell executes the same program in parallel on local neighborhoods in order to update their state. Since the program can change, genomes have the ability to duplicate or even to destroy themselves.

3 Experimental Setup

A genetic algorithm (GA) is designed to evolve solutions to 4 different problems:

✓ Development of given structure from a seed;
✓ Replication of given structures (minimum 3 replicas);
✓ Development of given structure from a seed followed by replication of developed structure (minimum 3 replicas);
✓ Re-evolution of solutions found for a given structure for the development and replication of a different structure.

The used morphologies are shown in Fig. 1, where different colors identify different cellular states. Structures of different sizes, complexities and number

Table 2. Instructions added to IBA instruction-set in order to allow Self Modification (SM).

Instruction	Parameters	Description
SKIP	$N[i_1] = Nskips$	Skip next $N[i_1]$ instructions. Not a SM instruction
MOVE	$N[i_3] = Start$ $N[i_4] = Insert$	Move instruction at line $N[i_3]$ just before $N[i_4]$
DUPE	$N[i_3] = Start$ $N[i_4] = Insert$	Copy instruction at line $N[i_3]$ just before $N[i_4]$
DEL	$N[i_3] = Start$	Delete instruction at line $N[i_3]$
CHF	$N[i_3] = Start$ $N[i_4] = Instr$	Change instruction at $N[i_3]$ to instruction at $N[i_4]$
CHP	$N[i_1] = Param$ $N[i_3] = Start$ $N[i_{2\|4}] = Value$	Change $N[i_1]$ parameter at $N[i_3]$ with value in $N[i_2]$ or $N[i_4]$ depending on $N[i_1]$

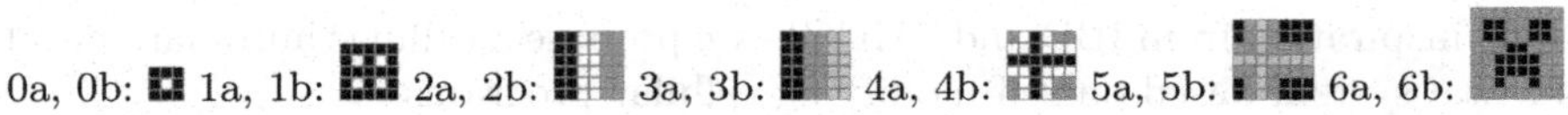

0a, 0b: 1a, 1b: 2a, 2b: 3a, 3b: 4a, 4b: 5a, 5b: 6a, 6b:

Fig. 1. Left to right; 0: simple square (2 states), 1: four squares (2 states), 2: three stripes flag (3 states), 3: generic four colors flag (4 states), 4: small Norwegian flag (3 states), 5: big Norwegian flag (4 states), 6: creeper (3 states). Note that **two versions** of each structure are used: (a) structure **with the necessary number states**, (b) structure **with an additional support state that is not required in the final structure**, but evolution is free to explore it. The different structures are referred in the text as structure 1a, structure 1b, etc. (Color figure online)

of states are used. The four problems are tested with two different CA genotype representations: traditional CA transition tables and CA using SMIBD. All the experiments are performed on 2-dimensional CA with von Neumann neighborhood (5 neighbors) with cyclic boundary conditions. The GA used a population of size 50, single point mutation with 2 % probability per genotype symbol, multi-point crossover with 10 % probability per genotype symbol, and fitness proportionate selection. Genotypes in the form of transition tables have a size of N^K, where N is the number of cell states and K is the neighborhood size (5 with von Neumann neighborhood). In case of SMIBD, the genotypes are composed by 10 instructions in the form *rule*, *op*1, *op*2, *op*3, *op*4, where *rule* identifies the rule number, *op*1 and *op*2 represent two neighbors, *op*3 and *op*4 are in the range *[0, number of instructions in program]*. CA development is executed for 40 steps. The fitness function for the development problem and for the replication problem is searching for matching structures, one in the case of development and three in the case of replication. For more information on similar fitness functions see [4,21]. For the developement and replication problem, the individual fitness functions are used, the one for development in the first 20 development steps and the one for replication in the 20 steps following the best developmental stage. Note that whether for the development problem the lattice size is the same as the wanted structure size, for the development and replication the lattice size is bigger as to guarantee enough space for the wanted replicas to emerge. As such, the initial development is considered to be a much harder problem because structures cannot rely on border conditions. For the re-evolution problem, the fitness function is modified by the different target structure. The same population is used as for the evolution of the first wanted structure.

4 Results and Analysis

4.1 Development

Results obtained for the morphogenesis of structures are summarized in Table 3. In particular, it is possible to notice that for structures 2a and 3a, the genomes that allow self-modification produce higher success rate in fewer generations on

Table 3. General results on the development problem using TT and SMIBD. Avg. over 100 runs.

	Transition table				Self modifying IBD			
Problem	2a	3a	4a	4b	2a	3a	4a	4b
Success rate %	86	34	0	0	100	96	5	5
Average (numGen)	13139	30676	x	x	1912	13309	57224	47042
StDev. (numGen)	18835	27660	x	x	9416	22687	42635	12950

average. For structures 4a and 4b, working solutions are found which were not achievable by transition table genomes. A general observation that emerged while inspecting the evolved solutions is that transition table representation tends to develop structures that quickly degenerate after few development steps and never recover. Self-modifying genomes often produce more stable solutions that retain the final structure, i.e., point attractors, or cycle a few steps before reaching the wanted structure again, i.e., short cyclic attractors. Another observation (see [6] for detailed results, not included here due to space limitations) is that a genotype representation that allows the program to modify itself, may allow a degree of control in developmental speed. As such, it may be able to evolve solutions that grow in different developmental times. This is of particular interest if one aims at developing given structures at specific points of the developmental time.

4.2 Replication

Table 4 presents the results for the replication problem on the tested structures. It is clearly visible that the proposed method outperforms the traditional CA transition table, both in terms of success rate and average number of generations needed to evolve a solution.

Table 4. Results on the replication problem using TT and SMIBD. Avg. over 100 runs.

	Transition table							Self modifying IBD						
Pattern	1a	2a	3a	4b	5a	6a	6b	1a	2a	3a	4b	5a	6a	6b
Success %	62	5	0	0	0	0	0	100	100	100	100	100	22	100
Avg. Gen	2116	4909	x	x	x	x	x	38	279	54	37	94	4737	54
StDev. Gen	2533	2009	x	x	x	x	x	24	344	53	25	72	2745	42

Continued Replication. It was observed that self-modifying genomes allowed solution to continue replication after the wanted number of replicas was achieved, whether transition tables often degenerated their behavior into a randomized pattern. As side experiment, solutions obtained with self-modification were re-developed in a bigger lattice of size 75×75 cells for a longer developmental time of 120 steps. In some cases, not only the replication process continued, but it produced a "massive" replication effect. Two examples are shown in Fig. 2, for structures 3a and 1a, respectively. Note: the original GA fitness required a perfect solution to produce at least 3 replicas (no additional award beyond 3).

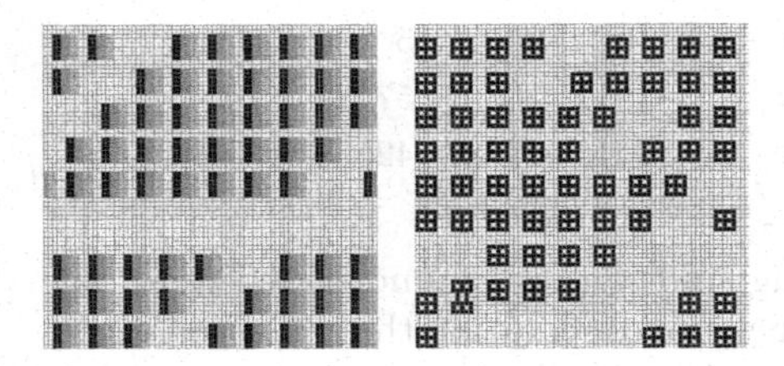

Fig. 2. Two examples of mass replications.

General Replicators. One of the aspects that we investigated was the "universality" of the evolved replicators, or in other words, whether the obtained solutions were able to generalize to other structures. As such, the evolved solutions obtained with self-modifying genomes were reused, i.e., once the solution was found for the replication of a structure, the structure in the lattice was replaced and the CA executed again. Quite surprisingly, in many cases the evolved programs could replicate any of the structures. Examples are shown in Figs. 3(a), (b), (c), and (d). In Fig. 3(a), a solution for structure 3a (4 states) is used on structure 2a (3 states). In this case, the supposedly unused state (yellow) is actually used as a support state. The two available support states are not used in Fig. 3(b), where structure 1a (2 states) is replicated with a solution for structure 3a (4 states). Figure 3(c) replicates structure 4a (4 states) using a solution for 3a (4 states), where the number of necessary states is the same. In Fig. 3(d), the spatial topology is preserved, whether the actual states (colors) are incorrect. Finally, Fig. 3(e) shows the replication of the Norwegian flag (4 states) using a solution for the French flag (4 states). Note that both solutions use a support state as in the final structures only 3 states are actually required. A transition table representation, on the other hand, would require an exponential scaling in the table size to encode for the neighborhood configurations resulting from the additional state. Also, solution evolved for a specific number of states are not practically usable with a different number of states, i.e., scale up or down. Self-modifying programs scale well in this regard.

4.3 Development and Replication

In this section, the described experiments target the development of a given structure first, followed by the replication of the grown structures. As such, the same genotype must encode both processes. This is a very interesting property that might be present in systems that target the development of self-replicating machines. In Table 5 numerical results are given for the tested problems, with a comparison of genotypes using transition tables and self-modification. For a simple structure as 0a (3×3 cells and 2 states), the results are fairly similar. More complex structures, as 2a and 3a did not produce any valid solution using transition tables. Self-modification allowed to produce some working solutions, able to both develop and replicate further. Solutions found by the two methods are inspected in Figs. 4 and 5. It is possible to notice that even if both examples

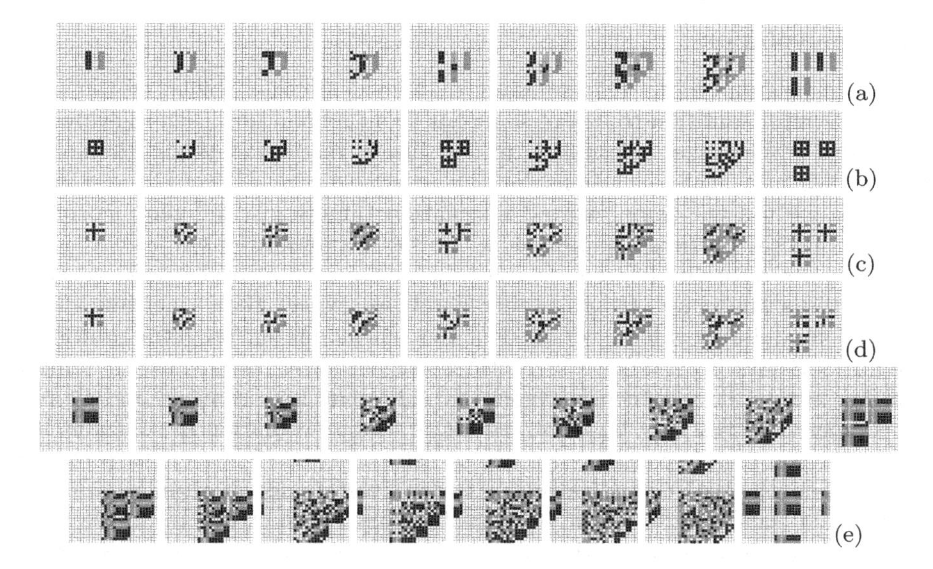

Fig. 3. Example of solution to the replication problem for (a) 3a used on 2a, (b) 3a used on 1a, (c) 3a used on 4a (4 state), (d) 3a used on 4a (3 state, it can no longer replicate the image, but the structure is perfectly replicated), (e) 3a used on the larger structure 5a (note that the lattice size had to be increased to accomodate the larger structure). (Color figure online)

Table 5. Results on development and replication problem using TT and SMIBD. Avg. over 100 runs.

	Transition table			Self modifying IBD		
Pattern	0a	2a	3a	0a	2a	3a
Success rate %	96	0	0	94	2	8
Avg. (NumGen)	877	x	x	1913	7652	5818
StDev. (NumGen)	1228	x	x	1944	2047	2973

produce valid solutions, there are clear differences. Using transition tables, the structure is replicated at step 3 and at step 7 four replicas of the given shape emerge. After, the genotype is not able to keep up the replication process and patterns soon disappear. The self-modifying genome example shows a different scenario (note that in Fig. 5, a bigger lattice is used to demonstrate the replication abilities, whether the original solution was evolved on a smaller lattice). At step 3 the wanted structure emerges and at step 7 five replicas have appeared. This process of replication never stops. In fact, at step 11, the 5 replicas are still present but with some additional space to allow new replicas to emerge, which finally appear at step 15. This process of making space and replicating continues indefinitely (at least for the observed developmental time).

4.4 Re-evolution

Taking inspiration from [12], we want to investigate the ability of the self-modifying representation to evolve solutions to a problem and then re-evolve to a different problem, i.e., adaptation to new fitness requirements/environmental change. We first evolve solutions for the morphogenesis of a given structure and then we use the same evolved population targeting a different morphology. This task is particularly difficult if transition tables are used, as evolution would require a totally different strategy, i.e., moving in a totally different area of the fitness landscape. On the other hand, we expect that a certain structure and modularity will be retained in the self-modifying program solutions. In addition, since both targeted structures are initialized from the same initial seed, i.e. zygote, developmental trajectories [20] may be visualized and shared developmental paths may be identified. Note that re-evolution of solutions for the replication task is considered a much easier task, as shown in the previous section where the evolved replicators presented a certain degree of generalization.

In Fig. 6 (Left) the structure 3a is evolved first, then structure 2b (2a with one additional available state) is used as a new target structure. A new solution is found in only 9 generations. Note that 2b is used as it has the same number of states but different arrangement of colors in the stripes pattern. The first five states in the trajectory are shared, then the paths split but retain the same "algorithmic" structure to reach the different solutions. In general, 100 evolutionary runs were performed. The GA using transition tables evolved the first solution 27 times. Out of those 27 times, 24 were re-evolved successfully to the second solution. On the other hand, with the usage of self-modification, 89 solutions were found to the first structure, which ended up in 84 successful re-evolution of the second structure. Note that with self-modifying genomes, loops (attractors) can be escaped as the regulation mechanism allows different parts of genomes to change or be active at different phenotypic stages.

In Fig. 6 (Right) the structure 4b (Norwegian flag with 4 states) was evolved first and then the structure 2b (French flag with 4 states) is targeted. The selected morphologies have different properties, e.g. horizontal symmetry vs. shifted symmetry. Using transition tables did not produce any result for the re-evolved structure. Self-modifying genomes allowed the 5 solutions found for the first structure to be successfully re-evolved to the second structure. In the shown example, some degree of trajectory modularity is present, indicating that the underlying "algorithmic" structure is retained.

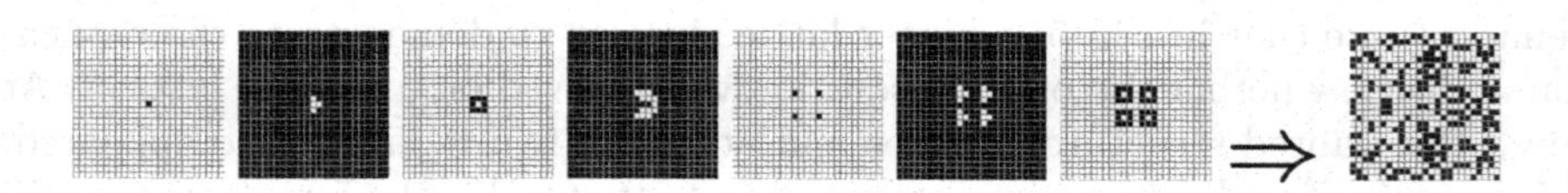

Fig. 4. TT solution to development and replication 0a. States 1 to 7 and 30.

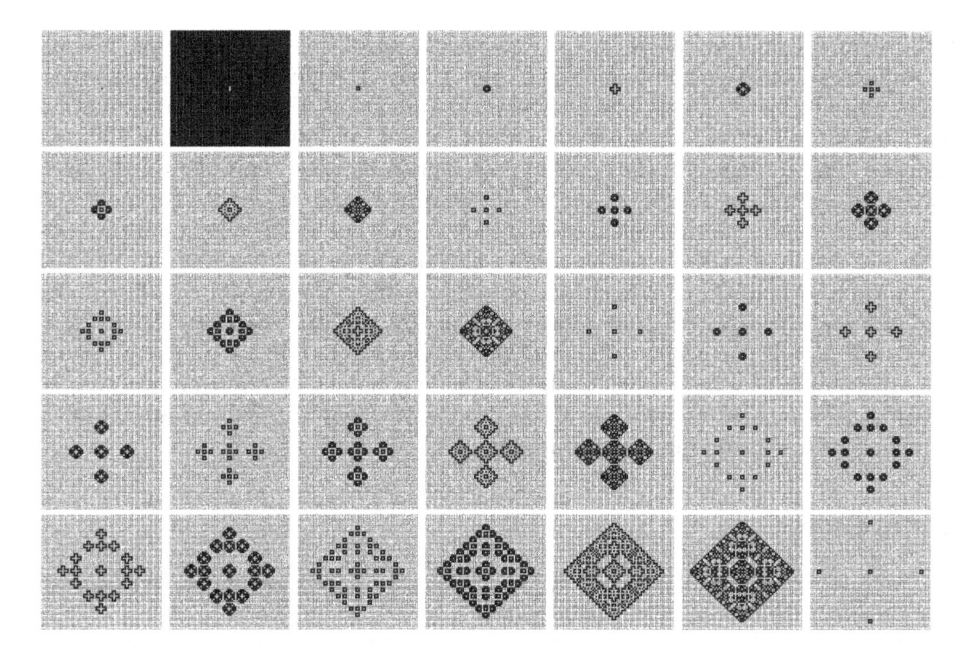

Fig. 5. SMIBD solution to the development and replication problem 0a, using a 75×75 lattice. At step nr. 31 a total of 61 replicas are present.

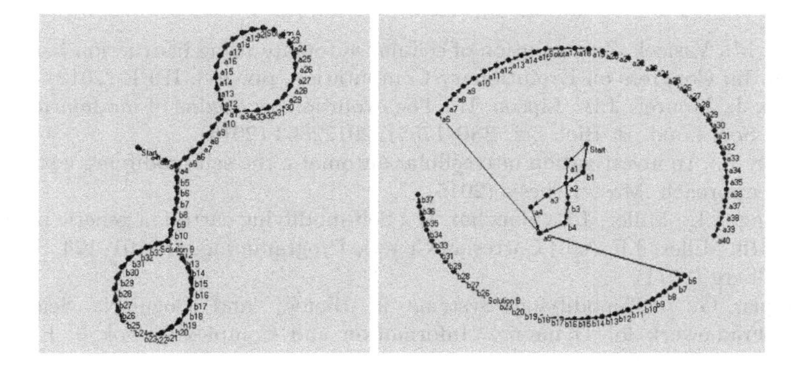

Fig. 6. Left: Developmental trajectories of a solution to problem 3a (black) and a re-evolved solution to 2b (blue), successfully found after 9 generations. The first 5 states are identical, then trajectories split but maintain similar topology. **Right**: Developmental trajectories of a solution to problem 4b (black) and a re-evolved solution to 2b (blue), with intertwined developmental trajectory. (Color figure online)

5 Conclusions and Future Work

In this paper we presented a novel method for regulation of gene expression for artificial evo-devo systems, namely Self-modifying Instruction Based Development.

Cellular automata have been used as experimental platform. Traditional CA transition tables have been compared to CA genotypes in the format of programs with self-modifying instructions, which allowed the emergence of a genome coordination mechanism. In fact, genomes have the ability to self-modify and activate different genes at different stages of the developmental process. Several problems have been solved successfully, as the development of structures, replication, development and replication combined, and reuse of an evolved genotype for development or replication of different structures than those initially targeted by evolution. SMIBD outperformed traditional transition tables, providing the possibility of evolving regulation mechanisms that are more modular. In the future we would like to analyze the scalability aspects of the proposed technique and, in particular, measure the evolvability of solutions with self-modifying genomes. In addition, we want to investigate reuse and re-evolution of solutions towards adaptivity to changing environments of lower and higher complexity.

References

1. Alon, U.: An Introduction to Systems Biology: Design Principles of Biological Circuits. CRC Press, Boca Raton (2006)
2. Altenberg, L., et al.: The evolution of evolvability in genetic programming. In: Advances in Genetic Programming, pp. 47–74 (1994)
3. Bidlo, M., Škarvada, J.: Instruction-based development: from evolution to generic structures of digital circuits. Int. J. Knowl.-Based Intell. Eng. Syst. **12**(3), 221–236 (2008)
4. Bidlo, M., Vasicek, Z.: Evolution of cellular automata using instruction-based approach. In: Congress on Evolutionary Computation, pp. 1–8. IEEE (2012)
5. Clune, J., Mouret, J.B., Lipson, H.: The evolutionary origins of modularity. Proc. Royal Soc. Lond. B: Biol. Sci. **280**(1755), 20122863 (2013)
6. Glover, T.: An investigation into cellular automata: the self-modifying instruction-based approach. Master thesis (2015)
7. Harding, S.L., Miller, J.F., Banzhaf, W.: Self-modifying cartesian genetic programming. In: Miller, J.F. (ed.) Cartesian Genetic Programming, pp. 101–124. Springer, Heidelberg (2011)
8. Kampis, G.: Self-modifying Systems in Biology and Cognitive Science: A New Framework for Dynamics, Information and Complexity, vol. 6. Elsevier, Amsterdam (2013)
9. Kashtan, N., Alon, U.: Spontaneous evolution of modularity and network motifs. Proc. Nati. Acad. Sci. U. S. A. **102**(39), 13773–13778 (2005)
10. Kitano, H.: Designing neural networks using genetic algorithms with graph generation system. Complex Syst. J. **4**, 461–476 (1990)
11. Kitano, H.: Building complex systems using developmental process: an engineering approach. In: Sipper, M., Mange, D., Pérez-Uribe, A. (eds.) ICES 1998. LNCS, vol. 1478, pp. 218–229. Springer, Heidelberg (1998)
12. Kovitz, B.: Experiments with cascading design. In: Proceedings of the 13th European Conference on Artificial Life (ECAL 2015), Workshop EvoEvo (2015)
13. Kowaliw, T., Banzhaf, W.: Augmenting artificial development with local fitness. In: Congress on Evolutionary Computation, pp. 316–323. IEEE (2009)

14. Koza, J.R.: Genetic Programming: On the Programming of Computers by Means of Natural Selection, vol. 1. MIT press, Cambridge (1992)
15. Kumar, S., Bentley, P.J.: Biologically inspired evolutionary development. In: Tyrrell, A.M., Haddow, P.C., Torresen, J. (eds.) ICES 2003. LNCS, vol. 2606, pp. 57–68. Springer, Heidelberg (2003)
16. Martin, A.P.: Increasing genomic complexity by gene duplication and the origin of vertebrates. Am. Nat. **154**(2), 111–128 (1999)
17. Miller, J.F., Thomson, P.: Cartesian genetic programming. In: Poli, R., Banzhaf, W., Langdon, W.B., Miller, J., Nordin, P., Fogarty, T.C. (eds.) EuroGP 2000. LNCS, vol. 1802, pp. 121–132. Springer, Heidelberg (2000)
18. Mitchell, M.: Complexity: A Guided Tour. Oxford University Press, Oxford (2009)
19. Nichele, S., Giskeødegård, A., Tufte, G.: Evolutionary growth of genome representations on artificial cellular organisms with indirect encodings. Artif. Life **22**(1), 76–111 (2016)
20. Nichele, S., Tufte, G.: Trajectories and attractors as specification for the evolution of behaviour in cellular automata. In: 2010 IEEE Congress on Evolutionary Computation (CEC), pp. 1–8. IEEE (2010)
21. Nichele, S., Tufte, G.: Evolutionary growth of genomes for the development and replication of multicellular organisms with indirect encoding. In: 2014 IEEE International Conference on Evolvable Systems (ICES), pp. 141–148. IEEE (2014)
22. Sipper, M.: Evolution of Parallel Cellular Machines. Springer, Heidelberg (1997)
23. Stanley, K.O., Miikkulainen, R.: Achieving high-level functionality through complexification. In: Proceedings of the AAAI-2003 Spring Symposium on Computational Synthesis, pp. 226–232 (2003)
24. Stanley, K.O., Miikkulainen, R.: Competitive coevolution through evolutionary complexification. J. Artif. Intell. Res. (JAIR) **21**, 63–100 (2004)
25. Taylor, J.S., Raes, J.: Duplication and divergence: the evolution of new genes and old ideas. Annu. Rev. Genet. **38**, 615–643 (2004)
26. Trefzer, M.A., Kuyucu, T., Miller, J.F., Tyrrell, A.M.: On the advantages of variable length grns for the evolution of multicellular developmental systems. IEEE Trans. Evol. Comput. **17**(1), 100–121 (2013)
27. Von Neumann, J., Burks, A.W., et al.: Theory of self-reproducing automata. IEEE Trans. Neural Netw. **5**(1), 3–14 (1966)

Evolution Under Strong Noise: A Self-Adaptive Evolution Strategy Can Reach the Lower Performance Bound - The pcCMSA-ES

Michael Hellwig(✉) and Hans-Georg Beyer

Vorarlberg University of Applied Sciences, Research Center PPE,
Hochschulstraße 1, 6850 Dornbirn, Austria
{michael.hellwig,hans-georg.beyer}@fhv.at
http://homepages.fhv.at/hemi
http://homepages.fhv.at/hgb

Abstract. According to a theorem by Astete-Morales, Cauwet, and Teytaud, "simple Evolution Strategies (ES)" that optimize quadratic functions disturbed by additive Gaussian noise of constant variance can only reach a simple regret log-log convergence slope $\geq -1/2$ (lower bound). In this paper a population size controlled ES is presented that is able to perform better than the $-1/2$ limit. It is shown experimentally that the pcCMSA-ES is able to reach a slope of -1 being the theoretical lower bound of *all* comparison-based direct search algorithms.

1 Introduction

In many real-world applications the problem complexity is increased by noise. Noise can stem from different sources such as randomized simulations or sensory disturbances. Evolutionary Algorithms (EAs) proved to be successful for optimization in the presence of noise [1,2]. However, the performance of the EAs degrades under strong noise and can even prevent the EA from converging to the optimizer.

Performance of EAs is usually measured by the amount of objective function evaluations n needed to reach a certain expected fitness compared to the non-noisy objective function value at the optimizer. This quantity is sometimes referred to as *simple regret* $\mathrm{SR}(n)$. It is defined in the case of minimization of the noisy function $\tilde{f}(\boldsymbol{y}), \boldsymbol{y} \in \mathbb{R}^N$ as

$$\mathrm{SR}(n) := \mathrm{E}[\tilde{f}(\boldsymbol{y}_n)] - f(\hat{\boldsymbol{y}}), \tag{1}$$

where the noisy fitness $\tilde{f}(\boldsymbol{y})$ is given by $\tilde{f}(\boldsymbol{y}) = f(\boldsymbol{y}) + \delta$ and $\boldsymbol{y}_n$ is the object vector recommended by the EA after n $\tilde{f}(\boldsymbol{y})$ evaluations. $f(\boldsymbol{y})$ is the deterministic objective function to be optimized which is disturbed by unbiased noise δ. The minimizer of $f(\boldsymbol{y})$ is denoted as $\hat{\boldsymbol{y}}$. The random variate δ describes the noise, which may or may not scale with the objective function value

$$\text{(a): } \delta \sim \sigma_{\epsilon_r} f(\boldsymbol{y}) \mathcal{N}(0,1) \qquad \text{and} \qquad \text{(b): } \delta \sim \sigma_\epsilon \mathcal{N}(0,1), \tag{2}$$

J. Handl et al. (Eds.): PPSN XIV 2016, LNCS 9921, pp. 26–36, 2016.
DOI: 10.1007/978-3-319-45823-6_3

and is assumed to be normally distributed with standard deviation $\sigma_{\epsilon_r}|f(\boldsymbol{y})|$ and σ_ϵ, respectively. The quantity σ_ϵ is referred to as noise strength.

There are different options to tackle the performance degradation of EAs that can be basically subdivided into two classes:

(i) reducing the noise observed by the EA by use of *resampling*, i.e. averaging over a number of κ objective function values (for a fixed $\boldsymbol{y}$), and
(ii) handling the noise by successively increasing the population size.

However, both methods implicate an increase of the required number n of fitness evaluations. In order to avoid a unnecessary excess of function evaluations, the question arises at which point to take the countermeasures (ii) or (i), i.e. to increase the population size or to use the $\tilde{f}$-averaging. As far as option (a) is concerned, there is a definite answer regarding the $(\mu/\mu_I, \lambda)$-Evolution Strategy (ES) on quadratic functions [3,4]: It is better to increase the population size than to perform resampling.

No matter whether one uses option (i) or (ii), in both cases techniques are required to detect the presence of noise. This can be easily done by resampling a candidate solution ($\kappa = 2$) because noise is reflected in changes of a candidate solution's measured fitness of two consecutive evaluations (for fixed $\boldsymbol{y}$). However, small noise strengths are usually well tolerated by the ES. That is, the ES can still approach the optimizer. In such cases there is no need to handle this noise. Another approach introduced in the UH-CMA-ES [5] considers the rank changes within the offspring individuals after resampling the population with $\kappa = 2$. If there are no or only a few rank changes, one can assume that the noise does not severely disturb the selection process. This approach is interesting, but seems still to be too pessimistic, i.e., even if there is a lot of rank changes, there may be still progress towards the optimizer due to the genetic repair effect taking place by the intermediate recombination operator. In [4] a population size control rule was proposed which is based on the residual error. The dynamics of the $(\mu/\mu_I, \lambda)$-ES in a noisy environment with constant noise strength σ_ϵ will usually approach a steady state in a certain distance to the optimizer. At that point, fluctuations of the parental fitness values around their mean value can be observed. The population size is then increased if the fitness dynamics on average does not exhibit further progress.

This paper presents a new detection method which is based on a linear regression analysis of the noisy fitness dynamics. Estimating the slope of the linear regression line, the direction of the trend can be determined. However, the estimated slope is a random variate. Therefore, a hypothesis test must be used to check the significance of the observed trend. If there is not a significant fitness decrease tendency, the population size will be increased. In the opposite case the population size can be decreased (up to a predefined limit). This approach is integrated into the covariance matrix self-adaptation evolution strategy (CMSA-ES) [6] yielding the *population controlled* (pc)CMSA-ES.

The applicability of the proposed algorithm is demonstrated on the noisy ellipsoid model. Investigating the SR(n) performance dynamics of the pcCMSA-ES in the *strong noise* scenario $\sigma_\epsilon =$ const. (i.e., the noise does *not* vanish at the optimizer), a remarkable observation can be made: $\mathrm{SR}(n) \approx c/n$. That is, the slope of the log-log plot reaches -1 approximately for sufficiently large number n of function evaluations. This is in contrast to a Theorem derived in [7]. There it is stated that *simple* ES can only reach a slope $\geq -1/2$, no matter whether one uses resampling or population upgrading. Remarkably, the -1 slope actually observed already represents the lower performance bound that cannot be beaten by any direct search algorithm as has been proven in [8].

The rest of this paper is organized as follows. The proposed noise detection technique by linear regression analysis is presented in Sect. 2. This technique is used to extend the CMSA-ES with a population control rule in Sect. 3 yielding the pcCMSA-ES. Empirical investigations are provided and discussed in Sect. 4. The paper closes with a summary and a outlook at future research questions.

2 Stagnation Detection by Use of Linear Regression Analysis

Stagnation or divergence behavior coincides with a non-negative trend within the observed fitness value dynamics of the ES (minimization considered). For trend analysis a regression model of the parental centroid fitness sequence of length L is used. If the slope of this model is *significantly* negative, the ES converges. In the opposite case, the population size must be increased. The decision will be based on statistical hypothesis testing.

Considering a not too long series of observed parental centroid fitness values, the observed time series can be approximated piecewise by a linear regression model. That is, a straight line is fitted through a set of L data points $\{(x_i, f_i), i = 1, \dots, L\}$ in such a manner that the sum of squared residuals of the model

$$f_i = ax_i + b + \epsilon_i \tag{3}$$

is minimal. Here ϵ_i models the random fluctuations. Determining the optimal a and b is a standard task yielding [9]

$$\hat{a} = \frac{\sum_{i=1}^{L}(x_i - \bar{x})(f_i - \bar{f})}{\sum_{i=1}^{L}(x_i - \bar{x})^2} \quad \text{and} \quad \hat{b} = \bar{f} - \hat{a}\bar{x}, \tag{4}$$

where $\bar{x}$ and $\bar{f}$ represent the sample means of the observations. Due to the ϵ_i random fluctuations the *estimate* $\hat{a}$ itself is a random variate. Therefore, the real (but unknown) a value can only be bracketed in a confidence interval. Assuming L sufficiently large, the central limit theorem guarantees that the estimator $\hat{a}$ of a is asymptotically normally distributed with mean a. Thus, the sum of squared residuals $\sum_{i=1}^{L}(f_i - b - ax_i)^2$ is distributed proportionally to χ^2_{L-2} with $L-2$ degrees of freedom and is independent of $\hat{a}$, cf. [9]. This allows to construct a test statistic

$$T_{L-2} = \frac{\hat{a} - a}{s_{\hat{a}}} \quad \text{with} \quad s_{\hat{a}} = \sqrt{\frac{\sum_{i=1}^{L}(f_i - b - ax_i)^2}{(L-2)\sum_{i=1}^{L}(x_i - \bar{x})^2}}, \tag{5}$$

where T_{L-2} is a t-distributed random variate with $L-2$ degrees of freedom [9].

Since $\hat{a}$ is a random variate, an observed $\hat{a} < 0$ does not guarantee convergence. Therefore, a hypothesis test will be used to put the decision on a statistical basis. Let $\mathrm{H}_0 : a \geq 0$ be the hypothesis that the ES increases the population size (because of non-convergence). We will only reject H_0 if there is significant evidence for the alternative $\mathrm{H}_1 : a < 0$. (In the latter case, the population size will not be increased.) That is, a left tailed test is to be performed with a significance level α (probability of wrongly rejecting H_0), i.e. $\Pr[\hat{a} < c|\mathrm{H}_0] = \alpha$, where c is the threshold (to be determined) below which the correct H_0 is rejected with error probability α. Resolving the left equation in (5) for $\hat{a}$ yields $\hat{a} = a + s_{\hat{a}} T_{L-2}$ and therefore $\Pr[a + s_{\hat{a}} T_{L-2} < c|\mathrm{H}_0] = \alpha$. This is equivalent to $\Pr[T_{L-2} < (c-a)/s_{\hat{a}}|\mathrm{H}_0] = \alpha$. Noting that $\Pr[T_{L-2} < (c-a)/s_{\hat{a}}] = F_{T_{L-2}}((c-a)/s_{\hat{a}})$ is the cdf of T_{L-2}, one can apply the quantile function yielding $(c-a)/s_{\hat{a}} = t_{\alpha;L-2}$, where $t_{\alpha;L-2}$ is the α quantile of the t-distribution with $L-2$ degrees of freedom. Solving for c one obtains $c = a + s_{\hat{a}} t_{\alpha;L-2}$. Thus, $c \geq s_{\hat{a}} t_{\alpha;L-2}$ and as threshold $(a = 0)$ one gets $c = s_{\hat{a}} t_{\alpha;L-2}$. That is, if

$$\hat{a} < s_{\hat{a}} t_{\alpha;\, L-2} \tag{6}$$

H_0 is rejected indicating a significant negative trend (i.e., convergence towards the optimizer, no population size increase needed).

3 The pcCMSA-ES Algorithm

Combining the convergence hypothesis test of Sect. 2 with the basic $(\mu/\mu_I, \lambda)$-CMSA-ES introduced in [6] an ES with adaptive population size control, the *population control* (pc)CMSA-ES, is presented in Algorithm 1. Until the algorithm has generated a list $\boldsymbol{\mathcal{F}}$ of L parental centroid function values an ordinary CMSA-ES run with truncation ratio ϑ is performed over L generations: In each generation the $(\mu/\mu_I, \lambda)$-CMSA-ES generates λ offspring with individual mutation strengths σ_l, see lines 4 to 10. The mutation strength σ_l can be interpreted as an individual scaling factor that is self-adaptively evolved using the learning parameter $\tau_\sigma = \frac{1}{\sqrt{2N}}$ (N – search space dimension). The mutation vector $\boldsymbol{z}_l$ of each offspring depends on the covariance matrix C which corresponds to the distribution of previously generated successful candidate solutions. The update rule can be found in line 30 where $\tau_c = 1 + \frac{N(N+1)}{2\mu}$ is used. After creation of the offspring, the objective function (fitness) values are calculated. Having completed the offspring population, the algorithm selects those μ of the λ offspring with the best (noisy) fitness values $\tilde{f}_{m;\lambda}$, $m = 1, \ldots, \lambda$. Notice, $m;\lambda$ denotes the mth best out of λ individuals. Accordingly, the notation $\langle . \rangle$ refers to the construction of the centroid of the respective values corresponding to the μ best offspring solutions. For example, the centroid of the mutation strengths is $\langle \sigma \rangle = \frac{1}{\mu} \sum_{m=1}^{\mu} \sigma_{m;\lambda}$. Subsequently, the pcCMSA-ES examines the list $\boldsymbol{\mathcal{F}}$ using the linear regression approach. The hypothesis test (6) is implemented within the program `detection`$(\boldsymbol{\mathcal{F}}_{int}, \alpha)$, line 19. Analyzing the fitness interval $\boldsymbol{\mathcal{F}}_{int}$, it returns the decision variable $td = 1$ if (6) is fulfilled, else $td = 0$. The parameter

Algorithm 1. pcCMSA-ES

1: Initialization: $g \leftarrow 0$; $wait \leftarrow 0$; $\langle\sigma\rangle \leftarrow \sigma^{(init)}$; $\langle\boldsymbol{y}\rangle \leftarrow \boldsymbol{y}^{(init)}$;
2: $\mu \leftarrow \mu^{(init)}$; $\mu_{\min} \leftarrow \mu^{(init)}$; $\boldsymbol{C} \leftarrow \boldsymbol{I}$; $adjC \leftarrow 1$
3: **repeat**
4: $\lambda \leftarrow \lfloor \mu/\vartheta \rfloor$
5: **for** $l \leftarrow 1$ to λ **do**
6: $\sigma_l \leftarrow \langle\sigma\rangle e^{\tau_\sigma \mathcal{N}(0,1)}$
7: $\boldsymbol{s}_l \leftarrow \sqrt{\boldsymbol{C}}\mathcal{N}(\boldsymbol{0}, \boldsymbol{I})$
8: $\boldsymbol{z}_l \leftarrow \sigma_l \boldsymbol{s}_l$
9: $\boldsymbol{y}_l \leftarrow \langle\boldsymbol{y}\rangle + \boldsymbol{z}_l$
10: $\tilde{f}_l \leftarrow \tilde{f}(\boldsymbol{y}_l)$
11: **end for**
12: $g \leftarrow g + 1$
13: $\langle\boldsymbol{z}\rangle \leftarrow \sum_{m=1}^{\mu} \boldsymbol{z}_{m;\lambda}$
14: $\langle\sigma\rangle \leftarrow \sum_{m=1}^{\mu} \sigma_{m;\lambda}$
15: $\langle\boldsymbol{y}\rangle \leftarrow \langle\boldsymbol{y}\rangle + \langle\boldsymbol{z}\rangle$
16: **add** $\tilde{f}(\langle\boldsymbol{y}\rangle)$ TO $\boldsymbol{\mathcal{F}}$
17: **if** $g > L \wedge wait = 0$ **then**
18: $\boldsymbol{\mathcal{F}}_{int} \leftarrow \boldsymbol{\mathcal{F}}(g - L : g)$
19: $td \leftarrow$ `detection`$(\boldsymbol{\mathcal{F}}_{int}, \alpha)$
20: **if** $td = 0$ **then**
21: $\mu \leftarrow \mu c_\mu$
22: $adjC \leftarrow 0$
23: **else**
24: $\mu \leftarrow \max(\mu_{\min}, \lfloor \mu/b_\mu \rfloor)$
25: **end if**
26: $wait \leftarrow L$
27: **else if** $wait > 0$ **then**
28: $wait \leftarrow wait - 1$
29: **end if**
30: $\boldsymbol{C} \leftarrow \left(1 - \frac{1}{\tau_c}\right)^{adjC} \boldsymbol{C} + \frac{adjC}{\tau_c}\langle \boldsymbol{s}\boldsymbol{s}^\top \rangle$
31: **until** termination condition
32: **return** $\langle\boldsymbol{y}\rangle$

α refers to the significance level of the hypothesis test. As long as a negative trend is detected the algorithm acts like the original CMSA-ES. Indication of a non-negative trend ($td = 0$) leads to an increase of the population size μ by multiplication with the factor $c_\mu > 1$, line 21, keeping the truncation ratio $\vartheta = \mu/\lambda$ constant by line 4. In order to prevent the next hypothesis test from being biased by old fitness values, the detection procedure is interrupted for L generations (line 26). Additionally the covariance matrix adaptation in line 30 is turned off, once the algorithm has encountered significant noise impact. For this purpose the parameter $adjC$ is set to zero in line 22. Stalling the covariance matrix update is necessary to avoid a random matrix process resulting in a rise of the condition number of $\boldsymbol{C}$ without gaining any useful information from the noisy environment.

In the case the hypothesis test returned $td = 1$, i.e. (6) is fulfilled, there is a significant convergence trend. In such a situation one can try to minimize the efforts and reduce the population size in line 24. Such a reduction can make sense in the distance dependent noise case (2a) where there is a minimal population size above which the ES converges without further population size increase. That is, the pcCMSA-ES increases first the population size aggressively and after reaching convergence, the population size is slowly decreased to its nearly optimal value. That is, the reduction factor b_μ should be related to that of c_μ, e.g. $b_\mu = \sqrt[k]{c_\mu}$ ($k = 2$, or 3), or can be chosen independently, but should fulfill $b_\mu < c_\mu$.

Regarding fitness environments where the ES has to deal temporarily with noisy regions, it might be beneficial to turn the covariance matrix adaptation on again once the ES has left the noisy region. That is, if a significant negative trend is present again the parameter $adjC$ should be reset to one in order to gain additional information about advantageous search directions. This can easily be obtained by inserting $adjC \leftarrow 1$ after line 24. However, as for the noisy fitness environments considered here, this adjustment is not able to provide significant improvements in terms of the ES's progress and therefore has not been implemented. It remains to be investigated in further studies.

4 Experimental Investigations and Discussion

The behavior of the proposed pcCMSA-ES algorithm is investigated on the ellipsoid model

$$f(\boldsymbol{y}) = \sum\nolimits_{i=1}^{N} q_i y_i^2 \tag{7}$$

with noise types (2b) and (2a). Especially, the cases $q_i = 1$ (sphere model) and $q_i = i, i^2$ have been considered. In the simulations the pcCMSA-ES is initialized with standard parameter settings and $\sigma^{(init)} = 1$ at $\boldsymbol{y}^{(init)} = \mathbf{1}$ in search space dimension $N = 30$. The initial population sizes are set to $\mu = 3$ and $\lambda = 9$ resulting in a truncation ratio $\vartheta = \frac{\mu}{\lambda} = \frac{1}{3}$ during the runs. The population size factors are $c_\mu = 2$ and $b_\mu = \sqrt{c_\mu}$. The significance level of the hypothesis test in line 19 of Algorithm 1 is $\alpha = 0.05$. The length L of the $\tilde{f}$-data collection phases must be chosen long enough to ensure a sufficient f improvement. As shown in [10], the effort to get an expected relative f improvement is proportional to the quotient of the trace of the Hessian of f and its minimal eigenvalue. Hence, for the sphere the effort is proportional to N and for the $q_i = i^2$ ellipsoid proportional to $\Sigma q := \sum_{i=1}^{N} q_i$. In the experiments $L = 5N$ and $L = \Sigma q$ are used.

Figure 1 shows the pcCMSA-ES dynamics for the (2b) case of constant $\sigma_\epsilon = 1$ noise. Considering the simple regret curves (blue), after a transient phase one observes that the ES on average continuously approaches the optimizer at a linear order in the log-log-plot. That means that $\mathrm{SR}(n) \propto n^a$ with $a < 0$. The parallelly descending dashed (magenta) lines $h(n) \propto n^{-1}$ indicate that the pcCMSA-ES actually realizes an $a \approx -1$. Fitting linear curves (solid magenta) to those $\mathrm{SR}(n)$

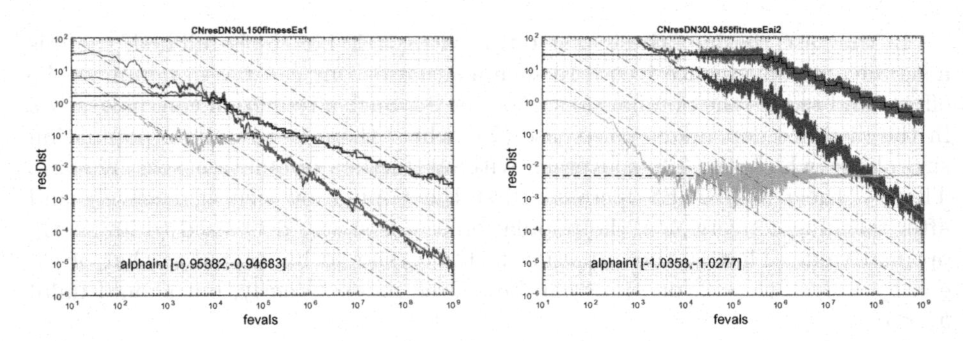

Fig. 1. The dynamical behavior of the pcCMSA-ES subject to additive fitness noise of strength $\sigma_\epsilon = 1$. Considering the sphere model as well as the ellipsoid model $q_i = i^2$ and search space dimensionality $N = 30$, four dynamics are plotted against the number of function evaluations n: the noise-free fitness of the parental centroid $\mathrm{SR}(n) = f(\langle \boldsymbol{y} \rangle)$ (blue), the corresponding weighted residual distance $R_q(n) = R_q(\langle \boldsymbol{y} \rangle)$ (red), and the mutation strength $\langle \sigma \rangle$ (green). The solid black step function predicts the residual steady state distance according to Eq. (8). In both cases, it is steadily reduced with each μ elevation. (Color figure online)

graphs, using the technique described by Eqs. (3)–(5), one can calculate the confidence intervals for a given confidence level, e.g. 95 %, which is displayed in Fig. 1. The observed $a \approx -1$ is remarkable since it apparently seems to violate a theorem by Astete-Morales, Cauwet, and Teytaud [7] that states that "Simple ES" can only reach an $a > -\frac{1}{2}$. The authors even supported their theorem with experiments regarding a tailored (1 + 1)-ES with resampling that came close to $-\frac{1}{2}$ and the UH-CMA-ES [5] that produced only a-values in the range of -0.1 to -0.3. Having a look at the assumptions made to prove the theorem, one finds the reason in the definition of "Simple ES". It contains a common assumption regarding the operation of ES – the scale invariance of the mutations. Roughly speaking, the expected value of the mutation strength should scale with the distance to the optimizer. That is, if one gets closer to the optimizer, the mutation strength should shrink. Looking at the (green) $\langle \sigma \rangle$ dynamics in Fig. 1 one sees that this assumption does not hold for the pcCMSA-ES. Remarkably, $\langle \sigma \rangle$ reaches a constant steady state value. Since theorems cannot be wrong, unlike the (1 + 1)-ES and the UH-CMA-ES, the pcCMSA-ES is *not* a "Simple ES".

While the pcCMSA-ES approaches a fixed mutation strength, on average it approaches the optimizer continuously as can be seen in Fig. 1 where the dynamics of the weighted residual distance R_q to the optimizer is displayed (red curves). This distance measure is defined as $R_q(\boldsymbol{y}) := \sqrt{\sum_{i=1}^N q_i^2 y_i^2}$. According to formula (22) in [3] the steady state expected value of $R_q(\boldsymbol{y})$ can be estimated for fixed population sizes

$$R_q^{\mathrm{ss}} = \sqrt{\frac{\sigma_\epsilon \Sigma q}{4\mu c_{\mu/\mu,\lambda}}}, \tag{8}$$

where $c_{\mu/\mu,\lambda}$ is the well-known progress coefficient [11]. This distance is reached by the CMSA-ES after a sufficiently long generation period (keeping μ and λ constant). Since the pcCMSA-ES changes the population size successively, the theoretical estimate (8) can be used to check whether the population size dynamics of the pcCMSA-ES works satisfactorily. The R_q dynamics follows closely the prediction of (8), which are displayed as (black) staircase curves.

As a second example, the case of distance dependent noise is considered in Fig. 2. The noise variance vanishes when approaching the optimizer. According to the progress rate theory for the noisy ellipsoid [12], one can derive an evolution condition

$$4\mu^2 c_{\mu/\mu,\lambda}^2 > \sigma^{*2} + \sigma_\epsilon^{*2} \tag{9}$$

that states that given upper values of normalized normalized noise and mutation strengths there is a parental population size μ ($\mu/\lambda = \text{const.}$) above which the ES converges to the optimizer. Here the normalized quantities are defined as $\sigma^* := \sigma \Sigma q / R_q$ and $\sigma_\epsilon^* := \sigma_\epsilon \Sigma q / (2R_q^2)$. Figure 2 shows the dynamics of the pcCMSA-ES on sphere and ellipsoid ($q_i = i^2$) model with normalized noise strengths $\sigma_\epsilon^* = 10$ and $\sigma_\epsilon^* = 4$, respectively. Taking a look at the solid blue lines representing the simple regret (being the noise-free fitness dynamics $f(\langle \boldsymbol{y} \rangle)$), one observes initially an *increase* of the parental simple regret. That is, the pcCMSA-ES departs from the optimizer. This is due to the choice of the initial population size of $\mu = 3$, $\lambda = 9$ being too small. However, after the first L generations, the first hypothesis test indicates divergence and the population size μ is increased by a factor $c_\mu = 2$. This increase repeats two or three times, as can be seen considering the (black) staircase curves displaying λ in Fig. 2, until a population size has been reached where the hypothesis test in line 19 of Algorithm 1 returns 1 indicating convergence, the SR-curves start to descend. This behavior is also reflected by the dynamics of the residual distance to the optimizer $R_q(\langle \boldsymbol{y} \rangle)$ (red). This attests that the pcCMSA-ES is able to adapt an appropriate population size needed to comply with Eq. (9) rather than simply increasing it arbitrarily. In contrast to the previous case of additive noise the mutation strength dynamics in Fig. 2 indicate a successive reduction of the noise strength σ. This is due to the decreasing influence of the distance dependent noise as the ES approaches the optimizer. In such cases the behavior of a "Simple ES" is desirable. The pcCMSA-ES behaves as such and demonstrates its ability to exhibit a linear convergence order similar to the non-noisy case. However, it has to be pointed out that the current population size reduction rule can result in interrupted convergence behavior in cases of very strong distance dependent noise. This can be inferred from the peaks in the right graph of Fig. 2. An attempt to address this disruption would be shortening both the test interval length L as well as the waiting time *wait* of the algorithm after each population size reduction and enlarging them again after a population size escalation, respectively. Also switching off the population size reduction might be a reasonable approach. Eventually, the population size control configuration under severe fitness proportional noise should be examined more closely in future investigations.

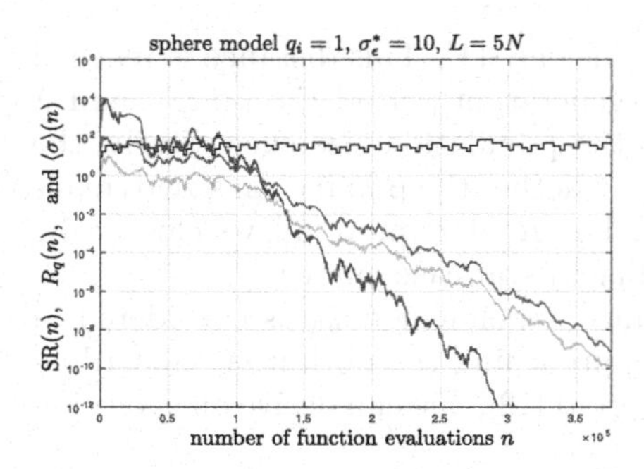

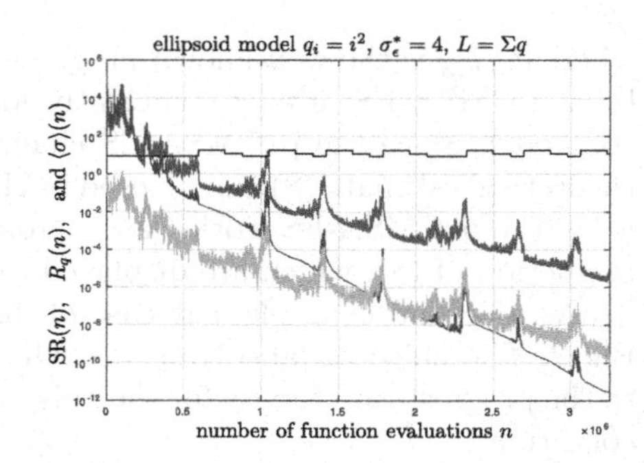

Fig. 2. The dynamical behavior of the pcCMSA-ES subject to distance dependent noise of normalized noise strength σ_ϵ^*. Considering the sphere model as well as the ellipsoid model $q_i = i^2$ with search space dimensionality $N = 30$, four dynamics are plotted against the number of function evaluations n: the simple regret of the parental centroid $\langle \boldsymbol{y} \rangle$ (blue), the corresponding residual distance $R_q(\langle \boldsymbol{y} \rangle)$ (red), and the mutation strength $\langle \sigma \rangle$ (green). The solid black staircase presents the offspring population size $\lambda = \lfloor \mu/\vartheta \rfloor$. According to Eq. (9), it will be increased up to a value where the strategy is able to establish continuous progress towards the optimizer. Afterwards the population size fluctuates around that specific value. (Color figure online)

5 Summary and Outlook

This paper presented an EA for the treatment of noisy optimization problems that is based on the CMSA-ES. Within its concept a mechanism for identification of noise-related stagnations or divergence behavior is integrated. Consequently, having identified noise related behavior the algorithm increases the size of the parental as well as the offspring population. This way it improves the likelihood to approach closer residual distances to the optimizer. Significant noise disturbances become noticeable by the absence of a clearly negative trend (minimization considered) within the noisy fitness dynamics. The slope of the respective trend can be deduced from the corresponding linear regression line. The estimated trend is used in a hypothesis test to decide whether there is convergence to the optimizer. If no further significant noise influences are discovered in subsequent tests the population size is again gradually reduced to avoid unnecessary function evaluations. This way the algorithm is capable to adapt the appropriate populations size. Accordingly, the adjusted CMSA-ES is denoted population control covariance matrix self-adaptation evolution strategy – pcCMSA-ES.

As a proof of concept, the pcCMSA-ES was tested on the noisy ellipsoid model considering two noise models, which obey different characteristics. The additive fitness noise case with constant noise strength σ_ϵ requires a permanent increase of the population size. On the other hand, the distance dependent noise case (which is equivalent to fitness proportionate noise in the case of the sphere model) requires only a limited population size increase. A well-crafted EA should be able to handle both cases (and of course, non-noisy optimization problems as well).

The empirical investigation of the strong noise case σ_ϵ = const. revealed a remarkable behavior of the pcCMSA-ES. The dynamics by which this ES approaches the optimizer seems to be already the fastest one can expect from a direct search algorithm on quadratic functions. The simple regret obeys an n^a dynamics with $a \approx -1$. This is remarkable since "Simple ES" should only allow for an $a \geq -1/2$ no matter how the noise is handled. The reason for this behavior is that unlike "Simple ES" the pcCMSA-ES does not scale the mutation strength σ in proportion to the distance to the optimizer in case of strong noise. This is different to other ESs such as $(1+1)$ or UH-CMA. However, if there is no strong noise, pcCMSA-ES behaves like a "Simple ES".

The pcCMSA-ES requires the fixing of additional exogenous strategy parameters. Particularly, the length L of the interval of observed fitness values that are considered in a single test decision has to be examined more closely. L should be large enough to ensure a sufficient evolution (convergence) of the fitness values. From the progress rate theory it is known that the number of generations needed for a certain fitness improvement scale with the quotient of the trace of the Hessian of f and its smallest eigenvalue. Therefore, L should be chosen proportional to N (search space dimensionality) in the sphere model case and to $\frac{N}{6}(N+1)(2N+1)$ in the case of the ellipsoid model $q_i = i^2$. However, in the black-box scenario the Hessian is not known. However, as long as the initial noise influence is small, the pcCMSA-ES transforms the optimization problem gradually into a local sphere model. In such cases, the $L \propto N$ choice should suffice. If, however, the noise is already strong in the initial phase, there is no definitive choice and the user has to make a guess regarding the trace vs. minimum eigenvalue ratio. Choosing L too large has a negative influence on the efficiency of the ES. It effects the lead time of the algorithm needed to establish an initial interval of fitness observations $\mathcal{F}_{int}$ as well as the waiting time *wait*. The parameter *wait* governs the length of the waiting period after a single population adjustment. After a transient phase of *wait* generations the algorithm starts again with the analysis of the fitness dynamics. It is not evident whether the parameter *wait* should depend on the length L of the fitness interval. The waiting time is essential to prevent wrong test decisions based on fitness dynamics resulting from different population specifications. A beneficial parameter setting has to be determined in future empirical investigations. There are also open questions regarding a profound choice of the population size change parameters c_μ and b_μ and the significance level $\alpha = 0.05$ used. These question should be also tackled by extended empirical investigations considering different test functions and noise scenarios.

Regarding theory, the analysis of certain aspects of the pcCMSA-ES seems to be possible using and extending the results presented in [12]. For example, the observed steady state σ in the strong noise case should be deducible from the self-adaptation response theory. Deriving the remarkable empirically observed $\mathrm{SR}(n) \propto n^{-1}$ law is clearly another task for future research.

Acknowledgements. This work was supported by the Austrian Science Fund FWF under grant P22649-N23 and by the Austrian funding program COMET (COMpetence centers for Excellent Technologies) in the K-Project Advanced Engineering Design Automation (AEDA).

References

1. Arnold, D.V.: Noisy Optimization with Evolution Strategies. Kluwer, Amsterdam (2002)
2. Jin, Y., Branke, J.: Evolutionary optimization in uncertain environments - a survey. IEEE Trans. Evolut. Comput. **9**(3), 303–317 (2005)
3. Beyer, H.-G., Arnold, D.V., Meyer-Nieberg, S.: A new approach for predicting the final outcome of evolution strategy optimization under noise. Genet. Progr. Evol. Mach. **6**(1), 7–24 (2005)
4. Beyer, H.-G., Sendhoff, B.: Evolution strategies for robust optimization. In: 2006 IEEE Congress on Evolutionary Computation CEC 2006, pp. 1346–1353 (2006)
5. Hansen, N., Niederberger, A.S.P., Guzzella, L., Koumoutsakos, P.: A method for handling uncertainty in evolutionary optimization with an application to feedback control of combustion. IEEE Trans. Evolut. Comput. **13**(1), 180–197 (2009)
6. Beyer, H.-G., Sendhoff, B.: Covariance matrix adaptation revisited – the CMSA evolution strategy. In: Rudolph, G., Jansen, T., Lucas, S., Poloni, C., Beume, N. (eds.) PPSN 2008. LNCS, vol. 5199, pp. 123–132. Springer, Heidelberg (2008)
7. Astete-Morales, S., Cauwet, M.-L., Teytaud, O.: Evolution strategies with additive noise: a convergence rate lower bound. In: Proceedings of the 2015 ACM Conference on Foundations of Genetic Algorithms XIII, FOGA 2015, pp. 76–84. ACM, New York (2015)
8. Shamir, O.: On the complexity of bandit and derivative-free stochastic convex optimization. In: COLT 2013 - The 26th Annual Conference on Learning Theory, NJ, USA, pp. 3–24 (2013)
9. Kenney, J.F.: Mathematics of Statistics: Parts 1–2. Literary Licensing, LLC, New York (2013)
10. Beyer, H.-G., Melkozerov, A.: The dynamics of self-adaptive multi-recombinant evolution strategies on the general ellipsoid model. IEEE Trans. Evolut. Comput. **18**(5), 764–778 (2014)
11. Beyer, H.-G.: The Theory of Evolution Strategies. Natural Computing Series. Springer, Heidelberg (2001)
12. Melkozerov, A., Beyer, H.-G.: Towards an analysis of self-adaptive evolution strategies on the noisy ellipsoid model: progress rate and self-adaptation response. In: GECCO 2015, Madrid, Spain, pp. 297–304. ACM (2015)

An Evolutionary Hyper-heuristic for the Software Project Scheduling Problem

Xiuli Wu[1(✉)], Pietro Consoli[2], Leandro Minku[3], Gabriela Ochoa[4], and Xin Yao[2(✉)]

[1] Department of Logistics Engineering, School of Mechanical Engineering, University of Science and Technology Beijing, Beijing, China
wuxiuli@ustb.edu.cn
[2] CERCIA, School of Computer Science, University of Birmingham, Birmingham, UK
{P.A.Consoli,X.Yao}@cs.bham.ac.uk
[3] Department of Computer Science, University of Leicester, Leicester, UK
leandro.minku@leicester.ac.uk
[4] Computing Science and Mathematics, University of Stirling, Stirling, UK
gabriela.ochoa@cs.stir.ac.uk

Abstract. Software project scheduling plays an important role in reducing the cost and duration of software projects. It is an NP-hard combinatorial optimization problem that has been addressed based on single and multi-objective algorithms. However, such algorithms have always used fixed genetic operators, and it is unclear which operators would be more appropriate across the search process. In this paper, we propose an evolutionary hyper-heuristic to solve the software project scheduling problem. Our novelties include the following: (1) this is the first work to adopt an evolutionary hyper-heuristic for the software project scheduling problem; (2) this is the first work for adaptive selection of both crossover and mutation operators; (3) we design different credit assignment methods for mutation and crossover; and (4) we use a sliding multi-armed bandit strategy to adaptively choose both crossover and mutation operators. The experimental results show that the proposed algorithm can solve the software project scheduling problem effectively.

Keywords: Software project scheduling · Hyper-heuristics · Adaptive operator selection · Sliding multi-armed bandit

1 Introduction

The Software Project Scheduling Problem (SPSP) relates to the decision of who does what task during a software project lifetime [1]. It plays an important role in reducing the duration and the cost of a software project [1, 15]. In China alone, it was reported that more than 40 % of unsuccessful software projects failed because of the inefficient planning of project tasks and human resources [8]. The SPSP, hence, is an important issue for IT companies.

However, the SPSP is particularly challenging when the project is large. The space of possible allocations of employees to tasks is enormous, and providing an optimal

J. Handl et al. (Eds.): PPSN XIV 2016, LNCS 9921, pp. 37–47, 2016.
DOI: 10.1007/978-3-319-45823-6_4

allocation of employees to tasks becomes a very difficult task [14]. It is impractical to use exact methods to solve medium or large SPSP instances. Evolutionary algorithms have been employed to solve the SPSP [1, 5, 12, 14, 16, 17]. Other metaheuristics have also been used, such as ant colony optimization and its variants [4, 6, 15]. A column generation approach was presented in [12], embedded within a branch-and-price procedure.

In those algorithms, different search operators (e.g., different types of crossover and mutation) may be good for different problem instances. However, little is known about which operators are most adequate for which types of instances. This motivates us to design an evolutionary algorithm capable of choosing the most effective operator automatically. Moreover, given a single problem instance, different search operators may be good at different stages of the search. As a result, it is very difficult to choose/design the operators to be used beforehand. Ideally, we would like an algorithm that can automatically choose which operators to use during the evolutionary process, and thus liberate practitioners from this difficult task [7]. This motivates our study of adaptive operator selection for the SPSP.

As a recent trend in optimization, hyper-heuristics search the space of heuristics rather than the space of solutions of the given problem, and use limited problem specific information to control the search process [9]. A hyper-heuristic is an automated methodology for selecting or generating heuristics to solve computational search problems [2]. We propose an evolutionary hyper-heuristic to solve the SPSP. Different from previous work on hyper-heuristics, our approach can be used to select both mutation and crossover operators, rather than being used to select only crossover or only mutation. We design different credit assignment methods for these two types of operators because mutation is typically used to exploit the solution space while crossover is typically used to explore it.

In summary, our novelty lies in the following: (1) this is the first work to adopt an evolutionary hyper-heuristic for SPSP; (2) this is the first work for adaptive selection of both crossover and mutation operators; (3) we design different credit assignment methods for the two types of operators: mutation and crossover; and (4) we use a sliding multi-armed bandit strategy to adaptively choose both crossover and mutation operators. We use a 3-sized crossover pool and a 3-sized mutation pool. Our experiments show that our approach is effective in selecting crossover and mutation operators for the SPSP.

The rest of this paper is organized as follows. Section 2 formulates the problem. Section 3 proposes an evolutionary hyper-heuristic for the SPSP. Section 4 reports the experimental results. Section 5 concludes the paper.

2 Formulation of SPSP

In this section, we explain the formulation of the SPSP [1, 11]. The notations adopted in the definitions are summarized in Table 1. A software project is composed of N tasks. A Task Precedence Graph (TPG) describes the precedence relations among tasks. It is used together with the decision variable and the task required efforts in order to determine the start and finishing time of each task (st_j and ed_j). This is done by creating a Gantt chart based on Algorithm 1 described in [11], which is omitted here due to space constraints. Each task t_j requires a set of skills req_j and has an estimated effort eff_j. There are M employees involved in the project. Each employee e_i can be described as a three-tuple array $(e_i^{skill}, e_i^{max}, e_i^{norm\_sal})$. A project requires a total of s skills. Each $skill^k$ (k = 1, 2, …, s) represents a kind of software development skill in the project, such as system analysis, designing, coding, algorithm, database, quality check, testing, etc.

Employees can work on several tasks simultaneously, as indicated by their dedication to certain tasks. The dedication $x_{ij} \in \{0/k, 1/k, \ldots, k/k\}$ of employee e_i to task t_j is the fraction of the employee's time devoted to that particular task. $k \in \mathbb{N}$ represents the granularity of the problem. A dedication of $x_{ij} = 1$ indicates that the employee e_i spends all his or her working time on task t_j. $x_{ij} = 0$ indicates that e_i does not spend any time on t_j.$0 < x_{ij} < 1$ indicates that e_i spends part of his or her working time on t_j. The matrix $\boldsymbol{X} = (x_{ij})$ of $M \times N$, where $x_{ij} \geq 0$, is the decision variable and represents a solution to the problem. This problem formulation [1, 11] assumes a static environment where employees will always be available during the lifetime of a project, i.e., they will not leave or be absent from work, and the task effort is fixed. As in [1, 11], we will also assume that $e_i^{max} = 1$ for all employees.

Table 1. SPSP notations

	Description
M	The number of the employees involved in the project
e_i	The i-th employee
e_i^{skill}	$e_i^{skill} = \{pro_i^1, pro_i^2, \ldots, pro_i^s\}$, pro_i^k (k = 1, 2, …, s) is a binary variable indicating whether the employee e_i possesses the skill $skill^k$
e_i^{max}	The max dedication of e_i to the project indicating the percentage of a full time employee e_i is able to dedicate to the project.
$e_i^{norm\_sal}$	The monthly salary for an employee e_i for his or her full normal working time
N	The software project is composed of N tasks
t_j	The j-th task
eff_j	The estimated effort for the task t_j
req_j	The required skills for the task t_j
TPG	The task precedence graph is an acyclic directed graph with tasks as nodes and task precedence as edges
x_{ij}	The decision variable to determine the degree of dedication of employee e_i to task t_j.
st_j	The starting time of task t_j
ed_j	The finishing time of task t_j

The SPSP is the problem of assigning employees to tasks in a software project so as to minimize the completing time (i.e., duration of the project as defined by Eq. (2)), and the cost (i.e., the total amount of salaries paid as defined by Eq. (3)). Equation (1) is the mixed objective, where w_1 and w_2 are the weights for the completing time and the cost, respectively. The assignment of employees to tasks is to determine the decision variable x.

The problem is subject to the aforementioned assumptions and the following two constraints: employees can only work on a task t_j if all employees working together have all the skills to perform the task (Eq. (4)); and employees should not exceed their maximum dedication to the tasks that are active at any given time moment t (Eq. (5)).

$$\text{Minimize}_x f(x) = w_1 f_1(x) + w_2 f_2(x) \tag{1}$$

$$f_1(x) = \max_j (ed_j), \tag{2}$$

where ed_j, $\forall j$, is obtained with Algorithm 1 from [11].

$$f_2(x) = \sum\nolimits_{i=1}^{M} \sum\nolimits_{j=1}^{N} \left(eff_j \middle/ \sum\nolimits_{k=1}^{M} x_{kj} \right) x_{ij}\, e_i^{norm\_sal}, \tag{3}$$

s.t.

$$\text{req}_j \subseteq \bigcup\nolimits_{i=1}^{n} \{skill_i | x_{ij} > 0\} \tag{4}$$

$$\sum\nolimits_{j \in \text{active\_tasks}(\tau)} x_{ij} \le e_i^{max}, \ \forall i, \tau, \tag{5}$$

where active_tasks(τ) are all tasks active at time τ according to the Gantt chart generated using Algorithm 1 from [11]

$$x_{ij} \in [0, 1] \tag{6}$$

It is worth noting that the SPSP is related to the Resource-Constrained Project Scheduling problem (RCPS), but there are some key differences [1]: (1) the SPSP has a cost associated to each employee; (2) SPSP has only one type of resource (employee); and (3) each activity in RCPS requires different quantities of different resources, whereas the SPSP requires different skills, which are not quantifiable entities.

3 The Evolutionary Hyper-heuristic

3.1 Hyper-heuristic Framework

The evolutionary hyper-heuristic (Fig. 1) chooses an operator to apply at each search stage. The high level algorithm is based on a (μ + λ)-EA, i.e. it maintains a population of μ candidate solutions and λ parents are selected at each generation. Before each evolutionary cycle, the adaptive operator selection function is called twice (lines 3 and 4) to choose the crossover and the mutation operator, respectively. At the end of each iteration, the two credit assignment functions: diversity-credit and improvement-credit, are called (lines 12 and 13) to assign a credit to the currently chosen crossover and mutation operator, respectively.

```
1: initialize a population pop with μ candidate solutions
2: repeat
3:    crossover= Operators-selection(crosscredit)
4:    mutation= Operators-selection(mutationcredit)
5:    for i=1:2:λ
6:        select 2 parents x(1) and x(2) from pop at random
7:        apply crossover to x(1) and x(2) to generate x'(1) and x'(2) with
probability Pc
8:        apply mutation to x'(1) and x'(2) with probability Pm
9:        pop=popU(x'(1),x'(2))
10:    end for
11:    Select the best μ solutions from pop to survive to the next gener-
ation
12:    Update crosscredit= diversity-credit(pop)
13:    Update mutationcredit= improvement-credit(pop)
14: until termination criteria are met
15: output the best candidate solution in pop.
```

Fig. 1. The evolutionary hyper-heuristic for SPSP

3.2 Adaptive Operator Selection

Adaptive operator selection (AOS) performs on-line selection of evolutionary operators to produce each new offspring, based on the recent known performance of each of the available operators. An adaptive operator selection is typically composed of a credit assignment and an operator selection rule. The former assigns a reward to an operator and the latter determines the operator to be chosen at each step. In the AOS framework, the performance of an operator in a very early stage may be irrelevant to its current performance [10]. More attention should be paid on the recent performance. We propose a sliding multi-armed bandit (SMAB) following the approach in [10]. The credit assignment and the operator selection rules adopted are as follows.

Credit Assignment of SMAB. To determine the credit assignment, one needs to make a decision on how to measure the impact in the search process caused by the application of an operator. We propose two credit assignment methods according to the main role each operator plays during the search process.

Considering that the main role of crossover is to explore the solution space, we employ the population diversity to evaluate the performance of one on-duty crossover operator. The diversity of the population is measured by the "population diversity", inspired by the entropy concept. It is calculated in Eq. (7) by computing the standard deviation of the same amount of dedication among solutions in the population.

$$ent = \sum\nolimits_{i=1}^{M} \sum\nolimits_{j=1}^{N} \sqrt{\frac{1}{\mu} \sum\nolimits_{k=1}^{\mu} (x_{ij}^{(k)} - \frac{1}{\mu} \sum\nolimits_{k=1}^{\mu} x_{ij}^{(k)})^2} \tag{7}$$

Considering that one mutation operator plays the role to guide a local search, we use the fitness improvements caused by the recent application of the operator under assessment. The fitness improvement is defined in Eq. (8).

$$r = (f_{old} - f_{new})/f_{old} \tag{8}$$

Where f_{old} and f_{new} is the fitness of the individual before and after the application of the operator respectively.

A sliding window with a fixed size W is used to store the fitness improvement values of the recently used operators. The sliding window is a two-dimensional list of 2 rows and W columns. The first row records the operator index number and the second row records the corresponding fitness improvement. It is organized as a first-in-first-out (FIFO) queue.

Selection Rule of SMAB. Based on the received credit values, the operator selection scheme selects one operator for generating new solutions. This paper uses a bandit-based operator selection scheme. Our scheme is similar to that in [10]. The major difference is that we use the entropy *ent* for the crossover operator and the fitness improvement value r as the quality $\hat{q}_{i,t-1}$ instead of the average of all the rewards received so far for an operator. The operator that maximizes Eq. (9) will be chosen as the on-duty operator.

$$argmax_{i=1,\ldots k}\left(\hat{q}_{i,t-1} + C\sqrt{\frac{2\log\sum_{k} n_{k,t}}{n_{i,t}}}\right) \tag{9}$$

where $\hat{q}_{i,t-1}$ is the empirical reward (the best result achieved by the operator in last W iterations) of the i-th arm (operator), C is a scale parameter, $n_{i,t}$ is the times that the i-th arm has been tried till the t-th iteration during the recent W applications.

3.3 The Low-Level Heuristics Pool

Crossover Operator Pool. There are 3 operators in the crossover operator pool.

Crossover 1: Swap-Row Crossover. For each employee, select its corresponding dedications to tasks from one randomly chosen parent to generate an offspring. This can be seen as changing some employees' dedication to tasks [11].

Crossover 2: Swap-Column Crossover. For each task, select its corresponding employees' dedications from one randomly chosen parent to compose an offspring. This can be seen as exchanging some tasks' resource assignment [11].

Crossover 3: Swap-Block Crossover. The Swap-Block Crossover [1] is a 2-D single point crossover applied to matrices. It randomly selects a row and a column (the same in the two parents) and then swaps the elements in the upper left quadrant and the lower right quadrant in both solutions.

Mutation Operator Pool. There are 3 operators in the mutation operator pool.

Mutation 1: Mutate-Position [11]. An individual is mutated by changing each entry x_{ij} of the dedication matrix to a random times of 1/7 with mutation probability, independently from other entries.

Mutation 2: Mutate-Row. An individual is mutated by swapping randomly dedications in each row. The selected positions to swap must be the nonzero dedication to keep the balance.

Mutation 3: Mutate-Column. An individual is mutated by swapping randomly dedications in each column.

4 Experimental Study

We run a number of experiments to determine whether the use of the proposed AOS technique is beneficial with respect to the adoption of a simple random selection of the crossover and mutation operators. We also compare the results of these algorithms with those achieved by a state-of-the-art approach [11]. We label the three algorithms for convenience GA-SMAB, GA-randAOS and GA.

We performed our experiments on a benchmark dataset of 48 instances used in [1]. For each instance, we provide the average results from 30 independent runs of the algorithms. In order to reduce the impact of different parameter settings on the results, we adopt the same parameter settings as those used in literature [1, 11] (Table 2) except for the mutation probability, which was set to a value that guaranteed the application of the mutation operator, necessary to observe its performance.

Table 2. Parameters setting

Parameter	Value	Description
μ	64	The size of the population
λ	64	The size of the offspring
Pc	0.75	The crossover probability
Pm	0.1	The mutation probability
maxg	200	The number of generations
w_1	10^{-1}	The weight of the duration
w_2	10^{-6}	The weight of the cost
winsize	7	The sliding window size
ScalingC	60	The scaling factor for crossover operators
ScalingM	110	The scaling factor for mutation operators

Table 3 reports the average results achieved by the GA-SMAB, GA-randAOS and GA algorithms on the 48 instances of the benchmark set. For every algorithm we report the average fitness, its standard deviation, the best result, the average cost and the average completion time. The results indicate how the fitness achieved by GA-SMAB is the lowest of the three. In particular, the average fitness of GA-SMAB is slightly lower than the best of GA. Similarly, GA-randAOS also shows a comparable improvement with respect to the GA. It is also worth noting that the cost has increased, while the average completing time has improved considerably. This is likely to be a result of the weights w_1 and w_2 used in the fitness function.

Table 4 summarizes the results of the Wilcoxon Rank-Sum (significance level of 0.05) test performed for each instance to determine the number of instances for which each algorithm yields statistically better (column W) results, comparable results (column T) and statistically worse results (column L). We also provide the p-values relative to the Wilcoxon Signed-Rank (significance = 0.05) test performed on the average fitness achieved by the three algorithms across the 48 instances of the benchmark set where instances with comparable results (according to the Wilcoxon Rank-Sum test) are treated as ties.

Table 3. Average results achieved by GA-SMAB, GA-randAOS and GA algorithms

	GA-SMAB	GA-RANDAOS	GA
Average fitness	4.5351	4.5552	4.6369
Standard dev of fitness	0.0856	0.1009	0.0509
Best fitness	4.3617	4.3709	4.5534
Average cost	1,830,377.1444	1,830,223.0918	1,830,037.0659
Average completing time	27.0460	27.2475	28.0686

Table 4. Win/tie/losses obtained based on Wilcoxon Rank-Sum tests for each instance, p-values of the Wilcoxon Signed-Rank test across instances and average Cohen's d effect size

	GA-SMAB			GA-randAOS			GA		
	w	t	l	w	t	l	w	t	l
GA-SMAB				25	23	0	44	4	0
				1.23E-05			7.62E-09		
				0.6368			1.5688		
GA-randAOS	0	23	25				37	11	0
	1.23E-05						1.14E-07		
	0.6368						0.9814		
GA	0	4	44	0	11	37			
	7.62E-09			1.14E-07					
	1.5688			0.9814					

The three algorithms produce statistically significantly different results across problem instances, as shown by the Wilcoxon Sign-Rank tests. The average Cohen's d effect sizes vary from 0.6368 (medium) to 1.5688 (large). Together with the win-tie-losses, this shows that it is beneficial to adopt GA-SMAB instead of GA-randAOS or GA. GA-SMAB achieved similar or better fitness than GA-randAOS and GA on all problem instances. In particular, GA-SMAB achieved statistically better fitness on 25 instances when compared to GA-randAOS, and similar fitness on 23 instances. This suggests that there are instances which do not require an AOS strategy, where GA-SMAB performs similarly to GA-randAOS. The improvement obtained through the use of GA-SMAB is also confirmed by the Cohen's d Effect size included in Table 4. However, an AOS strategy is needed for other problem instances.

When compared to the results achieved by GA, GA-SMAB outperforms its results on 44 instances. This difference can be explained by the interaction between the random parent selection and the SMAB strategy, as the increased exploration ability of the algorithm created more favorable conditions for the GA-SMAB.

In order to show the behavior of the algorithm, we provide the selection rates of the crossover and mutation operators for the instance *inst-employees20* respectively in Fig. 2. It is possible to notice trends in the search as one operator is preferred to the others during different periods of the search. This is particularly clear in the selection rates of the mutation operator, where operator Mutate-Column has a higher selection rate for most of the search, with the exception of some periods where the other two operators are preferred. In the plot relative to the crossover operator, on the other hand, it is possible to notice shorter trends over the course of the search, although Operator Swap-Block seems to be the one selected most of the times. This might be explained by the fact that the algorithm favors a frequent alternation of the crossover operators, as the repeated use of a single operator might cause a decrease of the population diversity.

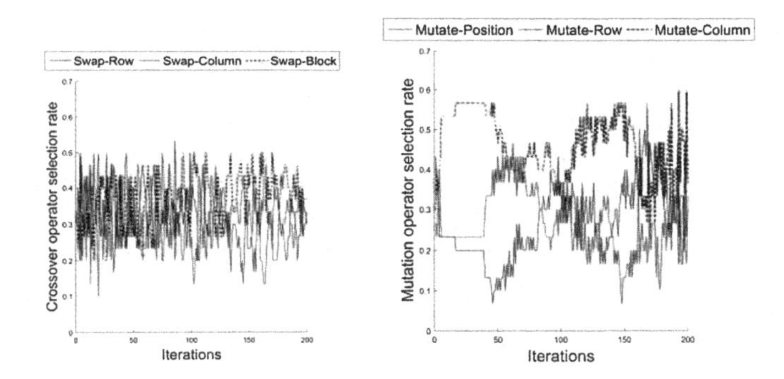

Fig. 2. Selection rates of the crossover (left) and selection rates of the mutation (right)

5 Conclusions

This paper proposes an evolutionary hyper-heuristic to address the SPSP. The hyper-heuristic uses an EA as a high level strategy and adapt automatically both mutation and crossover operators during evolution. A sliding window MAB strategy is used to adaptively select both operators during the search. The experiments performed on a set of 48 benchmark instances showed that the proposed algorithm can solve the SPSP effectively and outperform a strategy based on a simple random selection of the operators as well as a state-of-the-art approach from the literature. Future work includes a detailed analysis of the behavior of the proposed algorithm and the reasons for its ability to generate better solutions; an extension of the proposed algorithm in order to deal with the dynamic SPSP [13]; the use of alternative AOS strategies; and the inclusion of more aspects that could affect software projects into the problem formulation.

Acknowledgements. This paper was partly supported by the National Natural Science Foundation of China under Grant (Grants. 51305024 and 61329302) and EPSRC (Grant No. EP/J017515/1). Xin Yao was supported by a Royal Society Wolfson Research Merit Award.

References

1. Alba, E., Francisco, C.: Software project management with GAs. Inf. Sci. **177**, 2380–2401 (2007)
2. Burke, E., Hyde, M., Kendall, G., Ochoa, G., Ozcan, E., Woodward, J.: A classification of hyper-heuristic approaches. In: Gendreau, M., Potvin, J.-Y. (eds.) Handbook of Metaheuristics. International Series in Operations Research & Management, vol. 146, pp. 449–468. Springer, Berlin (2010)
3. Blazewicz, J., Lenstra, J., Rinnooy, K.: Scheduling subject to resource constraints: classification and complexity. Discret Appl. Math. **5**, 11–24 (1983)
4. Crawford, B., Soto, R., Johnson, F., Monfroy, E., Paredes, F.: A max-min ant system algorithm to solve the software project scheduling problem. Expert Syst. Appl. **41**, 6634–6645 (2014)
5. Chang, C., Jiang, H., Di, Y., Zhu, D., Ge, Y.: Time-line based model for software project scheduling with genetic algorithms. Inf. Softw. Tech. **50**, 1142–1154 (2008)
6. Chen, W., Zhang, J.: Ant colony optimization for software project scheduling and staffing with an event-based scheduler. IEEE Trans. Softw. Eng. **39**(1), 1–17 (2013)
7. Consoli, P.A., Minku, L.L., Yao, X.: Dynamic selection of evolutionary algorithm operators based on online learning and fitness landscape metrics. In: Dick, G., et al. (eds.) SEAL 2014. LNCS, vol. 8886, pp. 359–370. Springer, Heidelberg (2014)
8. Ding, R., Jing, X.: Five principles of project management in software companies. In: Project Management Technology, vol. 1 (2003). (in Chinese)
9. Jorge, A., Alcaraz, S., Ochoa, G., Swan, J., Carpio, M., Puga, H., Burke, E.: Effective learning hyper-heuristics for the course timetabling problem. Eur. J. Oper. Res. **238**, 77–86 (2014)
10. Li, K., Fialho, A., Kwong, S., Zhang, Q.: Adaptive operator selection with bandits for a multiobjective evolutionary algorithm based on decomposition. IEEE Trans. Evol. Comput. **18**(1), 114–130 (2014)
11. Minku, L., Sudholt, D., Yao, X.: Improved evolutionary algorithm design for the project scheduling problem based on runtime analysis. IEEE Trans. Softw. Eng. **40**(1), 83–102 (2014)
12. Montoya, C., Bellenguez-Morineau, O., Pinson, E., Rivreau, D.: Branch-and-price approach for the multi-skill project scheduling problem. Optim. Lett. **8**, 1721–1734 (2014)
13. Shen, X., Minku, L., Bahsoon, R., Yao, X.: Dynamic software project scheduling through a proactive-rescheduling method. IEEE Trans. Softw. Eng. 24 December 2015. doi:10.1109/TSE.2015.2512266
14. Penta, M., Harman, M., Antoniol, G.: The use of search-based optimization techniques to schedule and staff software projects: an approach and an empirical study. Softw. Pract. Exper. **41**, 495–519 (2011)
15. Xiao, J., Ao, X., Tang, Y.: Solving software project scheduling problems with ant colony optimization. Comput. Oper. Res. **40**, 33–46 (2013)

16. Xiao, J., Osterweil, L.J., Wang, Q., Li, M.: Dynamic resource scheduling in disruption-prone software development environments. In: Rosenblum, D.S., Taentzer, G. (eds.) FASE 2010. LNCS, vol. 6013, pp. 107–122. Springer, Heidelberg (2010)
17. Yannibelli, V., Amandi, A.: A knowledge-based evolutionary assistant to software development project scheduling. Expert Sys. Appl. **38**, 8403–8413 (2011)

The Multiple Insertion Pyramid: A Fast Parameter-Less Population Scheme

Willem den Besten[1], Dirk Thierens[1(✉)], and Peter A.N. Bosman[2]

[1] Institute of Information and Computing Sciences, Universiteit Utrecht, Utrecht, The Netherlands
d.thierens@uu.ne
[2] Centre for Mathematics and Computer Science, P.O. Box 94079, 1090 GB Amsterdam, The Netherlands

Abstract. The Parameter-less Population Pyramid (P3) uses a novel population scheme, called the population pyramid. This population scheme does not require a fixed population size, instead it keeps adding new solutions to an ever growing set of layered populations. P3 is very efficient in terms of number of fitness function evaluations but its run-time is significantly higher than that of the Gene-pool Optimal Mixing Evolutionary Algorithm (GOMEA) which uses the same method of exploration. This higher run-time is caused by the need to rebuild the linkage tree every time a single new solution is added to the population pyramid. We propose a new population scheme, called the multiple insertion pyramid that results in a faster variant of P3 by inserting multiple solutions at the same time and operating on populations instead of on single solutions.

1 Introduction

The recently introduced Parameter-less Population Pyramid evolutionary algorithm uses a novel population scheme, called the population pyramid [2,3]. A big advantage of the population pyramid is that there is no need for the user to set a fixed population size. Instead, P3 keeps adding new solutions to an ever growing set of layered populations. To explore new solutions, P3 uses the model-building technique and the Gene-pool Optimal Mixing operator from GOMEA [5]. The way it exploits these solutions however, is very different from the classical fixed size population method. P3 is very efficient in terms of number of fitness function evaluations but its run-time is significantly higher than that of GOMEA, this is because P3 rebuilds the linkage tree every time a single new solution is added to the population pyramid, while GOMEA only rebuilds the linkage tree every single generation. A typical GOMEA run only needs about 10 generations, so the difference between the number of linkage trees generated by P3 and GOMEA is huge. We propose a new population scheme, called the multiple insertion pyramid that results in a faster variant of P3 by significantly reducing the number of linkage trees built. Multiple insertion is a population scheme which basically combines P3 with ideas from the Exponential Population Scheme

J. Handl et al. (Eds.): PPSN XIV 2016, LNCS 9921, pp. 48–58, 2016.
DOI: 10.1007/978-3-319-45823-6_5

(EPS) by inserting growing populations instead of single solutions. The multiple insertion pyramid scheme drastically reduces the run-time overhead caused by maintaining a population pyramid. It also requires a lesser number of fitness evaluations than required by EPS [4]. Finally, the multiple insertion pyramid scheme yields a robust GOMEA that performs better in the number of fitness evaluations and on par with the run-time of the original GOMEA on the majority of the test problems. In the next Section, we discuss background information about the Gene-pool Optimal Mixing Evolutionary Algorithm, the Exponential Population Scheme, and the Parameter-less Population Pyramid. Section 3 introduces the Multiple Insertion Pyramid. Section 4 discusses experimental results, followed by our Conclusion.

2 Background

2.1 Gene-Pool Optimal Mixing Evolutionary Algorithm

The Gene-pool Optimal Mixing Evolutionary Algorithm (GOMEA) is designed for detecting and optimally mixing partial solutions into new solutions [5]. It uses a model, called the family of subsets (FOS), for representing problem variables whose values should be copied together during crossover. Each subset in the FOS is a set of problem variable indices. Each generation GOMEA learns a FOS from the population. Then each solution is subjected to the Gene-pool Optimal Mixing (GOM) crossover operator. First the solution is cloned into an offspring population and for each subset in the FOS, GOM will randomly select a donor from the population and copy the bits indicated by the subset into the current solution. If the donor and the parent solution have the same bits for the problem variables, then the offspring solution is equal to the parent solution and no fitness evaluation is performed. If the donation resulted in an offspring solution with a greater or equal fitness to its parent solution, then the changes are kept, otherwise the changes are undone. All algorithms in this paper use the linkage tree from the Linkage Tree GA as their FOS model [5].

2.2 Exponential Population Scheme

A simple parameter-less population scheme is to keep restarting a GA with an increased population size until the solution quality has become acceptable. The growth function determines the rate at which the population size grows in the number of converged runs. The scheme for doubling the population size is called the Exponential Population Scheme (EPS) [4]. EPS is comparable to the behavior of a user, who will often keep doubling the population size until the results are satisfying. Although there is only a constant overhead, EPS inevitably throws away several populations worth of fitness evaluations.

2.3 Parameter-Less Population Pyramid

The Parameter-less Population Pyramid (P3) uses a novel population scheme, called the population pyramid [3]. The population pyramid is a set of populations with two specific methods for inserting new solutions and performing crossover with the existing solutions. The set of populations expands slowly and forms a pyramid-like structure where the populations are arranged in increasing fitness and decreasing size. Each layer represents a population. Initially there is only one empty population P_0. Every iteration, a new random solution undergoes local search by a first-improvement hill climber and is then added to the bottom layer of the pyramid P_0. Whenever a population has changed due to the insertion of a solution, the linkage tree of that layer is rebuilt. Then the solution undergoes crossover with the layers of the population pyramid, starting with the bottom layer and working its way up. The crossover operator is the GOM operator from GOMEA. If crossover at layer P_i results in a fitness improvement, then the solution is added to the next population P_{i+1} (or a new population is created if there was none existing). The linkage tree at layer P_{i+1} is rebuilt and the solution is GOM crossed over with population P_{i+1}. This process is repeated until the top of the pyramid has been reached. To ensure diversity is preserved duplicate solutions are not added to the pyramid.

P3 selects the donor for the GOM crossover operator differently than GOMEA. GOMEA continues with the next FOS cluster when the solution and donor have equal values for the problem variables. P3 alters this procedure by searching the entire population in a random order until a donor with non-equal values has been found. The upper-bound on the run-time of the crossover operator increases by a factor linear in the population size. Another difference is that P3 sorts the clusters in smallest-first order instead of the randomized order used by GOMEA. However, this has only a minor, problem-dependent, effect on performance.

3 Multiple Insertion Pyramid

The exponential population scheme (EPS) is characterized by its efficiency in model-building, while the parameter-less population pyramid (P3) is characterized by its efficiency in fitness evaluations. From an algorithmic perspective the two algorithms look very different and share only the GOM operator and linkage tree model learning. We now show how to combine the two approaches into a new population scheme. A visual interpretation of the population schemes is given to more easily explain how they are similar and how they can be combined.

3.1 Visualizing Population Schemes

Deriving a generic representation for either population scheme requires a more careful look at how GOMEA works. Each of the schemes relies on the GOM operator for performing crossover between a solution and a population. Two key properties describe a call to GOM: the set of solutions that are available for donation, and what happens with the solution after crossover.

GOMEA. The grid in Fig. 1a represents GOMEA running for 5 generations (vertical axis) with a population size of 7 (horizontal axis). Each cell of the grid represents the state of solution x at generation y. Each row represents the state of the population at generation y, and each column is the progression of solution x over the generations. The solid arrows show whether the solution changed by GOM was accepted into the next generation. Because GOMEA always accepts the offspring solution into the new population, the progression is a chain of arrows in the visualization. GOMEA traverses the grid in left-to-right first (applying GOM to the population), bottom-to-top second order (initial generation until termination). Lastly, the dashed lines show which solutions from the previous generation have successfully donated bits to a solution.

The probability of a successful donation decreases over time, and this is seen in the uppermost row where the donations are more sparse. If a cell has no dashed line with the previous row, then there was no successful donation during crossover, hence the solution indicated by the cell was not even improved. As seen by the mess of dashed lines in the grid, each solution has the remainder of the population at its disposal for mixing. Let the age of a solution x be the number generations y it has existed, in other words the row index y of column x in the visualization. Two solutions live in the same population if they are improved under the same circumstances, meaning they are improved using the same linkage tree and have access to the same donor solutions. If two solutions live in the same population and are of the same age, then applying GOM crossover to them should be optimal as their fitness should be similar due the competition with each other over the span of multiple generations. This is automatically true for GOMEA, and together with the high availability of donors, the GOM operator is able to optimally mix solutions. GOMEA learns the model, namely the linkage tree, from the entire population (i.e. row) of generation y. Whenever a model is learned before GOM crossover is applied, the corresponding cell is colored gray. For GOMEA this is the start of the generation, i.e. the first column. Clearly GOMEA is minimal in the number of times a model is learned, because otherwise a model would have to be used over multiple generations. No parameter-less population scheme can learn only one model on a row, because then it would have perfectly predicted the population size. Only oracles can tell beforehand what the population size necessary to solve the problem instance is.

The visualization highlights the strengths and weaknesses of a population scheme. Given that the global optimum resides amongst solutions of similar fitness in the final generation, GOMEA has the weakness that it must fully evaluate every generation until the final generation has been reached. However, the sparsity of gray squares shows that relatively few linkage trees are constructed during a single run. The dashed arrows show how optimal mixing allows every solution to donate to every other solution over the entirety of the run.

EPS. Figure 1b shows the visualization of the exponential population scheme (EPS). EPS starts with a run of GOMEA of size one, and continuously restarts the population on convergence with a doubled population size. Each of the pink squares denotes one run of GOMEA and is traversed exactly as explained in the previous section. The pink squares illustrate how EPS is a rather simplistic wrapper around GOMEA for doubling the population size. An iteration of EPS is the complete evaluation of one run of GOMEA, traversed from left-to-right in the grid. Note that the horizontal axis no longer corresponds to a single population. Learning a linkage tree only uses the solutions of generation y inside the pink square instead of the entire row. The visualization shows the two major weaknesses of the EPS scheme. First, there is no optimal mixing between the runs of GOMEA as indicated by the lack of dashed arrows between the runs. Second, the gray squares are more abundant and show the wasted effort in building linkage trees for the previous generations. But it has already been shown that no parameter-less population scheme can get away with learning only one model per row, hence there being an additional cost of model learning is inevitable.

P3. Figure 1c shows the visualization of the parameter-less population pyramid. With P3 comes a slightly different interpretation which mostly changes how the grid traversed. Green squares indicate that a solution was already present in the population pyramid, hence it was not accepted into the next layer of the pyramid. There is no solid arrow leading into a green square for this very reason. There is a direct correspondence between a population and the layer of a pyramid. Both contain a set of solutions, but the latter cannot contain duplicates by definition. Almost every cell is colored gray to show just how often P3 learns a new model, with the only exception being when a solution is not accepted into the next population. That only occurs when the solution is a duplicate already present in the pyramid, or the solution did not improve in fitness after GOM crossover with the population. The latter happens more often for the higher layers of the pyramid, because improving the fitness of an already optimized solution is more difficult. The linkage tree must be updated every time a new solution is inserted into the corresponding pyramid layer. Each iteration of P3 a solution is drawn from the population generator, and is then improved until GOM crossover has been performed with each layer of the pyramid. A column x is then equivalent to x^{th} iteration of P3. Whereas GOMEA traverses the grid from the left-to-right first, P3 traverses the grid from bottom-to-top first. This behavior is the exact opposite of each other, as is illustrated further in the next subsection. Looking at the dashed lines in the visualization, the solutions receive only donations from the previously evaluated solutions. The mixing that occurs in P3 is unidirectional over the horizontal axis, in contrast to the bi-directional mixing in GOMEA and EPS. Not only that, a solution x at generation y only has access to other solutions that also reached generation y. Hence the number of donors available to a solution (x, y) is lower than the number of donors available to a similarly placed solution (x, y) in GOMEA. Age becomes an even more important factor when GOM is applied to solutions that were improved using different populations. How many the layers the solution has passed (the row) is equivalent to the age of the solution in GOMEA. The linkage trees learned by P3 become more precise as more solutions are added to the pyramid. However, a solution x was improved with a different linkage tree and population from any other solution. Thus when a solution is donated to, the age of the donor and the receiving solution will be the same, but they were raised in different populations. Hence there is a possible discrepancy in fitness and structure between the solution and the donor, which may lead to lesser performance when performing GOM crossover. The entropy in a population grows slower as more solutions are added, hence the linkage trees will change less over time. As a result, the discrepancy between solutions and donors is reduced over time, but is never truly eliminated. Again, the visualization helps in understanding the strengths and weaknesses of P3. The scheme is over-saturated with model learning at every step of the algorithm, even though the entropy in the populations changes increasingly slower over time. Crossover uses donors of equal age to the receiving solution, but the donors were constructed under different circumstances. This may lead to fitness discrepancies in the early stages of the algorithm, because the model and populations are changing significantly with the addition of a single solution. In exchange for these weaknesses, the population pyramid requires no population size parameter and it gains access to the older generations earlier on than GOMEA.

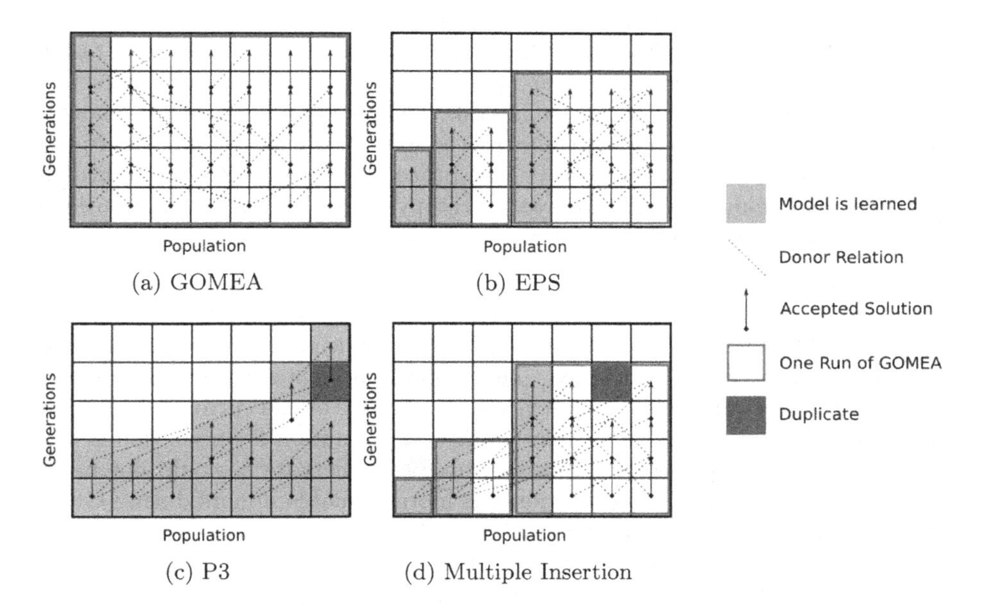

Fig. 1. Visual representation of the four population schemes.

3.2 Multiple Insertion

Multiple Insertion is a new population scheme designed to combine the efficiency in amount of linkage trees built of EPS with the efficiency in fitness evaluations of P3. The goal is to enhance EPS by using the population pyramid as a way to reuse the solutions from the previously converged populations. Similarly, the goal can be stated as using the strictly growing populations of EPS with P3 instead of one solution per iteration.

As in EPS, Multiple Insertion generates a new, larger population at each iteration. A growth function determines the size of the population generated at the start of iteration, this could be the exponential growth function or the linear growth function. Let P_i be the ith layer (or population) of the pyramid at the start of an iteration. First a population X is randomly generated with the population size determined by the growth function. Optionally the hill climber is invoked on each of the solutions in population X. The entire population X is checked for duplicates already in the pyramid, and the unique solutions are inserted into the bottom layer of the pyramid P_0 and a linkage tree model is learnt. Then GOM crossover is applied to the population X using the pyramid layer P_0. Those solutions whose fitnesses were improved by the GOM crossover are inserted into the set of accepted solutions. Solutions that are already present in the pyramid are filtered from the set of accepted solutions. Finally, the set of accepted solutions is inserted into the next pyramid layer P_{i+1}. This process repeats itself with X and the next pyramid layer P_{i+1} until the top of the pyramid has been reached. Note that the entire population X transitions to the next pyramid layer for GOM crossover, not just the solutions accepted into the next pyramid layer. As with P3, GOM crossover with the top layer may result in a new layer P_t. The new top layer P_t contains every solution from the set of accepted solutions at layer P_{t-1}. Unlike P3, the new layer may contain enough information for GOM crossover to improve the solutions, which would lead to the construction of a new top layer P_{t+1}. This process repeats itself until no fitness improvement is found in the top layer of the pyramid. If none of the solutions were improved by the GOM crossover, then nothing happens to the pyramid.

Visualization. Figure 1d shows the visualization of the multiple insertion population scheme. The density of the gray squares has been significantly reduced in contrast to P3, indicating that there should be a drastic gain in run-time because much less linkage trees are learned. The dashed lines show that solutions can receive donations from either the previous populations or the current population under construction. There is the constraint that a solution is only a donor if it was accepted in the previous generation, because otherwise it would not be in the population. The receiving solution has more available donors during GOM crossover than it would have in either EPS or P3. More importantly, the solutions in the current population are improved using the same model, hence they are of the same age and live in the same population. As mentioned for GOMEA, these conditions allow for optimal mixing between the solutions. When mixing occurs with a donor from a previous population, there may still be a fitness discrepancy as shown for P3. The linkage tree is no longer relearned for every solution inserted into the pyramid, so the population pyramid loses some of its immediate adaptiveness to a change in the pyramid. As more solutions are inserted into the pyramid layer at once, namely a subset of the population currently being improved, the entropy in the pyramid layer changes more significantly than it would have if one solution were inserted. This means that even though the model is relearned less frequently, the changes to the model are more substantial. The termination criterion for population convergence is a mix between EPS and P3, namely when the population has not been improved during an entire generation and when there are no more layers of the pyramid to perform crossover with.

4 Experimental Results

We have tested five, binary encoded, benchmark problems of increasing length.

1. **Deceptive Trap Function:** we consider the randomly linked DFT which tests the ability to detect and combine good building blocks ($k = 5$).
2. **Hierarchical If and Only If:** we consider the HIFF version where problem variables appear in randomized order.
3. **Ising Spin Glass:** the variables interact with neighboring vertices in a lattice graph. The problem instances are from the public repository hosting the source of the P3 algorithm [2].
4. **Nearest Neighbor NK Landscapes:** each variable depends on the $k = 4$ subsequent variables in the bit-string. The order of the variables in the representation are again assumed to be completely randomized.
5. **MaxCut:** find a maximum cut on a given graph. The instances are taken from [1], and were solved to optimality on the BIQMAC server.

Multiple insertion requires the specification of a growth function to determine how fast the population increases in size per iteration of the algorithm. Let $S(i)$ be the size of the population inserted into the population pyramid at iteration i. Three growth functions are tested: the linear function $S_l(i) = i$, the quadratic function $S_q(i) = i^2$, and the exponential function $S_e(i) = 2^{i-1}$. P3 uses the parameter configuration of the original P3 algorithm. The population pyramid scheme (PT) removes from P3 the hill climber, exhaustive donor searching, and it randomizes the FOS ordering. This way, PT is better comparable with the results from GOMEA. Multiple insertion is added to both population schemes. One can interpret both PT and P3 as a form of multiple insertion with a constant growth function $S(i) = 1$.

Figure 2 shows the relative average number of fitness evaluations for the P3, PT, and its multiple insertion variants. The labels in the legend follow a specific format: {Scheme} – MI – {Growth}, where Scheme is either PT or P3 and Growth is either the constant, linear, quadratic, or exponential growth function. The results are normalized using the results of the original P3 to better highlight the differences in the performance. As seen in Fig. 2 the quadratic and exponential growth functions diverge significantly from PT at several occasions. In contrast, the linear growth function is less prone to this erratic behavior. Especially the exponential growth function tends to have outliers in the number of fitness evaluations, to the point of exceeding EPS. For EPS it makes sense to scale exponentially, as a slower growth leads to more population being converged and rejected, which is terribly expensive. PT with multiple insertion is an extension of EPS that reuses every solution whereas EPS cannot do that. Because PT

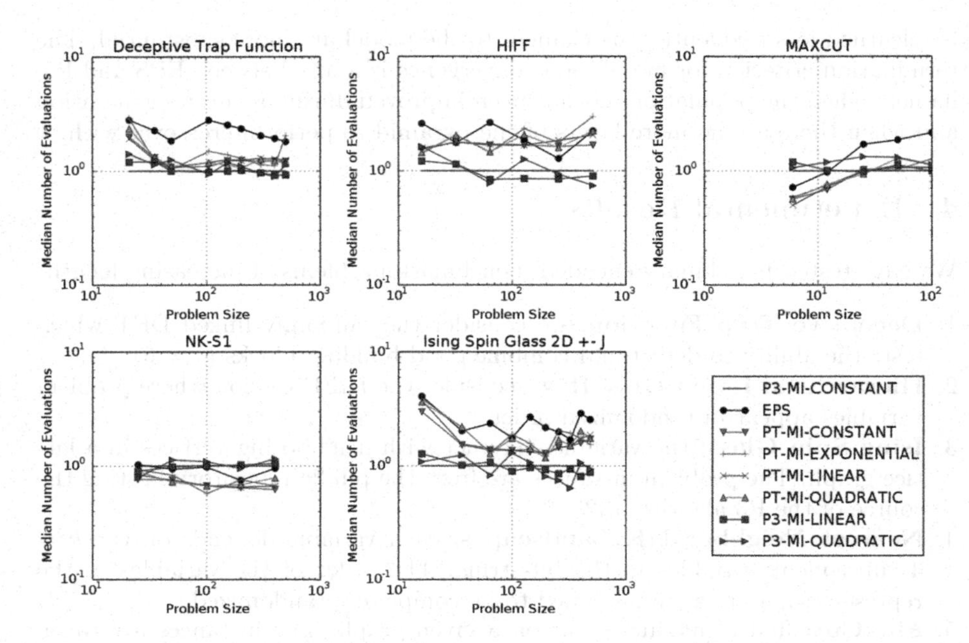

Fig. 2. Relative comparison to P3 of the median number of fitness evaluations for the multiple insertion variants of PT and P3.

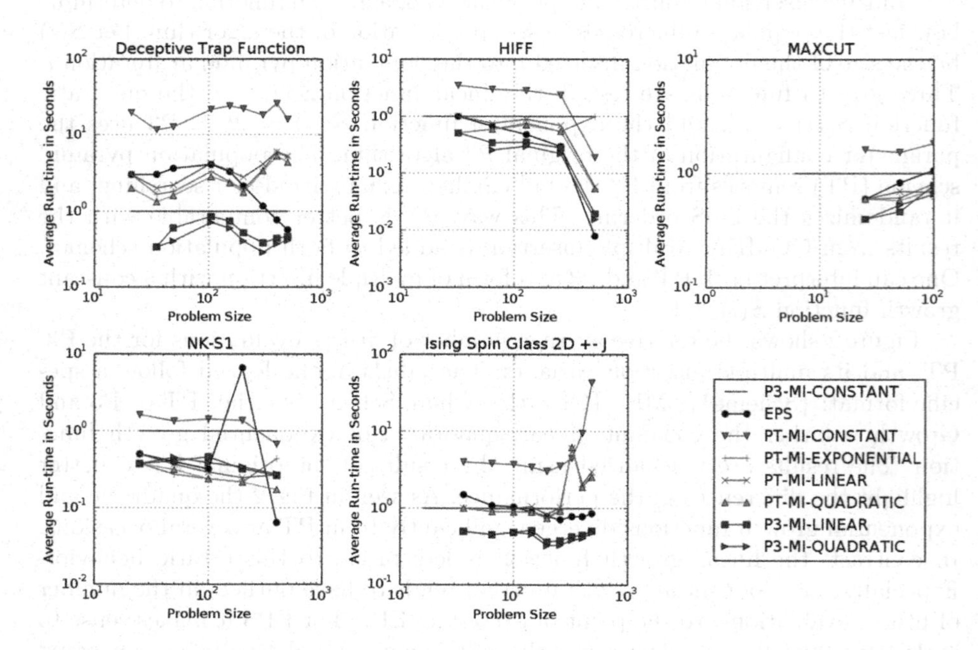

Fig. 3. Relative comparison to P3 of the average running time for the multiple insertion variants of PT and P3.

with multiple insertion is less prone to outliers with a slower growth function, there is little reason to use an exponential growth function.

Figure 3 shows the relative average run-time in milliseconds for P3, PT, and the multiple insertion variants. Using multiple insertion with PT reduces the run-time to the level of the original P3 which uses the hill climber. However, increasing the growth function beyond linear has no significant effect on the run-time, as seen by the tightly grouped lines per population scheme. Applying multiple insertion to the original P3 algorithm greatly speeds up the run-time. On any of the test problems the run-time of P3 matches or outperforms EPS, which is something it rarely achieved without multiple insertion. The exception to the rule are the NK-S1 problem instances, but that is due to the poor performance caused by the hill climber. The effectiveness of multiple insertion shows that P3 wastes a liberal amount of processing time on rebuilding linkage trees to little effect. The results also indicate that the performance of the linear growth function is on par with the quadratic function and more reliable than the exponential function, hence we recommend the use of the linear growth function.

5 Conclusion

We have proposed a novel parameter-less population scheme, called the Multiple Insertion Pyramid. We have discussed two populations schemes that do not require the user to set a fixed population size: the Exponential Population Scheme (EPS) and the Parameter-less Population Pyramid (P3). P3 is a model-based evolutionary algorithm that applies the linkage tree model building and the Gene-pool Optimal Mixing crossover operator from the Gene-pool Optimal Mixing Evolutionary Algorithm (GOMEA), but uses a novel population scheme as exploitation mechanism. P3's population pyramid scheme outperforms the linkage tree based GOMEA in terms of number of fitness evaluations at the cost of a significantly increased run-time. It is noted that P3's method to rebuilt the linkage tree whenever a single new solution is added to the population pyramid is too costly. By changing the population scheme to the Multiple Insertion Pyramid scheme the run-time is significantly reduced. The Multiple Insertion Pyramid combines EPS with the population pyramid, which allows solutions from previously converged populations to contribute to the current population. Because previous populations are retained for donation, the growth rate is reduced from exponential to a linear function. Multiple insertion can be applied to P3 to greatly reduce the time spent on relearning linkage trees.

References

1. Bosman, P.A., Thierens, D.: More concise and robust linkage learning by filtering and combining linkage hierarchies. In: Proceedings of the Genetic and Evolutionary Computation Conference, pp. 359–366. ACM (2013)
2. Goldman, B.W., Punch, W.F.: Parameter-less population pyramid. In: Proceedings of the Genetic and Evolutionary Computation Conference, pp. 785–792. ACM (2014)

3. Goldman, B.W., Punch, W.F.: Fast and efficient black box optimization using the parameter-less population pyramid. Evol. Comput. **23**(3), 451–479 (2015)
4. Harik, G.R., Lobo, F.G.: A parameter-less genetic algorithm. In: Proceedings of the Genetic and Evolutionary Computation Conference, pp. 258–267 (1999)
5. Thierens, D., Bosman, P.A.: Optimal mixing evolutionary algorithms. In: Proceedings of the Genetic and Evolutionary Computation Conference, pp. 617–624 (2011)

Doubly Trained Evolution Control for the Surrogate CMA-ES

Zbyněk Pitra[1,2,3(✉)], Lukáš Bajer[4,5], and Martin Holeňa[3]

[1] National Institute of Mental Health, Topolová 748, 250 67 Klecany, Czech Republic
z.pitra@gmail.com
[2] Faculty of Nuclear Sciences and Physical Engineering, Czech Technical University in Prague, Břehová 7, 115 19 Prague 1, Czech Republic
[3] Institute of Computer Science, Academy of Sciences of the Czech Republic, Pod Vodárenskou věží 2, 182 07 Prague 8, Czech Republic
{bajer,holena}@cs.cas.cz
[4] Faculty of Mathematics and Physics, Charles University in Prague, Malostranské nám. 25, 118 00 Prague 1, Czech Republic
[5] Unicorn College, V Kapslovně 2767/2, 130 00 Prague 3, Czech Republic

Abstract. This paper presents a new variant of surrogate-model utilization in expensive continuous evolutionary black-box optimization. This algorithm is based on the surrogate version of the CMA-ES, the Surrogate Covariance Matrix Adaptation Evolution Strategy (S-CMA-ES). Similarly to the original S-CMA-ES, expensive function evaluations are saved through a surrogate model. However, the model is retrained after the points in which its prediction was most uncertain have been evaluated by the true fitness in each generation. We demonstrate that within small budget of evaluations, the new variant of S-CMA-ES improves the original algorithm and outperforms two state-of-the-art surrogate optimizers, except a few evaluations at the beginning of the optimization process.

Keywords: Black-box optimization · Surrogate model · Evolution control · Gaussian process

1 Introduction

In many research and engineering tasks, optimization of real-world black-box functions that are costly to evaluate is a challenging problem of great importance. A single evaluation of the expensive function may require a great amount of resources in terms of time and performed experiments, measurements or simulations. In order to decrease the number of evaluations of the costly black-box function and still produce reasonably good solutions, a surrogate model can be employed [15]. Such models are built using the previous evaluations of the black-box function, and then are used to predict the values of new points instead of the original function.

J. Handl et al. (Eds.): PPSN XIV 2016, LNCS 9921, pp. 59–68, 2016.
DOI: 10.1007/978-3-319-45823-6_6

Nowadays, the *Covariance Matrix Adaptation Evolution Strategy* (CMA-ES) [5] is one of the most robust algorithms on real-world problems and is considered to be the state-of-the-art of continuous black-box optimization. In recent years, several surrogate-model approaches have been developed to increase the performance of the CMA-ES.

The ^{s*}ACM-ES [12] employs ordinal regression models based on SVM to estimate the ordering of the fitness function values. Furthermore, the parameters of the ordinal model are themselves optimized utilizing the CMA-ES algorithm during the optimization of the black-box function. In order to avoid premature convergence to local optima, the strategies ^{s*}ACM-ES-k [11] and BIPOP-^{s*}ACM-ES-k [13] improving ^{s*}ACM-ES by increasing the population size in generations evaluated by the model have been developed.

Another surrogate-assisted approach using continuous regression models to estimate the function values, called the Surrogate CMA-ES (S-CMA-ES), has been proposed in [1]. This approach employs models capable to predict the whole distribution of values of the objective function; however, the S-CMA-ES does not make use of that capacity of the models, and exploits only the means of the distribution of its values. On the other hand, several authors [10,14] have demonstrated the effective utilization of various criteria using the variances of predictions (e.g. expected improvement, probability of improvement) in optimization.

Different usage of surrogate modelling presents *Sequential Model-based Algorithm Configuration* method (SMAC) [8]. It fits surrogate models of algorithm settings in a parameter space and utilizes those models to make decisions about which settings to investigate. To make SMAC more useful in continuous optimization, random forest were replaced by Gaussian processes as a surrogate model in SMAC-BBOB [7].

The main contribution of this paper is to introduce Doubly Trained S-CMA-ES, the extension of the S-CMA-ES, using not only the means of the distributions predicted by the surrogate model, but also variances of those distributions. We experimentally evaluate different settings of this approach on the BBOB/COCO testing set [3,4] and compare it with the original version of the S-CMA-ES, the surrogate-assisted ^{s*}ACM-ES-k, and the SMAC method.

The remainder of the paper is structured as follows. Section 2 describes the S-CMA-ES and its model-training method. Section 3 defines its proposed extension, Doubly Trained S-CMA-ES. Section 4 contains the experimental part. Section 5 summarizes the results and draws conclusions.

2 Surrogate CMA-ES and Generation Evolution Control

The S-CMA-ES, introduced in [1], is a surrogate-model-based modification of the CMA-ES. After the initialization step, the following steps shown in Algorithm 1 are proceeded by the S-CMA-ES until the target fitness value is found: First, the population of one generation is sampled using the CMA-ES. Then, the evolution control is employed to evaluate sampled points. Finally, the CMA-ES strategy parameters (σ, $\mathbf{m}$, $\mathbf{C}$, etc.) are calculated using the original CMA-ES algorithm.

Algorithm 1. S-CMA-ES

Input: λ (population-size), y_{target} (target value), f (original fitness function), r (maximal distance between training points and $\mathbf{m}$), n_{MIN}, n_{MAX} (minimal and maximal number of points for model training), n_{orig} (number of original-evaluated points), $\mathcal{C}$ (uncertainty criterion), g_m (number of model generations)

1: $\sigma, \mathbf{m}, \mathbf{C}, g \leftarrow$ CMA-ES initialize
2: $\mathcal{A} \leftarrow \emptyset$
3: **while** $\min_{k\in\{1,\ldots,\lambda\}} y_k > y_{\text{target}}$ **do**
4: $\quad \mathbf{x}_k \sim \mathcal{N}(\mathbf{m}, \sigma^2\mathbf{C}) \quad k \in \{1, \ldots, \lambda\}$ *{CMA-ES sampling}*
5: $\quad (\{y_k\}_{k=1}^{\lambda}, \mathcal{A}) \leftarrow \text{evolutionControl}(\lambda, f, \mathcal{A}, \{\mathbf{x}_k\}_{k=1}^{\lambda}, \sigma, \mathbf{m}, \mathbf{C}, \ldots)$
6: $\quad \sigma, \mathbf{m}, \mathbf{C}, g \leftarrow$ CMA-ES update
7: **end while**
8: $\mathbf{x}_{\text{res}} \leftarrow \mathbf{x}_k$ where y_k is minimal

Output: $\mathbf{x}_{\text{res}}$

Algorithm 2. Generation evolutionControl in S-CMA-ES

Input: λ, σ, $\mathbf{m}$, $\mathbf{C}$, f, $\mathcal{A}$, $\{\mathbf{x}_k\}_{k=1}^{\lambda}$ (CMA-ES sampled population), g (generation), g_m (number of model generations), r, n_{MIN}, n_{MAX}

1: **if** g is original-evaluated **then**
2: $\quad y_k \leftarrow f(\mathbf{x}_k) \quad k = 1, \ldots, \lambda$ *{fitness evaluation}*
3: $\quad \mathcal{A} = \mathcal{A} \cup \{(\mathbf{x}_k, y_k)\}_{k=1}^{\lambda}$
4: $\quad f_{\mathcal{M}} \leftarrow \text{trainModel}(\mathcal{A}, \sigma, \mathbf{m}, \mathbf{C}, r, n_{\text{MIN}}, n_{\text{MAX}})$
5: **else**
6: $\quad y_k \leftarrow f_{\mathcal{M}}(\mathbf{x}_k) \quad k = 1, \ldots, \lambda$ *{model evaluation}*
7: $\quad$**if** g_m model generations passed **then** mark $(g+1)$ as original-evaluated
8: **end if**

Output: $(y_k)_{k=1}^{\lambda}$, $\mathcal{A}$

The *generation-based evolution control* (following Jin's terminology [9]) is used in S-CMA-ES as the evolution control step (Step 5 in Algorithm 1). This step is presented in more detail in Algorithm 2. At first, the population of one generation sampled using CMA-ES is evaluated by the original fitness function. Then, a surrogate model is constructed using the original-evaluated data. However, if the model has not enough training points, the original fitness function is utilized to evaluate sampled points. In the few subsequent generations, the function values of the samples are computed using the surrogate model; they are, consequently, used to calculate new CMA-ES parameters.

The phase of training the surrogate model is shown in Algorithm 3. In order to increase the accuracy of surrogate-model predictions (e.g. Gaussian process predictions), the points that have the Mahalanobis distance from the current CMA-ES mean $\mathbf{m}$ less than or equal to a specific bound r are selected for training. If the size of the training set is sufficient, k-NN clustering chooses n_{MAX} training points which are transformed to the basis defined by eigenvectors of CMA-ES' covariance matrix $\mathbf{C}$ through multiplication by $((\sigma^2\mathbf{C})^{-1/2})^{\top})$. Finally, the surrogate model is build using these transformed points. Naturally, the points for prediction of the model are transformed in the same way to ensure prediction with respect to the same base vectors.

3 Doubly Trained Evolution Control for the S-CMA-ES

In this section, an alternative to S-CMA-ES, called the *Doubly Trained* S-CMA-ES (DTS-CMA-ES), will be described. It uses not only model-predicted values of sampled points, but also their variances. Therefore, models capable to provide both values for each point have to be employed, in particular Gaussian processes [17] or random forests [2].

The DTS-CMA-ES differs from the S-CMA-ES through using *doubly trained evolution control* instead of the generation evolution control in Step 5 of Algorithm 1. The doubly trained evolution control is described in Algorithm 4 as follows: First, the values $\hat{y}$ and variances s^2 of CMA-ES sampled points are predicted by the surrogate model which is previously trained using the points evaluated by the original fitness function from previous generations. Second, the points are sorted according to the values of some uncertainty criterion $\mathcal{C}$ based on predicted $\hat{y}$ and s^2. Third, the n_{orig} most uncertain points are evaluated by the original fitness function. Next, the model is retrained using the points (chosen similarly to S-CMA-ES) evaluated by the original fitness function including the n_{orig} points from the previous step. Eventually, denoting λ as the population size, the $\lambda - n_{\text{orig}}$ points function values are predicted by the retrained model, and returned to the original S-CMA-ES to compute new parameters. Note that training the new model in step 1 differs from using the model from the previous generation since it uses updated CMA-ES state variables σ, $\mathbf{m}$ and $\mathbf{C}$.

3.1 Uncertainty Criteria

The following criteria $\mathcal{C}$, which determine the points for evaluation by the original fitness function, can be used in the DTS-CMA-ES (Algorithm 4).

Algorithm 3. S-CMA-ES trainModel

Input: σ, $\mathbf{m}$, $\mathbf{C}$, $\mathcal{A}$, r (maximal distance between training points and $\mathbf{m}$),
n_{MIN}, n_{MAX} (min. and max. number of points for training)
1: $(\mathbf{X}_{\text{tr}}, \mathbf{y}_{\text{tr}}) \leftarrow \{(\mathbf{x}, y) \in \mathcal{A} \,|\, (\mathbf{m}-\mathbf{x})^\top (\sigma^2\mathbf{C})^{-1/2}(\mathbf{m}-\mathbf{x}) \leq r\}$
2: **if** $|\mathbf{X}_{\text{tr}}| \geq n_{\text{MIN}}$ **then**
3: $\quad (\mathbf{X}_{\text{tr}}, \mathbf{y}_{\text{tr}}) \leftarrow$ choose n_{MAX} points by k-NN if $|\mathbf{X}_{\text{tr}}| > n_{\text{MAX}}$
4: $\quad \mathbf{X}_{\text{tr}} \leftarrow \{((\sigma^2\mathbf{C})^{-1/2})^\top \mathbf{x}_{\text{tr}} | \mathbf{x}_{\text{tr}} \in \mathbf{X}_{\text{tr}}\}$
5: $\quad f_{\mathcal{M}} \leftarrow$ buildModel$(\mathbf{X}_{\text{tr}}, \mathbf{y}_{\text{tr}})$
6: **else**
7: $\quad f_{\mathcal{M}} \leftarrow \emptyset$
8: **end if**
Output: $f_{\mathcal{M}}$

Algorithm 4. Doubly Trained evolutionControl in DTS-CMA-ES

Input: λ, σ, $\mathbf{m}$, $\mathbf{C}$, f, $\mathcal{A}$, $\{\mathbf{x}_k\}_{k=1}^{\lambda}$ (CMA-ES sampled population), $\mathcal{C}$ (uncertainty criterion), n_{orig} (number of original-evaluated points), r, n_{MIN}, n_{MAX}

1: $f_{\mathcal{M}} \leftarrow \text{trainModel}(\mathcal{A})$
2: $(\hat{y}_k, s_k^2) \leftarrow f_{\mathcal{M}}(\mathbf{x}_k)$ $\quad k \in \mathcal{I} := \{1, \ldots, \lambda\}$ $\quad$ {*model evaluation*}
3: $c_k \leftarrow \mathcal{C}(\hat{y}_k, s_k^2)$ $\quad k \in \mathcal{I}$ $\quad$ {*criterion evaluation*}
4: $\{c_{k_i}\}_{i=1}^{\lambda} \leftarrow \text{sort}\{c_k\}_{k=1}^{\lambda}$
5: $\mathcal{I}_{\text{orig}} = \{k_i \in \mathcal{I} \,|\, \{c_{k_i}\}_{i=1}^{n_{\text{orig}}}\}$
6: $y_k \leftarrow f(\mathbf{x}_k)$ $\quad k \in \mathcal{I}_{\text{orig}}$ $\quad$ {*fitness evaluation*}
7: $\mathcal{A} = \mathcal{A} \cup \{(\mathbf{x}_k, y_k)\}_{k \in \mathcal{I}_{\text{orig}}}$
8: $f_{\mathcal{M}} \leftarrow \text{trainModel}(\mathcal{A}, \sigma, \mathbf{m}, \mathbf{C}, r, n_{\text{MIN}}, n_{\text{MAX}})$
9: $y_k \leftarrow f_{\mathcal{M}}(\mathbf{x}_k)$ $\quad k \in \mathcal{I} \setminus \mathcal{I}_{\text{orig}}$ $\quad$ {*model evaluation*}

Output: $(y_k)_{k \in \mathcal{I}}$, $\mathcal{A}$

Variance. The variance s^2 of model-predicted function values $\hat{y}$:

$$\mathcal{C}_{s^2} = s^2. \tag{1}$$

The larger the variance, the higher the uncertainty of the predicted fitness.

Lower Confidence Bound (LCB). The lower confidence bound has been proposed in [14]:

$$\mathcal{C}_{\text{LCB}} = \hat{y} - 2s^2. \tag{2}$$

The points with lower values of the LCB criterion are considered more interesting for evaluation by the original fitness function than the points with higher values.

Probability of Improvement (PoI). The probability of improvement with respect to a given target $T \leq y_{\min}$ can be expressed as follows:

$$\mathcal{C}_{\text{PoI}} = P(f(\mathbf{x}) \leq T | y_1, \ldots, y_n) = \phi\left(\frac{T - \hat{y}}{s}\right), \tag{3}$$

where ϕ denotes the distribution function of $\mathcal{N}(0, 1)$ and $y_{\min}$ is the minimum value found so far.

Expected Improvement (EI). The expected improvement is described by [10]:

$$\mathcal{C}_{\text{EI}} = E((y_{\min} - f(\mathbf{x}))I(f(\mathbf{x}) < y_{\min}) | y_1, \ldots, y_n), \tag{4}$$

where

$$I(f(\mathbf{x}) < y_{\min}) = \begin{cases} 1 & f(\mathbf{x}) < y_{\min} \\ 0 & f(\mathbf{x}) \geq y_{\min}. \end{cases} \tag{5}$$

Similarly, $\mathcal{C}_{\text{EI}}$ can be expressed as [10]:

$$\mathcal{C}_{\text{EI}} = (y_{\min} - \hat{y})\,\phi\left(\frac{y_{\min} - \hat{y}}{s}\right) + s\varphi\left(\frac{y_{\min} - \hat{y}}{s}\right), \tag{6}$$

where φ denotes the density of $\mathcal{N}(0, 1)$.

4 Experimental Evaluation

We compared the performance of the DTS-CMA-ES to the original CMA-ES [5], two surrogate-model-based CMA-ES algorithms, the S-CMA-ES [1] and the BIPOP-^{s*}ACM-ES-k [13], and the SMAC algorithm [8] on the set of all 24 noiseless functions from the COCO/BBOB framework [3,4].

4.1 Experimental Setup

The considered algorithms were compared in dimensions $D = 2, 3, 5, 10$, and 20 using the standard BBOB settings, i.e. on 15 different function instances. The BBOB stopping criteria were reaching the distance from the function optimum $\Delta f_T = 10^{-8}$ and expending maximal number of evaluations per dimension (FE/D) which we have set to 100 due to our interest in expensive optimization where very few evaluations are available [6]. The parameters of the compared algorithms are summarized in the following paragraphs.

We have employed the original CMA-ES in its IPOP-CMA-ES version (Matlab code v. 3.61) with the following parameters: number of restarts = 4, IncPopSize = 2, $\sigma_{start} = \frac{8}{3}$, $\lambda = 4 + \lfloor 3 \log D \rfloor$. The remaining parameters were left default.

Loshchilov's ^{s*}ACM-ES-k was used in its bi-population version published in [13]. The BIPOP-^{s*}ACM-ES-k results have been downloaded from the BBOB results data archive[1] in its GECCO 2013 settings.

Gaussian processes (GP) have been employed in the S-CMA-ES as surrogate models for $g_m = 5$ model-evaluated generations. The distance r (see Algorithm 1) has been set to 10. The covariance function $K_{\text{Matérn}}^{\nu=5/2}$ with starting values $(\sigma_n^2, l, \sigma_f^2) = \log(0.01, 2, 0.5)$ were used for the GP model (see [1] for the details).

As opposed to [16], all the function values were normalized to zero mean and unit variance before training surrogate models in order to increase numerical accuracy. The CMA-ES parameter values have been set the same as in the original CMA-ES. All other settings were left default [1].

GP have been also employed in SMAC-BBOB [7], the continuous optimization version of the SMAC. The SMAC results were downloaded from the BBOB results data archive[2].

The DTS-CMA-ES was tested with multiple settings of parameters. First, all the uncertainty criteria from Sect. 3.1 (s^2, LCB, EI, PoI) were compared using $\lambda = 4 + \lfloor 3 \log D \rfloor$ and $n_{\text{orig}} = \lceil 0.1\lambda \rceil$ (see Algorithm 4) to find the most suitable one. For the remaining investigations, two different population sizes $\lambda_{\text{1pop}} = 4 + \lfloor 3 \log D \rfloor$ and $\lambda_{\text{2pop}} = 8 + \lfloor 6 \log D \rfloor$ and four n_{orig} values $\lceil 0.05\lambda \rceil$, $\lceil 0.1\lambda \rceil$, $\lceil 0.2\lambda \rceil$, $\lceil 0.4\lambda \rceil$ were used for comparison. The CMA-ES parameters, the distance r, and the GP model have been taken over from the S-CMA-ES.

---

[1] http://coco.gforge.inria.fr/data-archive/2013/BIPOP-saACM-k_loshchilov_noiseless.tgz.

[2] http://coco.gforge.inria.fr/data-archive/2013/SMAC-BBOB_hutter_noiseless.tgz.

4.2 Results

We have compared the performances of DTS-CMA-ES for four different uncertainty criteria described in Sect. 3.1. The results aggregated through the full set of benchmark functions show that the different criteria exhibit very similar convergence rate. However, the usage of $\mathcal{C}_{s^2}$ leads to a slightly better performances on most of BBOB functions, especially in 20D, and $\mathcal{C}_{\mathrm{EI}}$ performs the best on f16, f22, and f23 (if aggregated through dimensions).

Figure 1 presents comparison of DTS-CMA-ES employing criteria $\mathcal{C}_{s^2}$, $\mathcal{C}_{\mathrm{LCB}}$, $\mathcal{C}_{\mathrm{EI}}$, $\mathcal{C}_{\mathrm{PoI}}$ with $n_{\mathrm{orig}} = \lceil 0.1\lambda \rceil$ in 5D and 20D. Let Δ_f be the minimal distance found from the function optimum for the considered number of fitness function evaluations. The graphs depict a scaled logarithm of Δ_f depending on FE/D. Since all the algorithms ran for each function and dimension on 15 independent instances, only the empirical medians Δ_f^{med} over those 15 runs of Δ_f were taken for further processing. The scaled logarithms of Δ_f^{med} are calculated as

$$\Delta_f^{\log} = \frac{\log \Delta_f^{\mathrm{med}} - \Delta_f^{\mathrm{MIN}}}{\Delta_f^{\mathrm{MAX}} - \Delta_f^{\mathrm{MIN}}} \log_{10}\left(1/10^{-8}\right) + \log_{10} 10^{-8}$$

where Δ_f^{MIN} (Δ_f^{MAX}) is the minimum (maximum) $\log \Delta_f^{\mathrm{med}}$ found among all the compared algorithms for the particular function f and dimension D between 0 and 100 FE/D. Afterwards, graphs of $\Delta_f^{\log}$ can be aggregated across arbitrary number of functions and dimensions. Values in presented graphs are averages of $\Delta_f^{\log}$ across all 24 functions. More detailed results can be found on authors' webpage[3].

The graphs in Fig. 2 summarize the performance of four different n_{orig} values and two population sizes λ_{1pop} and λ_{2pop}. This and all the following experiments use the criterion $\mathcal{C}_{s^2}$ which performed best in the first set of experiments. We found that the lower the n_{orig}, the better the performance is observed. Moreover, testing showed overall best performance of $n_{\mathrm{orig}} = \lceil 0.05\lambda_{\mathrm{2pop}} \rceil$, which is in 2D, 3D, 5D equal to 1, and in 10D and 20D is equal to 2.

Table 1 illustrates the counts of the 1st ranks of the compared algorithms according to the lowest achieved Δ_f^{med} for 20, 40, and 80 FE/D respectively. These counts are for different dimensions summed across all 24 functions.

As can be seen in Fig. 3, DTS-CMA-ES provides the best average results among the tested algorithms during the middle part of the optimization process, i.e. between 30 and 80 FE/D. The SMAC excels at the very beginning of optimization progress (up to ca. 15 FE/D), and starting from ca. 80–130 FE/D (depending on dimension), the fastest converging algorithm is the ^{s*}ACM-ES-k.

The new algorithm demonstrates speed-up compared to the S-CMA-ES with the exception of f1; however, it still has problem with few multimodal functions (f17–f20). It can be interpreted as premature convergence in local optima.

[3] http://bajeluk.matfyz.cz/scmaes/ppsn2016/.

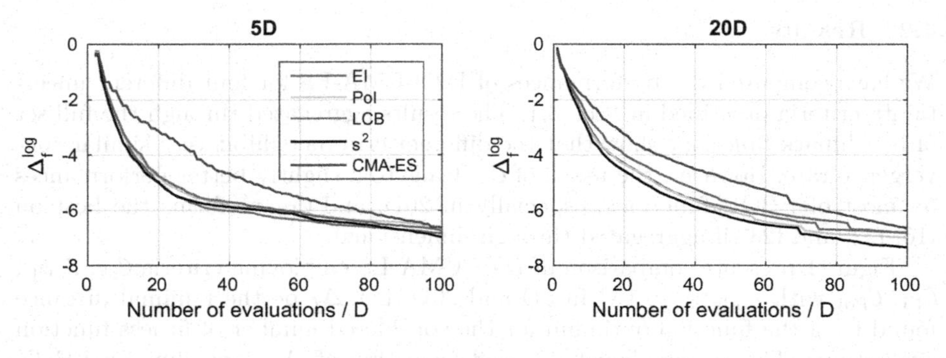

Fig. 1. Criterion comparison in 5D and 20D

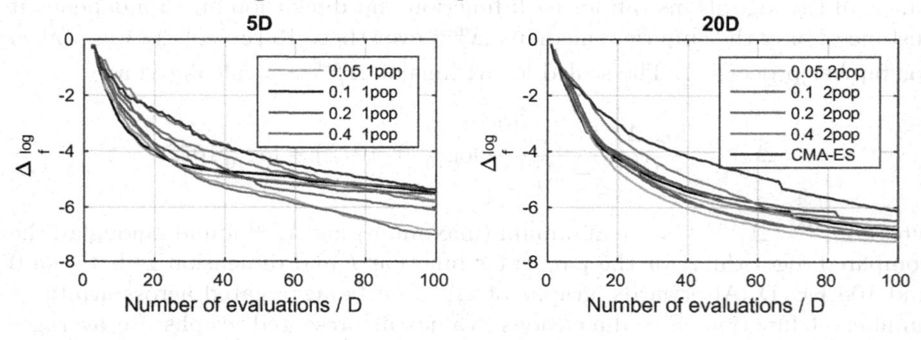

Fig. 2. Population-size and n_{orig} comparison in 5D and 20D

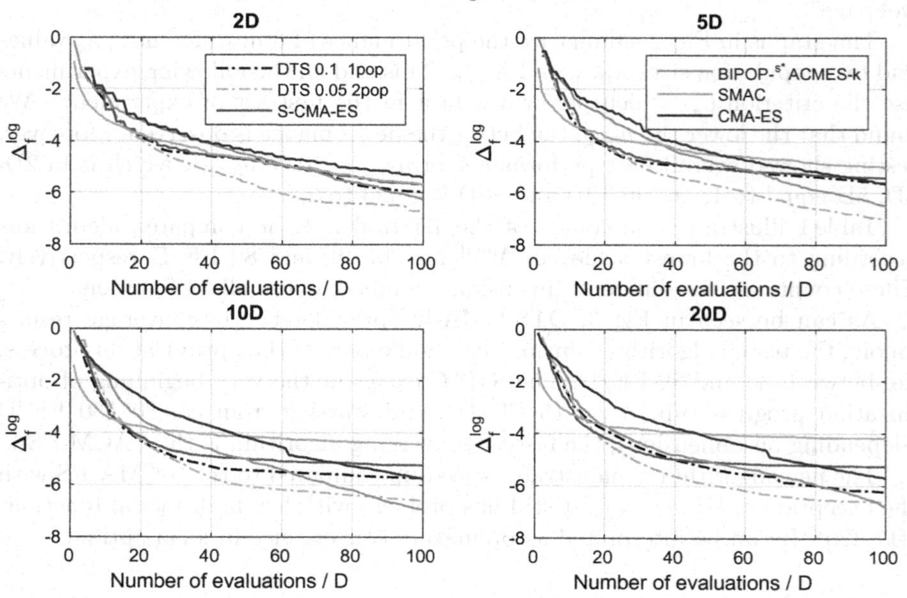

Fig. 3. Algorithm comparison in 2, 5, 10 and 20D

Table 1. Counts of the 1st ranks from 24 benchmark functions according to the lowest achieved Δ_f^{med} for different FE/D = {20, 40, 80} and dimensions D = {2, 3, 5, 10, 20}. Ties of the 1st ranks are counted for all respective algorithms. The ties often occure when $\Delta f_T = 10^{-8}$ is reached (mostly on f1 and f5).

FE/D	2D			3D			5D			10D			20D			$\sum$		
	20	40	80	20	40	80	20	40	80	20	40	80	20	40	80	20	40	80
DTS 0.1 1pop	6	3	2	**13**	6	3	**10**	4	3	**10**	7	3	2	4	5	**41**	24	16
DTS 0.05 2pop	**8**	**17**	**13**	7	**11**	**11**	9	**14**	**13**	6	**13**	**11**	**11**	**10**	8	**41**	**65**	**56**
S-CMA-ES	5	4	3	1	4	5	5	4	2	7	3	2	9	6	3	27	21	15
BIPOP-^{s*}ACM-ES-k	2	1	7	3	3	6	1	2	4	1	2	8	2	4	**9**	9	12	34
SMAC	5	4	4	4	4	5	3	4	5	4	2	2	3	4	3	19	18	19
CMA-ES	1	2	3	1	3	2	0	3	5	0	1	6	0	0	4	2	9	20

5 Conclusion and Future Work

This article presents a new version of the surrogate-based optimization algorithm S-CMA-ES. It further investigates the possibility to use surrogate models based on Gaussian processes in connection with the state-of-the-art black-box optimization algorithm CMA-ES. This improved algorithm introduces an additional model training within one generation, which shows a faster convergence to the global optima on many benchmark functions, independently of dimensions.

The choice of uncertainty criteria was not found as crucial in the speed of DTS-CMA-ES convergence. Furthermore, the comparison shows that the lower numbers of reevaluated points in each generation can lead to higher performance of the algorithm. We found that new approach usually reduces the number of necessary evaluations in expensive optimization more than other compared surrogate-model-based versions of the CMA-ES, namely BIPOP-^{s*}ACM-ES-k and S-CMA-ES, and except very early stages of the exploitation even more than SMAC-BBOB algorithm.

The main perspective of improving DTS-CMA-ES is to make the number of reevaluated points online adjustable, which should lead to more precise control of exploitation and facilitate escaping from the local optima. Another perspective is to additionally investigate different properties of surrogate models for better utilization of uncertainty criteria.

Acknowledgements. This work was supported by the Grant Agency of the Czech Technical University in Prague with its grant No. SGS14/205/OHK4/3T/14 by the Czech Health Research Council project NV15-33250A, by the project "National Institute of Mental Health (NIMH-CZ)", grant number ED2.1.00/03.0078 and the European Regional Development Fund, and by the project Nr.LO1611 with a financial support from the MEYS under the NPU I program. Further, access to computing and storage facilities owned by parties and projects contributing to the National Grid Infrastructure MetaCentrum, provided under the programme "Projects of Large Infrastructure for Research, Development, and Innovations" (LM2010005), is greatly appreciated.

References

1. Bajer, L., Pitra, Z., Holeňa, M.: Benchmarking Gaussian processes and random forests surrogate models on the BBOB noiseless testbed. In: Proceedings of the 17th GECCO Conference Companion. ACM, Madrid, July 2015
2. Breiman, L.: Classification and Regression Trees. Chapman & Hall/CRC, Boca Raton (1984)
3. Hansen, N., Auger, A., Finck, S., Ros, R.: Real-parameter black-boxoptimization benchmarking 2012: experimental setup. Technical report, INRIA (2012)
4. Hansen, N., Finck, S., Ros, R., Auger, A.: Real-parameter black-box optimization benchmarking 2009: noiseless functions definitions. Technical report RR-6829, INRIA (2009). Updated February 2010
5. Hansen, N.: The CMA evolution strategy a comparing review. In: Lozano, J.A., Larrañaga, P., Inza, I., Bengoetxea, E. (eds.) Towards a New Evolutionary Computation. Studies in Fuzziness and Soft Computing, vol. 192, pp. 75–102. Springer, Heidelberg (2006)
6. Holeňa, M., Linke, D., Bajer, L.: Surrogate modeling in the evolutionary optimization of catalytic materials. In: Soule, T. (ed.) Proceedings of the 14th GECCO, pp. 1095–1102. ACM, New York, Philadelphia (2012)
7. Hutter, F., Hoos, H., Leyton-Brown, K.: An evaluation of sequential model-based optimization for expensive blackbox functions. In: Proceedings of the 15th Annual Conference Companion on Genetic and Evolutionary Computation, GECCO 2013 Companion, pp. 1209–1216. ACM, New York (2013)
8. Hutter, F., Hoos, H.H., Leyton-Brown, K.: Sequential model-based optimization for general algorithm configuration. In: Coello, C.A.C. (ed.) LION 2011. LNCS, vol. 6683, pp. 507–523. Springer, Heidelberg (2011)
9. Jin, Y.: A comprehensive survey of fitness approximation in evolutionary computation. Soft Comput. **9**(1), 3–12 (2005)
10. Jones, D.R.: A taxonomy of global optimization methods based on response surfaces. J. Glob. Optim. **21**(4), 345–383 (2001)
11. Loshchilov, I., Schoenauer, M., Sebag, M.: Intensive surrogate model exploitation in self-adaptive surrogate-assisted CMA-ES (saACM-ES). In: Genetic and Evolutionary Computation Conference (GECCO), pp. 439–446. ACM Press, July 2013
12. Loshchilov, I., Schoenauer, M., Sebag, M.: Self-adaptive surrogate-assisted covariance matrix adaptation evolution strategy. In: Proceedings of the 14th GECCO, GECCO 2012, pp. 321–328. ACM, New York (2012)
13. Loshchilov, I., Schoenauer, M., Sebag, M.: BI-population CMA-ES algorithms with surrogate models and line searches. In: Genetic and Evolutionary Computation Conference (GECCO Companion), pp. 1177–1184. ACM Press, July 2013
14. Lu, J., Li, B., Jin, Y.: An evolution strategy assisted by an ensemble of local Gaussian process models. In: Proceedings of the 15th Annual Conference on Genetic and Evolutionary Computation, GECCO 2013, pp. 447–454. ACM, New York (2013)
15. Ong, Y.S., Nair, P.B., Keane, A.J.: Evolutionary optimization of computationally expensive problems via surrogate modeling. AIAA J. **41**(4), 687–696 (2003)
16. Pitra, Z., Bajer, L., Holeňa, M.: Comparing SVM, Gaussian process and random forest surrogate models for the CMA-ES. In: ITAT 2015: Information Technologies - Applications and Theory, pp. 186–193. CreateSpace Independent Publishing Platform, North Charleston (2015)
17. Rasmussen, C.E., Williams, C.K.I.: Gaussian Processes for Machine Learning. Adaptative Computation and Machine Learning Series. MIT Press, Cambridge (2006)

Efficient Global Optimization with Indefinite Kernels

Martin Zaefferer(✉) and Thomas Bartz-Beielstein

Faculty of Computer Science and Engineering Science,
Cologne University of Applied Sciences (TH Köln), Steinmüllerallee 1,
51643 Gummersbach, Germany
{martin.zaefferer,thomas.bartz-beielstein}@th-koeln.de

Abstract. Kernel based surrogate models like Kriging are a popular remedy for costly objective function evaluations in optimization. Often, kernels are required to be definite. Highly customized kernels, or kernels for combinatorial representations, may be indefinite. This study investigates this issue in the context of Kriging. It is shown that approaches from the field of Support Vector Machines are useful starting points, but require further modifications to work with Kriging. This study compares a broad selection of methods for dealing with indefinite kernels in Kriging and Kriging-based Efficient Global Optimization, including spectrum transformation, feature embedding and computation of the nearest definite matrix. Model quality and optimization performance are tested. The standard, without explicitly correcting indefinite matrices, yields functional results, which are further improved by spectrum transformations.

1 Introduction

When optimization requires time-consuming experiments, surrogate models are a well established approach to reduce the load of objective function evaluations [9]. Kernel-based models are a popular choice, e.g., Support Vector Machines (SVM) and especially Kriging. Often, kernels are required to be positive semi-definite (PSD), e.g., to allow for the existence of a map to a higher dimensional feature space (kernel trick) or to allow for interpretation of kernel matrices as a correlation matrices [5,21]. While ordinary kernels are PSD, users may have to apply uncommon kernels [20]. One example are distance-based kernels for combinatorial optimization problems, that may not be definite [18,26,27]. Even in real-valued search spaces, prior knowledge can be used to design promising, custom, indefinite kernels. While research on indefinite kernels with Kriging is sparse, the SVM field provides an useful starting point [20].

This study outlines existing techniques for dealing with indefinite distances and kernels. The issues of their application to Kriging are elaborated and possible solutions are explained. A comparative test-study with transparent, artificial test-functions is presented, with the goal of determining the benefit of different indefiniteness correction methods.

J. Handl et al. (Eds.): PPSN XIV 2016, LNCS 9921, pp. 69–79, 2016.
DOI: 10.1007/978-3-319-45823-6_7

2 Terms and Definitions

This study makes use of the following concepts.

Input space: The input space is a non-empty set $\mathcal{X}$.

Sample: A sample $x \in \mathcal{X}$ can be a vector (continuous or discrete), string, tree or some other object.

Kernel function: A symmetric function $\mathrm{k}(x, x')$ with $\mathrm{k} : \mathcal{X} \times \mathcal{X} \to \mathbb{R}$.

Distance function: A symmetric function $\mathrm{d}(x, x')$ with $\mathrm{d} : \mathcal{X} \times \mathcal{X} \to \mathbb{R}$, $\mathrm{d}(x, x') \geq 0$ and $\mathrm{d}(x, x') = 0$ if $x = x'$.

Distance metric: A distance function $\mathrm{d}(x, x')$ which a) is zero *iff* $x = x'$ and b) fulfills the triangle inequality $\mathrm{d}(x, x') + \mathrm{d}(x', x'') \geq \mathrm{d}(x, x'')$.

Kernel matrix: A matrix $\boldsymbol{K}$ with element $k_{ij} = \mathrm{k}(x_i, x_j)$.

Distance matrix: A matrix $\boldsymbol{D}$ with element $d_{ij} = \mathrm{d}(x_i, x_j)$.

Ill-conditioning: A symmetric matrix is ill-conditioned if $|\lambda_n|/|\lambda_1|$ is large. λ_n is the largest and λ_1 the smallest eigenvalue. Ill-conditioning is not in the focus of this paper, but may require related methods.

Definiteness: A symmetric $n \times n$ matrix $\boldsymbol{A}$ is positive definite (PD) iff $\boldsymbol{c}\boldsymbol{A}\boldsymbol{c}^T > 0$ for all $\boldsymbol{c} \in \mathbb{R}^n$. This is equivalent to all eigenvalues $\lambda_1 \leq \lambda_2 \leq \ldots \leq \lambda_n$ of $\boldsymbol{A}$ being positive. Respectively, a matrix is negative definite (ND) iff all eigenvalues are negative. The matrix is Positive or Negative Semi-Definite (PSD, NSD), iff all eigenvalues are non-negative (i.e., some are zero) or non-positive. A kernel matrix is usually required to be PSD. A broader class are Conditionally PSD or NSD (CPSD, CNSD) matrices, with the condition $\sum_{i=1}^{n} c_i = 0$. If a matrix matches none of these criteria, it is indefinite.

A function k is PSD (NSD) iff $\sum_{i=1}^{n} \sum_{j=1}^{n} c_i c_j \mathrm{k}(x_i, x_j) \geq (\leq)\ 0$, for all $n \in \mathbb{N}$ and $x \in \mathcal{X}$. It is conditionally definite if $\sum_{i=1}^{n} c_i = 0$. A distance measure $\mathrm{d}(x, x')$ is CNSD iff the Gaussian kernel $\mathrm{k}(x, x') = \exp(-\theta \mathrm{d}(x, x'))$ is PSD for all $\theta > 0$ [21, Proposition 2.28]. Also, the triangle inequality is a necessary condition for CNSDness [7, Corollary 1]. In case of SVM, PSD kernels guarantee that the mapping into some higher dimensional feature space exists (kernel trick) [21].

Correlation function: A special case of PSD kernels are correlation functions. Their values should be $-1 \leq k(x, x') \leq 1$, and $k(x, x') = 1$ if $x = x'$. Correlation matrices are required for statistical models like Kriging. The PSD requirement becomes clear when considering a linear combination of random variables. Indefinite matrices would imply negative variances of such combinations.

Kriging: This definition is based on [5] and some adaptations in [27]. Given a set of n samples $\boldsymbol{X} = \{x_i\}$, observations $\mathbf{y} = \{y_i\}$ and $i = 1 \ldots n$, Kriging interprets the observed responses $\mathbf{y}$ as realizations of a stochastic process. The set of random vectors $\boldsymbol{Y} = \{Y(x_i)\}$ is used to define this stochastic process. Correlations can, e.g., be modeled by the kernel

$$\mathrm{cor}\left[Y(x), Y(x')\right] = \mathrm{k}(x, x') = \exp(-\theta \mathrm{d}(x, x')). \tag{1}$$

with $\theta \in \mathbb{R}_+$. Both $\mathrm{k}(x, x')$ and $\mathrm{d}(x, x')$ can be chosen depending on the problem. For example, in case of combinatorial optimization $\mathrm{d}(x, x')$ can be a distance measures for binary strings, permutations, or trees [17,27].

Kriging predictor: The correlation matrix $\boldsymbol{K}$ is used in the predictor function

$$\hat{y}(x) = \hat{\mu} + \boldsymbol{k}^T \boldsymbol{K}^{-1}(\mathbf{y} - \mathbf{1}\hat{\mu}), \tag{2}$$

where $\hat{y}(x)$ is the predicted function value of a new sample x, $\hat{\mu}$ is the Maximum Likelihood Estimate (MLE) of the process mean, $\mathbf{1}$ is a vector of ones and $\boldsymbol{k}$ is the column vector of correlations between training samples $\boldsymbol{X}$ and the new sample x. Kernel parameters (e.g., θ) are determined by MLE. The MLE based on uncorrected, indefinite correlation matrices can be very misleading, producing unusable models. As an indefinite matrix can not be a correlation matrix, a basic assumption of the model is violated. Hence, indefiniteness requires correction.

Uncertainty estimate: The uncertainty of the prediction is estimated with

$$\hat{s}^2(x) = \hat{\sigma}^2(1 - \boldsymbol{k}^T \boldsymbol{K}^{-1} \boldsymbol{k}), \tag{3}$$

where $\hat{\sigma}^2$ is an estimate of the process variance, also determined by MLE.

Efficient Global Optimization: The uncertainty estimate $\hat{s}(x)$ is used in the Efficient Global Optimization (EGO) algorithm [11]. In EGO, a Kriging model is first built based on an initial set of observations $\mathbf{y}$ with elements $y_i = \mathrm{f}(x_i)$. Here, $\mathrm{f} : \mathcal{X} \to \mathbb{R}$ is an objective function to be minimized. It is assumed to be very expensive to evaluate (due to consumption of time or other resources). If $\hat{s}(x) > 0$, the Expected Improvement (EI) [15] of a sample is

$$\mathrm{EI}(x) = (\min(\mathbf{y}) - \hat{y}(x))\Phi\left(\frac{\min(\mathbf{y}) - \hat{y}(x)}{\hat{s}(x)}\right) + \hat{s}(x)\phi\left(\frac{\min(\mathbf{y}) - \hat{y}(x)}{\hat{s}(x)}\right),$$

else $\mathrm{EI}(x) = 0$, where Φ is the normal cumulative distribution function, and ϕ the normal probability density function. The sample x that maximizes $\mathrm{EI}(x)$ is evaluated with $\mathrm{f}(x)$. The resulting data is used to update the model. This repeats until a termination criterion is fulfilled (e.g., function evaluation budget).

3 Handling Indefinite Kernels

Several recent studies on SVMs (and related methods) dealt with indefinite kernels, cf. the survey in [20]. This topic has seen less attention in connection to Kriging [2,3,13]. Four types of methods can be identified. *Spectrum transformations* attempt to transform the matrix such that all eigenvalues have the desired sign. They have been used for SVMs [20] and, to some extend, for Gaussian Processes [2]. They are outlined and extended by repair methods in Sects. 3.1 to 3.3. *Nearest matrix* algorithms (Sect. 3.4) try to find matrices that are definite as well as close to the original matrices. *Feature embedding* (Sect. 3.5) understands the indefinite similarities (or distances) as features, and uses a standard, definite kernel to compute a surrogate similarity based on these features. *Method modifications* have been introduced to remove the necessity of definiteness in SVMs, e.g., by converting the quadratic programming problem to a linear one (LP-SVM or 1-norm SVM [12,14,28]). Method modifications are usually not transferable to Kriging and hence not considered here.

In the following, $\tilde{\boldsymbol{K}}$ denotes the definiteness-corrected variant of $\boldsymbol{K}$. Respectively, $\tilde{\boldsymbol{k}}$ will be the modified variant of $\boldsymbol{k}$ (cf. Eq. (2)). For distances, $\tilde{\boldsymbol{D}}$ and $\tilde{\boldsymbol{d}}$ are employed equivalently.

3.1 Spectrum Transformation: Kernel

The basis for the spectrum transformation is the decomposition of the kernel matrix $\boldsymbol{K} = \boldsymbol{U}\boldsymbol{\Lambda}\boldsymbol{U}^T$, where $\boldsymbol{U}$ is the matrix of eigenvectors of $\boldsymbol{K}$, $\boldsymbol{\Lambda} = \text{diag}(\boldsymbol{\lambda})$ the diagonal matrix containing the eigenvalues of $\boldsymbol{K}$. Following Chen et al. [4], the spectrum transformation can be written as a linear transformation based on some vector $\boldsymbol{a} \in \mathbb{R}^n$:

$$\tilde{\boldsymbol{K}} = \boldsymbol{A}\boldsymbol{K} \quad \text{with} \quad \boldsymbol{A} = \boldsymbol{U}\text{diag}(\boldsymbol{a})\boldsymbol{U}^T. \tag{4}$$

Several choices for $\boldsymbol{a}$ are available [20,24].

(I) Spectrum *flip* transforms the eigenvalues to their absolute values, with $\tilde{\lambda}_i = |\lambda_i|$ and $\boldsymbol{a}_{flip} = \text{sign}(\boldsymbol{\lambda})$. With Eq. (6) and using $\boldsymbol{a}_{flip}$, the resulting approach is very similar to the one described by Loosli et al. [12] for SVMs in Krein spaces.

(II) Spectrum *clip* removes negative eigenvalues by setting them to zero, with $\tilde{\lambda}_i = \max(\lambda_i, 0)$ and $\boldsymbol{a}_{clip} = \{\mathbb{I}(\lambda_1), \ldots, \mathbb{I}(\lambda_n)\}$, where $\mathbb{I}(\lambda_i) = 1$ if $\lambda_i \geq 0$ else $\mathbb{I}(\lambda_i) = 0$. Spectrum clip relates to the Moore–Penrose pseudoinverse [16], which is sometimes used in case of ill-conditioned $\boldsymbol{K}$.

(III) Spectrum *shift* uses $\tilde{\lambda}_i = \lambda_i + \eta$ with $\eta \in \mathbb{R}_+$ and $\tilde{\boldsymbol{K}} = \boldsymbol{K} + \eta \boldsymbol{I}_n$. Shifting is the same as the nugget effect that may be used in the Kriging model, where η is an additional parameter determined by MLE. It may be reasonable to combine it with some of the other transformations, e.g., to deal with numerical issues or noise. The nugget effect is often used to regularize ill-conditioned $\boldsymbol{K}$ [16].

(IV) Spectrum *square* uses $\tilde{\lambda}_i = (\lambda_i)^2$ and $\boldsymbol{a}_{sqr} = \boldsymbol{\lambda}$. Also: $\tilde{\boldsymbol{K}} = \boldsymbol{K}\boldsymbol{K}$.

(V) Spectrum *diffusion* uses $\tilde{\lambda}_i = \exp(\lambda_i)$ and $\boldsymbol{a}_{diff} = \exp(\boldsymbol{\lambda})/\boldsymbol{\lambda}$. This leads to the diffusion Kernel, with $\tilde{\boldsymbol{K}} = \exp(\boldsymbol{K})$ [24].
Of all these transformations, only shift (cf. nugget effect [5,16]) and clip (cf. pseudo-inverse [16] or multi dimensional scaling [3]) have been used with Kriging, although mostly for the purpose of dealing with noise or ill-conditioning.

The same transformation $\boldsymbol{A}$ has to be applied to $\boldsymbol{k}$ for prediction (see Eq. (2)):

$$\tilde{\boldsymbol{k}} = \boldsymbol{A}\boldsymbol{k}. \tag{5}$$

In case of spectrum shift, Eq. (5) is not required since the spectrum shift only affects self-similarities $\text{k}(x, x)$. While computing $\tilde{\boldsymbol{k}}$ is a consistent way to treat new test samples [4], it has been noted as a drawback due to the effort of (5) for each single prediction [12]. This issue can be remedied as follows. In the Kriging predictor given in Eq. (2), $\boldsymbol{k}^T\boldsymbol{K}^{-1}$ is computed. With the respective transformations we can prove that:

$$\tilde{\boldsymbol{k}}^T\tilde{\boldsymbol{K}}^{-1} = (\boldsymbol{A}\boldsymbol{k})^T\tilde{\boldsymbol{K}}^{-1} = \boldsymbol{k}^T\boldsymbol{A}^T\tilde{\boldsymbol{K}}^{-1}. \tag{6}$$

The computation of $\boldsymbol{A}^T\tilde{\boldsymbol{K}}^{-1}$ has to be performed only once after training, since it does not depend on the new sample. Afterwards, prediction requires only the usual computational effort of the Kriging predictor. Using Eq. (4) and $\boldsymbol{U}^T = \boldsymbol{U}^{-1}$ we can also prove that

$$\tilde{\boldsymbol{k}}^T\tilde{\boldsymbol{K}}^{-1} = \boldsymbol{k}^T\boldsymbol{K}^{-1}. \tag{7}$$

Similarly, the uncertainty estimate in Eq. (3) uses $\boldsymbol{k}^T\boldsymbol{K}^{-1}\boldsymbol{k}$. With Eqs. (4)–(6) this becomes:

$$\tilde{\boldsymbol{k}}^T\tilde{\boldsymbol{K}}^{-1}\tilde{\boldsymbol{k}} = \boldsymbol{k}^T\boldsymbol{A}^T\tilde{\boldsymbol{K}}^{-1}\boldsymbol{A}\boldsymbol{k}. \tag{8}$$

Hence, $\boldsymbol{A}^T\tilde{\boldsymbol{K}}^{-1}\boldsymbol{A}$ needs to be computed only once. In the following, *PSD-correction* refers to all methods that transform the spectrum of the kernel matrix, with $\tilde{\boldsymbol{K}} = \mathrm{SPEC_{PSD}}(\boldsymbol{K})$.

3.2 Spectrum Transformation: Distance

Spectrum transformations can also be applied to distances. PSD-correction can be applied directly via $\tilde{\boldsymbol{D}} = \mathrm{SPEC_{NSD}}(\boldsymbol{D}) = -\mathrm{SPEC_{PSD}}(-\boldsymbol{D})$. New data is handled accordingly, i.e., $\tilde{\boldsymbol{d}} = \boldsymbol{A}\boldsymbol{d}$. Equations (6)–(8) are not useful in this case. Thus, effort increases for prediction but decreases for MLE.

Alternatively, spectrum transformations can be used to generate CNSD matrices as described by Glunt et al. [6]. First, $\boldsymbol{Q} = \boldsymbol{I} - (2\mathbf{v}\mathbf{v}^T)/(\mathbf{v}^T\mathbf{v})$ with $\mathbf{v} = 1, 1, \ldots, 1, \sqrt{n}$ is computed and used to yield $\widehat{\boldsymbol{D}} = \boldsymbol{Q}(-\boldsymbol{D})\boldsymbol{Q}$. Then, $\widehat{\boldsymbol{D}}_{(-n,-n)}$ is extracted, which is $\widehat{\boldsymbol{D}}$ without last row and column. The matrix $\check{\mathbf{D}}$ is then constructed using $\mathrm{SPEC_{PSD}}(\widehat{\boldsymbol{D}}_{-n,-n})$ and the unchanged last row and column of $\widehat{\boldsymbol{D}}$. Finally, the matrix in the original form is $\tilde{\boldsymbol{D}} = -\boldsymbol{Q}\check{\mathbf{D}}\boldsymbol{Q}$. With spectrum clip, this approach is similar to Multi Dimensional Scaling, as used by Boisvert et al. [3] to correct indefiniteness in a Kriging model.

Due to the more complex transformations in the CNSD case, $\tilde{\boldsymbol{d}} = \boldsymbol{A}\boldsymbol{d}$ is no longer valid. Instead, the augmented distance matrix $\boldsymbol{D}_{aug}$ is computed, which includes distances between all training and new data. Then,

$$\boldsymbol{D}_{aug} = \begin{bmatrix} \boldsymbol{D} & \boldsymbol{d} \\ \boldsymbol{d}^T & 0 \end{bmatrix} \quad \text{and (after transformation)} \quad \begin{bmatrix} \tilde{\boldsymbol{D}} & \tilde{\boldsymbol{d}} \\ \tilde{\boldsymbol{d}}^T & \tilde{\delta} \end{bmatrix} = \tilde{\boldsymbol{D}}_{aug}, \tag{9}$$

where $\tilde{\delta}$ is the potentially non-zero self-distance of the transformed new data. The resulting $\tilde{\boldsymbol{d}}$ can be used in Eqs. (1) and (2). In the following, spectrum transformations of the distance matrix are denoted with *NSD-* or *CNSD-correction*.

3.3 Spectrum Transformations: Condition-Repair

The spectrum transformations may yield definite matrices that do not fulfill the additional conditions required for distance and correlation functions (cf. Sect. 2). One consequence is, that uncertainty estimates for observed samples (training data) become non-zero. This may stall the optimization progress (cf. a similar issue with the nugget effect described in [5]). Methods that mend this issue are referred to as *condition-repair*.

A correlation matrix can be repaired with $\tilde{k}^*_{ij} = \tilde{k}_{ij}/\mathrm{sqrt}(\tilde{k}_{ii}\tilde{k}_{jj})$ [19]. A CNSD distance matrix $\tilde{\boldsymbol{D}}$ can be repaired with $\tilde{d}^*_{ij} = 2\tilde{d}_{ij} - \tilde{d}_{ii} - \tilde{d}_{jj}$. The result is CNSD, non-negative and has zero diagonal [22]. In case of condition-repair, correlations $\tilde{\boldsymbol{k}}$ and distances $\tilde{\boldsymbol{d}}$ between training data and new samples have to be derived as outlined in Eq. (9). Spectrum shift only changes the diagonal of $\boldsymbol{K}$. Its influence on the uncertainty estimate can be remedied by re-interpolation [5].

3.4 Nearest Matrix Approach

Finding the nearest correlation matrix [8] or nearest euclidean distance matrix [6] is closely related to spectrum transformation. An alternating projections approach can be used to compute the nearest matrices. The first projection employs the spectrum clip. The second projection sets diagonals to one (correlation) or zero (distance). Thus, further condition-repair is not required. Unfortunately, these methods lack an efficient way of handling new data. Similarly to the condition-repair procedures, Eq. (9) can be used to derive $\tilde{\boldsymbol{d}}$ (or $\tilde{\boldsymbol{k}}$ analogously).

3.5 Feature Embedding

In feature embedding [12], non-CNSD distances can be used as input features for a CNSD distance function: $\tilde{d}_{ij} = \mathrm{d}_{\mathrm{def}}(\boldsymbol{d}_{i\cdot}, \boldsymbol{d}_{j\cdot})$, where $\boldsymbol{d}_{i\cdot}$ and $\boldsymbol{d}_{j\cdot}$ are the ith and jth rows of $\boldsymbol{D}$, and $\mathrm{d}_{def}(x, x')$ is a CNSD distance function (here: Euclidean). Distances $\boldsymbol{d}$ between training and new data have to be subject to $\tilde{d}_i = \mathrm{d}_{def}(\boldsymbol{d}, \boldsymbol{d}_{i\cdot})$.

4 Experimental Setup

Test-Problems: The samples x were restricted to be permutations, to enable a well understandable and controllable test case. Other object types are possible but were omitted for the sake of brevity. Different numbers of permutation elements were tested: $m = 5, 7, 10$. The experiments were performed with simple test-functions $\mathrm{f}(x) = \min_i \mathrm{d}(x, \gamma_i)$, where x is a sample (permutation), and the respective function value $\mathrm{f}(x)$ is the minimum distance to randomly chosen centers $\gamma_i \in \mathcal{X}$, with $i = 1, \ldots, w$. For the sake of this test, the function $\mathrm{f}(x)$ was assumed to be expensive. The number of centers w control the multi-modality of the function. In case of $w = 1$, $\mathrm{f}(x)$ is unimodal (as used in [18]). For the experiments, $w = 1, 3$ and 5 was tested. Two distance measures for permutations were used: The Interchange Distance is the minimal number of transpositions of arbitrary elements required to transform one permutation into another. It is metric, but not CNSD. As a more pathological (yet admittedly quite artificial) test-case, we chose the non-metric, non-CNSD distance $\mathrm{d}_{Lp}(x, x') = (\sum_{i=1}^{n} |x_i - x'_i|^p)^{1/p}$ with $p = 1/2$. Here, the permutations are interpreted as a vector of integers.

Performance Measures: Two sets of experiments were performed, (1) testing for modeling performance (including the quality of the uncertainty estimate) and (2) for optimization performance. The Root Mean Squared Error (RMSE) was used to estimate prediction accuracy. To assess the uncertainty estimate, standardized residuals $r = (y - \hat{y})/\hat{s}$ were computed, cf. [11,23]. These are used to calculate the Cramèr-von Mises (CVM) test statistic [1] (comparing against a normal distribution with zero mean and unit variance). 10-fold cross validation is used to receive statistically sound results. For the modeling experiments, the number of samples is $n = 20, 40$ and 60. For the optimization performance, best values found after 20 and 100 objective function evaluations are reported.

Model Settings: The Dividing Rectangles algorithm [10] was chosen to optimize the model parameters (θ, η) during MLE. For each parameter, 200 likelihood evaluations were allowed. A relative tolerance of 1e−6 was used to detect earlier convergence. For (uncorrected) indefinite matrices, the logarithmic likelihood evaluation was set to return a penalty of $-1e4 + \lambda_1$, to drive the search into the direction of PSD matrices. In all cases, PSD matrices could be established. However, the resulting matrix was sometimes numerically intractable in case of spectrum diffusion, which was hence excluded from further analysis (see Sect. 5). Re-interpolation [5] was employed to correct the uncertainty estimates in case of spectrum shift. Note, that η was always added to the diagonal of $\tilde{\boldsymbol{K}}$, i.e., *after* applying other correction methods. The models always used the same distance functions that were employed in the test function, combined with the kernel in Eq. (1). This simulates the case where an adequate distance is chosen by prior knowledge. All experiments were repeated 20 times.

Optimization Settings: For optimization, most settings remain unchanged. The budget of evaluations of $\mathrm{f}(x)$ was set to 100. Ten initial samples were chosen at random and evaluated with $\mathrm{f}(x)$. In each following step, the candidate that maximized EI (cf. Sect. 2) was determined by a Genetic Algorithm (GA). The GA had a budget of 2000 model evaluations for each step, except for $m = 5$, where brute force was used ($m! = 120$ model evaluations). The GA used interchange mutation (transposition of arbitrary elements) and cycle crossover. The population size was 20, the permutation rate $1/m$ and the recombination rate 0.5. As a baseline-comparison, a simple and model-free random search with 100 objective function evaluations was performed. All experiments were repeated 20 times.

5 Observations and Discussion

To summarize overall performance, statistical multiple-comparison tests were used. Since the data were non-normal and the variances inhomogeneous, a rank

Table 1. Ranks for RMSE (R), CVM values (C), best value after 20 evaluations (F_a) and 100 evaluations (F_b). Ranks are based on Tukey's HSD test, small values are better. P indicates percentage of cases where the optimum was found within 100 evaluations, large values are better. Table is sorted by $F_a + F_b$, with tie-breaker P. Color indicates a rank of 1, or $P \geq 0.9$. In the *names* columns, the leading boolean denotes whether condition-repair was used (T) or not (F). *CNSD/NSD/PSD*: the correction type, *feature*: feature embedding, *near*: nearest matrix approach, *standard*: no specific correction and *random*: random search. Other terms refer to the spectrum transformations.

names	R	C	F_a	F_b	P	names	R	C	F_a	F_b	P
T.flip.PSD	6	4	1	1	1	T.square.PSD.shift	6	5	2	2	0.91
F.clip.CNSD	1	6	1	1	0.99	standard.shift	7	4	2	2	0.91
F.clip.CNSD.shift	1	8	1	1	0.98	T.square.NSD	3	2	2	2	0.89
T.clip.NSD	1	2	1	1	0.96	T.square.CNSD	3	2	2	2	0.89
T.clip.CNSD	1	2	1	1	0.96	T.square.CNSD.shift	3	4	3	2	0.92
F.flip.CNSD.shift	1	10	1	1	0.95	F.square.PSD	4	4	2	3	0.86
near.CNSD	3	4	1	1	0.94	T.square.PSD	5	3	2	3	0.85
F.flip.CNSD	1	7	1	1	0.94	T.clip.PSD	4	3	2	3	0.84
T.flip.PSD.shift	3	5	2	1	0.98	F.clip.PSD	4	3	2	3	0.84
T.clip.CNSD.shift	2	3	2	1	0.98	standard	4	3	2	3	0.84
near.CNSD.shift	3	5	2	1	0.98	near.PSD	5	3	2	3	0.83
F.clip.PSD.shift	3	5	2	1	0.97	F.clip.NSD.shift	2	9	3	3	0.84
T.clip.PSD.shift	3	7	2	1	0.97	F.clip.NSD	2	9	3	3	0.84
T.clip.NSD.shift	2	3	2	1	0.97	F.flip.NSD.shift	4	8	3	3	0.83
F.flip.PSD	7	4	2	1	0.96	F.flip.NSD	4	6	3	3	0.82
F.flip.PSD.shift	6	4	2	1	0.95	near.PSD.shift	3	8	3	4	0.8
T.flip.NSD.shift	2	2	2	1	0.94	F.square.PSD.shift	3	6	3	4	0.76
feature	3	1	2	1	0.94	F.square.CNSD	4	8	3	5	0.7
T.flip.NSD	1	1	1	2	0.92	F.square.CNSD.shift	3	9	4	5	0.73
feature.shift	3	2	3	1	0.94	F.square.NSD.shift	2	9	4	5	0.69
T.flip.CNSD.shift	2	2	2	2	0.93	F.square.NSD	4	10	3	6	0.67
T.square.NSD.shift	3	4	2	2	0.92	random			5	7	0.35
T.flip.CNSD	1	1	2	2	0.92						

transformation was performed for each combination of n, w, m and distance function. Then, Tukey's Honest Significant Differences (HSD) test [25] was used with a significance level $\alpha = 0.05$. Results were largely confirmed by a non-parametric test, which disagreed in about 2 % of the cases. With the resulting pair-wise comparison, a ranking was computed. All methods that were not significantly worse than any other received rank 1 and were removed. From the remainder, every method that was not significantly worse than any other received rank 2, and so on. Results from the spectrum diffusion approach were excluded as it performed poorly and failed several times, due to numerical issues with excessively large numbers. Table 1 reports the respective ranks. Interestingly, the ranks for model accuracy and optimization performance disagree often. One reason may be, that optimization only requires a locally accurate model.

It could be observed, that usable models were achieved by the standard approach, as it outperformed the random search. That is because even a non-CNSD distance matrix may yield a PSD kernel matrix if θ is chosen large enough, but not too large. This becomes obvious with $\lim_{\theta \to \infty} \boldsymbol{K} = \boldsymbol{I}$, which is of course PD. However, if $\theta \to \infty$, Eq. (2) will just yield the mean of observations $\mathbf{y}$.

Enhancing the standard approach by spectrum shift improved optimization performance, but received the worst RMSE ranks. In general, a clear benefit of shift could not be observed. In combination with other indefiniteness-correcting methods, it either improved or deteriorated results. Due to the additional cost of fitting η, it may be undesirable for non-noisy data.

The simple feature embedding performed robustly, but not for smaller data sets. The performance after 20 evaluations (F_a in Table 1) was suboptimal. Feature embedding seemed to require larger data-sets to learn the embedding.

Spectrum transformations were among the best performers. Their main draw-back is the difficulty of deciding on (a) usage of condition-repair (b) type of transformation and (c) whether NSD-, CNSD- or PSD-correction should be used. For a), the results are not quite conclusive, but a large block of the worse performing methods ($F_b > 2$ in Table 1) does not employ condition-repair. CVM statistic values are often better if condition-repair is used. For (b), spectrum square is clearly worse than clip or flip, yet it may provide good results in combination with spectrum shift. Spectrum flip was not significantly different from spectrum clip. For (c), the results were mixed, but RMSE ranks seemed to better with NSD- and CNSD-correction compared to PSD-correction. Intuitively, this makes sense: NSD- and CNSD-correction correct the distance matrix, which was the source of the indefiniteness. If the kernel function is the source, only PSD-correction is applicable. Despite very similar performance, NSD- may be preferred to CNSD-correction due to higher computational complexity of the latter.

The nearest matrix approaches required the most computational effort, with tenfold run-times or more. This is due to the necessity of solving an optimization problem for each correction. Since they performed no better than the related spectrum clip methods, the nearest matrix approaches can be disregarded.

6 Conclusions and Outlook

This study dealt with indefinite kernels in the Kriging-based EGO algorithm. Working Kriging models could be derived, even when indefiniteness was not explicitly corrected (besides the penalty described in Sect. 4). Methods based on spectrum transformations improved the performance. The spectrum transformations were compared to feature embedding and computations of the nearest definite matrix. As some of the resulting matrices were no proper correlation matrices, further condition-repair mechanisms were included. In some cases, this additional condition-repair was beneficial. From the set of spectrum transformations, spectrum flip and clip performed best, while square and diffusion performed poorly, in the latter case producing numerically intractable results.

Overall, the results indicate that choosing an adequate method automatically may be problematic. Cross-validation is an option, but not ideal, due to the lack of agreement between model accuracy and optimization performance. Also, some of the worst performing models reported large likelihoods, hence disqualifying a selection based on likelihood. More extensive experiments or a theoretical analysis of the various approaches could help dealing with this issue. For theoretical considerations, it is promising to see that spectrum flip works so well, since it is theoretically well-founded for SVMs [12]. Furthermore, these results may also be of interest in the context of regularization or ill-conditioning, especially with respect to condition-repairing procedures and handling of new data samples.

References

1. Anderson, T.W.: On the distribution of the two-sample Cramer-von Mises criterion. Ann. Math. Stat. **33**(3), 1148–1159 (1962)
2. Ayhan, M.S., Chu, C.-H.H.: Towards indefinite gaussian processes. Technical report, University of Louisiana at Lafayette (2012)
3. Boisvert, J.B., Deutsch, C.V.: Programs for kriging and sequential Gaussian simulation with locally varying anisotropy using non-Euclidean distances. Comput. Geoscie. **37**(4), 495–510 (2011)
4. Chen, Y., Gupta, M.R., Recht, B.: Learning kernels from indefinite similarities. In: Proceedings of the 26th Annual International Conference on Machine Learning, ICML 2009, pp. 145–152. ACM, New York (2009)
5. Forrester, A., Sobester, A., Keane, A.: Engineering Design via Surrogate Modelling. Wiley, Hoboken (2008)
6. Glunt, W., Hayden, T.L., Hong, S., Wells, J.: An alternating projection algorithm for computing the nearest Euclidean distance matrix. SIAM J. Matrix Anal. Appl. **11**(4), 589–600 (1990)
7. Haasdonk, B., Bahlmann, C.: Learning with distance substitution kernels. In: Rasmussen, C.E., Bülthoff, H.H., Schölkopf, B., Giese, M.A. (eds.) DAGM 2004. LNCS, vol. 3175, pp. 220–227. Springer, Heidelberg (2004)
8. Higham, N.J.: Computing the nearest correlation matrix-a problem from finance. IMA J. Numer. Anal. **22**(3), 329–343 (2002)
9. Jin, Y.: A comprehensive survey of fitness approximation in evolutionary computation. Soft Comput. **9**(1), 3–12 (2005)
10. Jones, D.R., Perttunen, C.D., Stuckman, B.E.: Lipschitzian optimization without the lipschitz constant. J. Optim. Theory Appl. **79**(1), 157–181 (1993)
11. Jones, D.R., Schonlau, M., Welch, W.J.: Efficient global optimization of expensive black-box functions. J. Glob. Optim. **13**(4), 455–492 (1998)
12. Loosli, G., Canu, S., Ong, C.: Learning SVM in Krein spaces. IEEE Trans. Pattern Anal. Mach. Intell. **38**(6), 1204–1216 (2015)
13. Manchuk, J.G., Deutsch, C.V.: Robust solution of normal (kriging) equations. Technical report, CCG Alberta (2007)
14. Mangasarian, O.L.: Generalized support vector machines. In: Smola, A.J., Bartlett, P., Schölkopf, B., Schuurmans, D. (eds.) Advances in Large Margin Classifiers, pp. 135–146. MIT Press (2000)
15. Mockus, J., Tiesis, V., Zilinskas, A.: The application of bayesian methods for seeking the extremum. In: Towards Global Optimization 2, pp. 117–129. North-Holland, Amsterdam (1978)

16. Mohammadi, H., Le Riche, R., Durrande, N., Touboul, E., Bay, X.: An analytic comparison of regularization methods for Gaussian Processes. Research report, Ecole Nationale Supérieure des Mines de Saint-Etienne, LIMOS (2016)
17. Moraglio, A., Kattan, A.: Geometric generalisation of surrogate model based optimisation to combinatorial spaces. In: Merz, P., Hao, J.-K. (eds.) EvoCOP 2011. LNCS, vol. 6622, pp. 142–154. Springer, Heidelberg (2011)
18. Moraglio, A., Kim, Y.-H., Yoon, Y.: Geometric surrogate-based optimisation for permutation-based problems. In: Proceedings of the 13th Annual Conference Companion on Genetic and Evolutionary Computation, GECCO 2011, pp. 133–134. ACM, New York (2011)
19. Rebonato, R., Jäckel, P.: The most general methodology to create a valid correlation matrix for risk management and option pricing purposes. J. Risk **2**(2), 17–27 (1999)
20. Schleif, F.-M., Tino, P.: Indefinite proximity learning: a review. Neural Comput. **27**(10), 2039–2096 (2015)
21. Schölkopf, B., Smola, A.J.: Learning with Kernels: Support Vector Machines, Regularization, Optimization, and Beyond. MIT Press, Cambridge (2001)
22. Schölkopf, B.: The kernel trick for distances. In: Leen, T.K., Dietterich, T.G., Tresp, V. (eds.) Advances in Neural Information Processing Systems, vol. 13, pp. 301–307. MIT Press, Cambridge (2001)
23. Wagner, T.: Planning and multi-objective optimization of manufacturing processes by means of empirical surrogate models. Ph. D. thesis, TU Dortmund. Vulkan Verlag (2013)
24. Wu, G., Chang, E.Y., Zhang, Z.: An analysis of transformation on non-positive semidefinite similarity matrix for kernel machines. In: Proceedings of the 22nd International Conference on Machine Learning (2005)
25. Yandell, B.S.: Practical Data Analysis for Designed Experiments. Chapman and Hall/CRC, London (1997)
26. Zaefferer, M., Stork, J., Bartz-Beielstein, T.: Distance measures for permutations in combinatorial efficient global optimization. In: Bartz-Beielstein, T., Branke, J., Filipič, B., Smith, J. (eds.) PPSN 2014. LNCS, vol. 8672, pp. 373–383. Springer, Heidelberg (2014)
27. Zaefferer, M., Stork, J., Friese, M., Fischbach, A., Naujoks, B., Bartz-Beielstein, T.: Efficient global optimization for combinatorial problems. In: Genetic and Evolutionary Computation Conference, GECCO 2014, pp. 871–878. ACM (2014)
28. Zhu, J., Rosset, S., Hastie, T., Tibshirani, R.: 1-norm support vector machines. Adv. Neural Inf. Process. Syst. **16**(1), 49–56 (2004)

A Fitness Cloud Model for Adaptive Metaheuristic Selection Methods

Christopher Jankee[1], Sébastien Verel[1(✉)], Bilel Derbel[2], and Cyril Fonlupt[1]

[1] Univ. Littoral Côte d'Opale, EA 4491 - LISIC, Calais, France
verel@lisic.univ-littoral.fr
[2] Université Lille, CRIStAL – UMR 9189 – Inria, Lille, France

Abstract. Designing portfolio adaptive selection strategies is a promising approach to gain in generality when tackling a given optimization problem. However, we still lack much understanding of what makes a strategy effective, even if different benchmarks have been already designed for these issues. In this paper, we propose a new model based on fitness cloud allowing us to provide theoretical and empirical insights on when an on-line adaptive strategy can be beneficial to the search. In particular, we investigate the relative performance and behavior of two representative and commonly used selection strategies with respect to static (off-line) and purely random approaches, in a simple, yet sound realistic, setting of the proposed model.

1 Introduction

Context and Motivation. In the last decades, the optimization community has gained much expertise in the design of general purpose randomized heuristics to tackle hard optimization problems. Nonetheless, there cannot exist a universal solving method; which partially explains the plethora of available algorithms. We argue that the automatic choice of an effective algorithm, the smart combination of low level components and the proper tuning of their underlying parameters is one of the most challenging questions that the optimization community has to face in the next coming years. This issue is of interest both for its practical importance and also for the new research opportunities it opens for the design of novel high level techniques.

Two main approaches can be reported [3]: (i) off-line tuning (static choice of parameters before optimization) and (ii) on-line tuning (dynamic tuning of parameters). It is still an open issue to understand what makes these approaches act differently both at the practical level, and also at a more fundamental level with respect to the performance of solvers as a function of problem features. Generally speaking, this research aims at enhancing our understanding of such an issue by abstracting from a specific problem and instead proposing a high level model allowing us to provide both theoretical and empirical evidence on the expected behavior of algorithm configuration methods.

Background on Models for Adaptive Selection Strategies. In this article, we focus on on-line adaptive algorithm selection. From a portfolio of algorithms

J. Handl et al. (Eds.): PPSN XIV 2016, LNCS 9921, pp. 80–90, 2016.
DOI: 10.1007/978-3-319-45823-6_8

at each iteration of the search, a *selection strategy* aims at choosing the hopefully "best" algorithm to execute on the current set of solutions according to the previously observed performance of available algorithms in the portfolio. In [1,2], the authors use some specific benchmarks to study novel selection strategies and improving the underlying reward metrics. For instance, in [1], continuous benchmarks are experimented using a portfolio of variants of the well-established differential evolution operator. Alternatively, other works considered to directly define the rewards associated with the algorithms using particular stochastic distributions [4,11]. The purpose is to be able to study some specific properties of a given adaptive selection strategy such as its ability to detect and to learn the best algorithm from the portfolio. In [11], the set of possible rewards is defined by different uniform random distributions that are reassigned randomly to the portfolio at different time intervals. For instance, in [4], the so-called "Two-Values benchmarks" is used where two possible reward values and a probability of winning the highest is defined depending on pre-computed time intervals. Recently, a benchmark was proposed in [6] where the rewards depend on the number of times that an operator is applied during a time window in order to study a scenario where a number of operators providing different exploration/exploitation trade-offs are available. Several properties should be fulfilled by a relevant benchmark depending on the target issue to be studied. First, one has to take into account the stochasticity of most heuristic algorithms. Hence, the reward of each algorithm in the portfolio should typically be defined by choosing a relevant probability distribution. In order to appreciate the relative quality of the target selection strategies, the so-called "oracle", that is the optimal selection strategy, should be known. At last, since the performance of an algorithm in the portfolio could evolve during the optimization process, the reward distribution has to be tightly coupled with the state of the search. This aims at increasing generality and abstracting away specific algorithmic design issues. For instance, in [11] and related benchmarks, the reward depends on time, and not directly on the state of search; in [6] and related benchmarks, the state of the search is defined by the number of times an operator is used independently of the quality of current solutions. We argue that despite their skillful design, the existing benchmarks are not sufficient by their own to allow for a global fundamental understanding of the design of adaptive methods and the setting of relevant theory for them.

Contribution. In this work, we propose a new model called Fitness Cloud (FC) model inspired by *fitness cloud* [12]. The proposed model is to be viewed in a complementary manner to existing benchmarks. In the FC model, the state of the search is naturally defined by the fitness of the current solution, and the performance of a given metaheuristic is function of the current fitness value. The reward distribution is hence *not* controlled explicitly; but instead, kept as an implicit feature implied by the considered adaptive mechanisms or approaches to be designed independently and studied subsequently.

As a preliminary step we consider in this work a simple usage of the FC model with a portfolio composed by two metaheuristics having fixed performance qualities across two configurable fitness ranges. This setting allows however to

Algorithm 1. A single-solution single-operator basic metaheuristic.

```
1: x_0 ← initialization()
2: repeat
3:    for i = 1 ... λ_t do
4:       y_i ← operator(x_t)
5:    end for
6:    x_{t+1} ← selection(x_t, y_1, ..., y_{λ_t})
7: until stopping criterion is true
```

study two main issues. First, it allows us to provide theoretical evidence on when a static (off-line) selection strategy is more beneficial upon an adaptive (on-line) strategy. Second, through an empirical analysis, and by considering two widely used on-line adaptive strategies based on multi-armed bandits, we gain a more deep understanding on when and why such approaches could be effective with respect to baseline static or purely random strategies [5,9].

The rest of the paper is organized as follows. In Sect. 2, the fitness cloud model is defined with a simple theoretical analysis. In Sect. 3, different instantiations of the proposed scenario are considered and the relative performance and behavior of different selection strategies are elicited by a throughout empirical study. In Sect. 4, we conclude the paper and discuss future research directions.

2 Fitness Cloud Model and Theoretical Analysis

Before going into more details, and although the proposed model is independent of a particular metaheuristic, let us consider for the sake of clarity the template of Algorithm 1 rendering the design of a basic single-solution single-operator metaheuristic. The considered iterative algorithm has two parts. First, a stochastic local operator is applied to the current solution x_t to produce a set of λ_t candidate solutions y_i. Such an operator could be the random bit-flip mutation when the search space is the set of binary strings. Second, a new current solution x_{t+1} is selected. This is typically performed according to the fitness values, given by the fitness function f, of the newly generated solutions y_i, and the current solution x_t. A classical example of selection is the $(1+\lambda)$-EA which selects the best solutions so far. Notice that despite its simplicity such a template encompasses a wide range of algorithms.

2.1 Model Definition

The Fitness Cloud (FC) model informs about the fitness value of solutions after one iteration according to the fitness of the current solution. To make it simple, the FC model supposes that the state of the search is only given by the fitness $f_t = f(x_t)$ of the current solution x_t (see Algorithm 1). Assuming that the selection rule only takes into account the fitness values (which is a common practice for a wide range of metaheuristics), no particular model is required for the selection step. But a specific model is needed to capture the stochastic behavior of most evolutionary operators when generating new candidate solutions.

The basic idea behind the FC model is to assume that the fitness after applying a stochastic operator is given by a conditional probability distribution:

$$\Pr(f(y) = z' \mid f_t = z) \tag{1}$$

Being said, different choices of this probability distribution can be made such as discrete distributions (binomial, Poisson, etc.), or continuous distributions (normal, Weibull, etc.). Given its properties of convergence, we choose the use a normal distribution in this paper:

$$\Pr(f(y) = z' \mid f_t = z) \sim \mathcal{N}(\mu(z), \sigma^2(z)) \tag{2}$$

where $\mu(z)$ and $\sigma^2(z)$ are respectively the mean and the variance of the normal distribution which can depend on the fitness z of the solution and which are to be set to map a target setting. As a consequence, the evolution of the fitness during one iteration follows a conditional probability distribution which embeds the previous distribution. One important feature of the probability distribution is the *expected improvement* of one metaheuristic iteration, denoted by $E^+(z)$, which is the expected progress of the fitness given the current fitness value is z:

$$E^+(z) = \int_z^\infty \Pr(f_{t+1} = z' \mid f_t = z)\ z'\ dz' \tag{3}$$

2.2 Definition of a Simple Scenario with Two Fitness Ranges

The previous considerations are broad enough to allow us to define a more concrete and relevant simple scenario, where we are given a portfolio of two elitist metaheuristics (see Fig. 1). More precisely, we first assume that the possible fitness values are normalized in the range [0,1]. The search is assumed to start with fitness value 0 and stops when the fitness value 1 is reached. The whole range [0,1] is then divided into two fitness ranges: the first one from fitness 0 to $r \leqslant 1$, and the second range from r to 1. We then consider a portfolio of two heuristic algorithms having different relative performance in these two ranges. For this purpose, the relative behavior of each algorithm in the portfolio is modeled accordingly in each fitness range using the fitness cloud model. More precisely, at each fitness range, we shall fix the mean and variance of the conditional normal distribution in Eq. 2 as follows: $\mu_i(z) = z + K_{\mu_i}$ and $\sigma_i^2(z) = K_{\sigma_i}$ where for each metaheuristic M_i, $i \in \{1,2\}$, parameters K_{μ_i} and K_{σ_i} are different constant numbers at each fitness range. Therefore, we end up with 9 parameters to be fixed in this scenario: r, and the 8 parameters for the normal distributions for each metaheuristic and at each fitness range. However, as it will be shown in Sect. 2.3, the expected running time to reach the optimal value 1 depends on the expected improvement of each metaheuristic. Hence, only 5 parameters are free as illustrated by Fig. 1; where $E^+_{i,j}$ denotes the expected fitness improvement of metaheuristic M_i, $i \in \{1,2\}$, for the fitness range $j \in \{1,2\}$. Additionally, we assume that the best metaheuristic for the first fitness range is M_1, whereas it switches to M_2 in the second fitness range, i.e., $E^+_{2,1} < E^+_{1,1}$, and $E^+_{1,2} < E^+_{2,2}$.

Finally, like in many optimization problems, we assume that the expected improvement decreases when the fitness value increases: $E^+_{1,2} < E^+_{1,1}$, and $E^+_{2,2} < E^+_{2,1}$. It is important to notice that the relative performance of algorithms in the portfolio does not depend explicitly neither on time (number of iterations), nor on the number of times a metaheuristic is applied; but solely on the state of the search which is assumed to be implied by the current fitness value.

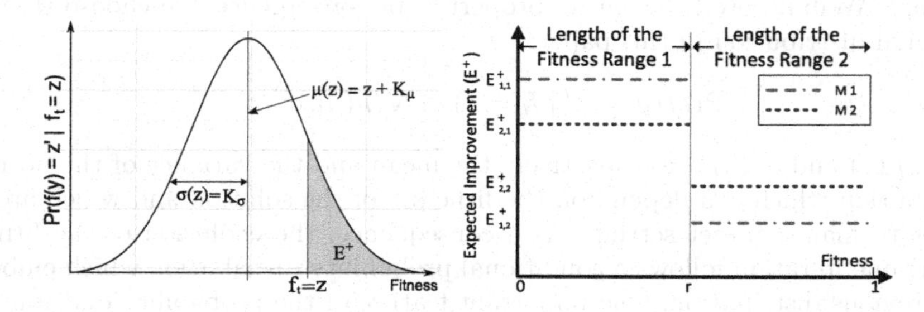

Fig. 1. Fitness cloud model: scenario with two metaheuristics and two fitness ranges.

2.3 Theoretical Insights

In this section, we assume an off-line static strategy that selects arbitrary one metaheuristic, denoted M, in the portfolio and executes it on the previously described scenario until reaching the target fitness value 1. We shall assume that the considered metaheuristic follows the template of Algorithm 1 initialized with a solution having fitness value 0 and implementing an elitist selection when deciding on the next solution, *i.e.* the best solution is retained for the next iteration. Notice that the expected improvement of the considered metaheuristic M at each iteration is by definition constant within each of the two fitness ranges defined in the considered scenario. We shall denote by E^+_1 (resp. E^+_2), the expected improvement within the first (resp. second) fitness range. Let us consider the running time of metaheuristic M, that is, the first hitting time (number of iterations) to reach the final fitness value: $T = \min\{t \mid F_t \geqslant 1\}$ where F_t is the random variable which gives the fitness value of the best solution found at iteration t. Notice that the total number of evaluations depends on λ_t and the number of evaluation in each operators. We then can prove the following:

Theorem 1. *The expected running time verifies:* $T_{up} - \frac{\delta}{E^+_2} \leqslant E[T] \leqslant T_{up}$ *with*

$$T_{up} = \frac{r}{E^+_1} + \frac{1-r}{E^+_2} \quad \textit{and} \quad \delta = \begin{cases} 1-r & \textit{if} 1-r \leqslant E^+_1, \\ E^+_1 & \textit{if} E^+_1 < 1-r \end{cases}$$

Proof. Let $T_1 = \min\{t \mid F_t \geqslant r\}$ and $T_2 = \min\{t - T_1 \mid F_t \geqslant 1\}$. By definition and by the linearity of expectation, we have that $E[T] = E[T_1] + E[T_2]$. Now we prove the following lemma:

Lemma 1. *Let $(k, \ell) \in [0,1]^2$ such that either $r \leqslant k \leqslant \ell$ or $k \leqslant \ell \leqslant r$. Let $T^{'} = \min\{t \mid F_t \geqslant \ell\}$ and $F_0 = k$. Let E^+ is the expected improvement in the fitness range $[k, \ell]$. Then, $E[T^{'}] = (\ell - k)/E^+$.*

The proof of the Lemma is an application of theorem 1 in [8] which can be stated in short by: if $E[X_t - X_{t+1}|X_t] = \delta$, then $E[T_0|X_0] = X_0/\delta$. In fact, by considering the random variable $X_t = \ell - F_t$, we have by definition of the fitness cloud model $E[X_t - X_{t+1} \mid X_t] = E^+$ which gives the necessary additive drift condition and the proof of the lemma follows immediately.

Since the expected improvement in the first fitness range is E_1^+, applying the previous lemma with $F_0 = 0$ provides: $E[T_1] = r/E_1^+$. Similarly, let $k \geqslant r$ the fitness of the (best) current solution just after a solution x_t with fitness greater than r is found for the first time. Since the expected improvement in the second fitness range is E_2^+, applying the previous lemma with $F_0 = k \geqslant r$ provides: $E[T_2] = (1-k)/E_2^+ \leqslant (1-r)/E_2^+$. The stated upper bound is hence proved. Let's now define $Y_{t'} = 1 - F_{t'}$ where $t^{'} = t - T_1$. Hence, $E[T_2|Y_0] = Y_0/E_2^+$. By applying the law of total expectation, we get $E[T_2] = E[Y_0]/E_2^+$. Let $F_{T_1} = F_{T_1-1} + \Delta_{T_1-1}$ where Δ_{T_1-1} is the random variable of the fitness difference between the iterations $T_1 - 1$ and T_1. By definition of T_1, $F_{T_1-1} < r$, and from the fitness cloud model we have that $E[\Delta_{t'-1}] = E_1^+ > 0$. It follows that: $E[Y_0] \geqslant 1 - r - E_1^+$. When $1 - r - E_1^+ \leqslant 0$, the algorithm is able to reach final fitness without any iteration in the second fitness range. Otherwise, for $1 - r - E_1^+ > 0$, the algorithm spends at least $(1 - r - E_1^+)/E_2^+$ iterations in the second fitness range 2. □

For example, with the (1 + 1)-EA which generates a single candidate solution and keeps it if it is better than the current one, the expected improvement $E^+(z)$ at fitness value z is given by: $E^+(z) = \int_z^\infty \frac{t}{\sigma\sqrt{2\pi}} \exp(\frac{-(t-\mu)^2}{2\sigma^2})dt = (\mu - z) \cdot (1 - \Phi(\frac{-(\mu-z)}{\sigma})) + \frac{\sigma}{\sqrt{2\pi}} \exp(\frac{-(\mu-z)^2}{2\sigma^2})$ where Φ is the cumulative distribution function of the standard normal distribution. Accordingly, the evolution of the expected running time for the two possible metaheuristics M_1 and M_2 as a function of the length r of the first fitness range, is illustrated in the right side of Fig. 2.

3 Experimental Analysis of Adaptive Selection Strategies

3.1 Experimental Design

From the scenario defined previously (see Fig. 1) where a portfolio of two metaheuristics are given; with the metaheuristic M_1 (resp. M_2) being better on the fitness range 1 (resp. 2), we were able to experiment several possible settings of the underlying FC model. Overall, and considering that M_1 and M_2 are implemented as an elitist (1 + 1)-EA, we only retain 3 cases corresponding to typical different instantiations that were found to be the most representative of the different challenges that this scenario allows to consider. In Fig. 2, we summarize these 3 experimental cases while providing the parameters used in the FC model.

Actually, as shown in the theoretical analysis, the expected improvement (EI for short) is crucially important in each fitness range. Hence, the mean and the standard deviation of the normal distributions are chosen to obtain the desired EI. By choosing K_μ negative, we emulate the behavior of a typical stochastic operator that decreases on average the fitness of current solution as it is the case very often in practice. In all the 3 cases, the EIs of M_1 and M_2 are the same in the first fitness range, while being different by a factor of 2. This actually does not penalize much metaheuristics when moving from one range to the other, which makes Case 2 a reference case with respect to the other cases. In fact, in Case 1, the EI values are much more closer in the second range, *i.e.*an oracle would act as in Case 2, but the performance of a static selection strategy that would always choose M_1 becomes much closer to the oracle than in Case 2. As for Case 3, the same factor of 2 is kept between the EIs in the second range; but the EIs has been reduced by a huge factor of 15, hence making the progress in the second range relatively much more difficult than in the first range compared to Case 2.

Values are given with a factor of 10^{-3}.

Cases	Meta.	Fitness range 1			Fitness range 2		
		$E^+_{i,1}$	K_{μ_i}	K_{σ_i}	$E^+_{i,2}$	K_{μ_i}	K_{σ_i}
Case 1	M_1	6	-1	16.27	1.8	-2	6.72
	M_2	3	-1	8.72	2	-2	7.24
Case 2	M_1	6	-1	16.27	1	-2	4.59
	M_2	3	-1	8.72	2	-2	7.25
Case 3	M_1	6	-1	16.27	0.2	-2	2.14
	M_2	3	-1	8.72	0.4	-2	2.84

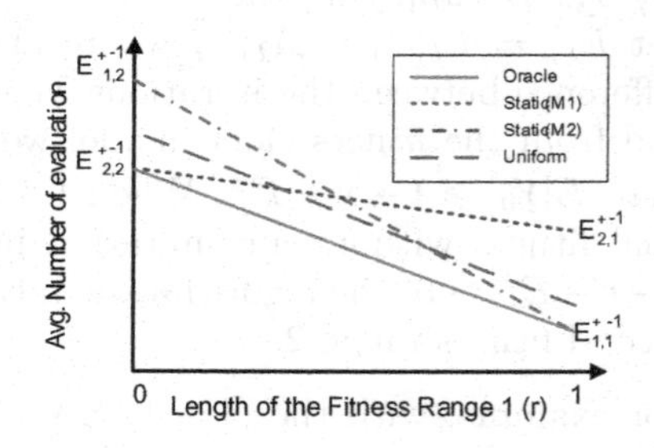

Fig. 2. Parameters values of the 3 cases with its corresponding sketch (upper bounds of expected running time): exp. impr. ($E^+_{i,j}$), mean difference (K_{μ_i}), and std. dev. (K_{σ_i}).

For our experimental investigations, we consider two selection strategies used in multi-armed bandits framework. Due to the lack of space, we only detail the experimented parameters without going into a technical discussion; the reader is referred to [6] for a review and a detailed description of the following adaptive selection strategies, namely Upper Confidence Bound (UCB) and Adaptive Pursuit (AP). The UCB strategy [4] estimates the upper confidence bound of the expected reward of each arm and selects the one with the higher bound. A parameter C tunes the exploitation/exploration trade-off. The AP selection strategy [10] uses exponential recency weighted average to estimate the average, tuned by an adaption rate α, and selects a metaheuristic according to probabilities updated by a learning rate β. For UCB, the set of studied parameters C is $\{0.0008, 0.01, 0.1, 0.75, 2, 4, 10, 20, 25, 50\}$, and for AP, both adaptation and learning rate parameters are in the set $\{0.1, 0.3, 0.5, 0.7, 0.9\}$. Additionally, we include three other strategies in our analysis. The *oracle strategy* selects the best metaheuristic in each fitness range, *i.e.* M_1 in the first range and M_2 in the second range. The *uniform strategy* selects at each iteration one metaheuristic uniformly at random among M_1 and M_2. Notice that in this case the expected improvement is the mean of expected improvements of M_1 and M_2. The *static strategy* selects always the same metaheuristic, that is either M_1 or M_2 before the execution is started. All results are averaged over 100 independent runs.

3.2 Empirical Analysis

In this section, we analyze the relative performance and the behavior of the considered strategies. The performance measure is the average number of evaluations to reach the target fitness value 1. For fairness, we consider the best parameter setting for each strategy. For each case, we compute the average rank of a setting over all values of r, and the best ranked setting is selected. The performance comparison is based on the Mann-Whitney test with a confidence level of 0.05. In Fig. 3, we show the performance obtained in the three test cases as a function of the length r of the first fitness range. Notice that for the considered Cases 1, 2, and 3, the average difference over the r-values of the performance between the oracle and the best static strategy (either with M_1 or M_2) is respectively 20, 57, and 100 evaluations. This is the maximum performance gap between an optimal adaptive method and an optimal off-line static strategy tuned for each value of r.

Adaptive Strategies *vs.* Uniform. As suggested by Theorem 1, the expected performance of static, oracle, and uniform selection strategies decreases linearly with the length of the first fitness range. The performance of UCB and AP strategies are also linear from our empirical data (Pearson correlation coefficients are very close to -1). For all test cases, and for any length r of the first range, UCB and AP strategies perform significantly better than the uniform random selection strategy (except for few values of r where no significant difference with AP is found). Interestingly, the average gap between UCB and uniform strategy in test cases 1, 2, and 3 is respectively around 26, 87, and 263 evaluations, which is much higher than the difference between the oracle and the best static strategy.

Adaptive Strategies *vs.* Static. The performance of adaptive strategies can be worse than a static strategy. For example, in the test case 2, UCB is better than any static strategy for $r \in [0.11, 0.99]$. Otherwise, when the length of the fitness range where the expected improvement of M_1 is the best, is short ($r < 0.11$), the static strategy choosing M_2 is better than UCB. At the opposite, for $r > 0.99$, the static strategy choosing M_1 is better than UCB. The performance of the adaptive selection strategies also depends on the expected improvements in the second range. Respectively for the cases 1 and 3, the r-values intervals where UCB strategy outperforms static strategies are $[0.09, 0.81]$ and $[0.61, 1.00]$. On average over the r-values, in test case 1 and 2, UCB outperforms the static strategy by 10 and 49 evaluations respectively. However, in the test case 3, when the expected improvements of algorithms in the portfolio in the second range are very small compared to the first one, a static strategy is preferred, indeed UCB requires 71 additional evaluations on average than static.

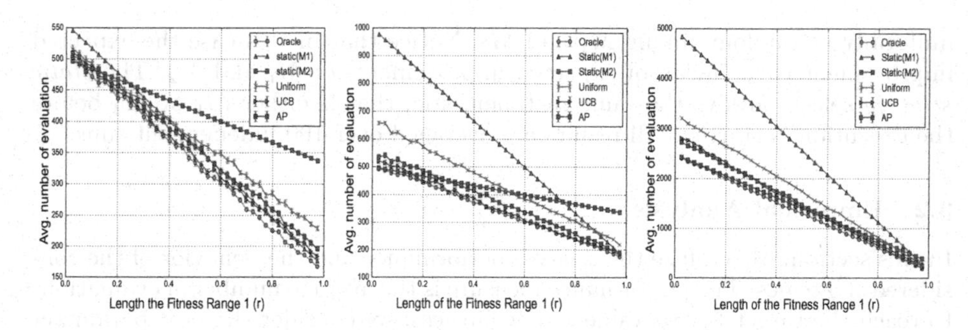

Fig. 3. Comparison of selection methods with the best parameters settings for UCB $C = 4$, and for AP $\alpha = 0.1$, $\beta = 0.1$. From left to right: test cases 1, 2, and 3.

UCB *vs.* AP Strategies. Overall, the AP strategy never outperforms UCB strategy significantly except in 3 minor exceptions for the lowest values of r in the case 3. In the case 1, UCB is better than AP for $r > 0.35$, and the difference between UCB and AP is 10 evaluations in average, which is half of the difference between best static strategy and oracle. In test case 2, UCB strategy is better than AP except for the 3 largest values of r, while being very close to the oracle, *i.e.* the average difference for UCB is only 8 evaluations compared to the 32 evaluations for AP, and 57 evaluations for a static strategy. In case 3, the performance difference between of UCB and AP is much closer (only 15 evaluations on average). Although UCB outperforms AP for r-values larger than 0.37, both strategies are on average worst than the optimal static strategy by a factor of 1.8. The Fig. 4 shows the selection frequency of the best metaheuristic according to the current fitness value for UCB and AP in Case 2 and when the expected improvement of metaheuristics is changed at fitness value $r = 0.5$. The UCB strategy converges at fitness 0.3 for $C = 0.004$, and AP around fitness 0.5 for $\alpha = 0.1$. In addition, when the best metaheuristic changes at fitness value 0.5, the UCB strategy recovers more quickly the best metaheuristic than AP.

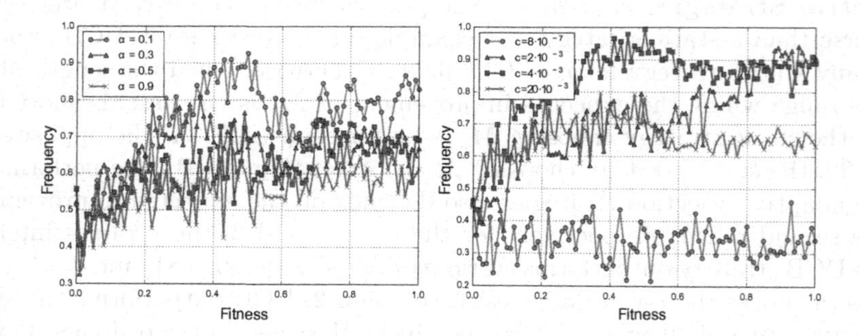

Fig. 4. Frequency of the best metaheuristic selection according to the fitness value for different parameter settings. Case 2 with $r = 0.5$ AP (left) with $\beta = 0.1$ UCB (right).

Uniform *vs.* Static. It is shown several times that random selection of parameters could outperformed a tuning method with static value of parameters [5,9]. The model with two fitness ranges scenario helps us to understand why and when random selection can be advantageous. The uniform strategy is better than the best static strategy when the length r belongs to the intervals $[0.14, 0.4]$, $[0.56, 0.88]$, and $[0.6, 1]$ respectively for cases 1, 2, and 3. Roughly speaking, a random uniform selection, and moreover an adaptive strategy, is more efficient when the performances of each metaheuristic in the portfolio are close.

Discussion. This two-fitness-range scenario allows us to fine-tune the performance of each metaheuristic at the two stages of the search. The comparison of case 2 and 3 shows that when the average expected improvements at the second stage is much lower than in the first stage, an adaptive method becomes less efficient except when the length of the first stage is very large. Indeed, the time that can be gained in the first stage becomes negligible and the main difficulty then turns out to be the final convergence to the optimum value. When the scale of the average expected improvements between the two stages is moderate like in test case 2, an adaptive method like UCB strategy is very effective. However, when the performance difference between metaheuristics at one stage of the search becomes small, like in test case 1, the problem is equally difficult for all metaheuristics in the portfolio, and the adaptive selection becomes rather useless besides the fact that it becomes more difficult to detect the best performing metaheuristic at a given iteration.

4 Conclusion

It is our hope that the fitness cloud model opens new research paths allowing to understand, to test and to design new adaptive portfolio methods in different settings. In this work, using a simple scenario, we provide properties of when and why an (on-line) adaptive selection strategy, or random selection could outperform an (off-line) static strategy. Indeed, the fitness cloud model goes beyond the intuition and allows to give a formal framework in order to analyze selection strategies in portfolio and to understand their behavior in different settings.

Following the natural question of Baudiš *et al.* in his conclusion [1] on *"the influence of portfolio size and composition on performance of various strategies"*, it would be possible to design relevant scenarios to deeply study those questions. It would also be possible to conduct a fine grained analysis of other selection strategies in a sequential as well as in parallel context [7]. It will also be interesting to extend the fitness cloud model to multi-objective optimization where the design of adaptive portfolio method is relatively in its infancy beginning.

References

1. Baudiš, P., Pošík, P.: Online black-box algorithm portfolios for continuous optimization. In: Bartz-Beielstein, T., Branke, J., Filipič, B., Smith, J. (eds.) PPSN 2014. LNCS, vol. 8672, pp. 40–49. Springer, Heidelberg (2014)
2. DaCosta, L., Fialho, A., Schoenauer, M., Sebag, M.: Adaptive operator selection with dynamic multi-armed bandits. In: GECCO 2008, p. 913 (2008)
3. Eiben, A.E., Michalewicz, Z., Schoenauer, M., Smith, J.E.: Parameter control in evolutionary algorithms. In: Lobo, F.G., Lima, C.F., Michalewicz, Z. (eds.) Parameter Setting in Evolutionary Algorithms. Studies in Computational Intelligence, vol. 54, pp. 19–46. Springer, Heidelberg (2007)
4. Fialho, A., Da Costa, L., Schoenauer, M., Sebag, M.: Analyzing bandit-based adaptive operator selection mechanisms. AMAI **60**, 25–64 (2010)
5. García-Valdez, M., Trujillo, L., Merelo-Guérvos, J.J., Fernández-de-Vega, F.: Randomized parameter settings for heterogeneous workers in a pool-based evolutionary algorithm. In: Bartz-Beielstein, T., Branke, J., Filipič, B., Smith, J. (eds.) PPSN 2014. LNCS, vol. 8672, pp. 702–710. Springer, Heidelberg (2014)
6. Goëffon, A., Lardeux, F., Saubion, F.: Simulating non stationary operators in search algorithms. Appl. Soft Comput. **38**, 257–268 (2016)
7. Jankee, C., Verel, S., Derbel, B., Fonlupt, C.: Distributed adaptive metaheuristic selection: comparisons of selection strategies. In: Bonnevay, S., et al. (eds.) EA 2015. LNCS, vol. 9554, pp. 83–96. Springer, Heidelberg (2016). doi:10.1007/978-3-319-31471-6_7
8. Lehre, P.K., Witt, C.: General drift analysis with tail bounds. Technical report (2013). arXiv:1307.2559
9. Tanabe, R., Fukunaga, A.: Evaluation of a randomized parameter setting strategy for island-model evolutionary algorithms. In: CEC 2013, pp. 1263–1270 (2013)
10. Thierens, D.: An adaptive pursuit strategy for allocating operator probabilities. In: GECCO 2005, pp. 1539–1546 (2005)
11. Thierens, D.: Adaptive strategies for operator allocation. In: Lobo, F.G., Lima, C.F., Michalewicz, Z. (eds.) Parameter Setting in Evolutionary Algorithms. Studies in Computational Intelligence, vol. 54, pp. 77–90. Springer, Heidelberg (2007)
12. Verel, S., Collard, P., Clergue, M.: Where are bottlenecks in NK fitness landscapes? In: CEC 2003, pp. 273–280 (2003)

A Study of the Performance of Self-⋆ Memetic Algorithms on Heterogeneous Ephemeral Environments

Rafael Nogueras and Carlos Cotta(✉)

Dept. Lenguajes y Ciencias de la Computación, Universidad de Málaga, ETSI Informática, Campus de Teatinos, 29071 Málaga, Spain
ccottap@lcc.uma.es

Abstract. We consider the deployment of island-based memetic algorithms (MAs) endowed with self-⋆ properties on unstable computational environments composed of a collection of computing nodes whose availability fluctuates. In this context, these properties refer to the ability of the MA to work autonomously in order to optimize its performance and to react to the instability of computational resources. The main focus of this work is analyzing the performance of such MAs when the underlying computational substrate is not only volatile but also heterogeneous in terms of the computational power of each of its constituent nodes. We use for this purpose a simulated environment subject to different volatility rates, whose topology is modeled as scale-free networks and whose computing power is distributed among nodes following different distributions. We observe that in general computational homogeneity is preferable in scenarios with low instability; in case of high instability, MAs without self-scaling and self-healing perform better when the computational power follows a power law, but performance seems to be less sensitive to the distribution when these self-⋆ properties are used.

1 Introduction

Population-based optimization algorithms are very well suited to parallel environments thanks to their flexibility and decentralized nature. This has been known and exploited since the late 80s. In contrast to the dedicated networks of computational resources that were typical in the past, recent years have witnessed the emergence of other kind of environments of a much more dynamic and unsteady nature though. This is the case of peer-to-peer (P2P) networks and volunteer computing networks, composed of volatile nodes whose availability usually responds to uncontrollable external factors. Such environments are particularly interesting in light of the increasingly pervasive abundance of computational devices which are permanently networked (think for example of smartphones, wearables, and any other kind of handheld devices) and whose computing power is often unused or at least under-exploited [6]. Capitalizing on such power can be a practical solution for solving many complex computational tasks but tackling the underlying dynamic computational landscape is not exempt of difficulties.

J. Handl et al. (Eds.): PPSN XIV 2016, LNCS 9921, pp. 91–100, 2016.
DOI: 10.1007/978-3-319-45823-6_9

Of course, intermediate layers can be constructed to hide the transient nature of computational nodes but it is not easy to have such an abstract layer making effective use of brief, ephemeral bursts of computing availability. While this direction is in any case interesting and valid, we consider here a much more direct approach in which the optimization algorithm is cognizant of the volatile environment.

Algorithms consciously running on computational environments with the features mentioned above must be resilient in order to withstand sudden node failures. In the case of evolutionary algorithms this resilience is partly provided by their inherent features [13,15], and can be further boosted by exploiting their capacity for adaptiveness and self-control [8,11]. This latter feature is essential to have the algorithm readjusting its behavior in response (or even in anticipation) to the fluctuations of the environment. Indeed, much work has been done in the area of self-adaptation in evolutionary algorithms, e.g., [5,24,25] in general, and in connection with unstable environments in particular. In this work we build on previous research [19–22] in order to tackle the potential heterogeneity of the environment [1] in terms of the computational power of individual nodes (which in a setting such as the one described before could range from tiny devices to desktop computers for example) and ascertain to which extent this can exert an influence in the performance of the algorithm. The underlying rationale for examining this matter lies in the potentially different impact than the failure of a node can have on the system as a whole depending on its computational power, and determining whether or in which conditions the system is sensitive to this environmental heterogeneity. To this end we consider island-based memetic algorithms (MAs) endowed with self-$\star$ properties [2] and use a simulated computational environment that allows experimenting with different scenarios both in terms of the volatility of computing nodes and the distribution of computing power of constituent nodes. A broad experimentation is done to assess the performance of the MA in these different scenarios.

2 Materials and Methods

2.1 Basic Algorithmic Setting

As stated in the introduction, the basic algorithm considered is an island-based MA. Let there be n_ι panmictic islands, each of them running a simple MA (using tournament selection, one-point crossover, bit-flip mutation, and replacement of the worst parent) on a different computing node. Such nodes are interconnected among them according to a certain topology $\mathcal{N}$ – see Sect. 2.3. In addition to standard selection, variation and local improvement, each island perform asynchronous migration: at the beginning of each cycle the island checks if migrants were received from any neighboring nodes and are stored in the input buffer. Were this the case, they would be inserted in the population following a certain migrant replacement policy. Later, at the end of each cycle, each island decides stochastically whether to send individuals to neighboring islands. If done, migrants are selected using a given migrant selection policy and sent to the neighbors. Following previous analysis of migration strategies in island-based MAs [18], we use random selection of migrants and deterministic replacement of the worst individuals in the receiving island.

2.2 Self-⋆ Properties

Self-⋆ properties [2] are those that enable a computational system to exert advanced control on its own functioning and/or structure. This goes beyond parameter control (its practical importance notwithstanding) and encompasses advanced capabilities such as, e.g., self-maintaining in proper state, self-healing externally infringed damage, or self-optimizing its behavior, just to cite a few. In the following we will describe the particular self-⋆ properties with which the MA considered is endowed.

Self-Generation. According to [4], a system is self-optimized if starting from an arbitrary initial configuration it is capable of improving a certain objective function of its global state. In the case of bioinspired optimization algorithms, this objective function does not directly refer to the fitness function to be optimized, but to the capability of the algorithm to optimize the latter. Such capability can be improved for example by tuning some parameters (self-parameterization) or even by adjusting qualitatively the way the search is done (self-generation). The latter approach amounts to have the algorithm adjusting the search strategy during runtime [12], and is related to the notion of memetic computing [23] whereby memes (understood as representations of problem solving strategies) are explicitly represented and evolved [17].

In the MA considered we follow the model by Smith [24] in which memes are attached to individuals and evolve alongside them. More precisely, these memes take the form of pattern-based rewriting rules $A \to B$, where A, B are variable-length strings taken from the same alphabet used to encode solutions plus a wildcard symbol. The action of the meme is finding an occurrence of pattern A in the solution and changing it by pattern B if it leads to a fitness improvement (otherwise the solution is left unchanged). Self-generation is attained due to the fact that memes are subject to mutation and are transferred from parent to offspring via local selection (offspring inherit the meme of the best parent).

Self-Scaling. This property involves the ability of the system to react efficiently to changes in its scale parameters, that is, changing its size or its structure in response to modifications in the size of the problem being solved, in the amount of computational resources available, or in any other circumstance of the computation, e.g., [9,28]. In this case, the main factor to be taken into account is the volatility of the environment that results in certain islands getting lost when the supporting node goes down. This implies that the overall size of the system will fluctuate, affecting genetic diversity and resulting in the loss of information. To cope with this, a self-balancing policy has been proposed [22]. This strategy is aimed to resize dynamically islands so that some of them increase their sizes when they detect a neighboring island has gone down, and analogously decrease their sizes when a new neighbor appears. This is done by communicating with neighbors at the beginning of each evolutionary cycle exchanging information on the size of their populations and the number of active neighbors, and performing a local balancing procedure [27] by transferring individuals. New islands can absorb this way a part of the existing population in neighboring nodes and, likewise, nodes detecting that a previously active neighbor is no longer available try to compensate this loss by increasing their own population sizes (using the recorded information on the size and number of active neighbors the lost island had in order to calculate the required increase). While simultaneous failures of neighboring nodes can still produce fluctuations, this strategy promotes the stabilization of the overall population size.

Self-Healing. This property focuses on the maintenance and restoration of system attributes that may have been affected by internal or external actions. In the context of evolutionary algorithms this property is not new since the use of ad-hoc procedures for repairing infeasible solutions produced by variation operators in constrained problems [16] can be regarded as a simple form of self-healing. More particularly for the case of ephemeral computational environments, the volatility of the system can be the source of at least two issues the algorithm needs to deal with: (i) node failures disrupt the connectivity of the network, limiting the flow of information and hindering the progress of the search, and (ii) forcing an island to increase its size can perturb the convergence of the search if the new information is simply random. To tackle the first issue, a self-rewiring strategy [19] is used: whenever an island detects that its number of active neighbors has fallen below a predefined threshold, it looks for additional neighbors to reach this minimum level, hence aiming to maintain a rich connectivity at all times. As to the second issue, it is dealt with by means of self-sampling [20], that is, the island keeps a probabilistic model of the current population and samples it (much like it is done in EDAs) when new individuals are required. This way, the latter are representative of the current state of the population. We consider here a tree-based bivariate probabilistic model.

2.3 Environmental Model

This island-based model runs on a simulated distributed system composed of n_ι nodes. These nodes are interconnected following a scale-free topology. This connectivity pattern is commonly observed in many real-world systems, particularly in P2P systems. To generate this topology we consider the Barabási-Albert (BA) model [3]: starting with a clique of $m + 1$ nodes (m being a parameter of the model), new nodes are added one at a time, selecting for each of them m neighbors among previous nodes; neighbor selection is driven by preferential attachment, whereby the probability of picking a certain node is proportional to the number of neighbors it already has.

To model the volatility of the nodes, we consider that failures/recoveries are Weibull distributed [14]. This distribution is described by a shape parameter η

and a scale parameter β: the probability of a node being available up to time t is $p(t, \eta, \beta) = \exp(-(t/\beta)^{\eta})$. Thus, for shape parameters larger than 1 (as we use in the experiments), failure/recovery probabilities increase with time.

Network heterogeneity is modeled by assuming each node i has a certain computing power $w_i \in \mathbb{N}^+$. These coefficients represent for simplicity a relative performance index and hence each node's power can be understood to be proportional to its coefficient. From the point of view of the MA, this computational power determines the number of evolutionary cycles (and hence the number of fitness function evaluations) each node can perform per unit of time. We have considered several scenarios regarding the distribution of values for these coefficients but in all cases, the overall computing power of the network $W = \sum_i w_i$ is the same so as to not introduce any bias towards any particular configuration:

- uniform: the overall computing power W is evenly distributed among nodes, meaning that $\lfloor W/n_\iota \rfloor \leqslant w_i \leqslant \lceil W/n_\iota \rceil$.
- random: each coefficient w_i can have a uniformly random value in $\{1, \ldots, W - n_\iota + 1\}$, subject to $W = \sum_i w_i$ as mentioned before. This is accomplished by having $w_i = 1$ initially, attributing random values in $(0, 1)$ to each node and then using D'Hondt's method to distribute $W - n_\iota$ additional units among nodes according to these values.
- binomial: coefficients can take values in $\{1, \ldots, W - n_\iota + 1\}$ and the probability of a certain value w is $p(w) = C(W - n_\iota, w - 1)q^{w-1}(1 - q)^{W - n_\iota - w + 1}$ where $q = 1/n_\iota$. As in the previous case, the boundary conditions ensure that each node has at least unit power.
- power law: coefficients are grouped in r levels, where $r \in \{0, \ldots, r_{\max}\}$ with $r_{\max} = \lfloor \log_2 n_\iota \rfloor - 1$, so that there is a single node with power $\lfloor n_\iota/2 \rfloor$ in the highest level, and in subsequent levels there are twice as many nodes, each with half as much power as in the previous upper level (although depending on the value of n_ι the lowest level can have additional nodes if these are not enough to create a new level).

The value W implied for the last configuration (power law) is used for the remaining distributions. Notice that depending on the configuration a node failure will have a different impact on the overall capacity of the system. For example, under the power law distribution around half of node failures will have a low impact in this overall capacity but larger disruptions are possible (albeit with a increasingly lower probability). On the opposite side of the spectrum, all failures have a priori the same moderate impact under a uniform distribution. Next section describes the experimentation conducted with these distributions to determine more quantitatively the effect they exert on performance.

3 Experimental Analysis

We consider $n_\iota = 32$ islands whose initial size is $\mu = 16$ individuals and a total number of evaluations *maxevals* $= 50000$. Meme lengths evolve within $l_{\min} = 3$ and $l_{\max} = 9$, mutating their length with probability $p_r = 1/9$.

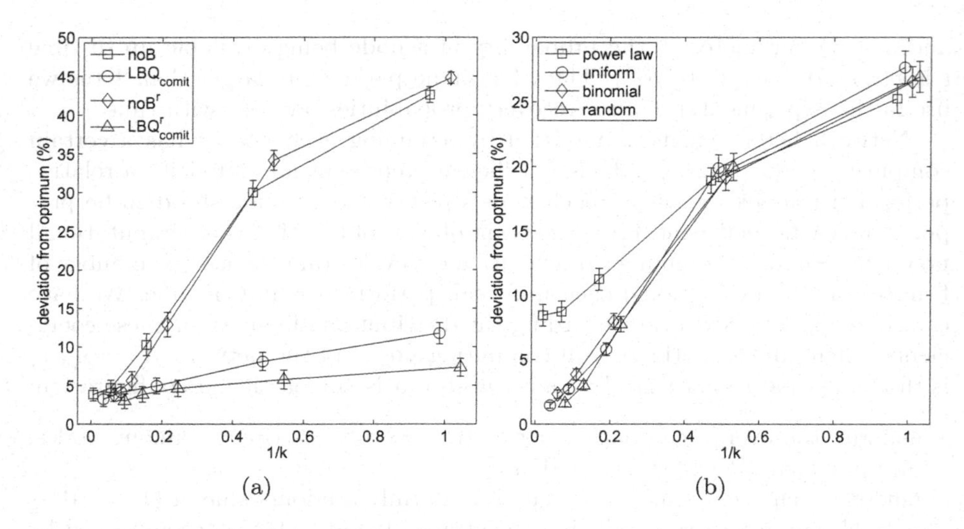

Fig. 1. Average deviation from the optimal solution across all problems. (a) According to algorithmic variant. (b) According to power distribution.

Table 1. Results of Holm test ($\alpha = 0.05$) using $\text{LBQ}^r_{\text{comit}}$ as control algorithm.

i	strategy	z-statistic	p-value	α/i
1	$\text{LBQ}_{\text{comit}}$	2.934e+00	1.670e−03	5.000e−02
2	noB	6.293e+00	1.554e−10	2.500e−02
3	noB^r	9.298e+00	7.125e−21	1.667e−02

We use crossover probability $p_X = 1.0$, mutation probability $p_M = 1/\ell$, where ℓ is the genotype length, and migration probability $p_{mig} = 1/80$. Regarding network topology, we use $m = 2$ in the Barabási-Albert model. This model is also used for self-rewiring when a node has less than m active neighbors. Regarding node deactivation/reactivation, we use the shape parameter $\eta = 1.5$ to have an increasing hazard rate, and scale parameters $\beta = -1/\log(p)$ for $p = 1 - (kn_\iota)^{-1}$, $k \in \{1, 2, 5, 10, 20\}$. These parameters can be interpreted as corresponding to an average of one island going down/up every k cycles if the failure rate was constant (it is not since $\eta > 1$ but this serves as a first approximation). This provides different scenarios ranging from low volatility ($k = 20$) to very high volatility ($k = 1$). We perform 25 simulations for each algorithm and volatility scenario. We consider four algorithmic variants (in parentheses the self-$\star$ properties involved – all variants use self-generation): $\text{LBQ}^r_{\text{comit}}$ (self-rewiring, self-sampling, self-scaling), $\text{LBQ}_{\text{comit}}$ (self-sampling, self-scaling), noB^r (self-rewiring) and noB. The experimental benchmark comprises three test functions, namely Deb's trap function [7] (concatenating 32 four-bit traps), Watson et al.'s Hierarchical-if-and-only-if function [26] (using 128 bits) and Goldberg et al.'s Massively Multimodal Deceptive Problem [10] (using 24 six-bit blocks).

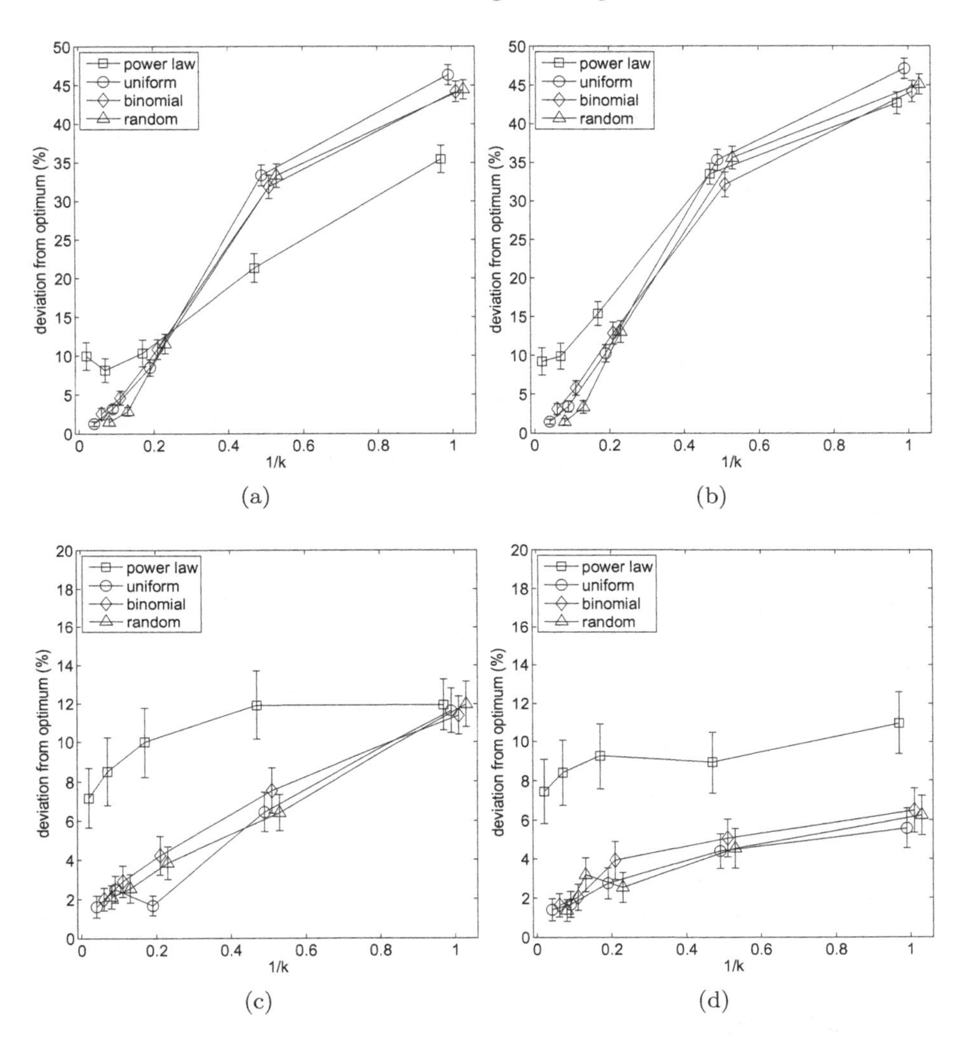

Fig. 2. Average deviation from the optimal solution across all problems and power distributions for each algorithmic variant. (a) noB (b) noB^r (c) LBQ$_{\text{comit}}$ (d) LBQ$^r_{\text{comit}}$.

Figure 1a shows the average deviation from the optimal for each of the four algorithmic variants across all problems and power distributions. The variants with self-scaling and self-sampling outperform variants without them, even in the presence of self-rewiring. This confirms the robustness of the former across different the scenarios considered. In fact, LBQ$^r_{\text{comit}}$ is significantly better than the remaining algorithms (Quade test p-value ≈ 0, Holm test passed at $\alpha = 0.05$ as shown in Table 1). Thus, from a global point of view endowing the MA with self-⋆ properties appears to be a advantageous option. If we now turn our attention to the overall results obtained under each different configuration by all four variants, the differences are not always large (Fig. 1b), although this is understandable in light of the great performance diversity of the algorithms involved that diminishes and conceals dissimilitudes among configurations. For this reason, it is more convenient to factorize the analysis and observe the impact that configurations have on each algorithm separately. This is shown in Fig. 2.

Comparing the behavior of the different algorithms, there is a common feature: in scenarios of low to moderate volatility ($k \geqslant 5$) the uniform distribution (corresponding to near-homogeneous nodes) provides better results. The differences are not always significant considering individual algorithms, but the trend is clear and yields a global significant difference (Quade test p-value $\approx 7.022e{-4}$, Holm test passed for all distributions except random at $\alpha = 0.05$ and for the latter as well at $\alpha = 0.1$). We interpret this result as indicating that in scenarios in which the environmental instability is not strong enough to pose a great handicap to the search process, device homogeneity contributes to balance the search, having all islands progressing at about the same rate. This is further vindicated by the comparatively worse results of the power law distribution in this range, suggesting that the decompensation of computational power (and thus the unbalance of search progress among islands) is not a cost-effective solution. It is also interesting to note the different behavior of noB and $\text{LBQ}^r_{\text{comit}}$ on the other part of the spectrum, that is, for moderate to high volatility ($k \leqslant 5$). In this case, it seems that noB benefits from heterogeneity. This can be due to the fact that having nodes with high computational power can push forward the search significantly during their short availability stint in a much more cost-effective way than less powerful nodes. The situation is different in the presence of self-healing and/or self-scaling, particularly for $\text{LBQ}^r_{\text{comit}}$. These self-$\star$ properties help to absorb the impact of the environment instability and hence $\text{LBQ}^r_{\text{comit}}$ is not in a so-markedly different scenario as before. In fact, power law is significantly worse than uniform for $\text{LBQ}^r_{\text{comit}}$ globally considering all values of k (although the performance is in that case still superior to the remaining algorithms).

4 Conclusions

Deploying population-based optimization algorithms on unstable environments require resilience to deal with the fluctuating computational landscape. Such fluctuations respond to the volatility of computing nodes but can also encompass the heterogeneity of the system and the corresponding variations in the computational power of nodes available at a certain moment. In this sense, endowing MAs with self-$\star$ properties has been shown as an effective solution. The combined use of self-scaling and self-healing seems robust under different configurations of the system, even in scenarios with extreme heterogeneity in which its performance is comparatively less favorable. Future work will be directed to confirm these findings, extending the range of scenarios considered both in terms of heterogeneity and of the volatility patterns of the system.

Acknowledgements. We acknowledge support from Spanish MinEco and FEDER under project EphemeCH (TIN2014-56494-C4-1-P), from Junta de Andalucía under project DNEMESIS (P10-TIC-6083), and from Universidad de Málaga, Campus de Excelencia Internacional Andalucía Tech.

References

1. Anderson, D.P., Reed, K.: Celebrating diversity in volunteer computing. In: Proceedings of the 42nd Hawaii International Conference on System Sciences HICSS 2009, pp. 1–8. IEEE Computer Society, Washington, DC (2009)
2. Babaoğlu, Ö., Jelasity, M., Montresor, A., Fetzer, C., Leonardi, S., van Moorsel, A., van Steen, M.: Self-star Properties in Complex Information Systems. Lecture Notes in Computer Science, vol. 3460. Springer, Heidelberg (2005)
3. Barabási, A.L., Albert, R.: Emergence of scaling in random networks. Science **286**(5439), 509–512 (1999)
4. Berns, A., Ghosh, S.: Dissecting self-⋆ properties. In: Third IEEE International Conference on Self-Adaptive and Self-Organizing Systems - SASO 2009, pp. 10–19. IEEE Press, San Francisco, CA (2009)
5. Caraffini, F., Neri, F., Picinali, L.: An analysis on separability for memetic computing automatic design. Inf. Sci. **265**, 1–22 (2014)
6. Cotta, C., Fernández-Leiva, A., de Vega, F.F., Chávez, F., Merelo, J., Castillo, P., Bello, G., Camacho, D.: Ephemeral computing and bioinspired optimization - challenges and opportunities. In: 7th International Joint Conference on Evolutionary Computation Theory and Applications, pp. 319–324, Lisboa, Portugal (2015)
7. Deb, K., Goldberg, D.: Analyzing deception in trap functions. In: Whitley, L. (ed.) Second Workshop on Foundations of Genetic Algorithms, pp. 93–108. Morgan Kaufmann Publishers, Vail (1993)
8. Eiben, A.E.: Evolutionary computing and autonomic computing: shared problems, shared solutions? In: Babaoğlu, Ö., Jelasity, M., Montresor, A., Fetzer, C., Leonardi, S., van Moorsel, A., van Steen, M. (eds.) SELF-STAR 2004. LNCS, vol. 3460, pp. 36–48. Springer, Heidelberg (2005)
9. Fernández, F., Vanneschi, L., Tomassini, M.: The effect of plagues in genetic programming: a study of variable-size populations. In: Ryan, C., et al. (eds.) Genetic Programming. LNCS, vol. 2610, pp. 317–326. Springer, Heidelberg (2003)
10. Goldberg, D., Deb, K., Horn, J.: Massive multimodality, deception and genetic algorithms. In: Männer, R., Manderick, B. (eds.) Parallel Problem Solving from Nature - PPSN II, pp. 37–48. Elsevier Science Inc., New York (1992)
11. Hinterding, R., Michalewicz, Z., Eiben, A.: Adaptation in evolutionary computation: a survey. In: Fourth IEEE Conference on Evolutionary Computation, pp. 65–69. IEEE Press, Piscataway, New Jersey (1997)
12. Krasnogor, N., Gustafson, S.: A study on the use of "self-generation" in memetic algorithms. Nat. Comput. **3**(1), 53–76 (2004)
13. Laredo, J., Castillo, P., Mora, A., Merelo, J., Fernandes, C.: Resilience to churn of a peer-to-peer evolutionary algorithm. Int. J. High Perform. Syst. Archit. **1**(4), 260–268 (2008)
14. Liu, C., White, R., Dumais, S.: Understanding web browsing behaviors through Weibull analysis of dwell time. In: 33rd International ACM SIGIR Conference on Research and Development in Information Retrieval - SIGIR 2010, pp. 379–386. ACM, New York (2010)

15. Lombraña González, D., Jiménez Laredo, J., de Vega, F.F., Guervós, J.M.: Characterizing fault-tolerance in evolutionary algorithms. In: de Vega, F.F., et al. (eds.) Parallel Architectures and Bioinspired Algorithms. SCI, vol. 415, pp. 77–99. Springer, Heidelberg (2012)
16. Michalewicz, Z.: Repair algorithms. In: Bäck, T., et al. (eds.) Handbook of Evolutionary Computation, pp. C5.4:1–5. Institute of Physics Publishing and Oxford University Press, Bristol (1997)
17. Neri, F., Cotta, C.: Memetic algorithms and memetic computing optimization: a literature review. Swarm Evolut. Comput. **2**, 1–14 (2012)
18. Nogueras, R., Cotta, C.: An analysis of migration strategies in island-based multimemetic algorithms. In: Bartz-Beielstein, T., Branke, J., Filipič, B., Smith, J. (eds.) PPSN 2014. LNCS, vol. 8672, pp. 731–740. Springer, Heidelberg (2014)
19. Nogueras, R., Cotta, C.: Self-balancing multimemetic algorithms in dynamic scale-free networks. In: Mora, A., Squillero, G. (eds.) Applications of Evolutionary Computing. LNCS, vol. 9028, pp. 177–188. Springer, Heidelberg (2015)
20. Nogueras, R., Cotta, C.: Self-sampling strategies for multimemetic algorithms in unstable computational environments. In: Vicente, J.M.F., Álvarez-Sánchez, J.R., López, F.P., Toledo-Moreo, F.J., Adeli, H. (eds.) Bioinspired Computation in Artificial Systems. LNCS, vol. 9108, pp. 69–78. Springer, Heidelberg (2015)
21. Nogueras, R., Cotta, C.: Self-healing strategies for memetic algorithms in unstable and ephemeral computational environments. Nat. Comput. (2016, in press)
22. Nogueras, R., Cotta, C.: Studying self-balancing strategies in island-based multimemetic algorithms. J. Comput. Appl. Math. **293**, 180–191 (2016)
23. Ong, Y., Lim, M., Chen, X.: Memetic computation -past, present and future. IEEE Comput. Intell. Mag. **5**(2), 24–31 (2010)
24. Smith, J.E.: Self-adaptation in evolutionary algorithms for combinatorial optimisation. In: Cotta, C., Sevaux, M., Sörensen, K. (eds.) Adaptive and Multilevel Metaheuristics. SCI, vol. 136, pp. 31–57. Springer, Heidelberg (2008)
25. Smith, J.: Self-adaptative and coevolving memetic algorithms. In: Neri, F. (ed.) Handbook of Memetic Algorithms. SCI, vol. 379, pp. 167–188. Springer, Heidelberg (2012)
26. Watson, R.A., Hornby, G.S., Pollack, J.B.: Modeling building-block interdependency. In: Eiben, A.E., Bäck, T., Schoenauer, M., Schwefel, H.-P. (eds.) PPSN 1998. LNCS, vol. 1498, pp. 97–106. Springer, Heidelberg (1998)
27. Zambonelli, F.: Exploiting biased load information in direct-neighbour load balancing policies. Parallel Comput. **25**(6), 745–766 (1999)
28. Zhao, W., Schulzrinne, H.: DotSlash: a self-configuring and scalable rescue system for handling web hotspots effectively. In: Chi, C.-H., van Steen, M., Wills, C. (eds.) WCW 2004. LNCS, vol. 3293, pp. 1–18. Springer, Heidelberg (2004)

Lyapunov Design of a Simple Step-Size Adaptation Strategy Based on Success

Claudia R. Correa[1], Elizabeth F. Wanner[2(✉)], and Carlos M. Fonseca[3]

[1] Post-Graduate Program in Mathematical and Computational Modeling, CEFET-MG, Belo Horizonte, Brazil
claudinharmc@yahoo.com.br

[2] Department of Computer Engineering, CEFET-MG, Belo Horizonte, Brazil
efwannerr@decom.cefetmg.br

[3] CISUC, Department of Informatics Engineering, University of Coimbra, Coimbra, Portugal
cmfonsec@dei.uc.pt

Abstract. A simple success-based step-size adaptation rule for single-parent Evolution Strategies is formulated, and the setting of the corresponding parameters is considered. Theoretical convergence on the class of strictly unimodal functions of one variable that are symmetric around the optimum is investigated using a stochastic Lyapunov function method developed by Semenov and Terkel [5] in the context of martingale theory. General expressions for the conditional expectations of the next values of step size and distance to the optimum under $(1 \overset{+}{,} \lambda)$-selection are analytically derived, and an appropriate Lyapunov function is constructed. Convergence rate upper bounds, as well as adaptation parameter values, are obtained through numerical optimization for increasing values of λ. By selecting the number of offspring that minimizes the bound on the convergence rate with respect to the number of function evaluations, all strategy parameter values result from the analysis.

Keywords: Step-size adaptation · Evolution strategy · Lyapunov function theory · Convergence rate

1 Introduction

Evolution strategies (ESs) are a particular class of Evolutionary Algorithms (EAs) that have attracted a significant amount of attention in the last decades. ESs traditionally emphasize the use of selection and mutation as search operators, where the mutation operator consists in creating an offspring by adding a random vector to the current solution, or individual. Adaptive methods that dynamically rescale mutation step-length parameters have proved to be rather effective [1–3].

Convergence analysis of ESs with step-size adaptation has deserved due attention in the research community, but it has proved to be a difficult task.

J. Handl et al. (Eds.): PPSN XIV 2016, LNCS 9921, pp. 101–110, 2016.
DOI: 10.1007/978-3-319-45823-6_10

The present work extends the methodology proposed in [4] to the analysis of both $(1, \lambda)$-ESs and $(1+\lambda)$-ESs with an arbitrary number of offspring, λ, and a success-based step-size adaptation rule. Based on the theoretical results developed in [4,5], general expressions for the expectation of the next individual's step size and distance to the optimum are analytically derived for both types of selection and, using a Lyapunov function, upper bounds on the convergence rate on the class of strictly unimodal functions of one variable that are symmetric around the optimum are determined. Moreover, the number of offspring that minimizes the bound on the convergence rate with respect to the number of function evaluations can be easily determined as a by-product, providing a useful guideline for the selection of this remaining strategy parameter.

2 Literature Review

Convergence analysis is a major topic of research in evolutionary algorithms, which has been addressed mostly separately depending on whether the optimization problems of interest are continuous or discrete. Drift analysis is a state-of-art technique for the study of the expected hitting time of randomized search heuristics on discrete problems. The use of drift analysis combined with Markov chain theory in order to obtain lower and upper bounds on the expected hitting time on discrete search spaces is presented, for example, in [6–10]. Proofs of convergence of the $(1+1)$-EA applied to pseudo-Boolean linear functions can be found in [6,9,10]. In [6], the author combines drift analysis and Markov-chains for the first time to state bounds on the expected optimization time. The upper bound on the expected runtime of the algorithm was improved in [9], using multiplicative drift analysis. Later, in [10], the author improves the upper bound further, also using multiplicative drift analysis.

For continuous optimization problems, works dealing with convergence analysis of self-adaptive $(1, \lambda)$-ES applied to sphere functions using martingale theory include [4,5,11]. In [5], a stochastic Lyapunov function method is developed in the context of martingale theory, but Monte Carlo simulations are used to verify the convergence of a mutative self-adaptive $(1, \lambda)$-ES. Based on [5], a Lyapunov synthesis procedure for the adaptation parameters of a simple derandomized self-adaptive $(1, 2)$-ES is proposed in [4]. The methodology is based on particular candidate functions which become stochastic Lyapunov functions through suitable choices of the algorithm adaptation parameters. Considering the class of strictly unimodal functions of one variable which are symmetric around the optimum, and through the appropriate setting of the algorithm parameters, it is proved that both the decision variable and the mutation step-size converge almost surely to the optimum and to zero, respectively. Moreover, an upper bound on the rate of convergence is derived, and suitable values for the ES parameters are determined numerically.

The parallel between the drift analysis used in [9] and the method used in [5] and [4] is worth noting. With both methods, the behavior of an evolutionary algorithm is analyzed through an auxiliary function, which must be chosen in such a way that convergence of the algorithm on the true objective function can be proved by verifying conditions on the auxiliary function only. In the case of drift analysis, the auxiliary function, also known as a potential function, is used to derive bounds on the expected runtime of the algorithm with respect to problem size. On the other hand, in [4,5], continuous optimization is considered and bounds on the convergence rate are provided instead.

3 The Proposed ES

The aim of the $(1 \overset{+}{,} \lambda)$-ES analyzed here is to minimize a real-valued function $f : \mathbb{R} \to \mathbb{R}$. A vector (x_t, d_t) describes an individual, the fitness of which depends on the value of the decision variable $x_t \in \mathbb{R}$. The step size which controls the variation is given by $d_t \in \mathbb{R}_+ \backslash \{0\}$. The evolution cycle consists of two steps: the mutation step, which creates λ offspring, $x_{i,t}$, $i = 1, \ldots, \lambda$, and the selection step, which selects the next parent, x_{t+1}, from the offspring (and possibly the current parent, x_t, depending on the selection scheme), and determines the associated step size, d_{t+1}.

Consider the following $(1, \lambda)$-ES, where $\mu_{i,t}$, $i = 1, \ldots, \lambda$, are random variables uniformly distributed in $[-1, 1]$, $0 < \alpha_f \leq 1$, and $\alpha_s \geq 1$:

$$
\begin{aligned}
x_{t+1} &= F(x_t, d_t, \mu_{1,t}, \ldots, \mu_{\lambda,t}) \\
&= \underset{x \in \{x_{i,t} = x_t + \mu_{i,t} d_t,\ i=1,\ldots,\lambda\}}{\arg\min} f(x) \\
d_{t+1} &= G(x_t, d_t, \mu_{1,t}, \ldots, \mu_{\lambda,t}) \\
&= \begin{cases} \alpha_f \cdot d_t \text{ if } f(x_{t+1}) > f(x_t) \\ \alpha_s \cdot d_t \text{ if } f(x_{t+1}) \leq f(x_t) \end{cases}
\end{aligned}
\tag{1}
$$

Note that, if f is any strictly unimodal function of one variable that is symmetric around the minimum, x_{t+1} is selected as the offspring $x_{i,t}$, $i = 1, \ldots, \lambda$, that is closest to the minimum point of f, even if that minimum point is not known. In fact, the selection process is translation invariant, meaning that the minimum of f can be considered to be located at zero after an appropriate translation, without loss of generality. Therefore, any even function with such properties could be chosen. Moreover, observe that d_{t+1} depends indirectly on $(\mu_{1,t}, \mu_{2,t}, \ldots, \mu_{\lambda,t})^T$ through a direct dependence on x_{t+1}. The search space $\mathbb{R}$ is equipped with the Borel σ-algebra.

The mutation step-size adaptation process can be described as follows: (i) if any offspring is at least as good as the parent, then the step size is increased, and (ii) if all offspring are worse than the parent, then the step-size is decreased. The only difference between this $(1, \lambda)$-ES and the corresponding $(1 + \lambda)$-ES is the selection scheme: in the latter, the best among the λ offspring *and* the parent itself, x_t, is selected to become x_{t+1}. As a consequence, the parent represents the all-time best individual at each iteration.

4 Theoretical Background

Consider a deterministic discrete dynamic system represented by $x_{t+1} = F(x_t)$ where $t \in \mathbb{N}$ is the time index, $x_t \in \mathbb{R}^n$ is the system state vector at time t, n is the number of system state variables, and $F : \mathbb{R}^n \to \mathbb{R}^n$ is a function in class C^p, with $p \geq 1$. An equilibrium point, x^*, of such a system is a point such that $F(x^*) = x^*$. A discrete version of the direct method of Lyapunov [12] essentially states that, if there is a function $V \in \mathcal{C}^1$ such that $V(x^*) = 0$ and $V(x) > 0$ $\forall\, x \neq x^*$, and, in addition, $V(x_{t+1}) < V(x_t)\ \forall\, x_t \neq x^*$, then the equilibrium is uniformly asymptotically stable [13]. Function V is known as a Lyapunov function.[1]

Since the proposed $(1 \stackrel{+}{,} \lambda)$-ESs are stochastic processes, the above result is not directly applicable, but convergence may still be studied through the conditional expectation $E^{A_t}[V(x_{t+1}, d_{t+1})] \triangleq E[V(x_{t+1}, d_{t+1})|x_1, \ldots, x_t, d_1, \ldots, d_t]$ of a suitable Lyapunov function $V(x_t, d_t)$ of the stochastic process (x_t, d_t). Theoretical results presented in [5] allow the convergence rate to be studied as well.

Consider that the inequality

$$|x_t| \leq \exp(-at), \tag{2}$$

holds asymptotically almost surely[2] for some scalar $a > 0$. Then, the convergence rate of x_t to the point of equilibrium $x^* = 0$ is $e^{-\bar{a}}$, where $\bar{a}$ is the supremum of the set of values of a for which (2) holds asymptotically almost surely. If only a lower bound a on the exponential decay constant $\bar{a}$ can be determined, with $0 < a \leq \bar{a}$, then $|x_t| \leq \exp(-\bar{a}t) \leq \exp(-at)$.

In [5], the asymptotic behavior of a supermartingale[3] V_t is analyzed under the following conditions: at each time step, V_t decreases on average by at least a constant $a > 0$, and the conditional variance of V_t is at most $b > 0$. The following result is proved therein:

Proposition 1. *Let V_t be a supermartingale and $V_0 = 0$. If the following conditions hold*

$$E^{A_t}(V_{t+1}) \leq V_t - a \tag{3}$$

$$E^{A_t}([V_{t+1} - E^{A_t}(V_{t+1})]^2) \leq b \tag{4}$$

where $a > 0$, $b > 0$, then $\forall\, \epsilon > 0$ the following inequality holds almost surely:

$$V_t \leq -at + o(t^{0.5+\epsilon}). \tag{5}$$

Proposition 1 is employed in the following result to establish the exponential convergence of the proposed $(1, \lambda)$-ES algorithm.

---

[1] Equivalently to $V(x^*) = 0$ and $V(x) > 0\ \forall\, x \neq x^*$, one may require that $V(x) \to -\infty$ only when $x \to x^*$.

[2] An event holds asymptotically almost surely if it holds with probability $1 - o(1)$, i.e. the probability of success goes to 1 in the limit as $n \to \infty$ [14].

[3] A stochastic process V_t is said to be a *supermartingale* if $E^{A_t}(V_{t+1}) \leq V_t$.

Proposition 2. *Consider the stochastic process* (x_t, d_t) *defined in Eq. (1). If* f *is a strictly unimodal function of one variable that is symmetric around its minimum, then this process converges to* $(0, 0)$ *almost surely, and the following inequalities*

$$|x_t| \leq \exp(-at) \quad \text{and} \quad d_t \leq \exp(-at) \tag{6}$$

hold asymptotically almost surely for some α_f, α_s, *and* $a > 0$.

Convergence of the corresponding $(1 + \lambda)$-ES can be stated (and proved) in the same way.

5 Convergence Rate Analysis

The following Lyapunov function is used in this work:

$$V_t = V(x_t, d_t) = \ln(|x_t| + wd_t) - k \ln(d_t) \tag{7}$$

where $w, k \in \mathbb{R}$, $w > 0$, and $0 \leq k < 1$. Observe that $V_t \to -\infty$ only when $x_t \to 0$ and $d_t \to 0$.

The proof of Proposition 2 consists of three steps:

1. Determining values for α_s, α_f, w, k and a such that inequality (3) is verified.
2. Proving that the conditional variance of V_t is bounded (inequality (4)).
3. Proving that, under Proposition 1, inequalities (6) hold asymptotically almost surely.

Since steps 2 and 3 depend only on the structure of the Lyapunov function and/or on the type of mutation and form of step-size adaptation considered, but not on whether step-size adaptation is based on success or on the length of the selected step, the corresponding proofs are identical to those presented in [4], where the same Lyapunov function and type of mutation are used together with a different two-point adaptation rule. Moreover, they apply to both $(1, \lambda)$ and $(1 + \lambda)$ selection.

In particular, step 2 can be accomplished by showing that V_{t+1} has support of bounded length for all x_t and d_t. By noting that the next value of the step size must satisfy $d_t\alpha_f \leq d_{t+1} \leq d_t\alpha_s$, is possible to show that both $\ln(d_{t+1})$ and $\ln(|x_{t+1}| + wd_{t+1})$ have supports of bounded length, and so does V_{t+1}.

Having this in mind, only step 1 is considered here for the proposed $(1 \overset{+}{,} \lambda)$-ESs. In order to derive (an upper bound on) $E^{A_t}(V_{t+1})$ for an arbitrary number of offspring, λ, expectations $E^{A_t}(|x_{t+1}|)$ and $E^{A_t}(d_{t+1})$ are calculated first for each type of selection. Actually, since the step-size adaptation scheme is exactly the same for $(1, \lambda)$ and $(1 + \lambda)$ selection, $E^{A_t}(d_{t+1})$ is the same in both cases. However, depending of the type of selection, $E^{A_t}(|x_{t+1}|)$ will result in different expressions.

The $(1, \lambda)$-ES is considered first. In determining $E^{A_t}(|x_{t+1}|)$, two cases must be considered: $0 \leq \frac{x_t}{d_t} < 1$, meaning that x_t is close to the origin, and $\frac{x_t}{d_t} \geq 1$, meaning that x_t is far from the origin.

In the case where x_t is close to the origin,

$$\begin{aligned} E^{A_t}(|x_{t+1}|) &= \tfrac{\lambda}{(2d_t)^\lambda}\left[\int_{x_t-d_t}^{0} -z(2z+2d_t)^{\lambda-1}\mathrm{d}z\right] \\ &+\tfrac{\lambda}{(2d_t)^\lambda}\left[\int_{0}^{-x_t+d_t} z(2d_t-2z)^{\lambda-1}\mathrm{d}z\right] \\ &+\tfrac{\lambda}{(2d_t)^\lambda}\left[\int_{-x_t+d_t}^{x_t+d_t} z(x_t+d_t-z)^{\lambda-1}\mathrm{d}z\right] \\ &= \left(\tfrac{1}{\lambda+1}\right)\left[\tfrac{x_t^{\lambda+1}}{d_t^{\lambda+1}}+1\right] d_t \end{aligned} \tag{8}$$

and in the case where x_t is far from the origin,

$$\begin{aligned} E^{A_t}(|x_{t+1}|) &= \tfrac{\lambda}{(2d_t)^\lambda}\left[\int_{x_t-d_t}^{x_t+d_t} z(x_t+d_t-z)^{\lambda-1}\mathrm{d}z\right] \\ &= \left[\tfrac{x_t}{d_t}-\left(\tfrac{\lambda-1}{\lambda+1}\right)\right] d_t. \end{aligned} \tag{9}$$

Considering (1+λ)-ES, three cases arise: $0 \leq \frac{x_t}{d_t} < \frac{1}{2}$, $\frac{1}{2} \leq \frac{x_t}{d_t} < 1$ and $\frac{x_t}{d_t} \geq 1$. In the case where $0 \leq \frac{x_t}{d_t} < \frac{1}{2}$,

$$\begin{aligned} E^{A_t}(|x_{t+1}|) &= \tfrac{\lambda}{(2d_t)^\lambda}\left[\int_{x_t-d_t}^{-x_t} x_t(2z+2d_t)^{\lambda-1}\mathrm{d}z\right] \\ &+\tfrac{\lambda}{(2d_t)^\lambda}\left[\int_{-x_t}^{0} -z(2d_t+2z)^{\lambda-1}\mathrm{d}z\right] \\ &+\tfrac{\lambda}{(2d_t)^\lambda}\left[\int_{0}^{x_t} z(2d_t-2z)^{\lambda-1}\mathrm{d}z\right] \\ &+\tfrac{\lambda}{(2d_t)^\lambda}\left[\int_{x_t}^{-x_t+d_t} x_t(2d_t-2z)^{\lambda-1}\mathrm{d}z\right] \\ &+\tfrac{\lambda}{(2d_t)^\lambda}\left[\int_{-x_t+d_t}^{x_t+d_t} x_t(x_t+d_t-z)^{\lambda-1}\mathrm{d}z\right] \\ &= \tfrac{d_t}{\lambda+1}-\left(1-\tfrac{x_t}{d_t}\right)^\lambda\left(\tfrac{d_t-x_t}{\lambda+1}\right). \end{aligned} \tag{10}$$

For $\frac{1}{2} \leq \frac{x_t}{d_t} < 1$,

$$\begin{aligned} E^{A_t}(|x_{t+1}|) &= \tfrac{\lambda}{(2d_t)^\lambda}\left[\int_{x_t-d_t}^{0} -z(2z+2d_t)^{\lambda-1}\mathrm{d}z\right] \\ &+\tfrac{\lambda}{(2d_t)^\lambda}\left[\int_{0}^{-x_t+d_t} z(2d_t-2z)^{\lambda-1}\mathrm{d}z\right] \\ &+\tfrac{\lambda}{(2d_t)^\lambda}\left[\int_{-x_t+d_t}^{x_t} z(x_t+d_t-z)^{\lambda-1}\mathrm{d}z\right] \\ &+\tfrac{\lambda}{(2d_t)^\lambda}\left[\int_{x_t}^{x_t+d_t} x_t(x_t+d_t-z)^{\lambda-1}\mathrm{d}z\right] \\ &= \left(\tfrac{x_t}{\lambda+1}\right)\left(\tfrac{x_t}{d_t}\right)^\lambda+\tfrac{d_t}{\lambda+1}-\tfrac{d_t}{2^\lambda(\lambda+1)}. \end{aligned} \tag{11}$$

Finally, for $\frac{x_t}{d_t} \geq 1$,

$$\begin{aligned} E^{A_t}(|x_{t+1}|) &= \tfrac{\lambda}{(2d_t)^\lambda}\left[\int_{x_t-d_t}^{x_t} z(x_t+d_t-z)^{\lambda-1}\mathrm{d}z\right] \\ &+\tfrac{\lambda}{(2d_t)^\lambda}\left[\int_{x_t}^{x_t+d_t} x_t(x_t+d_t-z)^{\lambda-1}\mathrm{d}z\right] \\ &= x_t+d_t\left(\tfrac{1-\lambda}{\lambda+1}\right)-\tfrac{d_t}{2^\lambda(\lambda+1)}. \end{aligned} \tag{12}$$

Regarding the expected step size, $E^{A_t}(d_{t+1}) = d_t\left[\alpha_f P_F + \alpha_s P_S\right]$, and all that is required is to derive expressions for the failure and success probabilities,

P_F and $P_S = 1 - P_F$, respectively. The probability of failure is simply $P_F = [P(|x_{i,t}| > |x_t|)]^\lambda = [P(x_{i,t} > x_t) + P(x_{i,t} < -x_t)]^\lambda$, leading to two distinct cases.

If $0 \leq \frac{x_t}{d_t} < \frac{1}{2}$, then $-x_t > x_t - d_t$, and thus $P(x_{i,t} < -x_t) \neq 0$. Therefore, $P_F = \left(1 - \frac{x_t}{d_t}\right)^\lambda$ and

$$E^{A_t}(d_{t+1}) = d_t \left[\left(1 - \frac{x_t}{d_t}\right)^\lambda (\alpha_f - \alpha_s) + \alpha_s \right]. \tag{13}$$

If $\frac{x_t}{d_t} \geq \frac{1}{2}$, then $-x_t \leq x_t - d_t$ and $P(x_{i,t} < -x_t) = 0$. Thus, $P_F = \frac{1}{2^\lambda}$ and

$$E^{A_t}(d_{t+1}) = d_t \left[\frac{1}{2^\lambda}(\alpha_f - \alpha_s) + \alpha_s \right]. \tag{14}$$

To complete step 1, inequality (3) can be rewritten as:

$$E^{A_t}[\ln(|x_{t+1}| + wd_{t+1})] - kE^{A_t}[\ln(d_{t+1})] - \ln(|x_t| + wd_t) + k\ln(d_t) \leq -a. \tag{15}$$

Since $\ln(\cdot)$ is a concave function, using Jensen's inequality,

$$E^{A_t}[\ln(|x_{t+1}| + wd_{t+1})] \leq \ln[E^{A_t}(|x_{t+1}| + wd_{t+1})] \tag{16}$$

and it is sufficient to prove that there exist α_f, α_s, w, k and a such that the following inequality holds:

$$\ln[E^{A_t}(|x_{t+1}|) + wE^{A_t}(d_{t+1})] - kE^{A_t}[\ln(d_{t+1})] - \ln(|x_t| + wd_t) + k\ln(d_t) \leq -a. \tag{17}$$

Due to the particular form of its left-hand side, inequality (17) may be rewritten as:

$$\Psi\left(\frac{|x_t|}{d_t}\right) = \ln\left[\frac{E^{A_t}(|x_{t+1}|) + wE^{A_t}(d_{t+1})}{|x_t| + wd_t}\right] - k(E^{A_t}[\ln(d_{t+1})] - \ln(d_t)) \leq -a. \tag{18}$$

Analytical expressions for function $\Psi(r)$, $r = |x_t|/d_t$, can be obtained analytically by considering the intervals (A) $0 \leq \frac{x_t}{d_t} < \frac{1}{2}$, (B) $\frac{1}{2} \leq \frac{x_t}{d_t} < 1$, and (C) $\frac{x_t}{d_t} \geq 1$, and combining expressions (8) to (14) as appropriate for each interval and each type of selection.

Then, it is necessary to show that, for all $r \geq 0$, $\Psi(r) \leq -a$ for some $a > 0$, particularly at the ends of each interval ($r = 0, 1/2, 1$ and $r \to +\infty$) and at any critical points inside those intervals ($r = r^*$ such that $\Psi'(r^*) = 0$).

Note that, regardless of the selection scheme, $\Psi(r)$ is represented in each interval by a sum of logarithms of rational fractions, and that $\Psi'(r)$ is represented by a rational fraction, in variable r. Therefore, the critical points within each interval, and the corresponding values of Ψ, can be determined numerically for given values of α_f, α_s, w and k, and the following constrained nonlinear optimization problem can be formulated:

$$(\alpha_s^*, \alpha_f^*, w^*, k^*, a^*) = \underset{\alpha_s,\alpha_f,w,k,a}{\arg\max}\ a$$

$$\text{subject to:} \begin{cases} \alpha_s \geq 1 \\ 0 < \alpha_f \leq 1 \\ w > 0 \\ 0 \leq k < 1 \\ \Psi(r) + a \leq 0, \quad r = 0, 1/2, 1, +\infty \\ \Psi(r^*) + a \leq 0, \; r^* : \Psi'(r^*) = 0 \end{cases} \tag{19}$$

Problem (19) can be solved numerically for either $(1 \overset{+}{,} \lambda)$-ES and any selected value of λ. To prove this statement, the principle of finite induction is used. Firstly, it is shown that Problem (19) can be solved both for the $(1+1)$-ES and for the $(1,2)$-ES (see Table 1). Then, it must also be shown that, if it can be solved for a generic λ, then it can also be solved for $\lambda + 1$.

In order to highlight the dependence of function Ψ on the number of offspring λ, the following notation is introduced:

$$\Psi_\lambda(r) = \ln\left[\frac{E_\lambda^{A_t}(|x_{t+1}|) + wE_\lambda^{A_t}(d_{t+1})}{|x_t| + wd_t}\right] - k\{E_\lambda^{A_t}[\ln(d_{t+1})] - \ln(d_t)\} \tag{20}$$

Clearly, $\forall\lambda \in \mathbb{N}$,

$$-k\{E_{\lambda+1}^{A_t}[\ln(d_{t+1})] - \ln(d_t)\} \leq -k\{E_\lambda^{A_t}[\ln(d_{t+1})] - \ln(d_t)\}, \tag{21}$$

so it is sufficient to show that $\forall\lambda \in \mathbb{N}, \exists\, w' > 0:$

$$E_{\lambda+1}^{A_t}(|x_{t+1}|) + w'E_{\lambda+1}^{A_t}(d_{t+1}) \leq E_\lambda^{A_t}(|x_{t+1}|) + wE_\lambda^{A_t}(d_{t+1}), \tag{22}$$

where $E_{\lambda+1}^{A_t}(|x_{t+1}|) \leq E_\lambda^{A_t}(|x_{t+1}|)$ and $E_{\lambda+1}^{A_t}(d_{t+1}) \geq E_\lambda^{A_t}(d_{t+1})$.

Suppose that there exist $0 < \alpha_f \leq 1$, $\alpha_s \geq 1$, $w > 0$, $0 \leq k < 1$, and $a > 0$ such that $\Psi_\lambda(r) \leq -a$ for all $r \geq 0$, and let $w_1 > 0$ and $w_2 > 0$ be defined as follows:

$$w_1 = w\,\frac{2^{\lambda+1}\alpha_f}{\alpha_f + \alpha_s(2^{\lambda+1} - 1)} < w\,\frac{(1-r)^\lambda(\alpha_f - \alpha_s) + \alpha_s}{(1-r)^{\lambda+1}(\alpha_f - \alpha_s) + \alpha_s} = w\,\frac{E_\lambda^{A_t}(d_{t+1})}{E_{\lambda+1}^{A_t}(d_{t+1})} \tag{23}$$

where $0 \leq \frac{x_t}{d_t} < \frac{1}{2}$, and

$$w_2 = 2w\,\frac{\alpha_f + \alpha_s(2^\lambda - 1)}{\alpha_f + \alpha_s(2^{\lambda+1} - 1)} = w\,\frac{E_\lambda^{A_t}(d_{t+1})}{E_{\lambda+1}^{A_t}(d_{t+1})} \quad \text{where} \quad \frac{x_t}{d_t} \geq \frac{1}{2}. \tag{24}$$

Letting $w' = \min\{w_1, w_2\}$, inequality (22) is shown to hold true, and there are indeed $0 < \alpha_f \leq 1$, $\alpha_s \geq 1$, $w = w' > 0$, $0 \leq k < 1$ and $a > 0$ such that $\Psi_{\lambda+1}(r) \leq -a$ for all $r \geq 0$, concluding the induction step.

By solving Problem (19) numerically for each type of selection and number of offspring, suitable step-size adaptation parameters α_f and α_s are obtained, and inequality (3) is shown to hold true, completing step 1 of the proof of Proposition 2. An upper bound e^{-a} on the convergence rate of such an ES on any strictly unimodal, symmetric function of one variable is also obtained as a by-product.

Table 1 shows the parameter and auxiliary values, as well as the convergence bounds, obtained for $(1, \lambda)$ and $(1 + \lambda)$ selection and several values of λ. The results were obtained using the sequential quadratic programming solver in the GNU Octave numerical package. From the table, it is possible to select the number of offspring λ which minimizes the convergence rate bound for each $(1 \overset{+}{,} \lambda)$-ES with respect to the number of function evaluations. For the $(1, \lambda)$-ES, $\lambda = 4$ can be seen to lead to the highest value of a/λ, whereas for the $(1 + \lambda)$-ES, the best number of offspring appears to be $\lambda = 1$.

Table 1. Values for α_s, α_f, w, k, a and a/λ obtained by solving optimization problem (19) for $(1, \lambda)$ and $(1 + \lambda)$-ESs and different numbers of offspring (λ).

$(1, \lambda)$-ES		$\lambda = 2$	$\lambda = 3$	$\lambda = 4$	$\lambda = 5$	$\lambda = 6$	$\lambda = 7$
α_s		1.15180	1.19847	1.19591	1.18729	1.18036	1.17567
α_f		0.72873	0.52779	0.42236	0.36531	0.33489	0.31895
w		2.51320	1.44610	1.09380	0.90415	0.77973	0.68860
k		0.31395	0.30306	0.29601	0.29158	0.28901	0.28761
a		0.00843	0.02379	0.03370	0.03932	0.04223	0.04362
a/λ		0.00422	0.00793	0.00842	0.00786	0.00704	0.00623
$(1 + \lambda)$-ES	$\lambda = 1$	$\lambda = 2$	$\lambda = 3$	$\lambda = 4$	$\lambda = 5$	$\lambda = 6$	$\lambda = 7$
α_s	1.89274	1.73948	1.64704	1.59411	1.56531	1.55027	1.54260
α_f	0.75675	0.60518	0.51496	0.46382	0.43603	0.42133	0.41370
w	0.11219	0.10567	0.09626	0.08560	0.07521	0.06598	0.05819
k	0.23988	0.24304	0.24488	0.24591	0.24648	0.24681	0.24701
a	0.04309	0.07039	0.08660	0.09569	0.10060	0.10318	0.10453
a/λ	0.04309	0.03518	0.02887	0.02392	0.02012	0.01720	0.01493

6 Conclusions

In this paper, a simple success-based $(1 \overset{+}{,} \lambda)$-ES with uniformly-distributed mutations for functions of one variable was proposed. Following the theoretical approach proposed in [5], the convergence of the ES was studied on the class of unimodal functions symmetric around the optimum using stochastic Lyapunov function and martingale theory. General expressions for the expectation of $|x_{t+1}|$ and d_{t+1} were derived considering both $(1 \overset{+}{,} \lambda)$ selection. Using an appropriate Lyapunov function, upper bounds on the convergence rate and specific ES parameter values were obtained via numerical optimization, for growing values of λ. The number of offspring that minimizes the bound on the convergence rate with respect to the number of function evaluations was also determined in this way. Future work includes an experimental study of the actual convergence rates achieved with the proposed parameter settings as well as extending the theoretical results to other function classes, other mutation distributions, and multiple decision variables.

Acknowledgment. This work was partially supported by national funds through the Portuguese Foundation for Science and Technology (FCT) and by the European Regional Development Fund (FEDER) through COMPETE 2020 – Operational Program for Competitiveness and Internationalization (POCI). The authors also would like to thank the Brazilian funding agencies, CAPES, CNPq and FAPEMIG.

References

1. Hansen, N., Ostermeier, A.: Completely derandomized self-adaptation in evolution strategies. Evol. Comput. **9**(2), 159–195 (2001)
2. Doerr, B., Doerr, C.: Optimal parameter choices through self-adjustment: applying the 1/5-th rule in discrete settings. In: Proceedings of the 2015 ACM-GECCO Genetic and Evolutionary Computation Conference, pp. 1335–1342. ACM (2015)
3. Doerr, B., Doerr, C.: A tight runtime analysis of the $(1 + (\lambda, \lambda))$ genetic algorithm on onemax. In: Proceedings of the 2015 ACM-GECCO Genetic and Evolutionary Computation Conference, pp. 1423–1430. ACM (2015)
4. Wanner, E.F., Fonseca, C.M., Cardoso, R.T.N., Takahashi, R.H.C.: Lyapunov stability analysis and adaptation law synthesis of a derandomized self-adaptive (1, 2)-ES. Under review
5. Semenov, M.A., Terkel, D.A.: Analysis of convergence of an evolutionary algorithm with self-adaptation using a stochastic Lyapunov function. Evol. Comput. **11**(4), 363–379 (2003)
6. Jägersküpper, J.: A blend of Markov-chain and drift analysis. In: Rudolph, G., Jansen, T., Lucas, S., Poloni, C., Beume, N. (eds.) PPSN 2008. LNCS, vol. 5199, pp. 41–51. Springer, Heidelberg (2008)
7. He, J., Yao, X.: Drift analysis and average time complexity of evolutionary algorithms. Artif. Intell. **127**, 57–85 (2001)
8. He, J., Yao, X.: A study of drift analysis for estimating computational time of evolutionary algorithms. Natural Comput. **3**, 21–35 (2004)
9. Doerr, B., Johannsen, D., Winzen, C.: Multiplicative drift analysis. Algorithmica **64**(4), 673–697 (2011)
10. Witt, C.: Tight bounds on the optimization time of a randomized search heuristic on linear function. Comb. Probab. Comput. **22**(02), 294–318 (2013)
11. Hart, W.E.: Rethinking the design of real-coded evolutionary algorithms: making discrete choices in continuous search domains. Soft Comput. J. **9**, 225–235 (2002)
12. Lyapunov, A.M.: The general problem of stability of motion (reprint of the original paper of 1892). Int. J. Control **55**(3), 531–773 (1992)
13. Hahn, W.: Stability of Motion. Springer, Heidelberg (1967)
14. Janson, S., Luczak, T., Rucinski, A.: Random Graphs. Wiley, Hoboken (2000)

Differential Evolution and Swarm Intelligence

TADE: Tight Adaptive Differential Evolution

Weijie Zheng[1,2(✉)], Haohuan Fu[1(✉)], and Guangwen Yang[1,2(✉)]

[1] Ministry of Education Key Laboratory for Earth System Modeling, Center for Earth System Science, Tsinghua University, Beijing, China
haohuan@mail.tsinghua.edu.cn

[2] Tsinghua National Laboratory for Information Science and Technology (TNList), Department of Computer Science and Technology, Tsinghua University, Beijing, China
zhengwj13@mails.tsinghua.edu.cn, ygw@mail.tsinghua.edu.cn

Abstract. Differential Evolution (DE) is a simple and effective evolutionary algorithm to solve optimization problems. The existing DE variants always maintain or increase the randomness of the differential vector when considering the trade-off of randomness and certainty among three components of the mutation operator. This paper considers the possibility to achieve a better trade-off and more accurate result by reducing the randomness of the differential vector, and designs a tight adaptive DE variant called TADE. In TADE, the population is divided into a major subpopulation adopting the general "current-to-*p*best" strategy and a minor subpopulation utilizing our proposed strategy of sharing the same base vector but reducing the randomness in differential vector. Based on success-history parameter adaptation, TADE designs a simple information exchange scheme to avoid the homogeneity of parameters. The extensive experiments on CEC2014 suite show that TADE achieves better or equivalent performance on at least 76.7 % functions comparing with five state-of-the-art DE variants. Additional experiments are conducted to verify the rationality of this tight design.

Keywords: Differential evolution · Differential vector · Adaptive

1 Introduction

Differential Evolution (DE), proposed in [1], is a simple and effective evolutionary algorithm to solve complex optimization problems. It has been shown to outperform some nature-inspired metaheuristics, such as genetic algorithm and particle swarm optimization over several benchmark functions [2], and has been adopted to various applications according to [3,4]. However, due to its stochastic nature, it suffers from long computing period and has the potential to improve the accuracy further. Since the mutation operator is the main engine that drives the population toward improvement, to achieve a more accurate and efficient DE algorithm, plenty of researches have been done based on its three components, the base vector, scaling factor and differential vector(s).

J. Handl et al. (Eds.): PPSN XIV 2016, LNCS 9921, pp. 113–122, 2016.
DOI: 10.1007/978-3-319-45823-6_11

The basic DE variants make the performance improvement mainly by reducing the randomness of base vector, such as DE/best/1, increasing the randomness of differential vector(s), such as DE/rand/2, or performing on both aspects, such as DE/best/2. The control parameters remain unchanged throughout the process. However, this certainty of the unchanged parameters makes it impractical due to high time cost of the required parameter tuning step, and this certainty somehow does harm to the performance since different parameter settings are fit for different stages. Therefore, many existing methods turned to introducing the randomness into the parameters and making them alterable and adaptable to different stages, such as jDE [6], NSDE [7], JADE [8] and SHADE [9] on a single mutation strategy, and CoDE [11], SaDE [10] and EPSDE [12] further combining multiple strategies. These adaptive (or self-adaptive) variants achieved more accurate result via increasing the randomness of search length, but still did not reduce the randomness of the differential vector.

These existing variants maintained the wide randomness on differential vector, which seems sensible since the randomness can ensure the possibility of reaching global optimum. However, it may waste time in the computation on the unpromising area, which results in its slow convergence. Introducing the certainty may result in a more accurate result in limited function evaluation times. Recently, our previous work [5] took the first step to reduce the diversity of the differential vector, and achieved a better result against DE/rand/1 and DE/best/1 on several benchmark functions. This shows the possibility of reducing the randomness of differential vector to achieve a more accurate result. However, although that work obtained a trade-off among the base vector and the differential vector, it maintained the certain scaling factor. As far as we know, there has not been any work considering the trade-off among all the three aspects while reducing the randomness of differential vector to get a competitive result.

In this paper, we firstly propose a novel DE mutation strategy. It takes current-to-*p*best as the base vector, maintains the random choice of the starting point of differential vector and adopts the current (target) vector as the ending point. Due to the reduced randomness in the search direction, this greedy mutation may lead to premature and is unfit to drive the whole population. Therefore, it is then utilized as the engine of a minor subpopulation. The major subpopulation is evolved via current-to-*p*best [8] which has the same base vector and can share the exploration information in time with the minor subpopulation. Both subpopulations adopt the success history based parameter adaption [9], and the exchange of the two subpopulations is designed to further enhance the diversity of the scaling factor. Extensive experiments are conducted to compare this Tight Adaptive DE (TADE) with five state-of-the-art DE variants (SHADE, JADE, CoDE, EPSDE and SaDE) on the CEC2014 benchmark suite.

The contributions of this paper can be summarized as follows:

- This paper designs a tight adaptive DE scheme, which includes a proposed mutation reducing the randomness in differential vector, the information exchange on mutation strategies to get out of the local optima, and the information exchange on control parameters to enhance the randomness

- TADE achieves better or equivalent performance on at least 76.7 % functions comparing with five state-of-the-art DE variants
- Verification for the rationality of our tight design is conducted by the experiments varying partition and parameter exchange rate

Organization of the rest paper is as follows. The brief introduction of Differential Evolution and an insight view are shown in Sect. 2. Section 3 discusses the motivation and detail of the proposed method. Extensive experiments and analysis are conducted on Sect. 4. Finally, Sect. 5 concludes the paper and discusses the future work.

2 Differential Evolution

This section briefly describes the framework and an insight view of DE algorithm. DE undergoes mutation, crossover and selection operators iteratively until satisfying the accuracy condition or reaching a predefined function evaluation times FES_{max}. The random change happens in mutation so that every candidate has the opportunity to enter the next generation and get itself inherited. Crossover operator generates the trial vector u_i^g that exchanges the information of the mutant vector v_i^g with the target vector x_i^g and further widen the diversity. Then the selection operator is employed to preserve the most promising vector entering the next generation, ensuring the non-degeneration evolution process. Obviously, mutation is the main engine to pull the population to improvement.

All mutation operators are composed of three parts: the base vector, differential vector(s) and the search length (scaling factor). Taking "current-to-*p*best" in JADE [8] as an example:

$$v_i^g = x_i^g + F_i(x_{pbest}^g - x_i^g) + F_i(x_{r_1}^g - \tilde{x}_{r_2}^g) \tag{1}$$

The mutation happens around the neighborhood of the base vector $x_i^g + F_i(x_{pbest}^g - x_i^g)$, giving a rough guess on where the promising search area is. Then the differential vector $x_{r_1}^g - \tilde{x}_{r_2}^g$ determines the search direction, and the scaling factor F_i controls how far it will search along the direction.

Throughout the DE developing history, many existing methods can be regarded as looking for a more accurate result via achieving a better trade-off between the randomness and the certainty. Randomness represents the huge diversity of the offspring candidates that can ensure the possibility of reaching the global minimum, while certainty carries the information which leads to a strategy or belief on where the promising area is. Therefore, although randomness maintains the possibility, the huge area to search may cause inefficiency, and although certainty provides relatively smaller area and processes faster, it may cause premature due to its myopia or greediness.

Specifically, as for the base vector, differential vector(s) and the search length (scaling factor) in mutation, the classical operators discuss the first two compositions but fix the scaling factor. DE/rand/1 maximizes the randomness of both

Algorithm 1. General framework of TADE

– *Initialization*

1: $g = 0, partition = 8/10, exR = 0.3$, Archive $A = \emptyset$;
2: Index counter $k_1 = k_2 = 1$;, $N_1 = partition * N$, $N_2 = N - N_1$;
3: $M_1 = \{(0.5, 0.5)_i | i = 1, ..., N_1\}$, $M_2 = \{(0.5, 0.5)_i | i = 1, ..., N_2\}$
4: $E_1 = \{N_1 - N_2 * exR + 1, ..., N_1\}, E_2 = \{N_1 + 1, ..., N_1 + N_1 * exR)\}$;
5: Initialize population $P^0 = \{x_1^0, ..., x_N^0\}$, evaluating P^0, $FES = N$
6: where major subpop $P_1^0 = \{x_1^0, ..., x_{N_1}^0\}$; minor subpop $P_2^0 = \{x_{N_1+1}^0, ..., x_N\}$;

– *Evolution*

1: **while** $FES < FES_{max}$ **do**
2: $S_1 = S_2 = \emptyset$;
3: **for** i = 1,2 **do**
4: **for** j in P_i^g **do**
5: generate (F_j^g, CR_j^g) from M_i;
6: generate *p*best from P^g;
7: generate mutator v_j^g by (eq.i);
8: generate trial vector u_j^g
9: **if** $f(x_j^g) > f(u_j^g)$ **then**
10: $x_j^{g+1} = u_j^g$; $x_i^g \rightarrow A$;
11: $(F_j^g, CR_j^g) \rightarrow S_i$;
12: **else**
13: $x_j^{g+1} = x_j^g$;
14: **end if**
15: **end for**
16: **end for**
17: **for** $i = 1, 2$ **do**
18: **if** $S_i \neq \emptyset$ **then**
19: $AS = \{(F_j^g, CR_j^g) \in S_{3-i} | j \in E_{3-i}\}$
20: Update M_i based on $S_i \cup AS$
21: $k_i = (k_i + 1) mod N_i$
22: **end if**
23: **end for**
24: $g = g + 1$, $FES = FES + N$
25: **end while**

parts, making it a most robust one, and DE/best/1 believes that the promising offsprings may be more likely to surround the best vector, and utilizes this certainty to design the best vector as the base vector and achieves a greedy but more rapid process that helps to reach a more accurate result in a limited time. Recently, our previous work further reduced the uncertainty of the differential vector, and get a competitive performance over DE/rand/1 and DE/best/1 on several functions. The basic operators hold the certainty that the control parameter is constant in the whole period. However, some randomness adding into this certainty indeed improves the performance, like jDE, JADE, SHADE and other adaptive methods.

3 TADE

Special attentions are given to JADE and SHADE. JADE increased the randomness to overcome the premature caused by greedy base vector. Firstly, JADE utilized a random vector from top $p\%$ best individuals to replace the greedy best vector in "current-to-best", which introduced the uncertainty of base vector. Moreover, JADE added an archive A to the population when generating the differential vector, which increased the randomness of candidate search directions. Besides, JADE used the current succeed parameters to partly influence the new ones, and added more uncertainty like Gaussian or Cauchy distribution. These strategies well alleviated the greediness of the base vector and enlarged the diversity. SHADE, the work based on JADE, further increased the diversity of the parameter. Instead of the current succeed parameters, SHADE maintained a historical succeed parameters, which can bring a wider randomness.

Since our previous work [5] found the possibility of reducing the randomness of the differential vector, this paper discussed whether it can be combined with the trade-off of other two aspects to achieve a more accurate result. We firstly propose a novel mutation strategy reducing the randomness of the differential vector. Different from [5] that selected the target vector as the starting point of the differential vector and maintained the randomness of the ending point, this mutation maintains the uncertainty of the starting point and takes the target vector as the ending. For the base vector, we adopt "current-to-*p*best" vector to increase the diversity and avoid the myopia of best vector. This mutation strategy is noted as "current-to-*p*best(half-rand)", and can be written as

$$v_i^g = x_i^g + F_i(x_{pbest}^g - x_i^g) + F_i(x_i^g - \tilde{x}_{r_2}^g) = x_i^g + F_i(x_{pbest}^g - \tilde{x}_{r_2}^g) \qquad (2)$$

The proposed mutation strategy largely reduces the number of possible candidate directions from $(N-1)(N+|A|-2)$ to $N+|A|-1$ for the current population. It is unfit to drive all or even the majority of the population to evolve since too much searching area has been cut. Therefore, this mutation strategy could act on a small part of the population as a pioneer soldier to rapidly explore on a narrow area. The majority of the population is controlled by a relatively farsighted "current-to-pbest(rand)" (that is the "current-to-pbest" in JADE and (rand) represents both randomness in differential vector). The reason of this choice is that "current-to-pbest(rand)" and "current-to-pbest(half-rand)" have the same base vector, the explored information from the pioneer soldier can easily feedback to the farsighted commanding officer in the next generation, and the officer can give a real-time and global-view new command with no extra cost to change the local search area the pioneer will explore in the next generation.

As for the parameter adaption, both parts adopt the way in SHADE due to its larger diversity. The basic thought of SHADE maintained the historical succeed parameters of several generations. Simply, the major and minor subpopulation update the success history merely by their own success parameters. However, with the iterative evolution, each element in the memory may become homoplastic, which hurts the diversity of search length and becomes harder to jump out. If the homoplasty can be postponed, larger area may be explored and a more accurate result may be achieved. Therefore, an exchange on the subpopulation to update the success memory is designed due to the inherently different mechanism of two mutation strategies. Specifically, an exchange rate exR is designed to determine the proportion of one subpopulation that will be used for the other subpopulation when updating the success memory.

The framework of this tight adaptive DE, called TADE, is shown in Algorithm 1. In *Evolution* phase, Lines 2–16 undergo the general process of DE for major and minor subpopulations. In line 6, the mutation information exchange happens since $p$best for each subpopultion is generated from the whole population P^g. Lines 17–23 update the success memory for both subpopulations. The success parameter exchange happens in Lines 19–20. Taking updating M_1 for major subpopulation as an example, the set AS is from minor success parameters and there are at most $|E_2| = N_1 * exR$ success parameters in the minor subpopulation that will join in the success memory update for the major subpopulation.

The Tight in TADE reflects on two aspects: the real-time information exchange on generating base vector, and the influence of success parameters in one subpopulation to the other. Next section shows the performance and rationality of this tight trade-off among base vector, scaling factor and differential vector.

4 Experiments

4.1 Settings

CEC2014 [13] benchmark suite is employed to demonstrate the performance comparison. Functions in this suite are all single objective optimization problems, containing both unimodal (F1-F3) and multimodal functions (F4-F30). More specifically, F4-F16 are simple multimodal functions, F17-F22 are hybrid functions and F23-F30 are composition functions. For each problem, 51 independent runs are conducted to obtain a reliable result. The dimension D is all set to be 30. The evolution process ends when the function evaluation time reaches $FES_{max} = 10000 * D$, and the error value smaller than 10^{-8} is taken as 0 [13].

4.2 Comparison and Analysis

Five state-of-the-art DEs, SHADE [9], JADE [8], CoDE [11], EPSDE [12] and SaDE [10], are utilized to compare with TADE. The settings of comparing methods are configured the same as the related papers. In TADE, the total population size N=100, the population partition is set to 8:2 for the major and minor subpopulation, and the success memory updating exchange rate $exR = 0.3$.

Wilcoxon's rank sum test at 5 % significance level is conducted between TADE and the comparing methods to measure whether TADE can obtain a significantly superior result or not. When TADE performs significantly better, "–" is marked, and "+" when TADE performs significantly poorer. "=" is marked when there is no significant difference. Due to the limited space, the detailed mean error value and the standard deviation of 51 independent runs are provided in http://thuhpgc.org/images/8/8e/Sup.jpg, but the number of cases on different function categories that TADE achieves better "–", equivalent "=" and worse "+" results against the comparing method are summarized in Table 1.

(a) Comparison with SHADE and JADE. From Table 1, TADE shows a quite competitive performance over SHADE, and it achieves better results on 12 functions and loses 7 functions. The attention should be paid to the unimodal as well

Table 1. Experimental results over 51 independent runs on unimodal functions of 30 variables with 300 000 FES

Function type	SHADE	JADE	CoDE	EPSDE	SaDE
Unimodal (F1-3)	$1^{-}2^{=}0^{+}$	$2^{-}1^{=}0^{+}$	$1^{-}2^{=}0^{+}$	$1^{-}2^{=}0^{+}$	$2^{-}1^{=}0^{+}$
Simple multimodal (F4-16)	$3^{-}6^{=}4^{+}$	$9^{-}3^{=}1^{+}$	$5^{-}5^{=}3^{+}$	$11^{-}2^{=}0^{+}$	$11^{-}1^{=}1^{+}$
Hybrid (F17-22)	$3^{-}2^{=}1^{+}$	$5^{-}1^{=}0^{+}$	$3^{-}1^{=}2^{+}$	$6^{-}0^{=}0^{+}$	$5^{-}1^{=}0^{+}$
Composition (F23-30)	$5^{-}1^{=}2^{+}$	$3^{-}4^{=}1^{+}$	$5^{-}2^{=}1^{+}$	$3^{-}0^{=}5^{+}$	$6^{-}1^{=}1^{+}$
Total	$12^{-}11^{=}7^{+}$	$19^{-}9^{=}2^{+}$	$14^{-}10^{=}6^{+}$	$21^{-}4^{=}5^{+}$	$24^{-}4^{=}2^{+}$

as the hybrid and composition functions. Indeed on the unimodal function F1, the mean error of SHADE is around $E+02$ and that value is $E-07$ for TADE, which demonstrates TADE's superior performance. For the hybrid and composition functions, the better cases TADE achieves are quite more than the better cases SHADE achieves. The difference between TADE and SHADE lies on the proposed mutation strategy for subpopulation and the mechanism preventing from homogeneity of the success memory. The real-time information exchange of two different mutations can prevent the whole population from being stuck in a local area, and the parameter information exchange prolongs the coming time of homogeneity, ensuring more time for exploration. These are the essential reasons why TADE performs better than SHADE. Also from Table 1, TADE shows an overwhelming superiority over JADE on every category. This advantage comes partly from the success history inherited from SHADE that widens the diversity, and partly from the tight major-minor scheme that makes a clear division of labour and improves the whole generation more effectively.

(b) Comparison with CoDE, EPSDE and SaDE. This comparison shows the performance against some state-of-the-art variants that combined the strengths of several mutation strategies. These methods employed a wide diversity of mutation strategies but with no tight information exchange. From Table 1, we see an overwhelming superiority of TADE over these methods, especially EPSDE and SaDE. A major contribution comes from the real-time information exchange in TADE, and the clear division and role of different methods help to achieve a win-win situation that pulls the generation towards improvement.

(c) Overall Comparison. From Table 1, TADE performs better or equivalently on at least 23(76.7 %) functions when compared with these five methods, and that number can reach 28(93.3 %) when compared with JADE and SaDE. The strength of the tight and cooperative scheme can be seen especially in the unimodal and the hybrid functions. The unimodal functions are relatively simple and the greedy "current-to-pbest(half-rand)" can rapidly explore the promising simple area. For hybrid functions that are usually the sum of several functions, two different mutations work for different components and search areas, and are tightly cooperated as well, thus resulting in the fast searching ability and enabling the possibility of reaching more promising area.

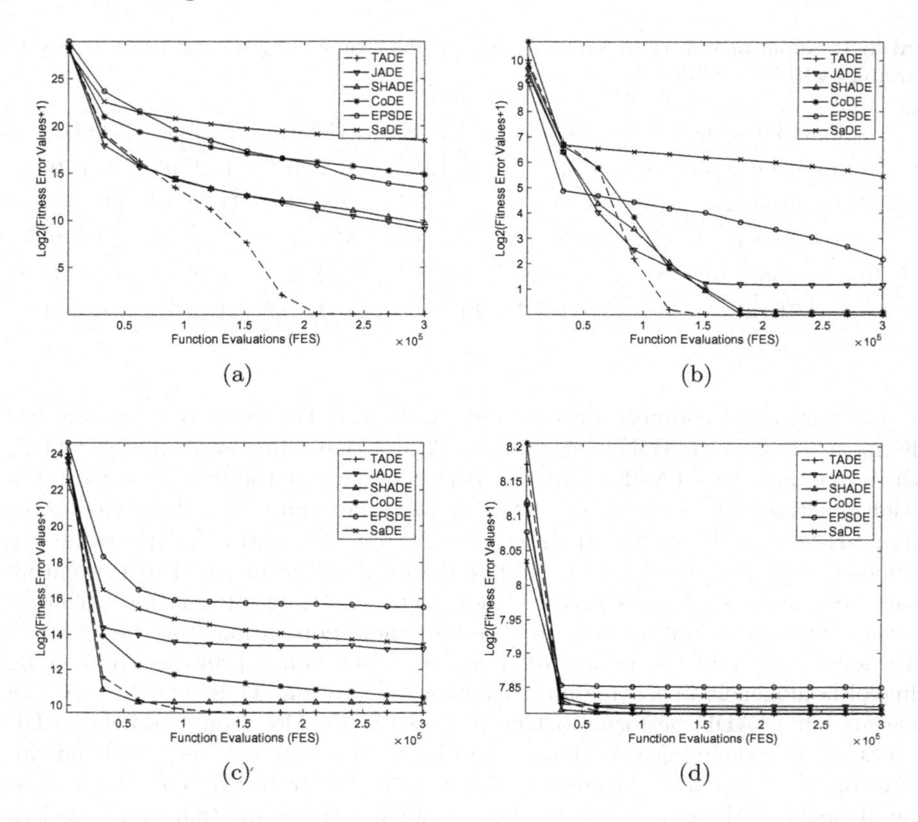

Fig. 1. Convergence curves of six DE variants on (a) F1, (b) F4, (c) F17, (d) F24.

More visually, Fig. 1 plots the convergence curves of these six methods on four functions selected from four categories. We can see the overwhelming superiority of TADE on Fig. 1(a), where TADE reaches several-magnitude better result with much rapid convergence speed than other methods. From Fig. 1(b), TADE and SHADE outperform the others, but TADE has a rapid speed and achieves the minimum ahead of SHADE. In Fig. 1(c), all six methods have almost the same speed at the early phase of the process, but TADE can reach a further accurate result due to the tight cooperative scheme that prevents the whole population from being all trapped into the local minima. TADE reaches competitive results in Fig. 1(d) against SHADE and CoDE. In a word, the strength of TADE is further shown from these curves.

4.3 Rationality of Our Tight Design

The experiments on different subpopulation partitions as well as different exchange rates are conducted to demonstrate the rationality of the tight designed scheme. Table 2 shows the results of 51 independent runs on different partitions

Table 2. Different partitions

Part	1:9	3:7	5:5	7:3	8:2	9:1
	21	18	13	4	−	7
	6	6	12	25	=	21
	3	6	5	1	+	2

Table 3. Different exchange rates

exR	0.0	0.1	0.2	0.3	0.4
	8	3	0	−	0
	20	27	30	=	30
	2	0	0	+	0

from 1:9 to 9:1 with fixed $exR = 0.3$, and Table 3 shows the comparison results on different exchange rates from 0.0 to 0.4 with 8:2 population partition.

From Table 2, when the proportion of the proposed "current-to-*p*best (half rand)" becomes larger (*Part* from 5:5 to 1:9), the performance becomes worse, showing the greediness of this mutation. Therefore, the design is reasonable that this greedy mutation is assigned with a smaller subpopulation. The effect of this greedy but fast pioneer is to get a fast feedback of its exploring area to the dominant role, "current-to-*p*best (rand)". This design can prevent the whole population from being trapped in the local place, and can improve the efficiency as well. The comparison result against 10:0 partition, that is the comparison with SHADE in Table 1, $12^{-} 11^{=}$ and 7^{+}, shows the actual effect of the pioneer discussed before and shows a win-win performance of both mutation strategies.

From Table 3, when $exR = 0.0$, which means no exchange between two subpopulations, this setting achieves 8 worse cases and only 2 better cases, which verifies TADE's parameter exchange can delay the homogeneity to some degree and results in the exploration of larger promising area. When the rate is from 0.2 to 0.4, there is no significant difference because almost every success setting of the minor subpopulation will join in the success history updating of the major one since there are only 20 individuals of minor subpopulation in this 8:2 partition.

5 Conclusion and Future Work

Generally, DEs maintain both randomness of the vectors that generate the differential vector to ensure the wide range of the candidate search directions. The competitive performance of our previous work on reducing the randomness of differential vector leads to a new thought on existing variants. The engine mutation operator of DE has three aspects, base vector, scaling factor and differential vector(s). The competitive performance and the behavior of existing variants can be explained by the trade-off of the randomness and certainty among these three components. However, existing methods only considered the trade-off without reducing the randomness of differential vector. This paper designed tight adaptive DE (TADE) that took the randomness-reduced differential vector into account. This proposed half-rand mutation was used to guide a minor subpopulation while the majority was leaded by the general "current-to-*p*best". The same base vector in both mutation helped to share the exploration information timely, and difference prevented from all being trapped in the local area. Moreover, based on success-memory parameter adaption, this paper designed a parameter information exchange scheme to delay the homogeneity and premature.

The extensive experiments in this paper were conducted on 30 CEC2014 benchmark functions. Firstly, TADE was compared with five state-of-the-art DE variants, SHADE, JADE, CoDE, EPSDE and SaDE. TADE showed a competitive performance against SHADE, and a superior performance against other four methods. Secondly, for the two tightness in this design, the different partitions and exchange rates were conducted. The results demonstrated the rationality of this design and reflected the characteristics of the proposed mutation strategy.

In the future, other basic designs and adaptive schemes on the randomness-reduced differential vector are encouraged to achieve a better performance.

Acknowledgments. This work is supported in part by National Natural Science Foundation of China (Grant Nos. 61303003, 41374113), by Tsinghua University Initiative Scientific Research Program (Grant No. 20131089356).

References

1. Storn, R., Price, K.: Differential evolution-a simple and efficient heuristic for global optimization over continuous spaces. J. Global Optim. **11**(4), 341–359 (1997)
2. Vesterstrom, J., Thomsen, R.: A comparative study of differential evolution, particle swarm optimization, and evolutionary algorithms on numerical benchmark problems. In: IEEE CEC, vol. 2, pp. 1980–1987 (2004)
3. Das, S., Suganthan, P.N.: Differential evolution: a survey of the state-of-the-art. IEEE Trans. Evol. Comput. **15**(1), 4–31 (2011)
4. Das, S., Mullick, S.S., Suganthan, P.N.: Recent advances in differential evolution – an updated survey. Swarm Evol. Comput. **27**, 1–30 (2016)
5. Zheng, W., Fu, H., Yang, G.: Targeted mutation: a novel mutation strategy for differential evolution. In: IEEE ICTAI, pp. 286–293 (2015)
6. Brest, J., Greiner, S., Boskovic, B., Mernik, M., Zumer, V.: Self-adapting control parameters in differential evolution: a comparative study on numerical benchmark problems. IEEE Trans. Evol. Comput. **10**(6), 646–657 (2006)
7. Yang, Z., Yao, X., He, J.: Making a difference to differential evolution. In: Siarry, P., Michalewicz, Z. (eds.) Advances in Metaheuristics for Hard Optimization. NCS, pp. 397–414. Springer, Berlin (2007)
8. Zhang, J., Sanderson, A.C.: JADE: adaptive differential evolution with optional external archive. IEEE Trans. Evol. Comput. **13**(5), 945–958 (2009)
9. Tanabe, R., Fukunaga, A.: Success-history based parameter adaptation for differential evolution. In: IEEE CEC, pp. 71–78 (2013)
10. Qin, A.K., Huang, V.L., Suganthan, P.N.: Differential evolution algorithm with strategy adaptation for global numerical optimization. IEEE Trans. Evol. Comput. **13**(2), 398–417 (2009)
11. Wang, Y., Cai, Z., Zhang, Q.: Differential evolution with composite trial vector generation strategies and control parameters. IEEE Trans. Evol. Comput. **15**(1), 55–66 (2011)
12. Mallipeddi, R., Suganthan, P.N., Pan, Q.K., Tasgetiren, M.F.: Differential evolution algorithm with ensemble of parameters and mutation strategies. Appl. Soft. Comput. **11**(2), 1679–1696 (2011)
13. Liang, J.J., Qu, B.Y., Suganthan, P.N.: Problem definitions and evaluation criteria for the CEC 2014 special session and competition on single objective real-parameter numerical optimization. Technical report, Zhengzhou University and Nanyang Technological University (2013)

An Extension of Algebraic Differential Evolution for the Linear Ordering Problem with Cumulative Costs

Marco Baioletti, Alfredo Milani, and Valentino Santucci(✉)

Department of Mathematics and Computer Science,
University of Perugia, Perugia, Italy
{marco.baioletti,alfredo.milani}@unipg.it,
valentino.santucci@dmi.unipg.it

Abstract. In this paper we propose an extension to the algebraic differential evolution approach for permutation based problems (DEP). Conversely from classical differential evolution, DEP is fully combinatorial and it is extended in two directions: new generating sets based on exchange and insertion moves are considered, and the case $F > 1$ is now allowed for the differential mutation operator. Moreover, also the crossover and selection operators of the original DEP have been modified in order to address the linear ordering problem with cumulative costs (LOPCC). The new DEP schemes are compared with the state-of-the-art LOPCC algorithms using a widely adopted benchmark suite. The experimental results show that DEP reaches competitive performances and, most remarkably, found 21 new best known solutions on the 50 largest LOPCC instances.

Keywords: Algebraic differential evolution · Linear ordering problem with cumulative costs · Permutations neighborhoods

1 Introduction and Related Work

Algebraic Differential Evolution (ADE) [13] is a recently proposed effective meta-heuristic for combinatorial optimization. ADE works on discrete search spaces by mimicking the behavior of the numerical Differential Evolution (DE) [16].

In the past, the numerical DE has been applied to combinatorial problems by adopting transformation techniques to decode a numerical vector (genotype) in the corresponding discrete solution (phenotype) in the evaluation step (see for example [2]). However, a single discrete solution can be represented by a potentially infinite number of continuous individuals, thus introducing a one-to-many mapping from the phenotypic to the genotypic space. As a consequence, large plateaus are very likely to be introduced in the search landscape and this is probably the main reason of the poor performances of these combinatorial applications of DE. Conversely, ADE allows to implement a discrete differential mutation operator (the key component of DE) that directly handles the discrete solutions of combinatorial problems.

J. Handl et al. (Eds.): PPSN XIV 2016, LNCS 9921, pp. 123–133, 2016.
DOI: 10.1007/978-3-319-45823-6_12

The only requirement of ADE is that the combinatorial search space at hand must be representable by means of a finitely generated group. This requirement is met in most of the combinatorial search spaces, e.g.: binary strings with the XOR and permutations with the usual composition operator. This algebraic structure cleanly establishes connections with the common solutions neighborhoods usually considered in combinatorial problems. In the case of permutations, the abstract differential mutation has been implemented using a generating set based on the adjacent swap moves [15]. The algebraic Differential Evolution for Permutations (DEP) has been applied to flowshop scheduling problems [12,13] and to the linear ordering problem [1,14] where, respectively, state-of-the-art and competitive results have been obtained.

In this paper, we extend our previous works on DEP in four directions: (i) we propose two new implementations of DEP by using generating sets based on exchange and insertion moves [15]; (ii) we extend the definition of the discrete differential mutation by allowing a scale factor parameter larger than 1; (iii) we apply DEP to a popular problem in the field of wireless communications systems, i.e., the Linear Ordering Problem with Cumulative Costs (LOPCC) [3]; (iv) we select other secondary components of DEP in order to tackle LOPCC.

LOPCC has been introduced in [3] as a cumulative variant of the linear ordering problem. Given a complete digraph of n nodes with node weights $d_i \geq 0$ and arc weights $c_{ij} \geq 0$, LOPCC aims to find a permutation of nodes $\pi = \langle \pi_1, \ldots, \pi_n \rangle$ that minimizes

$$f(\pi) = \sum_{i=1}^{n} \alpha_{\pi_i} \tag{1}$$

where the α-costs are backward recursively calculated as

$$\alpha_{\pi_i} = d_{\pi_i} + \sum\nolimits_{j=i+1}^{n} c_{\pi_i \pi_j} \alpha_{\pi_j} \qquad \text{for } i = n, n-1, \ldots, 1. \tag{2}$$

In [3], LOPCC has been proven to be NP-hard. Therefore, the exact algorithms available in literature [3,11] are effective only when $n \leq 16$. For larger instances, meta-heuristic approaches have been proposed. A Tabu Search (TS) scheme is described in [7]. EvPR is introduced in [8] and mainly consists in a GRASP procedure hybridized with an evolutionary path relinking technique. Finally, [17] proposes the so called Heterogeneous Cellular Processing Algorithm (HetCPA), a pseudo-parallel hybridization of a GRASP procedure with a scatter search scheme. As recently reported in [17], HetCPA and EvPR looks to be the state-of-the-art algorithms so far for LOPCC.

2 Algebraic Differential Evolution for Permutations

As described in [13], the design of the Algebraic Differential Evolution (ADE) mimics that of the classical DE. A population of N candidate solutions $\{x_1, \ldots, x_N\}$ is iteratively evolved by means of the three operators of differential mutation, crossover and selection. Differently from numerical DE, ADE addresses combinatorial optimization problems whose search space is representable by finitely generated groups. Since crossover and selection schemes for combinatorial spaces are widely available in literature, our proposal mainly focuses on the Differential Mutation (DM) operator. DM is widely recognized as the key component of DE [16] and, in its most common variant, generates a mutant v according to

$$v \leftarrow x_{r_0} \oplus F \odot (x_{r_1} \ominus x_{r_2}) \tag{3}$$

where $x_{r_0}, x_{r_1}, x_{r_2}$ are three randomly selected population individuals, while $F > 0$ is the scale factor parameter. In numerical DE, the operators $\oplus, \ominus, \odot$ are the usual vectorial operations of $\mathbb{R}^n$, while, in ADE, their definitions are formally derived using the algebraic structure of the search space.

The triplet $(X, \circ, G)$ is a finitely generated group representing a combinatorial search space if: (i) X is the discrete set of solutions; (ii) $\circ$ is a binary operation on X with the group properties, i.e., closure, associativity, identity (e), and invertibility (x^{-1}); and (iii) $G \subseteq X$ is a finite generating set of the group, i.e., any $x \in X$ has a (not necessarily unique) minimal-length decomposition $\langle g_1, \ldots, g_l \rangle$, with $g_i \in G$, and whose evaluation is x, i.e., $x = g_1 \circ \cdots \circ g_l$. For the sake of clarity, the length of a (minimal) decomposition of x is denoted with $|x|$. Using $(X, \circ, G)$ we can provide the formal definitions of the operators $\oplus, \ominus, \odot$ for ADE. Let $x, y \in X$ and $\langle g_1, \ldots, g_k, \ldots, g_{|x|} \rangle$ be a decomposition of x, then

$$x \oplus y := x \circ y \tag{4}$$

$$x \ominus y := y^{-1} \circ x \tag{5}$$

$$F \odot x := g_1 \circ \cdots \circ g_k \text{ with } k = \lceil F \cdot |x| \rceil \text{ and } F \in [0, 1]. \tag{6}$$

The algebraic structure on the search space naturally defines neighborhood relations among the solutions. Indeed, it induces a colored digraph whose nodes represent the solutions in X and two generic solutions $x, y \in X$ are linked by an arc with color $g \in G$ if and only if $y = x \circ g$. Hence, a one-step search move is directly encoded by a generator, while a composite move can be synthesized as the evaluation of a sequence of generators (a path on the graph). Analogously to what happens in $\mathbb{R}^n$, the elements of X can be dichotomously interpreted both as solutions (nodes on the graph) and as displacements between solutions (path colors on the graph). As detailed in [13], this allows to provide a rational interpretation to the discrete DM of definition (3). The key idea is that the difference $x \ominus y$ is the evaluation of the colors/generators on a shortest path from y to x. This geometric interpretation brings also to some connections with the Geometric DE proposed in [10].

Clearly, the definitions (4) and (5) do not depend on the generating set, thus they are uniquely defined. Conversely, the definition (6) requires a decomposition of x that is not unique in general, therefore a fair stochastic decomposition scheme has been suggested in [13].

The algebraic Differential Evolution for Permutations (DEP) [13] is an implementation of ADE for the search space of permutations. Indeed, permutations of the set $\{1, \ldots, n\}$, together with the usual composition operator, form the widely known symmetric group $\mathcal{S}(n)$, whose neutral element is the identity permutation e. In the previous series of works [1,12–14], the generating set ASW based on adjacent swap moves has been adopted. Formally, $ASW = \{\sigma_i : 1 \leq i < n\}$ where σ_i is the identity permutation with the items i and $i+1$ exchanged. The randomized decomposer for ASW, namely $RandBS$, has been devised by generalizing the classical bubble sort algorithm.

3 Exchange and Insertion Based Generating Sets

The generating sets based on exchange and insertion moves are respectively defined as $EXC = \{\epsilon_{ij} : 1 \leq i < j \leq n\}$ and $INS = \{\iota_{ij} : 1 \leq i, j \leq n\}$. ϵ_{ij} is the identity permutation with the items i and j exchanged, while ι_{ij} is the identity where the item i is shifted to position j. Their cardinalities are $|EXC| = \binom{n}{2}$ and $|INS| = (n-1)^2$ and both are proper supersets of ASW. The implementation of DM based on EXC and INS requires a stochastic decomposition algorithm for both the generating sets. Following the same idea used for ASW, we propose the two randomized decomposer for EXC and INS, respectively $RandSS$ and $RandIS$.

3.1 RandSS

Any permutation π can be decomposed in a sequence of generators in EXC by sorting π through successive exchange moves. Then, the decomposition is obtained by reversing the sequence of exchanges.

In order to identify the minimal sequence of exchange moves that sorts $\pi \in \mathcal{S}(n)$ we have to consider the cycle representation of π. Indeed, any permutation can be uniquely represented as a product of disjoint cycles [9]. A k-cycle of π is a sequence of k items $(\pi_{i_0}, \ldots, \pi_{i_{k-1}})$ such that, for any $0 \leq j < k$, the item π_{i_j} appears at position $\pi_{i_{(j-1) \bmod k}}$ in π. For example, $\langle 26745831 \rangle = (1268)(37)(4)(5)$. It is important to note that: (i) e is the only permutation with exactly n cycles, and (ii) an exchange of items belonging to the same cycle breaks the cycle into two new cycles, thus increasing the number of cycles by one. Therefore, a minimal decomposition can be obtained by iteratively choosing an exchange move that breaks a cycle.

The randomized decomposer for EXC, namely $RandSS$, is formally defined in Algorithm 1. The cycle weights w_i have been introduced in order to uniformly sample ϵ_{ij} among all the suitable exchanges (lines 7–8). Indeed, any k-length cycle can be broken with $\binom{k}{2} = k(k-1)/2$ different exchanges (line 5). The cycle representation (line 2) can be computed in $\Theta(n)$. The loop at lines 6–11 performs no more than $n-1$ iterations. The operations inside the loop can be performed in $\Theta(n)$. Therefore, the worst-case time complexity of $RandSS$ is $\Theta(n^2)$.

Finally, note that $RandSS$ generalizes the classical selection sort algorithm. Indeed, it can be shown that selection sort works similarly to $RandSS$ but with some limitations: it always breaks the cycle containing the smallest out-of-place item, and it divides the chosen k-length cycle in two cycles of lengths 1 and $k-1$, respectively.

Algorithm 1. RandSS - Randomized Decomposer for EXC

```
1: function RandSS(π ∈ S(n))
2:     s := ⟨ ⟩                          ▷ decomposition sequence of π incrementally built
3:     c := getCycles(π)          ▷ c_i is the ith cycle of π; c_ij is the jth item of cycle c_i
4:     for i := 1, len(c) do
5:         w_i := len(c_i)(len(c_i) − 1)/2                              ▷ weight of cycle c_i
6:     while len(c) < n do
7:         c_r := randomly choose a cycle through a roulette wheel basing on the weights w_i
8:         i, j := uniformly choose a pair of indexes from the cycle c_r
9:         π := π ∘ ϵ_ij
10:        append ϵ_ij to s
11:        update the cycles in c and their weights in w
12:    reverse the sequence s return s
```

3.2 RandIS

The INS decomposition of a permutation $\pi \in \mathcal{S}(n)$ can be obtained by sorting π using only insertion moves. Indeed, the decomposition is the sorting sequence of insertions reversed and inverted, i.e., every ι_{ij} is replaced with its inverse ι_{ji}.[1]

In order to compute the minimal sequence of insertions that sorts π we have to consider the longest increasing subsequence (LIS) of π. A LIS of π is not generally unique and it is defined as one of the longest monotonically increasing subsequence of (not necessarily consecutive) items of π [4]. It is important to note that: (i) e is the only permutation with exactly one LIS of maximal length n, and (ii) an insertion of a new item into a LIS increases the LIS length by one. Therefore, a minimal decomposition can be obtained by iteratively choosing an insertion that moves a new item in a LIS.

The randomized decomposer for INS, namely $RandIS$, is formally defined in Algorithm 2. At line 3 a random LIS L is obtained by modifying the LIS computation algorithm presented in [4][2]. The set U contains the items not in L (line 4). In order to uniformly sample ι_{ij} among all the suitable insertions (lines 9–11), any item in U is weighted by the number of suitable insertions in which it is involved (lines 5–7). The loop at lines 8–14 stops when $len(L) = n$, $U = \emptyset$ and $\pi = e$, therefore no more than $n - 1$ iterations are performed. The operations inside the loop have been implemented in $\Theta(n)$, thus the loop complexity is $\Theta(n^2)$. Moreover, since it is possible to show that the loop complexity dominates the rest, $RandIS$ requires time $\Theta(n^2)$ in the worst-case.

Finally, note that $RandIS$ generalizes the classical insertion sort algorithm. Indeed, classical insertion sort iteratively increases a sorted subsequence maintained at consecutive indexes on the left side of the permutation. Conversely, $RandIS$ allows to spread the sorted subsequence anywhere in the permutation.

4 Extended Differential Mutation

A limit of the discrete differential mutation previously introduced is that the definition of the multiplication operator does not allow to use a scale factor

[1] The inverting step is not considered in Sect. 3.1 because the exchange generators are self-invertible.

[2] For the sake of space, its description is not reported here.

Algorithm 2. RandIS - Randomized Decomposer for *INS*

```
1: function RandIS(π ∈ S(n))
2:     s := ⟨⟩                      ▷ decomposition sequence of π incrementally built
3:     L := get a random LIS of π
4:     U := {1, ..., n} \ L                                 ▷ set of unassigned items
5:     for all k ∈ U do
6:         P^L_{π,k} := set of positions in π where it is possible to shift item k in order to increase len(L)
7:         w_k := |P^L_{π,k}|                                    ▷ weight of item k
8:     while len(L) < n do
9:         r := randomly choose an item in U through a roulette wheel basing on the weights w_k
10:        i := position of r in π                              ▷ formally, π^{-1}(r)
11:        j := uniformly choose a position from P^L_{π,r}
12:        π := π ∘ ι_{ij}
13:        append ι_{ij} to s
14:        update L, U and, for any k ∈ U, update P^L_{π,k} and w_k
15:    reverse the sequence s
16:    invert the generators in s return s
```

parameter $F > 1$. Here, the abstract definition (6) is generalized by defining the properties that $z := F \odot x$ with any $F \geq 0$ has to respect, i.e.:

P1 $|z| = \lceil F \cdot |x| \rceil$,
P2 either a decomposition of x is a prefix of a decomposition of z (case $F \in [0, 1]$) or vice versa (case $F > 1$).

Clearly, when $F \in [0, 1]$, definition (6) meets both P1 and P2. For $F > 1$, P1 and P2 mean that, given the decomposition $\langle g_1, \ldots, g_{|x|} \rangle$ of x, a possible decomposition for $F \odot x$ is $\langle g_1, \ldots, g_{|x|}, g_{|x|+1}, \ldots, g_{\lceil F \cdot |x| \rceil} \rangle$ for a suitable choice of the generators $g_{|x|+1}, \ldots, g_{\lceil F \cdot |x| \rceil}$. Note that $F \odot x = x \circ g_{|x|+1} \circ \cdots \circ g_{\lceil F \cdot |x| \rceil}$. When the search space is finite, its diameter D constrains the maximum value allowed for F to $F_x^{max} = D/|x|$. Anyway, it is possible to extend the definition of $\odot$ by setting $F \odot x := F_x^{max} \odot x$ for any $F > F_x^{max}$. Geometrically, given two generic solutions x, y and $F > 1$, a decomposition of $F \odot (x \ominus y)$ can be interpreted in the search space graph as the sequence of arc colors in a shortest path starting from y, passing for x and extending beyond x. Unfortunately, when $F > 1$ there exist search spaces for which the multiplication operator is not always defined. An example is provided later on.

It is important to observe that, since $\mathcal{S}(n)$ has a finite diameter, the extended case of $F \odot x$ can be implemented by moving x away from e, i.e., towards a diametrically opposite permutation with respect to the identity.

For *ASW*, given $\pi \in \mathcal{S}(n)$, the extended multiplication operator $F \odot \pi$ can be implemented by sorting π in descending order, and composing π with the first $\lceil F \cdot |\pi| \rceil - |\pi|$ adjacent swap generators encountered during the sort. Denoting with r the "reverse" permutation $\langle n, \ldots, 1 \rangle$, the sorting step can be performed as $RandBS(r \circ \pi)$, thus reusing the randomized bubble sort proposed in [13]. Therefore, the worst-case complexity is $\Theta(n^2)$ as for $F \in [0, 1]$.

For *EXC*, an algorithm similar to *RandSS* is employed to compute $F \odot \pi$ with $F > 1$. We call it *MergeCycles* and it works by iteratively merging two cycles into one. Indeed, *MergeCycles* iteratively exchanges two items belonging to different cycles in order to merge the two cycles. The iteration stops when $\lceil F \cdot |\pi| \rceil - |\pi|$ exchanges have been performed. Then, the corresponding exchange generators are composed to the right of π to obtain $F \odot \pi$. Again, the worst-case complexity is $\Theta(n^2)$.

Finally, $\mathcal{S}(n)$ with the INS generating set is an example of a search space where the extended multiplication is not well defined. In order to satisfy the properties P1–P2 above, a necessary condition is that, for all $\pi \in \mathcal{S}(n)$ there must exist at least an insertion $\iota \in INS$ such that $LIS\_length(\pi \circ \iota) = LIS\_length(\pi) - 1$. However, an example can be used to show that this condition is not verified. Indeed, none of the 9 insertions of $\mathcal{S}(4)$ reduces the LIS length of $\langle 2413 \rangle$. Hence, in this paper, we do not consider the case $F > 1$ for the permutations search space generated by insertion moves.

5 Other Algorithmic Components

Though differential mutation is the core operator of DEP, its main scheme requires also a crossover and a selection operator.

In this work we have experimented two popular crossovers for permutation representations, namely, the two point crossover TPII adopted in [13] and the order based crossover OBX used in [1]. Given the parents $\rho', \rho'' \in \mathcal{S}(n)$, both TPII and OBX select a random subset of positions $P \subseteq \{1, \dots, n\}$ and build the offspring $v \in \mathcal{S}(n)$ by setting $v_i \leftarrow \rho'_i$ for any $i \in P$, and inserting the remaining items starting from the leftmost free place of v and following the order of appearance in ρ''. The difference between TPII and OBX is that TPII uses an interval of positions, while, in OBX, P can be any subset. Furthermore, TPII and OBX have been modified in order to consider the parameter $CR \in [0, 1]$. The modified variants, TPII^{CR} and OBX^{CR}, constrain the size of P to $|P| = \lceil CR \cdot n \rceil$. Therefore, for each pair of population and mutant individuals x_i, v_i, DEP generates an offspring u_i by applying $\text{TPII}^{\text{CR}}/\text{OBX}^{\text{CR}}$ to v_i and x_i respectively.

Regarding selection, the crowding scheme proposed in [18] is adopted. Each offspring u_j has a closest population individual $closest(u_j)$. Therefore, every population individual x_i is associated to the set of offsprings $U_i = \{u_j : closest(u_j) = x_i\}$. Then, for $1 \leq i \leq N$, the new population individual x'_i is selected to be the fittest among the solutions in $U_i \cup \{x_i\}$. Finally, the computation of the closest population individual has been implemented using the "position based distance" [15] because it is computed in $\Theta(n)$.

6 Experiments

Experiments have been held using the benchmark suites adopted in [7,8,17] (and available at http://www.optsicom.es/lopcc): UMTS (100 instances with $n = 16$), LOLIB (42 selected instances with $44 \leq n \leq 60$), RND (three sets of 25 instances of size, respectively, 35, 100 and 150).

DEP population size has been preliminarily set to $N = 80$, while F and CR are self-adapted using the popular jDE scheme [5] modified by introducing $\hat{F}$, i.e., a cap value for F. In order to test also the extended differential mutation, two choices have been considered for $\hat{F}$, i.e., $\hat{F} \in \{1, 1.2\}$. Hence, every possible combination of $\hat{F}$, generating set and crossover have been tested. Each DEP setting has been run ten times per instance and a run terminates if the best solution so far has not been updated during the last m evaluations[3]. m is set to 5 000, 100 000, 10 000 000 for, respectively, UMTS, LOLIB/RND35 and RND100/RND150 instances.

As in [7,8,17], the best result of every DEP setting on every instance has been used to analyze the performances and to compare DEP with the state-of-the-art algorithms HetCPA [17], EvPR [8] and TS [7]. The results of the competitors have been obtained from their respective papers. However, since the full results of EvPR are not available, we have considered optimistic lower bounds for EvPR as done in [17].

Table 1 provides a comparative analysis among the various DEP settings. The average ranks reported for every set of instances show that DEP/INS/1/TPII^{CR} is the best setting for DEP. Therefore, it has been selected for a further comparison with the state-of-the-art algorithms.

Table 1. Average Ranks among the DEP settings

Bench	ASW $\hat{F}=1$ OBX^{CR}	ASW $\hat{F}=1$ TPII^{CR}	ASW $\hat{F}=1.2$ OBX^{CR}	ASW $\hat{F}=1.2$ TPII^{CR}	EXC $\hat{F}=1$ OBX^{CR}	EXC $\hat{F}=1$ TPII^{CR}	EXC $\hat{F}=1.2$ OBX^{CR}	EXC $\hat{F}=1.2$ TPII^{CR}	INS $\hat{F}=1$ OBX^{CR}	INS $\hat{F}=1$ TPII^{CR}
UMTS	**5.50**	**5.50**	**5.50**	**5.50**	**5.50**	**5.50**	**5.50**	**5.50**	**5.50**	**5.50**
LOLIB	**5.33**	5.55	5.45	6.23	**5.33**	5.44	5.56	5.44	**5.33**	**5.33**
RND35	**5.50**	**5.50**	**5.50**	**5.50**	**5.50**	**5.50**	**5.50**	**5.50**	**5.50**	**5.50**
RND100	5.92	7.04	5.86	5.72	6.62	5.92	4.52	4.80	5.72	**2.88**
RND150	8.32	5.44	9.24	5.76	5.40	2.80	5.04	3.72	7.16	**2.12**
AVG	8.32	5.44	9.24	5.76	5.40	2.80	5.04	3.72	7.16	**2.12**

For each instance set and every algorithm, Table 2, provides both the average rank (R_{avg}) and the average relative percentage deviation (ARPD) from the best solution. By noting that R_{avg} values are more trustful than the ARPDs because they do not depend on the particular fitness distribution of the instance at hand, Table 2 clearly shows that DEP outperforms the state-of-the-art algorithms on most cases. Moreover, a Friedman+Finner statistical test [6] (with $\alpha = 0.05$) has been conducted. The R_{avg} values of the algorithms significantly outperformed by DEP are marked with a minus in Table 2. Interestingly, DEP significantly outperforms TS on every instance set except on the small UMTS instances, while the only competitor with similar performances is the hypothetic EvPR. However, since its results are an optimistic assumption, there are good chances that DEP significantly outperforms also EvPR.

[3] Additionally, a run terminates also if its CPU time exceeds one hour. However, this criterion has been sporadically met only on the RND150 instances.

Table 2. Average ranks and ARPDs among DEP and the state-of-the-art algorithms

Benchmark	HetCPA		Hyp_EvPR		TS		DEP	
	R_{avg}	ARPD	R_{avg}	ARPD	R_{avg}	ARPD	R_{avg}	ARPD
UMTS	**2.50**	*0.00*	**2.50**	*0.00*	**2.50**	*0.00*	**2.50**	*0.00*
LOLIB	2.29	*9.00*	**2.24**	14.20	3.24^{-}	12.59	**2.24**	14.20
RND35	2.44	0.37	**2.22**	*0.00*	3.12^{-}	0.49	**2.22**	*0.00*
RND100	1.98	2.27	2.10	2.42	4.00^{-}	15.00	**1.92**	*2.12*
RND150	2.64^{-}	10.28	1.90	*4.28*	3.66^{-}	27.13	**1.80**	5.56
OVERALL	2.41	4.38	2.30	*4.18*	3.02^{-}	11.04	**2.27**	4.38

Most importantly, DEP found 21 new best known solutions (according to [7,8,17]) on the 50 largest instances with $n \in \{100, 150\}$. These new upper bounds are provided in Table 3 together with the DEP setting that has obtained the corresponding result. Note that, though DEP/INS/1/TPII^{CR} obtained 10 new best known solutions, also many other DEP settings are present in Table 3. Therefore, another Friedman+Finner test has been conducted by considering a hypothetical DEP^{H} algorithm that produces, for every instance, the best result among all the settings. This test has shown that DEP^{H} would significantly outperform all the competitor algorithms.

Table 3. New best known solutions found by DEP

Instance	DEP setting	*Obj.Val.*	Instance	DEP setting	*Obj.Val.*
t1d100.1	DEP/ASW/1/OBX^{CR}	252.885	t1d100.25	DEP/INS/1/TPII^{CR}	632.586
t1d100.2	DEP/EXC/1.2/TPII^{CR}	286.888	t1d150.2	DEP/EXC/1/TPII^{CR}	163 274.856
t1d100.3	DEP/EXC/1.2/TPII^{CR}	1 288.298	t1d150.6	DEP/INS/1/TPII^{CR}	44 961.697
t1d100.8	DEP/EXC/1.2/OBX^{CR}	2 755.536	t1d150.7	DEP/EXC/1/TPII^{CR}	156 480.244
t1d100.9	DEP/INS/1/TPII^{CR}	61.772	t1d150.10	DEP/INS/1/TPII^{CR}	108 000.853
t1d100.10	DEP/ASW/1.2/TPII^{CR}	155.892	t1d150.12	DEP/INS/1/TPII^{CR}	65 708.550
t1d100.12	DEP/ASW/1.2/OBX^{CR}	231.347	t1d150.13	DEP/INS/1/TPII^{CR}	91 988.932
t1d100.17	DEP/ASW/1.2/TPII^{CR}	715.613	t1d150.16	DEP/INS/1/TPII^{CR}	16 231 674.691
t1d100.20	DEP/ASW/1.2/OBX^{CR}	236.088	t1d150.21	DEP/INS/1/TPII^{CR}	39 663.393
t1d100.22	DEP/ASW/1/OBX^{CR}	144.344	t1d150.22	DEP/INS/1/TPII^{CR}	683 618.275
t1d100.24	DEP/INS/1/TPII^{CR}	464.961			

7 Conclusion and Future Work

The algebraic differential evolution for permutations (DEP) has been extended by introducing the algorithmic implementations for two new generating sets based on exchange and insertion moves, and also by allowing a scale factor parameter larger than one. Moreover, two crossover operators and a crowding selection scheme have been adopted in order to tackle the linear ordering problem with cumulative costs (LOPCC). The proposed approach has been tested on a standard benchmark suite for LOPCC. The experimental results show that DEP reaches state-of-the-art performances by also producing 21 new best known solutions on the 50 largest instances. The notable results obtained by different DEP settings suggest that there is room for further improvement. Therefore, a future line of research will be the design of a meta-DEP scheme that considers all the different settings altogether, for example by using a self-adaptive scheme.

References

1. Baioletti, M., Milani, A., Santucci, V.: Linear ordering optimization with a combinatorial differential evolution. In: 2015 IEEE International Conference on Systems, Man, and Cybernetics (SMC), pp. 2135–2140 (2015)
2. Bean, J.C.: Genetic algorithms and random keys for sequencing and optimization. ORSA J. Comput. **6**(2), 154–160 (1994)
3. Bertacco, L., Brunetta, L., Fischetti, M.: The linear ordering problem with cumulative costs. Eur. J. Oper. Res. **189**(3), 1345–1357 (2008)
4. Bespamyatnikh, S., Segal, M.: Enumerating longest increasing subsequences and patience sorting. Inf. Process. Lett. **76**(1–2), 7–11 (2000)
5. Brest, J., Greiner, S., Boskovic, B., Mernik, M., Zumer, V.: Self-adapting control parameters in differential evolution: a comparative study on numerical benchmark problems. IEEE Trans. Evol. Comput. **10**(6), 646–657 (2006)
6. Derrac, J., Garca, S., Molina, D., Herrera, F.: A practical tutorial on the use of nonparametric statistical tests as a methodology for comparing evolutionary and swarm intelligence algorithms. Swarm Evol. Comput. **1**(1), 3–18 (2011)
7. Duarte, A., Laguna, M., Martí, R.: Tabu search for the linear ordering problem with cumulative costs. Comput. Optim. Appl. **48**(3), 697–715 (2009)
8. Duarte, A., Martí, R., Álvarez, A., Ángel-Bello, F.: Metaheuristics for the linear ordering problem with cumulative costs. Eur. J. Optim. Res. **216**(2), 270–277 (2012)
9. Herstein, I.N.: Abstract Algebra, 3rd edn. Wiley, New York (1996)
10. Moraglio, A., Togelius, J., Silva, S.: Geometric differential evolution for combinatorial and programs spaces. Evol. Comput. **21**(4), 591–624 (2013)
11. Righini, G.: A branch-and-bound algorithm for the linear ordering problem with cumulative costs. Eur. J. Oper. Res. **186**(3), 965–971 (2008)
12. Santucci, V., Baioletti, M., Milani, A.: Solving permutation flowshop scheduling problems with a discrete differential evolution algorithm. AI Commun. **29**(2), 269–286 (2016)
13. Santucci, V., Baioletti, M., Milani, A.: Algebraic differential evolution algorithm for the permutation flowshop scheduling problem with total flowtime criterion. IEEE Trans. Evol. Comput. (to be published– preprint online)
14. Santucci, V., Baioletti, M., Milani, A.: An algebraic differential evolution for the linear ordering problem. In: Proceedings of GECCO 2015, pp. 1479–1480 (2015)
15. Schiavinotto, T., Stützle, T.: A review of metrics on permutations for search landscape analysis. Comput. Oper. Res. **34**(10), 3143–3153 (2007)

16. Storn, R., Price, K.: Differential evolution – a simple and efficient heuristic for global optimization over continuous spaces. J. Global Optim. **11**(4), 341–359 (1997)
17. Terán-Villanueva, J.D., Fraire Huacuja, H.J., Carpio Valadez, J.M., Pazos Rangel, R., Puga Soberanes, H.J., Martínez Flores, J.A.: A heterogeneous cellular processing algorithm for minimizing the power consumption in wireless communications systems. Comput. Optim. Appl. **62**(3), 787–814 (2015)
18. Thomsen, R.: Multimodal optimization using crowding-based differential evolution. In: Proceedings of CEC 2004, vol. 2, pp. 1382–1389 (2004)

Analysing the Performance of Migrating Birds Optimisation Approaches for Large Scale Continuous Problems

Eduardo Lalla-Ruiz[1(✉)], Eduardo Segredo[2], Stefan Voß[1], Emma Hart[2], and Ben Paechter[2]

[1] Institute of Information Systems, University of Hamburg, Hamburg, Germany
{eduardo.lalla-ruiz,stefan.voss}@uni-hamburg.de
[2] School of Computing, Edinburgh Napier University, Edinburgh, Scotland, UK
{e.segredo,e.hart,b.paechter}@napier.ac.uk

Abstract. We present novel algorithmic schemes for dealing with large scale continuous problems. They are based on the recently proposed population-based meta-heuristics *Migrating Birds Optimisation* (MBO) and *Multi-leader Migrating Birds Optimisation* (MMBO), that have shown to be effective for solving combinatorial problems. The main objective of the current paper is twofold. First, we introduce a novel neighbour generating operator based on Differential Evolution (DE) that allows to produce new individuals in the continuous decision space starting from those belonging to the current population. Second, we evaluate the performance of MBO and MMBO by incorporating our novel operator to them. Hence, MBO and MMBO are enabled for solving continuous problems. A set of well-known large scale functions is used for comparison purposes.

Keywords: Continuous neighbourhood search · Migrating Birds Optimisation · Large scale continuous problems · Global optimisation

1 Introduction

Nature-inspired computing counts with an extensive variety of algorithms mimicking natural processes and events from the universe that are frequently used for tackling real-world optimisation problems. Along these algorithms, those inspired by the collective living and travelling of animals have attracted a considerable interest from the related research community [11]. In this regard, the collective behaviour and swarm intelligence of migratory birds and its algorithmic translation have been recently studied by Duman et al. [2], and Lalla-Ruiz et al. [4]. Authors exploit, by means of their corresponding proposed algorithmic approaches, the advantage of sharing information and cooperating among a group of individuals. While *Migrating Birds Optimisation* (MBO), which is inspired by the V-flight formation of migratory birds with one leader, was proposed in [2], in [4], based on field studies, *Multi-leader Migrating Birds Optimisation* (MMBO) was introduced, which allows different types of flight formation shapes, as well as several leading individuals, to be managed.

J. Handl et al. (Eds.): PPSN XIV 2016, LNCS 9921, pp. 134–144, 2016.
DOI: 10.1007/978-3-319-45823-6_13

Recently, MBO has shown its good performance for combinatorial problems, such as the *Quadratic Assignment Problem* (QAP) [2], the *Dynamic Berth Allocation Problem* (DBAP) [5], and *Hybrid Flow-shop Scheduling* [8], among others. In regard to continuous optimisation, an initial adaptation to low-dimensional problems, which uses sphere-shaped neighbourhoods, was developed in [1]. Results provided by said scheme, however, did not show a high performance. Regarding MMBO, it showed to provide better quality results than those achieved by MBO for the QAP [4]. Concerning its performance for continuous optimisation, as far as we know, this is the first time that MMBO is enabled for dealing with these types of problems, as well as the first time that MBO is assessed when solving large scale continuous problems.

The main goal of this work is to propose suitable adaptations of MBO and MMBO for tackling continuous optimisation problems. For doing that, we propose a novel neighbourhood structure based on the well-known *Differential Evolution* (DE) [10], which is able to generate solutions in a continuous decision space. The computational experimentation provided in this work, which involves the use of a set of well-known large scale continuous problems [7], indicates that our proposals are able to improve, for some cases, the results obtained by one of the best performing variants of DE considering that set of large scale functions [3].

The remainder of this paper is organised as follows. Section 2 describes our proposed MBO and MMBO approaches. Afterwards, in Sect. 3, the experimental evaluation carried out in this paper is exposed. Finally, Sect. 4 draws the main conclusions extracted from this work and provides some lines for further research.

2 Schemes Based on Migrating Birds Optimisation for Continuous Problems

This section focuses on describing our algorithmic proposals. Section 2.1 is devoted to describe the scheme MBO, while the approach MMBO is depicted in Sect. 2.2. Finally, in Sect. 2.3 we introduce our novel neighbour generating operator based on DE.

2.1 Migrating Birds Optimisation

Migrating Birds Optimisation (MBO) is a population-based algorithm based on the V-formation flight of migrating birds. It considers a population or flock, of individuals or birds, that are aligned in a V-flight formation. Following that formation, the first individual corresponds to the leader of the flock and the other ones define the rest of the flock. The birds maintain a cooperative relationship among them by means of sharing information. The way the flow of information is shared is unidirectional. Namely, one individual sends information and the other receives it. The direction of the information shared starts from the leader bird and goes to the rest of the flock by following the V-shape flight formation.

Algorithm 1 depicts the pseudocode of MBO. The input parameters are: (i) the number of birds in the flock (n), (ii) the maximum number of neighbours

Algorithm 1. Migrating Birds Optimisation pseudocode ([2])

Require: n, K, m, k, and x
1: Generate n initial birds in a random manner and place them on an hypothetical V-formation arbitrarily
2: $g = 0$
3: **while** $(g < K)$ **do**
4: **for** $(j = 1 : m)$ **do**
5: Try to improve the leader bird by generating k neighbours
6: $g = g + k$
7: **for all** (non-leader bird s in the flock) **do**
8: Try to improve non-leader bird s by using $k - x$ generated neighbours and x unused best neighbours from those birds in the front of it
9: $g = g + (k - x)$
10: **end for**
11: **end for**
12: Move the leader bird to the end of the V-formation and forward one of the birds following it to the leader position
13: **end while**
14: Return the best bird in the flock

generated by the flock of birds (K), (iii) the number of iterations performed before changing the leader bird (m), (iv) the number of neighbours generated by each bird (k), and (v) the number of best discarded neighbours to be shared among birds (x). The first step consists of generating n individuals or birds (line 1). The current number of neighbours generated by the flock of birds, i.e. g, is initially set to zero (line 2). During the search process, firstly, k neighbours are generated starting from the leader bird. In case the best neighbour leads to an improvement of the leader in terms of the objective function value, the latter is replaced by the former (line 5). Secondly, for each non-leader bird s, $k - x$ neighbours are generated. Additionally, the neighbourhood of s receives x unused best neighbours from those birds in front of it (lines 7–10). If s is improved by its best neighbour, then the former is replaced by the latter (line 8). The V-formation is maintained until a prefixed number of iterations $m > 0$ is reached. Once that, the leader bird becomes the last bird in the V-formation and one of its immediate successors becomes the new leader (line 12). The above steps are executed until a maximum number of neighbours, i.e. K, is generated (line 3). Finally, we should mention that, in our case, neighbours are created (lines 5 and 8) by using the operator described in Sect. 2.3.

2.2 Multi-leader Migrating Birds Optimisation

Multi-leader Migrating Birds Optimisation (MMBO) is a novel population-based meta-heuristic inspired by the flight formation of migratory birds which tries to improve its predecessor MBO. In MMBO, birds are distributed in a line formation mimicking the flight formation of migratory birds, which is determined according to given *relationship criteria*, e.g. by means of the objective function of the problem at hand. Depending on those criteria, we can have birds located at positions that are closer than others regarding the front of the migratory formation during the flight. Starting from each bird, a given number of feasible neighbours are generated through a predefined neighbourhood structure.

Algorithm 2. Multi-leader Migrating Birds Optimisation pseudocode ([4])

```
Require: n, K, k, and x
 1: Create the initial flock P by randomly generating n birds
 2: while (K neighbours have not been generated) do
 3:   Determine the interaction among birds of P and establish the formation
 4:   while (stopping formation criterion is not met) do
 5:     Generate k neighbours starting from each bird b ∈ P_L ∪ P_I
 6:     Replace each bird included into P_L by its best neighbour if
        the latter improves the former
 7:     Replace each bird included into P_I by its best neighbour
 8:     for all (bird f ∈ P_F) do
 9:       Generate k − x neighbours starting from f
10:       Get the best unused x neighbours from the previous birds of f in the group
11:       Replace f by its best neighbour if the latter improves the former
12:     end for
13:   end while
14: end while
15: Return the best bird in P
```

The neighbourhoods reflect the particular points of view about the solution space of each individual. As mentioned above, depending on the relationship criteria and how information is shared among individuals, different roles arise in MMBO:

- *Leader.* It is that bird with the best objective value when compared to the adjacent ones. Therefore, it does not receive information from any bird, but shares x neighbours with each adjacent one. Moreover, starting from a leader, k neighbours are generated. Since the objective value determines the position within the formation, a leader is the best performing bird, and consequently, the most advanced one in its corresponding group within the formation. The set of leaders is denoted as P_L.
- *Follower.* It is that bird which explores the search space considering its own information and the information received from the birds in front of it within the formation. It generates $k - x$ neighbours and receives x neighbours from the adjacent birds. The set of followers is denoted as P_F.
- *Independent.* It is that bird which is not included into any other of the above categories. It does not exchange information with any other individual, but generates k neighbours. The set of independent birds is denoted as P_I.

The pseudocode of MMBO is depicted in Algorithm 2. The first step is to obtain the initial flock P which consists of n birds generated at random (line 1). While the stopping criterion is not met, MMBO iterates (line 2). In this work, we consider a stopping criterion based upon a maximum number of neighbours to be generated (K). The relationship criteria among birds are based on the objective function value. This allows to recognise groups, as well as the formation (line 3). Then, the search process starts (lines 4–13) and it is executed until a stopping formation criterion is met. In case said criterion is satisfied, the search process is stopped in order to establish a new formation. During the search process, firstly, k neighbours are obtained starting from each bird $b \in P_L \cup P_I$ (line 5). Then, sets P_L and P_I are updated (lines 6–7). Secondly, for each follower $f \in P_F$, $k - x$ neighbours are generated (line 9), and it receives x neighbours from the adjacent birds according to the formation (line 10). Afterwards, if f is improved by its best neighbour, the former is replaced by the latter (line 11). The unused best neighbours of f are shared with the adjacent birds. In this work, neighbours are obtained (lines 5 and 9) by applying the operator introduced in Sect. 2.3.

2.3 Neighbour Generating Operator Based on Differential Evolution

This work presents a novel neighbour generating operator to be used with MBO and MMBO in order to enable their operation with continuous optimisation problems. This operator is based on the well-known *Differential Evolution* (DE), a search algorithm which was specifically proposed for global optimisation [10].

For encoding individuals, a vector of D real-valued decision variables or dimensions x_i is used, i.e. $\boldsymbol{X} = [x_1, x_2, \ldots, x_i, \ldots, x_D]$. The objective function $f(\boldsymbol{X})(f : \Omega \subseteq \mathbb{R}^D \rightarrow \mathbb{R})$ determines the quality of every vector $\boldsymbol{X}$. Hence, finding a vector $\boldsymbol{X}* \in \Omega$, where $f(\boldsymbol{X}*) \leq f(\boldsymbol{X})$ is satisfied for all $\boldsymbol{X} \in \Omega$, is the goal in a global optimisation problem. Considering box-constrained problems, the feasible region Ω is defined by $\Omega = \prod_{i=1}^{D}[a_i, b_i]$, where a_i and b_i represents the lower and upper bounds of variable i.

Regarding the most widely used nomenclature for DE [10], i.e. DE/x/y/z, where x is the vector to be mutated, y defines the number of difference vectors used, and z indicates the crossover approach, our neighbour generating operator is inspired by the scheme DE/rand/1/bin. We selected this variant due to its simplicity and popularity and because it was able to provide the best performance in previous work with the set of large scale problems we consider herein [3].

Given a particular individual $\boldsymbol{X}_{j=1 \ldots NP}$ (*target vector*) from a flock of either MBO or MMBO with size NP, a neighbour is obtained as follows. First, the *mutant generation strategy* rand/1 is applied for obtaining a *mutant vector* ($\boldsymbol{V}_j$). Thus, the mutant vector is generated as Eq. 1 shows. We should note that r_1, r_2, and r_3 are mutually exclusive integers randomly selected from the range $[1, NP]$, all of them different from the index j. Finally, F denotes the *mutation scale factor*.

$$\boldsymbol{V}_j = \boldsymbol{X}_{r_3} + F \times (\boldsymbol{X}_{r_1} - \boldsymbol{X}_{r_2}) \tag{1}$$

After obtaining the mutant vector, it is combined with the target vector to produce the *trial vector* ($\boldsymbol{U}_j$) through a crossover operator. The combination of the mutant vector generation strategy and the crossover operator is usually referred to as the *trial vector generation strategy*. One of the most commonly applied crossover operators, which is considered in this work, is the *binomial crossover* (*bin*). The crossover is controlled by means of the crossover rate CR, and uses Eq. 2 for producing a trial vector, where $x_{j,i}$ represents decision variable i belonging to individual $\boldsymbol{X}_j$. A random number uniformly distributed in the range $[0, 1]$ is given by $rand_{j,i}$, and $i_{rand} \in [1, 2, ..., D]$ is an index selected at random that ensures that at least one variable belonging to the mutant vector is inherited by the trial one. Variables are thus inherited from the mutant vector with probability CR. Otherwise, variables are inherited from the target vector.

$$u_{j,i} = \begin{cases} v_{j,i} \; if \; (rand_{j,i} \leq CR \; or \; i = i_{rand}) \\ x_{j,i} \; otherwise \end{cases} \tag{2}$$

It can be observed that the trial vector generation strategy may generate vectors outside the feasible region Ω. In this work, unfeasible values are reinitialised at random in their corresponding feasible ranges, being this approach one of the most frequently used in the related literature. Finally, we should note that the trial vector becomes the newly generated neighbour.

3 Experimental Evaluation

In this section we describe the experiments carried out with both algorithms depicted in Sect. 2. In addition to those schemes, we also considered the variant DE/rand/1/bin as an independent approach for comparison purposes.

Experimental Method. MBO and MMBO, as well as DE/rand/1/bin, were implemented by using the *Meta-heuristic-based Extensible Tool for Cooperative Optimisation* (METCO) [6]. Experiments were run on a Debian GNU/Linux computer with four AMD® Opteron™ processors (model number 6164 HE) at 1.7 GHz and 64 GB RAM. Every execution was repeated 30 times, since all experiments used stochastic algorithms. Bearing the above in mind, comparisons were carried out by applying the following statistical analysis [9]. First, a *Shapiro-Wilk test* was performed to check whether the values of the results followed a normal (Gaussian) distribution or not. If so, the *Levene test* checked for the homogeneity of the variances. If the samples had equal variance, an ANOVA *test* was done. Otherwise, a *Welch test* was performed. For non-Gaussian distributions, the non-parametric *Kruskal-Wallis* test was used. For all tests, a significance level $\alpha = 0.05$ was considered.

Problem Set. A set of scalable continuous optimisation functions proposed in the *2013* IEEE *Congress on Evolutionary Computation* (CEC'13) for its *Large Scale Global Optimization* (LSGO) competition [7] was considered as the problem set. We should note that this suite is the latest proposed for large scale global optimisation in the field of the CEC, and therefore, it was also used for the LSGO competitions organised in CEC'14 and CEC'15. The suite consists of 15 different problems (f_1–f_{15}) with different features: fully-separable functions (f_1–f_3), partially additively separable functions (f_4–f_{11}), overlapping functions (f_{12}–f_{14}), and non-separable functions (f_{15}). By following the suggestions given for different editions of the LSGO competition, we fixed the number of decision variables D to 1000 for all the above functions, with the exception of functions f_{13} and f_{14}, where 905 decision variables were considered due to overlapping subcomponents.

Parameters. Table 1 shows parameter values considered in this work for MBO and MMBO. They were selected by carrying out a previous parameter setting

Table 1. Configuration of the approaches MBO, MMBO, and DE/rand/1/bin

Parameter values for MBO and MMBO			
Parameter	Value	Parameter	Value
Stopping criterion (K)	3×10^6	Number of neighbors (k)	4
Flock size (n)	150	Number of flights (x)	1
Number of flights (m)	10		
Parameter values for DE/rand/1/bin			
Stopping criterion	3×10^6	Mutation scale factor (F)	0.5
Population Size (NP)	150	Crossover rate (CR)	0.9

study. As it can be observed in Sect. 2, parameter m is only considered by MBO. In past research, a configuration of the scheme DE/rand/1/bin, from among a candidate pool with more than 80 different parameterisations of said approach, was able to provide the best overall results for problems f_1–f_{15} [3]. This is the main reason why our neighbour generating operator is based on DE/rand/1/bin. Moreover, that best performing configuration, whose parameter values (NP, F, and CR) are also shown in Table 1, is considered herein as an independent method for measuring the performance attained by MBO and MMBO. Our operator also makes use of those parameter values. Finally, the stopping criterion was fixed to a maximum amount of 3×10^6 evaluations, following the recommendations provided by the LSGO competition.

Results. Figure 1 shows box-plots reflecting the results obtained by the considered schemes. It can be observed that, for some problems (f_2, f_3, f_5, f_9, and f_{11}) MBO and/or MMBO were able to obtain better solutions than those provided by the best performing variant of DE/rand/1/bin found for the large scale problems we consider in this work, thus showing the benefits that can be obtained from our hybridisation between MBO/MMBO and our novel neighbour generating operator based on DE. Since our neighbour generating operator is based on DE/rand/1/bin, it was expected that results obtained by MBO and MMBO were very similar to those provided by the former scheme executed independently. However, the features of MBO and MMBO for sharing information among individuals, as well as for establishing a structure among them, combined with the the exploration and exploitation abilities of our neighbour generating operator based on DE/rand/1/bin, were able to obtain even better results in 5 out of 15 problems. Taking into account the remaining functions, we should note that MBO and/or MMBO were able to achieve similar solutions than those attained by the best performing variant of DE, with except to some cases, such as f_1, where DE provided better solutions.

In order to give the aforementioned conclusions with statistical confidence, Table 2 shows, for each problem, the p-values obtained from the statistical comparison between the approach MBO and the rest of schemes, by following the statistical procedure explained at the beginning of the current section. It also

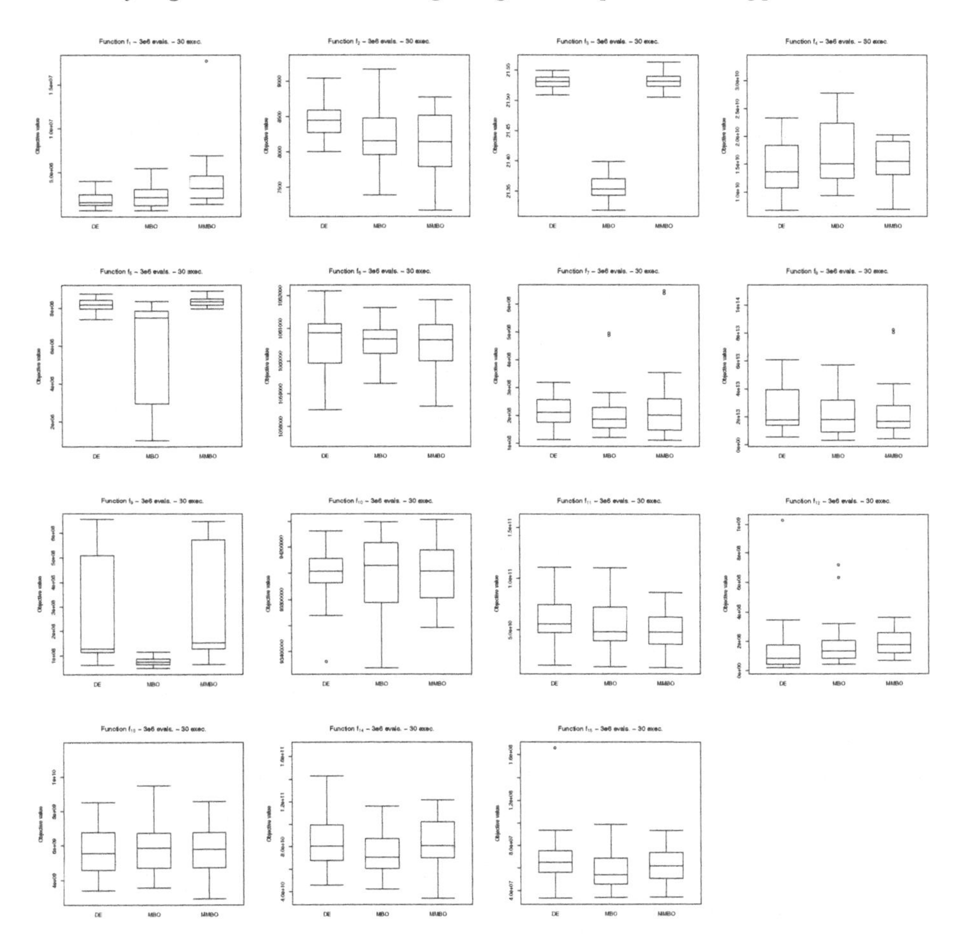

Fig. 1. Box-plots showing the results obtained by different schemes for functions f_1–f_{15}

shows cases for which MBO was able to statistically outperform other strategy ($\uparrow$), cases where other strategy outperformed MBO ($\downarrow$), and cases where statistically significant differences between MBO and the corresponding method did not arise ($\leftrightarrow$). Scheme A statistically outperforms method B if there exist statistically significant differences between them, i.e. if the p-value is lower than $\alpha = 0.05$, and if at the same time, A provides a lower mean and median of the objective value than B, since we are dealing with minimisation problems. Finally, Table 3 shows the same information, but regarding MMBO.

With respect to MBO, it is worth mentioning that it was able to outperform DE in 4 out of 15 problems (f_2, f_3, f_5, and f_9). Additionally, it was not outperformed by DE in any test case. For remaining problems, MBO and DE did not present statistically significant differences. Concerning MMBO, we should note that it

Table 2. Statistical comparison between MBO and remaining schemes considering problems f_1–f_{15}

f	Alg	p-value	Dif	f	Alg	p-value	Dif	f	Alg	p-value	Dif
f_1	DE	2.739e-01	$\leftrightarrow$	f_2	DE	3.331e-02	$\uparrow$	f_3	DE	4.665e-40	$\uparrow$
	MMBO	9.674e-03	$\uparrow$		MMBO	3.821e-01	$\leftrightarrow$		MMBO	5.700e-43	$\uparrow$
f_4	DE	2.550e-01	$\leftrightarrow$	f_5	DE	1.415e-07	$\uparrow$	f_6	DE	5.946e-01	$\leftrightarrow$
	MMBO	6.361e-01	$\leftrightarrow$		MMBO	1.229e-09	$\uparrow$		MMBO	9.101e-01	$\leftrightarrow$
f_7	DE	1.882e-01	$\leftrightarrow$	f_8	DE	3.912e-01	$\leftrightarrow$	f_9	DE	1.794e-06	$\uparrow$
	MMBO	7.227e-01	$\leftrightarrow$		MMBO	8.130e-01	$\leftrightarrow$		MMBO	8.701e-08	$\uparrow$
f_{10}	DE	5.742e-01	$\leftrightarrow$	f_{11}	DE	4.333e-01	$\leftrightarrow$	f_{12}	DE	5.277e-02	$\leftrightarrow$
	MMBO	9.053e-01	$\leftrightarrow$		MMBO	2.089e-01	$\leftrightarrow$		MMBO	1.433e-01	$\leftrightarrow$
f_{13}	DE	5.946e-01	$\leftrightarrow$	f_{14}	DE	8.367e-02	$\leftrightarrow$	f_{15}	DE	6.249e-02	$\leftrightarrow$
	MMBO	7.325e-01	$\leftrightarrow$		MMBO	5.809e-02	$\leftrightarrow$		MMBO	2.428e-01	$\leftrightarrow$

Table 3. Statistical comparison between MMBO and remaining schemes considering problems f_1–f_{15}

f	Alg	p-value	Dif	f	Alg	p-value	Dif	f	Alg	p-value	Dif
f_1	DE	8.401e-05	$\downarrow$	f_2	DE	1.402e-03	$\uparrow$	f_3	DE	8.786e-01	$\leftrightarrow$
	MBO	9.674e-03	$\downarrow$		MBO	3.821e-01	$\leftrightarrow$		MBO	5.700e-43	$\downarrow$
f_4	DE	4.965e-01	$\leftrightarrow$	f_5	DE	6.671e-02	$\leftrightarrow$	f_6	DE	8.130e-01	$\leftrightarrow$
	MBO	6.361e-01	$\leftrightarrow$		MBO	1.229e-09	$\downarrow$		MBO	9.101e-01	$\leftrightarrow$
f_7	DE	5.742e-01	$\leftrightarrow$	f_8	DE	5.543e-01	$\leftrightarrow$	f_9	DE	2.739e-01	$\leftrightarrow$
	MBO	7.227e-01	$\leftrightarrow$		MBO	8.130e-01	$\leftrightarrow$		MBO	8.701e-08	$\downarrow$
f_{10}	DE	9.176e-01	$\leftrightarrow$	f_{11}	DE	2.713e-02	$\uparrow$	f_{12}	DE	2.962e-03	$\downarrow$
	MBO	9.053e-01	$\leftrightarrow$		MBO	2.089e-01	$\leftrightarrow$		MBO	1.433e-01	$\leftrightarrow$
f_{13}	DE	7.901e-01	$\leftrightarrow$	f_{14}	DE	9.176e-01	$\leftrightarrow$	f_{15}	DE	3.912e-01	$\leftrightarrow$
	MBO	7.325e-01	$\leftrightarrow$		MBO	5.809e-02	$\leftrightarrow$		MBO	2.428e-01	$\leftrightarrow$

was able to beat DE in problems f_2 and f_{11}, it was beaten by DE considering functions f_1 and f_{12}, and both approaches did not show statistically significant differences when dealing with remaining test cases. Bearing the above in mind, MBO/MMBO were able to provide better solutions than those achieved by DE in 5 out of 15 problems. However, DE was able to outperform MBO/MMBO in 2 out of 15 functions. This means that MBO/MMBO were able to attain similar or even better solutions than DE in 13 out of 15 problems.

If we compare MBO with respect to MMBO we can make the following observations. MBO provided statistically better results than MMBO in 4 problems (f_1, f_3, f_5, and f_9), while the latter was not statistically better than the former in any case. Taking into account the remaining problems, statistically significant differences did not appear between both schemes.

4 Conclusions and Future Work

Algorithms inspired by the nature comprise an important type of solution approaches used for solving practical problems. Some of these approaches have been successfully applied to combinatorial problems, such as MBO and MMBO. Nevertheless, to our best knowledge, they had not been used for tackling large scale continuous problems. Hence, in this work we propose novel adaptations of both population-based meta-heuristics for solving relevant problems in this research area. For doing that, we developed a novel neighbour generating operator based on DE that allows new individuals to be generated in the continuous decision space. The experimental evaluation carried out indicates that our proposals are suitable and competitive for performing the optimisation of large scale continuous problems. In this regard, results demonstrate that MBO and MMBO are able to obtain similar solutions, and even better for some cases, than those provided by one of the best performing variants of DE considering the set of large scale continuous problems at hand.

Bearing in mind the contributions of this work, our research agenda will be focused on the assessment of the influence that the different parameters of MBO and MMBO have over their performance when solving continuous problems. Additionally, an analysis about the impact that different neighbourhood structures have over the behaviour of MBO and MMBO might also be of great interest.

References

1. Alkaya, A.F., Algin, R., Sahin, Y., Agaoglu, M., Aksakalli, V.: Performance of migrating birds optimization algorithm on continuous functions. In: Tan, Y., Shi, Y., Coello, C.A.C. (eds.) ICSI 2014, Part II. LNCS, vol. 8795, pp. 452–459. Springer, Heidelberg (2014)
2. Duman, E., Uysal, M., Alkaya, A.: Migrating birds optimization: a new metaheuristic approach and its performance on quadratic assignment problem. Inf. Sci. **217**, 65–77 (2012)
3. Kazimipour, B., Li, X., Qin, A.: Effects of population initialization on differential evolution for large scale optimization. In: 2014 IEEE Congress on Evolutionary Computation (CEC), pp. 2404–2411, July 2014
4. Lalla-Ruiz, E., de Armas, J., Expósito-Izquierdo, C., Melián-Batista, B., Moreno-Vega, J.M.: Multi-leader migrating birds optimization: a novel nature-inspired metaheuristic for combinatorial problems. Int. J. Bio-Inspired Comput. (2015, in press)
5. Lalla-Ruiz, E., Expósito-Izquierdo, C., de Armas, J., Melián-Batista, B., Moreno-Vega, J.M., et al.: Migrating birds optimization for the seaside problems at maritime container terminals. Inf. Sci. J. Appl. Math. **2015**, 1–12 (2015)
6. León, C., Miranda, G., Segura, C.: METCO: a parallel plugin-based framework for multi-objective optimization. Int. J. Artif. Intell. Tools **18**(4), 569–588 (2009)
7. Li, X., Tang, K., Omidvar, M., Yang, Z., Qin, K.: Benchmark functions for the CEC 2013 special session and competition on large scale global optimization. Technical report, Evolutionary Computation and Machine Learning Group, RMIT University, Australia (2013)

8. Pan, Q.K., Dong, Y.: An improved migrating birds optimisation for a hybrid flowshop scheduling with total flowtime minimisation. Inf. Sci. **277**, 643–655 (2014)
9. Segura, C., Coello, C.A.C., Segredo, E., Aguirre, A.H.: A novel diversity-based replacement strategy for evolutionary algorithms. IEEE Trans. Cybern. **PP**, 1–14 (2015)
10. Storn, R., Price, K.: Differential evolution - a simple and efficient heuristic for global optimization over continuous spaces. J. Glob. Optim. **11**(4), 341–359 (1997)
11. Yang, X.S.: Nature-Inspired Metaheuristic Algorithms. Luniver press (2010)

How Far Are We from an Optimal, Adaptive DE?

Ryoji Tanabe[1(✉)] and Alex Fukunaga[2]

[1] Institute of Space and Astronautical Science,
Japan Aerospace Exploration Agency, Kanagawa, Japan
rt.ryoji.tanabe@gmail.com
[2] Graduate School of Arts and Sciences, The University of Tokyo, Tokyo, Japan
fukunaga@idea.c.u-tokyo.ac.jp

Abstract. We consider how an (almost) optimal parameter adaptation process for an adaptive DE might behave, and compare the behavior and performance of this approximately optimal process to that of existing, adaptive mechanisms for DE. An optimal parameter adaptation process is an useful notion for analyzing the parameter adaptation methods in adaptive DE as well as other adaptive evolutionary algorithms, but it cannot be known generally. Thus, we propose a Greedy Approximate Oracle method (GAO) which approximates an optimal parameter adaptation process. We compare the behavior of GAODE, a DE algorithm with GAO, to typical adaptive DEs on six benchmark functions and the BBOB benchmarks, and show that GAO can be used to (1) explore how much room for improvement there is in the performance of the adaptive DEs, and (2) obtain hints for developing future, effective parameter adaptation methods for adaptive DEs.

1 Introduction

Differential Evolution (DE) is an Evolutionary Algorithm (EA) that was primarily designed for continuous optimization [17], and has been applied to many real-world problems [4]. A DE population $\boldsymbol{P} = \{\boldsymbol{x}^1, ..., \boldsymbol{x}^N\}$ is represented as a set of real parameter vector $\boldsymbol{x}^i = (x_1^i, ..., x_D^i)^{\mathrm{T}}$, $i \in \{1, ..., N\}$, where D is the dimensionality of the target problem and N is the population size.

After initialization of the population, for each generation t, for each $\boldsymbol{x}^{i,t}$, a mutant vector $\boldsymbol{v}^{i,t}$ is generated from the individuals in $\boldsymbol{P}^t$ by applying a mutation strategy. The most commonly used mutation strategy is the rand/1 strategy: $\boldsymbol{v}^{i,t} = \boldsymbol{x}^{r_1,t} + F_{i,t}\,(\boldsymbol{x}^{r_2,t} - \boldsymbol{x}^{r_3,t})$. The indices r_1, r_2, r_3 are randomly selected from $\{1, ..., N\}$ such that they differ from each other as well as i. The scale factor $F_{i,t} \in (0, 1]$ controls the magnitude of the differential mutation operator. Then, the mutant vector $\boldsymbol{v}^{i,t}$ is crossed with the parent $\boldsymbol{x}^{i,t}$ in order

Electronic supplementary material The online version of this chapter (doi:10.1007/978-3-319-45823-6_14) contains supplementary material, which is available to authorized users.

J. Handl et al. (Eds.): PPSN XIV 2016, LNCS 9921, pp. 145–155, 2016.
DOI: 10.1007/978-3-319-45823-6_14

to generate a trial vector $\boldsymbol{u}^{i,t}$. Binomial crossover, the most commonly used crossover method in DE, is implemented as follows: For each $j \in \{1, ..., D\}$, if $\mathrm{rand}[0,1] \leq C_{i,t}$ or $j = j_r$ (where, $\mathrm{rand}[0,1]$ denotes a uniformly generated random number from $[0,1]$, and j_r is a decision variable index which is uniformly randomly selected from $\{1, ..., D\}$), then $u_j^{i,t} = v_j^{i,t}$. Otherwise, $u_j^{i,t} = x_j^{i,t}$. $C_{i,t} \in [0,1]$ is the crossover rate. After all of the trial vectors $\boldsymbol{u}^{i,t}$, $i \in \{1, ..., N\}$ have been generated, each individual $\boldsymbol{x}^{i,t}$ is compared with its corresponding trial vector $\boldsymbol{u}^{i,t}$, keeping the better vector in the population, i.e., if $f(\boldsymbol{u}^{i,t}) \leq f(\boldsymbol{x}^{i,t})$, $\boldsymbol{x}^{i,t+1} = \boldsymbol{u}^{i,t}$ for minimization problems. Otherwise, $\boldsymbol{x}^{i,t+1} = \boldsymbol{x}^{i,t}$.

It is well-known that the performance of EAs is significantly influenced by control parameter settings [6,11], and DE is no exception [4]. Since identifying optimal control parameter values *a priori* is impractical, *adaptive* DE algorithms, which automatically adjust their control parameters online during the search process, have been studied by many researchers. Most of the well-known adaptive DEs [3,10,13,18,20] automatically adjust the F and C parameters. However, while many adaptive DEs have been proposed, their parameter adaptation methods are poorly understood. Previous work such as [3,10,13,18,20] only proposed a novel adaptive DE variant and evaluated its performance on some benchmark functions, but analysis of their adaptation methods have been minimal. The situation is not unique to the DE community – Karafotias et al. [11] have pointed out the lack of the analysis of adaptation mechanisms in EA. There are several previous work that try to analyze the parameter adaptation method in adaptive DE [3,5,13,16,20]. However, almost all merely visualized how F and C values change during a typical run on functions, and the analysis is limited to qualitative descriptions such as "a meta-parameter of C in adaptive DE quickly drops down to $[0, 0.2]$ after several generations on the Rastrigin function".

In this paper, we consider how an (almost) *optimal* parameter adaptation process might behave, and compare the behavior and performance of this approximately optimal process to that of existing, adaptive mechanisms for DE. We first define what we mean by an optimal parameter adaptation process, and propose a simulation process which can be used in order to greedily approximate the behavior of such an optimal process. We propose GAODE, which applies this methodology to DE and simulates an approximately optimal parameter adaptation process for a specific adaptive DE framework. We compare the behavior of GAODE to typical adaptive DE algorithms on six benchmark functions and the BBOB benchmarks [8], and discuss (1) the performance of current adaptive DE algorithms compared to GAODE, and (2) the implications of these results for developing more effective parameter adaptation method for adaptive DEs.

2 The Proposed GAO Framework for Adaptive DEs

First, note that this paper focuses on *parameter adaptation methods* for F and C in adaptive DEs such as jDE [3], EPSDE [13], JADE [20], MDE [10], SHADE [18]. In general, the term "adaptive DE" denotes a complex algorithm composed of multiple algorithm components. For example, "JADE" consists of three key components: (a) current-to-$p$best/1 mutation strategy, (b) binomial crossover, (c) JADE's parameter adaptation method of F and C. In this paper we want to focus on analyzing (c) *only*, rather than "JADE", the complex DE algorithm composed of (a), (b) and (c). Therefore, we extracted only (c) from each adaptive DE variant, and generalized it so that it can be combined with arbitrary mutation and crossover methods. This approach is taken in recent work [5,16].

Due to space limitations, the parameter adaptation methods in jDE, EPSDE, JADE, MDE, and SHADE cannot be described here (see Section A in the supplemental materials [1]), but the general framework can be described as follows: (i) At the beginning of each generation t, the $F_{i,t}$ and $C_{i,t}$ values are assigned to each individual $\boldsymbol{x}^{i,t}$. (ii) For each $\boldsymbol{x}^{i,t}$, a trial vector $\boldsymbol{u}^{i,t}$ is generated using a mutation strategy with $F_{i,t}$ and crossover method with $C_{i,t}$. (iii) At the end of each generation t, the F and C values used by successful individuals influence the parameter adaptation on the next generation $t+1$, where we say that an individual i is *successful* if $f(\boldsymbol{u}^{i,t}) \leq f(\boldsymbol{x}^{i,t})$.

2.1 Optimal Parameter Adaptation Process $\boldsymbol{\theta}^*$

We define the notion of an optimal parameter adaptation process in an adaptive DE. Below, DE-(a, m) denotes an adaptive DE algorithm using a and m, where a is a parameter adaptation method, and m is a DE mutation operator. Let L be the number of function evaluations (FEvals) until the search finds an optimal solution. An *adaptation process* $\boldsymbol{\theta}^a_m = (\{F_1, C_1\}, ..., \{F_L, C_L\})^{\mathrm{T}}$ is defined as the series of the F and C parameters generated when DE-(a, m) is executed with some adaptation mechanism a and some DE mutation operator m.

For some fixed m, an *optimal parameter adaptation process* $\boldsymbol{\theta}^*_m = (\{F^*_1, C^*_1\}, ..., \{F^*_L, C^*_L\})^{\mathrm{T}}$ is defined as an adaptation process which minimizes the expected value of L, i.e., there exists no a' such that $E[|\boldsymbol{\theta}^{a'}_m|] < E[|\boldsymbol{\theta}^*_m|]$. In the rest of the paper, we abbreviate this as $\boldsymbol{\theta}^*$. An *optimal parameter adaptation method* a^* is an adaptation method such that $\boldsymbol{\theta}^{a^*}_m = \boldsymbol{\theta}^*$.

a^* and $\boldsymbol{\theta}^*$ are useful notions for analyzing the parameter adaptation methods in adaptive DE. If $\boldsymbol{\theta}^*$ is known for some problem instance I, this by definition is a lower bound on the performance of DE-(a, m) (no other adaptation process can have a shorter expected length). This allows quantitative discussions regarding the performance of DE-(a, m) relative to the lower bound, e.g., "DE-(jDE, best/1) is 12.34 times slower than DE-(a^*, best/1)". We can also use such bounds in order to assess whether further improvements to a certain class of methods are worthwhile, e.g., "DE-(JADE, rand/2) performs worse than CMA-ES [9], but the performance of DE-(a^*, rand/2) is better than CMA-ES. Therefore, further improvements to the adaptation method a may result in a version of DE-(a, rand/2) which could outperform CMA-ES."

Besides providing a bound on the performance of DE-(a, m), $\boldsymbol{\theta}^*$ might be useful in guiding the development of more efficient parameter adaptation methods. For example, if for some problem instance I, the F values in $\boldsymbol{\theta}^*$ are relatively high at the beginning of the search while they are low at the end of the

Algorithm 1. GAODE (the DE with GAO)

```
1  t ← 1, initialize P^t = {x^{1,t}, ..., x^{N,t}}, l ← 1, θ^GAO ← ∅;
2  while The termination criteria are not met do
3      for i = 1 to N do
4          U^l ← ∅;
5          for j = 1 to λ do
6              F_{l,j} = rand(F^min, F^max], C_{l,j} = rand[C^min, C^max];
7              The (virtual) trial vector u^{l,j} is generated using an arbitrary mutation
               strategy with F_{l,j} and crossover method with C_{l,j}, then u^{l,j} → U^l;
8          Evaluates the (virtual) trial vectors in U^l by f, and select u^{l,best};
9          u^{i,t} = u^{l,best}, θ^GAO ← {F_{l,best}, C_{l,best}}, l ← l + 1;
10         If f(u^{i,t}) ≤ f(x^{i,t}), x^{i,t+1} = u^{i,t}. Otherwise, x^{i,t+1} = x^{i,t};
11     t ← t + 1;
```

search, then this suggests that we might be able to improve the performance of DE-(a, m) on problems similar to I by designing a so that the adaptation process of DE-(a, m) more closely resembles of $\boldsymbol{\theta}^*$ for I.

However, in practice, it is generally not possible to know $\boldsymbol{\theta}^*$. It is well-known that the appropriate parameter settings depend on the current search situation, and are not fixed values such as $F = 0.5$ and $C = 0.9$, i.e., there are different optimal parameter values as $(\{F_1^*, C_1^*\}, \{F_2^*, C_2^*\}, \{F_3^*, C_3^*\}, ...)$ for each FEvals $(1, 2, 3, ...)$. $\{F_l^*, C_l^*\}$ are also context-dependent, so we can not compute $\{F_l^*, C_l^*\}$ for some time step l in isolation – the search state at l depends on the control parameter settings used in steps $1, ..., l-1$.

2.2 Approximating an Optimal Adaptation Process $\boldsymbol{\theta}^*$

As discussed above, $\boldsymbol{\theta}^*$ would be very useful for analyzing the parameter adaptation methods, but it cannot be obtained in practice. Thus, we propose a *Greedy Approximate Oracle method* (GAO) in order to approximate $\boldsymbol{\theta}^*$, and apply the proposed GAO method to DE.

The basic idea is as follows: suppose that in step (i) of the adaptive DE framework described in the beginning of Sect. 2, we could enumerate all possible parameter settings $\{F, C\}$, and then retroactively select the $\{F, C\}$ pair which results in the best child – this would give us the optimal, 1-step adaptation process. Similarly, the optimal k-step adaptation process can be obtained by recursively simulating the execution of the DE for all possible k-step adaptation processes and then selecting the best k-step process. Of course, the number of possible adaptation processes grows exponentially in the number of steps, so in general, the k-step process can not be obtained, and in fact, fully enumerating *all* possible 1-step processes is impractical. We therefore obtain an approximation to the 1-step optimal process by randomly sampling $\{F, C\}$ values.

This is implemented as GAODE, shown in Algorithm 1. For each current FEvals l, let us consider that the individual $\boldsymbol{x}^l$ $(=\boldsymbol{x}^{i,t})$ generates the trial vector $\boldsymbol{u}^l$ $(=\boldsymbol{u}^{i,t})$ using parameter settings $\boldsymbol{\theta}^l = \{F_l, C_l\}$ $(=\boldsymbol{\theta}^{i,t})$. The optimal 1-step greedy parameter settings for step l is $\boldsymbol{\theta}^{g^*,l} = \{F_l^{g^*}, C_l^{g^*}\}$, and GAO seeks $\boldsymbol{\theta}^{\mathrm{GAO},l}$, which approximates values of $\boldsymbol{\theta}^{g^*,l}$, by random sampling of $\{F, C\}$ values.

For each $\boldsymbol{x}^l$, λ trial vectors $\boldsymbol{U}^l = \{\boldsymbol{u}^{l,1}, \ldots, \boldsymbol{u}^{l,\lambda}\}$ are generated (Algorithm 1, lines 4∼7). Parameter values $\boldsymbol{\theta}^{l,j} = \{F_{l,j}, C_{l,j}\}, j \in \{1, \ldots, \lambda\}$ used for generating $\boldsymbol{u}^{l,j}$ are uniformly randomly selected from $(F^{\min}, F^{\max}]$ and $[C^{\min}, C^{\max}]$ respectively (Algorithm 1, line 6). In DE, pseudo-random numbers are used for (a) parent selection in the mutation operator, and (b) the crossover operator. If two different virtual DE configurations which have different $\{F, C\}$ parameter values also use different random numbers for (a) and (b), it complicates the analysis because we cannot determine whether the configuration which generates the better trial vector did so because of its $\{F, C\}$ values or because of the random numbers used in (a) and (b). Therefore, in our experiments, we synchronized the pseudorandom generators for all of the virtual DE configurations so that they all used the same random numbers at both (a) and (b) for generating all trial vectors in $\boldsymbol{U}^l$ – this eliminates the possibility that a virtual DE configuration outperforms another due to fortunate random numbers used for (a) and (b).

The trial vectors in $\boldsymbol{U}^l$ are evaluated according to the function f, and the $\boldsymbol{u}^{l,\mathrm{best}}$ with the best (lowest) function value in $\boldsymbol{U}^l$ is selected (Algorithm 1, line 8). The selected $\boldsymbol{u}^{l,\mathrm{best}}$ is treated as $\boldsymbol{u}^l$ (=$\boldsymbol{u}^{i,t}$) of $\boldsymbol{x}^l$ (=$\boldsymbol{x}^{i,t}$). Note that, λ times evaluations according to f which are used to select $\boldsymbol{u}^{l,\mathrm{best}}$ in $\boldsymbol{U}^l$ (Algorithm 1, line 8), are *not counted* as the FEvals used in the search – this simulates a powerful oracle which "guesses" $\boldsymbol{u}^{l,\mathrm{best}}$ in one try. $\boldsymbol{\theta}^{l,\mathrm{best}} = \{F_{l,\mathrm{best}}, C_{l,\mathrm{best}}\}$ used for generating $\boldsymbol{u}^{l,\mathrm{best}}$ can be considered a approximation to $\boldsymbol{\theta}^{g^*,l} = \{F_l^{g^*}, C_l^{g^*}\}$, and is stored in $\boldsymbol{\theta}^{\mathrm{GAO}}$ (Algorithm 1, line 9).

Previous work has investigated optimal parameter values in adaptive EAs, especially in Evolution Strategies (ES) community [2,6,7]. For example, the optimal step size σ^* in (1 + 1)-ES on the Sphere function is $\sigma^* = 1.224\,\|\boldsymbol{x}^* - \boldsymbol{x}\|/D$ [7], where $\|\boldsymbol{x}^* - \boldsymbol{x}\|$ is the Euclidean distance between the optimal solution $\boldsymbol{x}^*$ and the current search point $\boldsymbol{x}$. The optimal mutation rate p_m schedule of (1 + 1)-GA on the one-max problem is also studied by Bäck [2]. While theoretically well-founded, these results are limited to a specific algorithm running on a specific problem, and have also been limited to one parameter value, e.g., σ and p_m. In contrast, the proposed GAO framework is more general. While we focus on applying GAO to DE for black-box optimization benchmarks in this paper, we believe the GAO approach can be straightforwardly generalized and applied to combinations of various problem domains (e.g., combinatorial problems, single/multi-objective problems, etc.), algorithms (e.g., GA, ES, MOEA, etc.), and parameters (e.g., crossover and mutation rate, crossover method, etc.).

3 Evaluating the Proposed GAO Framework

We compare GAODE, the DE with GAO, to the parameter adaptation methods used by representative adaptive DEs on six benchmark functions. We show that GAO can be used to (1) explore how much room for improvement there is in the performance of the adaptive DEs (Sect. 3.1), and (2) obtain hints for developing future, effective parameter adaptation methods (Sect. 3.2).

We used six benchmark functions: Sphere, Ellipsoid, Rotated-Ellipsoid, Rosenbrock, Ackley, Rastrigin functions. The first three are unimodal, and the last three are multimodal (the Rosenbrock function is unimodal for $D \leq 3$). The Rotated-Ellipsoid and Rosenbrock functions are nonseparable, and the (Rotated-) Ellipsoid functions are ill-conditioned functions. For details, see Table A.1 in [1].

The dimensionality D of each function was set to 2, 3, 5, 10, and 20. The number of runs per problem was 51. Random number seeds for parts of the DE are synchronized as explained in Sect. 2.2. Each run continues until either (i) $|f(\boldsymbol{x}^{\mathrm{bsf}}) - f(\boldsymbol{x}^*)| \leq 10^{-8}$, in which case we treat the run as a "success", or (ii) the number of fitness evaluations (FEvals) exceeds $D \times 10^5$, in which case the run is treated as a "failure". $\boldsymbol{x}^{\mathrm{bsf}}$ is the best-so-far solution found during the search process, and $\boldsymbol{x}^*$ is the optimal solution of the target problem. Following [9], we used the Success Performance 1 (SP1) metric, which is the average FEvals in successful runs divided by the number of successes, as a performance metric of the DE algorithms. SP1 represents the expected FEvals to reach the optimal solution, i.e., a small SP1 value indicates a fast and stable search.

We used five parameter adaptation methods in the representative adaptive DE variants: jDE [3], EPSDE [13], JADE [20], MDE [10], and SHADE [18]. For details, see Section A in [1]. The most basic rand/1/bin operator [17], described in Sect. 1, was used for all DEs. Following [14], the population size N was set to $5 \times D$ for $D > 5$, and $N = 20$ for $D = 2$ and 3. For each algorithm, we used the control parameter values that were suggested in the original papers as follows: $\tau_F = 0.1$ and $\tau_C = 0.1$ for jDE, F-pool $= \{0.4, ..., 0.9\}$ and C-pool $= \{0.1, ..., 0.9\}$ for EPSDE, $c = 0.1$ for JADE, and $H = 10$ for SHADE.

In the GAO framework, the parameter generation ranges $(F^{\min}, F^{\max}]$ and $[C^{\min}, C^{\max}]$ have to be set. In preliminary experiments, GAODE failed on some nonseparable functions when these ranges were set to $(0, 1]$ and $[0, 1]$ respectively. We believe this failure is due to small F values, so we also evaluated GAODE with $F^{\min} = 0.4$, where 0.4 is a lowest F value suggested by Rönkkönen et al. [15]. Unless explicitly noted, we denote GAODE with the former and later settings as GAODE00 and GAODE04 respectively, and a virtual DE algorithm that is a composition of GAODE00 and GAODE04 as GAODE (GAODE returns the best result obtained by running both GAODE00 and GAODE04). λ, the number of configurations sampled by GAODE at each individual, was set to 200.

3.1 Experiment 1: How Much Room Is There for Improvement with Adaptive DE Algorithms with the rand/1/bin Operator?

Figure 1 shows the results of GAODE, jDE, EPSDE, JADE, MDE, and SHADE on the six functions. For GAODE, instead of SP1, we show the lowest FEvals for reaching the optimal solution in the composed results of GAODE00 and GAODE04. The data of GAODE indicates an approximate bound on the performance that can be obtained by an adaptive DE using the rand/1/bin operator.

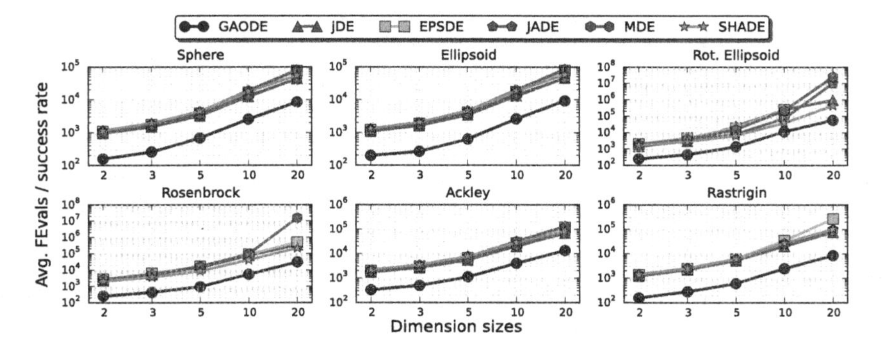

Fig. 1. Comparison of GAODE with the parameter adaptation methods in the adaptive DEs on each function. The horizontal axis represents the dimensionality D, and the vertical axis represents the SP1 values. Data with success rate = 0 is not shown.

As shown in Fig. 1, all runs of EPSDE fail on the Rotated-Ellipsoid function for $D = 20$. JADE also fails on all runs on the Rosenbrock function when $D \geq 10$. MDE can reach the optimal solution on the both functions, but its SP1 values are significantly worse than other methods. Consistent with the results in [16,20], adaptation methods tend to perform poorly when used with operators that are different from the operators used in the original papers where the adaptation methods were proposed. Although the performance rank among the methods depends on the functions and the dimensionality D, jDE and SHADE perform better than other compared methods in almost all cases. However, as shown in Fig. 1, jDE and SHADE converge to the optimal solution 4∼20 times slower than GAODE. This shows that even the best current adaptive methods perform poorly compared to an approximation of a 1-step greedy optimal process (and are therefore even worse compared to a k-step optimal process). *Thus, it appears that despite significant progress in recent years, there is still significant room for improvement in parameter adaptation methods for DE.*

3.2 Experiment 2: How Should We Adapt the Control Parameters?

Let us consider how the behavior of GAODE differs from existing adaptation methods. Figure 2 shows the frequency of appearance of $\{F, C\}$ value pairs during the search process for SHADE and GAODE on the 10-dimensional Rosenbrock and Rastrigin functions. Data from the best run out of 51 runs is shown. The results of jDE, EPSDE, JADE, and MDE can be seen in Fig. A.1 in [1].

As shown in Fig. 2, SHADE frequently generates F and C values in the range $[0.5, 0.7]$ and $[0.9, 1.0]$ on the Rosenbrock function, and $[0.9, 1.0]$ and $[0.1, 0.4]$ on the Rastrigin function respectively. These results are consistent with previous studies for DE [3,4] and adaptive DEs [3,20]. On the other hand, GAODE mainly generates F values in the range $[0.0, 0.1]$ on both functions. The C values frequently appear in $[0.0, 0.2]$ and $[0.8, 1.0]$ on the Rastrigin function, and GAODE mainly generates C values in *both* $[0.9, 1.0]$, and $[0.0, 0.1]$ on the Rosenbrock

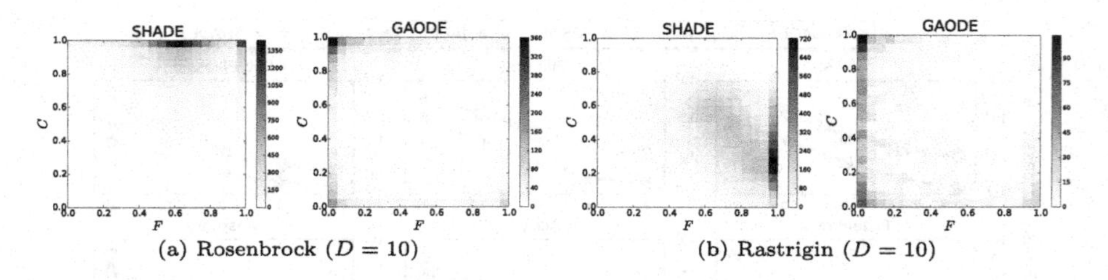

(a) Rosenbrock ($D = 10$) (b) Rastrigin ($D = 10$)

Fig. 2. Frequency of appearance of $\{F, C\}$ value pairs during the search process for SHADE and GAODE on the 10-dimensional (a) Rosenbrock and (b) Rastrigin functions. Darker colors indicate more frequent generation of the corresponding values by the parameter adaptation method.

function, i.e., the C values are bimodal. Interestingly, for the both functions, GAODE occasionally generates F and C values in the extreme regions $[0.9, 1.0]$ and $[0.0, 0.1]$ respectively (see bottom right in the figures).

In summary, GAODE frequently generates small F values, and C values in the range $[0, 0.2]$ and $[0.8, 1]$. Although CoBiDE [19], a recently proposed *non-adaptive* DE, generates the $F_{i,t}$ and $C_{i,t}$ values for each $\boldsymbol{x}^{i,t}$ according to a bimodal (two Cauchy) distribution, we are not aware of such a bimodal sampling approach in any previously proposed adaptive method. An adaptive DE algorithm using such sampling method may also perform better than the existing methods [3,10,13,18,20]. Thus, analysis of the approximate optimal parameter adaptation process obtained by GAO suggests that instead of unimodal sampling procedures implemented in previous adaptation methods, adaptive mechanisms using multimodal sampling may be a promising direction for future work.

4 Comparing GAODE with State-of-the-Art EAs

GAODE, which is an approximate simulation of an optimal, 1-step adaptation process, significantly performs better than the current state-of-the-art parameter adaptation methods for DE using the rand/1/bin operator, as described in Sect. 3.1 (again, we reemphasize that *GAODE is not a practical algorithm and is for analysis only* – the "performance" of GAODE ignores the $\lambda - 1$ samples which are discarded by GAODE at each iteration). It is interesting to compare GAODE with other state-of-the-art EAs. Here, we compare the adaptive DE variants including GAODE[1] with HCMA [12] and best-2009 on the BBOB benchmarks, consisting of 24 various functions [8]. HCMA, an efficient surrogate-assisted algorithm portfolio, represents the state-of-the-art on the BBOB benchmarks. Best-2009 is a *virtual* algorithm portfolio that is retrospectively constructed from the performance data of 31 algorithms participating in

[1] The BBOB benchmarks provide 15 instances for each function, i.e., there are $24 \times 15 = 360$ function instances. In this study, we applied GAODE00 and GAODE04 three times for each instance, and only the best result among them is used for GAODE.

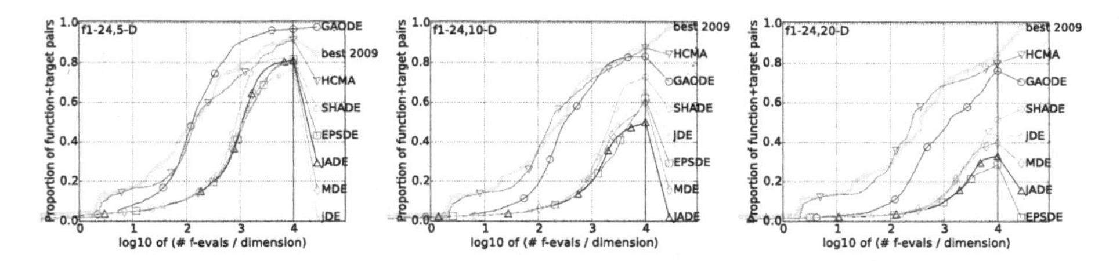

Fig. 3. Comparisons of GAODE with the adaptive DE variants, HCMA, and best-2009 on BBOB benchmarks ($D \in \{5, 10, 20\}$). These figures show the bootstrapped Empirical Cumulative Distribution Function (ECDF) of the FEvals divided by dimension for 50 targets in $10^{[-8..2]}$ for 5, 10, 20 dimensional all functions (higher is better). For details of the ECDF, see a manual of `COCO software` (http://coco.gforge.inria.fr/).

the GECCO BBOB 2009 workshop. Is it possible for an adaptive DE algorithm using the *classical* rand/1/bin operator to be competitive with these methods?

Figure 3 shows the Empirical Cumulative Distribution Function (ECDF) for each algorithm for 24 BBOB benchmark problems ($D = 5, 10, 20$) when maximum FEvals $= D \times 10^4$. The results for each function class and for $D = 2, 3$ can be found in Figs. A.2∼A.6 in [1]. As shown in Fig. 3, GAODE clearly outperforms jDE, EPSDE, JADE, MDE, and SHADE for all dimensions, in terms of both the quality of the best-so-far solution obtained during the search process and the anytime performance. GAODE also performs significantly better than HCMA and best-2009 for $D \leq 5$. This result suggests that if we can find a parameter adaptation method which performs similarly to the GAODE model, then an adaptive DE algorithm using the classical rand/1/bin could possibly outperform state-of-the-art algorithm portfolios such as HCMA for $D \leq 5$.

On the other hand, when the dimensionality increases, the performance of GAODE degrades compared to HCMA and best-2009. For $D = 20$, GAODE is outperformed by HCMA and best-2009. This may indicate that for high-dimensionality problems, it may not be possible to develop an adaptive DE using the rand/1/bin operator which is competitive with methods such as HCMA. However, this result may be due to the fact that GAODE only simulates an approximately optimal 1-step adaptation process – increasing the number of steps (i.e., a k-step optimal process) may result in better results, and is a direction for future work. In addition, different mutation operators (e.g., best/2, current-to-$p$best/1, etc.) may enable significantly better performance for adaptive DEs.

5 Conclusion

We proposed a Greedy Approximate Oracle method (GAO) which approximates an optimal parameter adaptation process $\boldsymbol{\theta}^*$. In GAO, λ parameter sets are randomly generated for each individual in the population, and the best parameter set with respect to the objective function value is used as a greedily approximated optimal parameter set (the other $\lambda - 1$ sets are discarded and are not counted). We evaluated GAODE, a DE algorithm with GAO, on 6 standard benchmark functions and the BBOB benchmarks [8], and compared it with the parameter adaptation methods in 5 adaptive DE variants. We showed that (1) current adaptive DEs are significantly worse than even an approximate, 1-step optimal adaptation, suggesting that there is still much work to be done in the development of adaptive mechanisms (Sect. 3.1), and (2) GAO can be used to identify promising directions for developing an efficient parameter adaptation method in adaptive DE (Sect. 3.2). We also compared GAODE with HCMA [12] and best-2009 on the BBOB benchmarks [8] in Sect. 4, and showed that a better adaptive mechanism may enable a DE using the classical rand/1/bin operator to achieve state-of-the-art performance.

The proposed GAO framework is a first attempt to approximate the optimal parameter adaptation process, and there is much room for improvement, as discussed in Sect. 2. In this paper, we applied GAO to the DE with the rand/1/bin operator, and evaluated its performance on single-objective continuous optimization problems. Future work will explore GAO as a general framework that can be applied to analyze the behavior of any adaptive EA (independent of specific operators and problem domains).

References

1. Supplement. https://sites.google.com/site/tanaberyoji/home/tf-ppsn16-sup.pdf
2. Bäck, T.: Optimal mutation rates in genetic search. In: ICGA, pp. 2–8 (1993)
3. Brest, J., Greiner, S., Bošković, B., Mernik, M., Žumer, V.: Self-adapting control parameters in differential evolution: a comparative study on numerical benchmark problems. IEEE TEVC **10**(6), 646–657 (2006)
4. Das, S., Mullick, S.S., Suganthan, P.N.: Recent advances in differential evolution - an updated survey. Swarm Evol. Comput. (2016, in press)
5. Drozdik, M., Aguirre, H., Akimoto, Y., Tanaka, K.: Comparison of parameter control mechanisms in multi-objective differential evolution. In: Jourdan, L., Dhaenens, C., Marmion, M.-E. (eds.) LION 9 2015. LNCS, vol. 8994, pp. 89–103. Springer, Heidelberg (2015)
6. Eiben, A.E., Hinterding, R., Michalewicz, Z.: Parameter control in evolutionary algorithms. IEEE TEVC **3**(2), 124–141 (1999)
7. Hansen, N., Arnold, D.V., Auger, A.: Evolution Strategies. Springer, Heidelberg (2015)
8. Hansen, N., Finck, S., Ros, R., Auger, A.: Real-parameter black-box optimization benchmarking 2009: noiseless functions definitions. Technical report, INRIA (2009)
9. Hansen, N., Kern, S.: Evaluating the CMA evolution strategy on multimodal test functions. In: Yao, X., et al. (eds.) PPSN 2004. LNCS, vol. 3242, pp. 282–291. Springer, Heidelberg (2004)
10. Islam, S.M., Das, S., Ghosh, S., Roy, S., Suganthan, P.N.: An adaptive differential evolution algorithm with novel mutation and crossover strategies for global numerical optimization. IEEE Trans. SMC B **42**(2), 482–500 (2012)
11. Karafotias, G., Hoogendoorn, M., Eiben, A.E.: Parameter control in evolutionary algorithms: trends and challenges. IEEE TEVC **19**(2), 167–187 (2015)
12. Loshchilov, I., Schoenauer, M., Sebag, M.: Bi-population CMA-ES agorithms with surrogate models and line searches. In: GECCO Com, pp. 1177–1184 (2013)

13. Mallipeddi, R., Suganthan, P.N., Pan, Q.K., Tasgetiren, M.F.: Differential evolution algorithm with ensemble of parameters and mutation strategies. ASC **11**, 1679–1696 (2011)
14. Pošík, P., Klema, V.: JADE, an adaptive differential evolution algorithm, benchmarked on the BBOB noiseless testbed. In: GECCO Com, pp. 197–204 (2012)
15. Rönkkönen, J., Kukkonen, S., Price, K.V.: Real-parameter optimization with differential evolution. In: IEEE CEC, pp. 506–513 (2005)
16. Segura, C., Coello, C.A.C., Segredo, E., León, C.: On the adaptation of the mutation scale factor in differential evolution. Opt. Lett. **9**(1), 189–198 (2015)
17. Storn, R., Price, K.: Differential evolution - a simple and efficient heuristic for global optimization over continuous spaces. J. Glob. Optim. **11**(4), 341–359 (1997)
18. Tanabe, R., Fukunaga, A.: Success-history based parameter adaptation for differential evolution. In: IEEE CEC, pp. 71–78 (2013)
19. Wang, Y., Li, H., Huang, T., Li, L.: Differential evolution based on covariance matrix learning and bimodal distribution parameter setting. ASC **18**, 232–247 (2014)
20. Zhang, J., Sanderson, A.C.: JADE: adaptive differential evolution with optional external archive. IEEE TEVC **13**(5), 945–958 (2009)

Feature Based Algorithm Configuration: A Case Study with Differential Evolution

Nacim Belkhir[1,2(✉)], Johann Dréo[1], Pierre Savéant[1], and Marc Schoenauer[2]

[1] Thales Research & Technology, Palaiseau, France
{nacim.belkhir,johann.dreo,pierre.saveant}@thalesgroup.com
[2] TAO, Inria Saclay Île-de-France, Orsay, France
marc.schoenauer@inria.fr

Abstract. Algorithm Configuration is still an intricate problem especially in the continuous black box optimization domain. This paper empirically investigates the relationship between continuous problem *features* (measuring different problem characteristics) and the *best parameter configuration* of a given stochastic algorithm over a bench of test functions — namely here, the original version of Differential Evolution over the BBOB test bench. This is achieved by learning an empirical performance model from the problem features and the algorithm parameters. This performance model can then be used to compute an empirical optimal parameter configuration from features values. The results show that reasonable performance models can indeed be learned, resulting in a better parameter configuration than a static parameter setting optimized for robustness over the test bench.

Keywords: Empirical study · Black-box continuous optimization · Problem features · Algorithm configuration · Empirical Performance Model · Differential Evolution

1 Introduction

Today, it is widely acknowledged that the quest of a universal black box optimization algorithm is vain, even if the *No Free Lunch Theorem* [17] has been questioned in the continuous framework [1]. However, many algorithms exist, more or less specific to different classes of optimization problems, and the new grail of optimizers has now turned toward Algorithm Selection, as formulated by Rice [15], or Algorithm Configuration, that can be considered as yet another (meta-)optimization problem [6]. In both cases, the choice (of an algorithm, or of the parameters of a given algorithm) is made w.r.t. the user's preference, aka a performance criterion (e.g., quality of the solution obtained in a given CPU cost, the smallest CPU cost to reach a given solution quality, the probability to reach a given quality, with given thresholds, etc.).

A first approach to Algorithm Configuration is to optimize this performance criterion once and for all using a specific algorithm, e.g., SMAC [8]. But this

J. Handl et al. (Eds.): PPSN XIV 2016, LNCS 9921, pp. 156–166, 2016.
DOI: 10.1007/978-3-319-45823-6_15

results in a single configuration, and even if several problems are used to compute the performance criterion, the generalization of the results to other problems might be problematic.

More recent approaches are based on a description of the objective function in some *feature space*, and try to learn a mapping from this feature space onto the space of parameter configurations of the algorithm at hand, based on examples of the behavior of several configurations on a training set of objective functions. And the most successful approach for learning such a mapping is to first learn an empirical model of the algorithm performance (that predicts the performance criterion for a given set of features and an algorithm configuration). When a new problem arises (i.e. a new set of features), finding the algorithm configuration that is predicted to have the best performance is then straightforward. This approach, initially proposed in [10], has demonstrated successful results in different combinatorial optimization domains [7,9,18].

In continuous domains, however, though several feature sets have been proposed [12,13], and successfully demonstrated to accurately classify problem instances [3,13], only Algorithm Selection problems have actually been tackled [3,12–14].

The present work addresses the Algorithm Configuration problem for continuous domains, building an Empirical Performance Model (EPM) based on the problem features in continuous search spaces cited above. The approach is experimentally validated with the original version of Differential Evolution [16], that has few hyper-parameters, but is known to be highly sensitive to their setting. The set of objective functions used for this validation is the well-known BBOB test bench [5].

The paper is organized as follows. Section 2 presents the general idea of feature-based Empirical Performance Model and subsequent algorithm configuration. Section 3 surveys the problem features in the case of continuous optimization. Section 4 then details the DE case study and the BBOB testbench used in this work. Section 5 describes the different experiments, and the corresponding results are detailed and discussed in Sect. 6, before the concluding Sect. 7.

2 Algorithm Configuration with an Empirical Performance Model

Context and Notations. The general context is the Black Box Optimization of *objective functions* $f : \Omega \mapsto \mathbb{R}$. An algorithm $\boldsymbol{\mathcal{A}}$ is given together with its control parameters $\theta \in \boldsymbol{\Theta}$. We assume that the objective functions can be described by some *features* $\boldsymbol{\psi} \in \boldsymbol{\Psi}$. The goal of Instance-based Algorithm Configuration is to find automatically, for a given objective function f described by its features ψ^f, the best possible *configuration* of $\boldsymbol{\mathcal{A}}$, i.e., values $\theta^*_f \in \boldsymbol{\Theta}$ such that running $\boldsymbol{\mathcal{A}}$ with parameters θ^*_f on f leads to optimal performances w.r.t. a given *performance measure* φ.

Empirical Performance Model. The first step is to build an *Empirical Performance Model* (EPM) $\widehat{\varphi}$ that approximate φ on $\boldsymbol{\Psi} \times \boldsymbol{\Theta}$. $\boldsymbol{\mathcal{A}}$ is run to optimize different functions f_i (described by their features ψ_{f_i}) using different parameter configurations θ_j. This allows to compute the exact values $\varphi(\psi_{f_i}, \theta_j)$ for different pairs (i, j)[1]. The set of all $((\psi_{f_i}, \theta_j), \varphi(\psi_{f_i}, \theta_j))$ is a training set that can be used as input to any standard regression method to learn a model $\widehat{\varphi}$ for φ. Note that building such a model is done once, and hence its computer cost is not a critical issue.

Empirical Optimal Configuration. When a new objective function g is to be optimized with $\boldsymbol{\mathcal{A}}$, its features ψ_g are computed, and the optimization of $\widehat{\varphi}(\psi_g, \theta_g)$ on parameter space $\boldsymbol{\Theta}$ leads to the empirical optimal parameters of $\boldsymbol{\mathcal{A}}$ for g. Here, the cost of computing the features ψ_g, in terms of number of calls to g, is here of utter importance. In particular, it should be compared to the cost of running a full 'ad hoc' meta-optimization of $\boldsymbol{\mathcal{A}}$ parameters for g (using e.g., SMAC [8]).

The remaining of the paper is concerned with objective functions defined on some continuous domain $\mathcal{D} \subset \mathbb{R}^d$ for a given dimension $d \in \mathbb{N}$. Different features, taken from the literature, will first be discussed, before the case study is detailed.

3 Features for Continuous Optimization

In this section, a single objective function $f : \mathbb{R}^d \mapsto \mathbb{R}$ is considered, and the feature space $\boldsymbol{\Psi}$ is a vector space of p real-valued features. The black-box context implies that features should be computed from samples of f[2], i.e. n pairs $(x_i, f(x_i))$), specifically gathered for that purpose. The set of values $\{f(x_i)|i = 1, \ldots, n\}$ is denoted $\mathcal{Y}$.

A first set of 55 features is taken from [12]. These features are grouped into six classes: the 3 *y–Distribution* features are related to the distribution of the values in $\mathcal{Y}$, the 18 *Levelset* features to the relative position of $\mathcal{Y}$ w.r.t a given threshold, the 9 *Meta-Model* features rely on meta-modeling of the sample set w.r.t linear and quadratic regression models, the 14 *Curvature* features on some numerical estimation of the Hessian and gradient of the problem, the 4 *Convexity* features on the empirical probability of convexity, and the 7 *Local Search* features on the ratio of local optima and global optima, estimated using some iterated local search procedure.

The *y–Distribution*, the *Levelset* and the *Meta-Model* features can all be evaluated on the same sample dataset, hence their cost altogether is n, the number of samples. However, some additional evaluations are required for the other feature classes, that depend on the previous samples. The orders of magnitudes are about $10^3 \times d$ for the *Convexity* features, around $10^4 \times d$ for the *Curvature* and the *Local Search* features.

[1] The same θ_j need not have been tried for all f_i.

[2] d, the dimension of the search space, can be considered as the only external feature — or the Algorithm Configuration can be conducted anew for each dimension (more in Sect. 5).

A set of 16 *Dispersion* features was originally proposed in [11]. They are based on comparisons of the distances between best samples from different percentiles of the overall sample (in terms of solution quality) to the mean or median distance between all samples. Finally, 5 *Information Content* features were proposed in [13], giving information about the global structure of the landscape.

Recent works [3,12,13] successfully demonstrated that these features could be used in order to classify the optimization problems w.r.t their classes in BBOB (that will be introduced in Sect. 5) for the Algorithm Selection Problem. More recently, in [4] a subset of these features were used in order to improve the process of a parameter tuning algorithm, relying on the SMBO method [8].

4 A Case Study in Continuous Domain: DE on BBOB

Differential Evolution and its Parameters. Differential Evolution [16] is a popular continuous optimization algorithm that encountered many successes. It is also known for its simplicity, at least in its original version, that comes at the price of a large sensitivity to its parameter setting: this is the reason why it has been chosen here, making it easier to see big differences of results for different parameter settings. Several advanced versions of DE exist, that clearly outperform the original version, but comparing our results with theirs is left for further work.

DE generates new individuals from the current population by adding to each individual in turn a difference vector between two other individuals, and recombining the result with another individual from the population. The original version of DE has only four static parameters:

- the population size $\mathbf{NP} \in \mathbb{N}$;
- the strategy $\boldsymbol{\mathcal{S}} \in \{\text{best1bin}, \text{randtobest1bin}, \text{best2bin}, \text{rand2bin}, \text{rand1bin}\}$ controls how to choose the endpoints of the difference vector;
- $\boldsymbol{F} \in [0, 2]$ controls the intensity of the difference vector;
- the crossover rate $\boldsymbol{CR} \in [0, 1]$.

In this work the population size $\boldsymbol{NP}$ is kept to the default value $15 \times d$ recommended by the authors[3]. Note that the recommendation for the other parameters is $\boldsymbol{\mathcal{S}} = best1bin$, $\boldsymbol{F} = 0.8$, and $\boldsymbol{CR} = 0.9$, and will be used as one of the baseline (Sect. 6).

Test Bench. The following experiments consider the noiseless test functions from the Black Box Optimization Benchmark (BBOB[4]) [5]. The BBOB test bench is made of 24 analytically defined functions defined on $[-5, 5]^d$, with known global optima and known difficulties (e.g., non-separability, multi-modality, etc.). They have been manually classified in five classes of problems. In this work, only three of the BBOB classes are considered: 5 separable functions, 4 uni-modal functions with low or moderate conditioning, and 5 uni-modal functions with

[3] http://www1.icsi.berkeley.edu/~storn/code.html.

[4] http://coco.gforge.inria.fr.

high conditioning (functions F1 to F14). Dimensions 2, 3, 5, 10 are considered for all functions. As advocated in the original framework, any independent run on a function is actually done on a variant, in order to get over a possible algorithm bias. Variants are obtained from the original function by a translation of the position of the optimum and — for the non-separable functions — a rotation of the coordinate system.

Performance Measure. Following the COCO/BBOB framework, the performance measure used here is the Expected Run Time[5] (ERT) needed to reach the optimal objective value with a given precision. Let $\overline{RT_s}$ be the average running time of successful runs, and p_s the empirical probability of success (out of the 15 independent runs). The ERT is defined as $ERT = ERT(f, \theta) = \overline{RT_s}/p_s$ if the results were obtained with DE configuration θ optimizing test function f.

Features. From the features briefly introduced in Sect. 3, different set of problem features are considered. All features have been computed using the R package kindly publicly provided by Pascal Kerschke[6]. Features are distinguished by their costs:

- $\psi^\star$ includes all features from Sect. 3, with an initial sample of size $k = 2000 \times d$. However, as discussed in Sect. 3, the actual cost is much larger because of features requiring additional evaluations.
- $\psi^\bullet_{\mathbf{k}}$: $k \times d$ is the size of the initial sample, and only features not requiring any additional function evaluation beyond those of the initial sample are considered (i.e., *Meta-Model*, *Information Content*, *Levelset*, *y−Distribution* and *Dispersion*). Results with $k = 500$ or $k = 2000$ are presented in the following.

Regression Model. Preliminary experiments, not discussed here, lead to consider a Random Forests regression model (in accordance with [9]). A grid search with ten-fold cross-validation has been run on the meta-parameters of the Random Forest. The implementation of the scikit-learn python library has been used throughout this work[7], using 10 trees of maximal depth 200.

5 Experiments

Dataset. A 40-steps discretization is used for $\mathbf{F} \in [0, 2[$ and $\mathbf{CR} \in [0, 1]$, resulting in $5 \times 40 \times 40$ different configurations. For each of the 14 functions of the test bench and for each dimension $d \in \{2, 3, 5, 10\}$, each one of its 15 variants is optimized with these 8000 DE configurations, and the ERT is computed. The initial dataset is hence made of 14×8000 entries per dimension, or 440 000 entries in total, considering all dimensions.

[5] Measured as the number of function evaluations.
[6] http://github.com/flacco.
[7] http://scikit-learn.org/.

Dimensions. As discussed, the dimension d can be considered as a particular feature, available "for free" (without any function evaluation), or as part of the problem definition — and there are as many problems as dimensions. These two points of view will be compared here: the EPM will be learned either using only the entries of the dataset of the same dimension, or all entries, and the dimension will then be used as an additional feature in the feature vector.

Cross-validation. All experiments are based on a leave-one-out procedure: one of the 14 functions in the test bench is completely removed from the dataset (all dimensions and, of course, all variants). An EPM is then learned, and the left-out function considered a "unknown". The only exception is the *robust* baseline described below.

Baselines. Different DE configurations are computed for each function, and used as a baseline for comparison with the results of the proposed approach. The *default configuration* recommended by DE authors (Sect. 4) is the first obvious baseline. However, it is likely to perform poorly across the whole test bench.

At the other extreme, the specific configuration found by some meta-optimizers for a given problem is likely to give the best overall results: two such meta-optimization were performed for each function: on the one hand, the best configuration encountered while computing the full dataset using the grid described above is saved; on the other hand, SMAC [8] is applied to each function, using the ERT performance measure as fitness. The best of both configurations is reported as *adhoc configuration.*

Finally, one single SMAC optimization is performed using the average ERT over all 14 functions as performance: the idea behind this is to try to find some robust configuration that would give good results on all functions simultaneously. The resulting configuration is termed *robust* and considered as the reference (Sect. 6).

Experiment Costs. All DE runs were allowed a maximum budget of $10^4 \times d$ function evaluations — though of course some runs did stop earlier, having reached the target precision. And all results have been averaged over 15 independent variants for each function and dimension.

Furthermore, both adhoc baseline configurations have the same cost, because the budget given to SMAC was purposely chosen to match that of the grid search, i.e., 8000×15 runs of at most $\times 10^4 \times dim$ evaluations each.

On the other hand, the robust configuration did cost 14 times more, as each of its iterations required to evaluate all the 14 functions — but has to be done only once.

Compared to that, the cost of finding the best empirical configuration using an EPM is the cost of the features: 500 or 2000 in the case of $\psi^{\bullet}_{500}$ and $\psi^{\bullet}_{2000}$ features, or around $2.10^4 \times d$ for the full set of features ψ^*.

6 Results

The series of experiments described above are presented in this section from two points of view: first, the different EPM are analyzed and compared to the ground truth — and the Empirical Optimal Configuration is compared to the true optimal configuration in parameter space. Then, the actual results of DE optimization using the Empirical Optimal Configuration are compared to those of the different baselines, keeping in mind the actual costs of the different approaches (see last paragraph above).

6.1 EPM Analyses

Due to space constraints, only few typical figures are displayed[8] (Fig. 1) and will be discussed here. There are indeed some strong similarities between top and bottom colormaps for Figs. 1a, c, and d, even though they correspond to different functions, dimensions, feature sets, and dimension handling modes. On the other hand, the two plots for F4 (Fig. 1b) are very different, and here the EPM fails to capture even an approximate shape of the true ERT landscape.

But beyond such comparisons, the optimal configurations of both plots are shown (on both plots too, to ease the comparison), displaying very different situations: in Fig. 1a and c, both optima are rather close (1c), or at least are both in the same color area of the true ERT; on the opposite, in Fig. 1b and d, both optima are far from one another, and the Empirical Optimal Configuration

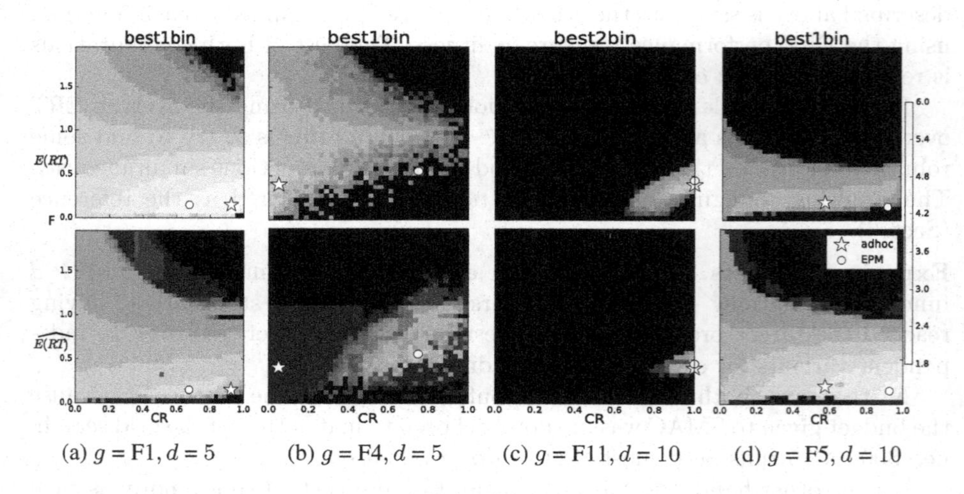

(a) $g = \text{F1}, d = 5$ (b) $g = \text{F4}, d = 5$ (c) $g = \text{F11}, d = 10$ (d) $g = \text{F5}, d = 10$

Fig. 1. Examples of comparisons between the true ERT (top) and the EPM (bottom) for 4 different functions and dimensions. Both EPMs of F1 (a) and F4 (b) have been learned only on samples from dimension 5, with features $\psi^{\bullet}_{k=2000}$ while those of F11 (c) and F5 (d) have been learned on samples of all dimensions, and with $\psi^{\bullet}_{k=500}$. Each subplot shows performances colormaps (without interpolation) of $\log_{10}(ERT/d)$, for one DE strategy, with the 2 other DE parameters F and CR on the axes. The true optimal configurations are plotted as white stars ☆ and the Empirical Optimal Configurations as white small circles ∘.

[8] Additional plots are available at https://drive.google.com/open?id=0B9GuQcCjvwtFdkotR1h1N3dlOG8.

lies in a region of very poor true ERT: the performance of these configurations used within DE for the corresponding function will be poor too (see forthcoming Sect. 6.2).

6.2 Empirical Optimal Configurations at Work

For each computed EPM (described in Sect. 5), an Empirical Optimal Configuration is obtained by optimization on the parameter space, and the ERT of this configuration is obtained by running DE on each of the 15 variants of the target function. Table 1 shows, for each function, the ratios of the ERTs of these different configurations against the *robust configuration* defined in Sect. 5 (the smaller the better). The first two columns are the other baselines, θ^d are the parameters values recommended by DE authors, and θ^L the adhoc configuration with best results (see Sect. 5). The two series of 3 columns correspond to the 3 different feature sets $\psi^{\bullet}_{500}$, $\psi^{\bullet}_{2000}$, and $\psi^{\star}$ (as defined in Sect. 4) and the two types of dimension handling (learning performed only on the same dimension as testing, or on all dimensions at once — see Sect. 5).

As it could be expected, the adhoc configuration is a clear winner, and the default values recommended by DE authors a clear loser. The results of dimension 5 (as well as those in dimensions 2 and 3, not shown due to space constraints)

Table 1. Percentages of the ERTs of different Empirical Optimal Configurations w.r.t. that of the *robust configuration* defined in Sect. 5, for dimensions 5 and 10. See text for details. Best results are printed in blue bold face when smaller than 100, in red italic when larger than 100; Worst results are printed in light gray; the $\star$ symbol indicates that the *robust configuration* never reached the target, and was artificially attributed an ERT of 15 times the maximum budget of one run, and ∞ indicates that the corresponding Empirical Optimal Configuration never reached the target.

g	θ^d	θ^L	train on dim 5			train on all dims		
			$\psi^{\bullet}_{500}$	$\psi^{\bullet}_{2k}$	$\psi^{\star}$	$\psi^{\bullet}_{500}$	$\psi^{\bullet}_{2k}$	$\psi^{\star}$
F1	156	*17*	34	29	44	**26**	**26**	2.78
F2	153	*27*	57	**54**	**54**	57	**54**	240
F3	2415	*25*	158	639	*123*	158	129	∞
F4	561	*08*	97	97	**93**	97	97	∞
F5	139	*28*	214	232	**48**	54	54	∞
F6	145	*25*	71	**42**	49	56	71	486
F7	123	*37*	78	84	**62**	78	**62**	∞
F8	146	*29*	57	78	**36**	57	109	190
F9	114	*28*	48	66	**41**	98	73	133
F10	120	*27*	**35**	**35**	53	**35**	**35**	∞
F11	120	*29*	**30**	32	**30**	**30**	**30**	134
F12	113	*39*	171	137	**55**	118	137	1431
F13	118	*28*	**52**	91	∞	**52**	225	460
F14	119	*26*	∞	255	∞	∞	**85**	240

Dimension 5

g	θ^d	θ^L	train on dim 10			train on all dims		
			$\psi^{\bullet}_{500}$	$\psi^{\bullet}_{2k}$	$\psi^{\star}$	$\psi^{\bullet}_{500}$	$\psi^{\bullet}_{2k}$	$\psi^{\star}$
F1	94	*4.9*	**12**	14	42	**1.2**	**1.2**	3.4
F2	100	*7.7*	39	71	179	45	71	**32**
F3$\star$	100	*0.2*	∞	∞	**1**	∞	∞	∞
F4$\star$	10	*0.3*	∞	∞	**2**	∞	∞	∞
F5$\star$	10	*0.2*	∞	∞	∞	∞	∞	∞
F6	157	*12.9*	33	**29**	46	56	82	56
F7	99	*0.22*	**24**	35	34	35	35	∞
F8	114	*17*	∞	∞	∞	∞	∞	**29**
F9	107	*15.4*	∞	28	29	27	29	**17**
F10	107	*12.9*	**17**	**17**	20	**17**	28	226
F11	114	*9.7*	**15**	16	**15**	**15**	16	16
F12	429	*7.8*	∞	21	**10**	∞	**10**	63
F13$\star$	10	*0.1*	∞	2	∞	∞	∞	**1**
F14	100	*14.4*	∞	∞	**16**	∞	∞	28

Dimension 10,

bring several good news: most proposed approaches perform better than the robust configuration, and at least one does, for all functions but F3. For some functions (F11, and also F8 and F10), some feature-based approaches even get close to the best adhoc configurations, never being worse than twice that best performance, except for F4 — as could be foreseen on Fig. 1b. When it comes to compare the different EPM settings, learning only from the single target dimension gives better results than learning for all dimensions together — and in the former case, using all available features does improve over only using the cheap features.

The situation is not so clear in dimension 10: in several cases, the Empirical Optimal Configuration cannot even reach the target in the allocated budget — a situation for which an example was given in Fig. 1d. However, when an optimum can be found, similar conclusions to the dimension 5 case can be drawn, though not as contrasted.

6.3 Discussion

Our results suggest that the learned EPM can be similar to the actual performance map, at least in a large part of the parameter space. Nonetheless, when a particular feature is not included in the learning set, it can be very hard for the EPM to achieve good accuracy and performance prediction, as witnessed with function F4 (Fig. 1b): only F3 and F4 are multi-modal functions, and they have very different structures (apart from being separable, and hence belong to the first BBOB class).

This was clear, too, with the F5 function (Fig. 1d): F5 is the only linear function of the test bench, hence the EPM was learned without any linear function in the training set. However, the global accuracy of the EPM for F5 is rather good — not worse than F1 for instance (Fig. 1a). But unfortunately, the small region of the parameter space where the EPM differs from the true ERT is the region that contains the optimal configuration.

This clearly demonstrates that a good accuracy over the parameter space of the EPM w.r.t. the true performances, such as the one being optimized by the learning algorithm (Random Forests here), is not required to reach the ultimate goal of the Algorithm Configuration process — find a quasi-optimal configuration for unknown instances. The only important property of the EPM is to be able to robustly identify good-performing regions of the parameter space. This opens several new possible research paths. At the level of the learning algorithm, the best regions of the parameter space could be weighted more than other parts of the space; at the extreme, rank-based learning could be used rather than regression of ERT values. At the level of the sampling, only good configurations could be used — e.g., the configurations encountered while running SMAC to find the true optimal configuration.

No clear differences can be seen between EPM learned using $\psi^{\bullet}_{500}$ or $\psi^{\bullet}_{2000}$, except for some functions, in dimension 10, where EPMs learned with $\psi^{\bullet}_{2000}$ solve the function while those learned from $\psi^{\bullet}_{500}$ don't. On the other hand, using the full set of features ψ^* does help, both in dimension 5 for the dimension-specific models, and in dimension 10 where it succeeds in reaching the target precision where the other models fail (e.g., F3 and F4, the only multi-modal functions of the test bench). While not surprising, this demonstrates that both the training set and the set of features should cover all the foreseeable difficulties of the unknown forthcoming instances. Any limited test bench (including BBOB) might hence be insufficient to learn a general-purpose configurator.

Finally, the fact that learning the EPM for a specific dimension leads to better results was to be expected. While this makes difficult to build an universal EPM, it does not prevent from any practical use of this approach, as the dimension is usually known (and constant) in most real world applications.

7 Conclusion

This paper has investigated the computation and use of an Empirical Performance Model (EPM) in the context of continuous black box optimization and has demonstrated that it is possible to learn a reasonable approximation of the real performance. More importantly, it was demonstrated that an efficient parameter configuration can be extracted from the learned EPM by optimizing the predicted performance, given a set of features on a new unknown function. In particular, it was possible to obtain empirical configurations that outperform a static parameter setting optimized for an average performance, over the whole test bench at the same overall cost. However, some open issues remain related to the robustness of the results, and deeper analyses are necessary to better understand (and avoid) some rare cases where the approach fails.

Several paths for further research are suggested by this work, both at the level of the learning algorithm and of the sampling of the parameter search space, as discussed in Sect. 6.3. A promising direction is to embed the EPM as a parameter control mechanism within the optimization process itself, assuming that the features can be efficiently approximated using a rather small number of samples, (e.g., w.r.t. an approximation of the objective function, as proposed in [2]). This would open a new perspective on the on-line parameter tuning grail.

References

1. Auger, A., Teytaud, O.: Continuous lunches are free plus the design of optimal optimization algorithms. Algorithmica (2009). https://hal.inria.fr/inria-00369788
2. Belkhir, N., Dréo, J., Savéant, P., Schoenauer, M.: Surrogate assisted feature computation for continuous problems. In: Proceedings of LION 10 (2016, to appear). https://hal.archives-ouvertes.fr/hal-01303320
3. Bischl, B., Mersmann, O., Trautmann, H., Preuß, M.: Algorithm selection based on exploratory landscape analysis and cost-sensitive learning. In: Proceedings of the 14th Annual Conference on Genetic and Evolutionary Computation, pp. 313–320. ACM (2012)

4. Bossek, J., Bischl, B., Wagner, T., Rudolph, G.: Learning feature-parameter mappings for parameter tuning via the profile expected improvement. In: Proceedings of the 2015 on Genetic and Evolutionary Computation Conference, pp. 1319–1326. ACM (2015)
5. Hansen, N., Auger, A., Finck, S., Ros, R.: Real-parameter black-box optimization benchmarking 2010: experimental setup. Technical report, RR-7215, INRIA (2010)
6. Hoos, H.H.: Programming by optimization. Commun. ACM **55**(2), 70–80 (2012)
7. Hutter, F., Hamadi, Y., Hoos, H.H., Leyton-Brown, K.: Performance prediction and automated tuning of randomized and parametric algorithms. In: Benhamou, F. (ed.) CP 2006. LNCS, vol. 4204, pp. 213–228. Springer, Heidelberg (2006)
8. Hutter, F., Hoos, H.H., Leyton-Brown, K.: Sequential model-based optimization for general algorithm configuration. In: Coello, C.A.C. (ed.) LION 5, 2011. LNCS, vol. 6683, pp. 507–523. Springer, Heidelberg (2011)
9. Hutter, F., Xu, L., Hoos, H.H., Leyton-Brown, K.: Algorithm runtime prediction: methods & evaluation. Artif. Intell. **206**, 79–111 (2014)
10. Leyton-Brown, K., Nudelman, E., Shoham, Y.: Learning the empirical hardness of optimization problems: the case of combinatorial auctions. In: Hentenryck, P. (ed.) CP 2002. LNCS, vol. 2470, pp. 556–572. Springer, Heidelberg (2002)
11. Lunacek, M., Whitley, D.: The dispersion metric and the cma evolution strategy. In: Proceedings of the 8th GECCO, pp. 477–484. ACM (2006)
12. Mersmann, O., Bischl, B., Trautmann, H., Preuss, M., Weihs, C., Rudolph, G.: Exploratory landscape analysis. In: Proceedings of the 13th GECCO, pp. 829–836. ACM (2011)
13. Munoz, M., Kirley, M., Halgamuge, S.K., et al.: Exploratory landscape analysis of continuous space optimization problems using information content. IEEE Trans. Evol. Comput. **19**(1), 74–87 (2015)
14. Muñoz, M.A., Kirley, M., Halgamuge, S.K.: A meta-learning prediction model of algorithm performance for continuous optimization problems. In: Coello, C.A.C., Cutello, V., Deb, K., Forrest, S., Nicosia, G., Pavone, M. (eds.) PPSN 2012, Part I. LNCS, vol. 7491, pp. 226–235. Springer, Heidelberg (2012)
15. Rice, J.R.: The algorithm selection problem. Adv. Comput. **15**, 65–118 (1976)
16. Storn, R., Price, K.: Differential evolution-a simple and efficient heuristic for global optimization over continuous spaces. J. Glob. Optim. **11**(4), 341–359 (1997)
17. Wolpert, D., Macready, W.: No free lunch theorems for optimization. IEEE Trans. Evol. Comput. **1**(1), 67–82 (1997)
18. Xu, L., Hutter, F., Hoos, H.H., Leyton-Brown, K.: Satzilla: portfolio-based algorithm selection for sat. J. Artif. Intell. Res. 565–606 (2008)

An Asynchronous and Steady State Update Strategy for the Particle Swarm Optimization Algorithm

C.M. Fernandes[1,2(✉)], J.J. Merelo[2], and A.C. Rosa[1]

[1] LARSyS: Laboratory for Robotics and Systems in Engineering and Science, University of Lisbon, Lisbon, Portugal
{cfernandes, acrosa}@laseeb.org

[2] Department of Computer Architecture, University of Granada, Granada, Spain
jmerelo@geneura.ugr.es

Abstract. This paper proposes an asynchronous and steady state update strategy for the Particle Swarm Optimization (PSO) inspired by the Bak-Sneppen model of co-evolution. The model consists of a set of fitness values (representing species) arranged in a network. By replacing iteratively the least fit species and its neighbors with random values (simulating extinction), the average fitness of the population tends to grow while the system is driven to a critical state. Based on these rules, we implement a PSO in which only the worst particle and its neighbors are updated and evaluated in each time-step. The other particles remain *steady* during one or more iterations, until they eventually meet the update criterion. The steady state PSO (SS-PSO) was tested on a set of benchmark functions, with three different population structures: *lbest* ring and lattice with von Neumann and Moore neighborhood. The experiments demonstrate that the strategy significantly improves the quality of results and convergence speed with Moore neighborhood. Further tests show that the major factor of enhancement is the selective pressure on the worst, since replacing the best or a random particle (and neighbors) yields inferior results.

1 Introduction

The Particle Swarm Optimization (PSO) is a population-based metaheuristics inspired by the social behavior of bird flocks and fish schools [8]. The search is carried out by a swarm of candidate solutions (called *particles*) that move around the fitness landscape of the target-problem, guided by mathematical rules that define their velocity at each time step. The most common configurations of PSO are synchronous: the fitness values of all particles are first computed and only then the particles update their velocity. Carlisle and Dozier [5] proposed a variant in which the velocity vector is updated immediately after computing the fitness of the corresponding particle. In this case, each particle is updated knowing the current best position found by half of its neighbors and the previous best found by the other half: the population of the asynchronous PSO (A-PSO) interacts with imperfect information about the global search. Asynchronous PSOs have been compared to the synchronous configuration (S-PSO) with contradictory results. While Carlisle and Dozier [5] suggested that A-PSO yields better results

J. Handl et al. (Eds.): PPSN XIV 2016, LNCS 9921, pp. 167–177, 2016.
DOI: 10.1007/978-3-319-45823-6_16

than S-PSO, Rada-Vilela *et al.* [13] reported that S-PSO is better than A-PSO in terms of the quality of the solutions and convergence speed.

One of the main motivations for investigating asynchronous update strategies for PSO is the possibility of parallelization [14]. Standard PSOs are easily parallelized (by assigning a particle or a set of particles to each processor, for instance) but due to load imbalances, synchronous update does not make an efficient use of the computational resource. For parallel PSO, asynchronicity is the logical approach. In addition, asynchronicity can also be useful in diversity maintenance and prevention of premature convergence [1], or to speed up convergence by skipping function evaluations [13]. In this paper, we follow an alternative approach. The goal is to design an asynchronous PSO that, unlike the standard A-PSO, significantly improves S-PSO in a wide range of problems. With that objective in mind, we propose a steady state PSO (SS-PSO). A system is said to be in steady state when some of its parts do not change for a period of time. In the SS-PSO, only a fraction of the population is updated and evaluated in each iteration.

The strategy is inspired by the Bak-Sneppen model of co-evolution between interacting species [3]. In order to investigate the dynamics of species extinction and coupled selection, Bak and Sneppen arrange a set of random fitness values (representing species) in a ring structure. Then, they replace the worst species and its neighbors by random values (extinction event), repeating the procedure during several iterations. After a long run, the system is driven to a critical state where most species have reached a fitness above a certain threshold and avalanches of extinction events produce non-equilibrium fluctuations in the configuration of the fitness values.

SS-PSO uses a similar scheme. However, here the worst particle and its neighbors are updated, instead of being replaced by random solutions. Like in the Bak-Sneppen model, the other particles remain steady until an update event hits them. For a proof of concept, the algorithm was tested on ten benchmark functions and compared to S-PSOs. The results show that SS-PSO significantly improves the performance of the S-PSO structured in a 2-dimensional square lattice with Moore neighborhood.

The remaining of the paper is structured as follows. Section 2 gives a background review on synchronous and asynchronous update strategies for the PSO. Section 3 describes the Bak-Sneppen model of co-evolution and introduces the proposed update strategy. Section 4 describes the experiments and discusses the results. Finally, Sect. 5 concludes the paper and outlines futures lines of research.

2 Synchronous and Asynchronous Particle Swarms

PSO is a population-based algorithm in which a group of solutions travels through a fitness landscape according to a set of rules that drives it towards optimal regions of the space. The algorithm is described by a simple set of equations that define the velocity and position of each particle. The position vector of the *i-th* particle is given by $\vec{X}_i = (x_{i,1}, x_{i,2}, \ldots x_{1,D})$, where D is the dimension of the search space. Velocity is given by $\vec{V}_i = (v_{i,1}, v_{i,2}, \ldots v_{1,D})$. The particles are evaluated with a fitness function $f(\vec{X}_i)$ in each time step and then their positions and velocities are updated by:

$$v_{i,d}(t) = \omega v_{i,d}(t-1) + c_1 r_1 (p_{i,d} - x_{i,d}(t-1)) + c_2 r_2 (p_{g,d} - x_{i,d}(t-1)) \quad (1)$$

$$x_{i,d}(t) = x_{i,d}(t-1) + v_{i,d}(t) \quad (2)$$

where p_i is the best solution found so far by particle i and p_g is the best solution found so far by the neighborhood. Parameters r_1 and r_2 are vectors of random numbers uniformly distributed in the range $[0, 1]$ and c_1 and c_2 are acceleration coefficients that tune the relative influence of each term of the formula. In order to prevent particles from stepping out of the limits of the search space, positions $x_{i,d}(t)$ of the particles are limited by constants that in general correspond to the domain of the problem: $x_{i,d}(t) \in [-Xmax, Xmax]$. Velocity may also be limited within a range in order to prevent the *explosion* of the velocity vector: $v_{i,d}(t) \in [-Vmax, Vmax]$. Usually, $Xmax = Vmax$. Parameter ω is the inertia weight, proposed by Shi and Eberhart [17] to help fine-tuning the balance between local and global search, and it is widely used in PSO implementations.

The neighborhood of the particle defines the value of p_g and is a key factor in the performance of PSO. Most of the PSOs use one of two simple sociometric principles for defining the neighborhood network. One connects all the members of the swarm to one another, and it is called *gbest* (or *star*), where g stands for *global*. The degree of connectivity of *gbest* is $k = n$, where n is the number of particles. The other typical configuration, called *lbest* (where l stands for local), creates a neighborhood that comprises the particle itself and its k nearest neighbors. The most common *lbest* topology is the *ring* structure, in which the particles are arranged in a ring (resulting in a degree of connectivity $k = 3$, including the particle).

Between the $k = 3$ connectivity of *lbest* ring and $k = n$ of *gbest*, there are several possibilities. Two of the most used are the 2-dimensional square lattices with von Neumann and Moore neighborhood. In [9], Kennedy and Mendes tested several social structures and concluded that when they are ranked by the quality of solutions the structures with $k = 5$ (like the von Neumann lattice) perform better, but when ranked according to the number of iterations needed to meet the criteria, configurations with higher degree of connectivity (like Moore neighborhood, with $k = 9$) perform better. These results are consistent with the premise that low connectivity favors robustness, while higher connectivity favors convergence speed (at the expense of reliability).

In the standard PSO, all particles are evaluated before updating their velocity. Therefore, they move with complete information about the state of the search. In the asynchronous variant, each particle is evaluated immediately after being updated. Independently of the social structure (assuming it is regular), A-PSO particles use the current best position found by half of its neighbors and the previous best found by the other half: the particles are guided by partial or imperfect information.

A-PSO was first discussed by Carlisle and Dozier [5]. Several reports claim that A-PSO outperforms S-PSO. Luo and Zhang [12], for instance, compared the algorithms and concluded that A-PSO is more accurate and faster. However, they tested the algorithms in only two functions and no statistical test is given. Perez and Basterrechea [15] tested the algorithm on six problems and concluded that A-PSO is faster and as accurate as S-PSO. Rada-Vilela et al. [16] compared S-PSO and A-PSO with ten functions, using a ring structure with number of neighbors k ranging from 2 to 30 and a population of 30 particles. They measured the quality of the solutions and speed of convergence and performed statistical tests on the results, concluding that S-PSO yields better results than A-PSO in unimodal functions. As for the multimodal, S-PSO yields similar or better results. These findings contradict the results of Carlisle and Dozier [5], Luo and Zhang [12] and Perez and Basterrechea [15].

As stated above, parallelization is one of the main motivations for investigating asynchronous PSOs, mainly because synchronous parallel implementations do not make an efficient use of computational resources when load imbalance exists. In this line of work, Koh *et al.* [10] compared parallel asynchronous and synchronous PSOs in homogeneous and heterogeneous environments. They concluded that the parallel performance of the asynchronous version is significantly better than that of asynchronous PSO for heterogeneous environments or heterogeneous computational tasks. Venter and Sobieszczanski-Sobieski [18] also studied and compared parallel synchronous and asynchronous PSOs. Their results indicate that the asynchronous PSO significantly outperforms the synchronous in terms of parallel efficiency.

In this paper, we are not yet concerned with parallelization. We follow a different approach from the works based on Carlisle and Dozier's seminal proposal. Our main objective is to evaluate the numerical results of the algorithm and validate it as an alternative to the standard synchronous approach. The strategy is based on a model of co-evolution that is described in the next section.

3 From a Model of Co-evolution to the Steady State PSO

Natural species in the same eco-system are related through several features (like food chains or symbiosis, for instance) and the extinction of one species affects the species that are related to them, in a chain reaction that can reach large proportions. Fossil records suggest that the size of extinctions events is in power-law proportion to its frequency. It is also known that the biological history of life on Earth is punctuated by catastrophic extinction events. The Bak-Sneppen model [3] was conceived with the objective of understanding the mechanisms underlying these mass extinctions. It consists of a number of species, each one with a fitness value assigned and connected to other species (neighbors). Every time step, the least fit species and its neighbors are eliminated from the system and replaced by individuals with random fitness.

This description may be translated to a mathematical model. The system is defined by n^d fitness numbers f_i arranged on a d-dimensional lattice (ecosystem) with n cells. At each time step, the smallest f value and its $2 \times d$ neighbours are replaced by uncorrelated random values drawn from a uniform distribution. With this simple rule applied iteratively, the system is driven to a critical state where most species have a fitness above a certain threshold. Complex behavior is observed even in the 1-dimensional case, where species are arranged in a ring and each one has two neighbors.

The Bak-Sneppen model is an example of a system with self-organized criticality (SOC) [2], a critical state formed by self-organization in a long transient period at the border of order and chaos. While *order* means that the system is working in a predictable regime where small disturbances have only local impact, *chaos* is an unpredictable state very sensitive to initial conditions or small disturbances. In complex adaptive systems, complexity and self-organization usually arise at that transition region between order and chaos, or *edge of chaos*, as it is sometimes stated. SOC systems are dynamical with a critical point at the region between order and chaos as an attractor. However, and unlike many physical systems, which have a parameter that needs to be tuned in order to obtain the critical state, SOC systems are able to self-tune to the critical point.

SOC and the Bak-Sneppen model inspired, for instance, a metaheuristic called *extremal optimization* (EO) [4]. In EO, a single solution to a problem is modified by local search. The algorithm removes the worst components of the solution and replaces them with randomly generated material. By plotting the fitness of the solution, distinct stages of evolution are observed, where improvement is disturbed by brief periods of dramatic decrease in the quality of the solution. Chen *et al.* [6] used EO to enhance the search abilities of PSO and prevent premature convergence to local optima. They tested the hybrid algorithm on a set of benchmark functions and compared it favorably with other metaheuristics.

Løvbjerg and Krink [11] applied SOC to PSO in order to control the convergence of the algorithm and maintain diversity. The authors introduce a *critical value* associated with each particle and define a rule that increments it when two particles are closer than a *threshold distance*. When the critical value of a particle exceeds a globally set *criticality limit*, the algorithm responds by dispersing the criticality of the particle within a certain surrounding neighborhood. In addition, the algorithm uses the critical value to control the inertia weight. The authors claim that their method is faster and attains better solutions than the standard PSO. However, the algorithm introduces five parameters that must be tuned or set to constant *ad hoc* values.

More recently, Fernandes et al. [7] used the Bak-Sneppen model to control the inertia weight and acceleration coefficients of each particle. An experimental setup demonstrates the validity of the algorithm and shows that the incorporation of each control mechanism improves its performance or at least reduces the tuning effort.

Like the Bak-Sneppen model, the population of PSO is structured by a network. With this likeness in mind, we devised an asynchronous and steady state update strategy for PSO in which only the least fit particle and its neighbors are updated and evaluated in each time step. The neighborhood is defined by the network: if the particles are connected by *lbest* with $k = 3$, only the worst particle and its two nearest neighbors are updated and evaluated; if a lattice with Moore neighborhood is used ($k = 9$), the least fit and its eight nearest neighbors are updated. Please note that local synchronicity is used here: the fitness values of the worst and its neighbors are first computed and only then their velocity is updated. For the remaining working mechanisms and parameters, the algorithm is exactly as standard PSO. Since part of the population remains steady in each time step, we named the algorithm steady state PSO (SS-PSO). SS-PSO is summarized in Algorithm 1.

1.Initialize velocity and position of each particle.
2.For (each particle j):Compute fitness.
3. For (each particle j):Compute $p_i$and p_g.
4. For (each particle j):if jis the least fit particle, update velocity and position of jand neighbors
5. Compute fitness of particles jand neighbors.
6. If (stop criteria not met) return to 3; else, end.

Algorithm 1: SS-PSO

4 Experiments and Results

The experimental setup was constructed with ten benchmark problems (Table 1). Functions f_1, -f_3 are unimodal; f_4, -f_8 are multimodal; f_9 is the shifted f_2 with noise and f_{10} is the rotated f_5 (f_9 global optimum and f_{10} matrix were taken from the CEC2005 benchmark). The dimension of the search space is $D = 30$ (except f_6, with $D = 2$). In order to construct square lattices with von Neumann and Moore neighborhood, population size μ is set to 49, a value that lies within the typical range [8]. Following [16], c_1 and c_2 were set to 1.4962 and ω to 0.7298. *Xmax* is defined as usual by the domain's upper limit and *Vmax* = *Xmax*. A total of 50 runs for each test were performed. Asymmetrical initialization is used (initialization ranges are in Table 1).

In order to assess the quality of solutions and convergence speed of the algorithms, two sets of experiments were conducted. First, the algorithms were run for a limited amount of iterations (3000 for f_1, f_3 and f_6, 20000 for the remaining) and the fitness of the best solutions found were recorded over the 50 runs. In the second set of experiments the algorithms were all run for 20000 iterations or until reaching a function-specific stop criterion (given in Table 1). The number of iterations required to meet the criterion was recorded and statistical measures were taken over the 50 runs. A success measure was defined as the number of runs in which an algorithm attains the stop criterion. The experimental setup is similar to those in [9, 16].

SS-PSO and S-PSO were implemented with three topologies: *lbest* with $k = 3$ and 2-dimensional square lattices with von Neumann ($k = 5$) and Moore neighborhood ($k = 9$). *Gbest* was not tested for two reasons. Firstly, it is fast but converges often to local optima. We have performed some tests with *gbest* and the success rates were very poor. Furthermore, SS-PSO uses the neighborhood structure to decide which particles to update, i.e., in the von Neumann ($k = 5$), five particles are updated. Since *gbest* has $k = n$, the proposed strategy would update the entire population and be equivalent to the S-PSO. Hence, we have restricted the tests to *lbest*, von Neumann and Moore. Please note that at this point of the research we are not primarily concerned in comparing the update strategy with state of the art PSOs. First, it is necessary to investigate in which situations the proposed algorithm is able to improve the convergence speed and accuracy of standard PSO and understand its underlying mechanisms. Only after the proof of concept we can compare it against other PSOs.

With *lbest*, the steady state strategy could not improve the standard synchronous update: SS-PSO$_{\text{lbest}}$ yields worse results than S-PSO$_{\text{lbest}}$ in most of the functions. As for the von Neumann neighborhood, the results are dual: SS-PSO$_{\text{VN}}$ yields better results in multimodal functions but it is outperformed in the unimodal by the S-PSO$_{\text{VN}}$.

Due to space restrictions, we omit the numerical results of the PSOs with *lbest* and von Neumann network and proceed to the PSOs with Moore neighborhood.

S-PSO$_{\mathrm{Moore}}$ attained the best results in most of the functions when compared to *lbest* and it is faster (considering both mean and median values of the evaluations required to meet the criteria) in every function. When compared to S-PSO$_{\mathrm{VN}}$, S-PSO$_{\mathrm{Moore}}$ is also faster in every function and attains better mean and median fitness values in unimodal functions. These results are consistent with the ones in Kennedy and Mendes [9]. Therefore, we believe that the Moore neighborhood structure is well suited for assessing the validity and relevance of our proposal.

Table 2 shows mean, standard deviation and statistical measures of the empirical distributions of best fitness values attained by S-PSO$_{\mathrm{Moore}}$ and SS-PSO$_{\mathrm{Moore}}$. The later yields better results in most of the functions: it attains lower mean and median fitness values in every unimodal function and in the multimodal f_4, f_6 and f_7. Mann-Whitney U tests were performed to compare the distribution of fitness values of each algorithm in each function. Results of the tests are significant at $p \leq 0.05$ for $f_1, f_2, f_3, f_4, f_6, f_7, f_9$, i.e., the null hypothesis that the two samples come from the same population is rejected. For the remaining functions (f_5, f_8, f_{10}), the null hypothesis is not rejected.

In terms of function evaluations (Table 3), SS-PSO$_{\mathrm{Moore}}$ is faster in the entire set of unimodal problems. In the multimodal problems, SS-PSO$_{\mathrm{Moore}}$ needs less evaluations in f_5, f_6, f_6 and f_8. Results of Mann-Whitney U tests are significant at $p \leq 0.05$ for functions $f_1, f_2, f_3, f_5, f_7, f_8, f_{10}$. Although S-PSO$_{\mathrm{Moore}}$ requires less evaluations in f_4, the result of the statistical test is not significant. Finally, the success rates (Table 3) are similar, except for f_7, in which SS-PSO clearly outperforms the synchronous version, and f_9. In conclusion: the empirical results, together with the statistical tests, show that SS-PSO$_{\mathrm{Moore}}$ outperforms S-PSO$_{\mathrm{Moore}}$ in most of the functions according to accuracy, speed and reliability, while not being outperformed in any case.

The previous tests demonstrate that the steady state update strategy in a PSO structured with Moore neighborhood significantly improves its performance. However, at this point, a question arises: what is the major factor for the performance enhancement, the steady state approach, or the set of particles that are updated? In order to shed light on this issue, a final test was conducted. Two variants of SS-PSO were implemented: one updates the best particle and its neighbors (*replace-best*); the second updates a randomly selected particle and its neighbors (*replace-random*). The algorithms were tested on the set of benchmark functions (see Table 4) and compared to results of the proposed SS-PSO$_{\mathrm{Moore}}$ (or *replace-worst*) given in Tables 2 and 3.

Replace-best update strategy is clearly inferior to *replace-worst*. With the exception of f_1 and f_3, the quality of solutions is degraded when compared to the proposed SS-PSO and even to S-PSO. Success rates are considerably lower in most functions. As for *replace-random*, it improves S-PSO in some functions, but in general it is not better than *replace-worst*: *replace-random* strategy is less accurate and slower in most of the functions. The test shows that selective pressure on the least fit individuals is a major factor in the performance SS-PSO.

Table 1. Benchmark functions.

	Mathematical representation	Range of search/initialization	Stop criterion
Sphere f1	$f_1(\vec{x}) = \sum_{i=1}^{D} x_i^2$	$(-100, 100)^D$ $(50, 100)^D$	0.01
Quadric f2	$f_2(\vec{x}) = \sum_{i=1}^{D} \left(\sum_{j=1}^{i} x_j \right)^2$	$(-100, 100)^D$ $(50, 100)^D$	0.01
Hyper Ellipsoid f3	$f_1(\vec{x}) = \sum_{i=1}^{D} i x_i^2$	$(-100, 100)^D$ $(50, 100)^D$	0.01
Rastrigin f4	$f_4(\vec{x}) = \sum_{i=1}^{D} \left(x_i^2 - 10\cos(2\pi x_i) + 10 \right)$	$(-10, 10)^D$ $(2.56, 5.12)^D$	100
Griewank f5	$f_5(\vec{x}) = 1 + \frac{1}{4000} \sum_{i=1}^{D} x_i^2 - \prod_{i=1}^{D} \cos\left(\frac{x_i}{\sqrt{i}} \right)$	$(-600, 600)^D$ $(300, 600)^D$	0.05
Schaffer f6	$f_6(\vec{x}) = 0.5 + \frac{\left(\sin \sqrt{x^2 + y^2} \right)^2 - 0.5}{(1.0 + 0.001(x^2 + y^2))^2}$	$(-100, 100)^2$ $(15, 30)^2$	0.00001
Weierstrass f7	$f_7(\vec{x}) = \sum_{i=1}^{D} \left(\sum_{k=0}^{kmax} \left[a^k cos\left(2\pi b^k (x_i + 0.5) \right) \right] \right)$ $-D \sum_{k=0}^{kmax} \left[a^k cos\left(2\pi b^k \cdot 0.5 \right) \right],$ $a = 0.5,\ b = 3,\ kmax = 20$	$(-0.5, 0.5)^D$ $(-0.5, 0.2)^D$	0.01
Ackley f8	$f_8(\vec{x}) = -20exp\left(-0.2\sqrt{\frac{1}{D} \sum_{i=1}^{D} x_i^2} \right)$ $-exp\left(\frac{1}{D} \sum_{i=1}^{D} \cos(2\pi x_i) \right) + 20 + e$	$(-32.768, 32.768)^D$ $(2.56, 5.12)^D$	0.01
Shifted Quadric with noise f9	$f_9(\vec{z}) = \sum_{i=1}^{D} \left(\sum_{j=1}^{i} x_j \right)^2 * (1 + 0.4\lvert N(0.1) \rvert),$ $\vec{z} = \vec{x} - \vec{o},$ $\vec{o} = [o_1, ..o_D]\ :\ shifted\ global\ optimum$	$(-100, 100)^D$ $(50, 100)^D$	0.01
Rotated Griewank f10	$f_{10}(\vec{z}) = 1 + \frac{1}{4000} \sum_{i=1}^{D} z_i^2 - \prod_{i=1}^{D} \cos\left(\frac{z_i}{\sqrt{i}} \right),$ $\vec{z} = \boldsymbol{M}\vec{x}$, **M**:ortoghonal matrix	$(-600, 600)^D$ $(300, 600)^D$	0.05

Table 2. Best fitness: mean, standard deviation, median, minimum and maximum.

	$S\text{-}PSO_{Moore}$					$SS\text{-}PSO_{Moore}$				
	Mean	St.dev	Median	Min	Max	Mean	St.dev.	Median	Min	Max
f_1	1.31e−11	1.12e−11	1.04e−11	1.39e−12	4.93e−11	**1.14e−14**	1.16e−14	**7.83e−15**	**2.06e−16**	**6.33e−14**
f_2	1.38e−29	6.36e−29	5.88e−31	4.71e−34	4.42e−28	**1.40e−46**	9.90e−46	**0.00e00**	**0.00e00**	**7.00e−45**
f_3	3.61e−11	3.50e−11	2.76e−11	2.64e−12	1.65e−10	**3.57e−14**	5.32e−14	**1.58e−14**	**9.43e−16**	**2.56e−13**
f_4	6.36e+01	1.73e+01	6.17e+01	3.78e+01	1.13e+02	**5.25e+01**	1.45e+01	**5.12e+01**	**2.19e+01**	**1.04e+02**
f_5	**5.86e−03**	7.22e−03	**1.08e−19**	**0.00e00**	2.95e−02	1.28e−02	2.06e−02	9.86e−03	**0.00e00**	1.20e−01
f_6	1.94e−04	1.37e−03	**0.00e00**	**0.00e00**	9.72e−03	**0.00e00**	0.00e00	**0.00e00**	**0.00e00**	**0.00e00**
f_7	1.94e−01	5.27e−01	2.27e−02	**2.86e−05**	3.16e00	**1.73e−02**	8.17e−02	**2.86e−05**	**2.86e−05**	**5.19e−01**
f_8	**1.01e−15**	2.01e−16	**8.88e−16**	**8.88e−16**	1.33e−15	1.07e−15	2.20e−16	**8.88e−16**	**8.88e−16**	**1.33e−15**
f_9	3.60e+01	2.32e+02	9.8e−05	6.44e−07	1.64e+03	**7.20e−05**	1.54e−04	**1.01e−05**	1.73e−08	7.11e−04
f_{10}	**6.70e−03**	9.20e−03	**1.08e−19**	**0.00e00**	3.68e−02	8.37e−03	1.01e−02	**1.08e−19**	0.00e00	3.70e−02

Table 3. Evaluations: mean, standard deviation, median, minimum and maximum.

	$S\text{-}PSO_{Moore}$						$SS\text{-}PSO_{Moore}$					
	Mean	St.dev.	Median	Min	Max	SR	Mean	St.dev.	Median	Min	Max	SR
f_1	20434.0	840.8	20433. 0	18326	22099	50	**17241.3**	716.2	**17320.5**	**14526**	**18594**	50
f_2	168599.0	12721.1	168119.0	140630	193501	50	**133140.6**	16854.2	**135828.0**	**105399**	**171702**	50
f_3	22987.9	1075.4	22956.5	20972	26019	50	**19519.6**	788.0	**19561.5**	**18045**	**21600**	50
f_4	**15635.0**	7771.5	**13524.0**	**7448**	49392	49	15902.8	8047.7	14256.0	7659	58248	49
f_5	18671.0	986.8	18595.5	16366	21952	50	**16419.2**	1300.7	**16060.5**	**14607**	**19683**	50
f_6	11443.0	9439.1	7105.0	3822	39788	49	**8049.0**	4852.6	**6381.0**	**2727**	**21744**	50
f_7	37272.7	1590.1	36970.5	34790	41846	24	**33192.0**	1184.8	**33340.5**	**30645**	**35685**	46
f_8	21029.8	1164.7	20923.0	19012	24794	50	**17723.6**	957.0	**17752.5**	**15750**	**19809**	50
f_9	704144.6	96262.6	706972	453201	922327	47	**653808.8**	95860.3	**671175**	**425655**	852786	50
f_{10}	18876.8	901.7	18963	**16954**	**20727**	**50**	**16140.8**	1122.9	**15975**	**13995**	**18612**	**50**

Table 4. Results of SS-PSO variants: median, min, max and success rates (SR)

	$SS\text{-}PSO_{Moore}$ (replace-best)							$SS\text{-}PSO_{Moore}$ (replace-random)						
	Fitness			Evaluations				Fitness			Evaluations			
	Median	Min	Max	Median	Min	Max	SR	Median	Min	Max	Median	Min	Max	SR
f_1	4.09e−29	2.50e−33	2.00e +04	9468	6714	24669	45	6.04e−14	7.86e−14	6.59e−12	18972	16425	20781	50
f_2	1.50e+04	0.00e00	4.50e +04	66717	65844	79443	3	8.33e−32	4.59e−34	5.00e+03	170091	136062	195498	47
f_3	3.01e−27	9.54e−34	1.00e +05	11718	8208	36000	35	1.66e−12	1.30e−13	2.25e−11	21118	19548	23283	50
f_4	1.30e+02	7.46e+01	2.00e +02	15192	8964	108495	9	5.62e+01	2.39e+01	8.76e+01	11052	5679	23571	50
f_5	4.17e−02	1.08e−19	9.05e +01	8014.5	6570	21186	28	7.40e−03	0.00e00	4.18e−02	17190	15570	19989	50
f_6	3.59e−04	0.00e00	9.72e −03	39811.5	1242	140247	38	0.00e00	0.00e00	9.72e−03	8460	3276	62091	50
f_7	7.35e00	2.51e00	1.38e +01	-	-	-	0	7.57e−04	2.86e−05	2.02e00	34168,5	31041	42507	32
f_8	2.28e00	8.86e−16	3.84e00	20898	13158	28764	6	1.11e−15	8,86e−16	1.33e−15	19822.5	18252	25416	50
f_9	1.06e−01	1.98e−03	1.53e +04	902407	812736	949590	12	1.64e−04	1.44e−06	6.01e+01	736713	546858	891432	49
f_{10}	4.17e−02	1.08e−19	9.05e +01	8014.5	6570	21186	27	7.40e−03	0.00e00	4.18e−02	17190 (50)	15570	19989	50

5 Conclusions and Future Work

This paper proposes an asynchronous and steady state update strategy for the PSO based on a model of co-evolution. Instead of the whole population, like in standard particle swarms (either synchronous or asynchronous), only the worst solution and its neighbors are updated and evaluated in each time step. The remaining particles are kept in a steady state. Accordingly, we have named it steady state PSO (SS-PSO).

The strategy was implemented with three social network structures (*lbest* and square lattices with von Neumann and Moore neighborhood) and tested on a set of ten unimodal, multimodal, shifted, noisy and rotated benchmark problems. Quality of solutions, convergence speed and success rates were compared. SS-PSO significantly improved the performance of S-PSO on a lattice with Moore neighborhood in every function. Since S-PSO$_{\text{Moore}}$ has been found to be the more accurate and faster POS in the set of benchmark functions, we believe that these results validate the proposal.

The strategy was tested with standard PSOs. In the future, and in order to assess the contribution of our proposal to the state of the art, we intend to test it with (efficient variants of the standard PSO and even compare it to other metaheuristics. Scalability of the steady state PSO regarding population size and problem dimension will also be studied. Finally, the emergent patterns of the algorithm (extension of events, stasis, critical values) will be compared to those of the Bak-Sneppen model.

Acknowledgements. First author wishes to thank FCT, *Ministério da Ciência e Tecnologia*, his Research Fellowship SFRH/BPD/66876/2009). This work was supported by FCT PROJECT [UID/EEA/50009/2013], EPHEMECH (TIN2014-56494-C4-3-P, Spanish Ministry of Economy and Competitivity), PROY-PP2015-06 (Plan Propio 2015 UGR), project CEI2015-MP-V17 of the Microprojects program 2015 from CEI BioTIC Granada.

References

1. Aziz, N.A.B., Mubin, M., Mohamad, M.S., Aziz, K.A.: A synchronous-asynchronous particle swarm optimisation algorithm. Sci. World J. **2014**, 1–17 (2014). Article ID 123019
2. Bak, P., Tang, C., Wiesenfeld, K.: Self-organized criticality: an explanation of 1/f noise. Phys. Rev. Lett. **59**(4), 381–384 (1987)
3. Bak, P., Sneppen, K.: Punctuated equilibrium and criticality in a simple model of evolution. Phys. Rev. Lett. **71**(24), 4083–4086 (1993)
4. Boettcher, S., Percus, A.G.: Optimization with extremal dynamics. Complexity **8**(2), 57–62 (2003)
5. Carlisle, A., Dozier, G.: An off-the-shelf PSO. In: Workshop on Particle Swarm Optimization (2001)
6. Chen, M.-R., Li, X., Lu, Y.-Z.: A novel particle swarm optimizer hybridized with extremal optimization. App. Soft Comput. **10**(2), 367–373 (2010)
7. Fernandes, C.M., Merelo, J.J., Rosa, A.C.: Controlling the parameters of the particle swarm optimization with a self-organized criticality model. In: Coello, C.A., Cutello, V., Deb, K., Forrest, S., Nicosia, G., Pavone, M. (eds.) PPSN 2012, Part II. LNCS, vol. 7492, pp. 153–163. Springer, Heidelberg (2012)

8. Kennedy, J., Eberhart, R.: Particle swarm optimization. In: Proceedings of IEEE International Conference on Neural Networks, vol. 4, pp. 1942–1948 (1995)
9. Kennedy, J., Mendes, R.: Population structure and particle swarm performance. In: Proceedings of the IEEE World Congress Evolutionary Computation, pp. 1671–1676 (2002)
10. Koh, B.-I., George, A.D., Haftka, R.T., Fregly, B.J.: Parallel asynchronous particle swarm optimization. Int. J. Numer. Meth. Eng. **67**(4), 578–595 (2006)
11. Løvbjerg, M., Krink, T.: Extending particle swarm optimizers with self-organized criticality. In: Proceedings of the 2002 IEEE Congress on Evolutionary Computation, vol. 2, pp. 1588–1593. IEEE Computer Society (2002)
12. Luo, J., Zhang, Z.: Research on the parallel simulation of asynchronous pattern of particle swarm optimization. Comput. Simul. **22**(6), 78–170 (2006)
13. Majercik, S.: GREEN-PSO: conserving function evaluations in particle swarm optimization. In: Proceedings of the IJCCI 2013, pp. 160–167 (2013)
14. McNabb, A.: Serial PSO results are irrelevant in a multi-core parallel world. In: Proceedings of the 2014 IEEE Congress on Evolutionary Computation, pp. 3143–3150 (2014)
15. Perez, R., Basterrechea, J.: Particle swarm optimization and its application to antenna farfield-pattern prediction from planar scanning. Microw. Opt. Technol. Lett. **44**(5), 398–403 (2005)
16. Rada-Vilela, J., Zhang, M., Seah, W.: A performance study on synchronous and asynchrounous updates in particle swarm. Soft. Comput. **17**(6), 1019–1030 (2013)
17. Shi, Y., Eberhart, R.C.: A modified particle swarm optimizer. In: Proceedings of IEEE International Conference on Evolutionary Computation, pp. 69–73. IEEE (1998)
18. Venter, G., Sobieszczanski-Sobieski, J.: A parallel particle swarm optimization algorithm accelerated by asynchronous evaluations. J. Aerosp. Comput. Inf. Commun. **3**(3), 123–137 (2006)

Dynamic, Uncertain and Constrained Environments

Augmented Lagrangian Constraint Handling for CMA-ES — Case of a Single Linear Constraint

Asma Atamna(✉), Anne Auger, and Nikolaus Hansen

Inria, Research centre Saclay–Île-de-France, TAO team, LRI (UMR 8623),
University of Paris-Saclay, Saint-Aubin, France
{atamna,auger,hansen}@lri.fr

Abstract. We consider the problem of minimizing a function f subject to a single inequality constraint $g(\mathbf{x}) \leq 0$, in a black-box scenario. We present a covariance matrix adaptation evolution strategy using an adaptive augmented Lagrangian method to handle the constraint. We show that our algorithm is an instance of a general framework that allows to build an adaptive constraint handling algorithm from a general randomized adaptive algorithm for unconstrained optimization. We assess the performance of our algorithm on a set of linearly constrained functions, including convex quadratic and ill-conditioned functions, and observe linear convergence to the optimum.

1 Introduction

Evolution strategies (ESs) are derivative-free continuous optimization algorithms that are now well-established to solve unconstrained optimization problems of the form $\min_{\mathbf{x}} f(\mathbf{x})$, $f : \mathbb{R}^n \to \mathbb{R}$, where n is the dimension of the search space. The state-of-the-art ES, the covariance matrix adaptation evolution strategy (CMA-ES) [7], is especially powerful at solving a wide range of problems and particularly ill-conditioned problems [5,8]. It typically exhibits linear convergence. The default CMA-ES algorithm implements comma selection where the best solution is not preserved from one iteration to the next one (contrary to plus selection). Comma selection is an important feature of CMA-ES that entails robustness of the algorithm to various types of ruggedness including noise.

Linear convergence being a central aspect of an ES in the unconstrained case, a $(1+1)$-ES using an adaptive augmented Lagrangian constraint handing—to deal with a single inequality constraint—has been introduced in [3] with the motivation to obtain a linearly converging algorithm. Empirical results show the linear convergence of the algorithm on the sphere and moderately ill-conditioned ellipsoid functions, subject to one linear constraint. In [4], the authors present a variant of the previous $(1+1)$-ES with augmented Lagrangian constraint handling and study theoretically its linear convergence using a Markov chain approach. In both mentioned works, the step-size is adapted using the $1/5$th success rule [10] while the covariance matrix is fixed to the identity. On ill-conditioned

J. Handl et al. (Eds.): PPSN XIV 2016, LNCS 9921, pp. 181–191, 2016.
DOI: 10.1007/978-3-319-45823-6_17

problems, however, adapting the covariance matrix is crucial. It is hence natural to wonder whether it is possible to design a CMA-ES variant with augmented Lagrangian constraint handling. The algorithms presented in [3,4], however, use plus selection and *can thus a priori not be used* directly to design such a variant.

In this context, we consider the constrained problem of minimizing $f : \mathbb{R}^n \to \mathbb{R}$ subject to a single inequality constraint $g(\mathbf{x}) \leq 0$, $g : \mathbb{R}^n \to \mathbb{R}$. More formally, we write

$$\min_{\mathbf{x}} f(\mathbf{x}) \quad \text{subject to} \quad g(\mathbf{x}) \leq 0. \tag{1}$$

We bring to light that the algorithms previously presented in [3,4] derive from a more general framework that seamlessly allows to build an adaptive constraint handling algorithm from a general adaptive stochastic search method. We then naturally apply this finding to build a $(\mu/\mu_\mathrm{w}, \lambda)$-CMA-ES variant with adaptive augmented Lagrangian constraint handling. We opted for using the median success rule step-size adaptation (MSR) [2] because it is an extension of the 1/5th success rule algorithm used in [3,4]. We then test the resulting algorithm—the $(\mu/\mu_\mathrm{w}, \lambda)$-MSR-CMA-ES with augmented Lagrangian constraint handling—on a set of functions, including convex quadratic as well as ill-conditioned functions, subject to one linear inequality constraint.

The rest of this paper is organized as follows: we introduce some basics about augmented Lagrangian in Sect. 2. Then, we define the general framework and apply it to the $(\mu/\mu_\mathrm{w}, \lambda)$-MSR-CMA-ES in Sect. 3. We present our empirical results in Sect. 4 and conclude with a discussion in Sect. 5.

Notations. We introduce here the notations that are not explicitly defined in the rest of the paper. We denote $\mathbb{R}^+$ the set of positive real numbers and $\mathbb{R}^+_>$ the set of strictly positive real numbers. $\mathbb{N}_>$ is the set of natural numbers without 0. $\mathbf{x} \in \mathbb{R}^n$ is a column vector, $\mathbf{x}^\intercal$ is its transpose, and $\mathbf{0} \in \mathbb{R}^n$ is the zero vector. $\|\mathbf{x}\|$ denotes the Euclidean norm of $\mathbf{x}$ and $\sim$ equality in distribution. $(\mu/\mu_\mathrm{w}, \lambda)$ denotes comma selection with weighted recombination and $(1+1)$ denotes plus selection with one parent and one offspring. $\mathbf{I}_{n\times n} \in R^{n\times n}$ is the identity matrix. $\mathbf{x}_i$ is the ith component of vector $\mathbf{x}$. The derivative with respect to $\mathbf{x}$ is denoted $\nabla_{\mathbf{x}}$. Finally, $\mathbf{1}_{\{A\}}$ returns 1 if A is true and 0 otherwise.

2 Augmented Lagrangian Methods

Augmented Lagrangian methods are constraint handling approaches that transform the constrained optimization problem into an unconstrained one where an augmented Lagrangian is optimized [9,12].

The augmented Lagrangian consists of a Lagrangian $\mathcal{L}$ and a penalty function, with $\mathcal{L} : \mathbb{R}^{n+1} \to \mathbb{R}$ defined as

$$\mathcal{L}(\mathbf{x}, \gamma) = f(\mathbf{x}) + \gamma g(\mathbf{x}) \tag{2}$$

for the objective function f subject to one constraint $g(\mathbf{x}) \leq 0$, where $\gamma \in \mathbb{R}$ is the Lagrange factor. The Lagrangian encodes the KKT stationarity condition which states that, given some regularity conditions are satisfied (constraint qualifications), if $\mathbf{x}^* \in \mathbb{R}^n$ is a local minimum of the constrained problem, then there exists a constant $\gamma^* \in \mathbb{R}^+$, called the Lagrange multiplier, such that

$$\underbrace{\nabla_{\mathbf{x}} f(\mathbf{x}^*) + \gamma^* \nabla_{\mathbf{x}} g(\mathbf{x}^*)}_{\nabla_{\mathbf{x}} \mathcal{L}(\mathbf{x}^*, \gamma^*)} = \mathbf{0},$$

where we assume here that f and g are differentiable at $\mathbf{x}^*$.

A penalty function is combined with the Lagrangian $\mathcal{L}$ to create the augmented Lagrangian h. There exist different ways to construct the augmented Lagrangian and we refer to [11] for a deeper discussion about this topic. In this work, we use the following augmented Lagrangian

$$h(\mathbf{x}, \gamma, \omega) = f(\mathbf{x}) + \begin{cases} \gamma g(\mathbf{x}) + \frac{\omega}{2} g^2(\mathbf{x}) & \text{if } \gamma + \omega g(\mathbf{x}) \geq 0 \\ -\frac{\gamma^2}{2\omega} & \text{otherwise} \end{cases}, \tag{3}$$

where $\omega > 0$ is a penalty factor. The same augmented Lagrangian was used for the first time within an ES in [3]. The function h is minimized successively with respect to $\mathbf{x}$, and γ and ω are updated so that γ approaches the Lagrange multiplier γ^* and ω favors feasible solutions. By adapting γ, the penalty factor ω does not have to grow to infinity to achieve convergence, unlike with quadratic penalty function methods [11].

Let $\mathbf{x}_{\text{opt}}$ be the optimum of the constrained problem in (1) and let γ_{opt} be the corresponding Lagrange multiplier. If f and g are differentiable at $\mathbf{x}_{\text{opt}}$, then for all $\omega > 0$,

$$\nabla_{\mathbf{x}} h(\mathbf{x}_{\text{opt}}, \gamma_{\text{opt}}, \omega) = \nabla_{\mathbf{x}} f(\mathbf{x}_{\text{opt}}) + \max(0, \gamma_{\text{opt}} + \omega g(\mathbf{x}_{\text{opt}})) \nabla_{\mathbf{x}} g(\mathbf{x}_{\text{opt}}) = \mathbf{0}.$$

3 A General Framework for Adaptive Augmented Lagrangian Constraint Handling

In [3,4], the authors present two $(1+1)$-ESs with an augmented Lagrangian constraint handling approach for the optimization problem in (1). The algorithms derive from a general framework for building a constraint handling adaptive algorithm. This framework starts with a randomized adaptive algorithm for minimizing an unconstrained function $f : \mathbb{R}^n \to \mathbb{R}$: the randomized adaptive algorithm can be identified by the sequence of its states $\mathbf{s}_t$ at iteration t that are iteratively computed from an update function $\mathcal{F}$ such that

$$\mathbf{s}_{t+1} = \mathcal{F}^f(\mathbf{s}_t, \mathbf{U}_{t+1}), \tag{4}$$

where the superscript indicates the function being minimized and where $(\mathbf{U}_t)_{t \in \mathbb{N}_>}$ is a sequence of independent identically distributed (i.i.d.) random vectors. For instance, in the case of a $(1+1)$-ES in [3,4], the state is a vector of the search space (current estimate of the optimum) and a step-size.

We assume that the state $\mathbf{s}_t$ of the algorithm includes a vector $\mathbf{X}_t \in \mathbb{R}^n$ which typically encodes the current estimate of the optimum at iteration t. Note that the transition function $\mathcal{F}$ above includes a step where candidate solutions are sampled from the current state $\mathbf{s}_t$ and the random vector $\mathbf{U}_{t+1}$, and evaluated on the objective function f.

From the adaptive algorithm above, we construct an algorithm with adaptive constraint handling to take into account a single constraint in the following way: we add to the state of the algorithm two scalars γ_t and ω_t that correspond respectively to the Lagrange factor and the penalty factor of the augmented Lagrangian h at iteration t. Therefore, the state at iteration t is $\mathbf{s}_t{}' = [\mathbf{s}_t, \gamma_t, \omega_t]$. The objective function used at each iteration to evaluate a candidate solution $\mathbf{X}_{t+1}^i$ is now

$$h_{(\gamma_t,\omega_t)}(\mathbf{X}_{t+1}^i) := h(\mathbf{X}_{t+1}^i, \gamma_t, \omega_t), \tag{5}$$

where h is the augmented Lagrangian defined in (3). Finally, the update of the state $\mathbf{s}_t{}'$ of the adaptive algorithm with augmented Lagrangian constraint handling takes place in two steps: first, $\mathbf{s}_t$ is updated via

$$\mathbf{s}_{t+1} = \mathcal{F}^{h_{(\gamma_t,\omega_t)}}(\mathbf{s}_t, \mathbf{U}_{t+1}), \tag{6}$$

where candidate solutions are now evaluated on $h_{(\gamma_t,\omega_t)}$ instead of f. Then, the parameters γ_t and ω_t of h are updated. In [3], γ_t is updated according to

$$\gamma_{t+1} = \max(0, \gamma_t + \omega_t g(\mathbf{X}_{t+1})), \tag{7}$$

while in [4], the authors use the following update

$$\gamma_{t+1} = \gamma_t + \omega_t g(\mathbf{X}_{t+1}). \tag{8}$$

For ω_t, the following update is used in both [3,4]

$$\omega_{t+1} = \begin{cases} \omega_t \chi^{1/4} & \text{if } \omega_t g^2(\mathbf{X}_{t+1}) < k_1 \frac{|h(\mathbf{X}_{t+1},\gamma_t,\omega_t) - h(\mathbf{X}_t,\gamma_t,\omega_t)|}{n} \\ & \text{or } k_2 |g(\mathbf{X}_{t+1}) - g(\mathbf{X}_t)| < |g(\mathbf{X}_t)| \\ \omega_t \chi^{-1} & \text{otherwise} \end{cases}, \tag{9}$$

for some constants $\chi > 1$, $k_1, k_2 \in \mathbb{R}^+$.

Based on these examples, we introduce some general update functions $\mathcal{G}_\gamma$ and $\mathcal{G}_\omega$ for the updates of γ_t and ω_t defined implicitly via

$$\gamma_{t+1} = \mathcal{G}_\gamma^g((\gamma_t, \omega_t), \mathbf{X}_{t+1}) \tag{10}$$

$$\omega_{t+1} = \mathcal{G}_\omega^{(f,g)}((\mathbf{X}_t, \gamma_t, \omega_t), \mathbf{X}_{t+1}). \tag{11}$$

The superscript in $\mathcal{G}_\gamma$ and $\mathcal{G}_\omega$ indicates that the function value is used in the update.

3.1 The $(\mu/\mu_w, \lambda)$-MSR-CMA-ES with Adaptive Augmented Lagrangian

We now apply the general framework sketched above to the covariance matrix adaptation evolution strategy (CMA-ES) with median success rule step-size adaptation (MSR). We start by presenting the algorithm for the unconstrained case then we give the updates of the augmented Lagrangian parameters γ_t and ω_t.

The (Unconstrained) CMA-ES with MSR. The original CMA-ES with MSR is given in Algorithm 1, without the highlighted parts. The algorithm proceeds iteratively: at each iteration t, λ candidate solutions (offspring) $\mathbf{X}_{t+1}^i$, $i = 1, \ldots, \lambda$, are sampled according to Line 5, where $\mathbf{X}_t \in \mathbb{R}^n$ is the current estimate of the optimum (mean vector), $\sigma_t \in \mathbb{R}^+$ is the step-size, and $\mathbf{U}_{t+1}^i \in \mathbb{R}^n$, $i = 1, \ldots, \lambda$, are i.i.d. random vectors sampled from the normal distribution $\mathcal{N}(\mathbf{0}, \mathbf{C}_t)$, with mean $\mathbf{0} \in \mathbb{R}^n$ and covariance matrix $\mathbf{C}_t \in \mathbb{R}^{n \times n}$. The offspring are ordered according to their fitness (f-value in the unconstrained case) in Line 6, where $i : \lambda$ is the index of the ith best offspring. The μ best offspring (parents) are then recombined (Line 7) to create the new mean vector $\mathbf{X}_{t+1}$, where the weights $w_i > 0$, $i = 1, \ldots, \mu$, satisfy $w_1 > \ldots > w_\mu$ and $\sum_{i=1}^{\mu} w_i = 1$.

The step-sized σ_t is adapted in Lines 8 to 11 using the MSR step-size adaptation [2]. MSR is a success-based step-size adaptation method which extends the well-known 1/5th success rule step-size adaptation [10], used with plus selection, to comma selection. The step-size is adapted depending on "success", where the success is defined as the median offspring $\mathbf{X}_{t+1}^{\boldsymbol{m}(\lambda)}$ (fitness-wise) of the current population being better than the jth best offspring $\mathbf{X}_t^{j:\lambda}$ of the previous population. In practice, we choose j to be the 30th percentile–the value for which the median success probability is roughly 1/2 on the sphere function with optimal step-size [2]. The number K_{succ} of offspring better than $\mathbf{X}_t^{j:\lambda}$ is computed in Line 8. Note that $K_{\text{succ}} \geq \lambda/2$ is equivalent to $h(\mathbf{X}_{t+1}^{\boldsymbol{m}(\lambda)}, \gamma_t, \omega_t) \leq h(\mathbf{X}_t^{j:\lambda}, \gamma_t, \omega_t)$. Therefore, we define the success measure z_t in Line 9 such that $z_t \geq 0$ if and only if $\mathbf{X}_{t+1}^{\boldsymbol{m}(\lambda)}$ is successful. z_t is cumulated in q_{t+1} (Line 10) and, finally, σ_t is updated in Line 11: it increases in the presence of success ($q_{t+1} > 0$) and decreases otherwise in order to increase the probability of success.

The covariance matrix $\mathbf{C}_t$ is adapted with CMA [7] in Lines 12 and 13. The update is a combination of the so-called rank-one-update and rank-μ-update. A detailed discussion on CMA can be found in [6].

Finally, the jth best offspring is updated in Line 17. Therefore, the state of the algorithm in the unconstrained case is

$$\mathbf{s}_t = (\mathbf{X}_t, \sigma_t, q_t, p_t, \mathbf{C}_t, \mathbf{X}_t^{j:\lambda}).$$

The constrained $(\mu/\mu_w, \lambda)$-MSR-CMA-ES with adaptive augmented Lagrangian. As explained in the general framework, the fitness f is replaced with the augmented Lagrangian h in the constrained case. The parameters γ_t and ω_t are adapted in Lines 15 and 16 in Algorithm 1, where changes in comparison to the unconstrained case are highlighted in gray.

The Lagrange factor γ_t is adapted in Line 15. It is increased when the new solution $\mathbf{X}_{t+1}$ is unfeasible and decreased otherwise, unless it is zero. The derivation of this update is discussed in details in [11].

For the penalty parameter ω_t, we use the original update proposed in [3] for the $(1+1)$-ES with augmented Lagrangian. The update rule is given in Line 16. ω_t is increased either when (i) the augmented Lagrangian h does not change "enough" after γ_t and ω_t are updated to avoid stagnation. This is translated by the first inequality where

$$\omega_t g^2(\mathbf{X}_{t+1}) \approx |h(\mathbf{X}_{t+1}, \gamma_t + \Delta\gamma_t, \omega_t + \Delta\omega_t) - h(\mathbf{X}_{t+1}, \gamma_t, \omega_t)|$$

is compared to the change in h due to the change in $\mathbf{X}_t$, $|h(\mathbf{X}_{t+1}, \gamma_t, \omega_t) - h(\mathbf{X}_t, \gamma_t, \omega_t)|$. ω_t is also increased when (ii) the change in the value of the constraint function is not large enough (second inequality in Line 16). To prevent an unnecessary ill-conditioning of the problem, ω_t is decreased whenever conditions (i) and (ii) are not satisfied.

4 Empirical Results

We evaluate Algorithm 1 on the sphere function (f_{sphere}), two ellipsoid functions (f_{elli}) with condition numbers $\alpha = 10^2, 10^6$, f^2_{sphere}, $f^{0.5}_{\text{sphere}}$, the different powers function ($f_{\text{diff\_pow}}$), and the Rosenbrock function (f_{rosen}), with one linear inequality constraint. The functions are defined in Table 1. We consider the case where the constraint is active at the optimum $\mathbf{x}_{\text{opt}}$, i.e. $g(\mathbf{x}_{\text{opt}}) = 0$. We choose the optimum to be at $\mathbf{x}_{\text{opt}} = (10, \ldots, 10)^\intercal$ and construct the constraint function, $g(\mathbf{x}) = \mathrm{b}^\intercal \mathbf{x} + c$, so that the KKT stationarity condition is satisfied at $\mathbf{x}_{\text{opt}}$ with $\gamma_{\text{opt}} = 1$. Therefore,

$$\mathrm{b} = -\nabla_{\mathbf{x}} f_\cdot(\mathbf{x}_{\text{opt}})^\intercal \quad \text{and} \quad c = \nabla_{\mathbf{x}} f_\cdot(\mathbf{x}_{\text{opt}}) \mathbf{x}_{\text{opt}},$$

for each function. Note that all considered functions are differentiable at $\mathbf{x}_{\text{opt}} = (10, \ldots, 10)^\intercal$.

For the step-size and the covariance matrix adaptation, we use the Python implementation of CMA-ES whose source code can be found at [1], with the default parameter setting detailed in [6]. We run the algorithm 11 times in $n = 10$, with $\mathbf{X}_0$ sampled uniformly in $[-5, 5]^n$, $\sigma_0 = 1$, $\gamma_0 = 5$, and $\omega_0 = 1$. The results are presented for one run in Figs. 1 (f_{sphere}, f^2_{sphere}, and $f^{0.5}_{\text{sphere}}$) and 2 (f_{elli} with $\alpha = 10^2, 10^6$, $f_{\text{diff\_pow}}$, and f_{rosen}). On the left column of each figure are graphs of the evolution of the distance to the optimum $\|\mathbf{X}_t - \mathbf{x}_{\text{opt}}\|$, the step-size σ_t, the distance to the Lagrange multiplier $\|\gamma_t - \gamma_{\text{opt}}\|$, and the penalty factor ω_t in log-scale. On the right column of the figures are graphs representing the evolution of the coordinates of the mean vector $\mathbf{X}_t$.

Graphs on the right column of Figs. 1 and 2 show the overall convergence of the algorithm to $\mathbf{x}_{\text{opt}}$. We also observe linear convergence of $\mathbf{X}_t$ to $\mathbf{x}_{\text{opt}}$, as well

---

Algorithm 1. $(\mu/\mu_{\mathrm{w}}, \lambda)$-MSR-CMA-ES with Augmented Lagrangian Constraint Handling

---

0 **given** $n \in \mathbb{N}_{>}$, $\chi = 2^{1/n}$, $k_1 = 3$, $k_2 = 5$, $\mu, \lambda \in \mathbb{N}_{>}$, $j = 0.3\lambda$, $0 \leq w_i < 1$, $\sum_{i=1}^{\mu} w_i = 1$,

$$\mu_{\mathrm{eff}} = 1/\sum_{i=1}^{\mu} w_i^2,\ c_\sigma = 0.3,\ d_\sigma = 2 - 2/n,\ c_c = \frac{4 + \mu_{\mathrm{eff}}/n}{n + 4 + 2\mu_{\mathrm{eff}}/n}$$

$$c_1 = \frac{2}{(n+1.3)^2 + \mu_{\mathrm{eff}}},\ c_\mu = \min\left(1 - c_1, 2\frac{\mu_{\mathrm{eff}} - 2 + 1/\mu_{\mathrm{eff}}}{(n+2)^2 + \mu_{\mathrm{eff}}}\right)$$

1 **initialize** $\mathbf{X}_0 \in \mathbb{R}^n$, $\sigma_0 \in \mathbb{R}^+_>$, $\mathbf{C}_0 = \mathbf{I}_{n \times n}$, $t = 0$, $q_0 = 0$, $p_0 = \mathbf{0}$,
constrained_problem // **true** if the problem is constrained, **false** otherwise

2 **if** constrained_problem

3 **initialize** $\gamma_0 \in \mathbb{R}$, $\omega_0 \in \mathbb{R}^+_>$

4 **while** stopping criteria not met

5 $\mathbf{X}^i_{t+1} = \mathbf{X}_t + \sigma_t \mathbf{U}^i_{t+1}$, $\mathbf{U}^i_{t+1} \sim \mathcal{N}(\mathbf{0}, \mathbf{C}_t), i = 1, \ldots, \lambda$ // sample candidate solutions

6 Extract indices $\{1:\lambda, \ldots, \lambda:\lambda\}$ of ordered candidate solutions such that

$$\begin{cases} h(\mathbf{X}^{1:\lambda}_{t+1}, \gamma_t, \omega_t) \leq \cdots \leq h(\mathbf{X}^{\lambda:\lambda}_{t+1}, \gamma_t, \omega_t) & \text{if constrained\_problem} \\ f(\mathbf{X}^{1:\lambda}_{t+1}) \leq \cdots \leq f(\mathbf{X}^{\lambda:\lambda}_{t+1}) & \text{otherwise} \end{cases}$$

7 $\mathbf{X}_{t+1} = \sum_{i=1}^{\mu} w_i \mathbf{X}^{i:\lambda}_{t+1} = \mathbf{X}_t + \sigma_t \sum_{i=1}^{\mu} w_i \mathbf{U}^{i:\lambda}_{t+1}$ // recombine μ best candidate solutions

8 $$K_{\mathrm{succ}} = \begin{cases} \sum_{i=1}^{\lambda} \mathbf{1}_{\{h(\mathbf{X}^i_{t+1}, \gamma_t, \omega_t) \leq h(\mathbf{X}^{j:\lambda}_t, \gamma_t, \omega_t)\}} & \text{if constrained\_problem} \\ \sum_{i=1}^{\lambda} \mathbf{1}_{\{f(\mathbf{X}^i_{t+1}) \leq f(\mathbf{X}^{j:\lambda}_t)\}} & \text{otherwise} \end{cases}$$

9 $z_t = \frac{2}{\lambda}\left(K_{\mathrm{succ}} - \frac{\lambda}{2}\right)$ // compute success measure

10 $q_{t+1} = (1 - c_\sigma) q_t + c_\sigma z_t$

11 $\sigma_{t+1} = \sigma_t \exp\left(\frac{q_{t+1}}{d_\sigma}\right)$ // update step-size

12 $p_{t+1} = (1 - c_c) p_t + \sqrt{c_c(2 - c_c)\mu_{\mathrm{eff}}}\left(\frac{\mathbf{X}_{t+1} - \mathbf{X}_t}{\sigma_t}\right)$ // cumulation path for CMA

13 $$\mathbf{C}_{t+1} = (1 - c_1 - c_\mu)\mathbf{C}_t + c_1 p_{t+1} p^{\mathsf{T}}_{t+1} + c_\mu \sum_{i=1}^{\mu} w_i \left(\frac{\mathbf{X}^i_{t+1} - \mathbf{X}_t}{\sigma_t}\right)\left(\frac{\mathbf{X}^i_{t+1} - \mathbf{X}_t}{\sigma_t}\right)^{\mathsf{T}}$$
// update covariance matrix

14 **if** constrained_problem

15 $\gamma_{t+1} = \max(0, \gamma_t + \omega_t g(\mathbf{X}_{t+1}))$ // update Lagrange factor

16 $$\omega_{t+1} = \begin{cases} \omega_t \chi^{1/4} & \text{if } \omega_t g^2(\mathbf{X}_{t+1}) < k_1 \frac{|h(\mathbf{X}_{t+1}, \gamma_t, \omega_t) - h(\mathbf{X}_t, \gamma_t, \omega_t)|}{n} \\ & \text{or } k_2 |g(\mathbf{X}_{t+1}) - g(\mathbf{X}_t)| < |g(\mathbf{X}_t)| \\ \omega_t \chi^{-1} & \text{otherwise} \end{cases}$$
// update penalty factor

17 $\mathbf{X}^{j:\lambda}_{t+1} = \mathbf{X}_t + \sigma_t \mathbf{U}^{j:\lambda}_{t+1}$ // update jth best solution

18 $t = t + 1$

---

as linear convergence of γ_t to γ_{opt} and σ_t to 0 (left column of Figs. 1 and 2). Moreover, $\|\mathbf{X}_t - \mathbf{x}_{\mathrm{opt}}\|$, $\|\gamma_t - \gamma_{\mathrm{opt}}\|$, and σ_t decrease at the same rate. On the other hand, the penalty factor ω_t is observed to converge to a stationary value after a certain number of iterations. We sometimes observe a stagnation in graphs of $\|\mathbf{X}_t - \mathbf{x}_{\mathrm{opt}}\|$ due to numerical precision.

The largest convergence rate (when excluding the initial adaptation phase) is observed on f_{sphere} and the smallest one on $f^{0.5}_{\mathrm{sphere}}$, where there is a factor of approximately 1.5 between the two convergence rates. However, there is some

Table 1. Definitions of the tested functions, where $f_{\text{sphere}} := f^1_{\text{sphere}}$.

Name	Definition	Name	Definition
$f^{\alpha}_{\text{sphere}}(\mathbf{x})$	$\left(\frac{1}{2}\sum_{i=1}^{n}\mathbf{x}_i^2\right)^{\alpha}$	$f_{\text{diff\_pow}}(\mathbf{x})$	$\sqrt{\sum_{i=1}^{n}\lvert\mathbf{x}_i\rvert^{2+4\frac{i-1}{n-1}}}$
$f_{\text{elli}}(\mathbf{x})$	$\frac{1}{2}\sum_{i=1}^{n}\alpha^{\frac{i-1}{n-1}}\mathbf{x}_i^2$	$f_{\text{rosen}}(\mathbf{x})$	$\sum_{i=1}^{n-1}\left(10^2(\mathbf{x}_i^2-\mathbf{x}_{i+1})^2+(\mathbf{x}_i-1)^2\right)$

variance in the empirical convergence rate. In particular, on 11 performed runs we observe the highest variance in the empirical convergence rate for f_{elli} with $\alpha = 10^6$, $f_{\text{diff\_pow}}$, and f_{rosen}.

On f_{elli} with $\alpha = 10^6$, $f_{\text{diff\_pow}}$, and f_{rosen}, we observe a stagnation of $\mathbf{X}_t$ in the early stages of the algorithm (left column in Fig. 2). The reason is that the adaptation of the covariance matrix takes longer on ill-conditioned problems.

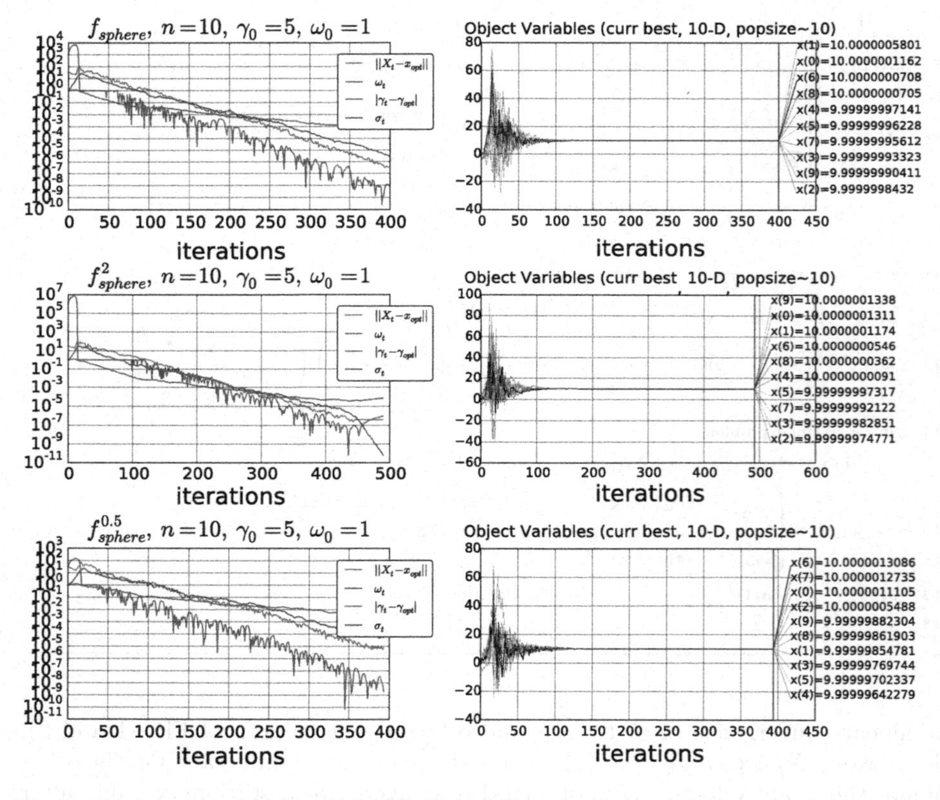

Fig. 1. Single runs of $(\mu/\mu_w, \lambda)$-MSR-CMA-ES with augmented Lagrangian on f_{sphere} (top row), f^2_{sphere} (middle row), and $f^{0.5}_{\text{sphere}}$ (bottom row) in $n = 10$. The optimum $\mathbf{x}_{\text{opt}} = (10, \ldots, 10)^{\intercal}$. Left: evolution of the distance to the optimum, the distance to the Lagrange multiplier, the penalty factor, and the step-size in log-scale. Right: evolution of the coordinates of $\mathbf{X}_t$.

This explains the slow convergence of some coordinates of $\mathbf{X}_t$ to 10 (right column in Fig. 2). Once the covariance matrix is adapted, the convergence occurs.

When comparing 11 single runs of Algorithm 1 to the $(1+1)$-ESs with augmented Lagrangian in [3,4] (not shown for space reasons) on constrained f_{sphere}, f_{elli} (in $n = 10$), it appears that on f_{sphere}, Algorithm 1 needs approximately

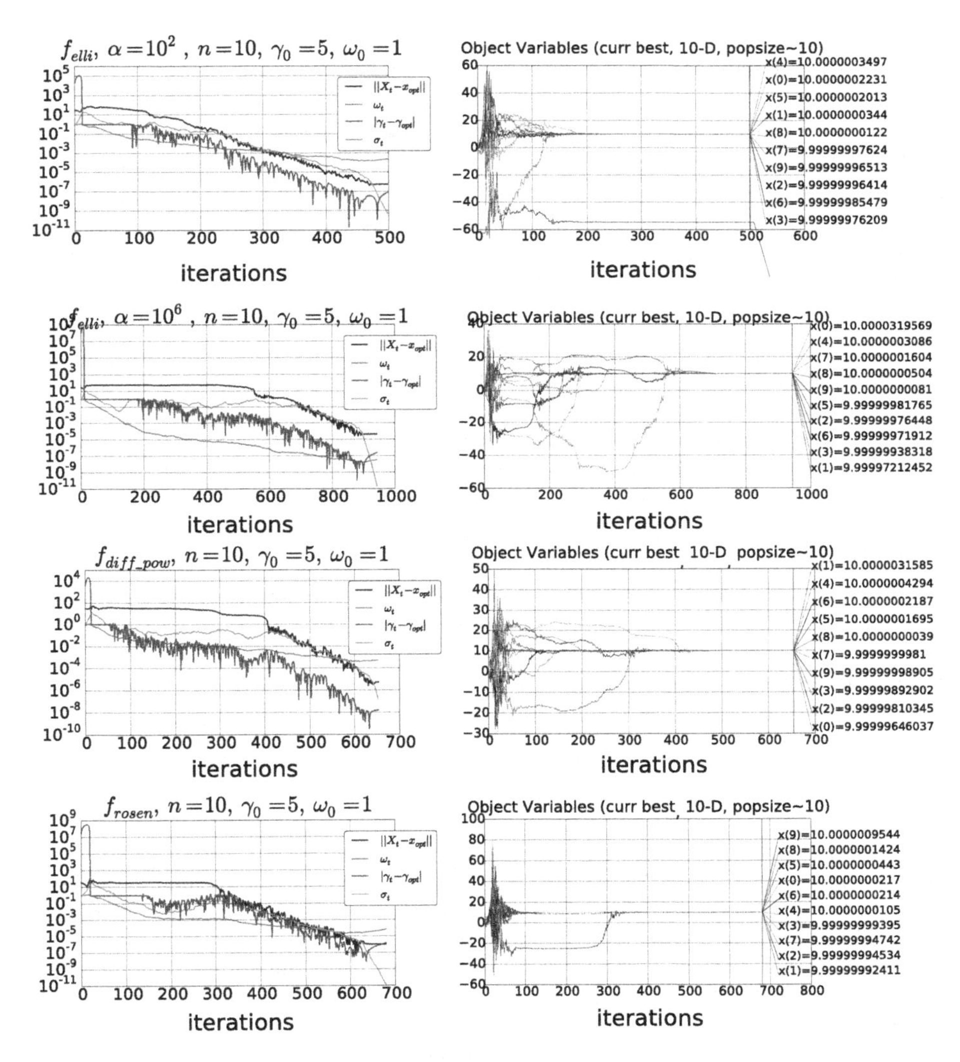

Fig. 2. Single runs of $(\mu/\mu_w, \lambda)$-MSR-CMA-ES with augmented Lagrangian on f_{elli} with $\alpha = 10^2$ (first row), f_{elli} with $\alpha = 10^6$ (second row), $f_{\text{diff\_pow}}$ (third row), and f_{rosen} (fourth row) in $n = 10$. The optimum $\mathbf{x}_{\text{opt}} = (10, \ldots, 10)^\top$. Left: evolution of the distance to the optimum, the distance to the Lagrange multiplier, the penalty factor, and the step-size in log-scale. Right: evolution of the coordinates of $\mathbf{X}_t$.

up to 1.5 times more function evaluations than algorithms in [3,4] to reach a distance to the optimum of 10^{-4}. On f_{elli} with $\alpha = 10^2$, however, Algorithm 1 is faster and needs approximately 1.3 times less function evaluations to reach the same distance, with $\alpha = 10^6$, Algorithm 1 is around 167 times faster to reach a target of 15 (this large difference is due to the adaptation of the covariance matrix).

5 Discussion

Linear convergence is a key aspect of ESs in both unconstrained and constrained optimization scenarios. As stated in [3], the minimum requirement for a constraint handling ES is to converge linearly on convex quadratic functions with a single linear constraint. On the other hand, an algorithm for constrained optimization should be able to tackle ill-conditioned problems. Having that in mind, we proposed a $(\mu/\mu_{\text{w}}, \lambda)$-CMA-ES with an augmented Lagrangian approach for handling one inequality constraint, where the choice of the augmented Lagrangian constraint handling was motivated by the promising results of its implementation for the $(1+1)$-ESs with 1/5th success rule in [3,4]. Moreover, we showed that our algorithm–as well as $(1+1)$-ESs with augmented Lagrangian constraint handling in [3,4]–is an instance of a more general framework for building an adaptive constraint handling algorithm from a general adaptive algorithm for unconstrained optimization.

Experiments on linearly constrained convex quadratic functions, as well as ill-conditioned functions (including the ellipsoid and Rosenbrock functions), showed linear convergence of our algorithm to the unique optimum of the constrained problem.

Acknowledgments. This work was supported by the grant ANR-2012-MONU-0009 (NumBBO) of the French National Research Agency.

References

1. Python source code of CMA-ES. https://pypi.python.org/pypi/cma
2. Ait Elhara, O., Auger, A., Hansen, N.: A median success rule for non-elitist evolution strategies: study of feasibility. In: Genetic and Evolutionary Computation Conference, pp. 415–422. ACM Press (2013)
3. Arnold, D.V., Porter, J.: Towards au augmented lagrangian constraint handling approach for the (1+1)-ES. In: Genetic and Evolutionary Computation Conference, pp. 249–256. ACM Press (2015)
4. Atamna, A., Auger, A., Hansen, N.: Analysis of linear convergence of a $(1+1)$-ES with augmented lagrangian constraint handling. To appear in the Proceedings of the Genetic and Evolutionary Computation Conference (2016)
5. Auger, A., Hansen, N., Perez Zerpa, J.M., Ros, R., Schoenauer, M.: Experimental comparisons of derivative free optimization algorithms. In: Vahrenhold, J. (ed.) SEA 2009. LNCS, vol. 5526, pp. 3–15. Springer, Heidelberg (2009)

6. Hansen, N.: The CMA Evolution Strategy: A Tutorial (2016). http://arxiv.org/pdf/1604.00772v1.pdf
7. Hansen, N., Ostermeier, A.: Completely derandomized self-adaptation in evolution strategies. Evol. Comput. **9**(2), 159–195 (2001)
8. Hansen, N., Ros, R., Mauny, N., Schoenauer, M., Auger, A.: Multiplier and gradient methods. Appl. Soft Comput. **11**, 5755–5769 (2011)
9. Hestenes, M.R.: Multiplier and gradient methods. J. Optim. Theory Appl. **4**(5), 303–320 (1969)
10. Kern, S., Müller, S.D., Hansen, N., Büche, D., Ocenasek, J., Koumoutsakos, P.: Learning probability distributions in continuous evolutionary algorithms - a comparative review. Nat. Comput. **3**(1), 77–112 (2004)
11. Nocedal, J., Wright, S.J.: Numerical Optimization, 2nd edn. Springer, New York (2006)
12. Powell, M.J.D.: A method for nonlinear constraints in minimization problems. In: Fletcher, R. (ed.) Optimization, pp. 283–298. Academic Press, New York (1969)

An Active-Set Evolution Strategy for Optimization with Known Constraints

Dirk V. Arnold(✉)

Faculty of Computer Science, Dalhousie University,
Halifax, Nova Scotia B3H 4R2, Canada
dirk@cs.dal.ca

Abstract. We propose an evolutionary approach to constrained optimization where the objective function is considered a black box, but the constraint functions are assumed to be known. The approach can be considered a stochastic active-set method. It labels constraints as either active or inactive and projects candidate solutions onto the subspace of the feasible region that is implied by rendering active inequality constraints equalities. We implement the approach in a $(1+1)$-ES and evaluate its performance using a commonly used set of test problems.

1 Introduction

Evolutionary algorithms are stochastic black box optimization strategies. They are commonly used in connection with optimization problems that do not admit a convenient mathematical representation of the objective, or if gradient estimates can be obtained only at a high cost or are necessarily inaccurate. Examples include scenarios where the evaluation of the quality of a candidate solution requires running a simulation model. In the context of constrained optimization with evolutionary algorithms, the constraint functions are often considered as black boxes as well. However, in many cases, including the case of bound constraints, it is not uncommon that the constraint functions are known and relatively inexpensive to evaluate. The objective of this paper is to develop a simple evolutionary algorithm for constrained optimization with known constraints.

Active-set methods are a common approach to solving constrained optimization problems [12]. They maintain a set of active inequality constraints and perform optimization in the subspace of the feasible region that is implied by rendering the active inequality constraints equalities. The algorithm we introduce in this paper can be considered a stochastic active-set approach implemented in a $(1+1)$-ES[1]. The step size of the $(1+1)$-ES is commonly controlled using the $1/5$th rule [13]. That rule can fail in the presence of small constraint angles (i.e., small angles between the gradient of the objective function and the normal vector of the constraint function) in cases as simple as a linear objective with a single linear constraint [2,15]. Small constraint angles result in low success rates

[1] See Hansen et al. [7] for an introduction to evolution strategy terminology.

J. Handl et al. (Eds.): PPSN XIV 2016, LNCS 9921, pp. 192–202, 2016.
DOI: 10.1007/978-3-319-45823-6_18

and thus a systematic reduction of the step size. Projection of infeasible candidate solutions onto the feasible region in connection with an active-set approach can potentially circumvent unwarranted decreases of the step size.

The remainder of this paper is organized as follows. In Sect. 2 we outline the class of optimization problems we strive to solve, formalize notation, and propose an active-set $(1+1)$-ES for optimization with known constraints. In Sect. 3, that algorithm is applied to commonly used test problems and its performance is discussed. Section 4 concludes.

2 Problem and Algorithm

We consider the problem of minimizing objective function $f : \mathbb{R}^n \to \mathbb{R}$ subject to constraints

$$\begin{aligned} g_i(\mathbf{x}) &\leq 0 \quad \text{for } i \in \{1, \ldots, l\} \\ h_j(\mathbf{x}) &= 0 \quad \text{for } j \in \{1, \ldots, m\}. \end{aligned} \tag{1}$$

The active set $\mathcal{A}(\mathbf{x})$ of a (feasible) candidate solution $\mathbf{x}$ is the set of indices of all those inequality constraints with $g_i(\mathbf{x}) = 0$. Assuming a single globally optimal solution $\mathbf{x}^*$ to the optimization problem, we refer to the active set $\mathcal{A}(\mathbf{x}^*)$ of that solution as the optimal active set $\mathcal{A}^*$. We refer to the subspace of the search space where all equality constraints and active inequality constraints in $\mathcal{A}(\mathbf{x})$ are satisfied as equalities as the reduced search space at $\mathbf{x}$. We write n^* for the dimension of the reduced search space at the optimal solution $\mathbf{x}^*$.

Our active-set $(1+1)$-ES evolves a feasible candidate solution $\mathbf{x} \in \mathbb{R}^n$ to the optimization problem at hand, adapting the step size $\sigma \in \mathbb{R}_+$ using the 1/5th rule. Offspring candidate solutions are usually projected onto the reduced search space at the parent. However, with a certain probability the use of the active set is suspended, allowing to break out of the reduced search space. Adapting the step size of the algorithm only in those steps where the active inequalities are enforced as equalities prevents unwarranted decreases of the step size. A single iteration of the algorithm is described in Fig. 1.

Boolean flag κ determines whether or not the active set of $\mathbf{x}$ is enforced for offspring candidate solution $\mathbf{y}$. If κ is false, then the search proceeds in the reduced search space. If it is true, then the inequality constraints active at the parental candidate solution will be enforced as inequalities rather than as equalities. If the dimension of the reduced search space at $\mathbf{x}$ is zero, then there is no use in enforcing the active constraints as they would repeatedly yield the same solution, and κ is thus set to true. Otherwise, the active inequality constraints are enforced as equality constraints with probability $1 - p$. Larger values of p decrease the likelihood that the algorithm will spend unproductive time in non-optimal reduced subspaces. However, once the optimal active set $\mathcal{A}^*$ has been found, smaller values of p are useful as unproductive steps beyond the optimal reduced search space are avoided. We use $p = 0.2$ throughout.

1. Compute the dimension $n' = n - \text{rank}(N)$ of the reduced search space at $\mathbf{x}$, where N is the matrix whose columns are the normal vectors at $\mathbf{x}$ of the equality constraints and the inequality constraints in $\mathcal{A}(\mathbf{x})$.
2. Repeat
   (a) Generate standard normally distributed $\mathbf{z} \in \mathbb{R}^n$ and let
   $$\mathbf{y} = \mathbf{x} + \sigma\mathbf{z}\,.$$
   (b) If $n' = 0$, let κ be true. If $n' > 0$, let κ be true with probability p and false otherwise.
   (c) If κ is true, project $\mathbf{y}$ onto the feasible region; otherwise, project $\mathbf{y}$ onto the intersection of the feasible region with the reduced search space at $\mathbf{x}$.

   until offspring candidate solution $\mathbf{y} \in \mathbb{R}^n$ is feasible.
3. If $f(\mathbf{y}) < f(\mathbf{x})$, then
   (a) Let
   $$\mathbf{x} \leftarrow \mathbf{y}\,.$$
   (b) If κ is false, let
   $$\sigma \leftarrow \sigma 2^{1/n'}\,.$$
   Otherwise, if κ is false, let
   $$\sigma \leftarrow \sigma 2^{-1/(4n')}\,.$$

Fig. 1. Single iteration of the active-set (1 + 1)-ES.

Projection of $\mathbf{y}$ onto the intersection of the feasible region with the reduced search space at $\mathbf{x}$ is accomplished by minimizing function $d(\mathbf{w}) = \|\mathbf{w} - \mathbf{y}\|^2$ subject to constraints

$$\begin{aligned} g_i(\mathbf{w}) &\leq 0 \quad \text{for } i \in \{1,\dots,l\} \setminus \mathcal{A}(\mathbf{x}) \\ g_i(\mathbf{w}) &= 0 \quad \text{for } i \in \mathcal{A}(\mathbf{x}) \\ h_j(\mathbf{w}) &= 0 \quad \text{for } j \in \{1,\dots,m\}. \end{aligned}$$

When the use of the active set is suspended, projection onto the feasible region is accomplished by minimizing that some function, but subject to the original set of constraints from Eq. (1). Notice that minimization does not make use of the objective f of the original optimization problem and that thus the algorithm performs only a single evaluation of f per iteration. Minimization of d can be accomplished using any algorithm for constrained optimization. We use the active-set method implemented in `fmincon` in *Matlab*'s optimization toolbox. Step 2 involves a loop as minimization of d may fail to yield a feasible solution.

The update of σ in Step 3 of the algorithm employs the implementation of the 1/5th rule due to Kern et al. [9]. The step size is updated only in those iterations

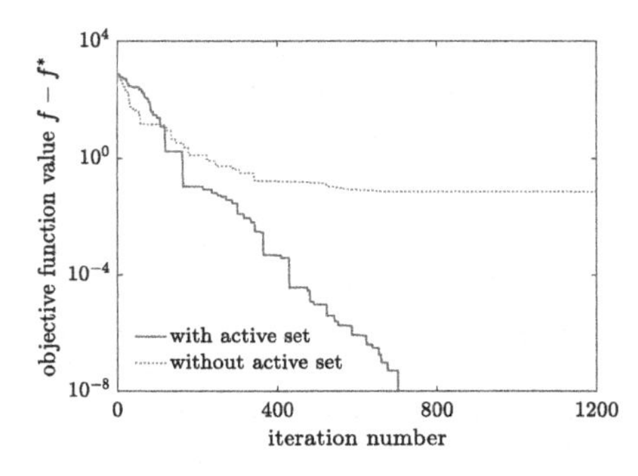

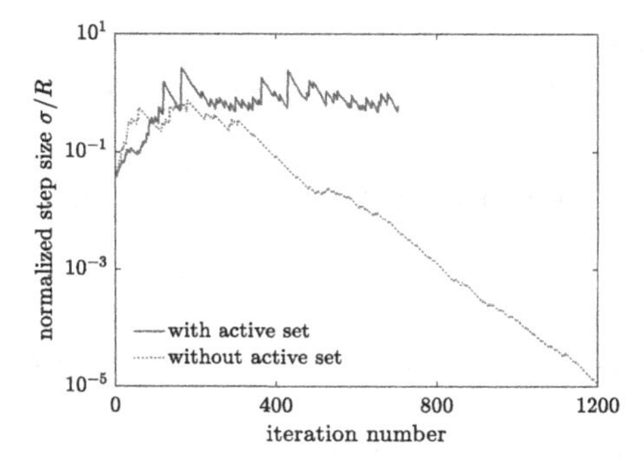

Fig. 2. Traces from runs of the active-set (1+1)-ES and a (1+1)-ES that projects infeasible candidate solutions onto the feasible region but does not enforce active inequalities as equalities applied to a 10-dimensional sphere with five mutually orthogonal linear inequality constraints active at the optimal solution. Shown are the evolution of the difference between $f(\mathbf{x})$ and the optimal objective function value f^* and the step size σ normalized by division by $R = \|\mathbf{x} - \mathbf{x}^*\|$.

where the search proceeds in the reduced search space, thus avoiding the issue of systematically decreasing σ when the use of the active set is suspended.

Throughout a run of the algorithm, we store the current active set. Constraints are added to the active set whenever a candidate solution is accepted for which those constraints are tight. Notice that it is straightforward from the output of `fmincon` to determine which constraints are tight by identifying those inequality constraints that have positive Lagrange multipliers. The active set is replaced in those iterations where a candidate solution replaces its parent that is generated with use of the active set suspended.

Figure 2 illustrates the advantage of the active-set based approach over a $(1+1)$-ES that simply projects infeasible candidate solutions onto the feasible region, without enforcing active inequality constraints as equalities, and that updates the step size in every iteration. We have conducted 21 runs of both strategies for objective function $f(\mathbf{x}) = \mathbf{x}^{\mathrm{T}}\mathbf{x}$ with $n = 10$ and constraints $\mathbf{x}^{\mathrm{T}}\mathbf{e}_i \geq 1$ for $i \in \{1, \ldots, 5\}$, where $\mathbf{e}_i$ is the unit vector in the direction of the ith coordinate axis. All runs are initialized to start at $\mathbf{x} = (9, \ldots, 9)^{\mathrm{T}}$ and with step size $\sigma = 1$. The active-set $(1+1)$-ES attained an objective function value within a factor of $(1 + 10^{-8})$ of the optimal objective function value $f^* = f(\mathbf{x}^*)$ in each of the runs; the run that required the median number of iterations is shown in the figure. It can be seen that the strategy converges linearly, and that after an initial increase, the step size σ is controlled to be approximately proportional to the distance R from the optimal solution $\mathbf{x}^*$. Not shown, in the run depicted, the five constraints become active between the 66th and the 100th iteration and remain active until the algorithm terminates. Without the use of the active set, none of the 21 runs obtained a solution with an objective function value within a factor of $(1+10^{-8})$ of f^* before terminating after 1,200 iterations. The corresponding trace shown in the figure is that of a random run. It can be seen that not enforcing active inequalities as equalities and updating σ in every iteration eventually results in very small step sizes and slow progress.

3 Evaluation and Discussion

We evaluate the active-set (1 + 1)-ES by applying it to the test problems `g01` through `g11` gathered by Michalewicz and Schoenauer [11] and summarized by Liang et al. [10]. We initialize the parental candidate solution by uniformly randomly sampling a point in the bound constrained search space and then projecting it onto the feasible region using the same approach as described for offspring candidate solutions in Sect. 2. The initial step size is set to $\sigma = 0.2 \min\{u_i - l_i \,|\, i = 1, \ldots, n\}$, where the l_i and u_i are the lower and upper bounds of the search space in dimension i for the respective problem. A run of the algorithm is terminated and considered a success if a candidate solution $\mathbf{x}$ with an objective function value $f(\mathbf{x}) < (1 + \epsilon)f^*$ is found, where f^* is the objective function value of the optimal solution to the problem. It is considered unsuccessful if after 1,200 iterations (and thus as many evaluations of the objective function) no solution satisfying the termination criterion has been found. We refer to ϵ as the target accuracy.

Table 1. Test function properties and results.

	g01	g02	g03	g04	g05	g06	g07	g08	g09	g10	g11
Dimension n	13	20	10	5	4	2	10	2	7	8	2
Reduced dimension n^*	0	19	9	0	1	0	4	2	5	2	1
Success rate:											
$\epsilon = 10^{-4}$	1.0	0.0	1.0	1.0	1.0	1.0	0.98	0.52	1.0	1.0	1.0
$\epsilon = 10^{-8}$	1.0	0.0	1.0	1.0	1.0	1.0	0.99	0.53	1.0	1.0	1.0
Median number of function evaluations:											
$\epsilon = 10^{-4}$	49	—	458	22	36	3	411	123	302	117	27
$\epsilon = 10^{-8}$	46	—	865	24	84	4	684	229	616	243	106

We have conducted 100 runs of the algorithm for each test problem and target accuracies $\epsilon \in \{10^{-4}, 10^{-8}\}$[2]. Table 1 summarizes the results. For eight of the eleven test problems, the globally optimal solution was found to the desired accuracy in all 100 runs conducted. The three exceptions are as follows:

[2] The `ConstraintTolerance` parameter of `fmincon` is set to its default value of 10^{-6} for the runs with target accuracy $\epsilon = 10^{-4}$. For target accuracy $\epsilon = 10^{-8}$ we used `ConstraintTolerance` 10^{-9} instead as some runs for problem `g03` terminate unsuccessfully if the default accuracy is used.

- For g02 not a single run successfully located the globally optimal solution. Problem g02 has a very large number of local minima, and the search space is such that the likelihood of starting in the basin of attraction of the global optimum is near zero. A stochastic hill climber, such as the $(1+1)$-ES, will almost always converge toward a merely local minimum.
- A small number of unsuccessful runs are observed for problem g07, even though the problem is unimodal. The search space is ten-dimensional, with six inequality constraints active at the optimal solution. In some runs, the upper bound constraint for variable x_8 is included in the active set of the algorithm at some point during the run. The likelihood of escaping the point that is optimal for this active set in one of the steps where the use of the active set is suspended can be observed to be no higher than 5 % for a large range of values of the step size. It may thus take hundreds of iterations before the upper bound constraint is rendered inactive, and either reaching the iteration limit or the step size becoming so small that the limits of numerical accuracy are reached prevents successfully escaping the non-optimal reduced search space.
- Problem g08 has four locally optimal solutions. No constraints are active at the globally optimal one of those. The $(1+1)$-ES converges to one of the merely locally optimal solutions with a likelihood of just under one half. Conducting multiple runs of the algorithm would allow locating the globally optimal solution with high probability.

Figure 3 shows histograms of the number of objective function evaluations required to solve problems g01 and g03 through g11 to both target accuracies. The ranges of the histograms are such that all successful runs are included. No data are shown for g02 as no successful runs were observed for that problem. It can be seen that the histograms for g01, g04, and g06 differ fundamentally from those for the other problems in that there is little difference between the data for $\epsilon = 10^{-4}$ and $\epsilon = 10^{-8}$. This is due to the dimension of the optimal reduced search space being zero. In that case, solving for the solution at the intersection of the active constraints yields the optimal solution (up to the limits of numerical accuracy). For the remaining problems, the gap between the histograms for $\epsilon = 10^{-4}$ and $\epsilon = 10^{-8}$ is due to the need for the $(1+1)$-ES to search a non-zero reduced search space, with a larger discrepancy for those cases where the dimension of that space is large.

Figure 4 shows traces from runs requiring the median number of iterations to reach target accuracy $\epsilon = 10^{-8}$ for test problems g04 and g10. The former is an example of a problem where the dimension of the optimal reduced search space is zero; for the latter, that dimension is $n^* = 2$. Plotted against the iteration number are the difference between the objective function value $f(\mathbf{x})$ and the optimal objective function value f^* as well as the step size σ of the algorithm. Those iterations where the algorithm suspends the use of the active set (i.e., κ is true) are marked with small circles.

In the run on g04, the evolution strategy generates the optimal active set $\mathcal{A}^*$ in iteration 24, at which point it terminates. Between iterations 18 and 21, the use

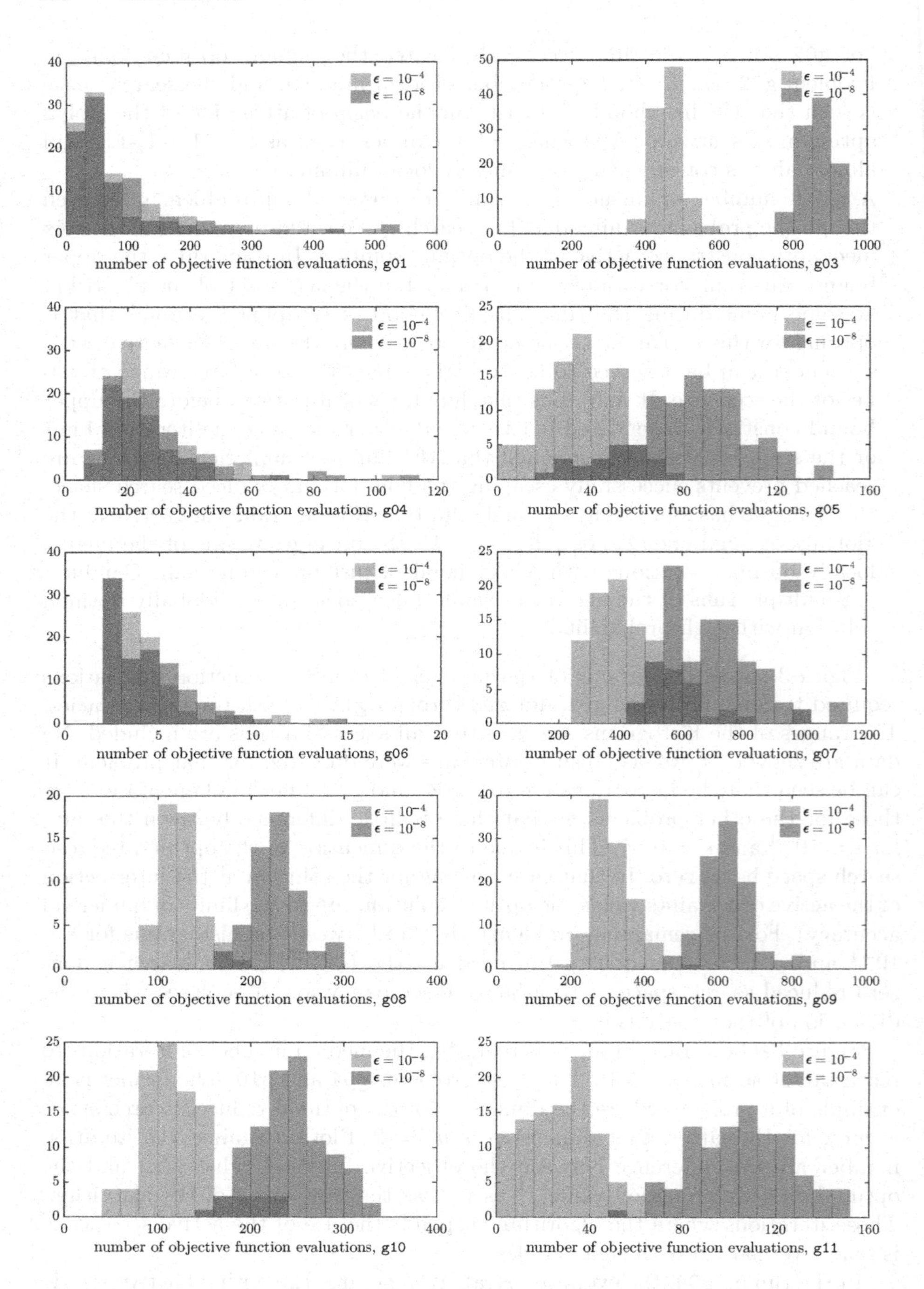

Fig. 3. Histograms showing the number of objective function evaluations required to solve problems g01 and g03 through g11.

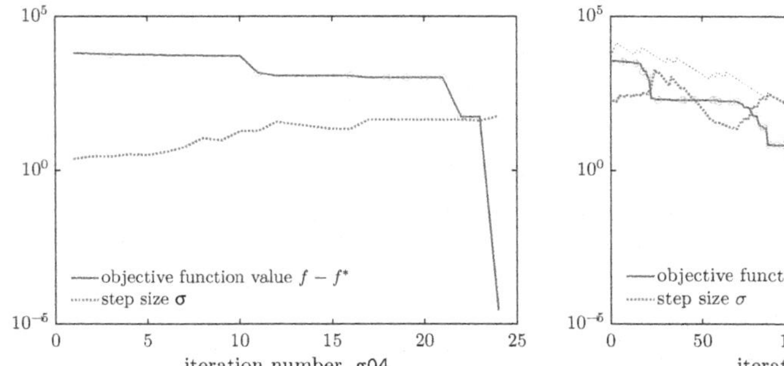

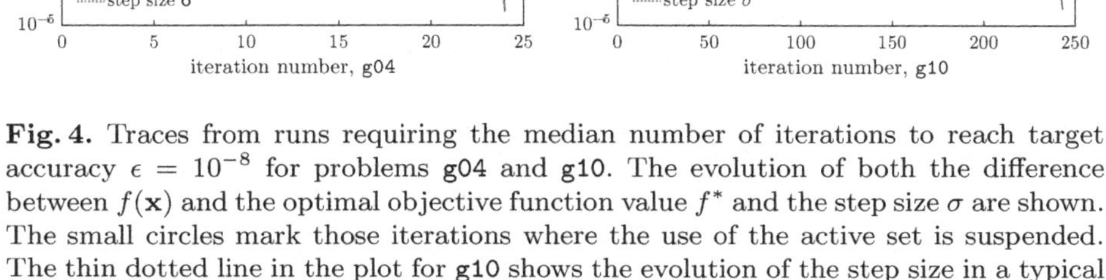

Fig. 4. Traces from runs requiring the median number of iterations to reach target accuracy $\epsilon = 10^{-8}$ for problems g04 and g10. The evolution of both the difference between $f(\mathbf{x})$ and the optimal objective function value f^* and the step size σ are shown. The small circles mark those iterations where the use of the active set is suspended. The thin dotted line in the plot for g10 shows the evolution of the step size in a typical run of the $(1+1)$-ES on an unconstrained, two-dimensional sphere function.

of the active set is suspended in each step as it is such that the dimension of the reduced search space is zero. The step size σ largely increases through most of the run.

In the run on g10, the algorithm arrives at the optimal active set $\mathcal{A}^*$, which consists of the six constraints active at the global optimum of the problem, at iteration 70. The active set remains stable after this, and optimization effectively proceeds in a reduced subspace of dimension $n^* = 2$, with the exception of those steps where the use of the active set is suspended, but which have no further effect on the sequence of successful candidate solution generated. It can be seen that the step size σ largely decreases as the search in the two-dimensional subspace progresses. Comparison with the thin dotted curve, which shows the rate of decrease of the step size of a $(1+1)$-ES on an unconstrained, two-dimensional sphere function, shows a similar rate of linear convergence.

A comparison of the performance of the active-set $(1+1)$-ES with that of other approaches to constrained evolutionary optimization is not straightforward as both initialization conditions and termination criteria often differ for approaches found in the literature. More significantly, most other approaches consider the constraint functions as black boxes and are thus not easily able to project candidate solutions onto the feasible region (or subspaces thereof). That said, some useful points of comparison do exist:

- The numbers of objective function evaluations required by the active covariance matrix adaptation based approach by Arnold and Hansen [3] to solve four of the problems[3] to target accuracy 10^{-8} range from 308 for g06 to 3,976 for g10. The corresponding figures from Table 1 range from 4 for g06 to 684

[3] Multimodal problems and those with equality constraints were not considered in that paper, leaving only problems g06, g07, g09, and g10.

for g07. It is important to keep in mind though that the active-set approach introduced here assumes knowledge of the constraint functions, whereas the algorithm from [3] considers the constraint functions as black boxes.
- The algorithm by Takahama and Sakai [16], which performed best among all entries submitted to the *CEC 2006 Special Session on Constrained Real-Parameter Optimization*, assumes knowledge of gradient vectors and thus does not treat the constraint functions as black boxes. It requires median numbers of objective function evaluations ranging from 1,182 for g08 to 105,799 for g10. The algorithm does successfully solve g02.
- Bagheri et al. [6] propose SACOBRA, a self-adaptive variant of the surrogate based COBRA algorithm by Regis [14], which solves most of the eleven problems (though not g02, and others only with limited accuracy) with fewer than 500 objective function evaluations. SACOBRA considers the constraint functions as black boxes and is thus applicable when the constraints are not known. It requires more function evaluations than the active-set $(1+1)$-ES for a number of those problems where the dimension of the optimal reduced search space is low. However, it appears to often converge faster where that dimension is not very small. This latter advantage is a consequence of the smooth nature of the test problems, which admit polynomial surrogate models that make it possible to converge superlinearly. As shown by Teytaud and Gelly [17], as a comparison-based algorithm that does not use objective function values other than in comparisons, the active-set $(1+1)$-ES cannot exhibit super-linear convergence.

4 Conclusions

To conclude, we have proposed an active-set $(1+1)$-ES for constrained numerical optimization with known constraints. The algorithm usually generates offspring candidate solutions constrained to the reduced subspace at their parents, but with a fixed probability samples offspring that do not necessarily fall into that space, thus allowing it to render active constraints inactive. Key to the functioning of the algorithm is to adapt the step size only in those iterations where the offspring are constrained to the reduced search space. The algorithm can be implemented in a few lines of *Matlab* code and performs very well when compared with related work.

It is of interest to apply the proposed active-set approach in evolutionary algorithms other than the $(1+1)$-ES. The $(\mu/\mu, \lambda)$-ES is an evolutionary algorithm less sensitive to noise and ruggedness of the objective than the $(1+1)$-ES as it is capable of proceeding with larger steps. Restart variants of that algorithm [5] are commonly used for multimodal optimization problems and may exhibit improved performance for problems g02 and g08. However, using the active-set approach with cumulative step size adaptation is less than straightforward, and it has been seen that care has to be taken when integrating constraint handling techniques with step size adaptation [1,8]. Also of interest is the problem of employing the active-set approach in combination with other constraint handling techniques, such as augmented Lagrangian methods [4], in case only a proper subset of the constraints is known explicitly. Finally, it is of interest to employ the active-set approach in evolutionary algorithms that use surrogate models of the objective function in order to reduce the time spent on optimization in reduced search spaces.

Acknowledgements. This research was supported by the Natural Sciences and Engineering Research Council of Canada (NSERC).

References

1. Arnold, D.V.: Resampling versus repair in evolution strategies applied to a constrained linear problem. Evol. Comput. **21**(3), 389–411 (2013)
2. Arnold, D.V., Brauer, D.: On the behaviour of the (1+1)-ES for a simple constrained problem. In: Rudolph, G., Jansen, T., Lucas, S., Poloni, C., Beume, N. (eds.) PPSN 2008. LNCS, vol. 5199, pp. 1–10. Springer, Heidelberg (2008)
3. Arnold, D.V., Hansen, N.: A (1 + 1)-CMA-ES for constrained optimisation. In: Genetic and Evolutionary Computation Conference – GECCO, pp. 297–304. ACM Press (2012)
4. Arnold, D.V., Porter, J.: Towards an augmented Lagrangian constraint handling approach for the (1+1)-ES. In: Genetic and Evolutionary Computation Conference – GECCO, pp. 249–256. ACM Press (2015)
5. Auger, A., Hansen, N.: A restart CMA evolution strategy with increasing population size. In: IEEE Congress on Evolutionary Computation – CEC, pp. 1769–1776. IEEE Press (2005)
6. Bagheri, S., Konen, W., Emmerich, M., Bäck, T.: Solving the G-problems in less than 500 iterations: improved efficient constrained optimization by surrogate modeling and adaptive parameter control (2015). arXiv:1512.09251
7. Hansen, N., Arnold, D.V., Auger, A.: Evolution strategies. In: Kacprzyk, J., Pedrycz, W. (eds.) Springer Handbook of Computational Intelligence, pp. 871–898. Springer, Berlin (2015)
8. Hellwig, M., Arnold, D.V.: Comparison of constraint handling mechanisms for the $(1, \lambda)$-ES on a simple constrained problem. Evol. Comput. **24**(1), 1–23 (2016)
9. Kern, S., Müller, S.D., Hansen, N., Büche, D., Ocenasek, J., Koumoutsakos, P.: Learning probability distributions in continuous evolutionary algorithms – a comparative review. Nat. Comput. **3**(1), 77–112 (2004)
10. Liang, J.J., Runarsson, T.P., Mezura-Montes, E., Clerc, M., Suganthan, P.N., Coello Coello, C.A., Deb, K., Problem definitions and evaluation criteria for the CEC special session on constrained real-parameter optimization. Technical report, Nanyang Technological University, Singapore (2006)
11. Michalewicz, Z., Schoenauer, M.: Evolutionary algorithms for constrained parameter optimization problems. Evol. Comput. **4**(1), 1–32 (1996)
12. Nocedal, J., Wright, S.: Numerical Optimization, 2nd edn. Springer, Berlin (2006)
13. Rechenberg, I.: Evolutionsstrategie - Optimierung technischer Systeme nach Prinzipien der biologischen Evolution. Friedrich Frommann Verlag, Stuttgart (1973)
14. Regis, R.G.: Constrained optimization by radial basis function interpolation for high-dimensional expensive black-box problems with infeasible initial points. Eng. Optim. **46**(2), 218–243 (2014)

15. Schwefel, H.-P.: Numerical Optimization of Computer Models. Wiley, Hoboken (1981)
16. Takahama, T., Sakai, S.: Constrained optimization by the ϵ constrained differential evolution with gradient-based mutation and feasible elites. In: IEEE World Congress on Computational Intelligence - WCCI, pp. 308–315. IEEE Press (2006)
17. Teytaud, O., Gelly, S.: General lower bounds for evolutionary algorithms. In: Runarsson, T.P., Beyer, H.-G., Burke, E.K., Merelo-Guervós, J.J., Whitley, L.D., Yao, X. (eds.) PPSN 2006. LNCS, vol. 4193, pp. 21–31. Springer, Heidelberg (2006)

Speciated Evolutionary Algorithm for Dynamic Constrained Optimisation

Xiaofen Lu[1,2(✉)], Ke Tang[2], and Xin Yao[1,2]

[1] CERCIA, School of Computer Science, University of Birmingham, Edgbaston, Birmingham B15 2TT, UK
{xxl332,x.yao}@cs.bham.ac.uk
[2] UBRI, School of Computer Science and Technology, University of Science and Technology of China (USTC), Hefei 230027, Anhui, China
ketang@ustc.edu.cn

Abstract. Dynamic constrained optimisation problems (DCOPs) have specific characteristics that do not exist in dynamic optimisation problems with bounded constraints or without constraints. This poses difficulties for some existing dynamic optimisation strategies. The maintaining/introducing diversity approaches might become less effective due to the presence of infeasible areas, and thus might not well handle with the switch of global optima between disconnected feasible regions. In this paper, a speciation-based approach was firstly proposed to overcome this, which utilizes deterministic crowding to maintain diversity, assortative mating and local search to promote exploitation, as well as feasibility rules to deal with constraints. The experimental studies demonstrate that the newly proposed method generally outperforms the state-of-the-art algorithms on a benchmark set of DCOPs.

Keywords: Evolutionary algorithm · Speciation · Deterministic crowding · Local search · Dynamic constrained optimisation problem

1 Introduction

In the real world, many optimisation problems are changing over time due to the dynamic environments [6,19]. These problems require an optimisation algorithm to quickly find the new optimum once the problem changes [20]. As a class of nature-inspired optimisation methods, evolutionary algorithms (EAs) can have good adaptation to the changing environments, and thus have been widely studied in the field of dynamic optimisation (DO). Many evolutionary DO approaches have been developed, which include maintaining/introducing diversity strategies, memory approaches, prediction approaches, multi-population approaches and so on [20]. As revealed in [20,23], most existing studies of DO focus on unconstrained or bounded constrained dynamic optimisation problems (DOPs), and few consider dynamic constrained optimisation problems (DCOPs) despite their high popularity in real-world applications. In a DCOP, a change may occur

J. Handl et al. (Eds.): PPSN XIV 2016, LNCS 9921, pp. 203–213, 2016.
DOI: 10.1007/978-3-319-45823-6_19

in either constraints or objective functions or both. Therefore, DCOPs have some specific characteristics compared to unconstrained or bounded constrained DOPs. For a DCOP, the distribution of infeasible/feasible solution might change, and the global optima might move to another disconnected feasible region, or appear in a new region without changing the current optima due to the dynamics of environments [19,23].

Addressing DCOPs poses difficulties for some existing DO strategies and constraint handling (CH) techniques due to their characteristics. As discussed in [19,23], maintaining/introducing diversity methods such as random-immigrants (RI) [12] and hyper-mutation (HyperM) [5] become less effective on DCOPs than on DOPs, when combined with penalty function that prefers feasible solutions to infeasible ones. This is because that they cannot maintain enough diversity to adapt to the new problem as the introduced random solutions are likely to be rejected by the penalty function if they are infeasible. Furthermore, without enough diversity, they may not deal well with the switch of the global optimum between disconnected feasible regions, as it needs to go through an infeasible path from the previous optimum to the current optimum. Moreover, some adaptive/self-adaptive CH techniques also face challenges in solving DCOPs as they need the knowledge of problem that is unavailable in a dynamic environment or historical information that might be outdated once the problem changes [23].

Researchers have recently carried out some studies trying to solve the challenges of DCOPs. To allow diversified infeasible solutions distributed in the whole search space, the authors in [19,25], and [2] applied the repair method [27] to handle constraints along with RI/HyperM to deal with dynamics. However, these methods need a lot of feasibility checkings, and thus cannot be applied to DCOPs in which the ratio of feasible solutions is very low and a feasibility checking is computationally costly. The authors in [1] employed simple feasibility rules [17] as the CH strategy along with RI and combined DE variants to introduce diversity after each change. However, this method may not maintain diversity well during the run as infeasible solutions are still likely abandoned by feasibility rules. Furthermore, it might be ineffective to make the partially converged population to re-diversify to track the switched global optima. Similarly, the proposed approach in [4] to deal with DCOPs might not quickly find the switched optima as the population tends to converge during the run.

In this paper, a speciation-based method, called speciated evolution with local search (SELS), is suggested to address the challenges of DCOPs. Speciation allows an EA to find multiple optima through making comparisons among similar individuals [7]. Thus, newly generated and promising infeasible solutions can be accepted in the new method. Furthermore, good solutions can be maintained in different feasible regions, and thus SELS should react quickly when the global optimal solution switches to another feasible region. In the literature, speciation has been utilized to solve dynamic unconstrained or bounded constrained optimisation problems [14,15], but no studies apply them to DCOPs. In addition to speciation, a local search strategy is employed in SELS to promote exploitation of the promising regions to quickly find the changed optimum.

SELS also uses a change detection method, and adds some random immigrants into the population to introduce diversity once a change is detected. Finally, to deal with constraints, the simple and parameter-free feasibility rules [17] are employed.

The remaining part of this paper is organized as follows. Section 2 summarizes the existing related work, and Sect. 3 details the new method. In Sect. 4, experimental results are presented, and conclusions and future work will be given in Sect. 5.

2 Related Work

Solving DCOPs does not only need DO strategies to deal with dynamics but also requires CH techniques to handle constraints. This section will introduce the efforts that researchers have done in combining DO strategies and CH techniques to solve DCOPs.

The RI and HyperM methods were combined in [23] with a penalty function proposed in [18]. When using RI, a fraction of the population is replaced by random solutions at every generation to maintain diversity. In hybrid with HyperM, the original mutation rate will change to a higher one to introduce diversity once change is detected. However, as most of the added random solutions are infeasible and they are likely rejected by the used penalty function, these combination methods can not maintain enough diversity to adapt to the new problem. Therefore, the authors suggested that a CH technique that can maintain diversified infeasible solutions is needed, when combined with RI and HyperM to solve DCOPs.

To maintain diversified infeasible solutions in solving DCOPs, the authors in [22] combined the repair method with RI and HyperM, respectively. By using the repair method, an infeasible solution was evaluated by the repaired feasible solution. Thus, the infeasible solutions that can make good feasible solutions are reserved. Other studies using the repair methods exist in [2, 24]. The former work used an improved repair method, and the latter applied the repair method and RI together in a DE context. However, using the repair method has a big disadvantage. That is, it requires a considerable number of feasibility checking during the repair process, and thus not suitable for problems with very small feasible area or expensive feasibility checking.

A simple ranking scheme in [13] was applied to handling constraints in [3]. To maintain population diversity, the method monitored the population diversity and switched between a global search and a local search operator according to whether the diversity degree is larger than a threshold. As a result, this method will highly depends on the setting of the threshold. To avoid the setting of the threshold, the authors in [4] used the Shannon's index of diversity as a factor to balance the influence of the global-best and local-best search directions. However, the population tends to converge in this method, and it might be ineffective to make the partially converged population to re-diversify to track the switched global optimum.

The authors in [1] employed simple feasibility rules in [17] as the CH strategy, and considered RI and combined DE variants to maintain and introduce diversity, respectively. However, this method may not well maintain diversity as infeasible solutions are likely abandoned according to feasibility rules, and thus might not effectively solve the switching global optima between disconnected feasible regions.

Except DO strategies to introduce/maintain diversity were considered, other DO strategies such as memory and prediction methods were also combined with CH techniques to deal with DCOPs. The study in [26] adapted abstract memory method, and the infeasibility driven evolutionary algorithm (IDEA) was combined with prediction method (to predict the future optima) in [10] and [11] to solve DCOPs. However, both memory and prediction methods are only applicable to particular dynamic problems (i.e., cyclic and predictable dynamic problems, respectively).

3 The Proposed Method

The proposed SELS method considers simple feasibility rules to handle constraints, which do not need to repair infeasible solutions. However, as feasibility rules prefer feasible solutions to infeasible ones, the diversity will decrease quickly. To avoid this, the speciation is employed, which makes comparison among similar individuals. Thus, SELS should maintain good diversity and respond quickly once the change happens. As speciation focuses on exploration, the new optima might not be found quickly as promising regions are not exploited sufficiently. Therefore, SELS uses a local search strategy to promote exploitation of the promising regions. In the following part of this section, the elements of SELS will be first described, and then the pseudo-code is given.

Speciation Method. This work uses deterministic crowding (DC) as the speciation method. Algorithm 1 gives the pseudo-code of DC. The DC method pairs all population elements randomly and generates two offspring for each pair based on EA operators. Selection is then operated on these four individuals, and a similarity measure is used to decide which offspring competes against which parent. The offspring will replace the compared parent and enter next generation if it is fitter.

In addition to DC, SELS employs an assortative mating (AM) [8] to induce speciation in the population. As DC can maintain good solutions on different peaks or in different feasible regions, intuitively, we would not like to operate crossover between solutions on different peaks or in different feasible regions as doing this will likely generate solutions in the valley or infeasible regions. To avoid this, assortative mating is used, which mates individuals with the most similar non-identical partner in the population. Through doing crossover between individuals in proximity, species will be automatically generated, and the exploitation of the corresponding search area will also be enhanced. In this work, Euclidean distance is used as the similarity measure.

Algorithm 1. Deterministic Crowding [16]

```
1: Randomly pair all individuals in the population
2: for each pair of individuals, p1 and p2, do
3:    Generate two offspring, o1 and o2, based on EA operators
4:    if dist(p1, o1) + dist(p2, o2) ≤ dist(p1, o2) + dist(p2, o1) then
5:       p1 = fitter(p1, o1)
6:       p2 = fitter(p2, o2)
7:    else
8:       p1 = fitter(p1, o2)
9:       p2 = fitter(p2, o1)
10:   end if
11: end for
```

Feasibility Rules. Feasibility rules have been commonly used to solve constrained optimisation problems due to its simple and parameter-free characteristics, which make comparison between individuals through the following three selection criteria:

1. If both are feasible solutions, the one with the highest fitness value is selected.
2. Between a feasible solution and an infeasible solution, the feasible one is preferred.
3. If both are infeasible, the one with the lowest sum of constraint violation wins.

In the proposed SELS, feasibility rules are employed to determine the fitter one in each pair of parent and offspring in the deterministic crowding.

Local Search (LS). In this work, the local evolutionary search enhancement by random memorizing (LESRM) [28] is applied to the best solution (x_{best}) at each generation. The LESRM uses an EA that has a step size control and adjusts the search direction based on individuals encountered before, and thus can do efficient exploitation in the area that the best solution is located in. In this paper, the EA used in LESRM is given in Algorithm 2, and the random memorizing part of LESRM is the same as in [28]. Here, ls_{num} denotes the maximum number function evaluations (FEs) permitted for LS at every generation, and D denotes dimension of the problem. The success ratio of δ denotes the ratio that x_{new} (generated by δ) is better than x_{best}.

Change Detection and Diversity Introduction. To detect the change, assume k is the number of individuals for detection and NP is the size of population, the (NP/k)-th, ($2*NP/k$)-th, ($(k-1)*NP/k$)-th, ..., (NP)-th (denoted as detection index) individual in the population are reevaluated at every generation to detect changes in time. Once a change is detected, the whole population will be re-evaluated, and NI random individuals will be generated to randomly replace the individuals of the population except the best individual. In our work, we

Algorithm 2. Evolutionary Search in LESRM in SELS (x_{best})

1: Find the closest non-identical solution x_{near} to x_{best}
2: Set $\delta =$ dist (x_{near}, x_{best}), and LS generation counter $ls_g = 0$
3: **repeat**
4: **repeat**
5: Set $ls_g = ls_g + 1$, and $x_{new} = x_{best} + \delta$*randn(1,D);
6: **if** mod(ls_g,2)==0 **then**
7: Calculate the success ratio of δ
8: Set $\delta = \delta/2$ if the success ratio of δ is less than 0.5
9: Set $\delta = \delta * 2$ if the success ratio of δ is larger than 0.5
10: **end if**
11: **until** x_{new} is fitter than x_{best}
12: Set $x_{best} = x_{new}$
13: Do random memorizing as in [28]
14: **until** ls_{num} function evaluations are used up

first give each individual in the population a rank based on the feasibility rules, and estimate the degree of change severity as the ratio of the number of reverse order after re-evaluation. We then set NI to max(($\lceil reverse\_ratio * NP \rceil$, 2).

The Framework of the Proposed Method. Algorithm 3 gives the whole process of SELS, which begins with a randomly generated population of candidate solutions and utilizes genetic algorithm (GA) to evolve this population.

Algorithm 3. The Framework of SELS

1: Evaluate a randomly generated population $\mathbf{P} = \{x_i | i = 1, 2, ..., NP\}$
2: **while** computational resources are not used up **do**
3: **repeat**
4: Randomly select an unpaired individual x from the population
5: Calculate Euclidean distance between x and each other unpaired individual
6: Pair x with the individual that has smallest positive Euclidean distance to x
7: **until** all individuals in the population are paired
8: **for** $i \leftarrow 1, NP$ **do**
9: Re-evaluate x_i or x_{i+1} if i or $i + 1$ is one detection index
10: **if** the change is detected **then**
11: Re-evaluate solutions in $\mathbf{P}$ and introduce diversity, go to Step 17
12: **else**
13: Generate two offspring o_1, o_2 from x_i and x_{i+1} with GA crossover and mutation
14: Do deterministic crowding(x_i, x_{i+1}, o_1, o_2), and set $i = i + 1$
15: **end if**
16: **end for**
17: Do local search to the best solution in $\mathbf{P}$
18: **end while**

4 Experimental Studies

4.1 Experimental Setup

To assess the efficacy of SELS, we conducted experiments on 11 DCOP benchmark test functions proposed in [21]. They are: G24-l(dF,fC), G24-2(dF,fC), G24-3(dF,dC), G24-4(dF,dC), G24-5(dF,dC), G24-6a(2DR,hard), G24-6c(2DR,easy), G24-7(fF,dC), G24-6d(2DR,hard), G24-8b(fC,OICB). We recorded the performance of SELS on each test function using the modified offline error [22] as the performance metric, then compared its performance to that of 6 state-of-the-art algorithms. They are, dGArepairRIGA [22], dGArepairHyperM [22], GSA + Repair [24], DDECV + Repair [2], DDECV [1] and EBBPSO-T [4]. In the experiments, the number of changes is set to 12, the change frequency is 1000 objective function evaluations, and the change severity is medium (i.e., $k = 0.5$, and $s = 20$). We use this setting as all of the 6 compared algorithms have presented complete experimental results only on this setting in their original papers.

Note that the first 4 of the compared algorithms use a repair scheme, so they need a lot of feasibility checking but they ignore the cost. To make a fair comparison, we run SELS only evaluating the feasibility for an infeasible solution and do not count in the number of used fitness evaluations. The resulted algorithm is denoted as eSELS and compared to the 4 repair algorithms. When compared to DDECV and EBBPSO, SELS evaluates both the feasibility and objective function value for every individual, no matter feasible or infeasible, which is the same to what DDECV and EBBPSO do.

In the experiments, for both SELS and eSELS, intermediate crossover with $p_c = 1.0$, and Guassian mutation with $p_m = 1/D$ and scale$=0.1$ are used, respectively. In the mutation, at least one variable is mutated every time. The number of LS objective function evaluations (ls_{num}) is set to 16, and 4 individuals are used for change detection every generation. Each algorithm is run 50 times on each test function.

4.2 Comparison Results with Existing Algorithms

Table 1 summarizes the mean and standard deviation of the modified offline error over 50 runs obtained by DDECV, EBBPSO-T and SELS as well as the performance rank of each algorithm on each test function (in case of ties, average ranks are assigned) based on Z-test with a level of 0.05. We applied the Friedman test and further a Holm's post-hoc procedure [9], which was used for multiple comparison of algorithms, to investigate whether SELS performed best on the set of test functions. The analysis shows that SELS has significant improvement than DDECV and EBBPSO-T at a level of 0.05.

Table 2 gives the performance rank of eSELS and the other 4 repair algorithms on each test function. We also applied the Friedman test and further a Holm's post-hoc procedure [9] to do a multiple-problem comparison among the 5 algorithms. The statistical test results show that eSELS performed significantly better than each other algorithm on the test function set at a level of 0.05.

Table 1. Comparison results between DDECV, EBBPSO-T and SELS based on experimental results of DDECV and EBBPSO-T in their original papers [1,4], respectively. The best result obtained on each function is marked in **bold**.

Func	DDECV[rank]	EBBPSO-T[rank]	SELS[rank]
G24-l	0.109 ± 0.033[3]	0.084 ± 0.041[2]	**0.025 ± 0.008**[1]
G24-2	0.126 ± 0.030[2]	0.136 ± 0.013[3]	**0.050 ± 0.015**[1]
G24-3	0.057 ± 0.018[3]	**0.032 ± 0.005**[1]	0.044 ± 0.022[2]
G24-3b	0.134 ± 0.033[3]	0.104 ± 0.015[2]	**0.052 ± 0.018**[1]
G24-4	0.131 ± 0.032[2.5]	0.138 ± 0.022[2.5]	**0.082 ± 0.021**[1]
G24-5	0.126 ± 0.030[2.5]	0.126 ± 0.019[2.5]	**0.054 ± 0.014**[1]
G24-6a	0.215 ± 0.067[3]	0.116 ± 0.099[2]	**0.055 ± 0.009**[1]
G24-6c	0.128 ± 0.025[2]	0.251 ± 0.061[3]	**0.052 ± 0.008**[1]
G24-6d	0.288 ± 0.055[2.5]	0.312 ± 0.203[2.5]	**0.041 ± 0.007**[1]
G24-7	0.106 ± 0.022[3]	**0.045 ± 0.009**[1]	0.087 ± 0.016[2]
G24-8b	0.151 ± 0.058[2]	0.312 ± 0.086[3]	**0.055 ± 0.022**[1]

Table 2. Comparison results among eSELS and the other 4 repair algorithms based on the experimental results in their original papers.

Func	dRepairRIGA [rank]	dRepairHyperM [rank]	GSA + Repair [rank]	DDECV + Repair [rank]	eSELS[rank]
G24-l	0.082 ± 0.015[3]	0.093 ± 0.023[4]	0.132 ± 0.015[5]	0.061 ± 0.010[2]	**0.013 ± 0.007**[1]
G24-2	0.162 ± 0.021[3.5]	0.171 ± 0.026[3.5]	0.182 ± 0.019[5]	0.062 ± 0.006[2]	**0.030 ± 0.008**[1]
G24-3	0.029 ± 0.004[4]	0.027 ± 0.005[2.5]	0.028 ± 0.004[2.5]	0.046 ± 0.006[5]	**0.018 ± 0.004**[1]
G24-3b	0.058 ± 0.007[2]	0.071 ± 0.014[3]	0.076 ± 0.009[4]	0.084 ± 0.006[5]	**0.021 ± 0.004**[1]
G24-4	0.140 ± 0.028[5]	0.059 ± 0.010[2]	0.073 ± 0.012[3]	0.088 ± 0.011[4]	**0.036 ± 0.009**[1]
G24-5	0.152 ± 0.017[4.5]	0.131 ± 0.019[3]	0.153 ± 0.013[4.5]	0.078 ± 0.008[2]	**0.027 ± 0.024**[1]
G24-6a	0.366 ± 0.033[4.5]	0.358 ± 0.049[4.5]	**0.033 ± 0.003**[1]	0.036 ± 0.005[2.5]	0.038 ± 0.006[2.5]
G24-6c	0.323 ± 0.037[4.5]	0.326 ± 0.047[4.5]	0.045 ± 0.004[3]	0.041 ± 0.010[1.5]	**0.040 ± 0.007**[1.5]
G24-6d	0.315 ± 0.029[5]	0.286 ± 0.035[4]	0.037 ± 0.007[2]	0.079 ± 0.006[3]	**0.029 ± 0.004**[1]
G24-7	0.154 ± 0.031[5]	0.067 ± 0.014[3]	**0.018 ± 0.002**[1]	0.107 ± 0.011[4]	0.035 ± 0.045[2]
G24-8b	0.341 ± 0.053[5]	0.257 ± 0.042[4]	0.192 ± 0.034[3]	0.074 ± 0.025[2]	**0.025 ± 0.006**[1]

4.3 The Performance Effect of AM, LS and Different Dynamics

We further conducted experiments to check whether the AM and LS can help in SELS, and comparisons were made among (1) SELS without AM or LS (SELS-am-ls in short), (2) SELS without LS (SELS-ls in brief), and (3) SELS. Table 3 summarizes the mean and standard deviation of the modified offline error for each of them along with the comparison results. It is shown that SELS-ls overall outperformed SELS-am-ls, and the used LS further improves SELS-ls. This demonstrates the benefits of using AM and LS.

To evaluate the performance of SELS on different dynamics, we also conducted experiments on small change severity (i.e., $k = 1.0$, and $s = 10$) and large severity (i.e., $k = 0.25$, and $s = 50$). Figure 1 gives the evolutionary curves

Table 3. Comparison results among SELS-am-ls, SELS-ls, and SELS based on experimental results implemented on DCOP test functions. Here, +, −, and ≈ denotes whether one algorithm is better, worse or equal to another according to Wilcoxon ranksum test with a level of 0.05.

Func	SELS-am-ls	SELS-ls vs SELS-am-ls	SELS vs SELS-ls
G24-1	0.133 ± 0.037	0.069 ± 0.019 +	0.025 ± 0.008 +
G24-2	0.186 ± 0.017	0.121 ± 0.021 +	0.050 ± 0.015 +
G24-3	0.124 ± 0.038	0.118 ± 0.025 ≈	0.044 ± 0.022 +
G24-3b	0.233 ± 0.039	0.144 ± 0.021 +	0.052 ± 0.018 +
G24-4	0.199 ± 0.027	0.171 ± 0.030 +	0.082 ± 0.021 +
G24-5	0.162 ± 0.022	0.117 ± 0.017 +	0.054 ± 0.014 +
G24-6a	0.318 ± 0.040	0.163 ± 0.020 +	0.055 ± 0.009 +
G24-6c	0.284 ± 0.030	0.152 ± 0.017 +	0.052 ± 0.008 +
G24-6d	0.198 ± 0.033	0.128 ± 0.023 +	0.041 ± 0.007 +
G24-7	0.121 ± 0.017	0.138 ± 0.018 −	0.087 ± 0.016 +
G24-8b	0.387 ± 0.044	0.242 ± 0.031 +	0.055 ± 0.022 +

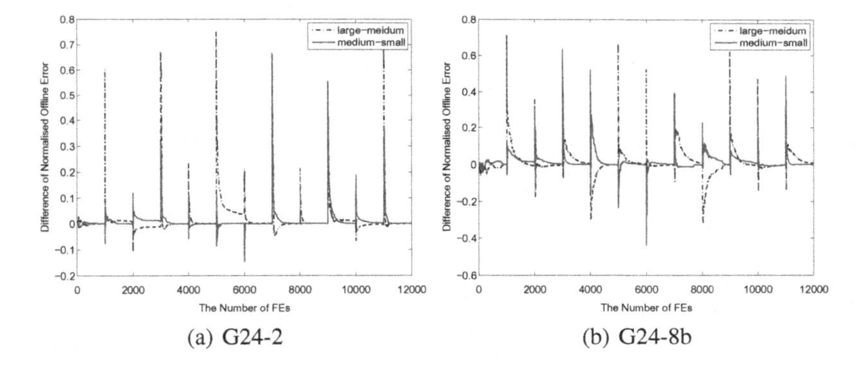

(a) G24-2 (b) G24-8b

Fig. 1. The Evolutionary difference curves of SELS between different change severity

of the normalised offline error differences between medium and small severity, and between large and small severity. The X axis denotes the number of objective function evaluations, and the Y axis denotes difference of the normalised offline error at each evaluation, which is normalised on each problem in one test function. In general, we found that SELS performed best on small severity, second best on medium, and worst on large severity.

5 Conclusion and Future Work

In this paper, a novel speciation-based method was proposed to solve DCOPs, which combines speciation methods as well as local search together.

Although the techniques used in SELS are not new, the experimental studies demonstrated the combination leads to an effective algorithm. In future work, we will study the performance effect of the choice of local search strategies on the proposed method, and will also evaluate this new method on more test functions.

Acknowledgments. This work was partially supported by NSFC (Grant No. 61329302), EPSRC (Grant No. EP/K001523/1), and Royal Society Newton Advanced Fellowship (Ref. no. NA150123). The authors thank Stefan Menzel for giving the valuable advice.

References

1. Ameca-Alducin, M.Y., Mezura-Montes, E., Cruz-Ramirez, N.: Differential evolution with combined variants for dynamic constrained optimization. In: 2014 IEEE Congress on Evolutionary Computation (CEC), pp. 975–982. IEEE (2014)
2. Ameca-Alducin, M.Y., Mezura-Montes, E., Cruz-Ramírez, N.: A repair method for differential evolution with combined variants to solve dynamic constrained optimization problems. In: Proceedings of 2015 on Genetic and Evolutionary Computation Conference, pp. 241–248. ACM (2015)
3. Campos, M., Krohling, R.: Bare bones particle swarm with scale mixtures of gaussians for dynamic constrained optimization. In: 2014 IEEE Congress on Evolutionary Computation (CEC), pp. 202–209. IEEE (2014)
4. Campos, M., Krohling, R.A.: Entropy-based bare bones particle swarm for dynamic constrained optimization. Knowl.-Based Syst. **000**, 1–21 (2015)
5. Cobb, H.G.: An investigation into the use of hypermutation as an adaptive operator in genetic algorithms having continuous, time-dependent nonstationary environments. Technical report, DTIC Document (1990)
6. Cruz, C., González, J.R., Pelta, D.A.: Optimization in dynamic environments: a survey on problems, methods and measures. Soft. Comput. **15**(7), 1427–1448 (2011)
7. Darwen, P., Yao, X.: Automatic modularization by speciation. In: Proceedings of IEEE International Conference on Evolutionary Computation, pp. 88–93. IEEE (1996)
8. De, S., Pal, S.K., Ghosh, A.: Genotypic and phenotypic assortative mating in genetic algorithm. Inf. Sci. **105**(1), 209–226 (1998)
9. Demšar, J.: Statistical comparisons of classifiers over multiple data sets. J. Mach. Learn. Res. **7**, 1–30 (2006)
10. Filipiak, P., Lipinski, P.: Infeasibility driven evolutionary algorithm with feed-forward prediction strategy for dynamic constrained optimization problems. In: Esparcia-Alcázar, A.I., Mora, A.M. (eds.) EvoApplications 2014. LNCS, vol. 8602, pp. 817–828. Springer, Heidelberg (2014)
11. Filipiak, P., Lipinski, P.: Making IDEA-ARIMA efficient in dynamic constrained optimization problems. In: Mora, A.M., Squillero, G. (eds.) EvoApplications 2015. LNCS, vol. 9028, pp. 882–893. Springer, Heidelberg (2015)
12. Grefenstette, J.J., et al.: Genetic algorithms for changing environments. In: PPSN, vol. 2, pp. 137–144 (1992)
13. Ho, P.Y., Shimizu, K.: Evolutionary constrained optimization using an addition of ranking method and a percentage-based tolerance value adjustment scheme. Inf. Sci. **177**(14), 2985–3004 (2007)

14. Kundu, S., Biswas, S., Das, S., Suganthan, P.N.: Crowding-based local differential evolution with speciation-based memory archive for dynamic multimodal optimization. In: Proceedings of 15th Annual Conference on Genetic and Evolutionary Computation, pp. 33–40. ACM (2013)
15. Li, C., Nguyen, T.T., Yang, M., Yang, S., Zeng, S.: Multi-population methods in unconstrained continuous dynamic environments: the challenges. Inf. Sci. **296**, 95–118 (2015)
16. Mahfoud, S.W.: Niching methods for genetic algorithms. Urbana **51**(95001), 62–94 (1995)
17. Mezura-Montes, E., Coello Coello, C.A., Tun-Morales, E.I.: Simple feasibility rules and differential evolution for constrained optimization. In: Monroy, R., Arroyo-Figueroa, G., Sucar, L.E., Sossa, H. (eds.) MICAI 2004. LNCS (LNAI), vol. 2972, pp. 707–716. Springer, Heidelberg (2004)
18. Morales, A.K., Quezada, C.V.: A universal eclectic genetic algorithm for constrained optimization. In: Proceedings of 6th European Congress on Intelligent Techniques and Soft Computing, vol. 1, pp. 518–522 (1998)
19. Nguyen, T.T.: Continuous dynamic optimisation using evolutionary algorithms. Ph.D. thesis, University of Birmingham (2011)
20. Nguyen, T.T., Yang, S., Branke, J.: Evolutionary dynamic optimization: a survey of the state of the art. Swarm Evol. Comput. **6**, 1–24 (2012)
21. Nguyen, T.T., Yao, X.: Benchmarking and solving dynamic constrained problems. In: 2009 IEEE Congress on Evolutionary Computation, pp. 690–697. IEEE (2009)
22. Nguyen, T.T., Yao, X.: Solving dynamic constrained optimisation problems using repair methods. IEEE Trans. Evol. Comput. (2010, submitted)
23. Nguyen, T.T., Yao, X.: Continuous dynamic constrained optimization-the challenges. IEEE Trans. Evol. Comput. **16**(6), 769–786 (2012)
24. Pal, K., Saha, C., Das, S.: Differential evolution and offspring repair method based dynamic constrained optimization. In: Panigrahi, B.K., Suganthan, P.N., Das, S., Dash, S.S. (eds.) Swarm, Evolutionary, and Memetic Computing. LNCS, vol. 8297, pp. 298–309. Springer, Heidelberg (2013)
25. Pal, K., Saha, C., Das, S., Coello, C., et al.: Dynamic constrained optimization with offspring repair based gravitational search algorithm. In: 2013 IEEE Congress on Evolutionary Computation (CEC), pp. 2414–2421. IEEE (2013)
26. Richter, H.: Memory design for constrained dynamic optimization problems. In: Di Chio, C., et al. (eds.) EvoApplicatons 2010, Part I. LNCS, vol. 6024, pp. 552–561. Springer, Heidelberg (2010)
27. Salcedo-Sanz, S.: A survey of repair methods used as constraint handling techniques in evolutionary algorithms. Comput. Sci. Rev. **3**(3), 175–192 (2009)
28. Voigt, H.M., Lange, J.M.: Local evolutionary search enhancement by random memorizing. In: The 1998 IEEE International Conference on Computational Intelligence, pp. 547–552. IEEE (1998)

On Constraint Handling in Surrogate-Assisted Evolutionary Many-Objective Optimization

Tinkle Chugh[1(✉)], Karthik Sindhya[1], Kaisa Miettinen[1], Jussi Hakanen[1], and Yaochu Jin[1,2]

[1] University of Jyvaskyla, Department of Mathematical Information Technology, PO Box 35 (Agora), FI-40014 University of Jyvaskyla, Finland
{tinkle.chugh,karthik.sindhya,kaisa.miettinen, jussi.hakanen,yaochu.jin}@jyu.fi

[2] Department of Computer Science, University of Surrey, Guildford, UK

Abstract. Surrogate-assisted evolutionary multiobjective optimization algorithms are often used to solve computationally expensive problems. But their efficacy on handling constrained optimization problems having more than three objectives has not been widely studied. Particularly the issue of how feasible and infeasible solutions are handled in generating a data set for training a surrogate has not received much attention. In this paper, we use a recently proposed Kriging-assisted evolutionary algorithm for many-objective optimization and investigate the effect of infeasible solutions on the performance of the surrogates. We assume that constraint functions are computationally inexpensive and consider different ways of handling feasible and infeasible solutions for training the surrogate and examine them on different benchmark problems. Results on the comparison with a reference vector guided evolutionary algorithm show that it is vital for the success of the surrogate to properly deal with infeasible solutions.

1 Introduction

Problems involving several conflicting objective functions are called multiobjective optimization problems. Such problems are typical e.g. in industrial applications. Because of the conflict, there typically does not exist a single solution but multiple so-called Pareto optimal solutions. The set of all Pareto optimal solutions in the objective space is called a Pareto front. Problems involving more than three objectives are sometimes referred to as many-objective optimization problems. In industrial optimization problems, computationally expensive functions are common, where function evaluations are time-consuming because of employing e.g. finite element methods. Such problems are usually handled using surrogates, which are approximate functions that replace the computationally expensive ones. For overviews of surrogate-assisted evolutionary algorithms (SAEAs) for single and multiobjective optimization, see [1,2]. Surrogate-assisted evolutionary algorithms for many-objective optimization have not received much attention but recently a novel Kriging-assisted evolutionary algorithm for many-objective optimization called K-RVEA [3] has been proposed.

J. Handl et al. (Eds.): PPSN XIV 2016, LNCS 9921, pp. 214–224, 2016.
DOI: 10.1007/978-3-319-45823-6_20

Although industrial problems involve constraints, they have received little attention in the literature. The constraints pose a challenge for evolutionary algorithms to generate feasible solutions. More importantly, the presence of both feasible and infeasible solutions within a population especially during early generations may cause problems in surrogate training. Usually, a feasible set of solutions is required to train the surrogate. In unconstrained problems, the solutions generated always lie in the feasible region while in constrained problems, it may not be the case. In many cases, an initial set of feasible solutions is not available or it may take a substantial number of function evaluations. In addition, infeasible solutions can also play a major role in updating the surrogates as it will affect the performance of the surrogates in subsequent generations.

Next, we present a summary of approaches used in the literature for constrained SAEAs. In [4–6], initial training of surrogates was performed without considering any information from infeasible solutions, while in [7] a prefixed number of feasible solutions was used to train Kriging models. For updating the surrogate, in [4], all feasible nondominated solutions from the latest generation were re-evaluated and added to the training data set. In [5], all nondominated solutions after using surrogates were reevaluated without considering the feasibility of the solutions. In [6,7], the probability of feasibility was used for selecting individuals to update the surrogates. However, all these algorithms were tested on biobjective optimization problems. Therefore, a detailed investigation has not been done for handling infeasible solutions in constrained many-objective optimization.

In this study, we focus on constrained SAEAs for many-objective optimization problems and investigate three different approaches for creating a training data set for surrogates. In the first approach, we neglect all infeasible solutions and the surrogate is trained only with feasible solutions. In the second approach, we consider some infeasible solutions close to the feasible region in addition to the feasible ones and in the third approach, we add a penalty to infeasible solutions to train the surrogates. In all of these cases, we also consider infeasible solutions for selecting individuals to update the surrogates and to limit the size of the training data set. To update the surrogates, we select individuals so that a maximum number of feasible solutions is used without a compromise in convergence and diversity. A similar strategy is used to limit the size of the training data set. As this can affect the training time, we eliminate individuals in such a way that the performance of the surrogate is not compromised.

We assume that constraint functions themselves are not computationally expensive. In other words, the computation time of evaluating constraints is significantly lower than evaluating objective functions and therefore, surrogates are not trained for constraint functions. Such a scenario where constraint functions are not computationally expensive can exist in different cases. For instance, if objective and constraint functions are independently evaluated or constraints are available as analytical functions of the decision variables e.g. thickness to height ratio while considering the design of some structural part of an aircraft. Regardless of this assumption, our major contribution is towards showing the effect of infeasible solutions in training surrogates.

To test different approaches for handling infeasible solutions, we use K-RVEA [3]. One of the main reasons and also an advantage to use K-RVEA is its ability to solve problems with more than three objectives. K-RVEA is based on the reference vector guided evolutionary algorithm RVEA [8], where the management of surrogates involves reference vectors. In RVEA, reference vectors are used to decompose the original problem into a number of subproblems. These subproblems are simultaneously solved and a set of solutions that approximate the entire Pareto front is finally obtained. Additionally, the balance of convergence and diversity of the solutions in the high-dimensional objective space is achieved by using a novel scalarization approach called angle penalized distance (APD) [8]. K-RVEA is an extension of this algorithm and it uses Kriging models as surrogates to approximate computationally expensive functions. A flowchart of K-RVEA is presented in Fig. 1.

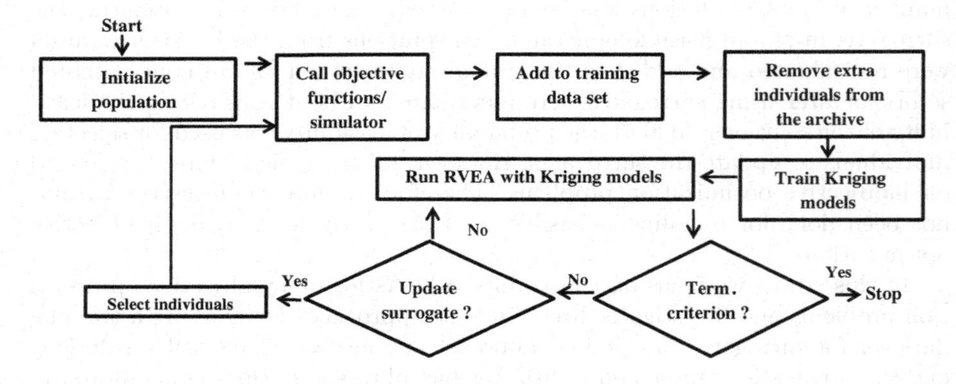

Fig. 1. Flowchart of K-RVEA

Initially, a population is initialized randomly or e.g. using Latin hypercube sampling [9]. Individuals of this population are then evaluated with the original objective functions and added to a training data set. If the size of this set exceeds a predefined limit, we eliminate individuals from it. Kriging models for each objective function are then trained and used to approximate objective function values. In any generation, if a termination criterion e.g. maximum number of function evaluations is not met, we update the surrogates after a prefixed number of generations. To update the surrogates, an efficient selection of individuals is performed with the help of reference vectors. Individuals are selected so that both convergence and diversity are managed while updating the surrogates. These individuals are then re-evaluated with the original functions and added to the training data set. If the termination criterion is met, nondominated solutions among all individuals evaluated with the original functions are obtained as the final solutions. For more details about K-RVEA, see [3].

In the next section, we provide the details of different approaches to handle infeasible solutions. In Sect. 3, we test and compare three approaches with the constrained RVEA [8]. Finally, in Sect. 4, we conclude the paper and discuss future research directions.

2 Approaches to Handle Infeasible Solutions

In this section, an extension of K-RVEA for constrained problems is presented, to be called cK-RVEA is given in Algorithm 1. cK-RVEA has three phases; initialization, using the surrogates and updating the surrogates.

Initialization. In the initialization phase, an initial set of feasible and/or infeasible solutions is used to train the surrogates. It may be difficult to obtain enough feasible solutions in the first generation therefore, in some cases, we first find feasible solutions by optimizing the constraint violation as an objective function. These individuals are stored in a training data set A_1. In addition, another set A_2 is used as the storage of nondominated feasible solutions.

Using the Surrogates. In the phase of using the surrogates, Kriging models are used to approximate objective function values. We use Kriging up to a predefined fixed number of generations (w_{max}) before updating the surrogates. We use the same parameter for the prefixed number of generations that was proposed for K-RVEA based on a sensitivity analysis. For the selection criterion in this phase, an individual from each subpopulation with minimum APD is selected if it is feasible. Otherwise, an individual with a minimum constraint violation is selected. Individuals thus selected are used as the population for the next generation.

Updating the Surrogates. The Kriging models are updated after using them for a fixed number of generations. The selection of individuals to be re-evaluated is very important for the performance of the surrogates especially when constraints are involved. For example, it may be possible that after re-evaluations, the number of infeasible solutions increases. Therefore, a maximum number of feasible solutions should be selected. In K-RVEA, a set of individuals U is selected based on the need of convergence or diversity. To this end, a fixed set of reference vectors (V_f) is generated in addition to the adaptive reference vectors (V_a). These reference vectors are used in the selection strategy to be described below.

Selection Strategy to Update the Surrogates. In K-RVEA, after using Kriging models for a fixed number of generations, individuals are assigned to the fixed reference vectors. Then the change in the number of inactive (or empty) fixed reference vectors from the previous update is calculated. If this change is smaller than a threshold, we select an individual with the minimum APD, otherwise with a maximum uncertainty (from the Kriging models). In cK-RVEA, we use APD or uncertainty if there is at least one feasible solution. Otherwise, we select an individual with a minimum constraint violation. Next, we provide a strategy to manage the training data set.

Managing the Training Data. In order to reduce the computation time to train the Kriging models, we limit the size (maximum size is N_I) of the training data set. For this purpose, we eliminate some individuals from the set after every time we update the surrogates. We first assign individuals other than the recently

Algorithm 1. cK-RVEA

Input: FE_{max}= max number of function evaluations; w_{max}= prefixed number of generations before updating Kriging models; N_I= max number of individuals in set A_1

Output: nondominated feasible solutions of all evaluated ones from A_2

/*Initialization*/

1: Initialize the number of function evaluations FE=0, the generation counter for using Kriging models w=1 and a counter for the number of updates, $t_u = 0$. Initialize set $A_2 = \phi$

2: Obtain solutions (all feasible OR feasible and infeasible) in the training data set A_1 and update $A_2 = A_1$

3: Train a Kriging model for each objective function by using individuals in A_1

4: **while** $FE \leq FE_{max}$ **do**

/*Using the surrogates*/

5: **while** $w \leq w_{max}$ **do**

6: Run RVEA with Kriging models and update $w = w + 1$

7: **end while**

/*Updating the surrogates*/

8: Select a set of individuals U using a selection strategy to update the surrogates and re-evaluate them with the original functions and update $FE = FE + |U|$

9: Add individuals from step 8 to A_1 and A_2 and update $|A_1| = |A_1| + |U|$ and $|A_2| = |A_2| + |U|$

10: Remove $|A_1| - N_I$ individuals from A_1 using management of the training data, update $w = 1$ and $t_u = t_u + 1$ and go to step 3

11: **end while**

evaluated ones to the adaptive reference vectors. These reference vectors are then clustered into a prefixed number of clusters and an individual either randomly (if feasible) or with a minimum constraint violation (if infeasible) is selected from each cluster. In this way, a fixed number of individuals is maintained in the training data set in order to improve the quality of Kriging models as much as possible while limiting the computation time.

In the following, we present three different approaches to handle infeasible solutions and variants of cK-RVEA using them are denoted by cK-RVEA1, cK-RVEA2 and cK-RVEA3.

Rejecting All Infeasible Solutions. In cK-RVEA1, surrogates are trained only with feasible solutions. Using feasible solutions to train the surrogates can help in increasing their performance, especially when the feasible region is very small. This is the case for example in problem C1-DTLZ1 [10], which also contains many locally Pareto optimal solutions. If the surrogate is trained with infeasible solutions, the approximated values from it may be far from the feasible region. Therefore, it is appropriate to find feasible solutions first and then use the surrogate for approximating objective functions.

Using Some Infeasible Solutions. In cK-RVEA2, we use some infeasible solutions close to the feasible region in addition to the feasible ones to train the surrogates. The main advantage of this is that when infeasible solutions are also used for training, the surrogates may be able to approximate a more diverse area without too much reduction in their performance. However, how close and how many infeasible solutions should be used are two important challenges.

For both cases mentioned above i.e. cK-RVEA1 and cK-RVEA2, a single objective genetic algorithm with niche based selection [11] is used for considering constraint violation as the objective function to obtain a fixed number of solutions in the feasible region. The termination criterion in this algorithm is to obtain adequate number of feasible solutions. A niche based selection ensures that a diverse set of feasible individuals in the decision space is obtained to train the surrogates. However, the diversity in the decision space does not guarantee diversity in the objective space and needs further attention.

Adding Penalty to Infeasible Solutions. In cK-RVEA3, we train surrogates with individuals generated randomly or e.g. with a Latin hypercube sampling and add a penalty to infeasible solutions. The main challenge in penalty based methods is to use an appropriate penalty parameter and we adopt here three methods from the study in [12]. In the first method, denoted by cK-RVEA3-I, a static penalty is added to each objective function value of f_i i.e.

$$f_i(x) = f_i(x) + R\sum_{j=1}^{m} |g_j(x)|, \tag{1}$$

where R is the penalty parameter and $|\ |$ denotes the absolute value of the constraint g_j. Note, however, that, we use $|\ |$ to represent the number of individuals in a set hereafter (except in (3)).

In the second method of using a penalty parameter denoted by cK-RVEA3-II, we adapt it with the number of feasible solutions obtained. After a certain number of function evaluations e.g. $FE \geq FE_{th}$, the penalty parameter R is decreased if the number of feasible solutions has increased from the previous generation and vice versa i.e.

$$R = \begin{cases} \dfrac{R}{c_1} & \text{if } FE \geq FE_{th}\ \Delta|P_f| > 0 \\ Rc_2 & \text{if } FE \geq FE_{th}\ \Delta|P_f| < 0 \end{cases} \tag{2}$$

where c_1 and c_2 are predefined parameters and $\Delta|P_f|$ denotes the change in the number of feasible solutions.

In the third method, cK-RVEA3-III, we use the method of parameter free penalty, where

$$f_i(x) = \begin{cases} f_i(x) & \text{if} \quad g_j(x) \geq 0,\ j = 1, \dots, m \\ f_i^{max} + \sum_{j=1}^{m} |g_j(x)| & \text{otherwise} \end{cases} \tag{3}$$

where f_i^{max} is the maximum value of f_i at the current generation. The main advantage of using this method is that no parameter is included and infeasible solutions are always penalized.

3 Numerical Experiments

In this section, results of experiments on the constrained versions of DTLZ problems [10] are presented. As mentioned, we consider three different approaches and compare them with each other and also with the constrained variant of RVEA [8]. Parameter values for niching are the same as used in [11] and values of parameters involved in K-RVEA are as follows: (a) number of individuals to train the surrogate in initialization phase = number of reference vectors, $N_I = 50$, (b) number of independent runs = 10, (c) maximum number of function evaluations = 300 and (d) number of generations before updating Kriging models, $w_{max} = 20$. In addition, we introduced the following parameters in cK-RVEA: (a) number of feasible solutions in cK-RVEA2 = 40, (b) static penalty used in cK-RVEA3-I, R=10000, (c) parameters used in cK-RVEA3-II (from [12]), $c_1 = 3$, $c_2 = 4$ and initial value of penalty parameter, $R = 1$.

The number of decision variables was set to 10 for all problems and the number of constraints varied from one to ten. Inverted generational distance (IGD) was used as the performance measure and a Wilcoxon rank sum test analysis with a significance level of 5% was adopted to compare the results. Results for cK-RVEA1, cK-RVEA2, cK-RVEA3-I and cRVEA for different numbers of objectives (denoted by k) are reported in Table 1, where ↑ represents that cK-RVEA1 performed better than the other, ↓ means that it performed worse, while ≈ means that statistically there is no significant difference between the two algorithms.

Table 1. Results for IGD values obtained by cK-RVEA1, cK-RVEA2, cK-RVEA3-I and cRVEA. The best results are highlighted

Prob.	k	cK-RVEA1				cK-RVEA2				cK-RVEA3-I				cRVEA		
		Min	Mean	Max		Min	Mean	Max		Min	Mean	Max		Min	Mean	Max
C1-DTLZ1	3	**0.098**	**0.154**	**0.166**	≈	0.147	0.159	0.168	↑	No feasible solution			↑	No feasible solution		
	6	0.148	0.176	**0.199**	≈	**0.107**	**0.174**	0.219	↑	No feasible solution			↑	No feasible solution		
	8	0.258	0.269	0.281	↓	**0.217**	**0.248**	**0.270**	↑	No feasible solution			↑	No feasible solution		
	10	0.197	**0.205**	**0.236**	≈	**0.166**	0.212	0.252	↑	0.309	0.359	0.420	↑	0.194	0.228	0.311
C2-DTLZ2	3	0.155	**0.213**	**0.271**	≈	0.189	0.215	0.283	↑	0.433	0.592	0.752	↑	0.205	0.260	0.291
	6	**0.373**	**0.388**	**0.407**	≈	**0.349**	0.406	0.443	↑	0.599	0.737	0.965	↑	0.389	0.435	0.530
	8	**0.387**	**0.479**	**0.598**	≈	0.424	0.542	0.755	↑	0.533	0.782	0.974	↑	0.522	0.601	0.703
	10	0.527	0.623	0.729	↑	0.571	0.727	0.878	↑	0.624	0.783	0.956	↑	0.571	0.615	0.673
C3-DTLZ4	3	**0.163**	**0.198**	0.256	≈	**0.160**	**0.187**	**0.216**	↑	0.183	0.249	0.386	≈	0.199	0.220	0.236
	6	**0.467**	**0.500**	**0.534**	≈	0.489	0.527	0.602	↑	0.537	0.587	0.646	↑	0.574	0.595	0.649
	8	**0.629**	**0.674**	**0.713**	≈	**0.602**	0.682	0.808	↑	0.713	0.801	0.856	↑	0.739	0.798	1.008
	10	0.779	0.860	0.903	↓	0.781	**0.824**	**0.897**	↑	0.891	0.991	1.299	≈	0.799	**0.836**	0.916

As can be seen, in C1-DTLZ1, cK-RVEA3-I and cRVEA were not able to find any feasible solutions in 300 function evaluations. The feasible region in C1-DTLZ1 is very small, therefore, using directly a surrogate or adding a penalty without finding feasible solutions was not useful for the surrogates as solutions are far from the feasible region. Therefore, it is important to find sufficiently many feasible solutions and then use surrogates. Both cK-RVEA1 and

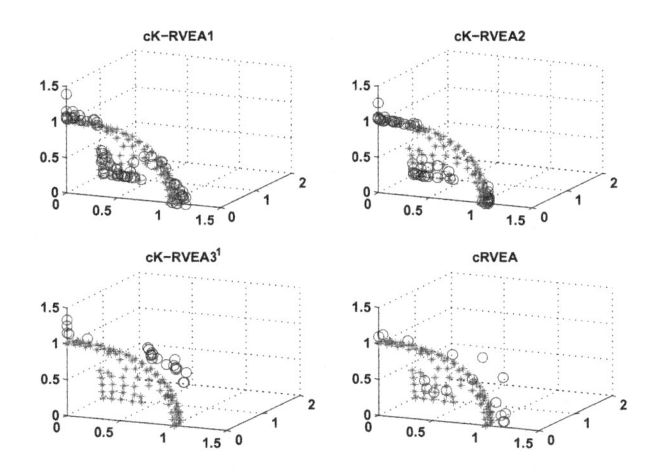

Fig. 2. Nondominated solutions obtained by cK-RVEA1, cK-RVEA2, cK-RVEA3-I and cRVEA denoted by circles of the run with the best IGD value for three-objective C2-DTLZ2 test problem. Here ∗'s represent the Pareto front.

cK-RVEA2 found feasible solutions using the single objective genetic algorithm with constraint violation as the objective function.

We also performed a sensitivity analysis for the parameters of cK-RVEA2 i.e. the number of infeasible solutions and how close to the feasible region they should be. As mentioned in the parameter settings, the number of solutions was 50 to train the surrogates. In this sensitivity analysis, we used 10, 20 and 30 infeasible solutions out of 50 and rest of them were feasible. For each case, we changed the distance of solutions from the feasible region. To do that, we used the normalized constraint violation of 0.5, 0.25, 0.1 and 0.001. Therefore, all together 12 studies were performed to analyze the number of infeasible solutions and their distance to the feasible region. Out of all these limited studies, the case with 10 infeasible solutions and the normalized constraint violation of 0.1 performed best and results from this case are shown in Table 1. However, self-adapting both the parameters is a future research topic.

Nondominated solutions of C2-DTLZ2 with three objectives of the run with the best IGD value from cK-RVEA1, cK-RVEA2, cK-RVEA3-I and cRVEA are shown in Fig. 2. As can be seen, cK-RVEA1 and cK-RVEA2 performed comparably and solutions from both variants got close to the Pareto front. In contrast, solutions of cK-RVEA3-I, where a penalty is added to infeasible solutions did not converge to the Pareto front. However, when infeasible solutions were used in addition to feasible ones in cK-RVEA2, they got closer to the Pareto front. Parallel coordinate plots of C3-DTLZ4 with 10 objectives of the run with the best IGD values are shown in Fig. 3. As can be seen, solutions from both cK-RVEA1 and cK-RVEA2 had large ranges in some of the objective values compared to other algorithms. Furthermore, as can be seen from the table, in most of the cases, the

Table 2. Results for IGD values obtained by cK-RVEA3-I, cK-RVEA3-II and cK-RVEA3-III. The best results are highlighted

Problem	k	cK-RVEA3-I				cK-RVEA3-II				cK-RVEA3-III		
		Min	Mean	Max		Min	Mean	Max		Min	Mean	Max
C1-DTLZ1	10	0.309	0.359	0.420	≈	**0.257**	**0.313**	**0.336**	≈	0.284	0.362	0.497
	3	0.433	0.592	0.752	↓	**0.206**	**0.332**	**0.461**	≈	0.318	0.637	0.961
C2-DTLZ2	6	0.599	0.737	0.965	↓	**0.528**	**0.636**	**0.913**	≈	0.570	0.850	1.036
	8	**0.533**	0.782	0.974	≈	0.579	**0.682**	**0.787**	≈	0.620	0.888	1.038
	10	0.624	0.783	0.956	≈	**0.575**	**0.704**	**0.867**	↑	0.632	0.903	0.998
	3	0.183	0.249	0.386	↓	**0.180**	**0.200**	**0.222**	≈	0.195	0.251	0.293
C3-DTLZ4	6	**0.537**	**0.587**	**0.646**	≈	0.554	0.599	0.723	↑	0.586	0.673	0.730
	8	**0.713**	**0.801**	**0.856**	≈	0.731	0.837	1.006	↑	0.772	0.938	1.052
	10	0.891	0.991	1.299	≈	**0.836**	**0.967**	**1.126**	↑	1.013	1.169	1.363

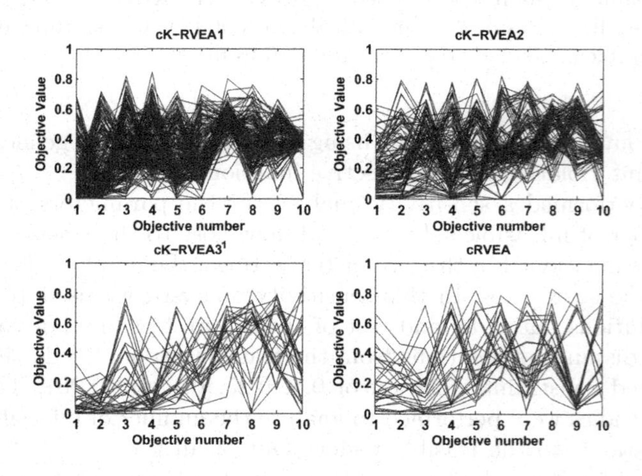

Fig. 3. Parallel coordinate plot of nondominated solutions obtained by cK-RVEA1, cK-RVEA2, cK-RVEA3-I and cRVEA of the run with the best IGD value on 10-objective C3-DTLZ4 test problem.

constrained variant of RVEA (i.e. without surrogates) performed worse than the others.

When comparing penalty based methods as detailed in Sect. 2, results are given in Table 2. As can be seen, the adaptive penalty method in most of the cases performed equivalently or better than the static penalty method. In any case, the method of the parameter free penalty was not able to outperform other methods. All these results show the influence of infeasible solutions on the performance of the surrogates. These results indicate that an adaptive way of handling infeasible solutions seems to be needed although more testing needs to be done on other benchmark problems.

4 Conclusions and Future Research

In this paper, we investigated the influence of different constraint handling strategies for collecting training data on the performance of surrogates. These strategies were investigated on the constrained DTLZ problems using K-RVEA. Results from the study show that handling infeasible solutions in selecting training data is very important. Using only feasible solutions i.e. cK-RVEA1 in most cases performed better than others because individuals approximated by the surrogates lie in the feasible region. However, it depends on the problem used as infeasible solutions may be helpful to increase the performance of the surrogates. Moreover, a hybrid approach to combine the different approaches e.g. how many and how close should be infeasible solutions to the feasible region, how to adapt the penalty parameter etc. can be beneficial. In addition, as few constrained many-objective optimization problems exist in the literature, developing and testing on new problems will also be our future work.

Acknowledgement. This work was supported by the FiDiPro project DeCoMo funded by TEKES, The Finnish Funding Agency for Innovation.

References

1. Jin, Y.: A comprehensive survey of fitness approximation in evolutionary computation. Soft. Comput. **9**, 3–12 (2005)
2. Chugh, T., Sindhya, K., Hakanen, J., Miettinen, K.: Handling computationally expensive multiobjective optimization problems with evolutionary algorithms - a survey. Reports of the Department of Mathematical Information Technology, Series B, Scientific Computing no. B 4/2015, University of Jyvaskyla (2015)
3. Chugh, T., Jin, Y., Miettinen, K., Hakanen, J., Sindhya, K.: K-RVEA: a Kriging-assisted evolutionary algorithm for many-objective optimization. Reports of the Department of Mathematical Information Technology, Series B, Scientific Computing no. B 2/2016, University of Jyvaskyla (2016)
4. Singh, H.K., Ray, T., Smith, W.: Surrogate assisted simulated annealing (SASA) for constrained multi-objective optimization. In: Proceedings of IEEE Congress on Evolutionary Computation, pp. 1–8. IEEE (2010)
5. Chen, G., Han, X., Liu, G., Jiang, C., Zhao, Z.: An efficient multi-objective optimization method for black-box functions using sequential approximate technique. Appl. Soft Comput. **12**, 14–27 (2012)
6. Singh, P., Couckuyt, I., Ferranti, F., Dhaene, T.: A constrained multi-objective surrogate-based optimization algorithm. In: Proceedings of IEEE Congress on Evolutionary Computation, pp. 3080–3087. IEEE (2014)
7. Martinez-Frutos, J., Herrero-Perez, D.: Kriging-based infill sampling criterion for constraint handling in multi-objective optimization. J. Glob. Optim. **64**, 97–115 (2016)
8. Cheng, R., Jin, Y., Olhofer, M., Sendhoff, B.: A reference vector guided evolutionary algorithm for many objective optimization. IEEE Trans. Evol. Comput. (2016, accepted). doi:10.1109/TEVC.2016.2519378
9. Mckay, M.D., Beckman, R.J., Conover, W.J.: A comparison of three methods for selecting values of input variables in the analysis of output from a computer code. Technometrics **42**, 55–61 (2000)

10. Jain, H., Deb, K.: An evolutionary many-objective optimization algorithm using reference-point-based nondominated sorting approach, part II: Handling constraints and extending to an adaptive approach. IEEE Trans. Evol. Comput. **18**, 602–622 (2014)
11. Deb, K.: An efficient constraint handling method for genetic algorithms. Comput. Methods Appl. Mech. Eng. **186**, 311–338 (2000)
12. Miettinen, K., Makela, M.M., Toivanen, J.: Numerical comparison of some penalty-based constraint handling techniques in genetic algorithms. J. Glob. Optim. **27**, 427–446 (2003)

Artificially Inducing Environmental Changes in Evolutionary Dynamic Optimization

Renato Tinós[1] and Shengxiang Yang[2](✉)

[1] Department of Computing and Mathematics, FFCLRP, University of São Paulo, Ribeirão Preto, São Paulo 14040-901, Brazil
rtinos@ffclrp.usp.br
[2] Centre for Computational Intelligence (CCI), School of Computer Science and Informatics, De Montfort University, Leicester LE1 9BH, UK
syang@dmu.ac.uk

Abstract. Biological and artificial evolution can be speeded up by environmental changes. From the evolutionary computation perspective, environmental changes during the optimization process generate dynamic optimization problems (DOPs). However, only DOPs caused by intrinsic changes have been investigated in the area of evolutionary dynamic optimization (EDO). This paper is devoted to investigate artificially induced DOPs. A framework to generate artificially induced DOPs from any pseudo-Boolean problem is proposed. We use this framework to induce six different types of changes in a 0–1 knapsack problem and test which one results in higher speed up. Two strategies based on immigrants, which are used in EDO, are adapted to the artificially induced DOPs investigated here. Some types of changes did not result in better performance, while some types led to higher speed up. The algorithm with memory based immigrants presented very good performance.

1 Introduction

In a recent work [11], Steinberg and Ostermeir investigated the hypothesis that environmental changes can help molecular evolution to cross fitness valleys. They experimentally tested four strategies for inducing environmental changes in the evolution of an antibiotic resistance gene (*TEM-15* β-lactamase). One particular strategy, where low antibiotic resistance individuals are selected in the initial steps, produced very interesting results. When the evolutionary pathways were analysed, it was observed that an initially deleterious mutation allowed to access a promising part of the sequence space. This part of the sequence space was very difficult to be reached when environmental changes had not occurred.

The idea that biological and artificial evolution can be speeded up by environmental changes is not new [6,9,12]. Kashtan *et al.* [6] compared two strategies for inducing environmental changes in the *in silico* evolution of five models: (i) logic circuits; (ii) feed-forward logic circuits; (iii) feed-forward artificial neural networks; (iv) feed-forward circuits; (v) RNA structure. The two strategies were modularly varying goals and randomly varying goals. Greater speed

J. Handl et al. (Eds.): PPSN XIV 2016, LNCS 9921, pp. 225–236, 2016.
DOI: 10.1007/978-3-319-45823-6_21

up was obtained for the first strategy, where subgoals are inserted or removed during the optimization process. Populations can spend long periods around metastable states. Environmental changes modify the fitness landscape and evolution dynamics [13], allowing populations to eventually escape from local optima and plateaus [6][1].

In the evolutionary computation (EC) perspective, the occurrence of environmental changes during artificial evolution generates dynamic optimization problems (DOPs). In recent years, there is an increasing interest in evolutionary dynamic optimization (EDO) [2,8]. However, to the best of the authors' knowledge, the works published in this area deal with DOPs where environmental changes are *intrinsic*. In other words, *artificially induced* DOPs are not considered. Here, we investigate artificially induced DOPs in a perspective of EDO. In [6], strategies for inducing environmental changes in *specific* DOPs were investigated. We propose a general framework to artificially induce environmental changes in *any* pseudo-Boolean optimization problem[2]. The proposed framework is based on the DOP benchmark generator introduced in [14], which will be presented in Sect. 2. The proposed framework will be presented in Sect. 3.

In the experiments used to test the proposed framework, environmental changes are artificially induced in the 0–1 knapsack problem in order to eventually speed up evolution. The experimental results are presented in Sect. 4. It is important to highlight that testing whether environmental changes can speed up evolution is only one of the possible motivations to artificially induced changes in EC. For example, we can artificially induce environmental changes in order to increase the robustness of the solutions [3]. Also, we can control when environmental changes can be inserted in some applications, e.g., those involving cooperation and competition [10]

From a programmer point of view, there are two main differences between artificially induced and intrinsic DOPs. In artificially induced DOPs, the programmer should decide *when* and *how* to change the problem, which is impossible in intrinsic DOPs. To this aim, one needs to answer two questions: (i) *When should the changes be inserted?* (ii) *How the fitness landscape should be modified?* We strongly believe that trying to answer these two questions opens new research possibilities in EDO. Researchers can investigate the best way to change the fitness landscapes from a theoretical point of view.

From a practical point of view, knowing beforehand when the changes occur, new algorithms and operators can be designed. For example, hypermutation re-introduces diversity by increasing the mutation rate after a change. Knowing when a change will occur allows to apply hypermutation some generations before the change. The development of new algorithms and operators is also important

[1] The idea of changing the static fitness landscape in order to make the optimization process easier is also present in other approaches. For example, in multi-objectivization, a single-objective problem is transformed into a multi-objective problem [7]. Another example is adding noise to the fitness function [5].

[2] In a pseudo-Boolean optimization problem P, the fitness function is $f_P(\mathbf{x}) \in \mathbb{R}$, where $\mathbf{x} \in \mathbb{B}^l$ is a candidate solution vector with dimension l.

because the goal in intrinsic and artificially induced DOPs can be different. In intrinsic DOPs, the goal is to track the moving optima, while in artificially induced DOPs we can be interested in finding the optima only for the static problem. Here, we test two very simple strategies to deal with artificially induced DOPs in Sect. 3.

2 DOP Benchmark Generator

Based on the analysis of fitness landscape changes in some DOPs, a benchmark generator for dynamic pseudo-Boolean optimization problems was proposed in [14]. The generator allows to create DOPs from any pseudo-Boolean optimization problem P with (static) fitness function $f_P(\mathbf{x})$, where $\mathbf{x} \in \mathbb{B}^l$. A DOP is considered as a sequence of static landscapes (environments) modified by changes [8]. In the DOPs created by the generator, the fitness function is given by:

$$f(\mathbf{x}, e) = f_P\big(g(\mathbf{x}, e)\big) + \Delta f\big(g(\mathbf{x}, e), e\big), \tag{1}$$

where e is the index of environment, i.e., it indicates a static fitness landscape between two consecutive changes [13]. Instead of computing the static fitness $f_P(.)$ at position $\mathbf{x}$, it is computed at position $g(\mathbf{x}, e)$. Besides, a deviation $\Delta f\big(g(\mathbf{x}, e), e\big)$ is added to $f_P(.)$. The generator allows to create 6 different types of DOPs based on the choice of $g(\mathbf{x}, e)$ and $\Delta f\big(g(\mathbf{x}, e), e\big)$, as described below.

2.1 DOP Type 1: DOP with Permutation

In this case, $\Delta f\big(g(\mathbf{x}, e), e\big) = 0$ and $g(\mathbf{x}, e)$ is given by a permutation of $\mathbf{x}$. In the generator, 3 different ways to permute $\mathbf{x}$ are employed.

DOP Type 1.1 (Permutation of the XOR Type): The candidate $\mathbf{x}$ is permuted according to: $g(\mathbf{x}, e) = \mathbf{x} \oplus \mathbf{m}(e)$, where:

$$\mathbf{m}(e) = \begin{cases} \mathbf{0}_l, & \text{for } e = 1 \\ \mathbf{m}(e-1) \oplus \mathbf{r}(e), & \text{for } e > 1 \end{cases} \tag{2}$$

where "$\oplus$" is the XOR operator and $\mathbf{r}(e)$ is a binary template that is randomly created in each environment e and contains $\lfloor \rho \cdot l \rfloor$ ones, where $0 \leq \rho \leq 1$. The change severity is controlled by ρ. DOP Type 1.1 produces the same type of change as the XOR DOP generator [14].

DOP Type 1.2 (Permutation Defined by a Permutation Matrix): The permutation is given by: $g(\mathbf{x}, e) = \mathbf{B}(e)\mathbf{x}$, where the permutation matrix $\mathbf{B}(e)$ is incrementally modified according to:

$$\mathbf{B}(e) = \begin{cases} \mathbf{I}_l, & \text{for } e = 1 \\ \mathbf{C}(e)\mathbf{B}(e-1), & \text{for } e > 1 \end{cases} \tag{3}$$

where $\mathbf{C}(e)$ is a permutation matrix obtained by randomly exchanging $\lfloor \rho \cdot l \rfloor$ lines of the l-dimensional identity matrix $\mathbf{I}_l$. In fact, the use of matrices would imply in a computational cost $O(l^2)$. This cost can be reduced to $O(l)$ by using an integer vector to record the positions of the permuted variables of $\mathbf{x}$. Similar strategies are adopted for other DOP types.

DOP Type 1.3 (Permutation According to a Set of Templates): The permutation is defined by:

$$g(\mathbf{x}, e) = \begin{cases} \mathbf{x} \oplus \mathbf{m}_j(e), \text{ if } \mathbf{x} \in \mathbf{s}_j(e), j = 1, \ldots, n_s \\ \mathbf{x}, \quad \text{otherwise} \end{cases} \tag{4}$$

where $\mathbf{s}_j(e)$ is a template defining a hyperplane in $\mathbb{B}^l$ and n_s is the number of templates. The templates, or schemata in the genetic algorithms terminology, are composed of digits 0, 1 and * (do not care) and can be associated with subsets of solutions. Each template $\mathbf{s}_j(e)$ is given by:

$$\mathbf{s}_j(e) = \begin{cases} \mathbf{0}_l, & \text{for } e = 1 \\ \mathbf{r}_j, & \text{for } e = 2 \\ \mathbf{D}(e)\mathbf{s}_j(e-1), & \text{for } e > 2 \end{cases} \tag{5}$$

where $\mathbf{r}_j$ is a random template with order equal to o_s, and $\mathbf{D}(e)$ is a permutation matrix obtained by randomly exchanging o_s lines of the l-dimensional identity matrix. The template $\mathbf{m}_j(e) \in \mathbf{s}_j(e)$ contains $\frac{l-o_s}{2}$ ones generated in random non-fixed positions of $\mathbf{s}_j(e)$. The order of the template $\mathbf{s}_j(e)$ is equal to o_s for $e > 1$. The following combinations $(o_s, n_s) \in \{(3,1), (2,1), (1,1), (1,2), (1,3)\}$, corresponding to $\rho \in \{0.125, 0.25, 0.5, 0.75, 0.875\}$, are used.

2.2 DOP Type 2: Copying Decision Variables

Here, $\Delta f\big(g(\mathbf{x}, e), e\big) = 0$ and $g(\mathbf{x}, e)$ is a transformation that produces decision variables that are copies of other decision variables. Two ways of copying the variables are considered: one where the variables in $\mathbf{x}$ are copied from other variables in $\mathbf{x}$ and another where the variables are copied from those in a template.

DOP Type 2.1 (Copying Decision Variables Using a Linear Transformation): The candidate solutions are linearly transformed by: $g(\mathbf{x}, e) = \mathbf{L}(e)\mathbf{x}$, where $\mathbf{L}(e)$ is a binary matrix generated according to:

$$\mathbf{L}(e) = \begin{cases} \mathbf{I}_l, & \text{for } e = 1 \\ \mathbf{Q}(e), & \text{for } e > 1 \end{cases} \tag{6}$$

where $\mathbf{Q}(e)$ is a matrix obtained by randomly copying $\lfloor \rho \cdot \frac{l}{2} \rfloor$ lines of the l-dimensional identity matrix into other lines.

DOP Type 2.2 (Copying Decision Variables from a Template): The transformation is given by:

$$g(\mathbf{x}, e) = \begin{cases} \mathbf{m}(e), & \text{if } \mathbf{x} \in \mathbf{s}(e) \\ \mathbf{x}, & \text{if } \mathbf{x} \notin \mathbf{s}(e) \end{cases} \tag{7}$$

where $\mathbf{s}(e)$ is a template given by:

$$\mathbf{s}(e) = \begin{cases} \mathbf{0}_l, & \text{for } e = 1 \\ \theta(e), & \text{for } e > 1 \end{cases} \tag{8}$$

The order of the random template $\theta(e)$ is $l - \lfloor \rho \cdot \frac{l}{2} \rfloor$. The binary template $\mathbf{m}(e) \in \mathbf{s}(e)$ is randomly generated at each environment e.

2.3 DOP Type 3: Adding Fitness Deviation by a Set of Templates

In this DOP type, $\mathbf{x}$ is not transformed, i.e., $g(\mathbf{x}, e) = \mathbf{x}$. The fitness deviation $\Delta f\big(g(\mathbf{x}, e), e\big) = \Delta f(\mathbf{x}, e)$ is given by:

$$\Delta f(\mathbf{x}, e) = \sum_{j=1}^{n_s} a(\mathbf{x}, \mathbf{s}_j(e), e), \tag{9}$$

where n_s is the number of templates. The order of each template $\mathbf{s}_j(e)$ is o_s. The parameters o_s and n_s are defined in the same way as in DOP Type 1.3. In Eq. (9), $a\big(\mathbf{x}, \mathbf{s}_j(e), e\big)$ is given by:

$$a\big(\mathbf{x}, \mathbf{s}_j(e), e\big) = \begin{cases} \Delta f_j(e), & \mathbf{x} \in \mathbf{s}_j(e) \\ 0, & \mathbf{x} \notin \mathbf{s}_j(e) \end{cases} \tag{10}$$

where $\Delta f_j(e)$ is the fitness deviation for $\mathbf{s}_j(e)$. Here, $\Delta f_j(e)$ is randomly generated from a uniform distribution in the range $[-\rho f_{range}, \rho f_{range}]$ in each environment e. The value of f_{range} is given by the difference between the best and mean fitness in the initial population (or, if this difference is too small, by the best fitness in the initial population).

3 Framework for Inducing Environmental Changes

Here, changes are artificially induced in order to test whether they can speed up evolution. Changes are inserted according to one of the 6 DOP types described in Sect. 2; we want to test which one produces the best results for speeding up evolution of a genetic algorithm applied to the 0–1 knapsack problem. In fact, a little modification is introduced in DOP types 1.3, 2.2, and 3, as described in Sect. 3.2. The framework for inducing environmental changes in EDO is described in Sect. 3.1. Variants of the standard genetic algorithm used in EDO are also tested (Sect. 3.3).

3.1 Framework

As the objective is to speed up evolution for problem P, we propose a framework where the static environment for problem P is modified every τ iterations of the algorithm (generations). The DOP is seen as a sequence of environments, where the type of each environment is indicated by $d(e)$. While $d(e) = 0$ indicates that $f(\mathbf{x}, e) = f_P(\mathbf{x})$ (i.e., the e-th environment is equal to the static environment for problem P), $d(e) = c \neq 0$ indicates an environment produced by DOP type c, where $c \in \{1.1, 1.2, 1.3, 2.1, 2.2, 3\}$. When e is odd, i.e., $\mathrm{mod}(e, 2) = 1$, the e-th environment is equal to the static environment for problem P, i.e., $d(e) = 0$. When e is even, i.e., $\mathrm{mod}(e, 2) = 0$, two strategies are compared: (i) Static, where $d(e) = 0$; (ii) Dynamic, where $d(e) = c$ and c identifies the DOP Type for environments where $\mathrm{mod}(e, 2) = 0$. Experiments with each one of the six DOP types will be presented in Sect. 4.

3.2 DOP Types

Some properties of the DOP types produced by the generator are described [14]:

- Neighbourhood relations: the transformation of the fitness landscapes for DOP Types 1.1 and 1.2 preserves the neighbourhood relations in the search space. In other words, instead of transforming the fitness landscape, we could move the population according to the respective transformation only one time after the change and compute $f(\mathbf{x}, e) = f_P(\mathbf{x})$ during τ generations. The neighbourhood relations are not preserved for the other DOP types.
- All solutions of the search space are changed for DOP Type 1.1. For DOP Types 1.3. and 3, the fractions of the search space affected by a change are equal to $\rho \in \{0.125, 0.25, 0.5, 0.75, 0.875\}$. For the remaining DOP types, the number of solutions of the search space affected by a change varies from 2^{l-1} to $2^l - 2$ for $\rho > 0$.

A consequence of the last property is that the change can have no effect in the dynamics of the population. For example, no effect will be observed when the solutions in the population are not among those affected by the fitness landscape modification. As we want to change the dynamics of the population here, we will use the knowledge about the best current solution in order to change the fitness landscape for DOP Types 1.3, 2.2 and 3. In this way, a small modification is introduced. In the DOP generator presented in [14], the first template $\mathbf{s}_j(e)$ for DOP Types 1.3, 2.2 and 3 is randomly chosen with no restriction. Here, the template is chosen assuring that $\mathbf{x_b}(e-1) \in \mathbf{s}_j(e)$, where $\mathbf{x_b}(e-1)$ is the best solution found in the e-th environment. Thus, DOPs produced by changes of types 1.3, 2.2 and 3 have the time-linkage property, i.e., knowing the current best solution influences the future dynamics of the problem [8].

3.3 Algorithms

The influence of artificially inducing changes in the 0–1 knapsack problem optimized by a standard genetic algorithm (GA) is investigated here. We also test variants of approaches used in EDO that replaces part of the population by immigrants. Two types of immigrants are tested:

- *Random immigrants* (RIs) [1]: when this strategy is used, 20 % of the population is replaced by randomly generated individuals.
- *Memory immigrants* (MIs) [15]: when this strategy is used, 10 % of the population is replaced by individuals stored in a memory population.

Instead of inserting immigrants in every generation, they are inserted only after a change. Also, as we want to optimize problem P, the memory is formed by the best individuals found in environments with $d(e) = 0$. As all the individuals in the memory were generated in environments with the same fitness landscape, it is not necessary to re-evaluate them when they are re-introduced in the population. The MIs are re-introduced in environments where $d(e) = 0$. The maximum size of the memory population is equal to the size of the GA population (*popsize*). When the maximum size is reached, a random individual of the memory population is replaced by the new individual, with exception for the best individual in the memory population. One can observe that we are using the knowledge about the changes in the problem in order to design the memory immigrants approach.

4 Experiments

4.1 Experimental Design

The fitness function for the 0–1 knapsack problem [4] is given by:

$$f_P(\mathbf{x}) = \sum_{i=1}^{l} p_i x_i - R(\mathbf{x}) \tag{11}$$

where $\mathbf{x} \in \mathbb{B}^l$ defines the subset of items in the knapsack and p_i is the profit of the i-th item. The penalty $R(\mathbf{x})$ is equal to zero if the sum of the weights in the knapsack is less than the knapsack capacity C. Otherwise, the penalty is:

$$R(\mathbf{x}) = \alpha \left(\sum_{i=1}^{l} w_i x_i - C \right) \tag{12}$$

where w_i is the weight of the i-th item and $\alpha = \max_{i=1,\ldots,l}(p_i/w_i)$. The profits and weights are integers randomly generated in the beginning of each run.

The profits are in the range $[40, 100]$, while the weights are in the range $[5, 20]$. The capacity C is equal to 50% of the sum of all weights. The objective is to maximize the fitness given by Eq. (11).

For all algorithms, the population size (*popsize*) is 100. Tournament selection, elitism, bit flip mutation, and uniform crossover are employed. The mutation rate is $1/l$, while the crossover rate is 0.6. In tournament selection, the best among 3 individuals randomly chosen is selected. Results for 4 algorithms, where RIs and MIs are inserted or not, are tested. The 2 strategies described in Sect. 3.1 are tested. For the dynamic strategy, results for runs with each one of the 6 DOP types are presented. The results of 50 runs for each combination of algorithm, dimension (l), change severity (ρ), and DOP strategy are presented. In the runs, the change period (τ) is equal to 500 generations. Each algorithm is run for l seconds. As the execution time is fixed and the number of evaluations for the algorithms can be different, the number of generations can also be different.

The best fitness obtained in each run is compared to the evaluation of the global optimum obtained by dynamic programming. The complexity of dynamic programming for the 0–1 knapsack problem is $O(lC)$, i.e., if C is polynomial, the algorithm runs in polynomial time. However, for the general case, the problem is NP-complete. In the experiments presented in the next section, the best fitness is stored only for the environments where $d(e) = 0$. In this way, the best from all generations are considered for the static strategy. However, for the dynamic strategy, only the results for environments where the index is odd are recorded.

4.2 Experimental Results

Table 1 shows the average error for the experiments. The average error is obtained by comparing, for each run, the static fitness of the global optimum with the static fitness of the best solution found by the algorithm. In order to test whether the best results are due to the use of immigrants (instead of due to changing the environment), results for runs of the static case with RIs and MIs are also presented. The results for the dynamic (with different DOP types) strategy are compared to the respective results for the static strategy. The Wilcoxon signed-rank test with the confidence level equal to 0.95 is used to test the statistical significance of the results.

Changing the environments resulted in better performance for some DOP types, but not for all. The worse results were obtained for DOP Types 1.1 and 1.2. As commented in Sect. 3.2, neighbourhood relations in the search space are preserved for changes in DOP Types 1.1 and 1.2. The changes produce the same effect that uniformly moving the solutions to other regions of the search space. Uniformly moving the individuals of the algorithm to new regions of the search space did not result in a better performance in the experiments.

The best results were obtained by DOP Type 2.2, followed by DOP Type 3. It is interesting to observe that, even directly optimizing $f_P(.)$ in approximately

Table 1. Average error (over 50 runs) for static and dynamic environments. The symbol *s* indicates that the results are statistically different according to the Wilcoxon signed-rank test. Bold face indicates that the results for the changing environments are statistically better than the respective results for the static environment. Italic face indicates the best result for each dimension.

RI	MI	ρ	DOP type						
			Static	1.1	1.2	1.3	2.1	2.2	3
l = 200									
No	No	0.125	0.8 ± 1.4	0.7 ± 1.1	0.7 ± 1.3	**0.1 ± 0.4 (s)**	**0.2 ± 0.4 (s)**	**0.1 ± 0.2 (s)**	**0.1 ± 0.4 (s)**
		0.500		2.3 ± 1.8 (s)	1.5 ± 1.6 (s)	**0.2 ± 0.5 (s)**	0.8 ± 1.1	**0.1 ± 0.3 (s)**	**0.2 ± 0.5 (s)**
		0.875		3.1 ± 2.2 (s)	1.9 ± 2.4 (s)	0.5 ± 1.1	1.3 ± 1.4	**0.1 ± 0.4 (s)**	**0.2 ± 0.5 (s)**
No	Yes	0.125	0.7 ± 1.1	1.4 ± 1.7 (s)	1.5 ± 2.0 (s)	**0.3 ± 0.5 (s)**	0.9 ± 1.2	**0.2 ± 0.6 (s)**	0.6 ± 1.1
		0.500		1.6 ± 1.7 (s)	1.6 ± 1.9 (s)	0.5 ± 0.9	1.5 ± 1.8 (s)	**0.3 ± 0.5 (s)**	0.7 ± 1.2
		0.875		1.4 ± 1.9 (s)	1.5 ± 2.0 (s)	0.7 ± 1.4	1.1 ± 1.5 (s)	**0.1 ± 0.5 (s)**	0.8 ± 1.2
Yes	No	0.125	1.1 ± 1.7	0.7 ± 1.4	0.9 ± 1.2	**0.2 ± 0.6 (s)**	**0.4 ± 0.7 (s)**	*0.0 ± 0.2 (s)*	**0.2 ± 0.5 (s)**
		0.500		1.9 ± 1.9 (s)	2.0 ± 2.5	**0.2 ± 0.5 (s)**	0.8 ± 1.1	**0.1 ± 0.4 (s)**	**0.2 ± 0.5 (s)**
		0.875		2.2 ± 1.9 (s)	1.6 ± 2.0	**0.4 ± 0.7 (s)**	1.2 ± 1.5	*0.0 ± 0.2 (s)*	**0.1 ± 0.4 (s)**
Yes	Yes	0.125	0.7 ± 1.2	1.2 ± 1.5	1.4 ± 1.9 (s)	**0.3 ± 0.5 (s)**	0.9 ± 1.4	**0.1 ± 0.4 (s)**	0.7 ± 1.3
		0.500		1.4 ± 1.7 (s)	1.8 ± 1.9 (s)	0.5 ± 1.0	1.2 ± 1.8	**0.2 ± 0.6 (s)**	0.7 ± 1.3
		0.875		1.2 ± 1.6 (s)	1.0 ± 1.5	0.7 ± 1.1	1.5 ± 1.8 (s)	**0.1 ± 0.3 (s)**	0.6 ± 1.0
l = 500									
No	No	0.125	4.4 ± 2.8	61.6 ± 9.6 (s)	60.8 ± 10.0 (s)	13.0 ± 3.5 (s)	27.8 ± 5.0 (s)	4.5 ± 2.0	7.1 ± 2.8 (s)
		0.500		80.7 ± 15.5 (s)	72.0 ± 18.6 (s)	9.1 ± 3.1 (s)	31.4 ± 7.4 (s)	4.5 ± 1.5	4.6 ± 1.7
		0.875		124.0 ± 15.5 (s)	65.9 ± 11.7 (s)	16.2 ± 5.6 (s)	75.1 ± 10.0 (s)	7.1 ± 2.4 (s)	6.8 ± 2.2 (s)
No	Yes	0.125	6.3 ± 3.4	**4.9 ± 2.3 (s)**	5.7 ± 3.0	5.7 ± 2.7	**4.9 ± 2.6 (s)**	**2.9 ± 1.9 (s)**	5.6 ± 3.1
		0.500		5.3 ± 2.8	8.2 ± 3.1 (s)	**4.1 ± 2.7 (s)**	5.4 ± 3.5	**2.0 ± 1.6 (s)**	**3.6 ± 2.3 (s)**
		0.875		7.7 ± 3.2 (s)	7.3 ± 3.5	**4.7 ± 2.7 (s)**	7.1 ± 3.9	**2.0 ± 1.6 (s)**	**4.2 ± 2.3 (s)**
Yes	No	0.125	5.2 ± 2.4	31.6 ± 7.4 (s)	76.1 ± 11.2 (s)	16.5 ± 4.3 (s)	29.5 ± 4.8 (s)	7.1 ± 2.0 (s)	**4.3 ± 2.2 (s)**
		0.500		75.3 ± 14.8 (s)	63.6 ± 13.3 (s)	19.8 ± 4.3 (s)	62.2 ± 9.4 (s)	**3.9 ± 1.6 (s)**	4.7 ± 2.1
		0.875		74.2 ± 16.5 (s)	71.6 ± 11.7 (s)	28.7 ± 6.9 (s)	83.6 ± 10.6 (s)	8.0 ± 2.6 (s)	6.9 ± 3.3 (s)
Yes	Yes	0.125	5.0 ± 2.6	5.5 ± 3.0	6.0 ± 3.3	**3.8 ± 2.2 (s)**	7.6 ± 4.6 (s)	**2.7 ± 1.9 (s)**	5.7 ± 2.6
		0.500		5.7 ± 3.4	4.8 ± 2.7	4.4 ± 2.5	7.5 ± 3.4 (s)	**3.2 ± 2.1 (s)**	6.2 ± 3.1 (s)
		0.875		5.7 ± ± 3.4	5.5 ± 2.8	4.7 ± 3.2	7.6 ± 4.1 (s)	***1.8 ± 1.6 (s)***	6.6 ± 3.3 (s)

half of the generations, the algorithms eventually obtained better results for the changing environments. With few exceptions, the dynamic strategy with DOP Type 2.2 resulted in better performance than the static strategy. Table 2 shows the percentage of successful runs, i.e., where the global optimum was found. For $l = 200$, the best result for the static case is 68%, while the global optimum was found in 96% of the runs of the algorithm with RIs for the dynamic strategy with DOP Type 2.2. The best results for the changing environments generally were obtained when the immigrants strategies were employed. However, immigrants generally did not result in better performance for the static environment. In particular, the best results for the changing environments for the experiments with $l = 500$ were obtained when MIs were inserted.

Table 2. Percentage of runs where the global optimum was found. Bold face indicates that the result for the dynamic environment is better than the respective result for the static environment. Italic face indicates the best result for each dimension.

RI	MI	ρ	DOP Type ($l = 200$)							DOP Type ($l = 500$)						
			Static	1.1	1.2	1.3	2.1	2.2	3	Static	1.1	1.2	1.3	2.1	2.2	3
No	No	0.125	62	62	62	**88**	**84**	**94**	**92**	4	0	0	0	0	2	0
		0.500		22	38	**88**	54	**94**	**82**		0	0	0	0	2	2
		0.875		16	40	**70**	40	**92**	**86**		0	0	0	0	0	0
No	Yes	0.125	64	46	50	**78**	54	**84**	**68**	2	2	2	2	2	**16**	0
		0.500		36	38	62	44	**76**	62		0	0	**8**	**4**	**22**	**6**
		0.875		44	48	**68**	52	**90**	56		0	2	2	0	**24**	2
Yes	No	0.125	56	**62**	54	**82**	**76**	***96***	**84**	0	0	0	0	0	0	**2**
		0.500		32	34	**80**	56	**88**	**84**		0	0	0	0	**2**	0
		0.875		24	42	**74**	42	***96***	**88**		0	0	0	0	0	0
Yes	Yes	0.125	68	44	48	**76**	58	**90**	62	0	**2**	0	**8**	**2**	**14**	0
		0.500		42	40	**72**	54	**86**	66		**2**	**4**	**2**	0	**6**	0
		0.875		44	54	58	46	**92**	66		**2**	0	**4**	0	***26***	0

5 Conclusions

We investigated artificially induced DOPs in this paper. Environmental changes can be artificially induced for different reasons, e.g., for speeding up evolution. A framework for generating artificially induced DOPs from any pseudo-Boolean problem was presented. Six different types of changes can be induced in the framework proposed here. The experiments with DOPs generated based on the 0–1 knapsack problem showed that better performance was obtained only for some change types and change severities.

Particularly, changes generated in DOP Type 2.2 resulted in better performance. For static environments, the best percentages of successful runs were: 68% ($l = 200$) and 4% ($l = 500$). For DOP Type 2.2, the best percentages of successful runs were: 96% ($l = 200$) and 26% ($l = 500$). Results not shown here for experiments with $l = 300$ and $l = 400$ also indicate better performance for the dynamic strategy[3]. The best results were obtained when random and memory immigrants were employed. The memory immigrants approach employed here makes use of the knowledge about the sequence of changes in the problem. This is an example of designing strategies to deal with artificially induced DOPs. The knowledge about the changes and their impact are usually not known in intrinsic DOPs. In artificially induced DOPs, the designer controls when and how to change the environments.

[3] The best percentages of successful runs for DOP Type 2.2 were 66% ($l = 300$) and 42% ($l = 400$), against 28% ($l = 300$) and 18% ($l = 400$) for the static environments.

Several future works are possible. Concerning the framework proposed here, it is necessary to better understand the impact of the different change types in different problems and state-of-art algorithms developed for static and dynamic optimization. In artificially induced DOPs, it is necessary to theoretically investigate how and when to change the environment according to the objectives of the programmer. Also, it is necessary to investigate new algorithms and operators that make use of the knowledge about the changes.

Acknowledgments. This work was funded partially by FAPESP under grant 2015/06462-1 and CNPq in Brazil, and partially by the Engineering and Physical Sciences Research Council (EPSRC) of U.K. under grant EP/K001310/1.

References

1. Cobb, H.G., Grefenstette, J.J.: Genetic algorithms for tracking changing environments. In: Proceedings of 5th International Conference on Genetic Algorithms, pp. 523–530 (1993)
2. Cruz, C., González, J., Pelta, D.: Optimization in dynamic environments: a survey on problems, methods and measures. Soft Comput. **15**, 1427–1448 (2011)
3. Fu, H., Sendhoff, B., Tang, K., Yao, X.: Robust optimization over time: problem difficulties and benchmark problems. IEEE Trans. Evol. Comp. **19**(5), 731–745 (2015)
4. Han, K.H., Kim, J.H.: Genetic quantum algorithm and its application to combinatorial optimization problem. In: Proceedings of the 2000 Congress on Evolutionary Computation, vol. 2, pp. 1354–1360 (2000)
5. Jin, Y., Branke, J.: Evolutionary optimization in uncertain environments-a survey. IEEE Trans. Evol. Comp. **9**(3), 303–317 (2005)
6. Kashtan, N., Noor, E., Alon, U.: Varying environments can speed up evolution. Proc. Natl. Acad. Sci. **104**(34), 13711–13716 (2007)
7. Knowles, J.D., Watson, R.A., Corne, D.W.: Reducing local optima in single-objective problems by multi-objectivization. In: Zitzler, E., Deb, K., Thiele, L., Coello Coello, C.A., Corne, D.W. (eds.) EMO 2001. LNCS, vol. 1993, p. 269. Springer, Heidelberg (2001)
8. Nguyen, T.T., Yang, S., Branke, J.: Evolutionary dynamic optimization: a survey of the state of the art. Swarm Evol. Comp. **6**, 1–24 (2012)
9. Parter, M., Kashtan, N., Alon, U.: Facilitated variation: how evolution learns from past environments to generalize to new environments. PLOS Comput. Biol. **4**(11), e1000206 (2008)
10. Richter, H.: Coevolutionary intransitivity in games: a landscape analysis. In: Mora, A.M., Squillero, G. (eds.) EvoApplications 2015. LNCS, vol. 9028, pp. 869–881. Springer, Heidelberg (2015)
11. Steinberg, B., Ostermeier, M.: Environmental changes bridge evolutionary valleys. Sci. Adv. **2**(1), e1500921 (2016)
12. Tan, L., Gore, J.: Slowly switching between environments facilitates reverse evolution in small populations. Evolution **66**(10), 3144–3154 (2012)
13. Tinós, R., Yang, S.: Analyzing evolutionary algorithms for dynamic optimization problems based on the dynamical systems approach. In: Yang, S., Yao, X. (eds.) Evolutionary Computation for Dynamic Optimization Problems. SCI, vol. 490, pp. 241–267. Springer, Heidelberg (2013)

14. Tinós, R., Yang, S.: Analysis of fitness landscape modifications in evolutionary dynamic optimization. Inf. Sci. **282**, 214–236 (2014)
15. Yang, S.: Genetic algorithms with memory-and elitism-based immigrants in dynamic environments. Evol. Comput. **16**(3), 385–416 (2008)

Efficient Sampling When Searching for Robust Solutions

Juergen Branke(✉) and Xin Fei

Warwick Business School, University of Warwick, Coventry, UK
juergen.branke@wbs.ac.uk, xin.fei.14@mail.wbs.ac.uk

Abstract. In the presence of noise on the decision variables, it is often desirable to find *robust* solutions, i.e., solutions with a good *expected* fitness over the distribution of possible disturbances. Sampling is commonly used to estimate the expected fitness of a solution; however, this option can be computationally expensive. Researchers have therefore suggested to take into account information from previously evaluated solutions. In this paper, we assume that each solution is evaluated once, and that the information about all previously evaluated solutions is stored in a memory that can be used to estimate a solution's expected fitness. Then, we propose a new approach that determines which solution should be evaluated to best complement the information from the memory, and assigns weights to estimate the expected fitness of a solution from the memory. The proposed method is based on the Wasserstein distance, a probability distance metric that measures the difference between a sample distribution and a desired target distribution. Finally, an empirical comparison of our proposed method with other sampling methods from the literature is presented to demonstrate the efficacy of our method.

1 Introduction

Many practical real-world problems involve uncertainty on decision variables. For example, in engineering, the actual product often does not correspond to the original design because of manufacturing tolerance. In such cases, the solutions should not only be good, but also *robust*. If $\xi \in \Xi$ are the possible disturbances to the decision variables, then the solution's expected fitness (which in the following, due to consistency with previous publications, we call *effective* fitness) is

$$f_{eff}(x) = \int_{\Xi} f(x+\xi)dP(\xi) \tag{1}$$

where $P(\xi)$ is the probability distribution of disturbance ξ. The effective fitness can be estimated by sampling as $\hat{f}_{eff}(x) = \sum_n f(x+\xi_n), \xi_n \in \Xi$. However, this is computationally expensive. Several researchers have thus attempted to speed up the search for robust solutions, surveys on these topics can be found in [2,8].

A previous study [11] has suggested that, for evolutionary algorithms (EAs), a single disturbed sample $f(x+\xi)$ that is used to evaluate a solution may actually

J. Handl et al. (Eds.): PPSN XIV 2016, LNCS 9921, pp. 237–246, 2016.
DOI: 10.1007/978-3-319-45823-6_22

be sufficient. The same study has reported that, in the case of infinite population size, an evolutionary algorithm with a single disturbed sample employed to evaluate each individual behaves in an identical manner to an evolutionary algorithm that operates directly on the effective fitness function. In the context of evolution strategies, [1] propose a mechanism to adaptively increase the population size over a run, along with a mechanism that adjusts mutation to account for the noise on the decision variables. To improve the estimate of an individual's effective fitness, previous studies have proposed to compute for the average of multiple samples, preferably based on Latin Hypercube Sampling [3,5].

EAs are population-based iterative search methods; hence, they usually converge to a promising region of the search space and then evaluate many samples in this area. Hence, at least towards the end of the optimisation run, when the EA would like to evaluate a solution, information about many other solutions in the neighbourhood is likely to be available if it is stored in a memory. This information can be exploited when estimating the robustness of a solution. Two questions arise:

1. How should the fitness values from the memory be weighted to yield a good (i.e., accurate and unbiased) estimate of the effective fitness of an individual?
2. If new information in terms of additional fitness evaluations can be collected, at what location(s) should this information be collected?

In [3], a new sample is taken at $\xi = 0$, and all the previous fitness values are weighted with the probability that a disturbance might actually result in the corresponding decision vector. However, this may result in a rather biased estimate if the distribution of memory samples is quite different from the distribution of expected disturbances. [9] propose to generate several candidate disturbances ξ_n, and then select the one that has the maximal minimum distance to any of the existing memory samples. This aims to fill in gaps in the distribution of memory samples; however, it is a rather simple heuristic and often results in extreme solutions being evaluated that are close to the disturbance boundary. [10] uses surrogate models to estimate the effective fitness.

In this paper, we propose a new method based on the Wasserstein distance to address the above two questions mentioned above. The Wasserstein distance measures the distance between two probability measures. The idea is to derive a large-sample target distribution from the known probability distribution of disturbances, and then collect new information and reallocate weight values such that the Wasserstein distance between the used samples and the large-sample target distribution is minimised.

The paper is structured as follows. Section 2 describes our proposed method and the mathematical foundation. Section 3 reports on several empirical experiments and a comparison with other methods from the literature. Finally, the paper concludes with a summary and some ideas for future work.

Algorithm 1. EA with ASA

Set $t \leftarrow 1$, initialise population P^t.
while Termination criterion is not met **do**
  Generate offspring population O from P^t.
  Generate N disturbance samples $x_n^t \in \Xi^t, n = 1, \ldots, N$
  for each solution $x^m \in O, m = 1, \ldots, \lambda$ **do**
    Compute approximate target $z_n = x^m + \xi_n^t$
    Identify memory solutions in neighbourhood $\mathcal{A}(x^m)$
    Construct N approximate set candidates $\mathcal{Y}^n(x^m) = \mathcal{A}(x^m) \cup z_n^t$
    Compute the Wasserstein distance value of each approximate set $\mathcal{Y}^n(x^m)$.
    Select the best approximate set $\mathcal{Y}^*(x^m)$ with the minimum distance value.
    Compute the optimal weight values $P(\mathcal{Y}^*(x^m))$.
    Compute $\hat{f}_{eff}(x^m) = \sum_{k=1} P(y_k) f(y_k), y_k \in \mathcal{Y}^*(x^m)$.
  end for
  Set $t \leftarrow t + 1$, update population P^t according to $\hat{f}_{eff}(x^m), m = 1, \ldots, \lambda$
end while

2 Proposed Method

This section describes our proposed method called archive sample approximation (ASA). The following notations are used throughout the paper:

- $\xi_n \in \Xi, n = 1, \ldots, N$: the underlying disturbance on decision variables.
- $z_n = x + \xi_n \in \mathcal{Z}(x), n = 1, \ldots, N$ (*approximation target*): one realisation of disturbed solution x.
- $y_k \in \mathcal{Y}(x), k = 1, \ldots, K (K \leq M)$ (*approximation set*): the set is used to approximate $\mathcal{Z}(x)$.
- $P(\mathcal{Z}), P(\mathcal{Y})$: the probability measure over approximation sets $\mathcal{Z}(x), \mathcal{Y}(x)$.
- $W(P(\mathcal{Y}), P(\mathcal{Z}))$: the Wasserstein distance between $P(\mathcal{Y})$ and $P(\mathcal{Z})$.
- $\mathcal{A} = \{a_1, \ldots, a_L\}$: the archive of previous fitness evaluations.
- $\mathcal{A}(x)$: a subset of $\mathcal{A}$ that contains locations in the "disturbance neighbourhood"of solution x.

Algorithm 1 describes how ASA can be integrated into an evolutionary algorithm. First, we generate a set of disturbances Ξ from the underlying noise distribution. Note that the disturbance set changes in every generation, but we use the same disturbances for all the individuals within one generation. With disturbance set Ξ, we can generate the approximation target set $\mathcal{Z}(x^m)$ for solution x^m that would, if evaluated, allow us to compute a good-enough estimate of the individual's effective fitness.

Next, we search $\mathcal{A}(x^m)$ in the "disturbance neighbourhood" of solution x^m from archive $\mathcal{A}$, and include all available memory solutions into approximation set $\mathcal{Y}(x^m)$. Meanwhile, we would like to add one additional disturbance realisation z_n for solution x^m into its approximation set. As candidates, we consider all the points in the target set $\mathcal{Z}(x)$, and try inserting each one, resulting in N approximation sets $\mathcal{Y}^n(x^m) = \mathcal{A}(x^m) \cup z_n$. The goal of the ASA procedure is to find the best approximation set $\mathcal{Y}^*(x^m)$ with probability measures $P(\mathcal{Y}^*(x^m))$ to well approximate the target $\mathcal{Z}(x^m)$.

Our algorithm uses the Wasserstein distance [6] to decide which approximation set is the best option (i.e., where to sample) and how to weigh the samples. This paper implements the L_1 Wasserstein distance to quantify the error in the approximation set, which can be computed by solving the Kantorovich-Rubinstein transportation problem, as follows.

$$\begin{aligned} W(P(\mathcal{Y}^n), P(\mathcal{Z})) = \min_{\mu} \quad & \sum_k \sum_n d(y_k, z_n)\mu(y_k, z_n) \\ \text{s. t.} \quad & \sum_k \mu(y_k, z_n) = P(z_n), \quad \forall n \\ & \mu \geq 0 \end{aligned} \tag{2}$$

Once we obtain the optimal "transportation plan" μ^* in (2) is obtained, the optimal weights (probability measure $P(\mathcal{Y}^n)$ can be determined immediately by using

$$P(y_k) = \sum_n \mu^*(y_k, z_n), \quad \forall k. \tag{3}$$

To identify the best candidate, we simply add, one by one, each target point to the set of relevant memory locations $\mathcal{A}(x)$, and compute their Wasserstein distance values $W(P(\mathcal{Y}^n), P(\mathcal{Z}))$. The ASA algorithm aims to find the approximation set $\mathcal{Y}^*$ with the minimum distance value given by

$$W(P(\mathcal{Y}^*), P(\mathcal{Z})) = \min_n (W(P(\mathcal{Y}^n), P(\mathcal{Z}))). \tag{4}$$

Finally, we discuss the computation issue of linear program (2) and its efficient solution method. For the Kantorovich-Rubinstein transportation problem, the computation complexity is significantly influenced by the size of the approximation target. In this paper, we apply duality theory to reduce the computational effort. We assume that η_n is the dual decision variable for the nth constraint in (2), then we have

$$\begin{aligned} W(P(\mathcal{Y}^n), P(\mathcal{Z})) = \max_{\eta} \quad & \sum_n P(z_n)\eta_n \\ \text{s.t.} \quad & \eta_n \leq d(y_k, z_n), \quad \forall n, \quad \forall k \end{aligned} \tag{5}$$

The optimal value is found if η_n satisfies

$$\eta_n^* = \min_n \ d(y_n, z_k) \tag{6}$$

Hence, the Wasserstein value can be computed by using

$$W(P(\mathcal{Y}^n), P(\mathcal{Z})) \leq \sum_n P(z_n)\eta_n^*. \tag{7}$$

The equality will hold if the linear program exhibits *strong duality.* By nature of the transportation problem, the optimal decision algorithm should select the closest starting point for each destination to minimise the total transportation cost. Therefore, the optimal weight value for each element in the approximation set can be computed as

$$P(y_k) = \frac{\sum_n \lambda^{kn} P(z_n)}{N} \tag{8}$$

with

$$\lambda^{kn} = \begin{cases} 1 & if\ y_k\ is\ the\ closest\ sample\ to\ z_n \\ 0 & otherwise \end{cases}$$

where λ^{kn} is an index function that is used to count the number of times y_k is the closest sample to z_n.

3 Numerical Experiments

3.1 Test Functions

We demonstrate the performance of our proposed method on three 5-D test problems listed in Table 1. A 1-D visualisation of each test function is shown in Fig. 1. TP 1 has a single asymmetric peak and has been adapted from [10]. It will allow to examine an algorithm's ability to precisely identify the location of the robust solution. TP 2 has been taken from [4] and is multi-modal. The original fitness function has its optimum at $x = 1$, whereas the optimum of the effective fitness is at $x = -1$, which allows to test whether an algorithm is able to correctly identify the robust optimum. TP 3 combines both characteristics and has been adapted from a function used in [10].

Table 1. Test function description

	Formulation	Noise
TP 1	$\min \quad 0.9d + \sum_{i=1}^{d} Q_1(x_i)$, $Q_1(x_i) = \begin{cases} -(8-x_i)^{0.1} e^{-0.2(8-x_i)} & x_i < 8 \\ 0 & otherwise \end{cases}$, $x \in [0, 10]$	$U(-1, 1)$
TP 2	$\min \quad \sum_{i=1}^{d} Q_2(x_i)$, $Q_2(x_i) = \begin{cases} -(x_i+1)^2 + 1.4 - 0.8\,\lvert sin(6.283x_i)\rvert & -2 < x_i < 0 \\ 0.6 \cdot 2^{-8\lvert x_i - 1\rvert} + 0.958887 - 0.8\,\lvert sin(6.283x_i)\rvert & 0 \leq x_i < 2 \\ 0 & otherwise \end{cases}$, $x \in [-2, 2]$	$U(-0.2, 0.2)$
TP 3	$\min \quad \sum_{i=1}^{d} Q_3(x_i)$, $Q_3(x_i) = 2sin(10e^{(-0.2x_i)}x_i)e^{(-0.25x_i)}$, $x \in [0, 10]$	$U(-1, 1)$

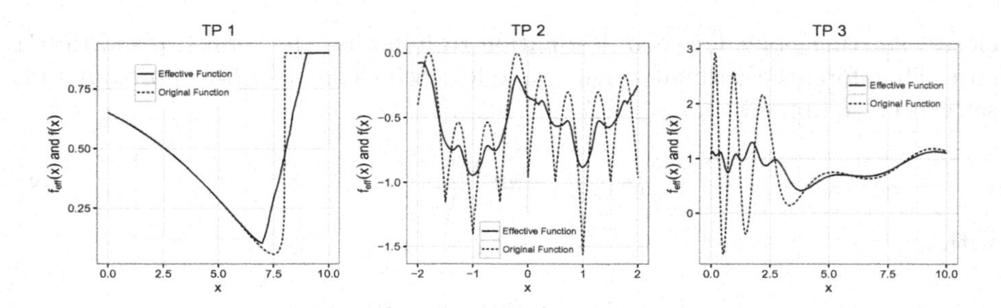

Fig. 1. 1-D visualisation of test functions

3.2 Evolutionary Algorithm

Our method can be combined with any metaheuristic. In this paper, we use a standard CMA-ES [7] for our experiments, with $\mu = 4, \lambda = 8$, initial $\sigma_0 = \frac{1}{4} Search\,Interval\,Width$ and equal weighting of the four individuals to determine the next centre of the mutation distribution.

3.3 Final Solution Selection and Performance Measure

Due to the noise, the effective fitness estimated by the algorithm is likely to deviate from the true effective fitness. For this reason, we use the barycenter of the selected parents as the solution that would be returned to the user.

$$x_{final} = \sum_{i=1}^{\mu} w_i x_i, \quad w_i = \frac{1}{\mu}$$

The effective fitness of final solution is evaluated using Monte-Carlo simulation with $N = 10,000$ samples.

$$f_{eff}(x_{final}) = \frac{1}{N} \sum_{i=1}^{N} f(x_{final} + \xi_i)$$

In order to better understand the quality of the effective fitness estimation, we furthermore report on the average absolute error AE_t by calculating the mean squared error between the true and approximate effective fitness as follows.

$$AE_t = \frac{1}{\lambda} \sum_{j=1}^{\lambda} \left| f_{eff}(x_m) - \hat{f}_{eff}(x_m) \right|, \quad x_m \in P_t$$

3.4 Target Samples Generation

We test the following three techniques for generating target samples.

1. Monte Carlo sampling (MC).
2. Latin hypercube sampling (LHS).
3. Equidistant sampling (ES) which places all samples on a regular grid with three points in each dimension.

3.5 Experimental Setup

We compare our method with three alternative approaches:

1. SEM: take one random sample for each solution. This is the approach proposed in [11].
2. SEM+AR: take one additional random sample for each solution, but also take into account all memory points in the area of disturbance. The new sample and all memory points are equally weighted. This is the approach proposed in [3].
3. ABRSS: A method that uses Latin Hypercube Sampling as reference points, and then includes for each reference point the closest memory point. To add a new sample, some random samples are generated and the one furthest from any memory point is selected. This method has been proposed in [9].

All methods are incorporated into CMA-ES, the size of target samples or reference points is $3^5 = 243$ for all methods. All reported results are averaged over 30 runs. The computational budget for each run was 2,500 fitness evaluations.

3.6 Results on Convergence Rate and Average Approximation Error

Figures 2 and 3 compare the convergence rate and approximation error of different methods. The effective fitness of the final solution is also reported in Table 2. As can be seen, ASA has the best convergence behaviour and smallest approximation error in all three test problems. The use of the Wasserstein distance effectively controls the approximation error, and thus it has a fast convergence rate. On TP 1, SEM+AR works almost as good as ASA. Because this problem is unimodal, all algorithms converge to the correct peak, and a lot of memory samples accumulate there, leading to a very small approximation error also for SEM+AR. ABRSS is much worse, probably because its sampling mechanism tries to sample away from existing memory samples, which in this case means at less relevant points and introducing a bias. The increase in approximation error for SEM and ABRSS can be explained by their focusing on the peak area, which has a very large gradient.

ABRSS is the second best method for TP 2 and TP 3. Since SEM and SEM+AR draw samples randomly, they are more prone to "lucky" over evaluations of individuals. As a consequence, they always discover new presumably good solutions, move there, and then realise after some time that the solution was actually not really very good, leading to a jumping behaviour from one local optimum to another.

3.7 Influence of the Target Sample Generation Mechanism

ASA requires a set of target samples to start with. To better understand the influence of the target sample generation mechanism, we compare the influence of different sample generation methods in Figs. 4 and 5. LHS (which we also

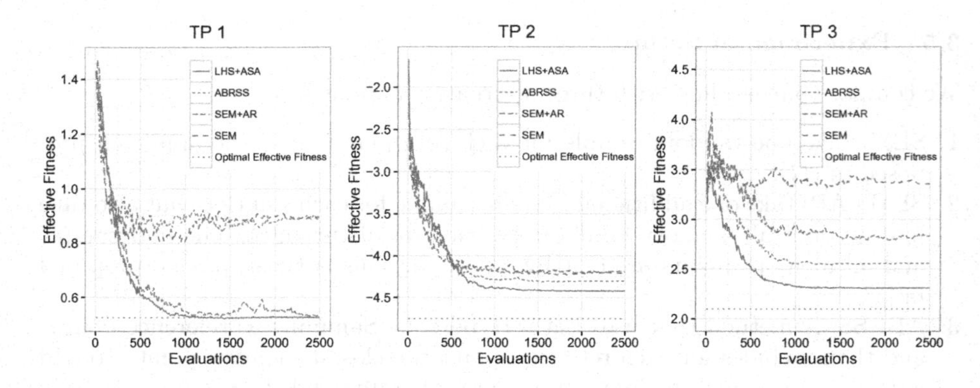

Fig. 2. Convergence rate of different methods

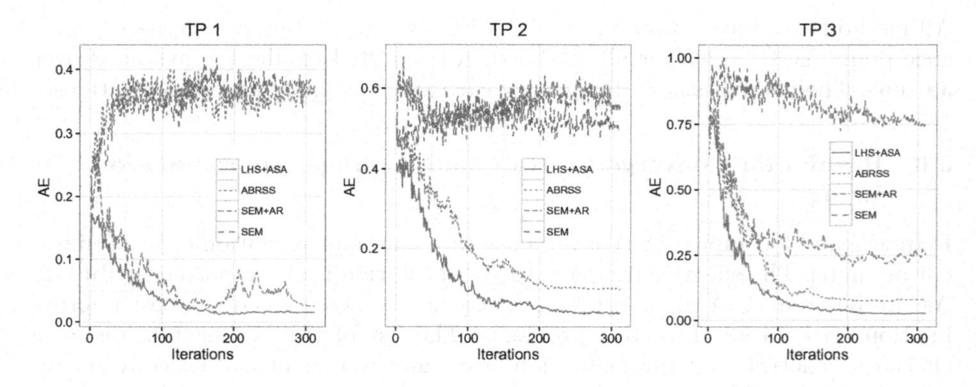

Fig. 3. Average absolute error of different methods

Table 2. Effective fitness after 2,500 evaluations

Method	*mean* ± *s.e.*		
	TP 1	TP 2	TP 3
SEM	0.8988 ± 0.0314	-4.1279 ± 0.0434	3.3569 ± 0.0945
SEM + AR	0.5388 ± 0.0069	-4.1845 ± 0.0420	2.8368 ± 0.0939
ABRSS	0.8893 ± 0.0451	-4.3293 ± 0.0388	2.5574 ± 0.0717
LHS + ASA	0.5269 ± 0.0013	-4.4231 ± 0.0362	2.3083 ± 0.0297

used in the previous experiment) performs well in all test functions. Interestingly, equidistant sampling outperforms other sampling methods in TP 3, but produces a bad solution in TP 1. This is probably because the boundary of the disturbance region has a particular large influence for TP 1, and the way we chose the grid structure that boundary region was never sampled.

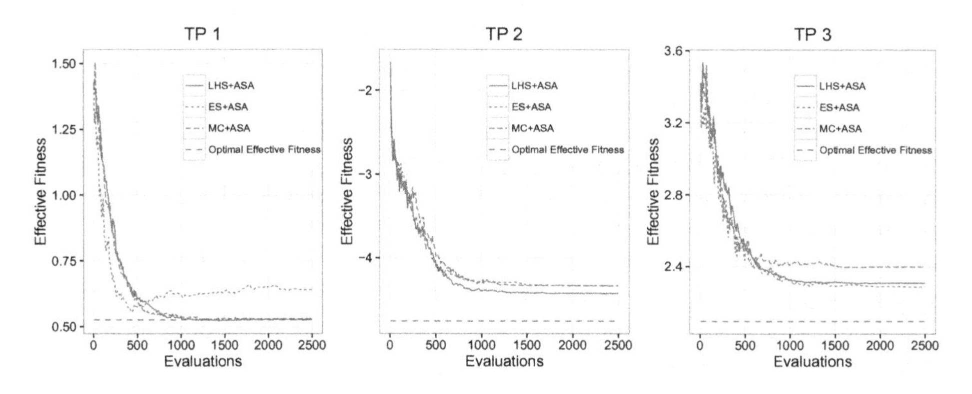

Fig. 4. Convergence rate of different target sample generation schemes

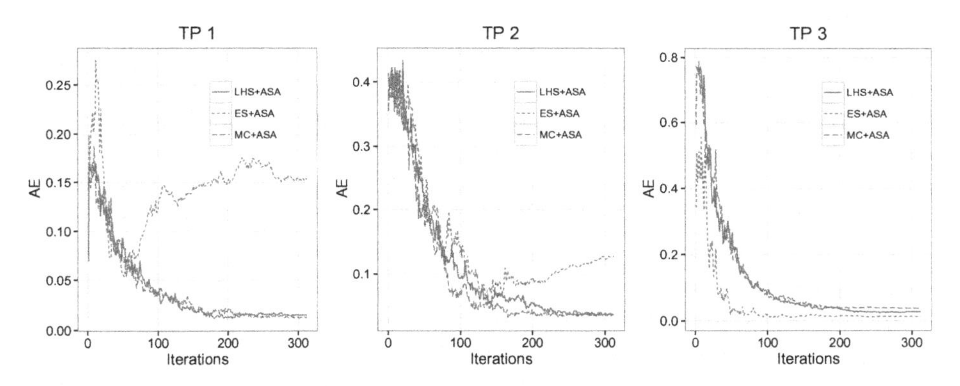

Fig. 5. Average absolute error of different target sample generation schemes

Table 3. Effective fitness after 2,500 evaluations

Target samples	*mean* ± *s.e.*		
	TP 1	TP 2	TP 3
MC+ASA	0.5304 ± 0.0030	−4.3308 ± 0.0625	2.3988 ± 0.0583
ES+ASA	0.6416 ± 0.0076	−4.3042 ± 0.0466	2.2872 ± 0.0439
LHS+ASA	0.5269 ± 0.0013	−4.4231 ± 0.0362	2.3083 ± 0.0297

4 Conclusion

We have looked at the problem of searching for robust solutions, where robust means a good expected fitness over a given distribution of disturbances to the decision variables. In particular, we have re-considered the idea of estimating a solution's effective fitness by making use of previous fitness evaluations stored in the memory. We proposed a methodology based on the Wasserstein distance to decide at what location we should evaluate the fitness in order to gain the most useful additional information about a solution's effective fitness, and also how to weigh the different samples in the neighborhood for estimating the effective fitness. Empirical comparisons with several previous methods for this problem on three test functions demonstrates the superiority of our new approach.

Future work will include developing other distance metrics and moving from an individual based view to a population based view when determining where to evaluate fitness.

References

1. Beyer, H.-G., Sendhoff, B.: Evolution strategies for robust optimization. In: World Congress on Computational Intelligence, pp. 4489–4496. IEEE (2006)
2. Beyer, H.-G., Sendhoff, B.: Robust optimization - a comprehensive survey. Comput. Methods Appl. Mech. Eng. **196**(33), 3190–3218 (2007)
3. Branke, J.: Creating robust solutions by means of evolutionary algorithms. In: Eiben, A.E., Bäck, T., Schoenauer, M., Schwefel, H.-P. (eds.) PPSN 1998. LNCS, vol. 1498, pp. 119–128. Springer, Heidelberg (1998)
4. Branke, J.: Evolutionary Optimization in Dyamic Environments. Kluwer, Boston (2001)
5. Branke, J.: Reducing the sampling variance when searching for robust solutions. In: Genetic and Evolutionary Computation Conference, pp. 235–242. Morgan Kaufmann, San Francisco (2001)
6. Dudley, R.M.: Real Analysis and Probability, vol. 74. Cambridge University Press, Cambridge (2002)
7. Hansen, N., Ostermeier, A.: Completely derandomized self-adaptation in evolution strategies. Evol. Comput. **9**(2), 159–195 (2001)
8. Jin, Y., Branke, J.: Evolutionary optimization in uncertain environments - a survey. IEEE Trans. Evol. Comput. **9**(3), 303–317 (2005)
9. Kruisselbrink, J., Emmerich, M., Bäck, T.: An archive maintenance scheme for finding robust solutions. In: Schaefer, R., Cotta, C., Kołodziej, J., Rudolph, G. (eds.) PPSN XI. LNCS, vol. 6238, pp. 214–223. Springer, Heidelberg (2010)
10. Paenke, I., Branke, J., Jin, Y.: Efficient search for robust solutions by means of evolutionary algorithms and fitness approximation. IEEE Trans. Evol. Comput. **10**(4), 405–420 (2006)
11. Tsutsui, S., Ghosh, A.: Genetic algorithms with a robust solution searching scheme. IEEE Trans. Evol. Comput. **1**(3), 201–208 (1997)

Genetic Programming

Optimising Quantisation Noise in Energy Measurement

William B. Langdon(✉), Justyna Petke, and Bobby R. Bruce

CREST, Department of Computer Science, University College London,
Gower Street, London WC1E 6BT, UK
w.langdon@cs.ucl.ac.uk
http://crest.cs.ucl.ac.uk/

Abstract. We give a model of parallel distributed genetic improvement. With modern low cost power monitors; high speed Ethernet LAN latency and network jitter have little effect. The model calculates a minimum usable mutation effect based on the analogue to digital converter (ADC)'s resolution and shows the optimal test duration is inversely proportional to smallest impact we wish to detect. Using the example of a 1 kHz 12 bit 0.4095 Amp ADC optimising software energy consumption we find: it will be difficult to detect mutations which an average effect less than 58 μA, and typically experiments should last well under a second.

Keywords: Theory · Genetic improvement · Genetic programming · Software engineering · SBSE · Parallel EC · Distributed power monitoring

1 Introduction

Evolutionary computing (EC) can be incorporated into product development either by inventing new designs or optimising existing ones. In both it is fundamentally important to be able to decide if a design is fit or not. The widespread adoption of fully functional mobile computers in the form of smartphones has thrust optimising software energy usage, and so battery life, into the limelight.

In many cases the quality of designs is calculated using simulators before manufacture. However, it is necessary that the simulation be detailed enough so that it can tell automatically a better design from an already good design. In the case of simple electronics such high quality simulator may exist. However even in the case of single chip devices, such simulators run several orders of magnitude slower that the software running on the chip and good simulators for the whole of a portable device may not be feasible. So for feasibility, cost, credibility and speed there is increasing interest in optimising portable electronic devices by using real devices and real power monitors (Fig. 1) to measure their true energy

W.B. Langdon—http://www.cs.ucl.ac.uk/staff/W.Langdon/.
J. Petke—http://www.cs.ucl.ac.uk/staff/J.Petke/.
B.R. Bruce—http://www.cs.ucl.ac.uk/staff/r.bruce/.

J. Handl et al. (Eds.): PPSN XIV 2016, LNCS 9921, pp. 249–259, 2016.
DOI: 10.1007/978-3-319-45823-6_23

Fig. 1. MAGEEC power measurement board ©http://mageec.org/

consumption and use it as part of the EC fitness function (Bruce 2015). With the advent of genetic improvement (GI) (Langdon 2015) it is increasingly common to view software as mutable and apply EC directly to it (White *et al.* 2008; Bruce 2015; Schulte *et al.* 2014a). There is great interest in using real measurements. Although our immediate use case is genetic improvement and the evolution of better software, here we are concerned with the practical limits of using real-world measuring devices in EC.

The next section presents a mathematical model of the accuracy of a single measuring device directly connected to single test device. Since fitness testing is usually the bottleneck in EC, it is common to consider running fitness tests in parallel. Section 3 expands the model of discretised measurement to a high speed Ethernet local area network based distributed system of dozens of computer hardware under test. Since Ethernet is a stochastic protocol, network delays are necessarily variable. Section 4 calculates that the best tests will be surprisingly short, under one second. This is in keeping with our view that often too much care is taken to get an accurate fitness value, where it is only necessary to be able to tell a good mutant from a less good one. Section 5 discusses the results in Sect. 4, ways to avoid EC degenerating into random search, three alternatives to LAN messages and concludes. To save space some of the intermediate mathematical steps and some of the discussion have be omitted. (The full text can be found in our technical report of the same name RN/16/01.)

2 Directly Connected Monitor

Figure 2 shows a system to automatically measure physical components of an EC fitness function. The "physical system" will be subject to mutations taken from the current population and the system will attempt to quantify the mutation's

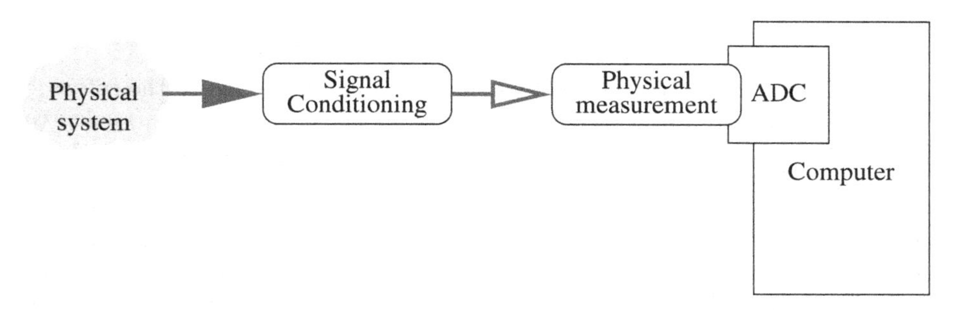

Fig. 2. Typical modern measuring and monitoring systems interface to the real world (physical system) via an analogue signal conditioning unit, a measuring device (e.g. a thermocouple) and an Analogue to Digital Converter (ADC). Although we consider optimising energy consumption, our mathematical framework can be generally applied. The conditioned signal is converted into an analogue electrical signal, which converted into a digital signal by the ADC, which is then read periodically at a fixed rate by the computer.

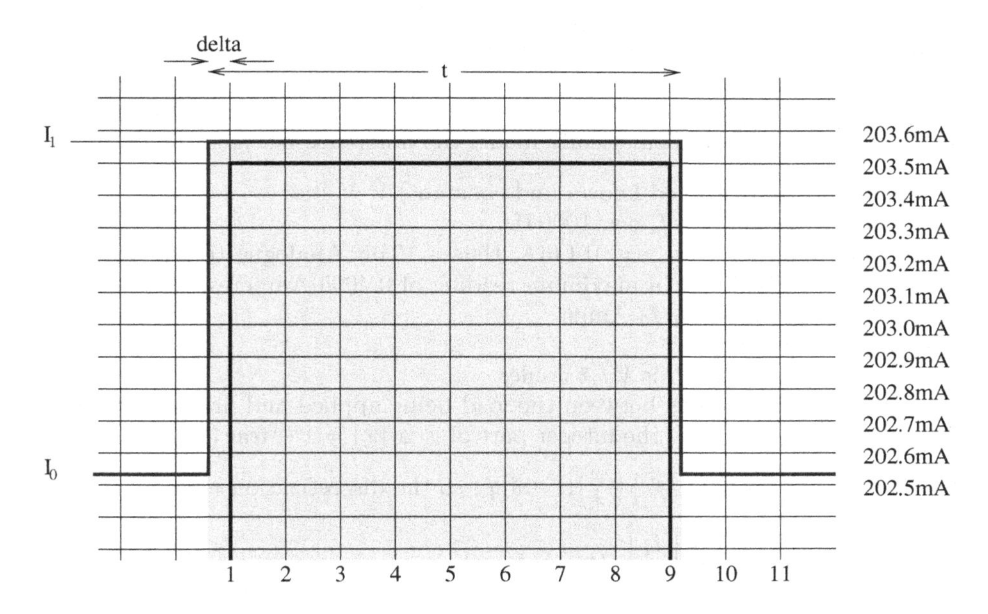

Fig. 3. Energy used is given by area of yellow rectangle times supply voltage (5 V) $E = 5I_1t = 5 \times 203.6\,\text{mA} \times 8.6\,\text{mS} = 8.753\,\text{mJ}$. Current resolution $a = 0.1\,\text{mA}$ (12 bit ADC full scale 0.4095 Amp). Sampling frequency $f = 1\,\text{kHz}$. Quantised energy $= 5 \times 203.5\,\text{mA} \times 8\,\text{mS} = 8.14\,\text{mJ}$. Noise $= 8.753 - 8.14 = 0.6134$. Relative noise $= 0.6134/8.753 \approx 7\,\%$.

effect. Our model applies generally to EC using physical measurement. It could deal with not just the power consumed by the CPU but also by other activities particularly the screen (Li *et al.* 2015), radio links and GPS.

In the case of genetic improvement, the mutation is applied to the software running on the physical devices (e.g. a smartphone) and the ADC (analog to digital converter) will measure its power consumption. Since phones operate at about 5 V little signal conditioning other than a fixed resistor is needed.

The simple model we present is potentially suitable for the very high frequency response that modern oscilloscopes are capable of. Since such oscilloscopes cost many thousands of pounds we will concentrate on automated power monitors costing a few tens of pounds each (such as the one in Fig. 1). Notice that although they cannot measure very high frequency (short duration) effects, they can still accurately measure average power consumption. Even if there is significant amounts of power at high frequency, it does not disappear when measured at lower frequencies and (assuming there are no serious aliasing effects) it simply contributes to the low frequency average.

The simple model presented in Fig. 3 assumes running the test causes the power consumption to rise but that the energy monitoring is quantised both into discrete time samples and that measurements of power consumption are also discrete. It assumes the power monitor is not synchronised to the start of the test software but that the start and end of the test are known. The actual energy used by the test is proportional to the area of the yellow rectangle in Fig. 3 but the reported (discretised) energy is proportional to the number of unit rectangles inside the rectangle bounded by the thick black lines and the x axis. Next we will mathematically model the difference between the two.

- Supply voltage (assumed known and constant) V Volts.
- Sampling frequency = f, e.g. 1000 Hz.
- Current resolution = a, e.g. 0.1 mA, thus a 12 bit Analogue to Digital Converter (ADC) will have a maximum reading of 0.4095 Amperes.
- Unloaded current draw I_0 Amps.
- Actual load I_1 Amps.
- The actual energy used is VI_1t Joules.
- δ is the time in seconds between the load being applied and first the sample.
- Assuming x is positive, the integer part of x is $\lfloor x \rfloor = x - \mathrm{frac}\,(x)$.

The measured energy is $\frac{Va}{f} \lfloor \frac{I_1}{a} \rfloor \lfloor (t-\delta)f \rfloor$ so the discretization noise is

$$
\begin{aligned}
&= VI_1t - \frac{Va}{f}\left\lfloor \frac{I_1}{a} \right\rfloor \lfloor (t-\delta)f \rfloor \\
&= VI_1t - \frac{Va}{f}\left(\frac{I_1}{a} - \mathrm{frac}\left(\frac{I_1}{a}\right)\right)\left((t-\delta)f - \mathrm{frac}\,((t-\delta)f)\right) \\
&= Vat\ \mathrm{frac}\left(\frac{I_1}{a}\right) + VI_1\delta - Va\delta\ \mathrm{frac}\left(\frac{I_1}{a}\right) \\
&\quad + \frac{V}{f} I_1 \mathrm{frac}\,((t-\delta)f) - \frac{Va}{f}\mathrm{frac}\left(\frac{I_1}{a}\right)\mathrm{frac}\,((t-\delta)f)
\end{aligned} \tag{1}
$$

Since the start of running the software is unrelated to the exact point in time measurements are taken, δ will be uniformly scattered in the range [0 to $1/f$] and so the expected value of δ is $1/2f$ (Fig. 3). Since I_1 is much bigger than a, it is reasonable to assume the fractional part of I_1/a, i.e. frac (I_1/a), is uniformly distributed across the interval [0–1]. (With a uniform distribution in [0–1], the expected value of frac $(\cdot)$ is $1/2$ and the standard deviation is $\sqrt{1/12} = 0.288675$). So the expected noise (Eq. 1) becomes:

$$\frac{Vat}{2} + VI_1\frac{1}{2f} - Va\,\frac{1}{2} \times \frac{1}{2f} + \frac{V}{f}I_1\frac{1}{2} - \frac{Va}{f}\frac{1}{2} \times \frac{1}{2} = \frac{1}{2}Vat + V\frac{I_1}{f} - \frac{1}{2}V\frac{a}{f}$$

$$\text{Fractional noise} = \frac{\text{noise}}{\text{true energy}} = \frac{\frac{1}{2}Vat + V\frac{I_1}{f} - \frac{1}{2}V\frac{a}{f}}{VI_1t} = \frac{1}{2}\frac{a}{I_1} + \frac{1}{ft} - \frac{1}{2}\frac{a}{fI_1t} \quad (2)$$

We can approximate the fractional noise by dropping the last term in Eq. 2. We can express Eq. 2 in terms of the current measurement resolution and number of samples $N = ft$. Each ADC raw value is $I_1/a = k$. (For a twelve bit resolution analogue to digital converter and I_1 near the middle of the range $k \approx 2048$.) So Eq. 2 becomes fractional noise $\approx 1/4096 + 1/N$. That is, with a coarse sampling the noise is dominated by the number of samples N but if we can either increase the sampling rate or run the experiment for longer, the $1/N$ term becomes less important and the noise tends to a limit given by the resolution of the ADC. Further, once the number of samples, N, exceeds the resolution of the ADC there is only marginal reduction in noise from increasing the number of samples. Using our 12 bit 1 kHz example ADC, there is only marginal gain in increasing the number of measurements above 4096. That is, greatly increasing the measurement time, t, above $4096/f \approx 4$ s, gives little further improvement. See also end of Sect. 4.

3 Distributed Power Measurement

In the previous section we assume that the onset of the load and when its finished are known exactly. In the case of distributed power monitoring, two commands are sent via a local area network (LAN). The first is to start the recording of energy consumption and the second to stop the recording. Initially we shall concentrate upon the variation introduced by the LAN and then include the energy measurement noise given by Eq. 1.

Measuring energy is initiated when the start message packet (p_1) reaches the monitoring computer at time s_1. (The LAN packets are shown by dotted arrows in Fig. 4.) When the acknowledgement packet (p_2) reaches the test computer (s_2), it starts the experiment, raising the current from rest (I_0) to I_1. t seconds later (e_1) the experiment finishes: the load drops back to I_0 and the test computer sends a message packet (p_3) stopping the measurement (e_2). In Fig. 4

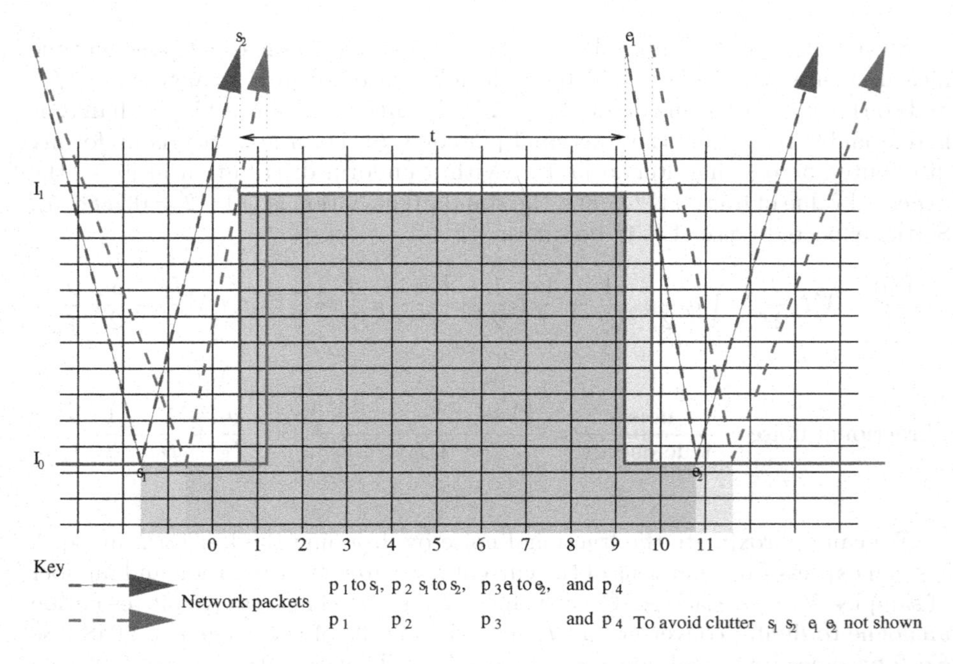

Fig. 4. Measuring energy is initiated when the start message (left arrow) reaches the monitoring computer s_1. When the acknowledgement reaches the test computer s_2, it starts the experiment, raising the current from rest (I_0) to I_1. t seconds later the experiment finishes: the load drops back to I_0 and the test computer sends a message e_1 ending the measurement e_2. The experiment is done twice (blue and red) but different results are obtained since the network delays are different. As in Fig. 3, energy used is given by area of under current curves times supply voltage (5 V). Left (blue) = 12.60 mJ, right (red) = 12.40 mJ, relative difference = 0.2/12.60 ≈ 1.6%. (Color figure online)

the experiment is done twice but different results are obtained since although the test computer starts at the same time and the experiment takes t seconds in both cases, the network delays are different.

The measured energy is $V\,(I_0(s_2 - s_1) + (I_1 - I_0)t)$. Where $(s_2 - s_1)$ is the observed duration. This is longer than t because of the transit times of the two network packets p_2 and p_3. (Figure 5 gives transit times for two LAN packets, there and back.) Now $(s_2 - s_1) = p_2 + t + p_3$ so measured energy $= V\,(I_0(p_2 + t + p_3) + (I_1 - I_0)t) = V\,(I_0(p_2 + p_3) + I_1 t)$.

We will assume that the transit times for the LAN packets are on average the same and that variations are independent. Thus the variance in the energy measurement due to network work variations (i.e. V, I_1 and t are assumed fixed):

$$V^2 I_0^2\,(\text{var}\,(p_2) + \text{var}\,(p_3)) = 2V^2 I_0^2\ \text{var}\,(p) \tag{3}$$

Since we assume that p_2 are p_3 are equally distributed and independent we drop their subscripts are refer to them both as p. So var (p) is the variance of LAN

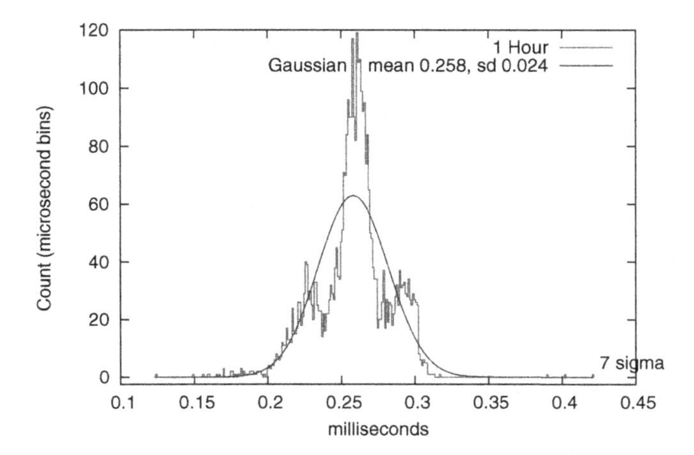

Fig. 5. Distribution of 3780 network delays. Notice approximate match of Normal distribution and also long tail of much longer delays.

packet transit times ($\mathrm{SD}\,(p) = \sqrt{\mathrm{var}\,(p)}$). The fractional variation in the energy measurement is

$$= \frac{\sqrt{2}\; VI_0\mathrm{SD}\,(p)}{V\,(2I_0p + I_1t)} = \frac{\sqrt{2}\;\mathrm{SD}\,(p)}{(2p + tI_1/I_0)}$$

Figure 5 suggests the mean of the two packet transit time ($2p$) is typically 0.258 mS and $\sqrt{2}\; SD(p)$ is 24 μs.

The variation in the discretization noise (given by Eq. 1) is due to variation in the duration t and size I_1 of the load. Treating these as independent gives the variance in the discretization noise. (Remember the variance of the product of two independent variables x and y (of means X and Y) is $\mathrm{var}\,(xy) = X^2\mathrm{var}\,(y) + Y^2\mathrm{var}\,(x) + \mathrm{var}\,(x)\mathrm{var}\,(y)$ (Goodman 1960 Eq. 2).)
Remember (Eq. 1) discretization noise/V

$$= at\,\mathrm{frac}\left(\frac{I_1}{a}\right) + I_1\!\left(\delta + \frac{1}{f}\mathrm{frac}\,((t-\delta)f)\right) - a\delta\;\mathrm{frac}\left(\frac{I_1}{a}\right) - \frac{a}{f}\mathrm{frac}\left(\frac{I_1}{a}\right)\mathrm{frac}\,((t-\delta)f)$$

We now calculate the variance of discretization noise/V one term at a time. Note the variance of the uniform distribution of the range [0-1] is 1/12. Starting with the first (depends on t) and last terms

$$\mathrm{var}\left(at\;\mathrm{frac}\left(\frac{I_1}{a}\right)\right) = a^2\mathrm{var}\,(t)/3 + a^2t^2/12 \qquad (4)$$

$$\mathrm{var}\left(-\frac{a}{f}\mathrm{frac}\left(\frac{I_1}{a}\right)\mathrm{frac}\,((t-\delta)f)\right) = 7/144\;\frac{a^2}{f^2} \qquad (5)$$

Now the middle terms (which depend on both I_1 and δ).

$$I_1\left(\delta+\frac{1}{f}\text{frac}\left((t-\delta)f\right)\right)-a\delta\ \text{frac}\left(\frac{I_1}{a}\right)=\delta\left(I_1-a\ \text{frac}\left(\frac{I_1}{a}\right)\right)+\frac{I_1}{f}\text{frac}\left((t-\delta)f\right)$$

Taking the variance of the first part (assuming that δ and I_1 are independent)

$$\text{var}\left(\delta\left(I_1-a\ \text{frac}\left(\frac{I_1}{a}\right)\right)\right)=\text{var}(\delta)\left(I_1-a/2\right)^2+\delta^2\left(\text{var}(I_1)+a^2/12\right)$$
$$+\ \text{var}(\delta)\left(\text{var}(I_1)+a^2/12\right)$$

and of the second part

$$\text{var}\left(\frac{I_1}{f}\text{frac}\left((t-\delta)f\right)\right)=\frac{\text{var}(I_1)}{f^2}/3+\frac{I_1^2}{f^2}/12 \qquad (6)$$

Combining formulae 4–6 gives var (discretization noise/V) as:

$$=a^2\text{var}(t)/3+a^2t^2/12+\text{var}(\delta)\left(I_1-a/2\right)^2+\delta^2\left(\text{var}(I_1)+a^2/12\right)$$
$$+\text{var}(\delta)\left(\text{var}(I_1)+a^2/12\right)+\frac{\text{var}(I_1)}{f^2}/3+\frac{I_1^2}{f^2}/12+7/144\ \frac{a^2}{f^2}$$
$$=\frac{a^2\text{var}(t)}{3}+\text{var}(\delta)\left(\left(I_1-\frac{a}{2}\right)^2+\frac{a^2}{12}\right)+\delta^2\text{var}(I_1)$$
$$+\text{var}(\delta)\text{var}(I_1)+\frac{\text{var}(I_1)}{3f^2}+\frac{a^2t^2}{12}+\frac{\delta^2a^2}{12}+\frac{I_1^2}{12f^2}+\frac{7a^2}{144f^2}$$

Referring to the end of Sect. 2 we have $t=N/f$ and $I_1=ka$. Since the load and measurement computers are not synchronised $\delta=1/2f$ and $\text{var}(\delta)=f^2/12$ (Fig. 3). So var (discretization noise/V) becomes

$$=\frac{a^2\text{var}(t)}{3}+\frac{a^2}{12f^2}\left(\left(k-\frac{1}{2}\right)^2+\frac{1}{12}\right)+\frac{a^2}{4f^2}\text{var}(k)$$
$$+\frac{a^2}{12f^2}\text{var}(k)+a^2\frac{\text{var}(k)}{3f^2}+\frac{a^2N^2}{12f^2}+\frac{a^2}{48f^2}+\frac{k^2a^2}{12f^2}+\frac{7a^2}{144f^2}$$
$$=\frac{a^2}{3}\text{var}(t)+\frac{2a^2}{3f^2}\text{var}(k)+\frac{a^2}{144f^2}\left(12N^2+24k^2-12k+14\right)$$

Assuming a 12 bit ADC and I_1 approx. half full scale var (discretization noise)

$$=\frac{V^2a^2}{3}\text{var}(t)+\frac{2V^2}{3f^2}\text{var}(I_1)+\frac{V^2a^2t^2}{12}+\frac{V^2a^2}{144f^2}\left(24k^2-12k+14\right) \qquad (7)$$
$$\approx\frac{V^2a^2}{3}\text{var}(t)+\frac{2V^2}{3f^2}\text{var}(I_1)+\frac{V^2a^2t^2}{12}+\frac{698880\ V^2a^2}{f^2}$$

We will assume t is long compared to both the sampling frequency f and the network variation. This allows us to assume that the variance in the energy reported is given by the sum of the variance due to network variation (Eq. 3) and that due noise in the measuring system (Eq. 7).

$$= 2V^2 I_0^2 \text{var}(p) + \frac{V^2 a^2}{3} \text{var}(t) + \frac{2V^2}{3f^2} \text{var}(I_1) + \frac{V^2 a^2 t^2}{12} + \frac{V^2 a^2}{144 f^2} \left(24k^2 - 12k + 14\right)$$

Assuming both t and I_1 are fixed so variance of energy measurement is

$$= 2V^2 I_0^2 \text{var}(p) + \frac{V^2 a^2 t^2}{12} + \frac{V^2 a^2}{144 f^2} \left(24k^2 - 12k + 14\right) \tag{8}$$

4 Maximising Beneficial Mutation Detection Rate

Suppose we run the original version of the software to be improved and record its use of energy. We then mutate the software. Suppose the mutation is beneficial, in that it reduces the energy consumed by Δ. (Here we assume the power consumption is spread uniformly across the time the software runs. Notice we are assuming the mutation changes the power consumption but the runtime t is not changed. See Sect. 5). If Δ^2 is large compared to the measurement variance (Eq. 8) then we can reasonably expect to measure that the mutation has been beneficial. If the difference is small, we may want to repeat the measurement to increase Δ. However, this would proportionately reduce the rate that we can test mutations. Equation 8 means we can ask: Is

$$\Delta^2 \text{ much bigger than } 2V^2 I_0^2 \text{var}(p) + \tfrac{V^2 a^2 t^2}{12} + \tfrac{V^2 a^2}{144 f^2} \left(24k^2 - 12k + 14\right) \text{ ?} \tag{9}$$

Let $\Delta I = \Delta / Vt$ be the beneficial effect of the mutation expressed in terms of energy divided by the length of the testing period. Notice that increasing the mutation testing time also increases the variance in the energy measurement. We divide by the supply voltage V so that ΔI can be expressed as the average reduction in current. Using $\Delta^2 = (\Delta I)^2 V^2 t^2$ in Question 9 and then dividing through by V^2 means Question 9 is the same comparison as: Is (the signal)

$$t^2 (\Delta I)^2 \text{ much bigger than} \tfrac{a^2 t^2}{12} + 2I_0^2 \text{var}(p) + \tfrac{a^2}{144 f^2} \left(24k^2 - 12k + 14\right)$$

Notice the last two terms do not depend on t and so for $\Delta I > a\sqrt{1/12}$ we can make the energy signal bigger than its variability by increasing t. However, we cannot effectively detect beneficial mutations with a proportionate effect less than $\Delta I = a\sqrt{1/12} \approx 0.3\ a$. If we require the signal to be at least twice the variability (4 times the variance) we can calculate the minimum time required.

$$t^2 (\Delta I)^2 = \frac{a^2 t^2}{3} + 8 I_0^2 \text{var}(p) + \frac{a^2}{36 f^2} \left(24k^2 - 12k + 14\right)$$

$$t = \sqrt{\frac{8 I_0^2 \text{var}(p) + \frac{a^2}{36 f^2} \left(24k^2 - 12k + 14\right)}{(\Delta I)^2 - a^2/3}}$$

Let $\Delta k = \Delta I/a$, assume $I_0 \approx I_1 = ka$

$$t \approx \sqrt{\frac{24k^2 \text{var}(p) + \frac{1}{12f^2}(24k^2 - 12k + 14)}{3(\Delta k)^2 - 1}} \tag{10}$$

Alternatively we can express this minimum time (Eq. 10) as a minimum number of number of samples using $N = ft$ (N was defined at the end of Sect. 2).

$$N \approx \sqrt{\frac{k^2 (24f^2 \text{var}(p) + 2) - k + 14/12}{3(\Delta k)^2 - 1}}$$

Again assuming a 1 kHz 12 bit ADC and noting that Fig. 5 suggests $\sqrt{2}\ SD(p)$ is 24 μs. i.e. $\text{var}(p) = 2.86 \times 10^{-10}\,\text{s}^2$. So $f^2\text{var}(p) = 2.86 \times 10^{-4}$.

$$N \approx k\sqrt{\frac{2}{3(\Delta k)^2 - 1}} \tag{11}$$

Δk is the mutation's impact on energy consumption, assumed constant over time, expressed as a current in units of the analogue to digital converter's resolution. If the average impact of the mutation is large compared to the resolution of the ADC, then $\Delta k \gg 0.58$. Therefore for our 1 kHz 12 bit ADC and mutations with a reasonably large impact the measurement need only last $1.7/\Delta k$ s.

5 Discussion and Conclusions

Experimental work suggests that the impact of software mutations is very non-uniform, with many mutations having no effect or being detrimental and only a small number being beneficial (Langdon and Petke 2015; Schulte *et al.* 2014b). Hence setting the experimental parameters to allow rapid detection of large impact mutations risks not detecting many small impact mutations. Where large mutations are rare this risks the EC degenerating into random search. Indeed if the impact of mutations is too small to be reliably detected (i.e. $\Delta I < 0.58a$) then we cannot expect miracles from EC.

We have modelled the energy consumption of software mutations by assuming their impact is spread uniformly throughout each test run. This is unlikely to be true and more sophisticated models might look at how the impact of mutations is distributed. However, for a mutation to be detected its effect will still need to be large compared to the ADC sensitivity. This suggests our present lower bound ($\Delta I = 0.58a$) might be improved at the cost of assuming more about software mutants, however, it appears that a critical lower bound will still exist.

If the test program is run repeatedly in order to integrate the mutation's effect, we would expect repeated patterns in the power monitor's signal. There are very sensitive algorithms which can reliably measure periodic differences even in the presence of sizeable noise.

Alternatively, it might be possible to use signal processing to recognise the onset and termination of the measurement period. Or, several low end test beds (e.g. the Raspberry Pie) have output pins which could be used to start and stop energy measurement. Finally, both the computer under test and the computer running the energy monitors have sophisticated clocks, which can be synchronised and thus absolute time (rather than explicit message passing) might be used to keep track of the start and end of energy consumption experiments.

1. It will be difficult to detect mutations which have on average an effect less than $\sqrt{(1/3)}\ a$ (a is the ADC's resolution) on the current consumed. For our example 12 bit 0.4095 Amp ADC this sets a lower limit of 58 μA.
2. On the other hand if the effect is much bigger than 58 μA, there is little to be gained by running measurement for longer than a second. Equation 11 suggests the ideal duration falls in proportion to the smallest effect size we wish our evolutionary system to detect.

References

Bruce, B.R.: Energy optimisation via genetic improvement a SBSE technique for a new era in software development. In: GECCO GI-2015 Workshop, pp. 819–820 (2015). http://www.cs.bham.ac.uk/%7Ewbl/biblio/gp-html/Bruce_2015_gi.html

Goodman, L.A.: On the exact variance of products. J. Am. Stat. Assoc. **55**(292), 708–713 (1960). http://dx.doi.org/10.2307/2281592

Langdon, W.B., Petke, J.: Software is not fragile. In: CS-DC 2015 Proceedings in Complexity. Springer (2015, Forthcoming). http://www.cs.bham.ac.uk/%7Ewbl/biblio/gp-html/langdon_2015_csdc.html

Langdon, W.B.: Genetically improved software. In: Gandomi, A.H., Alavi, A.H., Ryan, C. (eds.) Handbook of Genetic Programming Applications, Chap. 8, pp. 181–220. Springer, Heidelberg (2015). http://dx.doi.org/10.1007/978-3-319-20883-1_8

Li, D., et al.: Optimizing display energy consumption for hybrid Android apps. In: DeMobile 2015, Bergamo, Italy, 31 August, pp. 35–36. ACM, Invited Talk (2015). http://dx.doi.org/10.1145/2804345.2804356

Schulte, E., et al.: Post-compiler software optimization for reducing energy. In: ASPLOS 2014, Salt Lake City, Utah, USA, 1–5 March, pp. 639–652. ACM (2014a). http://www.cs.bham.ac.uk/%7Ewbl/biblio/gp-html/schulte2014optimization.html

Schulte, E., et al.: Software mutational robustness. GP&EM **15**(3), 281–312 (2014b). http://dx.doi.org/10.1007/s10710-013-9195-8

White, D.R., et al.: Searching for resource-efficient programs: low-power pseudorandom number generators. In: GECCO 2008, pp. 1775–1782. ACM (2008). http://www.cs.bham.ac.uk/%7Ewbl/biblio/gp-html/White2_2008_gecco.html

Syntactical Similarity Learning by Means of Grammatical Evolution

Alberto Bartoli, Andrea De Lorenzo, Eric Medvet(✉), and Fabiano Tarlao

Department of Engineering and Architecture, University of Trieste, Trieste, Italy
emedvet@units.it

Abstract. Several research efforts have shown that a similarity function synthesized from examples may capture an application-specific similarity criterion in a way that fits the application needs more effectively than a generic distance definition. In this work, we propose a similarity learning algorithm tailored to problems of syntax-based entity extraction from unstructured text streams. The algorithm takes in input pairs of strings along with an indication of whether they adhere or not adhere to the same syntactic pattern. Our approach is based on Grammatical Evolution and explores systematically a similarity definition space including all functions that may be expressed with a specialized, simple language that we have defined for this purpose. We assessed our proposal on patterns representative of practical applications. The results suggest that the proposed approach is indeed feasible and that the learned similarity function is more effective than the Levenshtein distance and the Jaccard similarity index.

Keywords: Distance learning · Entity extraction · String patterns

1 Introduction and Related Work

Many solutions to practically relevant applications are based on techniques that rely on a form of *similarity* between data items, i.e., on a quantification of the difference between any pair of data items in a given feature space. Although such a similarity may be quantified by many different generic functions, i.e., distances or pseudo-distances, a wealth of research efforts have advocated the usage of similarity functions that are *learned* from collections of data pairs labelled as being either "similar" or "dissimilar" [1–3]. Indeed, similarity functions constructed by a *similarity learning* algorithm have proven very powerful in many different application domains, as such functions may capture the application-specific similarity criterion described by the available *examples* in a way that fits the application needs more effectively than a generic distance definition.

In this work, we focus on the problem of learning a similarity function suitable for syntax-based *entity extraction* from *unstructured text* streams. The identification of strings which adhere to a certain syntactic pattern is an essential component of many workflows leveraging digital data and such a task occurs routinely

J. Handl et al. (Eds.): PPSN XIV 2016, LNCS 9921, pp. 260–269, 2016.
DOI: 10.1007/978-3-319-45823-6_24

in virtually every sector of business, government, science, technology. Devising a similarity function capable of capturing syntactic patterns is an important problem as it may enable significant improvements in methods for constructing syntax-based entity extractors from examples automatically [4–14]. We are not aware of any similarity definition capable of (approximately) separating strings which adhere to a common syntactic pattern (e.g., telephone numbers, or email addresses) from strings which do not.

We propose an approach based on GE, in which we explore systematically a similarity definition space including all functions that may be expressed with a specialized, simple *language* that we have defined for this purpose. The language includes the basic flow control, arithmetic and relation operators. It is expressive enough to describe important, existing similarity definitions, that we use as baseline in our experimental evaluation. A candidate solution, i.e., an individual, represents a program in the language which takes a pair of strings as input and outputs a number quantifying their similarity. Programs are executed with a *virtual machine* that we designed and implemented. The virtual machine is necessary only for assessing the quality of candidate solutions during the evolutionary search: the final solution can obviously be implemented in a more compact and more efficient way based on the specific technology in which the learned similarity function will be inserted.

We assessed our proposal on several tasks representative of practical applications, each task being a large text stream annotated with the strings following a task-specific pattern. We emphasize that we did not learn one similarity definition for each task: instead, we learned a single similarity function from all tasks except for one and then evaluated the behavior of the learned similarity function on the remaining task—i.e., on a syntactic pattern that was not available while learning. The results, averaged across all the tasks, demonstrate that the proposed approach is indeed feasible, i.e., it is able to learn a similarity function capable of (approximately) separating strings based on their adherence to a given syntactical pattern. Most importantly, the learned function is more effective than the Levenshtein distance and the Jaccard similarity index.

An evolutionary approach to metric learning can be found in [15]. The cited work proposes a general approach for multi-label clustering problems in a given feature space. We focus instead on a different and more specific problem: syntax-based entity extraction from unstructured text streams. Furthermore, we aim at learning a similarity function and do not insist in requiring that the learned function be a distance. Several proposals have advocated genetic approaches to similarity learning in the context of case-based reasoning [16–18]. In those cases, though, the problem was learning a meaningful similarity criterion between problem definitions, to enable effective comparison of a new problem to a library of known, already solved problems. We consider instead similarity between pairs of strings that are a small part of a problem instance. Our problem statement follows a common approach in similarity learning: input data consist of pairs of data points, where each pair is known to belong to either the same class (i.e., the same pattern) or to different classes [1]. An alternative framework is based on input data which consist of triplets of data points (a, b, c) labelled with the information regarding whether a is more similar to b or to c [19–21]. Such a *relative comparisons* framework has proven to be quite powerful, in particular, for clustering applications. A relative comparison approach could be applied also to our entity extraction problem and indeed deserves further investigation.

1.1 Problem Statement

The problem input consists of a set of tasks $\{T_1, \ldots, T_n\}$ where each task describes a *syntactic pattern* by means of *examples.* Task T_i consists of a pair of sets of strings (P_i, N_i): P_i contains strings which adhere to the ith pattern while N_i contains strings which do not adhere to that pattern. The problem consists in learning a *similarity function* $\hat{m}(s, s')$ which, given two strings s, s', returns a *similarity index* capable of capturing to which degree s and s' adhere to the same (unknown) syntactic pattern. That is, intuitively, pairs of strings in P_i should be associated with a "large" similarity index, while pairs consisting of a string in P_i and a string in N_i should be associated with a "small" similarity index. Furthermore, this requirement should be satisfied for all tasks by the same function $\hat{m}$.

In details, the ideal learned function should satisfy the following requirement:

$$\forall i \in \{1, \ldots, n\}, \forall x \in M(P_i, N_i), \forall y \in M(P_i, P_i), x < y \tag{1}$$

where $M(S, S') = \{m(s, s') : s \in S, s \in S'\}$. For a given problem input, a function satisfying Eq. 1 may or may not exist; and, even if it exists, a learning algorithm may or may not be capable of learning that function.

2 Our Approach

2.1 Search Space and Solution Quality

We consider a search space composed of functions that may be expressed with the language L described in Fig. 1 in the Backus-Naur Form (BNF). The available mathematical operators are defined in the rule concerning the ⟨ValueReturningFunction⟩ non-terminal while relation operators are defined in rule concerning the ⟨Condition⟩ non-terminal. The language includes basic flow control operators and allows defining numeric variables and arrays dynamically. Access to variables and array elements occur by index.

The language is expressive enough to describe commonly used similarity indexes: in particular, we described the Levenshtein distance and the Jaccard similarity index—which we used in our experimental evaluation as baselines—using this language.

We propose an evolutionary approach based on *Grammatical Evolution* (GE) [22, 23]. GE is an evolutionary framework where candidate solutions (*individuals*) are represented as fixed-length numeric sequences. Such sequences (*genotype*) are translated into similarity functions (*phenotype*) by means of a mapping procedure which uses the production rules in a grammar definition.

Rules

```
1.  ⟨BlockCode⟩ ::= ⟨RowOfBlockCode⟩
2.  ⟨Statement⟩ ::= ⟨Assign⟩ | ⟨CreateArray⟩ | ⟨CreateVariable⟩ | ⟨For⟩ | ⟨If⟩ |
    ⟨Return⟩ | ⟨SetArrayItem⟩
3.  ⟨ValueReturningFunction⟩ ::=⟨Constant⟩ | ⟨GetVariableValue⟩ | ⟨Add⟩ | ⟨Decrement⟩ |
    ⟨Maximum⟩ | ⟨Minimum⟩ | ⟨GetArrayItem⟩ | ⟨GetArrayLength⟩ | ⟨Division⟩ |
    ⟨Multiplication⟩
4   ⟨Assign⟩ ::= var[⟨ValueReturningFunction⟩] = ⟨ValueReturningFunction⟩
5.  ⟨CreateArray⟩ ::= newArray[⟨ValueReturningFunction⟩]
6.  ⟨CreateVariable⟩ ::= createVariable()
7.  ⟨Division⟩ ::= (⟨ValueReturningFunction⟩ / ⟨ValueReturningFunction⟩)
8.  ⟨For⟩ ::= for(index0 = 0; index0 < ⟨ValueReturningFunction⟩; index0++) ⟨BlockCode⟩
9.  ⟨If⟩ ::= if(⟨Condition⟩) ⟨BlockCode⟩ else ⟨BlockCode⟩
10. ⟨Return⟩ ::= return ⟨ValueReturningFunction⟩
11. ⟨SetArrayItem⟩ ::= array[⟨ValueReturningFunction⟩][⟨ValueReturningFunction⟩] =
    ⟨ValueReturningFunction⟩
12. ⟨Add⟩ ::= ⟨ValueReturningFunction⟩ + ⟨ValueReturningFunction⟩
13. ⟨Subtract⟩ ::= ⟨ValueReturningFunction⟩ - ⟨ValueReturningFunction⟩
14. ⟨Maximum⟩ ::= maximum(⟨ValueReturningFunction⟩,⟨ValueReturningFunction⟩)
15. ⟨Minimum⟩ ::= minimum(⟨ValueReturningFunction⟩,⟨ValueReturningFunction⟩)
16. ⟨Multiplication⟩ ::= ⟨ValueReturningFunction⟩ * ⟨ValueReturningFunction⟩
17. ⟨GetArrayItem⟩ ::= array[⟨ValueReturningFunction⟩][⟨ValueReturningFunction⟩]
18. ⟨GetArrayLength⟩ ::= array[⟨ValueReturningFunction⟩].length
19. ⟨Constant⟩ ::= 0 | 1 | ... | 255
20. ⟨GetVariableValue⟩ ::= var[⟨ValueReturningFunction⟩]
21. ⟨RowOfBlockCode⟩ ::= ⟨Statement⟩ | ⟨Statement⟩ \n ⟨RowOfBlockCode⟩
22. ⟨Condition⟩ ::= ⟨EqualCondition⟩ | ⟨NotEqualCondition⟩ |
    ⟨GreaterCondition⟩ | ⟨GreaterOrEqualCondition⟩
23. ⟨EqualCondition⟩ ::= ⟨ValueReturningFunction⟩ == ⟨ValueReturningFunction⟩
24. ⟨NotEqualCondition⟩ ::= ⟨ValueReturningFunction⟩ != ⟨ValueReturningFunction⟩
25. ⟨GreaterCondition⟩ ::= ⟨ValueReturningFunction⟩ > ⟨ValueReturningFunction⟩
26. ⟨GreaterOrEqualCondition⟩ ::= ⟨ValueReturningFunction⟩ >= ⟨ValueReturningFunction⟩
```

Alternative rules

```
2.  ⟨Statement⟩ ::= ⟨CreateVariable⟩
3.  ⟨ValueReturningFunction⟩ ::= ⟨Constant⟩
21. ⟨RowOfBlockCode⟩ ::= ⟨Statement⟩
```

Fig. 1. BNF grammar for the language L: below the set of alternative rules (see text).

After early experimentation, we chose to tailor several aspects of the general GE framework to our specific problem.

In our case, we represent an individual with a genotype consisting of a tuple $\boldsymbol{g} \in [0, 255]^{n_{\text{gen}}}$, where each g_i element is a positive 8-bit integer. We chose $n_{\text{gen}} = 350$ because with such value we were able to obtain, from two suitable genotypes, the phenotypes corresponding to the Levenshtein distance and the Jaccard similarity, according to the mapping procedure described below. Given a genotype, we obtain the corresponding phenotype, i.e., a similarity function expressed as a program l in the language L, according to an iterative *mapping procedure* which works as follows, starting with $l = \langle\text{BlockCode}\rangle$ and $i = 0$: (i) we consider the first occurrence of a non-terminal in l and the corresponding rule in the BNF grammar for L; (ii) among the $n_{\text{rule}} \geq 1$ alternatives (i.e., possible replacements separated by | in the rule), we choose the $(j+1)$th one, with j equals to the remainder between g_i and n_{rule}; (iii) we increment i by one: if i exceeds n_{gen}, we set to 1.The procedure is iterated until no more non-terminals exist in l: since it is not guaranteed that this condition is satisfied in a finite number of iterations, we implemented a mechanism to overcome this limitation. We associate a number c with each non-terminal x in l: the value of c is set to 0 for the starting non-terminal ⟨BlockCode⟩, or to $c' + 1$ otherwise, where c' is the number associated with the non-terminal whose replacement lead to the insertion of x in l. Whenever a non-terminal among ⟨Statement⟩, ⟨ValueReturningFunction⟩, and ⟨RowOfBlockCode⟩ has to be replaced, if its c exceeds a parameter $c_{\max} = 40$, we use the alternative rules shown at the bottom of Fig. 1 instead of the original ones for those non-terminals—in other words, with this mechanism we pose a depth limit on the derivation trees.

We quantify the quality of an individual encoding a similarity function m by its *fitness* $f(m)$, that we define as follows. Given a numeric multiset I, let $I_{p\%}$ indicate the smallest element $i \in I$ greater or equal to the p percentile of elements in I. Given a pair of numeric multisets (X, Y), we define the *overlapness* function $o(X, Y; p) \in [0, 1]$ as follows:

$$o(X, Y; p) = \frac{|\{x \in X : x \geq Y_{p\%}\}| + |\{y \in Y : y \leq X_{(100-p)\%}\}|}{|X| + |Y|} \tag{2}$$

Intuitively, $o(X, Y; p)$ measures the degree of overlapping between elements of X and Y, assuming that elements in X are in general smaller than elements of Y: when X and Y are perfectly separated, $o(X, Y; p) = 0, \forall p$. The value of p is used to discard extreme (greatest for X and smallest for Y) elements in the multisets. The fitness $f(m) \in [0, 1]$ of m is given by:

$$f(m) = \frac{1}{2n} \sum_{i=1}^{n} o\big(M(P_i, N_i), M(P_i, P_i); 10\big) + o\big(M(P_i, N_i), M(P_i, P_i); 0\big) \tag{3}$$

where $M(S, S')$ is defined as for Eq. 1. In other words, the fitness of m is the average overlapness over the tasks in $\{T_1, \ldots, T_n\}$: for each task, $f(m)$ takes into account the average between the overlapness of the two multisets $M(P_i, N_i)$ and $M(P_i, P_i)$ computed on the whole multisets and after discarding 10 % extreme values. The rationale for the latter design choice is to avoid giving too much importance to possible outliers in the data. Note that a similarity function satisfying Eq. 1 has zero fitness—i.e., fitness should be minimized.

During the evolutionary search, we evolve a fixed-size population of n_{pop} individuals for $n_{\text{iter}} = 200$ generations by means of the *mutation* and *two-point crossover* genetic operators, which are applied to individuals selected by means of a tournament of size 3.

2.2 Virtual Machine

We designed and implemented a *virtual machine* (VM) capable of executing programs in language L. A VM program execution takes a pair of strings (s, s') as input and returns the value $m(s, s')$, m being the similarity function represented by the program.

As described in Sect. 2.1, the language allows defining numeric variables and arrays dynamically with access occurring by index. VM provides a running program with a list of numeric variables and a list of numeric arrays. Indexes start from 0 and when a new variable is created the next free index is used: the actual variable/array being accessed is determined by the reminder of $\frac{i}{n_v}$. When execution starts, VM creates two arrays into the arrays list, one for s and the other for s': the ith element of each array contains the UTF-8 representation of the ith character in the corresponding string. The execution stops when a return statement is reached or when the last instruction has been executed: in the latter case, the returned value is $m(s, s') = 0$.

A VM program execution may fail, in which case execution terminates and the returned value is $m(s, s') = 0$. Failure occurs when one of the following conditions is met: division by zero; maximum number $n_{\max}$ of executed instructions exceeded; maximum array size n_{array} exceeded—we set $n_{\max} = 40000$ and $n_{\max}$ $=10\,\text{length}(s)\ \text{length}(s')$.

3 Experimental Evaluation

As described previously, a task describes a syntactic pattern by means of examples, i.e., each task consists of a pair of sets of strings (P_i, N_i): P_i contains strings which adhere to the pattern while N_i contains strings which do not adhere to the pattern. We assess our proposal on several datasets representative of possible applications of our similarity learning method (the name of each dataset describes the nature of the data and the type of the entities to be extracted): HTML-href [11,13,14], Log-MAC+IP [11,13,14], Email-Phone [7,8,11,13,14], Bills-Date [12,14], Web-URL [7,11,13,14], Twitter-URL [11,13,14]. Each dataset consists of a text annotated with all and only the snippets that should be extracted.

We constructed a task (P, N) for each such dataset, as follows. Let d denote the annotated text in the dataset. Set P contains all and only the strings that should be extracted from d. Set N contains strings obtained by splitting the remaining part of d. It follows that no pair of elements in $P \cup N$ overlap. The splitting procedure is based on a *tokenization* heuristics that (approximately) identifies the tokens that delimit P strings in d; those tokens are then used for splitting N strings in d as well. For example, if strings in P are delimited by a space, then we split the remaining part of d by spaces and insert all the resulting strings in N. The details of the heuristic are complex because different P strings could be delimited by different characters—we omit the details for ease of presentation.

We performed a cross-fold assessment of our proposed method, i.e., we executed one experiment for each of the 6 tasks resulting from the available datasets. In each ith experiment we executed our method on a *learning set* consisting of all but the ith task. We obtained the actual jth pair (P'_j, N'_j) of the learning set by sampling $2n_{\text{ex}}$ items of the corresponding (P_j, N_j), i.e., $|P'_j| = |N'_j| = n_{\text{ex}}$, with $P'_j \subseteq P_j$, $N'_j \subseteq N_j$, where n_{ex} is a parameter of the experiment which affects the amount of data available for learning.

We used the remaining task (P_i, N'_i) (i.e., all of the examples in P_i and a number $|N'_i| = |P_i|$ of examples sampled randomly from N_i) for quantifying the quality of the learned similarity function $m^\star$—$m^\star$ being the individual with the best fitness after the last generation. Note that we assessed $m^\star$ on a task *different* from the tasks that we used for learning it.

For each task, we repeated the experiment for 5 times, each time using a different random seed. We considered the following indexes for each experiment, which we averaged across the 5 repetitions: the learning fitness LF, i.e., the fitness of $m^\star$ on the learning set; the testing fitness TF, i.e., the fitness of $m^\star$ on (P_i, N'_i); the number #I of instructions in $m^\star$; the average number #S of executed instructions while processing pairs in (P_i, N'_i) with $m^\star$.

We explored two different values for the population size n_{pop}, 50 and 100 individuals, and three different values for the cardinality of sets of examples n_{ex}: 10, 25 and 50.

Table 1 provides the key results (with $n_{\text{ex}} = 50$ and $n_{\text{pop}} = 50$), separately for each dataset and averaged across all datasets. To place results in perspective, we provide all indexes (except for LF) also for two baseline definitions: the Levenshtein distance, which counts the minimum number of character insertions, replacements or deletions required to change one string into the other, and the Jaccard similarity index, which considers each string as a set of bigrams and is the ratio between the intersection and the union of the two sets. The key result is that, on average, the definitions synthesized by our method exhibit the best results. By looking at individual tasks, our synthesized definitions outperform Jaccard in three tasks, are nearly equivalent in one task and are worse or slightly worse in the two remaining tasks. Thus, the similarity functions synthesized by our method are more effective at separating strings based on their adherence at a certain syntactic pattern with respect to the traditional Levenshtein and Jaccard metrics.

Table 2 provides further insights into our method by providing results averaged across all tasks for various combinations of available examples n_{ex} and

Table 1. Results of our method, with $n_{\text{ex}} = 50$ and $n_{\text{pop}} = 50$, and the baselines. Best TF figure highlighted.

Task	LF	TF			#I			#S [$\times 10^6$]		
	GE	GE	Jac.	Lev.	*GE*	Jac.	Lev.	GE	Jac.	Lev.
HTML-href	0.45	**0.42**	0.64	0.91	1877	174	103	0.22	3.49	2.25
Log-MAC+IP	0.44	**0.08**	0.82	0.91	179	174	103	0.06	0.42	0.75
Email-Phone	0.43	0.64	**0.56**	0.90	352	174	103	0.41	4.62	3.64
Bills-Date	0.49	0.85	**0.59**	0.90	1116	174	103	1.56	2.71	5.19
Web-URL	0.40	**0.30**	0.43	0.92	151	174	103	0.72	23.8	10.00
Twitter-URL	0.48	0.30	**0.29**	0.90	147	174	103	0.84	6.28	8.10
Average	0.45	**0.43**	0.55	0.90	637	174	103	0.64	6.90	4.99

population size n_{pop}. It can be seen that, with a larger population ($n_{\text{pop}} = 100$), the amount of learning examples does not impact TF significantly, but more examples lead to more compact and more efficient solutions (smaller #I and #S, respectively). On the other hand, the configuration with smaller population ($n_{\text{pop}} = 50$) exhibits a slight but consistent improvement in TF when the amount of examples grows. It can also be observed that more examples lead to solutions with varying length but that tend to be more efficient (no clear trend in #I and decreasing #S, respectively). This observation suggests that our method might perhaps be improved further by a multiobjective optimization search strategy, where the fitness of an individual would take into account not only its ability of capturing similarity as specified in the learning examples (to be maximized) but also the length of the individual (to be minimized).

Table 2. Results (including learning time t_l) for different values of n_{pop} and n_{ex}.

n_{pop}	n_{ex}	LF	TF	#I	#S [$\times 10^6$]	t_l [s]
50	10	0.37	0.45	552	0.59	52
	25	0.43	0.44	3076	0.56	245
	50	0.45	**0.43**	637	0.64	715
100	10	0.34	0.50	1138	2.76	110
	25	0.40	0.48	1224	0.94	326
	50	0.38	0.49	**443**	**0.44**	1056

Table 2 also shows the learning time t_l, averaged across repetition: we performed the experiments on a platform equipped with an Intel Core i7-4720HQ (2.60 GHz) CPU and 16 GB of RAM.

4 Concluding Remarks

We have investigated the feasibility of learning a similarity function capable of (approximately) separating strings which adhere to a common syntactic pattern (e.g., telephone numbers, or email addresses) from strings which do not. We are not aware of any similarity function with this property, which could enable significant improvements in methods for constructing syntax-based entity extractors from examples automatically—in many application domains, similarity functions learned over labelled sets of data points have often proven more effective than generic distance definitions.

We have proposed a method based on Grammatical Evolution which takes pairs of strings as input, along with an indication of whether they follow a similar syntactic pattern. The method synthesizes a similarity function expressed in a specialized, simple language that we have defined for this purpose.

We assessed our proposal on several tasks representative of practical applications, with an experimental protocol in which we learned a similarity function on a given set of tasks (i.e., patterns) and we assessed the learned function on a previously unseen task. The results demonstrate that the proposed approach is indeed feasible and that the learned similarity function is much more effective than the Levenshtein distance and the Jaccard similarity index.

We plan to extend our investigation in two ways: first, synthesize a more powerful similarity function, by using a broader set of patterns and a larger amount of labelled data points; in this phase there may certainly be room for further improvements to our Grammatical Evolution method; next, take advantage of the learned similarity function in order to improve methods for syntax-based entity extraction.

Acknowledgements. We are grateful to Michele Furlanetto who contributed in the implementation of our proposed method.

References

1. Yang, L., Jin, R.: Distance metric learning: a comprehensive survey. Michigan State Universiy **2** (2006)
2. Kulis, B.: Metric learning: a survey. Found. Trends Mach. Learn. **5**(4), 287–364 (2012)
3. Bellet, A., Habrard, A., Sebban, M.: A survey on metric learning for feature vectors and structured data (2013). arXiv preprint arXiv:1306.6709
4. Fernau, H.: Algorithms for learning regular expressions from positive data. Inf. Comput. **207**(4), 521–541 (2009)
5. Cicchello, O., Kremer, S.C.: Inducing grammars from sparse data sets: a survey of algorithms and results. J. Mach. Learn. Res. **4**, 603–632 (2003)
6. Cetinkaya, A.: Regular expression generation through grammatical evolution. In: Proceedings of the 2007 GECCO Conference Companion on Genetic and Evolutionary Computation, pp. 2643–2646. ACM (2007)
7. Li, Y., Krishnamurthy, R., Raghavan, S., Vaithyanathan, S., Jagadish, H.: Regular expression learning for information extraction. In: Proceedings of the Conference on Empirical Methods in Natural Language Processing, pp. 21–30. Association for Computational Linguistics (2008)
8. Brauer, F., Rieger, R., Mocan, A., Barczynski, W.M.: Enabling information extraction by inference of regular expressions from sample entities. In: Proceedings of the 20th ACM International Conference on Information and Knowledge Management, pp. 1285–1294. ACM (2011)
9. Murthy, K., P., D., Deshpande, P.M.: Improving recall of regular expressions for information extraction. In: Wang, X.S., Cruz, I., Delis, A., Huang, G. (eds.) WISE 2012. LNCS, vol. 7651, pp. 455–467. Springer, Heidelberg (2012)
10. Bartoli, A., Davanzo, G., De Lorenzo, A., Mauri, M., Medvet, E., Sorio, E.: Automatic generation of regular expressions from examples with genetic programming. In: Proceedings of the 14th Annual Conference Companion on Genetic and Evolutionary Computation, pp. 1477–1478. ACM (2012)
11. Bartoli, A., Davanzo, G., De Lorenzo, A., Medvet, E., Sorio, E.: Automatic synthesis of regular expressions from examples. Computer **12**, 72–80 (2014)
12. Bartoli, A., De Lorenzo, A., Medvet, E., Tarlao, F.: Learning text patterns using separate-and-conquer genetic programming. In: Machado, P., et al. (eds.) Genetic Programming, vol. 9025, pp. 16–27. Springer, Cham (2015)

13. Bartoli, A., De Lorenzo, A., Medvet, E., Tarlao, F.: Active learning approaches for learning regular expressions with genetic programming. In: Proceedings of the 31st Annual ACM Symposium on Applied Computing, pp. 97–102. ACM (2016)
14. Bartoli, A., De Lorenzo, A., Medvet, E., Tarlao, F.: Inference of regular expressions for text extraction from examples. IEEE Trans. Knowl. Data Eng. **28**(5), 1217–1230 (2016)
15. Megano, T., Fukui, K.i., Numao, M., Ono, S.: Evolutionary multi-objective distance metric learning for multi-label clustering. In: 2015 IEEE Congress on Evolutionary Computation (CEC), pp. 2945–2952. IEEE (2015)
16. Stahl, A., Gabel, T.: Using evolution programs to learn local similarity measures. In: Ashley, K.D., Bridge, D.G. (eds.) ICCBR 2003. LNCS, vol. 2689, pp. 537–551. Springer, Heidelberg (2003)
17. Xiong, N., Funk, P.: Building similarity metrics reflecting utility in case-based reasoning. J. Intell. Fuzzy Syst. **17**(4), 407–416 (2006)
18. Xiong, N.: Learning fuzzy rules for similarity assessment in case-based reasoning. Expert Syst. Appl. **38**(9), 10780–10786 (2011)
19. Schultz, M., Joachims, T.: Learning a distance metric from relative comparisons. In: Advances in Neural Information Processing Systems (NIPS), p. 41 (2004)
20. Xiong, S., Pei, Y., Rosales, R., Fern, X.Z.: Active learning from relative comparisons. IEEE Trans. Knowl. Data Eng. **27**(12), 3166–3175 (2015)
21. Hao, S., Zhao, P., Hoi, S.C., Miao, C.: Learning relative similarity from data streams: active online learning approaches. In: Proceedings of the 24th ACM International on Conference on Information and Knowledge Management, pp. 1181–1190. ACM (2015)
22. Ryan, C., Collins, J.J., Neill, M.O.: Grammatical evolution: evolving programs for an arbitrary language. In: Banzhaf, W., Poli, R., Schoenauer, M., Fogarty, T.C. (eds.) EuroGP 1998. LNCS, vol. 1391, pp. 83–96. Springer, Heidelberg (1998)
23. O'Neill, M., Ryan, C.: Grammatical evolution. IEEE Trans. Evol. Comput. **5**(4), 349–358 (2001)

Hierarchical Knowledge in Self-Improving Grammar-Based Genetic Programming

Pak-Kan Wong[1(✉)], Man-Leung Wong[2], and Kwong-Sak Leung[1]

[1] The Chinese University of Hong Kong, Sha Tin, Hong Kong
{pkwong,ksleung}@cse.cuhk.edu.hk
[2] Lingnan University, Tuen Mun, Hong Kong
mlwong@ln.edu.hk

Abstract. Structure of a grammar can influence how well a Grammar-Based Genetic Programming system solves a given problem but it is not obvious to design the structure of a grammar, especially when the problem is large. In this paper, our proposed Bayesian Grammar-Based Genetic Programming with Hierarchical Learning (BGBGP-HL) examines the grammar and builds new rules on the existing grammar structure during evolution. Once our system successfully finds the good solution(s), the adapted grammar will provide a grammar-based probabilistic model to the generation process of optimal solution(s). Moreover, our system can automatically discover new hierarchical knowledge (i.e. how the rules are structurally combined) which composes of multiple production rules in the original grammar. In the case study using deceptive royal tree problem, our evaluation shows that BGBGP-HL achieves the best performance among the competitors while it is capable of composing hierarchical knowledge. Compared to other algorithms, search performance of BGBGP-HL is shown to be more robust against deceptiveness and complexity of the problem.

Keywords: Genetic Programming · Hierarchical knowledge learning · Estimation of distribution programming · Adaptive grammar · Bayesian network

1 Introduction

Computer program is a set of instructions which can be represented by a parse tree. The seminal work in Genetic Programming (GP) [8] showed that genetic operators can be used to automatically generate and evolve parse trees composed of elements given in a terminal set and a function set through evolution processes. A fitness function measures how well the evolved parse tree is. Later, Grammar-Based Genetic Programming (GBGP) [20,21] was proposed to effectively define a set of parse trees by a grammar (see [9] for a review). For example, context-free grammar explicitly models the closure relations of functions and is often adopted in the early works of GBGP.

J. Handl et al. (Eds.): PPSN XIV 2016, LNCS 9921, pp. 270–280, 2016.
DOI: 10.1007/978-3-319-45823-6_25

However, context-free grammar fails to model many types of dependencies in a program, such as data dependencies and control dependencies. For example, the sequence of reading from and writing to the same memory resource can alter the output of a program. Subsequently, some combinations are preferable and semantically sound in *good* parse trees which preserve the program dependencies. These dependencies affect how we derive non-terminals in a production rule. To address this issue, Probabilistic Model-Building Genetic Programming (PMBGP) [18] approaches weaken the context-free assumption using probabilistic models in the grammar. In every generation, parse trees are created from the current estimate of distribution which is improved iteratively by good parse trees. When a good solution is discovered, the final estimate of distribution reflects the principles (or the knowledge) to generate good parse trees and gives insights into the problem nature. Dependency in knowledge forms hierarchy. The combination of knowledge and its hierarchy is called the hierarchical knowledge, which gives insight into a search problem.

In this paper, we demonstrate how to construct and utilize the hierarchical knowledge in the Bayesian Grammar-Based Genetic Programming with Hierarchical Learning (BGBGP-HL) for improving the design of a grammar and automating the knowledge discovery. Our approach facilitates knowledge discovery while maintaining the search efficiency.

This paper is organized as follows. In the next section, relevant works on PMBGP and Grammatical Evolution are summarized. Next, we introduce the deceptive royal tree problem which is often used as a benchmark accepted by the PMBGP community. In Sect. 4, we show the hierarchical probabilistic context-sensitive grammar for the deceptive royal tree problem. After that, the workflow of the whole system are discussed in Sect. 5. The results are presented in Sect. 6. The final section is the conclusion.

2 Related Works

The related works of Probabilistic Model-Building Genetic Programming and Grammatical Evolution are summarized below.

Probabilistic Model-Building Genetic Programming. The approaches can be broadly classified into two categories. The probabilistic prototype tree (PPT) model-based methods operate on a fixed-length chromosome represented in a tree structure [3,4,17] so the PPT model is completely different from our approach. Another class of approaches utilizes probabilistic context-free grammar (PCFG) [1]. A complete review can be found at [7]. We only summarize the recent works directly related to our approach. PAGE [5] employs PCFG with Latent Annotations (PCFG-LA) to weaken the context-free assumption. The estimate of annotations can be learnt using expectation-maximization (EM) algorithm or variational Bayes (VB) learning as reported in PAGE-EM and PAGE-VB respectively, while PAGE-VB can also infer the total number of annotations needed from the learning data. Unsupervised PAGE (UPAGE) [6] utilizes PCFG-LA mixture model to deal with local dependencies and global contexts.

BAP [15] applies a learnt Bayesian network on top of a constant size chromosome and reproduces new individuals using the Bayesian network. Tanev's work [19] relies on probabilistic context-sensitive grammar (PCSG), which extends PCFG by assigning the probability according to the predefined contexts.

Grammatical Evolution. Methods to evolve grammar have also been studied in non-PMBGP approaches. One of the most prominent approaches is Grammatical Evolution (GE) [11,12]. The original version of GE adopted Backus-Naur Form for expressing the grammar. Unlike Genetic Programming (GP) [8], each individual is represented in a chromosome which can then be translated to a syntax tree and tells the system how to apply the rules using the grammar. GE has been extended and improved in terms of the genotype-phenotype mapping rules, search operators and search strategies [10,13,16].

Our Contributions. Contrary to the aforementioned PMBGP approaches, we employ a hierarchical PCSG model to maintain the syntactical correctness and guide the derivation of individuals by assigning a Bayesian network to each production rule. In contrast to the previous works [22,23], our approach can adaptively create new production rules. This is done by inserting complex and useful rules composed of the existing rules so that the syntactically correctness is preserved. In other words, the grammar is self-evolving and production rules are gradually specialized to adapt to the problem-specific fitness function. The new rules can capture the substructure frequently occurred in the good individuals. Our system learns a good strategy to derive an individual from the grammar. From the best of our knowledge, it is the first attempt to discover hierarchical derivation structure and capture substructure(s) in parse trees on self-improving hierarchical PCSG.

3 Deceptive Royal Tree Problem

We applied our approach on the bipolar version of deceptive royal tree (DRT) problem [24]. It is commonly used as a benchmark for comparing the effectiveness of PMBGP approaches. The goal of DRT problem is to find a tree with maximum score. Some deceptive royal trees and their scores are shown in Fig. 1. The DRT problem has a set of functions labeled by symbols from A to Z and two terminals x and y. These functions are binary and symmetric, i.e. $f(x, y) = f(y, x)$. There is a strong dependency between a node and its subtrees. Upon a different combination of the node and its two subtrees, the score varies greatly due to the extra global and local completion bonus. The scoring system is calculated recursively from the leaf nodes to the root node. Given a tree T, which has a root node r_T and two subtrees T_L and T_R, its score $S(T)$ is calculated recursively using the function $S(T) = k_{global}(r_T, T_L, T_R) \times ((k_{local}(r_T, T_L) \times S(T_L) + k_{local}(r_T, T_R) \times S(T_R)))$, where $S(x) = 1$, $S(y) = 0.95$, $k_{global}(r_T, T_L, T_R)$ is 1 by default, or 2 when T_L and T_R are perfect subtrees of the same type; $k_{local}(r_T, T_{subtree})$ is $\frac{1}{3}$ by default, 2 when $T_{subtree}$ is a perfect subtree, or 1 when $T_{subtree}$ has a correct root node but is not a perfect subtree. In short, k_{global} and k_{local} assign bonus or penalty

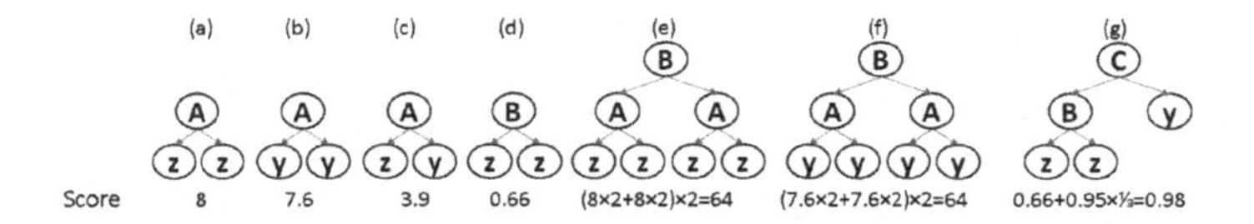

Fig. 1. Examples of the deceptive royal tree problem.

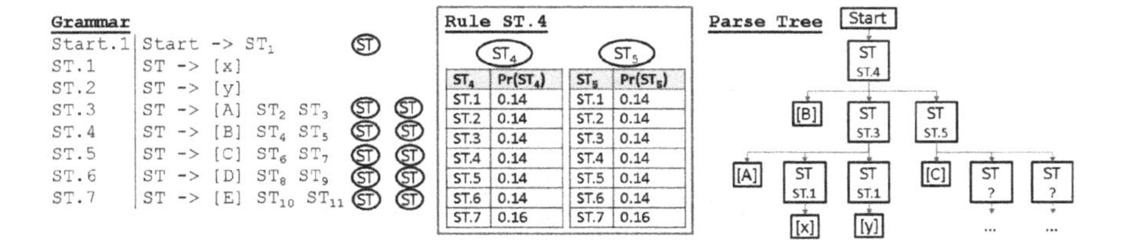

ST_4	$Pr(ST_4)$
ST.1	0.14
ST.2	0.14
ST.3	0.14
ST.4	0.14
ST.5	0.14
ST.6	0.14
ST.7	0.16

ST_5	$Pr(ST_5)$
ST.1	0.14
ST.2	0.14
ST.3	0.14
ST.4	0.14
ST.5	0.14
ST.6	0.14
ST.7	0.16

Fig. 2. Initial grammar for the deceptive royal tree problem.

to a tree and give rise to the *dependency among the subtrees and their immediate parent node*. Refer to Fig. 1, x and y are perfect subtrees only when their immediate parent node is A, so the trees in (a)–(c) score much higher than that in (d) due to the penalty for the incorrect immediate parent node B for the two terminal x nodes.

Another characteristics is that terminal y adds deceptiveness, i.e. importance of interactions among the nodes. These locally optimal royal trees are similar to the globally optimal royal trees in terms of structure and score. Refer to Fig. 1, the royal trees (a) and (b) are the global optimum and local optimum at level A respectively. At level B, the royal trees (e) and (f) are the global optimum and local optimum at level B respectively. The problem is challenging because of *deceptive attractors* which are local optimums deceiving the search system.

4 Grammar Model

A program is represented in a parse tree which is derived from the hierarchical PCSG to enforce the syntactic relations. Besides, the probabilistic dependency relations among the non-terminals of a production rule are captured by Bayesian networks. The hierarchical PCSG used in the DRT problem is shown in Fig. 2. For ease of reference, we label each rule and index each non-terminal on the right-hand side of the arrow (–>) with a subscript. Consider the rule $ST.4$ in the example. ST on the left-hand side of the arrow means the rule can be chosen when we derive any non-terminal ST. The right-hand side of the arrow gives the content when the rule is derived. Square brackets are used to enclose a terminal. Hence, $[B]$ is a terminal. ST_4 and ST_5 are the non-terminals. Furthermore, a Bayesian network is attached to each rule and captures the conditional dependencies among the non-terminals of that rule. For example, the box in

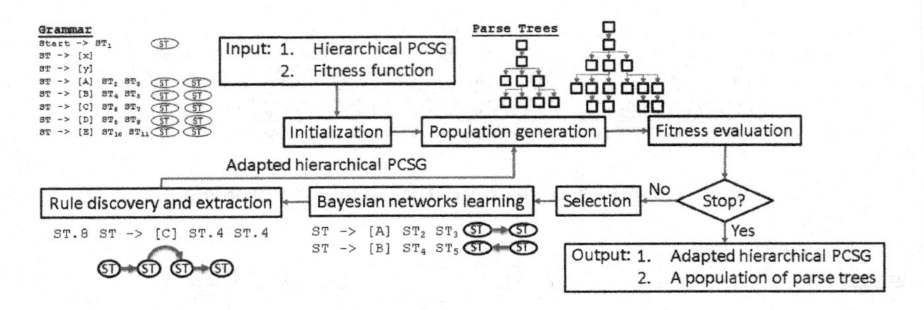

Fig. 3. System flowchart.

Fig. 2 shows the Bayesian network of rule $ST.4$. The Bayesian network models the conditional dependencies between non-terminals ST_4 and ST_5 while all terminals are constant and hence ignored. The ovals are the two random variables corresponding to ST_4 and ST_5. Non-terminal ST_4 has seven possible choices corresponding to the seven entries in the conditional probability table. Similarly, we have another table for ST_5. Initially, the probabilities are uniformly distributed and iteratively updated during evolution.

5 System Architecture

In this section, we present the details of BGBGP-HL. The entire process consists of seven steps (Fig. 3).

1. Initialize hierarchical PCSG to uniform distributions.
2. Create a new population of parse trees, some of which are derived from the hierarchical PCSG and the selected parse trees through elitism.
3. Calculate the fitness of the new parse trees in the population.
4. Select a set of parse trees based on their scores.
5. Learn the Bayesian networks in each production rule.
6. Discover and extract new rules (structure and parameters) if the probability converges.
7. Repeat from step 2 and the selected fitter parse trees survive to the next generation until meeting the stopping criteria.

This paper mainly focuses on capability of our system with the hierarchical grammar to discover new rules while the high-level ideas of the methods to identify and extract new rules are provided for completeness. In step 5, the Bayesian networks are learnt using K2 algorithm [2]. The samples come from the fitter parse trees, which are the knowledge of fitness function. After collecting the expansion choices of production rules among them, we obtain the statistics on how to derive *all* the non-terminals. Since these records are collected from fitter parse trees, each Bayesian network actually represents dependencies of the choices of non-terminals constituted in the fitter parse trees.

5.1 Derivation of a Parse Tree

To better understand the connection between hierarchical PCSG and parse trees, an example is present in this section. To derive a parse tree as shown in Fig. 2, the system starts from rule *Start.1* and picks a rule from $ST.1$ to $ST.7$ according to the conditional probability table (which is initially uniformly distribution). Let's say $ST.4$ is picked. Next, we derive the non-terminal ST_4 following the conditional probability distribution table (i.e. table on the left in the box in Fig. 2); and similarly for the non-terminal ST_5. Let's say $ST.3$ and $ST.5$ are picked. Again, we perform a random sampling recursively according to the probability distributions until all the leaf nodes in the parse tree are terminals.

5.2 Production Rules Discovery and Extraction

As previously mentioned, a Bayesian network captures the dependencies among the non-terminals in a production rule. In this section, we introduce the general idea of our method to discover and extract new production rules from the set of Bayesian networks of the rules.

Discovery of Interesting Assignments. The conditional probability table will be analysed. Using Fig. 5 as an example, we firstly identify choices which are almost certain and then compute the Shannon entropy (i.e. a measure of randomness of a random variable) of ST_4 for every parent configuration ST_5. When the entropy is below 0.5 bit, the randomness of random variable ST_4 is quite low. We have shaded the rows with entropy less than 0.5 bit in Fig. 5 (left).

Extraction of a Rule. Interesting expansion choices for conditional probability table $Pr(ST_4|ST_5)$ are marked with an asterisk (*) as shown in Fig. 5 (right). We want to predict where these interesting expansion choices are based on the choices of the non-terminals. In this paper, we apply C5.0, which is an advanced version of C4.5 [14] decision tree learning algorithm, to group and generalize the interesting expansion choices in *all* the conditional probability tables of the same production rule. C5.0 predicts the position of the entries marked with an asterisk using the choices of random variables as the feature attributes. Refer to the table on the right in Fig. 5, when $ST_4 = ST.3$ and $ST_5 = ST.4$, the entry is marked. Since this rule has two non-terminals, we will obtain a record with three attributes ST_4, ST_5 and *marked* (i.e. the attribute for a class label). These attributes contain the values $ST.3$, $ST.4$ and *marked* (meaning that there is an asterisk) respectively. We select branches with accuracy over 95 % and construct new hierarchical rules from the branches.

Importance of Rule Learning. During the evolution, we emphasized that the grammar not only specifies the valid individuals, but also the dependencies among the non-terminals. In order to generate more good individuals in the subsequent generations, the grammar is iteratively adapted by (1) adjusting the probabilities in the conditional probability tables, (2) identifying conditional dependencies among the non-terminals of the same rule, and (3) forming new

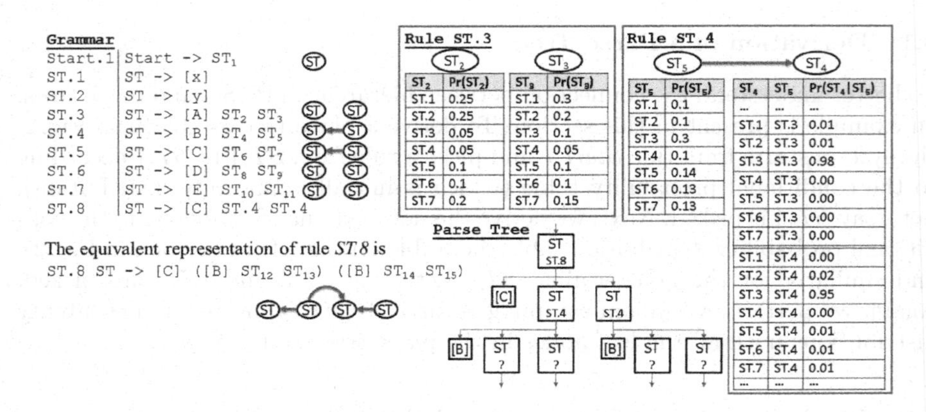

Fig. 4. Adapted grammar for the deceptive royal tree problem.

For rule ST.4, $ST_4 \rightarrow ST_5$

$Pr(ST_4 \mid ST_5)$	Random variable ST_4							
Random variable ST_5	ST.1	ST.2	ST.3	ST.4	ST.5	ST.6	ST.7	Entropy
ST.1	0.05	0.01	0.06	0.70	0.11	0.05	0.02	1.57
ST.2	0.02	0.00	0.02	0.93	0.03	0.00	0.00	**0.47**
ST.3	0.01	0.01	0.98	0.00	0.00	0.00	0.00	**0.16**
ST.4	0.00	0.02	0.95	0.00	0.01	0.01	0.01	**0.38**
ST.5	0.02	0.04	0.80	0.02	0.00	0.09	0.03	1.13
ST.6	0.14	0.24	0.18	0.19	0.05	0.01	0.19	2.52
ST.7	0.19	0.14	0.07	0.13	0.19	0.17	0.07	2.71

	Random variable ST_4						
Random variable ST_5	ST.1	ST.2	ST.3	ST.4	ST.5	ST.6	ST.7
ST.1							
ST.2				*			
ST.3			*				
ST.4			*				
ST.5							
ST.6							
ST.7							

Fig. 5. For rule $ST.4$, it conditional probability table $Pr(ST_4|ST_5)$ (left) and the table of interesting expansion choices (right).

production rules. An example is shown in Fig. 4. Firstly, we can see that the distribution of rule *ST.3* is no longer uniform and is updated to produce more good individuals. Secondly, it can be observed that there is an arrow in the Bayesian networks of both rule *ST.4* and *ST.5*. For instance, the arrow in the Bayesian network of rule *ST.4* means that ST_4 is conditional dependent to its parent ST_5. During the derivation of a non-terminal, its choice is dependent of the choices of other non-terminals. If ST_5 is sampled to $ST.3$, then the probability of ST_4 being $ST.3$ is 0.6 during the random sampling. Alternatively, if ST_5 is $ST.4$, then the probability of ST_4 being $ST.3$ drops to 0.1. Therefore, the conditional probability table in rule *ST.4* encodes the dependency. Thirdly, hierarchical PCSG enables the system to express the solutions in a recursive manner. A production rule is a composition of terminals, non-terminals and production rules. The recursive definition forms hierarchy among the rules. An example is rule $ST.8$ in Fig. 4. This rule is composed of rule $ST.4$ and rule $ST.3$. Its equivalent representation and the parse tree of $ST.8$ are also shown in the same figure. Note that the non-terminals of rule $ST.8$ can have a very different dependency structure (across multiple rules) when comparing with rule $ST.4$ and rule $ST.3$.

6 Evaluation

To demonstrate the search effectiveness and the capability of discovering hierarchical knowledge of our algorithm, we implemented BGBGP-HL and tested it with the DRT problem. Apart from the canonical GBGP, it was compared with two existing PMBGP methods: PAGE-EM [5], and PAGE-VB [5], which outperformed the univariate model (PIPE), adjacent model (EDP model) and simple GP [5]. The key parameters and their values of configuration are shown in Table 1. Due to time constraints, we declare an approach fails if it cannot obtain an optimal solution in 200,000 fitness evaluations.

We tested the four algorithms on DRT problems at level D, level E and level F. Table 2 shows the relative performance of the algorithms. At level D, the three PMBGP approaches performed well and found the optimal solutions over 88 % of runs. In contrast, GBGP did not perform well and the successful rate was below 10 %. At level E, the performance of PAGE-EM and PAGE-VB reduced by around 40 % and 20 % respectively. Their performance dropped continuously at level F. On average, BGBGP-HL needed roughly half of the number of fitness evaluations of PAGE-VB to find an optimum. The successful rate of BGBGP-HL deteriorated slowly and was less susceptible to the increase in tree depth. Among the successful runs at level E, the average number of fitness evaluations of PAGE-EM was similar to that of BGBGP-HL. We also applied t-test on the average

Table 1. System specific configuration.

GBGP		PAGE-EM		PAGE-VB		BGBGP-HL	
Parameter	Value	Parameter	Value	Parameter	Value	Parameter	Value
Population size	1,000	Population size	1,000	Population size	1,000	Population size	500
Generation	1,000	Annotation size	8, 16	Annotation set	{1, 2, 4, 8, 16}	Mutation	0.8
Crossover rate	0.9	Selection rate	0.1	Selection rate	0.1	Selection rate	0.4
Mutation rate	0.09	Elite rate	0.1	Elite rate	0.1	Elite rate	0.1
				Annotation exploration range	2	Accumulation size	100

Table 2. Results for Deceptive Royal Tree Problem at level D, E and F. It shows the statistics of the number of fitness evaluations in 50 runs. Column *Suc.*, *Mean*, *Std. dev.* and *t-test* are the number of successful runs, the average, the standard deviation of the number of fitness evaluations and the t-test results in the successful runs respectively. 'x' means the result is not applicable.

	Level D				Level E				Level F			
	Suc.	Mean	Std. dev.	t-test	Suc.	Mean	Std. dev.	t-test	Suc.	Mean	Std. dev.	t-test
BGBGP-HL	**50**	4,392	644	x	**48**	9,140	1,858	x	**45**	17,244	12,009	x
GBGP	9	18,972	2,260	+	0	x	x	x	0	x	x	x
PAGE-EM	**50**	10,660	4,543	+	27	9,704	3,220		4	1,704	4,205	−
PAGE-VB	44	9,395	2,896	+	33	21,212	5,689	+	27	39,296	11,293	+

numbers of fitness evaluations in the successful runs between our approach and the other approaches. The plus sign (+) indicates our approach is better than the compared approach with statistical significance (i.e. $p < 0.05$), and the minus sign (−) indicates the opposite. The table shows that our approach was better than the compared approaches with statistical significance. At level E, neither BGBGP-HL nor PAGE-EM is statistically significantly better than one another in respect to the average number of fitness evaluations. But the successful rate of BGBGP-HL is much higher than that of PAGE-EM. At level F, the successful rate of BGBGP-HL was 90 % and that of PAGE-EM was only 8 %. Although it seems that PAGE-EM used a smaller number of evaluations on average in the successful runs, the performance of PAGE-EM must be analyzed more carefully, as we have underestimated the average number of fitness evaluations of PAGE-EM because of its poor successful rate. The failed runs in GBGP was trapped in a local optima. PAGE-EM and PAGE-VB may require a larger population size to collect enough samples to estimate the dependencies when the tree depth grows. BGBGP-HL was more robust to the deceptiveness because the deceptiveness is localized to the leaf nodes y. New rules have learnt, say $ST \rightarrow ([A]\ [x]\ [x])$ and $ST \rightarrow ([B]\ ([A]\ ST\ ST)\ ([A]\ ST\ ST))$.

7 Discussion and Conclusion

BGBGP-HL can tackle the problems with strong dependence and deceptiveness. Meanwhile, it can construct hierarchical knowledge during evolution. Finding an optimal solution is a crucial step to understand an optimization problem. Grammar is an effective structural representation for describing the search space and hierarchical knowledge provided by the domain experts. Besides, contextual (and stochastic) knowledge can be expressed using hierarchical PCSG. The probabilistic distributions in Bayesian networks are transformed to explicit knowledge. Importantly, BGBGP-HL shows the steps of how knowledge is evolved (or derived). As a result, we can better understand how to composite the existing knowledge. While the knowledge produced is hierarchical, it remains comprehensible as it is expressed in the same language. We also argued that BGBGP-HL belongs to a new category of PMBGP approaches using *adaptive grammar*. It differentiates from current probabilistic grammar-based approaches since the adaptive grammar can also improve the structure (i.e. composition of the existing production rules) and the parameters (i.e. probability distributions) during evolution.

Acknowledgment. This research has been supported by General Research Fund LU310111 from the Research Grant Council of the Hong Kong Special Administrative Region.

References

1. Booth, T.L., Thompson, R.A.: Applying probability measures to abstract languages. IEEE Trans. Comput. **100**(5), 442–450 (1973)
2. Cooper, G.F., Herskovits, E.: A bayesian method for the induction of probabilistic networks from data. Mach. Learn. **9**(4), 309–347 (1992)
3. Hasegawa, Y., Iba, H.: Estimation of bayesian network for program generation. In: Proceedings of 3rd Asian-Pacific Workshop on Genetic Programming, p. 35 (2006)
4. Hasegawa, Y., Iba, H.: A bayesian network approach to program generation. IEEE Trans. Evol. Comput. **12**(6), 750–764 (2008)
5. Hasegawa, Y., Iba, H.: Latent variable model for estimation of distribution algorithm based on a probabilistic context-free grammar. IEEE Trans. Evol. Comput. **13**(4), 858–878 (2009)
6. Hasegawa, Y., Ventura, S.: Programming with annotated grammar estimation. In: Genetic Programming-New Approaches and Successful Applications, pp. 49–74 (2012)
7. Kim, K., Shan, Y., Nguyen, X.H., McKay, R.I.: Probabilistic model building in genetic programming: a critical review. Genet. Program Evolvable Mach. **15**(2), 115–167 (2014)
8. Koza, J.R.: Genetic Programming: On the Programming of Computers by Means of Natural Selection, vol. 1. MIT press, Cambridge (1992)
9. Mckay, R.I., Hoai, N.X., Whigham, P.A., Shan, Y., O'Neill, M.: Grammar-based genetic programming: a survey. Genet. Program Evolvable Mach. **11**(3–4), 365–396 (2010)
10. O'Neill, M., Brabazon, A.: Grammatical differential evolution. In: IC-AI, pp. 231–236 (2006)
11. O'Neill, M., Ryan, C.: Grammatical evolution. IEEE Trans. Evol. Comput. **5**(4), 349–358 (2001)
12. O'Neill, M., Ryan, C.: Grammatical Evolution: Evolutionary Automatic Programming in an Arbitrary Language, vol. 4. Springer, New York (2003)
13. O'Neill, M., Ryan, C.: Grammatical evolution by grammatical evolution: the evolution of grammar and genetic code. In: Keijzer, M., O'Reilly, U.-M., Lucas, S., Costa, E., Soule, T. (eds.) EuroGP 2004. LNCS, vol. 3003, pp. 138–149. Springer, Heidelberg (2004)
14. Quinlan, J.R.: C4.5: Programs for Machine Learning. Elsevier, Amsterdam (2014)
15. Regolin, E.N., Pozo, A.T.R.: Bayesian automatic programming. In: Keijzer, M., Tettamanzi, A.G.B., Collet, P., van Hemert, J., Tomassini, M. (eds.) EuroGP 2005. LNCS, vol. 3447, pp. 38–49. Springer, Heidelberg (2005)
16. Sabar, N.R., Ayob, M., Kendall, G., Qu, R.: Grammatical evolution hyper-heuristic for combinatorial optimization problems. IEEE Trans. Evol. Comput. **17**(6), 840–861 (2013)
17. Salustowicz, R., Schmidhuber, J.: Probabilistic incremental program evolution. Evol. Comput. **5**(2), 123–141 (1997)
18. Sastry, K., Goldberg, D.E.: Probabilistic model building and competent genetic programming. In: Riolo, R., Worzel, B. (eds.) Genetic Programming Theory and Practice, vol. 6, pp. 205–220. Springer, New York (2003)
19. Tanev, I.: Incorporating learning probabilistic context-sensitive grammar in genetic programming for efficient evolution and adaptation of snakebot. In: Keijzer, M., Tettamanzi, A.G.B., Collet, P., van Hemert, J., Tomassini, M. (eds.) EuroGP 2005. LNCS, vol. 3447, pp. 155–166. Springer, Heidelberg (2005)

20. Whigham, P.A.: Grammatically-based genetic programming. In: Proceedings of the Workshop on Genetic Programming: From Theory to Real-World Applications, vol. 16, pp. 33–41 (1995)
21. Wong, M.L., Leung, K.S.: Applying logic grammars to induce sub-functions in genetic programming. In: IEEE International Conference on Evolutionary Computation, vol. 2, pp. 737–740. IEEE (1995)
22. Wong, P.K., Lo, L.Y., Wong, M.L., Leung, K.S.: Grammar-based genetic programming with bayesian network. In: 2014 IEEE Congress on Evolutionary Computation, pp. 739–746. IEEE (2014)
23. Wong, P.K., Lo, L.Y., Wong, M.L., Leung, K.S.: Grammar-based genetic programming with dependence learning and bayesian network classifier. In: Proceedings of GECCO 2014, pp. 959–966. ACM (2014)
24. Yanase, T., Hasegawa, Y., Iba, H.: Binary encoding for prototype tree of probabilistic model building GP. In: Proceedings of GECCO 2009, pp. 1147–1154 (2009)

Parallel Hierarchical Evolution of String Library Functions

Jacob Soderlund[1], Darwin Vickers[1,2], and Alan Blair[1(✉)]

[1] School of Computer Science and Engineering, University of New South Wales, Sydney, Australia
j.soderlund@student.unsw.edu.au, d.vickers@unsw.edu.au, blair@cse.unsw.edu.au
[2] Data61, CSIRO, Sydney, Australia

Abstract. We introduce a parallel version of hierarchical evolutionary re-combination (HERC) and use it to evolve programs for ten standard string processing tasks and a postfix calculator emulation task. Each processor maintains a separate evolutionary niche, with its own ladder of competing agents and codebank of potential mates. Further enhancements include evolution of multi-cell programs and incremental learning with reshuffling of data. We find the success rate is improved by transgenic evolution, where solutions to earlier tasks are recombined to solve later tasks. Sharing of genetic material between niches seems to improve performance for the postfix task, but for some of the string processing tasks it can increase the risk of premature convergence.

1 Introduction

Evolutionary Computation typically involves a population of *agents* (individuals) undergoing repeated cycles of selection, crossover and mutation. It has long been recognized that a large-scale crossover would normally result in an initially inferior agent and that subsequent, smaller crossovers or mutations would be needed before the new agent becomes competitive in the general population. Methods have therefore been proposed to protect these young individuals for a period of time in an age-layered population structure or similar scheme [6].

Hierarchical evolutionary re-combination and the associated HERCL programming language were introduced as an alternative approach to this problem [1]. HERCL agents have a stack, registers and memory (Fig. 1), thus combining elements from linear GP [8] and stack-based GP [9,10]. Programs are divided hierarchically into *cells*, *bars* and *instructions*. Each *cell* is effectively a procedure or subroutine, containing a sequence of executable instructions. Cells are divided into smaller chunks called *bars*, delimited by the pipe symbol (|) – much like the bars in a musical score. Each instruction consists of a (single-character) command, optionally preceded by a sequence of dot/digits which form the argument for that command. The various commands are listed in Table 1 (see [1] for further details).

J. Handl et al. (Eds.): PPSN XIV 2016, LNCS 9921, pp. 281–291, 2016.
DOI: 10.1007/978-3-319-45823-6_26

```
    INPUT:  ickey
   OUTPUT:
   MEMORY:  Minnie..............................
REGISTERS:  .....[6]..[1].  [7]
    STACK:  MM
     CODE:  0[is|.<sy^5>};i|8{^s-~:+7=;wo8|-wo]
                                ^
```

Fig. 1. HERCL simulator, showing an evolved agent executing the `strcmp` task, to compare the strings "Minnie" and "Mickey". All items are floating point numbers, but the simulator prints them as a dot (zero), an ASCII character, or bracketed in decimal format, depending on their value.

Table 1. HERCL commands

	Input and Output
`i`	fetch INPUT to input buffer
`s`	SCAN item from input buffer to stack
`w`	WRITE item from stack to output buffer
`o`	flush OUTPUT buffer
	Registers and Memory
`<`	GET value from register
`>`	PUT value into register
`^`	INCREMENT register
`v`	DECREMENT register
`{`	LOAD from memory location
`}`	STORE to memory location
	Jump, Test, Branch and Logic
`j`	JUMP to specified cell (subroutine)
`|`	BAR line (RETURN on `.|` HALT on `8|`)
`=`	register is EQUAL to top of stack
`g`	register is GREATER than top of stack
`:`	if TRUE, branch FORWARD
`;`	if TRUE, branch BACK
`&`	logical AND
`/`	logical OR
`~`	logical NOT

	Stack Manipulation and Arithmetic	
`#`	PUSH new item to stack	$...... \mapsto ......x$
`!`	POP top item from stack	$......x \mapsto ......$
`c`	COPY top item on stack	$......x \mapsto ......x, x$
`x`	SWAP top two items	$...y, x \mapsto ...x, y$
`y`	ROTATE top three items	$z, y, x \mapsto x, z, y$
`-`	NEGATE top item	$......x \mapsto .....(-x)$
`+`	ADD top two items	$...y, x \mapsto ...(y+x)$
`*`	MULTIPLY top two items	$...y, x \mapsto ...(y*x)$
	Mathematical Functions	
`r`	RECIPROCAL	$..x \rightarrow ..1/x$
`q`	SQUARE ROOT	$..x \rightarrow ..\sqrt{x}$
`e`	EXPONENTIAL	$..x \mapsto ..e^x$
`n`	(natural) LOGARITHM	$..x \mapsto ..\log_e(x)$
`a`	ARCSINE	$..x \mapsto ..\sin^{-1}(x)$
`h`	TANH	$..x \mapsto ..\tanh(x)$
`z`	ROUND to nearest integer	
`?`	push RANDOM value to stack	
	Double-Item Functions	
`%`	DIVIDE/MODULO	$..y, x \mapsto ..(y/x), (y \bmod x)$
`t`	TRIG functions	$..\theta, r \mapsto ..r\sin\theta, r\cos\theta$
`p`	POLAR coords	$..y, x \mapsto ..\mathrm{atan2}(y,x), \sqrt{x^2+y^2}$

Hierarchical Evolution does not use a population in the usual sense, but instead maintains a stack or *ladder* of candidate solutions (agents), and a *codebank* of potential mates. At each step of the algorithm, the agent at the top rung of the ladder is selected and either mutated or crossed over with a randomly chosen agent from the codebank, or from an external library.

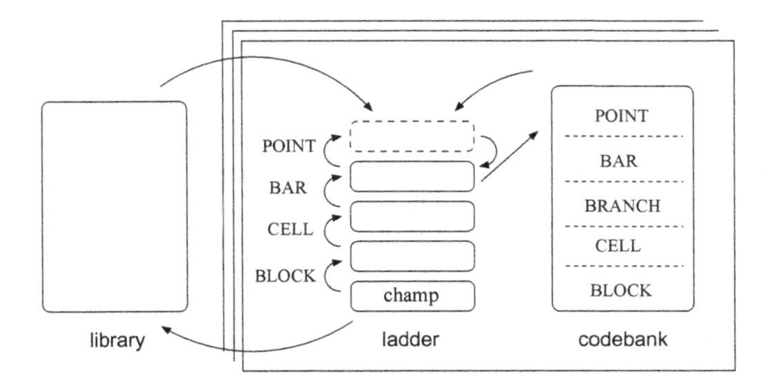

Fig. 2. Hierarchical evolutionary re-combination. If the top agent on the ladder becomes fitter than the one below it, the top agent will move down to replace the lower agent (which is transferred to the codebank). If the top agent exceeds its maximum number of allowable offspring without ever becoming fitter than the one below it, the top agent is removed from the ladder (and transferred to the codebank). When the algorithm is parallelized, each niche has its own ladder and codebank, but all of them may share a common library, containing the current champ from each niche.

Crossovers and mutations are classified into different *levels* (TUNE, POINT, BAR, BRANCH, CELL or BLOCK) according to what portion of code from the primary (ladder) parent is either mutated or replaced with code from the secondary (codebank or library) parent. A large crossover at the lowest rung of the ladder is followed up by a series of progressively smaller crossovers and mutations at higher rungs, concentrated in the vicinity of the large crossover.

In previous work, single-cell HERCL programs have successfully been evolved for coding tasks [1], dynamically unstable control problems [2] and classification tasks [3]. However, a number of drawbacks emerged:

(a) the algorithm sometimes experienced long periods of stagnation,
(b) for some of the more complex tasks, the single-cell programs became very long and difficult to evolve,
(c) the number of competing agents in a single ladder is rather low, potentially missing out on the benefits of parallelism inherent in other EC paradigms.

In the present work, we address these issues by introducing:

(a) incremental training, with reshuffling,
(b) multi-cell evolution,
(c) parallelized hierarchical evolution, on a multi-core architecture.

Aided by these enhancements, we test whether programs can be evolved to emulate a postfix calculator, and to perform ten string processing tasks modeled on functions from the standard C library.

2 HERCL Enhancements

(a) Incremental Training, With Reshuffling: Incremental training has previously been used to evolve HERCL programs for control problems such as the double pole balancing task [2]. We now extend this approach to supervised learning, with some additional modifications. The training items are shuffled into a random order and, initially, the fitness evaluation is based only on the first two items. Once a certain target cost has been achieved on the first k items, additional items are added, until the per-item target cost is no longer achieved. If the system runs for an entire *epoch* (100,000 offspring) without adding any new items, the last item on the list is swapped out and replaced with the next (in order) item – thus giving the system a chance to "move on", rather than getting stuck on a particularly difficult item. If two difficult items occur in succession, the algorithm will swap back and forth between them – up to a maximum of six attempts (three for each item). After the sixth failed attempt, a *reshuffle* event occurs: the training items are reshuffled into a new order, and training begins again with the first two items (according to the new ordering). Note that this is not the same as a random re-start, because the codebank and the champ are retained, and only the subset of the training data changes.

This reshuffling, combined with hierarchical search, gives rise to a process of *creative destruction*, where the components of agents evolved under previous orderings are re-combined, to optimize the fitness under the new ordering. Over time, code fragments that are advantageous for multiple sets of training items are more likely to survive to be incorporated into a global solution.

(b) Multi-cell Evolution: In order to evolve multi-cell programs we introduce new levels of crossover beyond the CELL level, labeled as BLOCK-1, BLOCK-2, BLOCK-4, etc. For a BLOCK-k mutation, a block of k cells from the secondary parent is transplanted into the primary parent (a JUMP instruction to the modified cell(s) may optionally be inserted elsewhere in the code). Subsequent (lower-level) mutations are concentrated in the vicinity of the transplanted block.

(c) Parallel Hierarchical Evolution: We parallelize the algorithm to run on a multi-core machine. Each core maintains a separate *niche* with its own ladder and codebank, but all of them may share a common library, comprised of the current best agent (champ) from each of the niches (Fig. 2). As soon as a global solution is found in one niche, a terminating signal is sent to the niches running on the other cores. Since competition between agents occurs only within a niche, data can be reshuffled independently in each niche, thus creating additional diversity in the system without compromising the "fairness" of the competition.

For comparison, we include some experiments where each niche is running completely independently, with no sharing of code between them. This allows us to examine to what extent improved performance is due to sharing of code, and to what extent it is due simply to a greater amount of searching.

3 String Processing Experiments

In our first set of experiments, we attempt to evolve solutions for a set of ten string processing tasks, adapted from functions in the standard C String Library `string.h` (listed in Table 2). The main motivation for choosing these tasks is to see whether certain general-purpose programming constructs could be evolved, and form a kind of "standard library" for HERCL, to facilitate the learning of more complex tasks such as those proposed in [5]. The right column of Table 2 gives a general indication of the kind of programming constructs that are required for each task. Using this information, along with preliminary experiments, we have tried to arrange the tasks roughly in order from easiest to hardest.

For each task, 1000 training and 1000 test cases are randomly generated. String contents are chosen uniformly from the set of all printable ASCII characters. For tasks involving a distinguished character or index, the number of

Table 2. String processing tasks and programming constructs.

TASK	DESCRIPTION	L	R	V	S	M	I
`strcpy`	input: a string output: the same string	✓					
`strcat`	input: a string followed by another string output: first string concatenated with second string	✓					
`strlen`	input: a string output: the length of that string	✓	✓				
`idxstr`	input: an index followed by a (non-empty) string output: the character at that index in the string	✓	✓				✓
`chrstr`	input: a character followed by a string output: index of first occurrence of that character (or an empty message, if it does not occur)	✓	✓	✓			
`stridx`	input: a (non-empty) string followed by a list of indices output: the list of characters at the specified indices	✓	✓	✓		✓	✓
`catstr`	input: a string followed by another string output: second string concatenated with first string	✓	✓	✓		✓	
`strchr`	input: a string followed by a character output: index of first occurrence of that character (or an empty message, if it does not occur)	✓	✓	✓		?	
`strrchr`	input: a string followed by a character output: index of last occurrence of that character (or an empty message, if it does not occur)	✓	✓	✓		?	
`strcmp`	input: a string followed by another string output: difference between characters at the first place where the two strings differ (or zero, if they are identical)	✓	✓	✓	✓	✓	

KEY: L = LOOP, R = REGISTER, V = compare VALUE, S = SUBTRACT, M = MEMORY, I = compare INDEX characters before and after it are geometrically distributed with a mean of 3 characters. This is equivalent to choosing the string lengths from a negative binomial distribution $\mathrm{NB}(2, \frac{3}{4})$ and then choosing the distinguished character uniformly within the string. Each run is conducted on a 16-core machine with 15 separate niches, plus one core dedicated to communication between the other cores (in a star network arrangement). For the cost function we use the Generalized Levenshtein Edit Distance [1], which is suitable for comparing outputs that may vary in length. The target cost is zero.

Table 3 shows the number of minutes to completion for the various evolutionary runs. In the first five runs (labeled Sa to Se) each task was evolved on its own, up to a maximum of 8 h. In the remaining runs, labeled as *transgenic*, the system attempts to evolve solutions for each task in turn, using a library consisting of the solutions evolved for all of the previous (successful) tasks on the list. Up to three attempts are made for each task, with each attempt running for a maximum of 8 h. As soon as one attempt is successful, the system adds the solution to its library and moves on to the next task. If all three attempts fail, the system moves to the next task without adding anything to the library.

For the runs labeled as *sharing*, code was shared between the 15 niches, with space reserved in the common library for the current best agent from each niche (updated asynchronously, at the end of each epoch). For the last three runs (labeled as non-sharing) the cores were run completely independently, with no genetic material transmitted between cores (and the library consisting only of the solutions to previous tasks).

Table 4 shows the evolved code from three selected runs (TSb, TSc and Ta). We see that almost all the runs succeeded in evolving solutions for `strcpy`, `strcat`, `strlen`, `idxstr`, `chrstr` and `stridx`. The first two tasks – `strcpy` and `strcat` – are easily solved within a few minutes and the resulting code is practically identical across all runs, as follows:

```
strcpy 0[i|sw;o]      strcat 0[i|sw;i|sw;o]
```

The solutions for `idxstr` all involve incrementing or decrementing a register, until the required index is reached. Those for `strlen` and `chrstr` involve counting items – either by incrementing an index, explicitly adding 1, or computing tanh of each character (which saturates to 1). The solutions for `stridx` all work by storing the string into successive memory locations and then accessing the value at each index in the list.

The last four tasks were solved considerably more often by the transgenic runs, and we can see several instances where code from previous tasks has been re-combined to solve later tasks. The solution for `catstr` in run Ta uses code from `stridx` to store the first string into memory, then transfers the second string to the output buffer, before retrieving the first string from memory. Indeed, `stridx` seems to be a kind of *bottleneck* task in the sense that failure on `stridx` in Run Tb has led to failure on all the subsequent tasks. Run TSc has found solutions for `strchr` and `strrchr` which invert the roles of the character and index, storing each index (plus 0.2) into the memory location specified by the character. If the same character occurs multiple times, the first or last occurrence (as appropriate) will overwrite the others and its index (rounded to the nearest integer) will be the one that prevails in the relevant memory location. We can recognize a fragment of code from `strrchr` in the solution for `strcmp`, which stores the first string into memory and then scans the second string, comparing it with the one in memory, one character at a time, until a non-zero difference is found.

Runs Sa, Sc, TSa, Ta and T2, which failed to evolve a solution for `strchr`, end up solving `strrchr` by storing each index and character alternately on the stack, then searching backwards. When the relevant character is found, the next number on the stack will be the correct index.

Run TSb provides a good example of how a suboptimal "gene" from one task can find its way into later tasks. For a string of length n, the solution for `strlen` is achieved not by incrementing a register but instead by computing the nearest integer to $\log\sqrt{(7.5)^n} \simeq 1.0075\,n$. This gives the correct answer for all strings up to length 67. The pattern then finds its way (with a slight twist) into the solution for `strchr` – which computes a formula based on the sum of the logs of the characters in the string, with an expected value of $1.04 + 0.985(k-1)$. In other words, the agent is exploiting the independent distribution of the characters, to find a solution which is good enough to give the correct answer for all 1000 training and test cases, but would not work for all possible inputs. The solution for `strrchr` similarly computes a function whose expected value is $0.79 + 1.02(k-1)$. It satisfies all the training data but makes an off-by-one error on 1 of the 1000 test items.

Table 3. Evolution time for string tasks.

Run	Sa	Sb	Sc	Sd	Se	TSa	TSb	TSc	TS2	Ta	Tb	T2
transgenic	← No →					← Yes →						
time limit	480 mins					3 × 480 = 1440 mins						
sharing	← Yes →									← No →		
cells	← 1 →								2	1		2
`strcpy`	2	3	1	4	2	2	4	1	4	1	3	3
`strcat`	15	6	19	15	15	15	15	16	17	16	8	11
`strlen`	2	1	190	2	12	109*	5	2	3	1	2	1
`idxstr`	8	2	275*	3	2	4	5	2	4	3	81	5
`chrstr`	122	44*	46	×	466	6	16	1	109	22	7	21
`stridx`	319	124	117	214	251	420	127	163	795	749	×	191
`catstr`	×	×	×	×	×	×	×	×	20	342	×	130
`strchr`	×	×	×	×	×	×	56	32	×	×	×	×
`strrchr`	97	×	37	×	×	272	5*	3	×	918	×	14
`strcmp`	×	×	×	×	×	×	×	174	×	×	×	×

KEY: × = failed to achieve zero cost on the training set
* = achieved zero cost on training set but not on test set

Table 4. Evolved code for selected runs of the string tasks.

<table>
<tr><th></th><th colspan="2">Run TSb (sharing)</th></tr>
<tr><td>strlen</td><td colspan="2">0[1#i|y*7.5#s;}qnzwo]</td></tr>
<tr><td>idxstr</td><td colspan="2">0[isi|=^:sx;|swo]</td></tr>
<tr><td>chrstr</td><td colspan="2">0[i{s>i|=h:+s;|!wo]</td></tr>
<tr><td>stridx</td><td colspan="2">0[i|s}^;is:o.|>{ws;o]</td></tr>
<tr><td>catstr</td><td colspan="2">–</td></tr>
<tr><td>strchr</td><td colspan="2">[i|ss;>is|=:x>;o.|!=h;eex=*;{e|6#**x.=x1;7><xgq;qnzwo]</td></tr>
<tr><td>strrchr</td><td colspan="2">[i|ss;>is|=:x>;o8|!he|7#**><xgq;qnzwo]</td></tr>
<tr><td>strcmp</td><td colspan="2">–</td></tr>
<tr><th></th><th>Run TSc (sharing)</th><th>Run Ta (no sharing)</th></tr>
<tr><td>strlen</td><td>0[i|<s^;wo]</td><td>0[is:{wo1:|}1.#+s;wo]</td></tr>
<tr><td>idxstr</td><td>0[is>2gi|<.=:s5^;|swo]</td><td>0[isi|%^g:s;|swo]</td></tr>
<tr><td>chrstr</td><td>0[i<s>is|=h:+s;|!wo]</td><td>0[i<s>is|=h:+s;|>wo]</td></tr>
<tr><td>stridx</td><td>0[i|s}^;is:o8|>{ws;o]</td><td>0[is|}^s;is|~:>{|ws1;o]</td></tr>
<tr><td>catstr</td><td>–</td><td>0[i|}^s;is|ws;1^{g:;|o]</td></tr>
<tr><td>strchr</td><td>0[.2#i>|<s^;}|>};is>{9=z~:o8|wo]</td><td>–</td></tr>
<tr><td>strrchr</td><td>0[.2#i>|.<s^5>};is>{3=z:w|o]</td><td>0[i|<^s;is>}|=:};o8|%wo]</td></tr>
<tr><td>strcmp</td><td>0[is|.<sy^5>};i|8{^s-~:+7=;wo8|-wo]</td><td>-</td></tr>
</table>

In three cases (Sb `chrstr`, Sc `idxstr` and TSa `strlen`) a very long suboptimal solution has been found which explicitly scans items one at a time, then prints out a hard-coded answer. In general, the non-sharing runs tend to produce code which is shorter and more robust than that of the sharing runs. The reason may be that a suboptimal solution can develop in one niche and then spread like a virus to the other niches. In that case, a system with all niches "quarantined" from each other may produce a global solution faster. For example, Run Ta (without sharing) has found a global solution for `catstr` using memory, whereas the three sharing runs all got stuck in suboptimal solutions which manage to store and retrieve up to 5 or 6 characters of the string using specific registers and stack manipualations, but fail for longer strings. The two-cell runs (TS2 and T2) were able to solve `catstr`, but their solutions avoid using memory and instead use recursion, with one recursive call for each item.

It is hard to say whether sharing makes a significant difference in the evolution times – although the sharing runs may finish slightly faster (or more often) than the non-sharing runs on tasks like `stridx` and `strrchr` where suboptimal solutions are unlikely to occur. It may be that some of these tasks – when provided with evolved solutions to the preceding tasks – effectively require only one new "trick" in order to succeed. Once this trick is discovered, the evolution rapidly proceeds to completion. Under this hypothesis, 15 niches running concurrently would be expected to find the vital "trick" in equal time, whether they are sharing code or evolving independently. Each trick then becomes a *stepping stone* to aid in the evolution of subsequent tasks [7].

4 Postfix Calculator

In order to further explore the issue of sharing vs. non-sharing evolution, we tried evolving on a different kind of task, whose solution is more likely to require a combination of separately evolved modalities. The task we chose is a `postfix` calculator. For this task, the input is a sequence of numbers and operators, forming an arithmetic expression in *Postfix* or *Reverse Polish* notation. The output is the numerical value to which the given expression evaluates.

The allowed operators are $+, -, *$ and $/$. Again, 1000 training and test cases are produced by a generative process, according to the probabilistic grammar shown in Table 5. Expressions involving divide by zero are excluded. For this task we again use the Generalized Levenshtein Edit Distance. The target cost is set at 10^{-6} in order to avoid failure due to roundoff errors.

Table 5. Probabilistic Grammar for generating postfix data.

S	→	Val	(0.1)	Op	→	+	(0.25)
S	→	Tree Tree Op	(0.9)	Op	→	−	(0.25)
Tree	→	Val	(0.6)	Op	→	*	(0.25)
Tree	→	Tree Tree Op	(0.4)	Op	→	/	(0.25)
Val	→	space, followed by numeric value from a Cauchy distribution, rounded to two decimal places					

For the `postfix` task, two-cell programs rather than single-cell programs were evolved in the first instance. Ten runs were performed with sharing of genetic material, and ten runs without sharing. All of the non-sharing runs failed to find a solution within the 8 h limit. Only two of the ten sharing runs managed to find a solution.

Ten additional runs were performed to see whether single-cell evolution, with sharing of code between niches, could produce solutions for the `postfix` task. For these runs, the time limit was extended to 24 h per run. Only one of these ten runs produced a solution. (In order to save computing time, single-cell experiments without sharing of code were not performed.) The exact evolution times and evolved solutions are shown in Table 6.

Note that the first of these solutions ultimately uses only one of the two cells. The code is quite short, and uses a serendipitous combination of trig functions and logs to distinguish the characters $+, -, *$ and $/$ and perform the appropriate operation. The other two solutions perform an explicit comparison against (the ASCII values of) these characters. Having two cells available during the evolution seems to free up the evolutionary process, and provide greater flexibility for targeted crossovers and mutations. But, later in the process the code from one cell may get randomly transplanted into the other cell, allowing it to perform the whole task on its own.

Table 6. Evolution times and evolved code for the postfix task.

```
MINS  CODE
314   0[]
      1[is|>39#g!:ss;|<ct>cnt!gn1:>g:r|*c{|hz-*+s3;wo]
395   0[43#>g~:!*s:wo8]
      1[isss:wc.|!ss|41#>g1;0j43#>g;47#>g:yr*<xy|.7#-t!z*+s1;wo]
721   0[isss:wo8|!s46#>s|g:!r*s;2;|vvg:}-<|vg:!+s:wo8|vg4;46#>g:
        3;|vvg:}-<|vg:!+s1:wo.|!*s:wo.|4;]
```

5 Conclusion and Further Work

We have parallelized the hierarchical evolutionary re-combination algorithm, and shown that HERCL programs can successfully be evolved to emulate a postfix calculator and perform ten different string processing tasks. In the process, fundamental programming constructs emerge such as incrementing and decrementing of registers, storing items into successive memory locations, looping until certain conditions are met and distinguishing between various arithmetic symbols to select between different computation paths.

This process can be compared to the primordial stages of biological evolution, where promiscuous recombinations in geographically separated regions over long periods of time eventually lead to beneficial fragments of genetic code, which can then be recombined into successively more complex organisms.

In future work we plan to package up the solutions for these string tasks into a "standard library" and test to what extent the inclusion of this library (and others like it) would speed up the learning of new tasks.

Our software is freely available on GitHub (via http://hercl.org).

Acknowledgment. This research was undertaken using computing resources from Intersect Australia's Orange HPC, provided through the National Computational Infrastructure (NCI), which is supported by the Australian Government.

References

1. Blair, A.: Learning the Caesar and Vigenere Cipher by hierarchical evolutionary re-combination. In: Congress on Evolutionary Computation, pp. 605–612 (2013)
2. Blair, A.: Incremental evolution of HERCL programs for robust control. In: Conference on Genetic and Evolutionary Computation Companion, pp. 27–28 (2014)
3. Blair, A.D.: Transgenic evolution for classification tasks with HERCL. In: Chalup, S.K., Blair, A.D., Randall, M. (eds.) ACALCI 2015. LNCS, vol. 8955, pp. 185–195. Springer, Heidelberg (2015)
4. Bruce, W.S.: The lawnmower problem revisited: stack-based genetic programming and automatically defined functions. In: Conference on Genetic Programming, pp. 52–57 (1997)

5. Helmuth, T., Spector, L.: General program synthesis benchmark suite. In: Genetic and Evolutionary Computation Conference, pp. 1039–1046 (2015)
6. Hornby, G.S.: ALPS: the age-layered population structure for reducing the problem of premature convergence. In: GECCO, pp. 815–822 (2006)
7. Lehman, J., Stanley, K.O.: Abandoning objectives: evolution through the search for novelty alone. Evol. Comput. **19**(2), 198–223 (2011)
8. Nordin, P.: A compiling genetic programming system that directly manipulates the machine code. Adv. Genet. Program **1**, 311–331 (1994)
9. Perkis, T.: Stack-based genetic programming. In: IEEE World Congress on Computational Intelligence, pp. 148–153 (1994)
10. Spector, L., Robinson, A.: Genetic programming and autoconstructive evolution with the push programming language. Genet. Program Evolvable Mach. **3**(1), 7–40 (2002)

On the Non-uniform Redundancy in Grammatical Evolution

Ann Thorhauer(✉)

University of Mainz, Mainz, Germany
thorhauer@uni-mainz.de
http://www.uni-mainz.de

Abstract. This paper investigates the redundancy of representation in grammatical evolution (GE) for binary trees. We analyze the entire GE solution space by creating all binary genotypes of predefined length and map them to phenotype trees, which are then characterized by their size, depth and shape. We find that the GE representation is strongly non-uniformly redundant. There are huge differences in the number of genotypes that encode one particular phenotype. Thus, it is difficult for GE to solve problems where the optimal tree solutions are underrepresented. In general, the GE mapping process is biased towards short tree structures, which implies high GE performance if the optimal solution requires small programs.

Keywords: Grammatical evolution · Redundant representation · Binary trees · Bias

1 Introduction

One core component of any evolutionary algorithm (EA) is the representation used [17]. Indeed, the choice of representation determines the success of a heuristic search method [17]. In general, there are two types of representations: a direct and an indirect representation [16]. When using a direct representation, no distinction between geno- and phenotype is made, just like in standard genetic programming (GP) [9], which uses tree structures to represent the individuals. Here, the search operators (e.g., crossover and mutation) are applied directly to these tree structures. An indirect representation distinguishes between geno- and phenotype. In this case, a representation describes how the genotypes (e.g., binary strings) are mapped to the phenotypes (e.g., expressions, trees) [16]. When using this type of representation, the search operators are applied to genotypes, but the actual effect of these operators is observed at the corresponding phenotypes.

Indirect representations may be biased due to redundant encodings [18]. If this is true, on average more than one genotype represents the same phenotype. A representation is uniformly redundant if every phenotype is represented by the same number of genotypes; it is non-uniformly redundant if one or more phenotypes are represented by a larger number of genotypes than others.

J. Handl et al. (Eds.): PPSN XIV 2016, LNCS 9921, pp. 292–302, 2016.
DOI: 10.1007/978-3-319-45823-6_27

Consequently, the use of redundant representations may be biased and could therefore influence the search process if the optimal solution or parts of it are underrepresented [18].

Grammatical evolution (GE) uses a redundant representation [15]. In contrast to standard GP, GE [19] uses variable-length binary strings to encode the programs/expressions and a grammar in Backus-Naur form (BNF) to map the binary genotypes to the tree phenotypes. Consequently, GE applies standard genetic search operators such as one-point crossover and mutation to linear bit strings.

The distribution of trees in the GP solution space is well studied for various problems [10] and can "[...] give an indication of problem difficulty for GP" [10]. For GE, there are no similar studies. In this paper, we study the non-uniform redundancy of the GE representation, especially the GE genotype to phenotype mapping. In our analysis, we focused on binary trees. We explored the entire solution space of GE by applying different grammars and different genotype lengths. We used two approaches to characterize trees: First, we used the tree size and tree depth; second, we took the shape of a tree into account since this property cannot be neglected when dealing with realistic programs. We showed that GE representation is strongly non-uniformly redundant. The number of different genotypes strongly exceeds the number of different phenotypes, and there are phenotypes that are encoded with higher probabilities than others. In general, short tree structures are represented most frequently.

In Sect. 2 we review bias and redundant representations. Section 3 reviews former studies of representation bias in GE. Our analysis and results are presented in Sect. 4. The paper ends with some concluding remarks.

2 Bias and Redundant Representations

A bias exists if some solutions or solution structures are visited more frequently than others during the run of a search procedure, or if certain actions are performed more frequently than others during the search [22]. The existence of a bias may be advantageous or disadvantageous for the search [17]. For example, a bias of search operators may be used to guide the search in a certain direction where promising solutions are presumed; this would be a desired bias. Unwanted bias occurs if there is an interaction between the search operators used and the chosen representation such that the problem becomes deceptive [1].

In heuristic search methods like GP, a desired bias results from the selection process. By selecting highly-fit individuals for the next generation, selection pushes a population in the direction of fitter individuals. In addition, by using problem-specific recombination or mutation operators, as well as suitable terminal and function sets, GP performance can be improved [21,22]. Indeed, Dignum and Poli showed that "[...] simple length biases can significantly improve the best fitness found during a GP run" [5].

Search bias can also be a result of redundant encodings [18]. Encodings are redundant if there are phenotypes that are encoded by more than one genotype.

A redundant encoding is biased (i.e., non-uniformly redundant) if not all phenotypes are encoded by the same number of genotypes; but rather some phenotypes are represented by a larger number of genotypes than others. When using GE, there is a redundant representation since there is more than one genotype that represents the same phenotype [15]. Non-uniform redundancy, where the phenotypes are non-uniformly represented by the genotypes, leads to a bias and causes structural difficulty since some solution structures are underrepresented. Rothlauf and Goldberg [18] examined the impact of redundant representations on the performance of evolutionary algorithms (EA). In general, redundant representations are less efficient because they use more alleles to store the same amount of information compared to non-redundant representations. In the case where a uniformly redundant representation is used, the general performance of the EA is neither decreased nor increased. In opposition to this, the performance of the algorithm can be increased or decreased if a non-uniformly redundant representation is used. In their experiments, Rothlauf and Goldberg [18] showed that performance can be increased when the optimal solution is overrepresented, and it can be decreased when the optimal solution is underrepresented.

3 Bias of Representation in Grammatical Evolution

GE [19] is a variant of GP that uses a complex genotype-phenotype mapping to create the phenotype programs/expressions from variable-length binary genotypes. A genotype consists of groups of eight bits (called codons) which encode an integer value that selects production rules from a grammar in BNF. These rules are used in the deterministic mapping process to create a phenotype.

Several studies have examined the bias of representation and grammar in GE. Most of them have focused on the impact of different representations or different search operators on performance rather than on the impact of the redundant representation. O'Neill and Ryan [12,14] examined the effect of genetic code degeneracy on genotypic diversity and the performance of GE. When a degenerate genetic code is used, each codon consists of more bits than actually necessary to encode a sufficient amount of integer values to select the rules from the grammar. They found that genetic diversity is higher when a degenerate genetic code is used. In addition, the amount of invalid individuals is lower. The impact on performance depends on the grammar used. Montes de Oca [11] focused on creating numerical values by concatenating digits and found that the most-commonly-used GE grammar induces a bias towards short-length numbers. Hemberg et al. [8] considered three different GE grammars (postfix, prefix, infix) and their influence on performance for various symbolic regression problems. They observed no differences between the grammars for small problem instances. However, for large problems, a postfix grammar was found to be advantageous. Fagan et al. [6] compared the performance of GE when using four different genotype-phenotype mappings (depth-first, breadth-first, random and πGE [13]) for four benchmark problems and measured the average best fitness, the average size of genotypes, and the average number of derivation tree nodes over the number of generations. The πGE mapper outperformed the other mapping strategies in three out of four problems. The breadth-first mapper produced larger trees in three out of four problems compared to the other mapping strategies. Harper [7] showed that standard GE random bit initialization produces trees that are non-uniformly distributed since "80 % of the trees have 90 % of their nodes on one side of the tree or the other" [7]. Trees are biased to be "tall and skinny" [7]. He distinguished between two types of grammars: explosive and balanced. He defined a grammar to "be explosive if the number of non-terminals (functions) exceeds the number of terminals" [7]. A grammar is balanced if the probability for expanding a non-terminal into a multiple of the same non-terminal is equal to the probability of expanding into a terminal. Therefore, when using an explosive grammar, the probability is high that the mapping process will run out of genes [7]. Daida and co-workers [2–4] studied the influence of the standard GP tree representation on the search process, and showed that not only crossover and selection does influence GP search behavior, but so does the representation. Inspired by their work, Thorhauer and Rothlauf [20] found that a random walk with GE using standard operators (one-point crossover, mutation and duplication) has a strong bias towards sparse tree structures.

Overall, GE literature is dominated by experimental studies on the impact of different search operators or representations on performance, but a fundamental mathematical analysis of the entire solution space has been omitted.

4 Analysis and Results

We studied the bias of representation in GE for binary trees. Using the definition from Harper [7], we used a balanced grammar (Fig. 1(a)) and an explosive grammar (Fig. 1(b)) to map the genotypes to the phenotype trees. We set the number of codons to 10 and 20, and created all possible binary genotypes by using 10 and 20 codons respectively, and studied their phenotypic properties. For the balanced grammar A (Fig. 1(a)), only the least significant bit in each codon (usually eight bits) determines which rule to choose during the mapping process. Therefore, we reduced the total number of possible binary genotypes from 2^{80} to $2^{10} = 1024$ when using 10 codons (each consists of one bit), and from 2^{160} to 2^{20} when using 20 codons. For the explosive grammar B (Fig. 1(b)), we needed the last two bits of each codon to define the rule to choose since we wanted to ensure that all rules were chosen with equal probability. Again, we got 2^{20} different genotypes. So the value of any bit directly affects the phenotype. We used standard depth-first mapping to create the phenotypes. No wrapping operator was used since the mapping process would never terminate and thus the corresponding individuals would be invalid anyway. This is a result of the structure of the chosen grammars. Before we analyzed the characteristics of all phenotype trees, the derivation trees were transformed into syntax trees to get binary trees that only consisted of functions (internal nodes) and terminals (leaf nodes). For all the tree structures, we measured the following properties: tree depth, tree size (internal nodes plus leaf nodes) and tree shape.

We started our analysis by distinguishing individuals according to their phenotype tree properties: size and depth. Figures 2(a)–(c) show the phenotype

```
<start> ::= <expr>
<expr>  ::= (<expr> + <expr>)
          | <var>
<var>   ::= X
```

(a) grammar A

```
<start> ::= <expr>
<expr>  ::= (<expr> + <expr>)
          | (<expr> + <expr>)
          | (<expr> + <expr>)
          | <var>
<var>   ::= X
```

(b) grammar B

Fig. 1. Production rules in BNF. Balanced (left) and explosive variant (right).

(tree) solution spaces that were created with 10 codons using grammar A, 20 codons using grammar A, and 10 codons using grammar B. We plotted the number of nodes (tree size) over the tree depth for all phenotype trees. The outer dashed lines show the boundaries for valid binary trees with a minimum and a maximum number of nodes. The circles between these lines represent valid trees, whereas the triangles represent invalid trees. In these cases, the mapping process could not be finished because the genotype ran out of genes. As a result, there are invalid binary trees, where not all internal nodes (+) have two child nodes or not all external nodes are terminals (X). When 10 codons and grammar A were used, we observe 7 different valid trees and 7 different invalid trees that can be created with 2^{10} different genotypes (Fig. 2(a)). Under these conditions, valid trees with a maximum size of 9 nodes can be created. Consequently, when using 20 codons and the same grammar, the phenotype solution space increases, and the limit of the maximum tree size rises to 19 nodes (Fig. 2(b)). Thus 2^{20} different genotypes encode 30 different valid and 16 invalid trees. Figure 2(c) shows the same result as seen in Fig. 2(a). Here, we used 10 codons and grammar B. In summary, the solution space of binary trees that can be covered with GE depends on the length of the genotype. The change from the balanced grammar A to the explosive grammar B does not modify the solution space of possible binary trees in GE.

To study the redundancy in GE representation, we had to examine the frequency of all trees. Figures 3(a)–(c) present the proportion of trees of a given size and depth in a 3D view. This clearly shows that the GE mapping process creates specific trees with higher probabilities than it does others. Therefore, representation in GE is non-uniformly redundant. Indeed, trees with a size of one and a depth of zero were created most frequently, independent from codon lenght or the grammar used. All trees of size 10 (Figs. 3(a) and (c)) and 20 (Fig. 3(b)) are the result of an unfinished mapping. As Harper [7] described, the use of the explosive grammar B strongly overrepresents these invalid individuals (Fig. 3(c)) compared to the use of the balanced grammar A (Figs. 3(a) and (b)). In summary, we observe an overrepresentation of short valid trees in all three figures. The GE mapping process has a strong bias towards short trees when grammar A is used (Fig. 3(a) and (b)); whereas when grammar B is used, the mapping process more frequently creates longer, but invalid, trees (Fig. 3(c)).

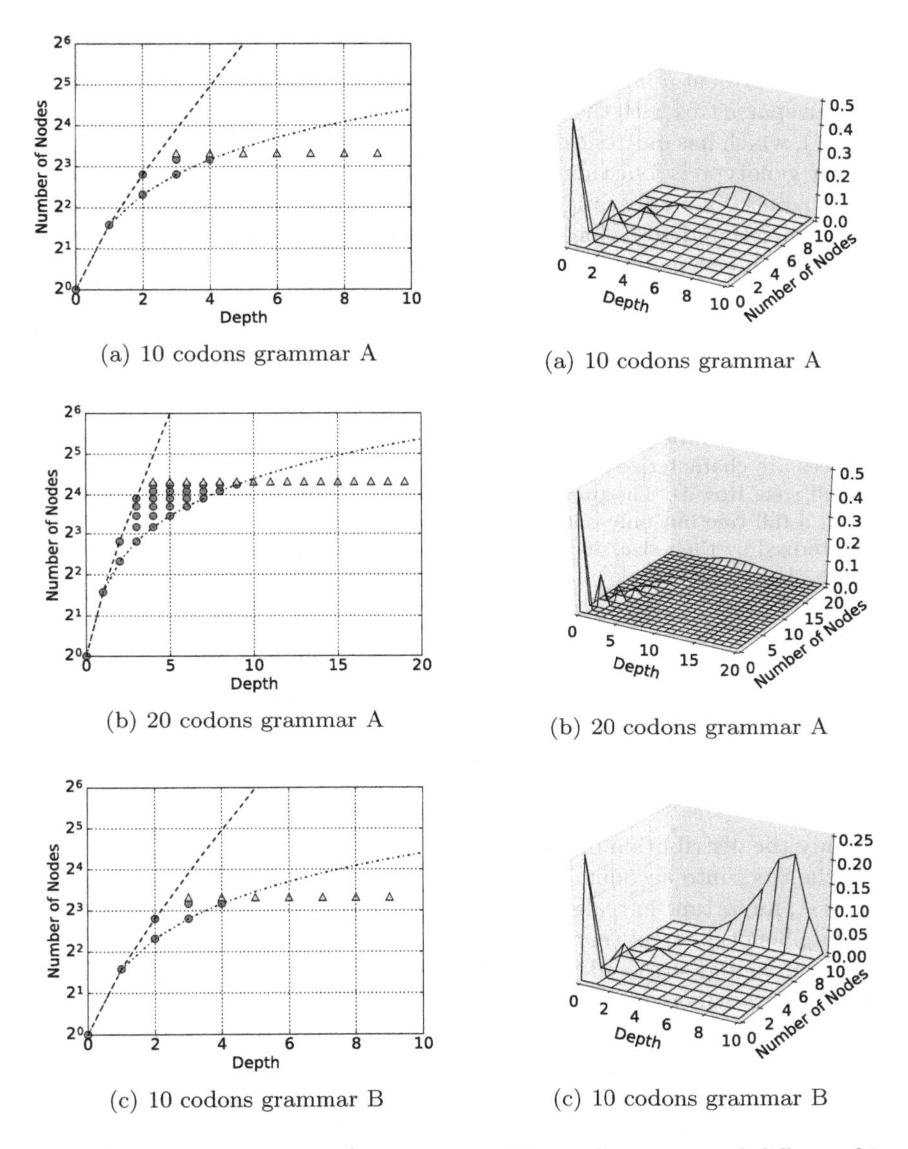

(a) 10 codons grammar A

(b) 20 codons grammar A

(c) 10 codons grammar B

Fig. 2. Phenotype (binary tree) solution spaces for different codon lengths and grammars.

(a) 10 codons grammar A

(b) 20 codons grammar A

(c) 10 codons grammar B

Fig. 3. Proportion of different binary trees for different codon lengths and grammars.

Figures 4(a)–(c) represent the proportion of trees that have the same size and depth (the proportion of invalid trees are not shown). Table 1 (rows A and B) describes how the probabilities to create particular trees can be calculated. The probabilities to represent specific trees in Fig. 4(a) when using 10 codons and grammar A cover a range between 0.5 and 0.0078125 ($\frac{1}{2^7}$). This implies 50 % of genotypes represent a tree that consists of one terminal X, whereas only 0.78125 % of genotypes represent a full tree of depth two (Fig. 4(a) d = 2, s = 7). The probability that a sparse tree with the same size of seven and a depth of three (Fig. 4(a) d = 3, s = 7), which has only one expanded node at any depth level, is represented by a binary genotype is four times higher (3.125 %). For 20 codons and grammar A, the probabilties actually range between 0.5 and approximately 0.000031. The use of grammar B and 10 codons hugely changes the probabilities to create each tree since it is three times more likely to choose the rule $< expr > + < expr >$ than to choose $< var >$ (see Table 1 column B). In this case, the probabilities range between 0.25 and about 0.00165. In summary, these results reveal a strong overrepresentation of short and rather sparse trees.

Table 1. The probabilities of representing particular trees for both grammars. Row A and B: Trees are characterized by their size and depth. Row A: The probabilities to create a full tree. Row B: The probabilities to create a sparse tree (i.e., same number of nodes as a full tree but only one expanded node at any depth level). Row C: Trees are characterized by their size, depth and shape; the probabilities to create any tree of a given size, depth and shape.

	grammar A	grammar B
A	$(\frac{1}{2})^{size}$	$(\frac{3}{4})^{(number\ internal\ nodes)} \times (\frac{1}{4})^{(number\ leaf\ nodes)}$
B	$(\frac{1}{2})^{size} \times 2^{(d-1)}$	$(\frac{3}{4})^{(number\ internal\ nodes)} \times (\frac{1}{4})^{(number\ leaf\ nodes)} * 2^{(d-1)}$
C	$(\frac{1}{2})^{size}$	$(\frac{3}{4})^{(number\ internal\ nodes)} \times (\frac{1}{4})^{(number\ leaf\ nodes)}$

To study the distribution of different trees in greater detail, we also took the shape of the trees into account. This implies that the exact ordering of nodes is relevant (i.e., important in realistic programs). Consequently, the number of different valid phenotype trees that can be encoded by 2^{10} different binary genotypes increases from 7 (Figs. 4(a) and (c)) to 23 (Figs. 5(a) and (c)). With 2^{20} genotypes, the number of trees increases from 30 (Fig. 4(b)) to 6918 (Fig. 5(b)). As an example, let's take the case of two trees of different shapes that have a size of five and a depth of two (Fig. 5(a) d = 2, s = 5). Table 1, row C shows how to calculate the probability that a specific tree will be created. The probabilities to encode the different phenotypes range between 0.5 and approximately 0.00195 ($\frac{1}{2^9}$) when grammar A and binary genotypes of 10 codons are used, and between 0.5 and about 0.000002 when grammar A and 20 codons are used. Again, the use of grammar B and 10 codons hugely changes the probabilty that a genotype creates a particular tree. In this case, the distribution of probabilites ranges between 0.25 and about 0.00031. In summary, the larger the size of a particular tree, the lower the probability that this tree is represented by a binary genotype (independently from the grammar used). The probability of creating a specific binary tree depends only on its size. The additional consideration of the tree shape increases the number of different trees that can be encoded by 2^{10} and 2^{20} different genotypes, but does not prevent the overrepresentation of particular binary trees.

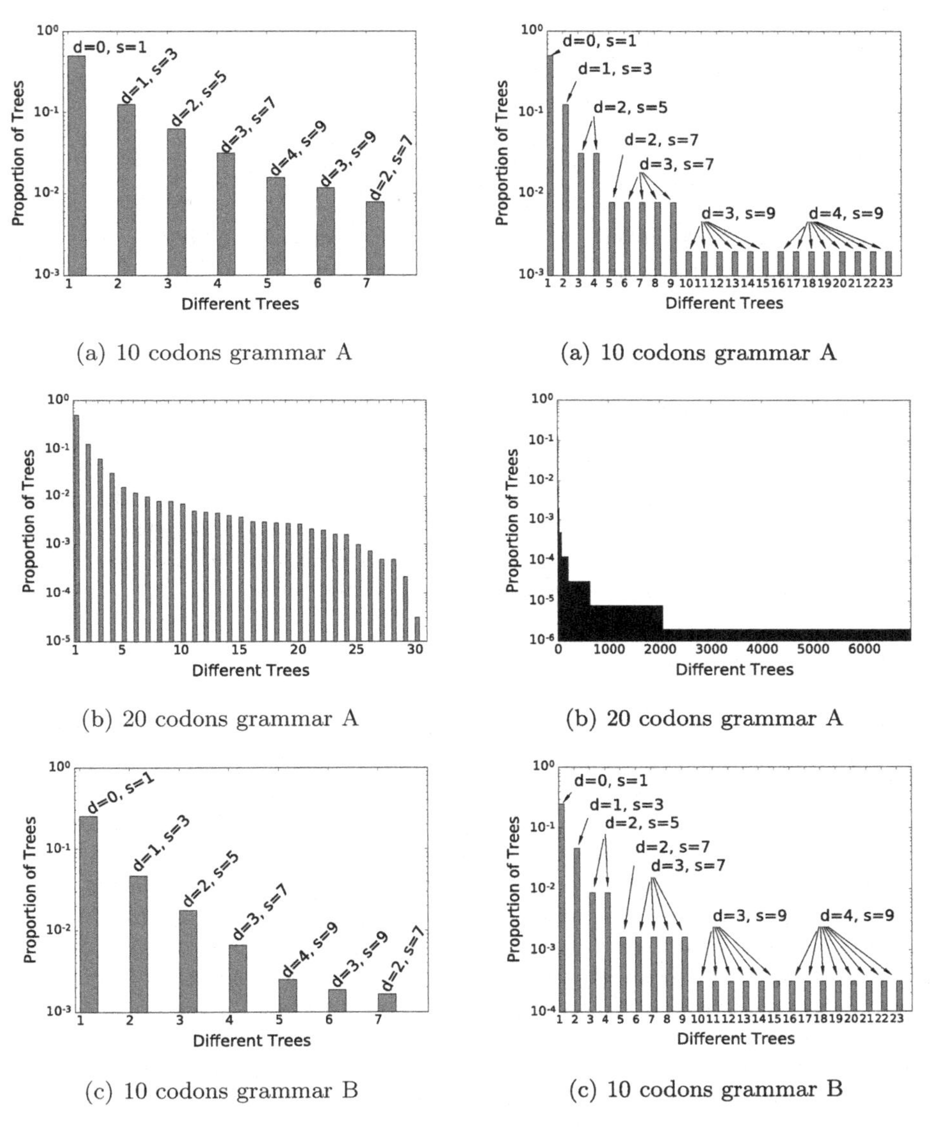

Fig. 4. Number of instances per valid tree (normalized over all tree instances) for different codon lengths and grammars. Trees are characterized by their size and depth.

Fig. 5. Number of instances per valid tree (normalized over all tree instances) for different codon lengths and grammars. Trees are characterized by their size, depth and shape.

Table 2 presents the number $|\Phi_p|$ of different valid phenotypes and the number $|\Phi_{p_{invalid}}|$ of different invalid phenotypes that can be encoded by genotypes of either 10 or 20 codons using grammars A and B. These results emphasize that

Table 2. Number $|\Phi_g|$ of different genotypes (genotype search space Φ_g), number $|\Phi_p|$ of different possible valid phenotype trees (phenotype solution space Φ_p), number $|\Phi_{p_{invalid}}|$ of different possible invalid phenotype trees ($\Phi_{p_{invalid}}$). Trees are characterized by their size, depth and shape.

	$\lvert\Phi_g\rvert$	$\lvert\Phi_p\rvert$	$\frac{\lvert\Phi_p\rvert}{\lvert\Phi_g\rvert}$	$\lvert\Phi_{p_{invalid}}\rvert$	$\frac{\lvert\Phi_{p_{invalid}}\rvert}{\lvert\Phi_g\rvert}$
grammar A, 10 codons	2^{10}	23	0.0225	252	0.2461
grammar A, 20 codons	2^{20}	6 918	0.0066	184 756	0.1762
grammar B, 10 codons	2^{20}	23	0.00002	252	0.00024

GE representation is strongly redundant since the phenotype solution space (Φ_p) is obviously smaller than the genotype search space (Φ_g). The number of different valid phenotype trees over the number of different genotypes for grammar A and 10 codons is 0.0225, whereas the proportion of different invalid phenotype trees is obviously larger (0.2461). Both values are even lower when grammar B is used. In general, $|\Phi_{p_{invalid}}|$ exceeds $|\Phi_p|$ for both grammars and codon lengths. During a GE run, these invalid trees are penalized with a mimimum fitness value, and provide no additional benefit since they will be sorted out.

5 Conclusions

We studied the redundancy in GE representation for binary trees. We used two different grammars (balanced and explosive) and two different genotype lengths. We explored the entire solution space of GE by creating all possible binary genotypes and mapped them to phenotype trees. When trees are characterized by their size and depth, sparse trees are more likely to be represented than full trees of the same size. If in addition the shape of a tree is relevant, the probability of creating a particular binary tree depends only on its size. In this case, the number of different invalid trees is larger than that of valid trees. Independent from the grammars used or codon lengths, the number of different binary genotypes strongly exceeds the number of different binary phenotypes. Moreover, there are large differences in the number of genotypes that encode one particular phenotype tree. Thus, it is difficult for GE to solve problems if the optimal tree solutions are underrepresented. In general, the GE mapping process is biased towards short tree structures.

A higher genotype length increases the number of different phenotypes that can be encoded. Moving from a balanced to an explosive grammar alters the probabilities for creating any tree. Furthermore, the probability that a genotype encodes an invalid tree is higher.

The focus of our study was on binary trees. Using more complex grammars and non-binary trees would create a greater variety of different trees and therefore reduce the probability of creating trees with the same characteristics. In the future, we will extend this analysis to different grammars that allow the creation of more complex non-binary trees.

References

1. Caruana, R.A., Schaffer, J.D.: Representation and hidden bias: Gray vs. Binary coding for genetic algorithms. In: Proceedings of the Fifth International Conference on Machine Learning, pp. 153–161. Morgan Kaufmann (1988)
2. Daida, J.M.: Limits to expression in genetic programming: lattice-aggregate modeling. In: CEC 2002, pp. 273–278. IEEE Press, NJ, USA (2002)
3. Daida, J.M., Hilss, A.M.: Identifying structural mechanisms in standard genetic programming. In: Cantú-Paz, E., et al. (eds.) GECCO 2003. LNCS, vol. 2724, pp. 1639–1651. Springer, Heidelberg (2003)
4. Daida, J.M., Li, H., Tang, R., Hilss, A.M.: What makes a problem GP-Hard? validating a hypothesis of structural causes. In: Cantú-Paz, E., et al. (eds.) GECCO 2003. LNCS, vol. 2724, pp. 1665–1677. Springer, Heidelberg (2003)
5. Dignum, S., Poli, R.: Operator equalisation and bloat free GP. In: O'Neill, M., Vanneschi, L., Gustafson, S., Esparcia Alcázar, A.I., De Falco, I., Della Cioppa, A., Tarantino, E. (eds.) EuroGP 2008. LNCS, vol. 4971, pp. 110–121. Springer, Heidelberg (2008)
6. Fagan, D., O'Neill, M., Galván-López, E., Brabazon, A., McGarraghy, S.: An analysis of genotype-phenotype maps in grammatical evolution. In: Esparcia-Alcázar, A.I., Ekárt, A., Silva, S., Dignum, S., Uyar, A.Ş. (eds.) EuroGP 2010. LNCS, vol. 6021, pp. 62–73. Springer, Heidelberg (2010)
7. Harper, R.: GE, explosive grammars and the lasting legacy of bad initialisation. In: CEC 2010, pp. 1–8. IEEE Press (2010)
8. Hemberg, E., McPhee, N., O'Neill, M., Brabazon, A.: Pre-, In- and postfix grammars for symbolic regression in grammatical evolution. In: IEEE Workshop and Summer School on Evolutionary Computing, pp. 18–22 (2008)
9. Koza, J.R.: Genetic Programming: On the Programming of Computers by Means of Natural Selection. MIT Press, Cambridge (1992)
10. Langdon, W., Poli, R.: Foundations of Genetic Programming. Springer, Berlin (2002)
11. Montes de Oca, M.A.: Exposing a bias toward short-length numbers in grammatical evolution. In: O'Neill, M., Vanneschi, L., Gustafson, S., Esparcia Alcázar, A.I., De Falco, I., Della Cioppa, A., Tarantino, E. (eds.) EuroGP 2008. LNCS, vol. 4971, pp. 278–288. Springer, Heidelberg (2008)
12. O'Neill, M., Ryan, C.: Genetic code degeneracy: implications for grammatical evolution and beyond. In: Floreano, D., Nicoud, J.D., Mondada, F. (eds.) Advances in Artificial Life. LNCS, vol. 1674, pp. 149–153. Springer, Berlin (1999)
13. O'Neill, M., Brabazon, A., Nicolau, M., Garraghy, S.M., Keenan, P.: π grammatical evolution. In: Deb, K., Tari, Z. (eds.) GECCO 2004. LNCS, vol. 3103, pp. 617–629. Springer, Heidelberg (2004)
14. O'Neill, M., Ryan, C.: Grammatical Evolution: Evolutionary Automatic Programming in an Arbitrary Language. Kluwer Academic Publishers, Norwell (2003)
15. O'Neill, M., Ryan, C., Keijzer, M., Cattolico, M.: Crossover in grammatical evolution. Genet. Program. Evolvable Mach. **4**(1), 67–93 (2003)
16. Rothlauf, F.: Representations for Genetic and Evolutionary Algorithms, 2nd edn. Springer, Berlin (2006)
17. Rothlauf, F.: Design of Modern Heuristics. Springer, Heidelberg (2011)
18. Rothlauf, F., Goldberg, D.E.: Redundant representations in evolutionary computation. Evol. Comput. **11**(4), 381–415 (2003)

19. Ryan, C., Collins, J.J., O'Neill, M.: Grammatical evolution: evolving programs for an arbitrary language. In: Banzhaf, W., Poli, R., Schoenauer, M., Fogarty, T.C. (eds.) EuroGP 1998. LNCS, vol. 1391, pp. 83–95. Springer, Heidelberg (1998)
20. Thorhauer, A., Rothlauf, F.: Structural difficulty in grammatical evolution versus genetic programming. In: GECCO 2013, pp. 997–1004. ACM (2013)
21. Whigham, P.A.: Grammatically-based genetic programming. In: Proceedings of the Workshop on Genetic Programming: From Theory to Real-World Applications, pp. 33–41. Morgan Kaufmann, San Mateo (1995)
22. Whigham, P.A.: Search bias, language bias and genetic programming. In: Proceedings of the First Annual Conference on Genetic Programming 1996, pp. 230–237. MIT Press, CA, USA (1996)

Tournament Selection Based on Statistical Test in Genetic Programming

Thi Huong Chu[1], Quang Uy Nguyen[1](✉), and Michael O'Neill[2]

[1] Faculty of IT, Le Quy Don Technical University, Hanoi, Vietnam
huongktqs@gmail.com, quanguyhn@gmail.com
[2] Natural Computing Research and Applications Group, UCD Business, University College Dublin, Dublin, Ireland
m.oneill@ucd.ie

Abstract. Selection plays a critical role in the performance of evolutionary algorithms. Tournament selection is often considered the most popular techniques among several selection methods. Standard tournament selection randomly selects several individuals from the population and the individual with the best fitness value is chosen as the winner. In the context of Genetic Programming, this approach ignores the error value on the fitness cases of the problem emphasising relative fitness quality rather than detailed quantitative comparison. Subsequently, potentially useful information from the error vector may be lost. In this paper, we introduce the use of a statistical test into selection that utilizes information from the individual's error vector. Two variants of tournament selection are proposed, and tested on Genetic Programming for symbolic regression problems. On the benchmark problems examined we observe a benefit of the proposed methods in reducing code growth and generalisation error.

Keywords: Genetic Programming · Tournament selection · Statistical test

1 Introduction

There are several factors that can effect the performance of Genetic Programming (GP) for given problems. These factors include the size of a population, the fitness evaluation of individuals, the selection mechanisms for reproduction, the encoding and genetic operations for modifying individuals. Amongst these, selection plays a critical role in GP performance [4]. To date, there have been many selection schemes proposed, and popular selection schemes in GP include fitness proportionate selection, ranking selection, and tournament selection [9]. Among these, the most widely used selection in GP is tournament selection [6].

Tournament selection is based on comparing the fitness values of sampled individuals. The individual with the best fitness is then selected as the winner. This implementation is simple and its effectiveness has been evidenced by a number of research [6]. However, the standard implementation used only fitness

J. Handl et al. (Eds.): PPSN XIV 2016, LNCS 9921, pp. 303–312, 2016.
DOI: 10.1007/978-3-319-45823-6_28

value while ignoring information from the error of individuals in all fitness cases. Consequently, some information that is potentially useful for GP search may be lost. Recent research has shown that significant benefit could be gained by using semantic information of GP individuals (e.g., [10,12]). Thus, it is attractive to examine whether using the error value of individuals on the fitness cases for selection can improve GP performance.

In this paper, the error vector of individuals is used in tournament selection. Two individuals are compared using a statistical test using their error vector. If the statistical test (using a Wilcoxon signed rank test in this case) shows that there is a significant difference, the individual with the better fitness is selected and tested against the others. This process is repeated for all individuals in the tournament sample with the winner selected based on the statistical test. We test the proposed selection technique on a set of benchmark regression problems, and observe that the proposed method helps to reduce GP code growth and generalisation error.

The remainder of this paper is organized as follows. In the next section, we briefly review the related work on improving tournament selection in GP. The two proposed tournament selection methods are presented in Sect. 3. Section 4 presents the experimental settings adopted, with the results presented and discussed in Sect. 5. Finally, Sect. 6 concludes the paper and highlights some future work.

2 Related Work

This section presents a brief review of previous research on tournament selection in GP. Tournament selection is the most popular selection operator in GP [17]. In standard tournament selection, a number of individuals (tournament size) are randomly selected from the population. These individuals are compared together and the winner (in terms of better fitness) is selected to go to the mating pool. This process is then repeated N times where N is the population size [4]. The advantage of tournament selection is that it allows the adjustment of the selection pressure by tuning the tournament size. Moreover, this method does not require a comparison of the fitness between all individuals that helps to save a large amount of processing time [19].

As the standard tournament selection consists of two steps: sampling and selecting. There is a large number of research focusing on different sampling and selecting strategies in tournament selection. Xie et al. [20] analysed the performance of no-replacement tournament selection in GP. In the no-replacement strategy, no individual can be sampled multiple times within the same tournament. Another problem in tournament selection is not-sampled problem, in which some individuals are not sampled at all if a too small tournament size is used. This problem was discussed by Xie et al. in [18]. Later, Gathercole et al. [7] analyzed the selection frequency of each individual and the likelihoods of not-selected and not-sampled individuals in tournament selection of different tournament sizes. Sokolov and Whitley proposed unbiased tournament selection [15] where all individuals have a fair chance to participate into the tournament.

Overall, previous research has shown that sampling strategies have minor impact to GP performance. Consequently, researchers paid more attention to the second step in tournament selection: selection. Baeck [3] introduced the selection probability of an individual of rank j in one tournament for a minimization task, with an implicit assumption that the population is wholly diverse. Blickle and Thiele [4] extended the selection probability model in [3] to describe the selection probability of individuals with the same fitness f_j. They defined the worst individual to be ranked 1st and introduced the cumulative fitness distribution, which denotes the number of individuals with fitness value f_j or worse.

In this paper, we propose a method for ranking individuals in tournament selection that is based on the use of a statistical test. To the best of our knowledge, this techniques has not been studied in GP. The detailed description of our method will be presented in Sect. 3.

3 Methods

This section describes two new tournament selection techniques. The first technique is called *Statistics-TS1*. Similar to the standard tournament selection, a number of individuals are randomly selected and compared. The winner is then chosen to go to the mating pool. However, instead of using the fitness value for comparing between individuals, a statistical test was applied to the error vector of these individuals. For a pair of individuals, if the test shows that they are different, then the individual with better fitness value is the winner. Conversely, if the test confirms that two individuals are not different, a random individual is selected from the pair. After that, the winner individual is tested against other individuals and the process is repeated for all individuals in the tournament size. The detailed description of Statistics-TS1 is presented in Algorithm 1.

Algorithm 1. Statistics test tournament selection 1

```
Input: Tour size, Population.
Output: The winner individual.
A ⟵ RandomIndividual();
for i ← 1 to TourSize do
    B ⟵ RandomIndividual();
    sample1 ⟵ Error(A);
    sample2 ⟵ Error(B);
    p − value ⟵ Testing(sample1, sample2);
    if p − value < alpha then
        A ⟵ GetBetterFitness(A, B);
    else
        A ⟵ GetRandom(A, B);
    end
end
TheWinnerIndividual ⟵A ;
```

Table 1. Problems for testing statistical tournament selection

Abbreviation	Name	Attributes	Training	Testing
A. **Benchmarking Problems**				
F1	korns-11	5	20	20
F2	korns-12	5	20	20
F3	korns-14	5	20	20
F4	vladislavleva-1	2	20	2025
F5	vladislavleva-2	1	100	221
F6	vladislavleva-4	5	500	500
F7	vladislavleva-5	3	300	2640
F8	vladislavleva-6	2	30	93636
F9	vladislavleva-7	2	300	1000
F10	vladislavleva-8	2	50	1089
B. **UCI Problems**				
F11	airfoil_self_noise	5	800	703
F12	casp	9	100	100
F13	Slump_test_Compressive	7	50	53
F14	slump_test_FLOW	7	50	53
F15	slump_test_SLUMP	7	50	53
F16	winequality-red	11	800	799
F17	winequality-white	11	1000	1000
F18	wpbc	31	100	98

In Algorithm 1, the function $RandomIndividual()$ returns a random individual from the GP population. Function $Error(A)$ calculates the vector error of individual A and function $Testing(sample1, sample2)$ performs a Wilcoxon signed rank test on two samples. Two last functions, $GetBetterFitness(A, B)$ and $GetRandom(A, B)$ aims at finding the better fitness individual among A and B or returning a random individual between two, respectively. Finally, $alpha$ is the critical value used to decide if the null hypothesis is rejected or accepted. If the output of the test ($p - value$) is smaller than the critical value, then the null hypothesis is rejected. This means that two individuals are significantly different and the better individual is selected as the winner. If the test can not reject the null hypothesis, then a random individual is selected from the pair.

The second tournament selection is called *Statistics-TS2*. Statistics-TS2 is similar to Statistics-TS1 but aims at reducing code grow in the GP population. In Statistics-TS2, if the statistical test can not reject the null hypothesis, then the individual with smaller size is selected from the pair. In other words, if two individuals involved in the test are not statistically different, then the smaller individual will be the winner.

Table 2. Evolutionary parameter values.

Parameter	Value
Population size	500
Generations	100
Selection	Tournament
Tournament size	7
Crossover probability	0.9
Mutation probability	0.1
Initial Max depth	6
Max depth	17
Max depth of mutation tree	15
Raw fitness	mean absolute error on all fitness cases
Trials per treatment	30 independent runs for each value

4 Experimental Settings

In order to measure the impact of the two new tournament selection to GP performance, we tested them on eighteen multivariate regression problems. Among these, ten problems are benchmark problems [16] and eight problems were taken from UCI machine learning dataset [2]. The tested problems are presented in Table 1.

The GP parameters used for our experiments are shown in Table 2. The terminal set for each problem includes N variables corresponding to the number of attributes of that problem. The function set include eight functions (+, -, *, /, sin, cos, log, exp) that are popularly used in GP. The raw fitness is the mean of absolute error on all fitness cases. The elitism technique was also used in which the best individual in the current generation is copied to the next generation. In the new tournament selection schemes, two critical values (0.05 and 0.1) were used for the Wilcoxon signed rank test to decide if the null hypothesis is rejected. For each problem and each parameter setting, 30 runs were performed.

5 Results and Discussion

This section analyses the performance of two new tournament selection methods and compares them with the standard tournament selection (Standard-TS). There metrics used for the comparison are: training error, testing error and solution size.

The first metric is the mean best fitness on the training data and this is presented in Table 3. This table shows that two new selection methods did not help to improve the performance of GP on the training data. By contrast, the training error of standard tournament selection is often better than that of statistical test based tournament selections. This result is not very surprising since the statistical based tournament selection techniques impose less pressure on the improving training error compared to standard tournament selection. Comparing between

Table 3. The mean best fitness on training data. If the result of Statistics-TS1 and Statistics-TS2 is significantly worse ($p - value < 0.05$) than the result of standard-TS, than its value is printed bold and italic faced.

Problems	Standard-TS	Statistics-TS1		Statistics-TS2	
		alpha = 0.05	alpha = 0.1	alpha = 0.05	alpha = 0.1
A. **Benchmarking Problems**					
F1	1.44	***2.33***	***2.24***	***3.59***	***2.84***
F2	0.24	***0.35***	***0.33***	***0.58***	***0.48***
F3	4.71	6.09	5.50	***6.74***	***6.83***
F4	0.01	0.01	0.01	***0.03***	***0.02***
F5	0.04	0.04	0.04	***0.06***	0.05
F6	0.12	0.12	0.12	0.12	0.12
F7	0.10	0.10	0.09	0.09	0.10
F8	0.37	0.49	***0.62***	***1.08***	***1.05***
F9	1.32	***1.53***	***1.53***	***1.81***	***1.60***
F10	0.42	0.45	0.41	***0.53***	***0.48***
B. **UCI Problems**					
F11	8.17	9.54	8.73	9.00	8.77
F12	3.48	***3.90***	***3.92***	***4.19***	***4.00***
F13	3.35	***4.79***	***4.62***	***7.20***	***6.29***
F14	8.05	***10.02***	***9.82***	***12.22***	***11.90***
F15	4.31	***5.95***	***5.53***	***7.28***	***6.90***
F16	0.49	0.50	***0.50***	***0.52***	***0.50***
F17	0.61	***0.62***	***0.63***	***0.64***	***0.63***
F18	25.04	***30.18***	***28.95***	***32.02***	***31.78***

Statistics-TS1 and Statistics-TS2, the table shows that Statistics-TS2 is often slightly worse than Statistics-TS1 on the training data.

We also conducted a statistical test to compare the training error of standard tournament selection with two new selection methods using a Wilcoxon signed rank test with the confident level of 95 %. If the test shows that the training error of statistical based tournament selection techniques is significantly worse than that value of standard tournament selection, this value is printed bold and italic faced in Table 3. It can be seen that, on most problem, the training error of statistical based selection is significantly worse compared to standard-TS.

The second metric used to compare the performance of the tested tournament techniques is their ability to generalize beyond the training data. In each run, the best solution was selected and evaluated on an unseen data set (the testing set). The testing error of the best individual was then recorded and the median of these values across 30 runs was calculated and presented in Table 4. This table shows that the testing error of two new tournament selection methods is often better than the value of standard tournament selection. This is very encouraging since the result in Table 3 shows that the training error of statistical based selection is often worse compared to standard tournament selection. The result on the testing error demonstrates that, using statistical test to only select the winner individual for the mating pool when the individual is statistically better than others help to improve the generalization of GP.

Table 4. The Median of test error. If the result of Statistics-TS1 and Statistics-TS2 is significantly better than the result of standard-TS, than their value is printed bold faced. Conversely, if their result is significantly worse than standard-TS, this value is printed bold and italic faced.

Problems	Standard-TS	Statistics-TS1		Statistics-TS2	
		alpha = 0.05	alpha = 0.1	alpha = 0.05	alpha = 0.1
A. **Benchmarking Problems**					
F1	10.75	**6.80**	**6.58**	**5.28**	**4.09**
F2	1.02	**0.89**	**0.90**	**0.82**	**0.82**
F3	38.70	**13.98**	**15.13**	**15.27**	**13.85**
F4	0.93	0.39	0.74	0.81	0.79
F5	0.03	0.04	0.05	0.07	0.05
F6	0.13	0.13	0.13	0.13	0.13
F7	0.22	0.23	0.24	**0.20**	0.18
F8	1.63	1.37	1.90	1.97	2.02
F9	1.86	***2.19***	***2.07***	***2.53***	2.25
F10	1.75	1.76	1.74	1.67	**1.44**
B. **UCI Problems**					
F11	24.85	20.99	27.06	28.23	31.57
F12	4.86	4.79	4.91	**4.65**	**4.64**
F13	7.49	6.86	6.80	8.40	8.03
F14	18.23	15.83	16.01	**13.11**	**14.51**
F15	8.97	8.31	**8.10**	8.57	**8.01**
F16	0.55	0.54	0.56	0.55	0.55
F17	0.66	0.65	0.66	0.66	0.65
F18	40.69	38.19	39.91	**37.03**	**37.06**

The statistical test on the testing error using a Wilcoxon signed rank test with the confident level of 95 % shows that two new tournament selection techniques are more frequently better than standard-TS on the testing error. Precisely, Statistics-TS1 is significantly better than standard-TS on three and four problems with $alpha = 0.05$ and $alpha = 0.1$ respectively while standard-TS is significantly better than Statistics-TS1 on one problem, $F9$. The testing error of Statistics-TS2 is significantly better than standard-TS on seven and eight problems with $alpha = 0.05$ and $alpha = 0.1$ respectively while standard-TS is significantly better than Statistics-TS2 on only one problem ($F9$) with $alpha = 0.05$. Comparing between Statistics-TS1 and Statistics-TS1, the statistical test shows that Statistics-TS2 is often slightly better than Statistics-TS1 on the unseen data.

Table 5. The average of solutions size of three selection methods. If the solutions found by Statistics-TS1 and Statistics-TS2 are more complex than those found by standard-TS, their value is printed bold and italic faced.

Problems	Standard-TS	Statistics-TS1		Statistics-TS2	
		alpha = 0.05	alpha = 0.1	alpha = 0.05	alpha = 0.1
A. **Benchmarking Problems**					
F1	295.1	263.2	258.7	97.4	118.0
F2	172.9	161.0	145.5	33.9	36.9
F3	260.8	***278.3***	***283.4***	90.0	89.2
F4	175.2	***176.1***	163.4	40.6	48.1
F5	207.8	***213.4***	***235.1***	55.6	61.3
F6	100.8	97.8	84.2	36.4	44.1
F7	120.6	***135.5***	***141.1***	57.0	57.2
F8	176.2	135.1	134.9	55.9	37.8
F9	143.1	***154.8***	***152.6***	50.4	71.2
F10	156.9	***157.2***	***166.4***	47.4	32.6
B. **UCI Problems**					
F11	264.7	***313.5***	***296.9***	185.6	211.6
F12	218.4	164.1	171.6	28.8	45.8
F13	216.4	143.0	152.9	22.8	31.5
F14	185.3	135.1	146.7	20.1	25.1
F15	212.1	164.7	152.1	24.2	30.4
F16	121.4	120.0	121.2	47.1	55.8
F17	147.8	125.6	145.7	40.1	46.1
F18	326.9	97.0	173.5	6.4	11.3

The last metric used to analyze the efficiency of statistics based tournament selection techniques is the size of their solutions. We recored the size of the best fitness individual in each runs. These values are then averaged over 30 runs and are presented in Table 5 In Table 5, when the solutions found of Statistics-TS1 and Statistics-TS2 are more complex than those obtained by standard-TS, their result is printed in bold and italic faced. It can be observed from this that the two new tournament selection techniques often help to find the solution of smaller size.

Apparently, the size of the solutions found by Statistics-TS1 is smaller than that of Standard-TS on most problem. Sometimes, Statistics-TS1 finds solutions that are more complex than the standard tournament selection and this happens on seven out of eighteen problems with $alpha = 0.05$ and on six out of eighteen problems with $alpha = 0.1$. For Statistics-TS2, the size of its solutions is much smaller than that of Standard-TS. It can be seen that the size of the solution obtained by Statistics-TS2 is often equal to one third of the solutions of Standard-TS on most problem. Overall, the results in this section show that statistical based tournament selection methods help GP to find simpler solution and generalize better on unseen data. This result is promising since finding simple solutions that achieve good performance on unseen data is the main objective of GP systems.

6 Conclusions and Future Work

In this paper, we introduced the idea of using a statistical test as part of selection step that utilizes information from fitness case error vectors of GP individuals. We proposed two variations of tournament selection that used statistical tests to select the winner for the mating pool. The effectiveness of the approach was examined on eighteen symbolic regression problems. In the experimental results we observe that the proposed techniques helped GP to reduce code growth and generalisation error.

There are a number of research areas for future work, which arise from this paper. First, we would like to study the approach to improve the performance of the statistical based tournament selection techniques on the training data. This may help the new techniques to perform better on a wider range of problems. One possible approach that can improve the performance of Statistics-TS1 and Statistics-TS2 is to combine them with local search techniques such as Soft Brood Selection [1]. Another approach is to implement these techniques with recent semantic based crossovers [11,13]. Second, at the theoretical level, it is still unclear while Statistics-TS1 and Statistics-TS2 perform well on unseen data though their performance on the training data is not as good as standard tournament selection. One possible reason is that they help to reduce code growth resulting in more parsimonious solutions. It is interesting to compare and analyze these tournament techniques with code bloat control methods like multi-objective GP [8] and operator equalisation [14]. Finally, a potential limitation of the proposed approach is the overuse of statistical tests without consideration for the increased probability of a significant difference being detected by chance [5]. Future research will include an exploration of the impact of different statistical tests and assessment of their suitability.

Acknowledgment. This research is funded by Vietnam National Foundation for Science and Technology Development (NAFOSTED) under grant number 102.01-2014.09. MON acknowledges the support of Science Foundation Ireland grant 13/IA/1850.

References

1. Altenberg, L.: The evolution of evolvability in genetic programming. In: Advances in Genetic Programming, pp. 47–74. MIT Press (1994)

2. Bache, K., Lichman, M.: UCI machine learning repository (2013). http://archive.ics.uci.edu/ml
3. Bäck, T.: Selective pressure in evolutionary algorithms: a characterization of selection mechanisms. In: Proceedings of the First IEEE Conference on Evolutionary Computation, pp. 57–62. IEEE Press, Piscataway (1994)
4. Blickle, T., Thiele, L.: A comparison of selection schemes used in evolutionary algorithms. Evol. Comput. **4**(4), 361–394 (1996)
5. Cumming, G.: Understanding The New Statistics: Effect Sizes, Confidence Intervals, and Meta-Analysis. Routledge, New York (2012)
6. Fang, Y., Li, J.: A review of tournament selection in genetic programming. In: Cai, Z., Hu, C., Kang, Z., Liu, Y. (eds.) ISICA 2010. LNCS, vol. 6382, pp. 181–192. Springer, Heidelberg (2010)
7. Gathercole, C.: An investigation of supervised learning in genetic programming. Ph.D. thesis. University of Edinburgh (1998)
8. Jong, E.D.D., Pollack, J.B.: Multi-objective methods for tree size control. Genet. Program. Evolvable Mach. **4**(3), 211–233 (2003)
9. Kim, J.J., Zhang, B.T.: Effects of selection schemes in genetic programming for time series prediction. Proc. Congr. Evol. Comput. **1**, 252–258 (1999)
10. Nguyen, Q.U., Nguyen, X.H., O'Neill, M., McKay, R.I., Galvan-Lopez, E.: Semantically-based crossover in genetic programming: application to real-valued symbolic regression. Genet. Program. Evolvable Mach. **12**(2), 91–119 (2011)
11. Nguyen, Q.U., Pham, T.A., Nguyen, X.H., McDermott, J.: Subtree semantic geometric crossover for genetic programming. Genet. Program. Evolvable Mach. **17**(1), 25–53 (2016)
12. Pawlak, T.P., Wieloch, B., Krawiec, K.: Review and comparative analysis of geometric semantic crossovers. Genet. Program. Evolvable Mach. **16**(3), 351–386 (2015)
13. Pawlak, T.P., Wieloch, B., Krawiec, K.: Semantic backpropagation for designing search operators in genetic programming. IEEE Trans. Evol. Comput. **19**(3), 326–340 (2015)
14. Silva, S., Dignum, S., Vanneschi, L.: Operator equalisation for bloat free genetic programming and a survey of bloat control methods. Genet. Program. Evolvable Mach. **13**(2), 197–238 (2012)
15. Sokolov, A., Whitley, D.: Unbiased tournament selection. In: Proceedings of the 7th Annual Conference on Genetic and Evolutionary Computation, pp. 1131–1138. ACM, New York (2005)
16. White, D.R., McDermott, J., Castelli, M., Manzoni, L., Goldman, B.W., Kronberger, G., Jaskowski, W., O'Reilly, U.M., Luke, S.: Better GP benchmarks: community survey results and proposals. Genet. Program. Evolvable Mach. **14**(1), 3–29 (2013)
17. Xie, H., Zhang, M.: Parent selection pressure auto-tuning for tournament selection in genetic programming. IEEE Trans. Evol. Comput. **17**(1), 1–19 (2013)
18. Xie, H., Zhang, M., Andreae, P., Johnston, M.: Is the not-sampled issue in tournament selection critical? In: IEEE World Congress on Computational Intelligence, pp. 3710–3717, June 2008
19. Xie, H., Zhang, M., Andreae, P.: Automatic selection pressure control in genetic programming. In: Yang, B., Chen, Y. (eds.) 6th International Conference on Intelligent System Design and Applications, pp. 435–440. IEEE (2006)
20. Xie, H., Zhang, M., Andreae, P., Johnson, M.: An analysis of multi-sampled issue and no-replacement tournament selection. In: Proceedings of the 10th Annual Conference on Genetic and Evolutionary Computation, GECCO 2008, pp. 1323–1330. ACM, New York (2008)

Kin Selection with Twin Genetic Programming

William B. Langdon(✉)

CREST, Department of Computer Science, University College London,
Gower Street, London WC1E 6BT, UK
w.langdon@cs.ucl.ac.uk
http://crest.cs.ucl.ac.uk

Abstract. In steady state Twin GP both children created by sub-tree crossover and point mutation are used. They are born together and die together. Evolution is little changed. Indeed fitness selection using the twin's co-conceived doppelganger is possible.

Keywords: Theory · Genetic programming · Artificial intelligence · Emergent behaviour · TinyGP · Steady state evolution

1 Introduction

In genetic programming it is very common to represent programs as trees (like Lisp s-expressions) and to use two point subtree crossover (Koza 1992) to create new programs. Although subtree crossover can be symmetric and can be used to create two new programs, it is common to it to use it to create a single child (Poli *et al.* 2008, p. 15). The offspring inherits its root node from one parent and a subtree from its other parent. Typically the second parent passes less genetic material to its offspring than the first. Also, being lower in the child's tree, typically the second parent's genes have less impact. Here we create and use both children of each two point crossover. The combined genetic contents of the two children is the same as the combined contents of their two parents. Although often subtree crossover creates a smaller and a larger child, it never changes the average size of programs. However, subtree crossover will typically mean the children are different from each other and from their parents. In many cases not only are they genetically different (their trees are different) but also the children are different sizes from each other and different from both parents.

For simplicity we restrict the GP function set to binary functions so that the trees have internal nodes with two outward facing edges. However theoretical results have also been provided for mix-arity as well as fixed arity trees (Dignum and Poli 2010). Without selection and using only crossover, GP populations of any initial distribution of sizes and shapes rapidly converge to a limiting distribution (Dignum and Poli 2010). The final size distribution depends on the initial total number of functions of each arity and the number of terminals (tree

W.B. Langdon—http://www.cs.ucl.ac.uk/staff/W.Langdon/.

J. Handl et al. (Eds.): PPSN XIV 2016, LNCS 9921, pp. 313–323, 2016.
DOI: 10.1007/978-3-319-45823-6_29

leafs). For typical GP populations, the limit distribution contains a few very small trees, the number of trees rises to a peak near the average tree size and then there is a long tail of rapidly decreasing frequency of bigger trees (see, e.g. Dignum and Poli 2010, Fig. 6).

The tendency for GP populations to bloat, i.e. for evolution to produce progressively larger programs with little increase in their ability, is well known (Poli *et al.* 2008, Sect. 11.3). It has been suggested (e.g. Dignum and Poli 2010) that bloat in tree genetic programming is due to subtree crossover producing small children with below average fitness. Such children are removed from the population by fitness selection. Since crossover does not change the average size, the remaining programs will tend to be bigger than average. Thus the remaining population after selection will on average be bigger than their parents. In the next round of crossovers, subtree crossover will again produce some small programs. It will keep doing this no matter how big the parent trees are. The limiting distribution still contains a sizeable fraction of small programs no matter how big its mean. This bloat theory says they will always have a tendency to be too small to be useful and so fitness will always cause selection to preferentially remove them, so the average size of trees will always increase.

Twin genetic programming was conceived with the idea (which failed) of foiling bloat. The idea being to force simultaneous fitness based removal of both the smaller and the larger child. Thus in twin GP, **two children are created** with the same average size as their parents. Although, whilst in the population, they can be independently selected to be parents, **when either child is deleted, she takes her twin with her**. (If the two parents were not selected independently but locked together, like the offspring, then the population would degenerate into independent lines and interbreeding would be impossible.)

The next section presents twin GP, Sect. 3 describes our experiments, whose results are given in Sect. 4. The failure to contain bloat and the success of kin selection are analysed in Sect. 5, including a mathematical model of the disruption caused by kin selection (in Sect. 5.3), before we summarise in Sect. 6.

2 Twin Genetic Programming

2.1 TinyGP

Our implementation of twin genetic programming is based on Riccardo Poli's C implementation of TinyGP (Poli *et al.* 2008) for Boolean problems. TinyGP provides a steady state (Syswerda 1989) fixed sized population evolutionary framework.

2.2 Two Offspring Sub-tree Crossover

TinyGP's subtree crossover essentially provides Koza's subtree crossover (Koza 1992) but without a bias in favour of choosing functions as crossover points. That is, the two crossover points (one per parent tree) are chosen independently at random from all internal and external (leafs) nodes in the tree.

In TinyGP once the initial population has been created, there are no size or depth restrictions on the evolving trees and after a few generations bloat is usually rampant.

The only change for twin GP is to use the two crossover points twice so creating two offspring (the twins).

2.3 Point Mutation

Children not created by crossover are created by applying point mutation (Poli *et al.* 2008, pp. 16–17) at 5 % per node to a copy of their parent. Thus larger trees are proportionately more likely to be changed at multiple places. Notice this type of tree mutation does not change the size or the shape of the program.

For twin genetic programming, two child programs are created by two independent mutations and then locked together as twins. Since the parents are chosen independently the twins (like those created by crossover) are typically of different sizes and shapes.

2.4 Fitness 6 Multiplexor

As a demonstration we use TinyGP's Boolean six multiplexor problem. The goal is to evolve a program which takes six Boolean inputs and outputs a Boolean corresponding to 6-Mux (Koza 1992, Sect. 7.4). I.e. two inputs correspond to two address lines (giving 4 combinations) which select one of the remaining four inputs and connect it to the output. There are $2^6 = 64$ possible tests. We use them all. An individual program's fitness is the number of test cases it for which it gives the correct answer. I.e. fitness is an integer between 0 and 64.

2.5 Twin Selection in Steady State Populations

In twin GP both selection to be a parent and deciding which two programs are to be removed from the population is on the basis of the fitness of the two twins. We looked at five ways of combining the twins' fitnesses: twin) the default, just use the individual program's fitness. MEAN) use the mean of the fitness of the program and its twin. MAX) use the best fitness of the two programs. MIN) use the worse fitness. KIN) use the fitness of the twin, i.e. kin selection. Finally, as a sanity check, we ignore fitness of the twins entirely and select randomly. As expected under random selection, evolution does not solve the problem at all and the populations do not bloat.

3 Experiments

We tried each of the five settings described in Sect. 2.5 and the original TinyGP. (Results for TinyGP "no twin" are at the top of Table 2). The parameters of twin GP are summarised in Table 1. Initial results for kin selection were disappointing and no solutions were found. Hence kin was re-run with a population ten times as big. We run each experiment 30 times. The results are summarised in Table 2. (For completeness, all GP runs were also made with the larger population size.)

Table 1. Twin genetic programming parameters for solving 6 multiplexor

Terminals:	6 Boolean inputs D0–D5
Functions:	AND, OR, NAND, NOR
Fitness:	All 64 fitness cases
Selection:	Binary tournaments used for both parent and replacement selection
Population:	1000 (or 10 000)
Initial pop:	Grow, max depth 6
Parameters:	80 % subtree crossover. Both crossover points chosen at random, i.e. no function bias
	20 % point mutation. 5 % chance of substitution with primitive of the same arity per primitive. Notice mutants are subjected to zero or more flips and larger programs have proportionately more changes
	No depth or size limits
Termination:	Problem is solved, or 100 generation equivalents

4 Results

Table 2 shows twin GP working surprisingly well.[1] Evolution of successful programs is even possible if we use the fitness of the worst of the twins. Although kin selection is obviously doing less well, if we increase the population size, evolution can proceed even if we totally ignore the fitness of the individual and always use instead the fitness of her twin. As the last pair of columns in Table 2 makes clear, twin GP has totally failed to address bloat.

Table 2. Twin genetic programming six multiplexor (30 runs each)

Experiment	Successful runs		Best fitness		Mean size generation 100	
pop size	1000	10000	1000	10000	1000	10000
no twin	21	30	64	64	1518.2	-
twin	13	29	64	64	1357.7	1654.5
MEAN	17	30	64	64	1167.8	-
MAX	19	30	64	64	1136.4	-
MIN	6	28	64	64	960.6	1103.9
KIN	0	11	61	64	259.4	327.0
RAND		0		50		13.6

[1] According to the binomial distribution the first three variants of twin GP, i.e. twin, mean and max, are not significantly worse than TinyGP without twins.

5 Discussion: Why Does Twin GP Bloat

5.1 Two Can Bloat Too

Figure 1 plots the change in total program size each time a new pair of programs is created (and so a twin is removed from the population). Figure 1 suggests a near symmetric distribution but notice the rapidly increase in variation as the population evolves to contain bigger trees. What Fig. 1 conceals is that the distribution is not exactly symmetric. Figure 2 plots the average change. In almost every generation, on average, smaller trees are replaced with bigger ones. I.e. binary fitness tournaments have a bias towards selecting larger trees to be parents than they select to be killed. This leads to bloat. (Binary tournaments have the lowest selection pressure or intensity of any simple tournament selection scheme (Blickle and Thiele 1996, Fig. 4).) That is, despite twin GP's careful control of the genetic operations of crossover and mutation, to ensure they do not change the total size, fitness selection is still bloating the population (Langdon and Poli 1997).

5.2 How Different Are Twins?

A possible explanation for all twin selection runs evolving fitter trees (including sometimes finding solutions), might have been that the twins are identical or at least very similar. However as Fig. 3 shows, there is more to it than that. Firstly we consider what do we mean by two trees are similar. Figure 3 considers four similarity metrics. Firstly we look at the trees themselves and then we look at two metrics based on their outputs.

We can consider if the trees are identical. (This is Koza 1992's population variety.) And secondly if they are the same size. As expected twin GP populations, like usual GP populations (Koza 1992), do not converge in terms of their genotypes. Even in the early generations there are almost no tournaments between identical trees and, in a typical run, none at all after generation nine. If we look at a much loose definition of tree similarity: are the trees the same size, we see a similar picture. (Of course identical trees must also have the same sizes, but not vice versa.) In the early generation about of 5 % of tournaments are between trees of the same size but this falls to less than 1 % after generation 13.

We also looked at similarity of behaviour. As expected (Langdon *et al.* 1999), the populations converge to some extent. Again Fig. 3 looks at two types of (phenotypic) convergence. Do two programs return identical answers on all the test cases and secondly do they have the same fitness. (Since fitness is define by the test cases, two trees which give the same answers on all 64 test cases must have the same fitness, but not vice versa.)

Phenotypes (i.e. behaviour) in twin GP populations do show some convergence (e.g. Fig. 3) and also twins' phenotypes (and so fitness) are slightly more similar than those of the population as a whole. For example at the end of the run in Fig. 3 30.6 % (+) of twins gave exactly the same answer on all 64 tests whilst the figure for random pairs was 29.4 % (×).

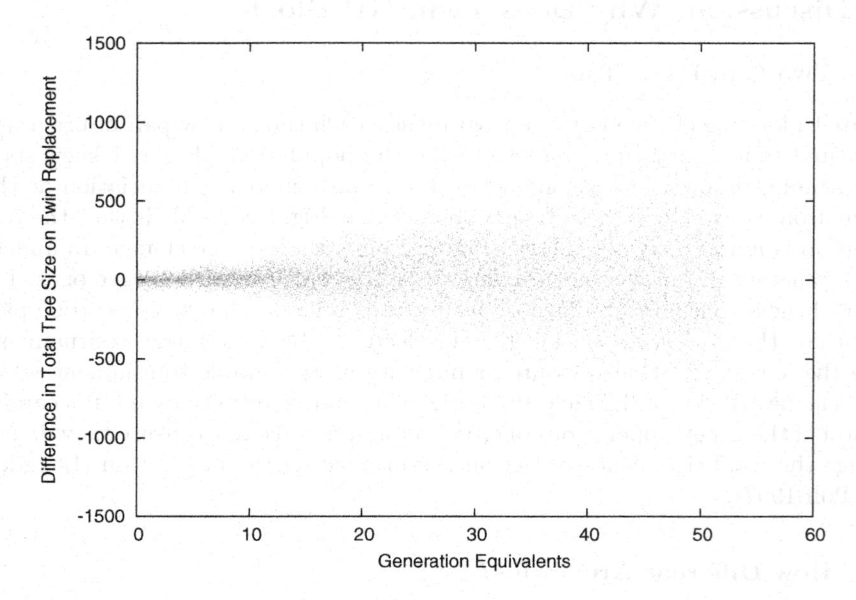

Fig. 1. Change in combined size (2×2 trees) per tournament in typical twin run (i.e. selection uses twin's own fitness). 6-Mux. Population 1000.

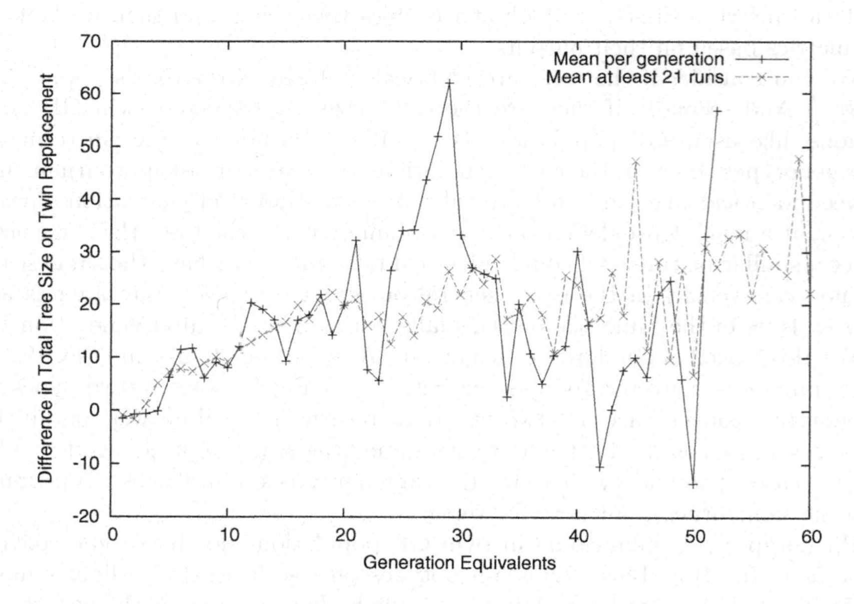

Fig. 2. Mean of size changes per generation. (+) same run as Figure 1

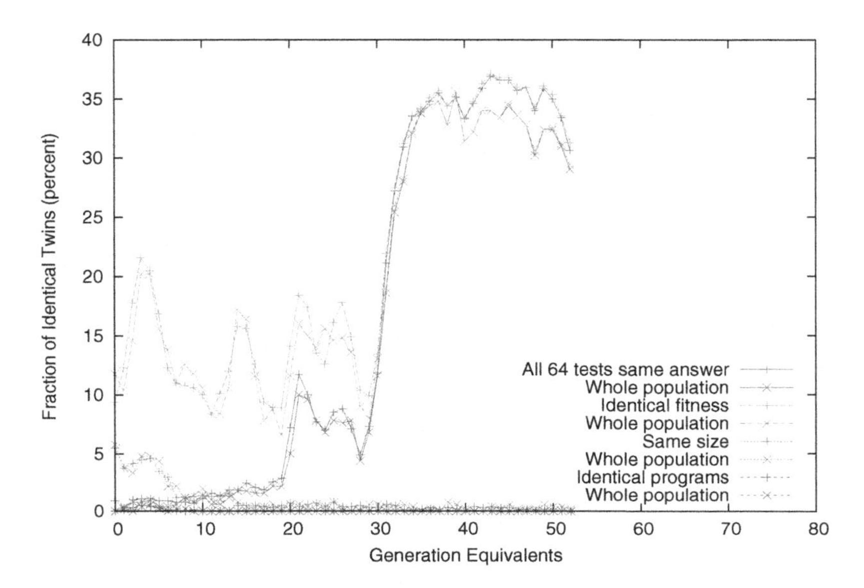

Fig. 3. Evolution of 4 measures of twin similarity in typical twin selection run (same run as Figure 1). + similarity of twins. × similarity of population. Twins (+) are slightly more similar than the population as a whole (×).

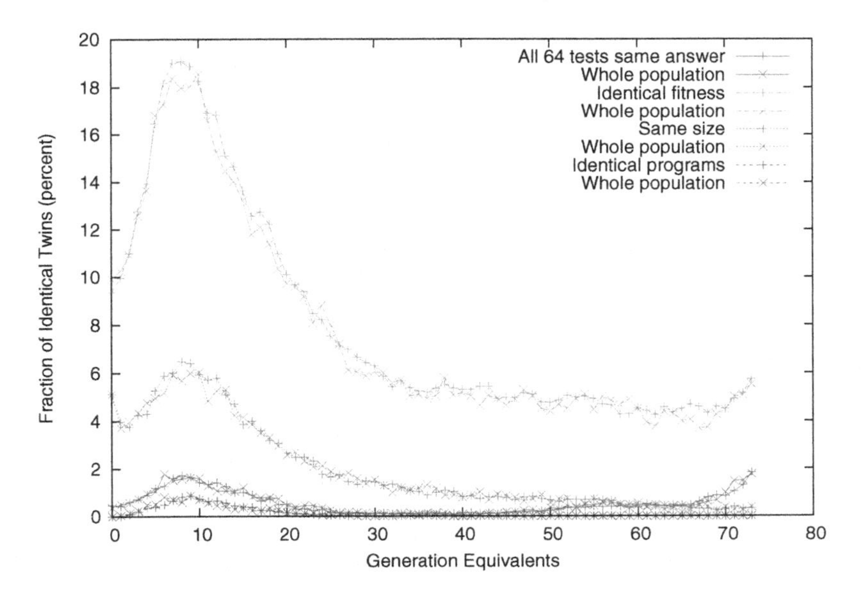

Fig. 4. Evolution of 4 measures of twin similarity in typical kin selection run with population of 10 000. Notice although the population convergence is smaller, otherwise population and twin similarity are much like selecting by the individuals fitness, Figure 3. Again twins are slightly more similar than the population.

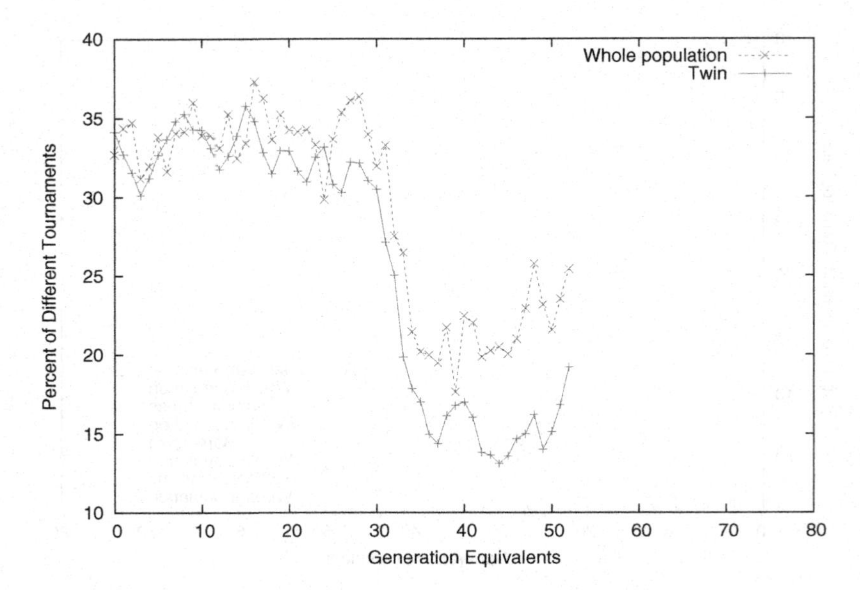

Fig. 5. Evolution of expected fraction of binary tournaments with different outcome in typical twin run with population of 1000 (+). (Same run as Figure 1) Expected fraction for whole population plotted × for comparison. Ties are broken at random.

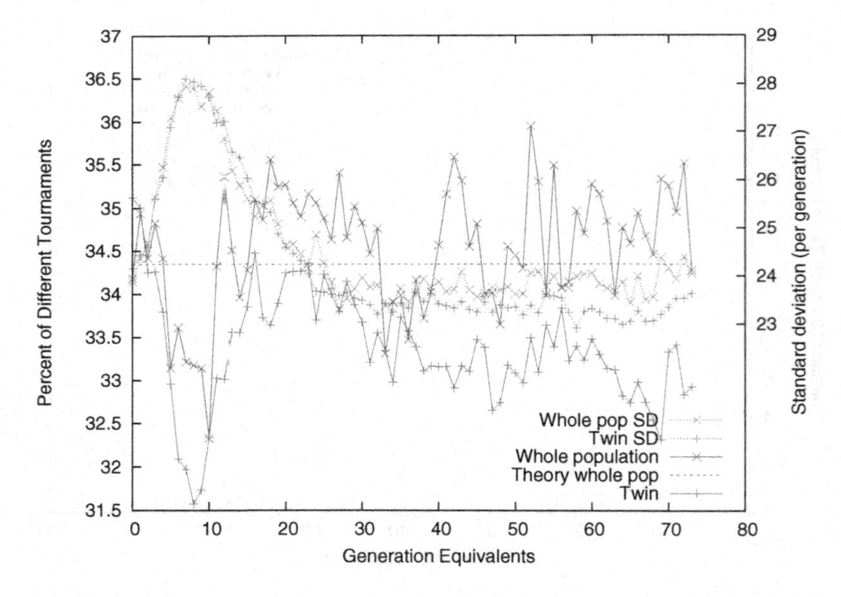

Fig. 6. Evolution of expected fraction of binary tournaments with different outcome in typical kin selection run with population of 10 000 (+). (Same run as Figure 4) Expected fraction for whole population plotted × for comparison. Twin's fitness gives same outcome slightly more often than using the population.

5.3 Expected Impact of Kin Selection in Boolean Problems

We next show in the limit in random populations kin selection impacts half fitness selection tournaments and calculate the ratio for $n = 6$ (34.3 % and show it agrees with experiment (see Fig. 6).

In GP it is common (as we do here, see Table 1) to use an unbiased set of primitives. Thus before fitness selection, in Boolean problems, like the 6 Multiplexor, the chance of getting any individual test right is 50 %. Therefore for an n bit problem, in large trees the initial random distribution of fitness follows the binomial distribution $2^{-N}C_i{}^N$ (where $N = 2^n$).[2] Notice $2^{-N}C_i{}^N$ is symmetric about $N/2$. In the limit of large n the binomial distribution can be approximated by a Gaussian distribution (The same holds when the functions are reversible (Langdon 2003).) The distribution's mean is 2^{n-1} and its variance is 2^{n-2} (standard deviation $2^{n/2-1}$). (E.g. for the 6-mux the mean is 32 and the standard deviation is 4.) In a random 6-mux population there is a reasonable chance of drawing two programs with the same fitness. In higher order problems (i.e. letting n increase) the width of the distribution grows. When it is large compared to 1.0 there is essentially no chance two random programs will have the same fitness. Thus for large n we need not consider the chance of tournaments having to consider a draw where individuals have the same fitness.

Consider a program with fitness i having a twin with fitness j in a binary tournament with a program of fitness k. It will win if $i > k$ and lose if $i < k$. Draws $i = k$ are resolved randomly. We can calculate the likelihood that substituting it with its twin's fitness will not change the outcome of the tournament. The twin still wins if $i > k \;\&\; j > k$ and still loses if $i < k \;\&\; j < k$ and half draws will yield the same answer as before, i.e. $\frac{1}{2}(i = k \mid j = k)$. Assuming fitness are randomly distributed, we can calculate the probability of the same outcome as:

$$\sum_{i=0}^{N} 2^{-N}C_i{}^N \sum_{j=0}^{N} 2^{-N}C_i{}^N \sum_{j=0}^{N} 2^{-N}C_i{}^N \begin{pmatrix} \delta(i > k \;\&\; j > k) + \\ \delta(i < k \;\&\; j < k) + \\ 1/2\delta(i = k \mid j = k) \end{pmatrix} \tag{1}$$

where $\delta(x)$ is 1 if x is true and 0 otherwise. If N is large we can ignore the draws (i.e. neglect the space occupied by $(i = k \mid j = k)$). $(i > k \;\&\; j > k)$ and $(i < k \;\&\; j < k)$ both partition the $(0..N)^3$ cube and allocate a quarter of it each. I.e. a half in total. Since the density function $2^{-3N}C_i{}^N C_j{}^N C_k{}^N$ is symmetric about the centre of the $(0..N)^3$ cube the total sum of probabilities will be a half. Thus in random populations of large Boolean problems kin selection which uses the twin's fitness will disrupt half of binary tournaments.

In the case of 6-mux, $N = 64$ and we evaluate Eq. 1 numerically using the actual random tree's fitness distribution as 65.7 %. (See horizontal line in Fig. 6.) Notice the close agreement with estimates drawn from a real run in Fig. 6 for the first generation. Some variation as the population moves away from its initial random distribution might be expected but the similarity in later generations

[2] For the small trees the distribution is only approximately binomial (Langdon 2009).

suggests although the population's average fitness has changed its variation (standard deviation) remains similar (see dotted lines in Fig. 6).

6 Conclusions

In twin genetic programming the combined size of the two children is always identical to that of their parents. Thus no genetic operation changes the average size of programs in the evolving population. Nonetheless a very small but sustained bias in fitness selection to kill smaller trees leads to cumulative increase in program size commonly known as bloat.

Twin genetic programming can be effective even when using elements of the fitness of the other twin, see Table 2. Evolution is adversely effected when either ignoring the best fitness of the twins or exclusively using the fitness of the other twin (kin selection). However we have demonstrated kin selection can evolve solutions. Surprisingly it is able to do this although using the twin's fitness disrupts almost as many selection tournaments as choosing at random from the population (Fig. 6) and yet evolution makes no progress at all with totally random selection (last row in Table 2). Section 5.3 presents a generic theoretical analysis of the impact of kin selection using binary tournaments for large programs in high order (large n) Boolean problems and applies numerical values for the special case of small trees and $n = 6$.

Implementation

C code for twin GP is available via anonymous FTP and via http://www.cs.ucl.ac.uk/staff/W.Langdon/ftp/gp-code/tiny_gp_twin.c

Acknowledgements. I am grateful for discussions with T.H. Westerdale.

References

Blickle, T., Thiele, L.: A comparison of selection schemes used in evolutionary algorithms. Evolut. Comput. **4**(4), 361–394 (1996). http://www.cs.bham.ac.uk/~wbl/biblio/gp-html/DBLP_journals_ec_BlickleT96.html

Dignum, S., Poli, R.: Sub-tree swapping crossover and arity histogram distributions. In: Esparcia-Alcázar, A.I., Ekárt, A., Silva, S., Dignum, S., Uyar, A.Ş. (eds.) EuroGP 2010. LNCS, vol. 6021, pp. 38–49. Springer, Heidelberg (2010). http://www.cs.bham.ac.uk/~wbl/biblio/gp-html/Dignum_2010_EuroGP.html

Koza, J.R.: Genetic Programming: On the Programming of Computers by Natural Selection. MIT press, Cambridge (1992). http://www.cs.bham.ac.uk/~wbl/biblio/gp-html/koza_book.html

Langdon, W.B., Poli, R.: Fitness causes bloat. In: Chawdhry, P.K., et al. (eds.) Soft Computing in Engineering Design and Manufacturing, pp. 13–22. Springer, London (1997). http://www.cs.bham.ac.uk/~wbl/biblio/gp-html/Langdon_1997_bloatWSC2.html

Langdon, W.B., Soule, T., Poli, R., Foster, J.A.: The evolution of size and shape. In: Spector, L., et al. (eds.) Advances in Genetic Programming 3, pp. 163–190. MIT Press, Cambridge (1999). Chapter 8. http://www.cs.bham.ac.uk/~wbl/biblio/gp-html/langdon_1999_aigp3.html

Langdon, W.B.: The distribution of reversible functions is Normal. In: Riolo, R.L., Worzel, B. (eds.) Genetic Programming Theory and Practice, pp. 173–187. Kluwer, Boston (2003). Chapter 11. http://www.cs.bham.ac.uk/~wbl/biblio/gp-html/langdon_2003_normal.html

Langdon, W.B.: Scaling of program functionality. Genet. Program. Evolvable Mach. **10**(1), 5–36 (2009). http://www.cs.bham.ac.uk/~wbl/biblio/gp-html/langdon_2007_gpem.html

Poli, R., Langdon, W.B., McPhee, N.F.: A Field Guide to Genetic Programming (2008). (With contributions by Koza, J.R.) http://lulu.com and freely http://www.gp-field-guide.org.uk, http://www.cs.bham.ac.uk/~wbl/biblio/gp-html/poli08_fieldguide.html

Syswerda, G.: Uniform crossover in genetic algorithms. In: David Schaffer, J. (ed.) Proceedings of the Third International Conference on Genetic Algorithms, pp. 2–9. George Mason University, Morgan Kaufmann, 4–7 June 1989

Using Scaffolding with Partial Call-Trees to Improve Search

Brad Alexander(✉), Connie Pyromallis, George Lorenzetti, and Brad Zacher

School of Computer Science, University of Adelaide, Adelaide 5005, Australia
bradley.alexander@adelaide.edu.au
http://www.cs.adelaide.edu.au/~brad

Abstract. Recursive functions are an attractive target for genetic programming because they can express complex computation compactly. However, the need to simultaneously discover correct recursive and base cases in these functions is a major obstacle in the evolutionary search process. To overcome these obstacles two recent remedies have been proposed. The first is Scaffolding which permits the recursive case of a function to be evaluated independently of the base case. The second is Call-Tree-Guided Genetic Programming (CTGGP) which uses a partial call tree, supplied by the user, to separately evolve the parameter expressions for recursive calls. Used in isolation, both of these approaches have been shown to offer significant advantages in terms of search performance. In this work we investigate the impact of different combinations of these approaches. We find that, on our benchmarks, CTGGP significantly outperforms Scaffolding and that a combination CTGGP and Scaffolding appears to produce further improvements in worst-case performance.

Keywords: Recursion · Genetic programming · Call-tree · Scaffolding · Grammatical evolution

1 Introduction

Recursive functions solve challenging problems by defining larger solutions in terms of sub-solutions. Recursive functions are often compact and expressive, which has made the evolution of recursive functions a popular target for genetic programming (GP) [5,6,9]. Unfortunately, the evolution of non-trivial recursive functions through GP has proven difficult in practice [1,6]. One cause of this difficulty is that fitness functions based solely on test cases are very sensitive to the correctness of the candidate code for the base case. Thus, when the base case's code is wrong, the fitness function will often give very low fitness to candidate solutions even when the major part of the code – the recursive case – is entirely correct [6]. This dependence results in the need to simultaneously evolve correct code for both the base and recursive cases [1,7] before a high fitness score is achieved.

Several approaches to improve search for recursive functions have been implemented. These have included: the use of niches to preserve diversity during search

J. Handl et al. (Eds.): PPSN XIV 2016, LNCS 9921, pp. 324–334, 2016.
DOI: 10.1007/978-3-319-45823-6_30

[7] and narrowing search-spaces using templates that express common patterns of recurrence [10,11]. Moraglio et al. [6] defined a simple and general approach, called *Scaffolding*, which employed existing test cases to always return a correct result for recursive calls. This allowed the fitness of the recursive case to be gauged independently of the base case. Later, Alexander and Zacher [2] defined Call-Tree-Guided Genetic Programming (CTGGP), which used information in user-defined partial call trees to separately deduce recursive calls. More recently, Chennupati et al. [4] implemented Multi-core Grammatical Evolution on parallel platforms to speed up the evolution of parallel recursive integer and list functions.

In this paper we benchmark both Scaffolding and CTGGP alone and in combination. The question of which configuration of these frameworks gives the most benefit is interesting, because both Scaffolding and CTGGP showed significant improvement over unassisted GP search on their respective recursive benchmarks. A-priori, Scaffolding and CTGGP have a complementary focus: Scaffolding allows the base case to be evolved without impacting the search for the recursive case and, in contrast, CTGGP separately evolves the parameter to recursive calls. To date, there has been no research that has compared the two approaches either in isolation or in combination. In this paper we compare the performance of Scaffolding and CTGGP on a range of recursive benchmarks. We show that CTGGP performs better than Scaffolding in isolation and the combination of the two approaches marginally improves average performance, but consistently improves worst-case performance. We also describe improvements to the CTGGP system and briefly examine their impact.

The remainder of this paper is structured as follows. In the next section we describe the conceptual frameworks for Scaffolding and CTGGP. In Sect. 3 we describe an implementation of our framework combining Scaffolding and CTGGP. In Sect. 4 we describe our experimental parameters and results and, finally, in Sect. 5 we present our conclusions and ideas for future work.

2 Conceptual Frameworks

In this section we describe the concepts of Scaffolding and CTGGP. Both concepts are enhancements to Genetic Programming (GP). Here we define a GP search framework as a function *GPSearch* which attempts to discover a target code fragment f. To define f's behaviour, *GPSearch* requires test-cases in the form of a list of inputs to f: $in = [i_1, \ldots, i_n]$ and a corresponding list of desired outputs from f: $out = [o_1, \ldots, o_n]$. Given these definitions, the *GPSearch* can be defined:

$$GPSearch(in, out) = f = \mathop{\mathbf{argmin}}_{f_x \in Genset} Error(f_x, in, out)$$

where *Genset* is the set of candidate functions that is generated by *GPSearch* and *Error* is an error function that returns a measure of how much the outputs of f_x deviate from the desired outputs in *out*. In GP, *Genset* is generated dynamically by an evolutionary process guided by the values of *Error*. In order to make search tractable, the elements of *Genset* are usually constrained to belong to a grammar relevant to the problem domain of interest.

In this paper, all of our target benchmarks f are recursive integer functions of one argument that return one value. We constrain the members of *Genset* to this form using Grammatical Evolution (GE) [8] – a form of GP which maps a binary string genotype to code via a BNF grammar provided by the user. In the following, we define *Error* as:

$$Error(f_x, in, out) = \Sigma_{k=1}^{n} |f_x(in_k) - out_k|$$

The aim of our search is to converge to target f such that $Error(f, in, out) = 0$. Both Scaffolding and CTGGP extend the framework just described. We describe these extensions in turn.

2.1 Scaffolding

The motivation for Scaffolding stems from the observation that any candidate recursive function f_x applied to test cases *in* and *out* will receive a very poor fitness evaluation unless the base case is correct. For example, in the `fib` function shown in Algorithm 1 a change of the base case guard on line 2 to `x` $<$ `1` will cause the program to fail on most foreseeable test cases. Likewise a change of the base case body on line 3 to `return 1` also causes most test cases to fail. This happens despite the remaining code being entirely correct.

Algorithm 1. Correct C implementation of the Fibonacci function

```
1 int correctRecurse(int x) {
2   if (x >= orig_x) {
3     return recurse(x);
4   }
5   for (int i = 0; i < n; i++) {
6     if (inp[i] == x) return out[i];
7   }
8   return recurse(x);
9 }
```

This sensitivity to errors in the base case has a significant impact on search. Moraglio proposed Scaffolding [6] to help address this problem. Under Scaffolding, every recursive call is replaced by a call to a non-recursive function that returns the correct value for that call. The correct value for each call is mined from the original test cases in *in* and *out*. Thus, for example, Scaffolding would replace the call fib(2), where $in = [0, 1, 2, 4, 6]$ and $out = [0, 1, 1, 3, 8]$, with the call correct-fib(2) which would then return $out_2 = 1$. If the input to the call for the recursive function doesn't correspond to a value in *in*, then recursion is allowed to progress as normal until a call with a value in *in* is encountered. After evolution has completed Scaffolding replaces all calls with the original function name.

The above scheme makes it possible to evaluate the correctness of the recursive case somewhat independently of the base case. The correctness of the base case still contributes to fitness to the extent that it is directly or indirectly exercised by *in*.

A final detail that must be addressed in Scaffolding is avoiding cases where the parameter of the recursive call matches the original parameter to the function. Without intervention, Scaffolding would find these cases to be correct while the actual program when run would make no progress. Scaffolding handles these cases by explicitly detecting them and avoiding replacement with the correct version of the call. This leads to infinite recursion and, thus, poor fitness.

2.2 CTGGP

CTGGP speeds up the GP search process by conducting a separate search for the parameters of the recursive call. Thus, if the target function is `fib` in Fig. 1, CTGGP would conduct a search for parameter expressions `x-1` and `x-2` separately from the rest of the code.

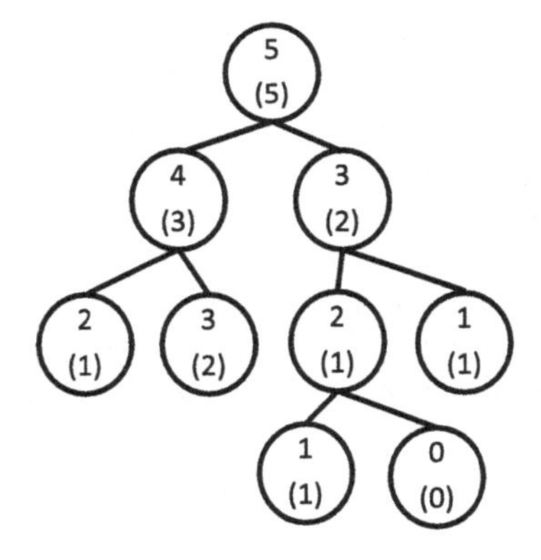

Fig. 1. An example partial call tree for a fib function

This search is guided by information embedded in a user-provided partial call tree. An example of a call for the `fib` function is shown in Fig. 1. The nodes of the call tree can contain up to two numbers. The top number is the parameter of the recursive call. The bottom number, in brackets, is the value returned by that call. Note that this return value is not required for every node in the tree. In addition, the user doesn't have to provide every call starting from the call at the root node. Moreover, the tree is allowed to be disjoint and overlapping calls from parent nodes can be shared, forming a directed-acyclic graph. The premise underpinning CTGGP is that in order to create the input and output values required to drive a GP process the user will often sketch an approximation of a call tree in order to obtain these input and output values. Thus, by using CTGGP we are merely exploiting information that the user already has in hand.

The partial call tree can be mined for information to guide CTGGP. The values in *in* are the top values in each node. The values in *out* are the bottom values in each node. In addition, CTGGP uses the structure of the tree to create a list of tree-segments: *segs* which contains a list of entries $[s_1, \ldots, s_m]$ where each s_i is a pair $(pnt_i, [c_{i1}, \ldots, c_{im_i}])$ which defines the relationship between the parent: pnt_i and its child nodes in the call tree. As an example of the correspondence between the call tree and the values above the values extracted from the call tree shown in Fig. 1 are shown in Fig. 2

$$\begin{aligned} in &= [5,4,3,2,3,2,1,1,0] \\ out &= [5,3,2,1,2,1,1,1,0] \\ segs &= [(5,[4,3]),(4,[2,3]),(3,[2,1]),(2,[]),(3,[]),(2,[1,0]),(1,[]),(0,[])] \end{aligned}$$

Fig. 2. Values extracted by CTGGP from the call tree in Fig. 1

The search process in CTGGP starts when the user specifies the call tree via a GUI interface. The interface application infers from the arity of tree fragments, with confirmation from the user, details of the grammars to be used. These details include the number of base cases and the inferred number of recursive calls [2]. From this step two grammars are produced. The first grammar, called *grammar1*, is used for searching for the parameters of the recursive calls. An example of this grammar for the call tree in Fig. 1, is shown in Fig. 3(a). A second grammar, called *grammar2*, is used for searching for the remainder of the function. The corresponding grammar2 is shown in Fig. 3(b). The candidate expressions generated from grammar1 are referenced in part (b) by the expressions **param1** and **param2**. The recursive calls in grammar2 are always denoted by the generic name "recurse".

Once the grammars have been defined, evolution in CTGGP proceeds in two concurrent phases of GE. Phase 1 produces individual expressions from grammar1 in order to produce parameter expressions for the recursive call(s). Phase 2 uses grammar2 to evolve the rest of the recursive function. During phase 2 evolution, the current best expressions from phase 1 are integrated into the candidate solutions. This simultaneous evolution is an improvement on our previous work [2] which ran phase 1 and phase 2 in sequence and thus required us to make an a-priori estimate of the amount of time phase 1 would require to find an acceptable solution.

In phase 2, individual candidate solutions f_x are evaluated by calling the $Error(f_x, in, out)$ function defined at the beginning of this section. In phase 1 the parameter expressions produced are evaluated by comparing the results of the candidate parameter expressions to the corresponding entries in *segs*. As an example of how this is done for the call tree in Fig. 1, consider a phase 1 search process for `fib` that produces the (correct) expressions: **param1** $= x - 1$ and

```
<expr_root> ::= <var> <op> <digit> |
  <digit> <op> <var>
<op> ::= - | * | + | /
<digit> ::= 0 | 1 | 2 | <big_digit>
<big_digit> ::= 3 | 4 | 5 | <bigger_digit>
<bigger_digit> ::= 6 | 7 | <huge_digit>
<huge_digit> ::= 8 | 9
<var> ::= x
```

(a)

```
<expr_root> ::=
 if(<var> < <digit>){
  return <lit>;
 }else{
  return <expr1> <op> <expr2>;
 }
<expr1> ::= <rec1> |
 (<rec1> <op> <lit>) |
 (<lit> <op> <rec1>)
<expr2> ::= <rec2> |
 (<rec2> <op> <lit>) |
 (<lit> <op> <rec2>)
<rec1> ::= recurse(param1)
<rec2> ::= recurse(param2)
<op> ::= - | * | +
<lit> ::= <digit> | <var>
... as per phase 1 grammar...
```

(b)

Fig. 3. Grammar1 (a) and grammar2 (b) generated by the call tree in Fig. 1.

param2 $= x - 2$. To test these we extract the first entry from *segs* from Fig. 2: $(5, [4, 3])$. This entry consists of the parent parameter 5 and the child parameters $[4, 3]$. We substitute the parent parameter 5 into each of **param1** and **param2**. These produce, respectively, the values 4 and 3. These are eliminated from the child parameter list leaving an empty list. The matching process then progresses other elements of *segs*. If all child parameter entries in *segs* are eliminated, then perfect fitness is given. If some items fail to match then fitness is penalised according to the distance between the output of the phase 1 expression and the closest match in the children of the relevant entry in *segs*.

The whole evolution progresses until either phase 2 evolution finishes with a perfect score or the maximum number of generations for phase 2 is reached.

3 Implementation and Experimental Setup

In this section we outline how Scaffolding is combined with CTGGP and how the experiments we use to measure their performance are set up.

To implement Scaffolding we need to embed code to access the correct answer to each recursive call in each candidate solution f_x. In our system this is done by implementing a function called `correctRecurse` that takes the place of the recursive calls: `recurse` in the phase 2 grammar. `correctRecurse` is implemented as part of the library code accessed by the candidate solution. The code for `correctRecurse` is shown in Algorithm 2. The variable `orig_x` is a global variable containing the value of the original call to `recurse`. The variables `in` and `out` are arrays representing *in* and *out* respectively. The variable `n` represents the length of both *in* and *out*. The if-statement on lines 2 to 4 checks to see if the parameter is the same as that of the original call. If so, it forces a call

Algorithm 2. Implementation of correctRecurse

```
1  int correctRecurse(int x) {
2    if (x >= orig_x) {
3      return recurse(x);
4    }
5    for (int i = 0; i < n; i++) {
6      if (inp[i] == x) return out[i];
7    }
8    return recurse(x);
9  }
```

back to `recurse` which, in this case, will usually result in infinite recursion and timeout thus giving a low fitness. The for-loop on lines 5 to 7 checks to see if `x` is part of *in*, if so, it will return the corresponding element of *out*. Otherwise the `recurse` function is called on line 8 with the possibility that it will, eventually, call `correctRecurse` again with a parameter that is in *in*.

The rest of the CTGGP framework remains the same as previously described. The framework itself is implemented in C++ and uses libGE version 0.26[1] to generate individuals and carry out search. Candidate functions are generated using C grammars and these are compiled into scaffold libraries using the Tiny-C-Compiler [3] (TCC). In phase 1 evolution, the test harness compares the output of the evolved parameter expressions with the elements of *segs*. In phase 2 evolution, which runs in a separate thread, the whole function is tested against the values in *in* and *out*. The phase 1 and phase 2 threads communicate via a C source file containing the parameter expressions generated by phase 1. This file is locked while being accessed so that our code remains thread-safe.

In our experiments we compare four different configurations on a range of target benchmarks. The four experiments are **Plain** - GE run against a grammar for each problem without Scaffolding or CTGGP; **Scaffolding** - GE run with Scaffolding but without separate evolution of parameters to recursive calls; **CTGGP** - GE running with CTGGP; and **Combined** combining CTGGP with Scaffolding as described above.

The target benchmarks are, for an integer parameter (n): *factorial* returns the factorial of n; *odd-evens* returns 0 if $n \bmod 2 = 0$ and 1 otherwise; *log2* finds $\lfloor \log_2 n \rfloor$; *fib* and *fib3* calculates the Fibonacci and Fibonacci-3 number for n; *lucas* calculates the nth Lucas number (this requires two base cases); and *pell* calculates the nth Pell number.

For the CTGGP runs, we use small call trees which provide information on the first five to six elements of the sequence. With our GUI these cases take less than 5 min each to draw. To enable a fair comparison for the non-CTGGP runs we simply integrate the phase1 grammar into the phase 2 grammar which defines a set of valid target programs. This setup allows the non-CTGGP

[1] This framework can be downloaded http://bds.ul.ie/libGE/.

experiments to take advantage of the specialised grammars generated with information from the structure of the call tree. This specialisation will positively influence their performance, making our relative estimates of the performance of CTGGP conservative.

In all experiments we use GE running on an underlying steady-state GA with tournament selection. The GE parameters remain same as the hand-tuned parameters from phase 2 from [2]. The replacement probability use is 0.25 and probabilities for crossover and mutation are, respectively, 0.9 and 0.01. In all experiments and both phases in the CTGGP experiments we use a population of 1000 running for 300 generations. The phases are set up to terminate early if an individual with perfect fitness is encountered. In all experiments using CTGGP phase 1 terminated faster than phase 2 so phase 2 statistics serve as the best indicator of the time the algorithm takes.

We ran our experiments on an AMD Opteron 6348 machine with 48 processors running at 2.8 GHz. When the load on the machine was light the average evaluation time per individual ranged from 2 ms to 20 ms depending on the complexity of the benchmark. All experiments were run for 50 trials.

4 Results and Discussion

Table 1 shows the results of our experiments. The columns show, respectively, the data for each experiment. The rows show the results for each benchmark – broken down into mean number of phase 2 evaluations ($\bar{\mathbf{x}}$), number of correct answers (**nc**) and worst case number of evaluations (**max**). Where not all runs in an experiment resulted in success the value of **max** simply indicates the time when the longest run terminated rather than maximum time-to-success. These cases are marked with an asterisk. We also mark bold an entry for $\bar{\mathbf{x}}$ if it is significantly better than corresponding value in the previous column (according to a log-rank test). As can be seen, on most benchmarks Scaffolding significantly outperforms plain GE. CTGGP significantly outperforms Scaffolding on all benchmarks – pointing to the advantage of utilising call tree information when it is available. In the last column Combined, only significantly outperforms CTGGP alone on the lucas benchmark. Exploring further, it can be seen that the value for $\bar{\mathbf{x}}$ is at least marginally better for combined than for CTGGP for all benchmarks. Moreover, the value of **max** is substantially lower for Combined and for CTGGP indicating that the combination may be a strategy for moderating worst-case performance.

As previously mentioned, in the latest implementation of CTGGP phase 1 and 2 evolution run concurrently. This means that on machines with spare processing capacity there is insignificant time overhead incurred from running phase 1. However, it is still interesting to observe how phase 1 and 2 interact over time. Figure 4 plots the best phase 1 fitness, best phase 2 fitness and and average phase 2 fitness against time for a long (75th percentile) run of the Combined framework on the fib3 benchmark.

Table 1. Mean number of evaluations ($\overline{\mathbf{x}}$), number of correct answers (**nc**) and worst case number of evaluations (**max**) for the four experimental configurations.

Problem		Plain	Scaffolding	CTGGP	Combined
factorial	$\overline{\mathbf{x}}$	4109	**2542**	**219**	157
	nc	50	50	50	50
	max	8942	7032	1403	366
oddeven	$\overline{\mathbf{x}}$	539	478	**269**	255
	nc	50	50	50	50
	max	2069	1845	1201	885
log2	$\overline{\mathbf{x}}$	21524	**9049**	**1404**	1206
	nc	50	50	50	50
	max	103783	22489	11527	3845
fib	$\overline{\mathbf{x}}$	53168	**31733**	**1189**	1081
	nc	40	49	50	50
	max	130923*	130107*	3698	3617
fib3	$\overline{\mathbf{x}}$	117875	**94818**	**12614**	10347
	nc	3	18	50	50
	max	125723*	124448*	84271	40771
lucas	$\overline{\mathbf{x}}$	105663	**35820**	**3081**	**1622**
	nc	8	49	50	50
	max	123455*	127936*	12288	7070
pell	$\overline{\mathbf{x}}$	56240	**28887**	**2127**	1879
	nc	41	49	50	50
	max	129823*	128358*	6186	4904

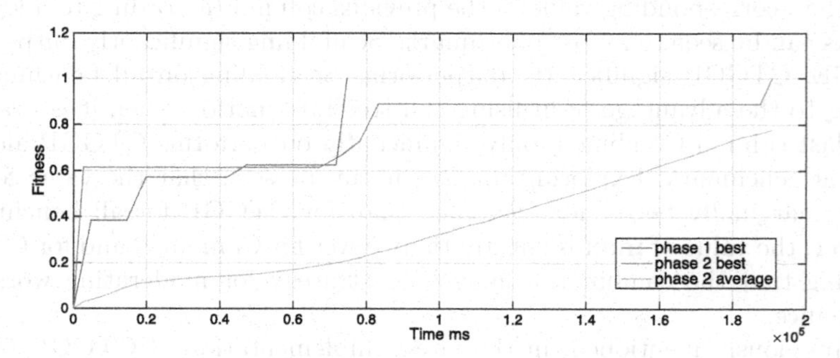

Fig. 4. Plot of fitness against time for the best individual in phase 1 and the best and average individual in phase 2 for the fib3 benchmark. (Color figure online)

As can be seen, phase 2 is able to make some progress, particularly in average fitness, while phase 1 is below perfect fitness - this indicates that even with incorrect parameter expressions search can progress. The speed of phase 2 search improves once phase 1 has produced the required parameters and terminated. This pattern of behavior is similar to that observed in other runs we have inspected.

5 Conclusions and Future Work

In this paper we have explored the effect of Scaffolding, CTGGP and a combination of these on GP search on a range of recursive benchmarks. We have shown that both Scaffolding and CTGGP significantly improve GP performance and there are indications that combining these is beneficial in terms of improving worst case performance. We have also shown that it is productive to run phase 1 and phase 2 evolution of CTGGP concurrently.

This work can be extended in several ways. We could further exploit the relationships between calling values and return values in the tree to help induct code that combines return values. We can extend the benchmarks for these experiments to include recurrences in loops. Finally we can conduct a more extensive study to confirm the effectiveness of the combined framework in reducing worst case times.

References

1. Agapitos, A., Lucas, S.: Learning recursive functions with object oriented genetic programming. In: Collet, P., Tomassini, M., Ebner, M., Gustafson, S., Ekárt, A. (eds.) EuroGP 2006. LNCS, vol. 3905, pp. 166–177. Springer, Heidelberg (2006)
2. Alexander, B., Zacher, B.: Boosting search for recursive functions using partial call-trees. In: Bartz-Beielstein, T., Branke, J., Filipič, B., Smith, J. (eds.) PPSN 2014. LNCS, vol. 8672, pp. 384–393. Springer, Heidelberg (2014)
3. Bellard, F., Tcc: Tiny C compiler (2003). http://fabrice.bellard.free.fr/tcc
4. Chennupati, G., Azad, R., Ryan, C.: Performance optimization of multi-core grammatical evolution generated parallel recursive programs. In: Proceedings of the 2015 on Genetic and Evolutionary Computation Conference, pp. 1007–1014. ACM (2015)
5. Koza, J.R., Andre, D., Bennett III, F.H., Keane, M.: Genetic Programming Darwinian Invention and Problem Solving. Morgan Kaufman, Burlington (1999)
6. Moraglio, A., Otero, F.E.B., Johnson, C.G., Thompson, S., Freitas, A.A.: Evolving recursive programs using non-recursive scaffolding. In: IEEE Congress on Evolutionary Computation, pp. 1–8 (2012)
7. Nishiguchi, M., Fujimoto, Y.: Evolution of recursive programs with multi-niche genetic programming (mnGP). In: Proceedings of the 1998 IEEE International Conference on Evolutionary Computation, pp. 247–252 (1998)
8. Ryan, C., Collins, J.J., Neill, M.O.: Grammatical evolution: evolving programs for an arbitrary language. In: Banzhaf, W., Poli, R., Schoenauer, M., Fogarty, T.C. (eds.) EuroGP 1998. LNCS, vol. 1391, pp. 83–96. Springer, Heidelberg (1998)
9. Spector, L., Robinson, A.: Genetic programming and autoconstructive evolution with the push programming language. Genet. Program. Evolvable Mach. **3**, 7–40 (2002)

10. Wong, M.L., Leung, K.S.: Evolving recursive functions for the even-parity problem using genetic programming. In: Advances in Genetic Programming, pp. 221–240. MIT Press (1996)
11. T. Yu and C. Clark. Recursion, lambda-abstractions and genetic programming. Cognitive Science Research Papers-University Of Birmingham CSRP, pp. 26–30 (1998)

Feature Extraction for Surrogate Models in Genetic Programming

Martin Pilát[1(✉)] and Roman Neruda[2]

[1] Faculty of Mathematics and Physics, Charles University in Prague, Malostranské Náměstí 25, 11800 Prague, Czech Republic
Martin.Pilat@mff.cuni.cz

[2] Institute of Computer Science, The Czech Academy of Sciences, Pod Vodárenskou Věží 271/2, 182 07 Prague, Czech Republic
roman@cs.cas.cz

Abstract. We discuss the use of surrogate models in the field of genetic programming. We describe a set of features extracted from each tree and use it to train a model of the fitness function. The results indicate that such a model can be used to predict the fitness of new individuals without the need to evaluate them. In a series of experiments, we show how surrogate modeling is able to reduce the number of fitness evaluations needed in genetic programming, and we discuss how the use of surrogate models affects the exploration and convergence of genetic programming algorithms.

Keywords: Surrogate model · Genetic programming · Random forest

1 Introduction

Evolutionary algorithms are great optimizers, however, they require a large number of objective function evaluations to find a suitable solution for a given problem. This large number of evaluations may be problematic in practice. Surrogate modeling [4] helps to reduce the number of fitness evaluations needed to find a solution of a given quality. Its main idea is to build an approximate model of the fitness function, which is used during the optimization as a cheap replacement of the expensive fitness. In the most common case, the model is built using various machine learning techniques from the individuals evaluated earlier in the evolution. The surrogate model is usually a standard regression model. So far, this technique is used almost exclusively in the field of continuous optimization, i.e. the optimization of functions $\mathbb{R}^n \rightarrow \mathbb{R}$.

The spread of surrogate modeling to different areas of evolutionary optimization is limited by the higher complexity of machine learning in these cases. Creating a surrogate model that maps, for example, a genetic program to a real number is a much more challenging task than running a regression algorithm on a vector of real numbers. In genetic programming, the surrogate modelling additionally includes, at least, the extraction of features from the genetic programs. So far, there are few applications of surrogate modeling outside the field

J. Handl et al. (Eds.): PPSN XIV 2016, LNCS 9921, pp. 335–344, 2016.
DOI: 10.1007/978-3-319-45823-6_31

of continuous optimization, one such example is provided by Li *et al.* [6] who use surrogate models to solve a problem in mixed integer programming.

Hildebrandt and Branke [3] proposed a method based on phenotypic features to predict the fitness values of an individual in genetic programming and used it to evolve dispatching rules for job shop scheduling. To create the feature vectors, they evaluate the individual on a few tasks and use the results of the individual as features. Then, they use a nearest neighbor model to predict the fitness of the individual (the fitness of the closest evaluated individual is used as a fitness of the new one). Such approach can be used in scenarios where the evolved program is used as a controller and it can be evaluated on a smaller task. However, in other scenarios the applicability of this method may be limited. For example, our motivation for this work is based on our previous work [5], where we attempted to evolve machine learning workflows with genetic programming. In such a scenario a partial evaluation of the individual does not make sense.

In this work we investigate the extraction of features directly from the tree individuals used in GP, with no need to evaluate the individuals. Our main goal is to create an algorithm, which can use these features to predict the quality of an individual based solely on it genotypic representation. To this end, we first extract as many features as possible from the individual and then train a surrogate model based on random forests. We also investigate the importance of individual features and how well the predictions match the real quality of individuals.

2 Surrogate-Based Genetic Programming

We propose a set of features, that can be extracted in a single pass through the tree without the need to evaluate the program. The features contain information of different kinds: general features regarding the tree, features concerning the primitives (i.e. functions) used in the tree, features on the arguments of the program, features regarding the constants used in the tree, and also the fitness of the parents of an individual. Particularly, the following features were used in the experiments in this paper:

- tree features – depth of the tree, size of the tree (number of nodes)
- constant features – maximum, minimum, and mean, number of constants and distinct constants divided by the size of tree
- argument features – average number of times an argument is used and proportion of arguments used
- for each terminal or primitive – the number of times it is used divided by the length of the individual
- parents' fitness – minimum, maximum and mean of the fitness of parents

Therefore, the number of features extracted from each individual equals the number of different non-terminals + the number of arguments of the program + 1 (for the constants as terminals) + 12 (the general features, the counts of arguments and constants, the statistics on the constants values, and the statistics

Algorithm 1. Surrogate-based Genetic Programming

Require: n: population size, τ: number of evaluations before surrogate modeling, A_m: maximum size of training set, w: the proportion of worst individuals to discard

1: $t \leftarrow 0$, $A = \emptyset$, $P_0 \leftarrow$ InitRandomPopulation(n)
2: **for** i in P_0 **do**
3: $\quad f_i \leftarrow$ Evaluate(i)
4: $\quad \varphi_i \leftarrow$ ExtractFeatures(i); $A \leftarrow A \cup \{(i, f_i, \varphi_i)\}$
5: **end for**
6: **while** termination criterion not met **do**
7: $\quad t \leftarrow t + 1$
8: $\quad S \leftarrow$ Selection(P_{t-1})
9: $\quad O_t \leftarrow$ GenerateOffspring(S)
10: $\quad$ **if** $|A| > \tau$ **then**
11: $\quad\quad T \leftarrow A$ ▷ training set
12: $\quad\quad$ **if** $|T| > A_m$ **then** $T \leftarrow$ RandomSample(T, A_m)
13: $\quad\quad M \leftarrow$ BuildModel(Features(T), Targets(T))
14: $\quad\quad I \leftarrow \{i \in O | \text{FitnessNotEvaluated}(i)\}$
15: $\quad\quad \hat{f}_i = \{\text{PredFit}(i, \text{ExtractFeatures}(i), M) | i \in I\}$
16: $\quad\quad W \leftarrow$ the indices of $w|I|$ worst individuals
17: $\quad\quad$ **for** i in I **do**
18: $\quad\quad\quad$ **if** $i \in W$ **then** $I \leftarrow$ replace i with its parent from S in I
19: $\quad\quad$ **end for**
20: $\quad$ **end if**
21: $\quad I \leftarrow \{i \in O | \text{FitnessNotEvaluated}(i)\}$
22: $\quad$ **for** i in P_t **do**
23: $\quad\quad f_i \leftarrow$ Evaluate(i)
24: $\quad\quad \varphi_i \leftarrow$ ExtractFeatures(i); $A \leftarrow A \cup \{(i, f_i, \varphi_i)\}$
25: $\quad$ **end for**
26: $\quad P_t \leftarrow$ Best(P_t) $\cup$ RemoveWorst(O)
27: **end while**

on fitness of parents). We investigate the importance of these features and also the performance of some models based on these features in Sects. 3.1 and 3.2.

We also considered a number of different features, i.e. the numbers times each tree of depth one is used. However, such structural features would increase the length of the feature vector significantly and would make the model training slower.

2.1 Baseline Algorithm

The main loop of the proposed algorithm (cf. Algorithm 1) is a relatively standard GP algorithm. It first generates a random initial population (line 1) and evaluates its fitness (line 3). Then, the evolution loop starts, the offspring are generated (line 9), those with unknown fitness are evaluated (line 23) and, finally, the environmental selection is performed (line 26). In the environmental selection, we use a weak elitism, i.e. the best individual from the parents replaces the worst offspring.

The part of the algorithm described in the paragraph above is also what we call the baseline algorithm in our experiments.

2.2 Surrogate Modeling

The surrogate version of the algorithm contains an archive of all individuals evaluated during the run which is initialized in the beginning (line 1) and updated after each evaluation of the real fitness with the value of the fitness and the features of the evaluated individual (lines 4 and 24).

The main part regarding the surrogate modeling lies between lines 10 and 20. First, there is a test, whether there are enough (at least τ) evaluated individuals in the archive, if not, the surrogate part is skipped and the algorithm works precisely as the simple algorithm described above. Otherwise, the training set is created. It contains either the whole archive, if there are less then A_m individuals, or a random sample of A_m individuals from the archive if there are more. The sampling step improves the speed of the surrogate model training.

The training of the model is performed on line 13. The features and the fitness of the individuals from the training set are collected and used for the training of the surrogate model. The output variable considered by the models is the fitness of the respective individual. Then, the individuals with unknown fitness, denoted by set I, are evaluated by the surrogate model. To this end, the features are extracted from each such individual and its fitness is predicted by the surrogate model. On line 16, the individuals are sorted by the estimated fitness and the indices of $w|I|$ worst individuals are put into set W. Each individual with its index in W is replaced by its parent. This ensures that such an individual does not need to be evaluated by a real fitness function, and it gives the parent another opportunity to generate new individual in the next generation (if it survives the mating selection).

Finally, the rest of the unevaluated individuals (now only $(1 - w)|I|$) is evaluated using the real fitness function. The newly evaluated individuals are added to the archive and the next generation begins.

2.3 Discussion

We use a rather unusual way of discarding the individuals predicted to be bad by the model – we replace them by their parents. In preliminary experiments, we have also tried a more traditional approach, i.e. discarding the individuals completely and replacing them by random or best parents. However, both these cases (random or best parents) lead to a fast loss of diversity in the population, as some of the parents get repeated. That in turn significantly slows down the convergence of the algorithm and can even lead to pre-mature convergence.

Another feature of the algorithm, which may be slightly unusual in genetic programming, is the use of weak elitism. We do not need the elitism from the point of view of preserving the best individual, as we already save the archive of all of them, however, preliminary experiments have shown, that it may slightly improve the performance of this simple algorithm.

3 Experiments

In order to evaluate the performance of our approach, we used four symbolic regression benchmarks described by White *et al.* [11] (keijzer-6, vladislavleva-4, nguyen-7, and pagie-1). The benchmarks represent a range of different functions with one to five arguments and each of the benchmarks uses a different set of primitives. For their precise definition refer to [8,11]. In preliminary experiements, we also used the korns-12 benchmark, however neither the baseline nor the surrogate algorithm were able to improve the random solutions from the initial population significantly, so these results are not presented here.

The objective is defined as the base 10 logarithm of the root mean square error (RMSE) of the prediction. The logarithm is used to make the values smaller and also easier for the surrogate model to train. As the algorithm uses only tournament selection (i.e. only comparisons of values), the objective is equivalent to RMSE.

The GP algorithm described above uses a population of 200 individuals and is run for a maximum of 15,000 objective function evaluations. The population is initialized by the ramped half-and-half method with maximum depth of trees set to 5. The mating selection is a tournament of 3 individuals, and the environmental selection uses a weak elitism (i.e. the best parent is added to the offspring population and the worst offspring is removed). The algorithm uses a simple one-point crossover (i.e. random subtrees are selected from each individual and swapped) and a uniform mutation which replaces a random subtree by a random subtree of a random depth between 1 and 4. The probability of crossover is 0.2 and the probability of mutation is 0.7. Those parents that do not undergo any of the operators are simply copied to the offspring population.

Additionally, the surrogate version of the algorithm starts using the surrogate model after $\tau = 1000$ evaluations and uses at most $A_m = 5000$ random individuals from the archive for the surrogate training. The surrogate model, unless otherwise noted, is a random forest for regression with 100 trees with the depth of at most 14 (the rest of the parameters of the trees uses the default values from scikit-learn [9]). The worst $w = 2/3$ of the individuals according to the model are discarded and replaced by their parents. The whole algorithm is implemented in Python using the deap framework [2] and the source codes are available as supplementary materials at the authors' webpage (www.martinpilat.com).

3.1 Model Quality

Before we use the surrogate model in the GP algorithm, we first test the performance of the features and different models for predicting the quality of solutions generated by the algorithm. To this end, we start the baseline GP and in each generation evaluate the new (therefore never before seen by the model) individuals with the surrogate model and also the real fitness function. Then, we compute the Spearman's rank coefficient, which expresses how similar the ranking provided by two methods is.

The Spearman's rank coefficient is used instead of correlation, as in the algorithm, we only use comparisons between pairs of individuals and not the actual values of the fitness (the algorithm uses tournament selection).

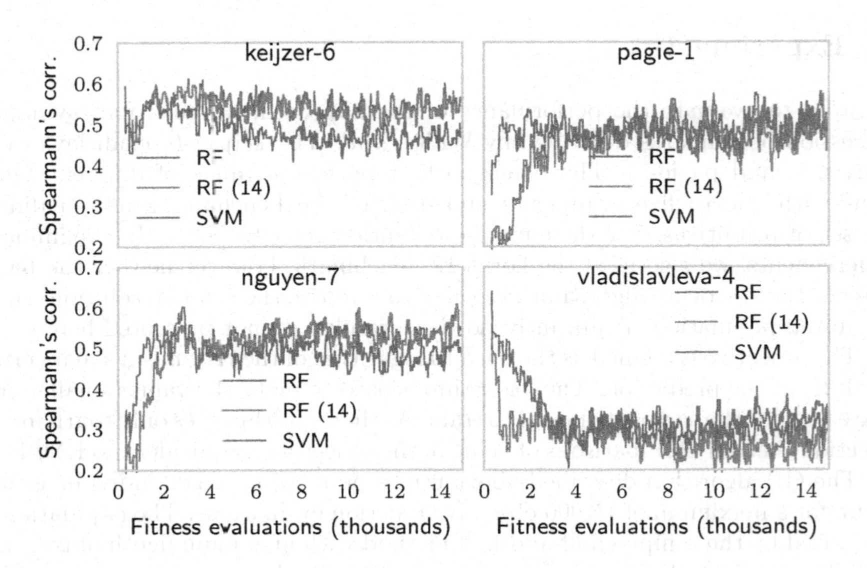

Fig. 1. The Spearman's correlation coefficient between the prediction provided by the model and the real fitness function. Median over 25 independent runs. The predictions made by the model were not used by the algorithm.

We compare support vector machines for regression (SVM) [10] and random forests [1] of 100 trees with two different depth limits - either unlimited (RF) or limited to 14 (RF (14)). Both methods use the default parameters from the scikit-learn framework and a standardization of the inputs is performed for the SVM. The limit on the depth of the random forest was set after preliminary experiments. It should be noted that the algorithm is quite sensitive to these settings and the provided values are a compromise, which seems to provide the best overall results. Different settings of the random forests were able to provide better results for some of the benchmarks than those presented here. In the preliminary experiments, we also used several types of linear models, however, their performance in this setting is unsatisfactory.

The results are presented in Fig. 1. The plots show the median of the Spearman's correlation coefficient computed over 25 independent runs as a function of the number of fitness evaluations. These runs were performed without the algorithm using the model in any way. In most cases, the model needs around 2,000 individuals in the training set before its performance becomes stable. After that, the Spearman's correlation coefficient is between 0.5 and 0.6 for most of the problems (except vladislavleva-4), which indicates a correlation of medium strength between the model and the observed fitness. For vladislavleva-4 the Spearmann's correlation drops to 0.3 after a thousand evaluations, this also coincides with the fact, that after thousand evaluations the convergence speed drops significantly (c.f. Fig. 2). This may be caused by the fact that the differences between the individuals get smaller and are thus harder to predict.

SVM provides the worst results of the three types of models used in this comparison, while the random forest with unlimited tree depth provides the best. However, the differences are rather small for some of the problems (mainly pagie-1 and vladislavleva-4). As expected, the performance of random forest with the depth of trees limited to 14 is slightly worse than that of unlimited random forest, but the difference is negligible. On the other hand, the smaller random forests are trained much faster, therefore, the RF (14) will be used in the rest of the paper.

3.2 Importance of Features

We have proposed a number of features. Naturally the importance of some of them is larger than the importance of others. In this section, we investigate which of the variables are the most important and therefore provide the most information about the quality of the programs. Such an information about the features is interesting not only for surrogate modeling in GP, but also for the design of better genetic programming operators. One can, for example, mutate more often those terminals that are more important for the quality of the program. We use the feature importance as computed by the random forest algorithm to judge the importance of the variables. This importance is computed [7] as the sum of the decreases in the performance metric used by the tree (mean square error in this case) caused by a given variable divided by the number of trees in the ensemble. Higher values indicate more important features.

To measure the importance, we run the same algorithm as in the previous section, i.e. the non-surrogate baseline with the model trained in each generation. This time, we log the importance of each variable provided by the model. The results of this experiment are summarized in Table 1. We report the four most important features for each of the problems after 5,000 and 10,000 evaluations together with their importance.

We can see that among the most important features are almost always some of the features which relate to the size of the tree – either the height of the tree or the size (len) of the tree. This should not be surprising, as too small trees can have a poor performance.

Other often important features are the counts of some of the primitives. Interestingly, among them the *exp* is the most important feature for the pagie-1 benchmark and one of the two most important features for the nguyen-7 dataset. In these cases, the features are important in the negative sense – programs with *exp* are inferior to programs without it, as none of the benchmarks use this function. Otherwise, the more simple primitives (multiplication, addition, subtraction) seem more important.

The vladislavleva-4 benchmark is interesting – it is the only one, where the importances of the different statistics on constants are among the top five features. The V-F1 feature in this benchmark corresponds to the special function in this benchmark defined as n^{ϵ}, where ϵ is a constant evolved by the algorithm. As this function is important to find the correct solution, it makes sense the number of times it is used is an important feature.

Table 1. The four most important features for each of the benchmarks after a given number of function evaluations. The importance is represented by median importance of each feature computed over 25 independent runs with the random forest regressor with 100 trees with maximum depth of 14. "depth" and "size" are the respective tree features, "const-mean" is the mean value of constants used in the individual and "arg-count" is the number of arguments used by the individual. The rest of the features represent the counts of the respective functions used by the individual.

Bench	Evals	Most important features
keijzer-6	5000	depth (0.397), safesqrt (0.097), mul (0.058), inverse (0.055)
keijzer-6	10000	depth (0.350), safesqrt (0.069), mul (0.050), size (0.050)
pagie-1	5000	exp (0.260), add (0.067), sub (0.060), size (0.060)
pagie-1	10000	exp (0.275), size (0.061), mul (0.058), arg-count (0.050)
nguyen-7	5000	depth (0.134), exp (0.096), safediv (0.068), mul (0.062)
nguyen-7	10000	depth (0.160), exp (0.085), size (0.065), mul (0.064)
vladisl.-4	5000	const-mean (0.150), V-F1 (0.076), depth (0.068), size (0.062)
vladisl.-4	10000	const-mean (0.159), size (0.062), V-F1 (0.059), depth (0.057)

3.3 Algorithm Performance

To test the algorithm, we made 25 runs on each of the benchmarks described above. The results are presented in Fig. 2. The plots show the dependence of the fitness (logarithm of the RMSE) on the number of fitness evaluations. Moreover, the red dotted line shows the p-value of one-sided Mann-Whitney U-test to test the statistical significance of the differences.

The best results were obtained for the keijzer-6 and pagie-1 benchmarks. Here, the surrogate version is able to decrease the number of evaluations needed to find a solution of given quality by almost 50the evolution (approx. between 2,000 and 7,000 evaluations). After this phase, the baseline version performs similarly on keijzer-6. For pagie-1, the surrogate is better during the whole 15,000 evaluations given as a budget to the algorithm, however, the differences get smaller. For the nguyen-7 benchmark, the median run of the surrogate algorithm is better between 2,000 and 6,000 fitness evaluations. However, the difference is rather small and the standard deviations are large, which means we can draw no definitive conclusion from this experiment.

The performance was the poorest for the vladislavleva-4 benchmark. There is no significant difference between the two algorithms in this case. We believe, there are two reasons for this behavior. First, most of the improvement happens before the surrogate modelling is even enabled – the baseline converges fast in the first 1,000 evaluations and does not improve much further. It may indicate that the problem becomes too difficult for the simple GP used in this work. Second, the performance of the model, as indicated by the tests in the previous sections was quite poor for this benchmark, which also may affect the results in a negative way.

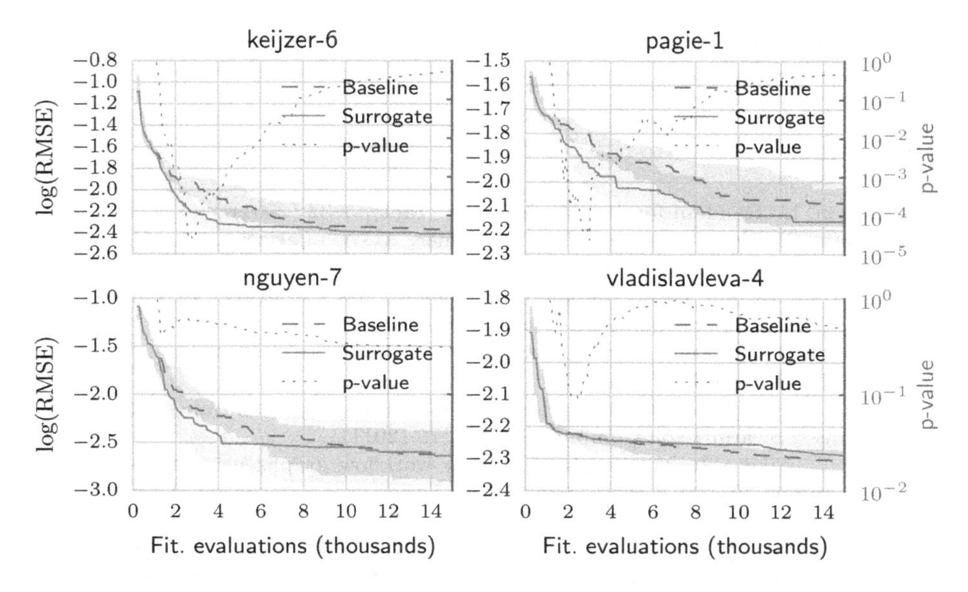

Fig. 2. The convergence rate of the algorithm on the four selected benchmarks. The lines represent the median of 25 runs, the shaded areas represent the first and third quartile. On the right axis, the red dotted line represents the p-value of the one-sided Mann-Whitney U-test computed after each 100 evaluations.

4 Conclusions and Future Work

We have proposed a simple surrogate-based genetic programming algorithm which provides promising results on the selected benchmark problems. We have shown that the surrogate models are capable of predicting the real fitness value without the need to evaluate the program, only by utilizing some static features. This may help to improve the effectiveness of genetic programming for problems with hard-to-evaluate fitness functions.

We also proposed a basic set of features which can be used for the building of surrogate models in genetic programming and we have evaluated the performance of these features. It seems that one of the most important features is the size of the tree, and among the more important features are the number of times each of the primitives is used. On the other hand, statistical features on the constants and features regarding the arguments actually used by the problem seem less important. In the case of the arguments, it may be caused by the fact that most of the benchmarks use all of the arguments, thus making these feature useless.

The presented approach should be considered mostly a proof of concept. There are definitely still many things that require more attention before it can be successfully used to solve practical tasks. The strategies for replacing the individuals deemed un-promising by the surrogate model can be refined, as well as the method for the selection of the training set from the archive – selecting a diverse set of samples instead of a random one may lead to better models.

Acknowledgments. This work is supported by Czech Science Foundation project no. P103-15-19877S.

References

1. Breiman, L.: Random forests. Mach. Learn. **45**(1), 5–32 (2001)
2. De Rainville, F.-M., Fortin, F.-A., Gardner, M.-A., Parizeau, M., Gagné, C.: DEAP: a python framework for evolutionary algorithms. In: Proceedings of the 14th Annual Conference Companion on Genetic and Evolutionary Computation, GECCO 2012, pp. 85–92. ACM, New York (2012)
3. Hildebrandt, T., Branke, J.: On using surrogates with genetic programming. Evol. Comput. **23**(3), 343–367 (2015)
4. Jin, Y.: Surrogate-assisted evolutionary computation: recent advances and future challenges. Swarm Evol. Comput. **1**(2), 61–70 (2011)
5. Křen, T., Pilat, M., Neruda, R.: Evolving workflow graphs using typed genetic programming. In: 2015 IEEE Symposium Series on Computational Intelligence, pp. 1407–1414, December 2015
6. Li, R., Emmerich, M., Eggermont, J., Bovenkamp, E., Bąck, T., Dijkstra, J., Reiber, J.: Metamodel-assisted mixed integer evolution strategies and their application to intravascular ultrasound image analysis. In: IEEE Congress on Evolutionary Computation, 2008. CEC 2008 (IEEE World Congress on Computational Intelligence), pp. 2764–2771, June 2008
7. Louppe, G., Wehenkel, L., Sutera, A., Geurts, P.: Understanding variable importances in forests of randomized trees. In: Burges, C., Bottou, L., Welling, M., Ghahramani, Z., Weinberger, K. (eds.) Advances in Neural Information Processing Systems, vol. 26, pp. 431–439. Curran Associates Inc., Red Hook (2013)
8. McDermott, J., White, D.R., Luke, S., Manzoni, L., Castelli, M., Vanneschi, L., Jaśkowski, W., Krawiec, K., Harper, R., Jong, K.D., O'Reilly, U.-M.: Genetic programming needs better benchmarks. In: Proceedings of the Fourteenth International Conference on Genetic and Evolutionary Computation Conference, pp. 791–798. ACM, Philadelphia (2012)
9. Pedregosa, F., Varoquaux, G., Gramfort, A., Michel, V., Thirion, B., Grisel, O., Blondel, M., Prettenhofer, P., Weiss, R., Dubourg, V., Vanderplas, J., Passos, A., Cournapeau, D., Brucher, M., Perrot, M., Duchesnay, E.: Scikit-learn: machine learning in Python. J. Mach. Learn. Res. **12**, 2825–2830 (2011)
10. Smola, A., Vapnik, V.: Support vector regression machines. Adv. Neural Inf. Process. Syst. **9**, 155–161 (1997)
11. White, D.R., McDermott, J., Castelli, M., Manzoni, L., Goldman, B.W., Kronberger, G., Jaśkowski, W., O'Reilly, U.-M., Luke, S.: Better GP benchmarks: community survey results and proposals. Genet. Prog. Evol. Mach. **14**, 3–29 (2013)

A General-Purpose Framework for Genetic Improvement

Francesco Marino[1], Giovanni Squillero[1], and Alberto Tonda[2](✉)

[1] Politecnico di Torino, Corso Duca Degli Abruzzi 24, 10129 Torino, Italy
francesco.marino@studenti.polito.it, giovanni.squillero@polito.it
[2] UMR GMPA, AgroParisTech, INRA, Université Paris-Saclay, 1 Av. Brétignières, 78850 Thiverval-Grignon, France
alberto.tonda@grignon.inra.fr

Abstract. Genetic Improvement is an evolutionary-based technique. Despite its relatively recent introduction, several successful applications have been already reported in the scientific literature: it has been demonstrated able to modify the code complex programs without modifying their intended behavior; to increase performance with regards to speed, energy consumption or memory use. Some results suggest that it could be also used to correct bugs, restoring the software's intended functionalities. Given the novelty of the technique, however, instances of Genetic Improvement so far rely upon ad-hoc, language-specific implementations. In this paper, we propose a general framework based on the software engineering's idea of mutation testing coupled with Genetic Programming, that can be easily adapted to different programming languages and objective. In a preliminary evaluation, the framework efficiently optimizes the code of the md5 hash function in C, Java, and Python.

Keywords: Genetic improvement · Genetic programming · Linear genetic programming · Software engineering

1 Introduction

The term "genetic improvement" has been commonly used to denote the science of applying genetic, breeding principles and biotechnology to *improve* plants and animals, that is, to maximize the expression of their genetic potential making them more productive for human use. While such techniques are dated back to 1700 s, the very same term has been quite recently renovated in a completely different context: computer science. Nowadays among evolutionary computation scholars, Genetic Improvement (GI) denotes the application of evolutionary, search-based optimization methods to the improvement of existing software.

The hope to automatically improve, let alone create, software has been a driving force of evolutionary computation. In 1992, John Koza asked "how can computers be made to do what is needed to be done, without being told exactly how to do it?", then tried to answer the question by introducing the paradigm of Genetic Programming (GP) [1]. Despite an unquestionable series of successes,

J. Handl et al. (Eds.): PPSN XIV 2016, LNCS 9921, pp. 345–352, 2016.
DOI: 10.1007/978-3-319-45823-6_32

GP cannot be used to evolve from scratch a full *computer program* able to solve a generic problem, yet.

With the more recent GI, practitioners are tackling an apparently easier problem: given the code of an existing program, GI strives to tweak it in order to reach a specific goal, such as: improve speed, reduce memory usage, reduce code length, remove bugs, etc. Such a technique raised the interest of the scientific community, and several works on the subject have appeared in literature, in the last few years [2–4]. Despite the interest, so far each application of GI has been developed in-house by researchers, with solutions designed for specific problems and computing languages, with little to none code re-usability.

In this paper, we propose a general-purpose framework for GI, able to tackle problems in any computer language and with user-definable goal. The approach exploits an existing general-purpose evolutionary algorithm (EA), and is tested on a simple but challenging test case, the md5 hash function. For three different languages, C, Java, and Python, the proposed methodology is proven able to reduce the code size of the function without introducing errors, given a target number of items for which to generate a hash value.

2 Background

2.1 Genetic Improvement

GI was introduced as a technique able to automatically modify the source code of existing software, optimizing its performance with regards to user-defined metrics [2]. GI was originally based on Genetic Programming (GP) [1]: individuals are encoded as linear graphs, each one representing a series of permutations on the code, ranging from commenting blocks, to swapping two lines, to change the initialization of a variable. Even if the fitness evaluation is specifically tailored for each application, the general idea is always to improve code behavior with respect to one or more objectives, all the while maintaining the software's functionalities.

GI has been successfully applied to different case studies, in order to decrease energy consumption [3], improve speed [2], specialize a program to optimize some specific functions [4], and minimize memory usage [5], respectively. Recent results, aiming at repairing the firmware of a router, prove that GI is also able to act on compiled code, correcting bugs without direct access to the source or to test suites [6]. The rising interest of the evolutionary research community for the topic culminated in a first workshop on the subject, organized during the GECCO conference in 2015[1].

As it is common for techniques in the early stages of research, case studies of GI use ad hoc tools, usually developed from scratch for the specific application. Now that GI is getting more and more adopted, a general-purpose framework could be extremely helpful to practitioners and researchers alike, speeding up prototyping and development of new ideas.

[1] http://geneticimprovementsoftware.com/.

2.2 μGP

μGP (MicroGP) [7,8] is an evolutionary toolkit. Originally devised to evolve assembly-language programs for test program generation [9], it was later expanded to a general-purpose open-source project[2] and exploited for several applications, ranging from Bayesian network structure learning [10] to analyzing the behavior of wireless network routing protocols [11], from adapting the number of cards in reactive pull systems [12] to the detection of power-related software errors in industrial verification processes [13].

What makes μGP suitable to tackle such a wide range of diverse problems is its design, based on a distinct separation between the description of individuals in the target application, the evolutionary core, and the fitness evaluator. In essence, the framework evolves a set of linear directed graphs, where each node represents a macro, that can in turn present several parameters. The description of the macros is specified by the user through a configuration file. When individuals are evaluated, the macros in each node are converted to text, and the resulting file is passed to a user-designed evaluator program. For a high-level depiction of the framework, see Fig. 1.

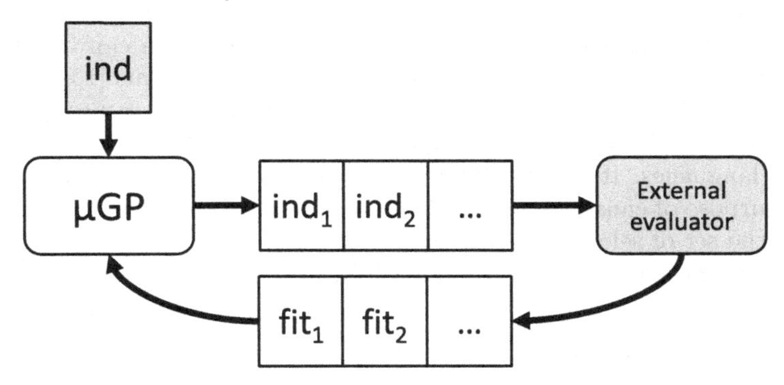

Fig. 1. High-level structure of the μGP toolkit. In order to prepare the framework to tackle a new application, only the parts in gray (XML description file and external evaluator) need to be modified by the user.

3 Proposed Approach

Given the rising interest around applications of GI, we propose a general-purpose framework, to ease prototyping and development. The framework is based on the μGP evolutionary toolkit, and can be quickly adapted to new GI applications, across different languages, without any need to recompile the source code, simply acting on configuration files and fitness evaluation.

[2] μGP is hosted on SourceForge http://ugp3.sourceforge.net/.

Table 1. Set of MT operations selected for the proposed framework.

Short-code	Description
CAR	Arithmetic operator replacement
CAS	Assignment operator replacement
CBI	Bitwise operator replacement
CCO	Logic connector replacement
CLO	Logic operator replacement
CST	Constant value replacement
CUN	Unary operator replacement
DEL	Statement deletion

An individual encodes a sequence of operations to be performed on the target code. Such operations are inspired by Mutation Testing (MT), a technique devised in the early 1970s to evaluate the quality of a test suite [14]. The basic idea is to slightly *mutate* a program, emulating developers' errors. All such mutants are eventually used to assess the effectiveness of a test suite in discriminating bug-free software.

As for GI, most approaches exploiting MT are either problem or language specific. However, being a well-established technique, one can find in literature lists of mutation operators that can be applied to programs [15–20].

Our approach exploits the possibility to mutate a source program. We select a compact list of standard MT operations that are both general and relevant for all languages. It is important to notice that some MT functions have the clear purpose of causing a fault, and they have not been considered here. Table 1 shows the set of selected operations.

The proposed approach is summarized in Fig. 2.

3.1 Evolutionary Core

An individual is a variable-length sequence of modifications. Each modification is encoded as an operation (see Table 1) and one or more operands. Possible operands, e.g., the list of used operators or the list of used constants, are pre-computed with static analysis.

The evolutionary core is the out-of-the-box μGP. μGP mutates and recombines individuals using classical genetic operators. In more details: an operation may be substituted with another operation, or its operands changed. Two individuals may be mixed using one-cut, two-cut, and uniform crossover operators.

Since the list of operations to be performed on the code is not language-specific, a parser is required to translate the generic high-level operations to language-specific ones. After the modified program has been generated, it is then tested on a set of test cases to ensure that the features are maintained and to evaluate the quality of the improvement reached, as for standard GI procedure.

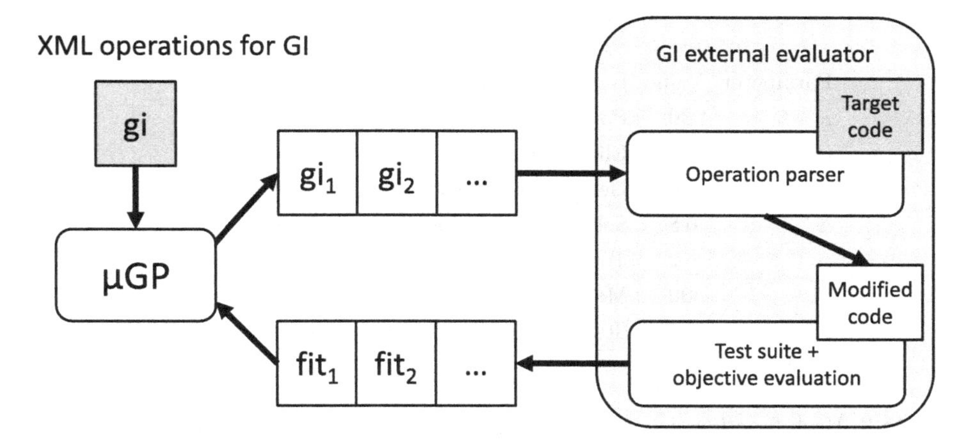

Fig. 2. Structure of the proposed approach. The μGP toolkit is used to generate a sequence of operations to be performed on the target code. The resulting modified program is then run through a test suite, in order to assess its functionality, and then evaluated with regards to a user-defined metric, for example speed, memory usage, etc.

4 Experimental Evaluation

In order to assess the suitability of our approach, we test the proposed framework on the MD5 function [21], a small, yet paradigmatic, case study. The MD5 message-digest algorithm is a widely used cryptographic hash function producing a 128-bit (16-byte) hash value, typically expressed in text format as a 32-digit hexadecimal number. MD5 has been utilized in a wide variety of cryptographic applications and is also commonly used to verify data integrity despite the fact that it is now considered unsuitable for further cryptographic use.

The classical MD5 implementation computes the value starting from the key and performing a series of arithmetic and binary operations on it. The experiments aim at *improving* the classical MD5 by reducing its size while still guaranteeing zero collision of the generated hashes on different fixed input set.

The necessity to optimize a general algorithm applied in a reduced scenario is not uncommon. And calculating hashes in an embedded system, thus for a known and fixed number of keys is a plausible scenario.

Experiments tackle the same function implemented in three different languages: C, Java and Python. Each function was improved in four scenarios, for a different number of keys: 8, 256, 1,024, and 4,096. The parameters used by the μGP in all the experiments are shown in Table 2. The evolution was stopped after 50 generations with no improvement in the fitness of the best individual.

In total, twelve different experiments were executed, producing four improved versions of the original function, in each language. Every improved program showed a reduced size, while guaranteeing to produce no collisions over the set of keys. Each run has been repeated 10 times. Results are shown in Table 3.

Table 2. μGP parameters

Parameter	Value	Description
μ	30	Population size
ϵ	1	Elite size
λ	20	Genetic operators applied in each generation
α	0.8	Self-adaptation inertia
τ	2	Tournament size
M	300	Maximum number of generations
St	50	Steady-state threshold

Table 3. Hash function size reduction compared among languages and test suites (10 repetitions)

# keys	C	Java	Python
8	46 %	40 %	36 %
256	41 %	37 %	39 %
1,024	41 %	36 %	36 %
4,096	37 %	38 %	44 %

Although preliminary, results are interesting. In both C and Java the achieved improvement is larger when the key set is smaller, as expected. The less different inputs are used, the more lines can be removed. Indeed, it must be noted that the tool is able to *modify* the original programs, tweaking constants or changing operators, and not only removing lines. For the Python implementation, on the other hand, the *type* of the keys and not their simple number, is the most important element. Thus, the tool is able to improve the original MD5, but improvement are not directly connected with the size of the key set.

Table 4 reports the average computational resources required to run the experiments. The tool was executed on a i7 computer with 16 GB of RAM, using a Linux-based operating system. Column **CPU** shows the total time required

Table 4. Time elapsed in each experiment (10 repetitions)

# keys	CPU (h:mm)			Generations		
	C	Java	Python	C	Java	Python
8	2:06	4:20	3:10	162	54	50
256	2:22	4:26	5:09	300	69	147
1,024	2:03	6:25	6:26	300	72	206
4,096	3:50	3:58	5:56	248	53	185

to run μGP, the tool for applying changes, and the evaluator. Column **Generations** reports the average number of generations before a steady-state is reached.

5 Conclusions

Genetic Improvement (GI) is a recently presented evolutionary technique for software engineering, able to automatically modify the source code of a program, increasing its performance with regards to energy/memory consumption or speed of execution. While the methodology has been proven to be rather promising, all solutions found in literature are ad-hoc implementations, often devised from scratch for a specific application. In this paper, we presented a generic framework for GI, able to target different programming languages and different objectives, requiring only minor tweaking on the part of the user. The proposed approach is experimentally tested on simple case studies in Python, C++ and Java, and the results show that it is able to satisfactorily perform in all instances. Future works will focus on providing a Graphical User Interface for the framework, and releasing a full test set of benchmarks for GI, in different languages and for different objectives.

References

1. Koza, J.R.: Genetic Programming: On the Programming of Computers by Means of Natural Selection, vol. 1. MIT Press, Cambridge (1992)
2. Langdon, W.B., Harman, M.: Optimising existing software with genetic programming. IEEE Transactions on Evolutionary Computation (99) (2013)
3. Bruce, B.R., Petke, J., Harman, M.: Reducing energy consumption using genetic improvement. In: Proceedings of the 2015 on Genetic and Evolutionary Computation Conference, pp. 1327–1334. ACM (2015)
4. Petke, J., Harman, M., Langdon, W.B., Weimer, W.: Using genetic improvement and code transplants to specialise a C++ program to a problem class. In: Nicolau, M., Krawiec, K., Heywood, M.I., Castelli, M., García-Sánchez, P., Merelo, J.J., Rivas Santos, V.M., Sim, K. (eds.) EuroGP 2014. LNCS, vol. 8599, pp. 137–149. Springer, Heidelberg (2014)
5. Wu, F., Weimer, W., Harman, M., Jia, Y., Krinke, J.: Deep parameter optimisation. In: Proceedings of the 2015 on Genetic and Evolutionary Computation Conference, pp. 1375–1382. ACM (2015)
6. Schulte, E.M., Weimer, W., Forrest, S.: Repairing COTS router firmware without access to source code or test suites: a case study in evolutionary software repair. In: Proceedings of the Companion Publication of the 2015 Annual Conference on Genetic and Evolutionary Computation. GECCO Companion 2015, pp. 847–854. ACM, New York (2015)
7. Squillero, G.: MicroGP - an evolutionary assembly program generator. Genet. Prog. Evol. Mach. **6**(3), 247–263 (2005)
8. Squillero, G.: Artificial evolution in computer aided design: from the optimization of parameters to the creation of assembly programs. Computing **93**(2–4), 103–120 (2011)

9. Corno, F., Sánchez, E., Squillero, G.: Evolving assembly programs: how games help microprocessor validation. IEEE Trans. Evol. Comput. **9**(6), 695–706 (2005)
10. Tonda, A., Lutton, E., Squillero, G., Wuillemin, P.-H.: A memetic approach to Bayesian network structure learning. In: Esparcia-Alcázar, A.I. (ed.) EvoApplications 2013. LNCS, vol. 7835, pp. 102–111. Springer, Heidelberg (2013)
11. Bucur, D., Iacca, G., Squillero, G., Tonda, A.: The impact of topology on energy consumption for collection tree protocols: an experimental assessment through evolutionary computation. Appl. Soft Comput. **16**, 210–222 (2014)
12. Belisário, L.S., Pierreval, H.: Using genetic programming and simulation to learn how to dynamically adapt the number of cards in reactive pull systems. Expert Syst. Appl. **42**(6), 3129–3141 (2015)
13. Gandini, S., Ruzzarin, W., Sanchez, E., Squillero, G., Tonda, A.: A framework for automated detection of power-related software errors in industrial verification processes. J. Electron. Test. **26**(6), 689–697 (2010)
14. DeMillo, R.A., Lipton, R.J., Sayward, F.G.: Hints on test data selection: help for the practicing programmer. Computer **4**, 34–41 (1978)
15. King, K.N., Offutt, A.J.: A FORTRAN language system for mutation-based software testing. Softw. Pract. Exp. **21**(7), 685–718 (1991)
16. Delamaro, M.E., Maldonado, J.C., Mathur, A.: Proteum-a tool for the assessment of test adequacy for C programs users guide. In: PCS, vol. 96, pp. 79–95 (1996)
17. Ma, Y.S., Offutt, J., Kwon, Y.R.: Mujava: an automated class mutation system. Softw. Test. Verif. Reliab. **15**(2), 97–133 (2005)
18. Derezińska, A., Rudnik, M.: Quality evaluation of object-oriented and standard mutation operators applied to C# programs. In: Furia, C.A., Nanz, S. (eds.) TOOLS 2012. LNCS, vol. 7304, pp. 42–57. Springer, Heidelberg (2012)
19. Derezińska, A., Hałas, K.: Operators for mutation testing of python programs. Research report (2014)
20. Jia, Y., Harman, M.: An analysis and survey of the development of mutation testing. IEEE Trans. Softw. Eng. **37**(5), 649–678 (2011)
21. Rivest, R.: The Md5 Message-digest Algorithm. Princeton, RFC (1992)

On the Use of Semantics in Multi-objective Genetic Programming

Edgar Galván-López[1(✉)], Efrén Mezura-Montes[2], Ouassim Ait ElHara[3], and Marc Schoenauer[3]

[1] School of Computer Science and Statistics, Trinity College Dublin, Dublin, Ireland
edgar.galvan@scss.tcd.ie
[2] Universidad Veracruzana, Xalapa, Veracruz, Mexico
emezura@uv.mx
[3] TAO, INRIA and LRI, CNRS & U. Paris-Sud, Université Paris-Saclay, Orsay, France
{ouassim.ait_elhara,marc.schoenauer}@inria.fr

Abstract. Research on semantics in Genetic Programming (GP) has increased dramatically over the last number of years. Results in this area clearly indicate that its use in GP can considerably increase GP performance. Motivated by these results, this paper investigates for the first time the use of Semantics in Muti-objective GP within the well-known NSGA-II algorithm. To this end, we propose two forms of incorporating semantics into a MOGP system. Results on challenging (highly) unbalanced binary classification tasks indicate that the adoption of semantics in MOGP is beneficial, in particular when a semantic distance is incorporated into the core of NSGA-II.

1 Introduction

Genetic Programming (GP) [9] has been successfully used in a range of different challenging problems (see Koza's article on human competitive results for a comprehensive review [10]). Despite its proven success, it also suffers from some limitations and researchers have been interested in making GP more robust by studying various elements of the search process, and also by e.g., considering other GP forms [7].

One of these elements that has relatively recently attracted the attention of researchers is the study of semantics in GP, resulting in a dramatic increase in the number of related publications (e.g., [2,8,11,12]).

Semantics is a broad concept that has been studied in different fields making it hard to give a precise definition of the concept. Moreover, the way semantics has been adopted in canonical GP varies significantly e.g., Beadle and Johnson [2] used reduced ordered binary decision trees on Boolean problems

Research conducted during Galván's stay at TAO, INRIA and LRI, CNRS & U. Paris-Sud, Université Paris-Saclay, France.

J. Handl et al. (Eds.): PPSN XIV 2016, LNCS 9921, pp. 353–363, 2016.
DOI: 10.1007/978-3-319-45823-6_33

to study semantics, whereas Uy's work on semantics has focused on repeatedly applying crossover to encourage semantic difference between parents and offspring (see [15] for a summary of works carried out in semantics).

This work uses a popular version of semantics GP, as originally proposed in [12], and used in recent works from the first author [8,13], in which the *semantics* of a (sub)tree is defined as the vector of output values computed by this (sub)tree for each set of input values in turn (a.k.a. each fitness case in most cases). Several semantic-based approaches have been proposed for GP which take semantics into account when e.g., choosing and modifying subtrees, such as the one that has been demonstrated beneficial in [13] and it is adopted in this work too.

To the best of our knowledge, however, there is no scientific study on the adoption of semantics in Evolutionary Multi-objective Optimisation at large [5], and in Multi-objective GP in particular and this paper intends to start filling this important research area.

The goal of this paper is to incorporate semantics into a Multi-objective GP paradigm by using the well-known NSGA-II. To this end, we adopted two different forms of incorporating semantics into NSGA-II: (a) one based on a relatively simple, efficient and straightforward semantic-based single-objective GP approach, and (b) one based on the adoption of a semantic distance into the core of the NSGA-II algorithm.

This paper is organised as follows. In Sect. 2, we introduce our proposed approaches. Section 3 provides details on the experimental setup used. The results presented in this paper are discussed in Sect. 4, and finally, conclusions and future work are drawn in Sect. 5.

2 Semantics in Multi-objective Genetic Programming

In this work, following [12], the *semantics* of a GP tree describes the behaviour of the tree when various values are given to the input variables. Two trees can be syntactically very different while behaving identically. What matters, as far as solving the problem at hand is concerned, is in fact the behaviour of the tree, i.e., its response to given inputs. These arguments support the use of semantics adopted here and at least partly explain the benefits of using semantics in GP as reported in [8,13].

In the case of a fitness based on the computation of several fitness cases, the semantics of a GP individual is a vector of size the number of fitness cases, one value for each fitness case. For instance, in the case of the problems used in this work (unbalanced data sets introduced in Sect. 3), the semantics of a GP tree is the vector of real-valued output by the tree for each of the examples in the e.g., training data set. In this work, the semantic distance between two trees is the number of outputs that are different between their semantics. Commonly, when computing the semantic distance, two outputs are considered different if their absolute difference is greater than a given threshold [8,13]. In this work, we set the threshold at 0.5.

2.1 Evolutionary Multi-objective Optimisation

Multi-objective optimisation (MO) is concerned with the simultaneous optimisation of several objectives. When these are in conflict, no single solution exists, and trade-offs between the objectives must be sought. The optimal trade-offs are the solutions for which no objective can be further improved without degrading another objective. This idea is captured in the Pareto dominance relation: a point x in the search space is said to *Pareto-dominate* another point y if x is at least as good as y on all objectives and strictly better on at least one objective.

The set of optimal trade-off solutions of a MO problem can then be defined as the set of points of the search space that are not dominated by any other point, and is called the *Pareto set* of the problem at hand. The goal of Pareto MO is to identify the Pareto set, or a good approximation of it. The *Pareto front* is the image of the Pareto set in the objective space.

Evolutionary multi-objective optimisation (EMO) [5] is based on the following: by replacing the single-objective selection steps, based on the comparison of fitness values, by some Pareto-based comparison, one turns a single-objective evolutionary optimisation algorithm into a multi-objective evolutionary optimisation algorithm, but because Pareto dominance is not a total order, some additional criterion must be used so as to allow the comparison of any pair of points of the search space.

In NSGA-II [6], the Pareto-based comparison uses the non-dominated sorting procedure: all non-dominated individuals in the population are assigned Rank 1 and removed from the population, the remaining non-dominated individuals are assigned Rank 2, and so on. The secondary criterion is the *crowding distance* that promotes diversity among the individuals having the same Pareto rank: in objective space, for each objective, the individuals in the population are ordered, and the partial crowding distance for each of them is the difference in fitness between its two immediate neighbours. The crowding distance is the sum over all objectives of these partial crowding distances. Intuitively, it can be seen as the Manhattan distance between the extremal vertices of the largest hypercube containing the point at hand and no other point of the population. Selecting points with the largest crowding distance amounts to favour the low-density regions of the objective space, thus favouring behavioural diversity.

The NSGA-II proceeds as follows. From a given population of size N, N offspring are created using standard variation operators (crossover and mutation). Parents and offspring are merged, and the resulting population, of size $2N$, is ordered using non-dominated sorting, and crowding distance as secondary criterion. The best N individuals according to this ranking are selected to survive at the next generation.

Because the underlying idea within NSGA-II is to favour behavioural diversity, but only considering the fitness as a whole, it can be hoped that introducing semantics in NSGA-II can only enforce this idea.

2.2 Incorporating Semantics in MOPG

In this work, we investigate two ways of incorporating semantics into a MOGP system (recall we use NSGA-II). One natural form to do so is to use semantics as commonly adopted in canonical GP (e.g., semantically-based crossover [13]). In our study, we adopted the semantics in the selection tournament mechanism [8] due to its simplicity and efficiency. Briefly, the idea is to create offspring that are semantically different from their parents when tournament selection is applied: the first parent is selected as usual and the second parent is selected if it is semantically different and fitter than the already selected parent, if this is not satisfied for any individual in the pool, one is chosen at random. We call this NSGA-II Semantics in Selection (SiS).

The second proposed way to add semantics to NSGA-II is to replace the crowding distance (see above) with a semantic-based indicator called Semantic-based Crowding Distance (SCD). This is computed the following way: a *pivot* is chosen, being the individual from the first Pareto front (Rank 1) that is the furthest away from the other individuals of this front using the crowding distance. For each point, its semantic distance with the pivot is computed. Similarly to the crowding distance, the SCD is computed as the average of the semantic distance differences with its closest neighbours in each direction. The higher values of this SCD are favored during the selection step of NSGA-II. This allows us to have a set of individuals that are spread in the semantic space, therefore, promoting semantic diversity, the same way NSGA-II promotes diversity ('spreadness') in the objective space. It is worth pointing out that this approach also works when there is only one front. This variant of NSGA-II will be called Distance-based Semantics (DBS) in the following.

3 MOGP Configuration and Experimental Design

To study the effects of semantics in MOGP, we used challenging binary (highly) unbalanced classification problems taken from the literature [1]. These problems are of different nature and complexity, e.g., they have from a few features up to dozens of them, these features include binary, integer, and real-valued features. Table 1 gives the details for all datasets. These have been used 'as is' (i.e., we did not try to balance the classes out). For each dataset, half of the data (with the same class balance than in the whole dataset) was used as a training set and the rest as a test set. All reported results are on the latter.

The terminal and function sets used in these experiments were the same than in [3]. The terminals are the problem features. The function set consists of the conditional IF function and the typical four standard arithmetic operators: $\mathcal{F} = \{if, +, -, *, /\}$, where the latter operator is the protected division, which returns the numerator if the denominator is zero. The IF function takes three arguments: if the first one is negative, the second argument is returned, otherwise the last argument is returned. These functions are used to build a classifier (e.g., mathematical expression) that returns a single value for a given input (data example to be classified). This number is mapped onto a set of class labels using

Table 1. Binary unbalanced classification data sets used in our research. Table adapted from [3].

Data set	Classes	Number of examples			Imb.	Features	
	Positive/Negative (Brief description)	Total	Positive	Negative	Ratio	No	Type
Ion	Good/bad (ionsphere radar signal)	351	126 (35.8 %)	225 (64.2 %)	1:3	34	Real
Spect	Abnormal/normal (cardiac tom. scan)	267	55 (20.6 %)	212 (79.4 %)	1:4	22	Binary
Yeast$_1$	mit/other (protein sequence)	1482	244 (16.5 %)	1238 (83.5 %)	1:6	8	Real
Yeast$_2$	me3/other (protein sequence)	1482	163 (10.9 %)	1319 (89.1 %)	1:9	8	Real

Table 2. Confusion matrix.

	Predicted positive	Predicted negative
Actual positive	True Positive (TP)	False Negative (FN)
Actual negative	False Positive (FP)	True Negative (TN)

zero as the class threshold. In our studies, an example is assigned to the minority class if the output of the classifier is greater or equal to zero. It is assigned to the majority class, otherwise.

The common way to measure the fitness of a classifier for classification tasks is the overall classification accuracy: for binary classification, the four possible cases are shown in Table 2. Assuming the minority class is the positive class, the accuracy is given by $Acc = \frac{TP+TN}{TP+TN+FP+FN}$. The drawback of using Acc alone is that it rapidly biases the evolutionary search towards the majority class [3]. A better approach is to treat each objective (class) 'separately' using a multi-objective approach: Two objectives are considered, the true positive rate $TPR = \frac{TP}{TP+FN}$, and the true negative rate $TNR = \frac{TN}{TN+FP}$, that measure the distinct accuracy for the minority and majority class, respectively.

The experiments were conducted using a steady state approach with tournament selection (of size 2 for NSGA-II and NSGA-II DBS, and of size 7 for NSGA-II SiS to encourage semantic diversity). Initialisation and sub-tree mutation used the ramped half-and-half method (initial and final depth set at 1 and 5, respectively). To control bloat, a maximum depth of 8 was specified (root is at depth 0), or a maximum number of 800 nodes was used. Crossover and mutation rates were set at 60 % and 40 %, respectively. To obtain meaningful results, we performed 50 independent runs for each of the MOGP approaches for each of the problems used in this work.

Table 3. Average (± standard deviation) hypervolume, where the reference point is (0,0), of evolved Pareto-approximated fronts, Pareto optimal (PO) front for the three MOGP used in this work: NSGA-II, NSGA-II SiS and NSGA-II DBS, over 50 runs.

Methods	Hypervolume	Ion	Spect	$Yeast_1$	$Yeast_2$
NSGA-II	Average	0.842 ± 0.070	0.542 ± 0.024	0.822 ± 0.041	0.944 ± 0.021
	PO Front	0.948	0.637	0.875	**0.978**
NSGA-II SiS	Average	0.858 ± 0.063	0.542 ± 0.020	0.827 ± 0.035	0.939 ± 0.048
	PO Front	0.960	0.642	**0.876**	0.977
NSGA-II DBS	Average	**0.856 ± 0.051**	0.548 ± 0.026	0.827 ± 0.015	**0.948 ± 0.011**
	PO Front	**0.977**	**0.664**	0.873	0.977

4 Results and Discussion

4.1 Front Hypervolume

As a measure of performance, in order to compare the different approaches, we use the hypervolume [4] of the evolved Pareto approximations. For bi-objectives problems, the hypervolume of a set of points in objective space (using reference point $(0,0)$) is easily computed as the sum of the areas of all trapezoids fitted under each point. Such measure was chosen as being the only known Pareto-compliant indicator to-date [16]: the larger the hypervolume, the better the performance. We also computed the Pareto-optimal (PO) front with respect to all 50 runs, i.e., the set of non-dominated solutions after merging all 50 Pareto-approximated fronts.

Table 3 reports, for each problem, both the average hypervolume over 50 runs, and the hypervolume of the PO. In this table, the best hypervolumes are highlighted in boldface. Furthermore, the statistical significance for the results on the average hypervolume was computed using Wilcoxon Test at 90 % level of significance, independently comparing each of the semantic-based approaches (NSGA-II SiS, NSGA-II DBS) against NSGA-II.

According to these results, in three out of the four problems, both semantic-based MOGP approaches achieve a higher hypervolume of the PO front compared to the NSGA-II. Moreover, the NSGA-II DBS is statistically better (indicated in boldface) than the NSGA-II on two classification problems, but not statistically different on the other two problems. On the other hand, NSGA-II SiS is not statistically different on any of the problems compared to NSGA-II. This suggests that the adoption of semantics into a MOGP approach should be in one of the pillars of the MO approach.

4.2 Evolved Solutions and Pareto-Optimal Front

Let us now focus on the coverage of the objective space achieved by the semantic variants of NSGA-II: Fig. 1 displays together on the same plot, for each problem (top to bottom), and for NSGA-II and NSGA-II DBS (left and center respectively), the 50 Pareto front approximations obtained in the 50 independent runs

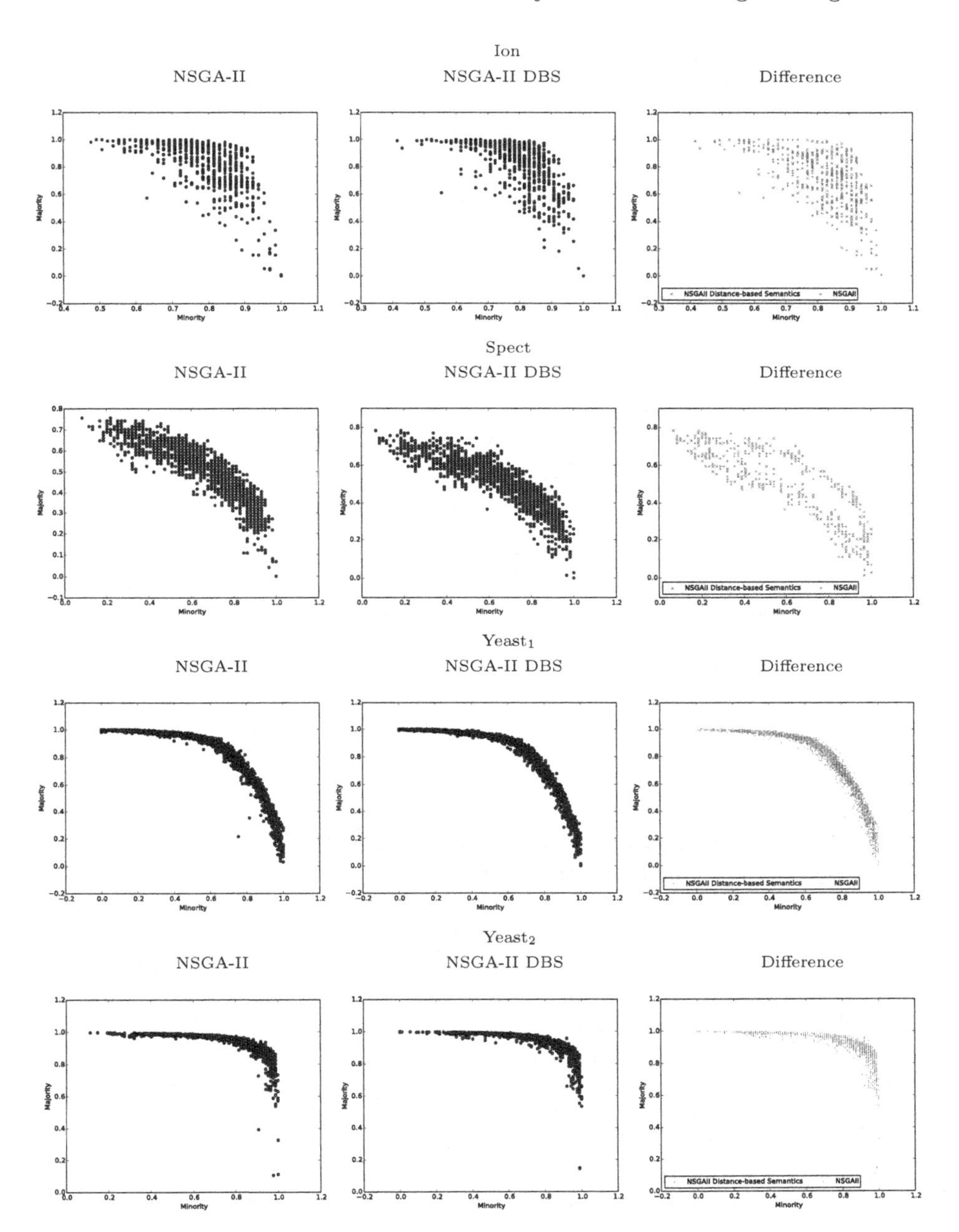

Fig. 1. Accuracy of all evolved solutions over 50 runs using the canonical NSGA-II and the NSGA-II DBS, shown in the left-hand side and centre of the figure, respectively. Plots in the the right-hand side of the figure show the evolved solutions that were exclusively found by either NSGA-II (indicated by a red plus '+' symbol) or NSGA-II DBS (indicated by a blue cross 'x' symbol). For clarity purposes, we reduced the size of the marker symbols in problems with denser areas (i.e., Yeast problems).

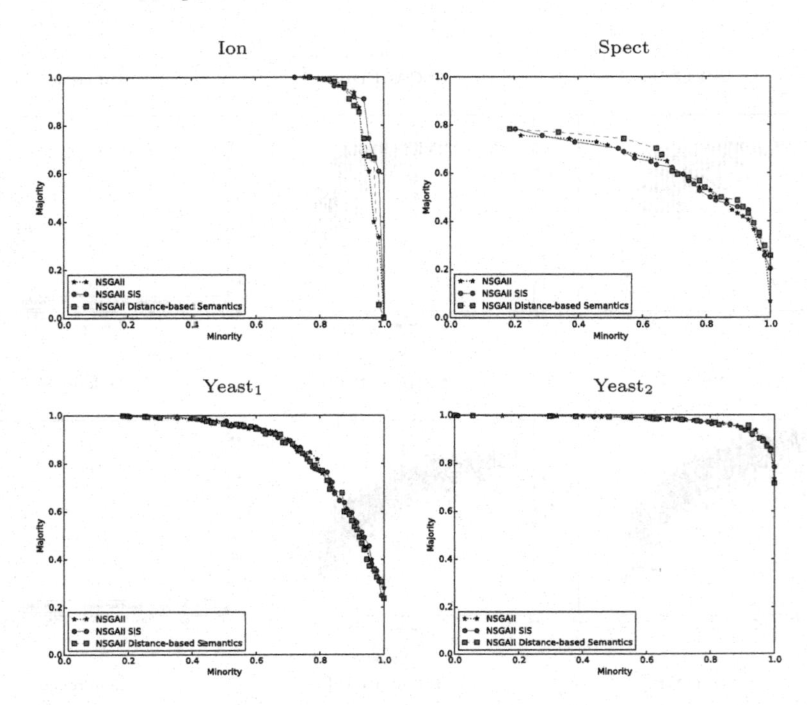

Fig. 2. Pareto-Optimal fronts each of the four problems for NSGA-II (black star symbols), NSGA-II SiS (blue circle symbols) and NSGA-II DBS (red square symbols). (Color figure online)

(NSGA-II SiS is omitted due to space constraints). For some problems (e.g., Ion), it is relatively easy to see that NSGA-II DBS has a better coverage of the objective space. A better look at the difference between NSGA-II and NSGA-II DBS is proposed on the right-hand side of the figure: only the points found by one of both algorithms are plotted, a red plus '+' symbol for NSGA-II, a blue cross 'x' symbol for NSGA-II DBS. More blue cross 'x' symbols are visible on top or right of the objective space, where the true Pareto front lies and explains why DBS has a better performance on the Ion data set.

Figure 2 shows, for each problem, the Pareto-Optimal fronts (POs) for each of the MOGP approaches used in this work. In accordance to the results reported in Table 3, little difference is observed among the three methods on the Yeast$_2$ problem, while NSGA-II SiS dominates on the Yeast$_1$ problem; and both semantic variants dominate parts of the front for Ion, while NSGA-II DBS is a clear winner for Spect.

4.3 Bloat

Bloat (dramatic increase of tree sizes as evolution proceeds) has always been an issue in GP, and should be monitored carefully when designing new GP variants.

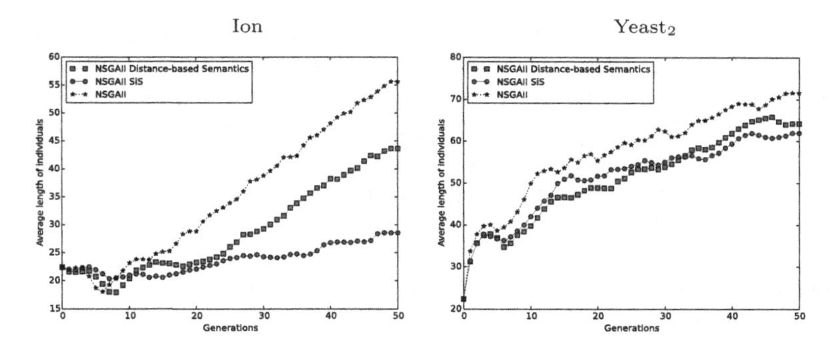

Fig. 3. Average length of evolved solutions vs generations, over 50 independent runs, for the Ion and $Yeast_2$ problems, for NSGA-II (black star symbols), NSGA-II SiS (blue circle symbols) and NSGA-II DBS (red square symbols). (Color figure online)

Contradictory results regarding bloat have been reported for semantic-based GP: semantics seems to prevent bloat in [8], while it exacerbates it in [13]. To shed some light on this issue here, Fig. 3 shows, for the Ion and $Yeast_2$ problems, the average length of evolved trees during evolution.

It is clear that the semantic-based approaches tend to produce slightly shorter programs compared to canonical NSGA-II on the $Yeast_2$ problem – and similar tendency was observed for the $Yeast_1$ and Spect problems (not shown here due to space constraints). Surprisingly, NSGA-II SiS is indeed able to produce much shorter programs compared to both other methods for the Ion problem. This is aligned to the results reported in [8], that indicates that SiS is capable of producing shorter programs compared to e.g., the well-known semantic-based crossover [13]. From this, we believe that researchers tend to report mixed results on bloat because its appearance is dependant on both: problem and approach used, and so, no general conclusions can be drawn on this.

5 Conclusions and Future Work

In Genetic Programming, semantics is commonly defined as the behaviour of syntactically correct programs. In canonical GP, semantics is represented by the output vector of the tree for different known inputs and the similarity between the semantics of two trees gives a much smoother idea of the similarity between the trees than either the syntactic description of the trees or their raw fitness.

This work proposed two ways to add semantics to Multi-Objective GP, more precisely NSGA-II for GP. The first one, Semantics in Selection (SiS), was adapted from canonical GP to NSGA-II. The second approach, named Distance-based Semantics (DBS), consists in using a semantic distance in lieu of the crowding distance at the heart of NSGA-II.

We have learned that semantic-based NSGA-II GP behaves better than plain NSGA-II GP on some well-known unbalanced binary classification problems. We also learned that NSGA-II DBS outperforms NSGA-II SiS. We believe that the reason behind this is because the concept of semantic distance is used into the very core of NSGA-II. There are multiple research areas that we will consider in the near future. An in-depth analysis is required to confirm, and understand why DBS outperforms SiS. Given the encouraging results, it is worth studying the effects of semantics in other parts of a MOGP algorithm (e.g., ranking system). It is also necessary to study the adoption of semantics and its impact in other well-known MO approaches.

Acknowledgements. EGL's research is funded by the Irish Research Council and co-funded by Marie Curie Actions. EGL would like to thank the TAO group at INRIA Saclay France for hosting him during the outgoing phase of the fellowship. The authors would like to thank the anonymous reviewers for their helpful comments.

References

1. Asuncion, A., Newman, D.J.: UCI Machine Learning Repository (2007)
2. Beadle, L., Johnson, C.: Semantically driven crossover in genetic programming. In: 2008 IEEE Congress on Evolutionary Computation CEC 2008. IEEE World Congress on Computational Intelligence, pp. 111–116, June 2008
3. Bhowan, U., Johnston, M., Zhang, M., Yao, X.: Evolving diverse ensembles using genetic programming for classification with unbalanced data. IEEE Trans. Evol. Comput. **17**(3), 368–386 (2013)
4. Coello, C.A.C., Lamont, G.B., Veldhuizen, D.A.V.: Evolutionary Algorithms for Solving Multi-objective Problems. Genetic and Evolutionary Computation. Springer, Secaucus (2006)
5. Deb, K.: Multi-Objective Optimization Using Evolutionary Algorithms. Wiley, New York (2001)
6. Deb, K., Pratap, A., Agarwal, S., Meyarivan, T.: A fast and elitist multiobjective genetic algorithm: NSGA-II. IEEE Trans. Evol. Comput. **6**, 182–197 (2002)
7. Galván-López, E.: Efficient graph-based genetic programming representation with multiple outputs. Intl. J. Autom. Comput. **5**(1), 81–89 (2008)
8. Galván-López, E., Cody-Kenny, B., Trujillo, L., Kattan, A.: Using semantics in the selection mechanism in genetic programming: a simple method for promoting semantic diversity. In: 2013 IEEE Congress on Evolutionary Computation, pp. 2972–2979, June 2013
9. Koza, J.R.: Genetic Programming: On the Programming of Computers by Means of Natural Selection. The MIT Press, Cambridge (1992)
10. Koza, J.R.: Human-competitive results produced by genetic programming. Genet. Prog. Evol. Mach. **11**(3–4), 251–284 (2010)
11. Krawiec, K., Pawlak, T.: Locally geometric semantic crossover: a study on the roles of semantics and homology in recombination operators. Genet. Prog. Evol. Mach. **14**, 31–63 (2013)
12. McPhee, N.F., Ohs, B., Hutchison, T.: Semantic building blocks in genetic programming. In: O'Neill, M., Vanneschi, L., Gustafson, S., Esparcia Alcázar, A.I., De Falco, I., Della Cioppa, A., Tarantino, E. (eds.) EuroGP 2008. LNCS, vol. 4971, pp. 134–145. Springer, Heidelberg (2008)
13. Uy, N.Q., Hoai, N.X., O'Neill, M., McKay, R.I., Galván-López, E.: On the roles of semantic locality of crossover in genetic programming. Inf. Sci. **235**, 195–213 (2013). Data-Based Control, Decision, Scheduling and Fault Diagnostics

14. Uy, N.Q., Hoai, N.X., ONeill, M., McKay, R., Phong, D.N.: On the roles of semantic locality of crossover in genetic programming: application to real-valued symbolic regression. Genet. Prog. Evol. Mach. **12**(2), 91–119 (2011)
15. Vanneschi, L., Castelli, M., Silva, S.: A survey of semantic methods in genetic programming. Genet. Prog. Evol. Mach. **15**(2), 195–214 (2014)
16. Zitzler, E., Brockhoff, D., Thiele, L.: The hypervolume indicator revisited: on the design of pareto-compliant indicators via weighted integration. In: Obayashi, S., Deb, K., Poloni, C., Hiroyasu, T., Murata, T. (eds.) EMO 2007. LNCS, vol. 4403, pp. 862–876. Springer, Heidelberg (2007)

Semantic Forward Propagation for Symbolic Regression

Marcin Szubert[1(✉)], Anuradha Kodali[2,3], Sangram Ganguly[3,4], Kamalika Das[2,3], and Josh C. Bongard[1]

[1] University of Vermont, Burlington, VT 05405, USA
Marcin.Szubert@uvm.edu
[2] University of California, Santa Cruz, CA 95064, USA
[3] NASA Ames Research Center, Moffett Field, CA 94035, USA
[4] Bay Area Environmental Research Institute, Petaluma, CA 94952, USA

Abstract. In recent years, a number of methods have been proposed that attempt to improve the performance of genetic programming by exploiting information about program semantics. One of the most important developments in this area is *semantic backpropagation*. The key idea of this method is to decompose a program into two parts—a subprogram and a context—and calculate the *desired* semantics of the subprogram that would make the entire program correct, assuming that the context remains unchanged. In this paper we introduce Forward Propagation Mutation, a novel operator that relies on the opposite assumption—instead of preserving the context, it retains the subprogram and attempts to place it in the semantically right context. We empirically compare the performance of semantic backpropagation and forward propagation operators on a set of symbolic regression benchmarks. The experimental results demonstrate that semantic forward propagation produces smaller programs that achieve significantly higher generalization performance.

Keywords: Genetic programming · Program semantics · Semantic backpropagation · Problem decomposition · Symbolic regression

1 Introduction

Standard tree-based genetic programming (GP) searches the space of programs using traditional operators of subtree-swapping crossover and subtree-replacing mutation [4]. These operators are designed to be generic and produce syntactically correct offspring regardless of the problem domain. However, their actual effects on the behavior of the program, and thus its fitness, are generally hard to predict. For this reason, many alternative search operators have been recently proposed that take into account the influence of syntactic modifications on program semantics [1,10,11,13].

Semantic backpropagation [12,15] is arguably one of the most powerful techniques employed by such semantic-aware GP operators. The two operators based

J. Handl et al. (Eds.): PPSN XIV 2016, LNCS 9921, pp. 364–374, 2016.
DOI: 10.1007/978-3-319-45823-6_34

on semantic backpropagation—Random Desired Operator (RDO) and Approximately Geometric Crossover (AGX) have proved to be successful on a number of symbolic regression and boolean program synthesis problems [11,12]. Both operators rely on semantic decomposition of an existing program into two parts—a subprogram and its context. Given a subprogram, both operators attempt to calculate its *desired semantics*, i.e., the values that it should return to make the entire program produce the desired output, assuming that the context remains unchanged. The desired semantics can be then used to find a replacement for the subprogram that improves the overall program behavior.

Despite their superior performance when compared to other GP search operators [11,12,15], backpropagation-based RDO and AGX face a few major challenges that can limit their practical applicability. First of all, they are much more computationally expensive than traditional syntactic operators. Indeed, in order to calculate desired semantics, the target program output needs to be *backpropagated* by traversing the tree and inverting the execution of particular instructions. The computational cost of this operation is similar to the cost of a single fitness evaluation (which is typically the most expensive component of GP). Moreover, using desired semantics to find a subprogram replacement usually requires even more computational effort. Finally, the results reported so far demonstrate that RDO and AGX tend to produce relatively large programs that are difficult to interpret and may suffer from overfitting.

In this paper, we introduce Forward Propagation Mutation (FPM), a novel semantic-aware operator that also relies on program decomposition but works in the opposite manner to semantic backpropagation. Instead of preserving the context and replacing the subprogram, forward propagation retains the subprogram and attempts to place it in the semantically right context. In contrast to semantic backpropagation, the FPM operator does not require an additional tree traversal and thus it incurs less computational overhead. Moreover, the experimental results obtained on a set of univariate and bivariate symbolic regression problems demonstrate that it achieves competitive performance in terms of the training error while producing much smaller programs that usually perform significantly better on the unseen test cases.

2 Semantic Genetic Programming

In order to incorporate semantic-awareness into genetic programming, most of the recently proposed methods adopt a common definition of program semantics, known as *sampling semantics* [13], which is identified with the vector of outputs produced by a program for a sample of possible inputs. In supervised learning problems considered here, where n input-output pairs are given as a training set $T = \{(\mathbf{x}_1, y_1), \ldots, (\mathbf{x}_n, y_n)\}$, semantics of a program p is equal to vector $\mathbf{s}(p) = [p(\mathbf{x}_1), \ldots, p(\mathbf{x}_n)]$, where $p(\mathbf{x})$ is a result obtained by running program p on input $\mathbf{x}$. Consequently, each program p corresponds to a point in n-dimensional semantic space and a metric d can be adopted to measure semantic distance between two programs. Furthermore, fitness of a program p can be calculated as a distance between its semantics $\mathbf{s}(p)$ and the target semantics $\mathbf{t} = [y_1, \ldots, y_n]$ defined by the training set, i.e., $f(p) = d(\mathbf{s}(p), \mathbf{t})$.

The information about program semantics and the structure of the semantic space endowed by a metric-based fitness function can be exploited in many ways to facilitate the search process carried out by GP. Apart from numerous semantic search operators [1,10,11,13], the knowledge about semantics can be used to maintain population diversity [3], to initialize the population [2] or to drive the selection process [7]. All such semantic-aware methods are collectively captured by the umbrella term of semantic genetic programming [14]. Recently, a paradigm of behavioral program synthesis [5] has been proposed, which extends semantic GP by using information not only about final program results but also about behavioral characteristics of program execution.

3 Semantic Backpropagation

One of the most important methods in semantic GP is semantic backpropagation [12]. The key concept behind this method is *program decomposition*: a program p is treated as a function (i.e., it is deterministic and has no side effects) that can be decomposed into two constituent functions (subprograms) p_1 and p_2 such that $p(\mathbf{x}) = p_2(p_1(\mathbf{x}), \mathbf{x})$. In particular, if a program is represented as a tree, such decomposition can be made at each node—the inner function p_1 is expressed by the subtree rooted at the given node, while the outer function p_2 corresponds to the rest of the tree (also termed *context* [9], see left part of Fig. 1).

Semantic backpropagation assumes that the desired program output $p^*(\mathbf{x})$ can be produced by retaining the outer function and replacing just the inner one by another subprogram p_s, i.e., $p^*(\mathbf{x}) = p_2(p_s(\mathbf{x}), \mathbf{x})$. Starting from the desired program output $p^*(\mathbf{x})$, the backpropagation algorithm heuristically inverts the program execution to calculate the desired semantics of the subprogram p_s, i.e., the values it should produce to make the entire program correct. This idea has been employed to design two operators, AGX and RDO, which differ with respect to what they use as the desired program output $p^*(\mathbf{x})$. In this study, we focus on RDO, a mutation operator that assumes that target semantics $\mathbf{t} = [y_1, \ldots, y_n]$ is given *a priori* and thus values $p^*(\mathbf{x}_i) = y_i$ can be used as an input for the backpropagation algorithm.

An example of a mutation performed by RDO is illustrated in Fig. 1 and proceeds as follows. First, a random mutation node is selected in the parent program (denoted as a circle with a double border in Fig. 1). The subtree p_1 rooted at this node is removed from the tree and the backpropagation algorithm is applied to calculate the desired semantics of the replacement p_s that would make the offspring program return desired values. The algorithm starts from the root of the tree, where desired semantics is given by $\mathbf{t}$, and follows the path to the removed subtree. For each node it calculates the desired semantics of its child by invoking the Invert function (a detailed description of this function and the RDO operator in general can be found in [12]).

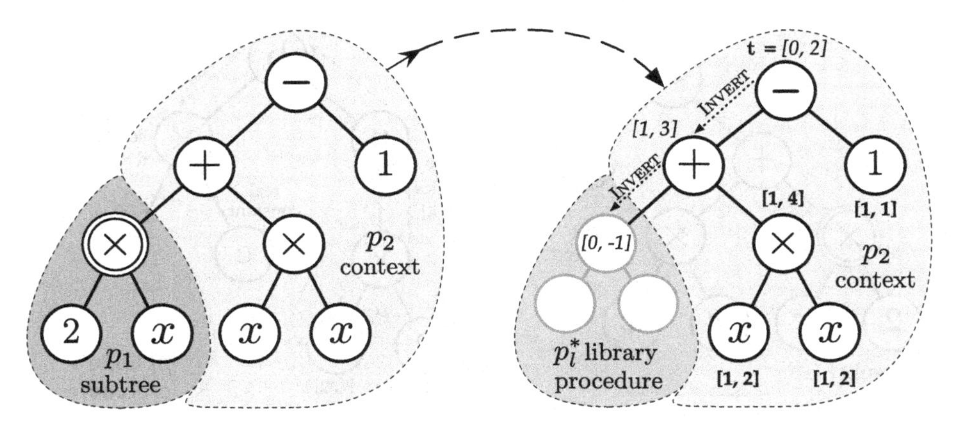

Fig. 1. A mutation performed by Random Desired Operator using semantic backpropagation. Desired semantics are denoted in italics.

For instance, let us assume that a training set contains just two cases with inputs $\mathbf{x} = [1, 2]$ and desired outputs $\mathbf{t} = [0, 2]$. As shown in Fig. 1, in the first step the algorithm finds out that to produce desired semantics at the root, knowing that outputs of its right child are equal to $[1, 1]$, the desired semantics of the left child must be equal to $[1, 3]$. This result is used in the subsequent step to calculate desired semantics for the next node. Finally, given desired semantics at the mutation node, the RDO operator attempts to replace the removed subtree with a subprogram that would produce such values. To this end, it employs a precomputed *library* of programs (procedures) that allows to efficiently retrieve a program p_l^* that has the smallest semantic distance to the desired semantics. Additionally, RDO also checks if a single constant real value would provide a better match to the desired semantics than p_l^*.

Importantly, in the process of semantic backpropagation, inverting certain functions can be ambiguous (if the function is not injective) or impossible (if the function is not surjective). As a result, the desired semantics may contain several values for each training case or special *inconsistent* elements. The library must be able to handle such queries efficiently [12,15].

4 Semantic Forward Propagation

Inspired by semantic backpropagation and RDO we propose an alternative mutation operator based on the complementary idea, which we term *semantic forward propagation*. Similarly to RDO, Forward Propagation Mutation (FPM) relies on decomposability of a program p into a subtree p_1 and a context p_2. However, while RDO assumes that a context can be preserved and attempts to replace the subtree, FPM makes the opposite assumption preserving the subtree and building a matching context for it.

The FPM operator starts by choosing a random mutation node in the parent program. The subtree p_1 rooted at this node is extracted from the tree and used

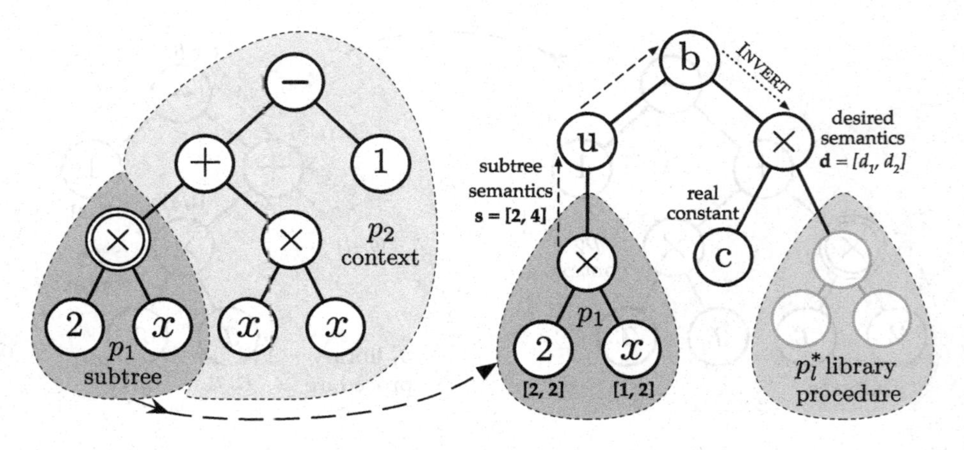

Fig. 2. An operation performed by Forward Propagation Mutation.

as a starting point for creating an offspring. In order to build a new context for this subtree, we assume a fixed structure of the context p_c containing 4 new nodes and a matching library procedure (see Fig. 2). We apply an exhaustive search to identify a context p_c^* of the assumed structure, that minimizes fitness of the entire offspring program $p_c^* = \arg\min_{p_c} f(p_c \circ p_1)$. To this end, we consider all pairwise combinations of the available unary (e.g., $\{\sin, \cos, \log, \exp\}$) and binary functions (e.g., $\{\times, +, -, /\}$) that could be placed directly above the selected subtree, as nodes u and b, respectively (cf. Fig. 2). Importantly, we extend the unary function set with the identity function $id(x) = x$. If the best found context p_c^* uses this function we skip adding the node u to the tree. For each pair of functions (u, b) placed above the subtree p_1, we *forward propagate* the semantics of the subtree up to the root of the new tree. Then, we apply just a single backpropagation step, using the same INVERT function as in RDO, to calculate desired semantics $\mathbf{d}$ of the other child of the node b, given the the target semantics $\mathbf{t}$ and the forward-propagated semantics $\mathbf{s}(u \circ p_1)$.

Since in this case the desired semantics is usually unambiguous, we can use a different method of searching the library, which could not be easily applied within the RDO operator. Here, we search for the library procedure which achieves highest cosine similarity. In other words, if we treat semantics as an n-dimensional vector, we return library procedure p_l^* that makes the smallest angle with the desired semantics $\mathbf{d}$, i.e.:

$$p_l^* = \arg\min_{p_l \in L} \arccos \frac{\mathbf{s}(p_l) \cdot \mathbf{d}}{\|\mathbf{s}(p_l)\| \|\mathbf{d}\|}.$$

Finally, we add a constant node c to scale the semantics of the library procedure making it closer to the desired semantics, i.e., $c = (\mathbf{s}(p_l^*) \cdot \mathbf{d}) \;/\; \|\mathbf{s}(p_l^*)\|^2$. An alternative, more computationally expensive approach, would be to run simple linear regression for each candidate program in the library, using its semantics as a single explanatory variable and desired semantics $\mathbf{d}$ as a response. This approach would require extending the context structure to accommodate both an intercept and a slope coefficient.

5 Experimental Setup

The main goal of the experiments is to compare the performance of RDO and FPM mutation operators on a suite of symbolic regression benchmarks. Additionally, as a control setup we employ traditional subtree-replacing mutation (SRM). All three mutation operators are used along with conventional subtree-swapping crossover in a standard generational tree-based GP algorithm with tournament selection. Each mutation operator is employed in five setups with different values of mutation and crossover probabilities (the source code of our experiments is available at https://github.com/mszubert/ppsn_2016).

Most of the GP parameters (summarized in Table 1) are adopted from the recent work on semantic backpropagation [12]. In particular, whenever a random mutation/crossover node needs to be selected, a *uniform depth node selector* is used. Given a program p, it first calculates program's height h, then draws uniformly an integer d from the interval $[0, h]$ and finally selects a random node from all nodes at depth d in program p. This technique has been recently shown to reduce bloat when compared to conventional Koza-I node selectors [6, 12].

Moreover, both RDO and FPM use population-based library which is constructed at each generation from all semantically unique subtrees (subprograms) in the current population. Since we impose an upper limit on the tree height (17), when searching the library we ignore all the procedures that would violate this constraint when inserted into the parent program.

We investigate training error, generalization performance (error on 1 000 unseen test cases) and the size of programs produced by using particular mutation operators on 11 symbolic regression benchmarks. We consider six univariate and five bivariate problems that are adopted from previous studies [4, 8, 12].

Table 1. Genetic programming parameters

Parameter	Value
Population size	256
Generations	100
Initialization	Ramped half-and-half with height range 2–6
	100 retries until accepting a syntactic duplicate
Instruction set	$\{+, -, \times, /, \exp, \log, \sin, \cos\}$ (log and/are protected)
Tournament size	7
Fitness function	Root-mean-square error (RMSE)
Node selection	Uniform depth node selector
Maximum tree height	17
Number of runs	30

Table 2. Symbolic regression benchmarks.

Benchmark name	Objective function	Variables	Training cases
P4 (QUARTIC)	$x^4 + x^3 + x^2 + x$	1	20
P7 (SEPTIC)	$x^7 - 2x^6 + x^5 - x^4 + x^3 - 2x^2 + x$	1	20
P9 (NONIC)	$\sum_1^9 x^i$	1	20
R1	$(x+1)^3/(x^2 - x + 1)$	1	20
R2	$(x^5 - 3x^3 + 1)/(x^2 + 1)$	1	20
R3	$(x^6 + x^5)/(x^4 + x^3 + x^2 + x + 1)$	1	20
K11 (KEIJZER-11)	$xy + \sin((x-1)(y-1))$	2	100
K12 (KEIJZER-12)	$x^4 - x^3 + \frac{y^2}{2} - y$	2	100
K13 (KEIJZER-13)	$6 \sin(x) \cos(y)$	2	100
K14 (KEIJZER-14)	$\frac{8}{2+x^2+y^2}$	2	100
K15 (KEIJZER-15)	$\frac{x^3}{5} + \frac{y^3}{2} - x - y$	2	100

Selected benchmarks (see Table 2) include polynomial, rational and trigonometric functions. For each problem, fitness was calculated as root-mean-square error on a number of training cases. The univariate problems use 20 cases distributed equidistantly in the $[-1, 1]$ range, while the bivariate ones use a grid of $10 \times 10 = 100$ points spaced evenly in the $[-1, 1] \times [-1, 1]$ square.

6 Results and Discussion

Table 3 presents detailed characteristics of the best-of-run individuals evolved with particular mutation operators. Each row of the table corresponds to a single combination of one of the five GP setups (with different crossover (**X**) and mutation (**M**) probabilities) and one of the three considered mutation operators (either FPM, RDO or SRM). We performed 30 independent GP runs for each of such 15 combinations on each of the 11 symbolic regression problems. To confirm statistically significant differences between the results obtained with particular mutation operators, for each problem and parameters setup we conducted the Kruskal-Wallis test followed by a post-hoc analysis using pairwise Mann-Whitney tests (with sequential Bonferroni correction). We set the level of significance at $p \leq 0.05$. Table 3 shows with an underline the results that were found significantly better than those achieved with the other operators.

The first part of Table 3 shows the average training errors. Although RDO achieves the best overall results for most univariate problems, for the bivariate ones FPM produces more competitive results. Regardless of the parameter settings, the traditional SRM operator leads to the highest training error. Noteworthy, the RDO and FPM operators obtain their best results under different crossover and mutation settings. While both of them benefit from using traditional crossover as an additional variation operator, the performance of FPM

Table 3. Detailed characteristics of best-of-run individuals produced by particular mutation operators (FPM, RDO, SRM), aggregated over 30 GP runs. Each operator was employed in 5 GP setups with different crossover (**X**) and mutation (**M**) probabilites. Bold marks the best results achieved under certain **X**/**M** settings on particular problems. Underline indicates statistically significant superiority.

	X	M	P4	P7	P9	R1	R2	R3	K11	K12	K13	K14	K15
			Average training error										
FPM	0.0	1.0	**0.0011**	0.0072	0.0153	**0.0064**	0.0049	**0.0024**	**0.0626**	**0.0418**	**0.0000**	**0.0031**	**0.0061**
FPM	0.5	0.5	**0.0001**	0.0018	**0.0025**	**0.0012**	0.0018	**0.0006**	**0.0299**	0.0111	**0.0000**	**0.0012**	**0.0007**
FPM	0.5	1.0	**0.0001**	0.0025	0.0037	0.0020	0.0025	0.0007	0.0358	0.0154	**0.0000**	**0.0021**	**0.0007**
FPM	1.0	0.5	**0.0000**	**0.0015**	0.0022	**0.0012**	**0.0013**	**0.0004**	**0.0283**	0.0086	**0.0000**	**0.0012**	**0.0004**
FPM	1.0	1.0	**0.0001**	0.0026	0.0029	0.0018	0.0019	0.0006	0.0334	0.0116	**0.0000**	**0.0017**	**0.0007**
RDO	0.0	1.0	0.0030	**0.0034**	**0.0147**	0.0071	**0.0043**	0.0030	0.0709	0.0444	0.0440	0.0587	0.0302
RDO	0.5	0.5	0.0004	**0.0017**	0.0029	0.0023	**0.0014**	0.0018	0.0455	**0.0090**	0.0038	0.0132	0.0024
RDO	0.5	1.0	**0.0001**	**0.0008**	**0.0004**	**0.0006**	**0.0004**	**0.0002**	**0.0294**	**0.0029**	0.0007	0.0044	0.0015
RDO	1.0	0.5	0.0003	0.0020	**0.0008**	0.0014	0.0015	**0.0004**	0.0504	**0.0063**	0.0015	0.0087	0.0041
RDO	1.0	1.0	**0.0001**	**0.0003**	**0.0003**	**0.0008**	**0.0004**	**0.0004**	**0.0294**	**0.0047**	0.0011	0.0033	0.0008
SRM	0.0	1.0	0.0518	0.0742	0.0758	0.0744	0.0811	0.0097	0.2025	0.3049	0.1552	0.2145	0.0723
SRM	0.5	0.5	0.0323	0.0968	0.0732	0.0834	0.0608	0.0156	0.1769	0.2328	0.1040	0.1138	0.0608
SRM	0.5	1.0	0.0449	0.0926	0.0638	0.0792	0.0880	0.0115	0.1781	0.2128	0.1267	0.1603	0.0748
SRM	1.0	0.5	0.0217	0.0882	0.0715	0.0663	0.0666	0.0078	0.1598	0.2005	0.0866	0.1690	0.0623
SRM	1.0	1.0	0.0282	0.0845	0.0792	0.0698	0.0754	0.0120	0.1942	0.2479	0.1437	0.1724	0.0628
			Median test error										
	X	M	P4	P7	P9	R1	R2	R3	K11	K12	K13	K14	K15
FPM	0.0	1.0	**0.0009**	**0.0084**	**0.0342**	**0.0071**	**0.0044**	**0.0026**	**0.0555**	**0.0529**	**0.0000**	**0.0028**	**0.0057**
FPM	0.5	0.5	**0.0000**	**0.0046**	**0.0256**	**0.0025**	**0.0123**	0.0030	**0.0425**	0.0581	**0.0000**	**0.0017**	**0.0008**
FPM	0.5	1.0	**0.0000**	**0.0045**	**0.0142**	**0.0037**	**0.0045**	**0.0015**	**0.0333**	**0.0290**	**0.0000**	**0.0032**	**0.0008**
FPM	1.0	0.5	**0.0000**	**0.0069**	0.0306	**0.0030**	**0.0042**	**0.0017**	**0.0260**	**0.0311**	**0.0000**	**0.0021**	**0.0005**
FPM	1.0	1.0	**0.0000**	**0.0055**	**0.0227**	**0.0025**	**0.0027**	**0.0009**	**0.0300**	**0.0295**	**0.0000**	**0.0024**	**0.0008**
RDO	0.0	1.0	0.0039	0.0593	0.0346	0.0087	0.0145	0.0071	0.1089	0.0988	0.0215	0.0774	0.0185
RDO	0.5	0.5	0.0025	0.5159	0.0469	0.0406	0.0148	**0.0028**	0.0738	**0.0374**	0.0070	0.0252	0.0036
RDO	0.5	1.0	0.0117	0.3084	0.0715	0.1522	0.0652	0.0618	0.0639	0.2124	0.0022	0.0556	0.0097
RDO	1.0	0.5	0.0006	0.0704	**0.0104**	0.0081	0.0319	0.0057	0.0445	0.0364	0.0030	0.0283	0.0014
RDO	1.0	1.0	0.0097	19.486	8E+3	0.0607	0.0466	0.0155	0.0402	0.3878	0.0023	0.0378	0.0026
SRM	0.0	1.0	0.0485	0.1170	0.1017	0.0836	0.0585	0.0123	0.2005	0.2649	0.1986	0.1458	0.0814
SRM	0.5	0.5	0.0240	0.0958	0.0810	0.0730	0.0592	0.0106	0.1770	0.1874	0.1122	0.0988	0.0525
SRM	0.5	1.0	0.0572	0.1865	0.0922	0.0800	0.0694	0.0105	0.1686	0.2037	0.1311	0.1207	0.0882
SRM	1.0	0.5	0.0191	0.0899	0.0785	0.0711	0.0641	0.0101	0.1493	0.1874	0.0998	0.1126	0.0446
SRM	1.0	1.0	0.0257	0.0734	0.0725	0.0739	0.0727	0.0142	0.1894	0.1853	0.1843	0.1723	0.0389
			Average program size										
	X	M	P4	P7	P9	R1	R2	R3	K11	K12	K13	K14	K15
FPM	0.0	1.0	172.6	179.0	195.9	162.2	187.9	161.0	210.9	172.4	**9.1**	204.9	207.7
FPM	0.5	0.5	150.4	341.4	322.1	325.3	352.8	347.7	328.3	305.5	**7.4**	326.5	260.6
FPM	0.5	1.0	**78.3**	292.4	271.7	287.9	283.3	265.9	286.4	264.5	**8.5**	258.9	239.6
FPM	1.0	0.5	**44.0**	346.6	354.4	327.2	339.2	311.0	328.0	311.1	**7.8**	298.6	300.0
FPM	1.0	1.0	**99.2**	283.4	271.0	255.7	253.3	270.6	244.8	230.6	**8.9**	239.8	264.4
RDO	0.0	1.0	537.6	690.6	550.8	777.5	2656.9	1203.7	418.6	434.8	85.0	147.2	250.2
RDO	0.5	0.5	503.6	637.9	686.0	493.9	529.6	485.7	358.4	482.4	497.1	346.4	1299.6
RDO	0.5	1.0	626.9	1004.3	934.1	906.7	854.0	747.2	654.2	841.2	464.3	548.6	1137.2
RDO	1.0	0.5	378.6	631.2	588.4	473.0	508.9	486.7	316.8	472.9	311.0	325.5	673.8
RDO	1.0	1.0	645.6	903.6	909.9	668.6	746.4	696.2	542.9	838.7	426.7	514.6	1034.4
SRM	0.0	1.0	**122.9**	**176.1**	**152.4**	**133.8**	**116.2**	**155.9**	**109.3**	**95.1**	95.7	**63.0**	**74.7**
SRM	0.5	0.5	**60.0**	**109.4**	**95.4**	**79.7**	**76.1**	**95.8**	**62.7**	**69.4**	59.6	**53.5**	**57.5**
SRM	0.5	1.0	111.8	**172.8**	**159.6**	**154.3**	**122.3**	**173.6**	**99.2**	**95.3**	96.1	**79.1**	**82.0**
SRM	1.0	0.5	97.9	**106.4**	**107.1**	**96.8**	**89.9**	**137.2**	**89.5**	**86.6**	81.1	**87.6**	**64.4**
SRM	1.0	1.0	119.0	**160.9**	**147.5**	**150.6**	**131.0**	**165.7**	**95.3**	**83.2**	95.9	**80.4**	**96.5**

decreases when mutation is performed too frequently (i.e., if $\mathbf{M} = 1.0$). To explain this phenomenon let us note that for a given subprogram, the FPM operator builds a context in a deterministic way. As a result, if two semantically equivalent subprograms are selected in the same generation, they will result in identical offspring. Consequently, FPM can lead to creating too many duplicated programs and thus losing diversity in the population. Importantly, although RDO is also deterministic, it is less susceptible to this problem because typically the number of distinct contexts is much larger than that of distinct subtrees.

In order to assess generalization performance of evolved programs, we calculate the root-mean-square error on 1000 test cases drawn uniformly from the same range as for the training cases. The median test errors committed by the best-of-run individuals are presented in the second part of Table 3. In most cases, the RDO operator (especially for setups that achieve the lowest training error) suffers from substantial overfitting resulting in large test error. Although the FPM operator is also vulnerable to overfitting (in particular on problem P9) it is not as severe as in the case of RDO. With a few exceptions, for each of the considered problems and parameter setups, the FPM operator obtains the highest generalization performance.

Finally, we investigate the average size of best-of-run individuals which is presented in the last part of Table 3. Not surprisingly RDO is the most bloating operator and this is one of the reasons for its poor performance on the unseen test data. On the other hand, in preliminary experiments with imposed program size limit of 300 nodes, we also observed overfitting of the RDO operator. The programs produced by FPM tend to be much smaller. In particular, on two relatively simple problems, P4 and K13, the FPM operator finds short programs that obtain zero test error. Apparently, employing FPM allows to discover solutions that are very close to the original function underlying the training data. However, on all the other problems, the programs produced by RDO and FPM are significantly larger than those created by the traditional SRM operator.

7 Conclusions

Semantic GP operators have proved to be effective on a number of symbolic regression problems [11,13,14]. In this study, we confirmed these observations by analyzing the performance of the RDO operator based on semantic backpropagation [12] and the FPM operator that employs a novel idea of semantic forward propagation. When applied to a suite of symbolic regression benchmarks, both operators significantly outperformed the subtree-replacing mutation operator conventionally applied in GP. However, while both considered semantic operators achieved competitive performance on the training data, the RDO operator was found much more susceptible to overfitting. The proposed FPM operator, on the other hand, consistently produced shorter programs that obtained significantly lower error on the unseen test data.

Despite achieving superior predictive accuracy and producing shorter programs than RDO, the programs constructed with the FPM operator are still too large to be easily understood. This is unfortunate since finding comprehensible solutions has been always considered as one of the primary benefits of using GP instead of black-box machine learning methods. As most semantic-aware operators tend to produce large or very large programs [10], the problem of bloat remains the major challenge that can limit the practical applicability of such methods. Therefore, one of the most important directions of future work is to investigate the performance of RDO and FPM operators combined with parsimony pressure mechanisms that control the complexity of evolved programs.

Acknowledgments. This work was supported by the National Aeronautics and Space Administration under grant number NNX15AH48G.

References

1. Beadle, L., Johnson, C.G.: Semantically driven crossover in genetic programming. In: Proceedings of the IEEE Congress on Evolutionary Computation, CEC 2008, pp. 111–116. IEEE (2008)
2. Beadle, L., Johnson, C.G.: Semantic analysis of program initialisation in genetic programming. Genet. Prog. Evol. Mach. **10**(3), 307–337 (2009)
3. Jackson, D.: Promoting phenotypic diversity in genetic programming. In: Schaefer, R., Cotta, C., Kołodziej, J., Rudolph, G. (eds.) PPSN XI. LNCS, vol. 6239, pp. 472–481. Springer, Heidelberg (2010)
4. Koza, J.R.: Genetic Programming: On the Programming of Computers by Means of Natural Selection. MIT Press, Cambridge (1992)
5. Krawiec, K.: Behavioral Program Synthesis with Genetic Programming, Studies in Computational Intelligence, vol. 618. Springer, Heidelberg (2016)
6. Krawiec, K., O'Reilly, U.M.: Behavioral programming: a broader and more detailed take on semantic GP. In: Proceedings of the 2014 Annual Conference on Genetic and Evolutionary Computation, GECCO 2014, pp. 935–942. ACM (2014)
7. Liskowski, P., Krawiec, K., Helmuth, T., Spector, L.: Comparison of semantic-aware selection methods in genetic programming. In: Proceedings of the Genetic and Evolutionary Computation Conference, pp. 1301–1307. ACM (2015)
8. McDermott, J., White, D.R., Luke, S., Manzoni, L., Castelli, M., Vanneschi, L., Jaskowski, W., Krawiec, K., Harper, R., De Jong, K., O'Reilly, U.M.: Genetic programming needs better benchmarks. In: Proceedings of the Genetic and Evolutionary Computation Conference, pp. 791–798. ACM (2012)
9. McPhee, N.F., Hopper, N.J.: Analysis of genetic diversity through population history. In: Proceedings of the Genetic and Evolutionary Computation Conference, vol. 2, pp. 1112–1120. Morgan Kaufmann (1999)
10. Moraglio, A., Krawiec, K., Johnson, C.G.: Geometric semantic genetic programming. In: Coello, C.A.C., Cutello, V., Deb, K., Forrest, S., Nicosia, G., Pavone, M. (eds.) PPSN 2012, Part I. LNCS, vol. 7491, pp. 21–31. Springer, Heidelberg (2012)
11. Pawlak, T.P., Wieloch, B., Krawiec, K.: Review and comparative analysis of geometric semantic crossovers. Genet. Prog. Evol. Mach. **16**(3), 351–386 (2015)
12. Pawlak, T., Wieloch, B., Krawiec, K.: Semantic backpropagation for designing search operators in genetic programming. IEEE Trans. Evol. Comput. **19**(3), 326–340 (2015)

13. Uy, N.Q., Hoai, N.X., O'Neill, M., Mckay, R.I., Galván-López, E.: Semantically-based crossover in genetic programming: application to real-valued symbolic regression. Genet. Prog. Evol. Mach. **12**(2), 91–119 (2011)
14. Vanneschi, L., Castelli, M., Silva, S.: A survey of semantic methods in genetic programming. Genet. Prog. Evol. Mach. **15**(2), 195–214 (2014)
15. Wieloch, B., Krawiec, K.: Running programs backwards: instruction inversion for effective search in semantic spaces. In: Proceedings of the 15th Annual Conference on Genetic and Evolutionary Computation, GECCO 2013, pp. 1013–1020. ACM, New York (2013)

Reducing Dimensionality to Improve Search in Semantic Genetic Programming

Luiz Otavio V.B. Oliveira[1(✉)], Luis F. Miranda[1], Gisele L. Pappa[1], Fernando E.B. Otero[2], and Ricardo H.C. Takahashi[3]

[1] Computer Science Department, Universidade Federal de Minas Gerais, Belo Horizonte, Brazil
{luizvbo,luisfmiranda,glpappa}@dcc.ufmg.br

[2] School of Computing, University of Kent, Chatham Maritime, UK
F.E.B.Otero@kent.ac.uk

[3] Mathematics Department, Universidade Federal de Minas Gerais, Belo Horizonte, Brazil
taka@mat.ufmg.br

Abstract. Genetic programming approaches are moving from analysing the syntax of individual solutions to look into their semantics. One of the common definitions of the semantic space in the context of symbolic regression is a n-dimensional space, where n corresponds to the number of training examples. In problems where this number is high, the search process can became harder as the number of dimensions increase. Geometric semantic genetic programming (GSGP) explores the semantic space by performing geometric semantic operations—the fitness landscape seen by GSGP is guaranteed to be conic by construction. Intuitively, a lower number of dimensions can make search more feasible in this scenario, decreasing the chances of data overfitting and reducing the number of evaluations required to find a suitable solution. This paper proposes two approaches for dimensionality reduction in GSGP: (i) to apply current instance selection methods as a pre-process step before training points are given to GSGP; (ii) to incorporate instance selection to the evolution of GSGP. Experiments in 15 datasets show that GSGP performance is improved by using instance reduction during the evolution.

Keywords: Dimensionality reduction · Semantic genetic programming · Instance selection

1 Introduction

Evolutionary computation methods have recently turned their attention to the semantics of the solutions represented by individuals instead of focusing only on their syntax [17]. Particularly, in the case of genetic programming, many methods are switching from the syntactic space to work on a n-dimensional semantic space, where n is the number of training instances we learn the function from.

J. Handl et al. (Eds.): PPSN XIV 2016, LNCS 9921, pp. 375–385, 2016.
DOI: 10.1007/978-3-319-45823-6_35

When applying any function—e.g., an individual—to the training set, the produced output corresponds to a point in the semantic space.

Given the definition above, the number of dimensions of the semantic space equals the number of training examples. In problems where this number is high—a common scenario in real-world applications—the search process can became harder as the number of dimensions increase, a problem well-known as the *curse of dimensionality* [5]. As the number of dimensions of the problem increases, the volume of the search space also increases exponentially.

One of the simplest ways to deal with the curse of dimensionality is to reduce the number of dimensions of the search space [5]. As in geometric semantic genetic programming each space dimension corresponds to a training instance, an alternative is to perform what in the machine learning literature is known as data instance selection. Data instance selection is a well-known problem within the context of data classification, but there is not extensive work regarding regression problems [2]. Instance selection methods are strongly based on distances between training instances from both the set of input and output features.

This paper evaluates the impact of reducing the number of dimensions of the semantic space in the context of geometric semantic genetic programming. This scenario is interesting since the crossover and mutation operators guarantee the semantic fitness landscape explored by GP is conic, which can be optimized by evolutionary algorithms with good results for virtually any metric, as indicated by [13]. Intuitively, a lower number of dimensions can make search more feasible, reducing the number of evaluations required to find a suitable solution while decreasing the chances of data overfitting.

Looking at how current instance selection methods work, we take advantage of previous knowledge and propose two approaches for instance selection: (i) apply current instance selection methods as a pre-process step before training points are given to GSGP; (ii) incorporate instance selection to the evolution of GSGP. In the first case, we use the methods Threshold Condensed Nearest Neighbor (TCNN) and Threshold Edited Nearest Neighbor (TENN) [9] to select instances, which are then given to GSGP. The second approach incorporates instance selection to the evolution of GSGP, through the proposed Probabilistic instance Selection based on the Error (PSE) method.

Computational experiments in 15 real-world and synthetic datasets, where the number of training instances varies from 50 to 4000 and instance number reduction (i.e. search space dimension reduction) of up to 68.50 %, show that results obtained by TCNN and TENN are no better than those generated by a random selection scheme. PSE, in turn, shows results statistically significantly better than GSGP with all instances in 5, and no statistical difference in 7 cases.

2 Related Work

Instance selection methods are commonly used in the classification literature [6], and play different roles in noisy and noise-free application scenarios. In noise-free scenarios, the idea is to remove points from the training set without degrading accuracy, such as improving storage and search time. In noisy application domains, the main idea is to remove outliers. In classification, these methods rely on the class labels of neighbour instances to determine the rejection/acceptance of an instance to the selected set. However, there are not many methods for instance selection in regression problems. A few works have extended well-known instance selection methods for classification to the context of regression [2].

The authors in [7] introduced a method based on mutual information, inspired by feature selection methods that rely on this criterion. The method focuses on noise-free scenarios, and has as its main objective to choose the best subset of instances to build a model. In this same direction, the authors in [16] propose Class Conditional Instance Selection for Regression (CCISR). It extends the Class Conditional Instance Selection method for classification, which uses a class conditional nearest neighbour relation to guide the search process. The authors in [9] proposed the Threshold Condensed Nearest Neighbor (TCNN) and Threshold Edited Nearest Neighbor (TENN) algorithms—regression versions of the ENN and CNN methods for classification, respectively. These algorithms will be discussed in the next section, as they are used in this paper.

Recently, the authors in [2] compared different strategies for instance selection in regression: discretization techniques—which transform the continuous outputs of the problem into discrete variables and then apply the traditional version of instance selection methods for classification—TCNN and TENN. They also proposed an ensemble method, namely bagging, to combine several instance selection algorithms. Each algorithm within the ensemble returns an array of binary votes (0 means the instance is not selected and 1 otherwise), and the relevance of an instance in the training set is considered proportional to the number of accumulated votes. The final instance selection is given by a threshold, which defines the percentage of votes an instance must have to be selected. As expected, the ensemble method presented the best results overall.

In our context, the use of an ensemble is not justifiable, as it is a time consuming task and would add too much time overhead to the search. For this reason, we adopted the threshold versions of TCNN and TENN, as the first assumes noise-free scenarios and the second focuses on outliers.

3 Strategies for Semantic Space Dimensionality Reduction

This section introduces two strategies to reduce the dimensionality of the semantic search space. First, we formally introduce the problem and motivation for instance selection in this scenario. Given a finite set of input-output pairs representing the training cases, defined as $T = \{(\mathbf{x_i}, y_i)\}_{i=1}^{n}$—where $(\mathbf{x_i}, y_i) \in \mathbb{R}^d \times \mathbb{R}$ $(i = 1, 2, \ldots, n)$—symbolic regression consists in inducing a model $p : \mathbb{R}^d \rightarrow \mathbb{R}$ that maps inputs to outputs, such that $\forall (\mathbf{x_i}, y_i) \in T : p(\mathbf{x_i}) = y_i$.

Let $I = \{\mathbf{x_1}, \mathbf{x_2}, \ldots, \mathbf{x_n}\}$ and $O = [y_1, y_2, \ldots, y_n]$ be the input set and output vector, respectively, associated to the training instances. The semantics of a program p represented by an individual evolved by GSGP, denoted by $s(p)$, is the vector of outputs it produces when applied to the set of inputs I, i.e.,

$s(p) = p(I) = [p(\mathbf{x_1}), p(\mathbf{x_2}), \ldots, p(\mathbf{x_n})]$. The semantics of any program can be represented as a point in a n-dimensional topological space $\mathcal{S}$, called semantic space, where n is the size of the training set.

GSGP introduces geometric semantic operators for GP that act on the syntax of the programs, inducing a geometric behaviour on the semantic level [14]. These operators guarantee the semantic fitness landscape explored by GP is conic, a property with positive effects on the search process. There is formal evidence that indicates evolutionary algorithms with geometric operators can optimise cone landscapes with good results for virtually any metric [13].

Algorithm 1. TENN

Input: $T = \{(\mathbf{x_i}, y_i)\}_{i=1}^n$, k, α
Output: Instance set $P \subset T$

1 Shuffle T;
2 $P \leftarrow T$;
3 **for** $i \leftarrow 1$ **to** n **do**
4 $\quad \hat{y} \leftarrow \mathit{regression}(\mathbf{x}_i, P \setminus (\mathbf{x}_i, y_i))$;
5 $\quad N \leftarrow \mathit{knn}(k, T)$;
6 $\quad \theta \leftarrow \alpha \cdot \mathit{sd}(N)$;
7 $\quad$ **if** $\theta = 0$ **then**
8 $\quad\quad \theta \leftarrow \alpha$
9 $\quad$ **if** $|y_i - \hat{y}| > \theta$ **then**
10 $\quad\quad P \leftarrow P \setminus (\mathbf{x}_i, y_i)$
11 **return** P;

Algorithm 2. TCNN

Input: $T = \{(\mathbf{x_i}, y_i)\}_{i=1}^n$, k, α
Output: Instance set $P \subset T$

1 Shuffle T;
2 $P \leftarrow (\mathbf{x_1}, y_1)$;
3 **for** $i \leftarrow 2$ **to** n **do**
4 $\quad \hat{y} \leftarrow \mathit{regression}(\mathbf{x}_i, P)$;
5 $\quad N \leftarrow \mathit{knn}(k, T)$;
6 $\quad \theta \leftarrow \alpha \cdot \mathit{sd}(N)$;
7 $\quad$ **if** $\theta = 0$ **then**
8 $\quad\quad \theta \leftarrow \alpha$
9 $\quad$ **if** $|y_i - \hat{y}| > \theta$ **then**
10 $\quad\quad P \leftarrow P \cup (\mathbf{x}_i, y_i)$
11 **return** P;

As the semantics in GSGP is defined as a point with a number of dimensions equivalent to the number of instances given as input to a candidate regression function, by reducing the number of input instances we automatically reduce the number of dimensions of the semantic space, which in turn reduces the complexity of the search space. Intuitively, the smaller the complexity the smaller the number of possible convex combinations, which may help the speed of convergence to the optimum. In this context, the first strategy we propose to reduce the number of dimensions of the search space is executed before data is given as input to GSGP, and depends only on the characteristics of the dataset. The second strategy, in turn, takes into account the median absolute error of an instance during GSGP evolution to select the most appropriate instances.

3.1 Pre-processing Strategies

We first introduce two methods for instance selection in regression. The Threshold Edited Nearest Neighbor (TENN) and Threshold Condensed Nearest Neighbor (TCNN) [9] adapt instance selection algorithms for classification problems—ENN [18] and CNN [8]—to the regression domain. They are presented in Algorithms 1 and 2.

These algorithms employ an internal regression method to evaluate the instances according to the similarity-based error. The decision of keeping or removing the i-th instance from the training set is based on the deviation of the instance prediction $\hat{y}_i$ and the expected output y_i, given by $|\hat{y}_i - y_i|$. If this difference is smaller than a threshold θ, $\hat{y}_i$ and y_i are considered similar and the instance is accepted or rejected, depending on the algorithm. The threshold θ is computed based on the local properties of the dataset, given by $\alpha \cdot sd(N)$, where α is a parameter controlling the sensitivity and $sd(N)$ returns the standard deviation of the outputs of the set N, composed by the k nearest neighbours of the instance.

The internal regression method adopted by TCNN and TENN—the procedure *regression* presented in Algorithms 1 and 2—can be replaced by any regression method. Our implementation uses the version of the kNN (k-nearest neighbour) algorithm for regression to infer the value of $\hat{y}$. Besides the training set T, these algorithms receive as input the number of neighbours to be considered and a parameter α, which controls how the threshold is calculated. At the end, the set P of instances selected to be used to train the external regression method is returned.

TENN is a decremental method, starting with all training cases in the set P and iteratively removing the instances diverging from their neighbours. An instance $(\mathbf{x}_i, y_i)$ is considered divergent if the output $\hat{y}$ inferred by the model learned without the instance is dissimilar from its output (y_i). TCNN, on the other hand, is an incremental method, beginning with only one instance from the training set in P and iteratively adding only those instances that can improve the search. The instance $(\mathbf{x}_i, y_i)$ is added only if the output $\hat{y}$ inferred by the model learned with P diverges from y_i.

3.2 GSGP Integrated Strategies

Both TENN and TCNN disregard any information about the external regression algorithm, since they are used in a pre-processing phase. In order to overcome this limitation, we propose a method to select instances based on their median absolute error, considering the output of the programs in the current population. The method, called Probabilistic instance Selection based on the Error (PSE), probabilistically selects a subset of the training set at each ρ generations, as presented in Algorithm 3. The higher the median absolute error, the higher the probability of an instance being selected to compose the training subset used by GSGP. The rationale behind this approach is to give higher probability to instances which are, in theory, more difficult to be predicted by the current population evolved by GSGP.

Given a GSGP population $P = \{p_1, p_2, \ldots, p_m\}$, the median absolute error of the i-th instance $(\mathbf{x}_i, y_i) \in T$ is given by the median value of the set $E = \{|p_1(\mathbf{x}_i) - y_i|, |p_2(\mathbf{x}_i) - y_i|, \ldots, |p_m(\mathbf{x}_i) - y_i|\}$. These values are used to sort T in descending order, and the position of the instance in T is used to calculate its probability of being selected to be part of the training set.

Algorithm 3. PSE method

```
   Input: Training set (T), population (pop), lower bound (λ)
   Output: Instance set P ⊂ T
 1 foreach inst = (x_i, y_i) ∈ T do        // Compute the median absolute error
 2 |  E ← [|p_1(x_i) − y_i)|, |p_2(x_i) − y_i)|, ..., |p_m(x_i) − y_i)|];
 3 |  inst.med ← median(E);
 4 Sort T by med value in descending order;
 5 P ← {};
 6 for i ← 1 to |T| do
 7 |  inst ← (x_i, y_i) ∈ T;
 8 |  r̃ ← (i−1)/(|T|−1) ;                   // Compute the normalized rank
 9 |  prob_sel ← 1 − (1 − λ) · r̃^2 ;        // Probability of selecting inst
10 |  if prob_sel ≥ rand() then           // Add inst to P with probability prob_sel
11 |  |  P ← P ∪ {inst};
12 return P;
```

In order to compute this probability, the method normalizes the rank of the instance in T to the range $[0, 1]$ by

$$\tilde{r} = \frac{(i-1)}{|T|-1}, \tag{1}$$

where i is the position of the instance in the ordered set T, $|.|$ denotes the cardinality of the set and $\tilde{r} \in [0, 1]$ is the normalized rank. The value of $\tilde{r}$ is used to calculate the probability of selecting the instance, given by

$$prob_{sel} = 1 - (1 - \lambda) \cdot \tilde{r}^2, \tag{2}$$

where λ is a parameter that determines the lower bound of the probability function. The higher the value of λ, the more instances are selected. The area under the function, equivalent to $\frac{2+\lambda}{3}$, corresponds to the proportion of instances selected from T.

4 Experimental Results

This section presents an experimental analysis of the instance selection strategies. The results obtained by GSGP with instance selection performed by TCNN and TENN (Sect. 4.1), and PSE (Sect. 4.2) are compared with GSGP with all instances.

The experiments were performed in a collection of datasets selected from the UCI machine learning repository [11], GP benchmarks [12] and a GSGP study from the literature [1], as presented in Table 1. For real-world datasets, we performed 5-fold cross-validations with 10 replications, and for synthetic ones, the data was sampled five times—according to Table 3 from [12]—and the algorithms were applied 10 times, both cases resulting in 50 executions. This sampling strategy justifies the adoption of the t-test in the statistical analysis performed in this section—the number of replications is larger than 30 [4] For compatibility purposes, we removed the categorical attributes of the datasets.

All executions used a population of 1,000 individuals evolved for 2,000 generations with tournament selection of size 10. The grow method [10] was adopted to generate the random functions inside the geometric semantic operators, and the ramped half-and-half method [10] used to generate the initial population, both with maximum individual depth equals to 6. The terminal set included the variables of the problem and constant values randomly picked from the interval $[-1, 1]$. The function set included three binary arithmetic operators $(+, -, \times)$ and the analytic quotient (AQ) [15] as an alternative to the arithmetic division. The GSGP method employed the crossover for Manhattan-based fitness function and mutation operators from [3] both with probability 0.5. The mutation step required by the mutation operator was defined as 10 % of the standard deviation of the outputs (O) given by the training data. All instances in the training set were used as input for the instance selection methods and GSGP.

Table 1. Datasets used in the experiments.

Dataset	Size	Nature	Source	Dataset	Size	Nature	Source
Airfoil	1503	Real	[1,11]	Keijzer-7	100	Synthetic	[12]
Bioavailability	359	Real	[1]	ppb	131	Real	[1]
Concrete	1030	Real	[1,11]	TowerData	4999	Real	[1]
cpu	209	Real	[1,11]	Vladislavleva-1	100	Synthetic	[1,12]
EnergyCooling	768	Real	[1,11]	WineRed	1599	Real	[1,11]
EnergyHeating	768	Real	[1,11]	WineWhite	4898	Real	[1,11]
Forestfires	517	Real	[1,11]	Yacht	308	Real	[1,11]
Keijzer-6	50	Synthetic	[1,12]				

4.1 Comparing Instance Selection Methods

In this section we compare the results obtained by GSGP with and without the instance selection performed before the evolutionary stage (pre-processing). The selection was performed by TCNN (GSGP-TCNN) and TENN (GSGP-TENN) methods, with $k = 9$ and 10 different values for α equally distributed in the intervals $[0.1, 1]$ and $[5.5, 10]$, respectively. Table 2 presents the median training and test RMSE's and the data reduction obtained with α resulting in the largest data reduction by TCNN and TENN methods—1 and 5.5, respectively.

In order to investigate the significance of instance selection methods in GSGP, we randomly selected l instances from each dataset, with no replacement, to compose a new training set used as input by GSGP. The value of l is defined as the smaller of the sizes of the sets resulting from TENN and TCNN. Table 2 presents the median training and test RMSE's of these experiments in the last

Table 2. Median training and test RMSE and reduction (*% red.*) achieved by the algorithms for each dataset. Values highlighted in bold corresponds to test RMSE statically worst than GSGP, according to a t-test with 95 % confidence.

Dataset	GSGP		GSGP-TCNN			GSGP-TENN			GSGP-Rnd	
	tr	ts	tr	ts	% red.	tr	ts	% red.	tr	ts
Airfoil	7.89	8.42	7.76	**8.74**	38.60	8.06	**8.60**	1.90	7.65	8.38
Bioavailability	9.89	30.74	4.95	**36.29**	46.30	9.84	31.38	0.90	4.55	**34.39**
Concrete	3.65	5.39	2.80	**6.40**	38.20	3.65	5.21	3.20	3.18	**5.95**
cpu	6.13	30.92	5.46	**33.61**	11.20	5.06	**51.54**	65.40	5.67	32.28
EnergyCooling	1.26	1.51	1.28	**2.49**	14.70	1.28	**1.83**	36.60	1.19	**1.71**
EnergyHeating	0.80	0.96	0.83	**1.87**	11.10	0.67	**1.84**	45.40	0.77	**1.11**
Forestfires	30.74	51.63	13.68	**101.90**	42.80	30.75	51.94	5.80	22.49	**57.57**
Keijzer-6	0.01	0.40	0.01	0.36	10.60	0.00	**1.25**	53.00	0.01	0.32
Keijzer-7	0.02	0.02	0.02	0.02	5.30	0.01	**0.40**	68.50	0.01	**0.05**
ppb	0.92	28.74	0.20	32.08	41.50	0.91	28.04	3.80	0.25	30.50
TowerData	20.44	21.92	19.82	**22.71**	12.60	20.44	**43.86**	41.90	20.40	**22.06**
Vladislavleva-1	0.01	0.04	0.01	**0.07**	20.90	0.01	**0.07**	43.40	0.01	**0.06**
WineRed	0.49	0.62	0.40	**0.73**	51.10	0.49	0.62	0.10	0.41	**0.66**
WineWhite	0.64	0.70	0.66	**0.78**	52.30	0.64	0.69	0.10	0.60	**0.71**
Yacht	2.12	2.52	2.20	**5.19**	36.90	2.11	**2.83**	24.30	2.01	**2.63**

two columns (denoted as 'GSGP-Rnd'). The results obtained show that using TCNN and TENN do not make any systematic improvement on GSGP results. Moreover, the results obtained by them are no better than those generated by a random selection scheme. Hence, the strategies used by these methods do not seem appropriate for the scenario we have.

4.2 Evaluating the Effects of PSE

In this section, we first investigate the sensitivity of PSE parameters and then compare the performance of GSGP with and without the PSE method. PSE parameters ρ and λ have a direct impact on the number of instances selected and how they are selected. In order to analyse their impact on the search, we fixed the GSGP parameters and focused on looking at the results as we varied these parameters. The values of ρ were set to 5, 10 and 15 while we varied the value of λ in 0.1, 0.4 and 0.7. Table 3 presents the median training RMSE obtained by the GSGP with these PSE configurations. The results show that higher values of ρ (15) with lower values of λ (0.1) tend to reduce the training RMSE.

The experiments with PSE adopt the values of ρ and λ resulting in the smallest median training RMSE, as presented in Table 3. Table 4 presents the median training and test RMSE's obtained by GSGP and by GSGP with PSE (GSGP-PSE). In order to identify statistically significant differences, we performed t-tests with 95 % confidence level, regarding the test RMSE of both methods in 50 executions. The symbol in the last column indicates datasets where the results present significant difference. Overall, GSGP-PSE performs better in terms of test RMSE than GSGP, being better in five datasets and worse in three.

Table 3. Median training RMSE of the GSGP-PSE with different values of λ and ρ for the test bed. The smallest RMSE for each dataset is presented in bold.

Dataset	$\rho = 5$			$\rho = 10$			$\rho = 15$		
	$\lambda = 0.1$	$\lambda = 0.4$	$\lambda = 0.7$	$\lambda = 0.1$	$\lambda = 0.4$	$\lambda = 0.7$	$\lambda = 0.1$	$\lambda = 0.4$	$\lambda = 0.7$
Airfoil	8.03	8.15	8.11	**7.97**	8.05	8.16	8.12	8.05	8.11
Bioavailability	9.66	9.66	9.88	9.70	9.69	9.83	**9.53**	9.77	9.81
Concrete	3.35	3.49	3.56	3.35	3.45	3.58	**3.34**	3.45	3.56
cpu	**4.93**	5.46	5.70	5.02	5.33	5.89	5.01	5.39	5.88
EnergyCooling	1.13	1.19	1.23	1.12	1.18	1.22	**1.12**	1.17	1.23
EnergyHeating	**0.66**	0.72	0.77	0.67	0.72	0.76	0.67	0.71	0.76
Forestfires	25.94	27.67	29.21	25.72	27.61	29.43	**25.55**	27.87	29.58
Keijzer-6	0.01	0.01	0.01	0.01	**0.01**	0.01	0.01	0.01	0.01
Keijzer-7	0.02	0.02	0.02	0.02	**0.02**	0.02	0.02	0.02	0.02
ppb	**0.50**	0.65	0.81	0.53	0.65	0.80	0.52	0.63	0.76
TowerData	19.22	19.74	19.98	19.22	19.61	20.09	**19.18**	19.61	19.92
Vladislavleva-1	0.01	0.01	0.01	0.01	0.01	0.01	**0.01**	0.01	0.01
WineRed	0.47	0.48	0.49	0.47	0.48	0.49	**0.47**	0.48	0.49
WineWhite	**0.63**	0.64	0.64	0.63	0.64	0.64	0.63	0.64	0.64
Yacht	1.94	2.02	2.09	**1.94**	2.01	2.08	1.94	2.00	2.09

Table 4. Median training and test RMSE's obtained for each dataset. The symbol ▲(▼) indicates GSGP-PSE is statistically better (worse) than GSGP in the test set according to a t-test with 95 % confidence.

	GSGP		**GSGP-PSE**	
Dataset	**tr**	**ts**	**tr**	**ts**
airfoil	7.88	8.42	7.97	8.55
bioavailability	9.89	30.74	9.53	32.16 ▼
concrete	3.65	5.39	3.34	5.24 ▲
cpu	6.13	30.92	4.93	33.44
energyCooling	1.26	1.51	1.12	1.38 ▲
energyHeating	0.80	0.96	0.66	0.84 ▲
forestfires	30.74	51.63	25.55	51.32
keijzer-6	0.01	0.40	0.01	0.32
keijzer-7	0.02	0.02	0.02	0.02
ppb	0.92	28.74	0.50	28.96
towerData	20.44	21.92	19.18	20.95 ▲
vladislavleva-1	0.01	0.04	0.01	0.05
wineRed	0.49	0.62	0.47	0.62 ▼
wineWhite	0.64	0.70	0.63	0.69 ▼
yacht	2.12	2.52	1.94	2.47 ▲

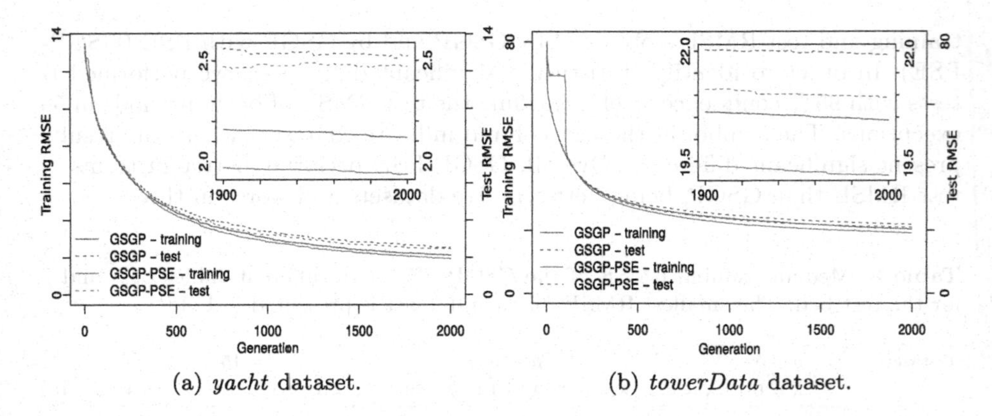

(a) *yacht* dataset.

(b) *towerData* dataset.

Fig. 1. Median RMSE in the training and test sets over the generations for GSGP with and without PSE for *yacth* and *towerData* datasets.

Figure 1 compares the evolution of the fitness of the best individual along the generations in the training and test sets for GSGP and GSGP-PSE, for two different datasets. Note that GSGP errors are overall higher than PSE. For instance, looking at the convergence of the dataset *towerData*, if we stop the evolution at generation 1,000, GSGP would have a test error of 25.02 and GSGP-PSE of 23.64. GSGP needs 293 more generations to reach that same error.

5 Conclusions and Future Work

This paper presented a study about the impact of instance selection methods on GSGP search. Two approaches were adopted: (i) selecting the instances in a pre-processing step; and (ii) selecting instances during the evolutionary process, taking into account the impact of the instance on the search.

Experiments were performed in a collection of 15 datasets in order to evaluate the impact of the instance selection. The first analysis showed GSGP fed with the whole dataset performs better in terms of test RMSE than when using subsets selected with TENN, TCNN or randomly. The second analysis showed that overall GSGP with PSE performs better in terms of test RMSE than the GSGP alone, and that instance selection to reduce the semantic space is worth further investigation.

Potential future works include analysing the effect of fitness functions that weight semantic space dimensions, exploring the impact of noise in the PSE method and studying approaches to insert information about the noisy instances during the selection.

Acknowledgements. The authors would like to thank CNPq (141985/2015-1), CAPES and Fapemig for their financial support.

References

1. Albinati, J., Pappa, G.L., Otero, F.E.B., Oliveira, L.O.V.B.: The effect of distinct geometric semantic crossover operators in regression problems. In: Proceedings of EuroGP, pp. 3–15 (2015)
2. Arnaiz-González, Á., Blachnik, M., Kordos, M., García-Osorio, C.: Fusion of instance selection methods in regression tasks. Inf. Fus. **30**, 69–79 (2016)
3. Castelli, M., Silva, S., Vanneschi, L.: A C++ framework for geometric semantic genetic programming. Genet. Prog. Evolvable Mach. **16**(1), 73–81 (2015)
4. Demšar, J.: Statistical comparisons of classifiers over multiple data sets. J. Mach. Learn. Res. **7**, 1–30 (2006)
5. Domingos, P.: A few useful things to know about machine learning. Commun. ACM **55**(10), 78–87 (2012)
6. Garcia, S., Derrac, J., Cano, J., Herrera, F.: Prototype selection for nearest neighbor classification: taxonomy and empirical study. IEEE Trans. Pattern Anal. Mach. Intell. **34**(3), 417–435 (2012)
7. Guillen, A., Herrera, L.J., Rubio, G., Pomares, H., Lendasse, A., Rojas, I.: New method for instance or prototype selection using mutual information in time series prediction. Neurocomputing **73**(10–12), 2030–2038 (2010)
8. Hart, P.: The condensed nearest neighbor rule (corresp.). IEEE Trans. Inf. Theor. **14**(3), 515–516 (1968)
9. Kordos, M., Blachnik, M.: Instance selection with neural networks for regression problems. In: Villa, A.E.P., Duch, W., Érdi, P., Masulli, F., Palm, G. (eds.) ICANN 2012, Part II. LNCS, vol. 7553, pp. 263–270. Springer, Heidelberg (2012)
10. Koza, J.R.: Genetic Programming: On the Programming of Computers by Means of Natural Selection, vol. 1. MIT Press, Cambridge (1992)
11. Lichman, M.: UCI Machine Learning Repository (2015). http://archive.ics.uci.edu/ml
12. McDermott, J., White, D.R., Luke, S., Manzoni, L., Castelli, M., Vanneschi, L., Jaskowski, W., Krawiec, K., Harper, R., De Jong, K., O'Reilly, U.M.: Genetic programming needs better benchmarks. In: Proceedings of GECCO, pp. 791–798 (2012)
13. Moraglio, A.: Abstract convex evolutionary search. In: Proceedings of the 11th FOGA, pp. 151–162 (2011)
14. Moraglio, A., Krawiec, K., Johnson, C.G.: Geometric semantic genetic programming. In: Coello, C.A.C., Cutello, V., Deb, K., Forrest, S., Nicosia, G., Pavone, M. (eds.) PPSN 2012, Part I. LNCS, vol. 7491, pp. 21–31. Springer, Heidelberg (2012)
15. Ni, J., Drieberg, R.H., Rockett, P.I.: The use of an analytic quotient operator in genetic programming. IEEE Trans. Evol. Comput. **17**(1), 146–152 (2013)
16. Rodrguez-Fdez, I., Mucientes, M., Bugarn, A.: An instance selection algorithm for regression and its application in variance reduction. In: 2013 IEEE International Conference on Fuzzy Systems (FUZZ), pp. 1–8, July 2013
17. Vanneschi, L., Castelli, M., Silva, S.: A survey of semantic methods in genetic programming. Genet. Program. Evolvable Mach. **15**(2), 195–214 (2014)
18. Wilson, D.L.: Asymptotic properties of nearest neighbor rules using edited data. IEEE Trans. Syst. Man Cybern. **2**(3), 408–421 (1972)

Multi-objective, Many-objective and Multi-level Optimisation

iMOACO$_\mathbb{R}$: A New Indicator-Based Multi-objective Ant Colony Optimization Algorithm for Continuous Search Spaces

Jesús Guillermo Falcón-Cardona(✉) and Carlos A. Coello Coello

Computer Science Department, CINVESTAV-IPN (Evolutionary Computation Group), Av. IPN No. 2508, Col. San Pedro Zacatenco, México D.F. 07300, Mexico
jfalcon@computacion.cs.cinvestav.mx, ccoello@cs.cinvestav.mx

Abstract. Ant colony optimization (ACO) is a metaheurisitc which was originally designed to solve combinatorial optimization problems. In recent years, ACO has been extended to tackle continuous single-objective optimization problems, being ACO$_\mathbb{R}$ one of the most remarkable approaches of this sort. However, there exist just a few ACO-based algorithms designed to solve continuous multi-objective optimization problems (MOPs) and none of them has been tested with many-objective problems (i.e., multi-objective problems having four or more objectives). In this paper, we propose a novel multi-objective ant colony optimizer (called iMOACO$_\mathbb{R}$) for continuous search spaces, which is based on ACO$_\mathbb{R}$ and the $R2$ performance indicator. Our proposed approach is the first specifically designed to tackle many-objective optimization problems. Moreover, we present a comparative study of our proposal with respect to NSGA-III, MOEA/D, MOACO$_\mathbb{R}$ and SMS-EMOA using standard test problems and performance indicators adopted in the specialized literature. Our preliminary results indicate that iMOACO$_\mathbb{R}$ is very competitive with respect to state-of-the-art multi-objective evolutionary algorithms and is also able to outperform MOACO$_\mathbb{R}$.

1 Introduction

In artificial intelligence, the social behavior of animals and insects has been a prominent source of inspiration for several metaheuristics which are part of the broad concept of Swarm Intelligence. Ant Colony Optimization (ACO), was originally proposed by Dorigo [1], and it is inspired by colonies of real ants that deposit a chemical substance (called pheromone) on the ground with the aim of tracing paths to a source of food. The ants tend to take, with a higher probability, those paths where there is a larger amount of pheromone. In fact, after some time, the shortest path is the one with the largest amount of pheromone [2]. Due to this property, ACO was originally applied to the solution of combinatorial optimization problems (COPs).

J.G. Falcón-Cardona—acknowledges support from CINVESTAV-IPN.
C.A. Coello Coello—acknowledges support from CONACyT project no. 221551.

J. Handl et al. (Eds.): PPSN XIV 2016, LNCS 9921, pp. 389–398, 2016.
DOI: 10.1007/978-3-319-45823-6_36

Over the years, the ACO metaheuristic has been extended to continuous search spaces, being the proposal of Bilchev and Parmee [3] the first of this sort. According to [4], there are several ACO-based optimizers for continuous domains although the ACO algorithm for continuous domains ($ACO_{\mathbb{R}}$) [5] is possibly the most remarkable. In spite of the relatively large amount of ACO-based algorithms currently available for continuous domains, there are just a few oriented to solve multi-objective optimization problems (MOPs). In [4] only two proposals are reported: the Population-based ACO Algorithm for Multi-Objective Function Optimization (PACO-MOFO) [6] and the Multi-Objective Ant Colony Optimizer ($MOACO_{\mathbb{R}}$) [7], both based on $ACO_{\mathbb{R}}$. Furthermore, in the specialized literature no multi-objective ant colony optimizer (MOACO) had been reported so far as being able to solve many-objective problems [8].

In this paper, we propose a novel indicator-based Multi-Objective Ant Colony Optimizer based on $ACO_{\mathbb{R}}$, called $iMOACO_{\mathbb{R}}$. To the authors' best knowledge, this is the first MOACO algorithm that is able to tackle many-objective optimization problems.

The remainder of this paper is organized as follows. Section 2 presents an overview of the previous work on ACO in continuous optimization problems. Section 3 briefly describes $ACO_{\mathbb{R}}$. The detailed description of our proposal is presented in Sect. 4. Then, we provide our experimental results in Sect. 5. Finally, Sect. 6 provides our conclusions and some possible paths for future research.

2 Previous Related Work

The first ACO algorithm designed for continuous search spaces was proposed by Bilchev and Parmee [3]. In this approach, each ant incrementally explores the search space from a single nest, defined as a promising point, trying different search directions at a radius not greater than R. At choosing a search direction, each ant's decision was biased by a trail quantity which was incremented if and only if the direction resulted in an improvement of the objective function; otherwise, the search direction was not taken into account. This process was repeated until a termination condition was met.

Socha and Dorigo proposed the $ACO_{\mathbb{R}}$ [5] algorithm whose fundamental idea is the use of a continuous probability density function (PDF) instead of a discrete one as in traditional ACOs. $ACO_{\mathbb{R}}$ uses a constant-size archive as its pheromone model where the best-so-far solutions are stored. For each dimension, a Gaussian-kernel PDF is defined using the corresponding elements of every stored solution. An ant incrementally constructs a new solution via the sampling of each Gaussian-kernel PDF. Once all ants have constructed a new solution, only the best ones are kept in the archive and the same number are removed from it. A detailed description of $ACO_{\mathbb{R}}$ will be provided in the next section.

The use of ACO in continuous MOPs has been scarcely explored [4]. We are only aware of two approaches. The first of them is PACO-MOFO, which is based on the Crowding Population-based ACO algorithm (CPACO) [9] and $ACO_{\mathbb{R}}$. PACO-MOFO applies a replacement operator based on crowding distance in order to maintain diversity and fitness sharing in furtherance of a uniform sampling of the objective space. The second proposal is MOACO$_\mathbb{R}$ [7], which is a direct extension of ACO$_\mathbb{R}$. The concept of *dominance depth* of NSGA-II [10] is used in this case to preserve at each iteration those solutions closer to the Pareto Front. Moreover, if the number of solutions exceed the size of the archive, those with a higher crowding distance value are removed in order to maintain constant the size of the archive.

3 ACO$_\mathbb{R}$ Overview

The pheromone model of ACO$_\mathbb{R}$ [5] is represented by an archive $\mathcal{T}$ that stores the k best-so-far solutions. For the ith dimension, a Gaussian-kernel PDF is defined using the corresponding components of all stored solutions as follows:

$$G^i(x) = \sum_{j=1}^{k} w_j g_j^i(x) = \sum_{j=1}^{k} w_j \frac{1}{\sigma_j^i \sqrt{2\pi}} \cdot e^{-\frac{(x-\mu_j^i)^2}{2{\sigma_j^i}^2}} \quad (1)$$

where $i = 1, \ldots, n$ and n is the number of decision variables. Each archive's solution j stores a vector of decision variables $s_j = (s_j^1, \ldots, s_j^n)$, an objective value $u(s_j)$ and the weight ω_j. The solutions are sorted by their quality, i.e., $u(s_1) \leq u(s_2) \leq \cdots \leq u(s_k)$, for a minimization problem.

Equation (1) depends on three vectors of parameters: μ^i is the vector of means, σ^i is the vector of standard deviations, and ω is the vector of weights. The vector of means μ^i is defined as follows:

$$\mu^i = \{\mu_1^i, \ldots, \mu_k^i\} = \{s_1^i, \ldots, s_k^i\} \quad (2)$$

The elements of σ^i have to be independently calculated for each Gaussian-kernel using the following formula:

$$\sigma_j^i = \xi \sum_{r=1}^{k} \frac{|s_r^i - s_j^i|}{k-1} \quad (3)$$

where $\xi > 0$ is a parameter of the algorithm that controls the way the long term memory is used, i.e., the speed of convergence. When ξ is large, the speed of convergence is slower and in case its value is close to zero, the speed of convergence is increased. Finally, each $\omega_j \in \omega$ is calculated as follows:

$$\omega_j = \frac{1}{qk\sqrt{2\pi}} \cdot e^{-\frac{(rank(s_j)-1)^2}{2q^2k^2}} \quad (4)$$

where $rank(\cdot)$ returns the solution's rank in $\mathcal{T}$ according to the established order and $q > 0$ is a parameter that controls the diversification of the search. As $q \to 0$, the best-ranked solutions are preferred to guide the search, and when it takes a large value, the weights tend to be more uniform.

In order to generate a new solution, first, each ant $a_i \in \mathcal{A}$ chooses, with probability $p_j = \omega_j / \sum_{r=1}^{k} \omega_r$, a guiding pheromone s_j from $\mathcal{T}$. Then, a_i samples $g_j^i(x)$, $i = 1, \ldots, n$, with the purpose of creating a new solution.

4 Our Proposed Approach

The hypervolume (HV) [18] and the $R2$ indicator [11] are two recommended unary performance indicators which simultaneously evaluate all the desired aspects of a Pareto Front approximation [11]. However, the $R2$ indicator requires less computational effort and it produces a more uniform distribution than HV. Given a Pareto Front approximation A, the unary version of the $R2$ indicator [11] is defined as follows:

$$R2(A, U) = \frac{1}{|U|} \sum_{u \in U} \min_{a \in A} \{u(a)\} \tag{5}$$

where U is a set of utility functions $u : \mathbb{R}^m \rightarrow \mathbb{R}$ that are a model of the decision maker's preference that maps each objective vector into a scalar value.

Motivated by the nice properties of the $R2$ indicator, Hernández and Coello proposed in [13] a ranking algorithm based on it, called $R2$-ranking. This mechanism groups solutions which optimize a set of utility functions, and place them on top, such that they get the first rank. Then, such points are removed and a second rank is assigned in the same way and so on until there are no more points left to be ranked. One of the advantages of this scheme is its good performance on many-objective problems.

Concerning the choice of the utility function u in Eq. (5), we use the achievement scalarizing function (ASF) [12] defined as:

$$u_{asf}(v \mid r, \lambda) = \max_{i \in \{1, \ldots, m\}} \frac{1}{\lambda_i} |v_i - r_i| \tag{6}$$

where r is a reference vector and λ is a convex weight vector, both of dimension m. The set $U = \{\lambda^i \mid i = 1, \ldots, N\}$ ($N = C_{m-1}^{H+m-1}$, H is a parameter of the algorithm) is computed using Simple-Lattice-Design (SLD). Moreover, we normalize each objective function $f_i(x)$ (the $R2$-ranking algorithm requieres this normalization) using the following formula:

$$f_i'(x) = \frac{f_i(x) - z_i^{min}}{z_i^{max} - z_i^{min}}, \forall i \in \{1, \ldots, m\} \tag{7}$$

where z^{min} and z^{max} are statistical approximations to the ideal and nadir vectors [12], respectively. These vectors are updated using a data structure called *RECORD*, which was proposed by Hernández and Coello [13].

When we deal with MOPs there is not a unique solution but a set of solutions which represent the best possible trade-offs among the objectives. Due to this fact, $\text{ACO}_{\mathbb{R}}$'s pheromone model has to be slightly modified in order to store the best solutions according to some criterion. A Pareto-based scheme is not a good choice if we aim to solve many-objective problems. Thus, we propose to use the $R2$-ranking algorithm because of its good performance in many-objective problems.

Each record of the archive $\mathcal{T}$ stores the same information as in ACO$_{\mathbb{R}}$, although in this case the objective value is treated as an objective vector $F(s_i)$. Additionally, it is added a field $rank(s_i)$. Once the solutions in $\mathcal{T}$ have been processed by the $R2$-ranking, the rank assigned to each solution s_j is stored in $rank(s_j)$. In order to create a new solution, we applied the standard process of ACO$_{\mathbb{R}}$ using Eqs. (1) to (4).

The underlying idea of the pheromone update is to promote a competition between the newly created solutions $\mathcal{A}$ and the pheromones in $\mathcal{T}$. Let $\Psi = \mathcal{A} \cup \mathcal{T}$. The union set is ranked by the $R2$-ranking and is immediately sorted, in increasing order, by the following criteria: (1) rank, (2) utility value, (3) L_2-norm. Finally, all pheromones in $\mathcal{T}$ are substituted by the first k solutions of Ψ.

In Algorithm 1, we describe our proposed iMOACO$_{\mathbb{R}}$.[1] The algorithm only requires three parameters: (1) the set of $N = C_{m-1}^{H+m-1}$ weight vectors, (2) the diversification parameter q, and (3) the convergence speed factor, ξ. The population size (M) and the archive size (k) are equal to N due to the optimal μ-distributions of the $R2$ indicator [11]. In lines 1 to 3, k random solutions are generated to initialize $\mathcal{T}$ and the RECORD structure is created. At each iteration, the $R2$-ranking is applied on $\mathcal{T}$ and afterwards every ant generates a new solution. Then, in line 8, the RECORD is updated using the ants' solutions with the aim of producing new values of z^{min} and z^{max}. From lines 9 to 13, the pheromone update is performed. This process is repeated until a termination condition is fulfilled and then the solutions in $\mathcal{T}$ are returned in line 14.

Algorithm 1. Main loop of iMOACO$_{\mathbb{R}}$.

Require: MOP, set of $N = C_{m-1}^{H+m-1}$ convex weight vectors, q, ξ
Ensure: Pareto front approximation
1: Randomly initialize archive $\mathcal{T}$
2: Initialize RECORD R
3: Initialize z^{min} and z^{max}
4: **while** termination condition is not fulfilled **do**
5: $\quad \mathcal{T} \leftarrow R2ranking(\mathcal{T})$
6: $\quad$ **for all** $ant \in \mathcal{A}$ **do**
7: $\quad\quad generateSolution(ant, z^{min}, z^{max})$
8: $\quad$ Update reference points (z^{min}, z^{max}) using R
9: $\quad \Psi \leftarrow \mathcal{A} \cup \mathcal{T}$
10: $\quad \Psi' \leftarrow R2ranking(\Psi)$
11: $\quad$ Remove all elements from $\mathcal{T}$
12: $\quad \Psi' \leftarrow sortReduction(\Psi')$
13: $\quad$ Copy the first k elements from Ψ' to $\mathcal{T}$
14: **return** $\mathcal{T}$

[1] The source code of our approach is available at: http://computacion.cs.cinvestav.mx/~jfalcon/iMOACOR/imoacor.html.

5 Experimental Results

In order to assess the performance of our proposed approach, we used the Zitzler-Deb-Thiele (ZDT) test suite, the Deb-Thiele-Laumanns-Zitzler (DTLZ) test suite, and the Walking-Fish-Group (WFG) test suite. However, due to space limitations, only the results for the ZDT and DTLZ test problems are included here. Our proposed approach was compared with respect to: NSGA-III[2] [15], MOEA/D [14], SMS-EMOA [16] (using HypE to estimate the hypervolume values [17]) and $MOACO_{\mathbb{R}}$[3] [7]. Results were compared using the hypervolume (HV), inverted generational distance plus (IGD+[4]) [19] and spacing (S) [18].

Attending the original papers, the common parameter settings for NSGA-III, MOEA/D and SMS-EMOA have been set as follows: $N_c = 20$, $P_c = 1.0$, $N_m = 20$ and $P_m = 1/n$. The neighborhood size of MOEA/D was set to 20. The number of samples in the HypE algorithm was set to 10,000. Based on an experimental study, the parameters (q, ξ) of $iMOACO_{\mathbb{R}}$ and $MOACO_{\mathbb{R}}$ were set as $(0.1, 0.5)$ for low and high dimensionality. In all cases, we performed a maximum number of 50,000 function evaluations. We used $N = 120$ weight vectors, which implies $H = 119$ and $H = 14$, for two and three dimensions, respectively.

For the scalability test, we employed DTLZ2 from four to nine objectives. All parameter values remained the same except for $N_c = 30$, as suggested in [15]. The maximum number of function evaluations remained the same as before. The experimental configurations $(m, N(H))$ are as follows: (4, 120(7)), (5, 126(5)), (6, 126(4)), (7, 84(3)), (8, 120(3)), (9, 165(3)) and (10, 220(3)).

5.1 Discussion of Results

This section compares $iMOACO_{\mathbb{R}}$ with three state-of-the-art MOEAs and a MOACO that was designed for continuous MOPs. The comparison is performed in terms of convergence and diversity of the solutions obtained. We perform 30 independent runs of each of the 5 algorithms on all the test instances adopted. Tables 1 and 2 show the average HV, IGD+ and S values, as well as the standard deviations (shown in parentheses) obtained by all the algorithms compared. The two best values among the algorithms are emphasized in gray scale, where the darker tone corresponds to the best value. A sharp symbol (#) is placed when a result is statistically different from $iMOACO_{\mathbb{R}}$'s result based on a single-tail Wilcoxon test (WT) using a significance level of 95 %.

Table 1 shows that NSGA-III yields the best HV results in three of four of the ZDT test problems and that $iMOACO_{\mathbb{R}}$ is the best in one of them. However, $iMOACO_{\mathbb{R}}$ obtained the second best HV value in the problems where NSGA-III wins. Moreover, $iMOACO_{\mathbb{R}}$ outperformed MOEA/D, SMS-EMOA (HypE)

[2] We used the implementation available at: http://web.ntnu.edu.tw/~tcchiang/publications/nsga3cpp/nsga3cpp.htm.

[3] The source code was provided by its author, Abel García Nájera.

[4] For each problem, the reference set is constructed joining the results from all algorithms and then applying the k-means clustering algorithm in order to reduce its cardinality to k.

and MOACO$_{\mathbb{R}}$ in all the ZDT problems and the differences are statistically significant. With respect to IGD+, iMOACO$_{\mathbb{R}}$ obtained the second place in 50 % of the problems and outperformed MOACO$_{\mathbb{R}}$ and SMS-EMOA(HypE) in 75 % of the problems (the differences were statistically significant).

Table 1. Comparison of iMOACO$_{\mathbb{R}}$ with respect to SMS-EMOA, MOEA/D, NSGA-III and MOACO$_{\mathbb{R}}$ in the ZDT test problems with two objectives. The symbol # is placed when the difference with respect to iMOACO$_{\mathbb{R}}$'s result is statistically significant, based on Wilcoxon's test. The two best values are shown in gray scale, where the darker tone corresponds to the best value. NC stands for a not computable result due to an algorithm's error.

Problem	Algorithm	HV	IGD+	S
ZDT1	iMOACO$_{\mathbb{R}}$	120.650592(0.002695)	0.006982(0.000262)	0.022642(0.000638)
	MOACO$_{\mathbb{R}}$	120.647992(0.001834)#	0.004849(0.000213)	0.005066(0.000462)
	SMS-EMOA	115.056548(0.204942)#	0.110323(0.006696)#	0.002271(0.002064)
	MOEA/D	120.556524(0.029634)#	0.002429(0.000194)	0.004075(0.000219)
	NSGA-III	120.662065(0.000361)	0.001986(0.000024)	0.008623(0.000175)
ZDT2	iMOACO$_{\mathbb{R}}$	120.319499(0.002065)	0.004579(0.000105)	0.022031(0.000424)
	MOACO$_{\mathbb{R}}$	NC	NC	NC
	SMS-EMOA	111.557974(0.298732)#	0.155868(0.009460)#	0.002802(0.000375)
	MOEA/D	120.303458(0.009442)#	0.002511(0.000320)	0.003736(0.000130)
	NSGA-III	120.328489(0.000541)	0.001919(0.000008)	0.003460(0.000067)
ZDT3	iMOACO$_{\mathbb{R}}$	128.746630(0.006519)	0.002151(0.000116)	0.017216(0.000907)
	MOACO$_{\mathbb{R}}$	128.718532(0.008340)#	0.003121(0.000213)#	0.005641(0.000496)
	SMS-EMOA	125.892282(1.903359)#	0.032616(0.025295)#	0.028416(0.019151)#
	MOEA/D	128.214272(0.946685)#	0.005800(0.006714)#	0.014828(0.001110)
	NSGA-III	128.774980(0.000181)	0.001413(0.000068)	0.011243(0.000922)
ZDT6	iMOACO$_{\mathbb{R}}$	117.381093(0.023285)	0.013011(0.001239)	0.023835(0.001014)
	MOACO$_{\mathbb{R}}$	NC	NC	NC
	SMS-EMOA	113.246727(1.377160)#	0.217514(0.070758)#	0.008383(0.016202)
	MOEA/D	116.763019(0.071014)#	0.050276(0.005115)#	0.003248(0.000604)
	NSGA-III	116.418956(0.002804)#	0.003126(0.000080)	0.001221(0.000018)

Table 2 shows the HV, IGD+ and S values in the DTLZ test problems with 3 objectives. MOACO$_{\mathbb{R}}$ obtained the best HV results in 40 % of the problems. iMOACO$_{\mathbb{R}}$ and NSGA-III performed similarly in HV with only one best value and in 40 % of the problems it ranked second. Moreover, iMOACO$_{\mathbb{R}}$ outperformed NSGA-III, MOEA/D and MOACO$_{\mathbb{R}}$ in 40 % of the problems and outperformed SMS-EMOA(HypE) in a statistically significant way, in all problems. On the other hand, both MOACO$_{\mathbb{R}}$ and MOEA/D outperformed, in terms of IGD+, the rest of the algorithms in 40 % of the problems in a statistically significant way. Finally, it is worth emphasizing that iMOACO$_{\mathbb{R}}$ obtained the best results in DTLZ6 for every indicator and outperformed the other MOEAs in a statistically significant way. However, iMOACO$_{\mathbb{R}}$ could not outperform MOACO$_{\mathbb{R}}$ in a statistically significant way.

It is worth indicating that, although SMS-EMOA(HypE) obtained the best S values and iMOACO$_{\mathbb{R}}$ the worst, we observed that the solutions obtained by SMS-EMOA(HypE) are not well spread and they tend to concentrate on a small region of objective function space. This is not reflected in the S values, because the solutions are all generated in the same small region. In contrast, iMOACO$_{\mathbb{R}}$ provides a better coverage along the Pareto front, but presents a non-uniform distribution in some cases, which is the explanation for its poor values.

Table 2. Comparison of iMOACO$_{\mathbb{R}}$ with respect to SMS-EMOA, MOEA/D, NSGA-III and MOACO$_{\mathbb{R}}$ in the DTLZ test suite with three objectives. The symbol # is placed when the difference with respect to iMOACO$_{\mathbb{R}}$'s result is statistically significant, based on Wilcoxon's test. The two best values are shown in gray scale, where a darker tone corresponds to the best value.

Problem	Algorithm	HV	IGD+	S
DTLZ2	iMOACO$_{\mathbb{R}}$	7.420386(0.000218)	0.020631(0.000148)	0.051706(0.000954)
	MOACO$_{\mathbb{R}}$	7.396275(0.005367)#	0.027855(0.001570)#	0.049300(0.004648)
	SMS-EMOA	4.096654(0.078739)#	0.327737(0.006341)#	0.015061(0.004850)
	MOEA/D	7.421695(0.000110)	0.019927(0.000004)	0.048915(0.000023)
	NSGA-III	7.421721(0.000480)	0.020182(0.000256)	0.048387(0.000899)
DTLZ4	iMOACO$_{\mathbb{R}}$	7.419849(0.000499)	0.031649(0.000353)	0.059475(0.003305)
	MOACO$_{\mathbb{R}}$	7.397087(0.004471)	0.037138(0.001348)	0.047430(0.004064)
	SMS-EMOA	4.540085(0.510681)#	0.244637(0.072285)#	0.020506(0.023548)
	MOEA/D	7.421583(0.000095)	0.029946(0.000007)	0.048923(0.000022)
	NSGA-III	7.219506(0.405047)	0.066563(0.073020)	0.040964(0.015405)
DTLZ5	iMOACO$_{\mathbb{R}}$	59.838732(0.006907)	0.002099(0.000280)	0.004906(0.003023)
	MOACO$_{\mathbb{R}}$	59.868424(0.001271)#	0.001168(0.000229)#	0.007250(0.000658)#
	SMS-EMOA	50.323056(0.664565)#	0.001539(0.000362)	0.006650(0.003698)
	MOEA/D	59.734700(0.001057)#	0.004707(0.000008)#	0.220014(0.005494)#
	NSGA-III	59.831769(0.008471)#	0.001708(0.000447)	0.011428(0.001740)#
DTLZ6	iMOACO$_{\mathbb{R}}$	1318.921707(0.019110)	0.006341(0.000682)	0.007476(0.003907)
	MOACO$_{\mathbb{R}}$	1315.603646(18.281311)	0.067062(0.338421)	0.028337(0.103803)
	SMS-EMOA	1179.647918(18.251463)#	0.260869(0.044008)#	0.012619(0.001664)#
	MOEA/D	1317.080995(0.438458)#	0.117785(0.030957)#	0.241082(0.002048)#
	NSGA-III	1317.572393(0.378312)#	0.081541(0.023995)#	0.060883(0.039139)#
DTLZ7	iMOACO$_{\mathbb{R}}$	1.481848(0.128161)	0.208907(0.064111)	0.110911(0.042411)
	MOACO$_{\mathbb{R}}$	1.955759(0.012511)	0.033899(0.002521)	0.066972(0.006164)
	SMS-EMOA	1.481553(0.151019)	0.192244(0.050467)	0.046280(0.016713)
	MOEA/D	1.827781(0.211815)	0.096353(0.150199)	0.166670(0.034483)#
	NSGA-III	1.937277(0.011787)	0.037603(0.001834)	0.058137(0.004748)

Table 3. Comparison of iMOACO$_{\mathbb{R}}$ with respect to three MOEAs in DTLZ2 with four to nine objectives. The symbol # is placed when the difference with respect to iMOACO$_{\mathbb{R}}$'s result is statistically significant, based on Wilcoxon's test. The two best values are shown in gray scale, where the darker tone corresponds to the best value.

Problem	Algorithm	HV	IGD+
DTLZ2 4D	iMOACO$_{\mathbb{R}}$	15.560885(0.000752)	0.050234(0.004783)
	SMS-EMOA	10.249730(0.677489)#	0.291283(0.009656)#
	MOEA/D	15.567068(0.000241)	0.037633(0.000019)
	NSGA-III	15.566456(0.000668)	0.038679(0.000926)
DTLZ2 5D	iMOACO$_{\mathbb{R}}$	31.650513(0.001900)	0.079425(0.004944)
	SMS-EMOA	21.358261(0.676573)#	0.363571(0.006165)#
	MOEA/D	31.667626(0.000250)	0.057564(0.000070)
	NSGA-III	31.665300(0.000589)	0.059863(0.000627)
DTLZ2 6D	iMOACO$_{\mathbb{R}}$	63.714682(0.002503)	0.091338(0.008945)
	SMS-EMOA	47.221717(1.575810)#	0.395902(0.009846)#
	MOEA/D	63.738154(0.000667)	0.048382(0.000030)
	NSGA-III	63.737999(0.001056)	0.051075(0.000791)
DTLZ2 7D	iMOACO$_{\mathbb{R}}$	127.695926(0.008977)	0.147653(0.008183)
	SMS-EMOA	82.448331(3.777842)#	0.501245(0.006396)#
	MOEA/D	127.747411(0.001454)	0.088905(0.000024)
	NSGA-III	127.749053(0.001358)	0.092568(0.001253)
DTLZ2 8D	iMOACO$_{\mathbb{R}}$	255.731810(0.060472)	0.166126(0.010043)
	SMS-EMOA	184.360111(8.860506)#	0.530043(0.005534)#
	MOEA/D	255.819317(0.001518)	0.093164(0.000159)
	NSGA-III	255.815238(0.001521)	0.099347(0.001020)
DTLZ2 9D	iMOACO$_{\mathbb{R}}$	511.711415(0.161806)	0.177908(0.009877)
	SMS-EMOA	414.480972(10.937714)#	0.548778(0.006587)#
	MOEA/D	511.866089(0.003034)	0.087360(0.000135)
	NSGA-III	511.870831(0.001247)	0.092867(0.001409)

Regarding our scalability test, in Table 3 we provide the HV and IGD+ values in DTLZ2 having from four to nine objectives. Clearly, MOEA/D and NSGA-III present better HV and IGD+ results than iMOACO$_{\mathbb{R}}$. However, the maximum observed difference, in relation to HV, is of order 10^{-1}, which is not very significant. iMOACO$_{\mathbb{R}}$ outperforms SMS-EMOA (HypE) in 100 % of the cases.

6 Conclusions and Future Work

In this paper, we have proposed a new ACO-based multi-objective optimizer for continuous search spaces, called iMOACO$_{\mathbb{R}}$. Our approach uses ACO$_{\mathbb{R}}$ as its search engine and employs a ranking algorithm based on the $R2$ indicator in order to define which solutions are better than the others. This allows our approach to tackle many-objective problems.

Our experimental results indicate that iMOACO$_{\mathbb{R}}$ had a competitive performance with respect to NSGA-III and MOEA/D and that is able to outperform SMS-EMOA (HypE) and MOACO$_{\mathbb{R}}$ in most of the test problems adopted. Therefore, we consider that iMOACO$_{\mathbb{R}}$ is a good starting point for having a highly competitive multi-objective optimizer based on ACO. However, one aspect that must be emphasized is the difficulty that iMOACO$_{\mathbb{R}}$ has on multi-frontal problems such as ZDT4, DTLZ1 and DTLZ3. Our proposed approach has difficulties to maintain diversity in these problems and more work in this direction is still required.

It is worth noticing that the solutions produced by iMOACO$_{\mathbb{R}}$ are similar to those generated by NSGA-III and MOEA/D in terms of distribution and it also achieves a competitive performance in terms of convergence. Furthermore, our proposed approach requires much less computational effort than SMS-EMOA.

As part of our future work, we are interested in studying different diversity mechanisms that allow us to maintain the biological metaphor of the ACO algorithm. Additionally, the pheromone structure still has a lot of room for improvement. Finally, we also aim to improve the performance of our approach in many-objective problems.

References

1. Dorigo, M.: Optimization, learning and natural algorithms. Ph.D. thesis, Politecnico di Milano, Italy (1992)
2. Dorigo, M., Stuetzle, T.: Ant Colony Optimization. MIT Press, Cambridge (2004)
3. Bilchev, G., Parmee, I.C.: The ant colony metaphor for searching continuous design spaces. In: Fogarty, T.C. (ed.) Evolutionary Computing. LNCS, vol. 993, pp. 232–244. Springer, Heidelgberg. (1995)
4. Leguizamón, G., Coello, C.A.C.: Multi-objective ant colony optimization: a taxonomy and review of approaches. In: Integration of Swarm Intelligence and Artificial, Neural Networks, pp. 67–94 (2011)
5. Socha, K., Dorigo, M.: Ant colony optimization for continuous domains. Eur. J. Oper. Res. **185**(3), 1155–1173 (2008)
6. Angus, D.: Population-based ant colony optimisation for multi-objective function optimisation. In: Randall, M., Abbass, H.A., Wiles, J. (eds.) ACAL 2007. LNCS (LNAI), vol. 4828, pp. 232–244. Springer, Heidelberg (2007)

7. Garcia-Najera, A., Bullinaria, J.A.: Extending $ACO_{\mathbb{R}}$ to solve multi-objective problems. In: Proceedings of the UK Workshop on Computational Intelligence (UKCI 2007), London, UK (2007)
8. Ishibuchi, H., Tsukamoto, N., Nojima, Y.: Evolutionary many-objective optimization: a short review. In: IEEE Congress on Evolutionary Computation (2008)
9. Angus, D.: Crowding population-based ant colony optimization for the multi-objective Travelling Salesman Problem. In: Proceedings of the 2007 IEEE Symposium on Computational Intelligence in Multicriteria Decision Making (MCDM 2007), pp. 333–340. IEEE Press, Honolulu (2007)
10. Deb, K., et al.: A fast and elitist multiobjective genetic algorithm: NSGA-II. IEEE Trans. Evol. Comput. **6**(2), 182–197 (2002)
11. Brockhoff, D., Wagner, T., Trautmann, H.: On the properties of the R2 indicator. In: Proceedings of the 14th Annual Conference on Genetic and Evolutionary Computation, pp. 465–472. ACM (2012)
12. Miettinen, K.: Nonlinear Multiobjective Optimization. Kluwer Academic Publisher, Boston (1999)
13. Hernández Gómez, R., Coello, C.A.C.: Improved metaheuristic based on the $R2$ indicator for many-objective optimization. In: Silva, S. (ed.) Proceedings of the 2015 Anual Conference on Genetic and Evolutionary Computation, pp. 679–686. ACM, Madrid (2015)
14. Zhang, Q., Li, H.: MOEA/D: a multiobjective evolutionary algorithm based on decomposition. IEEE Trans. Evol. Comput. **11**(6), 712–731 (2007)
15. Deb, K., Jain, H.: An evolutionary many-objective optimization algorithm using reference-point-based nondominated sorting approach, part I: solving problems with box constraints. IEEE Trans. Evol. Comput. **18**(4), 577–601 (2014)
16. Beume, N., Naujoks, B., Emmerich, M.: SMS-EMOA: multiobjective selection based on dominated hypervolume. Eur. J. Oper. Res. **181**(3), 1653–1669 (2007)
17. Bader, J., Zitzler, E.: HypE: an algorithm for fast hypervolume-based many-objective optimization. Evol. Comput. **19**(1), 45–76 (2011)
18. Coello, C.A.C., Van Veldhuizen, D.A., Lamont, G.B.: Evolutionary Algorithms for Solving Multi-objective Problems, vol. 242. New York Kluwer Academic, New York (2002)
19. Ishibuchi, H., Masuda, H., Tanigaki, Y., Nojima, Y.: Difficulties in specifying reference points to calculate the inverted generational distance for many-objective optimization problems. In: IEEE Symposium on Computational Intelligence in Multi-criteria Decision-Making (MCDM), pp. 170–177. IEEE (2014)

Variable Interaction in Multi-objective Optimization Problems

Ke Li[1](✉), Mohammad Nabi Omidvar[1], Kalyanmoy Deb[2], and Xin Yao[1]

[1] University of Birmingham, Edgbaston, Birmingham B15 2TT, UK
{likw,m.omidvar,x.yao}@cs.bham.ac.uk
[2] Michigan State University, East Lansing, MI 48864, USA
kdeb@egr.msu.edu

Abstract. Variable interaction is an important aspect of a problem, which reflects its structure, and has implications on the design of efficient optimization algorithms. Although variable interaction has been widely studied in the global optimization community, it has rarely been explored in the multi-objective optimization literature. In this paper, we empirically and analytically study the variable interaction structures of some popular multi-objective benchmark problems. Our study uncovers nontrivial variable interaction structures for the ZDT and DTLZ benchmark problems which were thought to be either separable or non-separable.

1 Introduction

Variable interaction is a major source of difficulty in numerical optimization, which hinders the performance of optimizers, especially on functions with complex variable interaction structures [7]. Variable interaction can be loosely defined as the extend to which the optimization of a variable is affected by the values taken by other variables. Complete lack of interaction between the decision variables is the simplest form of interaction structure in which case the variables can be optimized independently irrespective of the values taken by other variables. The other extreme is when each variable interacts with every other variable. However, most real-world problems fall in between these two extremes [8]. Such problems, which are often called partially separable, have a modular structure and contain several clusters of interacting variables. It is clear that if the variable interaction structure is known, the problem can be decomposed into a set of simpler problems which are easier to optimize. Decomposition-based optimization algorithms have been widely studied in the field of large-scale global optimization to alleviate the curse of dimensionality. Although there are numerous studies on both detecting and exploiting partial separability in global optimization [5,9], very limited studies have been dedicated to the analysis of variable interaction in the context of multi-objective optimization. It is worth noting that the multi-objective NK-landscape problems [1] consider variable interaction, but they are binary encoded and did

The first two authors, sorted alphabetically, make equal contributions to this work.

J. Handl et al. (Eds.): PPSN XIV 2016, LNCS 9921, pp. 399–409, 2016.
DOI: 10.1007/978-3-319-45823-6_37

not account for a modular design with respect to variable interaction. In this paper, by using the recently developed differential grouping method [5] and mathematical analysis, we empirically and theoretically analyze the variable interaction structures of two popular benchmark suites, ZDT [10] and DTLZ [3], from the evolutionary multi-objective optimization (EMO) literature. Contrary to the conventional wisdom [4], our analysis shows that most of the ZDT and DTLZ test problems exhibit nontrivial interaction structures which change with the number of objectives. A thorough understanding of variable interaction in the existing benchmarks can have implications on analyzing the behavior of existing algorithms, the design of new algorithms, and the design of future benchmark suites. The aim of this paper is to take a small step towards bridging this gap.

2 Preliminaries

The multi-objective optimization problem (MOP) considered in this paper is as:

$$\begin{array}{ll} \text{minimize} & \mathbf{F}(\mathbf{x}) = (f_1(\mathbf{x}), \cdots, f_m(\mathbf{x}))^T \\ \text{subject to} & \mathbf{x} \in \Omega \end{array} \quad (1)$$

where $\Omega = \prod_{i=1}^{n}[a_i, b_i] \subseteq \mathbb{R}^n$ is the feasible region of the decision (variable) space, and $\mathbf{x} = (x_1, \ldots, x_n)^T \in \Omega$ is a candidate solution. $\mathbf{F} : \Omega \rightarrow \mathbb{R}^m$ constitutes m objective functions, and $\mathbb{R}^m$ is the objective space.

Definition 1. *A function is partially additively separable if it takes the following general form [7]:*

$$f(\mathbf{x}) = \sum_{i=1}^{k} f_i(\mathbf{x}_i),\ k > 1, \quad (2)$$

where $\mathbf{x}_i$ are mutually exclusive decision variables of f_i, and k is the number of independent subcomponents.

This property makes it easy to optimize $f(\mathbf{x})$, because each subcomponent $\mathbf{x}_i$ can be optimized independently.

$$\underset{(\mathbf{x}_1,\cdots,\mathbf{x}_k)}{\operatorname{argmin}}\ f(\mathbf{x}) = \left[\underset{\mathbf{x}_1}{\operatorname{argmin}}\, f(\mathbf{x}), \cdots, \underset{\mathbf{x}_k}{\operatorname{argmin}}\, f(\mathbf{x})\right] \quad (3)$$

Definition 2. *Given a continuously differentiable function $f(\mathbf{x})$, for any pair of variables x_i and x_j, if $\frac{\partial^2 f}{\partial_{x_i} \partial x_j} \neq 0$, then x_i and x_j are said to interact with each other; otherwise, they are said to be independent from each other.*

The differential grouping method for detecting the variable interaction structure is derived from the following theorem [5].

Theorem 1. *For an additively separable function $f(\mathbf{x})$, $\forall a, b_1 \neq b_2, \delta \in \mathbb{R}, \delta \neq 0$, if the following condition holds:*

$$\Delta_{\delta,x_p}[f](\mathbf{x})|_{x_p=a,x_q=b_1} \neq \Delta_{\delta,x_p}[f](\mathbf{x})|_{x_p=a,x_q=b_2} \quad (4)$$

then x_p and x_q are non-separable where

Table 1. Mathematical definitions of ZDT and DTLZ benchmark suites

Name	Definition	Domain
ZDT1	$f_1 = x_1$ $g = 1 + 9 \cdot \sum_{i=2}^{n} x_i/(n-1)$ $h = 1 - \sqrt{f_1/g}$	$[0,1]$
ZDT2	as ZDT1, except $h = 1 - (f_1/g)^2$	$[0,1]$
ZDT3	as ZDT1, except $h = 1 - \sqrt{f_1/g} - (f_1/g)\sin(10\pi f_1)$	$[0,1]$
ZDT4	as ZDT1, except $g = 1 + 10 \cdot (n-1) + \sum_{i=2}^{n}(x_i^2 - 10\cos(4\pi x_i))$	$x_1 \in [0,1]$ $x_i \in [-5,5]$
ZDT6	$f_1 = 1 - \exp(-4x_1)\sin^6(6\pi y_1)$ $g = 1 + 9 \cdot (\sum_{i=2}^{n} x_i/(n-1))^{0.25}$ $h = 1 - (f_1/g)^2$	$[0,1]$
DTLZ1	$f_1 = (1+g)0.5\prod_{i=1}^{m-1} x_i$ $f_{j=2:m-1} = (1+g)0.5(\prod_{i=1}^{m-j} x_i)(1 - x_{m-j+1})$ $f_m = (1+g)0.5(1-x_1)$ $g = 100[n - m + 1 + \sum_{i=m}^{n}((x_i - 0.5)^2 -$ $\cos(20\pi(x_i - 0.5)))]$	$[0,1]$
DTLZ2	$f_1 = (1+g)0.5\prod_{i=1}^{m-1}\cos(x_i\pi/2)$ $f_{j=2:m-1} = (1+$ $g)0.5(\prod_{i=1}^{m-j}\cos(x_i\pi/2))(\sin(x_{m-j+1}\pi/2))$ $f_m = (1+g)\sin(x_1\pi/2)$ $g = \sum_{i=m}^{n}(x_i - 0.5)^2$	$[0,1]$
DTLZ3	as DTLZ2, except g is replaced by the one from DTLZ1	$[0,1]$
DTLZ4	as DTLZ2, except x_i is replaced by x_i^{α}, where $i \in \{1,\cdots,m-1\}, \alpha > 0$	$[0,1]$
DTLZ5	as DTLZ2, except x_i is replaced by $\frac{1+2gx_i}{4(1+g)}$, where $i \in \{2,\cdots,m-1\}$	$[0,1]$
DTLZ6	as DTLZ5, except the equation for g is replaced by $g = \sum_{i=m}^{n} x_i^{0.1}$	$[0,1]$
DTLZ7	$f_{j=1:m-1} = x_m$ $f_m = (1+g)(m - \sum_{i=1}^{m-1}[\frac{f_i}{1+g}(1+\sin(3\pi f_i))])$ $g = 1 + 9\sum_{i=m}^{n} x_i/(n-m+1)$	$[0,1]$

$$\Delta_{\delta,x_p}[f](\mathbf{x}) = f(\cdots, x_p + \delta, \cdots) - f(\cdots, x_p, \cdots) \quad (5)$$

refers to the forward difference of f with respect to variable x_p with interval δ.

Before the analysis, we describe the test problems used in this paper. ZDT benchmark suite [10] has been extensively used to benchmark numerous EMO algorithms for more than a decade and has the following general structure [2]:

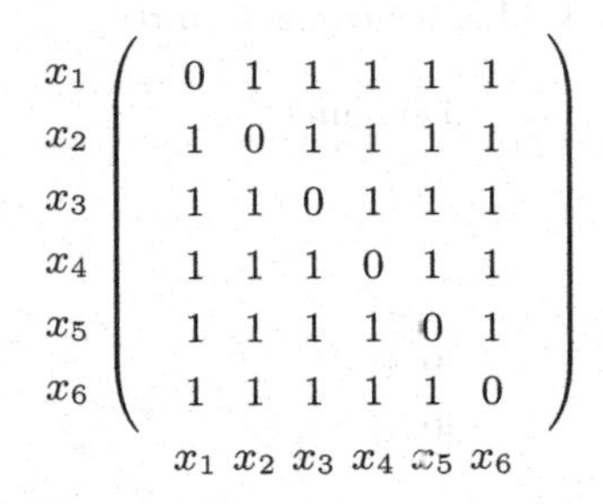

(a) variable interaction matrix

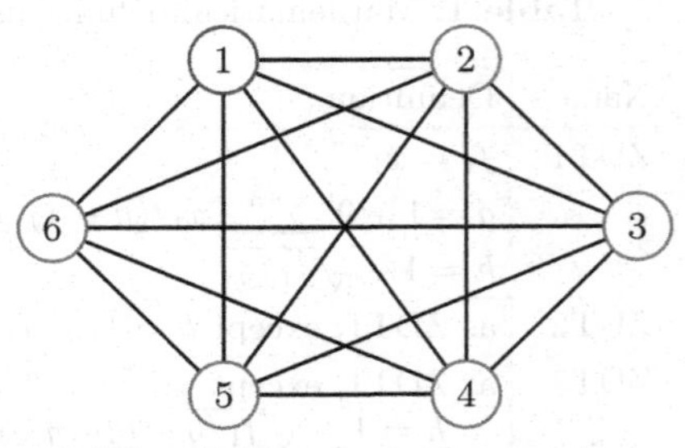

(b) variable interaction graph

Fig. 1. Variable interaction structures of the f_2 function of ZDT test suite.

$$\begin{aligned} \text{minimize} \quad & \mathbf{F}(\mathbf{x}) = (f_1(\mathbf{x}_I), f_2(\mathbf{x}_{II})) \\ \text{subject to} \quad & f_2(\mathbf{x}_{II}) = g(\mathbf{x}_{II}) \cdot h(f_1(\mathbf{x}_I), g(\mathbf{x}_{II})), \end{aligned} \tag{6}$$

where $\mathbf{x} = (\mathbf{x}_I, \mathbf{x}_{II})$ is partitioned into two non-overlapping sets. In particular, $\mathbf{x}_I = x_1$ and $\mathbf{x}_{II} = (x_2, \cdots, x_n)^T$ for all ZDT test problems. DTLZ [3] is another popular benchmark suite in the EMO literature. In essence, the DTLZ is developed based on the same principle as that of the ZDT. However, unlike ZDT, DTLZ test problems are scalable to any number of objectives. To help with the clarity of the analysis in the following section, the mathematical definitions of ZDT and DTLZ test problems are summarized in Table 1.

3 Variable Interaction Analysis via Differential Grouping

Differential grouping [5] is a function decomposition algorithm that can identify the underlying variable interaction structure of black-box continuous functions with a high accuracy. In this study, we employ its modified version (as shown in Algorithm 1) to analyze the ZDT and DTLZ benchmark suites. Due to the existence of multiple objective functions[1], Algorithm 1 applies differential grouping to each objective function independently, which results in m interaction structure matrices.

3.1 Variable Interaction Analysis on ZDT Benchmark Suite

Table 1 clearly shows that f_1 of all ZDT test problems is a fully separable function because it is only a function of x_1. Thus, we only need to analyze the variable interaction for the second objective function f_2. To keep the interaction matrices and the graphs within a manageable size, we set the number of variables to $n = 6$ which is large enough to reveal the patterns and regularities of the benchmark functions. The experimental results show that, by running Algorithm 1, f_2 of all ZDT test problems share the same variable interaction matrix, as shown

[1] The objective functions of ZDT and DTLZ test suites are genuinely independent.

Algorithm 1. Interaction Analysis via Differential Grouping

```
   Output: Interaction Structure Matrices I^(1)_{n×n}, ··· , I^(m)_{n×n}
1  for i ← 1 to m do
2      Initialize all entries of I^(i)_{n×n} to be 0;
3      for j ← 1 to n do
4          for k ← 1 to n ∧ k ≠ j do
5              p^1 ← rand(1, n), p^2 ← p^1 /*rand: random number generator */
6              repeat
7                  ξ_1 ← rand, ξ_2 ← rand;
8              until |ξ_1 − p^1_j| > ε_1 ∧ |ξ_2 − p^1_k| > ε_1;
9              p^2_j ← ξ_1;
10             Δ_1 ← f_i(p^1) − f_i(p^2);
11             p^1_k ← ξ_2, p^2_k ← ξ_2;
12             Δ_2 ← f_i(p^1) − f_i(p^2);
13             if |Δ_1 − Δ_2| > ε_2 then
14                 I^(i)_{jk} ← 1;
15 return I^(1)_{n×n}, ··· , I^(m)_{n×n}
```

in Fig. 1(a). The graphical representation of this interaction matrix is a fully connected graph which is shown in Fig. 1(b). This clearly shows that all the decision variables of f_2 interact with each other, making f_2 a fully non-separable function. In order to validate the correctness of this non-separability property, we use Definition 2 to prove Proposition 1.

Proposition 1. *f_2 of the ZDT benchmark suite is fully non-separable.*

Proof. Let us start from ZDT1. By taking the derivative of f_2 with respect to x_1, we have:

$$\frac{\partial f_2}{\partial x_1} = \frac{\partial(g - (1 - \sqrt{x_1/g}))}{\partial x_1} = \frac{\partial(g - \sqrt{x_1 g})}{\partial x_1}. \tag{7}$$

Since g is a function of x_2 to x_n, we can treat it as a constant in Eq. 7:

$$\frac{\partial f_2}{\partial x_1} = -0.5\sqrt{g/x_1}, \tag{8}$$

where $x_1 \neq 0$. According to Table 1, g is a summation of terms involving x_2 to x_n. Therefore:

$$\frac{\partial g}{\partial x_i} = 9/(n-1), \tag{9}$$

where $i \in \{2, \cdots, n\}$. Based on Eqs. 8 and 9, we have:

$$\frac{\partial^2 f_2}{\partial x_1 x_i} = -\frac{1}{4\sqrt{x_1 g}} \cdot \frac{\partial g}{\partial x_i} = -\frac{9}{4(n-1)\sqrt{x_1 g}}, \tag{10}$$

where $i \in \{2, \cdots, n\}$. Since $g > 0$, we have $\frac{\partial^2 f_2}{\partial x_1 x_i} \neq 0$. Based on Definition 2, we can see that x_1 interacts with all other variables, i.e., x_2 to x_n.

By taking the derivative of f_2 with respect to x_i for $i \in \{2, \cdots, n\}$, we have:

$$\frac{\partial f_2}{\partial x_i} = \frac{\partial g}{\partial x_i} - \frac{\partial \sqrt{x_1/g}}{\partial x_i} = \frac{9}{n-1}(1 - \frac{\sqrt{x_1}}{2\sqrt[4]{g}}). \tag{11}$$

By taking the derivative of Eq. 11 with respect to x_1, we have:

$$\frac{\partial^2 f_2}{\partial x_i \partial x_1} = -\frac{9}{4(n-1)\sqrt[4]{g}\sqrt{x_1}}, \tag{12}$$

where $x_1 \neq 0$. Since $g > 0$, we have $\frac{\partial^2 f_2}{\partial x_i \partial x_1} \neq 0$. Furthermore, by taking the derivative of Eq. 11 with respect to x_j, $j \in \{2, \cdots, n\}$ and $i \neq j$, we have:

$$\frac{\partial^2 f_2}{\partial x_i x_j} = \frac{81\sqrt{x_1}}{8(n-1)^2 g^{-5/4}}, \tag{13}$$

where $x_1 \neq 0$. Since $g > 0$, we have $\frac{\partial^2 f_2}{\partial x_i x_j} \neq 0$. In summary, we can see that all variables interact with each other, which means that the f_2 function of ZDT1 is fully non-separable. This agrees with the output of differential grouping. Since the other ZDT test problems share a similar form of h and g functions as that of ZDT1, we can use the above procedure to prove their non-separability. □

3.2 Variable Interaction Analysis on DTLZ Benchmark Suite

According to Table 1, the mathematical forms of DTLZ functions can be classified into three groups: DTLZ1 to DTLZ4, DTLZ5 to DTLZ6, and DTLZ7. Thus, we investigate the variable interaction structure of each group separately. Without loss of generality, we set $m = 4$ and $n = 6$ in the experiments. By running Algorithm 1 on DTLZ1 to DTLZ4, we can empirically verify that they share the same variable interaction matrices as shown in Fig. 2. Moreover, Fig. 3 is the graphical representation of the matrices in Fig. 2. To validate the correctness of this result, we again use Definition 2 to prove Proposition 2.

$$\begin{array}{c} x_1 \\ x_2 \\ x_3 \\ x_4 \\ x_5 \\ x_6 \end{array} \begin{pmatrix} 0 & 1 & 1 & 1 & 1 & 1 \\ 1 & 0 & 1 & 1 & 1 & 1 \\ 1 & 1 & 0 & 1 & 1 & 1 \\ 1 & 1 & 1 & 0 & 0 & 0 \\ 1 & 1 & 1 & 0 & 0 & 0 \\ 1 & 1 & 1 & 0 & 0 & 0 \end{pmatrix}$$
$$x_1 \; x_2 \; x_3 \; x_4 \; x_5 \; x_6$$

(a) f_1 and f_2

$$\begin{array}{c} x_1 \\ x_2 \\ x_3 \\ x_4 \\ x_5 \\ x_6 \end{array} \begin{pmatrix} 0 & 1 & 0 & 1 & 1 & 1 \\ 1 & 0 & 0 & 1 & 1 & 1 \\ 0 & 0 & 0 & 0 & 0 & 0 \\ 1 & 1 & 0 & 0 & 0 & 0 \\ 1 & 1 & 0 & 0 & 0 & 0 \\ 1 & 1 & 0 & 0 & 0 & 0 \end{pmatrix}$$
$$x_1 \; x_2 \; x_3 \; x_4 \; x_5 \; x_6$$

(b) f_3

$$\begin{array}{c} x_1 \\ x_2 \\ x_3 \\ x_4 \\ x_5 \\ x_6 \end{array} \begin{pmatrix} 0 & 0 & 0 & 1 & 1 & 1 \\ 0 & 0 & 0 & 0 & 0 & 0 \\ 0 & 0 & 0 & 0 & 0 & 0 \\ 1 & 0 & 0 & 0 & 0 & 0 \\ 1 & 0 & 0 & 0 & 0 & 0 \\ 1 & 0 & 0 & 0 & 0 & 0 \end{pmatrix}$$
$$x_1 \; x_2 \; x_3 \; x_4 \; x_5 \; x_6$$

(c) f_4

Fig. 2. Variable interaction matrices of DTLZ1 to DTLZ4.

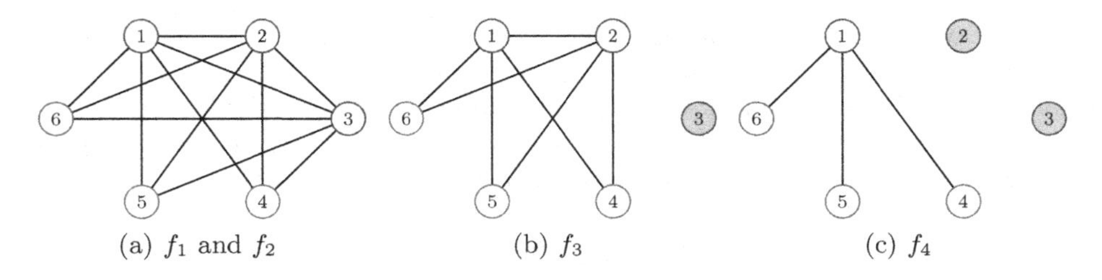

(a) f_1 and f_2 (b) f_3 (c) f_4

Fig. 3. Variable interaction graphs of DTLZ1 to DTLZ4.

Proposition 2. *For DTLZ1 to DTLZ4, $\forall f_i, i \in \{1, \cdots, m\}$, we divide the corresponding decision variables into two non-overlapping sets: $\mathbf{x}_I = (x_1, \cdots, x_\ell)^T$, $\ell = m - 1$ for $i \in \{1, 2\}$ while $\ell = m - i + 1$ for $i \in \{3, \cdots, m\}$; and $\mathbf{x}_{II} = (x_m, \cdots, x_n)^T$. All members of $\mathbf{x}_I$ not only interact with each other, but also interact with those of $\mathbf{x}_{II}$; all members of $\mathbf{x}_{II}$ are independent from each other.*

Proof. From Table 1 and Eq. 6, we re-write the objective functions of DTLZ1 to DTLZ4 in the following abstract form:

$$f_i(\mathbf{x}) = h(\mathbf{x}_I) \cdot g(\mathbf{x}_{II}), \tag{14}$$

where $i \in \{1, \cdots, m\}$. $\mathbf{x}_I = (x_1, \cdots, x_\ell)^T$, $\ell = m - 1$ for $i \in \{1, 2\}$ while $\ell = m - i + 1$ for $i \in \{3, \cdots, m\}$; and $\mathbf{x}_{II} = (x_m, \cdots, x_n)^T$. Notice that h function is a multiplication term of all individual variables of $\mathbf{x}_I$, while g function is some independent summations of terms involving all individual variables of $\mathbf{x}_{II}$.

Let us start from DTLZ1. By taking the derivative of f_i, where $i \in \{1, \cdots, m\}$, with respect to each member of $\mathbf{x}_I$, i.e., x_j, where $j \in \{1, \cdots, \ell\}$, we have:

$$\frac{\partial f_i}{\partial x_j} = 0.5(1 + g) \cdot \prod_{p=1, p \neq j}^{\ell} x_p. \tag{15}$$

Now by differentiating Eq. 15 with respect to x_k, where $k \in \{1, \cdots, n\}$ and $k \neq j$, we have:

$$\frac{\partial^2 f_i}{\partial x_j x_k} = \begin{cases} 0.5(1 + g) \cdot \prod_{p=1, p \neq i,j}^{m-1} x_p, & k \in \{1, \cdots, m - 1\} \\ 0.5 \frac{\partial g}{\partial x_k} \cdot \prod_{p=1, p \neq i}^{m-1} x_p, & k \in \{m, \cdots, n\}. \end{cases} \tag{16}$$

In particular, when $k \in \{m, \cdots, n\}$, we have:

$$\frac{\partial g}{\partial x_k} = 200(x_k - 0.5) + 2000\pi \sin(20\pi(x_k - 0.5)). \tag{17}$$

Note that both g and $\frac{\partial g}{\partial x_k}$ are not 0, when $x_k \neq 0.5, k \in \{m, \cdots, n\}$. In this case, we have $\frac{\partial^2 f_i}{\partial x_j x_k} \neq 0$, where $i \in \{1, \cdots, m\}$, $j \in \{1, \cdots, \ell\}$, $k \in \{1, \cdots, n\}$ and $k \neq j$. According to Definition 2, we can see that all members of $\mathbf{x}_I$ not only interact with each other, but also interact with those of $\mathbf{x}_{II}$. Note that since f_i, where $i \in \{3, \cdots, m\}$, is without of x_p, where $p \in \{m-i+2, \cdots, m-1\}$, we can treat x_p be independent/non-separable from the other variables for f_i.

In addition, by taking the derivative of f_i, where $i \in \{1, \cdots, m\}$, with respect to each member of $\mathbf{x}_{II}$, i.e., x_j, where $j \in \{m, \cdots, n\}$, we have:

$$\frac{\partial f_i}{\partial x_j} = 0.5 \prod_{p=1}^{\ell} x_p \cdot \frac{\partial g}{\partial x_j}. \tag{18}$$

According to Eq. 17, we can see that $\frac{\partial g}{\partial x_j}$ is a function of x_j. Thus, $\frac{\partial^2 f_1}{\partial x_j x_k} = 0$, where $k \in \{m, \cdots, n\}$ and $k \neq j$. According to Definition 2, we can see that all members of $\mathbf{x}_{II}$ are independent/non-separable from each other.

Since DTLZ2 to DTLZ4 have a similar form as DTLZ1, but are with some different exponentials. we can use the above proof procedure to derive the same variable interaction structure as DTLZ1. □

Then, by running Algorithm 1 on DTLZ5 and DTLZ6, we obtain the variable interaction matrices and graphs, as shown in Figs. 4 and 5, respectively. The correctness of this result is validated by the proof of Proposition 3.

$$\begin{array}{c} x_1 \\ x_2 \\ x_3 \\ x_4 \\ x_5 \\ x_6 \end{array} \begin{pmatrix} 0 & 1 & 1 & 1 & 1 & 1 \\ 1 & 0 & 1 & 1 & 1 & 1 \\ 1 & 1 & 0 & 1 & 1 & 1 \\ 1 & 1 & 1 & 0 & 1 & 1 \\ 1 & 1 & 1 & 1 & 0 & 1 \\ 1 & 1 & 1 & 1 & 1 & 0 \end{pmatrix}$$
$x_1\ x_2\ x_3\ x_4\ x_5\ x_6$

(a) f_1 and f_2

$$\begin{array}{c} x_1 \\ x_2 \\ x_3 \\ x_4 \\ x_5 \\ x_6 \end{array} \begin{pmatrix} 0 & 1 & 0 & 1 & 1 & 1 \\ 1 & 0 & 0 & 1 & 1 & 1 \\ 0 & 0 & 0 & 0 & 0 & 0 \\ 1 & 1 & 0 & 0 & 1 & 1 \\ 1 & 1 & 0 & 1 & 0 & 1 \\ 1 & 1 & 0 & 1 & 1 & 0 \end{pmatrix}$$
$x_1\ x_2\ x_3\ x_4\ x_5\ x_6$

(b) f_3

$$\begin{array}{c} x_1 \\ x_2 \\ x_3 \\ x_4 \\ x_5 \\ x_6 \end{array} \begin{pmatrix} 0 & 0 & 0 & 1 & 1 & 1 \\ 0 & 0 & 0 & 0 & 0 & 0 \\ 0 & 0 & 0 & 0 & 0 & 0 \\ 1 & 0 & 0 & 0 & 0 & 0 \\ 1 & 0 & 0 & 0 & 0 & 0 \\ 1 & 0 & 0 & 0 & 0 & 0 \end{pmatrix}$$
$x_1\ x_2\ x_3\ x_4\ x_5\ x_6$

(c) f_4

Fig. 4. Variable interaction matrices of DTLZ5 and DTLZ6.

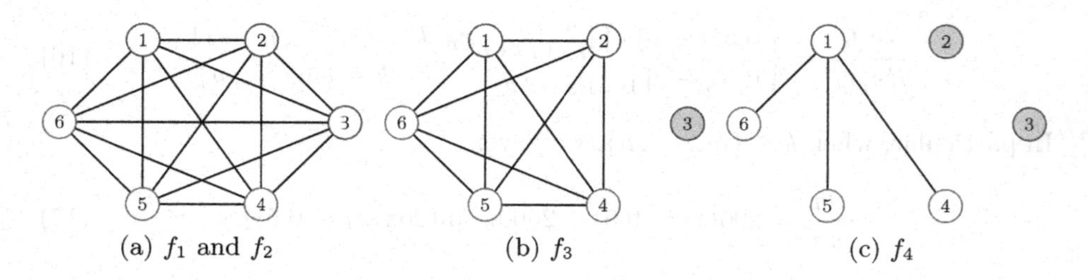

(a) f_1 and f_2 (b) f_3 (c) f_4

Fig. 5. Variable interaction graphs of DTLZ5 and DTLZ6.

Proposition 3. *For DTLZ5 and DTLZ6, $\forall f_i, i \in \{1, \cdots, m\}$, we divide the corresponding decision variables into two non-overlapping sets: $\mathbf{x}_I = (x_1, \cdots, x_\ell)^T$, $\ell = m - 1$ for $i \in \{1, 2\}$ while $\ell = m - i + 1$ for $i \in \{3, \cdots, m\}$; and $\mathbf{x}_{II} = (x_m, \cdots, x_n)^T$. For f_i, where $i \in \{1, \cdots, m-1\}$, all members of $\mathbf{x}_I$ and $\mathbf{x}_{II}$ interact with each other; for f_m, we have the same interaction structure as Proposition 2.*

Proof. From Table 1 and Eq. 6, we re-write the objective functions of DTLZ5 and DTLZ6 in the following abstract form:

$$f_i(\mathbf{x}) = h(\mathbf{x}_I, g(\mathbf{x}_{II})) \cdot g(\mathbf{x}_{II}), \tag{19}$$

where $i \in \{1, \cdots, m-1\}$. $\mathbf{x}_I = (x_1, \cdots, x_\ell)^T$, $\ell = m - 1$ for $i \in \{1, 2\}$ while $\ell = m - i + 1$ for $i \in \{3, \cdots, m\}$; and $\mathbf{x}_{II} = (x_m, \cdots, x_n)^T$. Comparing Eq. 19 with Eq. 14, the only difference lies on the h function which consists of both $\mathbf{x}_I$ and $\mathbf{x}_{II}$. Note that the objective functions of DTLZ5 and DTLZ6 have a similar form as that of DTLZ2, we can use the proof procedure of Proposition 2 to prove that all members of $\mathbf{x}_I$ not only interact with each other, but also interact with those of $\mathbf{x}_{II}$.

In addition, due to the additional term of $\mathbf{x}_I$ within the h function, we can derive that $\frac{\partial f_i}{\partial x_j}$, where $j \in \{m, \cdots, n\}$, should be a function of both x_j and members of $\mathbf{x}_I$. Thus, $\frac{\partial^2 f_i}{\partial x_j x_k} \neq 0$, where $k \in \{m, \cdots, n\}$ and $k \neq j$. This means that all members of $\mathbf{x}_{II}$ also interact with each other.

As for f_m, it still obeys the form of Eq. 14. According to the proof of Proposition 2, we can easily derive the same interaction structure as described in Proposition 2. □

At last, we run Algorithm 1 on DTLZ7 and find that all its objective functions are fully separable. This means that all entries of its interaction matrices should be 0, and the corresponding interaction graphs consist of n independent nodes. The proof of Proposition 4 validates the correctness of this result.

Proposition 4. *All objective functions of DTLZ7 are fully separable.*

Proof. From Table 1, we can see that f_i of DTLZ7 is a function of x_i for $i \in \{1, \cdots, m-1\}$. Thus, it is obvious that these objective functions are fully separable. As for f_m, we can re-write it as follows:

$$f_m = (1 + g)m - \sum_{i=1}^{m-1} (f_i + f_i \sin(3\pi f_i)) \tag{20}$$

In this case, f_m is the function of some independent summation terms involving x_1 to x_n. Therefore, it is also a separable function. □

4 Conclusions and Future Directions

We have seen that some of the ZDT and DTLZ test problems have complex variable interaction structures that change with the number of objectives.

More specifically, some objective functions are fully separable (e.g., f_1 of ZDT problems and all objectives of DTLZ7), some are fully non-separable (e.g., f_2 of ZDT problems and f_1 to f_{m-1} of DTLZ5 and DTLZ6), while the others are in between these two extreme cases, i.e., partially non-separable. This result is in contrast with the existing literature that coarsely classified the functions as separable or non-separable [4].

An interesting observation about the DTLZ functions is the existence of overlapping components within the objective functions. For example, in Fig. 3, at a first glance, the first two objective functions of DTLZ1 to DTLZ4 may be seen as a single non-separable component. However, upon a closer inspection, we can see that the variables form three components containing a set of shared decision variables. Concretely, $\{x_1, x_2, x_3, x_4\}$, $\{x_1, x_2, x_3, x_5\}$ and $\{x_1, x_2, x_3, x_6\}$ can be seen as three components with $\{x_1, x_2, x_3\}$ being the shared variables. This is analogous to functions with overlapping components in the large-scale global optimization literature [6]. Although differential grouping can discover the full variable interaction structure matrix, the optimal decomposition of functions with overlapping components is still an open question [6]. Based on the analysis in Sect. 3, it appears that objective functions with overlapping components are commonplace in multi-objective optimization. The analysis that we presented in this paper facilitates the study of this phenomenon with respect to both algorithm and benchmark designs.

Overall, variable interaction can affect various aspects of the EMO community, ranging from operator design to the choice of aggregation functions within decomposition-based EMO algorithms. We believe that variable interaction is an under-explored area in this literature, which might be due to extreme focus of the current research on small to medium sized problems. It is clear that when the dimensionality of a problem grows beyond a certain level, using a divide-and-conquer strategy becomes inevitable in which case considering variable interaction becomes a necessity. In the future, we plan to analyze a wider range of common benchmark suites within the EMO community. Additionally, similar to the large-scale global optimization [6], we plan to develop benchmark problems with challenging yet controllable variable interaction structures, which can better resemble the modular nature of real-world optimization scenarios.

Acknowledgement. This work was partially supported by EPSRC (Grant No. EP/J017515/1).

References

1. Aguirre, H.E., Tanaka, K.: Working principles, behavior, and performance of moeas on MNK-landscapes. Eur. J. Oper. Res. **181**(3), 1670–1690 (2007)
2. Deb, K.: Multi-objective genetic algorithms: Problem difficulties and construction of test problems. Evol. Comput. **7**(3), 205–230 (1999)
3. Deb, K., Thiele, L., Laumanns, M., Zitzler, E.: Scalable test problems for evolutionary multiobjective optimization. In: Abraham, A., Jain, L., Goldberg, R. (eds.) Evolutionary Multiobjective Optimization. AIKP, pp. 105–145. Springer, London (2005)

4. Huband, S., Hingston, P., Barone, L., While, R.L.: A review of multiobjective test problems and a scalable test problem toolkit. IEEE Trans. Evol. Comput. **10**(5), 477–506 (2006)
5. Omidvar, M.N., Li, X., Mei, Y., Yao, X.: Cooperative co-evolution with differential grouping for large scale optimization. IEEE Trans. Evol. Comput. **18**(3), 378–393 (2014)
6. Omidvar, M.N., Li, X., Tang, K.: Designing benchmark problems for large-scale continuous optimization. Inf. Sci. **316**, 419–436 (2015)
7. Salomon, R.: Re-evaluating genetic algorithm performance under coordinate rotation of benchmark functions. a survey of some theoretical and practical aspects of genetic algorithms. Biosystems **39**, 263–278 (1996)
8. Toint, P.L.: Test problems for partially separable optimization and results for the routine PSPMIN. Technical Report 83/4, Department of Mathematics, Facultés Universitaires de Namur, Namur, Belgium (1983)
9. Yang, Z., Tang, K., Yao, X.: Large scale evolutionary optimization using cooperative coevolution. Inf. Sci. **178**(15), 2985–2999 (2008)
10. Zitzler, E., Deb, K., Thiele, L.: Comparison of multiobjective evolutionary algorithms: empirical results. Evol. Comput. **8**(2), 173–195 (2000)

Improving Efficiency of Bi-level Worst Case Optimization

Ke Lu[1], Juergen Branke[1(✉)], and Tapabrata Ray[2]

[1] Warwick Business School, University of Warwick, Coventry, UK
Ke.Lu@warwick.ac.uk, Juergen.Branke@wbs.ac.uk
[2] University of New South Wales, Kensington, Australia
t.ray@adfa.edu.au

Abstract. We consider the problem of identifying the trade-off between tolerance level and worst case performance, for a problem where the decision variables may be disturbed within a set tolerance level. This is a special case of a bi-level optimization problem. In general, bi-level optimization problems are computationally very expensive, because a lower level optimizer is called to evaluate each solution on the upper level. In this paper, we propose and compare several strategies to reduce the number of fitness evaluations without substantially compromising the final solution quality.

1 Introduction

A typical problem in engineering is that manufacturing is not able to produce exactly to specification, but instead will introduce some deviations from the design variables. An engineer has to take this into account by allowing for manufacturing tolerances. In this paper, we use an evolutionary multi-objective (EMO) algorithm to determine the trade-off between tolerance level δ and worst case performance $f^{wc}(x) = \min_{x' \in [x-\delta, x+\delta]} f(x')$. This problem has first been addressed in [4], where an envelope-based (where the lower level is multi-objective) and a point-based algorithm (lower level is single-objective) were proposed. The point-based algorithm is a special case of a bi-level optimization algorithm. In general, a bi-level optimization problem can be formulated as follows.

$$\min \quad F(x_u, x_l) \tag{1}$$

$$\text{s.t. } x_l \in \operatorname{argmin}\{f(x_u, x_l) : g_i(x_u, x_l) \leq 0, j = 1, ..., J\} \tag{2}$$

$$G_k(x_u, x_l) \leq 0, k = 1, ...K \tag{3}$$

$$x_u \in X_U, x_l \in X_L \tag{4}$$

where x_u and x_l are the upper and lower level decision variables, F and f are the upper and lower level fitness function, and G and g are the upper and lower level constraints, respectively.

J. Handl et al. (Eds.): PPSN XIV 2016, LNCS 9921, pp. 410–420, 2016.
DOI: 10.1007/978-3-319-45823-6_38

The problem we want to solve can be described as

$$\max \quad f(x'), \delta \tag{5}$$
$$\text{s.t. } x' = \text{argmin}\{f(x'), x' \in [x-\delta, x+\delta]\}. \tag{6}$$

This is a special case of the above bi-level optimization problem with $x_u = \{x, \delta\}, x_l = x'$ and $g_1(x, x', \delta) = x' - x + \delta, g_2(x, x', \delta) = x - x' - \delta$. In particular, on the upper level we have a multi-objective problem with maximization of $f(x')$ and δ, and decision variables are x and δ, and for each upper level solution (x, δ) the worst case is identified by solving a lower level single objective problem with minimization of $f(x')$ and decision variable x'.

Because in bi-level optimization a lower level optimizer has to be run to evaluate each solution on the upper level, the procedure is computationally very expensive. In this paper, we re-visit the point-based algorithm and suggest and compare various strategies to reduce the necessary number of fitness function evaluations on the lower level.

The paper is structured as follows. First, we survey some related work in Sect. 2. Section 3 describes the baseline algorithm and presents six different strategies to reduce the necessary number of function evaluations. The test problems used are presented in Sect. 4, and the empirical results are discussed in Sect. 5. The paper concludes with a summary and some ideas for future work.

2 Related Work

As has already been mentioned in the introduction, the problem of looking at the trade-off between tolerance level and worst case performance has been introduced in [4] and the baseline algorithm we use here is the point-based algorithm from [4]. [7] considered a similar problem, looking at the trade-off between the nominal fitness $f(x)$ and the tolerance threshold that guarantees that the worst case fitness is by no more than a fixed value Δ worse than the nominal fitness. Since this is formulated as a three-level problem and computationally even more expensive than our bi-level problem, an approach using surrogates is introduced in [6]. Surrogate-assisted EAs have also been used in [8,9] for worst-case optimization. Multi-objective worst-case optimization is considered in [3]. Many papers that deal with uncertainty in the decision variables consider expected fitness optimization rather than worst-case optimization. A survey on robust design optimization can be found in [2].

There are a number of papers in evolutionary bi-level optimization. In particular [5] is also suggesting ways to improve the efficiency, including one of the methods we are testing in this paper, namely maintaining the population of a lower level EA so that it can be continued later. However, the special structure of the worst-case optimization problem allows us to exploit more ways to use upper level information to reduce the fitness evaluations on the lower level. In [1], surrogate models are combined with Differential Evolution in order to solve bi-level problems.

3 Algorithm and New Strategies

3.1 Worst Case Bi-level Evolutionary Algorithm

Algorithms 1 and 2 show the pseudocode for the upper and lower level EA, respectively. As can be seen, the upper level is a multi-objective NSGA-II type EA, the lower level is a standard single objective EA. Every individual on the upper level is evaluated by running a lower level EA. Line 10 of Algorithm 1 show that every individual surviving to the next generation is re-evaluated, and its fitness is updated if a new worst case is found. This is done to ensure that the worst case of solutions surviving over several generations is reliable. Otherwise, a solution for which the lower level was unable to find the true worst case might look deceivingly good on the upper level EA and misguide the search. However, re-evaluating each individual in every generation doubles the total number of evaluations and thus the computational cost.

Algorithm 1. Pseudocode for upper level MOEA

```
1: Initialize parent population P (x, δ)
2: Call lowerEA to evaluate each individual in P
3: for j=1 to g do                                   ▷ g is number of generations
4:     Non-dominate sort P
5:     Generate offspring population O by evolutionary operators
6:     Call lowerEA to evaluate each individual in O
7:     Get the union population U = P ∪ O
8:     Non-dominate sort U
9:     Select individuals to form the next generation parent population P
10:    Call lowerEA to re-evaluate each individual in the new next generation
       population P
11: end for
```

3.2 Strategies to Save Fitness Evaluations

Strategy I: Exploit Upper Level Information for Selection and Stopping Criteria at the Lower Level. In worst case optimization, the upper and lower level both use the fitness function f, but for the upper level it is an objective to maximize, whereas the lower level tries to minimize it. This can be exploited to prematurely stop lower level optimization runs if it is apparent that the corresponding upper level solution would not be interesting. The concept is visualized in Fig. 1. Let us assume the lower level is trying to find the worst case of solution E. Then, this solution's worst case objective moves down (lower f^w) during lower level optimization. If it enters the current dominated region of the upper level population (depicted by the red solutions $A-D$), then we know solution E will be dominated on the upper level, and not contribute significantly

Algorithm 2. Pseudocode for lower level EA

```
1: procedure LOWEREA(x, δ)
2:     Initialize parent population P′ such that each individual is within [x−δ, x+δ]
       ▷ (x, δ) is upper level individual
3:     Compute the fitness f(x′) of each individual in P′
4:     for j=1 to g’ do                           ▷ g’ is number of generations
5:         Generate offspring population O′ by evolutionary operators
6:         Compute the fitness of each individual in O′
7:         Get the union population U′ = P′ ∪ O′
8:         Sort U′ according to fitness
9:         Select best individuals from U′ to form the next generation parent popula-
           tion P′
10:    end for
11: return best individual in P′                                    ▷ lowest f
12: end procedure
```

to the upper level search, so we can choose to abort the lower level run prematurely. Note that we only do this for a solution's first lower level evaluation, not for re-evaluation, because the purpose of re-evaluation is to make sure we really found the worst case.

Strategy II: Use Neighbors to Update Worst Case Fitness. An important but computationally expensive step in the bi-level EA is the re-evaluation of all surviving individuals at the end of each upper level generation. In Strategy II, we propose to replace the re-evaluation by exploiting neighbors to update worst-case information instead. Consider the example depicted in Fig. 2 which

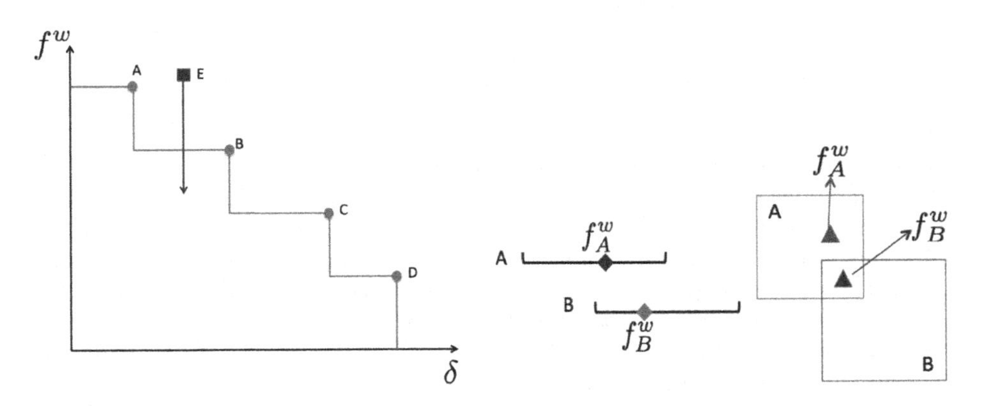

Fig. 1. Strategy I: if worst case found by lower level enters current dominated region of upper level, the lower level is aborted.

Fig. 2. Strategy II: update of worst cases using information from neighbors. Left part shows a one-dimensional example, right part shows a two-dimensional example.

shows two solutions A and B and their corresponding worst cases f_A^w and f_B^w as found by the lower level EA, respectively. The worst case f_B^w lies in the disturbance region of solution A. So, if $f_B^w < f_A^w$, then we know that f_A^w can not be the true worst case of solution A, and we can replace f_A^w by f_B^w as a more realistic estimate of A's worst case. Because EAs evaluate many solutions in promising regions of the search space, a solution's worst case estimate should quickly become accurate even without re-evaluation.

Strategy III: Skip Re-evaluation if It Does Not Improve Worst-Case Estimate. If re-evaluating a solution did not identify a new worst case, we can be more confident that the worst case we found previously is accurate. Strategy III chooses to skip re-evaluation for this solution in the following generation.

Strategy IV: Lower Level Smart Initialization. The purpose of re-evaluation is to make sure the lower level really correctly identified the worst case. In the baseline algorithm, the lower level is re-started from scratch for re-evaluation. Strategy IV suggests to keep the population at the end of a lower level run in memory, and re-start the search from this population if the corresponding upper level solution is to be re-evaluated. This has previously been proposed in [5]. Note that this may risk getting stuck in a local optimum. Since it does not make much sense to continue running the algorithm once it has converged, we abort the lower level EA in case the improvement over the past 5 generations was less than 0.001.

Strategy V: Make Lower Level Generations Adaptive to δ. For solutions with smaller tolerance level, the lower level search space is relatively smaller. Strategy V thus proposes to reduce the number of generations of the lower level EA for smaller δ. Specifically, we chose to set

$$Generations = \max\left\{10, \left\lceil maxGen \times \frac{2\delta}{\delta^{max} - \delta^{min}} \right\rceil\right\}$$

Strategy VI: Adjust Lower Level Population Size to δ. Instead of reducing the number of generations as in Strategy V, Strategy VI reduces the population size of the lower level EA for small δ. We chose a function that has a minimum population size of 1 for $\delta = 0$, and a maximum population size equal to the usual population size for $\delta = \delta^{max}$, which for the test functions considered later results in $popsize = \lceil (popsize^{max} - 1)/\delta^{max} \times \delta + 1 \rceil$ in case of a single decision variable, and $popSize = \lceil (popsize^{max} - 1)/(\delta^{max})^2 \times \delta^2 + 1 \rceil$ for the case of two dimensional problems.

4 Test Problems

Test Function 1. This is a simple 1 dimensional test function taken from [4] for which the optimal Pareto front is known. It is visualized in Fig. 3.

$$f_1(x_i) = \begin{cases} \frac{5}{3}x_i - \frac{2}{3} & \textbf{if } 1 < x_i \leq \frac{5}{2} \\ -\frac{5}{3}x_i + \frac{23}{3} & \textbf{if } \frac{5}{2} \leq x_i \leq 4 \\ \frac{2}{3}x_i - \frac{2}{3} & \textbf{if } 4 < x_i \leq \frac{11}{2} \\ -\frac{2}{3}x_i - \frac{20}{3} & \textbf{if } \frac{11}{2} < x_i \leq 7 \end{cases} \qquad 1.0 \leq x_i \leq 7.0$$

Test Function 2. This function, visualized in Fig. 4 is taken from [7]. It has more local optima which makes it more interesting, but we do not know the true Pareto front for this function.

$$\begin{aligned} f_2(x) &= 2e^{-(x-2)^2/0.32} + 2.2e^{-(x-3)^2/0.18} + 2.4e^{-(x-4)^2/0.5} \\ &+2.3e^{-(x-5.5)^2/0.5} + 3.2e^{-(x-7)^2/0.18} + 1.2e^{-(x-8)^2/0.18} \end{aligned}$$

where $0 \leq x \leq 10$.

Test Function 3. This is simply a 2D version of Test function 1, with $f_3(X) = \sum f_1(x_i)$. It is visualized in Fig. 5.

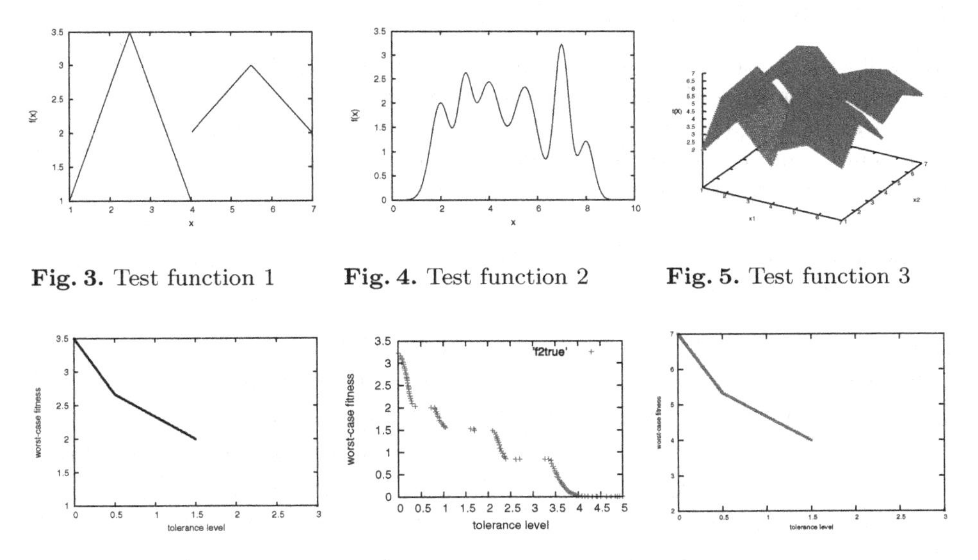

Fig. 3. Test function 1 **Fig. 4.** Test function 2 **Fig. 5.** Test function 3

Fig. 6. Pareto front of TF1 **Fig. 7.** Pareto front of TF2 **Fig. 8.** Pareto front of TF3

Table 1. Parameter setting for TF1 and 2

	UpperLevel	LowerLevel
Popsize	100	40
Max generations	500	100
Crossover prob.	0.9	0.9
Mutation prob	0.5	1.0

Table 2. Parameter setting for TF3

	UpperLevel	LowerLevel
Popsize	100	40
Max generations	500	100
Crossover prob.	0.9	0.9
Mutation prob	0.5	1.0

5 Empirical Results

5.1 Parameter Settings

The parameter settings are displayed in Tables 1 and 2.

For performance evaluation we use the Inverse Generational Distance (IGD) because, as explained in [4], it penalizes both, if solutions worse than the true Pareto front have been found, and if solutions seemingly better than the true Pareto front are found (which can happen because the lower level is not guaranteed to find the true worst case). For TF 2, the reference set is obtained by running our algorithm several times with a very large population size and very large number of generations. Note that we multiplied all IGD values by 100 to avoid small numbers. All results reported are averages over 20 runs.

5.2 Test Results and Analysis

We will look at the effect of each of the above six methods individually on different test functions, and then try combinations of different methods. In addition to figures showing the reduction of IGD over fitness evaluations, we show tables that report, for each method, the number of fitness evaluations needed (in percentage of the standard bi-level algorithm) and final IGD.

Table 3. Relative function evaluations and final IGD value when each of the six different strategies are used individually.

	TF1		TF2		TF3	
Strategy	Evals	IGD $\pm$ std.err	Evals	IGD $\pm$ std.err	Evals	IGD $\pm$ std.err
Standard	100.0 %	0.759 $\pm$ 0.012	100.0 %	2.827$\pm$ 0.089	100.0 %	1.265 $\pm$ 0.016
I	55.01 %	0.755 $\pm$ 0.011	54.68 %	2.870$\pm$ 0.093	56.43 %	1.265 $\pm$ 0.016
II	50.00 %	0.743 $\pm$ 0.014	50.00 %	3.001 $\pm$ 0.057	50.00 %	1.420 $\pm$ 0.021
III	74.96 %	0.745 $\pm$ 0.012	74.08 %	2.851 $\pm$ 0.080	72.30 %	1.431 $\pm$ 0.025
IV	62.27 %	0.779 $\pm$ 0.010	63.85 %	2.904 $\pm$ 0.063	52.57 %	1.477 $\pm$ 0.021
V	57.23 %	0.765 $\pm$ 0.011	91.01 %	2.918 $\pm$ 0.063	49.82 %	1.340 $\pm$ 0.085
VI	32.27 %	0.798 $\pm$ 0.009	50.87 %	3.179 $\pm$ 0.076	15.73 %	1.387 $\pm$ 0.023

Table 4. Combining Strategy I with any of the other five strategies, relative function evaluations and final IGD value.

	TF1		TF2		TF3	
Strategy	Evals	IGD±std.err	Evals	IGD±std.err	Evals	IGD±std.err
Standard	100.0 %	0.759 ± 0.012	100.0 %	2.827 ± 0.089	100.0 %	1.265 ± 0.016
I+II	6.04 %	0.726 ± 0.011	5.12 %	3.054 ± 0.062	6.26 %	1.511 ± 0.020
I+III	28.35 %	0.733 ± 0.009	28.56 %	2.927 ± 0.073	27.91 %	1.400 ± 0.018
I+IV	9.76 %	0.777 ± 0.008	9.70 %	2.775 ± 0.072	7.43 %	1.449 ± 0.018
I+V	39.94 %	1.056 ± 0.017	64.41 %	4.751 ± 0.121	32.46 %	1.283 ± 0.017
I+VI	20.70 %	0.783 ± 0.011	32.24 %	2.933 ± 0.085	10.84 %	1.471 ± 0.026

Individual Effects. The results of using each of the above 6 strategies individually is displayed in Table 3 and Figs. 9, 10, 11, 12 and 13. They are relatively consistent across test problems. Aborting lower level runs when they result in upper level dominated solutions saves almost 50 % of the evaluations. Replacing re-evaluation by neighborhood update (Strategy II) always saves

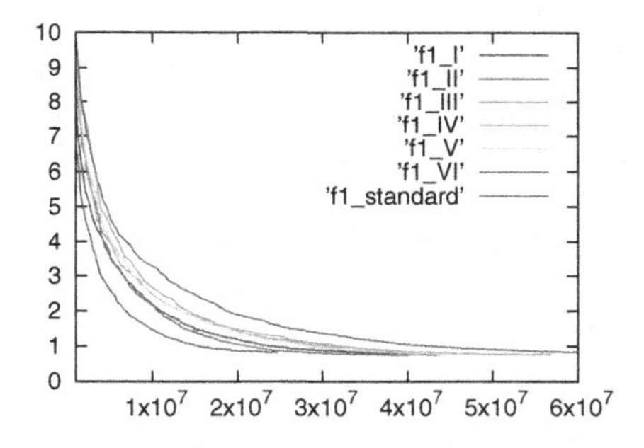

Fig. 9. Effects of six Strategies on TF1

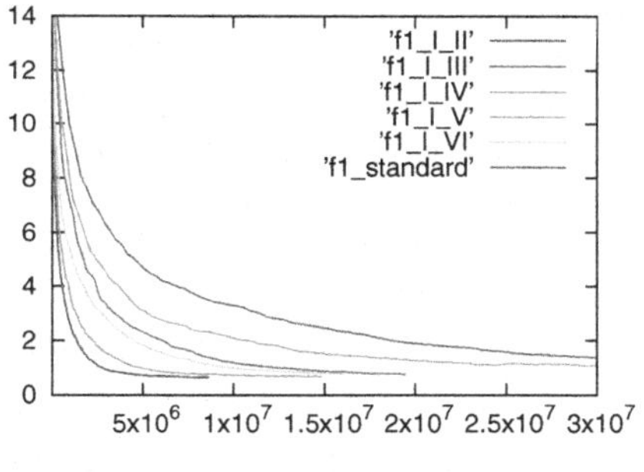

Fig. 10. Combinations with Strategy I on TF1

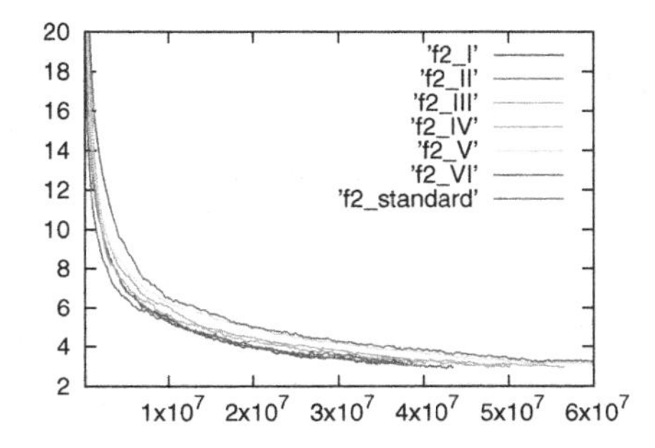

Fig. 11. Effects of six Strategies on TF2

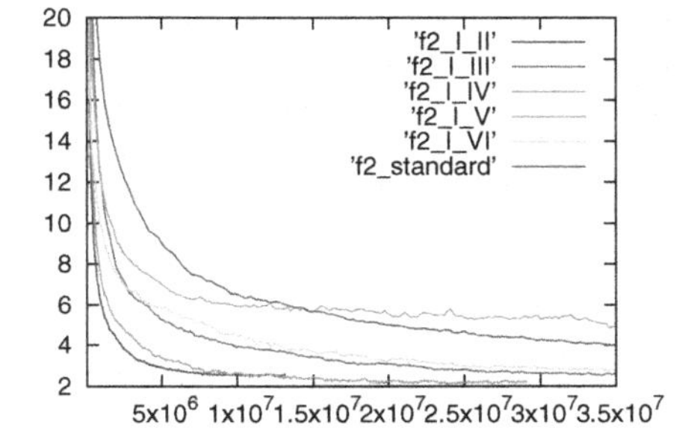

Fig. 12. Combinations with Strategy I on TF2

exactly 50 % of the evaluations. Strategy III (skipping some re-evaluations) saves only about 25 %. Smart initialization and early stopping in case of convergence yields between 36 % and 48 %. The biggest differences between test problems can be found in the techniques that adjust the number of generations or the population size to the size of the lower level search space. In TF3 this seems to yield the greatest savings, whereas savings in TF2 are relatively modest. Obviously this depends on how many Pareto optimal solutions with large δ exist in the upper level, and for TF1 and 3 there are relatively few, resulting in large savings. Reducing the population size seems to reduce the number of fitness evaluations by more than reducing the number of generations, although this certainly depends on the chosen parameter settings.

While the savings in terms of number of function evaluations are massive (up to 85 %), most of the methods do not suffer substantially in terms of obtained IGD. On TF1, all methods work quite well, although the last strategy (adapting the lower level population size to δ) performs worst. Reducing the number of generations seems to work better than reducing the population size across all three test functions. Strategy II (replacing re-evaluation by neighborhood update), despite saving 50 % of function evaluations, works better than the baseline algorithm on TF1. The reason is that in TF1, all the Pareto optimal solutions have one of three x values, so there are many solutions with overlapping neighborhoods which allows Strategy II to work particularly well. And because we can update fitness based on neighborhood *before* selection, whereas re-evaluation is done only *after* selection, it really can do better. Unfortunately, Strategy II is significantly worse than the baseline algorithm on TF2 which has substantially more peaks.

The convergence curves (Figs. 9, 10, 11, 12 and 13) confirm these observations. They also show that the acceleration in convergence by using the above strategies is generally more substantial in TF1 than in TF2.

Combinations of Strategies. Several of the strategies can be combined. Table 4 reports on results of combining the early abortion of unpromising runs

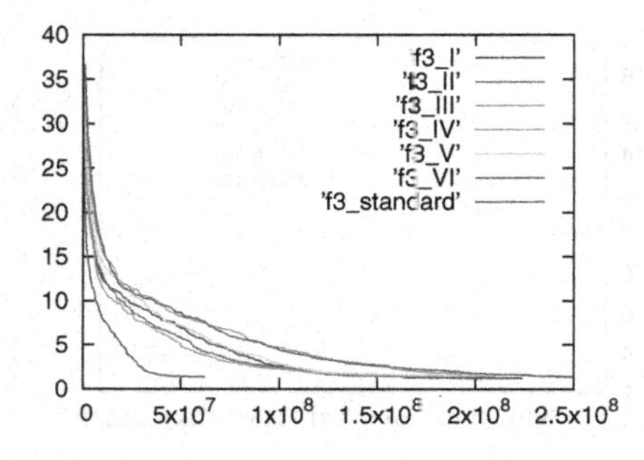

Fig. 13. Effects of six Strategies on TF3

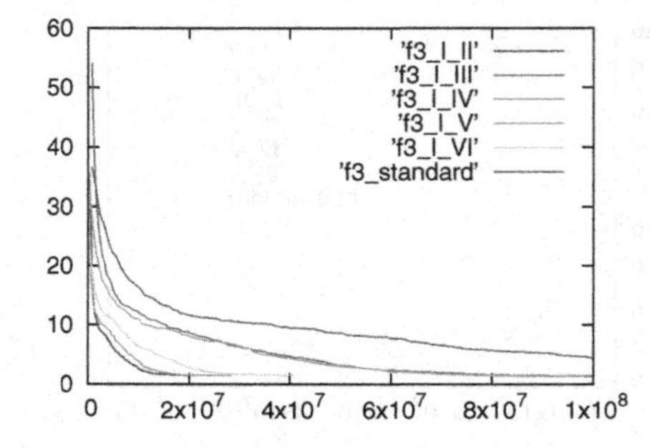

Fig. 14. Combinations with Strategy I on TF3

(Strategy I) with each of the other five strategies. The obtained savings in the number of fitness function evaluations is very remarkable and up to 95 % on TF2 for the case of combining Strategy I with Strategy II (abortion of lower level runs for unpromising solutions and replacing re-evaluation by neighborhood update). Note that the savings of Strategy I are independent of re-evaluation, as Strategy I is only applied during an individual's first evaluation, so combining Strategy I (only applied to first evaluation) and Strategy II (to get rid of re-evaluation) is complementary. Looking at the IGD of Strategy I+II, while the IGD is even lower than the standard algorithm for TF1, it is substantially higher for TF2 and TF3. A good alternative seems to be the combination of Strategies I+IV, as it also reduces fitness evaluations by more than 90 %, and works better on TF2 and TF3. Combination of Strategies I+V seems to work well on TF3, but poorly on TF1 and TF2 (has a very high standard error on TF2, and convergence plot in Fig. 12 indicates that it gets stuck in a local optimum).

6 Conclusion and Future Work.

This paper suggested and compared various strategies to reduce the necessary number of fitness function evaluations in bi-level worst case optimization, in particular when looking at the trade-off between worst case and tolerance level. The strategies obtained a reduction of fitness function evaluations of up to 95 %, with only modest decrease in performance. This is an important step towards making bi-level worst case optimization computationally feasible. In future research, we will consider whether the proposed strategies can be extended to the envelope based algorithm proposed in [4], and which of those strategies can be extended to general bi-level optimization problems. Also, a better understanding of when and why the different strategies work well would be helpful. Finally, it may be a good idea to vary the usage of these strategies over the run and, e.g., switch them off towards the end of the run.

References

1. Angelo, J.S., Krempser, E., Barbosa, H.J.: Differential evolution assisted by a surrogate model for bilevel programming problems. In: Congress on Evolutionary Computation, pp. 1784–1791. IEEE (2014)
2. Beyer, H.-G., Sendhoff, B.: Robust optimization-a comprehensive survey. Comput. Methods Appl. Mech. Eng. **196**(33), 3190–3218 (2007)
3. Branke, J., Avigad, G., Moshaiov, A.: Multi-objective worst case optimization by means of evolutionary algorithms. Evolutionary Computation (2013)
4. Branke, J., Lu, K.: Finding the trade-off between robustness and worst-case quality. In: Genetic and Evolutionary Computation Conference, pp. 623–630. ACM (2015)
5. Deb, K., Sinha, A.: An efficient and accurate solution methodology for bilevel multi-objective programming problems using hybrid evolutionary-local-search algorithm. Evol. Comput. **18**(3), 403–449 (2010)
6. Lim, D., Ong, Y.-S., Jin, Y., Sendhoff, B., Lee, B.S.: Inverse multi-objective robust evolutionary design. Genet. Program. Evolvable Mach. **7**(4), 383–404 (2007)

7. Lim, D., Ong, Y.-S., Lee, B.-S.: Inverse multi-objective robust evolutionary design optimization in the presence of uncertainty. In: Genetic and Evolutionary Computation Conference, pp. 55–62. ACM (2005)
8. Ong, Y.-S., Nair, P.B., Lum, K.: Max-min surrogate-assisted evolutionary algorithm for robust design. IEEE Trans. Evol. Comput. **10**(4), 392–404 (2006)
9. Zhou, A., Zhang, Q.: A surrogate-assisted evolutionary algorithm for minimax optimization. In: IEEE Congress on Evolutionary Computation (CEC), pp. 1–7. IEEE (2010)

Multi-objective Selection of Algorithm Portfolios: Experimental Validation

Daniel Horn[1](✉), Karin Schork[1], and Tobias Wagner[2]

[1] Computational Statistics, Technische Universität Dortmund, Vogelpothsweg 87, 44227 Dortmund, Germany
{daniel.horn,karin.schork}@tu-dortmund.de
[2] Institute of Machining Technology (ISF), Technische Universität Dortmund, Baroper Str. 303, 44227 Dortmund, Germany
wagner@isf.de

Abstract. The selection of algorithms to build portfolios represents a multi-objective problem. From a possibly large pool of algorithm candidates, a portfolio of limited size but good quality over a wide range of problems is desired. Possible applications can be found in the context of machine learning, where the accuracy and runtime of different learning techniques must be weighed. Each algorithm is represented by its Pareto front, which has been approximated in an a priori parameter tuning. Our approach for multi-objective selection of algorithm portfolios (MOSAP) is capable to trade-off the number of algorithm candidates and the respective quality of the portfolio. The quality of the portfolio is defined as the distance to the joint Pareto front of all algorithm candidates. By means of a decision tree, also the selection of the right algorithm is possible based on the characteristics of the problem.

In this paper, we propose a validation framework to analyze the performance of our MOSAP approach. This framework is based on a parametrized generator of the algorithm candidate's Pareto front shapes. We discuss how to sample a landscape of multiple Pareto fronts with predefined intersections. The validation is performed by calculating discrete approximations for different landscapes and assessing the effect of the landscape parameters on the MOSAP approach.

Keywords: Multi-objective optimization · Algorithm selection · Performance assessment · Benchmarking

1 Motivation

In algorithm selection tasks, it is still common practice to tune and compare competing algorithms or models with respect to a single performance measure. For instance, the mean error rate in classification or the best obtained function

D. Horn—We acknowledge partial support by the Mercator Research Center Ruhr under grant Pr-2013-0015 *Support-Vektor-Maschinen für extrem große Datenmengen.*

J. Handl et al. (Eds.): PPSN XIV 2016, LNCS 9921, pp. 421–430, 2016.
DOI: 10.1007/978-3-319-45823-6_39

value in optimization. The best algorithm can easily be selected based on the performance value. Often, however, additional performance measures are worth consideration. For instance, the budget of computation time or function evaluations could be considered as a second criterion. Since the performance measures are likely to be contradicting, both the tuning and the selection have to be adjusted. During the parameter tuning, the respective Pareto front has to be approximated for each algorithm. As a consequence, sets of solutions are compared in the selection step. As there is likely no single best candidate, the joint Pareto front is formed by a set of algorithms. In multi-objective selection of algorithm portfolios (MOSAP), we aim at approximating this subset of algorithms to allow selecting the best algorithm for a specific task a posteriori.

A possible application is the training of support vector machines (SVMs). Since the training of a single kernelized SVM scales at least quadratically with the number of observations, exact SVMs may be inapplicable for large datasets. Many approximative solvers have been introduced to compensate for this drawback. We conducted an exhaustive benchmark comparing the accuracy and the training time of some representative solvers in a multi-objective way [7].

To the best of our knowledge, there is no other work on the MOSAP topic. After the conceptual ideas of our MOSAP approach have been proposed and tested on the SVM application [6], we are now interested in benchmarking and validating MOSAP methods. In particular, we want to evaluate the performance of our own approach. To accomplish this, we propose a generator for constructing artificial data samples of candidate algorithms with known global Pareto fronts. Based on this generator, the performance of the resulting portfolios is evaluated for different properties of the generated data sets.

2 Multi-objective Selection of Algorithm Portfolios

In general, the performance of a set of r algorithms $\mathcal{A} = \{A_1, \ldots, A_r\}$ with respect to m objectives $(y_1, \ldots, y_m) \in Y^m$ shall be evaluated. Each algorithm A_i $(i = 1, \ldots, r)$ has its own set of parameters. We assume that a multi-objective parameter tuning has been performed in advance for each algorithm A_i. The resulting discrete approximation of the respective Pareto front is denoted as $PF(A_i)$. We focus on the common case of two objectives $(y_1, y_2) \in Y^2$.

Usually, there is stochasticity in the tuning results (e.g., random start designs in the optimization [8]). We assume that each tuning has been replicated $n > 1$ times, resulting in n independent approximations of $PF_j(A_i)$ $(j = 1, \ldots, n)$ for each algorithm. From these replications, we can compute the empirical attainment function [4]. In this paper, we use the median front (50 %-EAF) as representative of the outcome of each algorithm.

Our MOSAP approach is divided into three independent steps. In the first step, unnecessary candidate algorithms producing so-called interfering fronts are detected. Interfering fronts are completely dominated by the fronts of the other candidates and therefore do not contribute to the joint Pareto front. In our approach, we remove algorithms that are completely dominated in η of the replications. In the second step, we build a subset of algorithms with a reasonable trade-off between the size and the quality with regard to the joint Pareto front. This selection is a bi-objective decision making problem, as we aim to minimize the size of the subset and to maximize its quality. We define the quality of a given subset as the negative gap between its representative Pareto front and the joint Pareto front of all algorithms. The gap can be measured by any binary performance indicator, for example the hypervolume [12]. The decision making is implemented by optimizing the augmented Tschebyscheff norm [9] with a predefined weight vector w. In the third step, we aim at defining a decision rule for selecting the candidate algorithm for a specific problem. As we assume a bi-objective problem, we know that for non-dominated points the value of the second objective will decrease while the value of the first one increases. Hence, the solutions of the Pareto front can be indexed with regard to the first objective y_1. The domain of this objective is partitioned into intervals $[x_1, x_2], [x_2, x_3], \ldots, [x_{t-1}, x_t]$. Each interval is assigned to a specific algorithm A_i. For approximating this mapping, we calculate the joint non-dominated 50 %-EAF of all remaining algorithms and learn a decision tree [1] with input parameter y_1. To avoid that x_{k-1} and x_k $(k = 2, ..., t)$ are too close to each other, the decision tree is pruned with complexity control parameter cp.

In this paper, we aim at selecting an almost comprehensive portfolio of algorithms, therefore we parametrize our method as follows: $\eta = 0.5$, $cp = 0.1$ and $w = (0.01, 0.99)$ (it is more desirable to have a small gap than a small portfolio).

3 Test Case Generator

Our goal is the automatic construction of artificial test cases that resemble the real data we observed in the SVM benchmark [7], but are also able to take rather different shapes. In Fig. 1 an example of real data is displayed.

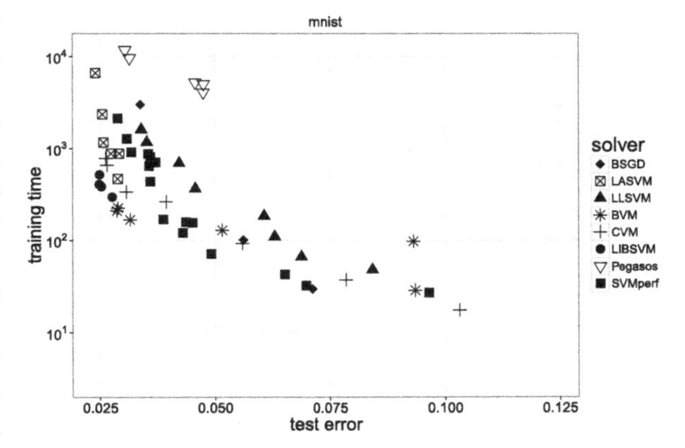

Fig. 1. Result of a biobjective parameter tuning (test error versus training time) of different approximative SVM solvers on the mnist dataset [7].

Our framework for creating the test cases consists of four steps. In the first step, we propose a flexible parametrized class of convex Pareto front shapes. By adjusting the parameters we are able to generate different Pareto fronts with predefined locations and shapes. In the second step, the sampling is extended to sets containing multiple Pareto fronts corresponding to r different algorithms. In this set, we differentiate between two types of Pareto fronts: active fronts that do have a contribution to the joint Pareto front and interfering fronts without contribution. The active fronts are constructed under consideration of predefined intersection points. In the third step, we describe how to generate discrete approximations from the continuously defined fronts in order to simulate the outcome of each algorithm. We propose four methods with different types of distributions and approximation error. In the fourth step, we discuss how to create n noisy replications of these discrete approximations.

Class of Functions. We define a parametrized function family $y = e^{-ax} - bx$ for $a, b > 0$ to construct convex functions which differ in the location of the knee (controlled by parameter b) and the curvature (controlled by a). We restrict the generator to convex functions based on our experiences with real-world data [6,7,10]. With this general formulation, we can only define Pareto fronts with a knee point skewed to lower values of the first objective y_1. For skewing the knee point towards lower values of y_2, the function is reflected on the angle bisector. To accomplish this, we utilize the Lambert W-function [2], which is the inverse function of xe^x. We assign this inverse function to negative values of b. A value of $b = 0$ results in a knee in the center of the front. The parameter a defines the curvature of the Pareto front, higher values of a result in a stronger severity of the knee. The effects of a and b are shown in Fig. 2. We normalize our fronts by subtracting $e^{-a} - b$ and dividing by $1 - e^{-a} + b$. After normalization, all functions of the function family intersect with the extreme points $(0, 1)$ and $(1, 0)$.

For preparing the next step of building defined sets of Pareto fronts, the parameters c and d are added to the function family. These parameters allow the Pareto fronts to be moved horizontally (c) and vertically (d). In addition, the parameter s is introduced for scaling the Pareto fronts.

The final class of functions is defined as

$$y = \begin{cases} \frac{1}{s}\left(\frac{e^{-a(x+c)} - b(x+c) - e^{-a} + b}{1 - e^{-a} + b} + d\right) & \text{if } b \geq 0 \\ \frac{1}{s}\left(\frac{1}{a|b|}\left[|b|W(u) + a(x+c-1)(e^{-a} - |b|) - a(x+c)\right] + d\right) & \text{if } b < 0 \end{cases}$$

with W the Lambert W-function [2] and

$$u = \frac{a}{|b|}\exp\left(\frac{a}{|b|}\left[e^{-a} - b + x(b - e^{-a} + 1)\right]\right).$$

Sets of Pareto Fronts. For generating a set of Pareto fronts, we want to sample N active and M interfering functions of the function family. The N active fronts are arranged according to predefined cut points $\{t_0 = 0, t_1, \ldots, t_{N-1}, t_N = 1\}$ of the joint Pareto front. Again, we want the joint Pareto front to lie in $[0, 1]^2$ with extreme points $(0, 1)$ and $(1, 0)$.[1] The M interfering fronts benchmark the ability of the algorithm selection approach to sort out unnecessary algorithms.

[1] If desired, an a posteriori scaling to arbitrary intervals is possible.

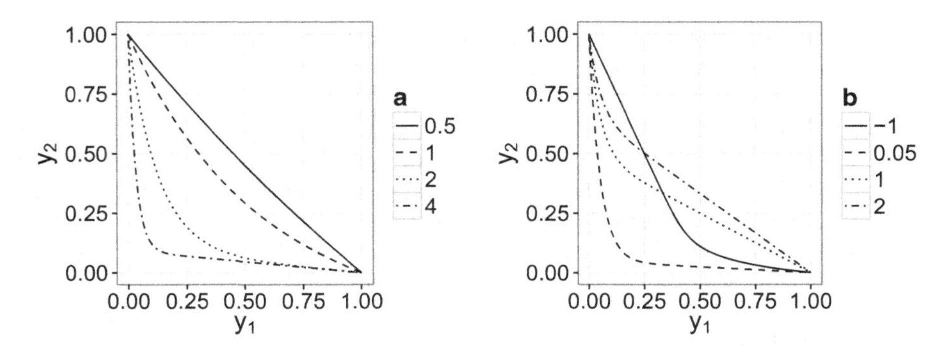

Fig. 2. Influence of the parameters a and b on the function family. In the left plot b is set to 0.1, in the right plot a is set to 20.

First the N fronts contributing to the joint Pareto front are sampled. The parameters a, b and c are drawn randomly as shown in Table 1. Due to numerical reasons the interval $[-0.05, 0.05]$ for parameter b is excluded. The value of c is slightly perturbed around the desired left cut point. For the first Pareto front, c is drawn with $\mu = 0$. This ensures that the knee of the front still lies in $(0, 1)^2$. The parameter d is automatically calculated based on the sampled values of a, b and c. It guarantees that the Pareto front intersects with the previous front or the extreme points in the predefined cut point.

Table 1. Types and parameters of the sampling distributions used in the generator.

Parameter	Distribution	Distr. parameters
$\log_2 a$	Uniform	$[-1, 5]$
b	Uniform	$[-5, -0.05] \cup [0.05, 5]$
c	Absolute	$\mu =$ left cut point t_{i-1}
	Normal	$sd = 0.05$

This procedure for generating the active Pareto fronts does not guarantee that a suitable joint Pareto front can always be constructed. To avoid infeasible data sets D, several checks are performed after each Pareto front part PF_j has been sampled. The front PF_j has to be dominated by the remaining fronts for $x \in [0, t_{j-1})$ and $x \in (t_j, 1]$. For $x \in [t_{j-1}, t_j]$, it has to be non-dominated. Furthermore, the front must not be quite similar to one of the other fronts. If one of this criteria is violated, a new value of the parameter c of the front is sampled. If no suitable front can be found after 10 samples, all its parameters are sampled again. This is done up to 100 times. If still no suitable front has been found, the next to last front is resampled, too.

In the next step, the M interfering fronts are generated. The parameters a and b are sampled according to Table 1. The parameters c and d are copied randomly from one of the active fronts and a positive noise value is added to make the function dominated by the corresponding active front. In addition, it is checked that the interfering front has no intersections with the joint front. If necessary, the parameters of the interfering front are sampled again. Up to now, only y_1 is scaled to $[0, 1]$. In the last step, this also done for y_2 by setting $d = d - \min y_2$ and $s = |\max y_2 - \min y_2|$.

Sampling Discrete Approximations. In reality, we have to deal with discrete approximations of the true Pareto front of an algorithm A_i. We propose four methods to construct those discrete approximations from the continuously defined Pareto fronts. In the first method called *deterministic* approximation, k points are distributed with a regular spacing along the front. To accomplish this, vectors $v_i = (v_{i,1}, v_{i,2})$ are generated with $v_{i,1} = \frac{i-1}{k-1}$ and $v_{i,2} = 1 - v_{i,1}$ $(i = 1, \ldots, k)$. The respective points on the front with $v_1 y_1 = v_2 y_2$ are calculated. The second method samples the weight vectors $v_{i,1} \in [0, 1]$ randomly from a uniform distribution (*random* approximation). The last two methods are based on actual approximations of the NSGA-II [3], where we use a population size of k. To construct the respective multi-objective problem, the continuous Pareto front is plugged as shape function h into the ZDT-concept [11] $\text{ZDT}(x_1, \ldots, x_l) = (x_1, g \cdot h(x)))$, where g is a function encoding the distance to the front with minimum 1. We consider an *NSGA-II* approximation, where g is fixed to its minimum value, and a *NSGA-II_g* approximation, where $g(x_2, \ldots, x_l) = 1 + \frac{9}{l-1} \sum_{i=2}^{l} x_i$ [11] is optimized for a few iterations to add a small approximation error to the front. In our experiments, we set $l = 10$ and fix the budget of the NSGA-II to 400 evaluations ($\frac{400}{k}$ generations).

Stochastic Replications. In the last step, we simulate n replications of potential tuning runs for a given joint Pareto front. We consider two practically motivated situations. In the first situation, the experiment is repeated under exactly the same circumstances. Hence, the only source of variation is the approximation quality of the tuning algorithm. This variation is simulated by adding noise to each point of the discrete approximation (*point-noise*). Following the idea of approximation error, we use absolute normal random variables with $\mu = 0$ and $\sigma = 0.02$. In the second situation, some details of the experiments change in the replications. In the context of machine learning, a different subset of learning instances may be considered during tuning [6], whereas a different rotation of the test function could be used for optimization [5]. For this situation, we create different instances by adding noise to the parameters of the Pareto fronts (*parameter-noise*). We use normally distributed random numbers with $\mu = 0$ and standard deviations according to Table 2. As a baseline method, we also consider a noiseless variant (*without-noise*), that simply replicates the discrete approximation n times. Examples of the possible combinations of approximation and replication methods are shown in Fig. 3. For reproducing our results, the generation of test data has been implemented in our MOSAP R-package[2].

Table 2. Standard deviations in the *parameter-noise* approach.

Parameter	sd
a	0.030
b	0.004
c	0.020
d	0.020

[2] https://github.com/danielhorn/multicrit_result_test.

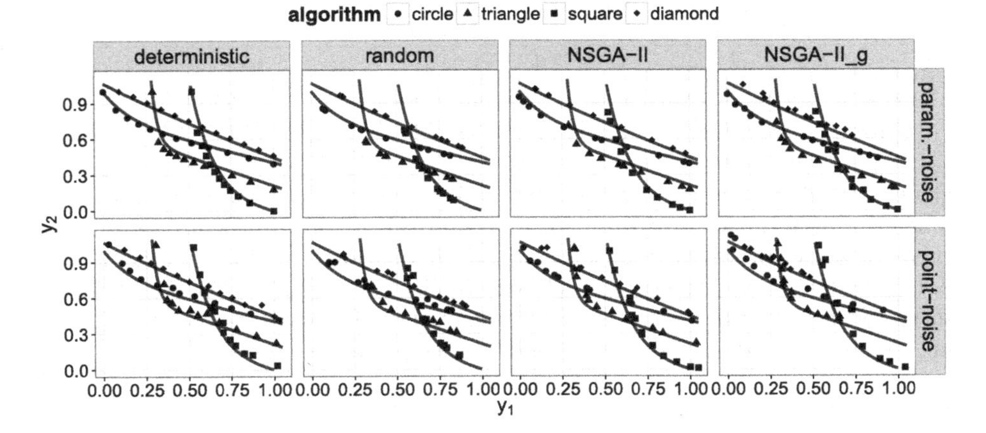

Fig. 3. Different types of generating discrete approximations and noisy replications.

4 Experimental Validation

A MOSAP method can make two types of errors: In the first type, it fails in predicting the correct subset of algorithms. A type 1 error occurs if active Pareto fronts are not selected, interfering fronts are selected, or the sequence of active fronts is swapped. The second type of error (type 2 error) is related to the accuracy of approximating the split points in the algorithm mapping.

Performance Measure. We propose an error measure that simultaneously takes both types of errors into account. Due to the construction principles of our test generator, there exists an oracle $f : [0, 1] \rightarrow \mathcal{A}$ which assigns the best algorithm A for a given value x of the performance measure y_1. Furthermore we define $\hat{f}$ as an estimator for f obtained by the MOSAP approach. The performance of $\hat{f}$ can be measured by

$$z(f, \hat{f}) = \int_0^1 \mathbb{1}(x)_{f(x)=\hat{f}(x)} dx.$$

z can be interpreted as the ratio of correct predictions of $\hat{f}$ over y_1. The optimum value is 1. In case of type 1 error, the z-value decreases by the length of the interval assigned to the wrong algorithm. In case of type 2 error, the z-value decreases by the approximation error of the split point. The integral can be easily computed in closed form because $\mathbb{1}(x)_{f(x)=\hat{f}(x)}$ is a piecewise constant function with known split points.

Benchmark. For benchmarking our MOSAP approach, we consider situations motivated by our practical applications in SVM tuning [7]. We consider $N \in \{2, 3, 5\}$ active Pareto fronts. The split points between these optimal fronts are

Table 3. Split points of the considered joint Pareto fronts.

N	2	2	3	3	5	5
Split.type	Unif.	Non-unif.	Unif.	Non-unif.	Unif.	Non-unif.
Split.points	$\{0.5\}$	$\{0.2\}$	$\{\frac{1}{3}, \frac{2}{3}\}$	$\{0.3, 0.5\}$	$\{0.2, 0.4, 0.6, 0.8\}$	$\{0.18, 0.2, 0.55, 0.75\}$

given in Table 3 and are chosen either *uniformly* (unif.) or *non-uniformly* (non-unif.). In addition, we add $M \in \{0, 2, 5\}$ interfering fronts and consider all four types of noise in creating the discrete approximation using $k \in \{4, 12, 40, 80\}$ points and all three types of stochastic replications resulting in 864 different setups with 100 replications each. For each experiment, we store the corresponding z–value. A higher error corresponds to decreasing z–values.

Hypotheses.

1. Even if a MOSAP method does not make any type 1 error, there will always be a type 2 error. This error should increase with the number of split points, the strength of the noise and decreasing quality of the coverage of the Pareto frontier (number of solutions k, spread and distribution).
2. The MOSAP-method should be able to eliminate the interfering fronts. Hence, M should not have a significant influence on the z–values.

All hypotheses are checked by means of a linear regression. The variables are coded as factors using dummy variables. The reference classes are set to: $N = 2$, $M = 0$, $k = 80$, split.type = *uniform*, discretize.type = *deterministic* and replications.type = *without-noise*. We do not report significance tests, as most observable results became significant due to the large number of replications.

5 Results

Due to space restriction we only provide the results of the linear regression in this paper, its coefficients are summarized in Table 4. The full results are available in the data-section of our MOSAP R-package.

The intercept of our model is slightly greater than 1. Hence, in the most easiest setting our method is able to reach a perfect result. With only $N = 2$ active and $M = 0$ inference fronts and no noise from both the discrete approximation and the replication, this case essentially measures how accurate the decision tree estimates the single split point.

The number of active fronts N has the largest effect. As expected, a higher number of active fronts or split points results in an increase of the error. For $N = 5$ we observe an effect size greater than 0.2. Hence, it is likely that some type 1 errors occur. This effect can be explained by the trade-off between the gap of the hypervolume and the number of algorithms which has to be found for deciding on the subset. Seemingly the gap is too small to be traded-off against the inclusion of another algorithm. This decision can be manipulated be using more extreme values of w in our method. Another option might be a human-in-the-loop, who reconsiders the parameter settings after looking at the results.

Table 4. Results of the linear regression.

Variable	Value	Estimator
Intercept		1.02
N	3	−9.47e-2
N	5	−2.35e-1
Split.type	Non-uniform	2.71e-2
M	2	−3.74e-3
M	5	−7.74e-3
k	40	9.48e-4
k	12	−1.43e-2
k	4	−1.26e-1
Replications.type	Point-noise	−2.90e-2
Replications.type	Parameter-noise	−2.89e-2
Discretize.type	NSGA-II	−1.58e-2
Discretize.type	Random	−7.69e-3
Discretize.type	NSGA-II_g	−1.12e-1

Both types of adding noise to the n replications do result in significant decreases of the z-value. This decrease is of the same level for both approaches. Compared to the noise added by the *NSGA-II_g* approach, however, the decrease of the z-values for both replication types is rather low.

The results of the discretization types are nearly as expected. The deterministic type is the easiest for the MOSAP approach. As expected, *NSGA-II_g* combining both approximation error and a non-uniform distribution results in a significant loss of performance. Surprisingly, *random* point sets result in better z-values than the ones of *NSGA-II* without approximation error. In fact, the *random* point sets are only slightly inferior to the deterministic ones. Hence, there is no need for perfectly spaced Pareto front approximations. This observation is confirmed by the effect of the approximation size. The reduction from $k = 80$ to $k = 40$ points even has a very small, positive effect, actually it is the only effect without a significant influence. Nevertheless, a further reduction to $k = 12$ or $k = 4$ points results in a either a slight or strong decrease of the z-value. Hence, k should not be set too small, but it is unnecessary to use very large discrete approximations. In conclusion, we can confirm our first hypothesis.

An increase of the number of interfering fronts M results in decreases of the z-value in the order of 10^{-3}. Compared to the effects of approximation error (*NSGA-II_g*) or additional active fronts ($N = 5$), these effects are small, but they indicate that an increasing amount of interfering fronts may have a negative effect on the result. Therefore, we can only partially confirm our second hypothesis.

As additional observation, the *non-uniform* cut points result in better z-values than the *uniform* ones. This seems meaningful, since some of the active fronts cover only a small portion of the joint front. Hence, type 1 errors will result in a smaller decrease of the z-value.

6 Conclusion and Outlook

In this paper, we present a validation framework for MOSAP methods. We applied the framework to evaluate to performance of our approach. As expected, the performance slightly decreases with an increasing number of active fronts and noisy approximation sets. Nevertheless, our method is capable of finding suitable portfolios and mappings even in the hardest cases considered.

In future work we are going to apply our MOSAP method to more practical test cases. One possibility would be to derive algorithm portfolios from the results of the Black-Box Optimization Benchmarking workshop (BBOB) [5]. In this workshop, the objectives are the number of function evaluations and the ratio of target levels attained over a set of functions. In this context, it would also be interesting whether the Pareto fronts can be merged over different instances instead of only replications on the same instance.

References

1. Breiman, L.: Random forests. Mach. Learn. **45**(1), 5–32 (2001)
2. Corless, R.M., Gonnet, G.H., Hare, D.E.G., Jeffrey, D.J., Knuth, D.E.: On the LambertW function. Adv. Comput. Math. **5**, 329–359 (1996)
3. Deb, K., Pratap, A., Agarwal, S., Meyarivan, T.: A fast and elitist multiobjective genetic algorithm: NSGA-II. IEEE Trans. Evol. Comput. **6**(2), 182–197 (2002)
4. da Fonseca, V.G., Fonseca, C.M.: The attainment-function approach to stochastic multiobjective optimizer assessment and comparison. In: Bartz-Beielstein, T., et al. (eds.) Experimental Methods for the Analysis of Optimization Algorithms, pp. 103–130. Springer, Heidelberg (2010)
5. Hansen, N., Auger, A., Finck, S., Ros, R.: Real-parameter black-box optimization benchmarking 2010: experimental setup. Technical report RR-7215, INRIA (2010)
6. Horn, D., Bischl, B., Demircioglu, A., Glasmachers, T., Wagner, T., Weihs, C.: Multi-objective selection of algorithm portfolios. Archives of Data Science (2016, under revision)
7. Horn, D., Demircioglu, A., Bischl, B., Glasmachers, T., Weihs, C.: A comparative study on large scale kernelized support vector machines. Advances in Data Analysis and Classification, S.I.: Science of Big Data: Theory, Methods and Applications (2016, under revision)
8. Hutter, F., Bartz-Beielstein, T., Hoos, H.H., Leyton-Brown, K., Murphy, K.: Sequential model-based parameter optimisation: an experimental investigation of automated and interactive approaches. In: Bartz-Beielstein, T., Chiarandini, M., Paquete, L., Preuß, M. (eds.) Empirical Methods for the Analysis of Optimization Algorithms, pp. 361–411. Springer, Heidelberg (2010)
9. Miettinen, K.: Nonlinear Multiobjective Optimization. International Series in Operations Research and Management Science, vol. 12, 4th edn. Kluwer Academic, Dordrecht (2004)
10. Weinert, K., Zabel, A., Kersting, P., Michelitsch, T., Wagner, T.: On the use of problem-specific candidate generators for the hybrid optimization of multi-objective production engineering problems. Evol. Comput. **17**(4), 527–544 (2009)
11. Zitzler, E., Deb, K., Thiele, L.: Comparison of multiobjective evolutionary algorithms: empirical results. Evol. Comput. **8**(2), 173–195 (2000)
12. Zitzler, Eckart, Thiele, Lothar: Multiobjective optimization using evolutionary algorithms - a comparative case study. In: Eiben, Agoston E., Bäck, Thomas, Schoenauer, Marc, Schwefel, Hans-Paul (eds.) PPSN 1998. LNCS, vol. 1498, p. 292. Springer, Heidelberg (1998)

Multi-objective Local Search Based on Decomposition

Bilel Derbel[1], Arnaud Liefooghe[1(✉)], Qingfu Zhang[2], Hernan Aguirre[3], and Kiyoshi Tanaka[3]

[1] University Lille, CNRS, UMR 9189 – CRIStAL/Inria Lille-Nord Europe, Villeneuve-d'ascq, France
bilel.derbel@univ-lille1.fr
[2] Computer Science Department, City University, Kowloon Tong, Hong Kong
[3] Faculty of Engineering, Shinshu University, Nagano, Japan

Abstract. It is generally believed that Local search (Ls) should be used as a basic tool in multi-objective evolutionary computation for combinatorial optimization. However, not much effort has been made to investigate how to efficiently use Ls in multi-objective evolutionary computation algorithms. In this paper, we study some issues in the use of cooperative scalarizing local search approaches for decomposition-based multi-objective combinatorial optimization. We propose and study multiple move strategies in the Moea/d framework. By extensive experiments on a new set of bi-objective traveling salesman problems with tunable correlated objectives, we analyze these policies with different Moea/d parameters. Our empirical study has shed some insights about the impact of the Ls move strategy on the anytime performance of the algorithm.

1 Introduction

Several single-objective approaches, ranging from problem-specific algorithms to more generic approaches such as meta-heuristics and evolutionary algorithms, have been designed, tuned and studied extensively in combinatorial optimization. Among many others, local search (Ls) heuristics [2] refer to algorithms where a solution is improved in an iterative search process by performing little perturbation on its vicinity. A common ingredient being at the basis of this class of algorithms is the so-called *neighborhood exploration* and *move strategy*. The specification of at least one neighborhood structure and its proper combination with a move strategy is in general a cornerstone in the design of advanced single-objective Ls-based algorithms. Actually, this statement holds also when turning to the multi-objective setting, where a whole set of solutions, optimizing simultaneously two or more objective functions, is to be computed. Ls components have been investigated to design effective aggregation-based [3,4,10,12] and dominance-based [9,10,12] multi-objective algorithms. In particular, within the class of dominance-based algorithms, it is shown in [9] how different move strategies can have a deep impact on search performance. In this paper, we are interested in studying the new opportunities offered by

J. Handl et al. (Eds.): PPSN XIV 2016, LNCS 9921, pp. 431–441, 2016.
DOI: 10.1007/978-3-319-45823-6_40

the so-called MOEA/D (multi-objective evolutionary algorithm based on decomposition) [14] framework in incorporating LS components. In fact, MOEA/D is a recently-proposed aggregation-based framework which was extensively studied for continuous problems. Interestingly, MOEA/D is a reference algorithm in multi-objective optimization, mainly due to its high flexibility in incorporating different search paradigms, and the high quality of the so-obtained algorithms. Nonetheless, very few investigations can be found on the proper incorporation of LS within MOEA/D for discrete domains. Some adaptations exist, but they are often based on genetic operators [1,11], and relatively few in-depth investigations [5,6] considering LS in MOEA/D were conducted against the large body of works in continuous domains.

In this paper, we provide a comprehensive study on incorporating basic LS move strategies into the MOEA/D framework. More precisely, our contribution is three-fold. Firstly, we revisit conventional single-objective move strategies and illustrate how they can be hybridized with MOEA/D. In particular, we highlight how the replacement flow of MOEA/D can be adapted to support such strategies. Secondly, we study the performance of the so-designed algorithms using a new set of bi-objective traveling salesman problem (TSP) instances with tunable objective correlations. Our thorough experimental analysis shows that different behaviors can be obtained depending on objective correlation, and more importantly on available budgets. Our findings are the byproduct of a running time analysis providing evidence on the importance of the LS move strategy in the design of anytime decomposition-based multi-objective algorithms. Thirdly, we provide a comprehensive study on the impact of MOEA/D common parameters. The research conducted in this paper is also to be viewed as establishing the first steps towards the design of more powerful decomposition-based multi-objective algorithms based on more advanced local search components. In fact, notwithstanding that we are not horse-racing against state-of-the-art algorithms for the considered optimization problems, and that we consider basic move strategies, our findings on the anytime performance of the designed algorithms suggests that incorporating LS into MOEA/D is still in its very infancy beginning, and hence, would deserve further research investigations in the future.

The rest of this paper is organized as follows. In Sect. 2, we recall some background on LS and MOEA/D. In Sect. 3, we describe in more details different strategies for incorporating LS components into MOEA/D. In Sect. 4, we give our experimental setup. In Sect. 5, we discuss our experimental findings. In Sect. 6, we conclude the paper and discuss some open research directions.

2 Background

A multi-objective optimization problem (MOP) can be defined by a solution set X and by an objective function vector $f = (f_1, \ldots, f_m)$ to be minimized.

The MOEA/D [14] **framework.** MOEA/D falls into the class of decomposition-based algorithms. It seeks good-performing solutions in multiple regions of the Pareto front by *decomposing* the original MOP into a number of *scalarized* single-objective *sub-problems*. Different scalarizing functions have been proposed so-far. In this paper, we use the common weighted Chebyshev function, to be minimized: $g(x \mid \lambda, z^\star) = \max_{k\in\{1,\dots,m\}} \lambda_k \cdot \left|z_k^\star - f_k(x)\right|$; where $x \in X$, $\lambda = (\lambda_1, \dots, \lambda_m)$ is a positive weighting coefficient vector, and $z^\star = (z_1^\star, \dots, z_m^\star)$ is a reference point. In this respect, the originality of the MOEA/D framework is to define a *T-neighborhood relation* between sub-problems. Let $(\lambda^1, \dots, \lambda^\mu)$ be a set of μ uniformly distributed weighting coefficient vectors defining μ sub-problems. MOEA/D maintains a population $P = (x^1, \dots, x^\mu)$, where every individual corresponds to one sub-problem. For each sub-problem $i \in \{1, \dots, \mu\}$, its T-neighbors, denoted $\mathcal{B}(i)$, are defined by considering the T closest weight vectors. Sub-problem solutions are evolved with respect to their neighbors. For every sub-problem, an offspring solution from the T-neighbors set $\mathcal{B}(i)$ is generated using some evolutionary operators. Then, the offspring can replace one or more T-neighbors if it improves the scalar (Chebyshev) value of the corresponding solution of the neighboring sub-problem. Different variants of this baseline MOEA/D flow exist. In the remainder, we consider the modifications introduced in [8], considered as a state-of-the-art variant in continuous domains, where (i) the T-neighbors of a sub-problem is the whole population with a small probability δ, or $\mathcal{B}(i)$ otherwise, and (ii) a newly generated offspring can replace at most nr other solutions, where nr and δ are two user-defined parameters. Other MOEA/D variants could be considered as well, but for the sake of analysis, we only consider the most common and widely-used variant from [8,14].

LS Move Strategies. LS is a single solution-based walk that iteratively improves the current solution by means of local transformations, and then moving to an improving close-by solution. Those transformations are usually based on a *neighborhood* function $\mathcal{N} : X \rightarrow 2^X$, which assigns a set of neighboring solutions $\mathcal{N}(x) \subset X$ to any solution $x \in X$. It should be clear for the reader that we differentiate between the T-neighborhood of MOEA/D and the neighborhood of a solution in LS. In the most simple LS variant, also referred to as *hill-climbing*, the search stops when the current solution is not outperformed by any neighbor. This means that a *local optimum* is reached. The move strategy, defining the transition rule to select an improving neighbor, is also a key ingredient in LS-based search. Typical strategies are as follows: (i) In a *best-improvement* (or steepest descent) move, the neighbor that improves the most is selected at each iteration. This means that the whole neighborhood is generated, which can be time-consuming for large neighborhoods. (ii) In a *first-improvement* move, the first improving neighbor is immediately selected. This avoids to systematically generate and evaluate the whole neighborhood. The exploration order of neighbors can remain unchanged, or instead can be randomly shuffled at each iteration. Additionally, the neighborhood structure can be used as a an evolutionary mutation operator when some few neighboring solutions are sampled at random. Hence, (iii) a *random* strategy can be considered as well, where a random neighbor is generated and replaces the current solution if there is an improvement.

3 The MLSD Scheme

Incorporating LS into MOEA/D can be viewed as a natural outcome since several single-objective sub-problems are to be improved cooperatively. Although the standard neighborhood exploration mechanisms of LS might not be very complicated to integrate into MOEA/D, still important design technicalities have to be explicitly and carefully specified, especially when exploring new neighboring solutions and when performing replacement in original MOEA/D.

In the high-level pseudo-code depicted in Algorithm 1, we provide a relatively detailed description of different possible ways of hybridizing MOEA/D with LS move policies. The proposed scheme is called MLSD-SR (Multi-objective Local Search based on Decomposition). One should notice that MLSD is parametrized by two elements, namely s (referring to the *selection* policy) and r (referring to the *replacment* policy). This allows us to differentiate between two stages: (i) the move selection stage (lines 10 to 21), and (ii) the replacement stage (lines 22 to 29). We thereby obtain four possible variants, as discussed in the following.

Algorithm 1. MLSD-SR: high-level pseudo-code

```
Input: μ: population size; T: neighborhood size; δ ∈ [0,1]; nr ∈ ⟦0,μ⟧; s ∈ {Best, First, Rnd};
       r ∈ {Min, Rnd}.
1  {λ^1,...,λ^μ} ← generate weight vectors w.r.t. μ sub-problems;
2  ∀i ∈ {1,...,μ} B(i) ← the T closest sub-problems w.r.t λ_i;
3  P={x^1,...,x^μ} ← generate the initial population;
4  evaluate P;
5  (update external archive with P;)                                  /* optional */
6  set z* from P;
7  while STOPPING CONDITION do
8      for i ∈ {1,...,μ} do
9          if rand{[0,1]} < δ then B_i ← B(i); else B_i ← P ;
           // Stage #1: Move selection
10         k ← rand{B_i};
11         I ← ∅;
                                   /* Check moves and record improved sub-problems */
12         for y ∈ N(x^k) do                               /* By default, s = Best */
13             evaluate y;
14             (update external archive with y;)                      /* optional */
15             update z* using y;
16             J_y ← {j ∈ B_i s.t. g(y | λ^j, z*) < g(x^j | λ^j, z*)};
17             if J_y ≠ ∅ then
18                 c_y ← 0;
19                 I ← I ∪ {(y, c_y, J_y)};
20                 if s = First then break;
21             if s = Rnd then break;                          /* go to line 22 */
           // Stage #2: Replacement
22         while ∃j ∈ B_i s.t. (∃(y, c_y, J_y) ∈ I s.t. j ∈ J_y and c_y < nr) do
23             if r = Min then
24                 y* ← arg min_{y s.t. (y,c_y,J_y)∈I} {g(y | λ^j, z*)}
25             else if r = Rnd then
26                 y* ← rand{y s.t (y, c_y, J_y) ∈ I};
27             x^j ← y*;
28             c_{y*} ← c_{y*} + 1;
29             B_i ← B_i \ {j};
```

The MLSD scheme iteratively loops over sub-problems until a stopping condition is satisfied. At each iteration w.r.t. sub-problem i, two stages are performed. The first stage consists in generating some new candidate solutions to be considered in the second stage. First, a parent solution x^k is selected randomly from the neighborhood of sub-problem i. The selected solution is then locally explored using the LS neighborhood structure $\mathcal{N}$. Three different move strategies can be considered. The first one ($\mathsf{s} = \mathsf{Best}$) consists in traversing all solutions $y \in \mathcal{N}(x^k)$ in an exhaustive manner while checking for any improvement. Notice that variable J_y (line 16) denotes the set of sub-problems improved by an incumbent solution y, and c_y is a counter initialized to 0. The tuple (y, c_y, J_y) is then saved into set I which contains all the records w.r.t any improving solution in $\mathcal{N}(x^k)$. In the second strategy ($\mathsf{s} = \mathsf{First}$), the exploration of neighbors $\mathcal{N}(x^k)$ stops as soon as an improving solution y is found. This strategy guarantees that if $\mathcal{N}(x^k)$ contains at least one improving solution, then it is selected and recorded in set I for the next stage. The last move strategy ($\mathsf{s} = \mathsf{Rnd}$) picks a single incumbent solution y uniformly at random from $\mathcal{N}(x^k)$, and records the tuple (y, c_y, J_y) in set I only if y is improving at least one neighboring sub-problem.

The second stage consists in replacing the solutions of neighboring sub-problems. If no improvement was observed, then the replacement stage is simply skipped. Otherwise, i.e. when $|I| \geq 1$, two possible strategies are considered. In the first one ($\mathsf{s} = \mathsf{Min}$), the solution of every sub-problem j in the T-neighborhood of sub-problem i is replaced by the best improving solution $y^\star$ found during the previous stage (if any). In the second one ($\mathsf{s} = \mathsf{Rnd}$), an improving solution (if any) is picked randomly to replace the current solution of j. Notice that in case the set I contains one single recorded tuple, the two previous replacement strategies are equivalent. Notice also that if a First or a Rnd policy is adopted in the selection stage, the designed replacement strategies are also equivalent. Hence, the two replacement strategies might imply different variants of MLSD only when a Best strategy is adopted in the first stage.

Finally, it is important to notice the role of the nr parameter in the replacement stage. In fact, since several candidate improving solutions can be considered in the case $\mathsf{s} = \mathsf{Best}$, each time a solution y is selected for the replacement in line 27, its associated counter c_y is incremented. Consequently, once this counter reaches the value nr, the corresponding solution cannot be selected anymore to replace any sub-problem, as specified by the condition of line 22.

4 Experimental Setup

For the sake of studying the behavior of the MLSD-SR framework, we consider the Traveling Salesman Problem (TSP) as a baseline benchmark problem. The motivation behind this choice is two fold. First, permutation-based optimization problems, like TSP, are of choice when evaluating the behavior of LS-based algorithms. Second, the TSP is a fundamental problem that appears at the bottleneck of many real-world applications and is representative of a wide range of more complex combinatorial optimization problems. We emphasize that this choice is to be understood from a purely benchmarking perspective. In particular, it is worth noticing that the multi-objective TSP has attracted a lot of interest in recent years and one can report several state-of-the-art algorithms, see e.g. [5,9,10,12]. This paper does not propose yet another algorithm for TSP, and we shall not consider to compare the MLSD-SR with those algorithms. Besides, designing TSP-specific algorithms is a whole piece of research that we are not targeting in this experimental study. Accordingly, we shall only focus on analyzing the relative performance of the different move strategies described previously.

Multi-objective TSP with Correlated Objectives. Given a complete graph $G = (V, E)$ with n nodes and non-negative edge costs, the symmetric single-objective TSP seeks a cyclic permutation that contains each node exactly once and such that the total cost is minimized. A solution can be represented as a permutation π of size n. Since multiple costs like distance or travel time can be considered, a multi-objective variant of the TSP can be formulated. Let $\{v_1, v_2, \ldots, v_n\}$ be the set of nodes, and $\{[v_i, v_j] \mid v_i, v_j \in V\}$ the set of edges. In the m-objective case, we have m cost matrices such that each edge $[v_i, v_j] \in E$ is assigned a *cost* c^k_{ij} for each objective function $k \in \{1, \ldots, m\}$. The objective functions can then be defined as follows: $f_k(\pi) = c^k_{\pi(n)\pi(1)} + \sum_{i=1}^{n-1} c^k_{\pi(i)\pi(i+1)}$. The multi-objective TSP is known to be NP-hard and intractable [10]. In this paper, we consider two-objective symmetric TSP instances ($m = 2$) with *correlated* random distance matrices. Following [12], edge costs are chosen from a uniform distribution in $[0, 4473]$. However, we additionally define a correlation coefficient $\rho \in [-1, 1]$ between the data contained in both cost matrices. The generation of correlated data follows a multivariate uniform distribution [13]. The positive (resp. negative) data correlation allows to decrease (resp. increase) the degree of conflict between the objective function values with a high accuracy. Notice than when $\rho = 0$, our instances are the same as [12].

Parameter Setting. We consider the *2-opt exchange* operator as the *neighborhood* $\mathcal{N}$ for TSP, i.e. given a candidate solution π, the sequence of nodes located between $\pi(i)$ and $\pi(j)$ is reversed. The neighborhood size is hence $\frac{n \cdot (n-1)}{2}$. We experiment instances of size $n = 100$ and correlation values: $\rho \in \{$-0.8, -0.4, 0.0, 0.4, 0.8$\}$. We consider a broad range for the other parameters, namely population size $\mu \in \{50, 100, 150, 200\}$, T-neighborhood size $T \in \{5, 10, 15, 20\}$, $nr \in \{1, 2, \infty\}$, and $\delta \in \{0.0, 0.1\}$. For every parameter combination, we consider the four variants of MLSD-SR as summarized in the table below, thus ending up with 1 920 configurations, each one independently executed 20 times. For s = First, neighboring solutions are explored in a random order. The stopping condition is a maximum budget of 10^8 function evaluations. The initial population is generated randomly and the weight vectors are generated as in [14].

r \ s	Best	First	Rnd
Rnd	✓(MLSD-BR)		
Min	✓(MLSD-BM)	✓(MLSD-FM)	✓(MLSD-RM)

5 Experimental Analysis

We follow the performance assessment protocol proposed in [7] by using the hypervolume relative deviation (I_{hv}) and the additive epsilon (I_{ε}^{+}) indicators. The hypervolume reference point is set to the worst objective-value, and the reference set is the best-found approximation over all tested configurations. Notice that we use an external archive recording all non-dominated solutions found so far.

High Budget Setting. We first report the descriptive statistics on the indicator-values, together with a Mann-Whitney non-parametric statistical test with a p-value of 0.05 and using a Bonferroni correction, for the highest budget of 10^8 calls of the evaluation function. In Table 1, we show the rank of different MLSD-SR variants with the rank being the number of variants that statistically outperform the one under consideration for each instance. The lower the rank, the better the algorithm. Both indicators agree that the best performing variant of MLSD over all considered instances is when a Best move strategy is adopted together with a Min replacement strategy. The objective correlation of considered instances appear to have a crucial impact. The gap between MLSD-BM and the other variants is substantial in the case of conflicting objectives whereas we found no significant differences for highly correlated objectives. Overall, the considered MLSD variants can be ranked as follows: MLSD-BM > MLSD-BR ≈ MLSD-FM > MLSD-RM. It is important to remark that combining a Best move strategy with an elitist replacement strategy is crucial, otherwise a First move strategy would be more appropriate. Notice that at this stage of the analysis, the MLSD-RM variant is overall the worst performing one, and the relative performance gap between different T-neighborhoods are not statistically significant. In the following, we shall show that these preliminary conclusions can only hold for a high computational budget.

Anytime Analysis. When analyzing the quality of the approximation with different budgets, we basically find that the relative performance of the considered variants is deeply impacted, independently of the parameter setting. This is illustrated in Fig. 1 for a particular parameter setting. Interestingly, the MLSD-BM and MLSD-BR variants can only outperform the other variants for a high budget. MLSD-RM, which was shown to be the worst-performing approach in such a setting, now appears to be the best anytime strategy. This might be surprising at a first glance. However, in the early stages of the search process, it is more likely that among few random samples, an improving solution for different sub-problems is found. In contrast, MLSD-BM would anyway explore all neighboring solutions (quadratic in n) and consider at most one solution for replacement. Hence, MLSD-RM is likely to progress faster and to save a significant number of evaluations. As the quality of the population gets better, it becomes more unlikely to find improving neighbors using random sampling. This can explain why MLSD-RM gets stuck and cannot improve the quality of the population anymore. It is also interesting to remark that MLSD-FM provides an intermediate trade-off, since it is relatively competitive against MLSD-RM while being able to catch MLSD-BM again on the latest stages. Interestingly, these

Table 1. Algorithm rank summary using 10^8 function evaluations, $\mu = 100$, $nr = 2$ and $\delta = 0.1$. The number in brackets stands for the average indicator-value.

		Hypervolume relative deviation ($I_{hv} \cdot 10^{-2}$)				Additive epsilon indicator ($I_{\epsilon}^{+} \cdot 10^{2}$)			
		s = B		MLSD-FM	MLSD-RM	s = B		MLSD-FM	MLSD-RM
ρ	T	MLSD-BM	MLSD-BR			MLSD-BM	MLSD-BR		
−0.8	5	0 (1.41)	4 (2.07)	4 (1.95)	12 (2.61)	0 (49.45)	5 (75.43)	4 (66.89)	5 (78.23)
	10	0 (1.38)	4 (2.05)	4 (2.02)	12 (2.57)	0 (51.21)	5 (85.53)	5 (86.38)	5 (76.68)
	15	0 (1.33)	4 (1.98)	6 (2.17)	12 (2.57)	0 (52.27)	5 (82.10)	10 (91.86)	5 (76.72)
	20	0 (1.39)	4 (2.04)	10 (2.28)	12 (2.47)	0 (53.95)	6 (86.20)	14 (103.3)	5 (77.53)
−0.4	5	0 (1.83)	1 (1.95)	8 (2.22)	12 (2.64)	0 (50.63)	2 (58.36)	4 (66.92)	8 (72.60)
	10	0 (1.78)	0 (1.84)	2 (2.03)	12 (2.50)	0 (50.92)	2 (60.35)	4 (65.97)	6 (68.70)
	15	0 (1.70)	0 (1.92)	5 (2.08)	12 (2.56)	0 (49.39)	2 (58.69)	6 (68.77)	8 (71.54)
	20	0 (1.78)	1 (1.95)	5 (2.06)	12 (2.51)	0 (52.14)	3 (60.60)	6 (69.52)	7 (69.94)
0.0	5	0 (2.42)	0 (2.30)	5 (2.67)	1 (2.69)	0 (45.08)	0 (41.62)	4 (51.59)	4 (52.27)
	10	0 (2.23)	0 (2.28)	0 (2.44)	5 (2.85)	0 (39.84)	0 (41.71)	0 (47.41)	6 (52.98)
	15	0 (2.32)	0 (2.25)	0 (2.52)	7 (2.71)	0 (42.15)	0 (42.22)	0 (49.12)	7 (50.31)
	20	0 (2.39)	0 (2.26)	0 (2.49)	7 (2.80)	0 (43.79)	0 (41.02)	0 (47.95)	7 (53.25)
0.4	5	0 (2.66)	0 (2.33)	0 (2.61)	0 (2.47)	1 (44.82)	0 (38.06)	0 (42.65)	0 (40.59)
	10	0 (2.51)	0 (2.43)	0 (2.44)	0 (2.50)	0 (42.17)	0 (39.45)	0 (38.80)	0 (39.44)
	15	0 (2.59)	0 (2.34)	0 (2.54)	0 (2.64)	0 (39.49)	0 (37.86)	0 (42.62)	0 (42.86)
	20	0 (2.54)	0 (2.30)	0 (2.68)	0 (2.52)	0 (39.23)	0 (38.48)	0 (42.14)	0 (41.33)
0.8	5	0 (2.54)	0 (2.15)	0 (2.08)	0 (2.10)	0 (33.76)	0 (29.78)	0 (28.00)	0 (28.25)
	10	0 (2.49)	0 (2.21)	0 (2.05)	0 (2.36)	0 (32.83)	0 (30.17)	0 (28.21)	0 (31.87)
	15	0 (2.56)	0 (2.22)	0 (2.14)	0 (2.31)	0 (32.78)	0 (28.68)	0 (27.56)	0 (30.62)
	20	0 (2.39)	0 (2.40)	0 (2.23)	0 (2.16)	0 (31.57)	0 (31.39)	0 (29.60)	0 (28.54)

results suggest that there is much room for future improvements in the anytime behavior of MLSD by considering hybrid move strategies.

Impact of the Population Size (μ). In Fig. 2, we show a subset of results on the impact of different population sizes on MLSD-BM and MLSD-RM (since no significant impact was found for MLSD-FM). The larger the population size, the better the final approximation set, independently of the considered strategy.

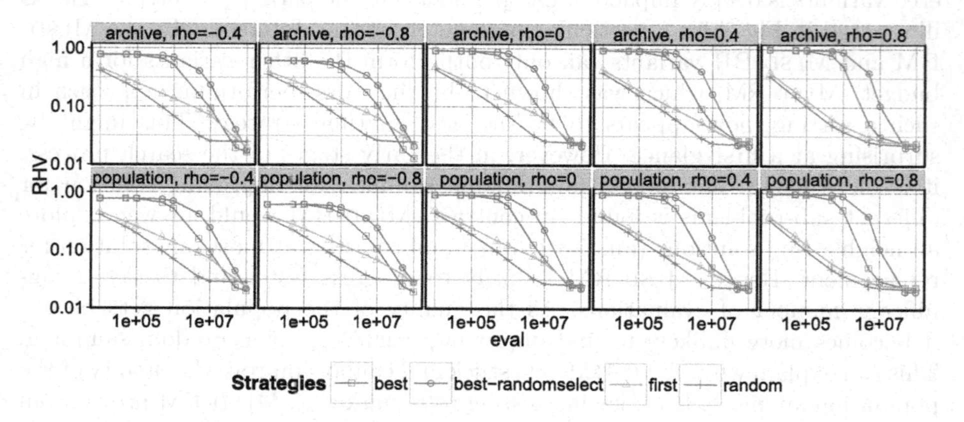

Fig. 1. Runtime analysis of the different algorithm variants. Error bars indicate 95 % confidence intervals. $\delta = 0$, $T = 10$, $nr = \infty$ and $\mu = 100$. Notice the log-scales.

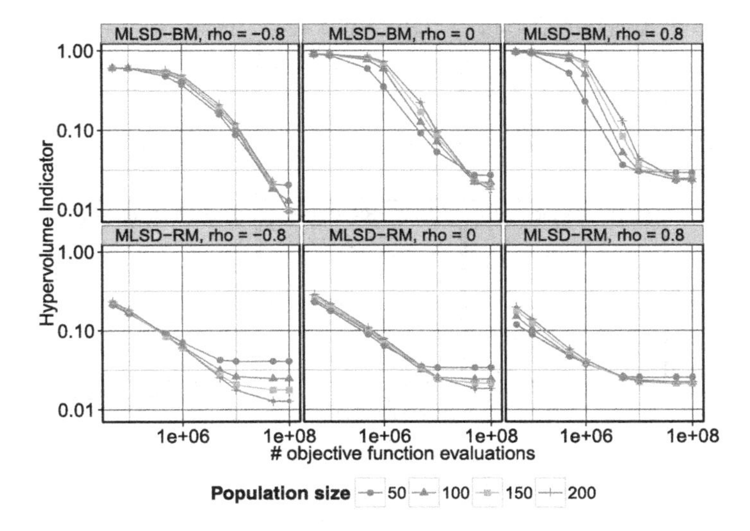

Fig. 2. Runtime analysis for different population sizes. $\delta = 0$, $T = 10$, $nr = \infty$.

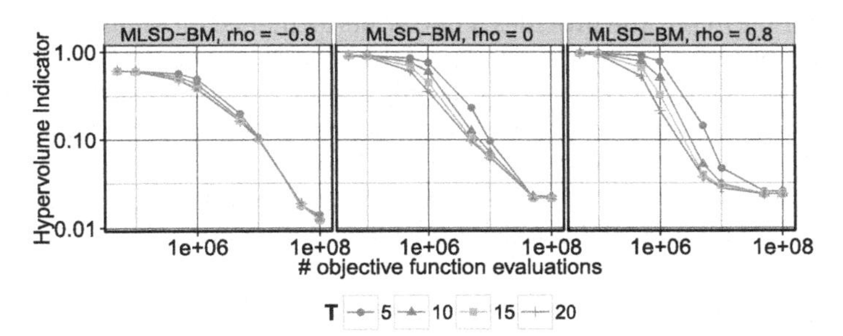

Fig. 3. Runtime analysis for different T−values. $\delta = 0$, $nr = \infty$ and $\mu = 100$.

However, smaller population sizes are better for smaller budgets, especially for instances with correlated objectives. We attribute this to the fact that a larger population size impacts the population diversity, and is thus more critical when the Pareto front is larger, which is the case for conflicting objectives.

Diversity Issues (T, nr and δ). We are able to report a significant impact of the T-neighborhood size only for the MLSD-BM variant, for highly correlated objectives and a small budget, as illustrated in Fig. 3. As for parameter nr, we found a significant impact only for MLSD-FM and MLSD-RM, as illustrated in Fig. 4. We recall that a larger nr−value allows a high-quality solution, possibly improving multiple sub-problems simultaneously, to replace all those solutions at once. Intuitively, the surviving solution has then more chance to improve the overall population quality in subsequent iterations, but at the price of decreasing diversity. we can see that smaller nr−values are better for convergence purposes,

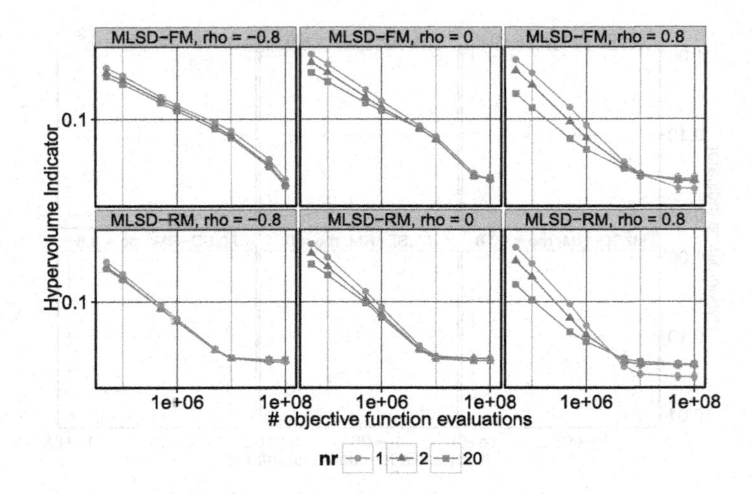

Fig. 4. Runtime analysis for different $nr-$values. $\delta = 0$, $T = 10$ and $\mu = 100$.

whereas a larger $nr-$value provides a better performance for small budgets. Interestingly, this observation holds only for highly-correlated objectives. As for parameter δ, the impact on performance was only significant when using MLSD-BM for correlated objectives with a small T-neighborhood size, but it was not helpful for improving the relative anytime performance. These empirical observations suggest that, contrary to the continuous case, the δ parameter might not be of great help when tackling combinatorial problems with conflicting objectives.

6 Conclusion

This paper investigates the foundations of the design of cooperative scalarizing local search approaches within decomposition-based algorithms for multi-objective combinatorial optimization. Our results revealed strong evidence on the need of adaptive algorithms that would enable to mix different move strategies and to better combine the neighborhood exploration with the replacement stage in order to properly balance the exploration/exploitation trade-off. It is our hope that our empirical study can enlighten our current understandings of decomposition-based approaches for multi-objective combinatorial optimization, and can stimulate new research paths towards the design of more powerful multi-objective randomized search heuristics based on local search and decomposition.

References

1. Chang, P.C., Chen, S.H., Zhang, Q., Lin, J.L.: MOEA/D for flowshop scheduling problems. In: CEC, pp. 1433–1438 (2008)
2. Hoos, H., Stützle, T.: Stochastic Local Search: Foundations and Applications. Morgan Kaufmann, Burlington (2004)

3. Ishibuchi, H., Murata, T.: A multi-objective genetic local search algorithm and its application to flowshop scheduling. IEEE Trans. Cyber. **28**(3), 392–403 (1998)
4. Jaszkiewicz, A.: Genetic local search for multi-objective combinatorial optimization. EJOR **137**(1), 50–71 (2002)
5. Ke, L., Zhang, Q., Battiti, R.: MOEA/D-ACO: a multiobjective evolutionary algorithm using decomposition and ant colony. IEEE Trans. Cyber. **43**(6), 1845–1859 (2013)
6. Ke, L., Zhang, Q., Battiti, R.: Hybridization of decomposition and local search for multiobjective optimization. IEEE Trans. Cyber. **44**(10), 1808–1820 (2014)
7. Knowles, J., Thiele, L., Zitzler, E.: A tutorial on the performance assessment of stochastic multiobjective optimizers. TIK report 214, Zurich, Switzerland (2006)
8. Li, H., Zhang, Q.: Multiobjective optimization problems with complicated Pareto sets, MOEA/D and NSGA-II. IEEE TEC **13**(2), 284–302 (2009)
9. Liefooghe, A., Mesmoudi, S., Humeau, J., Jourdan, L., Talbi, E.G.: On dominance-based local search. J. Heuristics **18**(2), 317–352 (2012)
10. Lust, T., Teghem, J.: Two-phase Pareto local search for the biobjective traveling salesman problem. J. Heuristics **16**(3), 475–510 (2010)
11. Palacios Alonso, J.J., Derbel, B.: On maintaining diversity in MOEA/D: application to a biobjective combinatorial FJSP. In: GECCO, pp. 719–726 (2015)
12. Paquete, L., Stützle, T.: Design and analysis of stochastic local search for the multiobjective traveling salesman problem. COR **36**(9), 2619–2631 (2009)
13. Verel, S., Liefooghe, A., Jourdan, L., Dhaenens, C.: On the structure of multiobjective combinatorial search space: MNK-landscapes with correlated objectives. Eur. J. Oper. Res. **227**(2), 331–342 (2013)
14. Zhang, Q., Li, H.: MOEA/D: a multiobjective evolutionary algorithm based on decomposition. IEEE TEC **11**(6), 712–731 (2007)

Analyzing Inter-objective Relationships: A Case Study of Software Upgradability

Zhilei Ren[1], He Jiang[1(✉)], Jifeng Xuan[2], Ke Tang[3], and Yan Hu[1]

[1] Key Laboratory for Ubiquitous Network and Service Software of Liaoning Province, School of Software, Dalian University of Technology, Dalian, China
{zren,jianghe,huyan}@dlut.edu.cn

[2] State Key Laboratory of Software Engineering, Wuhan University, Wuhan, China
jxuan@whu.edu.cn

[3] School of Computer Science and Technology, University of Science and Technology of China, Hefei, China
ketang@ustc.edu.cn

Abstract. In the process of solving real-world multi-objective problems, many existing studies only consider aggregate formulations of the problem, leaving the relationships between different objectives less visited. In this study, taking the software upgradability problem as a case study, we intend to gain insights into the inter-objective relationships of multi-objective problems. First, we obtain the Pareto schemes by uniformly sampling a set of solutions within the Pareto front. Second, we analyze the characteristics of the Pareto scheme, which reveal the relationships between different objectives. Third, to estimate the inter-objective relationships for new upgrade requests, we build a predictive model, with a set of problem-specific features. Finally, we propose a reference based indicator, to assess the risk of applying single-objective algorithms to solve the multi-objective software upgradability problem. Extensive experimental results demonstrate that, the predictive models built with problem-specific features are able to predict both algorithm independent inter-objective relationships, as well as the algorithm performance specific indicator properly.

Keywords: Pareto front · Meta-learning · Empirical analysis

1 Introduction

Many real-world multi-objective problems are solved with single-objective approaches [8,10], leaving the inter-objective relationships less studied. For example, the software upgradability problem is among the great challenges in the field of software engineering [10,14]. The problem aims to find the most suitable upgrade scheme that satisfies the users' upgrade requests. An upgrade scheme consists of a sequence of operations, including installing, removing,

J. Handl et al. (Eds.): PPSN XIV 2016, LNCS 9921, pp. 442–452, 2016.
DOI: 10.1007/978-3-319-45823-6_41

and/or upgrading packages. The software upgradability problem is inherently a multi-objective optimization problem, i.e., users may be interested in different upgrade objectives, such as software stability, package download size, etc. Even only considering single upgrade objective, the problem is reducible to the partial weighted MAXSAT problem [6], which is NP-hard. Moreover, the scalability of the software repositories poses great challenges for the upgrade process. Up to now, there are more than 43,000 packages in the Debian repository[1]. The intrinsic complexity and the scalability make the upgrade process a difficult problem. Meanwhile, in the literatures, most studies encode the upgrade requests into certain single-objective problem instances, such as partial weighted MAXSAT [6], Mixed Integer Linear Programming (MILP) [10], Pseudo Boolean Optimization [11], and Answer Set Programming [4]. Then, solvers are employed to resolve the encoded instances. However, despite the promising accomplishments these studies have achieved, there are still limitations to be improved. For example, in the existing approaches, multiple upgrade objectives are handled in aggregate ways, e.g., the weighted sum scalarization transformation or the lexicographic combination. Hence, a potential risk of such approaches is that, the relationships between different upgrade objectives may not be considered properly. If there exists drastic tradeoff between different objectives, the aggregation strategy has to be carefully chosen, e.g., the weight vector for the weighted sum approaches, or the objective order for the lexicographic approaches.

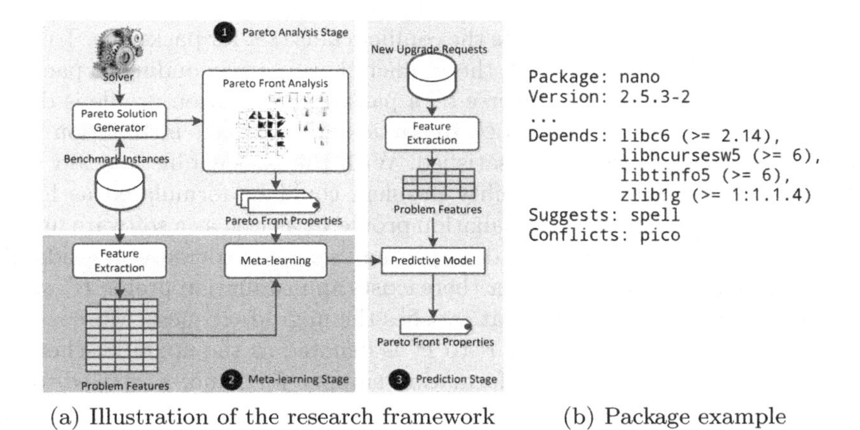

(a) Illustration of the research framework (b) Package example

Fig. 1. Background information

To face this challenge, we take the software upgradability problem as a case study, and intend to systematically investigate the relationships between different objectives. Motivated by the concept of the Pareto optimality, we are interested in the insights into the characteristics of the upgrade schemes that are not dominated by any other schemes (denoted as Pareto schemes). More specifically,

[1] http://www.debian.org.

Fig. 1(a) illustrates the research framework in this study, which comprises three stages. First, we intend to analyze the characteristics of the Pareto schemes. By uniformly sampling a set of Pareto schemes, we are able to analyze the relationships between different upgrade objectives. Then, for the meta-learning stage, we intend to capture the characteristics of the Pareto front by training a predictive model with features extracted from instances. Finally, we are interested in the possibility of predicting the relationships between objectives with the trained model, and leveraging the predicted indicator to evaluate the suitability of applying certain algorithms. More specifically, we consider the following Research Questions (RQs), which are listed as follows: **RQ1**: How are the different objectives correlated? **RQ2**: Are the inter-objective relationship of the Pareto schemes predictable with problem specific features? **RQ3**: Given an upgrade request, how to assess the suitability of applying single-objective optimization approaches?

2 Problem and Motivation

Let a universe U be a set of software packages, in which each package p is determined by the package name and a version number. Associated with each package, there exists a tuple (D, C), where D denotes the dependency clause set of p, in which each clause indicates a list of software packages. In the clause, at least one of the packages have to be installed so that package p could be installed properly. Accordingly, C represents the conflict clause set for package p. To install package p, none of the packages in the conflict clause corresponding to package p should be installed. Given a universe U, a package installation profile is defined as a subset of the packages within U. In particular, a package installation profile is valid if all the constraints are satisfied. With the package installation profile described, the software upgradability problem could be formulated as follows. Given a universe U, a package installation profile P, as well as a software upgrade request (install, remove, or upgrade a package set), the software upgradability problem aims to determine whether there exists an installation profile P', so that P' is a valid installation profile that satisfies the upgrade request. Moreover, the operation sequence that transfers P to P' is denoted as the upgrade scheme. In Fig. 1(b), we give the package information snippet for `nano`, a text editor. To install `nano` version 2.5.3-2, the constraints have to be met, e.g., packages tagged in the Depends and Conflicts fields have to be installed and removed accordingly.

In this paper, we focus on the optimization version of the problem, i.e., how to determine the most compact valid upgrade scheme for the request. In the literatures, there are 5 minimization upgrade objectives, which aim to minimize the number of packages removed in the solution (f_1: "*removed*"), the packages changed by the solution (f_2: "*changed*"), the number of outdated packages in the solution (f_3: "*notuptodate*"), the number of unsatisfiable package recommendations (f_4: "*unsat*"), and the number of extra packages installed (f_5: "*new*"), respectively. In the existing studies, there are mainly two types of aggregate criteria, both of which consider the lexicographic combination of multiple objectives, i.e., the objectives are handled in a hierarchical way, and the first objective has the highest priority. More specifically, the paranoid criterion first optimizes the "*removed*" objective, then the "*changed*" objective. Meanwhile, the trendy criterion considers the "*removed*", the "*notuptodate*", the "*unsat*", and the "*new*" objectives successively [11]. These lexicographic approaches do not have to enumerate all the combinations of the objectives. However, if there exists drastic tradeoff between these objectives, the search might be sensitive to the order of the objectives. Under such condition, analyzing the relationships between different upgrade objectives is necessary. Meanwhile, in the evolutionary computation literatures, a common resolution is to provide a set of Pareto optimal solutions. Such approaches do not only provide more choices for the decision maker, but also enable the analysis about the relationships between objectives, which might reveal insights into the problem [12]. Inspired by the concept of Pareto optimality, we are interested in deeper understanding of the software upgradability problem. In this process, the challenge lies in the fact that, obtaining the Pareto schemes for the software upgradability problem is very time consuming. Hence, we would adopt the meta-learning technique to tackle this challenge.

3 Experiments and Discussion

The experiments are conducted on an Intel Core i5 3.2 GHz CPU PC with 4 GB memory, running GNU/Linux with kernel 3.16. For the data set, we employ the benchmark from the Mancoosi International Solver Competition 2010–2012, in which the requests are generated from the Debian repository[2]. After filtering the infeasible upgrade requests[3], we obtain in total 350 upgrade requests. Then, we proceed to describe the Pareto scheme sampling procedure. The Pareto schemes could be defined as the upgrade schemes that are not dominated by any other upgrade schemes. In the literatures, there exist various mechanisms that could convert the problem of achieving the Pareto front into a number of scalar optimization problems, such as the weighted sum, the Tchebycheff aggregation, and the boundary intersection approaches [15]. Due to its simplicity and effectiveness, we adopt the weighted sum approach. More specifically, for each upgrade request, inspired by [15], the $\{5, 5\}$-simplex lattice design is employed to generate 126 weight vectors, which are then used to construct the weighted single-objective optimization problem. Then, for each weighted problem, we use a publicly available solver `mccs`[4] with Gurobi, which is a state-of-the-art MILP solver, to compute the optimal upgrade scheme in the corresponding direction.

3.1 RQ1: Conflict Analysis

First, we are interested in the comparisons between the two existing lexicographical criteria. With the trendy and the paranoid criterion, we apply `mccs`

[2] http://mancoosi.org/misc/.
[3] Note that these requests could be detected by the feature extraction phase, see RQ2.
[4] http://www.i3s.unice.fr/~cpjm/misc/mccs.html.

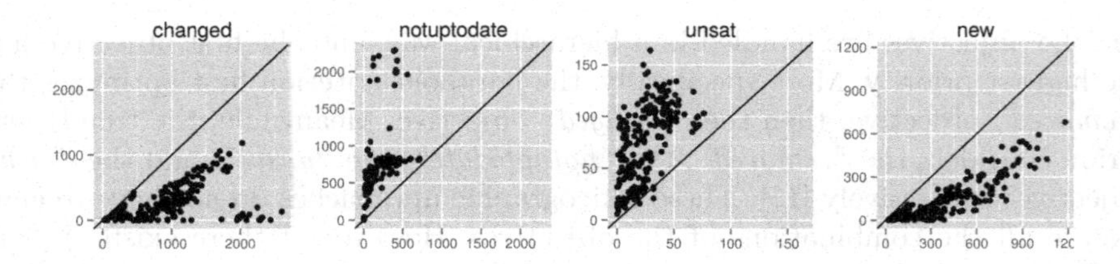

Fig. 2. Comparison between the trendy (x-axis) and the paranoid (y-axis) criteria

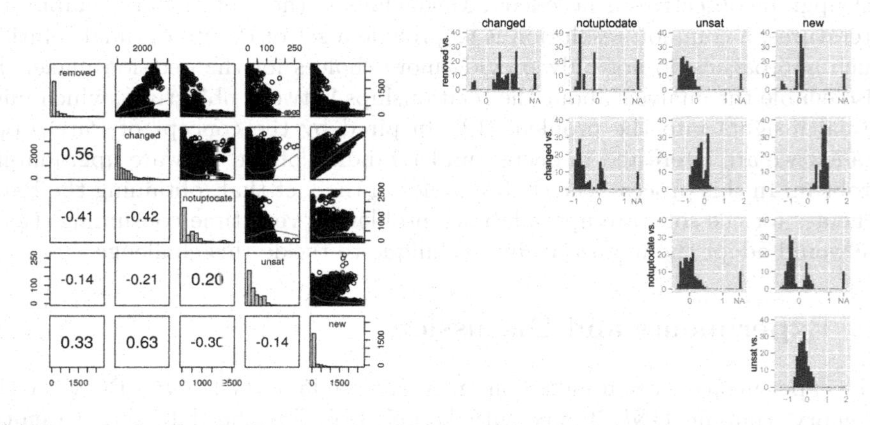

(a) Scatter plot for the Pareto schemes (b) Correlation distributions

Fig. 3. Properties of the Pareto schemes

over each upgrade request respectively, and plot the obtained single-objective optimal upgrade schemes in Fig. 2. Since the two criteria share the first objective "*removed*", we only plot the comparisons between the two upgrade criteria under the rest 4 objectives. In each subfigure, the x-axis and the y-axis represent the trendy and the paranoid criteria, respectively. From Fig. 2, the following observations could be drawn. (1) For the "*changed*" objective, `mccs` with the paranoid criterion outperforms `mccs` with the trendy criterion. This observation is as expected, in that the trendy criterion does not consider optimizing the "*changed*" objective. (2) Similarly, when we consider the "*notuptodate*" and the "*unsat*" objectives, which the paranoid criterion does not care, `mccs` with the trendy criterion could achieve better performance. (3) Surprisingly, for the "*new*" objective, which only the trendy criterion considers, `mccs` with the trendy criterion is outperformed by the same solver with the paranoid criterion. A possible reason is that, there exist certain correlation between objectives.

To examine the assumption, we plot the Pareto schemes in Fig. 3(a). Due to the dimensional issue of the problem, we adopt the pairwise scatter plot to illustrate the relationship between the 5 objectives. The figure comprises 3 components. First, the upper panel represents the scatter plot of the Pareto schemes projected on each specific plane. Second, the diagonal panel illustrates

Table 1. Feature category and examples

Domain	Feature category	Feature example
SAT (54 features)	Problem size features	Variable and clause numbers
	Variable–clause graph features	Variable and clause node degree statistics
	Variable graph features	Node degree statistics
	Clause graph features	Clause graph node degree statistics
	Balance features	Horn clauses fraction
MILP (141 features)	Problem size features	Number of variables and constraints
	Variable-constraint graph features	Variable and constraint node degree statistics
	Linear constraint matrix features	Cariable and constraint coefficient statistics
	Objective function features	Normalized objective coefficient statistics
	LP-based features	Mean value of integer slack vector
	Right-hand side features	Mean value of the right-hand side
	Probing based features	Mixed integer programming gap

the histograms that capture the distributions of the objective values for each objective. Finally, in the lower panel, we present the Spearman correlation coefficients between objectives. From Fig. 3(a), several interesting phenomena could be observed. First, under different objectives, the distributions of the objective values vary significantly. For example, for the "*unsat*" objective, the objective values of the Pareto schemes range within [100, 250]. Meanwhile for the "*new*" objective, the corresponding interval is [500, 2500]. Second, from the upper panel of Fig. 3(a), we observe that the relationships between objectives vary greatly as well. For example, when we consider the relationship between the "*changed*" and the "*new*" objectives, the points in the corresponding subfigure (row 2, column 5) exhibits a near linear pattern. This phenomenon conforms with what we observe in Fig. 2. Furthermore, the hypothesis that the two objectives are correlated is supported by the Spearman test, with a coefficient 0.63. When we consider the coefficients between other objectives, conflicts could be detected. For example, there exists a clear negative correlation between the "*removed*" and the "*notuptodate*" objectives.

More importantly, the correlation coefficients between objectives may also vary significantly over different upgrade requests. In Fig. 3(b), we plot the pairwise histograms for the 5 objectives. In Fig. 3(b), each subfigure corresponds to the distribution of correlations between objectives over the 350 upgrade requests, e.g., the upper left subfigure describes the distribution of the correlation between the "*removed*" and the "*changed*" objectives. From the figure, we can observe several phenomena that support our hypothesis. As expected, the correlation coefficients between objectives vary greatly. For example, the majority of correlation coefficients ranges within $[-1, 0]$, when we consider the relationship between the "*removed*" and the "*unsat*" objectives. This implies that the two objectives are negatively correlated. Meanwhile, the "*changed*" and the "*new*" objectives are positively correlated. Besides, in the figures, the "NA" indicates the upgrade request and the corresponding objectives, for which the objective value is constant.

Summary of RQ1: We detect conflict between different upgrade objectives. Moreover, the correlation coefficients between objectives vary greatly over different upgrade requests. Hence, more analysis is required, to reveal more insights.

3.2 RQ2: Correlation Prediction

In the previous experiment, we detect variation of the correlation coefficients when considering the relationships between objectives. In practice, the correlation coefficients between objectives could be helpful, such as the objective reduction in the field of many-objective optimization [13]. However, calculation of these coefficients requires sampling Pareto schemes, which further relies on exactly solving NP-hard problems. Consequently, this procedure is very time consuming. In our experiment setup, sampling Pareto schemes for the upgrade requests costs 225,070.4 s, which may not be tolerable in practice. As a solution, we adopt the meta-learning approach, to investigate the underlying linkages between problem specific features and the properties of Pareto scheme. In the literatures, meta-learning has been widely used for algorithm selection [7], and performance prediction [5]. These approaches share a commonality, i.e., a set of features are extracted from the instances, to characterize their properties.

To extract problem-specific features from benchmark instances, we first encode the upgrade requests with different formulations, and construct features with off-the-shelf feature extractors[5]. First, due to the close relationship between the software upgradability problem and SAT, we could transfer the upgradability requests into their decision version SAT instances, and obtain the problem-specific features accordingly. Second, the software upgradability problem could also be described as MILPs. Consequently, given an upgrade request, we generate a MILP instances considering the equally weighted sum of the 5 objectives. Then, feature extraction is conducted over the instance. In particular, we merge the features from the two problem domains together, which results in 195 features[6]. The feature categories and the examples for each category are listed in Table 1. For both problem formulations, the detail of the problem features could be found in [5]. Besides, a byproduct of the feature extraction is that, we could

[5] The code is obtained from http://www.cs.ubc.ca/labs/beta/Projects/EPMs/.

[6] We make the data publicly available at http://oscar-lab.org/upgradability/.

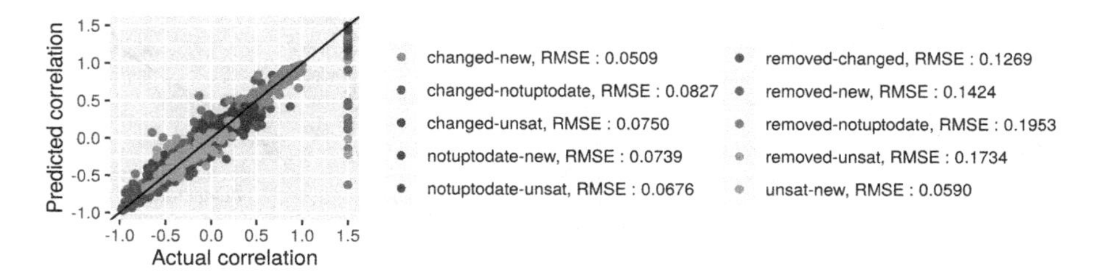

Fig. 4. Correlation prediction results (Color figure online)

detect infeasible upgrade request at an early stage. For example, if the transfered SAT instance is claimed to be unsatisfiable, satisfying the corresponding upgrade request will not be possible. For the sake of simplicity, with respect to the correlation associated with each pair of objectives, we build a regression model, namely random forest [1], due to its effectiveness. Given all the 350 upgrade request, we adopt the 10-fold cross-validation to evaluate the performance of the predictive model. In particular, those "NA" values are assigned with an exception value (1.5 in this study).

In Fig. 4, we present the prediction results of the regression. In the figure, the x-axis and the y-axis indicates the actual and the predicted correlation over the test set, within each fold of validation, respectively. Different objective combinations are denoted with different colors. Moreover, in the figure we present the Root Mean Square Error (RMSE) in the figure, to measure the accuracy of the prediction. From the figure, we observe that the trained model is able to estimate the correlation properly. The majority of the points lie closely around the reference line $y = x$. For the accuracy measure, all the RMSE values achieved by random forest lie below 0.2.

Summary of RQ2: With the problem specific features extracted from the upgrade requests, we could detect potential correlations between objectives.

3.3 RQ3: Tradeoff Assessment

As in RQ1, we observe that the performances of `mccs` with the trendy and the paranoid criteria may vary greatly over different upgrade requests. More importantly, using single-objective approaches such as lexicographic programming may pose risks within the problem solving process. If the search is overly concerned with certain objective, chances are that there may be other objectives over which the single-objective approaches may perform poorly [2]. Consequently, applying these methods may cause risks during the problem solving process.

In this experiment, we propose a measurements inspired by the reference based solution evaluation routine in the evolutionary computation literatures [9]. More specifically, the idea originates from the concepts of the ideal reference points, which are constructed by the best objective values with respect to each objective, considering all the Pareto schemes. With the ideal points described, we

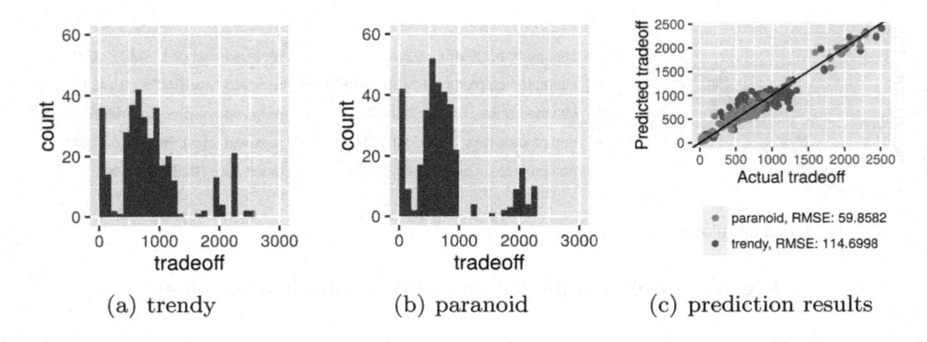

(a) trendy (b) paranoid (c) prediction results

Fig. 5. Distribution of *tradeoff* considering different criteria, and prediction results

define the risk of applying single-objective approaches for software upgradability problem as the maximum loss considering all the optimizing criteria:

$$tradeoff(s) = \max_{1 \leq i \leq 5} \{f_i(s) - f_i(ideal)\}, \quad (1)$$

where f_i indicates the objective function of each optimizing criteria, and s indicates the upgrade scheme achieved by certain algorithm. Given an upgrade scheme s achieved by certain lexicographic programming algorithm, if s is close to the ideal point, it would be suitable to accept the upgrade scheme to realize the upgrade process. Contrarily, a large *tradeoff* value implies that the corresponding scheme's quality is poor with respect to at least one objective. Accordingly, applying the corresponding upgrade scheme may be risky.

In Fig. 5, we present the distribution of the *tradeoff* measurements considering the two different criteria. From the figure, the following observations could be drawn. First, for both the criteria, the *tradeoff* value varies diversely over different upgrade requests. For example, for the paranoid criterion, the *tradeoff* value ranges within $[0, 2286]$, which means that there exists both easy upgrade requests (*tradeoff* $= 0$), and requests which leads to large *tradeoff*. Second, Similar as in RQ2, calculating the *tradeoff* indicator requires exactly solving the software upgradability problem. To make the measurement practical for guiding the problem solving process, we resort to the meta-learning mechanism again, to estimate the tradeoff with the problem specific features. The experimental setup is the same as in RQ2, except that we change the response variable from the correlation to *tradeoff* defined in Eq. 1. The prediction results are illustrated in Fig. 5(c), which is organized similarly with Fig. 4. From the figure, we can observe that the random forest model is able to predict the risk measurement *tradeoff* accurately. For both the trendy and the paranoid criteria, the RMSEs achieved by the random forest model are 59.8582 and 114.6998, respectively.

Summary of RQ3: In this experiment, we propose a reference based indicator *tradeoff*, to assess the suitability of applying aggregation based single-objective algorithms to solve the problem. Furthermore, we demonstrate that the measure is predictable, using the problem-specific features, which to some extent prevent the time consuming Pareto scheme sampling.

4 Conclusions and Future Work

In this study, we systematically investigate the relationships between multiple objectives of the software upgradability problem. The contributions of the paper could be summarized as follows. First, we design a series of experiments to analyze the inter-objective relationships for the software upgradability problem. Second, we apply the meta-learning technique to investigate the characteristics of the upgrade requests. Furthermore, the trained model enables the prediction of the properties of new upgrade requests. Finally, we propose a risk indicator, to measure the suitability of applying single-objective algorithms to solve the multi-objective software upgradability problem. However, there are still limitations that deserve future work. For example, the Pareto schemes are achieved by an exact solver. Due to the intrinsic complexity, sampling Pareto schemes for large scale upgrade request can be very time consuming. In the future, we intend to resort to multi-objective evolutionary algorithms [3,15] to approximate the Pareto schemes. Besides, in this study, we treat the off-the-shelf feature extractor as a black box, to capture the characteristics of the upgrade requests. Hence, deeper insights could be gained, if we further study the properties of the features mined from the upgrade requests.

Acknowledgement. This work is supported in part by the National Natural Science Foundation of China under Grants 61370144 and 61403057, in part by National Program on Key Basic Research Project under Grant 2013CB035906, and in part by the Fundamental Research Funds for the Central Universities under Grants DUT15TD37 and DUT16RC(4)62.

References

1. Breiman, L.: Random forests. Mach. Learn. **45**(1), 5–32 (2001)
2. Corne, D.W., Knowles, J.D.: Techniques for highly multiobjective optimisation: some nondominated points are better than others. In: Proceedings of the 9th Annual Conference on Genetic and Evolutionary Computation, pp. 773–780. ACM (2007)
3. Deb, K., Jain, H.: An evolutionary many-objective optimization algorithm using reference-point-based nondominated sorting approach, part I: solving problems with box constraints. IEEE Trans. Evol. Comp. **18**(4), 577–601 (2014)
4. Gebser, M., Kaminski, R., Schaub, T.: Aspcud: a linux package configuration tool based on answer set programming. In: Proceedings of Workshop on Logics for Component Configuration 2010 (2010)
5. Hutter, F., Xu, L., Hoos, H.H., Leyton-Brown, K.: Algorithm runtime prediction: methods & evaluation. Artif. Intell. **206**, 79–111 (2014)
6. Ignatiev, A., Janota, M., Marques-Silva, J.: Towards efficient optimization in package management systems. In: Proceedings of the 36th International Conference on Software Engineering, pp. 745–755 (2014)
7. Lindauer, M., Hoos, H.H., Hutter, F., Schaub, T.: Autofolio: an automatically configured algorithm selector. J. Artif. Intell. Res. **53**, 745–778 (2015)
8. Liu, Z., Yan, Y., Qu, X., Zhang, Y.: Bus stop-skipping scheme with random travel time. Transp. Res. Part C Emerg. Technol. **35**, 46–56 (2013)

9. Lokman, B., Köksalan, M.: Finding highly preferred points for multi-objective integer programs. IIE Trans. **46**(11), 1181–1195 (2014)
10. Michel, C., Rueher, M.: Handling software upgradeability problems with MILP solvers. In: Proceedings of Workshop on Logics for Component Configuration 2010, vol. 29, pp. 1–10 (2010)
11. Trezentos, P., Lynce, I., Oliveira, A.L.: Apt-pbo: solving the software dependency problem using pseudo-boolean optimization. In: Proceedings of the IEEE/ACM International Conference on Automated Software Engineering, pp. 427–436 (2010)
12. Verel, S., Liefooghe, A., Jourdan, L., Dhaenens, C.: Analyzing the effect of objective correlation on the efficient set of MNK-landscapes. In: Coello, C.A.C. (ed.) LION 2011. LNCS, vol. 6683, pp. 116–130. Springer, Heidelberg (2011)
13. Wang, H., Yao, X.: Objective reduction based on nonlinear correlation information entropy. Soft Comput. **20**(6), 2393–2407 (2016)
14. Xuan, J., Martinez, M., DeMarco, F., Clement, M., Marcote, S.L., Durieux, T., Berre, D.L., Monperrus, M.: Nopol: automatic repair of conditional statement bugs in java programs. IEEE Trans. Software Eng. (2016, online)
15. Zhang, Q., Li, H.: MOEA/D: a multiobjective evolutionary algorithm based on decomposition. IEEE Trans. Evol. Comp. **11**(6), 712–731 (2007)

Multicriteria Building Spatial Design with Mixed Integer Evolutionary Algorithms

Koen van der Blom[1(✉)], Sjonnie Boonstra[2], Hèrm Hofmeyer[2], and Michael T.M. Emmerich[1]

[1] Leiden Institute of Advanced Computer Science, Leiden University, Niels Bohrweg 1, 2333 CA Leiden, The Netherlands
{k.van.der.blom,m.t.m.emmerich}@liacs.leidenuniv.nl

[2] Department of the Built Environment, Eindhoven University of Technology, P.O. Box 513, 5600 MB Eindhoven, The Netherlands
{s.boonstra,h.hofmeyer}@tue.nl

Abstract. This paper proposes a first step towards multidisciplinary design of building spatial designs. Two criteria, total surface area (i.e. energy performance) and compliance (i.e. structural performance), are combined in a multicriteria optimisation framework. A new way of representing building spatial designs in a mixed integer parameter space is used within this framework. Two state-of-the-art algorithms, namely NSGA-II and SMS-EMOA, are used and compared to compute Pareto front approximations for problems of different size. Moreover, the paper discusses domain specific search operators, which are compared to generic operators, and techniques to handle constraints within the mutation. The results give first insights into the trade-off between energy and structural performance and the scalability of the approach.

Keywords: Evolutionary algorithms · Super-structure · Mixed integer optimisation · Multicriteria optimisation · Building spatial design · Building structural design · Building physics

1 Introduction

When designing buildings many disciplines have to be taken into account. For example structural design, because a building structure should have optimal strength, stiffness, and stability. Compliance is a specific measure of the stiffness of the building structure and will be subject to investigation in this paper. Another example is building physics, for which in this paper specifically climate control is used as objective, via the minimisation of the building outer surface, being a pre-cursor for future RC-network modelling obtaining minimal energy use for heating and cooling. This is an increasingly important objective due to unpredictable energy prices and climate protection. The built environment is responsible for about 40 % of the total use of energy and materials [1].

J. Handl et al. (Eds.): PPSN XIV 2016, LNCS 9921, pp. 453–462, 2016.
DOI: 10.1007/978-3-319-45823-6_42

Traditionally, energy efficiency and structural design objectives are dealt with in different engineering disciplines, and the same holds for various other objectives (e.g. architectural engineering, construction, etc.). Multidisciplinary optimisation aims to combine different disciplines in order to find building designs that perform well with respect to criteria from various disciplines. It has been used with great success in areas such as automotive and aerospace engineering [2], while in the building design domain its development is still somewhat limited.

This paper advances towards multidisciplinary optimisation of building designs, starting with finding building spatial designs based on criteria from structural design (compliance) and energy efficiency (total surface area). By proposing a multicriteria optimisation approach, the problem of conflicting objectives is discussed. In this case a Pareto front of building designs is computed that can be used in preparation of decision making, to understand design principles that lead to high performance in one discipline or the other discipline, and to find valid compromise solutions.

Traditional algorithms in (evolutionary) multicriteria optimisation, such as SMS-EMOA and NSGA-II, have been formulated for parametric design spaces. For such spaces they have been extensively tested and show a reliable performance. Recently a new super-structure for building spatial design was introduced by the authors [3,4] and here it is used for multicriteria optimisation for the first time. The super-structure encodes building spatial designs by means of a mixed integer representation. By changing discrete variables a large number of alternatives can be encoded. Continuous variables are used to change the dimensioning of these alternatives. Building spatial designs are viewed as configurations consisting of building spaces that do not overlap with each other. To enforce the feasibility of the structural designs generated for the building spatial designs, constraints on the variables are formulated by means of equations, which are checked before evaluation.

Given these preliminaries, this paper will provide the following research contributions: (1) first results on multicriteria optimisation of building spatial designs, including topology choices, (2) discussion of domain specific algorithm design aspects (search operators, constraint handling), and (3) interpretation and discussion of the evolved Pareto fronts in the multidisciplinary building design context. Another aspect discussed in this paper is the scalability of the approach in terms of the size and complexity of the building spatial design.

The remainder of this paper is structured as follows. Section 2 provides a brief summary of building design optimisation and the discipline-specific objectives. Then Sect. 3 discusses multicriteria optimisation techniques. The search space representation, constraints and objective functions are discussed in Sect. 4. Algorithm details are given in Sect. 5. Thereafter, in Sect. 6 numerical results are presented and Sect. 7 discusses these results and provides an outlook.

2 Building Spatial Design

Usually a building is designed by an architect and several engineers. They discuss their progress in project meetings, yet each discipline spends much effort on solving and optimising (discipline specific problems) at their own office. As such, fruitful interaction between disciplines is not guaranteed. This inefficiency of separated disciplines in the built environment gained acknowledgement [5], which gave rise to tools that allow more direct collaboration between engineers. One such tool is building information modelling (BIM) [6]. Through the modelling of data from various disciplines BIM allows information to be shared between engineers working on different building design aspects. Since choices made during the early stages of a design naturally propagate to the later stages, tighter collaboration by employing such tools may avoid one discipline disproportionally affecting performances in other disciplines.

An overview of optimisation tools in the built environment is provided by Palonen et al. [7]. Such tools generally parametrise components of the building design to enable the optimisation. Often these tools are limited to variation of the design through alteration of component variables, adding new components is rarely possible. Advances are made though, for example in the work by Hofmeyer and Davila Delgado [8], which focuses on optimisation via the simulation of a co-evolutionary preliminary building design process. Another interesting work is that of Hopfe et al. [9] where the significance of design variables on the building physics performance is predicted.

3 Multidisciplinary and Multicriteria Optimisation

Recently it has been recognised [5] that in order to help design teams consisting of experts from different disciplines in finding solutions, objectives and simulations from different disciplines have to be considered in concert. Multicriteria optimisation can be an important method in this context, as it allows to deal with conflicting objectives and can effectively support decision making.

In general, a multicriteria optimisation problem (MOP) is defined by a set of objective functions $f_i : X \rightarrow \mathbb{R}$, $i = 1, \ldots, m$ to be minimised (or maximised) for some search space X. Moreover, constraint functions $g_j(\boldsymbol{x})$ are usually considered, the value of which must be kept within a prescribed range.

For two feasible solutions $\boldsymbol{x}$ and $\boldsymbol{x}'$, it is said that $\boldsymbol{x}$ (Pareto) dominates $\boldsymbol{x}'$, if and only if $\forall i = 1, \ldots, m$: $f_i(\boldsymbol{x}) \leq f_i(\boldsymbol{x}')$ and there exists $j = 1, \ldots, m : f_j(\boldsymbol{x}) < f_j(\boldsymbol{x}')$. The efficient set X_E is the subset of X consisting of points that are not dominated by any point in X. The set $\{(f_1(\boldsymbol{x}), \ldots, f_m(\boldsymbol{x}))^T | \boldsymbol{x} \in X_E\} \subset \mathbb{R}^m$ is called the Pareto front (PF) of the MOP (given it exists). The PF provides valuable information about the space of all relevant solutions and their trade-offs. This paper aims to compute the PF for the real world problem of building spatial design and discuss the trade-offs between discipline specific objectives.

Recently, various powerful black box optimisation algorithms have been proposed for approximating Pareto fronts. Many of these belong to the class of evolutionary multicriterion optimisation, which use selection and variation (stochastic mutation, recombination) to steer a population of search points close to the Pareto front. The selection operator needs to take into account Pareto dominance, but diversity maintenance is also important in order to guarantee that all parts of the Pareto front are covered.

Two state-of-the-art evolutionary multicriterion optimisation algorithms, namely NSGA-II [10] and SMS-EMOA [11] are used as basic strategies in this paper. These algorithms will be instantiated for a domain specific search space.

4 Formal Problem Specification

4.1 Search Space Representation

The supercube representation, recently proposed by the authors [3,4], serves to represent the design space by means of continuous and discrete variables. The goal of the supercube representation was to formulate building design optimisation as a mixed integer nonlinear programming (MINLP) problem, an approach that in other domains is typically referred to as super-structure-based optimisation. Essentially, discrete variables encode the topology of spaces in the building spatial design and continuous variables determine the dimensioning of the spaces.

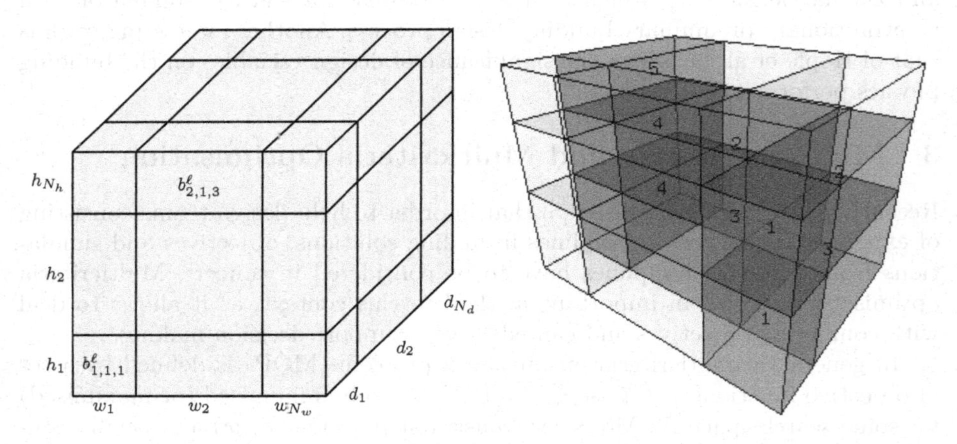

Fig. 1. Supercube grid representation (left) and building spatial design (right).

Building spatial designs consisting of N_{spaces} spaces are encoded in a cuboid (3D rectangle) grid of $N_w \times N_d \times N_h$ cells, these variables respectively refer to the number of cells in width, depth and height directions. In turn those same directions employ the indices $i \in \{1, \ldots, N_w\}$, $j \in \{1, \ldots, N_d\}$ and $k \in \{1, \ldots, N_k\}$, to determine their dimensioning with the variables w_i, d_j and h_k. Finally each cell may be turned on or off as being part of a space $\ell \in \{1, \ldots, N_{spaces}\}$ by the binary variable $b^{\ell}_{i,j,k}$. This is referred to as the supercube representation, Fig. 1 shows an example of a supercube and a derived building spatial design.

4.2 Topology Constraints

Four topology constraints are considered to disallow configurations of the supercube that are infeasible from an engineering point of view. All of these constraints can be described in mathematical form with just sums and products as presented by the authors in [3,4]. Textual explanations of the constraints and an example of the mathematical notation follow.

No Overlap ensures each cell is active for at most one space which can be defined mathematically with Eq. 1. Spaces should have a **Cuboid Shape**. This can be checked in two steps. Firstly it is ensured that for every space active cells appear at the same indices in all distinct rows, columns and beams. Secondly it is checked there are no gaps between the active cells of a space. **Vertical Gaps** between spaces, like archways and cantilevered parts, are disallowed in order to facilitate the check to determine whether a building stands on the ground by simple procedures. Finally a **Constant Number of Spaces** is enforced by making sure every space consists of at least one cell.

$$\forall_{i,j,k} : \sum_{\ell=1}^{N_{spaces}} b^{\ell}_{i,j,k} \leq 1 \tag{1}$$

4.3 Objective Functions

Energy performance is measured as the total outside surface area of the building spatial design, excluding the floor surface of the ground level. In the future, a RC-network model is planned to find heating and cooling energy per space.

For structural performance a black box simulator is used (meaning standard MINLP solvers cannot be used for optimisation) with the following settings. First the building spatial design is provided with a structural design via a so-called structural grammar. The grammar used here adds four concrete walls and a concrete roof (a slab) to every space, both with a thickness t=150 mm. Young's modulus of the concrete is set to E=30000 N/mm^2 and Poisson's ratio to $v = 0.3$. Live loads of 1.8 kN/m^2 are then applied to each slab, and wind loads from eight directions (N, NW, W, etc.) are applied to the building spatial design (with a pressure of 1.0 kN/m^2, a suction of 0.5 kN/m^2 and a shear of 0.4 kN/m^2) and transferred to the structural design. Using a finite element analysis (FEM), the compliance over all loads is calculated. For more details, see [8].

5 Algorithm Design

5.1 Volume Repair

A fixed volume V_0 for the building spatial design will be maintained during optimisation because otherwise objectives could possibly be optimised largely by taking extreme values for the continuous variables. The volume is taken as in Eq. 2 below. To exclude inactive cells $b_{i,j,k}$ is found by: $b_{i,j,k} = \sum_{\ell=1}^{N_{spaces}} b^{\ell}_{i,j,k}$, note that Eq. 1 needs to hold.

$$\sum_{i=1}^{N_w} \sum_{j=1}^{N_d} \sum_{k=1}^{N_h} b_{i,j,k} w_i d_j h_k = V_0 \tag{2}$$

When the volume of a new individual is not within a 1 % deviation of V_0 it is repaired by scaling the continuous variables. After scaling, continuous variables exceeding the lower bound are set to the lower bound. Variables exceeding the upper bound are multiplied by 0.95 until their value is within the bound. Naturally changes to variable values will also change the volume, therefore the process is repeated until the bound checks succeed without changes to the variables. Using the desired volume and the current volume V_c a factor $\alpha = V_0/V_c$ may be computed. Multiplying the dimensions of the supercube with the cubic root of α results in V_0. As such the scaling function is described by Eq. 3.

$$\forall_i : w_i = \sqrt[3]{\alpha w_i} \qquad \forall_j : d_j = \sqrt[3]{\alpha d_j} \qquad \forall_k : h_k = \sqrt[3]{\alpha h_k} \tag{3}$$

5.2 Optimisation and Constraint Checking

NSGA-II and SMS-EMOA are used with typical settings in the experiments below. In most cases they use the same settings and operators; otherwise it is indicated. A lower bound $lb = 3$ and upper bound $ub = 19.8$ are used for the continuous variables. Selection strategies are $(20 + 20)$ for NSGA-II and $(50 + 1)$ for SMS-EMOA. For the ease of notation $N_{cells} := N_w \times N_d \times N_h$ is defined. Binary variables have a probability of $1/N_{cells}$ to be initialised to one, or zero otherwise. Continuous variables are set to a value from $lb + (ub - lb) \times U$, where U is drawn uniformly at random from $]0, 1]$. Moreover, a fixed step size $0.05 \times (ub - lb)$ is used for the continuous variables. Following the initialisation the volume of the parent population is repaired as described in the previous subsection with a desired volume $V_0 = 4^3 \times N_{cells}$. Each individual is evaluated as follows. If any constraint is violated a penalty value *pen* is returned based on the number of violations CV such that $pen = 999,999,999 + CV - 1$. Here CV is an integer from one to five to indicate the number of violations. The five constraints relate to the four previously described constraints. The two parts of the cuboid shape constraint are counted separately. The objective functions are only evaluated when no constraints are violated. An evaluation budget of 2500 is used in the experiments. Note that constraint checks are not considered as evaluations here.

Each offspring is created by applying crossover and mutation. For crossover a parent $P1$ is selected uniformly at random from the population. A second parent $P2$ is then selected uniformly at random with a probability of 0.5, otherwise $P2 = P1$. Parents are either recombined with a probability of 0.5, or copied to the different children $C1$ and $C2$. Each binary variable is recombined as $C1 = P1$ and $C2 = P2$ with probability 0.5, or as $C1 = P2$ and $C2 = P1$ otherwise. Simulated binary crossover is applied to the continuous variables. When a variable exceeds a bound it is set to lb or ub as applicable. Finally either of the children is selected with probability 0.5. Mutation is applied with a probability of $1/N_{dims}$, where $N_{dims} := N_{cells} \times N_{spaces} + N_w + N_d + N_h$ is the total number of variables. Binary variables are mutated by bit flips. Polynomial mutation is applied to continuous variables above the lower bound, variables exactly at the boundary are reinitialised (as previously described). Following mutation variables exceeding their bounds are set to their appropriate boundary values. The volume of the produced offspring is repaired as previously described. NSGA-II then applies non dominated/crowding distance sorting to the population of size $\mu + \lambda$ before selecting the first μ individuals for the next parent population. SMS-EMOA selects based on the hypervolume contribution (reference point $(1.1e9, 1.1e9)$).

5.3 Smart Mutation

The general mutation and recombination operators used in NSGA-II and SMS-EMOA have difficulties navigating heavily constrained objective landscapes, such as considered here. A smart mutation operator is proposed which only produces mutants that do not violate the problem specific constraints. Since the algorithms have similar performance only SMS-EMOA is considered with smart mutation.

The smart mutation method works by extending or reducing spaces by either adding or removing a surface of cells. This is done by selecting one of the following faces of the space to make either an outward or an inward move: left, right, top, bottom, front or back. All moves are applied to all cells along the selected face of a space, such that the space remains cuboid when adding and removing cells. These moves are of size one, meaning that the width, depth or height (depending on the selected face) of a space grows or shrinks by a single cell. Whenever an outward move adds a cell to a space A that is already part of a space B the cell is set to inactive for space B. From all mutation steps that do not result in a constraint violation one is chosen uniformly at random.

A new offspring individual is then created as follows. A parent is selected uniformly at random. Smart mutation is applied with a 0.25 probability, otherwise a continuous variable that is relevant to at least one active cell is selected uniformly at random and mutated by polynomial mutation. No crossover is used.

Initialisation of binary variables is changed to ensure the initial population consists solely of valid individuals. For every space a non-fully occupied pillar is selected uniformly at random from the supercube and the first cell from the bottom that does not belong to any previously initialised space is set active for this space. To increase diversity in the initial population twenty smart mutations are applied to the initial individuals of single cell spaces.

Penalty values are no longer used since all offspring are now guaranteed to be valid. The remaining procedures are the same as in Subsect. 5.2.

6 Numerical Results

Problem configurations are denoted by four numbers. The first three indicate the dimensions of the supercube and the last indicates the number of building spaces that are considered. For example 2225 indicates a problem with a $2 \times 2 \times 2$ supercube and five spaces. Every experiment averages over five runs using average Pareto fronts (median attainment curves [12]). Tests were done for 222 and 333 configurations both with one, three and five spaces.

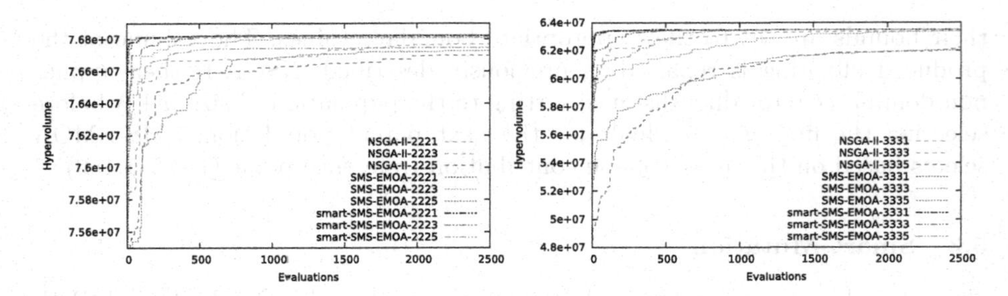

Fig. 2. Average hypervolume growth over five runs, reference point (35000, 2500), for one, three and five spaces in 222 (left) and 333 (right) configurations.

The various problem configurations show a quick convergence to a relatively stable hypervolume (taken with $log(1 + compliance), surface\ area)$ in Fig. 2, with more complicated configurations naturally taking move evaluations before stabilising. NSGA-II and SMS-EMOA produce similar attainment curves as may be observed in Fig. 3. This indicates the considered process works and results in a Pareto front approximation. The standard deviations of the hypervolume at the final generation are relatively small for most problem configurations and do not change the numerical result. Only for the 3335 configuration large deviations occur for the generic methods, but even their highest hypervolume solutions do not outperform the smallest hypervolume found by the method with smart mutation. A one sided Wilcoxon test between NSGA-II and SMS-EMOA results in $W = -1$, indicating there is no significant difference. Moreover, applying the one sided Wilcoxon test between either of those methods and smart SMS-EMOA results in $W = 15$, indicating the method with smart mutation is better with a statistical significance of 0.05.

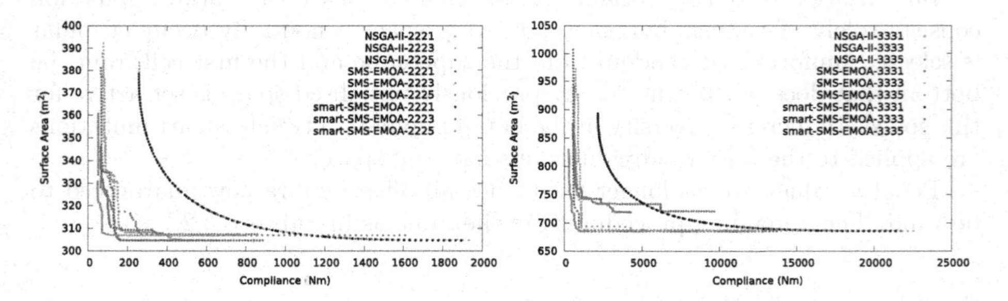

Fig. 3. Median attainment curves from five runs for one, three and five spaces in a 222 configuration (left) and a 333 configuration (right).

Smart SMS-EMOA produces similar results to the other two approaches for single space problems as can be observed in Fig. 2. For the problems with three spaces the method with smart mutation improves over the other two by a decent margin, and for five spaces it is clearly better both in terms of convergence speed and the final solution. The same behaviour can be observed in Fig. 3, where

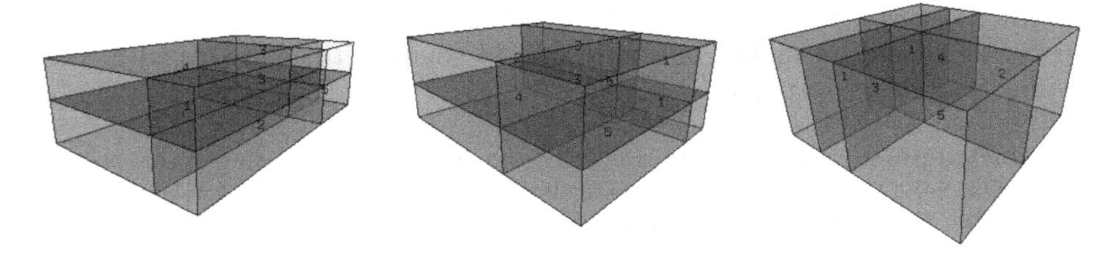

Fig. 4. Best spatial designs found with smart SMS-EMOA for the 3335 configuration. Minimal compliance (left), knee point (center) and minimal surface area (right).

differences in performance become more pronounced with larger problem sizes. Clearly, smart mutation produces a better Pareto front approximation.

Figure 4 shows the best found spatial designs in terms of each objective as well as a compromise solution at the knee point of the median attainment curve. As can be expected the optimal spatial design in terms of minimal surface area has a cuboid shape. The knee point solution is largely similar, but has a slightly lower structure and as a result is stretched in both width and depth to maintain the volume. Finally the minimal compliance solution has an L-shaped and elongated structure. The lower structure can be explained since it results in less strain on the structural elements, reducing the compliance.

Table 1. Average runtime over five runs, rounded to the closest whole minute.

Problem configuration	2221	2223	2225	3331	3333	3335
CPU time (minutes)	42	342	888	42	620	1008

Table 1 shows the CPU time used with smart mutation. The other methods performed similarly because the compliance computations used by far the most CPU time. Each experiment used a single core of an i7-3770 CPU @ 3.40 GHz processor and with 16 GiB DIMM DDR3 Synchronous 1600 MHz memory.

7 Discussion

Multicriteria optimisation algorithms for a building spatial design have been developed and tested for moderate size problems. The problem has been formulated as a mixed integer program. Moreover, the problem is characterised by a large number of constraints and a specific constraint handling mutation operator has been proposed. Pareto front approximations have been obtained. They always have a convex shape which makes it possible to find compromise solutions in knee points. The results show that smart mutations can be beneficial for exploring larger and more dense regions. However, in order to scale up the problem size further research in this direction is needed, including recombination operators. Moreover, surrogate modelling may allow for a more efficient exploration of the objective landscape. Finally, while statistical significant improvement was shown when using the method with smart mutation, parameter tuning should be applied in future work to compare the methods with their optimal parameter settings.

Acknowledgments. The authors gratefully acknowledge the financing of this project by the Dutch STW via project 13596 (Excellent Buildings via Forefront MDO, Lowest Energy Use, Optimal Spatial and Structural Performance).

References

1. European Commission: Challenging and Changing Europes Built Environment: A Vision for a Sustainable and Competetive Construction Sector By 2030. European Construction Technology Platform (2005)
2. Liao, X., Li, Q., Yang, X., Zhang, W., Li, W.: Multiobjective optimization for crash safety design of vehicles using stepwise regression model. Struct. Multi. Optim. **35**(6), 561–569 (2008)
3. van der Blom, K., Boonstra, S., Hofmeyer, H., Emmerich, M.T.M.: A super-structure based optimisation approach for building spatial designs. In: ECCOMAS 2016, 5–10 June, Greece (2016, accepted)
4. Boonstra, S., van der Blom, K., Hofmeyer, H., Amor, R., Emmerich, M.T.M.: Super-structure and super-structure free design search space representations for a building spatial design in multi-disciplinary building optimisation. In: EG-ICE 2016, 29 June–1 July, Poland (2016, accepted)
5. Martins, J.R., Lambe, A.B.: Multidisciplinary design optimization: a survey of architectures. AIAA J. **51**(9), 2049–2075 (2013)
6. Eastman, C., Eastman, C.M., Teicholz, P., Sacks, R.: BIM Handbook: A Guide to Building Information Modeling for Owners, Managers, Designers, Engineers and Contractors. Wiley, Hoboken (2011)
7. Palonen, M., Hamdy, M., Hasan, A.: MOBO a new software for multi-objective building performance optimization. In: Wurtz, E. (ed.) Proceedings of the 13th Internationcal Conference of the IBPSA, pp. 2567–2574. IBPSA c/o Miller-Thompson, Toronto (2013)
8. Hofmeyer, H., Davila Delgado, J.M.: Coevolutionary and genetic algorithm based building spatial and structural design. Artif. Intell. Eng. Des. Anal. Manuf. **29**(04), 351–370 (2015)
9. Hopfe, C.J., Emmerich, M.T.M., Marijt, R., Hensen, J.L.M.: Robust multi-criteria design optimisation in building design. In: Proceedings of Building Simulation and Optimization, Loughborough, UK, pp. 19–26. IBPSA, England (2012)
10. Deb, K., Pratap, A., Agarwal, S., Meyarivan, T.: A fast and elitist multiobjective genetic algorithm: NSGA-II. IEEE Trans. Evol. Comput. **6**(2), 182–197 (2002)
11. Emmerich, M.T.M., Beume, N., Naujoks, B.: An EMO algorithm using the hypervolume measure as selection criterion. In: Coello Coello, C.A., Hernández Aguirre, A., Zitzler, E. (eds.) EMO 2005. LNCS, vol. 3410, pp. 62–76. Springer, Heidelberg (2005)
12. Fonseca, C.M., da Fonseca, V.G., Paquete, L.: Exploring the performance of stochastic multiobjective optimisers with the second-order attainment function. In: Coello Coello, C.A., Hernández Aguirre, A., Zitzler, E. (eds.) EMO 2005. LNCS, vol. 3410, pp. 250–264. Springer, Heidelberg (2005)

The Competing Travelling Salespersons Problem Under Multi-criteria

Erella Matalon-Eisenstadt[1,2(✉)], Amiram Moshaiov[1], and Gideon Avigad[3]

[1] School of Mechanical Engineering, Tel Aviv University, Tel Aviv, Israel
[2] Department of Mechanical Engineering, ORT Braude College of Engineering, Karmiel, Israel
erella@braude.ac.il
[3] Robotics and Automation Department, Vineland Research and Innovation Centre, Lincoln, ON, Canada

Abstract. This paper introduces a novel type of a problem in which two travelling salespersons are competing, where each of them has two conflicting objectives. This problem is categorized as a Multi-Objective Game (MOG). It is solved by a non-utility approach, which has recently been introduced. According to this method all rationalizable strategies are initially found, to support posteriori decision on a strategy. An evolutionary algorithm is proposed to search for the set of rationalizable strategies. The applicability of the suggested algorithm is successfully demonstrated on the presented new type of problem.

1 Introduction

Studies on the Travelling Salesman Problem (TSP) have produced many versions of the original problem. Among such TSP versions are multi-objective problems (e.g., [1, 2]) and game problems (e.g., [3, 4]). This paper introduces a new version of a TSP, which amalgamates the aforementioned types of problems. The proposed TSP concerns a multi-objective game (MOG) between two salespersons that compete over markets, which are located in different cities. In such a game, each of the salespersons has multiple conflicting objectives (two in the current example).

The proposed MOG is examined based on the solution concept of rationalizable strategies' of MOGs. According to this concept, which has recently been introduced in [5], the rationalizable strategies are a result of a non-utility approach to MOGs. The set of the rationalizable strategies encloses all the strategies that may serve as the best response when considering a situation in which the players are undecided about their objective preferences. As shown in [5], finding the set of rationalizable strategies allows the players to select a preferred strategy, out of that set, using multi-criteria decision-making considerations. In order to find the set of rationalizable strategies, a modification to the evolutionary algorithm of [6] is suggested. The modified algorithm serves as an alternative search approach to the one presented in [7]. It is noted that a comparison of the proposed algorithm with that of [7] is left for future work.

The rest of this paper is organized as follows. Section 2 provides some background on MOGs under undecided objective preferences. Section 3 presents the proposed new

J. Handl et al. (Eds.): PPSN XIV 2016, LNCS 9921, pp. 463–472, 2016.
DOI: 10.1007/978-3-319-45823-6_43

type of problems and Sect. 4 describes the proposed algorithm. Next, an example is presented in Sect. 5. Finally, Sect. 6 concludes this paper.

2 Background

Many practical problems, in economics and engineering, can be modeled as games with multiple objectives or payoffs. Furthermore, each player might deal with her own conflicting objectives. In such games, which are termed MOGs, a vector of objective functions must be considered (e.g., [8]). If MOGs' players have objective preferences, then a utility function can be used to transform the MOG into a game with a single objective per each player. But, players may want to postpone the decision to a stage after all possible strategies are found and their performance trade-offs are examined. The idea of MOGs with undecided/postponed/undeclared objective preferences, was first presented in [6]. It was re-examined more recently in [5], where the concept of rationalizable strategies has been introduced.

As in [5], the considered game is zero-sum with respect to each component of the payoff vector. The game is pure strategy, single-act and non-cooperative. It is also of incomplete information because the players do not know what the opponent's objective preferences are. The ideas of [5] are summarized below.

2.1 The Players, Strategies and Payoffs

A MOG is considered between two players P_1 (maximizer), and P_2 (minimizer). Let S_1 and S_2 be the sets of all possible I and J strategies for P_1 and P_2 respectively, such that:

$$S_1 = \{s_1^1, s_1^2, \ldots, s_1^i, \ldots\ldots, s_1^I\},\ S_2 = \{s_2^1, s_2^2, \ldots, s_2^j, \ldots\ldots, s_2^J\} \tag{1}$$

Where I and J are the total number of strategies of P_1 and P_2, respectively.

The interaction between the ith strategy and the jth strategy played by P_1 and P_2, respectively, results in the following payoff vector:

$$\bar{f}_{i,j} = \left[f_{i,j}^{(1)}, f_{i,j}^{(2)}, \ldots, f_{i,j}^{(k)}, \ldots, f_{i,j}^{(K)}\right]^T \in \mathbb{R}^K \tag{2}$$

Where K is the number of objectives (payoffs) considered by each player.

The set of all interactions between a strategy s_1^i of player P_1 and each of the available strategies of the second player P_2 results with the following set of payoff vectors (performances) that are associated with strategy s_1^i:

$$F_{s_1^i} = \{\bar{f}_{i,1}, \ldots, \bar{f}_{i,j}, \ldots, \bar{f}_{i,J}\} \tag{3}$$

In the same way, the set of the associated payoff vectors (performances) of strategy s_2^j of the maximizer is:

$$F_{s_2^j} = \{\bar{f}_{1,j}, \ldots, \bar{f}_{i,j}, \ldots, \bar{f}_{I,j}\} \tag{4}$$

2.2 The Set of Rationalizable Strategies

As explained in [5], a general zero-sum MOG can be evaluated as a minimization or a maximization problem, depending on the players' viewpoint. Formally:

$$\text{For the maximizer } P_1 : max_{s_2^j \in S_2}\ min_{s_1^i \in S_1} \left(\bar{f}_{i,j}\right) \tag{5}$$

$$\text{For the minimizer } P_2 : min_{s_1^i \in S_1}\ max_{s_2^j \in S_2} \left(\bar{f}_{i,j}\right) \tag{6}$$

However, the above are ill-defined. When applied on a vector, the operators max and min require an interpretation. Here, following [5], the interpretation concerns the use of domination. When the maximizer is the optimizer, the representative set includes the Pareto-fronts (anti-optimal) as obtained by solving the minimization problem for each of the maximizer's strategies:

$$F_{s_1^i}^{-*} = \left\{\bar{f}_{i,j} \in F_{s_1^i} | \neg \exists \bar{f}_{i,j'} \in F_{s_1^i} : \bar{f}_{i,j'} \succ^{min} \bar{f}_{i,j}\right\} \forall j,j' = [1,J] \tag{7}$$

The set of all the anti-optimal fronts of P_1 is a set of sets, as defined below.

$$F_1^{-*} = \left\{F_{s_1^1}^{-*}, \ldots, F_{s_1^i}^{-*}, \ldots, F_{s_1^I}^{-*}\right\} \tag{8}$$

Sorting F_1^{-*} and selecting the dominating sets in the minimization problem (dominating in the inverse problem) will result in the maximizer's (P_1) *irrational* strategies. Hence, the set of irrational strategies of the maximizer are:

$$S_1^{irr} := \left\{s_1^i \in S_1 | \exists s_1^{i'} \in S_1 F_{s_1^i}^{-*} \succ^{min} F_{s_1^{i'}}^{-*} \forall i, i' \in [1,I]\right\} \tag{9}$$

The set S_1^{irr} includes all P_1's strategies that are associated with a dominating anti-optimal front in the maximization problem. The set of rationalizable strategies of P_1 is the relative complement of S_1 and S_1^{irr}:

$$S_1^R = S_1 - S_1^{irr} \tag{10}$$

The set S_1^R includes all P_1's strategies that are associated with a non-dominating anti-optimal front in the maximization problem where the cardinality of S_1^R is $|S_1^R| = I'$ and $1 \le I' \le I$. Each of the rationalizable strategies is represented in the objective space by its related anti-optimal front. The union of all the I' anti-optimal fronts of the rationalizable strategies $F_{s_1^i}^{-*}$ form the rationalizable layer of P_1:

$$F_1^R = \bigcup_{i=1}^{I'} F_{s_1^i}^{-*} \text{ where } F_{s_1^i}^{-*} \subseteq F_1^{-*} \tag{11}$$

2.3 Selecting a Strategy

The set of rationalizable strategies and their performances can be used to select a strategy. This, however, is not within the focus of this paper, and the interested reader is referred to [5] for some suggested ideas concerning such a selection.

3 Proposed TSP

The proposed type of a TSP concerns a MOG between two salespersons that compete over costumers that are located in several cities. Each of the salespersons has multiple conflicting objectives (taken as two in the studied case). The basic assumption is that neither of them knows what the chosen route of the opponent is.

The game arena is represented as undirected weighted graph. The considered graph contains N vertices (cities), where each vertex represents one member of the cities' set $C = \{c(1), c(2), \ldots, c(N)\}$. The arcs of the graph represent the roads between the cities. The game is between two competing salespersons (players), which are denoted by P_1 and P_2. A strategy of a player, which is a chosen route (path), is defined as a partial permutation of the cities' set C. Each player may visit a city no more than once. The routes of the players are described as the ordered sets $Path_1 = \left\{c_1^1, c_2^1, \ldots, c_{N_1}^1\right\}$ and $Path_2 = \left\{c_1^2, c_2^2, \ldots, c_{N_2}^2\right\}$ (where $1 \leq N_1, N_2 \leq N$) for the first and second players, respectively. Each element c_i^1 or c_j^2 of the routes denotes a city, such that the superscript points to which player visited that city, and the subscript indicates the order by which the player visited it.

Each interaction between two strategies of the players is assessed by each player using two payoffs. The first payoff is based on the difference between the lengths of players' routes. The second is based on a difference between the values of the cities visited by the players. Let $L_1 = Length(Path_1)$ and $L_2 = Length(Path_2)$ be the route length of the paths of P_1 (the maximizer) and P_2 (the minimize), respectively. It is assumed that each city $c(i)$ has a different market size, which is considered as the value of this city, $v(c(i)) \in [v_{min}, v_{max}]$. Then, the revenue of P_1 and P_2 is $V_1 = Value(Path_1) = \sum_{i=1}^{N_1} v(c_i^1)$ and $V_2 = Value(Path_2) = \sum_{i=1}^{N_2} v(c_i^2)$, respectively. If both routes contain common cities, namely $Path_1 \cap Path_2 \neq \varnothing$, then the first player arriving to a common city, earns the city value, whereas the other gets no added value for visiting this city. It is noted that the player with the shortest route to a common city is considered here as the first to arrive to that city.

In the considered game the first objective is $f^{(1)} = L_2 - L_1$ and the second is $f^{(2)} = V_1 - V_2$. Player P_1 aims at maximizing both objectives while P_2 aims at minimizing them. The reason for this objectives setting is to form a complete contradiction between the two players such that the game is a zero-sum game. In future work we intend to define the objectives such that each player concerns only for her own path length and path value.

4 Proposed One-sided Algorithm

The suggested evolutionary algorithm is a modified version of the one introduced in [6]. It searches for the rationalizable strategies of one player (optimizer) at a time. Hence, we refer to it as a one-sided algorithm, as opposed to the co-evolution algorithm of [7]. The principles of the algorithm are described below.

The proposed evolutionary algorithm includes an inner and an outer loop, in a manner similar to that of [6]. Note that in the current application the crossover operator is omitted. As in [6], the inner-loop finds the representative anti-optimal front of each strategy of the considered population (of the optimizer). Namely, at each inner iteration, the worst possible actions of the opponent are taken into considerations. When all the iterations of the inner-loop are completed, each of the optimizer's strategies of the current generation is represented by an anti-optimal front.

At the outer-loop, the optimizer implements a set-based optimization procedure, by applying a procedure similar to that of [6]. Namely, at the outer-loop, the optimizer is applying an optimization procedure based on the representative anti-optimal fronts of her strategies. However, there is a significant difference between the current work and [6]. This difference is in the process of selecting the strategies for the optimizer in the outer-loop. In [6], the optimizer *selects* the strategies with the *best* anti-optimal fronts while here, the optimizer *excludes* the strategies with the *worst* anti-optimal fronts. This difference stems from the definition of the rationalizable strategies in [5]. The outer-loop procedure to sort and select the strategies is described in the following.

4.1 Strategies' Fitness and Sorting

The outer-loop of the algorithm performs a lexicographic selection of strategies (similar to that in [9]). It is based on two indicators including: *rank* (r) and *in-rank* (ir) *grade*. When comparing two strategies, the one with a higher rank is preferred. If the ranks are equal, then the strategy with the higher in-rank grade is preferred.

The rank of a strategy r is calculated by sorting the strategies as follows. All strategies that their representative anti-optimal fronts do not dominate any other anti-optimal front (in the inverse optimization problem), receive the best rank ($rank = 1$). Next, these strategies and their representative anti-optimal fronts are removed from the list and the above step is repeated with $rank = rank + 1$. This procedure is repeated until all strategies are assigned with a rank.

The ir grade, of strategies of the same rank, is calculated based on [5], which follows [10, 11]. In order to assign a strategy q with its ir grade, the anti-optimal front F_q^{-*} of the strategy is compared with that of any p strategy within the same rank. Without loss of generality, in a maximization problem each such comparison results with a value that is calculated using the following indicator:

$$I_{\delta^+}\left(F_q^{-*}, F_p^{-*}\right) = \max_{\bar{f}_q \in F_q^{-*}} \min_{\bar{f}_p \in F_p^{-*}} \max_{k \in [1,K]} \left(f_q^{(k)} - f_p^{(k)}\right) \tag{12}$$

Namely, the indicator I_{δ^+} of the first anti-optimal front is the minimal distance that the second anti-optimal front needs to be moved (in both directions) until it becomes dominated by the first front in the inverse optimization problem. The in-rank-grade (*ir*) of a strategy is the minimal value of all its I_{δ^+} indicators, obtained out of all the pair-wise comparisons of its anti-optimal front with all the anti-optimal fronts of the other strategies within the same rank. Namely, if there are N strategies in the rank of the strategy with the anti-optimal front F_q^{-*} then the in-rank grade ($ir(q)$) of this strategy is:

$$ir(q) = \min_{p\in[1,N]\wedge p\neq q}\left(I_{\delta^+}\left(F_q^{-*}, F_p^{-*}\right)\right) \tag{13}$$

4.2 The Route Coding

Each route is represented by two vectors of size N, where N is the number of cities. The first vector is a permutation of the N cities. The second vector is a random binary vector. The route is a list of cities from the first vector which is created by eliminating from the permutation vector all the cities that are associated with a zero in the second vector. Consider a TSP with six cities, $C = \{c(1), c(2), c(3), c(4), c(5), c(6)\}$ or in short = {1, 2, 3, 4, 5, 6}. Within the considered case, a possible example of a permutation vector is {3, 1, 6, 4, 2, 5}. Assuming that the binary vector is {1, 0, 0, 1, 1, 0}, then the coded path is $path = \{c(3), c(4), c(2)\}$. It is noted that at the end of the tour the salesperson returns to the starting city. Hence, the actual path is $h = \{c(3), c(4), c(2), c(3)\}$, or in short $path = \{3,\ 4,\ 2,\ 3\}$.

5 Case Study

This section provides an example of the proposed competitive TSP. It is designed to illustrate the effectiveness of the proposed algorithm and the type of information that it provides to the salespersons.

5.1 The Arena

The considered arena includes fourteen cities. Twelve cities are arranged in two groups (metropolises), as shown in Fig. 1.

The index of each city is given in brackets, and the adjacent number is the city value. The groups are cities [1–6] and [7–12]. The two remaining cities (13 and 14) are located far from both metropolises (metros). These cities are hereby termed countryside cities. The countryside city 14 is closer to the 1st metro and the countryside city 13 is closer to the 2nd metro.

Each salesperson starts the tour from a different metro. The first salesperson starts at city 5 and the second one at city 8. The starting city of P_1 and P_2 is denoted in Fig. 1 by light and dark gray squares, respectively. It is noted that the values of the two countryside cities are taken higher than those of the metro cities. The higher values give the

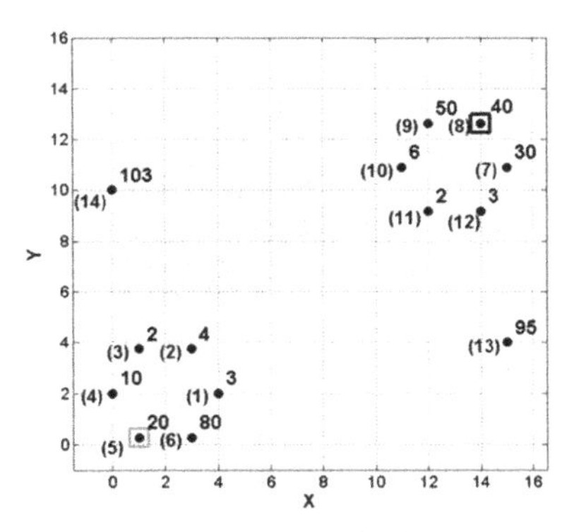

Fig. 1. Arena

salespersons a "motivation" to deviate from the metros toward the countryside cities. In this arena it is expected that each salesperson will not travel to the other metro. When traveling towards the other metro, the objective of a minimal path length is negatively affected. Moreover, by traveling to the other metro the value of the tour is not necessarily improved, as the other salesperson is expected to already visit the cities of that metro. On the other hand, it is not as clear if the incentive to go to the countryside cities, due to their higher values, is sufficient. This issue can be examined using the proposed algorithm.

It is noted that the current TSP problem is a selective one, namely the tour may hold any number of cities, such that $1 \leq |path| \leq N$. In the current problem setting, all the paths start from a fixed city, therefore the number of all possible paths is: $\sum_{n=1}^{N-1} n!$, which is the sum of all possible partial permutations of a set of $N - 1$ cities. Hence, in the given problem of = 14, the number of possible paths is 6.75×10^9 and therefore the number of possible interactions of the considered game is $(6.75 \times 10^9)^2 = 4.56 \times 10^{19}$.

5.2 Results

The following results were obtained using 100 individuals and 100 generations for the inner-loop and 50 individuals and 50 generations for the outer-loop. The employed mutation rate was 20 % and no crossover was used. The mutation is a random switching between two genes at each vector of the genotype. It is noted that the term rationalizable is hereby shorten to rational. Figure 2 shows an entire set of payoff vectors of all fronts, from a typical run. Clearly, from the maximizer perspective, the fronts of the rational strategies are better than those of the irrational strategies.

Figure 3 describes typical results as obtained for the maximizer. Two anti-optimal fronts are shown. The upper and lower ones correspond to the rational and irrational strategy of the middle and left panel, respectively. At each of these strategies, the starting city of the salesperson is denoted by a square and the second city by a star. The value of each city is indicated as well. Examining the objective space (right panel of Fig. 3) it can be seen that the anti-optimal front of the irrational path dominates the anti-optimal front of the rational path (in a minimization problem). The shown rational strategy of P_1 is path {5, 14,

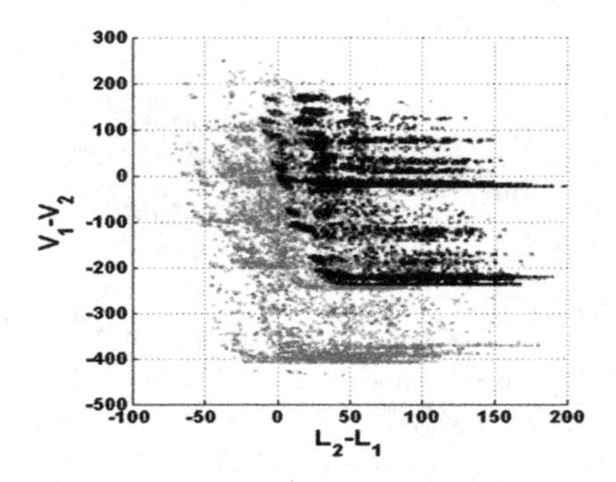

Fig. 2. Fronts of rational (black) and irrational (gray) strategies

4, 6, 3, 5}, while the irrational one is path {5, 11, 4, 14, 3, 5}. The main difference between the two paths is that the irrational includes city 11 of the remote metro, while the rationalizable one does not. Moreover, P_1 visits this city right after living the initial city, while in the rational path, the salesperson goes directly to city 14 and does not visit city 11 at all. It seems obvious that visiting city 11 is an irrational choice for P_1. Not only that this city is in the remote metro, but it is also the city with the lowest value there. Therefore, visiting this city is an irrational choice as by doing so the length of the path increases significantly while the value is hardly improved. Moreover, by visiting city 11, P_1 risks losing city 14 of the high value, to the opponent. Another difference between the rational and irrational paths is that in the irrational path, the salesperson skips city 6. This city is in the metro of P_1 and has the highest value (80) there, so skipping this city is also an irrational choice. On the other hand, in the rational path, the salesperson travels directly to the countryside city (14) with the highest value. After visiting city 14, the salesperson returns to her metro and visit two more cities with high values (cities 4 and 6) and one city with a low value (city 3).

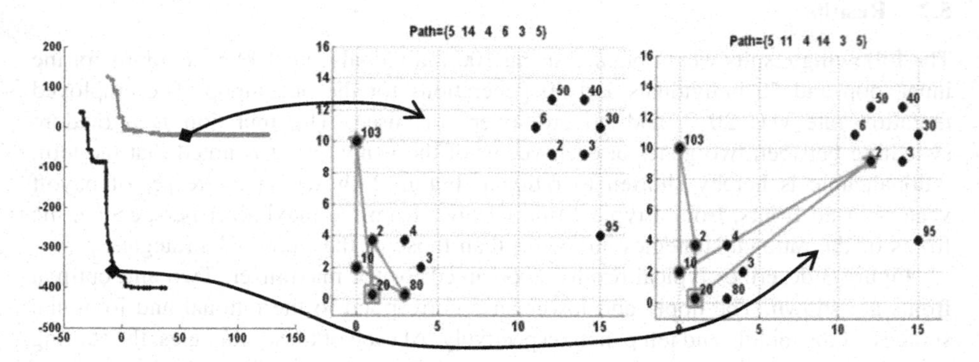

Fig. 3. Example of a rational and an irrational strategy of the maximizer, P_1 (middle and right panels respectively). The associated anti-optimal fronts (left panel).

Sixty-two runs were performed, where each run produced a multiset of (approximated) rationalizable strategies. To allow statistical analysis, several preliminary steps are carried out as follows. First, a set Q is created of all different paths, as obtained from all the 65 multisets. It turns out that the resulting set includes 64 different paths (64 out of 6.75×10^9 possible paths). These are denoted as p_r, where $r \in \{1, 62\}$. Second, the 62 multisets are united into a multiset, which is denoted as M. Third, the number of repetitions of p_r in M is denoted as q_r. Finally, the frequency of a solution p_r is defined as $100q_r/|M|$. The paths with the six highest frequencies are shown in Fig. 4, which also indicates (in brackets) their frequencies. It is noted that most of the 64 paths appear in a very low frequency (less than 1 %). Moreover, the frequencies are decaying fast such that the sixth solution is already with only 1.24 %.

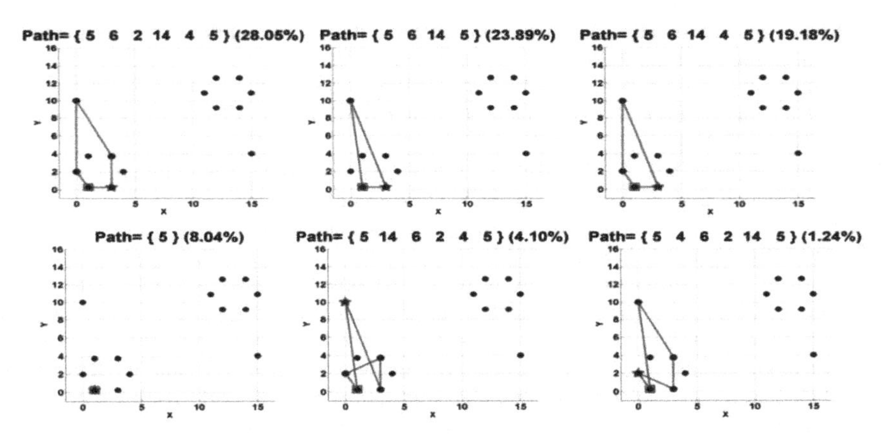

Fig. 4. The paths of P_1 with the six highest frequencies. (The 1st city is designated by a square and the 2nd by a star)

Almost all (94.5 %) of the rationalizable solutions are paths within the salesperson metro and the nearby city. Most (over 88.5 %) of the rationalizable solutions are paths that include city 14 (the countryside city with the highest value). Among the most frequent solutions is path {5} which corresponds to the shortest path length on the expense of a low value. The duplication of the most frequent solutions may indicate that the diversity is poor, and therefore convergence may be too fast. We suspect that this phenomenon is due to the use of the in-rank grade, which is measured within the objective space rather than using diversification in the genotypic (strategy) space.

6 Summary, Conclusions and Future Work

The current paper presents and defines a novel type of a TSP in which two salespersons are competing. In contrast to previous studies on competing TSP, here the problem is posed as a multi-objective game under undecided objective preferences. In addition, this work proposes a new algorithm to find the rationalizable strategies in such games. The proposed algorithm is demonstrated on the aforementioned TSP. Future work will include: (a) speciation within the strategy space by a measure of dissimilarity between paths, (b) a comparison of the results of the proposed algorithm with those obtained by the algorithm of [7], (c) extension to a non-zero-sum game, (d) solving more general cases of competitive TSP. Finally, it should be noted that some initial work on evaluating the algorithms using test functions and specially devised measures can be found in [12].

References

1. Ehrgott, M., Gandibleux, X.: An annotated bibliography of multiobjective combinatorial optimization, pp. 1–60 (2000)
2. Manthey, B., Ram, L.S.: Approximation algorithms for multi-criteria traveling salesman problems. In: Erlebach, T., Kaklamanis, C. (eds.) WAOA 2006. LNCS, vol. 4368, pp. 302–315. Springer, Heidelberg (2007)
3. Fekete, S.P., Fleischer, R., Fraenkel, A., Schmitt, M.: Traveling salesmen in the presence of competition. Theor. Comput. Sci. **313**(3), 377–392 (2004)
4. Kendall, G., Li, J.: Competitive travelling salesmen problem: a hyper-heuristic approach. J. Oper. Res. Soc. **64**, 208–216 (2012)
5. Eisenstadt, E., Moshaiov, A., Avigad, G., Branke, J.: Rationalizable strategies in multi-objective games under undecided objective preferences (2016). www.eng.tau.ac.il/~moshaiov/MOGJOTA.pdf
6. Avigad, G., Eisenstadt, E., Weiss-Cohen, M.: Optimal strategies for multi objective games and their search by evolutionary multi objective optimization. In: 2011 IEEE Conference on Computational Intelligence and Games (CIG), pp. 166–173 (2011)
7. Eisenstadt, E., Moshaiov, A., Avigad, G.: Co-evolution of strategies for multi-objective games under postponed objective preferences. In: 2015 IEEE Conference on Computational Intelligence and Games (CIG), pp. 461–468 (2015)
8. Zeleny, M.: Games with multiple payoffs. Int. J. Game Theory **4**(4), 179–191 (1975)
9. Deb, K., Pratap, A., Agarwal, S., Meyarivan, T.: A fast and elitist multiobjective genetic algorithm: NSGA-II. IEEE Trans. Evol. Comput. **6**(2), 182–197 (2002)
10. Avigad, G., Branke, J.: Embedded evolutionary multi-objective optimization for worst case robustness. In: Proceedings of the 10th Annual Conference on Genetic and Evolutionary Computation, pp. 617–624 (2008)
11. Zitzler, E., Thiele, L., Laumanns, M., Fonseca, C.M., Da Fonseca, V.G.: Performance assessment of multiobjective optimizers: an analysis and review. IEEE Trans. Evol. Comput. **7**(2), 117–132 (2003)
12. Eisenstadt, E., Moshaiov, A., Avigad, G.: Testing and comparing multi-objective evolutionary algorithms for multi-payoff games. https://www.eng.tau.ac.il/~moshaiov/TEC.pdf

A Parallel Multi-objective Memetic Algorithm Based on the IGD+ Indicator

Edgar Manoatl Lopez(✉) and Carlos A. Coello Coello

Departamento de Computación,
CINVESTAV-IPN (Evolutionary Computation Group),
07300 Mexico D.F., Mexico
emanoatl@computacion.cs.cinvestav.mx,
ccoello@cs.cinvestav.mx

Abstract. The success of local search techniques in the solution of combinatorial optimization problems has motivated their incorporation into multi-objective evolutionary algorithms, giving rise to the so-called multi-objective memetic algorithms (MOMAs). The main advantage for adopting this sort of hybridization is to speed up convergence to the Pareto front. However, the use of MOMAs introduces new issues, such as how to select the solutions to which the local search will be applied and for how long to run the local search engine (the use of such a local search engine has an extra computational cost). Here, we propose a new MOMA which switches between a hypervolume-based global optimizer and an IGD+-based local search engine. Our proposed local search engine adopts a novel clustering technique based on the IGD+ indicator for splitting the objective space into sub-regions. Since both computing the hypervolume and applying a local search engine are very costly procedures, we propose a GPU-based parallelization of our algorithm. Our preliminary results indicate that our MOMA is able to converge faster than SMS-EMOA to the true Pareto front of multi-objective problems having different degrees of difficulty.

1 Introduction

Most practical real-world problems have several objectives (these objectives are often in conflict) which need to be optimized at the same time. They are called Multi-objetive Optimization Problems (MOPs). Contrary to a Single-objective Optimization Problem (SOP), a MOP does not result in a single optimal solution. Instead, it results in a set of solutions which represent the best trade-offs among all the objectives. These solutions are known as *Pareto optimal* and their image is called the *Pareto Optimal Front* (POF). Most researchers are interested in finding Pareto fronts, which have a good (e.g., uniform) distribution.

E. Manoatl Lopez—Author acknowledges support from CONACyT and CINVESTAV-IPN to pursue graduate studies in Computer Science.
C.A. Coello Coello—Author gratefully acknowledges support from CONACyT project no. 221551 and from a Cátedra Marcos Moshinsky 2014 in Mathematics.

J. Handl et al. (Eds.): PPSN XIV 2016, LNCS 9921, pp. 473–482, 2016.
DOI: 10.1007/978-3-319-45823-6_44

There are several methods for solving MOPs such as Memetic Algorithms (MAs) which combine a global optimizer (e.g., an evolutionary algorithm) with a Local Search engine (LS). Recently, MAs have shown to efficiently solve MOPs (see for example [16, 17]). In general, LS techniques use decision variable space neighborhoods whose selected points generate vectors in the objective function space.

It is worth mentioning that combining a global optimizer with a local search technique for specific MOPs is critical to achieve good results, if the fitness function computation in real-world MOPs takes a considerable amount of running time. Likewise, there exist computational trade-offs between local and global search. Thus, some researchers, such as [4], have raised some specific questions related to the effectiveness and efficiency of local search engines:

- How often should the LS be applied based upon a probability, PLS?
- On which k solutions should LS be used given a neighborhood $N(x)$ where x is a current solution?
- How long should LS be run defined by a time period T?
- How efficient does LS need to be versus its effectiveness?

These questions involve some difficulties for designing new multi-objective memetic algorithms (MOMAs).

Here, we propose a new MOMA which uses a LS technique based on the modified inverted generational distance (IGD+) (this indicator was recently proposed by Ishibuchi [12, 13]) combined with a hypervolume-based global optimizer [3]. We want to combine different properties of each indicator for improving the performance of the overall MOMA. This is possible, since these indicators have nice properties (i.e., hypervolume is Pareto compliant and IGD+ is weakly Pareto compliant). However, the main drawback of the hypervolume is the high computational cost associated with its computation. So, this limits the use of this indicator, particularly in problems having many objectives. On the other hand, IGD+ has a very low computational cost, even in high dimensional problems. In spite of the fact that this hybridization is possible, there are still some drawbacks which limit the use of this type of combination, since computing the exact hypervolume contribution is highly costly.

Nowadays, this sort of limitations can be addressed by using massive parallel processors such as a Graphic Processing Unit (GPU). There is plenty of evidence that indicates that GPU-based approaches can reduce the running time without losing the advantages of CPU-based approaches (for more details see [2, 14, 18]). For this reason, we develop here a parallel implementation of our MOMA and illustrate its performance when using both indicators (hypervolume and IGD+).

The remainder of this paper is organized as follows. Section 2 provides some basic concepts related to multi-objective optimization. Our MOMA is described in Sect. 3. Section 4, presents our methodology and a brief discussion of our preliminary results. Finally, conclusions and some possible paths for future research are provided in Sect. 5.

2 Basic Concepts

We are interested in solving problems of the type:

$$\text{minimize} \quad \boldsymbol{f}(\boldsymbol{x}) := [f_1(\boldsymbol{x}), f_2(\boldsymbol{x}), \ldots, f_m(\boldsymbol{x})]^T \tag{1}$$

subject to:

$$g_i(\boldsymbol{x}) \leq 0, \quad i = 1, 2, \ldots, p \tag{2}$$

$$h_j(\boldsymbol{x}) = 0, \quad j = 1, 2, \ldots, q \tag{3}$$

where $\boldsymbol{x} = [x_1, x_2, \ldots, x_n]$ is the vector of decision variables, $f_i : \mathbb{R}^n \rightarrow \mathbb{R}$, $i = 1, \ldots, m$ are the objective functions and $g_i, h_j : \mathbb{R}^n \rightarrow \mathbb{R}$, $i = 0, \ldots, p$, $j = 1, \ldots, q$ are the constraint functions of the problem.

In order to describe our LS technique, we have to provide more details about the IGD+ indicator before presenting our proposed algorithm. According to [13], the IGD+ indicator can be described as follows:

$$IGD^{+}(\mathcal{A}, \mathcal{Z}) = \frac{1}{|\mathcal{Z}|} \left(\sum_{j=1}^{|\mathcal{Z}|} d_j^{+}(\boldsymbol{z}, \boldsymbol{a})^p \right)^{1/p} \tag{4}$$

where $\boldsymbol{a} \in \mathcal{A} \subset \mathbb{R}^m$, $\boldsymbol{z} \in \mathcal{Z} \subset \mathbb{R}^m$, $\mathcal{A}$ is the Pareto front set approximation and $\mathcal{Z}$ is the reference set. $d^{+}(\boldsymbol{a}, \boldsymbol{z})$ is defined as:

$$d^{+}(\boldsymbol{z}, \boldsymbol{a}) = \sqrt{(\max\{a_1 - z_1, 0\})^2, \ldots, (\max\{a_m - z_m, 0\})^2}. \tag{5}$$

Therefore, we can see that the set $\mathcal{A}$ represents a better approximation to the real $\mathcal{PF}$ when we obtain a lower $IGD+$ value, if we consider the reference set as $\mathcal{PF}_{True}$.

3 Our Proposed Multi-objective Memetic Algorithm

3.1 Global Optimizer

Our MOMA consists of two different approaches. The first one is a global optimizer which is based on SMS-EMOA [3]. The second method is our local search technique which uses an IGD+-based search technique. The global optimizer starts with an initial population of N individuals. Then, a new individual is created through the use of evolutionary operators. This new individual will become a member of the next population, if replacing an existing individual leads to a higher quality of the population with respect to the hypervolume contribution. Afterwards, one individual is discarded from the worst ranked front in order to maintain the same population size. If the cardinality of this front is larger than 1, the individual which minimizes the hypervolume contribution is eliminated. The LS technique is launched when a certain percentage of the total number of generations is reached. Next, we will provide more details of the way in which our LS works.

3.2 Local Search Engine

We focused on how to select the kth solution to which LS should be applied. A straightforward solution is to apply LS to all the individuals in the population. Although this involves a higher computational cost, in our case, this sort of scheme is possible because of our GPU-based implementation. Thus, our proposal is to apply several local search engines on different regions of the search space, which are specified by a clustering technique, based on the IGD+ indicator. It is worth noting that this indicator requires a reference set $\mathcal{Z}$. Our proposed approach creates different neighborhoods for each point in the reference set. The ith neighborhood is created by N points from the population. Such points have the nearest distance with respect to the ith reference point in terms of the $d+$ distance (see Eq. (5)). Our LS technique starts with a population $\mathcal{P}$ which contains N individuals obtained by our global search engine. The new i^{th} offspring is created by choosing three different parents from its neighborhood. The parents are recombined using the differential evolution operator, where the first parent is selected by the nearest distance in terms of the $d+$ distance and the rest of the parents are randomly chosen. The second step is to combine the parents and the offspring of each neighborhood to form the so-called Q set. The new population at generation $t+1$ is generated by finding the nearest point from Q for each z reference point in $\mathcal{Z}$. This process is repeated until the stopping criterion is satisfied (we use the maximum number of iterations).

3.3 Reference Set

We can approximate the geometrical shape of certain types of Pareto Fronts (PFs) using superspheres. A γ-supersphere is a type of curve which is described as follows:

$$\{(y_1, \dots, y_m) \in \mathbb{R}^m_+ | \quad y_1^\gamma + \cdots + y_m^\gamma = 1\} \tag{6}$$

where $\gamma \in \mathbb{R}_+$ is an arbitrary and fixed value. We only consider the "positive" parts of the γ-superspheres. According to [8], we can view the positive parts of the γ-superspheres as concave if $\gamma > 1$ or as convex if $0 < \gamma < 1$. Clearly, we can see that a set of weight vectors satisfies Eq. (6) when $\gamma = 1$, since a weight vector is defined as:

Definition 1 *Let* $\boldsymbol{w} = [w_1, \dots, w_m] \in \mathcal{R}^m$. *We say that* $\boldsymbol{w}$ *is a weight vector if* $\sum_{j=1}^m w_j = 1$ *and* $w_j \geq 0$.

In order to build the reference set, we assume that we have a set of weight vectors which is used to construct the reference set. We need to find the γ-value which will be used to transform the weights set into the reference set. Clearly, in order to find the γ-value, Eq. (6) would become a root-finding problem and we can say that the γ-value needs to satisfy:

$$y_1^\gamma + \cdots + y_m^\gamma - 1 = 0 \tag{7}$$

For solving Eq. (7), we use Newton's method for approximating the γ-value. Now, we can see that the next approximation to the root is defined as:

$$\gamma_{k+1} = \gamma_k - \frac{(\sum_{j=1}^{m} y_j^{\gamma_k}) - 1}{\sum_{j=1}^{m} y_j^{\gamma_k} \log(y_j)} \tag{8}$$

Let $\mathcal{Q}$ be the current set which was created combining the parent and offspring population. Thus, the reference set is created by Algorithm 1.

Algorithm 1. Computation of the reference set which is based on supersphere curves

Require: A current set $\mathcal{Q} \subset \mathbb{R}^m$, a set of weighted vectors $W \subset \mathbb{R}^m$, where m is the number of objectives, expand value $e \subset \mathbb{R}$ and translate value $t \subset \mathbb{R}$
Ensure: The reference set $\mathcal{Z}$ which is the best approximation of the set Q
1: Find the nondominated points from $\mathcal{Q}$ and save to $\mathcal{Q}'$
2: **for** each $\boldsymbol{p} \in \mathcal{Q}'$ **do**
3:     **for** each $\boldsymbol{w} \in \mathcal{W}$ **do**
4:         Compute $d^{\perp}(\boldsymbol{p}, \boldsymbol{w}) = \| \boldsymbol{p} - \boldsymbol{w}^T \boldsymbol{p} \boldsymbol{w} / \| \boldsymbol{w} \|^2 \|$
5:     **end for**
6:     Assign $r(w) = \operatorname{argmin}_{\boldsymbol{p} \in \mathcal{Q}'} \; d^{\perp}(\boldsymbol{p}, \boldsymbol{w})$
7: **end for**
8: $j \leftarrow 0$
9: **for** each $\boldsymbol{w} \in \mathcal{W}$ **do**
10:     $stepsize \leftarrow \boldsymbol{p}_{r(\boldsymbol{w})} \cdot \boldsymbol{w} / \| \boldsymbol{w} \|^2$
11:     $\boldsymbol{y} \leftarrow stepsize * \boldsymbol{w}$
12:     Approximate the γ value using equation (7)
13:     Compute the supersphere point as $z_{j,k} \leftarrow e(w_{j,k}^{\gamma}) - t$ for all $j = 1, \ldots, m$
14:     $j \leftarrow j + 1$
15: **end for**

In the first step of the algorithm, we find the non-dominated points from set $\mathcal{Q}$ which will establish the non-dominated region. After that, in the first loop, we search the nearest perpendicular distance between each weighted vector $\boldsymbol{w}$ and the non-dominated points (we find the best relationship between each weighted vector and each non-dominated point). In order to construct the reference surface, we project the nearest non-dominated point to a specific weighted vector $\boldsymbol{w}$. Once this is done, we can search the γ-value using Newton's method, which is described by Eq. (8). Finally, the reference point is computed using the γ-value. After that, we apply the expand and translate operations. These operations transform the surface for spreading the reference set along of objective space. We can see that this process is considered as a generation and is repeated for each weighted vector. For generating the weighted vectors, we adopted Das and Dennis' approach [5] and the number of weighted vectors was set to N.

3.4 GPU Implementation

The main idea of our parallel implementation is to use all the available hardware resources for improving the performance of our proposed MOMA. For this reason, in order to simplify the parallelization we focused only on the most time-consuming parts of the algorithm. Our implementation is based on two different parallel implementations, one for handling the local search technique, and another one, which is responsible of computing the hypervolume contribution from the global search engine. As we indicated in Sect. 3.2, the LS procedure is composed by a clustering technique, a procedure for generating new offspring as well as the evaluation of the objective functions. This process is repeated for a certain number of iterations. The main idea is to apply the LS technique to all the individuals of the population. For this reason, the parallelization of this procedure is done in the following way: we adopt a SIMD[1] model to apply the clustering technique to create each sub-region on the objective space at the same time. Thus, this procedure creates different blocks of threads, where each thread computes the $d+$ value (see Eq. (5)) between each reference point in $\mathcal{Z}$ and each current point in the population $\mathcal{Q}$. After that, each block searches the nearest distance (this process is repeated until having b elements for building the clustering region). After this is done, we create m offspring, each of them residing in a specific sub-region (the ith neighborhood) using a thread of the GPU for each of them. Thus, each thread in the block can assess the new offspring in the neighborhood. It is worth mentioning that this process needs to normalize all points for each generation of the local search technique, in order to handle objectives having different units.

In [14], the use of a GPU-based approach showed that it is possible to find a good approximation of MOPs using the hypervolume indicator as a selection mechanism without losing the advantages of a sequential approach. For this reason, we adopted this approach for implementing the second part of our MOMA.[2]

4 Experimental Results

We compare the performance of our memetic algorithm with respect to SMS-EMOA which has two different variants. The first version uses exact calculation of the hypervolume contribution for each generation of the search process. The second version incorporates the algorithm proposed in [1] for estimating the hypervolume using Monte Carlo sampling, instead of the exact hypervolume calculations adopted in the original implementation of SMS-EMOA. Our MOMA

[1] SIMD (Single Instruction Multiple Data) is a computer architecture which can handle only one instruction but applies it to many data streams simultaneously [9].

[2] The GPU-based approach computes in a faster way the hypervolume contribution of a point.

was compared with respect to its GPU-based implementation. Our proposed approach was implemented in CUDA-C.[3]

4.1 Test Problems

For our comparative study, we adopted two benchmarks: (1) the Deb-Thiele-Laumanns-Zitzler (DTLZ) test suite [7] and (2) the Walking-Fish-Group (WFG) test suite [10,11]. These problems include different aspects which make them more difficult to solve (for more details see [7,11]).

4.2 Methodology

For our comparative study, we decided to adopt the hypervolume indicator, which assesses both convergence and maximum spread along the Pareto front. Mathematically, if Λ denotes the Lebesgue measure, the hypervolume can be described as:

$$I_H(\mathcal{A}, \boldsymbol{y}_{ref}) = \Lambda\left(\bigcup_{\boldsymbol{y}\in\mathcal{A}} \{\boldsymbol{x} | \quad \boldsymbol{y} \prec \boldsymbol{x} \prec \boldsymbol{y}_{ref}\}\right) \tag{9}$$

where $\mathcal{A}$ is the approximation of the Pareto front optimal set and $\boldsymbol{y}_{ref} \in \mathbb{R}^k$ denotes the reference point. In order to compute I_H, we use different reference points for each test suite, which were set to $(1, \ldots, 1)$ for DTLZ1, $(2, \ldots, 2)$ for DTLZ2 to DTLZ6, $(2, \ldots, 2, 7)$ for DTLZ7 and $(3, 5, \ldots, 2m + 1)$ for the WFG test problems. Additionally, we also compared the running time of each MOEA, which was measured in minutes.

4.3 Parameterization

For the DTLZ test suite, the total number of decision variables is given by $n = m + k - 1$, where m is the number of objectives and k was set to 5 for DTLZ1, to 10 for DTLZ2 to DTLZ6 and to 20 for DTLZ7. The number of decision variables in the WFG test problems was set to 24, and the position-related parameter was set to $m - 1$. Instances with two and three objectives were adopted.

The parameters of each MOEA used in our study were chosen in such a way that we could do a fair comparison among them. The distribution indexes for the SBX and polynomial-based mutation operators [6] were set as: $\eta_c = 20$ and $\eta_m = 20$, respectively. The crossover probability was set to $p_c = 0.9$ and the mutation probability was set to $p_m = 1/L$, where L is the number of decision variables. In the SMS-EMOA-HyPE, the number of samples was set to 50,000.

[3] The GPU platform and API developed by Nvidia called CUDA [15] (Computer Unified Device Architecture), which is the one adopted in this work, is based on the CUDA-C language, which is an extension of C that allows the development of GPU routines called *kernels*. Each kernel defines instructions that are executed on the GPU by many threads at the same time.

The number of generations of the LS technique was set to 50 for the DTLZ test problems and to 80 for the WFG test problems, where each generation the LS is applied for each reference point. The control parameter F was set to 0.5 for the differential evolution operator. The total number of function evaluations was set in such a way that it did not exceed 30,000 for the DTLZ test problems and 50,000 for the WFG test suite. All the implementations were tested on the same computer which has the following characteristics: An Intel Core i7-3930k CPU running at 3.20 GHz, with 8 GB of RAM 1600 MHz DDR3. Our GPU was a Geforce GTX 680, and we ran our experiments in Fedora 18 (64-bit version).

4.4 Discussion of Results

Table 1 provides the average hypervolume over the 30 independent executions of each approach for each test suite. Additionally, we show the average time, which was measured in minutes, needed to perform the maximum number of function evaluations in each case and the speed up achieved (in parentheses). The best results are presented in **boldface**.

Table 1. Comparison of results for each test suite, using the average hypervolume indicator.

	Test Suite 1						Test Suite 2				
Problem	m	SMS-EMOA	SMS-EMOA-HYPE	IGD+-MA	IGD+-MA(GPU)	Problem	m	SMS-EMOA	SMS-EMOA-HYPE	IGD+-MA	IGD+-MA(GPU)
DTLZ1	2	**0.8732805**	0.8725481	0.8726563	0.8709863	WFG1	2	7.0150395	6.6915866	**7.4293168**	7.3778004
	3	**0.974249**	0.9666142	0.9737887	0.9731913		3	**62.566032**	53.1739159	62.4667318	62.4431766
DTLZ2	2	3.2109678	3.2095071	**3.2109715**	3.2109601	WFG2	2	11.4297746	11.4126833	**11.4304887**	11.429895
	3	**7.4313536**	7.4260692	7.4312298	7.4312795		3	100.9053244	100.3897934	**100.9423556**	100.8915152
DTLZ3	2	1.9302655	2.7552999	2.8205047	**2.8444797**	WFG3	2	10.9301202	10.8957265	**10.9346344**	10.9320503
	3	6.8129142	5.0821156	**7.0065842**	6.9233977		3	76.0218553	74.3412533	76.0301204	**76.0650755**
DTLZ4	2	**2.9687737**	2.8466417	2.9082368	2.9478005	WFG4	2	**8.6759874**	8.6474796	8.6749735	8.6751788
	3	6.927273	6.9031298	6.9659859	**7.0188906**		3	**77.3490714**	76.1356581	77.2287144	77.236047
DTLZ5	2	3.2109635	3.2095599	3.2109646	**3.210965**	WFG5	2	8.2444335	8.2422967	**8.2653013**	8.2702657
	3	**6.1052922**	6.1009119	6.1050065	6.1050063		3	**74.1569177**	73.3959324	74.1328251	74.1289772
DTLZ6	2	**3.0727714**	3.0898	2.9075032	2.8973674	WFG6	2	**8.3786401**	8.3522619	8.3785062	8.3762176
	3	**5.6964296**	5.2659912	5.2550039	5.2817297		3	74.5010368	73.4821868	74.5165829	**74.6646198**
DTLZ7	2	**4.4180206**	4.3527739	4.4174787	4.417529	WFG7	2	8.685331	8.6549507	**8.6863782**	8.6863691
	3	**12.8437627**	12.7603802	7.878233	7.5641662		3	**77.6304566**	76.4916201	77.5775899	77.5752613
						WFG8	2	8.3184115	8.2791368	**8.3251368**	8.3208976
							3	**73.6151505**	72.5266533	73.5156236	73.5167815
						WFG9	2	**8.5957132**	8.4786182	8.5693595	8.5555043
							3	76.279433	73.9882086	76.3432385	**76.3733424**

It is clear that the winner in this experimental study is our GPU-based MOMA in terms of CPU time. We are also able to obtain the same results as the sequential version, which verifies that our parallel implementation is working as expected (see Table 2). We can see that our MOMA is able to converge faster than SMS-EMOA on some test problems (e.g., in the multi-frontal problems) and it outperforms SMS-EMOA-HYPE in all instances. This confirms that our proposed IGD+-based LS is an effective way to solve MOPs. It is worth noting, however, that for DTLZ5, DTLZ6 and DTLZ7, SMS-EMOA performs better than our MOMA. The reason is probably that the true Pareto front of these problems is linear and disconnected, which makes the approximations produced by our approach to converge to a single region of the search space.

Table 2. Computational time (measured in minutes) required by each execution of the MOEAs compared. In the parentheses show the speed up.

	Test Suite 1						Test Suite 2					
Problem	m	SMS-EMOA	SMS-EMOA-HYPE	IGD+-MA	IGD+-MA(GPU)	Problem	m	SMS-EMOA	SMS-EMOA-HYPE	IGD+-MA	IGD+-MA(GPU)
DTLZ1	2	0.2353 (2.22x)	1.1378 (10.74x)	0.1571 (1.48x)	**0.1059**	WFG1	2	0.4365 (2.03x)	1.9709 (9.17x)	0.3661 (1.7x)	**0.2149**
	3	1.3434 (2.54x)	0.7481 (1.41x)	0.9396 (1.78x)	**0.5290**		3	6.1682 (3.25x)	17.4731 (9.2x)	4.5742 (2.41x)	**1.8995**
DTLZ2	2	0.3145 (2.36x)	5.1541 (38.69x)	0.2720 (2.04x)	**0.1332**	WFG2	2	0.6315 (2.64x)	3.8164 (15.94x)	0.4441 (1.86x)	**0.2393**
	3	3.5239 (2.26x)	14.3901 (9.21x)	2.9234 (1.87x)	**1.5616**		3	7.1067 (3.46x)	4.9679 (2.42x)	5.1835 (2.52x)	**2.0555**
DTLZ3	2	0.1349 (1.33x)	0.2886 (2.85x)	0.1488 (1.47x)	**0.1014**	WFG3	2	0.6760 (2.51x)	4.9836 (18.51x)	0.6065 (2.25x)	**0.2692**
	3	1.2725 (1.95x)	1.2460 (1.91x)	0.8001 (1.22x)	**0.6534**		3	6.4021 (2.75x)	17.4964 (7.53x)	5.7263 (2.46x)	**2.3238**
DTLZ4	2	0.2803 (2.11x)	3.6041 (27.11x)	0.2221 (1.67x)	**0.1329**	WFG4	2	0.7430 (2.78x)	7.4647 (27.92x)	0.5865 (2.19x)	**0.2673**
	3	2.9593 (2.67x)	10.6125 (9.59x)	2.0486 (1.85x)	**1.1070**		3	8.5252 (3.22x)	13.9963 (5.29x)	5.8787 (2.22x)	**2.6452**
DTLZ5	2	0.3151 (2.38x)	5.1250 (38.66x)	0.2710 (2.04x)	**0.1325**	WFG5	2	0.7245 (2.49x)	9.0330 (31.02x)	0.7122 (2.45x)	**0.2912**
	3	2.2238 (2.4x)	10.8695 (11.71x)	1.4279 (1.54x)	**0.9283**		3	8.1814 (3.15x)	14.4043 (5.55x)	5.9147 (2.28x)	**2.5967**
DTLZ6	2	0.1501 (1.56x)	0.6505 (6.76x)	0.1114 (1.16x)	**0.0962**	WFG6	2	0.5864 (2.31x)	6.1397 (24.17x)	0.5381 (2.12x)	**0.2540**
	3	1.7579 (2.51x)	4.2780 (6.12x)	1.2497 (1.79x)	**0.6993**		3	6.0747 (2.8x)	12.14762 (5.6x)	5.1019 (2.35x)	**2.1701**
DTLZ7	2	0.3111 (2.66x)	3.5508 (30.4x)	0.2026 (1.74x)	**0.1167**	WFG7	2	1.1146 (3.32x)	12.5224 (37.31x)	0.8211 (2.45x)	**0.3356**
	3	2.8511 (3.31x)	11.2552 (13.07x)	1.6282 (1.89x)	**0.8611**		3	8.4301 (2.53x)	19.5875 (5.89x)	7.6072 (2.29x)	**3.3255**
						WFG8	2	0.5485 (2.43x)	3.9599 (17.56x)	0.4466 (1.98x)	**0.2255**
							3	4.6612 (2.69x)	8.9829 (5.18x)	4.6498 (2.68x)	**1.7358**
						WFG9	2	0.9392 (2.91x)	10.4772 (32.51x)	0.7952 (2.47x)	**0.3222**
							3	8.8769 (2.67x)	18.8293 (5.67x)	7.8688 (2.37x)	**3.3232**

5 Conclusions and Future Work

We have proposed a new Multi-Objective Memetic Algorithm which has an IGD+-based local search engine. The core idea of our proposed algorithm is to combine properties of two different performance indicators. Our proposal includes a GPU-based implementation which makes it possible to launch multiple local search processes at the same time. Our preliminary results indicate that it is possible to improve the convergence of a hypervolume-based approach in multi-frontal problems.

Our proposed GPU-based multi-objective memetic algorithm is able to achieve a significant speed up (of up to 38x) with respect to SMS-EMOA. As part of our future work, we would like to improve the method for building the reference set, which is used for computing the IGD+ value, since it has a few drawbacks on some test problems. Additionally, we would like to test our approach in many-objective problems.

References

1. Bader, J., Zitzler, E.: HypE: An Algorithm for Fast Hypervolume-Based Many-Objective Optimization. Evolutionary Computation, 19(1): 45–76, Spring, 2011
2. de Oliveira, F.B., Davendra, D., Guimarães, F.G.: Multi-objective differential evolution on the GPU with C-CUDA. In: Snášel, V., Abraham, A., Corchado, E.S. (eds.) SOCO 2012. AISC, vol. 188, pp. 123–132. Springer, Heidelberg (2013)
3. Beume, N., Naujoks, B., Emmerich, M.: SMS-EMOA: multiobjective selection based on dominated hypervolume. Eur. J. Oper. Res. **181**(3), 1653–1669 (2007)
4. Coello Coello, C.A., Lamont, G.B., Van Veldhuizen, D.A.: Evolutionary Algorithms for Solving Multi-Objective Problems, 2nd edn. Springer, New York (2007). ISBN 978-0-387-33254-3
5. Das, I., Dennis, J.E.: Normal-boundary intersection: a new method for generating the pareto surface in nonlinear multicriteria optimization problems. SIAM J. Optim. **8**(3), 631–657 (1998)

6. Deb, K., Pratap, A., Agarwal, S., Meyarivan, T.: A fast and elitist multiobjective genetic algorithm: NSGA-II. IEEE Trans. Evol. Comput. **6**(2), 182–197 (2002)
7. Deb, K., Thiele, L., Laumanns, M., Zitzler, E.: Scalable test problems for evolutionary multiobjective optimization. In: Abraham, A., Jain, L., Goldberg, R. (eds.) Evolutionary Multiobjective Optimization. Theoretical Advances and Applications, pp. 105–145. Springer, New York (2005)
8. Emmerich, M.T.M., Deutz, A.H.: Test problems based on Lamé superspheres. In: Obayashi, S., Deb, K., Poloni, C., Hiroyasu, T., Murata, T. (eds.) EMO 2007. LNCS, vol. 4403, pp. 922–936. Springer, Heidelberg (2007)
9. Flynn, M.J.: Some computer organizations and their effectiveness. IEEE Trans. Comput. **21**(9), 948–960 (1972)
10. Huband, S., Barone, L., While, L., Hingston, P.: A scalable multi-objective test problem toolkit. In: Coello Coello, C.A., Hernández Aguirre, A., Zitzler, E. (eds.) EMO 2005. LNCS, vol. 3410, pp. 280–295. Springer, Heidelberg (2005)
11. Huband, S., Hingston, P., Barone, L., While, L.: A review of multiobjective test problems and a scalable test problem toolkit. IEEE Trans. Evol. Comput. **10**(5), 477–506 (2006)
12. Ishibuchi, H., Masuda, H., Nojima, Y.: A study on performance evaluation ability of a modified inverted generational distance indicator. In: 2015 Genetic and Evolutionary Computation Conference (GECCO 2015), 11–15 July 2015, Madrid, Spain, pp. 695–702. ACM Press (2015). ISBN 978-1-4503-3472-3
13. Ishibuchi, H., Masuda, H., Tanigaki, Y., Nojima, Y.: Modified distance calculation in generational distance and inverted generational distance. In: Gaspar-Cunha, A., Henggeler Antunes, C., Coello, C.C. (eds.) EMO 2015. LNCS, vol. 9019, pp. 110–125. Springer, Heidelberg (2015)
14. Lopez, E.M., Antonio, L.M., Coello Coello, C.A.: A GPU-based algorithm for a faster hypervolume contribution computation. In: Gaspar-Cunha, A., Henggeler Antunes, C., Coello, C.C. (eds.) EMO 2015. LNCS, vol. 9019, pp. 80–94. Springer, Heidelberg (2015)
15. NVIDIA Corporation. Cuda zone (2014)
16. Pilát, M., Neruda, R.: Hypervolume-based local search in multi-objective evolutionary optimization. In: 2014 Genetic and Evolutionary Computation Conference (GECCO 2014), 12–16 July 2014, Vancouver, Canada, pp. 637–644. ACM Press (2014). ISBN 978-1-4503-2662-9
17. Tan, Y.-Y., Jiao, Y.-C., Li, H., Wang, X.-K.: MOEA/D-SQA: a multi-objective memetic algorithm based on decomposition. Eng. Optim. **44**(9), 1095–1115 (2012)
18. Wong, M.L., Cui, G.: Data mining using parallel multi-objective evolutionary algorithms on graphics processing units. In: Tsutsui, S., Collet, P. (eds.) Massively Parallel Evolutionary Computation on GPGPUs, pp. 287–307. Springer, Heidelberg (2013). ISBN 978-3-642-37958-1

Towards Automatic Testing of Reference Point Based Interactive Methods

Vesa Ojalehto[1(✉)], Dmitry Podkopaev[2], and Kaisa Miettinen[1]

[1] University of Jyvaskyla, Department of Mathematical Information Technology, P.O. Box 35 (Agora), FI-40014 University of Jyvaskyla, Finland
{vesa.ojalehto,kaisa.miettinen}@jyu.fi

[2] Systems Research Institute, Polish Academy of Sciences, Newelska 6, 01-447 Warsaw, Poland
dmitry.podkopaev@ibspan.waw.pl

Abstract. In order to understand strengths and weaknesses of optimization algorithms, it is important to have access to different types of test problems, well defined performance indicators and analysis tools. Such tools are widely available for testing evolutionary multiobjective optimization algorithms.

To our knowledge, there do not exist tools for analyzing the performance of interactive multiobjective optimization methods based on the reference point approach to communicating preference information. The main barrier to such tools is the involvement of human decision makers into interactive solution processes, which makes the performance of interactive methods dependent on the performance of humans using them. In this research, we aim towards a testing framework where the human decision maker is replaced with an artificial one and which allows to repetitively test interactive methods in a controlled environment.

Keywords: Multiobjective optimization · EMO · Testing framework · Decision maker's preferences · Preference information · Aspiration level

1 Introduction

Many real-life problems of decision making and support are tackled by multiobjective optimization. A solution of a multiobjective optimization problem can be defined as a feasible solution which is the most preferred for a decision maker (DM). Therefore, multiobjective optimization methods that aim at supporting a DM rely on information about the DM's preferences (*preference information* for short) and incorporate mechanisms of communication with the DM. In the methods where such communication is organized in an interactive way (i.e. *interactive methods*), the solution process is carried out in iterations. In each iteration, the DM provides preference information and, as feedback, obtains information about Pareto optimal solutions derived based on this preference information [2,8,9]. Interactive methods are very suitable for solving practical problems due

J. Handl et al. (Eds.): PPSN XIV 2016, LNCS 9921, pp. 483–492, 2016.
DOI: 10.1007/978-3-319-45823-6_45

to several advantages [2]. First, the DM gets the possibility to learn progressively about the set of Pareto optimal solutions of a complex problem, which reduces cognitive load. Secondly, applying interactive methods does not necessitate generating many Pareto optimal solutions, which is essential in the case of computationally complex problems. Instead, only solutions that are interesting for the DM are generated.

Many interactive methods have been developed so far, see e.g. [2,8,9]. Naturally, the problem of testing and comparing different methods arises [7,8]. Making tests and comparisons of interactive multiobjective optimization methods is hampered by the necessity of involving DMs in tests. First of all, this involvement makes method testing much more costly than testing by computational means, taking into account that many problems of industry, management, engineering, etc. require DMs being experts in corresponding fields. Secondly, it is hard to conduct good quality experiments due to various barriers related to human nature: the difficulty of creating proper motivation of DMs if tested e.g. by students using artificial problems; the inconsistency of human nature and variability among humans; difficulties of accounting for improving DM's capabilities in time due to learning[1].

As noted in [7,8], only few interactive multiobjective optimization methods have been extensively tested, which means that information about the quality of most of the methods cannot be called reliable. The main sources of such information are intuitive conclusions of the authors of the methods and results of employing the methods for solving limited numbers of real-world or hypothetical problems. In order to overcome the deficiency of tests and comparisons of interactive methods, one can use artificial DMs understood as techniques of generating preference information. Because interactive methods vary significantly in approaches to preference information modeling [9,10], different artificial DMs should be created for different preference information types.

Compared to the diversity of interactive methods, the number of approaches to creating artificial DMs is very limited. In [8], some examples of testing methods by using artificial DMs were described. Since 1999, only few new works have appeared where actions of DMs have been simulated using artificial mechanisms. Among them, a DM was represented as an additive value function in [13], and that representation was used for generating goals in a simulated goal programming problem with a discrete number of alternatives. When generating goals, judgment errors and biases of the DM were simulated and then effects on the performance of goal programming algorithms were studied. In [15], a universal mechanism of generating DM's preference information was proposed based on minimizing the distance of the corresponding Pareto optimal solution to a given "goal solution". However, that mechanism has a limited application area. The work in [7] aimed at the same goal as our research, except that a DM was modelled via a value function, which does not allow generating reference point, but

[1] Humans learn, therefore, it is not easy to employ the same DMs to test different methods, as they have learnt about the problem while solving the problem, which affects the quality of a long series of experiments.

provides preference information as rankings of given sets of alternatives. Some mechanisms of modeling the imperfection of humans' judgments was incorporated into that model and used for testing the BC-EMOA algorithm [7].

More approaches to creating artificial DMs have been developed for enhancing existing methods, and they may be adoptable for method testing. The approach in [1] is an example of such a study (see also references therein), where the DM's preference model based on a fuzzy inference system was trained during the interactive solution process and used for providing additional preference information on behalf of the DM.

Clearly, each artificial DM created for testing methods should be tailored to the preference information expected by these methods. A popular way of modeling DM's preferences is via value functions (often referred to as utility functions). The advantages are a theoretically proved completeness [5], and the simplicity of representation. From a value function, one can easily obtain such preference information as pairwise comparisons or rankings of given sets of alternatives (as e.g. in [7]). Note that methods where the DM can be replaced by a value function are called non ad hoc methods [8,12,13]. However, in many interactive methods which are popular in practice, the preference information is provided in the form of reference points [2,9] representing desirable objective function values. Such methods are regarded as ad hoc, e.g. methods where the DM cannot be replaced by a value function [8,12,13]. To our knowledge, there are no artificial DMs developed for testing methods based on reference points.

In this paper, we develop an artificial DM for testing interactive methods, which involve preference information as a reference point. It is the first development of this kind. We mimic the behavior of a human DM who adjusts preferences based on obtained information about derived solutions, and demonstrates randomness in the behavior in responses to the uncertainty about the Pareto optimal set.

The paper is organized as follows. In Sect. 2, we describe the concept of an artificial DM and in Sect. 3 incorporate it into a framework for testing interactive methods. In Sect. 4, we present results of testing two methods: R-NSGA-II [4] and minimizing an achievement scalarizing function [14]. We conclude in Sect. 5.

2 Artificial Decision Maker

We propose to employ an artificial DM to replace the real DM. Our concept of an artificial DM and its interaction with an interactive method comprises the following three components:

- *Steady part*: the complexity of knowledge possessed by the DM and related to solving the considered class of problems which does not change during the solution process. This includes accumulated experience and the core preferences which do not change in time.
- *Current context*: the current situation as perceived by the DM, which may change in time. This includes: the knowledge about the problem accumulated

by the DM during the solution process, level of tiredness which can affect concentration, and the probability of making mistakes.
- *Preference information*: the method-specific information expressed by the DM during the solution process to guide the method toward solutions that are more preferred by the DM.

The artificial DM should be defined by the steady part which does not change in time, a mechanism of representing and updating the current context as the solution process continues, and the mechanism of generating the preference information based on the steady part and the current context. By varying the parameters of the steady part, one can obtain different artificial DMs for conducting multiple experiments.

It is tempting to describe the steady part as a classical model of DM's preferences (e.g. choice function, binary relation or utility function). However, as said, there are no studies describing how to generate preference information in terms of reference points from such models. Therefore, we construct the steady part in the form of some general preference information which cannot be called a preference model in the classical sense. We propose a procedure of generating the current preference information based on the steady preference information and taking into account the current context. The latter is represented by the current solution or the set of derived solutions available for the DM.

3 Testing Framework

The aim of this research is to create a framework for comparing different interactive methods with an artificial DM. The proposed framework is compatible with interactive methods where the DM provides one's preferences in each iteration of the method as a reference point. In what follows, we first give basic notions of multiobjective optimization, then describe the artificial DM used in the framework, and finally proceed with details on how the artificial DM is utilized.

Multiobjective optimization problems are formulated as follows:

$$\begin{aligned} \text{minimize} \quad & \mathbf{f}(\mathbf{x}) = (f_1(\mathbf{x}), \ldots, f_k(\mathbf{x}))^T \\ \text{subject to} \quad & \mathbf{x} = (x_1, \ldots, x_n)^T \in S, \end{aligned}$$

meaning that the DM wishes to simultaneously minimize k ($k \geq 2$) objective functions $f_i : S \to \mathbb{R}$ on the set S of *feasible solutions* (*decision vectors*) which is a nonempty compact subset of $\mathbb{R}^n$. The image of S is denoted by $\mathbf{f}(S)$. Its elements $\mathbf{z} = \mathbf{f}(\mathbf{x}) = (f_1(\mathbf{x}), \ldots, f_k(\mathbf{x}))^T$ in the objective space $\mathbb{R}^k$ consisting of objective (function) values are called *objective vectors*.

The set of Pareto optimal solutions of the problem (the *Pareto optimal set*) is defined by $E = \{\mathbf{x} \in S :$ there is no $\mathbf{x}' \in S$ such that $f_i(\mathbf{x}') \leq f_i(\mathbf{x}')$ for all $i = 1, \ldots, k$ and $\mathbf{f}(\mathbf{x}') \neq \mathbf{f}(\mathbf{x}')\}$.

Let us also introduce an *ideal objective vector* and a *nadir objective vector* defined, respectively, as $\mathbf{z}^{\star} = (z_1^{\star}, \ldots, z_k^{\star})^T$ where $z_i^{\star} = \min_{\mathbf{x} \in E} f_i(\mathbf{x})$ for $i = 1, \ldots, k$, and $\mathbf{z}^{\text{nad}} = (z_1^{\text{nad}}, \ldots, z_k^{\text{nad}})^T$ where $z_i^{\text{nad}} = \max_{\mathbf{x} \in E} f_i(\mathbf{x})$ for

$i = 1, \ldots, k$. Note that the nadir objective vector is, in general, more difficult to obtain than the ideal objective vector and, therefore, approximations are often used (see e.g. [3,8] and references therein).

The *current context* is defined as follows. We assume that the artificial DM is aware of the bounds of the objective functions, that is, the objective vectors $\mathbf{z}^\star$ and $\mathbf{z}^{\text{nad}}$. These vectors can either be known or estimated [8]. Their components give bounds on the aspiration levels which constitute a reference point. In addition, the set of derived Pareto optimal (or non-dominated) solutions, which is updated during the solution process, provides the DM with information about what combinations of objective function values are achievable.

As for *the steady part* of preference representation, for each objective f_i, $i = 1, \ldots, k$, we introduce a "ranking coefficient" w_i which determines the priority of the objective function f_i over the other objective functions. That is, the DM prefers more to obtain smaller values for those objective functions whose $w_i \in (0, 1]$ is larger. We assume that all objective functions are relevant to the problem and, therefore, each of them should have a ranking coefficient.

In addition to ranking coefficients, we utilize *initial aspiration levels*, $\text{asp}_i \in (z_i^\star, z_i^{\text{nad}}]$, $i = 1, \ldots, k$, that is, objective values the artificial DM would like to achieve. If $asp_i = z_i^{\text{nad}}$, we assume that the artificial DM initially does not have any preferences regarding objective f_i.

The probability $p \in (0, 1]$ determines how willing the artificial DM is to give up on the initial preferences. With larger p and larger w_i values, the artificial DM is more probable to consider the f_i objective relevant, i.e., to use asp_i as the reference point component (otherwise, it uses the component of the nadir vector). Alongside with the constant probability p, we introduce the varying probability p_λ which is initialized with p and decreased in the process of consecutive consideration of objective functions in the order defined by their priority (for details, see the scheme of the decision process below). Finally, the *preference information* is represented as a reference point $\text{ref} = (\text{ref}_1, \ldots, \text{ref}_k)$.

Now we can describe how the artificial DM interplays with a method. This process has the following parameters: θ – tolerance value which controls when an objective function value is considered to be acceptable; $t_{\max}$ – maximum number of iterations; t – iteration counter. Furthermore, we denote a uniformly distributed random number in the interval $[0, 1]$ by rand.

At the beginning of the solution process, when the set of derived solutions P is empty, we generate preference information as described below.

1. For each objective function f_i:
   (a) if asp_i is not defined, set $\text{asp}_i = z_i^\star$;
   (b) Set initial components of the reference point ref_i:
      i if $\text{rand} < p \cdot w_i$, set $\text{ref}_i = \text{asp}_i$,
      ii else, set $\text{ref}_i = z_i^{\text{nad}}$.

Here, if the aspiration level for an objective function is not defined, we set it to the ideal value of this objective in step (a), as each objective should have the opportunity to be improved, even if the artificial DM does not have a notion of what the values should be. Otherwise, the component of the initial reference point for each objective function is set at step (b) either as the aspiration level or the component of the nadir objective vector. The latter choice depends on the ranking coefficient of the objective function as well as the probability, if the aspiration level of the objective function should be used, in order to increase the priority of improving values of those objective functions.

After the artificial DM has been initialized with preference information, it is utilized with an interactive multiobjective optimization method as follows.

1. Set
   $p_\lambda = p$ – the varying probability of using the aspiration level as the component of the reference point,
   $F = \emptyset$ – the current index set of relevant objective functions,
   $t = 0$ – the iteration counter,
   P – the set of derived Pareto optimal (non-dominated) solutions,
   and generate initial preferences (reference point ref) as described above.
2. t++
3. Provide the interactive method with the current reference point ref to generate a new Pareto optimal solution x, and add this solution to the set P.
4. For each objective function f_i ordered by ranking coefficients w_i in a decreasing order:
   a. if $\text{asp}_i - f_i(x) < \theta$ and rand $< p$, add i to the set F,
      else, if rand $< p_\lambda$, add i to the set F.
   b. Set $p_\lambda = p_\lambda - \frac{p_\lambda}{i} \cdot |F|$.
5. If $|F| = k$, go to step 10.
6. For each objective function f_i with $i \in F$:
   a. Set the new component of the reference point ref_i:
      $\text{ref}_i = \text{asp}_i - (\text{asp}_i - f_i(x))/2$.
7. For each objective function f_l with $l \notin F$:
   a. Construct predict_l using a decision tree trained with previously obtained Pareto optimal solutions;
   b. set $\text{ref}_l = \min(\text{predict}_l, z_l^{\text{nad}})$.
8. If the new reference point is identical to the previous one, go to step 10.
9. If $t < t_{\max}$ go to step 2.
10. **STOP**. Select x as the solution to the problem.

In the beginning of the solution process, the interactive method is used to generate a new objective vector using the current reference point ref. Then we select and add to set F those objective functions which are considered to be relevant during this iteration. Firstly, if the objective vector value has achieved the desired aspiration level, it is selected with a high probability (but not equal to one), as we are assuming that the DM is not certain that the aspiration level is the best possible which could be achieved.

On the other hand, if the component of the objective vector has not achieved the desirable value, it is selected with a lower probability. This selection probability is then decreased based on the number of objective functions selected so far. The described scheme gives a strong preference on selecting most preferred objectives that have achieved desired values, while decreasing chances to select less desired objective functions. When all objective functions are selected to be relevant, the current Pareto optimal solution is considered to be the final solution of the problem. This means that it is possible that the artificial DM will end the solution process prematurely.

Next, in step 4. a new reference point ref is created. In order to take into account that the aspiration level might not be reachable, the reference point components for the selected objectives are set between the current objective function value and the aspiration level. Then the remaining reference point components are set either to the nadir value or to the value predicted by a decision tree [11] trained with previously obtained Pareto optimal solutions. A decision tree is built for each objective, using the values of the other objectives as a training data for predicting which values should be selected for other objectives in order to obtain the preferred value for the considered objective function.

The solution is finally accepted either when all initial aspiration levels have been achieved, artificial DM could not create a new reference point or after the maximum number of iterations has been conducted.

4 Numerical Experiments

Next we give some computational results to demonstrate application of the artificial DM. For this demonstration, we use two different methods for generating new Pareto optimal solutions: R-NSGA-II algorithm [4] and minimizing the achievement scalarizing function (ASF) of a reference point method [14] to project a reference point to the Pareto optimal set, where the differential evolution algorithm is used to minimize the ASF. As the latter method produces only a single Pareto optimal solution, while R-NSGA-II produces several ones, among the Pareto optimal (nondominated) solutions generated by R-NSGA-II, the one nearest to the reference point is selected. The R-NSGA-II algorithm had the population size 100 and was allowed to have maximum of 200 generations totaling to maximum of 20000 evaluations. The differential evolution method had the stopping criterion of maximum of 20000 evaluations.

Each method was used to solve four different problems: DTLZ1 – DTLZ4 [6] with the number of objectives (k) ranging from 2 to 6, totaling 24 different problems. Each problem was solved ten times using both methods, with ten different, randomly generated sets of initial preference information. The maximum number of iterations was set to 11.

Examples of two test runs when solving the ZDLT2 problem with three objective functions with both R-NSGA-II and the ASF methods can be seen in Figs. 1 and 2, respectively. In these figures, the search path taken by the artificial DM is shown as a continuous line, with x marking as each reference point constructed by the artificial DM. The diamond represents the initial reference point and the square represents the final reference point of the solution process. In Fig. 1, it can be seen that the artificial DM constructed seven reference points in the case of

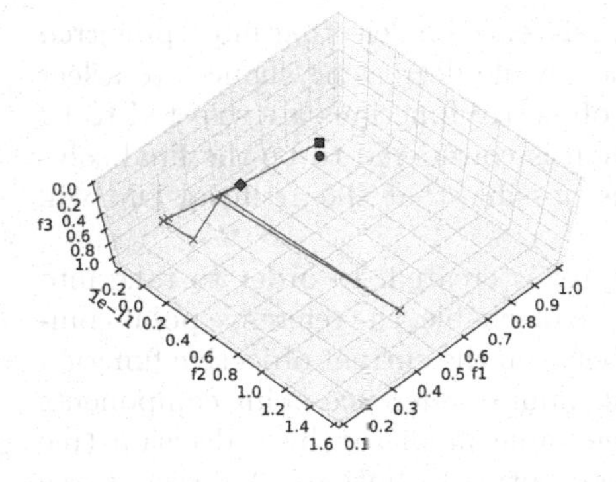

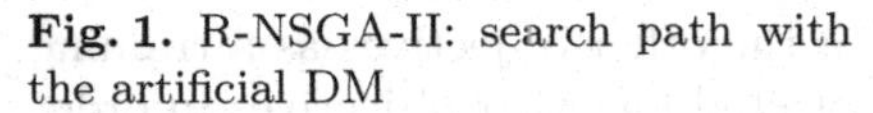
Fig. 1. R-NSGA-II: search path with the artificial DM

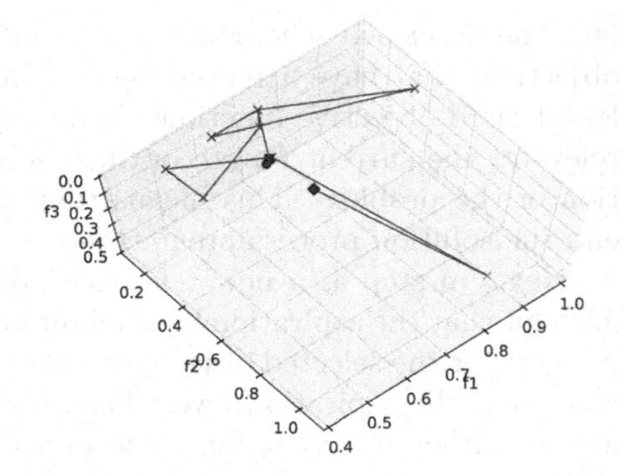

Fig. 2. ASF method: search path with the artificial DM

R-NSGA-II and nine for the ASF method (Fig. 2). The circle outside the search path shows the aspiration level, asp_i, $i = 1, \ldots, k$, that the artificial DM was aiming at. The Pareto optimal solutions generated during the test runs are not shown here.

At the beginning of each test run, the artificial DM does not have any knowledge of the Pareto optimal set of the problem being solved and as described in Sect. 3, the first steps taken are random. This means that all search paths are different for each test run, and the first steps can lead away from the aspiration levels. But as can be seen in figures, after the artificial DM has accumulated enough knowledge of the problem, the solution process converges towards the aspiration levels.

The obtained results are detailed in Table 1 for R-NSGA-II and in Table 2 for the ASF method, where for brevity we limit to the cases $k \in \{2, 4, 6\}$. In the tables, we give the name of the problem, the number of objectives (k), the mean and minimum distances to the initial aspiration levels and the standard deviation of the distances. Finally, the tables report how many iterations the artificial DM used the on average during the test runs.

As can be seen, a distinction between the two methods can be drawn, while both methods behave in a somewhat similar manner. The ASF method could find final solutions that are consistently closer to the initial preferences, i.e., mean values and deviations of distances are smaller than with R-NSGA-II. The ASF method was also able to achieve the initial aspiration levels, even though this did not happen in all runs. For problems with four and more objectives, the differences in the performance were slightly smaller, as the ASF method did not achieve initial aspiration levels consistently, but it should be noted that the performance of R-NSGA-II also deteriorated.

The latter result can be considered as somewhat surprising, taking into account that the NSGA-II algorithm underlying the R-NSGA-II algorithm does not typically perform well with problems having more than three objective functions, and it could be expected that the performance of R-NGSA-II would

Table 1. R-NSGA-II results

Problem	k	mean	dev	min	iter
DTLZ1	2	0.709	0.447	0.000	10
DTLZ2	2	0.040	0.044	0.002	11
DTLZ3	2	2.427	1.600	0.863	5
DTLZ4	2	0.237	0.278	0.006	9
DTLZ1	4	0.224	0.185	0.012	11
DTLZ2	4	0.273	0.185	0.061	11
DTLZ3	4	3.579	0.954	1.671	5
DTLZ4	4	0.439	0.355	0.038	11
DTLZ1	6	0.432	0.270	0.148	11
DTLZ2	6	0.365	0.192	0.024	11
DTLZ3	6	6.000	2.238	3.865	11
DTLZ4	6	0.411	0.269	0.026	11

Table 2. ASF method results

Problem	k	mean	dev	min	iter
DTLZ1	2	0.606	0.327	0.000	10
DTLZ2	2	0.002	0.002	0.000	11
DTLZ3	2	2.112	1.870	0.000	4
DTLZ4	2	0.002	0.002	0.000	11
DTLZ1	4	0.239	0.257	0.000	11
DTLZ2	4	0.004	0.002	0.001	11
DTLZ3	4	1.819	1.340	0.000	6
DTLZ4	4	0.069	0.147	0.001	11
DTLZ1	6	0.386	0.247	0.001	10
DTLZ2	6	0.102	0.207	0.001	10
DTLZ3	6	1.725	1.287	0.005	8
DTLZ4	6	0.115	0.155	0.001	11

deteriorate more. However, it should be noted that the aim of the interactive solution processes is not to obtain best possible coverage of the Pareto frontier, but to concentrate on the area which is the most interesting for the DM. As R-NSGA-II generates several solutions in that area in contrast to a single solution obtained by the ASF method, the former algorithm provides the artificial DM more information, i.e., Pareto optimal solutions to construct more suitable reference points. This implies that the comparison of population based and non-population based methods should be made fair by paying attention to the amount of information the artificial DM is trained with.

5 Conclusions

In this research, we proposed to build an automated framework for testing interactive multiobjective optimization methods, without utilizing a value function to represent the DM's preferences. This was achieved by replacing the human DM with an artificial DM constructed from two distinct parts: the steady part and the current context. With the steady part the artificial DM tries to maintain the search towards its preferences, while at the same time the current context allows changing the direction as well as ending the solution process prematurely, mimicking actions of a human DM. With the proposed framework, it is possible to carry out repeatable tests of interactive methods in a controlled environment.

The numerical experiments performed with the proposed testing framework indicate that the algorithm can identify differences between different interactive methods. In the experiments, two interactive methods were compared solely based on the distance between the final solution and the steady preference information. In addition to this distance, it would be interesting to construct new performance metrics specific for interactive methods, such as characteristics of the trajectory of the solution process in the objective space.

Acknowledgments. This work was supported on the part of Vesa Ojalehto by the Academy of Finland (grant number 287496).

References

1. Babbar-Sebens, M., Minsker, B.S.: Interactive genetic algorithm with mixed initiative interaction for multi-criteria ground water monitoring design. Appl. Soft Comput. **12**(1), 182–195 (2012)
2. Branke, J., Deb, K., Miettinen, K., Słowiński, R. (eds.): Multiobjective Optimization: Interactive and Evolutionary Approaches. Springer, Heidelberg (2008)
3. Deb, K., Miettinen, K., Chaudhuri, S.: Towards an estimation of nadir objective vector using a hybrid of evolutionary and local search approaches. IEEE Trans. Evol. Comput. **14**(6), 821–841 (2010)
4. Deb, K., Sundar, J., Udaya Bhaskara Rao, N., Chaudhuri, S.: Reference point based multi-objective optimization using evolutionary algorithms. Int. J. Comput. Intell. Res. **2**(3), 273–286 (2006)
5. Debreu, G.: Theory of Value: An Axiomatic Analysis of Economic Equilibrium. Cowles Foundation for Research in Economics at Yale University, New Haven (1959). Monograph 17
6. Huband, S., Hingston, P., Barone, L., While, L.: A review of multiobjective test problems and a scalable test problem toolkit. IEEE Trans. Evol. Comput. **10**(5), 477–506 (2006)
7. López-Ibáñez, M., Knowles, J.: Machine decision makers as a laboratory for interactive EMO. In: Gaspar-Cunha, A., Henggeler Antunes, C., Coello, C.C. (eds.) EMO 2015. LNCS, vol. 9019, pp. 295–309. Springer, Heidelberg (2015)
8. Miettinen, K.: Nonlinear Multiobjective Optimization. Kluwer Academic Publishers, Boston (1999)
9. Miettinen, K., Hakanen, J., Podkopaev, D.: Interactive nonlinear multiobjective optimization methods. In: Greco, S., Ehrgott, M., Figueira, J. (eds.) Multiple Criteria Decision Analysis: State of the Art Surveys, pp. 931–980. Springer, New York (2016)
10. Purshouse, R., Deb, K., Mansor, M., Mostaghim, S., Wang, R.: A review of hybrid evolutionary multiple criteria decision making methods. In: Proceedings of IEEE Congress on Evolutionary Computation (CEC), pp. 1147–1154 (2014)
11. Quinlan, J.R.: Induction of decision trees. Mach. Learn. **1**(1), 81–106 (1986)
12. Steuer, R.E.: Multiple Criteria Optimization: Theory, Computation, and Application. Wiley, New York (1986)
13. Stewart, T.J.: Goal programming and cognitive biases in decision-making. J. Oper. Res. Soc. **56**(10), 1166–1175 (2005)
14. Wierzbicki, A.: A mathematical basis for satisficing decision making. Math. Model. **3**, 391–405 (1982)
15. Zujevs, A., Eiduks, J.: New decision maker model for multiobjective optimization interactive methods. In: Proceedings of the Information Technologies, pp. 51–58. Kaunas: Technologija (2011)

Towards Many-Objective Optimisation with Hyper-heuristics: Identifying Good Heuristics with Indicators

David J. Walker(✉) and Ed Keedwell

University of Exeter, Exeter, UK
{D.J.Walker,E.C.Keedwell}@exeter.ac.uk

Abstract. The use of hyper-heuristics is increasing in the multi-objective optimisation domain, and the next logical advance in such methods is to use them in the solution of *many-objective* problems. Such problems comprise four or more objectives and are known to present a significant challenge to standard dominance-based evolutionary algorithms. We incorporate three comparison operators as alternatives to dominance and investigate their potential to optimise many-objective problems with a hyper-heuristic from the literature. We discover that the best results are obtained using either the favour relation or hypervolume, but conclude that changing the comparison operator alone will not allow for the generation of estimated Pareto fronts that are both close to and fully cover the true Pareto front.

1 Introduction

As the field of hyper-heuristic research matures, attention is moving from solving problems requiring the optimisation of a single objective to those comprising two or more objectives. A recent paper [14] proposed a multi-objective extension of a single-objective algorithm that identifies good sequences of heuristics to apply to a given problem. The original single-objective algorithm updates the transition probabilities that govern the selection of the next heuristic using the raw fitness value, and in the multi-objective extension this was replaced with a dominance-based approach.

Though such *multi-objective* problems are prevalent, it is well known that optimisation problems often comprise a large set of objectives that must be simultaneously optimised [10]. Problems with four or more objectives are often called *many-objective* problems. Using dominance-based multi-objective algorithms to solve many-objective problems is generally problematic, as the dominance relation does not scale well to even relatively small numbers of objectives; solutions quickly become incomparable, as they are considered equivalent under dominance. A considerable amount of research in the evolutionary computation field has been devoted to the investigation of evolutionary algorithms that are capable of solving many-objective problems. These generally take one of two approaches: either some of the problem objectives must be discarded so that a

J. Handl et al. (Eds.): PPSN XIV 2016, LNCS 9921, pp. 493–502, 2016.
DOI: 10.1007/978-3-319-45823-6_46

standard multi-objective EA can be employed, or an alternative to the dominance relation must be found. To our knowledge, no work in the hyper-heuristic field has considered many-objective problems. In this paper, we begin to investigate many-objective test problems [4], considering problems comprising four, five and six objectives. We take inspiration from work on dominance alternatives and consider indicators to investigate how useful they are when incorporated into a recent multi-objective hyper-heuristic, such that the indicator replaces the dominance relation for comparing solutions.

The remainder of this paper is organised as follows. Some relevant background material is presented in Sect. 2 before the indicators we examine are introduced in Sect. 3. Section 4 presents our experimental setup, and results are discussed in Sect. 5. We discuss our conclusions and future work in Sect. 6

2 Background

2.1 Many-Objective Optimisation

In the last decade research on *many*-objective optimisation has increased rapidly. A solution $\mathbf{x}$ to an arbitrary many-objective optimisation problem is described by an M-dimensional objective vector $\mathbf{y}$, such that $M \geq 4$:

$$\mathbf{y} = (f_1(\mathbf{x}), \ldots, f_M(\mathbf{x})). \tag{1}$$

Evolutionary algorithms are known to generate good solution sets to multi-objective problems. Such algorithms often use the *dominance* relation to compare the relative quality of two solutions. With the advent of research into many-objective optimisation it has been known that dominance does not scale well to compare many-objective solutions. As the number of objectives increases, so does the likelihood that two solutions will be equivalent; given just a 5-objective problem, and a uniform distribution of solutions, solutions residing in approximately 95 % of objective space will be incomparable under dominance.

Various approaches have been taken to address the inability of dominance-based MOEAs to optimise many-objective problems. Generally, these approaches either involve finding an approach that can compare solutions described by a large number of objectives [3] or identifying redundant objectives that can be discarded so that a standard dominance-based MOEA can be used. This work takes the former option, and we consider three approaches to comparing many-objective solutions; these are described in Sect. 3.

2.2 Hyper-heuristics

Hyper-heuristics are techniques that identify low-level heuristics that generate good solutions to optimisation problems. They operate above the *domain barrier*, meaning that they optimise the heuristics, rather than the solutions to a given optimisation problem, and require no problem-specific information to function.

They have been applied in a wide range of problem domains, often solving combinatoric problems but also in the continuous domain, to great success. Hyper-heuristics are either *generative* or *selection*-based. A generative hyper-heuristic creates novel low-level heuristics, such as mutation or crossover operators, that are tailored to work on a specific type of problem. In this work we only consider selection-based methods, which operate with a pre-defined pool of low-level heuristics and identify those that are well suited to a specific problem domain.

A central part of a selection hyper-heuristic is the mechanism by which the next low-level heuristic to apply is chosen. Common methods are random selection, selection with choice function, and more recently Markov-based methods. In this work we employ an algorithm based on a hidden Markov model [14], which is described later. A recent survey of hyper-heuristic approaches is provided in [2].

The use of hyper-heuristics within many-objective optimisation has received very little attention. Some studies have used them to solve multi-objective problems. One recent example was [13], which employed a reinforcement learning-based Markov chain approach to solving continuous multi-objective problems. Another approach incorporated the hypervolume indicator [8] into the move acceptance strategy of a hyper-heuristic and applied it to solve multi-objective test problems [11,12]. [9] presented a multi-objective hyper-heuristic designed to operate in the realm of search-based software engineering; their algorithm was based on NSGA-II, and used choice function in concert with a multi-armed bandit to select low-level heuristics. With the exception of the hypervolume example, these methods rely heavily on dominance; as discussed earlier, we hypothesise that these approaches will not scale well to deal with many-objective problems, and we discuss potential ways of addressing this issue in the next section.

3 Indicators

Since the discovery that standard, dominance-based, evolutionary optimisers do not provide sufficient selective pressure to locate an acceptable estimate of a many-objective problem's Pareto front [7] significant research efforts have been spent investigating alternatives to dominance. Three that are considered in this study are *hypervolume* [8], the *favour relation* [5], and an indicator based on the average rank method [1].

3.1 Hypervolume

An early contribution, which remains one of the principle indicators of solution quality is the hypervolume [8]. The hypervolume is the dominated space between a solution (or solutions) and a pre-defined reference point. Hypervolume has been used as an indicator in a range of studies, including a recent work in which it was incorporated into the acceptance strategy of a hyper-heuristic [11]. That work considered continuous multi-objective test problems, however was restricted to 2-objective problems only. Though the hypervolume scales to any number of objectives, it is often restricted to problems with low numbers of objectives because of its computational complexity when calculated exactly for a population (though it can be estimated accurately with Monte Carlo sampling [6]). This work is not hindered by such complexity issues, as the calculation is trivial for a single solution.

3.2 Favour Relation

Given two solutions $\mathbf{y}_i$ and $\mathbf{y}_j$, the favour relation [5] determines which is the fitter solution in terms of which is dominant on the most objectives. More formally:

$$\mathbf{y}_i <_f \mathbf{y}_j \Leftrightarrow |\{m : y_{im} < y_{jm}\}| > |\{m : y_{jm} < y_{im}\}|. \quad (2)$$

We incorporate the favour relation as a direct replacement for dominance, such that transition probabilities and the parent solution for the next generation are updated if the current parent solution does not favour its child.

3.3 Average Rank

Many of the dominance alternatives that have been proposed in the literature are based on population ranking. The MOSSHH algorithm is a point-based approach, and thus has no population that can be ranked. That said, it does have an external archive of non-dominated solutions that can be ranked, we use it in combination with the average rank method [1].

Given a population $\mathbf{Y} = \{\mathbf{y}_i\}_{i=1}^N$ of solutions, the solutions are ranked M times, once according to each objective such that r_{im} is the rank of the i-th solution on the m-th objective. The average rank $\bar{r}_i$ is then computed with:

$$\bar{r}_i = \frac{1}{M} r_{im}. \quad (3)$$

In order to use this formulation as an indicator we calculate the average rank of the elite archive. The indicator returns 1 if the rank of the new solution is superior to its parent, and 0 otherwise. It is necessary to ensure that the child has been added to the archive, however due to the formulation of the algorithm described shortly, it is not possible to evaluate the indicator if the child has not been added to the archive.

4 Experiments

In order to determine the usefulness of the indicators outlined in Sect. 3 we now incorporate them into a hyper-heuristic to compare them against the dominance-based approaches that has been shown to work well for multi-objective problems comprising two or three objectives. We employ a selection hyper-heuristic called MOSSHH [14], in which sequences of low-level heuristics that lead to good solutions are identified. A transition probability matrix is maintained, which governs

Algorithm 1. MOSSHH

```
 1: x, h_c, A, B = initialise()
 2: E = initialise_archive()                          Initialise the archive.
 3: repeat
 4:    h_p = h_c
 5:    h_c = select(A, h_p)                            Choose the next heuristic
 6:    AS = select(B, h_c)                             Set the acceptance strategy
 7:    record(h_p, h_c, AS)                            Record the current heuristic
 8:    x' = apply(x, h_c)                Apply the heuristic to the current solution
 9:    if AS == 1 then
10:      E = update_archive(E, f(x'))  If acceptance strategy was met update archive
11:      if ¬I(f(x), f(x')) then
12:        x = x'               If the indicator is true replace the parent with the child
13:        if archived(f(x')) then
14:          update_probabilities()
15:        end if
16:      end if
17:      clear_records()
18:    end if
19: until termination criterion met
```

the transition from one low-level heuristic to another, and each heuristic has an *acceptance strategy*, which determines the likelihood that the solution generated by a given heuristic will be accepted. Both transition probabilities and acceptance strategies are learned using an online learning process.

The indicator-based multi-objective sequence-based hyper-heuristic (MOSSHH) [14] algorithm is described in Algorithm 1. The algorithm begins by initialising a random parent solution, choosing a starting low-level heuristic, and initialising the transition probability and acceptance strategy matrices uniformly (Line 1). An empty elite archive is initialised (Line 2). The first stage in each iteration of the iterative process is to select the next low-level heuristic and acceptance strategy using the current low-level heuristic (Lines 4–6). The chosen values are recorded (Line 7), in case the current sequence of low-level heuristics is identified as being useful, and the new low-level heuristic is applied to generate a new solution (Line 8). If the acceptance strategy is met ($AS == 1$) then the solution's objective values are evaluated and the archive is updated. Any solutions dominated by the new solution are discarded, and if the solution itself is not dominated by the archive then it is added to it (Line 10). At this point, the parent and child solutions are compared using one of the indicators. If the indicator deems that the child is superior to the parent, then the child solution succeeds the parent solution as the parent in the next generation. Otherwise, the current parent solution is retained. If the solution was added to the archive, then the sequence of low-level heuristics that led to it is complete, and transition probability and acceptance strategies are updated accordingly (Line 14).

The problems we examine are drawn from the DTLZ suite of test problems [4]. Specifically, we investigate 4-, 5- and 6-objective instances of the DTLZ1,

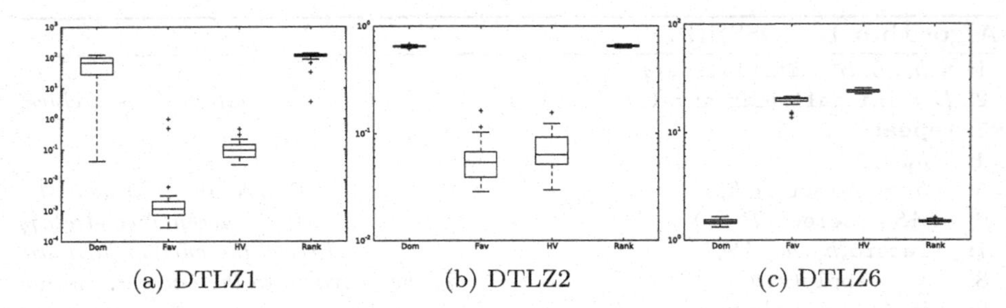

(a) DTLZ1 (b) DTLZ2 (c) DTLZ6

Fig. 1. Generational distance results for 6-objective instances of DTLZ1, DTLZ2 and DTLZ6.

DTLZ2 and DTLZ6 test problems. Each problem has been selected to demonstrate the hyper-heuristic's ability to cope with specific problem features (such as deceptive fronts and a discontinuous Pareto front). The problems are parametrised as suggested in [4]. In the case of DTLZ1 and DTLZ6, the algorithm is run for 50,000 function evaluations. DTLZ2 is known to be a easier problem, and as such is run for just 5,000 function evaluations due to computational time constraints.

The following set of low-level heuristics is employed:

Ruin and recreate two versions; in the first, the entire solution is destroyed and replaced with a random feasible solution. In the second, a single parameter is chosen and replaced.

Mutation three additive mutation operators. In each, a parameter is chosen at random and mutated with an additive mutation drawn from one of three probability distributions (uniform, in the region $(-0.05, 0.05)$; Gaussian, with 0 mean and standard deviation 0.1; beta, in the region $(-0.05, 0.05)$).

Archive selection two versions; one in which the entire solution is replaced with a solution drawn at random from the archive, and a second in which a parameter is replaced with an archived solution's corresponding parameter.

In total, $H = 7$ low-level heuristics are employed. To begin with, each has an equal probability of selection (transition probabilities are initialised to $1/H$).

Each problem is optimised with MOSSHH using each of the four indicators. In order to analyse the results, each instance of the problem is optimised 30 times for each of the four problems. We compare the results using the generational distance to examine the convergence properties of the algorithm, as well as using inverted generational distance to consider diversity.

5 Results

Figure 1 illustrates the generational distance results for 6-objective instances of DTLZ1, DTLZ2 and DTLZ6, while Fig. 2 shows the corresponding inverted generational distance results. The corresponding 4- and 5-objective results are

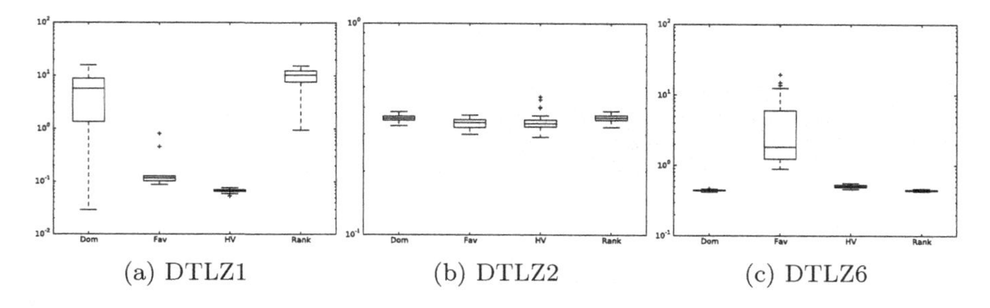

(a) DTLZ1 (b) DTLZ2 (c) DTLZ6

Fig. 2. Inverted generational distance results for 6-objective instances of DTLZ1, DTLZ2 and DTLZ6.

omitted for space. The results were generated by computing the mean distance from the final archive of solutions for each run to a set of Pareto optimal samples, in the case of generational distance, and the corresponding distance from the sample sets to the optimised solution sets, in the case of inverted generational distance.

In the case of DTLZ1, the results show that the dominance-based indicator has failed to converge to the Pareto front. Both the favour and hypervolume indicators have performed significantly better, converging much closer to the Pareto front, and covering it more extensively as can be seen from the IGD results. The rank-based indicator has not performed well, with at best comparable performance to that of the dominance indicator. The difference is less clear in the DTLZ2 case; though the favour and hypervolume indicators have again converged closer to the Pareto front, the difference is less significant. In terms of diversity, there is little to chose between the four alternatives; this is not surprising, as DTLZ2 is designed to be an easier problem for optimisers to solve.

Figure 3 shows representative estimated Pareto fronts obtained by optimising a 6-objective instance of DTLZ2 using the four indicators. To colour the solutions, the population was ranked to identify the objective on which each solution has the best rank. This information is used to colour the line representing each solution. As can be seen, the dominance and rank indicators have a spread of preferred objectives, whereas the favour and hypervolume indicators have optimised a specific objective (objective 6). The improved performance of these two indicators can be explained by this, as large numbers of solutions that optimise this objective (and objective 5) have been included in the estimated Pareto set, which means the overall mean distance between the estimated front and the true front is reduced.

The algorithm has managed to optimise DTLZ6, though, interestingly, the generational distance results are the reverse of those for DTLZ1 and DTLZ2. This is likely to be because of the available heuristics; given the propensity for the indicators to optimise specific regions of the Pareto front, as discussed above, a problem with discontinuities will present difficulties for an optimiser that does not have crossover heuristics available to it. Once the algorithm has converged to

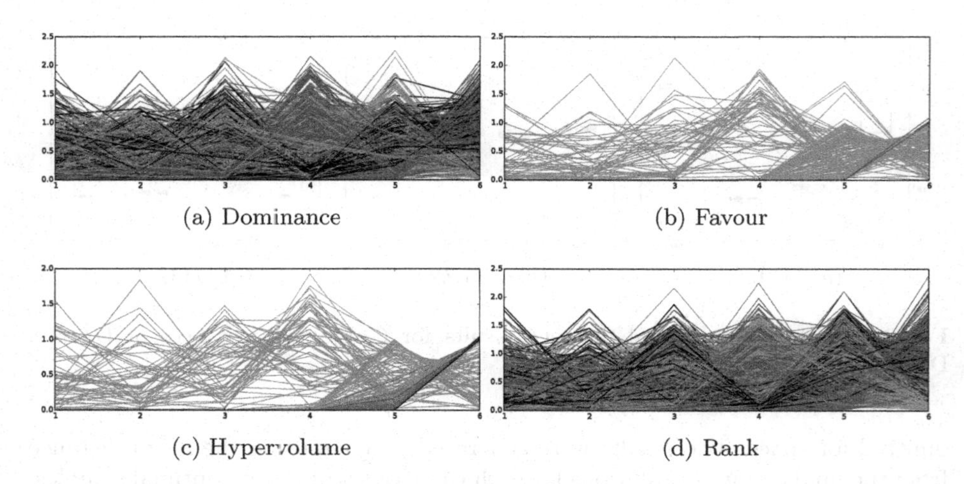

(a) Dominance (b) Favour

(c) Hypervolume (d) Rank

Fig. 3. Parallel coordinate plots showing the estimated Pareto front obtained by optimising a 6-objective instance of DTLZ2 with various indicators. (Color figure online)

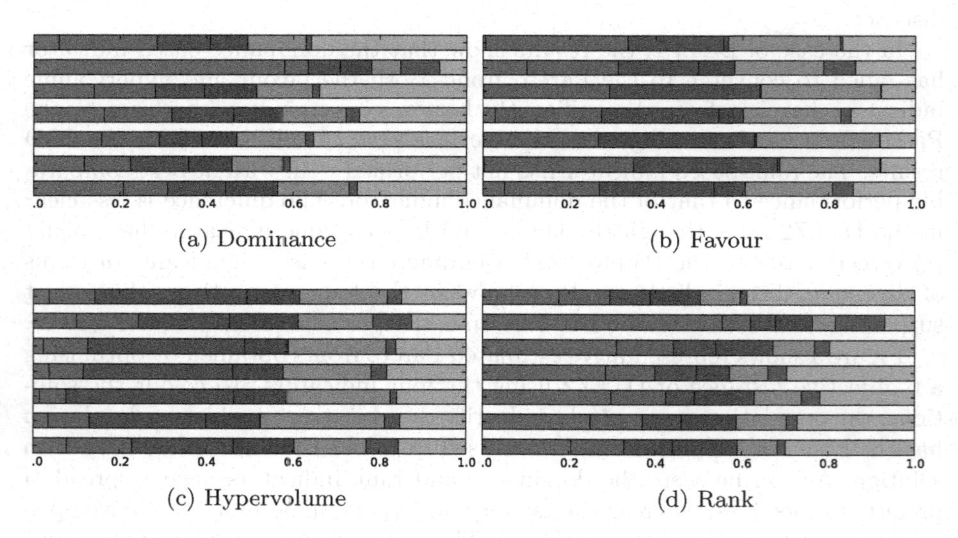

(a) Dominance (b) Favour

(c) Hypervolume (d) Rank

Fig. 4. Transition probability matrices for 6-objective instances of DTLZ1 (top). Key to low-level heuristics (bottom-top) – grey: ruin and recreate (solution); red: ruin and recreate (parameter); green: uniform mutation; blue: Gaussian mutation; cyan: beta mutation; magenta: archive replacement (solution); yellow: archive replacement (parameter). A large block indicates a large probability of transitioning to that heuristic from the current heuristic. (Color figure online)

a specific region of the Pareto front, the mutation heuristics used herein appear to lack the ability to cross discontinuities and, though the archive heuristics were intended to ameliorate this lack of crossover, they only allow the algorithm to return to areas of the space that have previously been explored. Crossover heuristics will be required to make this algorithm scale to problem features such as discontinuous Pareto fronts.

Figure 4 presents the transition probability matrices for the 6-objective instances of DTLZ1. In each case, the transition probability matrices from the thirty runs have been averaged. As can be seen, in the case of the dominance and rank-based indicators, which performed less well, there is a higher propensity to use the ruin and recreate and archive heuristics. In contrast, the more successful favour and hypervolume indicators have preferred mutational heuristics; this aligns with known results for multi-objective instances of these problems [14].

6 Conclusion

This paper has presented an analysis of the use of a selection hyper-heuristic to solve many-objective optimisation problems. This is, as far as we are aware, the first study of its type, and as such have evaluated the algorithm's performance on a small number of test problems with relatively small numbers of objectives; future work will expand this approach to a wider range of problems and consider many more objectives. The work presented three approaches to comparing many-objective solutions. Of the three, we consider the favour and hypervolume indicators to be the most successful, though we note that these were less useful when optimising discontinuous Pareto fronts. A wider range of low-level heuristics, including crossover heuristics, would likely address this issue, though that would require conversion to a population-based approach.

As we move toward optimising problems comprising larger numbers of objectives, we expect the conditions experienced in this work to become more pronounced. The dominance relation will become less able to provide selection pressure, and the favour and hypervolume indicators will likely continue to optimise specific regions of the Pareto front well, at the expense of other regions. It is therefore unlikely that considering alternative comparison methods alone will allow us to successfully optimise many-objective problems, and as such we are currently investigating additional ways in which hyper-heuristics can be modified so that they can be used to optimise such problems.

Acknowledgements. This work was funded under EPSRC grant EP/K000519/1.

References

1. Bentley, P.J., Wakefield, J.P.: Finding acceptable solutions in the pareto-optimal range using multiobjective genetic algorithms. In: Chawdhry, P.K., Roy, R., Pant, R.K. (eds.) Soft Computing in Engineering Design and Manufacturing, pp. 231–240. Springer, London (1998)
2. Burke, E.K., Gendreau, M., Hyde, M., Kendall, G., Ochoa, G., Ozcan, E., Qu, R.: Hyper-heuristics: a survey of the state of the art. J. Oper. Res. Soc. **64**(12), 1695–1724 (2013)

3. Corne, D., Knowles, J.: Techniques for highly multiobjective optimisation: some nondominated points are better than others. In: Genetic and Evolutionary Computation Conference, London, UK, pp. 773–780 (2007)
4. Deb, K., Thiele, L., Laumanns, M., Zitzler, E.: Scalable multi-objective optimization test problems. In: Proceedings of IEEE Congress on Evolutionary Computation, vol. 1, pp. 825–830 (2002)
5. Drechsler, N., Drechsler, R., Becker, B.: Multi-objective optimisation based on relation favour. In: Zitzler, E., Thiele, L., Deb, K., Coello Coello, C.A., Corne, D. (eds.) EMO 2001. LNCS, vol. 1993, pp. 154–168. Springer, Heidelberg (2001)
6. Everson, R.M., Fieldsend, J.E., Singh, S.: Full elite sets for multi-objective optimisation. In: Parmee, I.C. (ed.) ACDM 2002, vol. 5, pp. 343–354. Springer, Heidelberg (2002)
7. Farina, M., Amato, P.: On the optimal solution definition for many-criteria optimization problems. In: 2002 Annual Meeting of the North American Fuzzy Information Processing Society Proceedings, pp. 233–238 (2002)
8. Fleischer, M.: The measure of pareto optima. In: Fonseca, C.M., Fleming, P.J., Zitzler, E., Deb, K., Thiele, L. (eds.) EMO 2003. LNCS, vol. 2632, pp. 519–533. Springer, Heidelberg (2003)
9. Guizzo, G., Fritsche, G.M., Vergilio, S.R., Pozo, A.T.R.: A hyper-heuristic for the multi-objective integration and test order problem. In: Proceedings Genetic and Evolutionary Computation Conference, GECCO 2015, pp. 1343–1350 (2015)
10. Ishibuchi, H., Tsukamoto, N., Nojima, Y.: Evolutionary many-objective optimization: a short review. In: IEEE Congress on Evolutionary Computation, CEC 2008 (IEEE World Congress on Computational Intelligence), pp. 2419–2426, June 2008
11. Maashi, M., Kendall, G., Ozcan, E.: Choice function based hyper-heuristics for multi-objective optimization. Appl. Soft Comput. **28**, 312–326 (2015)
12. Maashi, M., Ozcan, E., Kendall, G.: A multi-objective hyper-heuristic based on choice function. Expert Syst. Appl. **41**(9), 4475–4493 (2014)
13. McClymont, K., Keedwell, E.: Markov chain hyper-heuristic (MCHH): an online selective hyper-heuristic for multi-objective continuous problems. In: Proceedings of the Genetic and Evolutionary Computation Conference (GECCO 2011), pp. 2003–2010. ACM (2011)
14. Walker, D.J. Keedwell, E.: Multi-objective optimisation with a sequence-based selection hyper-heuristic. In: Proceedings of the Genetic and Evolutionary Computation Conference (GECCO 2016) (2016)

Use of Piecewise Linear and Nonlinear Scalarizing Functions in MOEA/D

Hisao Ishibuchi(✉), Ken Doi, and Yusuke Nojima

Department of Computer Science and Intelligent Systems,
Graduate School of Engineering, Osaka Prefecture University,
1-1 Gakuen-cho, Naka-ku, Sakai, Osaka 599-8531, Japan
{hisaoi,nojima}@cs.osakafu-u.ac.jp,
ken.doi@ci.cs.osakafu-u.ac.jp

Abstract. A number of weight vector-based algorithms have been proposed for many-objective optimization using the framework of MOEA/D (multi-objective evolutionary algorithm based on decomposition). Those algorithms are characterized by the use of uniformly distributed normalized weight vectors, which are also referred to as reference vectors, reference lines and search directions. Their common idea is to minimize the distance to the ideal point (i.e., convergence) and the distance to the reference line (i.e., uniformity). Each algorithm has its own mechanism for striking a convergence-uniformity balance. In the original MOEA/D with the PBI (penalty-based boundary intersection) function, this balance is handled by a penalty parameter. In this paper, we first discuss why an appropriate specification of the penalty parameter is difficult. Next we suggest a desired shape of contour lines of a scalarizing function in MOEA/D. Then we propose two ideas for modifying the PBI function. The proposed ideas generate piecewise linear and nonlinear contour lines. Finally we examine the effectiveness of the proposed ideas on the performance of MOEA/D for many-objective test problems.

Keywords: Evolutionary multi-objective optimization (EMO) · Many-objective optimization · Decomposition-based evolutionary algorithm · MOEA/D

1 Introduction

In the EMO (evolutionary multi-objective optimization) community, many-objective optimization has been a hot topic in the last decade [9, 10]. The difficulty of many-objective optimization for EMO algorithms is explained as follows [9]: When a Pareto dominance-based EMO algorithm such as NSGA-II [4] and SPEA [14] is applied to a multi-objective problem with many objectives, all solutions in a population become non-dominated with each other in a very early stage of evolution. As a result, no strong selection pressure towards the Pareto front can be generated by its Pareto dominance-based fitness evaluation mechanism.

Recently a number of weight vector-based algorithms were proposed for many-objective problems in the framework of MOEA/D [13] such as I-DBEA [1], RVEA [2], NSGA-III [3] and MOEA/DD [11]. Those algorithms are characterized by

J. Handl et al. (Eds.): PPSN XIV 2016, LNCS 9921, pp. 503–513, 2016.
DOI: 10.1007/978-3-319-45823-6_47

the use of uniformly distributed normalized weight vectors. They also have similar fitness evaluation mechanisms. In Fig. 1, we show the angle a between the solution $f(x)$ and the nearest reference line l, the distance d_1 from $f(x)$ to the ideal point z^* along l, and the distance d_2 from $f(x)$ to l. Each solution is usually assigned to the nearest reference line l using the angle a or the distance d_2. Then the fitness of the assigned solution is evaluated by the closeness to the nearest reference line (i.e., a or d_2) and the closeness to the ideal point (i.e., d_1).

An important issue is how to strike a balance between the convergence (i.e., minimization of d_1) and the uniformity (i.e., minimization of d_2 or a). In MOEA/D [13], this balance was handled by the penalty parameter θ for the distance d_2 in the following PBI (penalty-based boundary intersection) function:

$$\text{Minimize}\, f^{PBI}(\mathbf{x}|\mathbf{w}, \mathbf{z}^*) = d_1 + \theta d_2, \tag{1}$$

where the penalty parameter θ is a non-negative real number. This parameter is used to handle the balance between the convergence d_1 and the uniformity d_2.

In this paper, we first discuss the difficulty of the penalty parameter specification in MOEA/D in Sect. 2. We also discuss a desired shape of the contour lines of a scalarizing function in MOEA/D. Next we propose two ideas for modifying the PBI function in Sect. 3. One is a piecewise linear function, and the other is a non-linear function. Then the performance of MOEA/D with each function is examined in Sect. 4. Finally we conclude this paper in Sect. 5.

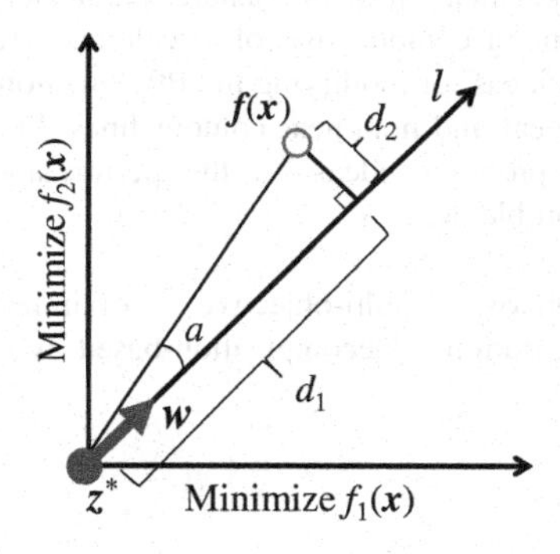

Fig. 1. The weight vector w, the reference line l, and the solution $f(x)$.

2 Parameter Specification in the PBI Function

In Fig. 2, we show the relation between the contour lines of the PBI function and the optimal solution for a concave Pareto front. When θ is not small, the optimal solution is on the intersection of the reference line and the Pareto front as shown in Fig. 2(b) and (c). However, when θ is small, the optimal solution is far from the reference line. For example, the red circle on the f_2 axis in Fig. 2(a) is the optimal solution for the reference line with the direction (0.8, 0.2). Moreover, when θ is small, it is difficult to find a solution on the concave region of the Pareto front as shown in Fig. 2(a).

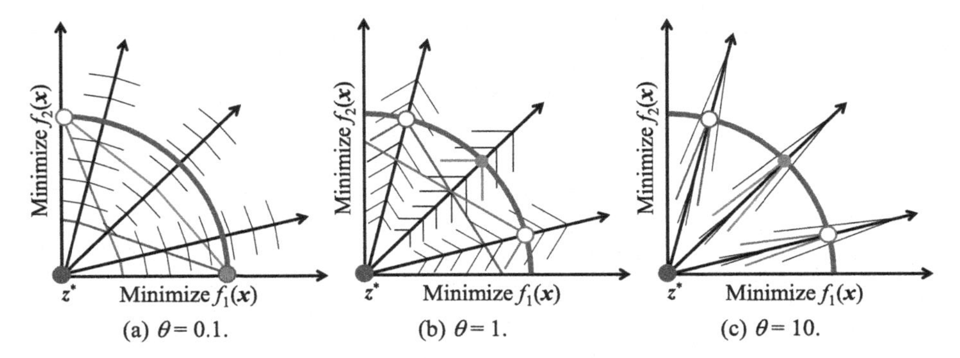

Fig. 2. Relation between the contour lines of the PBI function for three directions ((w = (0.2, 0.8), (0.5, 0.5), (0.8, 0.2)) and the optimal solution for the case of a concave Pareto front. (Color figure online)

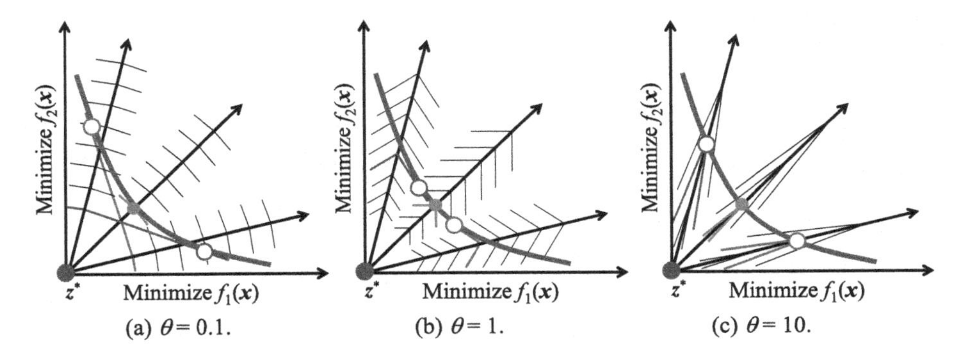

Fig. 3. Relation between the contour lines of the PBI function for three directions ((w = (0.2, 0.8), (0.5, 0.5), (0.8, 0.2)) and the optimal solution for the case of a convex Pareto front.

Well-distributed solutions are obtained from a small value of θ in Fig. 3(a) and a large value of θ in Fig. 3(c). However, in Fig. 3(b) with $\theta = 1$, the three solutions are close to each other around the center of the Pareto front (i.e., well-distributed solutions are not obtained). Thus an intermediate value of θ is not a good choice.

From these discussions, one may think that a large value of θ is a good choice. The use of a large value of θ is also consistent with the emphasis of the uniformity in the above-mentioned weight vector-based algorithms. However, a large value of θ degrades the convergence property of the PBI function in the same manner as the performance deterioration of Pareto dominance-based EMO algorithms for many-objective problems [7]. In Fig. 4, we show the region of solutions which are evaluated as being better than the red circle by the PBI function. When θ is small in Fig. 4(a), the solution has a large improved region. However, when θ is large in Fig. 4 (c), the improved region is very small. So, it is not likely that a better solution is easily found by crossover and mutation. The increase in the number of objectives exponentially decreases the ratio of this improved region in the neighborhood of the solution.

This exponential decrease explains poor performance of the PBI function with a large value of θ for many-objective knapsack problems [7]. Similar discussions were given about the specification of p in the weighted L_p scalarizing function in [12].

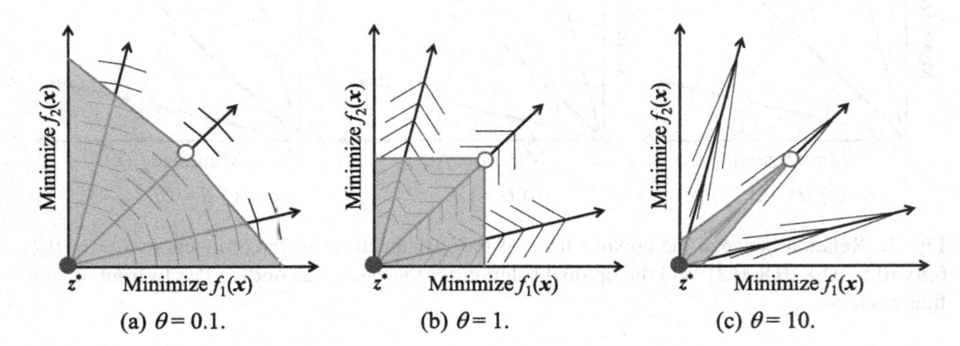

Fig. 4. Improved region for a solution (red circle) with respect to the PBI function. (Color figure online)

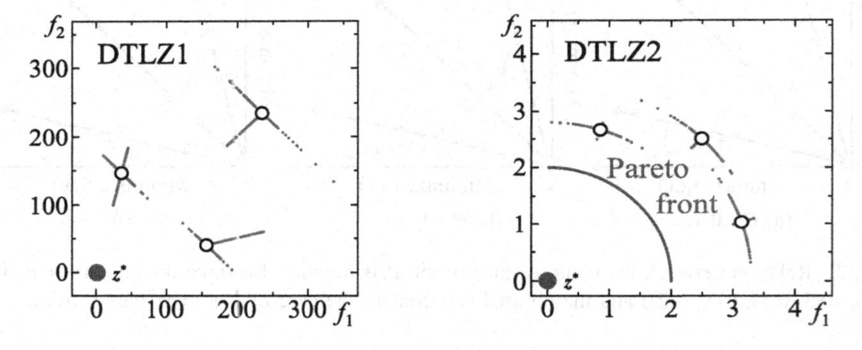

Fig. 5. 100 solutions generated by polynomial mutation of a randomly selected variable [8].

These discussions, however, are not consistent with experimental results in our former study [8] where good results were obtained from a large value of θ for many-objective DTLZ1 and DTLZ2 problems. This inconsistency can be explained by the special features of DTLZ 1-4 [5]. In DTLZ 1-4, the decision variable vector $\boldsymbol{x}$ is separable into the distance variable vector $\boldsymbol{x}_M$ and the position variable vector $\boldsymbol{x}_{\text{pos}}$. The objective vector $\boldsymbol{f}(\boldsymbol{x})$ is written as $\boldsymbol{f}(\boldsymbol{x}) = (1 + g(\boldsymbol{x}_M))\boldsymbol{h}(\boldsymbol{x}_{\text{pos}})$. Pareto optimal solutions are obtained by minimizing the scalar function $g(\boldsymbol{x}_M)$ to $g(\boldsymbol{x}_M) = 0$. Thus, the convergence improvement can be viewed as separate single-objective optimization.

In Fig. 5, we show 100 solutions generated by the polynomial mutation with the distribution index 20 to a randomly selected single variable from each of three solutions (open circles). When a distance variable in $\boldsymbol{x}_M$ is mutated, only the distance from the ideal point z^* is decreased or increased without changing any value of $\boldsymbol{h}(\boldsymbol{x}_{\text{pos}})$. Thus improved solutions are obtained on the line between the ideal solution z^* and the current solution in Fig. 5. If the current solution is on the reference line, those solutions are evaluated as being better than the current solution by the PBI function independent of the value of θ. When the mutation is applied to a position variable in $\boldsymbol{x}_{\text{pos}}$, the location of the solution is changed without changing the value of $g(\boldsymbol{x}_M)$ as shown in Fig. 5. Thus the uniformity can be improved separately from the convergence. Thanks to these special features, good experimental results were reported when a large value of θ was used for many-objective DTLZ 1-4 test problems. WFG 4-9 test problems [6] also have similar special features.

Discussions on the specification of θ in this section are summarized as follows.

(a) Small values of θ: The PBI function has high convergence ability even for many-objective problems. Its main difficulty is the handling of concave Pareto fronts.
(b) Values between (a) and (c): The diversity of solutions can be very small for values around $\theta = 1$ when the shape of the Pareto front is convex.
(c) Large values of θ: Uniformly distributed solutions are likely to be obtained. However, the convergence is degraded by the increase in the number of objectives.

These discussions may suggest two directions for improving the PBI function. One is to improve the uniformity for the PBI function with a small value of θ. This direction is illustrated in Fig. 6(a). The other is to improve the convergence for the PBI function with a large value of θ as illustrated in Fig. 6(b). The contour lines after the modification are similar between Fig. 6(a) and (b). That is, the convergence is emphasized only when a solution is close to the reference line. The uniformity is emphasized when a solution is far from the reference line.

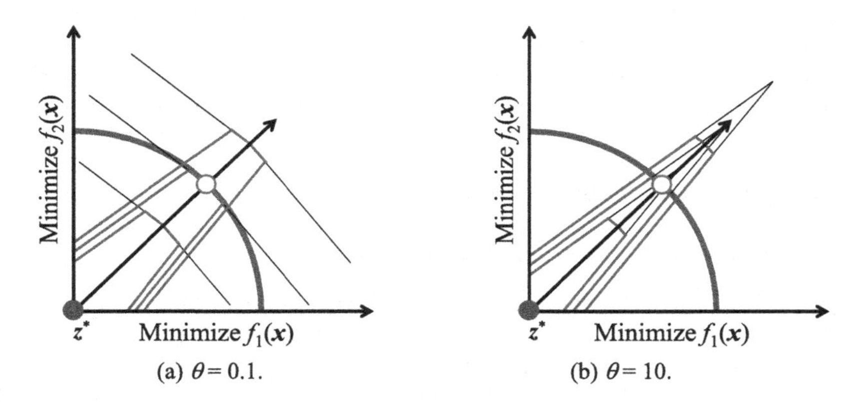

Fig. 6. Modifications of the contour lines of the PBI function.

3 Modifications of the PBI Functions

The PBI function after the modifications in Fig. 6 can be formulated using two penalty parameters θ_1 and θ_2 as the following two-level PBI function:

$$\text{Minimize} f_{2-Level}^{PBI}(\mathbf{x}|\mathbf{w},\mathbf{z}^*) = \begin{cases} d_1 + \theta_1 d_2, & \text{if } d_2 \leq d^*, \\ d_1 + \theta_1 d^* + \theta_2(d_2 - d^*), & \text{if } d_2 > d^*, \end{cases} \quad (2)$$

where $\theta_1 < \theta_2$ and d^* is a parameter to switch the penalty value between θ_1 and θ_2. If d_2 is smaller than d^*, a small penalty value θ_1 is used. If d_2 is larger than d^*, a large penalty value θ_2 is used for the amount of the violation: $d_2 - d^*$. In this paper, we specify the two penalty parameters θ_1 and θ_2 as $\theta_1 = 0.1$ and $\theta_2 = 10$.

The value of d^* is specified by solutions in the current population as follows:

$$d^* = \alpha \frac{1}{H} \frac{1}{m} \sum_{i=1}^{m} (f_i^{Max}(\mathbf{x}) - f_i^{Min}(\mathbf{x})), \quad (3)$$

where α is a parameter, H is an integer parameter used for generating uniformly distributed weight vectors in MOEA/D, m is the number of objectives, and $f_i^{Max}(x)$ and $f_i^{Min}(x)$ are the maximum and minimum values of the ith objective in the current population, respectively. In (3), the average width of the domain of each objective is divided by H to obtain a rough estimation for the distance between adjacent solutions. The parameter α is used to examine the validity of the formulation (3) through computational experiments with various values of α.

Our idea in (2) is to use a small penalty value only when a solution is close to the reference line. This idea can be also implemented as the following quadratic function.

$$\text{Minimize} f_{Quadratic}^{PBI}(\mathbf{x}|\mathbf{w},\mathbf{z}^*) = d_1 + \theta d_2 \frac{d_2}{d^*}, \quad (4)$$

where d^* is the same parameter as in (2), which is calculated by (3). The effect of the penalty parameter θ is decreased by the factor (d_2/d^*) when d_2 is small (i.e., $d_2 < d^*$) and increased by (d_2/d^*) when d_2 is large (i.e., $d_2 > d^*$). When $d_2 = d^*$, this formulation is the same as the PBI function in (1). The value of θ is specified as $\theta = 1$ in (4).

4 Computational Experiments

4.1 Experimental Results of the PBI Function

We applied MOEA/D with the PBI function to DTLZ 1-2 with four and eight objectives. Various values of θ between 0.01 and 100 were examined. The total number of examined solutions was used as the termination condition: $m \times 10{,}000$ solutions for m-objective problems. We examined various settings of the population size. The neighborhood size in MOEA/D was specified as 10 % of the population size. The number of decision variables (n) was $5 + m - 1$ (DTLZ1) and $10 + m - 1$ (DTLZ2). We used the SBX crossover with the distribution index 15 and the crossover probability 0.8, and the polynomial mutation with the distribution index 20 and the mutation probability $1/n$. The average hypervolume was calculated over 50 runs for the reference point (0.6, …, 0.6) of DTLZ1 and (1.1, …, 1.1) of DTLZ2.

In the same manner, we applied MOEA/D to 500-item 0/1 knapsack problems with four and eight objectives [7] except for the following settings: 400,000 solution evaluations, uniform crossover with the probability 0.8, bit-flip mutation with the probability 2/500, the reference point (0, …, 0) for the hypervolume calculation, and the reference point z^* for the PBI function as $z_i^* = 1.1 \times \max\{f_i(\mathbf{x})\}$ for $i = 1, 2, \ldots, m$ where $\max\{f_i(\mathbf{x})\}$ is the maximum value of $f_i(\mathbf{x})$ among all the examined solutions [7].

The average hypervolume value over 50 runs is shown in Figs. 7, 8 and 9. Each circle shows the average result from the corresponding setting of the population size (e.g., 56) and the value of θ (e.g., 0.01). The range of appropriate values of θ in each figure are as follows: $5 \leq \theta \leq 20$ in Fig. 7, $2 \leq \theta \leq 100$ in Fig. 8, and $0.01 \leq \theta \leq 0.1$ in Fig. 9. The PBI function with a small values of θ cannot handle the concave Pareto front of DTLZ2 in Fig. 8. Large values for θ deteriorate the convergence performance of the PBI function for many-objective knapsack problems in Fig. 9(b). Clear performance deterioration is also observed around $\theta = 1$ in Figs. 7 and 9. Figures 7, 8 and 9 show the difficulty and the importance of an appropriate parameter specification of θ.

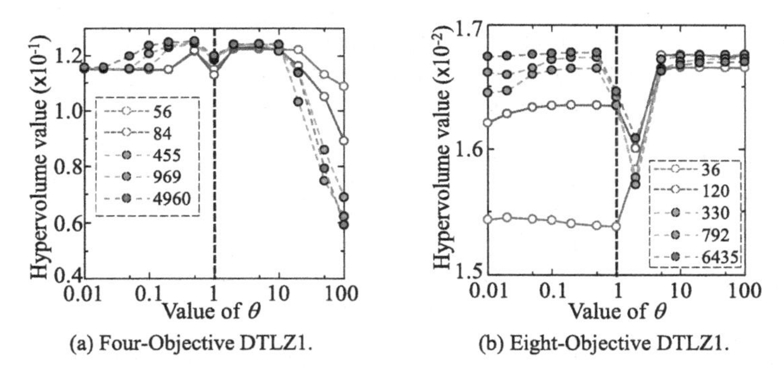

(a) Four-Objective DTLZ1.

(b) Eight-Objective DTLZ1.

Fig. 7. Results of the PBI function on DTLZ1 (Linear Pareto front).

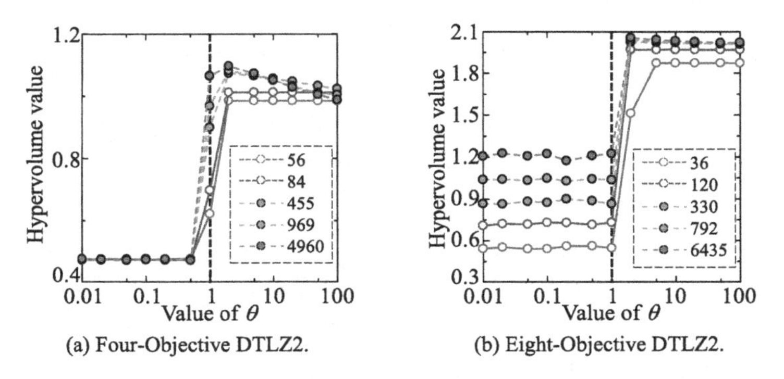

(a) Four-Objective DTLZ2.

(b) Eight-Objective DTLZ2.

Fig. 8. Results of the PBI function on DTLZ2 (Concave Pareto front).

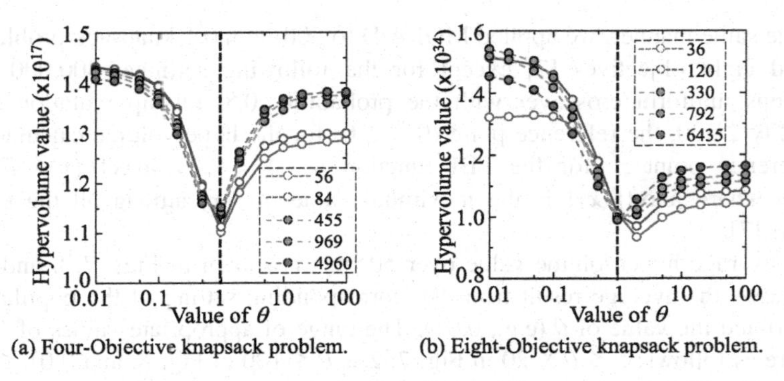

(a) Four-Objective knapsack problem. (b) Eight-Objective knapsack problem.

Fig. 9. Results of the PBI function on the knapsack problems (Convex Pareto front).

4.2 Experimental Results of the Two-Level PBI Function

Experimental results of the two-level PBI function are shown in Figs. 10, 11 and 12 where the horizontal axis is $1/\alpha$. At the leftmost (rightmost) point of each figure with a small (large) value of $1/\alpha$, $\theta_1 = 0.1$ ($\theta_2 = 10$) is mainly used. Thus the obtained results at the leftmost (rightmost) point of each figure are almost the same as those by $\theta = 0.1$ ($\theta = 10$) in Subsect. 4.1. Only for the knapsack problems, we use larger values of α (see the horizontal axis of each figure in Figs. 10, 11 and 12).

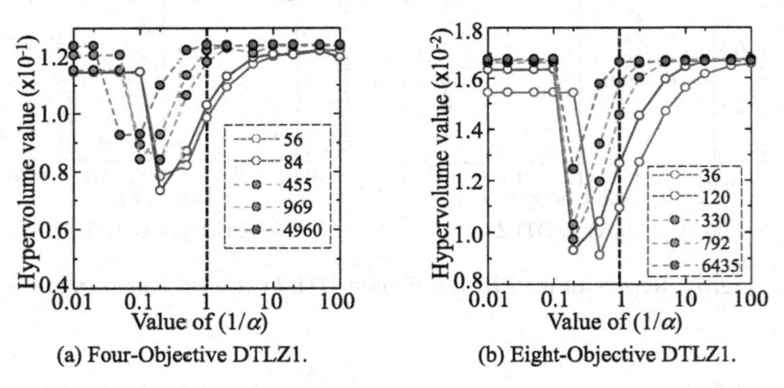

(a) Four-Objective DTLZ1. (b) Eight-Objective DTLZ1.

Fig. 10. Results of the two-level PBI function on DTLZ1.

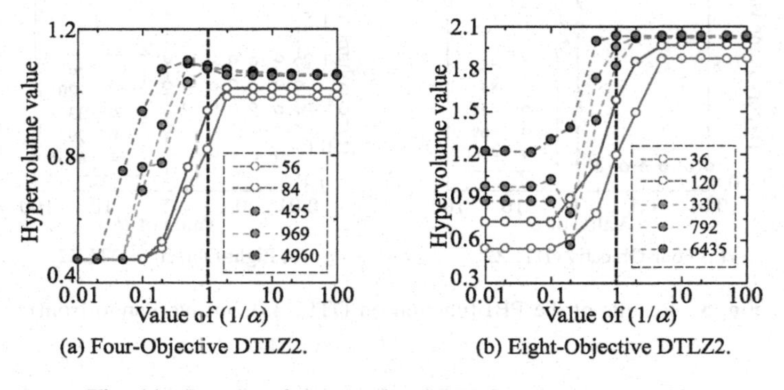

(a) Four-Objective DTLZ2. (b) Eight-Objective DTLZ2.

Fig. 11. Results of the two-level PBI function on DTLZ2.

4.3 Experimental Results of the Quadratic PBI Function

Experimental results of the quadratic PBI function are shown in Figs. 13, 14 and 15. The obtained results at the leftmost (rightmost) point of each figure are similar to those by $\theta = 0.01$ ($\theta = 100$) in Subsect. 4.1. This is because the penalty value is very small (very large) on average at the leftmost (rightmost) point in Figs. 13, 14 and 15. We can

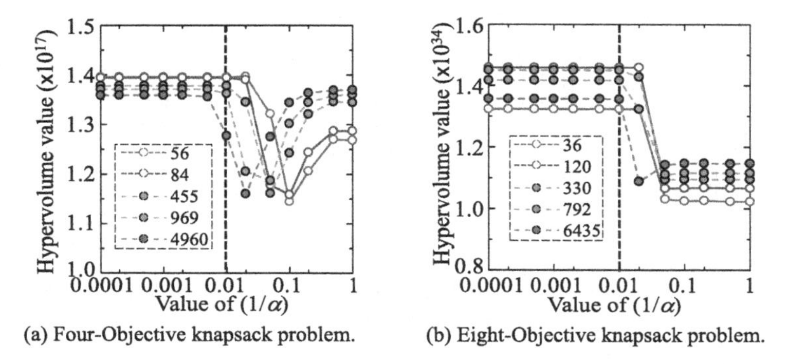

(a) Four-Objective knapsack problem. (b) Eight-Objective knapsack problem.

Fig. 12. Results of the two-level PBI function on the knapsack problems.

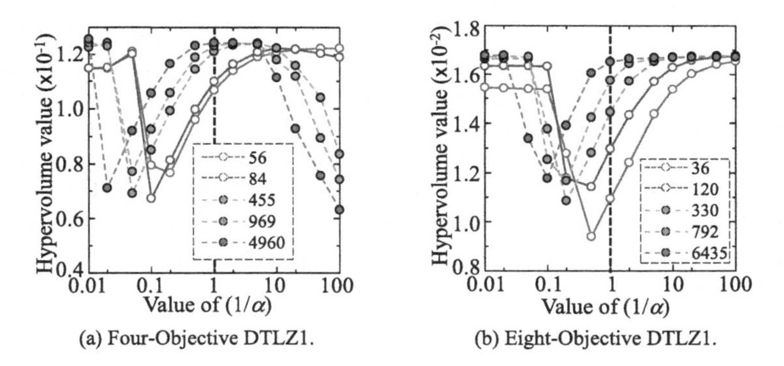

(a) Four-Objective DTLZ1. (b) Eight-Objective DTLZ1.

Fig. 13. Experimental results of the quadratic PBI function on DTLZ1.

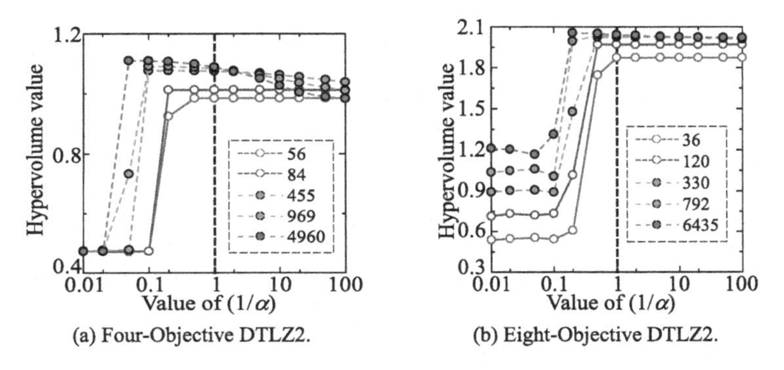

(a) Four-Objective DTLZ2. (b) Eight-Objective DTLZ2.

Fig. 14. Experimental results of the quadratic PBI function on DTLZ2.

also observe some similarity among the obtained results on each test problem in the three subsections such as the V-shape results in Figs. 9(a), 12(a) and 15(a).

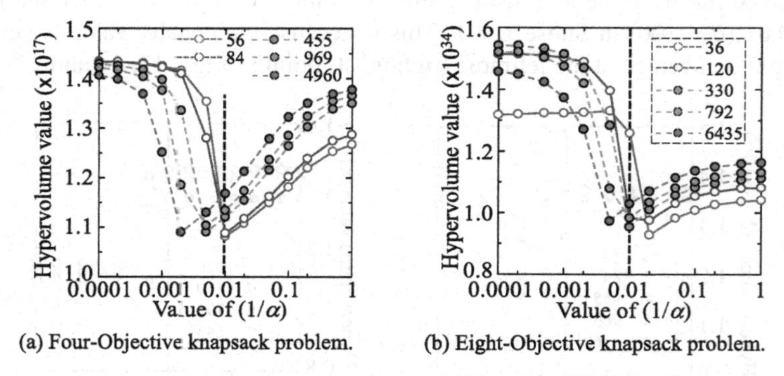

Fig. 15. Experimental results of the quadratic PBI function on the knapsack problems.

5 Conclusions

We first explained why the specification of the penalty value θ is difficult in the PBI function of MOEA/D. Then we proposed an idea of modifying the shape of the contour lines of the PBI function to strike a convergence-uniformity balance. This idea was implemented as two-level and quadratic PBI functions. By the proposed idea, we obtained interpolative results between small and large penalty value cases in Figs. 11 and 14 for the DTLZ2 problems. In Figs. 11(a) and 14(a), improvement was observed by the proposed idea from the interpolative results when $0.1 < 1/\alpha < 1$. However, for the DTLZ1 and knapsack problems, clear performance deterioration was observed from the interpolative results, which was similar to the performance deterioration by θ around 1.0 in the original PBI function.

References

1. Asafuddoula, M., Ray, T., Sarker, R.: A decomposition-based evolutionary algorithm for many objective optimization. IEEE Trans. Evol. Comput. **19**, 445–460 (2015)
2. Cheng, R., Jin, Y., Olhofer, M., Sendhoff, B.: A reference vector guided evolutionary algorithm for many-objective optimization. IEEE Trans. Evol. Comput. (in press). doi:10.1109/TEVC.2016.2519378
3. Deb, K., Jain, H.: An evolutionary many-objective optimization algorithm using reference-point-based non-dominated sorting approach, part I: solving problems with box constraints. IEEE Trans. Evol. Comput. **18**, 577–601 (2014)
4. Deb, K., Pratap, A., Agarwal, S., Meyarivan, T.: A fast and elitist multiobjective genetic algorithm: NSGA-II. IEEE Trans. Evol. Comput. **6**, 182–197 (2002)
5. Deb, K., Thiele, L., Laumanns, M., Zitzler, E.: Scalable multi-objective optimization test problems. In: Proceedings of IEEE CEC 2002, pp. 825–830

6. Huband, S., Hingston, P., Barone, L., While, L.: A review of multiobjective test problems and a scalable test problem toolkit. IEEE Trans. Evol. Comput. **10**, 477–506 (2006)
7. Ishibuchi, H., Akedo, N., Nojima, Y.: Behavior of multi-objective evolutionary algorithms on many-objective knapsack problems. IEEE Trans. Evol. Comput. **19**, 264–283 (2015)
8. Ishibuchi, H., Doi, K., Nojima, Y.: Characteristics of many-objective test problems and penalty parameter specification in MOEA/D. In: Proceedings of IEEE CEC 2016, pp. 1115–1122 (2016)
9. Ishibuchi, H., Tsukamoto, N., Nojima, Y.: Evolutionary many-objective optimization: a short review. In: Proceedings of IEEE CEC 2008, pp. 2424–2431 (2008)
10. Li, B., Li, J., Tang, K., Yao, X.: Many-objective evolutionary algorithms: a survey. ACM Comput. Surv. **48**(1), 13 (2015)
11. Li, K., Deb, K., Zhang, Q., Kwong, S.: An evolutionary many-objective optimization algorithm based on dominance and decomposition. IEEE Trans. Evol. Comput. **19**, 694–716 (2015)
12. Wang, R., Zhang, Q., Zhang, T.: Decomposition based algorithms using pareto adaptive scalarizing methods. IEEE Trans. Evol. Comput. (in press). 10.1109/TEVC.2016.2521175
13. Zhang, Q., Li, H.: MOEA/D: a multiobjective evolutionary algorithm based on decomposition. IEEE Trans. Evol. Comput. **11**, 712–731 (2007)
14. Zitzler, E., Thiele, L.: Multiobjective evolutionary algorithms: a comparative case study and the strength pareto approach. IEEE Trans. Evol. Comput. **3**, 257–271 (1999)

Pareto Inspired Multi-objective Rule Fitness for Noise-Adaptive Rule-Based Machine Learning

Ryan J. Urbanowicz(✉), Randal S. Olson, and Jason H. Moore

Institute for Biomedical Informatics, Perelman School of Medicine,
University of Pennsylvania, Philadelphia, PA, USA
{ryanurb,olsonran,jhmoore}@upenn.edu
http://www.epistasis.org/

Abstract. Learning classifier systems (LCSs) are rule-based evolutionary algorithms uniquely suited to classification and data mining in complex, multi-factorial, and heterogeneous problems. The fitness of individual LCS rules is commonly based on accuracy, but this metric alone is not ideal for assessing global rule 'value' in noisy problem domains and thus impedes effective knowledge extraction. Multi-objective fitness functions are promising but rely on prior knowledge of how to weigh objective importance (typically unavailable in real world problems). The Pareto-front concept offers a multi-objective strategy that is agnostic to objective importance. We propose a Pareto-inspired multi-objective rule fitness (PIMORF) for LCS, and combine it with a complimentary rule-compaction approach (SRC). We implemented these strategies in ExSTraCS, a successful supervised LCS and evaluated performance over an array of complex simulated noisy and clean problems (i.e. genetic and multiplexer) that each concurrently model pure interaction effects and heterogeneity. While evaluation over multiple performance metrics yielded mixed results, this work represents an important first step towards efficiently learning complex problem spaces without the advantage of prior problem knowledge. Overall the results suggest that PIMORF paired with SRC improved rule set interpretability, particularly with regard to heterogeneous patterns.

Keywords: Data mining · Classifier systems · Fitness evaluation · Multi-objective optimization · Machine learning

1 Introduction

Rule-based machine learning (RBML) algorithms learn a set of 'IF:THEN' association rules capturing piece-wise local patterns to map the problem. Learning classifier systems (LCS) are a well-studied type of RBML predominantly applied to supervised and reinforcement learning tasks [1]. LCSs evolve a set of rules that collectively comprise a solution/prediction model. This distributed solution varies from the standard machine learning paradigm of a single model solution, which has made LCS particularly well suited to complex, multifactorial, and

J. Handl et al. (Eds.): PPSN XIV 2016, LNCS 9921, pp. 514–524, 2016.
DOI: 10.1007/978-3-319-45823-6_48

heterogeneous problems such as the n-bit multiplexer machine learning benchmarks [2]. While most early LCS research has focused on reinforcement learning, supervised learning has become a major focus in recent years, particularly with regards to real-world applications [2–5]. One major area includes biomedical data mining and prediction. These types of problems are typically characterized as 'noisy', can include a large number of variables, and can involve complex underlying patterns of association such as epistatic interactions and heterogeneity. In 2015, [2] introduced ExSTraCS 2.0, a more scalable Michigan-style supervised LCS. This approach was able to detect and characterize epistatic and heterogeneous patterns in noisy simulated genetic data, and was the first algorithm to report solving the 135-bit multiplexer directly. However additional emphasis on accuracy in the fitness function was necessary to efficiently solve the set of multiplexer problems (i.e. the ν parameter, which controls the influence of accuracy on fitness, was set to 10 rather than the default of 1). Having prior knowledge that these problems were 'clean' (i.e. the problem could be optimally solved with 100 % prediction accuracy) was an important part of choosing an appropriate objective weight. In that case, accuracy was overemphasized as the only explicit objective. The same logic is true for being able to solve noisy problems. In [2,6] it was found that having ν set above 1 reduced performance in noisy domains. This is because noisy problems can not be solved with 100 % prediction accuracy, and 'optimal' rules for these problems will have an accuracy below 1. Overemphasizing accuracy in a noisy problem leads to dramatic over-fitting, and a loss of generalization, prediction accuracy, and interpretability.

Only a handful of studies have explored a multi-objective fitness functions in LCS. Implicit and explicit multi-objective learning approaches for Michigan and Pittsburgh-style LCS algorithms were reviewed in [7]. Multi-objective research in Pittsburgh-style LCSs has focused on balancing rule-set accuracy with parsimony [8,9]. The MOLeCS algorithm was introduced as the first explicitly multi-objective Pittsburgh LCS [10], applying competing objectives of rule-accuracy and coverage, where coverage refers to the number of training instances that were matched, and thus 'covered' by the rule. MOLeCS was the first LCS to consider a Pareto-front based rule fitness. Two different Pareto-front approaches were proposed in [10] to determine rule fitness ranking each generation of the genetic algorithm. Each involved the formation of a non-dominated rule-fitness front from rules in the current population. The first strategy gave all rules on the front the same 'best' fitness score, and all beneath, the same lower fitness score. The second strategy gave all rules on the front the best set of overall scores but rules on the non-dominated front with the highest accuracy also had the highest fitness. These approaches are not applicable to Michigan-style LCSs, which perform online rather than batch learning. Seeking to improve performance in noisy problems, a weighted-sum approach to multi-objective fitness function for Michigan-style LCS rules was recently proposed in ExSTraCS 2.1 [11] to avoid the overfitting issues that persist even when ν was set to 1 as seen in ExSTraCS 2.0 [2]. This new fitness function improved the interpretability and power to automatically characterize underlying complex patterns in the evolved rule set without sacrificing accuracy [11]. However this approach relies on the assumption that the data is noisy. Case in point, ExSTraCS 2.1 was no longer able to solve clean multiplexer problems beyond the 20-bit version since accuracy was now being undervalued.

In this study we present preliminary results for a Pareto-inspired multi-objective rule fitness (PIMORF). Our goal was to see if we could implement a Pareto-based Michigan-style LCS and determine whether we could identify Pareto-front properties that could be used to switch the objective weighting in favor of accuracy (in clean problems), and coverage (in noisy ones), without the advantage of prior knowledge. Also, building off work in [12], we introduce a fast rule compaction strategy that takes advantage of the multi-objective fitness function to globally rank rules for efficient rule set reduction that preserves performance. This proposed PIMORF was implemented and tested within the ExSTraCS 2.1 algorithm and evaluated over the 6-bit to 135-bit multiplexer problems, as well as a spectrum of complex, noisy simulated genetic datasets concurrently modeling epistatic and heterogeneous patterns of association. We expect that this work will (1) demonstrate the feasibility of adapting the Pareto-front concept to the Michigan-style LCS architecture, (2) improve knowledge extraction, and (3) pave the way for other data-driven fitness function adaptations to encourage assumption-free automated machine learning and data mining.

2 Methods

In this section we briefly (1) introduce the ExSTraCS algorithm, (2) describe how the PIMORF is updated and applied, (3) describe our proposed rule compaction strategy, and (4) outline the evaluation strategy.

2.1 Algorithm

The ExSTraCS algorithm [2] is a Michigan-style LCS algorithm, that has been expanded and adapted to better suit the needs of real-world supervised learning problems wherein classification, prediction, data mining, and/or knowledge discovery is the goal. Most recently in version 2.1, it was expanded to include a multi-objective fitness function that utilized a balanced weighting for the accuracy and coverage objectives. The accuracy and coverage metrics used in the present study were calculated as described in [11]. In short, the accuracy objective is the accuracy above what would be expected by random chance (based on the ratio cases to controls), transformed with an exponential function so that accuracy improvement beyond random chance were highly valued, but less emphasis was being placed on achieving 100 % accuracy. The coverage metric is a state-frequency adjusted measure of the proportion of instances correctly (i.e. accurately) covered by the given rule. For rules that have not yet seen all of the training instances (i.e. so called 'Not Epoch Complete' (NEC) rules), we extrapolate this proportion up to the expected correct coverage once all

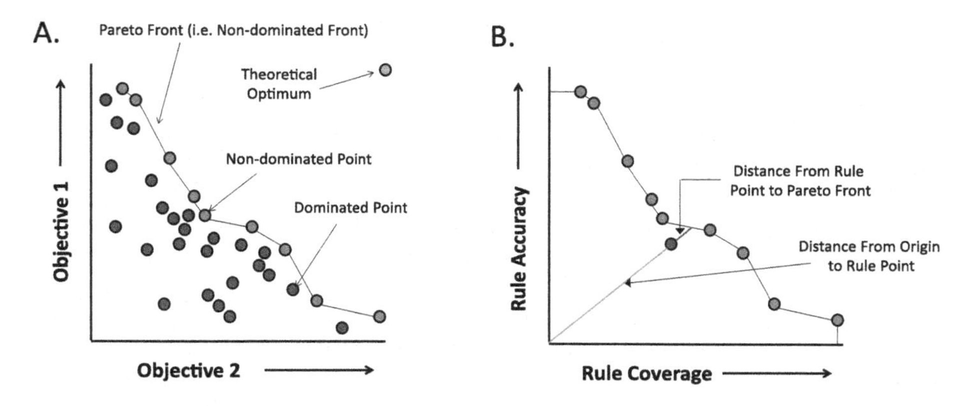

Fig. 1. Pareto front illustrations: (A) General representation of a 2-objective Pareto-front. (B) Application of the Pareto front concept to the calculation of rule-fitness in PIMORF.

data has been observed. For a detailed description of the ExSTraCS algorithm see [2,11]. For comparison we also evaluate ExSTraCS 2.0.1.2, which employs the typical accuracy-based LCS fitness [2]. All implementations are available on sourceforge.com or by request.

2.2 PIMORF for LCS

The Pareto-front is part of the Pareto-optimization approach popularized for multi-objective learning in genetic algorithms [13]. Figure 1A illustrates components of a general Pareto-front as it might be applied to any evolutionary modeling approach. Typically, a population of models are generated and objective performance is evaluated (often accuracy and parsimony). Each model appears as a point in the objective space (see Fig. 1A). The 'front' (i.e. non-dominated front) is the set of all non-dominated points. A point is non-dominated if at least one of its objective values is the maximum observed given the value of the second objective. Next, the set of non-dominated points/models are chosen as the parents for the next generation of models, while dominated models can be discarded. Over multiple generations, the goal is to evolve the front closer to the theoretical optimum. The benefit of this approach is that evolution takes both objectives into account without making any assumptions about objective weighting (i.e. all points on the non-dominated front are treated with equal priority). Our Pareto adaptation to Michigan-style supervised LCS algorithms, (PIMORF) is differently designed to calculate rule-front-relative multi-objective rule fitness values. Instead of points representing models in the multi-objective front space, they represent LCS rules, that are each only part of the overall LCS 'model'. In PIMORF, the rule-fitness front is updated during the course of learning i.e. every time a new rule is generated and added to the rule-population, we check to see if the rule is non-dominated. If it is, the rule-fitness front is updated accordingly. The PIMORF rule-front constitutes the current standard for optimal multi-objective rule fitness, and rules that do not fall on the front (i.e. dominated rules) can be preserved since they might be important contributors to the overall solution despite not possessing a non-dominated combination of objective values. Implementation of PIMORF involves the following: (1) Scaling the 'correct coverage' objective using the maximum observed rule coverage [11]. (2) Learning and updating two separate rule-fitness fronts: one for 'epoch complete' (EC) rules that have been around long enough to have trained on the entire dataset, and another for NEC rules which have seen at least 1000 instances in the training set. To allow for a fair coverage comparison, NEC rule coverage values are extrapolated as described in [11] up to the total training set size. (3) In the first 1000 learning iterations, prior to either front being established, accuracy alone is applied as a surrogate for multi-objective fitness. (4) Rule-fitness is calculated as the relative distance between the origin (where accuracy and coverage objectives are both 0) and the rule point vs. the origin and the intercept point on the rule-fitness front (see Fig. 1B). This is an agnostic approach to multi-objective fitness weighting since any rule on the front has the maximum fitness value. We also explored averaging this agnostic fitness value with a linear accuracy or coverage bias, to be applied in the case that we wanted to apply prior knowledge assuming a clean or noisy problem, or utilize characteristics of the rule-fitness front to detect this automatically. This PIMORF implementation, combining the relative parato distance with a coverage gradient bias will be referred to as ExSTraCS 2.1.1.

2.3 Rule Compaction

Rule compaction is a form of post-processing applied to the evolved LCS population following training. Its goal is to remove poor or redundant rules from the population and yield a more compact rule-set that is easier to interpret (i.e. extract knowledge), and ideally that preserves or improves power and predictive accuracy. In previous work, a variety of LCS rule compaction strategies were implemented and compared [12]. These strategies relied on an accuracy-based fitness function, and therefore has the drawback of being poor for globally ranking rules in the context of noisy problems. This is because highly accurate rules in the population consistently over-fit the training data. In this study, we introduce a simple rule compaction (SRC) scheme which we contrast with QRC, a rapid scheme from [12], that preserves or improves performance, but minimally reduces the overall rule-set size by removing clearly poor or inexperienced rules. SRC complements PIMORF which yields a more globally reliable rule-ranking metric than accuracy or rule-numerosity (i.e. the number of copies of a rule in the population). Numerosity had previously been applied as a rough estimator of global rule-value with mixed success [12]. SRC is implemented as follows: (1) Rank all rules in the population by PIMORF. (2) Progress through the rule set by descending PIMORF. (3) For each rule, identify and remove any instances in the training data that the rule correctly covers. If no remaining instances can be correctly covered, or the rule has an accuracy below the probability of randomly selecting the class specified by the rule, or the rule has not yet had the opportunity to train on the whole dataset (i.e. a NEC rule), this rule is excluded from the final rule-set. SRC stops once the training set is empty (i.e. it has been completely covered), or it has gone through the entire rule set.

2.4 Evaluation

In the present study we compare and evaluate ExSTraCS with and without the proposed PIMORF as well as compare QRF to SRC in the case where PIMORF is applied. Both implementations were run over the same set of 960 noisy (i.e. heritabilities of 0.1, 0.2, or 0.4), complex simulated genetic datasets with 20 discrete-valued attributes that were described and applied in [11] and generated using GAMETES [14]. Each dataset concurrently modeled patterns of epistasis and heterogeneity concurrently where four of the attributes were predictive and 16 were non-predictive. 20 replicates of each dataset were analyzed and 10-fold cross validation (CV) was employed to measure average testing accuracy and account for over-fitting. ExSTraCS was run up to 200,000 learning iterations. Pair-wise statistical comparisons were made using the Wilcoxon signed-rank tests. All statistical evaluations were completed using R. Comparisons were considered to be significant at $p \leq 0.05$. All analyses were performed using 'Discovery', a 2400 core Linux cluster available to the Dartmouth College research community. These comparisons are performed over a set of key performance metrics [2]. Both accuracy metrics were calculated as a respective 'balanced accuracy' to account for imbalanced datasets as the default output of ExSTraCS. 'Both Power' is the ability to correctly identify both two-locus heterogeneous models. 'Single Power' is the ability to have found at least one. 'Both Co-occur. Power' indicates the ability to detect both correct heterogeneous patterns, while 'Single Co-occur. Power' is to detect at least one. Macro Population refers to the number of unique classifiers in the classifier population. Additionally we generated 18 toy simulated genetic datsets each with 20 attributes and 1600 training instances. These included datasets with either (1) a single locus linear model, (2) a two-locus XOR interaction model, or (3) a three-locus XOR interaction model each with varying degrees of noise (0–100 %). Another 6 clean datasets with increasing sample sizes were generated for respective multiplexer benchmarks of (6-bit through 135-bit) [2]. This secondary analysis was designed to explore rule front properties that may serve as a 'switch' to automatically direct ExSTraCS to adopt an accuracy or coverage objective bias in a problem dependent manner.

3 Results and Discussion

Table 1 summarizes the statistical results comparing ExSTraCS with a multi-objective fitness function (v2.1) to ExSTraCS with a simpler accuracy based fitness (v2.0.2.1), as well as to our proposed implementation of PIMORF in

Table 1. Average performance over all 960 datasets.

Rule population performance after 200,000 iterations									
Performance	ExSTraCS								
Statistics	v2.1	v2.0.2.1	p	v2.1.1	p	+QRF	p	+SRC	p
Training Accuracy	.7472	.7975	↑ **	.7519	↑ **	.7485	↓ *	.7648	↑ **
Test Accuracy	.6215	.6177	↓ **	.6123	↓ **	.6130	-	.6192	↑ *
Both Power	.4104	.4031	-	.3895	↓ **	.3901	-	0.3875	↓ *
Single Power	.7802	.7542	↓ **	.7635	↓ **	.7710	↑ *	.7740	↑ **
Both Co-Occur. Power	.2292	.0333	↓ **	.2656	↑ **	.2675	-	.3542	↑ **
Single Co-Occur. Power	.8271	.7688	↓ **	.8260	-	.8266	-	.8375	↑ **
Macro Population	1248.5	1351.5	↑ **	875.6	↓ **	810.2	↓ **	192.7	↓ **
Run Time (min)	52.57	50.56	↓ **	35.56	↓ **	35.61	-	35.58	-

– No significant change
* $p < 0.05$ (Direction of change given by arrows)
** $p < 6.94 \times 10^{-4}$ (Cutoff assumes Bonferroni multiple test correction based on 72 comparisons)

ExSTraCS (v2.1.1). This table further presents statistical comparisons between v2.1.1 following the application of QRF rule compaction, and differently with the application of the proposed SRC approach. As expected, preliminary testing applying SRC to ExSTraCS with accuracy-based fitness yielded a much smaller rule-set but with large performance losses (not shown). As can be reiterated from this table, a multi-objective fitness function (in v2.1) globally improved or maintained average performance measures when compared to accuracy based fitness (in v2.0.2.1) over a spectrum of noisy datasets. Closer inspection of these results, replicating findings in [11], suggest some data set specific trade offs for accuracy and power metrics, enforcing the suboptimality of a multi-objective fitness function with constant equal objective weights. With the substitution of PIMORF as the fitness metric in ExSTraCS (in v2.1.1), we do observe significant performance losses in testing accuracy, Both Power and Single Power, but on the other hand observe a significant increase in Both Co-Occurence Power, which reflects the ability of the algorithm to accurately detect and interpret both underlying heterogeneous models, a critical advantage of LCS algorithms in comparison to other machine learning approaches. Closer inspection of the v2.1.1 results yielded similar dataset specific trade offs in performance, suggesting that when averaged over all datasets this new implementation was not ideal in terms of some key performance metrics, but universal performance metric improvements could be expected if the dataset could be paired to the proper objective weights. Furthermore, v2.1.1 significantly and dramatically reduced the macro-population size (i.e. number of unique rules in the final population), and significantly reduced algorithm run time. While PIMORF performance is not yet optimal without proper objective weighting, the results are promising and support the importance of a multi-objective fitness in noisy rule-based machine learning. Next we examine the effect of our proposed rule compaction strategy

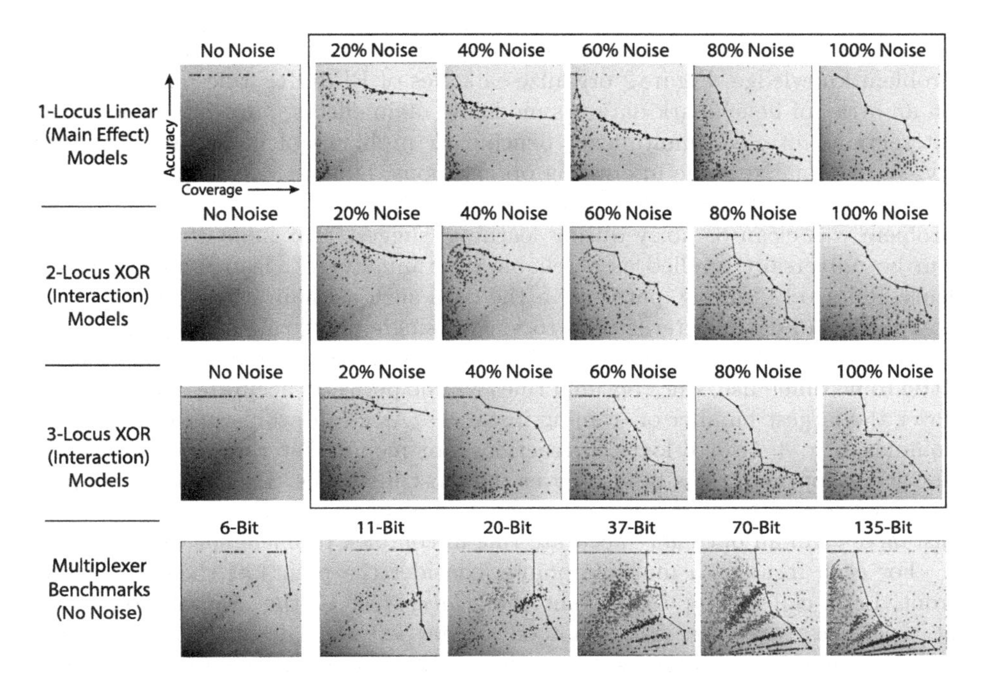

Fig. 2. Pareto inspired rule front comparisons. Each box gives the respective rule-fitness front, with accuracy and coverage axes each scaled between 0 and 1. Points represent rules in the final rule population. The background shading under the curve represents a basic illustration of underlying relative rule-fitness. Note that points found above the front are NEC rules with likely overestimates of objectives. The large black box groups all analyses involving noisy data.

(SRC) in comparison with no rule compaction and QRF. The results for v2.1.1 in Table 1 represent no rule compaction, and the following two columns present the results of QRF and SRC being independently applied to the same rule populations summarized in the column for v2.1.1. This comparison reveals that while QRF does indeed further reduce the rule population size while preserving if not slighly improving some performance metrics, SRC, benefiting from multi-objective fitness that better captures a global sense of rule value, significantly improves testing accuracy as well as all other power metrics with the exception of Both Power which yields a relatively small loss. Using SRC, we observe the largest significant increase in Both Co-Occurrence Power observed for any implementation of LCS or ExSTraCS on this array of simulated genetic benchmarks [2,5,6,11]. This performance metric has been by far the most difficult to improve. Given that SRC dramatically reduces the population size, while simultaneously improving performance relative the population without compaction, this strategy is an improvement over QRF and other strategies evaluated in [12].

In a related analysis, we sought to characterize evolved rule-fitness fronts learned under different conditions of problem complexity and noise. The goal was to see if properties of the front could be applied to appropriately adapt the fitness function to include a more appropriate objective bias without prior problem knowledge. Figure 2 organizes a series of PIMORF rule-fronts learned on an array of benchmark datasets modeling main effects, pure 2-way, or 3-way interactions, or clean multiplexer benchmark problems of increasing complexity. We summarize some interesting observations, but concede that preliminary analyses seeking to apply these characteristics to predict whether the underlying problem was clean or noisy during learning, suggest that none of these trends can be universally applied as a reliable discriminator of clean vs. noisy problems. Relatively 'simple' patterns in the data such as main effects or relatively complex clean data patterns tend to yield a single point rule front (1, 2, and 3 locus models without noise). In such problems, objective weighting likely makes little to no difference, since optimal rules will be perfectly accurate and correctly cover the largest number of training instances. As clean problems become more complex (e.g. 4, 5, or 6 locus interactions), or include heterogeneity, we would not expect optimal rules to also cover the most instances. This is because over-general rules, with sub-optimal accuracy, can correctly cover a larger number of instances than an optimally accurate rule in complex problem spaces.

For each front with multiple points, consider the points at the ends of the front. Let's call the far right point the '*CoverMax*' or the accuracy observed at the largest coverage. The point on the far left we will call the '*AccuracyMax*', or the largest coverage observed at the maximum rule accuracy. One interesting trend is that in partially noisy problems, *CoverMax* tends to be not only large, but larger than *AccuracyMax*. A more general way to view this trend is to notice that partially noisy problems tend to have a shallow overall slope. Alternatively, in clean, complex problems, such as the set of increasingly complex multiplexer problems, *AccuracyMax* tends to be both large and larger than *CoverMax*, or more generally, the slope of these fronts are steep. Unfortunately, these trends become unreliable indicators when (A) there is insufficient signal, or (B) problem complexity increases but the noise level fixed, or (C) the complexity/dimensionality of a problem become so great that the magnitude of *AccuracyMax* makes it difficult to distinguish a complex clean rule-front from a completely noisy signal. This makes the implementation of an automated 'switch', shifting from accuracy to coverage bias problematic. In a clean but complex problem, until at least one optimal rule is found, the characteristics of the front might suggest that the problem is noisy and add a coverage bias. The addition of the wrong bias makes it even more unlikely that optimal rules will be identified, and that the rule front will be correctly updated to an accurately characteristic shape. One final observation for the multiplexer problems, is that we can see clusters of rules forming linear patterns. These groups turned out to correspond with the number of attributes specified in respective rules. Here we can effectively observe the different linear relationships between the accuracy and coverage within candidate rules that have not specified all of the necessary attributes to correctly cover the underlying multiplexer problem (e.g. in the 135-bit problem the 5 clearly identifiable groups correspond to 1–5 attributes having been specified in those rules.

4 Conclusions

The initial results presented in this paper demonstrate the potential benefits of a Pareto-front inspired LCS rule-fitness and support taking an agnostic approach to objective weighting in the likely absence signal to noise ratio prior knowledge in real-world problems. Therefore to promote effective modeling (i.e. accurate prediction and interpretable solutions) in problem domains that are not known to be 100 % signal, a key goal should be to identify or properly estimate the signal to noise ratio, and apply this information to correctly weight accuracy and coverage objectives in the rule fitness function. Despite observing some interesting trends comparing simulated datasets with clean to noisy signals, we have not identified a reliable 'switch' that could be employed to automatically adapt the algorithm to employ the proper objective bias. Future work will explore a purely agnostic Pareto-based rule-fitness to evolve rules and rely on a rule compaction scheme to test different objective weight ratios, and select the best one as the final rule-set. While this work focuses on the adaptation of rule-based machine learning to problems with unknown noise properties, multi-objective fitness could still benefit performance on clean problems, where a small explicit generalization pressure, has the potential to speed up learning beyond the underlying implicit generalization pressures and the use of subsumption.

Acknowledgments. The computations in this work were performed on the Discovery cluster supported by the Research Computing group, ITS at Dartmouth College. This work was supported by NIH grants AI116794, LM009012, LM010098, EY022300, LM011360, CA134286, and GM103534.

References

1. Urbanowicz, R.J., Moore, J.H.: Learning classifier systems: a complete introduction, review, and roadmap. J Artif. Evol. Appl. **2009**, 25 (2009). Article no. 1
2. Urbanowicz, R.J., Moore, J.H.: ExSTraCS 2.0: description and evaluation of a scalable learning classifier system. Evol. Intel. **8**(2–3), 89–116 (2015)
3. Bernadó-Mansilla, E., Garrell-Guiu, J.M.: Accuracy-based learning classifier systems: models, analysis and applications to classification tasks. Evol. Comput. **11**(3), 209–238 (2003)
4. Orriols, A., Bernadó-Mansilla, E.: Class imbalance problem in UCS classifier system: fitness adaptation. In: The 2005 IEEE Congress on Evolutionary Computation, vol. 1, pp. 604–611. IEEE (2005)
5. Urbanowicz, R.J., Bertasius, G., Moore, J.H.: An extended michigan-style learning classifier system for flexible supervised learning, classification, and data mining. In: Bartz-Beielstein, T., Branke, J., Filipič, B., Smith, J. (eds.) PPSN 2014. LNCS, vol. 8672, pp. 211–221. Springer, Heidelberg (2014)
6. Urbanowicz, R., Moore, J.: The application of Michigan-style learning classifier systems to address genetic heterogeneity and epistasis in association studies. In: Proceedings of the 12th Annual Conference on Genetic and Evolutionary Computation, pp. 195–202. ACM (2010)
7. Bernadó-mansilla, E., Llorà, X., Traus, I.: Multiobjective learning classifier systems: an overview. Technical report 2005020 (2005)

8. Llorà, X., Goldberg, D.E.: Bounding the effect of noise in multiobjective learning classifier systems. Evol. Comput. **11**(3), 279–298 (2003)
9. Bacardit, J., Garrell, J.M.: Bloat control and generalization pressure using the minimum description length principle for a pittsburgh approach learning classifier system. In: Kovacs, T., Llorà, X., Takadama, K., Lanzi, P.L., Stolzmann, W., Wilson, S.W. (eds.) IWLCS 2003. LNCS (LNAI), vol. 4399, pp. 59–79. Springer, Heidelberg (2007)
10. Mansilla, E.B., Guiu, J.M.G.: MOLeCS Using multiobjective evolutionary algorithms for learning. In: Zitzler, E., Thiele, L., Deb, K., Coello, C.A.C., Corne, D. (eds.) Evolutionary Multi-criterion Optimization, pp. 696–710. Springer, Heidelberg (2001)
11. Urbanowicz, R., Moore, J.: Retooling fitness for noisy problems in a supervised Michigan-style learning classifier system. In: Proceedings of the 2015 on Genetic and Evolutionary Computation Conference, pp. 591–598. ACM (2015)
12. Tan, J., Moore, J., Urbanowicz, R.: Rapid rule compaction strategies for global knowledge discovery in a supervised learning classifier system. In: Advances in Artificial Life, ECAL, vol. 12 pp. 110–117 (2013)
13. Deb, K., Pratap, A., Agarwal, S., Meyarivan, T.: A fast and elitist multiobjective genetic algorithm: NSGA-II. IEEE Trans. Evol. Comput. **6**(2), 182–197 (2002)
14. Urbanowicz, R.J., Kiralis, J., Sinnott-Armstrong, N.A., Heberling, T., Fisher, J.M., Moore, J.H.: GAMETES: a fast, direct algorithm for generating pure, strict, epistatic models with random architectures. BioData Min. **5**(1), 1 (2012)

Decomposition-Based Approach for Solving Large Scale Multi-objective Problems

Luis Miguel Antonio(✉) and Carlos A. Coello Coello

Departamento de Computación, CINVESTAV-IPN
(Evolutionary Computation Group), 07300 Mexico D.F., Mexico
lmiguel@computacion.cs.cinvestav.mx, ccoello@cs.cinvestav.mx

Abstract. Decomposition is a well-established mathematical programming technique for dealing with multi-objective optimization problems (MOPs), which has been found to be efficient and effective when coupled to evolutionary algorithms, as evidenced by MOEA/D. MOEA/D decomposes a MOP into several single-objective subproblems by means of well-defined scalarizing functions. It has been shown that MOEA/D is able to generate a set of evenly distributed solutions for different multi-objective benchmark functions with a relatively low number of decision variables (usually no more than 30). In this work, we study the effect of scalability in MOEA/D and show how its efficacy decreases as the number of decision variables of the MOP increases. Also, we investigate the improvements that MOEA/D can achieve when combined with coevolutionary techniques, giving rise to a novel MOEA which decomposes the MOP both in objective and in decision variables space. This new algorithm is capable of optimizing large scale MOPs and outperforms MOEA/D and GDE3 when solving problems with a large number of decision variables (from 200 up to 1200).

1 Introduction

Although in real-world applications, many MOPs have hundreds or even thousands of decision variables, the effect of the scalability of decision variables space over modern MOEAs has not been properly addressed. In fact, scalability in decision variables space is a topic that has been only scarcely studied in the context of multi-objective optimization using MOEAs. This is perhaps motivated by the fact that most researchers assume that the currently available MOEAs should be able to work properly with a large number of decision variables. Nevertheless, there exists empirical evidence that indicates that most of the currently available MOEAs significantly decrease their efficacy as the number of decision variables of a MOP increases [4,5]. The work reported here tries to narrow the gap in this important topic.

We are interested in improving the MOEA/D [14] framework in order to make it capable to deal with large scale (in decision variables space) MOPs.

C.A.C. Coello—Author Gratefully acknowledges support from CONACyT project no. 221551.

J. Handl et al. (Eds.): PPSN XIV 2016, LNCS 9921, pp. 525–534, 2016.
DOI: 10.1007/978-3-319-45823-6_49

Thus, we study here the effect of scalability in MOEA/D and we investigate the improvements that this algorithm can achieve. For this purpose, we propose to combine the MOEA/D framework with Cooperative Coevolutionary techniques (which have shown to be very effective for large scale single-objective optimization [8,13]), giving rise to a novel MOEA based on a double decomposition (both in objective and decision variables space).

The remainder of this paper is organized as follows. The previous related work is discussed in Sect. 2. Section 3 describes our proposed approach and the experiments carried out to validate it. Finally, our conclusions and some possible paths for future work are drawn in Sect. 4.

2 Previous Related Work

Regarding studies on scalability in MOEAs, to the authors' best knowledge, the most significant ones are those reported by Durillo et al. [4,5], in which the behavior and effect of decision variables scalability over eight multi-objective metaheuristics (representatives of the state-of-the-art) are analyzed. For this sake, the authors adopted a benchmark of scalable problems (the Zitzler-Deb Thiele (ZDT) [16] test suite) using a number of decision variables that ranged from 8 up to 2048. The study paid particular attention to the computational effort required by each algorithm for reaching the true Pareto front of each problem. These papers provide empirical evidence of the decrease in efficacy and efficiency that multi-objective metaheuristics have when dealing with MOPs with a large number of decision variables, as it is shown in their results.

Another work in this direction is a small study presented in [14], where ZDT1 is solved with up to 100 decision variables using MOEA/D. They analyze how the computational cost, measured in terms of the number of function evaluations, increases as the number of decision variables of the problem increases. This is shown using a number of decision variables that ranges from 10 up to 100 variables. They used as a performance index the average number of function evaluations spent by MOEA/D for reducing the *D*-metric [17] and concluded that the average number of function evaluations linearly scales up, as the number of decision variables increases. They attribute these results to two facts: (i) the number of scalar optimization sub-problems in MOEA/D is fixed to be 100, regardless of the number of decision variables of the problem. (ii) the complexity of each single-objective optimization could scale linearly with the number of decision variables. However, this study is too small to show a general behavior of MOEA/D over large scale (in decision variables space) MOPs.

Although scalability in decision variables space is a topic that has been only scarcely studied in the evolutionary multi-objective optimization field, large-scale optimization has been the focus of an important amount of research in global (single-objective) optimization using evolutionary algorithms. The currently available approaches for large-scale global optimization can be roughly divided in two groups: those that decompose a high-dimensional decision variables vector into small subcomponents which can then be handled by conventional EAs (see for example [13]) and the ones that approach the problem by disturbing the population of the EA or by combining different evolutionary methods (see for example [9]). From these methods, cooperative coevolution has been found to be one of the most successful approaches for solving large and complex problems, through the use of problem decomposition.

3 Our Proposed Approach

The main idea of our proposed approach is to make use of the divide-and-conquer technique, adopted by the cooperative coevolutionary framework for large scale single objective optimization, and incorporate such concept into MOEA/D. Our motivation is that it is very natural to use scalar optimization methods in MOEA/D, since each solution is associated with a scalar optimization problem, in contrast with non-decomposition MOEAs where in most cases there is no easy way for them to take advantage of scalar optimization methods. Next, we give a brief description of both MOEA/D and cooperative coevolution.

3.1 MOEA/D

The *multi-objective evolutionary algorithm based on decomposition* (MOEA/D) [14] has attracted growing interest from the community, due to its simplicity and to its effectiveness when applied to a broad range of MOPs. MOEA/D decomposes the MOP into a set of single-objective subproblems and solves these subproblems simultaneously using an evolutionary algorithm. It adopts a set of weights each of which corresponds to a single subproblem. Each weight vector is used as a search direction to define a scalar function. For this sake, the so called Tchebycheff decomposition is the most widely used approach. Given a weight vector $\lambda = [\lambda_1, \ldots, \lambda_n]^T$ the corresponding subproblem is defined as:

$$\text{minimize} \quad g^{te}(x|\lambda, z^*) = \max_{1 \leq i \leq n} \lambda_i |f_i(x) - z_i^*| \tag{1}$$

where z^* is the reference point chosen as the minimum of objective function values found during the evolution. The main advantage of the Tchebycheff approach is that it works regardless of the shape of the Pareto front, while other decomposition approaches (like the weighted sum approach) only work for convex Pareto fronts. The weights are also used to define neighborhoods of the subproblems. The neighborhood relations among these subproblems are defined based on the distances between their aggregation coefficient vectors. At each generation, a new individual is generated and evaluated using its own neighborhood of weights, with the idea that any information about these closest weight vectors should be helpful for optimizing the current individual's subproblem. Once this new individual is created, it is compared to its parent and in case it is better, it replaces its parent. Moreover, it is also compared to other individuals in its neighborhood and is allowed to replace some of them. Therefore, at each generation, the population is composed of the best solution found so far (i.e., since the start of the run of the algorithm) for each subproblem.

3.2 Cooperative Coevolution

In nature, coevolution is the process of reciprocal genetic change in one species, or group, in response to another. That is, coevolution refers to a reciprocal evolutionary change between species that interact with each other [6]. A coevolutionary search involves the use of multiple species as the representation of a solution to an optimization problem. In the case of cooperative algorithms, which are the focus of this work, individuals are rewarded when they work well with other individuals and punished when they perform poorly together [11].

The first framework of cooperative coevolution (CC) utilized within evolutionary algorithms was originally introduced by Potter and De Jong [10], with their *Cooperative Coevolutionary Genetic Algorithm* (CCGA). This framework uses a divide-and-conquer approach to split the decision variables into subpopulations of smaller size, so that each of these subpopulations is optimized with a separate EA. The main idea was to decompose a *high-dimensional* problem into several low-dimensional subcomponents and evolve these subcomponents cooperatively. So, instead of evolving a population (global or spatially distributed) of similar individuals representing a global solution, the cooperative coevolutionary framework coevolves subpopulations of individuals representing specific parts of the global solution.

After this work, there were many more *cooperative coevolutionary* approaches, most of them for large scale global optimization since this showed to be a good framework for solving high-dimensional problems [8,13]. In general, the most common cooperative coevolutionary framework for high-dimensional global (single-objective) optimization can be summarized as follows:

1. Decompose a vector of decision variables into m low dimensional subcomponents.
2. Set $j = 1$ to start a new cycle.
3. Optimize the jth subcomponent with a certain EA for a predefined number of fitness evaluations (FEs).
4. If $j < m$ then $j++$, and go to Step 3.
5. Stop if the stopping criteria are satisfied; otherwise, go to Step 2 for the next cycle.

3.3 Description of Our Proposed Approach

If we are to extend the basic computational model of cooperative coevolution into an approach that already uses a decomposition strategy as the one adopted by MOEA/D, we must address the issues of a second problem decomposition, as well as other issues such as the interdependencies among subcomponents, credit assignment, and the maintenance of diversity. In order to do so and to provide reasonable opportunities for the success of co-adapted subcomponents and an increase in efficiency when dealing with large scale MOPs, we can not use the whole model of cooperative coevolution as we did in our previous work presented in [1], since it is much more costly (due to the use of multiple subpopulations) than the use of MOEA/D as a standalone algorithm. Instead, we only incorporate into MOEA/D a coevolutionary step where we make use of the divide-and-conquer technique that splits the MOP to be solved, but in *decision variables* space.

Our proposed approach divides the vector of decision variables into S subcomponents (species), each one representing a subset of all the decision variables at a time rather than taking only one variable per subcomponent. We assign each decision variable to its corresponding subcomponent in a random way, trying to increase the chance of optimizing some interacting variables together. However, it is important to note that the cooperative coevolutionary adaptation presented here does not work as in the original framework, since we do not intend to use several subpopulations for each subcomponent of the problem and we will not need individuals from the other species to assemble a complete solution in order to perform a fitness evaluation. Here, we only use decision variable decomposition to make operations (crossover and mutation) more effective and with this, we can manage in a better way the *curse of dimensionality* (the performance of an evolutionary algorithm deteriorates rapidly as the dimensionality of the search space increases [12]) present in MOEAs. So, individuals will still be representing a whole solution, but operators will be applied based on the corresponding species, and not based on the individuals. The algorithm of our proposed MOEA based on double decomposition (MOEA/D^2) works as follows:

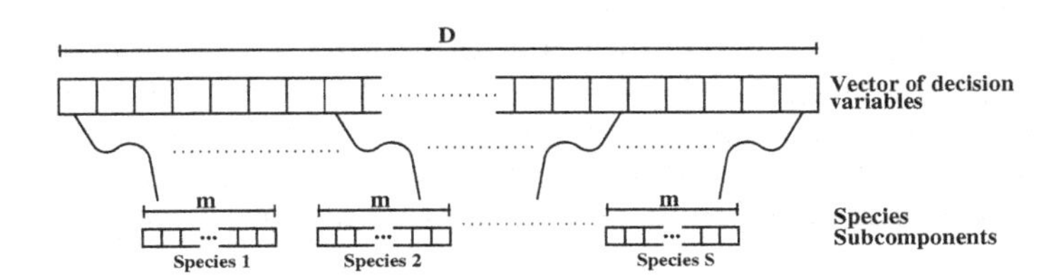

Fig. 1. Graphical representation of the subcomponents (species) creation. Here, we assume a vector of decision variables of dimension D which is divided into S subcomponents of dimension m, created in a random way from the original vector of decision variables and assigned to the S existing species, where $D = m * S$.

Input:
- The MOP with k objective functions
- N: The number of subproblems considered in MOEA/D
- S: The number of species for decision variables decomposition
- A set of N uniform spread weight vectors:
  $\lambda^1, \ldots, \lambda^N$
- T: The neighborhood size

Output:
- PS: the final solutions found during the search

Step (1) Initialization:

Step (1.1) Set the external population of final solutions $PS = \emptyset$.

Step (1.2) Find the T closest weight vectors to each weight vector. For each $i = 1, \ldots, N$, set $B(i) = \{i_1, \ldots, i_T\}$, where $\lambda^{i_1}, \ldots, \lambda^{i_T}$ are the T closest weight vectors to λ^i.

Step (1.3) Generate an initial population $x^1, \ldots x^N$ randomly or by a problem-specific method. Set $FV^i = f(x^i)$.

Step (1.4) Divide the problem into S subcomponents $c^1, \ldots, c^S$ each of dimension m, created in a random way from the original vector of decision variables x of dimension D (as shown in Fig. 1), where $D = m * S$, such that, for each $j = 1, \ldots, N$, $x^j = [c^1_j, \ldots, c^S_j]$.

Step (1.5) Initialize $z = [z_1, \ldots, z_k]^T$, where z_i is the best value found so far for objective f_i.

Step (2) Update:

For $i = 1, \ldots, N$ do

Step (2.1) Crossover and Mutation:

For $j = 1, \ldots, S$ do

Step (2.1.1) Randomly select two indexes p, q from $B(i)$, and then generate a new solution y^j_c from c^j_p and c^j_q using crossover.

Step (2.1.2) Apply a problem-specific repair improvement heuristic on y^j_c to produce y'^j_c.

Step (2.2) Assemble y' from $[y'^1_c, \ldots, y'^S_c]$, sorting the subcomponents to form the original vector of decision variables.

Step (2.3) For each $j = 1, \ldots, k$, if $z_j > f_j(y')$, then set $z_j = f_j(y')$.

Step (2.4) Update of Neighboring Solutions: For each index $j \in B(i)$ use (1) such that, if $g^{te}(y'|\lambda^j, z^*) < g^{te}(x^j|\lambda^j, z^*)$, then $FV^j = f(y')$.

Step (2.5) Remove from the external population PS all the vectors dominated by $f(y')$. Add $f(y')$ to PS if no vectors in PS dominate it.

Step (3) Stopping Criterion: Stop if the termination criterion is satisfied. Otherwise, go to Step 2.

Since c^j_p and c^j_q in Step 2.1.1 are the current best subcomponent (in decision variables space) solutions to neighbors of the ith subproblem (in objective function space) and their dimensions are less than the original vector of decision variables x, their offspring y'^j_c (already improved by mutation) should be a good contribution to the complete assemble of the new final solution y'. Therefore, the resultant solution is very likely to have a lower (improved) function value for the neighbors of the ith subproblem. Also, by using only the decomposition nature of the cooperative coevolutionary framework, there is no need for extra function evaluations. Therefore, the efficiency of MOEA/D is not lost.

3.4 Experimental Results

We validated MOEA/D^2 comparing its performance with respect to that of the original MOEA/D and with respect to GDE3 [7]. Although GDE3 is not a decomposition based algorithm, in the studies presented in [4] this differential evolution based MOEA obtained the best overall results, which is the reason why we decided to include it in our comparative study.

Methodology. For the purposes of this study, we adopted the Deb-Thiele-Laumanns-Zitzler (DTLZ) test suite [3] with instances of three objectives with a number of decision variables that ranges from 200 to 1200. In order to assess the performance of each approach, we selected the hypervolume indicator [15], since this measure can differentiate between degrees of complete outperformance of two sets. The hypervolume is defined as the n-dimensional space that is contained by an n-dimensional set of points. When applied to multi-objective optimization, the n-dimensional objective values for solutions are treated as points for the computation of such space. That is, the hypervolume is obtained by computing the volume (in objective function space) of the nondominated set of solutions Q that minimize a MOP. For every solution $i \in Q$, a hypercube v_i is generated with a reference point W and the solution i as its diagonal corner of the hypercube:

$$\mathcal{S} = Vol\left(\bigcup_{i=1}^{|Q|} v_i\right) \tag{2}$$

The aim of this study is to identify which of the algorithms being compared is able to get closer to the true Pareto front using the same number of objective function evaluations and how they behave as the dimensionality of the MOP increases.

Parameterization. The parameters of each algorithm used in our study were chosen in such a way that we could do a fair comparison among them. For MOEA/D^2 and MOEA/D, we adopted SBX and polynomial-based mutation [2] as the crossover and mutation operators, respectively. The mutation probability was set to $p_m = 1/l$, where l is the number of decision variables; the distribution indexes for SBX and the polynomial-based mutation were set as: $\eta_c = 20$ and $\eta_m = 20$. For the case of MOEA/D^2, different numbers of species were used for each problem instance, in order to have 2 decision variables per species. So, for problems with 200 decision variables, 100 species were used, for problems with 400 decision variables, 200 species were used, and so on. The maximum number of iterations adopted for all problems and MOEAs was set to 1000, regardless of their dimensionality. The F and CR values for GDE3 were set to 0.5. Finally, the population size for all algorithms in all problems instances was set to 100.

Discussion of Results. In our experiments, we obtained the hypervolume value over the 25 independents runs performed. Table 1 shows the average hypervolume of each of the MOEAs being compared for each test problem adopted, as well as the results of the statistical analysis that we made to validate our experiments, for which we used Wilcoxon's rank sum. Also, we show the improvement on the hypervolume value that our approach was able to obtain with respect to that of the other algorithms. GDE3 presented the poorest performance in all problem instances. MOEA/D produced competitive results for DTLZ2 and DTLZ4, although it could not outperform our approach in any problem instance.

Table 1. Average of the hypervolume indicator. The cells containing the best hypervolume value for each problem have a grey colored background. The improvement columns show the improvement on the hypervolume value that our approach was able to get against that of the other MOEAs. The P(H) columns shows the results of Wilcoxon's rank sum test. P is the probability of observing the given result (the null hypothesis is true). Small values of P cast doubt on the validity of the null hypothesis. $H = 0$ indicates that the null hypothesis ("medians are equal") cannot be rejected at the 5% level. $H = 1$ indicates that the null hypothesis can be rejected at the 5% level.

		MOEA/D^2	MOEA/D	MOEA/D^2-MOEA/D	MOEA/D^2-MOEA/D	GDE3	MOEAD^2-GDE3	MOEAD^2-GDE3
Function	No. Vars	HV	HV	Improvement	P(H)	HV	Improvement	P(H)
DTLZ1	200	124999991998953.0000	124999970289543.0000	21709410.2344	0.000000 (1)	124923656113375.0000	76335885578.2031	0.000000 (1)
	400	124999755014274.0000	124999298073533.0000	456940740.9063	0.000000 (1)	124181465622337.0000	818289391936.5000	0.000000 (1)
	600	124998478272908.0000	124996030413092.0000	2447859816.0625	0.000000 (1)	122039040800943.0000	2959437471964.5000	0.000000 (1)
	800	124995060011719.0000	124985536236730.0000	9523774988.8594	0.000000 (1)	117702084444497.0000	7292975567222.5300	0.000000 (1)
	1000	124986970597954.0000	124955356063479.0000	31614534475.2500	0.000000 (1)	110256296271387.0000	14730674326567.2000	0.000000 (1)
	1200	124970550659900.0000	124894718579624.0000	75832080276.4531	0.000000 (1)	99191612716078.9000	25778937943821.2000	0.000000 (1)
DTLZ2	200	728999.3904	728999.3862	0.0043	0.712386 (0)	728989.2937	10.0967	0.000000 (1)
	400	728999.3808	728999.3680	0.0128	0.043602 (1)	728321.7458	677.6350	0.000000 (1)
	600	728999.3605	728999.3085	0.0520	0.000000 (1)	721831.2296	7168.1309	0.000000 (1)
	800	728999.2870	728999.1289	0.1581	0.000000 (1)	698874.5807	30124.7063	0.000000 (1)
	1000	728999.0954	728998.4267	0.6687	0.000000 (1)	653788.8012	75210.2941	0.000000 (1)
	1200	728998.4393	728994.9746	3.4647	0.000000 (1)	571671.4472	157326.9921	0.000000 (1)
DTLZ3	200	1727999970755560.0000	1727999849624200.0000	121131355.7500	0.000000 (1)	1727222040269000.0000	777930486563.2500	0.000000 (1)
	400	1727996400439630.0000	1727991944817710.0000	4455621923.0000	0.000000 (1)	1716392835963730.0000	11603564475898.3000	0.000000 (1)
	600	1727970985655110.0000	1727945403715830.0000	25581939288.2500	0.000000 (1)	1679379508439150.0000	48591477215964.8000	0.000000 (1)
	800	1727890440027340.0000	1727805158813020.0000	85281214323.2500	0.000000 (1)	1597662758376130.0000	130227681651214.0000	0.000000 (1)
	1000	1727715593620590.0000	1727460212199730.0000	255381420857.2500	0.000000 (1)	1463152563598520.0000	264563030022069.0000	0.000000 (1)
	1200	1727363193497010.0000	1726773363259150.0000	589830237867.0000	0.000000 (1)	1259639566256750.0000	467723627240265.0000	0.000000 (1)
DTLZ4	200	728999.4140	728999.4078	0.0062	0.277231 (0)	728991.3901	8.0240	0.000000 (1)
	400	728999.4065	728999.3896	0.1154	0.000000 (1)	728201.9349	434.6057	0.000000 (1)
	600	728999.3788	728999.3464	0.0324	0.000000 (1)	720704.2965	8295.0824	0.000000 (1)
	800	728999.3150	728999.1945	0.1206	0.000000 (1)	696046.7161	32952.5989	0.000000 (1)
	1000	728999.1477	728998.6192	0.5285	0.000000 (1)	644641.3868	84357.7609	0.000000 (1)
	1200	728998.5780	728994.9171	3.6609	0.000000 (1)	559363.2154	169635.3626	0.000000 (1)
DTLZ5	200	1727866.0538	1727865.9384	0.1154	0.013007 (1)	1727431.4481	434.6057	0.000000 (1)
	400	1727865.5061	1727864.6278	0.8784	0.000000 (1)	1721336.0150	6529.4911	0.000000 (1)
	600	1727863.4153	1727859.1967	4.2187	0.000000 (1)	1697620.6971	30242.7183	0.000000 (1)
	800	1727857.2346	1727840.7092	16.5255	0.000000 (1)	1635056.7976	92800.4370	0.000000 (1)
	1000	1727837.7148	1727760.3047	77.4101	0.000000 (1)	1510727.5139	217110.2010	0.000000 (1)
	1200	1727773.5677	1727524.1382	249.4296	0.000000 (1)	1313095.5122	414678.0555	0.000000 (1)
DTLZ6	200	999967922.4861	999899450.4162	68472.0699	0.000000 (1)	999330750.1795	637172.3066	0.000000 (1)
	400	998891441.7157	996820925.0570	2070516.6587	0.000000 (1)	987305237.5147	11586204.2010	0.000000 (1)
	600	990768252.6193	981381295.3432	9386957.2761	0.000000 (1)	948509739.2585	42258513.3608	0.000000 (1)
	800	964064335.0353	937687686.8457	26376648.1896	0.000000 (1)	874883514.5826	89180820.4527	0.000000 (1)
	1000	906131328.8124	855654049.3429	50477279.4695	0.000000 (1)	744576613.9787	161554714.8337	0.000000 (1)
	1200	803763312.2166	712087777.8538	91675534.3628	0.000000 (1)	551828946.6234	251934365.5932	0.000000 (1)
DTLZ7	200	2203.4849	2203.4656	0.0193	0.000000 (1)	2055.0598	148.4252	0.000000 (1)
	400	2193.4627	2192.7324	0.7303	0.000000 (1)	1699.4438	494.0190	0.000000 (1)
	600	2090.3379	2067.9036	22.4343	0.000413 (1)	1338.1314	752.2065	0.000000 (1)
	800	1842.2642	1815.7768	26.4875	0.013007 (1)	1059.6383	782.6260	0.000000 (1)
	1000	1605.8489	1526.3840	79.4649	0.000001 (1)	855.6279	750.2210	0.000000 (1)
	1200	1398.8865	1352.2878	46.5987	0.006223 (1)	718.3771	680.5094	0.000000 (1)

According to Wilcoxon's test, we cannot reject the null hypothesis in only two cases when comparing our approach to MOEA/D, in DTLZ2 and DTLZ4 with 200 decision variables, which means that in these cases both algorithms have a similar behavior. This shows that our approach has a similar performance to MOEA/D in multi-frontal problems. The best overall performance of our approach was in DTLZ1, DTLZ3 and DTLZ6, where our approach significantly outperformed MOEA/D and GDE3, and as the results show, as the dimensionality of the problems grows, the improvement obtained by our approach on the hypervolume value increases. So, we can confirm that our approach can handle in a better way problems with degenerate Pareto optimal fronts, as is the case of DTLZ6. Decomposition is very effective when solving non-separable problems such as DTLZ1 and DTLZ3. For DTLZ5 and DTLZ7, the improvement was more remarkable as the dimensionality of the problems increased. However, our approach was also able to outperform both MOEA/D and GDE3 in all instances. Based on the results of Wilcoxon's test, we can confirm that the null hypothesis can be rejected, so MOEA/D^2 produced the best overall results.

4 Conclusions and Future Work

Here, we developed a novel decomposition-based MOEA called MOEA/D^2, which adopts decomposition based techniques used by cooperative coevolutionary algorithms. MOEA/D^2 uses a double decomposition of the MOP, one in objective functions space, as done by MOEA/D, and another one in decision variables space. Our experimental results indicate that MOEA/D^2 clearly outperforms MOEA/D and GDE3 in MOPs having from 200 up to 1200 decision variables. Our approach was able to deal with all the difficulties presented in the DTLZ test suite, even in high dimensionality. The results confirmed that our proposed approach is very effective and efficient in tackling large scale MOPs. As part of our future work, we intend to study other decomposition techniques for decision variable space. We are also interested in studying the possible use of other (computationally inexpensive) methods to generate a set of weight vectors more uniformly distributed for MOEA/D^2.

References

1. Antonio, L.M., Coello Coello, C.A.: Use of cooperative coevolution for solving large scale multiobjective optimization problems. In: 2013 IEEE Congress on Evolutionary Computation (CEC 2013), Cancún, México, 20–23 June 2013, pp. 2758–2765. IEEE Press (2013). ISBN 978-1-4799-0454-9
2. Deb, K., Pratap, A., Agarwal, S., Meyarivan, T.: A fast and elitist multiobjective genetic algorithm: NSGA-II. IEEE Trans. Evol. Comput. **6**(2), 182–197 (2002)
3. Deb, K., Thiele, L., Laumanns, M., Zitzler, E.: Scalable test problems for evolutionary multiobjective optimization. In: Abraham, A., Jain, L., Goldberg, R. (eds.) Evolutionary Multiobjective Optimization. Theoretical Advances and Applications, pp. 105–145. Springer, Heidelberg (2005)

4. Durillo, J., Nebro, A., Coello Coello, C., Garcia-Nieto, J., Luna, F., Alba, E.: A study of multiobjective metaheuristics when solving parameter scalable problems. IEEE Trans. Evol. Comput. **14**(4), 618–635 (2010)
5. Durillo, J.J., Nebro, A.J., Coello Coello, C.A., Luna, F., Alba, E.: A Comparative study of the effect of parameter scalability in multi-objective metaheuristics. In: 2008 Congress on Evolutionary Computation (CEC 2008), Hong Kong, June 2008, pp. 1893–1900. IEEE Service Center
6. Ehrlich, P.R., Raven, P.H.: Butterflies and plants: a study in coevolution. Evolution **18**(4), 586–608 (1964)
7. Kukkonen, S., Lampinen, J.: GDE3: the third evolution step of generalized differential evolution. In: 2005 IEEE Congress on Evolutionary Computation (CEC 2005), Edinburgh, Scotland, September 2005, vol. 1, pp. 443–450. IEEE Service Center
8. Omidvar, M.N., Li, X., Yao, X., Yang, Z.: Cooperative co-evolution for large scale optimization through more frequent random grouping. In: 2010 IEEE Congress on Evolutionary Computation (CEC), vol. 1, pp. 1–8, September 2010
9. Noman, N., Iba, H.: Enhancing differential evolution performance with local search for high dimensional function optimization. In: Proceedings of the 2005 Conference on Genetic and Evolutionary Computation, GECCO 2005, pp. 967–974. ACM, New York (2005)
10. Potter, M.A., De Jong, K.A.: A cooperative coevolutionary approach to function optimization. In: Davidor, Y., Männer, R., Schwefel, H.-P. (eds.) PPSN 1994. LNCS, vol. 866. Springer, Heidelberg (1994)
11. Potter, M.A., De Jong, K.A.: Cooperative coevolution: an architecture for evolving coadapted subcomponents. Evol. Comput. **8**(1), 1–29 (2000)
12. van den Bergh, F., Engelbrecht, A.P.: A cooperative approach to particle swarm optimization. Trans. Evol. Comput. **8**(3), 225–239 (2004)
13. Yang, Z., Tang, K., Yao, X.: Multilevel cooperative coevolution for large scale optimization. In: 2008 IEEE Congress on Evolutionary Computation, pp. 1663–1670. IEEE Press (2008)
14. Zhang, Q., Li, H.: MOEA/D: a multiobjective evolutionary algorithm based on decomposition. IEEE Trans. Evol. Comput. **11**(6), 712–731 (2007)
15. Zitzler, E., Brockhoff, D., Thiele, L.: The hypervolume indicator revisited: on the design of pareto-compliant indicators via weighted integration. In: Obayashi, S., Deb, K., Poloni, C., Hiroyasu, T., Murata, T. (eds.) EMO 2007. LNCS, vol. 4403, pp. 862–876. Springer, Heidelberg (2007)
16. Zitzler, E., Thiele, L.: Multiobjective evolutionary algorithms: a comparative case study and the strength pareto approach. IEEE Trans. Evol. Comput. **3**(4), 257–271 (1999)
17. Zitzler, E., Thiele, L., Laumanns, M., Fonseca, C.M., da Fonseca, V.G.: Performance assessment of multiobjective optimizers: an analysis and review. IEEE Trans. Evol. Comput. **7**(2), 117–132 (2003)

Parallel Algorithms and Hardware Issues

An Evolutionary Framework for Replicating Neurophysiological Data with Spiking Neural Networks

Emily L. Rounds[1(✉)], Eric O. Scott[2], Andrew S. Alexander[3], Kenneth A. De Jong[2], Douglas A. Nitz[3], and Jeffrey L. Krichmar[1]

[1] Department of Cognitive Sciences, University of California, Irvine, Irvine, CA, USA
{roundse,jkrichma}@uci.edu

[2] Department of Computer Science, George Mason University, Fairfax, VA, USA
{escott8,kdejong}@gmu.edu

[3] Department of Cognitive Science, University of California, San Diego, La Jolla, CA, USA
{aalexander,dnitz}@ucsd.edu

Abstract. Here we present a framework for the automatic tuning of spiking neural networks (SNNs) that utilizes an evolutionary algorithm featuring indirect encoding to achieve a drastic reduction in the dimensionality of the parameter space, combined with a GPU-accelerated SNN simulator that results in a considerable decrease in the time needed for fitness evaluation, despite the need for both a training and a testing phase. We tuned the parameters governing a learning rule called spike-timing-dependent plasticity (STDP), which was used to alter the synaptic weights of the network. We validated this framework by applying it to a case study in which synthetic neuronal firing rates were matched to electrophysiologically recorded neuronal firing rates in order to evolve network functionality. Our framework was not only able to match their firing rates, but also captured functional and behavioral aspects of the biological neuronal population, in roughly 50 generations.

Keywords: Spiking neural networks · Evolutionary algorithms · Indirect encoding · Neurophysiological recordings · Plasticity · Data matching · Parallel computing

1 Introduction

As the power and availability of high-performance computing resources grows, large and biologically realistic networks of spiking neurons are becoming increasingly relevant as a computational modeling tool. Networks consisting of on the order of hundreds or thousands of neurons allow researchers to formulate models that can represent how neural circuits give rise to cognition and behavior [12], and they allow engineers to prototype novel mechanisms that may prove useful in applications of neuromorphic hardware [9].

J. Handl et al. (Eds.): PPSN XIV 2016, LNCS 9921, pp. 537–547, 2016.
DOI: 10.1007/978-3-319-45823-6_50

An important step in the design of these networks is the selection of parameter values that enable the model to perform a desired target function. Simulations of spiking neural networks (SNNs) tend to be very computationally expensive, and involve a large number of free parameters. For instance, even after a model of a neurological system has been constrained with the best available physiological data, it is not uncommon for an SNN to exhibit tens or hundreds of thousands of unknown synaptic weight parameters that must be specified by the model designer. Furthermore, SNN applications are often based on recurrent network topologies, where gradient-based optimization methods (such as backpropagation) are inapplicable. For these reasons, the task of parameterizing an SNN to solve a particular task, or to accurately model particular biological data, is an especially difficult kind of neural network optimization problem.

In this paper, we propose a two-pronged framework for tuning the parameters of spiking neural networks. First, we achieve a drastic reduction in the dimensionality of the parameter space by using a learning mechanism as an *indirect encoding* method for automatically adapting the weights of neural connections. This allows us to use an evolutionary algorithm (EA) to tune only the coarse-grained structure of the network and the global parameters of the learning method itself. Second, we use a GPU-based SNN simulator to accelerate fitness evaluation. This allows us to compensate for the increased computational effort that is required to train the networks through learning. To learn the synaptic weights, we apply a standard nearest neighbor implementation of spike-timing-dependent plasticity (STDP) [11], a widely-used and biologically realistic model of synaptic plasticity which has been studied experimentally [4] as well as computationally.

We demonstrate the functionality of this framework by applying it to a case study in which an SNN is tuned to match neural recordings from the rat retrosplenial cortex (RSC) [1]. To our knowledge, this is the first attempt to apply search algorithms to train SNNs to replicate neurophysiological data from awake, behaving animals. Existing work in the area of SNN synthesis has either trained recurrent networks to match high-level animal behavior in cognitive tasks [7,13,17], or it has focused on tuning the parameters of individual neuron models to match electrophysiological data [8,14–16]. However, in order to better understand the mechanisms underlying neurological circuits and to verify theoretical models of cognition, it is important that they are able to match neurological data in terms of neuronal firing rates as well as population functionality and behavior. Sometimes the choice of these parameters can be constrained by high-quality physiological data [20], but even with the best-understood brain regions we almost never know the precise value that these parameters should assume to best mimic nature. We show that this can be done effectively through the use of the present evolutionary parameter-tuning framework.

In general, neural networks have been successfully evolved using both direct and indirect encoding schemes. The NEAT and HyperNEAT algorithms [18,19] utilize an indirect encoding scheme in order to evolve increasingly complex network topologies, while Carlson et al. [5] used a similar approach to ours to evolve SNNs whose neuronal responses gave rise to receptive fields similar to those found in neurons from the primary visual cortex. However, this study used artificial data and did not perform a behavioral task. Asher et al. [2] used a direct encoding scheme to train an artificial neural network (ANN) to perform visually- and memory-guided reaching tasks. However, this approach took thousands of generations to evolve, and yielded a network that had less biologically realistic neuronal units. To our knowledge, evolutionary algorithms utilizing indirect encoding have not been used to tune the parameters of networks containing realistic spiking neurons in order to perform a cognitive task.

To summarize, our approach is novel in three key ways: (1) We use biologically plausible spiking SNNs with realistic neural dynamics that not only reproduce the behavior of neural circuits, but also match empirical data at the neuron level while simultaneously capturing the holistic behavior of the circuit, (2) we use an indirect encoding approach evolutionary algorithm to tune SNNs, and (3) we use GPUs to run populations of SNNs simultaneously, thus speeding up the search process. This approach may be useful for replicating other neural datasets, and for creating biologically plausible SNNs.

2 Methodology

We test our STPD-based encoding method by fitting the activity of a network of 1,017 neurons to neurophysiological and behavioral data that have been previously collected by Alexander and Nitz from six male Long-Evans rats [1]. In neuroscience models, this topology is often loosely, manually specified based on the known, somewhat incomplete properties of a real structure in the brain. In the present case, we begin with a pre-specified network topology that defines the coarse-grained connectivity structure among several groups of neurons (Fig. 1). The goal of parameter tuning is to adjust the details of the network—such as synaptic weights, the number of connections between groups, and/or the behavioral parameters of the neurons in each group—such that the network successfully produces the desired target behavior.

2.1 RSC Model

In the current study, each SNN contained three groups of neurons, shown in Fig. 1: 417 excitatory input neurons, which handled the encoding of the behavioral inputs; 480 regular-spiking excitatory Izhikevich neurons and 120 fast-spiking inhibitory Izhikevich neurons [10]. The network had four types of connections: inputs to excitatory (`Inp→Exc`), inputs to inhibitory (`Inp→Inh`), recurrent excitatory (`Exc→Exc`), and inhibitory to excitatory (`Inh→Exc`). All synaptic projections were random with a 10 % chance of connectivity. No topology was enforced. To train the network, a learning rule known as STDP was used to update the weights of the network [4]; specifically, a standard nearest-neighbor implementation [11]. Homeostatic synaptic scaling was incorporated into the STDP rule in order to keep the neuronal firing rates within a reasonable regime by scaling to a target firing rate (for more details see Carlson et al. [6]).

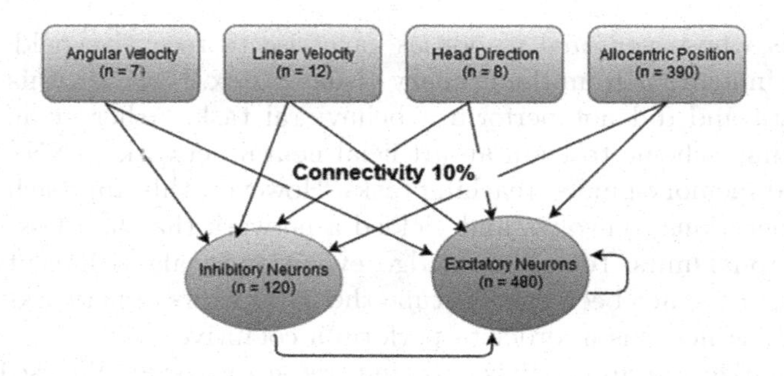

Fig. 1. The network topology used in the current study included four input groups, excitatory and inhibitory neuron populations, feedforward inhibition, and recurrent excitation, with 10 % connectivity between neurons.

2.2 Parameters and Training

The automated tuning framework was used to evolve a total of 18 parameters, which were related to plasticity, overall firing rates and weight ranges (see Table 1). These parameters were used as inputs to the CARLsim GPU-based simulation framework, which we used to run the SNN models [3,5]. Twelve parameters related to STDP were evolved, which correspond to three types of STDP curves. The remaining parameters control the homeostatic target base firing rates for the excitatory and inhibitory populations, and the initial and maximum weight values for each set of inter-group connections.

Table 1. Parameters initialized via the ECJ framework

Parameter	A^+	A^-	τ^+	τ^-	Base FR (exc)	Base FR (inh)	Inp-Exc Init.	Inp-Inh Init.	EE Init.	IE Init.
Minimum	−0.0002	−0.0002	5.0	5.0	5.0	5.0	0.01	0.01	0.001	0.001
Maximum	0.004	0.004	100.0	100.0	20.0	20.0	0.5	0.5	0.5	0.5
Std. dev	−0.00042	−0.00042	9.5	9.5	1.5	1.5	0.049	0.049	0.0499	0.0499

Each network was trained on a subset of trials from the Alexander and Nitz experiments [1]. In the present study, each generation underwent a training session and a testing session. Both consisted of 150 behavioral trials, which were drawn randomly from separate subsets of the total number of trials recorded to ensure that there was no overlap. In the testing phase, STDP was disabled in order to keep the synaptic weights fixed following training.

Following testing, the population was evaluated by summing the best correlations between the experimentally observed and simulated neurons for each SNN. The best correlations were found by first correlating every simulated neuron ($n = 600$) against every experimentally observed neuron ($n = 228$). Next, a match was chosen based on highest correlation value between each experimentally observed neuron and the corresponding simulated neuron (a neuron could only be chosen once). After all experimentally observed neurons had a match, the fitness score for that individual SNN was computed by summing the correlations ρ between each pair (1). A maximum mean firing rate threshold was also incorporated into the fitness function to ensure that simulated firing rates were reasonable and realistic. The firing rate of each neuron in the network was averaged across all trials, and the highest observed value was considered the maximum mean. If the observed maximum mean firing rate maxFR exceeded the threshold, then the fitness score was penalized by subtracting the difference between the threshold and the observed firing rate (2):

$$f(\boldsymbol{x}) = \begin{cases} \sum_{i=1}^{n} \rho(\mathtt{realFR}_i, \mathtt{synFR}_{\mathtt{match}}) & \text{if } \mathtt{maxFR} < \mathtt{FR}_{\mathtt{target}}, \\ \sum_{i=1}^{n} \rho(\mathtt{realFR}_i, \mathtt{synFR}_{\mathtt{match}}) - \mathtt{FR}_{\mathtt{error}} & \text{otherwise,} \end{cases} \tag{1}$$

where

$$\mathtt{FR}_{\mathtt{error}} = \mathtt{FR}_{\mathtt{max}} - \mathtt{FR}_{\mathtt{target}}, \tag{2}$$

and $\mathtt{FR}_{\mathtt{target}} = 250\,\text{Hz}$ was the maximum mean firing rate allowed for any given neuron.

After a generation, the fitness scores were sent to ECJ via the PTI for evaluation and constructing a new population. The simulations proceeded for 50 generations. The complete process was repeated 10 times to ensure repeatability. It is important to reiterate that the use of GPU processing speeds up the fitness function significantly. In this case, the fitness function runs 136,800 Pearson's r correlations (600 synthetic neurons multiplied by 228 neurophysiological neurons) per each individual, which is computationally very expensive. This complexity could increase considerably with the size of the dataset being replicated, the size of the network being run, and/or the number of individuals in the population, making it very important that the fitness function can be calculated in parallel on GPU.

2.3 Evolutionary Algorithm

We represented the parameter space of the RSC model as vectors in $\mathbb{R}^{18}$, and then applied a $(\mu + \lambda)$-style, overlapping-generations EA with truncation selection to maximize $f(\boldsymbol{x})$. We used a mutation operator that takes each parameter and adds 1-dimensional Gaussian noise with probability 0.5. The width of the Gaussian mutation operator was fixed at 10 % of the range that each parameter was allowed to vary within. The values of μ and λ were fixed at 3 and 15, respectively. It was straightforward to combine the SNN simulator with the ECJ evolutionary computation system [21] to create a unified parameter-tuning framework.

These decisions result in an algorithm with a small population and a strong selection pressure. This simple EA proved sufficient for our purpose, which is provide a proof of the feasibility of evolving SNNs with an STDP-based indirect encoding. We leave the problem of customizing EA design decisions to maximize performance for future work.

3 Results

3.1 Fitness Values and Firing Rate Correlations

Each of the 10 independent runs of the EA were executed for a small number of generations. Thanks to the indirect encoding, the best fitness found tended to be very high after just 50 generations (see Fig. 2(a)), with a mean of 105.93 ± 0.91. The highest observed fitness was 107.79. A total of 228 experimentally correlated neurons were matched, thus the average firing rate correlation was about 0.47 per neuron. The lowest observed fitness was 104.7, resulting in a correlation of about 0.46 per neuron (strong correlations by experimental standards). At the start of each evolutionary run, the average maximum fitness score was 84.57 ± 19.78.

Each of the ten evolutionary runs took 3.13 ± 1.26 days to complete. A breakdown of how long a generation took can be seen in Table 2. In the beginning, the population ran very slowly, taking approximately four hours to complete (slightly under two hours for training, and slightly more for testing). By the tenth generation, the population took roughly an hour to complete, which stayed relatively constant across the remaining generations (breaking down to about 20 min for training and 30 for testing). However, there was considerable variance in how long a generation could take at each point (each generation had a standard deviation of about one hour) because of the different firing rates of individual SNNs. Although the fitness increased during the evolutionary run, the selection strategy tended to include high and low firing SNNs in the population, which affects runtime performance in CARLsim.

The tuning framework was able to closely match experimental neural activity to simulated neural activity. Figure 2(c) shows two representative examples of matched neurons with high correlations. Note that the correlation values in the figure are not much higher than the average correlation value, suggesting that they are typical examples of matched neurons indicative of the network's overall fitness. Thus the EA was able to generate networks whose neuronal firing rates were able to match those of the experimental dataset.

Table 2. Average runtimes in minutes (mean/std. dev)

Generation	1	10	20	30	40	50
Training	115.09/24.86	21.85/20.71	22.23/21.54	20.29/17.62	18.5/11.83	21.17/16.79
Testing	126.28/39.51	42.43/40.55	42.91/30.89	39.39/28.34	32.1/17.32	39.65/37.13
Total	240.48/59.24	64.42/59.57	65.3/50.4	59.94/45.25	50.66/28.38	60.91/49.86

3.2 Replicating Empirical Data

The evolved networks were also able to capture functional aspects of the neurons observed in the electrophysiological data. In agreement with Alexander and Nitz [1], we found neurons that were active when the animal was turning left or right (turn, no mod. in Fig. 2(b)) and turn cells that were route modulated (i.e.,

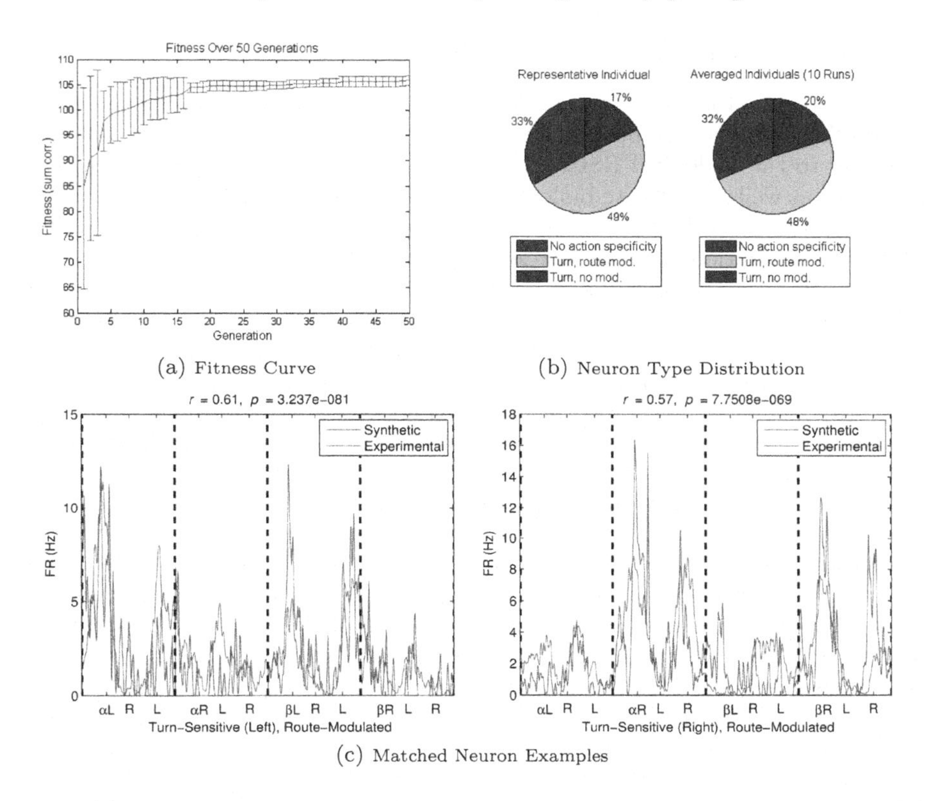

(a) Fitness Curve

(b) Neuron Type Distribution

(c) Matched Neuron Examples

Fig. 2. (a) Network fitness rapidly and consistently converged over 50 generations. (b) All evolved networks yielded consistent distributions of neuron types. (c) Two representative matched neuron examples are provided, demonstrating that firing rate correlations between synthetic and experimentally-observed neurons were generally quite high.

preferring one turn over another on the same route; e.g., the first left instead of the second left on an LRL route; see turn, mod. in Fig. 2(b)), as well as neurons that were turn-insensitive (see no action specificity in Fig. 2(b)). In agreement with the experimental data, we found that approximately 20 % of the population were turn-sensitive, but were not route modulated. The ratios of turn-sensitive and route modulated cells found in the evolved SNNs were comparable to those found experimentally (compare our Fig. 2(b) with Fig. 3(a) in Alexander and Nitz [1]). However, we found a higher proportion of neurons that were route modulated (48 % as opposed to 26 %), and surprisingly, fewer of our neurons were turn-insensitive (32 % as opposed to 50 %). This may be because our SNN inputs, which were derived from recorded behavioral metrics, were less noisy than the sensorimotor signals that the rat uses to navigate.

3.3 Replicating Population Functionality

Lastly, the EA produced network behavior that was quite similar to the empirical findings, which is important because it suggests that the network functions similarly to the biological RSC, and thus has the ability to capture population dynamics as well as replicate biological neuronal firing rates. The evolved agent's position along a track could be discerned from the simulated population activity (see Fig. 3). Positional reconstruction was computed by cross-correlating mean neural activity across even vs. odd trials. Similar to that observed experimentally, the positional ensemble firing rate reconstructions from a representative evolved SNN clearly showed that the neural activity at positions on even trials was highly correlated with neural activity at the same positions on odd trials (Fig. 3(a)), thus very accurate reconstructions could be determined from population activity when the subject was in the same environment. That is, the highest correlation values occurred between the same bin numbers across even and odd trials, as shown by the white dashed line, from the top left corner to the bottom right corner. The reconstruction process was also applied to trials when the tracks were in different locations (α and β). Figure 3(b) shows a correlation matrix between positions in β and positions in α for the LRL trajectory. These reconstructions indicated that the position of the agent could be inferred between the track positions as well, but with less accuracy than for the even vs. odd reconstructions. This is consistent with the results reported in [1] (compare our Fig. 3(a) and (b) with Fig. 3(e) and 6(a) in Alexander and Nitz [1]), suggesting that the evolved simulated network is capable of conjunctively encoding allocentric and route-centric information similar to the biological RSC. These results were consistent across all evolved SNNs.

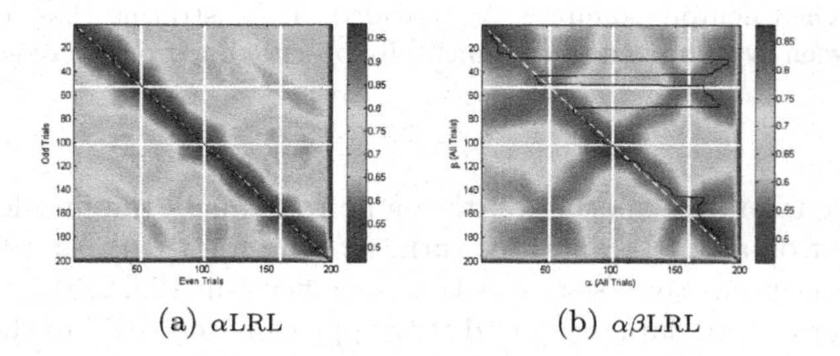

(a) αLRL (b) $\alpha\beta$LRL

Fig. 3. (a) Positional ensemble firing rate correlations (even vs. odd trials) which were highest fell along a 'perfect prediction line' suggesting that the network was able to infer its position along any given route so long as that route was in the same allocentric position. (b) Positional ensemble firing rate correlations for all trials at position α vs. position β deviated from the perfect prediction line, suggesting that the network discriminated routes that existed in different allocentric positions.

4 Discussion

In the present study, we introduced an automated tuning framework that leverages the search power of evolutionary algorithms combined with the parallelization of GPUs, which can result in a speedup of up to 60 times faster than a CPU in CARLsim [3]. This results in an efficient method for searching the SNN parameter space by drastically reducing its dimensionality via an indirect encoding scheme in which a learning rule, STDP, was used to specify the synaptic weights of each network. Performing fitness evaluation on each network in parallel further reduced the time necessary to tune the SNNs, even though every individual in the population was subjected to both a training and a testing phase. We successfully applied this framework to a case study in which it was used to evolve a model of the brain region RSC using electrophysiologically recorded neurons. Rather than altering the synaptic weights of each SNN directly, an evolutionary algorithm was used to alter the learning parameters of each SNN until a close match between synthetic and recorded neuronal firing rates was found, which resulted in a reduction of the number of parameters to be tuned from thousands to only 18. Furthermore, the evolutionary algorithm took only 50 generations to converge, demonstrating the framework was able to efficiently evolve a solution. This is in stark contrast to direct encoding methods of evolving neural networks, which can take thousands of generations to converge [2].

The phenomenological results of this case study suggest that the approach of using STDP as an indirect encoding scheme will generalize to other types of SNN tuning problems, and can be used to match other neurophysiological datasets, since many electrophysiological recordings are collected under conditions similar to the present dataset. First, the SNNs successfully captured the underlying network activity, which was reflected in the fitness score of each evolved network. Secondly, the SNNs captured neuronal function observed in the data, which was reflected in empirically observed distributions of non-route modulated turn-sensitive neurons and route modulated turn-sensitive neurons, respectively. Thirdly, the ensemble activity of the synthetic neurons captured behavioral functionality, such as position and route reconstruction.

The capacity to efficiently synthesize networks that reproduce neuron and network functionality across these three levels is of considerable importance as we attempt to move toward a greater understanding of brain function. We have demonstrated that we have created a powerful tool with this capacity by applying our framework to this case study of the RSC, which may be applied to a variety of modeling efforts and tuning problems involving SNNs. Further experiments are underway to investigate how the network responds to manipulations of its inputs, and to predict how neural activity in the retrosplenial cortex might change depending on environmental context. These predictions can then be tested by conducting new electrophysiological experiments, the results of which could lead to a better understanding of how neural responses give rise to behavior.

Acknowledgments. Supported by the National Science Foundation (Award IIS-1302125).

References

1. Alexander, A.S., Nitz, D.A.: Retrosplenial cortex maps the conjunction of internal and external spaces. Nat. Neurosci. **18**(8), 1143–1151 (2015)
2. Asher, D.E., Krichmar, J.L., Oros, N.: Evolution of biologically plausible neural networks performing a visually guided reaching task. In: Proceedings of the 2014 Annual Conference on Genetic and Evolutionary Computation (GECCO 2014), pp. 145–152. ACM, New York (2014)
3. Beyeler, M., Carlson, K.D., Chou, T.-S., Dutt, N., Krichmar, J.L.: CARLsim 3: a user-friendly and highly optimized library for thecreation of neurobiologically detailed spiking neural networks. In: 2015 International Joint Conference on Neural Networks (IJCNN 2015), pp. 1–8. IEEE (2015)
4. Bi, G.-Q., Poo, M.-M.: Synaptic modifications in cultured hippocampal neurons: dependence on spike timing, synaptic strength, and postsynaptic cell type. J. Neurosci. **18**(24), 10464–10472 (1998)
5. Carlson, K.D., Nageswaran, J.M., Dutt, N., Krichmar, J.L.: An efficient automated parameter tuning framework for spiking neuralnetworks. Front. Neurosci. **8** (2014)
6. Carlson, K.D., Richert, M., Dutt, N., Krichmar, J.L.: Biologically plausible models of homeostasis and STDP: stabilityand learning in spiking neural networks. In: The 2013 International Joint Conference on Neural Networks (IJCNN 2013), pp. 1–8. IEEE (2013)
7. Carnevale, F., deLafuente, V., Romo, R., Barak, O., Parga, N.: Dynamic control of response criterion in premotor cortex during perceptual detection under temporal uncertainty. Neuron **86**(4), 1067–1077 (2015)
8. Fountas, Z., Shanahan, M.: GPU-based fast parameter optimization for phenomenological spikingneural models. In: 2015 International Joint Conference on Neural Networks (IJCNN 2015), pp. 1–8, July 2015
9. Hu, M., Li, H., Chen, Y., Wu, Q., Rose, G.S., Linderman, R.W.: Memristor crossbar-based neuromorphic computing system: a case study. IEEE Trans. Neural Networks Learn. Syst. **25**(10), 1864–1878 (2014)
10. Izhikevich, E.M.: Simple model of spiking neurons. IEEE Trans. Neural Networks **14**(6), 1569–1572 (2003)
11. Izhikevich, E.M., Desai, N.S.: Relating STDP to BCM. Neural Comput. **15**(7), 1511–1523 (2003)
12. Krichmar, J.L., Coussy, P., Dutt, N.: Large-scale spiking neural networks using neuromorphic hardwarecompatible models. ACM J. Emerging Technol. Comput. Syst. (JETC), **11**(4) (2015). Article no. 36
13. Mante, V., Sussillo, D., Shenoy, K.V., Newsome, W.T.: Context-dependent computation by recurrent dynamics in prefrontal cortex. Nature **503**(7474), 78–84 (2013)
14. Prinz, A.A., Billimoria, C.P., Marder, E.: Alternative to hand-tuning conductance-based models: construction and analysis of databases of model neurons. J. Neurophysiol. **90**(6), 3998–4015 (2003)
15. Prinz, A.A., Bucher, D., Marder, E.: Similar network activity from disparate circuit parameters. Nat. Neurosci. **7**(12), 1345–1352 (2004)
16. Rossant, C., Goodman, D.F.M., Fontaine, B., Platkiewicz, J., Magnusson, A.K., Brette, R.: Fitting neuron models to spike trains. Front. Neurosci. **5**(9) (2011)

17. Song, H.F., Yang, G.R., Wang, X.J.: Training excitatory-inhibitory recurrent neural networks forcognitive tasks: a simple and flexible framework. PLoS Comput. Biol. **12**(2), e1004792 (2016)
18. Stanley, K.O., D'Ambrosio, D.B., Gauci, J.: A hypercube-based encoding for evolving large-scale neural networks. Artif. Life **15**(2), 185–212 (2009)
19. Stanley, K.O., Miikkulainen, R.: Evolving neural networks through augmenting topologies. Evol. Comput. **10**(2), 99–127 (2002)
20. Tripathy, S.J., Savitskaya, J., Burton, S.D., Urban, N.N., Gerkin, R.C.: Neuroelectro: a window to the world's neuron electrophysiology data. Front. Neuroinf. **8** (2014)
21. White, D.R.: Software review: the ECJ toolkit. Genet. Program. Evolvable Mach. **13**(1), 65–67 (2012)

A Cross-Platform Assessment of Energy Consumption in Evolutionary Algorithms

Towards Energy-Aware Bioinspired Algorithms

F. Fernández de Vega[1], F. Chávez[1(✉)], J. Díaz[1], J.A. García[1], P.A. Castillo[2], Juan J. Merelo[2], and C. Cotta[3]

[1] Universidad de Extremadura, Badajoz, Spain
{fccfdez,fchavez,mjdiaz,jangelgm}@unex.es
[2] ETSI Informática, Universidad de Granada, Granada, Spain
{pacv,jmerelo}@ugr.es
[3] ETSI Informática, Campus de Teatinos, Universidad de Málaga, Málaga, Spain
ccottap@lcc.uma.es

Abstract. Energy consumption is a matter of paramount importance in nowadays environmentally conscious society. It is also bound to be a crucial issue in light of the emergent computational environments arising from the pervasive use of networked handheld devices and wearables. Evolutionary algorithms (EAs) are ideally suited for this kind of environments due to their intrinsic flexibility and adaptiveness, provided they operate on viable energy terms. In this work we analyze the energy requirements of EAs, and particularly one of their main flavours, genetic programming (GP), on several computational platforms and study the impact that parametrisation has on these requirements, paving the way for a future generation of energy-aware EAs. As experimentally demonstrated, handheld devices and tiny computer models mainly used for educational purposes may be the most energy efficient ones when looking for solutions by means of EAs.

Keywords: Green computing · Energy-aware computing · Performance measurements · Evolutionary algorithms

1 Introduction

In the analysis of single or multi-processor algorithm performance, an important feature is frequently forgotten: energy consumption, which largely correlates with performances provided by new processors. That is why, in an environment where raw processor speed is no longer doubling at an accelerated pace, reducing energy consumption and taking it into account when evaluating algorithms becomes an issue, to the point that latest HPC benchmarks also include this measurement in their reports and there are calls for *energy-proportional computing* [5] and *green computing* [10], a term that was born in the last decade to refer to problems associated to energy consumption in computing environments, particularly

J. Handl et al. (Eds.): PPSN XIV 2016, LNCS 9921, pp. 548–557, 2016.
DOI: 10.1007/978-3-319-45823-6_51

in large data centers. But this energy-aware and proportional point of view is equally applicable to desktop computers and any kind of algorithms that may be run.

When dealing with evolutionary algorithms (EAs), big efforts have been applied to improve performances while applying parallel and distributed systems [13]. Improvements have tried to analyze global quality of solutions when compared with time required to find them. But similarly as the traveler considering not only speed but also price when selecting means of transport, we should also consider energy consumption when running an algorithm, and not just the time to solution.

To the best of our knowledge, the influence of this important parameter has not been analyzed yet in the context of EAs, although its importance has already been recognized [6]. This is the main goal of this work, to make a preliminary analysis of the impact of energy consumption when running a well know EA, Genetic Programming, on different hardware architectures, so that we may in the future be aware of the importance, and even design energy-aware EAs; we will also measure the impact of a particular feature, population size, in the energy consumption, so that these parameters can be taken into account in an energy-aware design of evolutionary algorithms.

The rest of the paper is organized as follows: Sect. 2 describes previous works on the area; Sect. 3 describes the experiments performed and Sect. 4 shows the results obtained. Finally we summarize our conclusions in Sect. 5.

2 Evolutionary Algorithms and Energy Consumption

Computer science took interest in energy efficiency a number of years ago, and a new research topic was born, *Green Computing* [10], together with the *energy-aware* [5,14] concept. Even processor makers offered new processors providing *dynamic frequency scaling*, which adapts energy consumption as well as heat dissipation to the need of the processes to be run [1,2,4].

On the other hand, EAs have already been applied as optimization algorithms in the context of energy management. We can thus find optimization problems associated to HVAC (Energy management of heating, ventilating and air-conditioning) [8,12]. We can also find EAs applied to energy dispatch [7]. But any of the above referred problems are only tangentially connected to the problem we are interested in: how to include energy consumption as one of the main features of EAs to be considered when looking for solutions, and its relationship with the main parameters of the algorithm.

The main concept discussed in this paper is the capability of an EA to adapt to dynamic environments in which energy consumption is one of the main components to be optimized [6]. This capability, which is one of the self-$\star$ features of a given algorithm, including EAs [6], has already been considered by researchers in other kind of algorithms and computer architectures [3], in some cases an essential part of them [14]. Energy-awareness is considered a key component in infrastructures of any size, from large data centers to processor architectures for mobile devices where battery life must be optimized. Also in this context, EAs have been employed to design cache hierarchies reducing energy consumption and heat dissipation [16].

Yet, to the best of our knowledge, EAs have never been studied from the point of view of their own energy needs. Given their stochastic nature, the number of parameters regulating the way they perform the search process and the plethora of hardware platforms available to run them, we consider it of interest to study the energy consumed when looking for a solution, so that in the future they may become energy-aware and capable of self-regulating when progressing towards the solution of the problem faced. This is what we have set out to do in this paper.

3 Methodology

This preliminary study tries to measure energy consumption for the Genetic Programming (GP) algorithm. In the following subsections both the algorithm setting (Sect. 3.1) and computational platforms (Sect. 3.2) are presented.

3.1 Algorithmic Setting

Given the stochastic nature of this kind of algorithms, we firstly decided to run each of the experiment 30 times so that the average can be computed as an estimation of the algorithm behavior. In order to establish a fair comparison among the different hardware platforms considered, these runs are done with the same 30 random seeds so that all of the runs are exactly the same in every platform, when considering high level operations defined in the high level programming language.

On the other hand, and given the influence of computing time in the total amount of energy consumed by the algorithm, we configured the main loop of the algorithm to finish when the optimal solution is found. We are thus mainly interested in the average computing time for the 30 runs, together with the energy consumed along that time. The only differences that may arise are due to hardware differences: instruction set architecture, processor speed and operating system; features that are not the focus of this work. Nevertheless, these differences may influence future decision on the preferred hardware and operating system for the algorithms.

We must also mention the interest in studying some of the main parameters of the algorithm: they have a well-known impact on the time to find solutions, and may thus also directly, or indirectly influence the energy consumed to reach that solution. In this preliminary study we have focused on population size and have tested several values for the problem selected. Although we are working with GP and a well-known problem, this first analysis will be helpful to see that energy consumption is an important issue when working with EAs.

Table 1. Main GP parameters for multiplexer 6.

Max number of generations	500
Population sizes	100, 200, 400, 500, 1000
Crossover probability	0.9
Mutation probability	0.1

Table 2. Devices

Device	Processor	Cores	RAM	OS
raspberry pi	Cortex-A7 900 MHz	4	1 GB	Raspbian GNU/Linux 7
tablet	Samsung Galaxy Tab 3 SM-T311, Exynos 4212 1.5 GHz	2	1.5 GB	Android 4.4.2 (kernel 3.0.31)
laptop	Intel(R) Core(TM) i5-2450M 2.5 GHz	4	8 GB	Ubuntu 12.04.5 LTS
iMac	Intel(R) Core(TM) i5 2.7 GHz	4	4 GB	OSX 10.11.4
blade	Virtual Machine (on IBM 8CPUs x 2 GHz Intel Xeon CPUE5504 @ 2.00 GHz (x2), 16 Gb RAM)	4	4 GB	Debian 6.0, 64 bits

In order to ease the compilation processes in every hardware platform, we have selected a well known implementation of GP in the C programming language: lilgp[1]. Regarding the problem selected for the experimental stage, we have selected one of the test problems provided by lilgp: the multiplexer problem. To be precise, we have set up to work with 6 bits. The main parameters of the algorithm are described in Table 1. Function and terminal sets are the standard ones as described by Koza [11].

3.2 Computational Platforms

Several computational platforms have been tested, i.e., raspberry pi, tablet, laptop, iMac and blade. Table 2 provides the details for the hardware architectures and operating systems used. Given the differences among hardware devices, we have employed different ways for measuring energy consumed by each of the algorithms.

Laptop and Raspberry Pi. Regarding the laptop and raspberry pi, we have employed a multimeter for measuring total power delivered to the device in two different scenarios: (i) when the algorithm is not running and (ii) when the algorithm is running. Starting with an initial measurement at rest (first scenario) in both cases, our multimeter is able to measure the watts delivered,

[1] http://garage.cse.msu.edu/software/lil-gp/.

which remains constant in the case of raspberry pi while the algorithm is running. So we can obtain watts delivered by the algorithm (second scenario) by simply subtracting both values. In the laptop the watts delivered vary continuously, so we record a video in order to register all possible values and how long it lasted each one. Once all data are collected, we analyze how long each value stays with respect to the total execution time. Thus, we can accurately compute the average power delivered and, finally by subtracting the initial value, we get the power delivered by the algorithm.

Tablet. In order to collect data about energy consumption to Android devices, such as smartphones or tablets, *PowerTutor*[2] [15] has been used. This app is a diagnostic tool for analyzing system and app energy consumption. In order to obtain energy measurements of the EA, PowerTutor runs in the background and logs data on energy utilization for each app, summarizing the info in an intuitive user interface (reporting the number of Joules the app has consumed during the run).

iMac. Data collection in the iMac was done using *HardwareMonitor*[3]. This application suite includes a command-line tool that provides readings of the internal hardware sensors built on the computer. In order to obtain power measurements, a shell script is run in parallel to each run of the EA. This script gathers sensor data periodically (we use a sampling frequency of 1s) and goes to sleep state between measurements. During the experimentation, no other application is run, apart from background processes under OS control. To gauge the data, the same data-collection process is run for 100 s before each batch of runs, thus providing an indication of the system basal consumption at that moment which is in turn used to compute the excess power delivered due to the EA (and hence accounting for eventual hysteretic phenomena).

Blade. Ecosystems such as clusters are often used to process big data sets. This kind of systems allow us to optimize, by sharing, resources like Ethernet, storage devices, power supply, etc. The ecosystem we use employs VMWare Esxi 5.0[4] (see Table 2) whose hypervisor provides us with information about energy consumed by both the hardware platform as well as any of the virtual machines running on it. Thus we can obtain specific data for the virtual machine running the algorithm, and thus we can compute the difference between energy consumption when the algorithm is running and when it is not running.

4 Results

As described in the previous section, computational times and power delivered for each of the devices are reported in Table 3; algorithms tested using

[2] http://ziyang.eecs.umich.edu/projects/powertutor/documentation.html.
[3] http://www.bresink.com/osx/HardwareMonitor.html.
[4] https://my.vmware.com/web/vmware/details?productId=229&downloadGroup=ESXI50.

Table 3. Time (in seconds) for lilgp-multiplexer-6 run on each system depending on the population size. The numbers denote the mean and the standard error of the mean for the 30 runs performed.

population size					
System	100	200	400	500	1000
raspberry pi	7.77 ± 1.31	19.91 ± 2.41	46.22 ± 4.01	61.10 ± 7.19	116.80 ± 13.55
laptop	1.73 ± 0.31	4.43 ± 0.54	10.60 ± 0.97	13.89 ± 1.68	27.13 ± 3.36
iMac	1.38 ± 0.28	3.69 ± 0.48	8.98 ± 0.84	11.74 ± 1.44	22.95 ± 2.85
tablet	4.43 ± 0.75	4.85 ± 0.78	35.68 ± 4.15	36.17 ± 4.17	68.70 ± 7.89
blade	2.59 ± 0.53	6.88 ± 0.91	16.78 ± 1.59	22.28 ± 2.77	43.53 ± 5.48

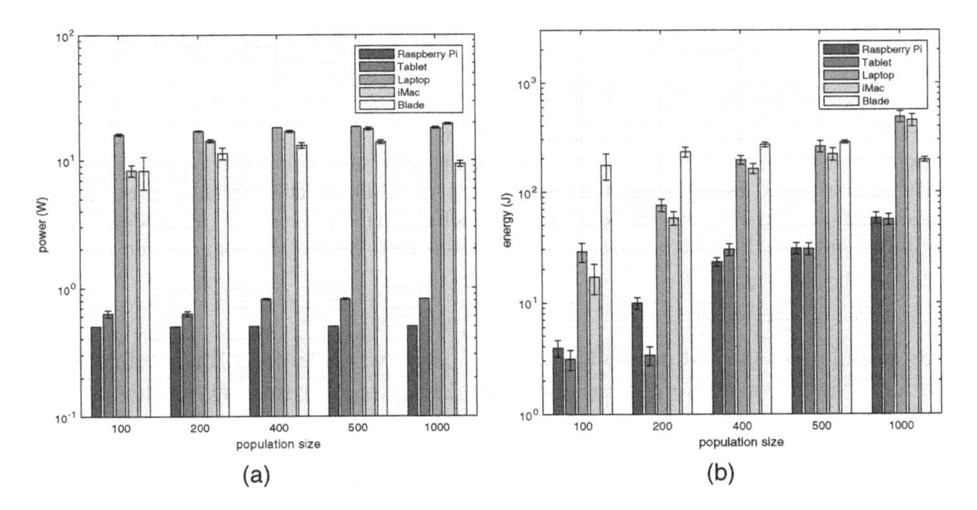

Fig. 1. (a) Average power delivered in each run. (b) Energy consumption per run. In both cases the bars (corresponding to raspberry pi, tablet, laptop, iMac and blade from left to right in each group) indicate mean values and the error bars span the standard error of the mean. Notice the logarithmic scale in the Y axis.

different population sizes for each of the experiments are shown in Fig. 1a. The first thing that may be noticed is the difference in computing time among devices analyzed, which corresponds with expectations: small devices (raspberry pi and tablet) require quite a long time to reach solutions and this is typically the reason why they are not frequently used as the hardware platform to run EAs – although they are useful when non-standard distributed models are analyzed (such as pool-based models, [9]).

Nevertheless, we are considering a different point of view, and will not just focus on computing time. Different features can thus be analyzed: (i) device behavior when considering energy consumed or power delivered by the algorithm per unit of time; (ii) total energy required to find a solution and (iii) the

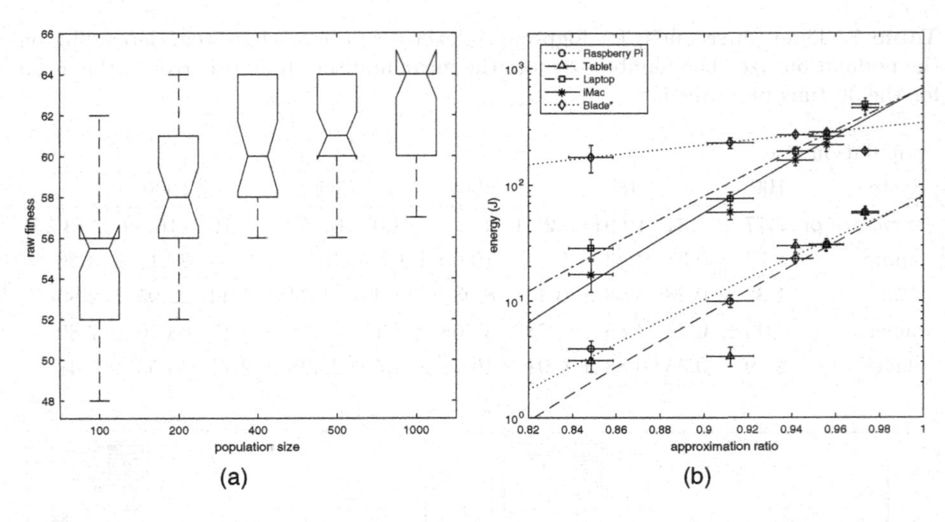

Fig. 2. (a) Distribution of fitness values attained for each population size (platform independent). (b) Trade-off between fitness attained and energy consumption. The data points mark mean values and the error bars indicate the standard error of the mean. A fit to a power-law $E \propto f^k$ is also included (in the case of the blade system, we have omitted the last point since it is an outlier). Notice the logarithmic scale in the Y axis.

way population size influences the energy needs when looking for a solution. If we focus firstly on the energy required by `lilgp` to be run in every device (see Fig. 1b), we notice that handheld devices (raspberry pi and tablet) are the least demanding ones, requiring an order of magnitude less energy to run the same algorithm when compared with more standard computers: iMac, laptop and blade system. Secondly, when considering the total energy required to reach the solution for the problem, the tablet running Android is the device that provides the *cheapest* solution, according to energy consumed, while blade and laptop are the most *expensive* ones in every case. Yet, if we only focus on computing time, the opposite is the case, and iMac, laptop and blade would be the preferred ones. But given that we are looking for energy-efficient ways of finding solutions by means of EAs, then raspberry pi and tablet might be preferred. Of course, this may look as a somewhat expected result a posteriori but it is nevertheless important to have obtained experimental confirmation of this fact, since different devices do not necessarily have to yield analogous trade-offs between energy consumption and speed.

In addition to absolute energy consumption values, it is also interesting to analyze how the energy requirements vary for a given device when the parameterization is changed. In this case, we have focused on population sizes for two reasons: firstly, it is an essential parameter that greatly influences the search process and can have an energy impact due to both the different behavior of the algorithm and memory management issues that might appear; secondly, its use in conjunction with a stopping condition based on generations influences the total work done as well as the quality of the solutions attained. This fact is used as a proxy to study energy consumption as a function of the attained quality of solutions in devices in which online energy measurements are not possible (and hence only the total energy consumed in a run can be measured).

We see in Fig. 1b that there is a general trend of increasing energy cost for increasing population sizes[5] which has a twofold cause: the longer computational time needed to complete the runs and the slightly larger power delivered in each case (i.e., energy spent per unit time is influenced by the population size). The latter effect can be due to issues related to memory management and is most remarkable in non-handheld devices (quite interestingly, no change is found for raspberry pi and quite small changes for the tablet). This increased energy toll does not always pay off as we can see by inspecting Fig. 2a; except for the largest population size, there does not seem to be a significant difference between the median fitness for a certain size and the immediately smaller size. A more focused perspective on this issue is shown in Fig. 2b in which we depict the energy required by each device to attain a certain fitness. The order of growth of this cost can be modeled as a power law as a first approximation. Such a power law is consistent with the superlinear cost of obtaining increasingly better approximations to the optimal solution and –while the fit can be obviously improved– it provides the means for a first comparison of these different devices. Thus, we can see how the general trends are not dramatically different for raspberry pi, tablet, laptop and iMac except for the offset of order of magnitude between the two former and the two latter. The blade system provides a more stable energy profile and can be preferred to laptop and iMac if a tight approximation to the optimum is sought. However, the smaller devices remain the best option in terms of absolute energy cost.

5 Conclusions

We have presented a preliminary study on the energy consumed by a well known EA: the Genetic Programming algorithm. We have analyzed the behavior of a benchmark GP problem, the multiplexer-6, and have run it using `lilgp` on different hardware devices running a number of operating systems, from blade systems using different Linux distributions to tablet devices running Android.

One of the first things we have learned is that although devices with better processors can run the algorithm faster, they spend larger amounts of energy, and the total energy required to find a solution is also larger. This means that although the standard preference for better hardware platforms and processors allows to find solutions more quickly, it incurs in waste of energy that should be considered: it is much more energy efficient to run the algorithm on a Raspberry

[5] In the case of the blade, we can observe that a population with 1000 individuals consumes less energy than with 500 individuals. This phenomenon is due to the processor frequency decreases because more memory is needed.

Pi or a tablet device. Of course, it is expected that some very computationally-expensive problems could not be fully solved in one of these devices. Then again, they can provide a valuable, energy-efficient contribution to the collective resolution of such problems when used within a larger ephemeral network of computing devices [6].

Secondly, we have seen the influence of one of the main parameters of the algorithm may have on the energy consumed: changing population sizes automatically produce a change in the amount of energy required to reach solutions, and this is a hint on future analysis. We should thus carefully consider how each of the parameters of the algorithm may also influence the amount of energy consumed when looking for solutions.

As a future line of work, it would be interesting to consider different devices in the same class exploring the space of solutions in a multiobjective way: which devices manage to find the solution faster for the least amount of energy? In principle the analysis is generic and does not rely on any feature that could be said to be GP-specific, so we believe the conclusions extracted are applicable to any EA. This said, it would be also interesting to further confirm this. We will thus expand the study to other evolutionary algorithms to check whether these energy profiles are exclusive of GP or there are variations among them. Energy profiling the algorithms will also allow us to find out where the energy expenses actually come from, allowing us to optimize the algorithm itself making it energy-aware.

Acknowledgements. We acknowledge support from Spanish Ministry of Economy and Competitiveness and European Regional Development Fund (FEDER) under project EphemeCH (TIN2014-56494-C4-{1,2,3}-P), from University of Granada, PROY-PP2015-06 (Plan Propio 2015 UGR), from Junta de Andalucía under project DNEMESIS (P10-TIC-6083), from Universidad de Málaga, Campus de Excelencia Internacional Andalucía Tech, from Junta de Extremadura FEDER, project GR15068 and FP7-PEOPLE-2013 IRSES Grant 612689 ACoBSEC.

References

1. Albers, S.: Energy-efficient algorithms. Commun. ACM **53**(5), 86–96 (2010)
2. Albers, S.: Algorithms for dynamic speed scaling. In: Schwentick, T., Dürr, C. (eds.) Leibniz International Proceedings in Informatics (LIPIcs). 28th International Symposium on Theoretical Aspects of Computer Science (STACS 2011), vol. 9, pp. 1–11. Schloss Dagstuhl-Leibniz-Zentrum fuer Informatik, Dagstuhl, Germany (2011). http://drops.dagstuhl.de/opus/volltexte/2011/2995
3. Almeida, F., Blanco, V., Cabrera, A., Ruiz, J.: Modeling energy consumption for master-slave applications. J. Supercomput. **65**(3), 1137–1149 (2013)
4. Bansal, N., Kimbrel, T., Pruhs, K.: Dynamic speed scaling to manage energy and temperature. In: 45th Annual IEEE Symposium on Foundations of Computer Science, 2004. Proceedings, pp. 520–529, October 2004
5. Barroso, L.A., Hölzle, U.: The case for energy-proportional computing. Computer **12**, 33–37 (2007)

6. Cotta, C., Fernández-Leiva, A., de Vega, F.F., Chávez, F., Merelo, J., Castillo, P., Bello, G., Camacho, D.: Ephemeral computing and bioinspired optimization - challenges and opportunities. In: 7th International Joint Conference on Evolutionary Computation Theory and Applications, pp. 319–324. SCITEPRESS, Lisboa (2015)
7. Fadaee, M., Radzi, M.: Multi-objective optimization of a stand-alone hybrid renewable energy system by using evolutionary algorithms: a review. Renew. Sustain. Energy Rev. **16**(5), 3364–3369 (2012). http://www.sciencedirect.com/science/article/pii/S1364032112001669
8. Fong, K., Hanby, V., Chow, T.: HVAC system optimization for energy management by evolutionary programming. Energy Build. **38**(3), 220–231 (2006). http://www.sciencedirect.com/science/article/pii/S0378778805000939
9. García-Valdez, M., Trujillo, L., Merelo, J.J., Fernández de Vega, F., Olague, G.: The evospace model for pool-based evolutionary algorithms. J. Grid Comput. **13**(3), 329–349 (2015). http://dx.doi.org/10.1007/s10723-014-9319-2
10. Hooper, A.: Green computing. Commun. ACM **51**(10), 11–13 (2008)
11. Koza, J.R.: Genetic Programming: On the Programming of Computers by Means of Natural Selection. MIT Press, Cambridge (1992)
12. Lee, W.S., Chen, Y.T., Kao, Y.: Optimal chiller loading by differential evolution algorithm for reducing energy consumption. Energy Build. **43**(23), 599–604 (2011). http://www.sciencedirect.com/science/article/pii/S0378778810003804
13. de Vega, F.F., Pérez, J.I.H., Lanchares, J.: Parallel Architectures and Bioinspired Algorithms, vol. 122. Springer, Heidelberg (2012)
14. Yoo, C.M., Sungjoo, K.: Energy-Aware System Design. Springer, Houten (2011)
15. Zhang, L., Tiwana, B., Dick, R.P., Qian, Z., Mao, Z.M., Wang, Z., Yang, L.: Accurate online power estimation and automatic battery behavior based power model generation for smartphones. In: 2010 IEEE/ACM/IFIP International Conference on Hardware/Software Codesign and System Synthesis (CODES+ISSS), pp. 105–114, October 2010
16. Álvarez, J.D., Risco-Martín, J.L., Colmenar, J.M.: Multi-objective optimization of energy consumption and execution time in a single level cache memory for embedded systems. J. Syst. Softw. **111**, 200–212 (2016). http://www.sciencedirect.com/science/article/pii/S0164121215002241

Comparing Asynchronous and Synchronous Parallelization of the SMS-EMOA

Simon Wessing[1(✉)], Günter Rudolph[1], and Dino A. Menges[2]

[1] Computer Science Department, Technische Universität Dortmund, Otto-Hahn-Str. 14, 44221 Dortmund, Germany
{simon.wessing,guenter.rudolph}@tu-dortmund.de
[2] Adept Technology GmbH, Revierstr. 5, 44379 Dortmund, Germany
dino.menges@adept.com

Abstract. We experimentally compare synchronous and asynchronous parallelization of the SMS-EMOA. We find that asynchronous parallelization usually obtains a better speed-up and is more robust to fluctuations in the evaluation time of objective functions. Simultaneously, the solution quality of both methods only degrades slightly as against the sequential variant. We even consider it possible for the parallelization to improve the quality of the solution set on some multimodal problems.

Keywords: Asynchronous · Synchronous · Parallel · Multiobjective · Evolutionary · Optimization

1 Introduction

With the rise of multi-core systems in all device classes from smartphones to desktops, parallel algorithms become more and more important. Parallelization is especially beneficial in optimization, where a high number of objective function evaluations should be enabled. An asynchronous parallelization appears preferable as it gets around the inevitable idle times caused by synchronous parallelization, but it must be precluded that this advantage is bought at the expense of solution quality.

Formerly, sophisticated algorithms containing message passing, master/slave concepts, or island models were often necessary to distribute execution on a cluster of nodes [3,5,6]. On present-day integrated multi-core architectures, simple shared memory communication may already be sufficient, especially for the application area of population-based optimization. Here, we focus on an evolutionary algorithm (EA) for multiobjective optimization, namely the S-metric selection evolutionary multiobjective optimization algorithm (SMS-EMOA) [1]. In its original form, the algorithm follows a steady-state scheme, which means that only one offspring solution is created per generation. This approach is in a sense optimal with regard to the exploitation of information in the current population, but unsuitable to synchronous parallelization. With synchronous parallelization, we mean the creation of $\lambda > 1$ offspring per generation, evaluating

J. Handl et al. (Eds.): PPSN XIV 2016, LNCS 9921, pp. 558–567, 2016.
DOI: 10.1007/978-3-319-45823-6_52

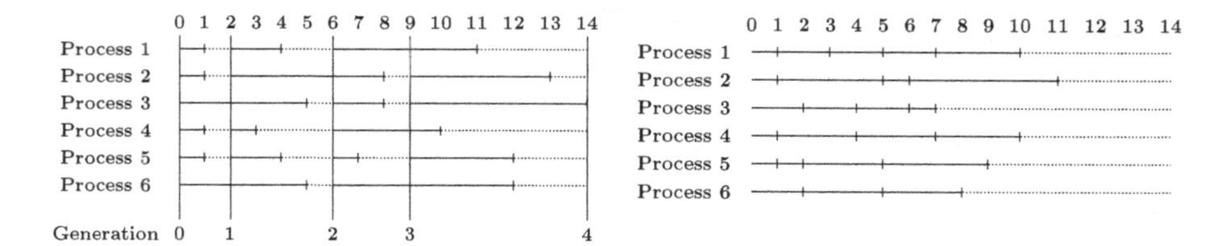

Fig. 1. Illustration of synchronous (left) and asynchronous parallelization (right). Idle times are visualized as dotted lines. In the asynchronous case, three time steps can be saved by giving up the generation concept.

them in parallel, and carrying out selection after all evaluations have finished. This approach requires an extension of the selection scheme, to be able to select μ individuals from $\mu + \lambda$. This $(\mu + \lambda)$-selection may have been avoided for SMS-EMOA to some extent, because the exact calculation of the hypervolume contributions probably requires algorithmically more complex code, or will take longer than the conventional $(\mu + 1)$-selection. Alternatively, one could sidestep the problem by taking a greedy approach of λ selections in the $(\mu + 1)$-scheme, removing one individual after another. However, Bringmann and Friedrich advise against this approach [2].

Unfortunately, the synchronization is unfavorable in case of fluctuating evaluation times, because it creates idle times (see Fig. 1). An alternative is asynchronous parallelization, which allows us to keep using $(\mu + 1)$-selection, but means giving up the concept of generations. Klinkenberg et al. [5] were the first to propose asynchronous parallelization of the SMS-EMOA. Their implementation was a master/slave approach using message passing on a cluster with twelve nodes. They combined it with metamodeling to save expensive function evaluations and applied it to a molecular control problem. The results indicated a nearly linear speed-up and only a slight decrease of solution quality between one and twelve processors. However, the analysis was not taken further because the parallelization was not the sole topic of the work. Another successful real-world application of the asynchronous SMS-EMOA is due to Menges et al. [7], who optimized the motion planning of a mobile robot.

Depolli et al. [3] investigate the asynchronous parallelization of a multiobjective differential evolution algorithm (AMS-DEMO) on a steel casting problem and benchmarks. They identify the *selection lag* as an important measure for the performance of asynchronous EMOAs. It is defined as "the number of solutions that undergo selection in the time between the observed solution's creation and selection" [3]. In their experiments, a linear speed-up with four processors could already be observed for evaluation times of only 0.01 seconds. While these results are encouraging, the question arises if the SMS-EMOA with its relatively expensive survivor selection achieves similar values. We try to answer this question with our experiment in Sect. 3. Before, we present the asynchronous SMS-EMOA in detail (Sect. 2), and afterwards we draw conclusions in Sect. 4.

Algorithm 1. Asynchronous SMS-EMOA

Input: *mutex*, population P_0

```
1: t ← 0
2: while stopping criterion not fulfilled do
3:    x ← createOffspring(P_t, mutex)                      // create 1 offspring
4:    evaluate(x)                                           // calculate objective values
5:    enter(mutex)                                          // lock out other processes
6:    Q_t ← P_t ∪ {x}
7:    {F_1, ..., F_w} ← nondominatedSort(Q_t)               // sort in w fronts
8:    r ← createReferencePoint(F_w)                         // calculate reference point for last front
9:    x* ← argmin_{x∈F_w}(Δ_s(x, F_w, r))                   // determine x* with smallest contribution
10:   P_{t+1} ← Q_t \ {x*}                                  // remove worst individual
11:   t ← t + 1
12:   leave(mutex)                                          // release lock for other processes
13: end while
```

2 The Asynchronous SMS-EMOA

Large differences in the evaluation times of individuals result in idle times in synchronously parallelized algorithms, because selection cannot start before all objective values are available. These idle times are of course a waste of resources. Figure 1 (left) shows an example of such a generational approach with a $(\mu+6)$-EA. The six offspring are evaluated in parallel. The time during which a process calculates a function value is indicated by a horizontal solid line. Dotted lines mark idle times. The vertical lines mark the synchronization points between the generations. In this example, 29 of the 84 time steps are unused, so the system is idle 35 % of the time. The asynchronous approach is more efficient, finishing the execution three time steps earlier and idling only 11 time steps at the end (Fig. 1, right).

Thus, we implement the SMS-EMOA as an asynchronous algorithm to minimize the idle time. The pseudocode is shown in Algorithm 1. The idea is to have several processes working on a shared population, with the additional benefit that the selection scheme can stay a $(\mu + 1)$. To make this work, all read and write operations involving the population must be protected by a lock, allowing only one process to access at a time. Entering and leaving the critical section is illustrated in lines 5 and 12. As the required functionality is provided by virtually all modern programming languages, these modifications are extremely simple. The objective function evaluation, where the most time is spent according to the common black-box optimization assumptions, may happen in parallel. After the process finishes an evaluation, it waits until it can enter the critical section to carry out the survivor selection. Naturally, the new individual may either replace another one or be rejected. Then, the section is left and the next generation starts with creating the next individual by variation. This loop continues until a stopping criterion is fulfilled. So, every single process executes all tasks of an SMS-EMOA, not only a subset as in a master/slave scenario.

Table 1. Experimental factors

Factor	Type	Symbol	Levels
Problem instances	Non-observable		{WFG1, ..., WFG9}
#Objective functions	Observable	m	$\{2, 3\}$
Evaluation base time	Observable	t_b	$\{0.01, 0.1, 1\}$
Evaluation time behavior	Observable		{fixed, random, proportional}
Parallelization	Control		{sync, async}
Parallelization degree	Control	p	$\{1, 4, 16, 64\}$
Population size	Control	μ	$\{10, 100\}$

Also the function createOffspring does a read access to the population for parent selection. It has to protect it by a critical section to avoid modifications of the population during this time.

3 Experiment

Research Question. How do the properties of multiobjective problems and parallelization settings of the SMS-EMOA influence performance?

Pre-experimental Planning. When the overhead of the SMS-EMOA is negligible in comparison to the evaluation time, an almost linear speed-up can be expected for the asynchronous variant [3,5]. The synchronous variant should only achieve the same run time with constant evaluation times, and is expected to suffer from fluctuations, as explained in Sect. 2. The hypervolume calculations of the SMS-EMOA become a bottleneck with increasing number of objectives m and increasing population size μ. Thus, it is expected that the speed-up deteriorates when m and μ are large and evaluation time is low. The experimental setup is chosen to enable quantification of this behavior.

Selection lag is recorded for both synchronous and asynchronous variants. Depolli et al. [3] identify the selection lag of the asynchronous variant (without queues) as $p - 1$ for p processors. We presume that in the synchronous case, a value of $\frac{1}{p}\sum_{i=0}^{p-1} i = (p-1)/2$ would be the expected value for the selection lag, because only individuals in the same generation can be selected during one individual's evaluation.

Task. We calculate the dominated hypervolume and the averaged Hausdorff distance (AHD, [8]) of the final population with respect to a Pareto-optimal reference set, after running the SMS-EMOA for a fixed number of function evaluations. The reference set contains 500 points for two objectives and 1000 for three. The reference points for the hypervolume are $(3, 5)^\top$ and $(3, 5, 7)^\top$. To assess running time, wall-clock time is measured and the weak speed-up in comparison to the sequential variant is computed [3]. The term *weak* speed-up means that we divide the sequential time by the parallel time without taking into account the potential quality differences of the results, which are regarded separately.

Setup. We implemented the algorithm described in Sect. 2 and the synchronous variant in the language Python (version 3.4). The code is publicly available in the packages evoalgos[1] and optproblems [9,10]. As variation operators, simulated binary crossover and polynomial mutation are used. The parameters of these operators are set to $\eta_m = 20$, $\eta_c = 20$, $p_m = 0.1$, and $p_c = 0.7$.

Table 1 contains the experimental factors for this experiment, which are combined in a full-factorial design. We carry out five stochastic replications per configuration. As test problems, the set from the walking fish group (WFG, [4]) is used with two and three objectives. The number of decision variables is set to 24, with $k = 4$ position-related parameters. The feasible region of the problems is normalized to the unit hypercube. The evaluation time of the objective functions is determined as follows. We assume a base value of t_b seconds and define three different ways to obtain the actual evaluation time t_e. The first alternative is to use the t_b value as it is, $t_e = t_b$. The second variant takes two random uniform numbers $u_1, u_2 \sim \mathcal{U}[0, 1]$ and sets $t_e = (u_1 + u_2) \cdot t_b$. This way, t_e has a triangular distribution between zero and $2t_b$. The last approach uses $t_e = t_b \cdot f_1(\boldsymbol{x})$, where f_1 is the first objective function of the WFG problems, whose image is always $[0, 2]$. This setup is motivated by different real-world scenarios. The fixed evaluation time corresponds to homogeneous hardware and constant simulation time. Random fluctuations will appear if the optimization is running on a heterogeneous cluster. Simulation time may also be solution-dependent, leading to correlated evaluation times, as in [7].

The number of function evaluations is set to 10000 for each algorithm run. We exclude the population initialization and only measure the time spent in the optimization loop. By using *sleep* system calls to spend the t_e seconds, we can simulate a parallelized SMS-EMOA run on a single core, because the other algorithm parts are in critical sections anyway. The experiment is run on AMD Opteron 6276 processors with 2.3 GHz; operating system is Ubuntu Linux.

Results and Observations. Figs. 2 and 3 show the weak speed-up. The solid and dashed lines depict median values of the 45 runs on the nine WFG problems. Error bars mark 95 % confidence intervals for the median. The grey diagonal represents the maximally possible linear speed-up. In most cases, the asynchronous variant obtains a better speed-up than the synchronous one, with a few exceptions for high parallelization degrees and low evaluation times on two objectives. The speed-up of the synchronous variant sometimes even drops below one for three objectives. Figure 4 illustrates the selection lag values for the different configurations. For this figure, we first calculated the mean selection lag for each run. The lines in the figure are the median of 135 runs, due to the number of remaining configurations per panel. Generally, the predicted values seem to be accurate, except that the selection lag of the asynchronous variant drops off when evaluation times are small compared to selection times. Figure 5 shows some selected indicator values for hypervolume and averaged Hausdorff distance. The random noise in these values is much higher than for the run times. However, in Fig. 5a

[1] With a runnable example in the documentation at https://ls11-www.cs.tu-dortmund.de/people/swessing/evoalgos/doc/algo.html.

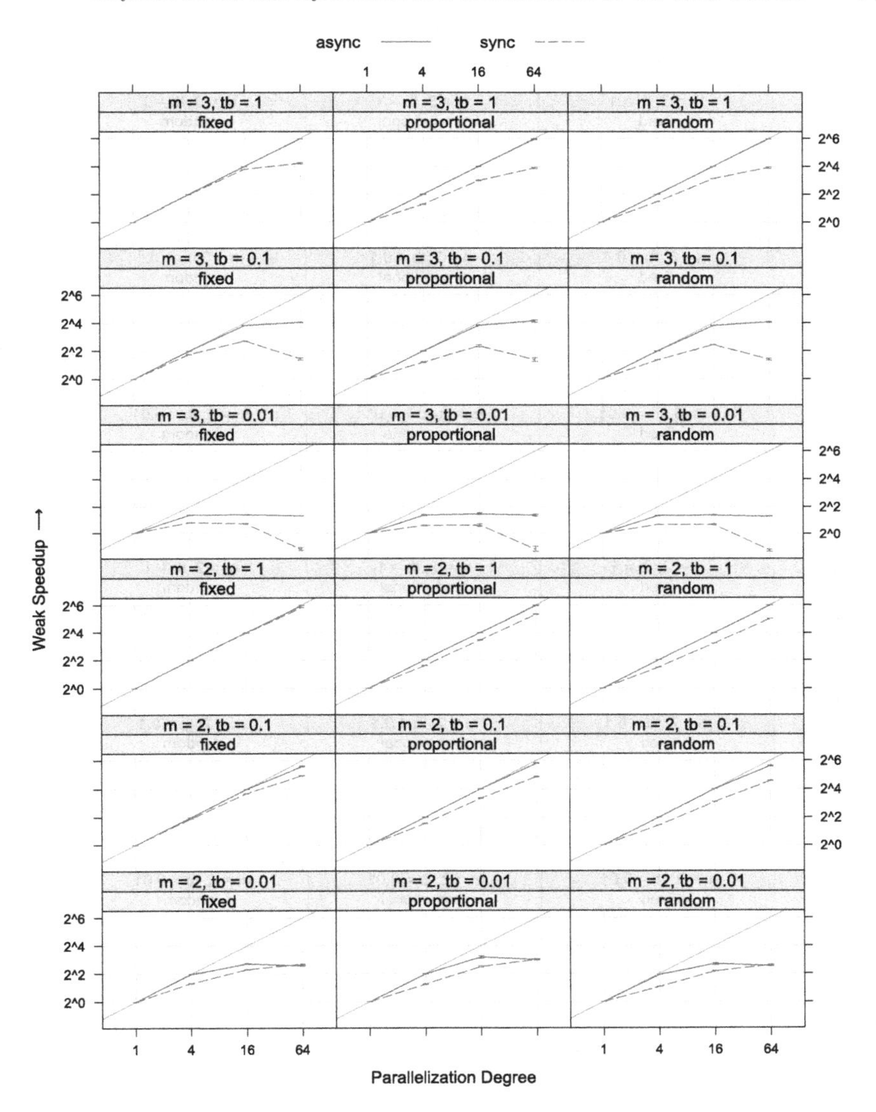

Fig. 2. Weak speed-up versus parallelization degree for $\mu = 10$.

there seems to be a positive effect on both indicators, while in Fig. 5b and c the results seem mixed.

Discussion. The cases where the synchronous beats the asynchronous variant regarding speed-up may be caused by differences in the overhead of the implementations. The asynchronous one especially makes more function calls, because each solution is processed individually. On the other hand, the decline of the synchronous variant for $m = 3$ may be because it has to compute hypervolume for

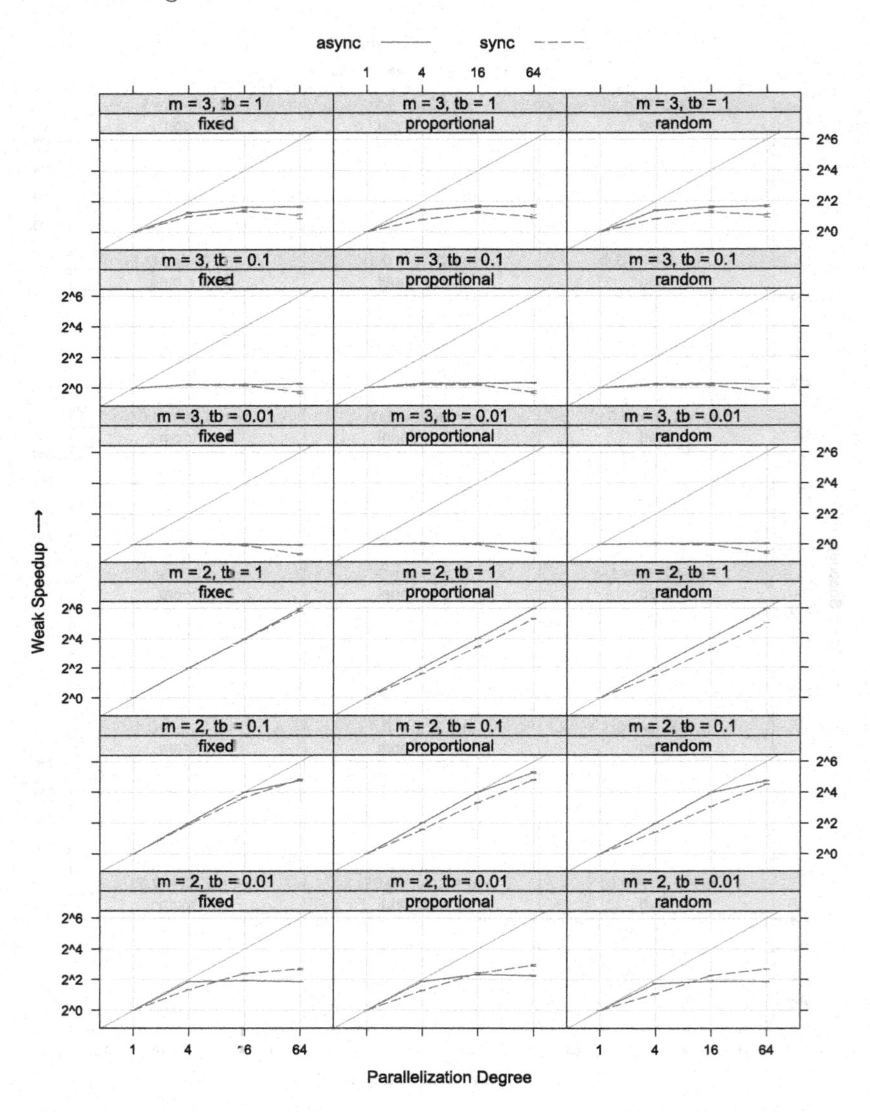

Fig. 3. Weak speed-up versus parallelization degree for $\mu = 100$.

$\mu + p$ solutions, while the asynchronous only ever does $\mu + 1$. Generally, the measured speed-ups should be seen as rather conservative estimates, if we consider that the SMS-EMOA, including the hypervolume computation, was implemented in pure Python. The speed-up could be further improved by implementing it in a lower-level language such as C++.

We are not entirely sure why the selection lag of the asynchronous variant is sometimes lower than expected. One explanation could be that the distribution of the budget on the worker processes becomes uneven with decreasing t_b. This

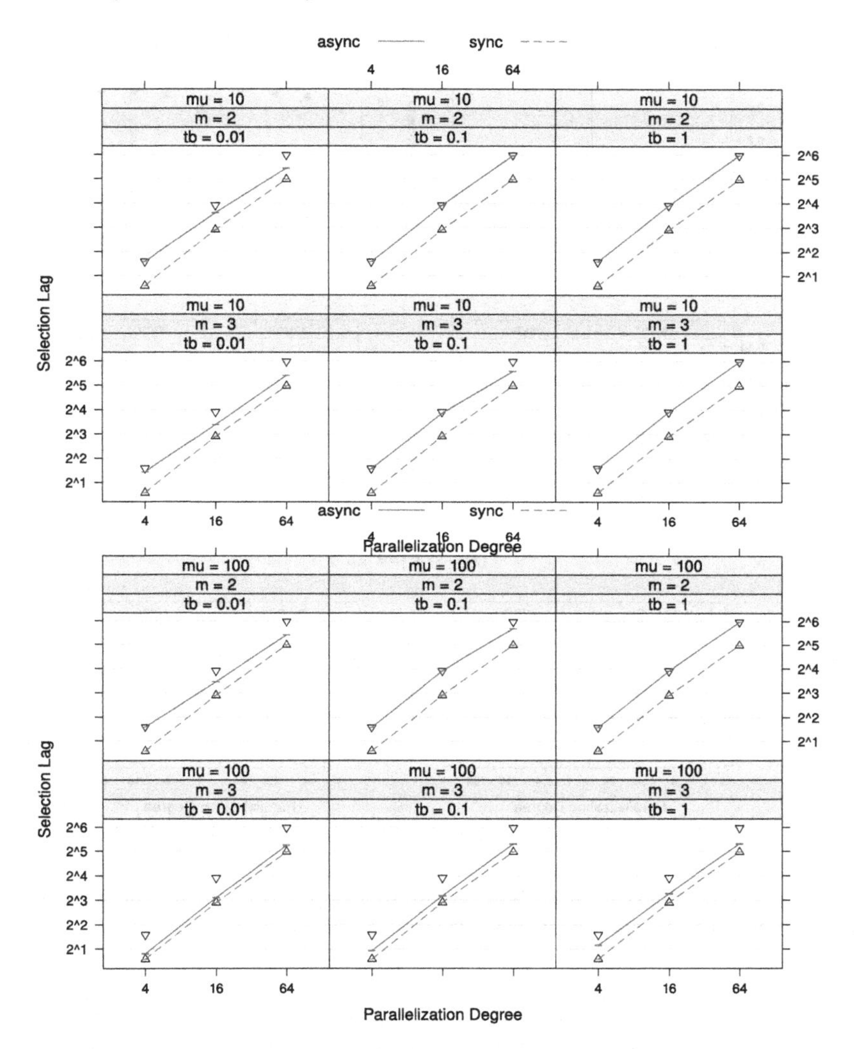

Fig. 4. Measured selection lag versus parallelization degree. The upper triangles ($\triangledown$) mark $p-1$, the lower ones ($\vartriangle$) $(p-1)/2$.

hypothesis stems from the observation that the scheduler only ran a single thread when we switched off the delay completely. If it is true, then the selection lag could also drop below $(p-1)/2$. To look into this issue, we recorded the partition of the budget and calculated its standard deviation. However, the effect could only be observed for fixed evaluation times, and the standard deviation did not exceed the values of the two fluctuating cases. In any case, we recommend to record this data on the actual parallelization also in future experiments.

The results in Fig. 5 can be explained by the fact that for multimodal problems it is usually beneficial to spend a larger part of the budget on exploration

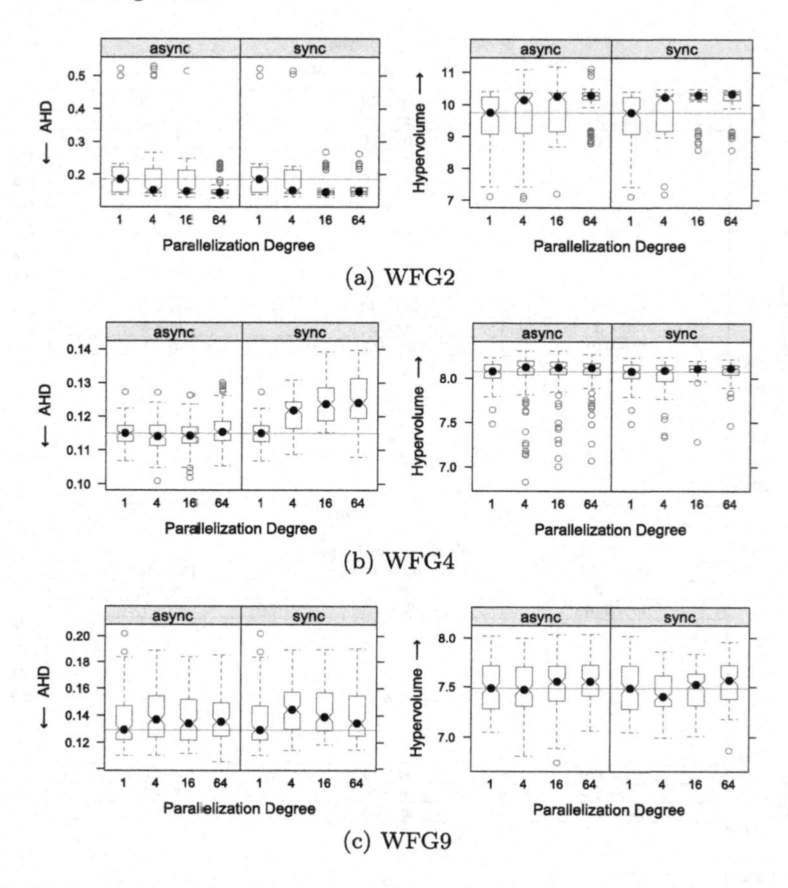

(a) WFG2

(b) WFG4

(c) WFG9

Fig. 5. AHD and hypervolume values of the final populations for configurations with $m = 2$ and $\mu = 10$. All three test problems are multimodal.

than for unimodal problems. The asynchronous variant additionally can sample objective space regions with lower evaluation times with a higher density than expensive regions, due to the missing synchronization. Thus, proportional evaluation times may cause a drift of the population towards a small part of the Pareto-front. This may either be beneficial, if simulation time is to be minimized [7], or detrimental, if we want to simulate interesting solutions with higher fidelity. Naturally, the assessments of AHD and hypervolume do not necessarily have to agree.

4 Conclusion

Both the synchronous variant with greedy selection and the asynchronous variant obtain an almost linear speed-up in a scenario of expensive function evaluations and moderate parallelization. The synchronous variant falls off more sharply under less favorable circumstances. The experiments have shown that the quality of solutions may even increase with more parallelism for multimodal problems whereas it may decrease for unimodal problems. We conjecture that this behavior is caused by a higher selection lag preventing a rapid movement to local optima. In theory, selection lag should only depend on the parallelization degree (and the queue size, if queues are used). However, our experiment discovered that for asynchronous parallelization, reality can somewhat deviate from theory. It is our impression that the measured selection lag especially deviates from the expectation in the cases where speed-up is low and parallelization does not work well. Thus, it might be used as a tool to assess the usefulness of asynchronous parallelization even when speed-up cannot be computed due to missing data on sequential performance.

References

1. Beume, N., Naujoks, B., Emmerich, M.: SMS-EMOA: multiobjective selection based on dominated hypervolume. Eur. J. Oper. Res. **181**(3), 1653–1669 (2007)
2. Bringmann, K., Friedrich, T.: Don't be greedy when calculating hypervolume contributions. In: Proceedings of the Tenth ACM SIGEVO Workshop on Foundations of Genetic Algorithms, FOGA 2009, pp. 103–112. ACM (2009)
3. Depolli, M., Trobec, R., Filipič, B.: Asynchronous master-slave parallelization of differential evolution for multi-objective optimization. Evol. Comput. **21**(2), 261–291 (2012)
4. Huband, S., Hingston, P., Barone, L., While, L.: A review of multiobjective test problems and a scalable test problem toolkit. IEEE Trans. Evol. Comput. **10**(5), 477–506 (2006)
5. Klinkenberg, J.W., Emmerich, M.T.M., Deutz, A.H., Shir, O.M., Bäck, T.: A reduced-cost SMS-EMOA using kriging, self-adaptation, and parallelization. In: Ehrgott, M., Naujoks, B., Stewart, J.T., Wallenius, J. (eds.) Multiple Criteria Decision Making for Sustainable Energy and Transportation Systems. Lecture Notes in Economics and Mathematical Systems, vol. 634, pp. 301–311. Springer, Heidelberg (2010)
6. Märtens, M., Izzo, D.: The asynchronous island model and NSGA-II: study of a new migration operator and its performance. In: Proceedings of the 15th Annual Conference on Genetic and Evolutionary Computation, GECCO 2013, pp. 1173–1180. ACM (2013)
7. Menges, D.A., Wessing, S., Rudolph, G.: Asynchrone Parallelisierung des SMS-EMOA zur Parameteroptimierung von mobilen Robotern. In: Hoffmann, F., Hüllermeier, E. (eds.) Proceedings 25, Workshop Computational Intelligence. Schriftenreihe des Instituts für Angewandte Informatik/Automatisierungstechnik, vol. 54, pp. 47–65. KIT Scientific Publishing (2015). (in German)
8. Schütze, O., Esquivel, X., Lara, A., Coello Coello, C.A.: Using the averaged Hausdorff distance as a performance measure in evolutionary multiobjective optimization. IEEE Trans. Evol. Comput. **16**(4), 504–522 (2012)
9. Wessing, S.: evoalgos: modular evolutionary algorithms (2016). Python package version 0.3. https://pypi.python.org/pypi/evoalgos
10. Wessing, S.: optproblems: infrastructure to define optimization problems and some test problems for black-box optimization (2016). Python package version 0.8. https://pypi.python.org/pypi/optproblems

A Parallel Version of SMS-EMOA for Many-Objective Optimization Problems

Raquel Hernández Gómez[1(✉)], Carlos A. Coello Coello[1], and Enrique Alba[2]

[1] Computer Science Department, CINVESTAV-IPN
(Evolutionary Computation Group), 07360 Mexico City, Mexico
rhernandez@computacion.cs.cinvestav.mx, ccoello@cs.cinvestav.mx
[2] Universidad de Málaga, 29071 Malaga, Spain
eat@lcc.uma.es

Abstract. In the last decade, there has been a growing interest in multi-objective evolutionary algorithms that use performance indicators to guide the search. A simple and effective one is the $\mathcal{S}$-Metric Selection Evolutionary Multi-Objective Algorithm (SMS-EMOA), which is based on the hypervolume indicator. Even though the maximization of the hypervolume is equivalent to achieving Pareto optimality, its computational cost increases exponentially with the number of objectives, which severely limits its applicability to many-objective optimization problems. In this paper, we present a parallel version of SMS-EMOA, where the execution time is reduced through an asynchronous island model with micro-populations, and diversity is preserved by external archives that are pruned to a fixed size employing a recently created technique based on the Parallel-Coordinates graph. The proposed approach, called $\mathcal{S}$-PAMICRO (PArallel MICRo Optimizer based on the $\mathcal{S}$ metric), is compared to the original SMS-EMOA and another state-of-the-art algorithm (HypE) on the WFG test problems using up to 10 objectives. Our experimental results show that $\mathcal{S}$-PAMICRO is a promising alternative that can solve many-objective optimization problems at an affordable computational cost.

1 Introduction

Numerous real-world problems can be formulated as Multi-Objective Optimization Problems (MOPs), which involve several (often conflicting) objectives to be optimized at the same time. In general, a MOP is formally described as follows:

$$\text{Minimize} \quad \boldsymbol{F}(\boldsymbol{x}) := (f_1(\boldsymbol{x}), f_2(\boldsymbol{x}), \ldots, f_m(\boldsymbol{x})) \tag{1}$$
$$\text{subject to} \quad \boldsymbol{x} \in \mathcal{S}, \tag{2}$$

C.A. Coello Coello—Author gratefully acknowledges support from CONACyT project no. 221551.
E. Alba—Author is partially funded by the Spanish MINECO and FEDER project TIN2014-57341-R (http://moveon.lcc.uma.es).

J. Handl et al. (Eds.): PPSN XIV 2016, LNCS 9921, pp. 568–577, 2016.
DOI: 10.1007/978-3-319-45823-6_53

where $\boldsymbol{x}$ is the *vector of decision variables*, $\mathcal{S} \subset \mathbb{R}^n$ is the *feasible region set* and $\boldsymbol{F}(\boldsymbol{x})$ is the vector of m ($\geq$2) *objective functions* ($f_i : \mathbb{R}^n \rightarrow \mathbb{R}$). The aim is to seek from among the set of all values, which satisfy the constraint functions defined in Eq. (2), the particular set $\boldsymbol{x}^*$ that yields the optimum values for all the objective functions.

Multi-Objective Evolutionary Algorithms (MOEAs) are stochastic, population-based, search techniques; which are well-suited for solving a wide variety of complex MOPs. In the last decades several MOEAs have been proposed (see, for example, [4, Chap. 2] and [18]), with the vast majority relying on two concepts: *Pareto dominance*[1] as their primary selection mechanism, followed by a *density estimator*.[2] The former favors non-dominated solutions over dominated ones, whereas the latter induces a total order of *incomparable solutions*,[3] preserving *diversity*[4] at the same time.

One of the main concerns is that Pareto-based MOEAs face difficulties to reach the *Pareto optimal front*[5] when dealing with many-objective optimization problems ($m \geq 4$) [9,11,13]. This is due to the fact that most or all solutions in the population quickly become non-dominated with respect to the rest, and the best individuals are identified only by the density estimator. Thus, in some cases good locally non-dominated solutions in terms of convergence might be discarded at the expense of keeping good solutions in terms of diversity, in spite of the fact that they may be distant from the Pareto optimal front [1]. To address this issue, a new trend is the incorporation of *performance indicators*[6] into the selection mechanism of a MOEA [2,6,19]. The *hypervolume indicator* [4, p. 257] is, with no doubt, a natural choice, (see for example [6,19]) since it is the only unary indicator that is known to be Pareto compliant. Also, it has been proven that maximizing the hypervolume is equivalent to reaching the Pareto optimal set [7]. However, the main drawback of this sort of approach is its computational cost, which increases exponentially with the number of objectives [3], making it prohibitive for many-objective optimization problems.

In this work, we focus on the $\mathcal{S}$-Metric Selection Evolutionary Multi-Objective Algorithm (SMS-EMOA) [6], due to its simplicity and superiority over Pareto- and Aggregation-based algorithms [6,10,16]. This optimizer is a steady state evolutionary algorithm that ranks individuals according to Pareto dominance and uses the hypervolume as its density estimator. The worst-case complexity of SMS-EMOA is $\mathcal{O}(|P|^m)$ [17]. Parallelizing SMS-EMOA arises as a possible alternative to reduce its computational cost, where at least two strate-

[1] A solution $\boldsymbol{x} \in \mathcal{S}$ dominates a solution $\boldsymbol{y} \in \mathcal{S}$ ($\boldsymbol{x} \prec \boldsymbol{y}$) if and only if $\forall i \in \{1, \ldots, m\}$, $f_i(\boldsymbol{x}) \leq f_i(\boldsymbol{y})$ and $\exists j \in \{1, \ldots, m\}$, $f_j(\boldsymbol{x}) < f_j(\boldsymbol{y})$.

[2] A density estimator models the distribution of a population, by measuring the similarity degree among individuals.

[3] Two solutions $\boldsymbol{x}, \boldsymbol{y} \in \mathcal{S}$ are incomparable if neither $\boldsymbol{x} \prec \boldsymbol{y}$ nor $\boldsymbol{y} \prec \boldsymbol{x}$ holds.

[4] *Diversity* refers to achieving a uniform distribution of solutions covering all regions of the objective function space.

[5] $POF := \{\boldsymbol{F}(\boldsymbol{x}) \in \mathbb{R}^m : \boldsymbol{x} \in \mathcal{S}, \nexists \boldsymbol{y} \in \mathcal{S}, \boldsymbol{y} \prec \boldsymbol{x}\}$.

[6] A performance indicator, defined as $I : \mathbb{R}^m \rightarrow \mathbb{R}$, measures the quality of an approximation set (the final population of a MOEA).

gies are possible [14]: (1) parallelization of the computations, in which the operations applied to an individual are performed in parallel, and (2) parallelization of the population, in which the population is partitioned and each subpopulation evolves in semi-isolation (individuals can be exchanged between subpopulations). Klinkenberg et al. [10] and Lopez et al. [12] have studied the first approach. In [10], a variation of SMS-EMOA parallelized the evaluations of individuals using a surrogate model, whose purpose was to approximate the function values. In [12], the exact hypervolume contributions of SMS-EMOA were parallelized through the use of Graphics Processing Units (GPUs). To the best of our knowledge, our work is the first attempt to incorporate the second sort of approach (parallelization of the population) into SMS-EMOA.

In order to get a better grasp of the variability of the execution time of SMS-EMOA, we sampled several points on DTLZ1 [4, p. 200], varying the number of objective functions and the population size on a PC Intel(R) Core(TM) i7 CPU 950 @ 3.07 GHz × 8 with 3.8 GB memory, using the same parameters in all experiments [6]. The average resulting surface is shown in Fig. 1. An interesting observation is that, regardless of the number of objectives, time was almost negligible when using small populations (less than 20 individuals). This fact is considered in our proposal, where we use micro-populations in an asynchronous island model [15]. Furthermore, diversity is improved by external archives that are kept to a constant size by a recently proposed density estimator [8], which is scalable in objective space.

The remainder of this paper is organized as follows. Section 2 is devoted to the description of our proposed parallel MOEA. In Sect. 3 we present our experimental results. Finally, Sect. 4 provides our conclusions and some potential lines of future research.

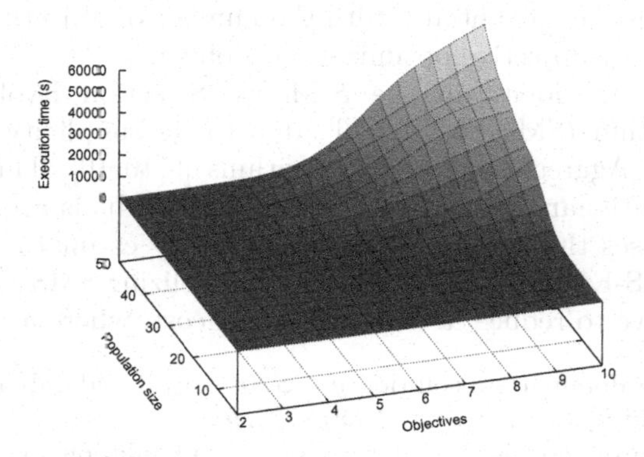

Fig. 1. Average execution time of SMS-EMOA.

2 Our Proposed Approach

The PArallel MICRo Optimizer based on the $\mathcal{S}$ metric ($\mathcal{S}$-PAMICRO) draws ideas from the island model, where the overall population is split into l micro-populations, called *islands*. Every island evolves independently a serial SMS-EMOA with an external archive of size $l \times |P|$, where $|P|$ corresponds to the micro-population size. In this approach, the islands are connected in a logical unidirectional ring, exchanging $nmig$ solutions occasionally[7] in an asynchronous fashion. The goal of $\mathcal{S}$-PAMICRO is to reduce the execution time of SMS-EMOA, hopefully also improving the quality of solutions in high dimensional spaces, because of the separated search of the islands, which changes the behavior of the serial version and yields a new kind of algorithm [14,15].

Algorithm 1. Outline of an island in $\mathcal{S}$-PAMICRO

Input: MOP, stopping criterion, island identification i, number of islands l, number of migrants $nmig$, and frequency of migration $fmig$.
Output: Final sub-population A

1: $A \leftarrow \emptyset$ {initialize external archive}
2: $n \leftarrow l|P|$ {archive size limit}
3: Initialize micro-population P at random
4: **while** the stopping criterion is not satisfied **do**
5: $P \leftarrow$ SMS-EMOA(MOP, $fmig$, P) {execute during $fmig$ evaluations of the objective vector}
6: $R \leftarrow$ Check the arrival of migrants from $(l + i - 1) \pmod{l}$ island
7: $A \leftarrow A \cup P \cup R$
8: **if** $|A| > n$ **then**
9: $A \leftarrow$ Pruning(A,n) {see Algorithm 2}
10: $S \leftarrow$Uniform_Random_Selection(A,$nmig$) {$nmig$ random solutions are selected from A}
11: Send copies of S to the $(i + 1) \pmod{l}$ island
12: $P \leftarrow$ Elitist_Ranking_Replacement($P \cup R$) {dominated individuals are likely to be discarded}
13: **return** A

In Algorithm 1, we present the pseudocode of an island in $\mathcal{S}$-PAMICRO. First, the external archive A and its maximum size are specified. Next, the micro-population P is initialized at random. In line 5, SMS-EMOA is executed during $fmig$ function evaluations. Then, an island receives, without blocking, the immigrants R from the source island, according to the adopted topology. In line 7, the external archive is updated, adding the current micro-population as well as the immigrants. In lines 8 and 9, the external archive is truncated if it exceeds its limits, using the technique described in the next paragraph. In the following two lines, the candidates to be migrated are selected by using the policy of uniform-random migration [15], in which $nmig$ individuals are randomly

[7] This is known as *migration*.

selected from the archive and a copy of them is sent to the destination island. In line 12, the micro-population is updated, replacing $|R|(<|P|)$ individuals with the immigrants. Here, we employed elitist-ranking replacement [15], where immigrants are combined with the current population, and then they are ranked using Pareto dominance, and the worst solutions are removed. This elitist mechanism preserves the currently best solutions for the next iteration, assuring proximity to the Pareto optimal front. At the end, the final sub-populations of all islands $i \in \{0, 1, \ldots, l-1\}$ are collected and adjusted to the size $l \times |P|$, using the same pruning technique. This operation is performed by a designated island.

Algorithm 2. Pruning

Input: Population P, desired size n
Output: Reduced population P
1: $\{F_1, \ldots, F_k\} \leftarrow$ Rank population P in k fronts according to Pareto dominance.
2: Calculate $\boldsymbol{z}^{min}$ and $\boldsymbol{z}^{max}$
3: Normalize population $p.\boldsymbol{y} \leftarrow \frac{p.\boldsymbol{y} - \boldsymbol{z}^{min}}{\boldsymbol{z}^{max} - \boldsymbol{z}^{min}}$, $\forall p \in P, p.\boldsymbol{y} \in \mathbb{R}^m$
4: **while** $|P| > n$ **do**
5:    **if** $|F_k| \leq |P| - n$ **then** {Remove members of the k-th front}
6:       $r \leftarrow F_k$
7:       $k \leftarrow k - 1$
8:    **else**
9:       $D \leftarrow$ Calculate density of P based on the Parallel-Coordinates graph
10:       $r \leftarrow \arg\max_{p \in F_k} D[p]$
11:       $F_k \leftarrow F_k \setminus \{r\}$
12:    $P \leftarrow P \setminus \{r\}$
13: **return** P

Our pruning technique is explained in Algorithm 2. First the population is ranked using the well-known non-dominated sorting procedure [4, p. 93]. In lines 2 and 3, the population is normalized in the objective space by means of two reference points: z^{min}, composed of the best objective values found so far, and z^{max}, formed with those vectors parallel to the axes with the lowest Euclidean norm. Next, all members of the worst current k-th front are removed if the size of this front is less or equal than the number of individuals to be removed (lines 5–7). Otherwise, the individual with the highest density value is eliminated from the current front (lines 9–11) until the desired size is achieved.

The density estimator, originally proposed in [8], is based on a visualization technique, called Parallel Coordinates. In this technique, a graph is built in the 2-dimensional plane where m copies of the real line $\mathbb{R}$ are placed perpendicular to the x-axis and a solution in $\mathbb{R}^m$ is represented by a series of connected line segments with vertices on the parallel axes. The core idea in the density estimator is to represent the Parallel Coordinates of each distinct pair of objective functions as a 2D matrix, where the $m(m-1)/2$ graphs are attached next to each other and only normalized individuals are considered. The dimension of this matrix depends on a resolution parameter (γ). An element of the matrix identifies the level of overlapping line segments and those individuals covering a wide area of the matrix have a better density estimator. Interested readers are referred to [8] for more details.

$\mathcal{S}$-PAMICRO was developed in the *EMO Project*,[8] our framework for Evolutionary Multi-Objective Optimization. This software is implemented in C language and MPICH.[9]

3 Experimental Results

In this section, we investigate the effectiveness of $\mathcal{S}$-PAMICRO on the Walking-Fish-Group (WFG) test suite [4, p. 209]. In this benchmark, properties, such as non-separability, multi-modality, deceptiveness and bias, are preserved as we increase the number of objectives, making these problems harder to solve for a MOEA. The decision variables (n) and the position-related parameter (k) are specified in Table 1.

We compared the results of our proposed algorithm with respect to SMS-EMOA, its parallel version using the asynchronous island model without external archives (pSMS-EMOA), and the Hypervolume Estimation Algorithm (HypE) [2] for 2, 3, 5 and 10 objectives. HypE ranks the population by means of Pareto dominance and its secondary selection criterion is based on the estimation of the hypervolume contributions using Monte Carlo sampling (for 2 and 3 objectives, the exact value is computed). All the MOEAs were implemented in the EMO Project, using real-numbers encoding.

The variation operators were polynomial-based mutation and simulated binary crossover (SBX) [5]. The crossover rate and its distribution index were set to 0.9 and 20, for 2 and 3 objectives, and 1.0 and 30 for many-objective problems. The mutation rate and its distributed index was set to $1/n$ and 20, respectively. For HypE, the number of sampling points was fixed to 20,000 and the resolution parameter of $\mathcal{S}$-PAMICRO (γ), as suggested in [8], is shown in Table 1.

Table 1. Parameters adopted in our experiments

m	WFG		MOEAs	pMOEAs		$feval$	$\mathcal{S}$-PAMICRO
	n	k	$\|P\|$	$\|P\|$	l		γ
2	24	4	100	10	10	40,000	3
3	24	4	120	10	12	50,000	2
5	47	8	196	11	18	50,000	2
10	105	18	276	11	25	80,000	2

The stopping criterion consisted of reaching a maximum number of objective function evaluations ($feval$), limiting the execution time to no more than two hours for each run. For fair comparisons, the parameters were similar in the

[8] Available at http://computacion.cs.cinvestav.mx/~rhernandez.
[9] https://www.mpich.org.

sequential and parallel cases. The population size $|P|$ of the sequential algorithms (SMS-EMOA/HypE) and the parallel MOEAs (pSMS-EMOA/$\mathcal{S}$-PAMICRO) are defined in Table 1, as well as the number of islands or processors (l) in the latter case. Here, l is equivalent to the division of the overall population size among the micro-population size. Experiments were carried on a Cluster of 10 PCs Intel(R) Core(TM) i7 CPU 950 @ 3.07 GHz × 8 with 3.8 GB memory. The frequency of migration, $fmig$, was set to 80 function evaluations and the number of migrants $nmig$ was set to 2 (these values were empirically determined). We performed 30 independent runs for all scenarios. For comparing results, we adopted the hypervolume indicator, bounded by the reference points $(3, 5, 7, \ldots)$ for the instances WFG1 and WFG3; and $(2.2, 4.2, 6.2, \ldots)$ for the rest of the problems. We applied the Wilcoxon rank sum test (one-tailed) to the mean hypervolume indicator values, in order to determine whether $\mathcal{S}$-PAMICRO performed better than the other MOEAs at the significance level of 5 %.

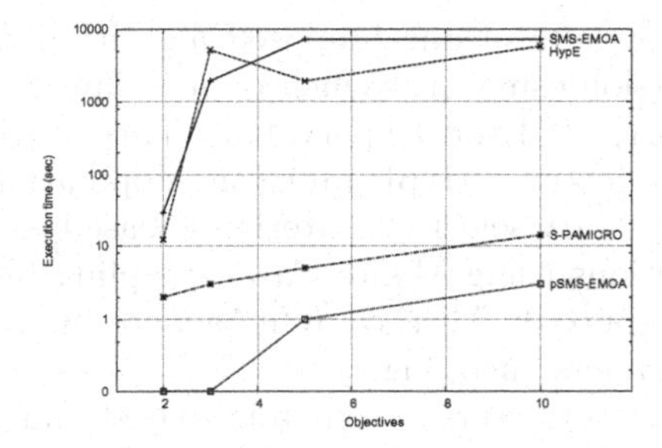

Fig. 2. Average execution time of optimizers.

The average execution time, using a logarithmic scale for the y-axis, is shown in Fig. 2. As it can be observed, $\mathcal{S}$-PAMICRO spent considerably less time than SMS-EMOA and HypE. For example, with 10 objectives, a run of our proposed approach took only 16 s out of the two hours that were allowed to the other MOEAs. Using 5 objective functions, $\mathcal{S}$-PAMICRO ended in 5 s, in contrast to the 26 min spent by HypE. Even in low dimensionality, our algorithm could reduce the run time a little bit. Furthermore, the overhead of handling the external archive in $\mathcal{S}$-PAMICRO is relatively low, compared to pSMS-EMOA that was the fastest optimizer.

On the other hand, interesting results with respect to the quality of solutions were obtained. In Table 2, we present the hypervolume indicator values of all the experiments. An arrow pointing upwards (↑) means that our algorithm outperformed in a significantly better way, the other MOEAs compared. Conversely, an arrow pointing downwards (↓) means that our algorithm was significantly beaten. An asterisk ($*$) means that the algorithm was interrupted because the allowed execution time was exceeded. In the majority of the cases for 5 and 10 objectives, $\mathcal{S}$-PAMICRO obtained the best results, outperforming SMS-EMOA,

Table 2. Median and standard deviation of the hypervolume indicator on the WFG benchmark. The two best values are shown in gray scale, where a darker tone corresponds to the best value.

m	HypE	SMS-EMOA	pSMS-EMOA	$\mathcal{S}$-PAMICRO
		WFG1		
2	5.17e+00 4.11e-1 ↑	4.45e+00 3.63e-1 ↑	3.66e+00 2.59e-1 ↑	6.61e+00 9.65e-1
3	5.66e+01 1.62e+0 ↓	5.28e+01 2.50e+0 ↑	4.23e+01 3.08e+0 ↑	5.56e+01 3.71e+0
5	2.82e+03 1.17e+2 ↑	3.18e+03 7.20e+1 * ↑	3.91e+03 4.83e+1 ↑	5.16e+03 3.88e+2
10	4.19e+09 1.81e+8 ↑	1.88e+09 2.62e+8 * ↑	5.28e+09 5.76e+7 ↑	5.87e+09 2.33e+8
		WFG2		
2	5.46e+00 2.79e-2 ↑	5.47e+00 1.25e-1 ↑	5.39e+00 1.71e-1 ↑	5.49e+00 4.00e-2
3	5.34e+01 4.21e+0 ↓	4.47e+01 4.47e+0	5.18e+01 2.00e+0 ↑	5.32e+01 2.50e-1
5	4.24e+03 3.00e+2 ↑	4.41e+03 3.32e+2 * ↑	4.66e+03 1.52e+1 ↑	4.75e+03 2.00e+1
10	4.66e+09 3.22e+8 ↑	3.80e+09 2.86e+8 * ↑	4.91e+09 1.75e+8 ↑	4.93e+09 1.96e+8
		WFG3		
2	1.09e+01 3.06e-2 ↑	1.09e+01 2.09e-2 ↑	1.08e+01 3.23e-2 ↑	1.09e+01 4.50e-2
3	7.59e+01 2.19e-1 ↑	7.60e+01 1.52e-1	7.48e+01 1.06e-1 ↑	7.61e+01 3.61e-1
5	5.55e+03 1.55e+2 ↑	6.84e+03 5.88e+1 * ↑	6.93e+03 3.11e+1 ↑	7.22e+03 5.86e+1
10	8.37e+09 1.38e+8 ↓	7.64e+09 1.95e+8 * ↑	5.91e+09 3.30e+8 ↑	8.19e+09 1.98e+9
		WFG4		
2	2.91e+00 3.46e-3 ↓	2.90e+00 1.08e-2	2.77e+00 2.05e-2 ↑	2.90e+00 2.10e-2
3	2.96e+01 5.19e-2 * ↓	2.97e+01 5.43e-2 ↓	2.66e+01 2.41e-1 ↑	2.88e+01 4.45e+0
5	1.69e+03 9.10e+1 ↑	2.50e+03 6.71e+1 * ↑	3.13e+03 7.15e+1 ↑	3.47e+03 1.16e+2
10	1.86e+09 1.03e+8 * ↓	1.37e+09 6.15e+7 * ↓	2.00e+09 4.38e+8 ↓	1.22e+09 5.81e+8
		WFG5		
2	2.59e+00 2.40e-3 ↑	2.58e+00 2.82e-3 ↑	2.53e+00 1.21e-2 ↑	2.59e+00 8.62e-3
3	2.74e+01 7.07e-1 * ↓	2.73e+01 1.38e-1 ↓	2.52e+01 1.92e-1 ↑	2.70e+01 1.46e-1
5	1.96e+03 1.33e+2 ↑	2.47e+03 5.10e+1 * ↑	2.75e+03 1.50e+2 ↑	3.31e+03 9.51e+1
10	1.95e+09 1.06e+8 * ↑	1.04e+09 3.14e+7 * ↑	1.04e+09 3.47e+8 ↑	3.99e+09 6.24e+8
		WFG6		
2	2.65e+00 5.79e-2 ↑	2.64e+00 5.43e-2 ↑	2.56e+00 3.93e-2 ↑	2.68e+00 2.11e-2
3	2.77e+01 2.68e-1	2.79e+01 2.12e-1 ↓	2.52e+01 3.86e-1 ↑	2.77e+01 4.05e-1
5	1.80e+03 1.37e+2 ↑	2.08e+03 7.00e+1 * ↑	2.93e+03 6.19e+1 ↑	3.39e+03 6.23e+1
10	1.83e+09 1.28e+8 ↑	9.82e+08 3.55e+7 * ↑	2.02e+09 2.55e+8 ↑	3.83e+09 5.36e+8
		WFG7		
2	2.92e+00 1.60e-3 ↓	2.91e+00 1.05e-2 ↓	2.84e+00 1.25e-2 ↑	2.91e+00 3.05e-1
3	2.97e+01 2.72e-2 * ↓	2.99e+01 1.35e-2 ↓	2.73e+01 2.64e-1 ↑	2.93e+01 1.95e-1
5	1.82e+03 1.10e+2 ↑	2.66e+03 7.07e+1 * ↑	3.20e+03 7.84e+1 ↑	3.55e+03 4.62e+1
10	2.22e+09 1.08e+8 ↓	1.26e+09 5.23e+7 *	1.12e+09 2.77e+8	8.52e+08 7.72e+8
		WFG8		
2	2.25e+00 1.46e-2 ↓	2.24e+00 1.13e-2 ↓	2.10e+00 2.99e-2 ↑	2.24e+00 3.37e-2
3	2.34e+01 2.82e-1 ↑	2.52e+01 8.04e-2 ↓	2.19e+01 4.28e-1 ↑	2.43e+01 5.25e-1
5	1.52e+03 1.20e+2 ↑	2.26e+03 5.62e+1 * ↑	2.55e+03 1.16e+2 ↑	2.86e+03 3.62e+2
10	1.84e+09 1.29e+8 ↓	1.06e+09 4.60e+7 * ↓	1.53e+09 3.69e+8 ↓	4.64e+08 7.71e+8
		WFG9		
2	2.30e+00 2.61e-1 ↑	2.78e+00 2.34e-1 ↑	2.63e+00 2.09e-1 ↑	2.81e+00 4.88e-1
3	2.16e+01 1.56e+0 * ↑	2.82e+01 1.77e+0 ↓	2.25e+01 1.10e+0 ↑	2.74e+01 6.78e+0
5	1.75e+03 1.65e+2 ↑	2.36e+03 1.12e+2 * ↑	2.57e+03 6.33e+1	2.61e+03 8.93e+2
10	1.66e+09 1.10e+8 ↑	1.12e+09 6.31e+7 * ↑	1.87e+09 3.46e+8 ↑	2.31e+09 9.27e+8

HypE and pSMS-EMOA. While with 2 and 3 objectives, our proposal only surpassed pSMS-EMOA, being competitive with SMS-EMOA and HypE.

In summary, we observed that $\mathcal{S}$-PAMICRO could achieve much better results than SMS-EMOA and HypE in high dimensionality, spending much less computational time. For this reason, we claim that our proposed approach is a promising alternative for solving many-objective optimization problems.

4 Conclusions and Future Work

This paper presented a parallel version of the $\mathcal{S}$-Metric Selection Evolutionary Multi-Objective Algorithm (SMS-EMOA). The new approach, called PArallel MICRo Optimizer based on the $\mathcal{S}$ metric ($\mathcal{S}$-PAMICRO), draws ideas from the asynchronous island model with relatively small populations. Diversity is preserved through external archives that are pruned to a limit size, using a recently proposed technique that is based on automatic image analysis. We compared our proposal with respect to HypE (Hypervolume Estimation Algorithm), and with respect to the serial version of SMS-EMOA and another parallel version of it. We observed that $\mathcal{S}$-PAMICRO is a viable alternative for solving many-objective optimization problems at an affordable computational time. In fact, the execution time seems to be dominated by polynomial terms and not the exponential terms when using micro-populations. The model of the execution time of $\mathcal{S}$-PAMICRO is $1.526m-1.632$, using least-squares approximation. Further studies are nevertheless required, adopting more benchmarks and comparing to other state-of-the-art MOEAs. We are also interested in studying the effects of the additional parameters related to the migration operator.

References

1. Adra, S.F., Fleming, P.J.: Diversity management in evolutionary many-objective optimization. IEEE Trans. Evol. Comput. **15**(2), 183–195 (2011)
2. Bader, J., Zitzler, E.: HypE: an algorithm for fast hypervolume-based many-objective optimization. Evol. Comput. **19**(1), 45–76 (2011)
3. Bringmann, K., Friedrich, T.: Don't be greedy when calculating hypervolume contributions. In: FOGA 2009: Proceedings of the Tenth ACM SIGEVO Workshop on Foundations of Genetic Algorithms, Orlando, Florida, USA, pp. 103–112. ACM, January 2009
4. Coello Coello, C.A., Lamont, G.B., Van Veldhuizen, D.A.: Evolutionary Algorithms for Solving Multi-objective Problems, 2nd edn. Springer, New York (2007). ISBN 978-0-387-33254-3
5. Deb, K., Agrawal, R.B.: Simulated binary crossover for continuous search space. Complex Syst. **9**, 115–148 (1995)
6. Emmerich, M.T.M., Beume, N., Naujoks, B.: An EMO algorithm using the hypervolume measure as selection criterion. In: Coello Coello, C.A., Hernández Aguirre, A., Zitzler, E. (eds.) EMO 2005. LNCS, vol. 3410, pp. 62–76. Springer, Heidelberg (2005)

7. Fleischer, M.: The measure of Pareto optima applications to multi-objective metaheuristics. In: Fonseca, C.M., Fleming, P.J., Zitzler, E., Deb, K., Thiele, L. (eds.) EMO 2003. LNCS, vol. 2632, pp. 519–533. Springer, Heidelberg (2003)
8. Hernández Gómez, R., Coello Coello, C.A.: A multi-objective evolutionary algorithm based on parallel coordinates. In: Proceedings of the 2016 Genetic and Evolutionary Computation Conference (GECCO 2016), New York, NY, USA. ACM Press (2016, in press)
9. Ishibuchi, H., Tsukamoto, N., Nojima, Y.: Evolutionary many-objective optimization: a short review. In: IEEE Congress on Evolutionary Computation, CEC 2008 (IEEE World Congress on Computational Intelligence), pp. 2419–2426, June 2008
10. Klinkenberg, J.-W., Emmerich, M.T.M., Deutz, A.H., Shir, O.M., Bäck, T.: A reduced-cost SMS-EMOA using Kriging, self-adaptation, and parallelization. In: Ehrgott, M., Naujoks, B., Stewart, T.J., Wallenius, J. (eds.) Multiple Criteria Decision Making for Sustainable Energy and Transportation Systems, vol. 634, pp. 301–311. Springer, Heidelberg (2010)
11. Li, B., Li, J., Tang, K., Yao, X.: Many-objective evolutionary algorithms: a survey. ACM Comput. Surv. **48**(1), 13:1–13:35 (2015)
12. Lopez, E.M., Antonio, L.M., Coello Coello, C.A.: A GPU-based algorithm for a faster hypervolume contribution computation. In: Gaspar-Cunha, A., Henggeler Antunes, C., Coello Coello, C.A. (eds.) EMO 2015. LNCS, vol. 9019, pp. 80–94. Springer, Heidelberg (2015)
13. Lücken, C., Barán, B., Brizuela, C.: A survey on multi-objective evolutionary algorithms for many-objective problems. Comput. Optim. Appl. **58**(3), 707–756 (2014)
14. Luna, F., Alba, E.: Parallel multiobjective evolutionary algorithms. In: Kacprzyk, J., Pedrycz, W. (eds.) Springer Handbook of Computational Intelligence, pp. 1017–1031. Springer, Heidelberg (2015)
15. Van Veldhuizen, D.A., Zydallis, J.B., Lamont, G.B.: Considerations in engineering parallel multiobjective evolutionary algorithms. IEEE Trans. Evol. Comput. **7**(2), 144–173 (2003)
16. Wagner, T., Beume, N., Naujoks, B.: Pareto-, aggregation-, and indicator-based methods in many-objective optimization. In: Obayashi, S., Deb, K., Poloni, C., Hiroyasu, T., Murata, T. (eds.) EMO 2007. LNCS, vol. 4403, pp. 742–756. Springer, Heidelberg (2007)
17. While, L., Bradstreet, L., Barone, L.: A fast way of calculating exact hypervolumes. IEEE Trans. Evol. Comput. **16**(1), 86–95 (2012)
18. Zhou, A., Bo-Yang, Q., Li, H., Zhao, S.-Z., Suganthan, P., Zhang, Q.: Multiobjective evolutionary algorithms: a survey of the state of the art. Swarm Evol. Comput. **1**(1), 32–49 (2011)
19. Zitzler, E., Künzli, S.: Indicator-based selection in multiobjective search. In: Yao, X., Burke, E.K., Lozano, J.A., Smith, J., Merelo-Guervós, J.J., Bullinaria, J.A., Rowe, J.E., Tiňo, P., Kabán, A., Schwefel, H.-P. (eds.) PPSN 2004. LNCS, vol. 3242, pp. 832–842. Springer, Heidelberg (2004)

Real-World Applications and Modelling

Evolution of Active Categorical Image Classification via Saccadic Eye Movement

Randal S. Olson[1(✉)], Jason H. Moore[1], and Christoph Adami[2]

[1] Institute for Biomedical Informatics, University of Pennsylvania, Philadelphia, PA, USA
{olsonran,jhmoore}@upenn.edu

[2] Department of Microbiology and Molecular Genetics, Michigan State University, East Lansing, MI, USA
adami@msu.edu

Abstract. Pattern recognition and classification is a central concern for modern information processing systems. In particular, one key challenge to image and video classification has been that the computational cost of image processing scales linearly with the number of pixels in the image or video. Here we present an intelligent machine (the "active categorical classifier," or ACC) that is inspired by the saccadic movements of the eye, and is capable of classifying images by selectively scanning only a portion of the image. We harness evolutionary computation to optimize the ACC on the MNIST hand-written digit classification task, and provide a proof-of-concept that the ACC works on noisy multi-class data. We further analyze the ACC and demonstrate its ability to classify images after viewing only a fraction of the pixels, and provide insight on future research paths to further improve upon the ACC presented here.

Keywords: Active categorical perception · Attention-based processing · Evolutionary computation · Machine learning · Supervised classification

1 Introduction

Pattern recognition and classification is one of the most challenging ongoing problems in computer science in which we seek to classify objects within an image into categories, typically with considerable variation among the objects within each category. With *invariant* pattern recognition, we seek to develop a model of each category that captures the essence of the class while compressing inessential variations. In this manner, invariant pattern recognition can tolerate (sometimes drastic) variations within a class, while at the same time recognizing differences across classes that can be minute but salient. One means of achieving this goal is through invariant feature extraction [1], where the image is transformed into feature vectors that may be invariant with respect to a set of transformations, such as displacement, rotation, scaling, skewing, and lighting changes. This method can also be used in a hierarchical setting, where subsequent layers extract compound features from features already extracted in lower

J. Handl et al. (Eds.): PPSN XIV 2016, LNCS 9921, pp. 581–590, 2016.
DOI: 10.1007/978-3-319-45823-6_54

levels, such that the last layer extracts features that are essentially the classes themselves [2]. Most of these existing methods have one thing in common: they achieve invariance either by applying transformations to the image when searching for the best match, or by mapping the image to a representation that is itself invariant to such transformations.

In contrast to these "passive" methods where transformations are applied to the image, we propose an active, attention-based method, where a virtual camera roams over and focuses on particular portions of the image, similar to how our own brain controls the focus of our attention [3]. In this case, the camera's actions are guided by what the camera finds in the image itself: In essence, the camera searches the image to discover features that it recognizes, creating in the process a time series of experiences that guides further movements and eventually allows the camera to classify the image. We call this camera an "active categorical classifier," or ACC for short.

Broadly speaking, the problem of classifying a spatial pattern is transformed into one of detecting differences within and between time series, namely the temporal sequence that the virtual camera generates in its sensors as it navigates the image. The method we propose here is inspired by models of visual attention [4], where attention to "salient" elements of an image or scene is guided by the image itself, such that only a small part of the incoming sensory information reaches short-term memory and visual awareness. Thus, focused attention overcomes the information-processing bottleneck imposed by massive sensory input (which can easily be $10^7 - 10^8$ bits per second in parallel at the optic nerve [4]), and serializes this stream to achieve near-real-time processing with limited computational requirements.

In previous work, we have shown that it is possible to evolve robust controllers that navigate arbitrary mazes with near-perfect accuracy [5] and simulate realistic animal behavior [6]. Independently, we have shown that we can evolve simple spatial classifiers for hand-written numerals in the MNIST data set [7]. Here we use the same technology to evolve active categorical classifiers that "forage" on images and respond to queries about what they saw in the image without needing to examine the image again.

2 Methods

In this section, we describe the methods used to evolve the active categorical classifiers (ACCs). We begin by describing the simulation environment in which the ACC scans and classifies the images. Next, we outline the structure and underlying neural architecture of an ACC. Finally, we provide details on the evolutionary process that we used to evolve the ACCs and the experiments that we conducted to evaluate them.

2.1 Simulation Environment

We evaluate the ACC on the MNIST data set, which is a well-known set of handwritten digits commonly used in supervised image classification research [8].

The MNIST data set contains 28 × 28 pixel images of hand-written digits—all with corresponding labels indicating what digit the image represents (0–9)—and comes in two predefined sets of training and testing data (60,000 and 10,000 images, respectively). In this project, we binarize the images such that any pixels with a grayscale value >127 (out of the range [0, 255]) are assigned a value of 1, and all other pixels are assigned a value of 0.

When we evaluate an ACC, we place it at a random starting point in the 28 × 28 image and provide it a maximum of 40 steps to scan the image and assign a classification. (The 40-step maximum is meant to limit each simulation to a reasonably short amount of time.) Every simulation step, the ACC decides (1) what direction to move, (2) what class(es) it currently classifies the image as, and (3) whether it has made its final classification and is ready to terminate the simulation early. The ACC is evaluated only on its final classification for each image in the training set, with a "fitness" score (F_{ind}) assigned as:

$$F_{\text{ind}} = \frac{1}{1000} \times \sum_{i=1}^{1000} \frac{\text{CorrectClass}_i}{\text{NumClassesGuessed}_i} \tag{1}$$

where i is the index of an individual image in the training set, $\text{CorrectClass}_i = 1$ if the correct class is among the $\text{NumClassesGuessed}_i$ guesses that the ACC offers (it is allowed to guess more than one), and $\text{CorrectClass}_i = 0$ otherwise. Thus, an ACC can achieve a minimum fitness of 0.1 by guessing *all* classes for all images, but only achieves a maximum fitness of 1.0 by determining only the correct class for every image. We note that due to computational limitations, we subset the MNIST training set to the first 100 images of each digit, such that we use only 1,000 training images in total (1/60th of the total set).

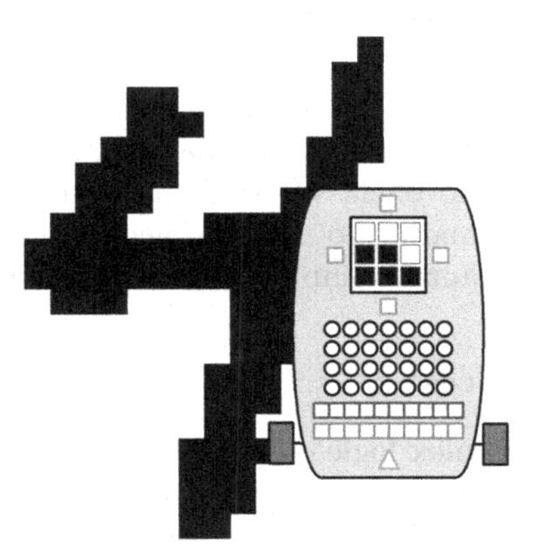

Fig. 1. Active categorical classifier (ACC) configuration. The ACC brain has 64 binary states that either fire or are quiescent, and represent sensory input from the image, internal memory, or decisions about how to interact with the image (described in the text).

2.2 Active Categorical Classifier (ACC)

We show in Fig. 1 the ACC in its natural habitat, roaming a digitized MNIST numeral. Each ACC has a brain that consists of 64 Markov neurons ("states") that either fire (state = 1) or are quiescent (state = 0), and represent sensory input from the image, internal memory, and decisions about how to interact with the image. The ACC uses nine of these states to view nine pixels of the image in a 3×3 square, and four of the states to probe for activated pixels outside of its field of view with four raycast sensors that project across the image from the 0°, 90°, 180°, and 270° angles of the 3×3 square (green squares in Fig. 1). The raycast sensors activate only when they intersect with an activated pixel, and allow the ACC to find the numeral even if its starting position is far from it.

We also provide the ACC two actuator states ("motor neurons") that allow it to "saccade" three pixels up/down and left/right, or any combination thereof (red rectangles denoted as wheels in Fig. 1). In addition, the ACC has 20 states dedicated to classifying the image: 10 states that can be activated to guess each digit class (blue squares), and 10 states to *veto* an activated guess for each digit class (purple squares), e.g., "this is definitely not a 4." This configuration allows the ACC to guess multiple classes at once, and combine its internal logic to veto any of those guesses if it believes them to be incorrect. Finally, the ACC has a "done" state (orange triangle), which allows it to end the simulation early if it has already decided on its final guess(es) for the current image. The remaining 28 neurons are "memory" states (black circles) used to process and store information, and integrate that information over time.

The "artificial brain" for the ACC in these experiments is a *Markov Network* (MN, see, e.g., [5,7,9]) that deterministically maps the 64 states (described above) at time t to a corresponding series of output states that we interpret to determine the ACC's movement actions and classifications at time $t + 1$. The combination of output states and sensory inputs from time $t + 1$ are then used to determine the output states for the ACC at time $t + 2$, and so on. Every MN must therefore usefully combine the information provided over time in the 64 states to decide where to move, classify the image, and finally to decide when it has gathered enough information to make an accurate classification. Making all these decisions at once requires complex logic that is difficult to design.

2.3 Optimization Process

In order to create the complex logic embodied by a Markov Network, we *evolve* the MNs to maximize classification accuracy on the training images. We use a standard Genetic Algorithm (GA) to stochastically optimize a population of byte strings [10], which deterministically map to the MNs that function as the ACC's "artificial brains" in the simulation described above. Due to space limitations, we cannot describe MNs in full detail here; a detailed description of MNs and how they are evolved can be found in [11].

In our experiments, the GA maintains a population of 100 byte strings ("candidates") of variable length (maximum = 10,000 bytes) and evaluates them according to the fitness function in Eq. 1. The GA selects the candidates to reproduce into the next generation's population via tournament selection, where it shuffles the population and competes every byte string against only one other byte string. In each tournament, the byte string with the highest fitness produces one exact copy of itself as well as one mutated copy of itself into the next generation, while the "loser" produces no offspring. We note that the GA applies only mutations to the offspring (no crossover/recombination), with a per-byte mutation rate of 0.05 %, a gene duplication rate of 5 %, and a gene deletion rate of 2 %.

2.4 Experiments

According to the evolutionary optimization process, the GA selects ACCs that are capable of spatio-temporal classification of MNIST digits. We first ran 30 replicates of the GA with random starting populations and distinct random seeds and allowed these replicates to run for 168 hours on a high-performance compute cluster. From those 30 replicates, we identified the highest-fitness ACC (the "elite"), and seeded another set of 30 replicates with mutants of the elite ACC. We allowed this second set of replicates to run for another 168 hours. In the following section, we report on the results of these experiments.

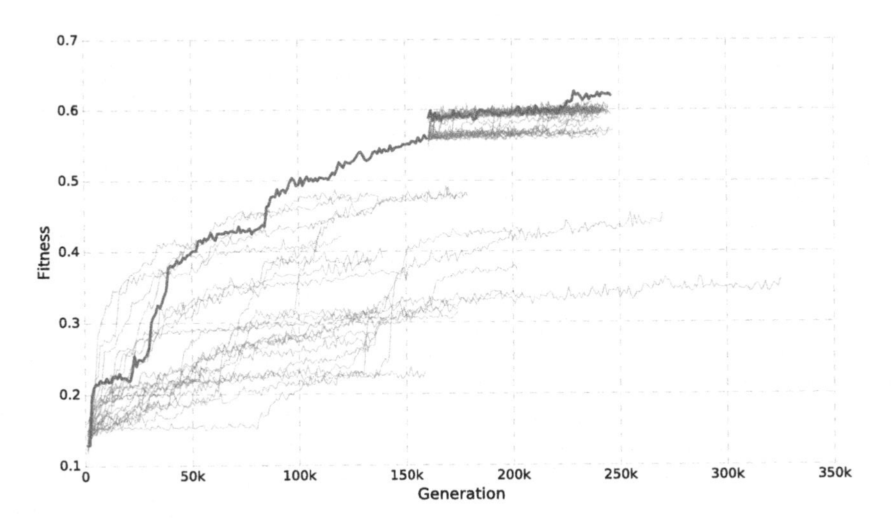

Fig. 2. Fitness over time on the MNIST training set. Each line represents a replicate of the evolutionary process that trains the active categorical classifiers. The lines represent the highest-fitness individual every 1,000 generations, where the blue line traces the lineage that led to the highest-fitness individual out of all replicates. After running all 30 replicates for one week, we took the best individual from the first set of runs and seeded another set of evolutionary runs with it, which is represented by the cluster of lines following the top lineage of first set.

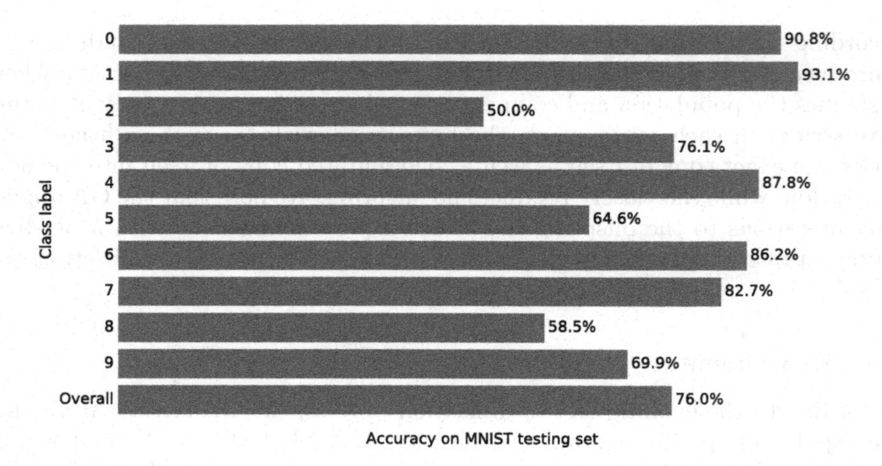

Fig. 3. Active categorical classifier (ACC) accuracy on the binarized MNIST testing set. We report per-digit accuracy (labeled 0–9) of the ACC as well as the average accuracy across all digits (labeled "Overall").

3 Results

At the completion of the second set of replicates, the remaining active categorical classifiers (ACCs) had been optimized for 336 hours and roughly 250,000 generations. Shown in Fig. 2, the ACCs experienced the majority of their improvements within the first 150,000 generations, and minimal improvements occurred in the second set of replicates, indicating that the ACCs had reached a plateau—either because the scan pattern required to improve was too complex, or because improving the classification accuracy on poorly classified digits compromised the ability to classify those digits the ACC was already proficient at. Such trade-offs are likely due to insufficient brain size, and investigations with larger brains are currently underway.

Instead of continuing the optimization process for a third set of replicates, we identified the highest-fitness ACC from replicate set 2 (highlighted in blue, Fig. 2) and analyzed its spatio-temporal classification behavior to gain insights into its functionality. For the remainder of this section, we focus on the best ACC evolved in replicate set 2, which we will simply call "the ACC." Shown in Fig. 3, the ACC achieved respectable but not state-of-the-art performance on the MNIST testing set: It managed to classify most of the 0s and 1s correctly for example, but failed to classify many of the 2s. Overall, the ACC achieved a macro-averaged accuracy of 76 %, which provides a proof-of-concept that the ACC works, but still has room for improvement on noisy multi-class data sets. We note that we have optimized ACCs on a set of hand-designed, non-noisy digits, where they managed to achieve 100 % accuracy. Thus, it is clear that the ACC architecture requires additional experimentation to fully adapt to noisy data, much like other methods currently in use. In Fig. 4B, we analyze the movement patterns of the

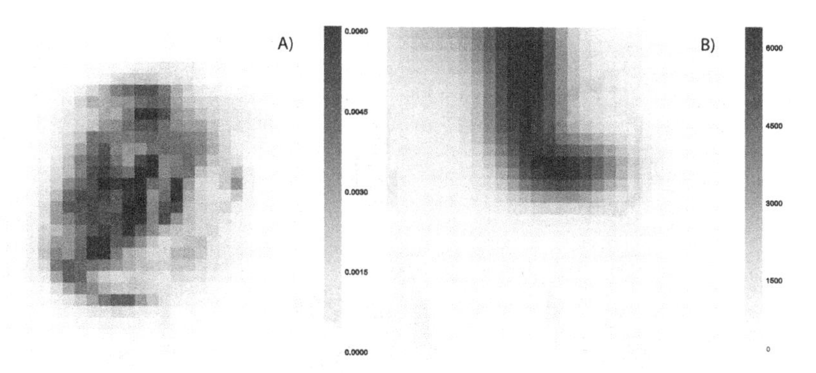

Fig. 4. Analysis of informative pixels in the MNIST training set. Panel A shows the most informative pixels in the MNIST training set according to feature importance scores from a Random Forest (i.e., Gini importance [12]), whereas Panel B shows the pixels that the best active categorical classifier visited most frequently when classifying the MNIST data set. In both cases, darker colors represent higher values.

ACC by counting how many times each pixel is viewed in the ACC's 3×3 visual grid when classifying the MNIST data set. Even though the ACC always starts at a random location in the image, we find that it follows a stereotypical scanning patterns of the digits: the ACC lines itself up to the top-left of the digit, then executes an L-shaped scanning pattern.

In contrast, Fig. 4A depicts the most informative pixels for differentiating the classes in the binarized MNIST data set with a Random Forest classifier as implemented in scikit-learn [13]. Here, we find that the most informative pixels exist in the center of the images, with several less-informative pixels on the image edges. Importantly, we note that the ACC never scans some of the most informative pixels in the lower half of the MNIST images (Fig. 4A vs. B). We believe that this behavior is the reason that the ACC is rarely able to classify any of the 2s, for example, because some of the most critical pixels for differentiating 2s from the rest of the digits are never visited.

We provide examples of the ACC scanning patterns in Fig. 5. Shown again is the stereotypical L-shaped scanning pattern starting at the upper-left corner of every digit. (We note that we trimmed the agent paths to only the final scanning pattern because the initial phase of ACC movements are simply lining up to the upper-left corner of the digit.) Interestingly, the ACC scans only a fraction of the available pixels to make each classification, and appears to be integrating information about the digit over space and time to identify distinctive sub-features of the digits. Furthermore, the ACC completes the majority of its scans within 5–10 steps and then immediately activates the "done" state, indicating that the ACC also learned when it knows the correct digit.

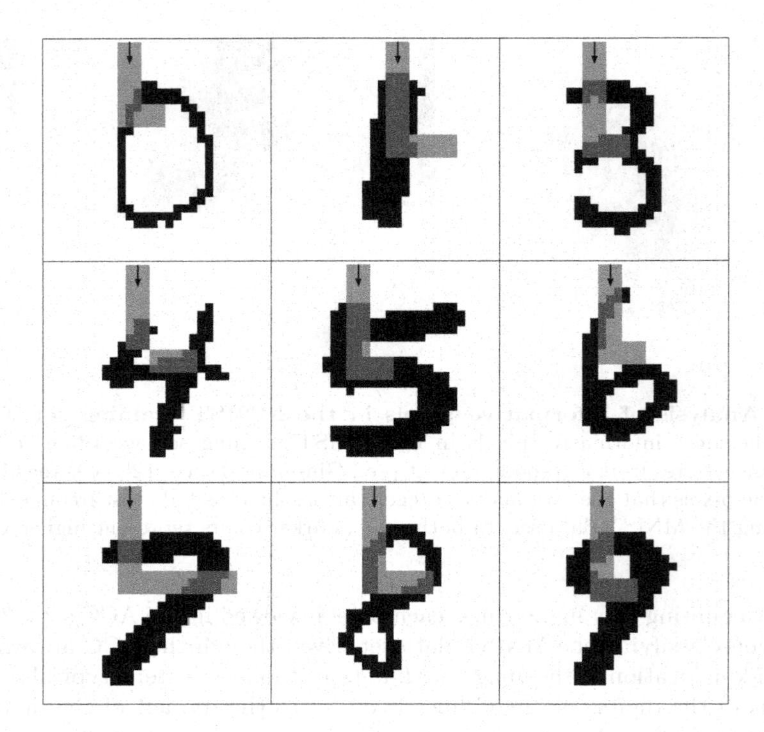

Fig. 5. Example trajectories of the best active categorical classifier (ACC). The arrows indicate the direction that the ACC followed, whereas the dark grey areas indicate the pixels that it scanned. Although the ACC starts all evaluations at random spots in the grid, it aligns itself to the digit to a common starting point and executes and L-shaped scan of the digit. We note that we excluded an example of digit 2 because the ACC rarely classifies it correctly, although it follows a similar L-shaped trajectory.

4 Discussion

The results that we display here show that it is possible to optimize an active categorical classifier (ACC) that scans a small portion of an image, integrates that information over space and time, and proceeds to perform an accurate classification of the image. Although the ACC does not achieve competitive accuracy on the MNIST data set compared to many modern techniques (76 % testing accuracy, Fig. 3), we believe that this result is due to the lack of training data rather than any particular limitation of ACCs: Due to computational limitations, we were only able to use a fixed set of 1,000 training images (100 of each class) to optimize the ACCs, while modern techniques use much larger training sets that even include additional variations of the training images [14]. Indeed, when we trained a scikit-learn Random Forest with 500 decision trees [13] on the same binarized training set of 1,000 images, it achieves only 88.5 % accuracy on the MNIST testing set as compared to 97.5 % when it is trained on the full training set. Thus, in future work we will focus on integrating methods that expose the ACCs to all training images in an efficient manner.

From the point of view of embodied artificial intelligence, the challenge presented to the ACC in the image classification task is remarkably difficult. For one, these experiments challenged a single artificial brain to simultaneously perform several complex tasks, including to line itself up to a consistent starting point regardless of where it randomly starts in the image, decide where it needs to move to complete the scan based on limited information about the image, determine what pixels are important to consider, *and* integrate that information over space and time to classify the image into 1 of 10 classes. We furthermore challenged the ACC to evolve something akin to a "theory of mind" such that it knows when it has guessed the correct class for the image and to end the simulation early. In future work, it will be illuminating to analyze the underlying neural architecture of the evolved ACCs to provide insight into the fundamentals of active categorical perception [15].

Unlike many modern image classification techniques that must analyze an entire static image to determine an image's class, the ACC instead integrates information from a small subset of the pixels over space and time. This method naturally lends itself to video classification, where feature compression will play a crucial role in overcoming the massive data size challenge for real-time classification of moving objects [3]. Lastly, recent work has shown that modern deep learning-based image classification techniques tend to be easily fooled because they are trained in a supervised, discriminative manner: They establish decision boundaries that appropriately separate the data they encounter in the training phase, but these decision boundaries also include (and thus mis-classify) many inappropriate data points never encountered during training [16]. Although most deep learning researchers respond to this challenge by creating additional "adversarial" training images to train the deep neural networks [17], we believe that the findings in [16] highlight a critical weakness in deep learning: the resulting networks are trained to precisely map inputs to corresponding target outputs, without generalizing far beyond the training data they are exposed to [18].

Due to their nature, deep neural networks are highly dependent on the training data, and only generalize to new challenges if they are similar to those encountered in the training data [17]. In contrast, heuristic-based machines such as the ACC learn simple, generalizable heuristics for classifying images that encode the conceptual representation [9] of the objects, and should not be so easily fooled. As such, even if the ACC in the present work does not achieve competitive accuracy when compared to modern deep learning techniques, we believe that further development of heuristic-based image classification machines will lead to robust classifiers that will eventually surpass deep neural networks in generalizability without the need for adversarial training images. We further believe that it is precisely those machines that carry with them complex representations of the world that will become the robust and sophisticated intelligent machines of the future. Whether the embodied evolutionary approach we describe here will succeed in this is, of course, an open problem.

Acknowledgments. We thank David B. Knoester, Arend Hintze, and Jeff Clune for their valuable input during the development of this project. We also thank the Michigan State University High Performance Computing Center for the use of their computing resources. This work was supported in part by the National Science Foundation BEACON Center under Cooperative Agreement DBI-0939454, and in part by National Institutes of Health grants LM009012, LM010098, and EY022300.

References

1. Trier, O.D., Jain, A.K., Taxt, T.: Feature extraction methods for character recognition - a survey. Pattern Recogn. **29**, 641–662 (1999)
2. LeCun, Y., Boser, B., Denker, J.S., et al.: Backpropagation applied to handwritten zip code recognition. Neural Comput. **1**, 541–551 (1989)
3. Mnih, V., Heess, N., Graves, A., Kavukcuoglu, K.: Recurrent models of visual attention. In: Advances in Neural Information Processing Systems, NIPS 2009, pp. 2204–2212 (2014)
4. Itti, L., Koch, C.: Computational modelling of visual attention. Nat. Rev. Neurosci. **2**, 194–203 (2001)
5. Edlund, J., Chaumont, N., Hintze, A., et al.: Integrated information increases with fitness in the evolution of animats. PLoS Comput. Biol. **7**, e1002236 (2011)
6. Olson, R., Hintze, A., Dyer, F., Knoester, D., Adami, C.: Predator confusion is sufficient to evolve swarming behaviour. J. Roy. Soc. Interface **10**, 20130305 (2013)
7. Chapman, S., Knoester, D., Hintze, A., Adami, C.: Evolution of an artificial visual cortex for image recognition. In: Liò, P., et al. (eds.) Advances in Artificial Life (ECAL 2013), pp. 1067–1074. MIT Press, Cambridge (2013)
8. LeCun, Y., Bottou, L., Bengio, Y., Haffner, P.: Gradient-based learning applied to document recognition. Proc. IEEE **86**, 2278–2324 (1998)
9. Marstaller, L., Hintze, A., Adami, C.: Cognitive systems evolve complex representations for adaptive behavior. Neural Comput. **25**, 2079–2105 (2013)
10. Eiben, A., Smith, J.: Introduction to Evolutionary Computing. Springer, Berlin (2003)
11. Olson, R.S., Knoester, D.B., Adami, C.: Evolution of swarming behavior is shaped by how predators attack. Artif. Life **22** (2016)
12. Breiman, L., Cutler, A.: Random forests - classification description, March 2016. http://www.stat.berkeley.edu/~breiman/RandomForests/cc_home.htm
13. Pedregosa, F., Varoquaux, G., Gramfort, A., et al.: Scikit-learn: machine learning in Python. J. Mach. Learn. Res. **12**, 2825–2830 (2011)
14. Wan, L., Zeiler, M., Zhang, S., LeCun, Y., Fergus, R.: Regularization of neural networks using DropConnect. In: Proceedings of the 30th International Conference on Machine Learning, ICML 2013 (2013)
15. Beer, R.D.: The dynamics of active categorical perception in an evolved model agent. Adapt. Behav. **11**, 209–243 (2003)
16. Nguyen, A., Yosinski, J., Clune, J.: Neural networks are easily fooled: high confidence predictions for unrecognizable images. In: Computer Vision and Pattern Recognition (CVPR 2015). IEEE Press (2015)
17. Goodfellow, I.J., Shlens, J., Szegedy, C.: Explaining and harnessing adversarial examples. In: International Conference on Learning Representations (2015). http://research.google.com/pubs/ChristianSzegedy.html
18. Szegedy, C., Zaremba, W., Sutskever, I., et al.: Intriguing properties of neural networks. In: International Conference on Learning Representations (2014). http://research.google.com/pubs/ChristianSzegedy.html

Cooperative Coevolution of Control for a Real Multirobot System

Jorge Gomes[1,2,3(✉)], Miguel Duarte[1,2,4], Pedro Mariano[3], and Anders Lyhne Christensen[1,2,4]

[1] BioMachines Lab, Lisbon, Portugal
[2] Instituto de Telecomunicações, Lisbon, Portugal
[3] BioISI, Faculdade de Ciências da Universidade de Lisboa, Lisbon, Portugal
jgomes@di.fc.ul.pt
[4] Instituto Universitário de Lisboa (ISCTE-IUL), Lisbon, Portugal

Abstract. The potential of cooperative coevolutionary algorithms (CCEAs) as a tool for evolving control for heterogeneous multirobot teams has been shown in several previous works. The vast majority of these works have, however, been confined to simulation-based experiments. In this paper, we present one of the first demonstrations of a real multirobot system, operating outside laboratory conditions, with controllers synthesised by CCEAs. We evolve control for an aquatic multirobot system that has to perform a cooperative predator-prey pursuit task. The evolved controllers are transferred to real hardware, and their performance is assessed in a non-controlled outdoor environment. Two approaches are used to evolve control: a standard fitness-driven CCEA, and novelty-driven coevolution. We find that both approaches are able to evolve teams that transfer successfully to the real robots. Novelty-driven coevolution is able to evolve a broad range of successful team behaviours, which we test on the real multirobot system.

Keywords: Cooperative coevolution · Evolutionary robotics · Novelty search · Reality gap · Heterogeneous multirobot systems

1 Introduction

Cooperative coevolutionary algorithms (CCEAs) allow for the evolution of solutions that consist of coadapted, interacting components [16,17]. CCEAs are a natural fit for the evolution of heterogeneous multiagent systems [18], as each agent can be represented as an independent component of the solution, and can therefore evolve a specialised behaviour (see for instance [12,18,22]). The classic CCEA architecture [17] operates with two or more populations, where each agent evolves in a separate population. Populations are isolated from one another, meaning that individuals only compete and reproduce with members of their own population. The individuals in each population are evaluated by forming teams with representative individuals from the other populations. These teams are

J. Handl et al. (Eds.): PPSN XIV 2016, LNCS 9921, pp. 591–601, 2016.
DOI: 10.1007/978-3-319-45823-6_55

evaluated in the problem domain, and the individual under evaluation receives the fitness score obtained by the team as a whole.

Previous works that have applied CCEAs to the evolution of agent behaviours can be divided in three main categories [14]: (i) *game-theoretic environments*, essentially strategy games where each agent is rewarded according to a payoff matrix [15,21]; (ii) *abstract embodied agents*, where the evolved agents are situated in an environment that they sense and act in, but the agents are abstract and unrelated to any real robotic platform [7,18,22]; and (iii) *simulated robotics tasks*, in which the evolved agents are modelled closely after a real robotic platform and a real task environment [10,11].

One notable category is missing from this list, namely *real robotics tasks* – tasks in which behavioural control is evolved in simulation, and then transferred to a real robot team. While this reality gap has been crossed using other evolutionary algorithms [19], in both single [8] and multirobot systems [3], to the best of our knowledge, CCEAs have been confined to simulation-based experiments up until now. The potential of CCEAs to evolve control for robot teams has been shown in simulation in tasks such as: predator-prey pursuit [7,11], herding [18], collective construction [13], multirobot foraging [5,12], and keepaway soccer [4].

In this paper, we evolve control for an aquatic surface multirobot system that must perform a cooperative predator-prey pursuit task. Predator-prey pursuit is one of the most commonly studied tasks in multiagent coevolution. In the cooperative version of this task [7,11,22], a team of predators must cooperate to capture an escaping prey. The predator-prey task is especially interesting in CCEA studies because behavioural heterogeneity and close cooperation in the predator team is required to effectively catch the prey [22]. After evolving the controllers offline in simulation, we transfer the controllers to the real robotic platform, and systematically evaluate them in an outdoor environment. The natural unpredictability associated with the aquatic environment (caused by inaccurate robot motion, waves, and currents) allow us to study transferability in a realistic scenario, and understand how controllers evolved by CCEAs are able to cope with noisy and stochastic conditions.

We evolve control using two cooperative coevolution approaches: a standard fitness-driven CCEA [17], and novelty-driven cooperative coevolution [7] – a recently proposed algorithm that aims at mitigating the premature convergence issues that commonly plague CCEAs [15,16]. Novelty-driven coevolution is based on novelty search [9], an evolutionary approach that rewards individuals displaying novel behaviours, rather than exclusively rewarding the individuals that display the highest performance with respect to a fitness function. Novelty-driven coevolution (*NS-Team*) relies on team-level behaviour characterisations, and rewards behaviourally novel teams in addition to high-fitness ones, as it is typically done in CCEAs. The team-level characterisations capture how the team as a whole behaves, without discriminating between the behaviours of the individual agents. Both the fitness and the novelty scores of the teams are used to reward the individuals, via a multi-objective algorithm. By rewarding agents that lead to novel team behaviours, an evolutionary pressure towards novel equilibrium states is created. Besides the ability to overcome premature convergence, and thus reach higher quality solutions, it has also been shown that *NS-Team* can evolve a diverse set of solutions for a given task [4,5,7].

2 Experimental Setup

2.1 Cooperative Predator-Prey Task

In our predator-prey pursuit task, a team of three predators must cooperate to capture one escaping prey. Only the controllers of the team of predators are evolved, while the prey has a pre-specified fixed behaviour.

In each trial, the three predators are placed in the centre of the arena, with random positions and orientations (Fig. 1). The prey is placed in a random location, ranging from 20 m to 35 m from the centre of the arena. A trial ends if a predator gets closer than 2 m from the prey (prey is *captured*), if the prey escapes the arena, or if the time limit (75 s) is reached. The prey moves in the opposite direction of the closest predator, if that predator is closer than 10 m, otherwise it does not move. The prey can move up to the maximum possible speed of the predators, meaning that the predators typically cannot outrun it. Cooperation among the predators is therefore essential to capture the prey.

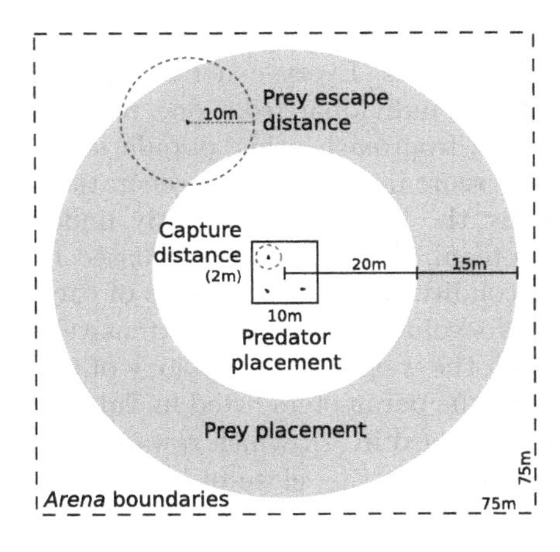

Fig. 1. Task setup used for the evolutionary process.

2.2 Robotic Platform

For our experiments, we use an aquatic multirobot system [1] that has been used in other evolutionary robotics studies in the past [3]. Each robot is a small (65 cm in length) differential drive mono-hull robot. The robots can move at speeds of up to 1.7 m/s, turn at a maximum rate of 90°/s, and are equipped with GPS and compass. The robots broadcast information (such as their position) to their neighbours up to a range of 40 m using Wi-Fi, which is then used to calculate the robots' sensory inputs. The same robotic platform is used for both the predator robots and prey robot.

Each predator is controlled by an artificial neural network, which receives the sensory inputs – the distance to other predators and the relative position of the prey – and has two outputs that control the linear speed and the angular velocity of the robot. The two output values are converted to left and right motor speeds and applied to the robot's motors. The network relies on the following sensory inputs, which are limited to a range of 40 m, and are normalised to $[-1, 1]$:

Predator Sensing: Six inputs for detecting the other predators, corresponding to six equally-sized circular sectors around the robot. Each input returns the normalised distance to the closest predator in the corresponding sector, or the maximum value if no predator is present there.

Prey Location: Two inputs returning (i) the relative angle from the predator to the prey (zero corresponds to straight ahead), and (ii) the normalised distance from the predator to the prey. If the prey is not within sensing range, the sensors return an angle of zero and the maximum distance.

2.3 Evolutionary Setup

Both fitness-driven and novelty-driven cooperative coevolution were implemented over the same standard coevolutionary architecture [17]. There are three coevolving populations, one for each of the predators. Every generation, each population is evaluated in turn. To evaluate an individual from one population, a team is formed with one representative from each other population – the individual that obtained the highest fitness score in the previous generation, or a random one in the first generation. Only the individual currently under evaluation receives the score obtained by the team. Every team is evaluated in 10 simulation trials, with randomized initial conditions. The controllers of each population are evolved by NEAT [20], a neuroevolution algorithm extensively used in evolutionary robotics, that evolves both the weights and topology of the networks. The three coevolving populations use the parameters listed in Table 1.

The fitness function F is the same as the one used in [7], which rewards the teams for capturing the prey as soon as possible, or getting close to it:

$$F = \begin{cases} 2 - \tau/T & \text{if prey captured} \\ max(0, (d_i - d_f)/size) & \text{otherwise} \end{cases}, \tag{1}$$

where τ is the time to capture the prey, T is the maximum trial length, d_i and d_f are, respectively, the average initial and final distance from the predators to the prey, and $size$ is the side length of the arena.

Novelty-driven coevolution is implemented as proposed in [7], using the *NS-Team* technique, which computes the individuals' novelty scores based on the behavioural novelty displayed by the team in which the individual participated. To calculate the novelty score of each team, we rely on four features to characterise team behaviour, all normalised to [0,1]: (i) whether the prey was captured

Table 1. Parameters used for NEAT and novelty search (last row).

Population size	150	Target species count	5	Crossover prob	20 %
Recurrency allowed	true	Mutation prob	25 %	Prob. add link	5 %
Prob. add node	3 %	Prob. mutate bias	30 %	Num. generations	250
Novelty k-nearest	15	Add archive prob	2.5 %	Max. archive size	2000

or not; (ii) average final distance of the predators to the prey; (iii) average distance of each predator to the other predators over the trial; and (iv) trial length. The novelty search algorithm was configured according to [6], see Table 1 (last row). The novelty score of each individual is combined with its fitness score using the NSGA-II [2] multiobjective ranking, as advocated in previous works [6,7].

2.4 Simulation

For the evolutionary process, we used a two-dimensional simulation environment, where the robots are abstracted as circular objects with a certain heading and position[1]. The robot motion model was implemented based on simple measurements taken on the real robots, and did not include complex physics simulation or fluid dynamics. In order to facilitate the transfer from simulation to reality and promote general behaviours, noise was applied to the sensors and actuators [3,8] based on measurements taken from real robots, and the initial task conditions were varied in every simulation trial. The following parameters were varied during the simulated trials:

Set for Each Trial: random individual motor speed offsets of up to 10 % of maximum speed; compass offset up to ±9°; the prey's escape speed varied between [75 %,100 %] of the predators's maximum speed; and the initial positions and orientations of all robots were varied according to Sect. 2.1.

Set at Each Time Step: GPS error up to 1.8 m; compass error up to ±10°; motor output varied up to 5 %; and the prey's escape direction randomly varied up to 50 % from the optimal direction.

3 Evolving and Identifying Diverse Behaviours

For both fitness-driven (*Fit*) and novelty-driven cooperative coevolution (*NS-Team*), we followed a methodology that allowed us to identify a set of diverse and high-quality solutions, that were then evaluated in the real multirobot system.

Evolutionary Process: Each evolutionary approach was repeated in ten independent evolutionary runs. To obtain a more accurate estimate of the evolved teams' quality and behaviour, all the *best-of-generation* teams (the teams that obtained the highest fitness score in each generation, in each evolutionary run) were re-evaluated a posteriori in 50 simulation trials. On average, the evolutionary runs of *Fit* achieved a highest fitness score of 1.09 ± 0.10, and *NS-Team* achieved 0.96 ± 0.20. While this difference is significant ($p = 0.043$, Mann-Whitney U test), both approaches managed to evolve high-quality solutions.

Behaviour Mapping: To visualise the diversity of behaviours evolved by each evolutionary approach, we mapped the *best-of-generation* teams according to their behaviour characterisation vector (see Sect. 2.3), as done in previous works [7]. The four dimensions of the behaviour characterisation were reduced

[1] https://github.com/BioMachinesLab/drones/tree/master/JBotAquatic.

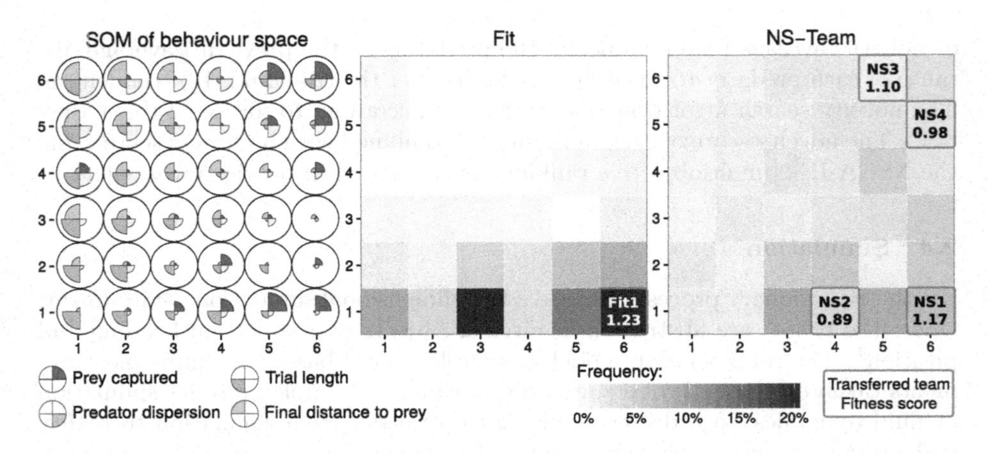

Fig. 2. Left: trained Kohonen map, where each node represents a region of the team behaviour space. Middle and right: team behaviour exploration by the two evolutionary approaches. The darker a region, the more of the evolved teams belonged to it.

to two dimensions using a Kohonen self-organising map in order to obtain a visual representation of the team behaviour space exploration, see Fig. 2 (left). The teams evolved by each evolutionary approach were then mapped: each team is assigned to the node (map region) with the closest weight vector, see Fig. 2. The results show that *NS-Team* explored the behaviour space much more uniformly, and could reach behaviour regions that were never reached by *Fit*, which is consistent with the results reported in previous works [7].

Selection of Solutions: We then proceeded to select a diverse set of solutions to be tested in the real robots. We selected different regions of the behaviour space where the prey capture rate was high, and identified the team belonging to each of those regions that obtained the highest fitness score, see Fig. 2. We chose one team evolved by *Fit*, as all the high-quality teams were found in the bottom-right corner of the map, and four solutions evolved by *NS-Team*, from different regions of the map with high *prey capture* values.

4 Transferring the Teams to Real Robots

The selected teams were then evaluated in the real multirobot system. The experiments were performed in a semi-enclosed water body, see Fig. 3. The task setup was similar to the simulation setup (see Sect. 2.1): the three predator robots were placed close to the centre of the arena, and the prey was placed at approximately 25, 30, and 35 m away from the centre, in each of the three trials that were used to assess the performance of the teams. Each trial lasted for at most 100 s, the arena boundaries were 100×100 m, and the prey moved at the maximum speed. To compare the results of the real-robot experiments with simulation, the chosen teams were re-evaluated in 500 simulation trials, using

Fig. 3. Photo of the real-robot experiments, at Parque das Nações, Lisbon, Portugal, in a semi-enclosed area in the margin of the Tagus river.

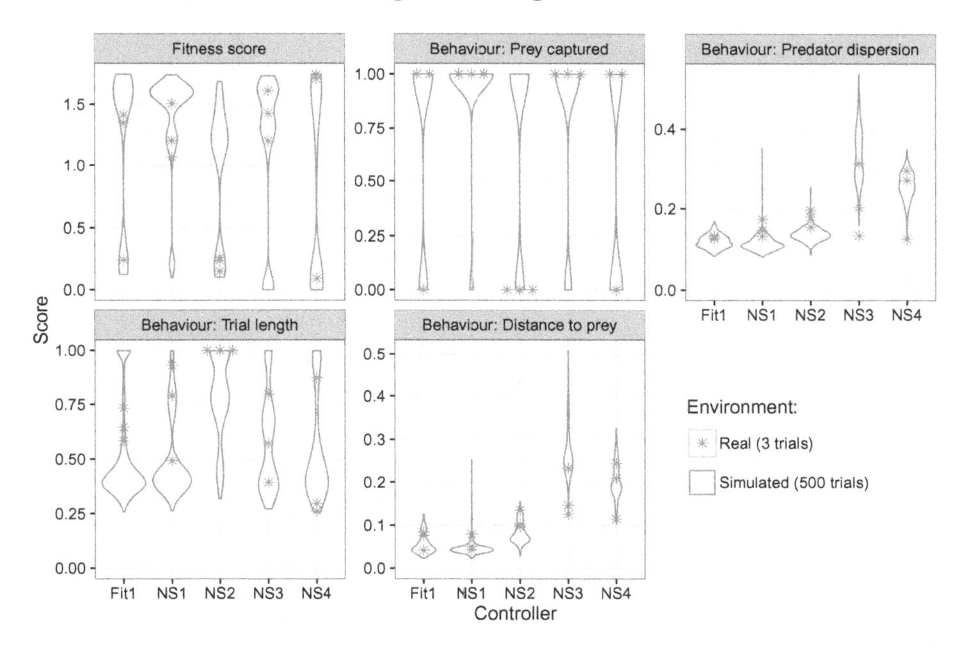

Fig. 4. Comparison of the fitness score and behaviour features obtained in the real-robot experiments (asterisks) and in simulation (violin plots) in similar conditions.

the same initial conditions as the real-robot experiments. The fitness scores and behaviour features of the teams operating in the real environment were computed using logged GPS data.

In Fig. 4 (Fitness), we compare the fitness scores obtained by the teams in simulation and in the real robots. We additionally explore the diversity of team behaviours by comparing the controllers' performance in reality and in simulation according to the behaviour features that were used in novelty-driven coevolution (Sect. 2.3). The results show that all teams except *NS2* were able to capture the prey in the majority of the trials. The fitness scores obtained in the

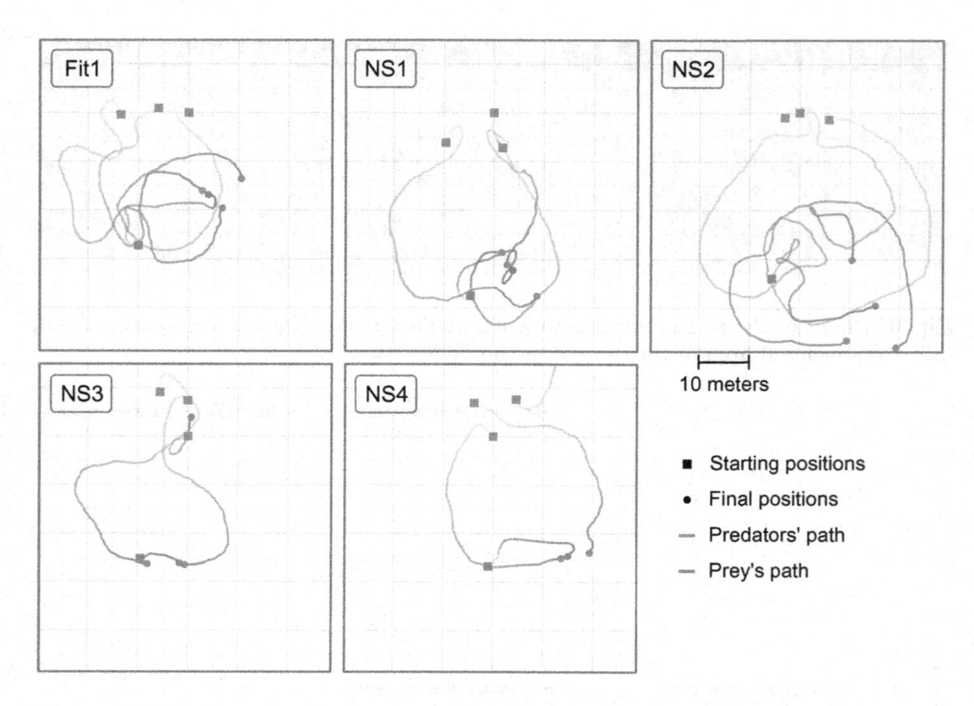

Fig. 5. Traces of one experimental trial (out of three) for each of the teams evaluated in the real robots. Traces and videos of all real-robot experiments are available online.[2]

real experiments are similar to the scores obtained in simulation, fitting in the distribution obtained in simulation. These results are a first indication that the evolved controllers were generally able to cross the reality gap successfully.

The effectiveness of the team behaviours was confirmed by analysing the traces of the real-robot experiments, shown in Fig. 5. The *Fit1* and *NS1* teams displayed a behaviour where the three predators would initially spread and move towards the prey, each approaching the prey from a different direction. The behaviour of *NS2* was similar to *Fit1* and *NS1*, but the predator team dispersed more. The teams *NS3* and *NS4* displayed a significantly different behaviour: only two predators chased the prey, approaching it from opposite directions, while the remaining predator would move away from the group. The observed robot traces are consistent with the measured behavioural features (Fig. 4), and confirm that novelty-driven coevolution was able to achieve a wide diversity of team behaviours. For instance, it is possible to observe that *NS3* and *NS4* display a higher dispersion and final distance to prey, which is explained by the fact that in these teams, only two predators chase the prey. The differences and similarities between the team behaviours observed in the real-robot experiments are consistent with the behaviour map obtained in simulation (Fig. 2).

Overall, despite the stochastic conditions of the aquatic environment, the predators displayed effective cooperation, and were consistently able to solve the task. The team of predators would often fail to capture the prey in the first attempt, but the team would then spread out and try to encircle the prey again. Moreover, robots sometimes displayed temporary motor failures (see supplementary videos[2]), which did not compromise the effectiveness of the team. These behaviours suggest that the teams were not overfitted to the simulation environment, and could effectively adapt to different scenarios.

5 Conclusion

In this paper, we employed cooperative coevolutionary algorithms (CCEAs) to evolve control for an aquatic multirobot system. Our experiments relied on a cooperative predator-prey task, where a heterogeneous team of three predators was evolved to capture one reactive prey. Two evolutionary approaches were applied: traditional fitness-driven cooperative coevolution, and novelty-driven cooperative coevolution. The evolutionary processes were conducted exclusively in simulation, and a number of high-fitness teams were then systematically evaluated in real robots operating in a non-controlled outdoor environment.

The evolved teams generally transferred well to the real robots, successfully crossing the reality gap. Out of the five teams tested, four teams could consistently capture the prey, and obtained fitness scores very similar to those obtained in simulation. The cooperation between robots that was exhibited in simulation was also observed in real robots, and the teams displayed robust behaviours that did not appear to be overfitted to the simulation environment. The successful transfer is especially notable given that we used low-fidelity simulator during evolution, and given the stochastic nature of the real task environment. We encouraged the evolution of robust and transferable controllers by introducing conservative amounts of noise and variations in the sensors and actuators of the robots in simulation, and by using multiple trials to evaluate each solution, with different initial conditions.

Novelty-driven cooperative coevolution was able to produce a good diversity of high-quality team behaviours for solving the task, which were identified following a systematic approach. The diversity of behaviours that was observed in simulation was also present in the real multirobot system.

In summary, we demonstrated that CCEAs can be successfully used to synthesise control for a real multirobot system, operating in an environment outside controlled laboratory conditions. Despite the large number of previous works that have showed the potential of CCEAs for evolving heterogeneous multirobot systems, our work stands amongst the first to demonstrate this potential in real robots and in a realistic environment. Our experiments also validated, for the first time, the potential of novelty-driven cooperative coevolution in real robots, and confirmed it as a valuable approach to evolve diverse team behaviours.

[2] Videos and logs of the experiments: http://dx.doi.org/10.5281/zenodo.49582.

Acknowledgements. This work was supported by centre grant (to BioISI, Centre Reference: UID/MULTI/04046/2013), from FCT/MCTES/PIDDAC, Portugal, and by grants SFRH/BD/89095/2012 and UID/EEA/50008/2013.

References

1. Costa, V., Duarte, M., Rodrigues, T., Oliveira, S.M., Christensen, A.L.: Design and development of an inexpensive aquatic swarm robotics system. In: OCEANS 2016-Shanghai, pp. 1–7. IEEE Press (2016)
2. Deb, K., Pratap, A., Agarwal, S., Meyarivan, T.: A fast and elitist multiobjective genetic algorithm: NSGA-II. IEEE Trans. Evol. Comput. **6**(2), 182–197 (2002)
3. Duarte, M., Costa, V., Gomes, J., Rodrigues, T., Silva, F., Oliveira, S.M., Christensen, A.L.: Evolution of collective behaviors for a real swarm of aquatic surface robots. PLoS ONE **11**(3), e0151834 (2016)
4. Gomes, J., Mariano, P., Christensen, A.L.: Avoiding convergence in cooperative coevolution with novelty search. In: International Conference on Autonomous Agents and Multiagent Systems (AAMAS), pp. 1149–1156. IFAAMAS (2014)
5. Gomes, J., Mariano, P., Christensen, A.L.: Cooperative coevolution of morphologically heterogeneous robots. In: European Conference on Artificial Life, pp. 312–319. MIT Press (2015)
6. Gomes, J., Mariano, P., Christensen, A.L.: Devising effective novelty search algorithms: a comprehensive empirical study. In: Genetic and Evolutionary Computation Conference (GECCO), pp. 943–950. ACM Press (2015)
7. Gomes, J., Mariano, P., Christensen, A.L.: Novelty-driven cooperative coevolution. Evol. Comput. (2016, in press)
8. Jakobi, N.: Evolutionary robotics and the radical envelope-of-noise hypothesis. Adapt. Behav. **6**(2), 325–368 (1997)
9. Lehman, J., Stanley, K.O.: Abandoning objectives: evolution through the search for novelty alone. Evol. Comput. **19**(2), 189–223 (2011)
10. Nitschke, G.: Designing emergent cooperation: a pursuit-evasion game case study. Artif. Life Robot. **9**(4), 222–233 (2005)
11. Nitschke, G.S., Eiben, A.E., Schut, M.C.: Evolving team behaviors with specialization. Genet. Program. Evolvable Mach. **13**(4), 493–536 (2012)
12. Nitschke, G.S., Schut, M.C., Eiben, A.E.: Collective neuro-evolution for evolving specialized sensor resolutions in a multi-rover task. Evol. Intell. **3**(1), 13–29 (2010)
13. Nitschke, G.S., Schut, M.C., Eiben, A.E.: Evolving behavioral specialization in robot teams to solve a collective construction task. Swarm Evol. Comput. **2**, 25–38 (2012)
14. Panait, L., Luke, S.: Cooperative multi-agent learning: the state of the art. Auton. Agent. Multi-Agent Syst. **11**(3), 387–434 (2005)
15. Panait, L., Luke, S., Wiegand, R.P.: Biasing coevolutionary search for optimal multiagent behaviors. IEEE Trans. Evol. Comput. **10**(6), 629–645 (2006)
16. Popovici, E., Bucci, A., Wiegand, R.P., De Jong, E.D.: Coevolutionary principles. In: Rozenberg, G., Back, T., Kok, J.N. (eds.) Handbook of Natural Computing, pp. 987–1033. Springer, Heidelberg (2012)
17. Potter, M.A., De Jong, K.A.: Cooperative coevolution: an architecture for evolving coadapted subcomponents. Evol. Comput. **8**(1), 1–29 (2000)
18. Potter, M.A., Meeden, L.A., Schultz, A.C.: Heterogeneity in the coevolved behaviors of mobile robots: the emergence of specialists. In: International Joint Conference on Artificial Intelligence (IJCAI), pp. 1337–1343. Morgan Kaufmann (2001)

19. Silva, F., Duarte, M., Correia, L., Oliveira, S.M., Christensen, A.L.: Open issues in evolutionary robotics. Evol. Comput. **24**(2), 205–236 (2016)
20. Stanley, K., Miikkulainen, R.: Evolving neural networks through augmenting topologies. Evol. Comput. **10**(2), 99–127 (2002)
21. Wiegand, R.P., Liles, W.C., De Jong, K.A.: Analyzing cooperative coevolution with evolutionary game theory. In: Congress on Evolutionary Computation (CEC), vol. 2, pp. 1600–1605. IEEE Press (2002)
22. Yong, C.H., Miikkulainen, R.: Coevolution of role-based cooperation in multiagent systems. IEEE Trans. Auton. Ment. Dev. **1**(3), 170–186 (2009)

Replicating the Stroop Effect Using a Developmental Spatial Neuroevolution System

Amit Benbassat(✉) and Avishai Henik

Ben-Gurion University of the Negev, Beer-Sheva, Israel
amitbenb@post.bgu.ac.il, henik@bgu.ac.il

Abstract. We present an approach to the study of cognitive phenomena by using evolutionary computation. To this end we use a spatial, developmental, neuroevolution system. We use our system to evolve ANNs to perform simple abstractions of the cognitive tasks of color perception and color reading. We define these tasks to explore the nature of the Stroop effect. We show that we can evolve it to perform a variety of cognitive tasks, and also that evolved networks exhibit complex interference behavior when dealing with multiple tasks and incongruent data. We also show that this interference behavior can be manipulated by changing the learning parameters, a method that we successfully use to create a Stroop like interference pattern.

1 Introduction

Much research in cognitive psychology has been devoted to goal directed behavior or to the mental processes involved in focusing on relevant information and declining or ignoring irrelevant information. One of the paradigmatic tasks in cognitive psychology is the Stroop task in which people are presented with words in color (e.g., RED in green) and asked to pay attention to the color and ignore the meaning of the word. The current work applies evolutionary algorithms (EAs) to study the mechanisms involved in the Stroop task.

1.1 The Stroop Effect

In his original work Stroop [14] presented participants with lists of stimuli on a card and asked them to name the color of the ink as fast as possible. He measured the time to name 100 stimuli on each card. Stroop used two conditions, incongruent (e.g., RED in green) and neutral (i.e., patches of colors). Responding was slower to the incongruent condition than to the neutral condition. Stroop suggested that the difference between incongruent and neutral conditions was an indication for the automaticity of word reading. Importantly, when he asked participants to read the words and ignore their color, word reading was not hampered by the incongruent colors.

J. Handl et al. (Eds.): PPSN XIV 2016, LNCS 9921, pp. 602–612, 2016.
DOI: 10.1007/978-3-319-45823-6_56

With the introduction of computers to psychology laboratories, the task changed to a trial-by-trial task. Vocal response time of participants was measured in milliseconds. These single trial experiments enable experimenters to include congruent trials (i.e., Green in green). As computer presentations mix all conditions, participants are unable to predict the appearance congruent inputs and cannot adopt a reading strategy for those inputs.

Research on goal directed behavior use not only the Stroop task but other tasks also [2,5,12].

1.2 Neuroevolution

Neuroevolution is the subfield of evolutionary computation concerned with growing *Artificial Neural Networks* (ANNs) via artificial evolution. The field has attracted much research effort. Stanly and Miikkulainen [13] created the NEAT system for the explicit purpose of evolving complex networks from simple initial networks. NEAT uses direct encoding where evolved genes relate to specific parts of the network. HyperNEAT is a neuroevolution system created on the basis of NEAT that uses indirect encoding and is widely used to evolve ANNs that perform various tasks [10,11,15]. HyperNEAT works by evolving Meta-networks with NEAT, that in turn decide on edge weights in the ANN that is meant to perform the task. Some neuroevolution systems take a cue from nature and use a developmental scheme. For example, Kitano [9] presented a method of evolving grammars that generate ANN connectivity maps. Another example is Gruau [4]. Gruau suggested the concept of *Cellular Encoding* (CE) where the individual neurons act as cells during the developmental process. In Gruau's system development of ANNs is dictated by the genome (a tree genome in [9]'s case. A linear genome in ours). Many other implementations are presented in a review by Floreano et al. [3]

1.3 The Current Work

In this work we present an evolutionary learning based approach to the study of cognitive phenomena. We employ an EA on populations of randomly generated ANNs in order to evolve networks that perform cognitive tasks. This allows us to explore the specific conditions under which certain phenomena may occur. Specifically, in the first phase of the project we wanted to create a neuroevolution system that features natural qualities. Next, we attempted to generate a Stroop effect. Specifically, we aimed to generate interference and facilitation and also the asymmetry between word reading and color naming. That is, significant interference and facilitation in color naming with small or null effects in word reading.

2 Spatial Developmental Neuroevolution System

As suggested earlier, we designed our evolutionary system with an eye towards nature. We cannot emulate all natural traits but we focused on three important traits which we integrated as design features. These important attributes of our system are listed below:

1. ANN based: We chose neuroevolution because the artificial neuron is an abstraction of the biological neuron. Though the two are by no means identical an ANN is similar to brain systems in being a decentralized computation system made of simple computation units that share some of the same attributes.
2. Developmental: A gene in the individual's genome does not map directly to a specific simple element in the final network. Rather, it is seen as a command that is to be performed by the developing network or a subset of its artificial neurons, during the developmental process.
3. Spatial: Every artificial neuron in our system is located in some point in a virtual space. Developmental steps are space related, placing, moving, and connecting neurons to each other using spatial coordinates.

2.1 The Spatial ANN

The ANNs in our system consist of three distinct layers: An input layer, an output layer, and a hidden layer. Each one of the layers exists in its own space defined by the user. The user defines the number of dimensions each layer has and the size of each dimension. The spaces contain a discrete grid of locations where a neuron may reside (e.g. A 3D layer of size $3 \times 4 \times 5$ contains exactly $3 \times 4 \times 5 = 60$ possible neuron locations).

In the input and output layers every location contains an artificial neuron. The hidden layer's content depends on the individual's genome. The genome defines all hidden neurons as well as all network edges. The input values are real numbers in the [−1,1] range (we typically use the extreme values −1 and 1 but the system supports using other values as well). Outputs are limited to the [−1,1] range.

2.2 Genome Structure

Our encoding is based on a genome in the form of a linear array of genome atoms (or genes). Each gene is a set of integers and real numbers denoted by

Table 1. Fields in the genome atom

Field name	Field type
`Read Mode`	Integer
`New Read Mode`	Integer
`Opcode`	Integer
`Weight`	Real
`Threshold`	Real
`Location Offset`	Integer array

field names. Table 1 presents the names of the fields. The last line shows the Location offset array of fields. This is an integer array that signifies a location offset in one of the network grids (the identity of which depends on the type of gene being read). The length of the array is determined by the number of spatial dimensions which the user controls using run parameters.

2.3 Run Parameters

The particulars of each run are controlled by the user with a series of run parameters. Evolutionary parameters control the evolutionary process, fitness parameters control the calculation of individual fitness, ANN parameters control the basic attributes of the networks and encoding parameters control the genome's encoding rules into the ANN phenotype.

Evolutionary parameters include number of generations, population size, crossover probability, mutation probability, type of selection and diversity maintenance parameters. Our system utilized single-point crossover variant that allows genome size to change by up to 20 % in each crossover event. Mutation is uniform (mutation probability is per gene and not per individual). When a spot in the genome is chosen for mutation, one of two actions is performed each with a probability of 0.5: Either the atom itself is randomly changed or a small genome segment beginning with the chosen atom is copied to another random location in the genome. Our system uses standard tournament selection. In the experiments described in this work we use a tournament size of 4.

The diversity maintenance measure limits the number of individuals with similar behavior profiles. An individual's behavior profile is an array made of all the output values its ANN gets for all the fitness tests (the values in the behavior profile are rounded to values in $\{-1,0,1\}$). The *distance ratio* between two behavior profiles is the number of locations where the profiles differ divided by the length of the profiles. We say that two individuals are neighbors if the distance ratio between their behavior profiles is lower than a value controlled by the user (which we typically set at 0.3). Our diversity maintenance system allows an individual to be selected only if the number of its neighbors already selected is lower than a certain threshold (typically 30). We used this method to encourage diversity not only in genome but also in behavior.

Fitness parameters include the test inputs used to calculate the fitness score and the test inputs used to calculate the benchmark score. The two calculations differ in that the fitness score calculates a much smoother function. When calculating fitness the individual is rewarded slightly for every output neuron that generates a correct output, as well as being given a bonus for the whole network giving the correct answer. The benchmark score is only affected by whether the network's answer to the test input is correct or not. In both the fitness and the benchmark score cases we normalize the scores to the [0,1000] range to make them easier to assess.

ANN network size parameters include a limit on the number of hidden layer neurons and of network links, which we set to 400 and 4000, respectively. Initial genome size is set to 80 (limited to a maximum of 150). The number of different read codes is set to 31. The read codes determine which neuron reads which gene.

The read encoding scheme for i codes is based on a complete binary tree with i nodes that are tagged in level order starting at 1. In order to decide whether or not a given neuron reads a given gene the read codes of the neuron and the gene have to match. Two read codes match if one is an ancestor of the other in the tree.

2.4 Spatial Developmental Encoding

Our system supports multiple encoding schemes. There are several different types of actions that a gene can cause. The probability of a gene encoding a certain action is controlled by the user, who chooses how much weight to assign to each of the possible gene types. Table 2 contains the encoding weights we chose for our experiments below. We chose these values empirically and with use of common sense.

Table 2. Different action types and their weights in our experiments. an action can have a weight of 0 or higher assigned to it. An action assigned a weight of 0 is impossible to encode with any gene. The probability that an action will be encoded by a randomly generated gene is proportionate to the weight of that action.

Action name	Weight	Action description
`New Node`	10	Create a new neuron
`Move`	2	Move neuron
`Connect`	4	Connect neuron to another neuron
`Connect output`	4	Connect neuron to an output
`Connect input`	4	Connect an input to neuron
`Connect all output`	0	Connect neuron to all outputs
`Connect all input`	4	Connect all inputs to neuron
`Mutate Threshold`	2	Change neuron threshold and factor
`Split`	8	Split existing neuron, creating a new neuron next to it
`Power Split`	4	Split existing neuron, creating a new neuron with all the same connections
`Sleep`	4	Neuron sleeps, no longer performing actions
`Awaken`	4	Sleeping neuron wakes up, and resumes performing actions
`Die`	2	Neuron dies and is removed from network

Of the actions described in Table 2 `New Node` stands out as working on the entire network (by adding a neuron to it). All other actions are activated by individual neurons for which the read encoding matches.

3 The Problem Domains

In this research, we interpret our outputs in a way analogous to nature. The human brain often makes decisions when a plurality of neurons signal together rather than relying on just one neuron. In our experiments we followed this principle. The domains explored in this work (i.e., reading words or naming colors) are classification tasks where the ANN is expected to tell a number of different classes apart (e.g., distinguishing between three colors; blue, red, green). In these tasks, the output is a two dimensional 4×5 grid, and each one of the 4 rows stands for one of the possible classes (e.g., blue). Decision is made by plurality rule, with the network choosing class i if and only if row i in the output has more 1's in it than any other row.

3.1 Color Perception

In the Color Perception task (or *CP*) the ANNs are required to identify the color of an input. In this work we define the CP task to work with a 3-dimensional input grid of size $4 \times 5 \times 5$. We see the input as made up of 4 2-dimensional grids: 3 colored "visual field" grids (*red*, *green* and *blue*) and 1 "task definition" grid that is used to differentiate between color perception and Color Reading tasks. In the CP task we expect the forth grid of the input to contain all -1's. The output is a 2-dimensional grid of size 4×5 with a correct output being one that contains a plurality of 1's in the row representing the right answer. Our convention is that the first row stands for *red*, the second stands for *green*, the third stands for *blue* and the forth is reserved for future use for inputs that do not have one dominant color. In Figs. 1 and 2 there are two examples of inputs for the CP task. In these figures, and all other figures further on that show examples of inputs, we assume that an empty square represents an input of -1 and a square containing a black circle represents an input of 1.

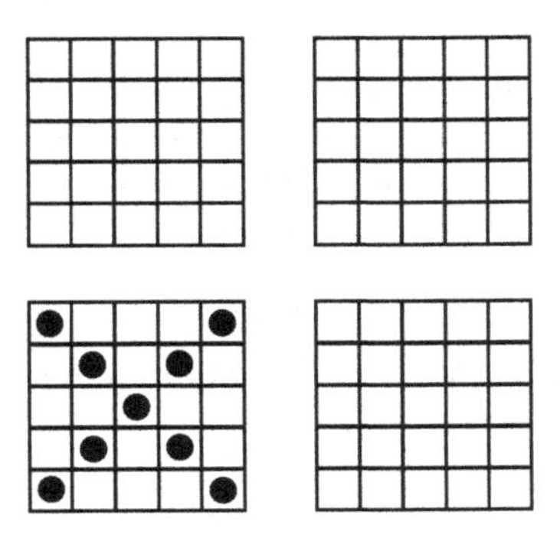

Fig. 1. Neutral CP input of $\times$ sign in blue

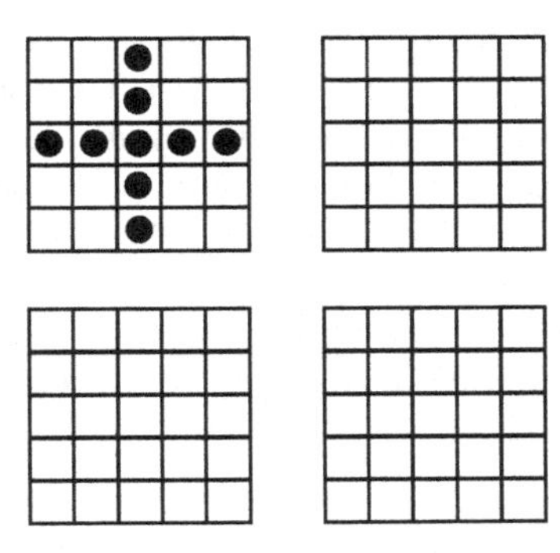

Fig. 2. Congruent CP input of $+$ sign in red

3.2 Color Reading

In the Color Reading task (or *CR*) the ANNs are required to read a colored symbol in the input. The input and output dimensions are identical to the CP task. In the CR task we expect the forth grid of the input to contain all 1's. Our convention for the output is similar to CR. We chose symbols to stand for the three base colors. The + symbol stands for *red*, the sideways `H` stands for *green* and the `H` stands for *blue*. In Figs. 2, 3 and 4 we see the symbols for red, green and blue, respectively.

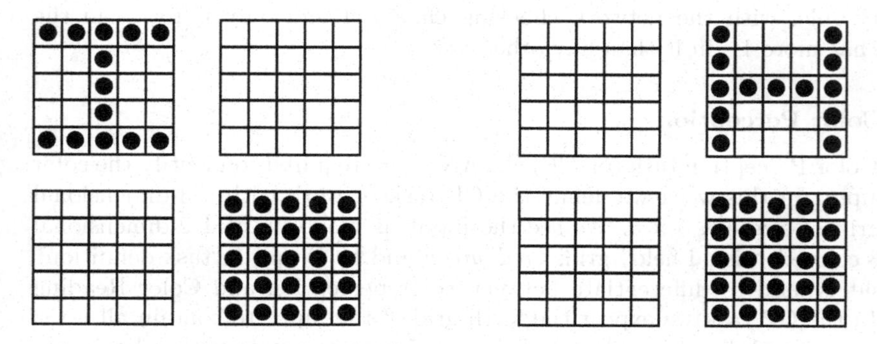

Fig. 3. Incongruent CR input of green sign (the sideways `H`) in red

Fig. 4. Incongruent CR input of blue sign (the `H`) in green

4 Experiments

We ran several experiments in the different tasks. Each experiment was designed to check another phenomenon. In each experiment we ran the same simulation 100 times in order to get sufficient data. For some experiments we had to compare two or more types of runs. In those cases each run type got its own set of 100 simulations. Each 100 simulation experiment set took at most about a day (running on a laptop computer with Intel Core i7-4700MQ 2400 MHz 4-core processor). In all the runs below we used an elitism rate of 0.02, a mutation rate of 0.02 and a crossover rate of 0.8.

The runs generate a lot of data. For brevity we do not present all generated data here. We will present data which we think is relevant to the experiments presented.

4.1 Evolving Color Perception and Reading

First we attempted to evolve ANNs that perform each task separately. For brevity's sake we do not go into much detail as far as these runs are concerned.

To evolve ANNs for the CP task we used a population of 200 individuals, running for 120 generations. In 81 out of 100 simulations the ANNs reached a perfect benchmark score on the CP task. To evolve ANNs for the CR task we used a population of 500 individuals, running for 900 generations. Though only 7 of the simulations resulted in individuals that had a perfect benchmark score on the CR task, they were not far off the mark. The best solution in a simulation had a mean benchmark score of 909.0457 ($\sigma = 80.3975$).

4.2 Mixed

We used a population of 400 individuals, running for 400 generations. We calculated the fitness score and the benchmark score using the 12 test inputs from The CP Experiment and the 21 test inputs from the CR Experiment. After the runs terminated, we checked the best individuals on congruent and incongruent inputs separately in both tasks.

Looking at congruent inputs the best solution in a simulation had a mean benchmark score of 949.999 ($\sigma = 119.6258$) in the CP task. Looking at incongruent inputs the best solution in a simulation had a mean benchmark score of 801.6637 ($\sigma = 173.5871$) in the CP task. Looking at congruent inputs the best solution in a simulation had a mean benchmark score of 893.3315 ($\sigma = 176.5132$) in the CR task. Looking at incongruent inputs the best solution in a simulation had a mean benchmark score of 446.6643 ($\sigma = 149.5968$) in the CR task.

We conducted one-way ANOVA on the 4 score types ($F(3, 396) = 208.3780$). The difference between the congruent and incongruent is significant in the CP task ($p < 0.0001$), and also in the CR task ($p < 0.0001$).

In both tasks there are no significant differences in other attributes of the individuals, which is to be expected as the individuals being compared are taken from the same population pools. These results show that interference does occur in our system and that the congruent inputs are easier for our evolved networks. However there is not clear directionality as in the Stroop case. Networks do better on congruent inputs in both the CP and CR tasks.

4.3 Weighted

We attempted to generate directionality by weighting the fitness function. In this experiment we used the same parameters as in the mixed experiment described in Sect. 4.2 except for the fitness function, which was weighted to bias evolution in favor of the CR task. Each fitness test-case from the CR test suite affected the fitness result as if it appeared 30 times in the suite.

Looking at congruent inputs the best solution in a simulation had a mean benchmark score of 749.9962 ($\sigma = 219.0449$) in the CP task. Looking at incongruent inputs the best solution in a simulation had a mean benchmark score of 493.3311 ($\sigma = 141.9752$) in the CP task. Looking at congruent inputs the best solution in a simulation had a mean benchmark score of 746.6633 ($\sigma = 246.7312$) in the CR task. Looking at incongruent inputs the best solution in a simulation had a mean benchmark score of 738.3301 ($\sigma = 191.3615$) in the CR task.

We conducted one-way ANOVA on the 4 score types ($F(3, 396) = 38.2965$). The difference between the congruent and incongruent is significant in the CP task ($p < 0.0001$), but it is insignificant in the CR task ($p = 0.9915$). This approach successfully creates the desired asymmetry that is an attribute of the Stroop effect.

4.4 Phased and Weighted

In this experiment we added a phasing element to our simulations. We ran a population of 400 for 150 generations evaluating fitness on both CP and CR tests, using the weighted mixed test suite we used in Sect. 4.3, then we allowed the population to evolve for 400 generations on CR inputs only, and then let it evolve for 150 more generations the weighted mixed test suite.

Looking at congruent inputs the best solution in a simulation had a mean benchmark score of 716.6629 ($\sigma = 243.3185$) in the CP task. Looking at incongruent inputs the best solution in a simulation had a mean benchmark score of 458.3311 ($\sigma = 142.8719$) in the CP task. Looking at congruent inputs the best solution in a simulation had a mean benchmark score of 819.9976 ($\sigma = 234.1278$) in the CR task. Looking at incongruent inputs the best solution in a simulation had a mean benchmark score of 794.9971 ($\sigma = 186.3103$) in the CR task.

We conducted one-way ANOVA on the 4 score types ($F(3, 396) = 64.7027$). Again the difference between the congruent and incongruent is significant in the CP task ($p < 0.0001$), while it is insignificant in the CR task ($p = 0.8256$). This approach also creates an asymmetry effect similar to Stroop, and also results in a higher score on the CR task (however, this difference is not statistically significant).

4.5 Neutral CP Input

Among our 12 inputs for the CP task there are 3 `X` shaped inputs that are neither congruent nor incongruent. These are considered *Neutral* inputs (see Fig. 1 for an example of a neutral CP input).

In the experiments we ran in which a Stroop like effect appeared the results on neutral CP inputs fell somewhere in between the congruent and incongruent scores. In the weighted fitness experiment the best solution in a simulation had a mean benchmark score of 596.6617 ($\sigma = 202.648$) on neutral inputs. In the phased weighted fitness experiment the best solution in a simulation had a mean benchmark score of 569.9951 ($\sigma = 202.6477$) on neutral inputs. In light of these results we can say that our experiments are Stroop like also in the classical sense using neutral inputs. On the other hand this is also where our results differ somewhat from the Stroop effect as it appears in humans (the difference in performance between congruent and neutral tests is small to negligible).

5 Concluding Remarks

Our system employs various measures to make the developmental process more like natural development.

We presented a new developmental spatial neuroevolution system for cognitive science research and used it to explore the Stroop effect and evolve ANNs that show some Stroop like behaviors. We successfully replicated, in our evolved networks, the phenomenon of interference due to conflict between the two tasks. We also succeeded in establishing that this conflict can be directional, by biasing the fitness function in favor of the reading task.

There is still much to be done and we want to explore these issues further and also expand our system and look into some new areas. Listed below are some avenues for future research which we believe are promising and we plan to pursue:

- We plan to examine numerical cognition, checking to see if using simple tasks as an evolutionary stepping stone improve the evolution of counting ability. We plan to follow up on work by Katz et al. [8] and Cantlon et al. [1] that suggests counting ability may have evolved from a simpler cognitive system for size perception.
- We plan to expand our inquiry into the evolutionary dynamics of evolving ANNs to perform cognitive tasks. Specifically we are interested in the effects of changing task and environment in mid run on the resulting population. Some of these effects have been demonstrated in the past in several simple domains [6,7].
- We plan to explore the Stroop effect and other similar effects such as Numerical Stroop and the Simon Effect.

Our system itself is still a work in progress, more functionality is needed in order to make it more flexible so it can cover more complex behavior. An obvious extension would be to allow for the evolution of recurrent networks that can handle domains with multiple instances that require the network to react according to new input as well as its own output (such as navigation tasks, and tasks that require networks to have memory capabilities).

Acknowledgments. The research leading to these results has received funding from the European Research Council under the European Union's Seventh Framework Programme (FP7/2007-2013)/ERC Grant agreement number 295644.

References

1. Cantlon, J.F., Platt, M.L., Brannon, E.M.: Beyond the number domain. Trends Cogn. Sci. **13**(2), 83–91 (2009)
2. Eriksen, B.A., Eriksen, C.W.: Effects of noise letters upon the identification of a target letter in a nonsearch task. Percept. Psychophys. **16**(1), 143–149 (1974)
3. Floreano, D., Dürr, P., Mattiussi, C.: Neuroevolution: from architectures to learning. Evol. Intel. **1**(1), 47–62 (2008)
4. Gruau, F.: Automatic definition of modular neural networks. Adapt. Behav. **3**(2), 151–183 (1994)
5. Henik, A., Tzelgov, J.: Is three greater than five: the relation between physical and semantic size in comparison tasks. Mem. Cogn. **10**(4), 389–395 (1982). http://dx.doi.org/10.3758/BF03202431

6. Kashtan, N., Alon, U.: Spontaneous evolution of modularity and network motifs. Proc. Natl. Acad. Sci. USA **102**(39), 13773–13778 (2005)
7. Kashtan, N., Noor, E., Alon, U.: Varying environments can speed up evolution. Proc. Natl. Acad. Sci. **104**(34), 13711–13716 (2007)
8. Katz, G., Benbassat, A., Diesendruck, L., Sipper, M., Henik, A.: From size perception to counting: an evolutionary computation point of view. In: Proceedings of 15th Annual Conference Companion on Genetic and Evolutionary Computation, pp. 1675–1678. ACM (2013)
9. Kitano, H.: Designing neural networks using genetic algorithms with graph generation system. Complex Syst. J. **4**, 461–476 (1990)
10. Lehman, J., Stanley, K.O.: Abandoning objectives: evolution through the search for novelty alone. Evol. Comput. **19**(2), 189–223 (2011)
11. Secretan, J., Beato, N., Ambrosio, D., Rodriguez, D.B., Campbell, A., Stanley, K.O.: Picbreeder: evolving pictures collaboratively online. In: Proceedings of SIGCHI Conference on Human Factors in Computing Systems, pp. 1759–1768. ACM (2008)
12. Simon, J.R., Rudell, A.P.: Auditory sr compatibility: the effect of an irrelevant cue on information processing. J. Appl. Psychol. **51**(3), 300 (1967)
13. Stanley, K.O., Miikkulainen, R.: Evolving neural networks through augmenting topologies. Evol. Comput. **10**(2), 99–127 (2002)
14. Stroop, J.R.: Studies of interference in serial verbal reactions. J. Exp. Psychol. **18**(6), 643 (1935)
15. Yosinski, J., Clune, J., Hidalgo, D., Nguyen, S., Zagal, J., Lipson, H.: Evolving robot gaits in hardware: the hyperneat generative encoding vs. parameter optimization. In: Proceedings of 20th European Conference on Artificial Life, pp. 890–897 (2011)

Evolving Cryptographic Pseudorandom Number Generators

Stjepan Picek[1(✉)], Dominik Sisejkovic[2], Vladimir Rozic[1], Bohan Yang[1], Domagoj Jakobovic[2], and Nele Mentens[1]

[1] KU Leuven, ESAT/COSIC and iMinds, Kasteelpark Arenberg 10, bus 2452, 3001 Leuven-Heverlee, Belgium
stjepan@computer.org

[2] Faculty of Electrical Engineering and Computing, University of Zagreb, Zagreb, Croatia

Abstract. Random number generators (RNGs) play an important role in many real-world applications. Besides true hardware RNGs, one important class are deterministic random number generators. Such generators do not possess the unpredictability of true RNGs, but still have a widespread usage. For a deterministic RNG to be used in cryptography, it needs to fulfill a number of conditions related to the speed, the security, and the ease of implementation. In this paper, we investigate how to evolve deterministic RNGs with Cartesian Genetic Programming. Our results show that such evolved generators easily pass all randomness tests and are extremely fast/small in hardware.

Keywords: Random number generators · Pseudorandomness · Cryptography · Cartesian Genetic Programming · Statistical tests

1 Introduction

Random number generators are used in a range of applications spanning from producing simple values and adding randomness to programs, over online betting to various cryptographic applications. Accordingly, they are important components in many real world scenarios. In cryptographic applications one relies on a source of randomness that can produce truly random numbers when generating seeds, nonces, initialization vectors (IVs), etc. However, for many of today's applications like generating masks or padding messages, true randomness is not needed, only statistical quality is required [1]. There, it is sufficient to use PRNGs that produce good results and yet are realized with deterministic methods. One example is the Blum Blum Shub generator [2] that produces numbers that are indistinguishable from true random numbers by means of standard statistical

This work has been supported in part by Croatian Science Foundation under the project IP-2014-09-4882. In addition, this work was supported in part by the Research Council KU Leuven (C16/15/058) and IOF project EDA-DSE (HB/13/020).

J. Handl et al. (Eds.): PPSN XIV 2016, LNCS 9921, pp. 613–622, 2016.
DOI: 10.1007/978-3-319-45823-6_57

testing. This example shows that it is possible to obtain numbers of a sufficient statistical quality with a deterministic method.

In this paper, we investigate the efficiency of Cartesian Genetic Programming (CGP) when evolving deterministic PRNGs. In order to do so, we present a new framework capable of generating many PRNGS that pass statistical tests and are fast and small when implemented in hardware.

Motivation and Contributions. One real-world application of PRNGs in cryptography is to use them for masking [3] as a countermeasure against side channel attacks. When used for such a purpose, we want those generators to be extremely fast and small when implemented in hardware. To obtain a generator with such characteristics, we cannot use "expensive" operations like multiplication or addition.

Therefore, in this paper we evolve PRNGs that pass all statistical tests and use only cheap operations, which is also an important difference between our approach and related work, since there multiplication and addition operations appear without constraints [4,5]. We emphasize that if we can use the multiplication operation, then there exist much smaller PRNGs than those presented in related work and it is a trivial task to design a PRNG that passes all statistical tests. Next, we consider some of the fitness functions that are used in related work actually inappropriate since they do not mimic the inner working of a PRNG. Therefore, in this paper we use a fitness function that we believe describes the PRNG behavior better. Finally, we are the first to apply CGP to this problem, since that paradigm is the most natural for PRNG structures because of its multiple input – multiple output configuration.

In Sect. 2, we give the necessary introduction to random numbers, PRNGs, and testing methods. Then, in Sect. 3, we give an overview of related work. In Sect. 4, we discuss the model of the PRNG we design and the obtained results. Section 5 offers a short discussion on the results, their applicability and some possible future research avenues. Finally, Sect. 6 offers a short conclusion.

2 (Pseudo) Random Number Generators

In this paper, we follow the terminology as given in the AIS 20/31 proposals [6]. An **ideal random number generator** is a mathematical construct that generates sequences of independent and uniformly distributed random numbers. A **random number generator** (RNG) is any group of components or an algorithm that outputs sequences of discrete values. RNGs can be divided into true random number generators and pseudorandom number generators.

A **true random number generator** (TRNG) is a device for which the output values depend on some unpredictable source that produces entropy. **Pseudorandom number generators** (PRNGs) or **deterministic random number generators** (DRNGs) represent mechanisms that produce random numbers by performing a deterministic algorithm on a randomly selected seed. **Cryptographically secure pseudorandom number generators** are PRNGs with properties suitable for use in cryptography.

A **seed** is a random value used to initialize the internal state of the generator. A **state** is an instantiation of a random number generator. Note that PRNGs can accept additional input data besides the seed value. In Fig. 1, we give a model of a PRNG.

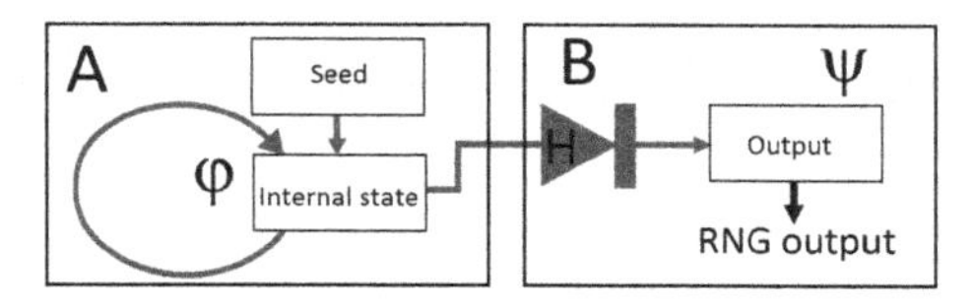

Fig. 1. Model of a PRNG.

In the rectangular A, we depict the input value for the generator (seed) and the PRNG (some function φ) and in the rectangular B the one-way function (H) with the RNG output. Here, ψ is the output function, φ is the state transition function, and $x_{n+1} = \varphi(x_n)$. In this work, we concentrate only on part A and we assume that the one-way function H does not exist. This diagram conforms to the model called DRNG.2 [6], but we note that here we are not interested in forward and backward secrecy requirements. Forward and backward secrecy ensure that it will not be possible to determine the successor and predecessor values from a known subsequence of output values.

2.1 Testing Randomness

The quality of PRNGs is evaluated using statistical tests which follow the same procedure as any other hypothesis testing. The hypothesis under test is that the PRNG produces a perfectly random output. Tests are applied on the output bit sequences. Each test defines a metric called *the test statistic* which can be computed from the sequence under test. A simple example of the test statistic is the bias of the sequence defined as:

$$\varepsilon = \left| \frac{N_{ones}}{N_{ones} + N_{zeroes}} - 0.5 \right|, \tag{1}$$

with N_{ones} the number of ones and N_{zeros} the number of zeros in the sequence. The next step is to compute the P_{value} which is the probability that an ideal RNG produces a sequence which is more extreme with respect to the defined metric than the sequence under test. For example, if the sequence under test contains 70 zeros and 30 ones, then the P_{value} is equal to the probability that an ideal RNG produces a 100-bit sequence with bias higher than 0.2. The final step is comparing the P_{value} with the predetermined constant, for example $\alpha = 0.01$. If the P_{value} is higher than the cut-off value α, than the sequence passes the test, otherwise it fails.

Each statistical test checks for a different statistical defect, therefore more extensive testing results in a more reliable outcome. Several batteries of statistical tests have been proposed, where the most famous ones are the NIST [7] and DIEHARD [8] test suites. In general, the tests applied on longer sequences are more likely to detect statistical weaknesses. In this work, we use the NIST battery of 15 tests applied on sequences of 10^6 bits. Statistical tests can produce both false-positive and false-negative errors. The probability of a false-positive error (truly random sequence fails the test) is equal to the chosen parameter α. Here, we use the value $\alpha = 0.01$ as recommended by the NIST standard [7].

3 Related Work

John Koza used genetic programming (GP) to evolve programs that output random numbers [9]. As a fitness function he used the notion of information entropy as defined by Shannon and the end result was a program that was able to accept a sequence of consecutive integer values and transform it into random binary digits. Hernandez, Seznec, and Isasi used GP to evolve random number generators where they used the strict avalanche criterion (SAC) as a fitness function [10]. Martinez et al. designed a pseudorandom number generator suitable for cryptographic usage by means of GP [4]. The obtained generator – Lamar was tested with a number of tests where the input values were obtained via a counter function. We consider this work the most serious attempt on evolving PRNGs for cryptographic usage although in our opinion this work has some potential drawbacks since it does not follow the structure a PRNG should have as discussed in Sect 5. Lopez et al. focused on the evolution of PRNGs that could be used in low cost RFID tags [5]. They followed an approach similar to previous work where the fitness function was based on the notion of the SAC and the testing on values obtained via a counter function.

4 The Proposed Model of a PRNG and Results

As a design choice, we decided to work with PRNGs that have four input terminals. Each terminal is represented with a 32-bit unsigned integer variable. This means that the state of our generator has 128 bits. Since we assume that the input and output sizes of the variables are of the same size, it means that our PRNG should output 128 bits of random data in every iteration.

We use the function set (inner nodes) that consists of binary Boolean primitives: rotate right/left for one position (RR/RL), shift right/left for one position (SR/SL), AND, NOT, XOR, and P(x). The function P(x) is a basic perfect outer shuffle where the bits are interleaved into two halves of a word and the outer (end) bits remain in the outer positions [11]. Note that the hardware implementations of RR, RL, SR, SL, and P consist only of rearranging the signal wires and therefore come without the cost of additional logic gates. Such an architecture is suitable for both hardware and software implementations because it utilizes a simple FIFO buffer. To handle the cases where the input value to a PRNG is all zeros or all ones, we use one constant that we select randomly and it equals $4E2D93A6_{16}$. Although we work with generators that have four variables and a 128-bit state, we note that we could have chosen any number of variables and any variable size that is available in ANSI C (for the results to be platform independent).

4.1 Cartesian Genetic Programming Approach

In Cartesian Genetic Programming (CGP) a program is represented as an indexed graph. The terminal set (inputs) and node outputs are numbered sequentially. Node functions are also numbered separately. CGP has three parameters that are chosen by the user; number of rows n_r, number of columns n_c, and levels-back l [12]. In our experiments, for the number of rows we use a value of one and for the levels-back parameter we use the same value as for the number of columns. The number of node input connections n_n is two, the number of node output connections n_o is one, and the population size equals five in all our experiments. To obtain 128 bits of random values we set the number of output nodes to 4, so we need only one iteration to obtain the full generator state. For the CGP individual selection, we use a $(1+4)$-ES strategy in which offspring are favored over parents when they have a fitness less than or equal to the fitness of the parent. The mutation type is probabilistic.

To evaluate the performance of the evolved PRNGs, we use the following procedure. In every generation we randomly create four 32-bit input values and assign them to the terminal set. Next, we run the PRNG with those input values and we obtain output values (which we informally call "initial output"). Then, we check how a small change at the input propagates to the output. To do so, we XOR the original input values with all vectors of 128 bits and Hamming weight equal to one. For each of those modified input values, we again run the PRNG and save the output values (called "modified outputs"). Then, we do a pairwise XOR between the "initial output" and the "modified output" values and we send the result to the Test function (see Eq. (2)).

The goal of this part of the evaluation process is to check the impact of a single bit change. One way to do this is by checking the avalanche criterion (AC), which indicates how many output bits change when a single input bit changes. The ideal generator would have on average 50 % of output bits changed for every input bit change. It is possible to enforce an even stricter criterion – SAC, where the demand is that *exactly* 50 % of output bits are changed for every input bit change. However, SAC does not necessarily imply better statistical quality. It is possible to construct a simple array of XOR gates which satisfies the SAC and yet has very poor performance as a PRNG. Instead, we choose to evaluate the entropy of the change caused by a single input bit flip.

We developed a test function based on the NIST approximate entropy test [7]. This choice was guided by two facts. First, the statistic of this test is the function that always results in a value between 0 and 1, with a higher value corresponding to better randomness. This is very suitable for computing the fitness function. Second, the approximate entropy test is applicable to very short bit sequences.

Note that when computing the fitness function we apply the test function on sequences of 128 bits.

The approximate entropy test uses the estimation of the entropy-per-bit as the test statistic. First, the relative frequencies of all 4-bit and 3-bit patterns are estimated. Based on these frequencies, the entropy is estimated as:

$$Test(output\_value) = \sum_{i(m=3)} \frac{\nu_i}{n} log \frac{\nu_i}{n} - \sum_{i(m=4)} \frac{\nu_i}{n} log \frac{\nu_i}{n}, \quad (2)$$

where ν_i is the number of occurrences of each bit pattern and n is the length of the sequence. Since the higher the value of the Test function the better, we aim to *maximize* the fitness value, where the maximal result equals 128 (scoring 1 on all 128 tests).

Finally, to check how a generator deals with the all-zeros and all-ones input vector, we run it for those values and the result is again sent to the Test function (now there are no pairs of output values to XOR before invoking the Test function). Finally, the resulting fitness function equals:

$$fitness = \sum_{1}^{130} Test(output\_value) - missing * 130, \quad (3)$$

where *output_value* is either an XOR between two output values (as is the case for the first 128 tests) or is a single output value (as is the case for the all-zeros and all-ones input vectors). The variable *missing* represents the number of missing terminals in the generator. The parameter 130 is selected so to enforce that every solution that is missing a terminal is worse than any correct solution (i.e. with all terminals).

Finally, we run our evaluation procedure for n "rounds". The round process is very simple and it just repeats the whole procedure described above, but instead of randomly selecting input variables for every round, we use the output variables from the previous round. By this technique we aim to mimic the mechanism of a PRNG since there the input of iteration $t+1$ is the output from the iteration t. When using a mechanism with multiple rounds, the cumulative fitness equals the *smallest fitness value* over all rounds. With this criterion, we ensure that the generator behaves at least as good as for the worst evaluation round. Indeed, when we work with only one round, it is hard to predict how the generator behaves when it takes the previously generated values as an input. We experimented with several round number values, but we did not observe any improvement when the number of rounds is greater than two. Consequently, in all our experiments we use two rounds.

4.2 Experimental Results

As one of our goals is to evolve generators that are as small and fast as possible in hardware, that intuitively means we want to restrict the size of the graph we have. Therefore, here we investigate what is the necessary population size and graph size to evolve PRNGs that pass statistical tests. In all the experiments we set the termination criterion to 20 000 generations. We emphasize that our experiments showed that it is already possible to evolve good PRNGs with a significantly lower number of generations.

The results in Table 1 show the obtained mean fitness values of the best individuals out of 30 runs for every combination of parameters. However, although the average value can help us to deduce which parameter combination works the best, it can also be misleading. Therefore, we additionally give the best obtained values for every combination of parameters.

Table 1. Average/max results for CGP.

Genotype/p_m	1	4	7	10
15	60.8925/74.8561	66.8327/73.576	69.4915/77.7739	69.284/73.6348
30	74.2445/80.5798	72.8384/80.4259	72.5254/76.4385	71.0447/77.7706
50	72.0271/75.471	74.8202/81.6425	76.9252/82.7421	75.1236/80.4513
100	70.9798/76.7845	76.7837/82.3453	78.4326/83.7166	76.057/**84.7544**
200	74.7316/82.4207	77.4329/83.157	79.2718/84.6614	78.9486/82.1012
500	75.3872/82.5726	80.1348/84.3552	80.0269/83.7311	80.3652/83.4072
1 000	78.3797/83.895	79.9121/84.458	**80.6124**/84.1686	79.1203/83.0524

On the basis of the results, we select a genotype size equal to 100 and a mutation probability of 10 % as the best performing parameters.

4.3 Evaluation of the Results

After we obtain the results from CGP, we need to test whether they actually pass the statistical tests. In order to do so, we first use a parser that takes as an input the CGP encoding of a solution and produces a C source code as given below. Note that this example of PRNG passes all statistical test, but otherwise we do not impose any other criterion in the choice of PRNG (i.e. we did not select it on the basis of the size or specific operations). Here, *uint* represents the unsigned int variable type and *const* is the value we chose as a constant ($4E2D93A6_{16}$). It is important to note that the full set of NIST statistical tests are applied only in this phase, and not during the evolution since they are relatively slow.

```
void CGP (uint x0,uint x1,uint x2,uint x3,uint*z0,uint*z1,uint*z2,uint*z3)
{ uint y4 = x0 & x1; uint y5 = x2 ^ x3; uint y6 = (y5 >> 1) | (y5 << 31);
uint y7 = p1(y6); uint y8 = x3 ^ y7; uint y9 = p1(y8); uint y10 = y6 ^ y9;
uint y11 = (y9 << 1) | (y9 >> 31); uint y12 = const; uint y13 = p1(y10);
uint y14 = y12 ^ y11; uint y15 = y12 ^ y13;
uint y16 = (y15 >> 1) | (y15 << 31); uint y17 = y10 ^ y16;
uint y18 = p1(y17); uint y19 = y18 >> 1; uint y20 = y18 ^ y4;
uint y21 = p1(y20); uint y22 = y18 ^ y21; uint y23 = p1(y18);
uint y24 = y19 ^ y18; uint y25 = y23 ^ y19; uint y26 = y22 ^ y14;
*z0 = y18; *z1 = y25; *z2 = y26; *z3 = y24; }
```

The source code is then automatically run until it outputs a string of bits of length n, with n equal to 1 000 000. That string serves as an input for the NIST statistical test suite [7]. Only if a generator passes all the tests, we consider it good enough to be used in practice. Our results showed that on average 80 % of evolved PRNGs pass statistical tests.

Finally, we implemented our CGP example solution and Lamar in Verilog HDL and then we compared them with the Mersenne Twister generator which is a widely used general-purpose PRNG [13]. These algorithms were synthesized using Xilinx ISE14.7 on a Virtex4 xc4vfx100-10ff1152 to draw a fair comparison with the reference implementation of the Mersenne Twister. The implementation result is given in Table 2. With the utilization of 188 slices, our algorithm achieves a maximum working frequency of 286 MHz. The Lamar can reach a working frequency of 43 MHz with 645 slices. Designs can be parallelized to obtain a higher throughput. Therefore we use throughput per slice as the metric for implementation efficiency. As shown in the table, given the same footprint on FPGA, our CGP implementation could be **90 times faster** than the Lamar and **3 times faster** than the Mersenne Twister design [13].

Table 2. Comparison of the hardware implementation results

	Slices	LUTs/FFs/BRAMs	Throughput/slice
CGP	188	317/128/0	**195Mbps/slice**
Lamar [4]	645	1045/238/0	2.16 Mbps/slice
Mersenne twister [13]	128	213/193/4	65.7 Mbps/slice

5 Discussion and Future Work

If comparing our approach with the one followed for the Lamar PRNG [4], we see there is a number of important differences. In Lamar, the authors run independent tests (rounds) for a number of times (usually repeated 16 384 times with an explanation that it is experimentally proven to be enough). Our first objection is that the number of repetitions is an additional parameter one needs to tune and there is no background knowledge one can use. Second, since they use independent input values to create new output values, this does not mimic the working of a PRNG, but rather resembles the procedure one would use when testing a number of Boolean functions. Although this does not necessarily lead to bad results, we find it potentially problematic since in general we do not aim to evolve PRNGs that output an extremely short sequence before needed to be reseeded with a new value. Next, Lamar uses operations like addition and multiplication that we believe are not suitable for small and fast PRNGs to be implemented in hardware. Besides those operations, there are also rotations and shift operations where the number of shift positions is huge and therefore results in zero values for shift operations and a number of unnecessary rotations in rotation operations. Considering the hardware implementation on both ASIC and FPGA, the fixed point multiplier usually has a larger footprint than other logical functions and elementary arithmetic functions, as addition and subtraction. The on-chip DSP slices on FPGAs can be used to implement the multiplication without occupying reconfigurable fabric. However, the number of these dedicated DSPs is relatively small.

Finally, the authors of Lamar consider using it as a stream cipher [1], which we believe is unrealistic. Since it is not possible to automatically test all the properties a stream cipher should have (since it would mean a fully automatic cryptanalysis, which is not possible) it is also not possible to write an appropriate fitness function. Therefore, although we do not categorically claim it is not possible to successfully evolve a stream cipher, we state that such a cipher would be good by accident since we cannot evolve it specifically for that purpose.

In future work, we plan to work with a variable number of rounds, where in the beginning we would use a smaller number of rounds and as the evolution progresses we would increase the number of rounds to increase the selection pressure and ensure our PRNGs have higher chances passing a posteriori testing. Next, we could use longer sequences as inputs for the fitness function since then the fitness function will be able to better discriminate PRNGs and consequently, the success rate of PRNGs after the statistical tests will be higher. To obtain such longer sequences, we could use a concatenation of results for several rounds. Additionally, recall Fig. 1 where we said that in this paper we disregard the rectangular B that includes the one-way function. The simplest solution in adding a one-way function would be to simply combine several bits of the output string via an XOR function. If one decides to go with the EC approach, then he could evolve one or more Boolean functions with high nonlinearity [14].

We believe more experiments are necessary to determine the limits of the evolutionary approach. Since we aim to find generators that not only pass all the statistical tests, but are also fast and small when implemented in hardware, we could improve the fitness function in an effort to reduce the number of nodes in CGP. Finally, we propose a setting where we believe that the evolutionary approach would display its full benefits. Consider an FPGA board that also has an ARM processor (e.g. Zynq). Then one can put on the ARM the CGP that evolves PRNGs. Such evolved PRNGs can be sent to the FPGA to be partially reconfigured. Therefore, with this approach we would effectively use evolvable hardware [15] to increase the security of a system.

6 Conclusions

In this paper, we address the issue of evolving pseudorandom number generators that are suitable for cryptography. The results obtained show that CGP can be used as a viable choice to evolve PRNGs. To define a real-world application for such generators, we discuss the limitations of PRNGs and where they could be used and consequently what properties they need to have. Furthermore, we present a fitness function that in our opinion ensures better results than those used before. We emphasize that we aimed to evolve PRNGs that are extremely small and fast in hardware and therefore do not rely on expensive operations like multiplication or addition.

References

1. Katz, J., Lindell, Y.: Introduction to Modern Cryptography, 2nd edn. Chapman and Hall/CRC, Boca Raton (2014)
2. Blum, L., Blum, M., Shub, M.: A simple unpredictable pseudo random number generator. SIAM J. Comput. **15**(2), 364–383 (1986)
3. Danger, J.L., Guilley, S., Barthe, L., Benoit, P.: Countermeasures against physical attacks in FPGAs. In: Badrignans, B., Danger, L.J., Fischer, V., Gogniat, G., Torres, L. (eds.) Security Trends for FPGAS: From Secured to Secure Reconfigurable Systems, pp. 73–100. Springer, Dordrecht (2011)
4. Lamenca-Martinez, C., Hernandez-Castro, J.C., Estevez-Tapiador, J.M., Ribagorda, A.: Lamar: a new pseudorandom number generator evolved by means of genetic programming. In: Runarsson, T.P., Beyer, H.-G., Burke, E.K., Merelo-Guervós, J.J., Whitley, L.D., Yao, X. (eds.) PPSN 2006. LNCS, vol. 4193, pp. 850–859. Springer, Heidelberg (2006)
5. Peris-Lopez, P., Hernandez-Castro, J.C., Estevez-Tapiador, J.M., Ribagorda, A.: LAMED - a PRNG for EPC class-1 generation-2 RFID specification. Comput. Stand. Interfaces **31**(1), 88–97 (2009)
6. Killmann, W., Schindler, W.: A proposal for: functionality classes for random number generators. Bundesamt für Sicherheit in der Informationstechnik (BSI), Germany (2011)
7. Bassham, III, Lawrence, E., Rukhin, A.L., Soto, J., Nechvatal, J.R., Smid, M.E., Barker, E.B., Leigh, S.D., Levenson, M., Vangel, M., Banks, D.L., Heckert, N.A., Dray, J.F., Vo, S.: A Statistical Test Suite for Random and Pseudorandom Number Generators for Cryptographic Applications, SP 800-22 Rev. 1a. National Institute of Standards & Technology, Gaithersburg, MD, USA (2010)
8. Marsaglia, G.: The Marsaglia Random Number CDROM including the Diehard Battery of Tests of Randomness (1995). http://www.stat.fsu.edu/pub/diehard/
9. Koza, J.R.: Evolving a computer program to generate random numbers using the genetic programming paradigm. In: Proceedings of the Fourth International Conference on Genetic Algorithms, pp. 37–44. Morgan Kaufmann (1991)
10. Hernandez, J., Seznec, A., Isasi, P.: On the design of state-of-the-art pseudorandom number generators by means of genetic programming. In: Congress on Evolutionary Computation, CEC2004, vol. 2, pp. 1510–1516, June 2004
11. Warren, H.S.: Hacker's Delight. Addison-Wesley Longman Publishing Co., Inc., Boston (2002)
12. Miller, J.F., Thomson, P.: Cartesian genetic programming. In: Poli, R., Banzhaf, W., Langdon, W.B., Miller, J., Nordin, P., Fogarty, T.C. (eds.) EuroGP 2000. LNCS, vol. 1802, pp. 121–132. Springer, Heidelberg (2000)
13. Tian, X., Benkrid, K.: Mersenne twister random number generation on FPGA, CPU and GPU. In: NASA/ESA Conference on Adaptive Hardware and Systems, AHS 2009, pp. 460–464, July 2009
14. Picek, S., Jakobovic, D., Miller, J.F., Batina, L., Cupic, M.: Cryptographic boolean functions: one output, many design criteria. Appl. Soft Comput. **40**, 635–653 (2016)
15. Sekanina, L.: Virtual reconfigurable circuits for real-world applications of evolvable hardware. In: Tyrrell, A.M., Haddow, P.C., Torresen, J. (eds.) ICES 2003. LNCS, vol. 2606, pp. 186–197. Springer, Heidelberg (2003)

Exploring Uncertainty and Movement in Categorical Perception Using Robots

Nathaniel Powell(✉) and Josh Bongard

Department of Computer Science, University of Vermont, Burlington, USA
nvpowell@uvm.edu

Abstract. Cognitive agents are able to perform categorical perception through physical interaction (active categorical perception; ACP), or passively at a distance (distal categorical perception; DCP). It is possible that the former scaffolds the learning of the latter. However, it is unclear whether ACP indeed scaffolds DCP in humans and animals, nor how a robot could be trained to likewise learn DCP from ACP. Here we demonstrate a method for doing so which involves uncertainty: robots are trained to perform ACP when uncertain and DCP when certain. We found evidence in these trials that suggests such scaffolding may be occurring: Early during training, robots moved objects to reduce uncertainty as to their class (ACP), but later in training, robots exhibited less action and less class uncertainty (DCP). Furthermore, we demonstrate that robots trained in such a manner are more competent at categorizing novel objects than robots trained to categorize in other ways.

Keywords: Uncertainty · Active categorical perception · Robotics

1 Introduction

The embodied approach to cognitive science holds that the body is a necessary component for the acquisition of adaptive—and, ultimately, cognitive—behavior [3,7]. Since the establishment of this approach, much work has been dedicated to investigating how the body can do so [8], and quantifying its contribution [4,6]. A common approach for doing so is to employ robots, in which all aspects of their morphology, control structure, and task environment can be observed and experimentally modified.

A common skill investigated from an embodied perspective is categorical perception: how an agent makes use of its body to generate the requisite stimuli to learn appropriate categories. Initially, Beer evolved minimally cognitive agents to achieve this 'active' form of categorical perception (ACP) [1]: the agents interacted with their environment in a way that reduces intracategorical differences and magnifies intercategorical ones. Subsequent studies explored this phenomenon using more complex robot morphologies [2,11,12]; those robots physically manipulated the objects to be categorized. However, sophisticated cognitive agents typically employ distal categorical perception (DCP)—categorizing an

J. Handl et al. (Eds.): PPSN XIV 2016, LNCS 9921, pp. 623–632, 2016.
DOI: 10.1007/978-3-319-45823-6_58

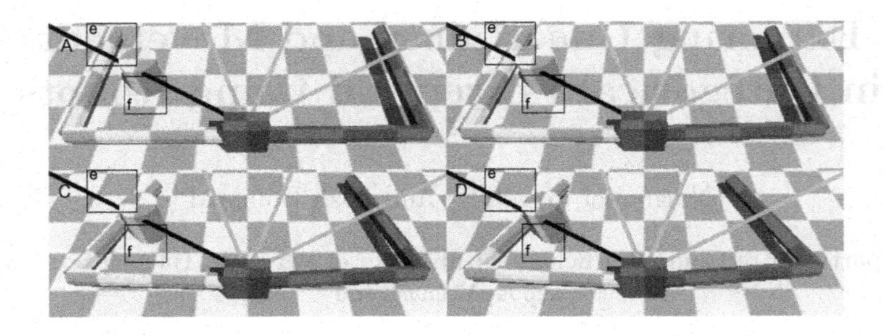

Fig. 1. Evolved behavior for a robot that interacts with its environment when uncertain to perform categorical perception. Time is tracked through panels A through D. Movement of the arm is tracked in box e. Movement of the object is tracked in box f.

object by sight and/or sound from a distance—as it has obvious advantages over ACP, such as rapidity, avoidance of potentially dangerous contact, and success even when physical contact is not possible. This raises questions regarding how animals learn (and how robots should learn) DCP, ACP, and how and when to switch between them. One hypothesis is that ACP scaffolds the learning of DCP: interactions with objects can structure perception in such a way as to facilitate learning of non-embodied skills [5].

The question remains however as to the conditions under which ACP or DCP should be employed. We hypothesize that such switching should be modulated by uncertainty: unfamiliar stimuli should trigger internal uncertainty, which in turn should trigger appropriate action resulting in ACP, which, finally, provides scaffolding for the learning of DCP when next presented with this object. Over a lifetime, this should result in an agent that exhibits more instances of DCP and fewer of ACP. Here, we demonstrate the usefulness of this particular mechanism by training simulated robots to perform ACP when uncertain and DCP when certain. We show that, when exposed to novel stimuli, these robots categorize better than robots trained to categorize in other ways.

2 Methods

We conducted a series of experiments in which a simulated yet embodied robot attempts to categorize objects in its environment (Fig. 1). It is embodied in the sense that, despite its virtual surroundings, actions it performs can impact the environment, and the agent immediately detects the sensory repercussions of those effects. An evolutionary algorithm was used to train the robot to perform ACP, DCP, or a combination of the two when exposed to a number of environments. The robot's task was to correctly categorize large objects as large, and small objects as small. When training concludes, the best robot's categorization abilities were tested by exposing it to novel environments, and its categorization error in those novel situations was measured.

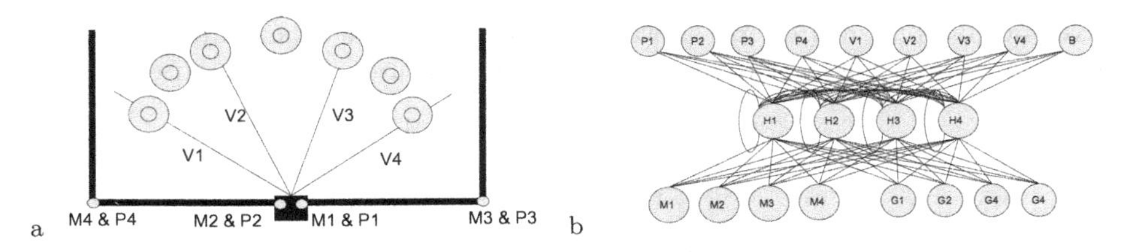

Fig. 2. (a) The robot body is constructed from two arms. The upper limbs are attached to a central body. The limbs are attached to each other and the body with four motorized joint, each containing an angle sensor. Thin lines represent line of sight for sensing distance. The robot can be exposed to any one of 14 objects, which are placed either in the line of sight or 'blind' to the robot. Objects are either large (large circles) or small (small circles). (b) The controller of the robot is instantiated as an artificial neural network. The input layer is made up of four proprioceptive neurons (P), four vision neurons (V) and a bias neuron (B). Hidden layer consists of 4 recurrent hidden neurons (H). Output layer: four motor neurons (M) and four guess neurons (G)

The Robot Body. The robot's body was constructed from four equal length cylinders and a small central body constructed from a rectangular solid. The arms adjacent to the main body are connected by supporting motorized joints to it, as are the forearms connected to the upper arms (Fig. 2a). There are a total of four motorized joints, yielding four mechanical degrees of freedom. Each of the motorized joints enabled the connected objects to rotate relative to one another through the robot's coronal plane, which, given its morphology, corresponds to the horizontal plane. This results in the arms flexing in horizontally toward the main body and extending outward from the main body, also horizontally. The arms were placed at a particular height so that they do not end up breaking the visual beams emanating from the robot (black and gray lines in Fig. 1).

The Robot Controller. The controller of the robot is instantiated as a partially-recurrent artificial neural network. There are three layers that make up the neural network: the input layer, a hidden layer, and an output layer (Fig. 2b). The input layer consists of two types of sensory neurons: vision neurons (V) and proprioceptive neurons (P). At each time step during which a robot is simulated, the angles of the four motorized joints are computed, normalized to real values in the range $[-1, +1]$, and supplied to the four proprioceptive neurons. Likewise, four beams are sent out from the main body such that they span the range $[-60^o, +60^o]$ in front of the robot. The angles between each pair of neighboring beams was set to 40°. While the robot may move its arms, it cannot move the visual beams. However, the beams may be broken by coming into contact with an external object. The length of each beam at each time step is computed and scaled to a real value in $[0, 1]$ such that zero indicates the beam is unbroken, while one indicates that an object is in contact with the base of the beam. The hidden layer consists of four fully recurrent hidden neurons (H): each hidden neuron receives input from each of the sensors (in addition to a fixed-output bias neuron B) as well as values from the other hidden neurons, including itself. The new value of the ith hidden neuron h_i is computed as

$$h_i = \tanh((\sum_{j=1}^{4} s_j w_{ji}) + w_{bi} + (\sum_{k=1}^{4} h_k w_{ki})) \quad (1)$$

where s_j is the value of the jth sensor neuron, w_{ji} is the weight of the synapse connecting the jth sensor neuron to the ith hidden neuron ($w_{ji} \in [-1, +1]$), w_{bi} is the weight of the synapse connecting the bias neuron (value clamped to one) to the ith hidden neuron, h_k is the value of the kth hidden neuron, w_{ki} is the weight of the synapse connecting the kth hidden neuron to the ith hidden neuron, and $\tanh(x)$ brings the hidden neuron values back into the range $[-1, 1]$.

The output layer is comprised of two different types of neurons: motor neurons (M) and guess neurons (G). The value of the ith output neuron is computed at each time step using $o_i = \tanh(\sum_{j=1}^{4} h_j w_{ji})$ regardless of whether it is a motor or guess neuron. The value of each of the four motor neurons is scaled to the range $[-45^o, +45^o]$ and then supplied, as a desired angle, to each of the four joints. A proportional-derivative (PD) controller is effected by supplying torque to the joint proportional to the difference between the current angle and the desired angle. The outputs arriving at the guess neurons are employed by the robot to perform categorical perception and are described in more detail below.

The Task Environment. When evaluated, a robot is equipped with a neural network labeled with a particular set of synaptic weights as described above. The robot is then exposed to one of 14 environments as shown in Fig. 2. There are seven possible positions. At each location a cylinder with a large or small radius may appear. Objects are placed in such a way that they are either initially unseen by the robot or in its direct line of sight.

The Evolutionary Algorithm. An evolutionary algorithm was used to train neural networks in the robot as described above. For each evolutionary trial, seven environments from the total set of 14 were chosen as random and fixed as the training set for that trial. In a different evolutionary trial, seven different environments may be chosen. At the outset of the trial, an initial population of 20 random neural networks were created. Each of these neural networks contained random synaptic weights drawn from $[-1, +1]$ with a uniform distribution. Each neural network was then evaluated on the robot seven times, in the seven environments chosen for that trial. During each of these seven evaluation periods, the robot was allowed to move for 25 time steps in the simulator. This population of neural networks was then evolved for 100 generations using a common evolutionary algorithm that balances increasing fitness over time while also maintaining genetic diversity in the population [9]. When the 100 generations complete, the ANN with highest fitness in the population is re-evaluated on the robot seven times, in the seven novel environments that were unseen during evolution.

The Fitness Functions. The robots in this experiment were evaluated against four different fitness functions, leading to four experimental conditions. In the first condition, robots were evolved simply to categorize correctly (C). In the second condition they were evolved to categorize correctly while minimizing movement (CnM). In the third condition they were evolved to categorize correctly while maximizing movement (CM). In the fourth and final condition they were evolved to categorize correctly, and to do so by moving when uncertain about the object's category and remaining still when certain about the object's category. This condition was referred to as CR, as the robot should establish a correlation (R) between movement and uncertainty. Fifteen independent evolutionary trials, each starting with a different randomly chosen set of seven objects and 20 random ANNs, were performed for each of the four conditions, yielding a total of 60 independent trials.

The fitness functions for these conditions were constructed from combinations of the following terms

$$C = \sum_{t=1}^{T}\sum_{e=1}^{E}\sum_{i=1}^{G}(g_{ei}^{(t)} - s_e)^2/TEG \tag{2}$$

$$K = \sum_{t=1}^{T}\sum_{e=1}^{E}\sum_{m=1}^{M}|(a_{em}^{(t)} - a_{em}^{(t-1)})|/TEM \tag{3}$$

$$R = \sum_{e=1}^{E}\mathrm{Corr}(\boldsymbol{G}_e, \boldsymbol{m}_e)/E \tag{4}$$

$$\boldsymbol{G}_e = [\sigma(\boldsymbol{g}_e^{(2)}), \sigma(\boldsymbol{g}_e^{(3)}), \ldots, \sigma(\boldsymbol{g}_e^{(25)})] \tag{5}$$

$$\boldsymbol{m}_e = [\sum_{m=1}^{M}|a_{em}^{(2)} - a_{em}^{(1)}|, \ldots, \sum_{m=1}^{M}|a_{em}^{(25)} - a_{em}^{(24)}|] \tag{6}$$

where

- C denotes how well a given neural network categorizes, averaged over all $T = 25$ time steps, $E = 7$ training objects, and $G = 4$ guess neurons ($C = 0$ indicates perfect categorization and $C = 1$ the worst possible categorization);
- K denotes the average amount of motion over all T time steps, E training environments, and $M = 4$ motors;
- R denotes the amount of correlation (Corr) between uncertainty ($\boldsymbol{G}_e$) and amount of movement ($\boldsymbol{m}_e$), averaged over all E environments ($R = 1$ indicates the robot moves maximally when uncertain and minimally when certain);
- $\boldsymbol{G}_e$ represents a vector containing the uncertainties of the ANN when exposed to the eth environment, at each time step of the exposure (with the exception of the first time step);
- $\boldsymbol{m}_e$ represents a vector containing the amount that the robot moved when exposed to the eth environment, at each time step of the exposure (with the exception of the first time step);
- $g_{ei}^{(t)}$ represents the output of the ith guess neuron during the tth time step of exposure to the oth object;

- s_e represents the size of the object in the eth environment (small object=-0.5, large object=+0.5);
- $a_{em}^{(t)}$ represents the angle of the mth motorized joint during the tth time step of exposure to the eth environment; and
- $\boldsymbol{g}_e^{(t)}$ represents a vector containing the values of the four guess neurons generated during the tth time step when exposed to the eth environment, and $\sigma(\boldsymbol{g}_e^{(2)})$ represents the variance within that vector.

Condition C: In the first condition, robots were only evolved to categorize, regardless of the amount or type of movement they employed to do so. This was accomplished by evolving robots that maximized the fitness function

$$F_C = 1/(1 + C). \tag{7}$$

Condition CnM: In the second condition, DCP was explicitly favored by evolving robots that successfully categorize while also minimizing movement:

$$F_{CnM} = (\frac{1}{1 + C})(\frac{1}{1 + K}). \tag{8}$$

Condition CM: In this third condition, ACP was explicitly favored by evolving robots to successfully categorize while maximizing movement:

$$F_{CM} = K/(1 + C). \tag{9}$$

Condition CR: Finally, robots were evolved in the fourth condition to employ DCP when uncertain as to the object's size and to employ ACP when they were certain. This was accomplished using

$$F_{CR} = R/(1 + C). \tag{10}$$

Prediction Variance and Uncertainty. Here, we employ the variance among the values of the guess neurons to denote a controller's uncertainty about the current object's category. In the machine learning literature, prediction variance is often employed as a proxy for uncertainty [10]. This is because, as long as individual units in a predictive model (here, the guess neurons) are independent, they are likely only to converge on the same prediction when that prediction is correct. This is not unlike a group of people with very different backgrounds generating diverse—and thus mostly wrong—answers to questions that touch on an area of their mutual ignorance, but who only generate similar responses when the question touches on an area of their common knowledge. The guess neurons here are independent because each guess neuron has its own synaptic weights connecting it to the input layer.

3 Results

At the termination of each run, for each condition, the robot with the best fitness is extracted from the population. Each of these controllers is then re-instantiated

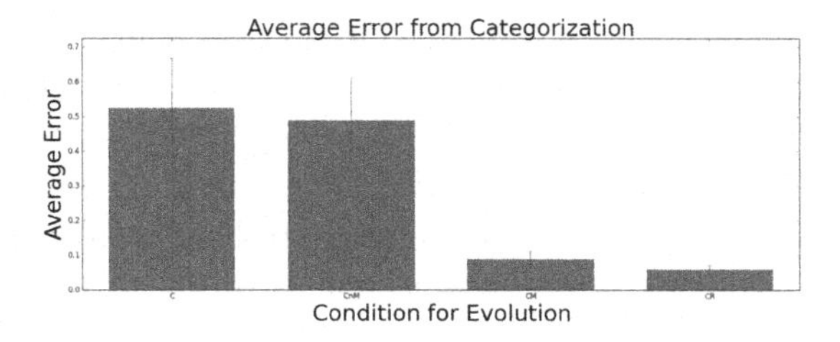

Fig. 3. The average ability of the best 15 controllers to correctly categorize unseen objects, for the four experimental conditions tested. The bars are in reference to the standard error for each condition.

in the robot and evaluated a further seven times, in the seven environments that the controller did not experience during training. We employed C (Eq. 2) to compute the robot's average ability to categorize these $E = 7$ novel environments. A controller that obtains lower values of C when exposed to these novel environments is thus exhibiting a better ability to generalize its ability to categorize, compared to another controller with a higher value of C. Figure 3 reports the average generalization abilities of the 15 evolved controllers extracted from the four conditions. Testing for significance was computed using multiple Mann-Whitney U tests. The tests looked for significant differences between the CR condition and the remaining three conditions (C, CnM, and CM). Significant P-values were found in each comparison. After correcting for multiple comparisons using the Bonferroni method of adjustment, the P-values found between CR and C, CR and CnM, and CR and CM are respectively 1.52×10^{-5}, 1.52×10^{-5}, and 1.18×10^{-3}. Significant P-values indicate that the relative average amount of categorization error performed by individuals in the CR condition is significantly lower than the error of individuals in the remaining three conditions. Therefore, it can be concluded that individuals evolved under the CR condition are better suited for tasks involving categorization of objects than individuals from the other three evolutionary conditions.

4 Discussion

Why is C worse than CR? The controllers evolved in the C condition performed worse than those evolved for CR for several reasons. Such controllers may have generated little to no motion, enabling rapid and successful categorization of seen objects, while sacrificing the ability to categorize unseen objects, in the training environments. This may have led to overfitting such that novel, seen objects in the testing environments were categorized incorrectly. Conversely, controllers may have evolved to perform more motion. Such a strategy might cause correct, instantaneous categorization which is subsequently lost when the robot comes into contact with the object.

Why is CnM worse than CR? Controllers evolved in the CnM condition are likely incentivized to memorize the categories of seen objects, and ignore the categories of unseen objects. This results in overfitting: the robot will not only poorly categorize novel, unseen objects, but novel, seen objects as well. In other words, the robots are deprived of the ability to reduce spurious differences between intracategory objects through motion.

Why is *CM* worse than *CR*? Conversely, controllers evolved in the *CM* condition may suffer from two disadvantages compared to controllers evolved using *CR*. Controllers from *CM* may not be able to afford to hold still when they are certain of a novel object's class. Moreover, they may have to exhibit so much motion that they end up magnifying spurious intracategory differences, rather than generating less—yet appropriate—movement that reduces those differences.

How did *CR* Succeed? Controllers produced in the *CR* condition presumably outperformed the controllers produced by the other three conditions because they can better employ ACP, or DCP, when that form of categorization is most appropriate. When the robot is certain about an object's category, it should employ DCP without moving: the robot does not need to wait for physical contact with the object to categorize the object. Moreover, if it does contact the object, the resulting sensor data may affect the guess neurons and thus draw the robot away from the already correctly-predicted category. Conversely, when the robot is uncertain as to the object's category, it should initiate movement as rapidly as possible: that is, exactly at the moment that it is uncertain. Another interesting result from the CR condition is that when a CR-generated controller's uncertainty is high, it also tends to exhibit higher category error At such times, CR-generated controllers are encouraged to move as much as possible. However, once in contact with the object and the amount of uncertainty starts to decrease, less motion is necessary: the robot is free to perform whatever actions are appropriate to reduce intracategory differences. In contrast, *CM* generated robots are more restricted in the kinds of actions they can employ to reduce these differences: they must generate high-magnitude movements.

Evidence of Scaffolding. Within the CR trials, there is some evidence that suggests robots may initially evolve ACP, which scaffolds the subsequent evolution of DCP in later generations. Figure 4 reports evolutionary changes in uncertainty and movement generated by the best robots extracted from 13 trials using the CR fitness function. Uncertainty and movement is only recorded when these robots encounter the leftmost large and small cylinders (Fig. 2 shows one such robot hitting the large cylinder). Movement among these robots increases during the first 30 generations: this movement may cause the robot to hit the large but not the small object, and thus may exaggerate the initially small yet intercategorical difference registered by the leftmost distance sensor when exposed to these two objects. This in turn may enable evolution to discover controllers that successfully use this magnified difference to correctly categorize these two objects. Later still, mutations may alter the controller to predict correctly using smaller and smaller differences from this visual sensor, earlier and earlier during the evaluation period, ending ultimately with a robot able to immediately categorize these two objects correctly using the leftmost visual sensor, thus not

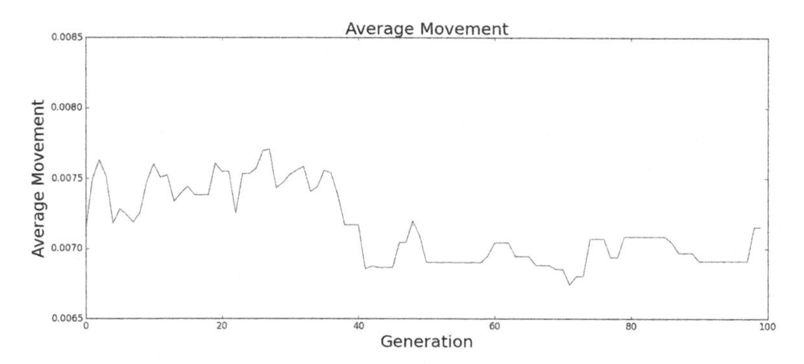

Fig. 4. Evolutionary increase, and subsequent decrease, in average movement among the best robots evolved using the CR fitness function.

requiring physical interaction with these objects. Indeed, motion does seem to decrease after generation 30 in these trials. If ACP had not been useful for scaffolding DCP, we should have seen evolved robots that only used DCP only in the presence of seen objects (thus no observed motion in the presence of these two leftmost objects) and ACP only in the presence of unseen objects. However, more analysis is required to determine whether the scaffolding of DCP by ACP did indeed occur in these trials.

Possible Sources of Error. One potential source of error is that only 100 generations of evolution were performed with a relatively small initial population of 20 individuals. This may not have allowed for significant optimization to occur in all conditions. The very short evaluation period may also be a source of error, because the CM and C conditions may allow for the discovery of exaggerated movements that yet, given enough time, eventually reduce intracategory differences. More generally, longer evaluation periods will allow for a greater range of movements that may then better help clarify the relationship between movement, categorization, and uncertainty.

5 Conclusion

Here we have shown that it is possible to explicitly train robots to exhibit ACP when uncertain and DCP when certain, and that such robots outperform other robots trained to perform ACP or DCP at will but without the ability to do so based on uncertainty; trained to always perform DCP (categorize without moving); or trained to always perform ACP (categorize via movement). Further, this approach does not require us to dictate how the robot should interact with its environment; it is free to discover its own strategies for reducing intracategory differences through physical interaction. This work helps to clarify the relationship between three competencies necessary for any embodied agent that wishes to categorize rapidly and successfully: the ability to categorize, the ability to interact with the world, and the ability to decide when such interactions are and are not needed for categorization. In future work we wish to employ more sophisticated optimization methods, such as multiobjective optimization, which enable better tradeoffs between multiple fitness terms. We would also like to investigate the kinds of physical interactions generated by the controllers, and how such actions differ based on different circumstances. Further, we wish to investigate how the successful controllers described here gradually acquired DCP, ACP, and/or whether the acquisition of one scaffolded the subsequent acquisition of the other. Finally, we wish to explore whether such dynamics relate to how humans categorize.

Acknowledgments. This work was supported by the National Science Foundation under projects PECASE-0953837 and INSPIRE-1344227.

References

1. Beer, R.D.: The dynamics of active categorical perception in an evolved model agent. Adapt. Behav. **11**(4), 209–243 (2003)
2. Bongard, J.: The utility of evolving simulated robot morphology increases with task complexity for object manipulation. Artif. Life **16**(3), 201–223 (2010)
3. Brooks, R.A.: Elephants don't play Chess. Robot. Auton. Syst. **6**(1), 3–15 (1990)
4. Hauser, H., Ijspeert, A.J., Füchslin, R.M., Pfeifer, R., Maass, W.: Towards a theoretical foundation for morphological computation with compliant bodies. Biol. Cybern. **105**(5–6), 355–370 (2011)
5. Lungarella, M., Sporns, O.: Information self-structuring: key principle for learning and development. In: Proceedings of the International Conference on Development and Learning, pp. 25–30. IEEE (2005)
6. Paul, C.: Morphological computation: a basis for the analysis of morphology and control requirements. Robot. Auton. Syst. **54**(8), 619–630 (2006)
7. Pfeifer, R., Bongard, J.: How the Body Shapes the Way We Think: A New View of Intelligence. MIT Press, Cambridge (2006)
8. Pfeifer, R., Iida, F., Lungarella, M.: Cognition from the bottom up: on biological inspiration, body morphology, and soft materials. Trends Cogn. Sci. **18**(8), 404–413 (2014)
9. Schmidt, M., Lipson, H.: Age-fitness pareto optimization. In: Riolo, R., McConaghy, T., Vladislavleva, E. (eds.) Genetic Programming Theory and Practice VIII, vol. 8, pp. 129–146. Springer, New York (2011)
10. Seung, H.S., Opper, M., Sompolinsky, H.: Query by committee. In: Proceedings of the Fifth Annual Workshop on Computational Learning Theory, pp. 287–294. ACM (1992)
11. Tuci, E., Massera, G., Nolfi, S.: Active categorical perception of object shapes in a simulated anthropomorphic robotic arm. IEEE Trans. Evol. Comput. **14**(6), 885–899 (2010)
12. Zieba, K., Bongard, J.: An embodied approach for evolving robust visual classifiers. In: Proceedings of the Genetic and Evolutionary Computation Conerence, pp. 201–208. ACM (2015)

Community Structure Detection for the Functional Connectivity Networks of the Brain

Rodica Ioana Lung[1], Mihai Suciu[1(✉)], Regina Meszlényi[2,3], Krisztian Buza[3], and Noémi Gaskó[1]

[1] Babeş-Bolyai University, Cluj-napoca, Romania
mihai-suciu@cs.ubbcluj.ro
[2] Department of Cognitive Science, Budapest University of Technology and Economics, Budapest, Hungary
[3] Brain Imaging Center, Research Center for Natural Sciences, Hungarian Academy of Sciences, Budapest, Hungary

Abstract. The community structure detection problem in weighted networks is tackled with a new approach based on game theory and extremal optimization, called Weighted Nash Extremal Optimization. This method approximates the Nash equilibria of a game in which nodes, as players, chose their community by maximizing their payoffs. After performing numerical experiments on synthetic networks, the new method is used to analyze functional connectivity networks of the brain in order to reveal possible connections between different brain regions. Results show that the proposed approach may be used to find biomedically relevant knowledge about brain functionality.

Keywords: Community structure · Weighted networks · Game theory · Brain functional connectivity networks

1 Introduction

During the last years, more and more computational methods for community structure detection focus on dealing with very large datasets [8], while small sets with more challenging structures are often ignored. However, many applications require 'sensible' algorithms that can reveal the inner structure in networks for which the architecture is not obvious and for which there is no available information about the real structure. An example of such networks are the functional connectivity networks of the brain, usually constructed from raw fMRI data. A growing interest in brain research is reflected by recent American and European large scale research projects that are dedicated to study the brain and its disorders[1]. As the expected impact of these projects may be compared to that of the

[1] Such projects include the BRAIN Initiative (http://www.braininitiative.nih.gov/, April, 2016) and the European Human Brain Project (https://www.humanbrainproject.eu/, April, 2016).

J. Handl et al. (Eds.): PPSN XIV 2016, LNCS 9921, pp. 633–643, 2016.
DOI: 10.1007/978-3-319-45823-6_59

celebrated Human Genome Project, we anticipate an increased need for methods that allow exploratory analysis and predictions based on datasets describing the dynamics of the brain, such as fMRI data. As different parts of the brain work in collaboration, community detection has a high potential to reveal relevant information about brain functionality which may contribute to better understanding the mechanisms of the brain and brain disorders.

In this context we are proposing exploring such networks with a new game theoretic tool that uses the concept of Nash equilibria within an extremal optimization algorithm to identify possible communities. We show that this method is capable to identify inner network connections that are not grasped by other methods.

2 Weighted Nash Extremal Optimization

The community structure detection problem consists in finding groups of nodes in a network that are more linked to each other than to the rest of the network [4]. In spite of the fact that there are many computational approaches to this problem, there still does not exist a formal definition for the community structure that is universally accepted to encompass the simple description from above. In this paper we explore the use of the Nash equilibrium concept from game theory as a possible characterization for the community structure with non-overlapping nodes in weighted, undirected, networks.

2.1 The Community Structure Detection Game

Consider a weighted graph $G = (V, E)$ where V is the set of nodes, $V = \{i\}_{i=\overline{1,n}}$, and E the set of edges. Let $W = \{w_{ij}\}_{i,j \in V}$ be the set of weights w_{ij} associated to each edge $e_{ij} = (i, j)$ from E. In this work we will consider positive weights.

Let game $\Gamma = (N, S, U)$ be composed of:

- the set of players $N = V$, i.e. each node is the network G is a player in game Γ;
- the set of strategy profiles $S = S_1 \times S_2 \times \ldots \times S_n$, where $\times$ represents the cartesian product, and S_i is the set of strategies of player i. In Γ, S_i represents communities in G, i.e. each node has to chose a community; an element $s \in S$ is called a strategy profile having the form $s = (s_1, s_2, \ldots, s_n)$, where s_i represents the community chosen by player i.
- the payoff functions $U = \{u_i\}_{i \in N}$, where $u_i : S \to \mathbb{R}$, computed as the contribution of a node to its community [16]:

$$u_i(s_1, s_2, \ldots, s_n) = f(s_i) - f(s_i \backslash \{i\}), \quad (1)$$

where

$$f(C) = \frac{\sum_{i,j \in C} w_{ij}}{\sum_{i \in C, j \in V} w_{ij}} \quad (2)$$

is the fitness of community C. Thus, the payoff of a node i depends on its strategy, as well as the strategies of the other nodes that have chosen the same community as i, and the nodes that did not choose the community of i.

A strategy profile s^* is a Nash Equilibrium if $u_i(s_i, s^*_{-i}) \leq u_i(s^*_i)$, $\forall i \in N$ and $\forall s_i \in S_i$, where $(s_i, s^*_{-i}) = (s^*_1, \ldots, s_i, \ldots, s^*_n)$ is the strategy profile in which all players chose their strategies from s^*, except player i that chooses s_i. A Nash equilibrium (NE) of game Γ is a partition over the set of nodes $N = V$ such that no node can increase its payoff by unilateral deviation. We can consider this also as an alternate definition for the community structure of a network; to test this hypothesis we use numerical experiments performed on benchmarks with known community structures.

NEs of a game can be computed with heuristic methods by using the Nash ascendancy relation between strategy profiles [9] that counts the number $t_N(s, q)$ of players that can improve their payoffs by unilateral deviation from one strategy profile s to q:

$$t_N(s, q) = card\{i \in N | u_i(s) < u_i(q_i, s_{-i}), q_i \neq s_i\}. \tag{3}$$

Strategy s is better in Nash sense than strategy q (or strategy s Nash ascends strategy q) if $t_N(s, q) < t_N(q, s)$. A strategy profile s^* is non-dominated with respect to the Nash ascendancy relation if $\nexists q \in S$ such that q Nash ascends s. It is known that the set of Nash non-dominated solutions is equal to the set of Nash equilibria of the game [9].

2.2 Method

The community detection problem in unweighted networks has been previously approached by an extremal optimization algorithm based on the game theoretic approach described above [16]. Another similar extremal optimization approach that maximizes the modularity function [12] can be found in [10]. In this paper we present a new extremal optimization variant, called Weighed Nash Extremal Optimization (W-NEO), designed to capture the community structure in weighted networks.

An extremal optimization (EO) algorithm [2] typically uses one individual $s = (s_1, \ldots, s_n)$ to search the space and preserves each iteration the best solution found up to that moment, s_{best}. A fitness value is assigned to each component s_i in s, $i = \overline{1, n}$. Each iteration, the component s_j having the worst fitness value is randomly re-initialized. If the new individual is better than s_{best}, it will replace it. If not, the search continues from the new value of s.

W-NEO extends EO by evolving a population of pairs (s, s_{best}) that search independently for the Nash equilibria of game Γ by using the Nash ascendancy relation.

Encoding. Individuals s (and s_{best}) are represented as strategy profiles of game Γ, i.e. integer vectors; a component i represents the community of node i. Communities are numbered from 0 to a maximum value n_{comm}. The value of n_{comm} differs between EO pairs (s, s_{best}), it is set at the beginning of the search, and

Algorithm 1. W-NEO step

1: For current configuration s evaluate $u_i(s)$, the payoff function corresponding to each node $i \in \{1, \ldots, n\}$.
2: find the k worst components in s and replace them with a random value;
3: **if** (s ***Nash ascends*** s_{best}) **then**
4: set $s_{best} := s$.
5: **end if**

Algorithm 2. *Weighted Nash Extremal Optimization*

1: Randomly initialize and evaluate *popsize* pairs of configurations (s, s_{best}).
2: Compute k_{Nash};
3: Set $k_1 = k_{Nash}$;
4: **repeat**
5: Update $k = \min\{k_{Nash}, [k_1 + 2\frac{nr.it}{MaxGen}(1 - k_1)]\}$;[2]
6: Apply a W-NEO step on each (s, s_{best}) pair;
7: Update k_{Nash};
8: **until** the maximum number of generation is reached;
9: Return s_{best} with highest fitness Φ.

[2] *nr.it* is the iteration number, and $[\cdot]$ represents the integer part.

takes values between a minimum and maximum expected number of communities c_{min} and c_{max}.

Fitness Assignment. For each individual, the payoff functions u_i, $i = \overline{1,n}$ (1) are computed and used to compare nodes within a W-NEO iteration. To compare s and s_{best}, a different fitness function, Φ, is used, computed as:

$$\Phi(s) = \sum_{i=1}^{n} u_i(s) \cdot w_i^{(in)}, \tag{4}$$

where $w_i^{(in)}$ is the sum of the weights of the links node i has with other nodes in its community. W-NEO uses fitness Φ as an alternative to the modularity function [12].

Extremal Optimization. Several EO variants proposed for the community structure problem in unweighted networks extend the typical EO by modifying more than one node during an iteration. This number, denoted by k, can be fixed, or set adaptively. In the first half of the search, Noisy EO [10] linearly decreases the value of k from a given value to 1, whereas in its second phase, k is kept constant. MNEO [16] decreases k exponentially throughout the search. The recommended initial value for k is 10 % of the number of nodes, which is a parameter for both methods.

W-NEO uses an adaptive mechanism to update k values by combining the linear decrease of NoisyEO with the $t_N(s, s_{best})$ operator used by the Nash ascendancy relation (3). Each iteration, the number k_{Nash} is computed as the maximum value of $t_N(s, s_{best})$ in that iteration. The number k of nodes changed in one iteration is computed as the minimum value between k_{Nash} and the one

corresponding to NoisyEO. The first k_{Nash} value, denoted by k_1, is computed immediately after the initialization of the population and it is used to set the initial value in the equation that decreases k linearly. Thus, W-NEO does not need a parameter for the initial value of k. The outline of W-NEO is presented in Algorithm 2 and a W-NEO step is detailed in Algorithm 1.
Parameters. W-NEO uses the following parameters:

- Population size - $popsize$;
- Maximum number of generations - $MaxGen$;
- Expected minimum and maximum number of communities.

2.3 Numerical Experiments - Synthetic Benchmarks

The performance of W-NEO is tested on a set of synthetic benchmarks and compared with the results obtained by other methods that compute the community structure in weighted networks.

Benchmark. The LFR benchmark [5] is used to evaluate the performance of W-NEO in a first phase. Three sets of networks and corresponding community structures were generated[2], with parameters presented in Table 1. The most important parameters are μ, representing the ratio of links a node has outside its community. A μ value of 0.5 indicates that the node has an equal number of links in its community and outside, and a μ value of 0.6 that the node has more links outside than inside. The μ_w parameter is similar, taking weights into account. $\mu_w = 0.6$ means that the sum of weights of the links the node has in its community is 0.4 of the total strength of that node.

Table 1. LFR benchmarks. 30 networks were generated for each μ and μ_w value. κ is the average node degree, κ_{max} is the maximum degree, and τ_1 and τ_2 are the minus exponents for the degree sequence and for the community size distribution, respectively.

Name	N	κ	κ_{max}	τ_1	τ_2	μ	μ_w	Comm. size
LFR 128	128	20	50	2	1	0.3,0.4,0.5,0.6	0.1–0.6	[10,50]
LFR 1000 S	1000	20	50	2	1	0.3,0.4,0.5,0.6	0.1–0.6	[10,50]
LFR 1000 B	1000	20	50	2	1	0.3,0.4,0.5,0.6	0.1–0.6	[20,100]

The most challenging sets in this benchmark are the small ones (128 nodes), with μ and μ_w values above 0.4, having the least well defined structures. The bigger networks may seem more challenging because of their size, but they all present a well defined community structure even for $\mu, \mu_w = 0.5$, because of the greater number of communities in which the outside links of a node can be distributed, making the difference between the number of links inside its

[2] By using the code available at https://sites.google.com/site/andrealancichinetti/software, accessed May, 2015.

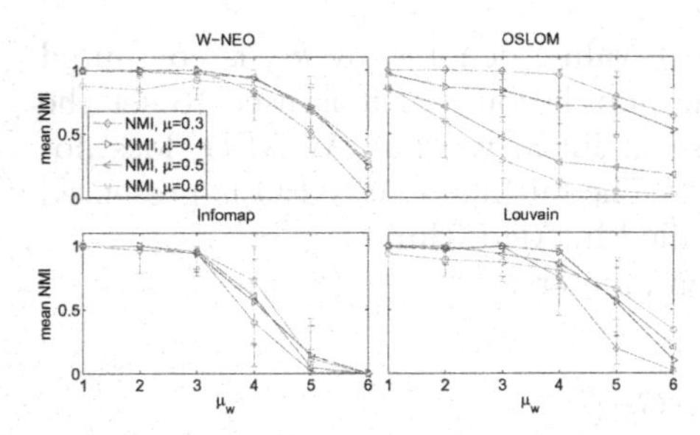

Fig. 1. Average NMI values for the LFR sets with 128 nodes. Wilcoxon sign-rank tests results are presented in Table 2

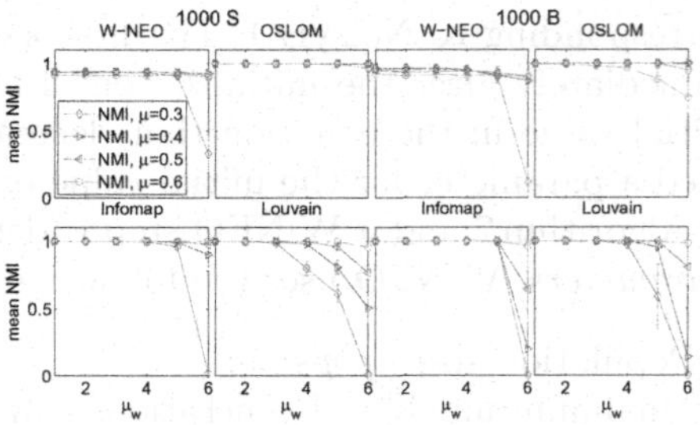

Fig. 2. Average NMI values for the 1000 nodes sets. Wilcoxon sign-rank tests results are presented in Table 2.

community and the number of links in any other community bigger than in the case of networks with 128 nodes and smaller number of communities.

Performance Evaluation. Results are evaluated by using the normalized mutual information indicator (NMI) [6]. A NMI of 1 indicates identical community structures. When two different community structures are compared to the real structure, the one having the higher NMI value is considered better.

Comparisons with Other Methods. The results obtained by W-NEO are compared with those obtained by three state of art methods: Oslom [7], Infomap [14], and Louvain [1]. Differences in median NMI values obtained by each method for each set of 30 networks are evaluated by using the Wilcoxon sign-rank test with a confidence level of 0.05.

Parameter Settings. W-NEO parameters are: population size, minimum and maximum expected number of communities, and maximum number of generations. Considering that (s, s_{best}) pairs evolve independently, the effect of size of the of the population is the usual one, in this case using a larger population being equivalent with performing multiple independent runs with smaller populations. The expected number of communities influences the results in a similar

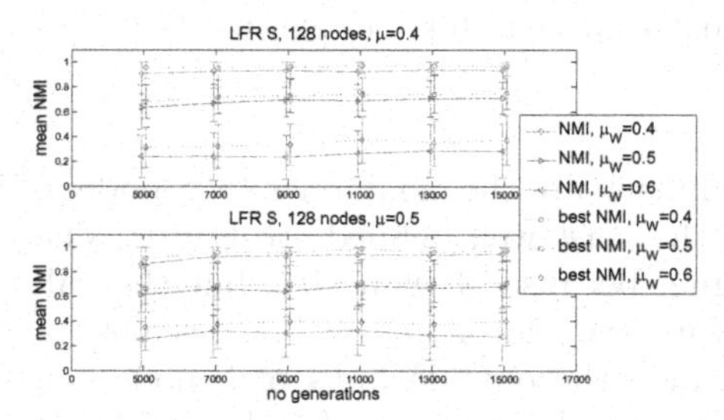

Fig. 3. NMI values of W-NEO for the LFR 128 nodes set, and different $MaxGen$ values.

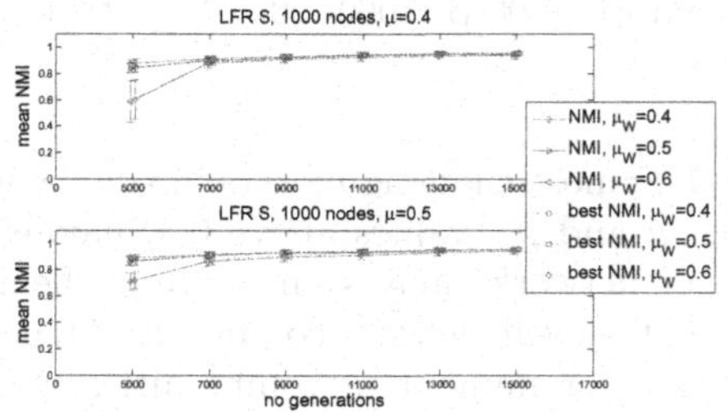

Fig. 4. NMI values of W-NEO for two LFR sets with 1000 nodes, and different $MaxGen$ values.

Table 2. Wilcoxon sign -rank test results. A • indicates that the corresponding method provided the best results. If there are more methods with results that are not statistically different from the best one, they are also marked with a •.

		128 nodes				1000 nodes S				1000 nodes B			
μ	μ_W	W -NEO	Oslom	Infomap	Louvain	W -NEO	Oslom	Infomap	Louvain	W -NEO	Oslom	Infomap	Louvain
0.3	0.1	•	•	•	•	-	•	•	•	-	•	•	•
	0.2	•	•	•	•	-	•	•	•	-	•	•	•
	0.3	•	•	-	•	-	•	•	-	-	•	•	•
	0.4	-	•	-	-	-	•	•	-	-	•	•	-
	0.5	-	•	-	-	-	•	-	-	-	•	-	-
	0.6	-	•	-	-	-	•	-	-	-	•	-	-
0.4	0.1	•	-	•	•	-	•	•	•	-	•	•	•
	0.2	•	-	•	•	-	•	•	•	-	•	•	•
	0.3	•	-	-	•	-	•	•	•	-	•	•	•
	0.4	•	-	-	•	-	•	•	-	-	•	•	•
	0.5	•	•	-	-	-	•	-	-	-	•	•	-
	0.6	-	•	-	-	-	•	-	-	-	•	-	-
0.5	0.1	•	-	•	•	-	•	•	•	-	•	•	•
	0.2	•	-	•	•	-	•	•	•	-	•	•	•
	0.3	•	-	•	•	-	•	•	•	-	-	•	•
	0.4	•	-	-	-	-	•	•	•	-	-	•	•
	0.5	•	-	-	-	-	•	•	-	-	-	•	•
	0.6	•	-	-	-	-	•	-	-	-	-	•	•
0.6	0.1	-	-	•	-	-	•	•	•	-	•	-	-
	0.2	-	-	•	-	-	-	•	•	-	-	•	•
	0.3	•	-	•	-	-	-	•	-	-	-	•	•
	0.4	•	-	-	-	-	-	•	•	-	-	•	-
	0.5	•	-	-	•	-	-	•	•	-	-	•	-
	0.6	•	-	-	•	-	-	•	-	-	-	•	•

manner. For these numerical experiments, the minimum and maximum number of communities was set such that approx. 20 % of the population has assigned the real number of communities. The population size was set to 30. Because the maximum number of generations indirectly influences the results, as it related to the value of k (Algorithm 2), several values are tested for this parameter.

Results and Discussion. Numerical results obtained on the synthetic benchmarks are presented as error-bars in Figs. 1 and 2 ($MaxGen = 10\,000$). The results of the Wilcoxon sign-rank test are presented in Table 2. For the small networks, the results provided by W-NEO are in some cases the best compared with the other methods, and in most cases as good as the others. For the 1000 nodes sets, W-NEO results are statistically different than all the others, but with NMI values greater than 0.9 in almost all cases (Fig. 2).

W-NEO Parameters. Figures 3 and 4 illustrate the variation of average NMI values with the maximum number of generations. For each set two values are represented: the average NMI of the individuals having the best Φ value in each run and the average NMI of the individual with the best NMI in the final population. The small differences between the two values indicate the the function Φ can be considered as an efficient fitness function for assessing the quality of a community structure.

3 Brain Functional Connectivity Networks

In order to examine if W-NEO can detect communities of the brain, we used a public resting-state fMRI database from the 1000 Functional Connectomes Project, Addiction Connectome Preprocessed Initiative. In our study we used the MTA 1 dataset with the ANTS registered, no scrubbing, no global signal regression preprocessing pipeline.[3] The dataset contains 126 subjects' resting-state data, based on which, and an atlas of 90 functional regions of interest (ROI) [15], we calculated the Pearson correlation between the activities of the ROIs.[4] We calculated an "averaged" network, in which nodes correspond to ROIs, denoted as $r_1, r_2, \ldots$, and the weight of each connection $\{r_i, r_j\}$ is the average of the correlations between r_i and r_j over all the subjects. Only positive correlations with values above 0.35 were considered.

For each subject, information about cannabis usage and the childhood diagnosis for Attention Deficit Hyperactivity Disorder (ADHD) is available. Therefore, additionally to the "averaged" network, we considered four disjoint groups of subjects: (A) the healthy subjects (no cannabis usage, no ADHD), (B) cannabis users without ADHD, (C) ADHD patients who do not use cannabis, and (D) subjects with childhood diagnoses of ADHD who regularly use cannabis. For each of these groups, we obtained a network of ROIs. In each of these networks, we calculated the weight of the connection $\{r_i, r_j\}$ as the average of the correlations between r_i and r_j for the subjects belonging to the group.

From the community structure detection point of view, the brain functional connectivity networks proved to be challenging; performing multiple runs with the four algorithms led to different results for each run and each algorithm, with Oslom, Infomap and Louvain finding structures with maximum 3 communities. However, by setting the values for the minimum and maximum number of communities to 10 and 20, W-NEO provides structures with more communities that can be further analyzed.

Thus, after performing 30 independent runs ($MaxGen = 3000$) for each network, the resulting community structures were aggregated in the following manner: each node was placed in the same community with the node with which it was placed in the same community most of the times in the 30 runs. If there are several such nodes, one of them is selected at random. Because the resulting community structure contained many communities formed only by two nodes, a further step consisted in uniting the communities having the smallest fitness values with those with which they have the strongest link. The strength of the link between two communities is computed as the ratio between the sum of weights of the links that connect the communities and the number of nodes that link them. Communities are merged until their number equals the recommendation of the domain experts, i.e. 14.

---

[3] See http://fcon_1000.projects.nitrc.org/indi/ACPI/html/ for details.

[4] One ROI (Basal Ganglia 4) did not include meaningful measurement for any of the 126 subjects, therefore we ignored this ROI in the subsequent analysis.

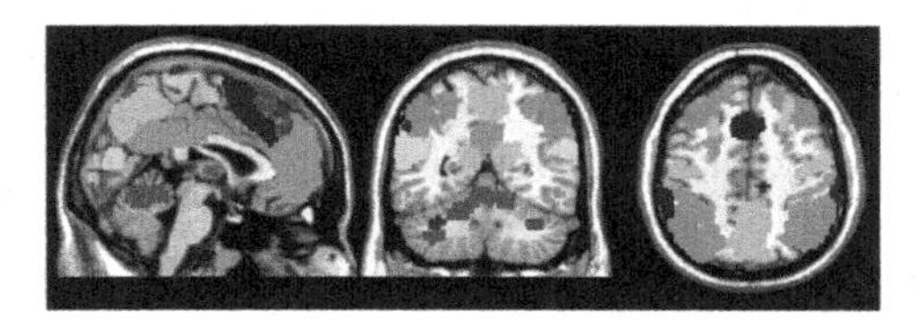

Fig. 5. Community structure of the averaged whole brain functional connectivity network.

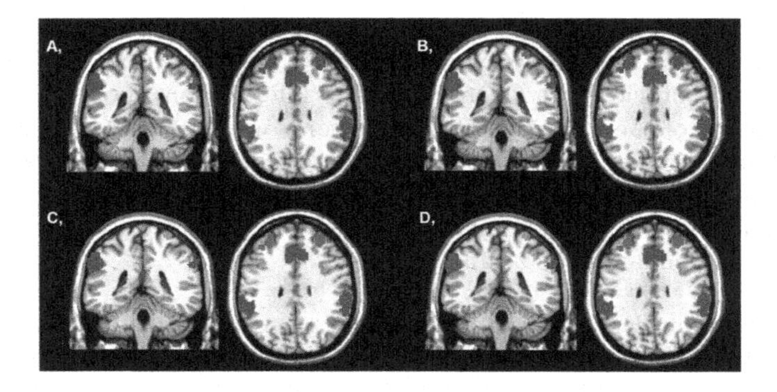

Fig. 6. Community structure of the anterior and posterior salience network in case of (A) healthy subjects, (B) cannabis users without ADHD, (C) subjects with childhood diagnoses of ADHD who does not use cannabis, (D) subjects with childhood diagnoses of ADHD who regularly use cannabis.

Results. The detected community structures (Figs. 5 and 6) are consistent with domain knowledge and they illustrate that the proposed community detection approach may be applicable to discover new insights about brain functionality and brain disorders. In particular, we examined the structure of two large communities of brain regions, the so called default mode network[5] (anterior and posterior default mode networks), and the salience network (anterior and posterior salience networks).

The role of the default mode network (DMN) in drug addiction has been shown by several studies [11,13]. In our community structures we found that the DMN is more intact (more ROIs are in the same community) in non-addicted subjects. In healthy subjects, 13 ROIs of the DMN belong to the same community, whereas we observed 11 ROIs of the DMN to be highly connected in ADHD patients. In contrast, in case of cannabis addicts, both with and without ADHD, the DMN is decomposed to several smaller communities (with less than 7 ROIs).

[5] We note that in the brain research community, the phrases *default mode network* and *salience network* are used to refer to two specific sets of strongly interconnected regions of the brain. Therefore, the *default mode network* and the *salience network* are *communities* according to the terminology used throughout this paper.

The salience network has a critical role in attention, therefore it is expected to be related to ADHD [3]. In healthy subjects and cannabis addicts without ADHD, the salience networks were found to be intact, in particular 11 and 12 ROIs were observed within the same community. However, in subjects diagnosed with ADHD, the salience network's largest community has only 7 ROIs, see Fig. 6.

4 Conclusions

The analysis of brain functional connectivity networks from the community structure point of view can offer important information about the structure and functioning of the brain. The brain networks are relatively small, with very unclear structure, not detected by existing algorithms. In this paper we propose a game theoretic approach capable to identify strong connections in these networks and construct community structures that can offer relevant knowledge about the functioning of the brain.

Acknowledgment. K. Buza was supported by the grant of the National Research, Development and Innovation Office - NKFIH PD 111710 and the János Bolyai Research Scholarship of the Hungarian Academy of Sciences. This work was also supported by a grant of the Romanian National Authority for Scientific Research and Innovation, CNCS - UEFISCDI, project number PN-II-RU-TE-2014-4-2332.

References

1. Blondel, V.D., Guillaume, J.L., Lambiotte, R., Lefebvre, E.: Fast unfolding of communities in large networks. J. Stat. Mech.: Theory Exp. **2008**(10), P10008 (2008)
2. Boettcher, S., Percus, A.: Nature's way of optimizing. Artif. Intell. **119**, 275–286 (2000)
3. Castellanos, F.X., Proal, E.: Large-scale brain systems in ADHD: beyond the prefrontal-striatal model. Trends Cogn. Sci. **16**(1), 17–26 (2012)
4. Fortunato, S.: Community detection in graphs. Phys. Rep. **486**, 75–174 (2010)
5. Lancichinetti, A., Fortunato, S.: Benchmarks for testing community detection algorithms on directed and weighted graphs with overlapping communities. Phys. Rev. E **80**, 016118 (2009)
6. Lancichinetti, A., Fortunato, S., Kertész, J.: Detecting the overlapping and hierarchical community structure in complex networks. New J. Phys. **11**(3), 033015 (2009)
7. Lancichinetti, A., Radicchi, F., Ramasco, J.J., Fortunato, S.: Finding statistically significant communities in networks. PloS One **6**(4), e18961 (2011)
8. Leskovec, J., Krevl, A.: SNAP datasets: stanford large network dataset collection, June 2014. http://snap.stanford.edu/data
9. Lung, R.I., Dumitrescu, D.: Computing nash equilibria by means of evolutionary computation. Int. J. Comput. Commun. Control **III**(Suppl. issue), 364–368 (2008)
10. Lung, R., Suciu, M., Gasko, N.: Noisy extremal optimization. Soft Comput. 1–18 (2015). http://dx.doi.org/10.1007/s00500-015-1858-3

11. Ma, N., Liu, Y., Fu, X.M., Li, N., Wang, C.X., Zhang, H., Qian, R.B., Xu, H.S., Hu, X., Zhang, D.R.: Abnormal brain default-mode network functional connectivity in drug addicts. PloS One **6**(1), e16560 (2011)
12. Newman, M.E.J.: Modularity and community structure in networks. Proc. Natl. Acad. Sci. **103**(23), 8577–8582 (2006)
13. Roberts, G.M., Garavan, H.: Evidence of increased activation underlying cognitive control in ecstasy and cannabis users. Neuroimage **52**(2), 429–435 (2010)
14. Rosvall, M., Bergstrom, C.T.: Maps of random walks on complex networks reveal community structure. Proc. Natl. Acad. Sci. **105**(4), 1118–1123 (2008)
15. Shirer, W., Ryali, S., Rykhlevskaia, E., Menon, V., Greicius, M.: Decoding subject-driven cognitive states with whole-brain connectivity patterns. Cereb. Cortex **22**(1), 158–165 (2012)
16. Suciu, M., Lung, R.I., Gaskó, N.: Mixing network extremal optimization for community structure detection. In: Ochoa, G., Chicano, F. (eds.) EvoCOP 2015. LNCS, vol. 9026, pp. 126–137. Springer, Heidelberg (2015)

Data Classification Using Carbon-Nanotubes and Evolutionary Algorithms

E. Vissol-Gaudin(✉), A. Kotsialos(✉), M.K. Massey, D.A. Zeze, C. Pearson, C. Groves, and M.C. Petty

School of Engineering and Computing Sciences, Durham University, Stockton Road, Durham DH1 3LE, UK

{eleonore.vissol-gaudin,apostolos.kotsialos,m.k.massey, d.a.zeze,christopher.pearson,chris.groves,m.c.petty}@durham.ac.uk

Abstract. The potential of Evolution in Materio (EiM) for machine learning problems is explored here. This technique makes use of evolutionary algorithms (EAs) to influence the processing abilities of an un-configured physically rich medium, via exploitation of its physical properties. The EiM results reported are obtained using particle swarm optimisation (PSO) and differential evolution (DE) to exploit the complex voltage/current relationship of a mixture of single walled carbon nanotubes (SWCNTs) and liquid crystals (LCs). The computational problem considered is simple binary data classification. Results presented are consistent and reproducible. The evolutionary process based on EAs has the capacity to evolve the material to a state where data classification can be performed. Finally, it appears that through the use of smooth signal inputs, PSO produces classifiers out of the SWCNT/LC substrate which generalise better than those evolved with DE.

1 Introduction and Background

Evolution-in-materio (EiM) is an Unconventional Computing (UC) technique which focuses on exploiting the underlying properties of materials to bring them to a computation inducing state [12]. Contrary to traditional computing with Metal-Oxide-Silicon-Field-Effect-Transistor (MOFSET) technology, where everything is designed, produced and programmed very carefully, EiM uses a bottom up approach where computation is performed by the material without having explicit knowledge of its internal properties [13].

The idea of EiM can be found in early work of Pask [2] which was concerned with growing an electrochemical ear. More recent work [21], is based on observations made when evolutionary algorithms (EAs) were used for designing electrical circuits on Field-Programmable-Gate-Arrays (FPGAs). The resulting circuit topologies were influenced by the material of the board used. Because of feedback provided by the iterative nature of stochastic optimisation interacting with the material, the identified solutions were based on the specific FPGA's properties that were unaccounted for during the board's design. EiM replaces

J. Handl et al. (Eds.): PPSN XIV 2016, LNCS 9921, pp. 644–654, 2016.
DOI: 10.1007/978-3-319-45823-6_60

the FPGAs with un-configured material systems favouring exploitation of some physical property by a search algorithm [12].

Here, using an iterative process, the material is configured until it reaches a state where a pre-specified scheme of interaction is uniquely translated as a computational input/output relationship. Viewing this iterative process as material training, this type of EiM requires the selection of finite training and verification datasets. Since the problem is about a computation, the datasets consist of known input/output pairs from its domain of definition and range, respectively. The training process requires the repetitive application of computation inputs sent to the material and measurements of its corresponding response. Measured responses are translated into computation outputs, which allows the definition of an error function. The physical property measured and the interpretation scheme of the material's response used for translating it into a computation output are pre-specified and fully known before the training process starts.

There are two types of incident signals on the material. Computation inputs, which are used to represent the arguments of a computation, and configuration inputs, which are used for changing the material's properties. Modulation of the incident signals is controlled by an error minimising optimisation algorithm, which explores the problem's search space. The search space itself, is a hybrid of the material's physical state and the subspace spanned by the independent configuration inputs. Hence, the optimisation algorithm aims at configuring the material at a particular state by finding the optimal configuration inputs producing that material state, the response of which can be uniquely translated into a computation. In effect, EiM is a bottom up approach for producing a computing device where the exact architecture, or material state, remains unknown. Reservoir computing is based on similar notions [5,10].

EiM has a broad scope and can be divided in four inter-dependant dimensions: (a) the type of material used, (b) the physical property manipulated to obtain a computation, (c) the computational problem itself and (d) the optimisation algorithm used for solving the corresponding problem. Figure 1 illustrates the basic concept.

An algorithm selects a set of configuration inputs. Computation inputs from the training dataset are sent to the material; its response is recorded for each input and is translated into a computation output. For each input/output pair, an error is calculated to allow an objective function evaluation. This objective function is minimised by a derivative-free optimisation algorithm.

In our implementation, the evolvable material is connected to a computer via an *mbed* micro-controller fixed on a custom-made motherboard. Configuration and computation input signals are constant voltage charges applied by the *mbed* to the material and outputs are direct current measurements. Voltages are sent to the material through a set of Digital-to-Analogue-Converters (DACs) fixed on the motherboard. They are connected to the array of gold micro-electrodes shown in Fig. 1 deposited on a glass slide using etch-back photolithography. The material blend is drop-deposited within a nylon washer (2.5 mm internal diameter) fixed to this platform for material/electronics interaction.

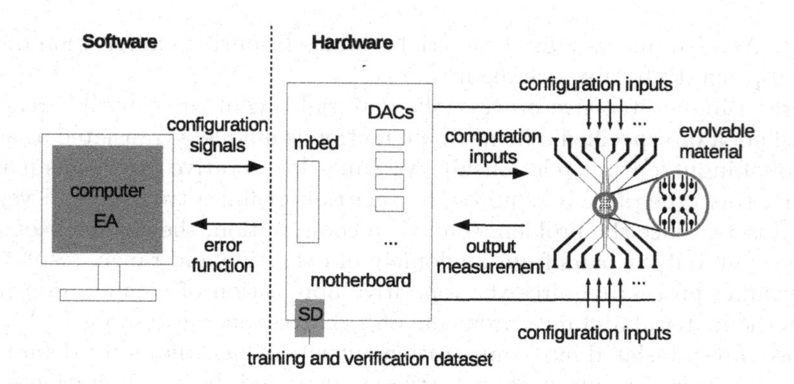

Fig. 1. EiM concept and electrode array (50 μm contacts, 100 μm pitch)

Different organic and inorganic media have been used as materials, such as slime moulds [7], bacterial consortia [1], cells (neurons) [18], liquid crystals (LC) panels [6] and nano-particles [3]. Single walled carbon nanotubes (SWCNT) based materials have shown the potential to solve computational problems [8,11,14,15,22]. In [19] it is argued that inorganic materials make a better medium for unconventional computing exploration. Following this argument, as well as results in [22], a mixture of SWCNT and LC in liquid form is used here.

These types of materials have a very complex structure and the development of analytical or stochastic models of their behaviour is very difficult. In their absence, EiM treats them as black boxes, leading to the use of derivative free population based stochastic search algorithms for solving the training problem. Here, a particle swarm optimisation (PSO) [9] and an implementation of differential evolution (DE) [17] are used, which will be referred to as EAs.

Several candidate computational problems can be used in the context of EiM. A more comprehensive review of potential problems can be found in [16]. The problem considered here is a simple binary data classification with different degrees of separation and data distributions.

2 Evolved Material

A mixture of SWCNT and LC, where nanotubes are dispersed in liquid crystals at varying concentrations, is used. SWCNT are both semiconducting and conducting; the samples used contain less than 15 % impurities (according to vendor specifications) as residual from the catalytic growth process.

It is shown in [22] that SWCNTs tend to bundle under an applied electric field, establishing a percolation path between electrodes. The greater length of these bundles or "ropes" with respect to the dimensions of LC molecules suggests that they are not highly influenced by movement of the latter. The purpose of a LC matrix is therefore to provide a fluid medium in which the SWCNTs can move in response to the field. Formation of percolation paths is variable and reconfigurable allowing the creation of complex electrical networks. This adds an extra dimension to the problem, compared to previous experiments where SWCNTs were mixed with a solid polymer [8,11].

3 The Classification Problem

Three variants of a binary data classification problem are considered based on different 2-dimensional datasets. A typical training and verification procedure is followed and the corresponding datasets have $K_t = 800$ and $K_v = 4000$ members. Figure 2(a) shows the training datasets for the separable (SC) and merged (MC) classes and (b) for the V1 class (V1C). The units of the two computation inputs are in Volts. When a particular pair is used, the two electrodes reserved to receive computation inputs are charged with the corresponding voltages. SC and MC data are organised into two different squares; the SC ones are not overlapping and are placed at a distance, whereas the MC ones overlap slightly. V1C's data are completely separable, but they are arranged diagonally so as to increase the problem's difficulty. After training using those datasets, the material must be in such a state so as to infer the class (C_1 or C_2) for any input pair randomly selected from the verification dataset. Effectively the objective is to evolve an analogue machine, capable of distinguishing the class an input belongs to.

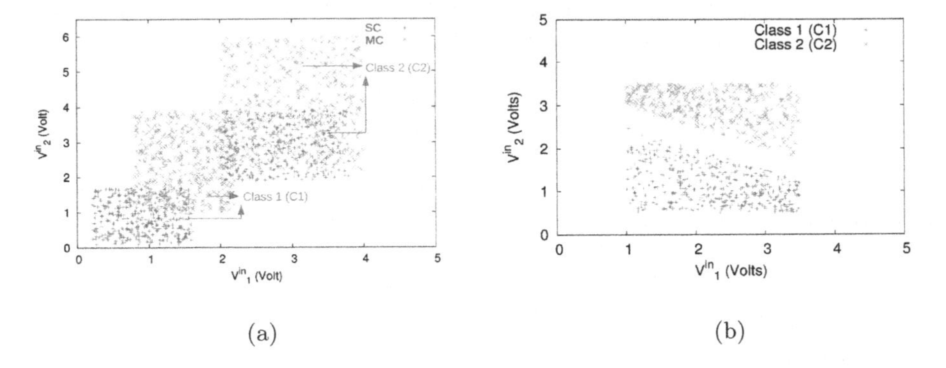

Fig. 2. (a) SC and MC and (b) V1C training datasets.

4 Problem Formulation

Evolution of such a device is formulated as an optimisation problem. There are sixteen connections on the micro-electrode array twelve of which are used. Two of those are used for sending computation inputs as voltage pulses of amplitude $\mathbf{V}^{in} = (V_1^{in}, V_2^{in})$ and eight are used for sending configuration voltages as pulses within the range $V_j \in [V_{\min}, V_{\max}]$, $j = 1, \ldots, 8$. The remaining two connections are reserved for measuring outputs currents $\mathbf{I} = (I_1, I_2)$ (A) when the material has been sent $\mathbf{V}^{in}$ and is under charge of the V_j's.

By considering as a decision variable only the possible locations where the two components of $\mathbf{V}^{in}$ are applied and using a simple increasing index scheme for assigning configuration voltages (e.g. if V_1^{in} is assigned to electrode 3 and V_2^{in} is assigned to 5, then the following assignment for the configuration inputs takes place: $V_1 \rightarrow 1$, $V_2 \rightarrow 2$, $V_3 \rightarrow 4$ $V_4 \rightarrow 6$, $V_5 \rightarrow 7$, $V_6 \rightarrow 8$, $V_7 \rightarrow 9$ $V_8 \rightarrow 10$) then there are ${}^{10}P_2 = 90$ possible connection assignments. A continuous variable $p \in [1, 90]$ is defined and updated by the EA used rounded to the nearest integer during the iterations.

The optimisation problem's vector of decision variables is defined as

$$\mathbf{x} = [V_1 \dots V_8 \; R \; p]^T \tag{1}$$

where R is a scaling factor. It is for a specific electrode assignment p and set of configuration voltages V_j, that the material's response to an input $\mathbf{V}^{in}$ is recorded. The response is a pair of measurements $\mathbf{I} = (I_1, I_2)$ (A) of the direct current at the two output locations, which are the basis of a comparison scheme using R for deciding the class $\mathbf{V}^{in}$ belongs to.

Let $\mathbf{I}^{(k)}$ denote the pair of direct current measurements taken when input data $\mathbf{V}^{in}(k)$ from class C_i, $i = 1$ or $i = 2$, are applied *while* the material is subjected to configuration voltages $V_j^{(k)}$. $\mathbf{V}^{in}(k)$ and $V_j^{(k)}$ are applied according to electrode assignment number $p^{(k)}$ and scaling factor $R^{(k)}$ is used. Also, let $C(\mathbf{V}^{in}(k))$ denote $\mathbf{V}^{in}(k)$'s real class and $C_M(\mathbf{V}^{in}(k), \mathbf{x})$ the material's assessment of it calculated according to the following rule:

$$C_M(\mathbf{V}^{in}(k), \mathbf{x}) = \begin{cases} C_1 \text{ if } I_1(k) > R I_2(k) \\ C_2 \text{ if } I_1(k) \leq R I_2(k). \end{cases} \tag{2}$$

For every training pair of data $\mathbf{V}^{in}(k)$, $k = 1, \dots, K_t$ the error from translating the material response according to rule (2) is

$$\epsilon_{\mathbf{x}}(k) = \begin{cases} 0 \text{ if } & C_M(\mathbf{V}^{in}(k), \mathbf{x}) = C(\mathbf{V}^{in}(k)) \\ 1 \text{ otherwise.} \end{cases} \tag{3}$$

The mean total error is given by

$$\Phi_e(\mathbf{x}) = \frac{1}{K_t} \sum_{k=1}^{K_t} \epsilon_{\mathbf{x}}(k). \tag{4}$$

Two penalty terms are added to (4), H and U. $H(\mathbf{x})$ penalises solutions with high configuration voltages and is given by

$$H(\mathbf{x}) = \frac{\sum_{j=1}^{8} V_j^2}{8 V_{max}^2}. \tag{5}$$

The rationale behind this penalisation is that incremental and generally low levels of configuration voltages are preferable. Solutions where high $V_j^{(k)}$ are applied can destroy material structures favourable to the problem formed during evolution. On the other hand, solutions that render the material unresponsive need to be avoided. A measure of such unresponsiveness is calculated at the end of each search iteration ι, where a sample equal to the population size S of error function evaluations is available. Let $\sigma^2_{o,\iota}$ denote the variance of $\Phi(\mathbf{x})$ and $\sigma^2_{V,\iota}$ the variance of $\sum_{j=1}^{8} V_j^2$ at iteration ι. A value of $\sigma^2_{o,\iota}$ close to zero indicates a non-responsive material and the penalty term takes the form

$$U_\iota = \left(1 - \frac{\sigma^2_{o,\iota}}{\sigma^2_{V,\iota}}\right)^2. \tag{6}$$

Hence, the total objective function $\Phi_s(\mathbf{x})$ for an arbitrary individual s at iteration ι is given by

$$\Phi_s(\mathbf{x}) = \Phi_e(\mathbf{x}) + H(\mathbf{x}) + U_\iota. \tag{7}$$

U_ι aims at leading the optimisation away from material states where the same response is given for different inputs.

The optimisation problem to be solved is that of minimising (7) for a population of size S, subject to voltage bound constraints $V_j \in [V_{\min}, V_{\max}]$, $R > 0$, electrode assignment p and classification rule (2). $V_{\min} = 0$ Volts and for the SC problem $V_{\max} = 4$ Volts whereas for the MC and V1C $V_{\max} = 7$ Volts.

Two different stochatic optimisation algorithms are used for solving this problem, differential evolution (DE) [20] and particle swarm optimisation (PSO) [4]. A constricted version of PSO with parameters taken from [9] is implemented. The DE algorithm implementation uses the parameters suggested in [17]. A population size of $S = 10$ is used for DE and PSO.

5 Results and Discussion

The first column of Table 1 presents the minimum error Φ^*_e achieved during training. Once training is terminated, verification is performed on the trained material by applying back the optimal solution achieved along with the previously unused verification data. The same verification procedure is repeated ten times. The other four columns of Table 1 refer to results of these runs. $\Phi^*_{e,v}$ is the minimum error, $\Phi^w_{e,v}$ the worst, $\overline{\Phi}_{e,v}$ the average and $\sigma^2_{\Phi_{e,v}}$ the variance. For both DE and PSO, the penalty terms $H(\mathbf{x})$ and U_ι are not included and the classification error for Φ_e is given, for the sake of brevity.

Table 1 shows that for all problems, except outliers, and both algorithms, the observed error increase at the verification phase is $\Phi^*_e - \Phi^*_{e,v} < 2.125\,\%$. This indicates that the material's behaviour is consistent and generalises well as a classifier. Solutions obtained during training using DE can be better than those of PSO, especially for the SC and V1C datasets. However, PSO outperforms DE with respect to consistency across experiments and generalisation of the solution. This can also be observed for the MC dataset where DE obtains both the smallest and largest error for verification ($\Phi^*_{e,v} = 4.37\,\%$ and $18.25\,\%$ respectively) whilst variance of PSO verification tests tends to be lower.

Table 1. Training and verification errors for SC, MC and V1C problems.

SC experiments	Φ_e^* (%)	$\Phi_{e,v}^*$ (%)	$\Phi_{e,v}^w$ (%)	$\overline{\Phi}_{e,v}$ (%)	$\sigma^2_{\Phi_{e,v}}$
PSO 1SC	1.3	1.675	2.35	2.0375	0.0527
PSO 2SC	1.6	2.125	3.2	2.6175	0.1277
PSO 3SC	1.3	1.975	2.45	2.25	0.0305
DE 1SC	0.7	1.05	1.625	1.3975	0.03193
DE 2SC	10.4	16.325	18.5	17.4035	0.3652
DE 3SC	1.6	1.675	2.55	2.185	0.06565
MC experiments	Φ_e^* (%)	$\Phi_{e,v}^*$ (%)	$\Phi_{e,v}^w$ (%)	$\overline{\Phi}_{e,v}$ (%)	$\sigma^2_{\Phi_{e,v}}$
PSO 1MC	5.8	6.6	7.075	6.815	0.0171
PSO 2MC	5.2	6.325	8.8	7.7225	0.648
PSO 3MC	5.7	7.825	9.025	8.5975	0.1184
DE 1MC	3.4	3.975	4.625	4.38	0.0439
DE 2MC	6.4	7.525	8.95	8.145	0.1739
DE 3MC	5.7	18.25	19.425	18.8375	0.1321
V1C experiments	Φ_e^* (%)	$\Phi_{e,v}^*$ (%)	$\Phi_{e,v}^w$ (%)	$\overline{\Phi}_{e,v}$ (%)	$\sigma^2_{\Phi_{e,v}}$
PSO 1V1C	2.7	3.975	5.175	4.6525	0.1318
PSO 2V1C	2.6	3.5	4.25	3.8625	0.0559
PSO 3V1C	1.1	2.525	3.375	2.915	0.063
DE 1V1C	1.3	2.325	2.725	2.4975	0.016
DE 2V1C	1.7	3.125	4.00	3.4975	0.071
DE 3V1C	0.007	4.55	6.2	5.575	0.2617

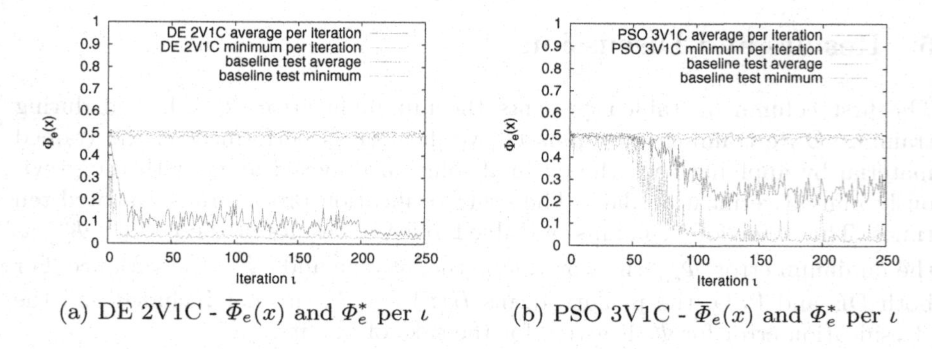

(a) DE 2V1C - $\overline{\Phi}_e(x)$ and Φ_e^* per ι (b) PSO 3V1C - $\overline{\Phi}_e(x)$ and Φ_e^* per ι

Fig. 3. Convergence patterns for training the material based on the V1C data.

Figure 3 shows the convergence pattern of the error for DE and PSO and is representative of the 18 experiments in Table 1. The baseline tests were performed using samples containing only LCs as material, without any SWCNTs. When DE is used, the material adapts within few iterations; subsequently the

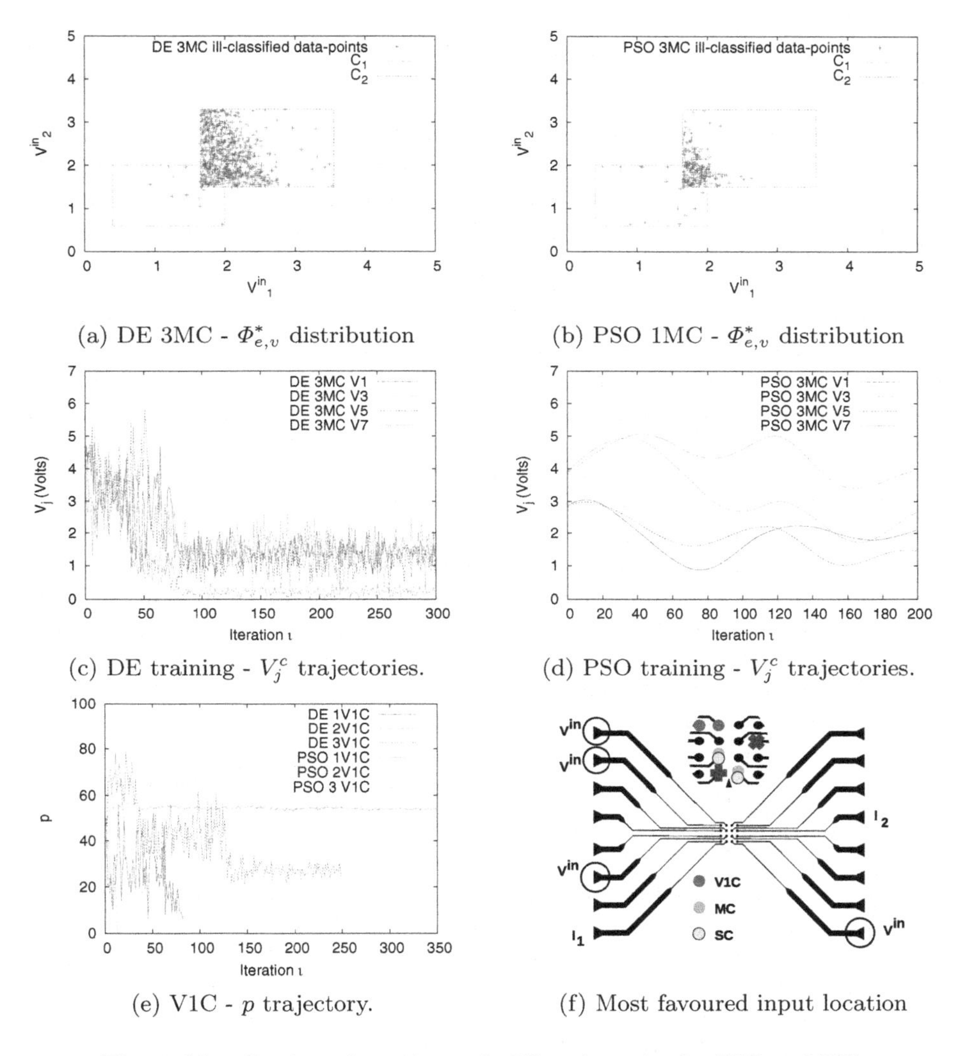

(a) DE 3MC - $\Phi^*_{e,v}$ distribution

(b) PSO 1MC - $\Phi^*_{e,v}$ distribution

(c) DE training - V_j^c trajectories.

(d) PSO training - V_j^c trajectories.

(e) V1C - p trajectory.

(f) Most favoured input location

Fig. 4. Visualisation of p and sample V_j^c trajectories for PSO and DE.

algorithm spends more iterations exploiting the minimum found. On the other hand, the PSO algorithm achieves better results at a later stage exploring more the search space.

Figures 4(a) and (b) present the verification error distribution of the 3rd runs of DE and PSO, respectively, using the MC dataset. Both converged to solutions with the same training error $\Phi^*_e = 5.7\,\%$, but with different $\overline{\Phi}_{e,v}$. The overlapping area of the two classes forms the core of the points that are erroneously classified. However, the better generalisation property of the PSO solution compared to that produced by DE is evident, as the errors outside the overlap are fewer.

When the DE solution is applied, the errors are far more widely dispersed and densely distributed into the area of C_2, making a poor classifier out of the material.

Figures 4(c) and (d) show the distinctive difference between the two algorithm's configuration voltages' trajectories, averaged over S, per iteration. It can be seen that the search performed by DE is more noisy even when the algorithm aims to exploit a minimum. On the other hand, PSO's exploration of the search space is based on smoother inputs. Figure 4(e) depicts the convergence trajectory of p for all experiments using the V1C dataset. Convergence is not towards the same value of p, but resulting input pin location is similar. Figure 4(f) presents the corresponding mapping of p with regard to input location on the micro-electrode array for the optimal solutions of the three problems. Experiments resulting to errors between 4–10 % for Φ_e^* and $\Phi_{e,v}^*$ tend to have a p corresponding to the most favoured locations shown in Fig. 4(f).

Our current hypothesis is that the poorer generalisation of the solutions obtained by DE is due to the pattern of average configuration voltages per iteration. The PSO algorithms smoother trajectories of V_j^c build structures inside the material, reinforcing responses minimising the classification error. The noisy V_j^c applied by DE appears to make the formation of such structures more difficult. Over the different experiments, DE is less consistent in its performance; exploration of the search space by the PSO algorithm results in better conductive circuit formation within the material. This hypothesis needs to be supported by more experiments and evidence, such as image analysis of the material before and after training.

6 Conclusion

This paper has presented the results of an investigation on evolution in materio for a mixture of single walled carbon nanotubes and liquid crystals. Under the influence of different levels of voltage applied at various locations of its body, conductive networks are formed by the nanotubes. Three simple classification problems are considered and training of the material as a data classifier is formulated as an optimisation problem. Results obtained with training and verification datasets are reported, showing that the solution can perform classification for similar problems with different instances. The stronger exploration element of PSO and the smoother input signals sent appear to result to classifiers that generalise better.

This is quite a new area of research and many issues need to be addressed. A more detailed investigation needs to be performed on the optimisation algorithms used and the impact of their search pattern on the solutions' quality. More recent variants of evolution-inspired algorithms need to be implemented as well. The impact of the SWCNT and LC concentration in the mix needs to be evaluated. Finally, more complicated problems will be considered and it would be very interesting to observe the material structure patterns formed for this purpose in each particular case.

References

1. Amos, M., Hodgson, D., Gibbons, A.: Bacterial self-oranisation and computation. Int. J. Unconventional Comput. **3**(3), 199–210 (2007)
2. Bird, J., Di Paolo, E.: Gordon pask and his maverick machines (Chap. 8). In: Husbands, P., Holland, O., Wheeler, M. (eds.) The Mechanical Mind in History, pp. 185–211. The MIT Press, Cambridge (2008)
3. Bose, S., Lawrence, C., Liu, Z., Makarenko, K., van Damme, R., Broersma, H., van der Wiel, W.: Evolution of a designless nanoparticle network into reconfigurable boolean logic. Nat. Nanotechnol. (2015)
4. Eberhart, R.C., Kennedy, J.: A new optimizer using particle swarm theory. In: Proceedings of the 6th International Symposium on Micro-Machine and Human Science, New York, NY, vol. 1, pp. 39–43 (1995)
5. Goudarzi, A., Lakin, M.R., Stefanovic, D.: Reservoir computing approach to robust computation using unreliable nanoscale networks. In: Ibarra, O.H., Kari, L., Kopecki, S. (eds.) UCNC 2014. LNCS, vol. 8553, pp. 164–176. Springer, Heidelberg (2014)
6. Harding, S.L., Miller, J.F.: Evolution in materio: computing with liquid crystal. Int. J. Unconventional Comput. **3**(4), 243–257 (2007)
7. Jones, J., Whiting, J.G., Adamatzky, A.: Quantitative transformation for implementation of adder circuits in physical systems. Biosystems **134**, 16–23 (2015)
8. Kotsialos, A., Massey, M.K., Qaiser, F., Zeze, D., Pearson, C., Petty, M.C.: Logic gate and circuit training on randomly dispersed carbon nanotubes. Int. J. Unconventional Comput. **10**(5–6), 473–497 (2014)
9. Laskari, E.C., Parsopoulos, K.E., Vrahatis, M.N.: Particle swarm optimization for integer programming. In: WCCI, pp. 1582–1587. IEEE (2002)
10. Lukoševičius, M., Jaeger, H.: Reservoir computing approaches to recurrent neural network training. Comput. Sci. Rev. **3**(3), 127–149 (2009)
11. Massey, M.K., Kotsialos, A., Qaiser, F., Zeze, D.A., Pearson, C., Volpati, D., Bowen, L., Petty, M.C.: Computing with carbon nanotubes: optimization of threshold logic gates using disordered nanotube/polymer composites. J. Appl. Phys. **117**(13), 134903 (2015)
12. Miller, J.F., Downing, K.: Evolution in materio: looking beyond the silicon box. In: Proceedings of the 2002 NASA/DoD Conference on Evolvable Hardware, pp. 167–176. IEEE (2002)
13. Miller, J.F., Harding, S.L., Tufte, G.: Evolution-in-materio: evolving computation in materials. Evol. Intel. **7**(1), 49–67 (2014)
14. Miller, J.F., Mohid, M.: Function optimization using cartesian genetic programming. In: Proceedings of the 15th Annual Conference Companion on Genetic and Evolutionary Computation, pp. 147–148. ACM (2013)
15. Mohid, M., Miller, J.F., Harding, S.L., Tufte, G., Lykkebø, O.R., Massey, M.K., Petty, M.C.: Evolution-in-materio: solving machine learning classification problems using materials. In: Bartz-Beielstein, T., Branke, J., Filipič, B., Smith, J. (eds.) PPSN 2014. LNCS, vol. 8672, pp. 721–730. Springer, Heidelberg (2014)
16. NASCENCE project (ICT 317662). Report on suitable computational tasks of various difficulties. Deliverable D4.2 (2013)
17. Pedersen, M.E.H.: Good parameters for differential evolution. Technical report, Hvass Computer Science Laboratories (2010)
18. Prasad, S., Yang, M., Zhang, X., Ozkan, C.S., Ozkan, M.: Electric field assisted patterning of neuronal networks for the study of brain functions. Biomed. Microdevices **5**(2), 125–137 (2003)

19. Stepney, S.: The neglected pillar of material computation. Physica D **237**(9), 1157–1164 (2008)
20. Storn, R., Price, K.: Differential evolution-a simple and efficient heuristic for global optimization over continuous spaces. J. Global Optim. **11**(4), 341–359 (1997)
21. Thompson, A.: An evolved circuit, intrinsic in silicon, entwined with physics. In: Higuchi, T., Iwata, M., Liu, W. (eds.) ICES 1996. LNCS, pp. 390–405. Springer, Heidelberg (1996)
22. Volpati, D., Massey, M.K., Johnson, D., Kotsialos, A., Qaiser, F., Pearson, C., Coleman, K., Tiburzi, G., Zeze, D.A., Petty, M.C.: Exploring the alignment of carbon nanotubes dispersed in a liquid crystal matrix using coplanar electrodes. J. Appl. Phys. **117**(12), 125303 (2015)

WS Network Design Problem with Nonlinear Pricing Solved by Hybrid Algorithm

Dušan Hrabec[1(✉)], Pavel Popela[2], and Jan Roupec[2]

[1] Faculty of Applied Informatics, Tomas Bata University, Nad Stráněmi 4511, 760 05 Zlín, Czech Republic
hrabec@fai.utb.cz

[2] Faculty of Mechanical Engineering, Brno University of Technology, Technická 2896/2, 616 69 Brno, Czech Republic
{popela,roupec}@fme.vutbr.cz

Abstract. The aim of the paper is to introduce a wait-and-see (WS) reformulation of the transportation network design problem with stochastic price-dependent demand. The demand is defined by hyperbolic dependency and its parameters are modeled by random variables. Then, a WS reformulation of the mixed integer nonlinear program (MINLP) is proposed. The obtained separable scenario-based model can be repeatedly solved as a finite set of MINLPs by means of integer programming techniques or some heuristics. However, the authors combine a traditional optimization algorithm and a suitable genetic algorithm to obtain a hybrid algorithm that is modified for the WS case. The implementation of this hybrid algorithm and test results, illustrated with figures, are also discussed in the paper.

Keywords: Stochastic transportation model · Network-design problem · Nonlinear pricing · Wait-and-see approach · Genetic algorithm · Hybrid algorithm

1 Introduction

The transportation network design problem (TNDP) remains a challenging research topic in transportation planning. From constructing new roads, pipelines, power lines, etc. to determining the optimal road toll, TNDP has provided valuable information for capital investment in transportation [1,7,18]. Various approaches have been used to solve TNDP. Steenbrink [17] and Magnanti and Wong [8] reviewed a number of the network design problems (NDP's) and some earlier algorithms. LeBlanc [7] proposed a branch-and-bound procedure to solve the problem but the algorithm did not perform well in large-scale problems. For a detailed review of solution techniques see, e.g., [1,11].

This paper presents a hybrid algorithm for the solution of a scenario-based wait-and-see (WS) stochastic mixed integer nonlinear program (MINLP), which models the design of a transportation network under price-sensitive stochastic demand. Regarding the solution technique, we mention our direct approach

J. Handl et al. (Eds.): PPSN XIV 2016, LNCS 9921, pp. 655–664, 2016.
DOI: 10.1007/978-3-319-45823-6_61

derived from modeling ideas (e.g., [13]). Due to the growing popularity of pricing strategies development and further applications in industry, we follow up on our previous modeling ideas presented in [4], where we modeled a mixed integer linear program with linearly price-dependent stochastic demand. So, we extend our previous model from [4] into a more complex case with a nonlinear (hyperbolic) price-demand dependency and, therefore, we also modify the previously used algorithm [4,13].

2 Stochastic TNDP with Pricing Solved by WS Approach

In this section, we develop the above mentioned MINLP which represents the design of a transportation network under price-sensitive stochastic demand. Note, that in our case, the network consists of three components: supply, demand, and transition parts of the system, see [2]. Before we deal with the stochastic problem and its WS reformulation, we shortly review the hyperbolic pricing function [10].

2.1 Pricing

Consider a price-setting firm that faces a price-dependent demand function, $b_i(p_i)$, describing the dependency between price p_i and demand b_i for each customer denoted by i. To capture real-world situations, we will further define the demand function as $b_i(p_i) = \alpha_i p_i^{-\beta_i}$, where $\alpha_i > 0$ and $\beta_i > 1$, see Fig. 1.

This means that the selling prices are decision variables, and so we want to find the optimal price p_i^* for each customer i.

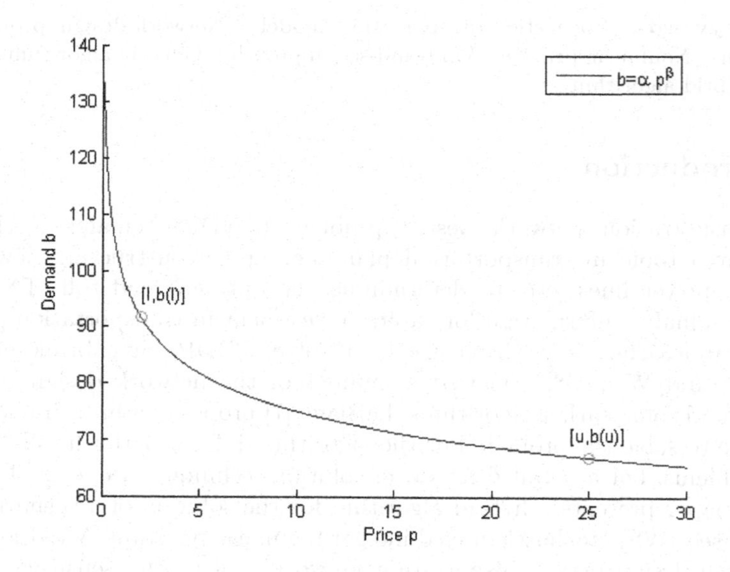

Fig. 1. Example of a hyperbolic demand-pricing function.

2.2 Stochastic Demand and the WS Approach

In real-world problems, the customer demand information is often uncertain and varying. This situation is usually modeled by one of the following deterministic reformulations: (a) the here-and-now (HN) approach, which means that the decisions are made before the demand is observed, see [15] and, specifically, [13]; (b) the wait-and-see (WS) approach, which means that the demand is known at the decision point. An interested reader can also find useful references to fundamental concepts of stochastic programming, e.g., in [6,15].

In this paper, we approach the stochastic TNDP with pricing using the WS scenario-based approach. The scenario-based approach assumes that we have enough observations of the parameters $\alpha_{i,s}$ and $\beta_{i,s}$ (one combination of the observations represents one particular scenario for each customer). In order to develop the mathematical model, we define the following (decision) variables, index sets and parameters.

- The decision variables:
  $x_{e,s}$: amount of the product to be transported on edge e in scenario s,
  $\delta_{e_n,s} \in \{0,1\}$: 1 if new edge e_n is built in scenario $s, 0$ otherwise,
  $p_{i,s}$: unit selling price for customer i in scenario s,
- second-stage variables:
  $y^+_{i,s}$: shortages for customer i in scenario s,
  $y^-_{i,s}$: leftovers for customer i in scenario s,
- index sets:
  E : set of edges, $e \in E$,
  E_n : set of new (built) edges, $e_n \in E_n,\ E_n \subset E$,
  i : set of customers (or locations with a non-zero demand), $i \in I$,
  j : set of production locations (or warehouses), $j \in J$,
  k : set of traffic nodes, $k \in K$,
  V : set of all nodes (vertices) in the network, $v \in V$,$V = I \cup J \cup K$,
  S : set of all possible scenarios, $s \in S$, $s = 1, 2, \ldots, m$,
- and parameters:

$A_{v,e}$: incidence matrix, $A_{v,e} \begin{cases} 1 & \text{if edge } e \text{ leads to node } v, \\ -1 & \text{if edge } e \text{ leads from node } v, \\ 0 & \text{otherwise,} \end{cases}$

$b_{v,s}$: the demand in node v for scenario s,
c_e : unit transporting cost on edge e,
d_{e_n} : cost of building of a new edge e_n,
r^+_i, r^-_i : unit penalty cost for shortages/leftovers at customer node i,
l, u : lower and upper bound for selling prices,
$\alpha_{i,s}, \beta_{i,s}$: scenario-based (and demand-related) parameters.

Then, we formulate the stochastic TNDP with nonlinear pricing, which we reformulate using WS approach, and so, we solve the model repeatedly, i.e., once for each scenario:

$\forall s \in S:$

$$\max \sum_{i \in I} (\sum_{e \in E} A_{i,e} x_{e,s}) p_{i,s} - \sum_{e \in E} c_e x_{e,s} - \sum_{e_n \in E_n} d_{e_n} \delta_{e_n,s} - \sum_{i \in I} (r_i^- y_{i,s}^- + r_i^+ y_{i,s}^+) \quad (1)$$

$$\sum_{e \in E} A_{i,e} x_{e,s} = b_{i,s} - y_{i,s}^+ + y_{i,s}^-, \quad \forall i \in I, \quad (2)$$

$$\sum_{e \in E} A_{j,e} x_{e,s} = b_{j,s}, \quad \forall j \in J, \quad (3)$$

$$\sum_{e \in E} A_{k,e} x_{e,s} = b_{k,s}, \quad \forall k \in K, \quad (4)$$

$$x_{e_n,s} \leq \delta_{e_n,s} \sum_{j \in J} (-b_j), \quad \forall e_n \in E_n, \quad (5)$$

$$y_{i,s}^+ \leq b_{i,s}, \quad \forall i \in I, \quad (6)$$

$$x_{e,s} \geq 0, \quad \forall e \in E, \quad (7)$$

$$\delta_{e_n,s} \in \{0,1\}, \quad \forall e_n \in E_n, \quad (8)$$

$$y_{i,s}^+, \ y_{i,s}^- \geq 0, \quad \forall i \in I, \quad (9)$$

$$p_{i,s} \geq l, \quad \forall i \in I, \quad (10)$$

$$p_{i,s} \leq u, \quad \forall i \in I, \quad (11)$$

$$b_{i,s} = \alpha_{i,s} p_{i,s}^{-\beta_{i,s}}, \quad \forall i \in I. \quad (12)$$

The objective function (1) maximizes the total profit, which is the revenue minus all the costs (transportation, network design and penalties for leftovers and shortages). Equations (2–4) are balance constraints, i.e. amount entering a node is equal to the demand plus the amount leaving; in addition, in the constraint (2) we consider quantities presenting leftovers and shortages, respectively. (5) guarantees that there will be no transported amount on non-built edges. (6) is a constraint on shortages, i.e. any shortage can not be higher than related demand. (7)–(11) state domains of decision variables, while Eq. (12) states the hyperbolic dependency between price and demand (see Fig. 1).

Obviously, the problem (1)–(12) is nonlinear, but it seems that the exact solvers deal with a linearized (MILP) version of it. Such nonlinear problems often requires a heuristic approach, especially large scale problems. Therefore, we further propose a hybrid algorithm in Sects. 3 and 4.

3 Hybrid Algorithm for the WS Approach

The above-mentioned model was coded in GAMS and solved by the BARON, MINOS and CPLEX solvers for suitable test instances. The obtained results are considered acceptable. The next solution attempt targeted large test problems using the same techniques; however, this led to an increase of the required computational time.

Due to the above, the decision to utilize previous experience was made, see [3,13]. This resulted in the implementation of a modified hybrid algorithm combining the GAMS code with a selected genetic algorithm (GA). The C++ implementation concentrating on the GAMS-GA interface is developed for the updated GA, as it was discussed in [12]. This can also be replaced by other GAs [9]. The principles of the following algorithmic scheme follow the papers [3,13].

1. Initialize the computer environment for parallel computations.
2. Define the scenario-based GAMS model and load the model and data into *.gms files for each scenario. Specify control parameters for the GA so that one instance is created for each scenario. The parameters can be defined either by the user (e.g., the population size) or inherited from the GAMS code (e.g., how many edges in the network should be taken into account).
3. Build an initial population for each GA instance. Specifically, the initial values of 0–1 variables must be generated and copied in the `$INCLUDE` files, from which they are read by the GAMS code.
4. The GAMS model is repeatedly solved (in parallel, two loops, one for scenarios and one by population size) by using the MINOS solver. Each run solves the program for the fixed values of 0–1 variables. The profit (or, alternatively, cost) function values are computed (initially in 3. and then in 8.).
5. The best results obtained from GAMS in 4. are saved for comparisons.
6. The termination conditions for the algorithm are tested (in parallel) and the algorithm is terminated if they are met. Otherwise the algorithm proceeds until the last scenario solution is obtained.
7. Input values for the GA from GAMS results are generated, see step 4. Specifically, the profit function values for each member of population of the GA are received from results of the GAMS runs in 4.
8. The GA run leads to an update of the set of 0–1 variables (population), see [12] for details.

Broadly speaking, the GA works with 0–1 variable $\delta_{e_n,s}$ for each scenario s, while MINOS solves the remaining nonlinear problem (NLP) for the fixed binary variable δ, i.e. MINOS computes optimal $x_{e,s}$, $p_{i,s}$ as well as value of the objective function. Afterwards, the value of objective/fitness function (1) is sent back for the solution assessment and then, according to 6., the algorithm continues.

4 Description of the Utilized Genetic Algorithm

This section shortly reviews key ideas of the utilized GA that works as the main part of the hybrid algorithm, see Sect. 5. It follows the previous ideas of one of the authors [12]; see also [13] for its extension.

In general, we consider a set of genetic operators containing: the crossover operator, the mutation operator, and eventually other problem dependent or implementation dependent operators. All these operators generate descendants from parents. The parent selection operator and the genetic operators have a probabilistic character and the deletion operator is usually deterministic. The fitness value f is a non-negative number which captures a relative measure of the quality of every individual in the current population. The run of our GA can be described using the following steps: (1) Generation of the initial population (random generation is often used) composed of individuals. (2) Computation of fitness function values related to 1). (3) Parent selection and generation of offspring. (4) Creation of the new population by using deletion operator and addition of offspring generated in the previous step. (5) Mutation. (6) If the stopping rule is not satisfied, go to step 3), otherwise continue to 7). (7) The result is the best individual in the population. It is usually advantageous to use some redundancy in genes, and then the physical length of the genes can be greater than one bit. Such a type of redundancy by shades was introduced by Ryan [14]. To prevent degeneration and the deadlock in a local extreme, a limited lifetime of individuals can be used. This limited lifetime is implemented via a death operator [12], which represents something like a continual restart of the GA. Many GAs are implemented on a population consisting of haploid individuals (each individual contains one chromosome). However, in nature, many living organisms have more than one chromosome and there are mechanisms used to determine dominant genes. Sexual recombination generates an endless variety of genotype combinations that increases the evolutionary potential of the population. Since it increases the variation among the offspring produced by an individual, this improves the probability that some of them will be successful in varying and often unpredictable environments. The modeling of sexual reproduction is quite simple. The population is divided into two parts - males and females. One parent from each part is selected for crossover. The sex of the individual is stored in the special gene; this gene is not mutated. The sex of the descendant is determined by a crossover of the sexual genes of parents, the descendant is placed into the corresponding part of population. The replacement scheme is associated with another problem. To ensure monotonous behavior the incremental replacement (steady-state replacement) was introduced. We can use least-fit member replacement where one (or more) elements with the worst fitness is replaced, or we can replace randomly chosen element(s). Therefore, the elitism brings a way to keep monotony while generational replacement is used. One or several best individuals represent the elite. The whole elite is directly taken into the next iteration.

So, the GA used in the paper for problem related computations uses ranking selection, haploid chromosomes, shadows and limited lifetime, as described above. We used uniform crossover and the probability of mutation of every gene was 5 %. Every 01 variable was stored in one gene having length of 3 bits. This redundant coding uses the shades technique mentioned above. The population size was 20 individuals; such a low value was chosen in relation to the computational complexity of evaluation of the fitness. The maximum number of iterations was limited to 50. The maximum lifetime of individual was set to 5 iterations.

5 Computations and Results

Figure 2 represents an initial visualization of an example. The example shows a distribution network: bold lines are existing edges and dash lines are possible

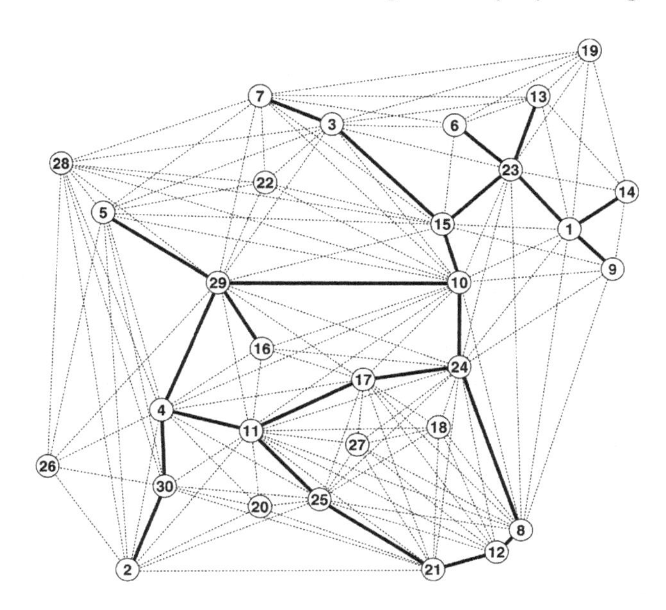

Fig. 2. Input network structure for the WS case [4].

edges that can be switched on by 0–1 variables, nodes 1–14 present customers, 15,16 production nodes, 17–30 transition nodes.

The main idea of the hybrid algorithm is based on the solution of a stochastic program for various sequences of the fixed $0-1$ variables repeatedly for each scenario. This extends the idea of [13] with modifications of the hybrid algorithm in Sect. 3. So, the optimal objective function values are obtained together with these sequences of zeros and ones. They serve as the input fitness value plus elements of the populations for the GA instances that utilizes its own above mentioned steps that are hidden within the GA structure. Updated sequences of zeros and ones are generated by the GA and sent to the GAMS through the updated `$INCLUDE` file and the computational loop continues until a satisfactory improvement of the network design is obtained. For the purpose of future comparison, we have utilized the test examples from [4]. The comparison between MINOS and of the proposed hybrid solution will be subject of our future research, but we have already shown on other MINLP problems that usage of exact solvers is not applicable in real (large) problems due to a huge computational time [4]. Therefore, using of the hybrid approach has one more reason in the MINLP's.

Results are described in Fig. 3 where the thicknesses of lines represent frequencies of usage in m scenarios, and hence, probabilities that variables x_e related to edges are non-zeros. The fixed lines are drawn as dash lines to emphasize the role of edges generated by the WS computations. We may also see that the stochastic demand usually requires new edges to bring the necessary adaptation in the results. In comparison with the HN solutions (cf. [13]) it can be done in a more flexible and cheaper way. Figure 3 also shows that only

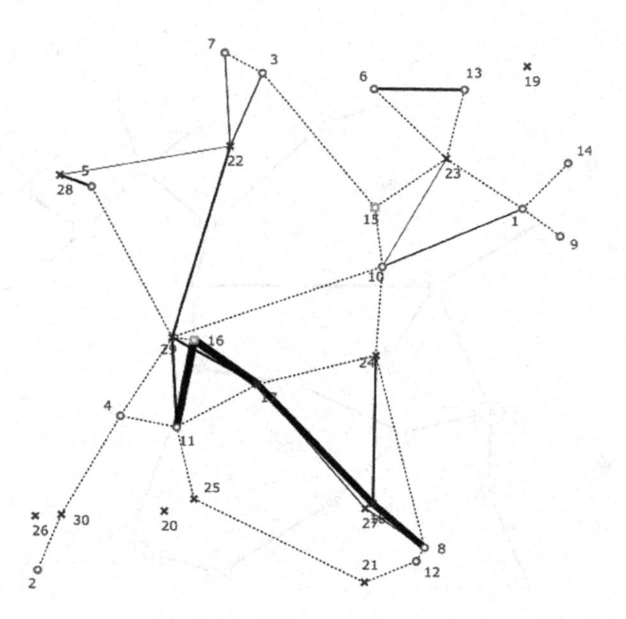

Fig. 3. Visualization of results for the hybrid algorithm for 100 scenarios.

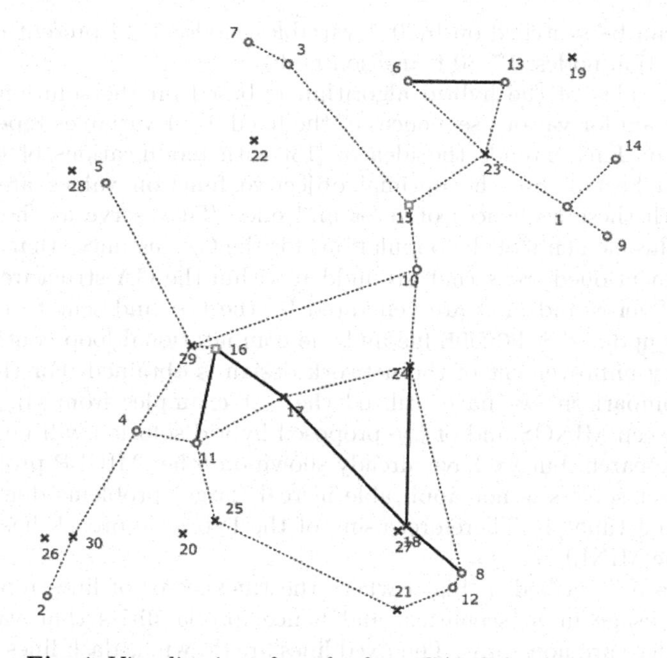

Fig. 4. Visualization of results from GAMS for 1 scenario.

suboptimality has been reached by computations for some scenarios, as extra unnecessary edges are switched on by the GA runs (e.g., 5–28).

To compare the obtained results, due to extreme time requirements of finding a traditional GAMS MINLP solution, we utilized one scenario case and provide a visualization of the result in Fig. 4. We leave further comparison of time requirements as well as values of objective functions for our further research.

6 Conclusions and Further Research

The paper presents a WS reformulation of a TNDP with stochastic price-dependent demands. The proposed mixed-integer nonlinear model is solved with the original hybrid algorithm involving GA for the solution of the WS network design problem. The previously introduced hybrid algorithm (see [4,13]) has been modified and successfully tested. This reconfirms our conclusions in [13] about the portability of the approach to other problems.

In our further research work, we plan to compare (or improve) the proposed hybrid algorithm with similar ideas dealing with differential evolution, specifically multi-chaotic success-history based parameter adaptation for differential evolution [5], which is a novel version of the standard GA that, hopefully, may achieve better computational results for our MINLP problems. Moreover, some obvious suboptimalities (see, e.g., Fig. 3) produced by the GA can easily be eliminated by appending a local search procedure to the GA run.

Similar mixed integer (nonlinear) stochastic programs may appear in many application areas, including NDP [11], traffic networks [3] or waste management problems [16]. Therefore, the suggested hybrid algorithm can be modified and widely applied.

Acknowledgments. This work was supported by the EEA Programme and Norway Grants, within the institutional cooperation project Nr. NF-CZ07-ICP-4-345-2016.

References

1. Babazadeh, A., Poorzahedy, H., Nikoosokhan, S.: Application of particle swarm optimization to transportation network design problem. J. King Saud Univ. Sci. **23**, 293–300 (2011)
2. Ghiani, G., Laporte, G., Musmanno, R.: Introduction To Logistic Systems Planning And Control. Wiley: Interscience Series in Systems and Optimization. Wiley, Chichester (2004)
3. Holešovský, J., Popela, P., Roupec, J.: On a disruption in congested networks. In: Proceedings of the 18th International Conference of Soft Computing MENDEL, pp. 191–196, Brno, Czech Republic (2013)
4. Hrabec, D., Popela, P., Roupec, J., Jindra, P., Novotný, J.: Hybrid algorithm for wait-and-see transportation network design problem with linear pricing. In: Proceedings of 21st International Conference on Soft Computing, MENDEL 2015, pp. 183–188, Brno, Czech Republic (2015)

5. Hrabec, D., Viktorin, A., Šomplák, R., Pluháček,M., Popela, P.: A heuristic approach to facility location problemfor waste management: a case study. In: Proceedings of 22nd International Conference on Soft Computing, MENDEL 2016, pp. 61–66, Brno, Czech Republic (2016)
6. Kall, P., Wallace, S.W.: Stochastic Programming, 2nd edn. Wiley, Chichester (1994)
7. LeBlanc, L.J.: An algorithm for discrete network design problem. Transp. Sci. **9**, 183–199 (1975)
8. Magnanti, T.L., Wong, R.T.: Network design and transportation planning: models and algorithms. Transp. Sci. **18**, 1–55 (1984)
9. Matoušek, R.: HC12: the principle of CUDA implementation. In: Proceedings of the 16th International Conference on Soft Computing MENDEL 2010, pp. 303–308, Brno, Czech Republic (2010)
10. Petruzzi, N.C., Dada, M.: Pricing and the newsvendor problem: a review with extensions. Oper. Res. **47**(2), 183–194 (1999)
11. Poorzahedy, H., Rouhani, O.M.: Hybrid meta-heuristic algorithms for solving network design problem. Eur. J. Oper. Res. **182**, 578–596 (2007)
12. Roupec, J.: Advanced genetic algorithms for engineering design problems. Eng. Mech. **17**(5–6), 407–417 (2011)
13. Roupec, J., Popela, P., Hrabec, D., Novotný, J., Olstad, A., Haugen, K.K.: Hybrid algorithm for network design problem with uncertain demands. In: Proceedings of the World Congress on Engineering and Computer Science, WCECS 2013, vol. 1, pp. 554–559, San Francisco, USA (2013)
14. Ryan, C.S.: Polygenic inheritance scheme. In: Proceedings of the 3rd International Conference on Soft Computing MENDEL 1997, pp. 140–147, Brno, Czech Republic (1997)
15. Shapiro, A., Dentcheva, D., Ruszczynski, A.: Lectures on Stochastic Programming: Modeling and Theory. SIAM, Philadelphia (2009)
16. Šomplák, R., Procházka, V., Pavlas, M., Popela, P.: The logistic model for decision making in waste management. Chem. Eng. Trans. **35**, 817–822 (2013)
17. Steenbrink, P.A.: Optimization of Transport Network. Wiley, New York (1974)
18. Tiratanapakhom, T., Kim, H., Nam, D., Lim, Y.: Braess' Paradox in the uncertain demand and congestion assumed stochastic transportation network design problem. KSCE J. Civil Eng. 1–10 (2016)

A Novel Efficient Mutation for Evolutionary Design of Combinational Logic Circuits

Francisco A.L. Manfrini[1,2], Heder S. Bernardino[1], and Helio J.C. Barbosa[1,3(✉)]

[1] Universidade Federal de Juiz de Fora (UFJF), Juiz de Fora, MG, Brazil
heder@ice.ufjf.br

[2] Instituto Federal do Sudeste de Minas Gerais, Juiz de Fora, MG, Brazil
francisco.manfrini@ifsudestemg.edu.br

[3] Laboratório Nacional de Computação Científica (LNCC), Petrópolis, RJ, Brazil
hcbm@lncc.br
http://www.hedersb.uk.to
http://www.lncc.br/hcbm

Abstract. In this paper we investigate evolutionary mechanisms and propose a new mutation operator for the evolutionary design of Combinational Logic Circuits (CLCs). Understanding the root causes of evolutionary success is critical to improving existing techniques. Our focus is two-fold: to analyze beneficial mutations in Cartesian Genetic Programming, and to create an efficient mutation operator for digital CLC design. In the experiments performed the mutation proposed is better than or equivalent to traditional mutation.

Keywords: Cartesian genetic programming · Point mutation operator · Circuit design · Combinational circuits

1 Introduction

The design of circuits is an important research field and the corresponding optimization problems are complex and computationally expensive. The design of a Combinational Logic Circuits (CLC) is based on the data from a truth table that lists all possible combinations of input logic levels with the corresponding output logic level. Given a certain truth table, it is possible to identify a CLC that meets the conditions prescribed by the truth table using traditional techniques and/or metaheuristics [5,9,14,16].

Several strategies for the design of combinational circuits have been reported [2,5,6,9,14,16]. The aim of these approaches is to find a functional solution, and to minimize the number of gates. Nowadays, CGP (Cartesian Genetic Programming) [15] is one of the most efficient methods for evolutionary design and optimization of digital combinational circuits [16,21]. CGP is a genetic programming technique in which the programs are modeled as directed acyclic graphs (DAG) and, thus, a large number of computational structures can be easily represented, such as CLCs [17]. That graph is represented by a matrix of potentially connected elements.

J. Handl et al. (Eds.): PPSN XIV 2016, LNCS 9921, pp. 665–674, 2016.
DOI: 10.1007/978-3-319-45823-6_62

The literature shows that different function sets are used in the evolutionary design. Koza [12] designed circuits using a small set of gates $\Gamma = \{and, or, not\}$. Miller *et al.* [15,17] and, recently, Goldman and Punch [10,11] used 4 types of gates $\Gamma = \{and, or, nand, nor\}$. Coello *et al.* [2–4,9] used 5 types of gates $\Gamma = \{and, not, or, xor, wire\}$. In [8], Gajda expanded the set of functions and used 9 types of gates $\Gamma = \{and, or, not, nand, nor, xor, wire, c_0, c_1\}$ where *not* and *wire* are unary functions (taking the first input of the gate) and c_k is a constant generator with the value k.

Understanding how search operators interact with solution representation is a critical step in order to design new techniques for improved search. There have been a number of previous studies into various aspects of GP evolution. For instance, [10,11] created methods to prevent wasted CGP evaluations and methods to overcome CGP's search limitations imposed by genome ordering [13].

The remainder of this paper is organized as follows: Sect. 2 summarises Cartesian Genetic Programming while Sect. 3 describes the proposed ideas. The computational experiments are presented in Sect. 4, where the obtained results are compared to those from the literature. Section 5 presents some discussions and, finally, Sect. 6 concludes the paper.

2 Cartesian Genetic Programming

In 1999, Miller [15] proposed a new form of Genetic Programming, called Cartesian Genetic Programming, in which the programs are modeled as directed acyclic graphs (DAG). Recently, [19] presented a CGP method that encodes programs via cyclic graphs. CGP provides a great generality enabling the representation of neural networks, circuits, and other computational structures [17]. Some features can be highlighted:

- CGP represents an individual using a matrix of processing nodes.
- Nodes contain genes describing what function they perform and how they are connected to other nodes.
- DAGs are represented by a collection of nodes connected by directed edges.
- CGP has three parameters associated to the representation and mapping process: the number of columns, the number of rows, and *levels-back*. *levels-back* controls the connectivity of the graph by constraining which columns a node can get its inputs from.
- Offspring are created by means of mutation.
- Offspring replace parents when they are better or have the same fitness value.
- The most common form of CGP uses a $(\mu + \lambda)$ reproduction strategy, where μ parents generate λ offspring and then, from the $(\mu + \lambda)$ individuals, the top μ are taken to be parents in the next generation.

Figure 1a shows an example of the matrix representation adopted by CGP, where I_1, I_2, I_3 are the primary inputs, O_1, O_2 are outputs, and each node represents an operation or its function (*or, if, switch, ...*). Figure 1b shows an example where *number of columns* = 4, *number of rows* = 2, and *levels-back* = *number of columns*. The nodes can have their inputs connected to the outputs of any nodes in the columns to the left of the current one or to a primary input. In this case, nodes 5, 6, 9, 10, and 11 are neutral, having no influence on the phenotype. These nodes are referred to as *inactive*. When a node is connected to an output (directly or indirectly), it is called *active*, as nodes 4, 7, and 8. Inactive nodes allow for genetic drift, as individuals can be mutated without changing their fitness. Figures 1c and d show phenotypes associated with the outputs O_1, and O_2, respectively.

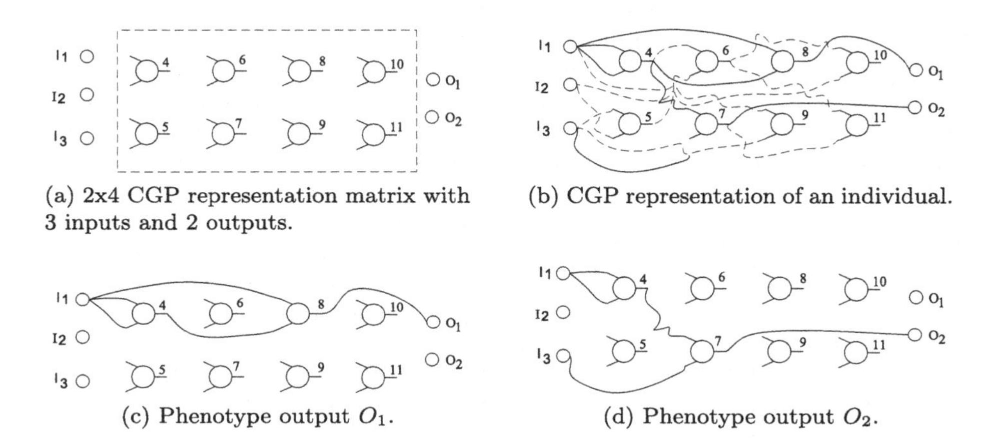

(a) 2x4 CGP representation matrix with 3 inputs and 2 outputs.

(b) CGP representation of an individual.

(c) Phenotype output O_1.

(d) Phenotype output O_2.

Fig. 1. CGP representation

Modern CGP practice has mostly done away with *rows* and *levels-back* in favor of *rows* = 1 and *levels-back* as the sum of the number of columns with the number of inputs.

2.1 Single Active Mutation (SAM)

CGP's usual variation operator is a point mutation. However, different implementations can be found in the literature, making it hard to define a standard version of the algorithm. For instance, in some papers [15] this operator chooses a set number of genes at random to be mutated, while in other papers [20] each gene can be mutated with a certain probability, allowing any number of genes to be mutated at once.

When mutations occur in non-coding sections of the genotype, no modification will appear in the phenotype and, consequently, both individuals (mutated and non-mutated) have the same fitness. In order to avoid this situation Goldman and Punch [10] proposed a method in which a single active gene is modified every time an offspring is generated. This alternative will be referred to as "SAM" here. SAM's iterative process generates an offspring by mutating randomly selected genes until an active gene is changed. When this mutation operator is used, one can see that: (i) one active gene is mutated, (ii) inactive genes may be changed, and (iii) no mutation rate needs to be specified by the user. As SAM achieved the best results in the hardest test-problem from [10], here we adopted this mutation operator as a baseline for all the computational experiments.

3 Description of the Biased Single Active Mutation

Traditionally, the mutation operator used in CGP is a point mutation operator, in which a randomly chosen position in the matrix representation is replaced by another randomly selected value. As the elements of the matrix are composed by a function/operation and its inputs, two different modifications can occur. When a function is chosen for gene mutation, then a valid value is the address of any function in the function set (here called mutation type gate), whereas if an input gene is selected to be modified, then a valid value is the address of the output of any previous node in the genotype, or of any program input.

The proposed approach is based on the idea of analyzing the behavior of the genotype during the evolutionary process for a given set of problems. Based on this analysis, we create a bias to help direct gene mutation when applied to other problems. For each run, every time the child has a fitness value better than that of its parent (the child proceeds to the next generation), we say a beneficial mutation occurred. Every time such improvement occurs, we check whether this mutation occurred on the function executed by the parent. In this case, we store the new and beneficial transition, from the previous (in the parent) to the new function (in the child).

At the end of the evolutionary process, we create the frequency table of all transitions, giving rise to a probability distribution. The creation of the probabilities transition matrix is illustrated in Fig. 2. This probability distribution is utilized to guide the evolutionary process. Every time a function is to be changed, this probability distribution will be used. This new mutation operator proposed here is referred to as *biased* SAM.

Figure 3 shows an example of the biased mutation, where a gate *and* is selected to be mutated. The new value of that node is chosen according to the probabilities present in the transition probabilities matrix. In this example, the new gene in the child is a *nor* gate.

4 Case Study

4.1 Analysis of the Evolution

Initially four benchmark problems, taken from [1], were chosen where the success of the mutations applied during the evolutionary process is studied as explained in Sect. 3. We used the expanded set of functions as in [8]: $\Gamma = \{and, or, not, nand, nor, xor, wire, c_0, c_1\}$, We also used $\mu = 1$, $\lambda = 4$, *number of rows*=1, *number of columns*=100, and *levels-back* = *number of columns*, as in [16].

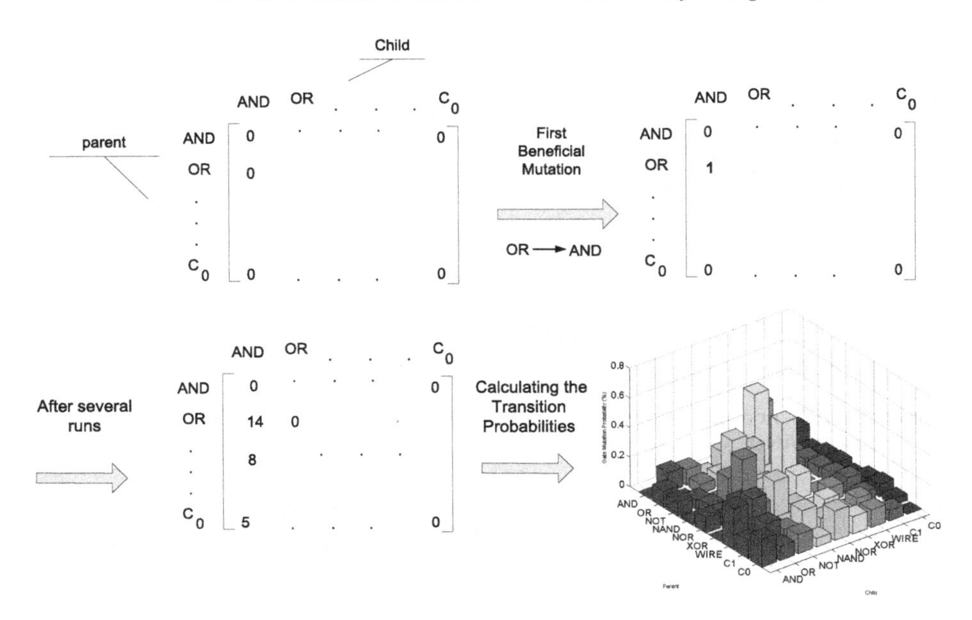

Fig. 2. Illustration of the generation of the transition probabilities matrix.

The test-problems are defined as:

Circuit 1: The first problem has four inputs and one output. The set F indicates the rows of the truth table in which the outputs are equal to one: $F = \{0, 1, 3, 6, 7, 8, 10, 13\}$.

Circuit 2: The second problem has five inputs, one output, and $F = \{2, 3, 6, 7, 10, 11, 13, 15, 18, 19, 21, 23, 25, 27, 29, 31\}$.

Circuit 3: The third problem has four inputs, three outputs, and $F_1 = \{0, 5, 10, 15\}$; $F_2 = \{1, 2, 3, 6, 7, 11\}$; $F_3 = \{4, 8, 9, 12, 13, 14\}$.

Circuit 4: The fourth problem has five inputs, three outputs, and
$F_1 = \{0, 1, 2, 3, 4, 5, 6, 7, 8, 9, 10, 11, 12, 13, 14, 15, 28, 29, 30, 31\}$;
$F_2 = \{0, 2, 4, 6, 7, 8, 10, 12, 14, 15, 16, 18, 20, 22, 23\ 24, 26, 28, 30, 31\}$;
$F_3 = \{4, 5, 12, 13, 20, 21, 28, 29\}$.

A hundred independent runs were performed and the algorithm is terminated when a correct circuit is found or the maximum number of evaluations is reached; here, 100000 evaluations are allowed.

For each beneficial mutation, the exchanges are stored and frequency of occurrence of each exchange will be used in order to build a matrix of transition probabilities. Figure 4 presents a bar plot of the values stored in that matrix at the end of the analysis. Notice that Fig. 3 shows one particular case: the probabilities for the *and* gate of a parent. The matrix of transition probabilities will be used to guide mutation (biased mutation) in other problems as will be seen in Sect. 4.2.

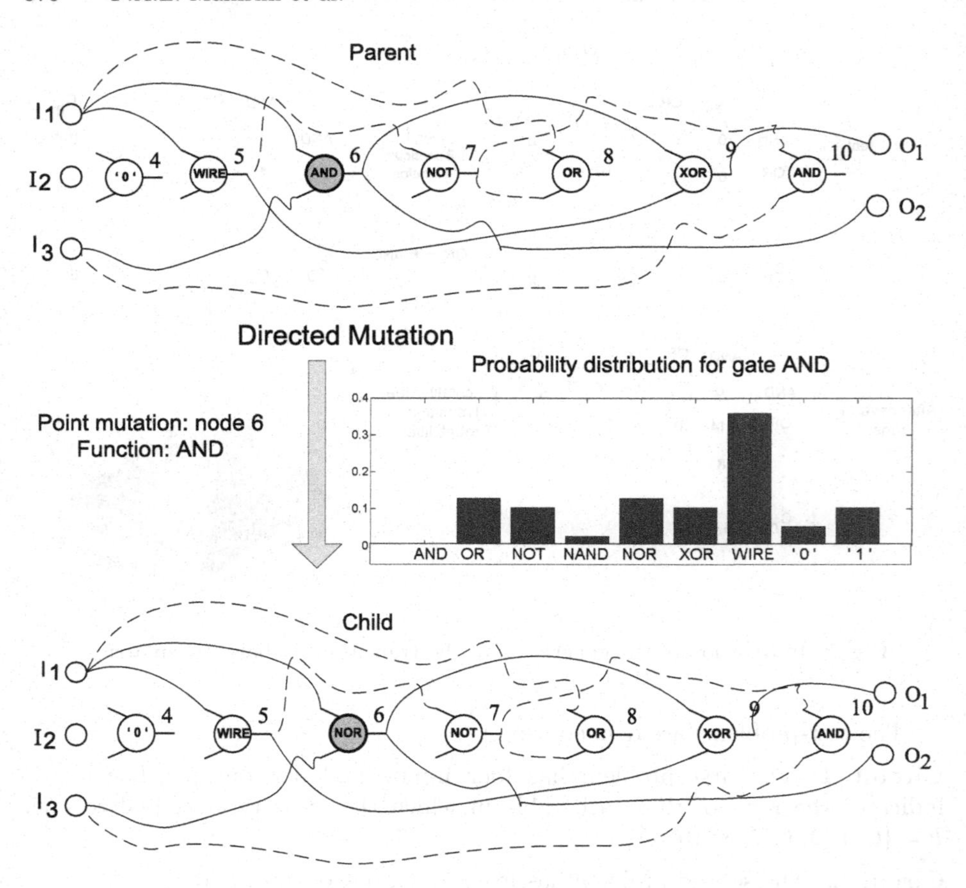

Fig. 3. Example of the biased mutation operator. When a gate (function) is selected to be mutated, then it is replaced by another one using a roulette wheel defined by the transition probabilities matrix obtained by counting the beneficial mutations.

4.2 Designing a Combinational Logic Circuit

For a comparative study four benchmark problems studied by Goldman and Punch [11] and widely used in the electronics literature [7,18] were chosen to verify the effectiveness of our approach. All experiments were implemented in MATLAB and for statistical analysis we used SPSS. The following values were calculated and used in the comparisons: the number of times a feasible solution is found (we call it a "hit"), the median of the number of objective function evaluations required to obtain a feasible solution (called here "MES"), and the number of beneficial mutations per thousand evaluations performed. Notice that larger values of this ratio indicate a smaller number of objective function evaluations unnecessarily wasted and, consequently, an increase in the performance of the method.

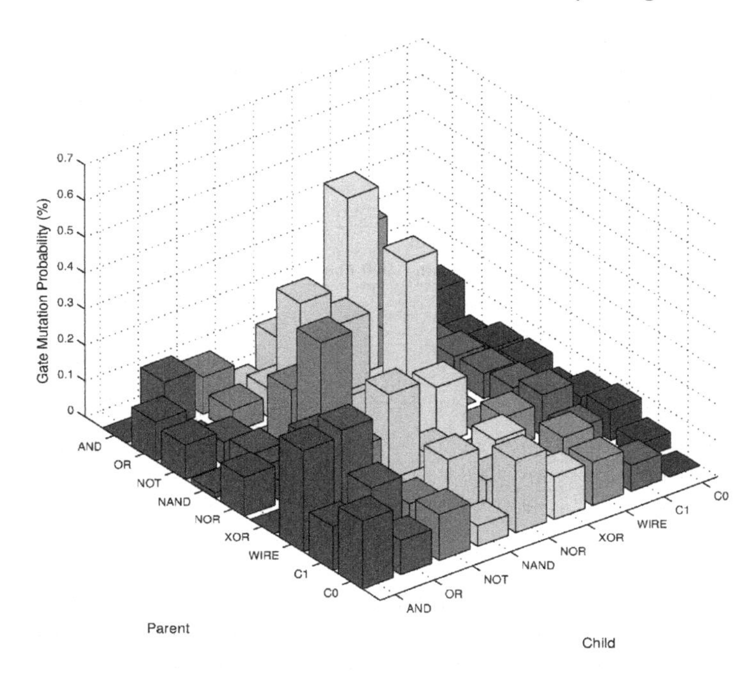

Fig. 4. Values in the transition probabilities matrix obtained using the four benchmark problems shown in Sect. 4.1, considering the beneficial mutations from parent to child gates.

For all test-problems, we used $\mu = 1$, $\lambda = 4$, performed 51 independent runs, and adopted the function set $\Gamma = \{and, or, not, nand, nor, xor, wire, c_0, c_1\}$. Also, for each problem, we employed the same number of nodes used by Goldman and Punch [11]. To ensure that a feasible solution is always found at the end of a run, a sufficiently large number of function evaluations (5000, 500000, 500000, 1000000, respectively for problems 1, 2, 3, and 4) is pre-defined for each problem. Goldman and Punch [11] were able to solve those problems in 95 % of the runs with 1487, 42278, 74939, and 611034, evaluations respectively. The maximum number of evaluations allowed here are at least 60 % higher than those required in [11] to solve the problem.

The results of each type of mutation are shown in Table 1. The first line for each problem shows the control configuration corresponding to SAM, as described in Sect. 2.1.

Problem 1: The first problem, 3-Bit Parity, is considered very simple, but is the most common test-problem in the CGP literature [11,17,22,23] and may help understand how the mutation changes affect the results. The topology configuration was *number of rows* = 1, *number of columns* = 500, as chosen in [11]. It can be seen that the biased mutation converges to a feasible solution with a smaller number of evaluations. The proposed technique obtained MES = 269 while the control configuration has MES = 413.

Table 1. Comparison between standard and biased SAM for the four test-problems used here. "Hits" represents the number of times that a given approach found a feasible solution. The number of times that a beneficial mutation occurs every 1000 evaluations is denoted by "Beneficial Mutation/1000 *evals*". "MES" is Median Evaluations to Success. The *p*-values calculated with the Mann-Whitney U test over the MESs are also presented.

Circuit	Single active mutation	Hits (%)	Beneficial mutation/1000 *evals*	MES MES	Confidence interval	*p*-value
Bit	Standard	98	15.7	413	285 .. 469	–
Parity	Biased	100	15.9	269	189 .. 393	0.049
16 to 4 bit	Standard	100	1.6	19473	16673 .. 22953	–
encoder	Biased	100	1.9	16153	14525 .. 20837	0.103
4 to 16 bit	Standard	90	1.0	332813	293501 .. 360637	–
decoder	Biased	100	2.1	165665	145681 .. 184161	0
3-Bit	Standard	76	1.3	559385	464745 .. 744909	–
Multiplier	Biased	88	1.6	435781	382125 .. 550645	0

Problem 2: The second problem is the 16-4 bit encoder, which can be found in [10,11]. The topology configuration was *number of rows* = 1, *number of columns* = 2000, as chosen in [11]. The proposed technique obtained MES = 15951 while the control configuration has MES = 19469.

Problem 3: The third problem, is the 16-4 bit decoder proposed in [10,11]. The topology configuration was *number of rows* = 1, *number of columns* = 1000, as chosen in [11]. The proposed technique obtained MES = 161889 while the control configuration has MES = 325193.

Problem 4: The fourth problem is the 3-bit multiplier, which can be found in [10,11]. The topology configuration was *number of rows* = 1, *number of columns* = 5000, as in [11]. This problem is very difficult by comparison to the other ones. The proposed technique obtained MES = 430487 while the control configuration has MES = 559385.

5 Discussion of the Results

Analyzing the transition probability matrix extracted from the four benchmark problems in Sect. 4.1, one can see that the most important modification during mutations is to change a *nand* gate in the parent to a *xor* gate. On the other hand, relatively fewer cases were observed in which a beneficial mutation arises from changing a *not* gate into an *or* gate. Thus, the reinforcement of the occurrence of the first exchange, and the avoidance of the second one can potentially improve the performance of the algorithm.

It is interesting to note in Fig. 4 that there is no gate to be preferable in the mutation for all cases. The probability of the transition varies with the gate to be modified. For instance, the *xor* gate has a higher probability value when *nand* or *or* gates are being modified while it presents a lower probability of improvement when replacing *wire* or *not* gates.

When the entire transition probabilities matrix is considered in the search, the results (presented in Sect. 4.2) are better than those obtained by the baseline (SAM) version, for all test-problems used here. Thus, one can see that the extraction of knowledge is possible and useful in improving the performance of CGP.

Finally, notice that beyond the decrease in the number of objective function evaluations to reach a feasible solution, the ratio between the average number of beneficial mutations and the average number of evaluations to success increased, showing that the efficiency of the search is also improved for all problems tested.

6 Conclusions

This paper proposed a new mutation for automatic design of combinational logic circuits via Cartesian Genetic Programming. Through the analysis of the evolutionary process in a given set of problems it was possible to gain knowledge and then use it to guide the search. The incorporated knowledge about the performance of the mutation operator constitutes an important step towards increasing the power of CGP as a design tool.

Experimental results confirmed the superiority of the new biased mutation operator over a standard mutation in reducing the number of fitness evaluations in the design of combinational logic circuits.

Nevertheless, a more in-depth study of the evolutionary mechanisms and beneficial mutations remains as a promising research area. The rationale behind the design of the biased Single Active Mutation applied here to circuit design, is not restricted to this type of application; other design domains can be investigated.

Acknowledgment. The authors would like to thank the support provided by CNPq (grant 310778/2013-1), FAPEMIG (grants APQ-03414-15 and PEE-00726-16), and PPGMC/UFJF.

References

1. Alba, E., Luque, G., Coello Coello, C.A., Hernández Luna, E.: Comparative study of serial and parallel heuristics used to design combinational logic circuits. Optim. Methods Softw. **22**(3), 485–509 (2007)
2. Coello, C.A.C., Aguirre, A.H., Buckles, B.P.: Evolutionary multiobjective design of combinational logic circuits. In: Proceedings of 2nd NASA/DoD Workshop on Evolvable Hardware, pp. 161–170. IEEE (2000)
3. Coello, C.A.C., Alba, E., Luque, G.: Comparing different serial and parallel heuristics to design combinational logic circuits. In: Proceedings of NASA/DoD Conference on Evolvable Hardware, pp. 3–12. IEEE (2003)
4. Coello, C.A.C., Christiansen, A.D., Aguirre, A.H.: Use of evolutionary techniques to automate the design of combinational circuits. Int. J. Smart Eng. Syst. Des. **2**, 299–314 (2000)

5. Coello, C.A.C., Luna, E.H., Hernández-Aguirre, A.: Use of particle swarm optimization to design combinational logic circuits. In: Tyrrell, A.M., Haddow, P.C., Torresen, J. (eds.) ICES 2003. LNCS, vol. 2606, pp. 398–409. Springer, Heidelberg (2003)
6. Coello, C.A.C., Zavala, R.L., García, B.M., Hernández-Aguirre, A.: Ant colony system for the design of combinational logic circuits. In: Miller, J.F., Thompson, A., Thompson, P., Fogarty, T.C. (eds.) ICES 2000. LNCS, vol. 1801, pp. 21–30. Springer, Heidelberg (2000)
7. Ercegovac, M.D., Moreno, J.H., Lang, T.: Introduction to Digital Systems. Wiley, Hoboken (1998)
8. Gajda, Z., Sekanina, L.: An efficient selection strategy for digital circuit evolution. In: Tempesti, G., Tyrrell, A.M., Miller, J.F. (eds.) ICES 2010. LNCS, vol. 6274, pp. 13–24. Springer, Heidelberg (2010)
9. García, B.M., Coello, C.A.C.: An approach based on the use of the ant system to design combinational logic circuits. Mathw. Soft Comput. **9**(3), 235–250 (2002)
10. Goldman, B.W., Punch, W.F.: Reducing wasted evaluations in cartesian genetic programming. In: Krawiec, K., Moraglio, A., Hu, T., Etaner-Uyar, A.Ş., Hu, B. (eds.) EuroGP 2013. LNCS, vol. 7831, pp. 61–72. Springer, Heidelberg (2013)
11. Goldman, B.W., Punch, W.F.: Analysis of cartesian genetic programming's evolutionary mechanisms. IEEE Trans. Evol. Comput. **19**(3), 359–373 (2015)
12. Koza, J.R.: Genetic Programming: On the Programming of Computers by Means of Natural Selection, vol. 1. MIT Press, Cambridge (1992)
13. Luke, S., Panait, L.: A comparison of bloat control methods for genetic programming. Evol. Comput. **14**(3), 309–344 (2006)
14. Manfrini, F., Barbosa, H.J.C., Bernardino, H.S.: Optimization of combinational logic circuits through decomposition of truth table and evolution of sub-circuits. In: IEEE Congress on Evolutionary Computation (CEC), pp. 945–950 (2014)
15. Miller, J.F.: An empirical study of the efficiency of learning Boolean functions using a cartesian genetic programming approach. In: Proceedings of Genetic and Evolutionary Computation Conference, vol. 2, pp. 1135–1142 (1999)
16. Miller, J.F.: Cartesian genetic programming. Springer, Berlin (2011)
17. Miller, J.F., Smith, S.L.: Redundancy and computational efficiency in cartesian genetic programming. IEEE Trans. Evol. Comput. **10**(2), 167–174 (2006)
18. Tocci, R.J., Widmer, N.S., Moss, G.L.: Digital Systems. Pearson, Upper Saddle River (2011)
19. Turner, A.J., Miller, J.F.: Recurrent cartesian genetic programming. In: Bartz-Beielstein, T., Branke, J., Filipič, B., Smith, J. (eds.) PPSN 2014. LNCS, vol. 8672, pp. 476–486. Springer, Heidelberg (2014)
20. Turner, A.J., Miller, J.F.: Neutral genetic drift: an investigation using cartesian genetic programming. Genet. Program. Evol. Mach. **16**(4), 531–558 (2015)
21. Vasicek, Z.: Cartesian GP in optimization of combinational circuits with hundreds of inputs and thousands of gates. In: Machado, P., et al. (eds.) EuroGP 2015. LNCS, vol. 9025, pp. 139–150. Springer, Berlin (2015)
22. Walker, J.A., Miller, J.F.: The automatic acquisition, evolution and reuse of modules in cartesian genetic programming. IEEE Trans. Evol. Comput. **12**(4), 397–417 (2008)
23. Yu, T., Miller, J.F.: Neutrality and the evolvability of boolean function landscape. In: Miller, J., Tomassini, M., Lanzi, P.L., Ryan, C., Tetamanzi, A.G.B., Langdon, W.B. (eds.) EuroGP 2001. LNCS, vol. 2038, p. 204. Springer, Heidelberg (2001)

Fast and Effective Multi-objective Optimisation of Submerged Wave Energy Converters

Dídac Rodríguez Arbonès[1], Boyin Ding[2], Nataliia Y. Sergiienko[2], and Markus Wagner[3(✉)]

[1] Datalogisk Institut, University of Copenhagen, Copenhagen, Denmark
didac@di.ku.dk
[2] School of Mechanical Engineering, The University of Adelaide, Adelaide, Australia
[3] School of Computer Science, The University of Adelaide, Adelaide, Australia
markus.wagner@adelaide.edu.au

Abstract. Despite its considerable potential, wave energy has not yet reached full commercial development. Currently, dozens of wave energy projects are exploring a variety of techniques to produce wave energy efficiently. A common design for a wave energy converter is called a buoy. A buoy typically floats on the surface or just below the surface of the water, and captures energy from the movement of the waves.

In this article, we tackle the multi-objective variant of this problem: we are taking into account the highly complex interactions of the buoys, while optimising the energy yield, the necessary area, and the cable length needed to connect all buoys. We employ caching-techniques and problem-specific variation operators to make this problem computationally feasible. This is the first time the interactions between wave energy resource and array configuration are studied in a multi-objective way.

Keywords: Wave energy · Multi-objective optimisation · Simulation speed-up

1 Introduction

Global energy demand is on the rise, and finite reserves of fossil fuels, renewable forms of energy are playing a more and more important role in our energy supply [11]. Wave energy is a widely available but largely unexploited source of renewable energy with the potential to make a substantial contribution to future energy production [3,9]. There are currently dozens of ongoing wave energy projects at various stages of development, exploring a variety of techniques [10].

A device that captures and converts wave energy to electricity is often referred to as a wave energy device or wave energy converter (WEC). One common WEC design is called a point absorber or buoy. A buoy typically floats on the surface or just below the surface of the water, and it captures energy from the movement of the waves [9]. In our research, we consider three-tether WECs (Fig. 1) as a technological alternative to the common single-tether WECs. While their

J. Handl et al. (Eds.): PPSN XIV 2016, LNCS 9921, pp. 675–685, 2016.
DOI: 10.1007/978-3-319-45823-6_63

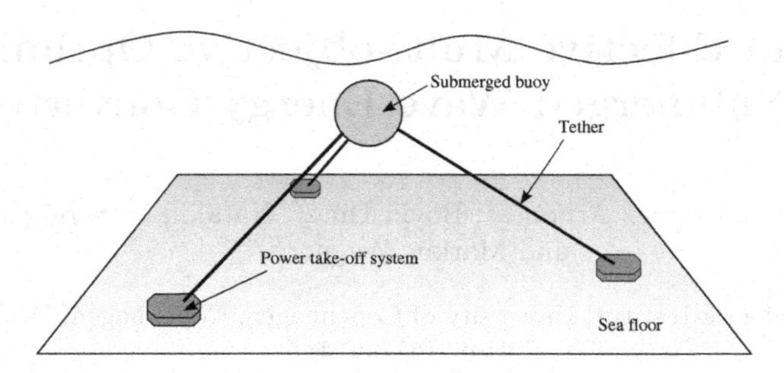

Fig. 1. Schematic representation of a three-tether WEC [18].

capital cost are higher than of conventional single-tether heaving buoys, they can extract significantly more energy from the waves [16]. In our case, the buoys are fully submerged and tethered to the seabed in an offshore location. They use the motion of the waves to drive a hermetically sealed hydraulic line to drive hydroelectric turbines to generate electricity, or to power a reverse osmosis desalination plant to create potable water.

A single wave energy converter can only capture a limited amount of energy alone, which is why it is essential to deploy wave energy devices in large numbers. A group of wave energy devices is commonly referred to as a wave energy farm or array [2]. In order to evaluate our arrays, we use a recently developed frequency domain model for arrays of fully submerged three-tether WECs [15]. This model allows us to investigate different parameters, such as number of devices and array layout. In addition to the objective of producing energy, we consider the following two objectives: the cable length needed to connect all buoys as given by the minimum spanning tree, and the area of the convex hull needed to place all buoys. The ideal choice of parameters leads to an optimisation problem: what are the best trade-offs of the buoys' locations, the area needed, and the cable length needed? To the best of our knowledge, this study is the first to investigate this question to reduce costs and to increase efficiency.

We proceed as follows. In Sect. 2, we introduce the multi-objective buoy placement problem and the different objectives that are subject to our investigations. Then, we present in Sect. 3 our speed-ups, the operators, and the constraint handling used. We report on our computational study in Sect. 4 before we conclude with a summary.

2 Preliminaries

In the following, we outline the different objectives and constraints that we consider for the WEC array optimisation.

Let $X = \{x_1, \ldots, x_n\}$ and $Y = \{y_1, \ldots, y_n\}$ be the set of x and y coordinates of n WECs in the plane. The goal is to find a set of coordinates such that the energy output of the whole wave farm is maximised. At the same time, the total length of the cable or pipe necessary to interconnect the buoys, as well as the area necessary for the wave farm, should be minimised.

The WEC design that we consider is a fully submerged spherical body connected to three tethers that are equally distributed around the buoy hull (Fig. 1). Each tether is connected to the individual power generator at the sea floor, which allows to extract power from surge and heave motions simultaneously [14].

2.1 Power Output Prediction

In the following, we briefly outline the model of this kind of WECs arrays as it was derived by Sergiienko et al. [15] and used by Wu et al. [18].

The dynamic equation of the WECs array is derived in the frequency domain using linear wave theory, where a fluid is inviscid, irrotational and incompressible [4]. This model considers three dominant forces that act on the WECs:

(i) excitation force includes incident and diffracted wave forces;
(ii) radiation force acts on the oscillating body due to its own motion;
(iii) power take-off force that exerts on the WEC from machinery through tethers.

The key point in the array performance is the hydrodynamic interaction between buoys that can be constructive or destructive depending on the array size and geometry.

Assuming that the total number of devices in the array is n and p is the body number, then the dynamics of the p-th WEC in time domain is described as:

$$\mathbf{M}_p \ddot{\mathbf{x}}_p(t) = \mathbf{F}_{exc,p}(t) + \mathbf{F}_{rad,p}(t) + \mathbf{F}_{pto,p}(t), \tag{1}$$

where $\mathbf{M}_p$ is a mass matrix of the p-th buoy, $\ddot{\mathbf{x}}_p(t)$ is a body acceleration vector in surge, sway and heave, $\mathbf{F}_{exc,p}(t)$, $\mathbf{F}_{rad,p}(t)$, $\mathbf{F}_{pto,p}(t)$ are excitation, radiation and power take-off (PTO) forces respectively. The power take-off system is modelled as a linear spring and damper for each mooring line with two control parameters, such as stiffness K_{pto} and damping coefficient B_{pto}.

In case of multiple bodies, where $p = 1 \ldots n$, Eq. (1) can be extended to include all WECs and expressed in frequency domain:

$$\left((\mathbf{M}_\Sigma + \mathbf{A}_\Sigma(\omega)) j\omega + \mathbf{B}_\Sigma(\omega) - \frac{\mathbf{K}_{pto,\Sigma}}{\omega} j + \mathbf{B}_{pto,\Sigma} \right) \hat{\dot{\mathbf{x}}}_\Sigma = \hat{\mathbf{F}}_{exc,\Sigma}, \tag{2}$$

where subscript Σ indicates a generalised vector/matrix for the array of N bodies, $\mathbf{A}_\Sigma(\omega)$ and $\mathbf{B}_\Sigma(\omega)$ are radiation added mass and damping coefficient matrices that include hydrodynamic interaction between buoys, $\mathbf{K}_{pto,\Sigma}$, $\mathbf{B}_{pto,\Sigma}$ are the stiffness and damping block-matrices of the PTO system.

The total power absorbed by the array of WECs can be calculated as:

$$P_\Sigma = \frac{1}{4}(\hat{\mathbf{F}}^*_{exc,\Sigma} \hat{\dot{\mathbf{x}}}_\Sigma + \hat{\dot{\mathbf{x}}}^*_\Sigma \hat{\mathbf{F}}_{exc,\Sigma}) - \frac{1}{2} \hat{\dot{\mathbf{x}}}^*_\Sigma \mathbf{B} \hat{\dot{\mathbf{x}}}_\Sigma, \tag{3}$$

where * denotes the conjugate transpose.

For more details on the model, we refer the interested reader to [15, 18].

2.2 Constraints and Assumptions

We have the following constraints placed on our optimisation. The first one enforces an upper bound on the area of the farm. This constraint ensures that we can only place a buoy i within a certain area, which is a realistic constraint for most layout problems. For a rectangular wave farm with length l and width w, this constraint is satisfied *iff*

$$0 \leq x_i \leq l \text{ and } 0 \leq y_i \leq w, 1 \leq i \leq n. \quad (4)$$

Because buoys can damage each other if they get too close, and also maintenance ships need to be able to navigate between them, the second constraint regulates the spacial proximity. It is satisfied *iff*

$$\sqrt{(x_i - x_j)^2 + (y_i - y_j)^2} \geq 50 \text{ meters}. \quad (5)$$

In addition to the above constraints, we assume that all WECs have the same power take-off characteristics.

2.3 Euclidean Minimum Spanning Tree

We use the Euclidean minimum spanning tree (MST) to calculate the minimum length of cable or pipe required to connect all buoys in a particular array configuration. It is computed by first constructing the complete graph on the set of points that represent the buoys and edge costs given by the Euclidean distance between any pair of buoys. Then, the minimum spanning tree for this graph is computed and used as an objective representing the costs of the cable or pipe length. Figure 2 displays an array layout as, as well as the minimum spanning tree, represented by lines joining each buoy.

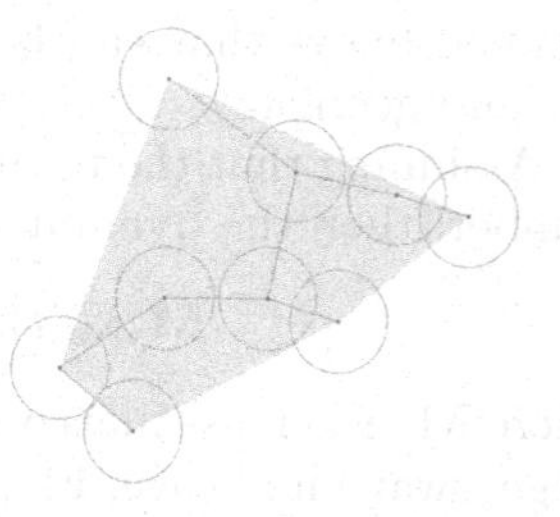

Fig. 2. An example WEC array. The circles visualise the safety distance.

2.4 (Cost of the) Convex Hull

In our study, the cost of the convex hull is defined as the area contained by the set of points forming the convex hull. This value is the minimum land area that is required for a wave farm layout. Figure 2 displays a buoy layout, as well as the area (cost) of the convex hull shaded in grey.

3 Computational Speed-Up, Operators, and Constraint Handling

3.1 Speed-Up of Simulation

In order to make the simulations computationally feasible, and to make the best use of the available hardware, we reimplemented the PTO system in C++ and

Table 1. Runtime per evaluation in seconds (median of 20 runs). Ω is the set of frequencies ω used. In each cell, single-thread results are on the left and multi-thread ones on the right. *Laptop* denotes a computer with a Intel Core i7-4910MQ CPU (up to 3.9 GHz, used with 4 threads) and 32 GB RAM. *Server* is a compute server with four AMD Opteron 6348 CPUs (up to 3.4 GHz, used with 48 threads) and 128 GB RAM.

n	$\|\Omega\|$	MATLAB		C++	
		Laptop	Server	Laptop	Server
4	50	29.28/11.64	56.15/3.54	2.83/0.98	5.48/0.38
9	25	80.96/31.14	153.75/9.48	8.32/3.05	16.28/0.98
16	25	262.86/97.58	508.63/29.88	29.31/9.94	55.42/3.23
25	25	658.21/239.37	1265.97/72.46	71.92/26.20	141.16/8.95

parallelised it with OpenMP [13]. Because the system is defined as a series, it is inherently parallelisable and a linear speed-up possible. Furthermore, OpenMP's framework allows for nested paralellisation, which further decreases the overall running time. The integrals are calculated with the GNU Scientific Library [5], which has support for integrals with singularities. After comparing the performance of complex linear system solving in C++ and MATLAB [12], the result from Eq. (2) is obtained from MATLAB. Note that the evaluation of Eq. (3) can only be parallelised for each of the frequencies ω considered.

The evaluation of a WEC array is time consuming even with parallelisation. The integral calculation is the bottleneck consuming upwards of 95 % of the running time. Wu et al. [18] used caching of integral computations because a large portion of these integrals are repeated, achieving a factor 7 speed-up in running time. We use the caching approach even more comprehensively, by caching results not only within a single layout evaluation like Wu et al., but by reusing them across multiple evaluations. This additional improvement can help if an optimisation algorithm modifies only part of a solution at each iteration, therefore reusing integrals computed in previous iterations.

In Table 1 we list the achieved time needed to compute the intra-buoy interactions. As we can see, the speed-ups (up to 142-fold) allow us to run significantly more evaluations if the overall available time is limited.

3.2 Problem-Specific Operators

As the problem is highly constrained due to a large number of buoys and the given safety margin around each buoy, the operators have to ensure that feasible placements are produced. We investigate the benefit of the two variation operators MovementMutation and BlockSwapCrossover by Tran et al. [17] over the commonly used Polynomial Mutation and Simulated Binary Crossover. The former pair was designed for wind turbine placement optimisation, where safety distance constraints and area constraints also need to be considered.

MovementMutation is an operator that does a local change to the current solution. For a randomly picked WEC, MovementMutation moves it to a randomly selected spot along a selected direction to a feasible location.

BlockSwapCrossover is designed to implant a randomly selected rectangular "block" of WECs from each of the two parents to produce two children, each with a varying degree of information from each parent. A repair operator is applied in case the number of WECs does not match to the target number.

Note that the fundamental difference between both the WEC positioning and the wind turbine positioning is that "shading" is the primary inter-turbine effect, while the primary inter-buoy effect is "phase shifting". However, as the operators do not consider these effects directly, we can apply them to our problem as well.

3.3 Constraint Handling

As described in Sect. 2.2, we consider area constraints and safety distance constraints in this study.

The area (box) constraint is enforced by applying a sinusoidal-shaped function that maps any value to a closed range. The function used has the form [6]:

$$x = a + (b - a) * (1 + cos(\pi * x/(b - a) - \pi))/2 \qquad (6)$$

The advantage of this function is twofold. First, the boundaries of the region are automatically enforced without the need for a check on each iteration. Second, it provides a smooth transition of the movements of buoys close to the boundaries, contributing to the performance of the optimisation algorithm.

The inter-buoy distance is enforced by applying a penalty to the objectives. This penalty is proportional to the distance that the buoys lie outside the safety margin. The resulting objectives O' are used in the optimisation process:

$$O' = O\left(1 - K\sum_{i}^{n}\sum_{j\neq i}^{n}\text{MAX}(M - d(i,j), 0)\right) \qquad (7)$$

where n is the number of buoys, M is the safety distance to keep between buoys, $d(i,j)$ is the Euclidean distance between the buoys, and $K \in \mathbb{R}^+$ is the penalty regularisation parameter. This parameter is meant to control the slope of the penalty applied, acting as a trade-off between discouraging solutions that lie far into the infeasible region, and allowing the exploration of boundary regions.

4 Experimental Study

In this section, we describe our experimental setup and report on the results of different multi-objective evolutionary algorithms using our speed-ups and variation operators for the multi-objective buoy placement problem.

4.1 Experimental Setup

For the basis of our study, we utilise the algorithms SMS-EMOA [1] and MO-CMA-ES [7], as implemented in the optimisation framework Shark 3.0 [8]. We use SMS-EMOA in two variants: (i) the default SMS-EMOA with Polynomial Mutation and Simulated Binary Crossover, and (ii) the problem-specific SMS-EMOA* with MovementMutation and BlockSwapCrossover (see Sect. 3.2).

We use a population of size $\mu = 50$ for all experiments, and the evaluation budget for each run is 6000 evaluations. All other parameters are used with their default values in the Shark library. Unless stated otherwise, we report the results of 20 independent runs.

For all runs, we initialise the first population with regular grids that are scaled from the tightest grid to the most generous one where buoys are placed on the boundaries as well. In Fig. 3 we show an example, which also shows the non-linear effect that arises from the constraint handling of the box constraints (Eq. 4). The side-effect of this initialisation is that we already achieve right from the beginning a population of solutions that is guaranteed to be diverse in the size of the convex hull and in the length of the minimum spanning tree. In preliminary experiments, we observed that this approach performed better than one with random initial layouts.

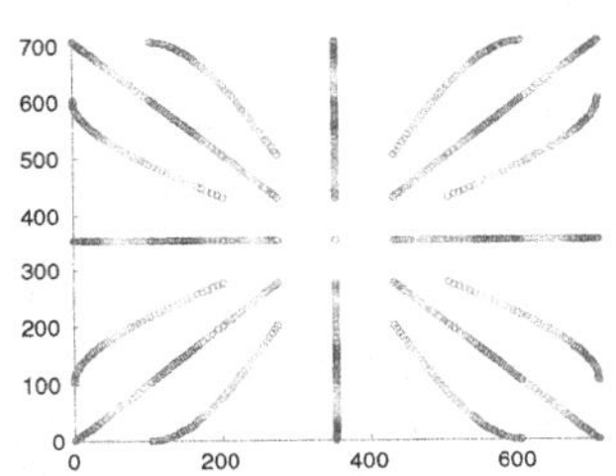

Fig. 3. Initial population in the 25-buoy scenario. The $\mu = 50$ layouts are shown in different colours. (Color figure online)

The scenarios are defined as follows. The goal is to place 4, 9, 16, and 25 buoys subject to the three objectives in a quadratic area. We scale the area available with the number (considering an area of 20,000 m^2 per buoy), which results in squares with sides of length 283 m, 424 m, 566 m, and 707 m respectively.

To compare the performance of the different setups, we inspect the sets of trade-offs visually, and we employ the hypervolume indicator. To compute the latter, we rescale the final solution set into the unit cube that is defined by the extreme values (of feasible layouts) observed for each scenario.

In each scenario, we use the sea state, i.e. the wave frequency distribution, and the features of the buoys as defined in the single-objective investigations in [18], which allows us to compare results for 25 buoys. The WEC radius is $a = 5$ meters, and their power take-off characteristics are kept static. The mass of each buoy is equal to 0.85 times the mass of the displaced water. Ocean depth is chosen to be 30 m and all WECs are submerged 6 m to the centre of buoy.

To compute the power output of a solution, we need to choose a number of discrete wave frequencies from the wave spectrum. While Wu et al. [18] observed that a single frequency from the entire spectrum of waves can be used with reasonable accuracy during buoy placement optimisation, we prefer to use a significantly more time-consuming approach with 25 or 50 wave frequencies. This provides us with very accurate power output predictions. Also, this greatly reduces the risk of unrealistic exploitation of local optima due to ill-conditioned scenarios, which we have observed in the one-frequency case.

4.2 Experimental Results

In the following, we compare the performance of the different multi-objective approaches. Figure 4 summarises the hypervolumes achieved by the final populations for the different scenarios. While the standard version of MO-CMA-ES outperforms the standard version of SMS-EMOA, both are easily outperformed (in terms of achieved hypervolume) by SMS-EMOA*. It appears that even though the latter employs operators previously used in wind turbine placement optimisation, they are also beneficial in our case.

Note that MO-CMA-ES hardly benefits from the caching of simulation results, as it tends to sample new coordinates for all buoys every time. While this behaviour is typically an advantage, it was infeasible for us to apply MO-CMA-ES to the optimisation of the larger scenarios.

Exemplarily, we show in Fig. 5 how the average objective scores across the populations as they evolve over time. Interestingly, SMS-EMOA and SMS-EMOA* behave quite differently, even though they differ only in their variation operators. For example, the standard SMS-EMOA performs best in terms of convex hull and length of the minimum spanning tree, but it produces on average

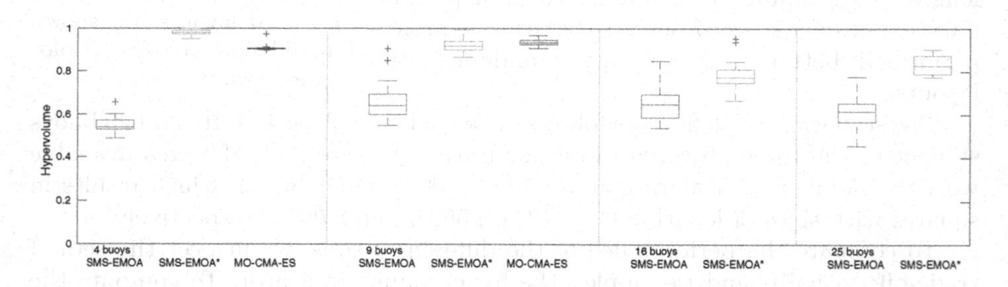

Fig. 4. Hypervolumes covered by the final populations. MO-CMA-ES results are missing for 16 and 25 buoys due to the unacceptable run-times.

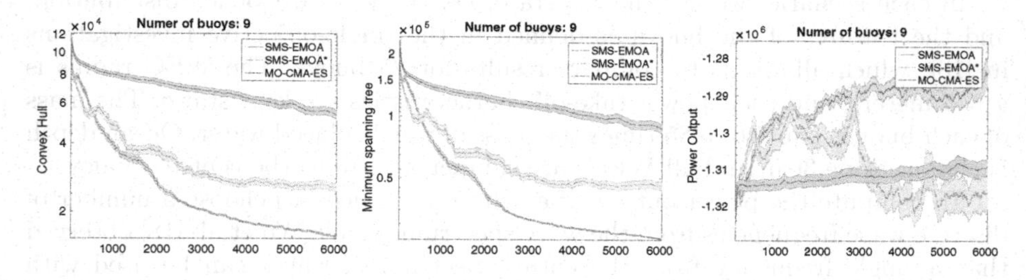

Fig. 5. Evolution of objective scores over time in case of the 9-buoy scenario. Shown are the averages and the 95 % confidence intervals. Note that we are minimising the negative power output, meaning that smaller values are better.

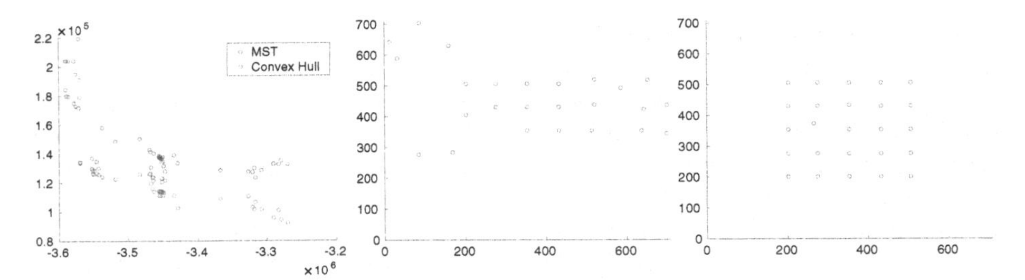

Fig. 6. Population with the highest power output layout for 25 buoys (left, computed by SMS-EMOA*), with the three-dimensional space being projected twice into the two-dimensional space. The layout with the highest power output (3.590 MW) is shown in the middle, and it uses more than twice the area of the layout with the lowest output (3.270 MW) in this population. Note that the waves arrive from the top of the layouts. The circles in the middle and in the right show the 5 m-buoys (to scale).

the layouts with the worst power output. It appears that all three approaches explore quite different parts of the objective space across all runs. This is not only reflected by the final mean values, but also by the spread across different runs by the same algorithm. In summary, this shows to us that a decision maker should not blindly trust multi-objective performance indicators, but inspect the solution sets as well. In practice, a single trade-off layout has to be chosen for implementation, and even though the nature of the problem is multi-objective, the decision eventually boils down to hidden preferences or economic factors.

Lastly, we briefly compare our results for 25 buoys with the ones by [18]. We take their best result and recompute the power prediction more accurately using 25 frequencies instead of the single frequency they used. As a result, their best layout has a predicted power output of 3.459 MW. Compared to this, our best layout (Fig. 6) has an output of 3.590 MW, which is 3.8 % better.

To conclude, we can see that the multi-objective optimisation of arrays of wave energy converters is feasible, if software engineering tricks are employed to speed-up the simulations, and if problem-specific variation operators are used. Also, it is important for engineers to explore different algorithms as these explore the objective spaces with different biases—and the consequence of this bias might matter to the decision maker.

5 Conclusions

Wave energy plays an increasing role in the energy supply world-wide. We have investigated the problem of placing wave energy converters on a given offshore area using different conflicting objective functions.

In a first step, we speeded up and parallelised the computations of the buoys' interactions, which resulted in a speed-up by a factor of up to 142 for 25 buoys. In order to improve the actual optimisation, we employed variation operators from the loosely related wind turbine problem. Interestingly, these problem-specific operators proved to be effective in our setting as well, despite the intra-device interactions being fundamentally different.

The computational study shows that multi-objective evolutionary algorithms can be used for the multi-objective buoy placement problem; in particular, our best performing configuration even improves the power output upon a previous single-objective results by 3.8 %. This improvement can translate into millions of additional dollars of income per year during the lifetime of the wave farm.

In the future, we will extend our research to optimising the individual power take-off characteristics of the buoys in addition to their position, as the effective sea state within the WEC arrays differs significantly from the state outside.

We made the code and results available online: http://cs.adelaide.edu.au/~optlog/research/energy.php

Acknowledgments. This work has been supported by the ARC Discovery Early Career Researcher Award DE160100850.

References

1. Beume, N., Naujoks, B., Emmerich, M.: SMS-EMOA: multiobjective selection based on dominated hypervolume. Eur. J. Oper. Res. **181**(3), 1653–1669 (2007)
2. de Andrés, A., Guanche, R., Meneses, L., Vidal, C., Losada, I.: Factors that influence array layout on wave energy farms. Ocean Eng. **82**, 32–41 (2014)
3. Drew, B., Plummer, A., Sahinkaya, M.N.: A review of wave energy converter technology. Proc. Inst. Mech. Eng. Part A: J. Power Energ. **223**(8), 887–902 (2009)
4. Falnes, J.: Ocean Waves, Oscillating Systems: Linear Interactions Including Wave-Energy Extraction. Cambridge University Press, Cambridge (2002)
5. GNU Scientific Library: Version 1.16 (2013). http://www.gnu.org/software/gsl/. Accessed 7 Apr 2016
6. Hansen, N.: CMA-ES Source Code: Practical Hints. https://www.lri.fr/~hansen/cmaes_inmatlab.html. Accessed 7 Apr 2016
7. Igel, C., Hansen, N., Roth, S.: Covariance matrix adaptation for multi-objective optimization. Evol. Comput. **15**(1), 1–28 (2007)
8. Igel, C., Heidrich-Meisner, V., Glasmachers, T.: Shark. J. Mach. Learn. Res. **9**, 993–996 (2008)
9. Lagoun, M., Benalia, A., Benbouzid, M.: Ocean wave converters: state of the art and current status. In: IEEE International Energy Conference, pp. 636–641 (2010)
10. López, I., Andreu, J., Ceballos, S., de Alegría, I.M., Kortabarria, I.: Review of wave energy technologies and the necessary power-equipment. Renew. Sustain. Energ. Rev. **27**, 413–434 (2013)
11. Lynn, P.A.: Electricity from Wave and Tide: An Introduction to Marine Energy. Wiley, Hoboken (2013)
12. MATLAB: Version 8.6.0 (R2015b). The MathWorks Inc., Natick (2015)
13. OpenMP Architecture Review Board: OpenMP Application Program Interface, Version 3.0, May 2008
14. Scruggs, J.T., Lattanzio, S.M., Taflanidis, A.A., Cassidy, I.L.: Optimal causal control of a wave energy converter in a random sea. Appl. Ocean Res. **42**(2013), 1–15 (2013)

15. Sergiienko, N.Y., Cazzolato, B.S., Ding, B., Arjomandi, M.: Frequency Domain Model of the Three-Tether WECs Array (2016). https://www.researchgate.net/publication/291972368_Frequency_domain_model_of_the_three-tether_WECs_array. Accessed 7 Apr 2016
16. Srokosz, M.A.: The submerged sphere as an absorber of wave power. Fluid Mech. **95**(4), 717–741 (1979)
17. Tran, R., Wu, J., Denison, C., Ackling, T., Wagner, M., Neumann, F.: Fast and effective multi-objective optimisation of wind turbine placement. In: Genetic and Evolutionary Computation, pp. 1381–1388. ACM (2013)
18. Wu, J., Shekh, S., Sergiienko, N., Cazzolato, B., Ding, B., Neumann, F., Wagner, M.: Fast and effective optimisation of arrays of submerged wave energy converters. In: Genetic and Evolutionary Computation Conference (2016). Accepted for publication

Evolution of Spiking Neural Networks Robust to Noise and Damage for Control of Simple Animats

Borys Wróbel[1,2(✉)]

[1] Systems Modeling Group, IO PAN, Sopot, Poland
[2] Evolutionary Systems Group,
Uniwersytet im. Adama Mickiewicza, Poznań, Poland
wrobel@evosys.org
http://www.evosys.org

Abstract. One of the central questions of biology is how complex biological systems can continue functioning in the presence of perturbations, damage, and mutational insults. This paper investigates evolution of spiking neural networks, consisting of adaptive exponential neurons. The networks are encoded in linear genomes in a manner inspired by genetic networks. The networks control a simple animat, with two sensors and two actuators, searching for targets in a simple environment. The results show that the presence of noise on the membrane voltage during evolution allows for evolution of efficient control and robustness to perturbations to the value of the neural parameters of neurons.

Keywords: Spiking neural networks · Adaptive exponential integrate-and-fire model · Genetic algorithm · Robustness to noise · Robustness to damage

1 Introduction

One of the central mysteries of biology is the enormous robustness of complex biological systems to perturbations [7]. This robustness is paradoxical because large complexity suggests fragility. And yet biological systems are robust not only to the fluctuations of the external environment, malfunctions of internal parts, but also the steady bombardment, over generations, of genetic disturbances (mutations) resulting in slight changes in structure of these systems. For example, biological genetic networks are robust to transcriptional noise, point mutations, deletions and duplications of genes. Perhaps the most complex systems known, biological neural networks, are robust to changes at several scales—developmental variability from one generation to the next, influencing the number of cells and theirs connectivity, fluctuations over individual life in the number of cells resulting from their death of cells and formation of new ones, and at a scale smaller still—destruction and formation of synapses, changes of the neurophysiological properties of individual neurons, etc.

J. Handl et al. (Eds.): PPSN XIV 2016, LNCS 9921, pp. 686–696, 2016.
DOI: 10.1007/978-3-319-45823-6_64

As the components of artificial computational systems get smaller, these systems become more difficult to build (resulting in the variability of structure) and more unreliable (with more noise and more fragility of each part). Hence the interest in building artificial systems inspired by biology, such as artificial genetic networks and artificial neural networks, which promise large computational resources, low power consumption, and robustness to silicon mismatch (for example [6]).

In this paper I investigate the interplay between the robustness to noise and to other perturbations in evolved artificial spiking neural networks. The networks are evolved to control the behavior of a simple animat, with and without the noise on membrane potential of neurons. I then analyse the robustness of these networks to changes in parameters of neurons and the functioning of the animats' actuators.

The model of evolution of networks used in this paper was built originally for artificial genetic networks. We called this approach a 'mixed paradigm' [8,9], because the encoding in the artificial genomes is inspired by the encoding of biological genetic networks, but the functioning of the networks is inspired by the networks of biological neurons in the brain. In biology, the encoding of the neural structures in the genome is much more indirect, with the number of neurons in large mammalian brains vastly larger than the number of genes. However, the computational task faced by the networks investigated here is quite simple, consisting of directional movement toward target of an animat with two sensors and two actuators, so a simple encoding is more than sufficient.

2 Model

2.1 Evolving Spiking Neural Networks

The network model used in this paper does not restrict the number of nodes or connections in the networks (more precisely, the restrictions imposed by the limited computer memory are never reached in practice; however, the task considered here does not require large networks). Each internal node is encoded in the genome as a series of *cis genetic elements* followed by a series of *trans elements* (Fig. 1). Each element in the genome has several fields (four in the version of the model used in this paper): the *type* (cis, trans, and input or output), *sign*, and two *coordinates*. Three types of connections are allowed between the nodes: input-cis—encoded by one input and one cis element—and, similarly, trans-cis, and trans-output. The signs determine if a particular connection is inhibitory (when the signs of two elements are different) or excitatory (when the signs are the same).

A connection is formed if the coordinates of two elements are such that the Euclidean distance between the corresponding points in an abstract 2-dimensional space is below a predefined threshold (5.0). The smaller the distance, the higher the weight of the connection (using the positive part of the function $\frac{10-2d}{d+1}$, where d is the distance between the elements). If more than one connection is formed between any two nodes, the weights are added, giving in the end either a positive synaptic weight (an excitatory connection) or a negative one (an inhibitory connection). For example, to calculate the weight between the second sensory neuron in the network for the animat in Figs. 1 and 3 (marked as S) to the first interneuron (which has 5 cis elements), we need to add the weights coming from the interactions between input-cis pairs $(3, 4)$, $(3, 5)$, and $(3, 8)$, the other pairs have distances higher than 5.0. This gives the weight $-0.16 - 3.07 + 2.04 = -1.19$.

Input nodes have one state variable, determined by the sensors on the animat; internal and output nodes have four state variables: membrane potential v, adaptation current w, excitatory conductance g_E, and inhibitory conductance g_I. They are governed by four differential equations, according to the adaptive exponential integrate-and-fire model of spiking neurons (AdEx, EIF) [2,5]:

$$\frac{dv}{dt} = \frac{g_L(E_L - v) + g_L\delta e^{\frac{v-V_T}{\delta}} + w + g_E(E_E - v) + g_I(E_I - v) + I_{offset}}{C} \quad (1)$$

$$\frac{dw}{dt} = \frac{a(V - E_L) - w}{\tau_w} \quad (2)$$

$$\frac{g_E}{dt} = \frac{-g_E}{\tau_E} \quad (3)$$

$$\frac{g_I}{dt} = \frac{-g_I}{\tau_I} \quad (4)$$

Euler integration was used, with 1-ms steps. The exponential term gives an upswing of the action potential (when the input current, $g_E(E_E - v) + g_I(E_I - v) + I_{offset}$, drives the membrane potential beyond V_T), which is stopped when the potential reaches 0 mV, and the downswing (in the next simulation step) results from the reset condition: v is given the value of V_r, and w is incremented by b. If the neuron has a negative (positive) connection, the inhibitory (excitatory) conductance of the postsynaptic neuron is incremented by the synaptic gain (0.003 μS) multiplied by the weight.

The values of the parameters used in this paper give tonic spiking when the input current is constant (above about 0.2204 nA): leak conductance $g_L = 0.01\,\mu S$, rest potential $E_L = -70\,mV$, slope factor $\delta = 2\,mV$, threshold potential $V_T = -50\,mV$, excitatory reversal potential $E_E = 0\,mV$, inhibitory reversal potential $E_I = -70\,mV$, offset current $I_{offset} = 0$ for internal neurons and $I_{offset} = 0.5\,nA$ for output nodes, membrane capacitance $C = 0.2\,nF$, adaptation coupling $a = 0.002\,\mu S$, adaptation time constant $\tau_w = 30\,ms$, synaptic time constants $tau_E = tau_I = 5\,ms$, reset voltage $V = -58\,mV$, and adaptation increment $b = 0\,nA$.

2.2 Animats and Their Environment

The animat has two sensors and two actuators. The state of the sensors depends on the amount of the signal received from the targets (which can be seen as,

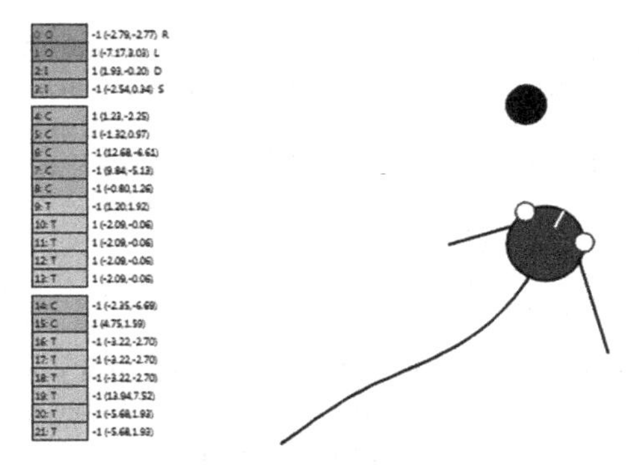

Fig. 1. The genome and the animat. The genome of animat shown in Fig. 3 is shown as an example (*left*); it consists of two sets of a series of cis (dark green) and trans (pink) elements, thus encoding two internal nodes. The four elements on top encode the outputs (green) and inputs (gray). The animat (*right*) has two sensors in front (light circles) and two actuators behind them (the direction of the thrust is shown by black straight lines); the trace of the movement for the first 500 ms is indicated by the curvy line; as the animat approaches the target on its left, the right actuator is slightly more active. (Color figure online)

for example, light intensity or scent [3]). The signal coming from a given target decreases with the Euclidean distance (d_{Euc}) from this target (as $\frac{1}{1+0.2d_{Euc}}$), and reaches maximum (1.0) at zero distance. The signal coming from all the targets is summed. When a target is reached, it disappears, and the signal field changes instantaneously. The activation of the sensor (S_L and S_R, Fig. 1) is equal to the value of signal at the sensor's location.

The sensory information is provided to the neurons that connect to the input nodes. The state of one of these nodes (S, for sum) depends on the average activation of both sensors on the animat ($\frac{2}{1+e^{-\gamma_{avg}(S_R+S_L)}} - 1$), and the state of the other (D, for difference)—on the difference in sensors' activation ($\frac{1}{1+e^{-\gamma_{dif}(S_R-S_L)}}$). In other words, the state of the node D is 0.5 when the activation of the left and right sensor on the animat is the same, and it decreases towards 0 (or increases towards 1) when the right-left difference decreases (or increases). The steepness of the sigmoid functions is set to amplify small differences or to allow for a dynamic response even when the animat is close to several targets ($\gamma_{dif} = 10$, $\gamma_{avg} = 0.5$). At each simulation step, the state of each input node is determined, rounded to the largest previous hundredth (to simulate sensors with limited precision), and this value, multiplied by the synaptic gain and the weight of the connection to a postsynaptic internal neuron to which the input node connects, is added to the excitatory (or inhibitory, if the weight is negative) conductance of this postsynaptic neuron.

The thrust forces (Fig. 1) generated by the actuators are proportional to number of spikes of the output neurons in the previous 120 ms. The directions of the forces are such that when the activations of the actuators differ, the animat turns, but even when only one actuator is active, the animat moves in a loop rather than turning on the spot. When both previously active actuators become inactive, the motion continues due to inertia until the animat is brought to a stop by drag (proportional to velocity). This drag also imposes a maximum velocity possible.

2.3 Genetic Algorithm

Each evolutionary run consisted of 250 generations of a genetic algorithm with a constant population size of 300 individuals, with binary tournament selection (draw two, keep the better one), and elitism (10 individuals). The genomes of the animats in the initial population had, apart from two input and two output elements, three series of cis and trans elements (the number of cis and trans in each series was drawn from the normal distribution with mean and standard deviation both equal to 3.0, rounded to the largest smaller integer; all numbers below 1.0 were binned to 1). Coordinates in genetic elements were determined by drawing a random direction and a random distance from $(0, 0)$ using a uniform distribution. Genetic operators were changes of coordinates (with probability 0.005 per gene; coordinate change causes the point corresponding to the element to move in the abstract 2-dimensional space by a distance drawn from a normal distribution), deletions, and duplications of individual elements. The probabilities of deletions and duplications were 0.00375 and 0.0025, respectively, creating a mutational pressure for short genomes.

The genetic algorithm aimed to minimize the average value of the fitness function over 5 random maps with 20 targets each, $f_{fit} = 1 - \frac{c_{targets}}{20})$, where $c_{targets}$ is the amount of targets reached. The animats were allowed to move for 24000 ms during evolution, but when analysing the champions after each run (and testing robustness), the fitness was re-evaluated by averaging for 1000 random maps with 20 targets and 48000 ms for each map.

Since the output nodes spike at a constant frequency without input from internal neurons (because of the positive offset current), most animats in the initial population moved, although not directionally.

3 Results and Discussion

Although the foraging task considered here is quite simple, corresponding to Braitenberg vehicle 2b [1], it is not completely trivial. This is because the number of targets gets smaller with each find, changing the activation of the sensors on the animat, and the control has to be tuned to the physics of the environment (the drag forces) and the animat (the thrust of the actuators). In fact, only about 6 % of the independent runs resulted in networks that allowed the animat to find nearly all the targets on any map (fitness function below 0.1, meaning that at most about 2 targets out of 20 were left on average). Only these champions were analysed further (9 from 159 independent runs conducted with Gaussian noise on the membrane potential with standard deviation 5 mV; 13 from 209

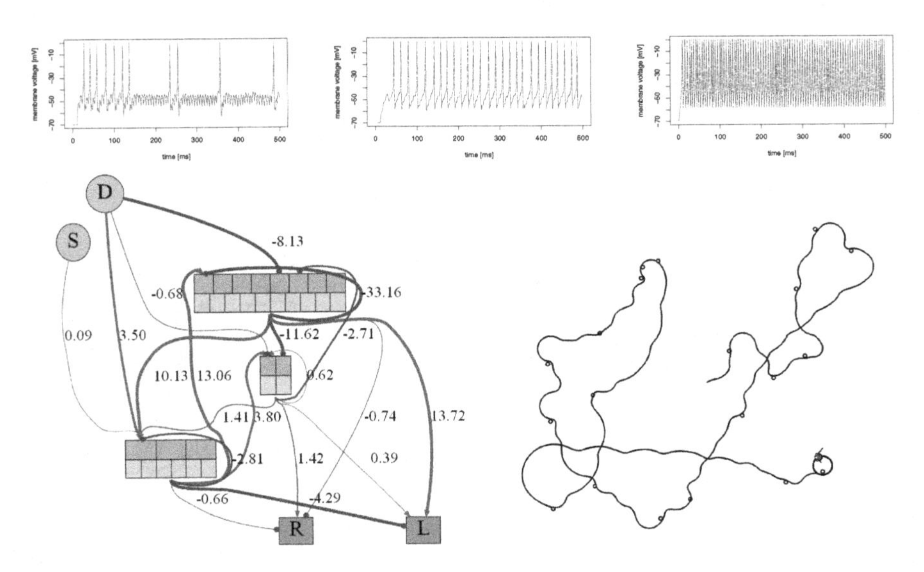

Fig. 2. The behavior of the best animat in the cohort of 13 champions evolved without noise. See text for details.

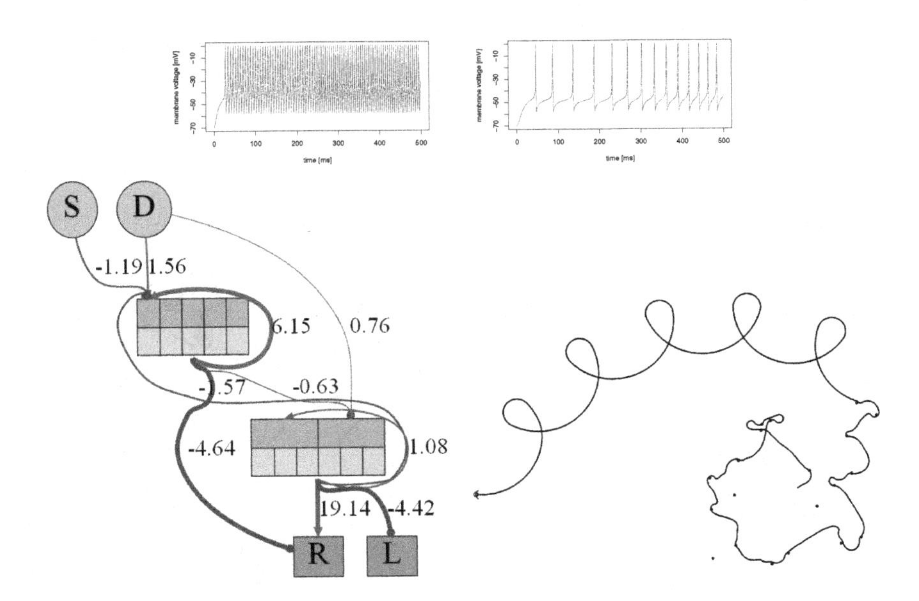

Fig. 3. The behavior of the worst animat in cohort of 13 champions evolved without noise. See text for details.

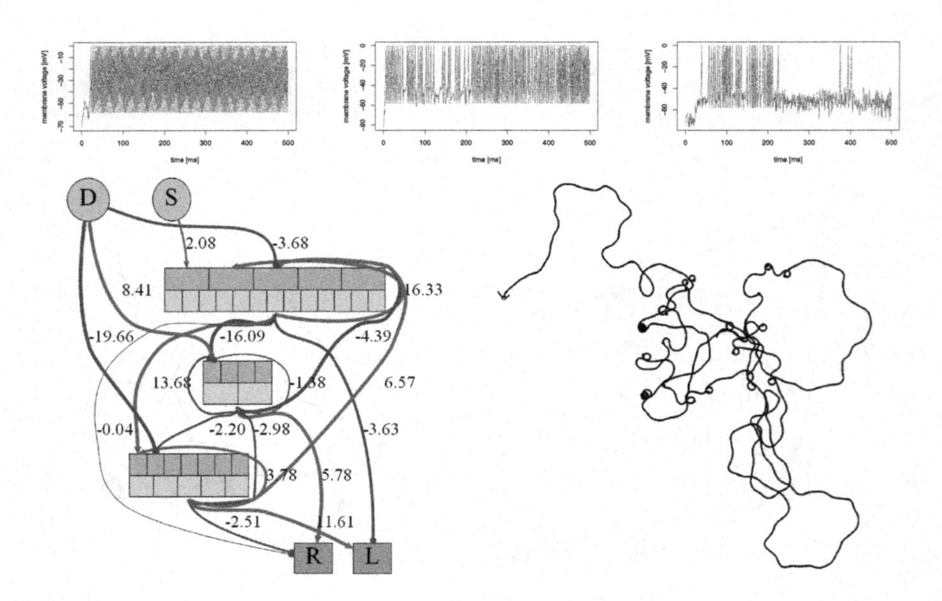

Fig. 4. The behavior of the best animat in the cohort of 9 champions evolved with noise. See text for details.

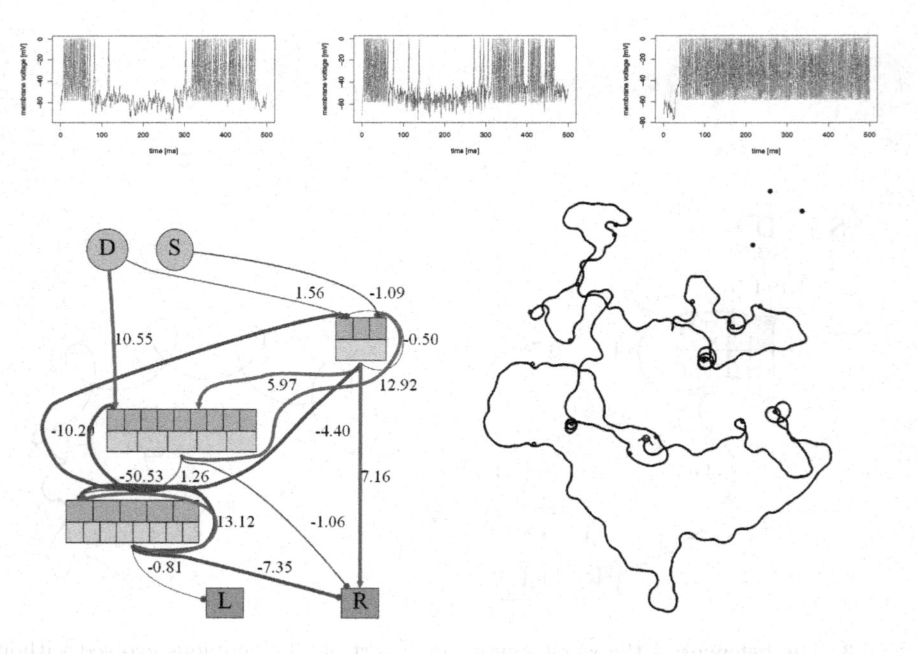

Fig. 5. The behavior of the worst animat in the cohort of 9 champions evolved with noise. See text for details.

Table 1. Comparison of the robustness of the 9 champions evolved with noise on voltage and 13 evolved without. The fitness function values (all the columns but the last 2) have been obtained for 1000 random maps and 48000 ms; the last 2 columns indicate the number of cis and the trans elements in the genomes. The values of the fitness function below 0.25 are highlighted in red, the values below 0.33 in green, and below 0.40 in grey. The last row shows the results of the comparison between 9 and 13 values in each column using the Anderson-Darling k-samples test, with values below 0.05 highlighted in red. See text for more details.

original f_fit	Min A 15%	Max A 15%	Int I -0.2	Int I 0.2	Out I 0.2	Out I 0.8	gain 0.0025	gain 0.0035	Vr -68	Vr -48	EL -100	EL -40	Sum cis	Sum trans
Evolved with noise on voltage (SD 5 mV)														
0.015	0.293	0.308	0.123	0.135	0.012	0.028	0.304	0.353	0.183	0.168	0.140	0.194	16	20
0.017	0.042	0.151	0.062	0.169	0.043	0.024	0.130	0.267	0.520	0.211	0.093	0.245	15	22
0.030	0.144	0.201	0.170	0.097	0.088	0.033	0.361	0.224	0.086	0.032	0.325	0.164	20	33
0.035	0.154	0.184	0.112	0.109	0.231	0.089	0.139	0.257	0.186	0.328	0.093	0.082	14	15
0.064	0.141	0.262	0.259	0.184	0.388	0.181	0.123	0.046	0.175	0.291	0.292	0.127	10	18
0.086	0.444	0.467	0.623	0.106	0.071	0.199	0.869	0.087	0.315	0.376	0.802	0.362	16	8
0.088	0.244	0.287	0.214	0.416	0.383	0.124	0.328	0.416	0.133	0.125	0.228	0.388	18	17
0.089	0.163	0.220	0.434	0.442	0.363	0.141	0.412	0.235	0.389	0.058	0.391	0.408	18	21
0.095	0.183	0.263	0.135	0.143	0.375	0.166	0.321	0.186	0.805	0.825	0.260	0.119	17	14
Evolved without noise on voltage														
0.000	0.001	0.211	0.995	0.935	0.669	0.497	0.911	0.952	0.829	0.784	0.999	0.934	13	18
0.008	0.176	0.284	0.991	0.991	0.998	0.632	0.565	0.556	0.427	0.870	0.991	0.992	14	8
0.048	0.250	0.651	0.997	0.983	0.998	0.664	0.564	0.341	0.414	0.972	0.982	0.994	8	13
0.053	0.437	0.587	0.994	0.927	0.999	0.722	0.949	0.995	0.946	0.995	0.992	0.854	14	8
0.062	0.218	0.491	0.447	0.952	0.999	0.224	0.632	0.281	0.677	0.986	0.998	0.982	10	21
0.066	0.400	0.464	0.997	0.998	0.995	0.571	0.992	0.810	0.402	0.765	0.997	0.998	15	20
0.074	0.287	0.392	0.995	0.988	0.999	0.727	0.765	0.836	0.233	0.981	0.995	0.978	15	13
0.087	0.430	0.461	0.744	0.527	0.997	0.584	0.622	0.122	0.831	0.989	0.973	0.792	16	14
0.088	0.431	0.450	0.997	0.995	0.998	0.501	0.940	0.766	0.636	0.992	0.999	0.994	7	9
0.090	0.294	0.353	0.995	0.993	0.998	0.779	0.907	0.739	0.475	0.970	1.000	0.991	11	9
0.096	0.224	0.384	0.913	0.996	0.287	0.357	0.896	0.778	0.268	0.994	0.996	0.923	18	14
0.098	0.280	0.299	0.998	0.970	0.998	0.443	0.998	0.925	0.997	0.926	0.999	0.980	11	18
0.099	0.116	0.481	0.987	0.997	0.998	0.427	0.766	0.161	0.236	0.943	0.988	0.995	7	11
AD test p-value	0.18	0.0052	5.4E-5	3.6E-5	0.00025	3.6E-5	0.0002	0.006	0.014	7.9E-5	3.8E-5	3.6E-5	0.019	0.055

independent runs without noise, Table 1; all the re-evaluations shown in the table were performed using the noise levels at which the networks were evolved).

The results clearly indicate that networks evolved with noise were much more robust to other perturbations (Table 1). Nearly all of such networks (with the exception of one) allowed to find at least two thirds of the targets when the activity of one of the actuators was increased by 15 %, and the activity of the other actuator was decreased by 15 % (Table 1, column 2 and 3). Some animats performed better when the left actuator had higher and the right lower activity; for the other animats the inverse situation resulted in better performance, so Table 1 shows the results for the better scenario in column 2 and for the worse scenario in column 3. The networks evolved without noise did not fare well in the worse scenario (see Table 1 for *p*-values of all the statistical tests; all tests were done using the Anderson-Darling k-sample test; I also used, when appropriate, the two-sample Kolmogorov-Smirnov test, the *p*-values for the two tests were very close in value).

Similarly, most of networks evolved with noise performed well when the internal neurons were given either positive or negative offset current of 0.2 nA (instead of 0; Table 1; column 4 and 5), when the offset current for the output nodes was decreased or increased by 0.3 nA (from 0.5 to 0.2 or 0.8, respectively; column 5 and 6), when synaptic gain was increased or decreased by 0.0005 μS (from 0.003 to 0.0025 or 0.0035; column 7 and 8), reset voltage decreased or increased by 10 mV (from 58 to −68 or −48; column 9 and 10), or rest potential decreased or increased by 30 mV (from 70 to −100 or −40; column 11 and 12). None of the networks evolved without noise performed well in these scenarios.

Two networks evolved with noise were not affected much even by a change from tonic spiking to tonic spiking with adaptation (by setting the adaptation increment $b = 0.1$ nA), the fitness function changed in these two cases from 0.017 to 0.119 and from 0.035 to 0.089, and two other networks could perform even with Gaussian noise with standard deviation 7 mV (the fitness function changed from 0.015 to 0.227, and from 0.030 to 0.109), while only one network without noise gave good behaviour with noise up to 2 mV (the fitness function changed from 0.074 to 0.198).

In general, the networks evolved with noise were encoded by slightly longer genomes than the networks evolved without noise (although all the networks in Table 1 had 3 internal nodes, with the exception of the one in the last row, shown in Figs. 1 and 3). This difference can be attributed mostly to a higher number of cis elements in the genomes evolved with noise (Table 1; column 13 and 14).

The detailed analysis of the topology of the networks indicates that all networks have a neuron spiking with high frequency and inhibiting one of the actuators (Figs. 2, 3, 4 and 5, the best and worst in each cohort is shown; the *top* panels in all four figures show the voltage in the first 500 ms of the three, or two, internal neurons; the order from left to right on the *top* corresponds to the order from top to bottom in the graph on the *bottom left* panel of each figure, D is for input node whose state corresponds to the difference in animat sensors, S is for the node whose state corresponds to the sum/average of the sensors, R and L indicate output nodes regulating right and left actuators, respectively, red edges correspond to excitatory, and blue to inhibitory connections; the *bottom right* shows the trajectory of the animat over 48000 ms).

For example, the network of the best animat in the cohort of 13 evolved without noise (Fig. 2) this high-frequency neuron is internal neuron 3, the one inhibiting stronger the left actuator. The network has also a self-sustaining loop between neurons 1 and 3, a weaker loop between 2 and 3, the neuron 2 activates itself very weakly, and neuron 1 inhibits itself strongly. The sustained activity of neurons 2 and 3 results in a continued circular counterclockwise movement after all the targets are reached (and the sensors are quiet); since the left actuator is inhibited (by neuron 3), the animat turns to the left.

The small network in Fig. 3 is a less efficient controller; the trajectory shows that three targets have been missed on this map (filled dark circles in the top right). With small activation of the sensors, and driven mostly by self-sustained spiking of both neurons (neuron 1 with higher frequency), the animat continues to move with the left actuator more activated than the right, and will not reach the targets if the simulation is continued.

In contrast, the animat in Fig. 4, evolved with noise, reaches all the targets, and—with no activation of the sensors—continues to move without any clear pattern, driven by noise, and sustained activity of the network (with neuron 1 at the highest frequency, much higher than the spiking frequency of the other neurons, but with neuron 3 having the most complex spiking pattern, with periods of activity separated by relative quiescence, apparent also in the first 500 ms, Fig. 4 *top*). Note that the noise prevents the animat to reach the target as precisely as is possible for the animats evolved and acting without noise (most targets are reached after an approach along a circular local trajectory).

The same is true for the worst animat in the cohort of 9 evolved with noise. This animat is relatively less efficient than the best one, but on this map it manages to collect all the missing 3 targets when the simulation is continued beyond 48000 ms shown (and used to reevaluate fitness); after all the targets are collected, the sustained activity of all neurons (with the highest frequency of neuron 3, and complex spiking pattern of neuron 2) continues.

4 Conclusion and Future Work

The main conclusion is that both an efficient control and robustness to perturbations to the value of the parameters of neurons can be evolved in the presence of noise, similarly to our previous results showing that robustness to noise in evolving genetic networks promotes robustness to damage [4]. The noise used in this paper was on one of the state variables of the spiking neural network; it will be interesting to see if other models of noise and perturbations (for example, relevant to neuromorphic hardware [6] will give similar results. Importantly, the networks in the model used here were encoded in an artificial genome in a way that does not limit in principle, and allows the evolutionary process to vary, the number of nodes and connections in the network. It remains to be seen if the evolution for more complex tasks and behaviors, possibly requiring more complex networks, is possible in this system.

Acknowledgments. This work was supported by Polish National Science Centre (project EvoSN, UMO-2013/08/M/ST6/00922). I am grateful to Volker Steuber for discussions, and to Ahmed Abdelmotaleb and Michal Joachimczak for their involvement in the development of GReaNs software platform.

References

1. Braitenberg, V.: Vehicles. Experiments in Synthetic Psychology. MIT Press, Cambridge (1986)
2. Brette, R., Gerstner, W.: Adaptive exponential integrate-and-fire model as an effective description of neuronal activity. J. Neurophysiol. **94**(5), 3637–3642 (2005)
3. Joachimczak, M., Wróbel, B.: Evolving gene regulatory networks for real time control of foraging behaviours. In: Proceedings of the Alife XII Conference, pp. 348–358. MIT Press, Cambridge (2010)

4. Joachimczak, M., Wróbel, B.: Evolution of robustness to damage in artificial 3-dimensional development. Biosystems **109**(3), 498–505 (2012)
5. Naud, R., Marcille, N., Clopath, C., Gerstner, W.: Firing patterns in the adaptive exponential integrate-and-fire model. Biol. Cybern. **99**(4–5), 335–347 (2008)
6. Stromatias, E., Neil, D., Pfeiffer, M., Galluppi, F., Furber, S.B., Liu, S.C.: Robustness of spiking deep belief networks to noise and reduced bit precision of neuro-inspired hardware platforms. Front. Neurosci. **9** (2015). paper number 222
7. Wagner, A.: Robustness and Evolvability in Living Systems. Princeton University Press, Princeton (2013)
8. Wróbel, B., Abdelmotaleb, A., Joachimczak, M.: Evolving networks processing signals with a mixed paradigm, inspired by gene regulatory networks and spiking neurons. In: Di Caro, G.A., Theraulaz, G. (eds.) International Conference on Bio-Inspired Models of Network, Information, and Computing Systems, BIONETICS 2012. LNICST, vol. 134, pp. 135–149. Springer, Heidelberg (2014)
9. Wróbel, B., Joachimczak, M.: Using the genetic regulatory evolving artificial networks (GReaNs) platform for signal processing, animat control, and artificial multicellular development. In: Kowaliw, T., Bredeche, N., Doursat, R. (eds.) Growing Adaptive Machines. SCI, vol. 557, pp. 187–200. Springer, Heidelberg (2014)

Anomaly Detection with the Voronoi Diagram Evolutionary Algorithm

Luis Martí[1,2(✉)], Arsene Fansi-Tchango[3], Laurent Navarro[3], and Marc Schoenauer[1]

[1] TAO Team, CNRS/INRIA/LRI, Université Paris-Saclay, Paris, France
luis.marti@inria.fr
[2] Universidade Federal Flumnense, Niterói, RJ, Brazil
[3] Thalés Research, Paris, France

Abstract. This paper presents the Voronoi diagram-based evolutionary algorithm (VorEAl). VorEAl partitions input space in abnormal/normal subsets using Voronoi diagrams. Diagrams are evolved using a multi-objective bio-inspired approach in order to conjointly optimize classification metrics while also being able to represent areas of the data space that are not present in the training dataset. As part of the paper VorEAl is experimentally validated and contrasted with similar approaches.

1 Introduction

Anomalous Internet traffic detection is a major question of computer network security. Intrusion detection systems (IDSs) [9] have proposed with the intention of tackling this issue. They are meant to protect a network by providing a line of defense that is able to detect and react to network attacks. Two main approaches are used when building an IDS: (i) misuse-based and (ii) anomaly-based detection. While the former focuses on detecting attacks that follow a known pattern or signature, the latter is interested in building a model representing the system's normal behavior while assuming all deviated activities to be anomalous or intrusions. Because of that fact anomaly detection has received increasing attention in the recent past.

Anomaly detection has been addressed with different approaches (see [2] for a survey). Among nature-inspired approaches artificial immune systems (AISs) [7] have received an special attention.

This paper proposes the Voronoi diagram-based evolutionary algorithm (VorEAl). VorEAl is inspired on AISs and the representations that had been proposed for evolutionary shape design consolidating previous progresses made in this direction [8]. Its main distinctive feature is that it evolves Voronoi diagram-based representations for normal/abnormal regions of the search space. Such representation offers a flexible and compact alternative to some common representations used in AIS such as hyper-spheres and hyper-rectangles. VorEAl applies a multi-objective approach that takes into account the detection accuracy and other especially devised volume-based methods that promotes the emergence of solutions that also adequately represent areas of the input space where

J. Handl et al. (Eds.): PPSN XIV 2016, LNCS 9921, pp. 697–706, 2016.
DOI: 10.1007/978-3-319-45823-6_65

no normal data has been received and, therefore, should represent anomalies. As in any multi-objective approach, the algorithm produces a set of trade-off solutions. VorEAl applies a committee approach that is based on the best (in term of *a priori* given set of preferences) subset of those solution.

The paper is organized as follows. Section 2 presents the context of AIS and some existing approaches to anomaly detection. Section 3 introduces the Voronoi representation for abnormal and normal input subsets, together with the variation operators and objective functions used to evolve it and VorEAl as a whole. Section 4 introduces our methodology for the experimental validation of VorEAl, also presenting the results of the study and comparing them with other approaches from the literature. Finally, Sect. 5 discusses the results and sketches some further research directions.

2 Foundations

There has been a consistent interest by the community on proposing nature-inspired approaches to anomaly detection. In this context, AISs have attracted attention as they embody an analogy to the biological immune system. They are particularly appealing for anomaly detection problems as they capture the ability of the biological system of telling apart normal body cells from pathogens. That is, from a computational perspective, they create a model that is able to discriminate between normal (self) and abnormal (non-self) objects. This feature make AISs specially suited to be applied in the context of anomaly-based IDSs.

In order to extend AISs' performance it is necessary to apply algorithms that combine a powerful representation capacity as well as the possibility of adequately adapting that capacity to meet the problem characteristics.

Voronoi diagrams are geometrical constructs that were known by the ancient Greeks. Any set of points (aka *Voronoi sites*) in some n-dimensional Euclidean space $\mathcal{E}$ defines a *Voronoi diagram*, i.e., a partition of that space into *Voronoi cells*: the cell corresponding to a given site S is the set of points whose closest site is S. The boundaries between Voronoi cells are the medians of the $[S_iS_j]$ segments, for neighbor Voronoi sites S_i and S_j. Though originally defined in two or three dimensions, there exist several algorithmic procedures to efficiently compute Voronoi diagrams in any dimension.

Voronoi diagrams offer a compact representation for shapes (surfaces in 2D, volumes in 3D, for instance), by attaching to each Voronoi cell (or, equivalently, to the corresponding Voronoi site), a Boolean label. The resulting Voronoi diagram is a partition of the space into 2 subsets: the "true" cells are the shape/volume, and the "false" cells are the outside of the shape/volume. The *genotype* is here a (variable length) list of labeled Voronoi sites, and the phenotype is the corresponding partition in the space into two subsets. More generally, any piece-wise constant function on the underlying space can be represented by a similar representation by using real-valued labels. Such representation has been successfully used in the context of Evolutionary Optimum Design [5,10]. In particular, it has been demonstrated that the local complexity of the representation can also be adjusted by evolution: in regions of the space where the shape has a complex boundary, several Voronoi sites will be used, whereas only a few of them will be necessary elsewhere.

In the context of classification, the target phenotypes are partitions of the parameter space into positive and negative examples (in the case of 2 classes), and can hence also be represented by Voronoi diagrams with Boolean labels —or with labels taken from a finite alphabet in the case of more than 2 classes.

3 The Voronoi Diagrams-Based Evolutionary Algorithm

We now discuss the building blocks of VorEAl. In particular, we present variation operators, the possible strategies used for evaluating the individuals and how these elements are assembled together to form the algorithm.

3.1 Variation Operators

The genotypes of Voronoi representations is a variable length list of Voronoi sites $(S_1, \ldots, S_p)$, with $p \in [P_{\min}, P_{\max}]$, where each site is defined by its n coordinates in $\mathcal{E}$. Each site S has an associated label $S.\ell$ that determines how a point that falls within the corresponding cell is classified.

```
function mutate_voronoi(I, p_s, p_f, p_t, p_+, p_-, η)
  ▷ I, individual to be mutated.
  ▷ p_s ∈ [0, 1], prob. of mutating a site.
  ▷ p_f ∈ [0, 1], prob. of mutating a site feature (coordinate).
  ▷ p_t ∈ [0, 1], prob. of changing the label of a site.
  ▷ p_+ ∈ [0, 1], prob. of adding a new site.
  ▷ p_- ∈ [0, 1], prob. of removing a site.
  ▷ η ∈ (0, ∞], learning rate.
  for all S ∈ I do
    if U[0, 1) < p_s then
      for all x ∈ S do
        if U[0, 1) < p_f then
          x ← mutate_log_normal(x, η)
      if U[0, 1) < p_t then
        S.ℓ ← switch_label(S.ℓ).
  if U[0, 1) < p_+ then
    I ← I ∪ {random_site}.
  if U[0, 1) < p_- then
    i ← U[1, |I|); I ← I \ {I(i)}.
  return I, mutated individual.
```

Fig. 1. Mutation of a Voronoi diagram.

Mutation Operator. Several mutation operators can be designed for such a variable-length representation.

- At the individual level, a Voronoi site can be added, at a randomly chosen position, with a random label; or a randomly chosen Voronoi site can be removed.
- At the site level, Voronoi sites can be moved around in the space – and the well-known self-adaptive Gaussian mutation has been chosen here, inspired by Evolution Strategies (see (1) below); or the label of a Voronoi site can be changed.

In the self-adaptive Gaussian mutation [11], each coordinate x of each Voronoi site also "carries" its own variance σ that is used for its Gaussian mutation. Coordinate x undergoes Gaussian mutation with variance σ while σ undergoes a log-normal mutation with learning rate η as follows:

$$x \leftarrow \sigma\mathcal{N}(x,1) \text{ and } \sigma \leftarrow \sigma e^{\eta\mathcal{N}0,1} \tag{1}$$

The different mutation operators are applied according to different probabilities, following the procedure described in Fig. 1.

function $\texttt{crossover\_voronoi}(\mathcal{I}_1, \mathcal{I}_2)$
$\triangleright$ $\mathcal{I}_1$ and $\mathcal{I}_2$, individuals to mate.
repeat
$\boldsymbol{P} \leftarrow \texttt{random\_hyperplane}(\mathcal{I}_1 \cup \mathcal{I}_2)$.
$\xi_1^{(1)}, \xi_2^{(1)} \leftarrow \texttt{split\_individual}(\mathcal{I}_1, \boldsymbol{P})$.
$\xi_1^{(2)}, \xi_2^{(2)} \leftarrow \texttt{split\_individual}(\mathcal{I}_2, \boldsymbol{P})$.
until $\xi_k^{(i)} \neq \varnothing, \forall i, k$
$\mathcal{O}_1 = \xi_1^{(1)} \cup \xi_2^{(2)}$; $\mathcal{O}_1.\ell = \mathcal{I}_1.\ell$;
$\mathcal{O}_2 = \xi_1^{(2)} \cup \xi_2^{(1)}$; $\mathcal{O}_2.\ell = \mathcal{I}_2.\ell$.
return $\mathcal{O}_1, \mathcal{O}_2$, offspring.

(a) Crossover of two Voronoi diagrams, applied with probability p_c.

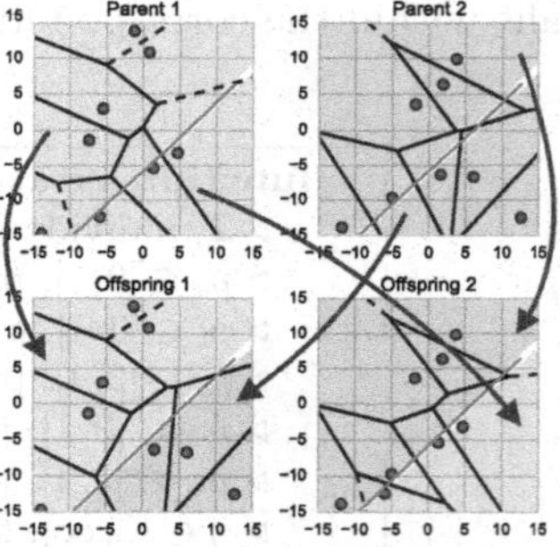

(b) Example of crossover of two Voronoi genotype individuals in 2D.

Fig. 2. Crossover operator for Voronoi diagrams.

Crossover Operator. The crossover operator for Voronoi representation should not simply exchange some Voronoi sites between both parents, but should respect the locality of the representation. Voronoi sites that are close to each other should have more chance to stay together than Voronoi sites that are far apart. This is achieved by the geometric crossover that operates on two (randomly selected) parents by creating a random cutting hyperplane, and exchanges the Voronoi sites from both sides of the hyperplane. The Voronoi diagrams are of course reconstructed after the crossover. This procedure is described in detail in Fig. 2a. A two-dimensional example is given in Fig. 2b.

3.2 Objectives and Fitness Assignment

Anomaly detection can be posed as a particular case of classification problem where data items must be tagged either as "normal" or "anomalous". That relying on a dataset $\Psi = \{\boldsymbol{x}^{(i)}, y^{(i)}\}$ where, without loss of generality we can state that $\boldsymbol{x} \in \mathbb{R}^n$ and $y^{(i)} \in \{\text{normal}; \text{anomaly}\}$ obtain a classifier that correctly detects instances that correspond to each of the two categories. Because of this fact the existing metrics devised to assess the quality of a classification algorithm are also applicable in this context. For this particular problem, the most relevant metrics are accuracy, recall and specificity, although many more could also be of use. Accuracy seems the best choice in the general case, as one wants to correctly identify all examples. But when dealing with anomalies, the dataset is generally highly imbalanced, as normally there are fewer anomalous instances than 'normal' ones. If only the classification accuracy is used, the error contribution of the anomalies will be reduced and hence the model will be biased to not regard them.

Furthermore, as already mentioned, the anomaly detection problem requires that the classifier is not only able to correctly classify the "normal" and "anomalous" instances present in the training dataset but is also capable of detecting when a given input falls in an area that was not covered by data of the training set and, therefore, also can be interpreted as an anomaly.

It is possible to prompt the Voronoi diagrams (individuals) to represent the known data in a form as compact as possible by expressing that as the relation between the volumes of the Voronoi cell and the convex hull of the training data that it contains. Let $\mathcal{I} = \{S_i, i = 1 \dots n_{\mathcal{I}}\}$ be a Voronoi diagram, and, for each cell C_i, let $v_i \in \mathbb{R}$ be its volume and $\mathcal{D}_i$ the set of data points it contains, i.e., $\mathcal{D}_i = \{\boldsymbol{x} \in \Psi; d(x, S_i) \leq d(x, S_j) \forall i \neq j\}$, d being the n-dimensional Euclidian distance. We can then define the individual compactness as the sum, for each cell, of the ratio of the volume of the convex hulls of $\mathcal{D}_i$ and the volume of the cell,

$$\mathrm{C}(\mathcal{I}) = \begin{cases} \sum_i \frac{\text{volume}(\text{convex\_hull}(\mathcal{D}_i))}{v_i} & \text{if } |\mathcal{D}_i| > n, \\ 0 & \text{in other case.} \end{cases} \tag{2}$$

It could be hypothesized that the previous formulation can be improved by adding a multiplicative term that counts the number of elements in $\mathcal{D}_i$, resulting in the multiplicative compactness

$$\mathrm{C}_{\text{mult}}(\mathcal{I}) = \begin{cases} \sum_i (|\mathcal{D}_i| - n) \frac{\text{volume}(\text{convex\_hull}(\mathcal{D}_i))}{v_i} & \text{if } |\mathcal{D}_i| > n, \\ 0 & \text{in other case.} \end{cases} \tag{3}$$

In both cases, maximizing the compactness will produce cells that contain the data in a form as tight as possible. Those compactness objectives can be complemented by one that promotes the existence of empty cells that represent areas of the input domain that are now present in the training data. Such objective would take care of sites with small $\mathcal{D}_i$'s and promote that they become empty as the evolution takes place. A form of representing this is by computing the total volume of cells with an anomaly label of an individual and rate it by the number of elements it contains,

$$\mathrm{EV}(\mathcal{I}) = \sum_{i, S_i.\ell=\text{anomaly}} \frac{v_i}{1 + 2\ln(|\mathcal{D}_i| + 1)}. \quad (4)$$

Consequently, it is obvious that it is necessary to jointly address all of those objectives. Therefore, a multi-objective optimization approach will empower the algorithm with the capacity to address all the requirements of the task at the same time.

3.3 Algorithm Description

VorEAl consolidates the previous components as an algorithm that constructs a classification model. The algorithm starts by creating an initial random population $\mathcal{P}_0$ of n_{pop} individuals. At a given iteration t, individuals in the population $\mathcal{P}_t$ are then mutated and mated using operators described above and thus producing an offspring population $\mathcal{P}_{\text{off}}$ that consists of n_{off} individuals. At this point, individuals that have not yet been evaluated are presented with the dataset and the values of the different objective functions are calculated. In this work, we compute the accuracy, recall and specificity, but it should be noted that others are available. From the union of $\mathcal{P}_t$ and $\mathcal{P}_{\text{off}}$, the best n_{pop} are selected using the non-dominated sorting selection of NSGA-II [3].

This process repeats until the stopping criterion of the algorithm is met. When that happens, the algorithm has a final population $\mathcal{P}_{\text{final}}$ from which the best individual(s) can be selected to represent the 'self' of the AIS. This a non-trivial task as it implies taking into account the different conflictive objectives. In this work, we select a committee of individuals $\mathcal{P}_{\text{committee}} \subseteq \mathcal{P}_{\text{final}}$ that contains the ρ-percent of $\mathcal{P}_{\text{final}}$ with the highest accuracy. Hence, the classifier returns the most voted decision among the members of $\mathcal{P}_{\text{committee}}$.

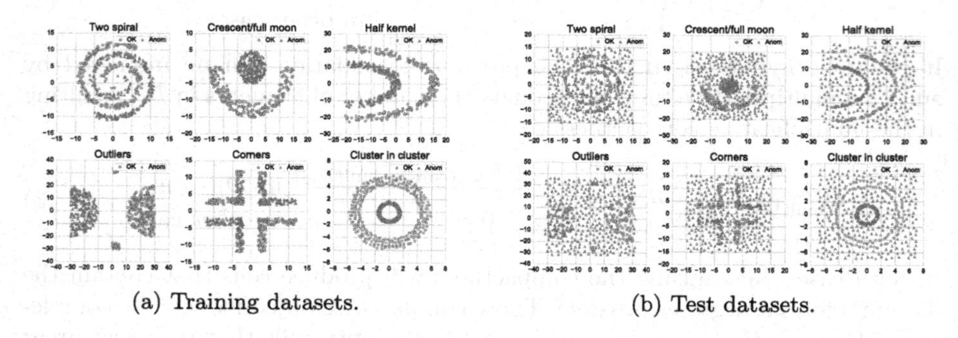

(a) Training datasets. (b) Test datasets.

Fig. 3. Training and testing datasets. Test set anomalies present in the test datasets are generated using the procedure described in Sect. 4.

4 Experimental Study

The previous discussion and proposal must be complemented by a set of experiments that establish the validity of VorEAl and studies the impacts of the different components presented. That is the focus of this section.

One of the main questions regarding VorEAl is at what point a multi-objective affinity function would actually generate better results at an admissible cost. It could be argued that there exists the possibility that adding more objectives would just make the search process more complex and resource demanding.

An important matter to be clarified was the impact of each of the objectives presented in previous section. For that reason different combinations were tested. In particular, we tested accuracy and compactness (a/c); accuracy, compactness and total empty volume (a/c/t); accuracy and multiplicative compactness (a/m) and accuracy, multiplicative compactness and total empty volume (a/m/t).

In order to provide grounds for comparison with similar approaches as well as well-known approaches, other methods were included in the experiments. In particular, we included the negative selection algorithm (NSA) [6] using both variable-sized hyper-spheres and hyper-rectangles. For fair comparisons, we applied the NSA^{+}_{sp} and NSA^{+}_{re} in which non-self training samples are subsequently used to enrich the detector library generated by NSA.

Similarly, we have included in the experiments two well-known classifiers: one-class vector machines (SVMs) [12] and the naïve Bayes classifier.

Table 1. Summary of the outcome of the statistical hypothesis tests for each problem and performance indicator. When an algorithm in the row has been significantly better than the one in the column the corresponding cell is marked with a "+". If it has been worst then the cell contains a "−". Cases where no significant difference was established are identified with a "∼".

Two Spiral

		VorEAl (a/c)	VorEAl (a/c/t)	VorEAl (a/m)	VorEAl (a/m/t)	NSA-re	NSA-sp	one-class SVM	naïve Bayes
Accuracy	VorEAl (a/c)	·	−	−	−	+	+	−	+
	VorEAl (a/c/t)	+	·	−	−	+	+	−	+
	VorEAl (a/m)	+	+	·	−	+	+	∼	+
	VorEAl (a/m/t)	+	+	+	·	+	+	+	+
	NSA-re	−	−	−	−	·	−	−	−
	NSA-sp	−	−	−	−	+	·	−	+
	one-class SVM	+	+	∼	−	+	+	·	+
	naïve Bayes	−	−	−	−	+	−	−	·
Recall	VorEAl (a/c)	·	−	−	−	+	+	−	+
	VorEAl (a/c/t)	+	·	∼	−	+	+	−	+
	VorEAl (a/m)	+	∼	·	−	+	+	∼	+
	VorEAl (a/m/t)	+	+	+	·	+	+	+	+
	NSA-re	−	−	−	−	·	−	−	−
	NSA-sp	−	−	−	−	+	·	−	−
	one-class SVM	+	+	∼	−	+	+	·	+
	naïve Bayes	−	−	−	−	+	+	−	·
Specificity	VorEAl (a/c)	·	−	−	−	−	−	−	+
	VorEAl (a/c/t)	+	·	∼	−	+	−	∼	+
	VorEAl (a/m)	+	∼	·	−	+	−	∼	+
	VorEAl (a/m/t)	+	+	+	·	+	−	+	+
	NSA-re	+	−	−	−	·	−	−	+
	NSA-sp	+	+	+	+	+	·	+	+
	one-class SVM	+	∼	∼	−	+	−	·	+
	naïve Bayes	−	−	−	−	−	−	−	·

Crescent/full moon

		VorEAl (a/c)	VorEAl (a/c/t)	VorEAl (a/m)	VorEAl (a/m/t)	NSA-re	NSA-sp	one-class SVM	naïve Bayes
Accuracy	VorEAl (a/c)	·	∼	−	−	+	+	+	∼
	VorEAl (a/c/t)	∼	·	−	−	+	+	+	∼
	VorEAl (a/m)	+	+	·	−	+	+	+	+
	VorEAl (a/m/t)	+	+	+	·	+	+	+	+
	NSA-re	−	−	−	−	·	−	−	−
	NSA-sp	−	−	−	−	+	·	−	−
	one-class SVM	−	−	−	−	+	+	·	−
	naïve Bayes	∼	∼	−	−	+	+	+	·
Recall	VorEAl (a/c)	·	∼	−	−	+	+	+	+
	VorEAl (a/c/t)	∼	·	−	−	+	+	+	+
	VorEAl (a/m)	+	+	·	∼	+	+	+	+
	VorEAl (a/m/t)	+	+	∼	·	+	+	+	+
	NSA-re	−	−	−	−	·	−	−	−
	NSA-sp	−	−	−	−	+	·	−	−
	one-class SVM	−	−	−	−	+	+	·	−
	naïve Bayes	−	−	−	−	+	+	+	·
Specificity	VorEAl (a/c)	·	∼	−	−	+	+	+	+
	VorEAl (a/c/t)	∼	·	−	−	+	+	+	+
	VorEAl (a/m)	+	+	·	−	+	+	+	+
	VorEAl (a/m/t)	+	+	+	·	+	+	+	+
	NSA-re	−	−	−	−	·	+	−	−
	NSA-sp	−	−	−	−	−	·	−	−
	one-class SVM	−	−	−	−	+	+	·	∼
	naïve Bayes	−	−	−	−	+	+	∼	·

Half kernel

		VorEAl (a/c)	VorEAl (a/c/t)	VorEAl (a/m)	VorEAl (a/m/t)	NSA-re	NSA-sp	one-class SVM	naïve Bayes
Accuracy	VorEAl (a/c)	·	−	−	−	+	−	−	−
	VorEAl (a/c/t)	+	·	∼	−	+	∼	+	+
	VorEAl (a/m)	+	∼	·	∼	+	∼	+	+
	VorEAl (a/m/t)	+	+	∼	·	+	+	+	+
	NSA-re	−	−	−	−	·	−	−	−
	NSA-sp	+	∼	∼	−	+	·	+	+
	one-class SVM	+	−	−	−	+	−	·	−
	naïve Bayes	+	−	−	−	+	−	+	·
Recall	VorEAl (a/c)	·	−	−	−	+	−	−	−
	VorEAl (a/c/t)	+	·	∼	∼	+	+	+	+
	VorEAl (a/m)	+	∼	·	∼	+	+	+	+
	VorEAl (a/m/t)	+	∼	∼	·	+	+	+	+
	NSA-re	−	−	−	−	·	−	−	−
	NSA-sp	+	−	−	−	+	·	+	+
	one-class SVM	+	−	−	−	+	−	·	∼
	naïve Bayes	+	−	−	−	+	−	∼	·
Specificity	VorEAl (a/c)	·	−	−	−	−	−	−	−
	VorEAl (a/c/t)	+	·	−	−	+	−	+	+
	VorEAl (a/m)	+	+	·	∼	+	−	+	+
	VorEAl (a/m/t)	+	+	∼	·	+	−	+	+
	NSA-re	+	−	−	−	·	−	−	−
	NSA-sp	+	+	+	+	+	·	+	+
	one-class SVM	+	−	−	−	+	−	·	+
	naïve Bayes	+	−	−	−	+	−	−	·

Outliers

		VorEAl (a/c)	VorEAl (a/c/t)	VorEAl (a/m)	VorEAl (a/m/t)	NSA-re	NSA-sp	one-class SVM	naïve Bayes
Accuracy	VorEAl (a/c)	·	∼	∼	∼	+	+	+	+
	VorEAl (a/c/t)	∼	·	−	∼	+	+	+	+
	VorEAl (a/m)	∼	+	·	+	+	+	+	+
	VorEAl (a/m/t)	∼	∼	−	·	+	+	+	+
	NSA-re	−	−	−	−	·	−	−	−
	NSA-sp	−	−	−	−	+	·	−	−
	one-class SVM	−	−	−	−	+	+	·	+
	naïve Bayes	−	−	−	−	+	+	−	·
Recall	VorEAl (a/c)	·	∼	∼	∼	+	+	+	+
	VorEAl (a/c/t)	∼	·	∼	∼	+	+	+	+
	VorEAl (a/m)	∼	∼	·	∼	+	+	+	+
	VorEAl (a/m/t)	∼	∼	∼	·	+	+	+	+
	NSA-re	−	−	−	−	·	−	−	−
	NSA-sp	−	−	−	−	+	·	−	−
	one-class SVM	−	−	−	−	+	+	·	∼
	naïve Bayes	−	−	−	−	+	+	∼	·
Specificity	VorEAl (a/c)	·	∼	∼	∼	+	+	+	+
	VorEAl (a/c/t)	∼	·	∼	∼	+	+	+	+
	VorEAl (a/m)	∼	∼	·	∼	+	+	+	+
	VorEAl (a/m/t)	∼	∼	∼	·	+	+	+	+
	NSA-re	−	−	−	−	·	+	+	+
	NSA-sp	−	−	−	−	−	·	−	−
	one-class SVM	−	−	−	−	−	+	·	+
	naïve Bayes	−	−	−	−	−	+	−	·

Cluster in cluster

		VorEAl (a/c)	VorEAl (a/c/t)	VorEAl (a/m)	VorEAl (a/m/t)	NSA-re	NSA-sp	one-class SVM	naïve Bayes
Accuracy	VorEAl (a/c)	·	−	−	−	+	+	−	−
	VorEAl (a/c/t)	+	·	−	−	+	+	−	∼
	VorEAl (a/m)	+	+	·	−	+	+	∼	+
	VorEAl (a/m/t)	+	+	+	·	+	+	+	+
	NSA-re	−	−	−	−	·	−	−	−
	NSA-sp	−	−	−	−	+	·	−	−
	one-class SVM	+	+	∼	−	+	+	·	+
	naïve Bayes	+	∼	−	−	+	+	−	·
Recall	VorEAl (a/c)	·	−	−	−	+	+	−	−
	VorEAl (a/c/t)	+	·	−	−	+	+	−	−
	VorEAl (a/m)	+	+	·	∼	+	+	+	∼
	VorEAl (a/m/t)	+	+	∼	·	+	+	+	∼
	NSA-re	−	−	−	−	·	−	−	−
	NSA-sp	−	−	−	−	+	·	−	−
	one-class SVM	+	+	−	−	+	+	·	−
	naïve Bayes	+	+	∼	∼	+	+	+	·
Specificity	VorEAl (a/c)	·	−	−	−	∼	−	−	−
	VorEAl (a/c/t)	+	·	−	−	+	−	−	−
	VorEAl (a/m)	+	+	·	−	+	−	−	∼
	VorEAl (a/m/t)	+	+	+	·	+	∼	+	+
	NSA-re	∼	−	−	−	·	−	−	−
	NSA-sp	+	+	+	∼	+	·	+	+
	one-class SVM	+	+	+	−	+	−	·	+
	naïve Bayes	+	+	∼	−	+	−	−	·

Corners

		VorEAl (a/c)	VorEAl (a/c/t)	VorEAl (a/m)	VorEAl (a/m/t)	NSA-re	NSA-sp	one-class SVM	naïve Bayes
Accuracy	VorEAl (a/c)	·	∼	+	+	+	+	∼	+
	VorEAl (a/c/t)	∼	·	+	∼	+	+	∼	+
	VorEAl (a/m)	−	−	·	−	+	+	−	+
	VorEAl (a/m/t)	−	∼	+	·	+	+	∼	+
	NSA-re	−	−	−	−	·	−	−	+
	NSA-sp	−	−	−	−	+	·	−	+
	one-class SVM	∼	∼	+	∼	+	+	·	+
	naïve Bayes	−	−	−	−	−	−	−	·
Recall	VorEAl (a/c)	·	∼	+	+	+	+	∼	+
	VorEAl (a/c/t)	∼	·	+	+	+	+	∼	+
	VorEAl (a/m)	−	−	·	∼	+	+	−	+
	VorEAl (a/m/t)	−	−	∼	·	+	+	−	+
	NSA-re	−	−	−	−	·	−	−	+
	NSA-sp	−	−	−	−	+	·	−	+
	one-class SVM	∼	∼	+	+	+	+	·	+
	naïve Bayes	−	−	−	−	−	−	−	·
Specificity	VorEAl (a/c)	·	∼	+	+	+	+	∼	+
	VorEAl (a/c/t)	∼	·	+	+	+	∼	∼	+
	VorEAl (a/m)	−	−	·	∼	+	−	−	+
	VorEAl (a/m/t)	−	−	∼	·	+	−	−	+
	NSA-re	−	−	−	−	·	−	−	+
	NSA-sp	−	∼	+	+	+	·	−	+
	one-class SVM	∼	∼	+	+	+	+	·	+
	naïve Bayes	−	−	−	−	−	−	−	·

The experiments involved six classification benchmarks problems: the 'two spiral', 'crescent and full moon', 'half densities', 'corners', 'outliers' and 'cluster in cluster' problems. They have the advantage that they can be visualized in 2D while still posing a substantial challenge to the algorithm. One key element that must be addressed is the ability of the method to detect anomalies that were present in the original dataset and also those that were not present. Six tests were prepared with that goal in mind by adding random anomaly data in the areas that did not had any data in the training dataset. The resulting training and test datasets can be observed in Fig. 3. Besides fixing these parameters we limited the population to 100 individuals and ran the algorithms for 500 generations. The rest of the parameters are tuned using a grid search procedure on a reduced-size problem. The same parameters were used for all problems. The mutation of the parameters were $p_s = 0.5$, $p_f = 0.5$, $p_t = 0.1$, $p_+ = 0.2$, $p_- = 0.1$ and $\eta = 0.5$, while the mating probability was 0.5, the minimum and maximum number of sites in an individual was set to 20 and 100, respectively and the committee selection percentile (ρ) was set to 5 % of the population.

The stochastic nature of the algorithms being analyzed calls for the use of an experimental methodology that relies on statistical hypothesis tests. Using those tests, we are able to determine in a statistical sound way if one algorithm instance outperforms another. The topic of assessing stochastic classification algorithms is studied in depth in [4]. There, it is shown that the Bergmann–Hommel procedure is the most suitable for our class of problem. In all cases, we have used a base level of significance of 0.05 and we run the same experiment instances 50 times. The results of this experiments are shown as box plots in Fig. 4. It can be inferred from those plots that the three-objective form of with accuracy, multiplicative compactness, and total empty volume VorEAl yielded the best results.

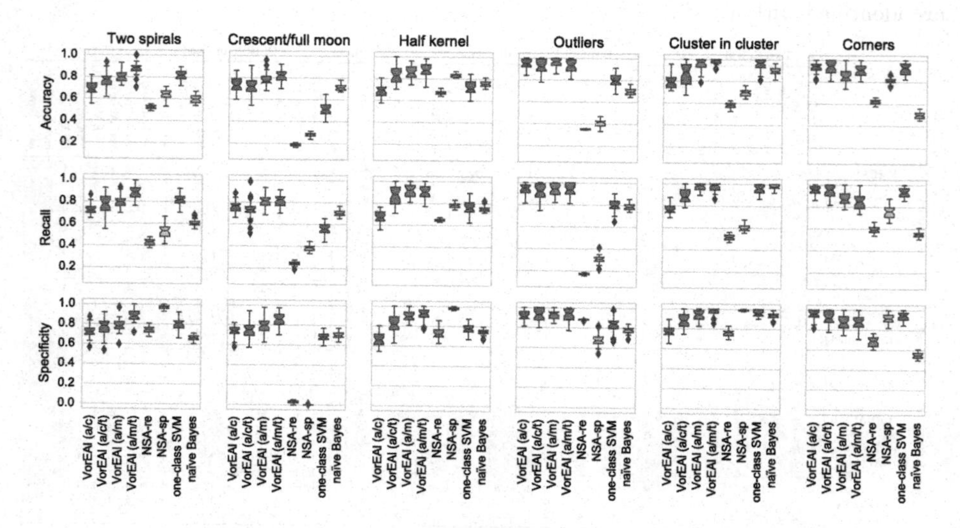

Fig. 4. Box plots of the experimental evaluations on the anomaly detection test sets.

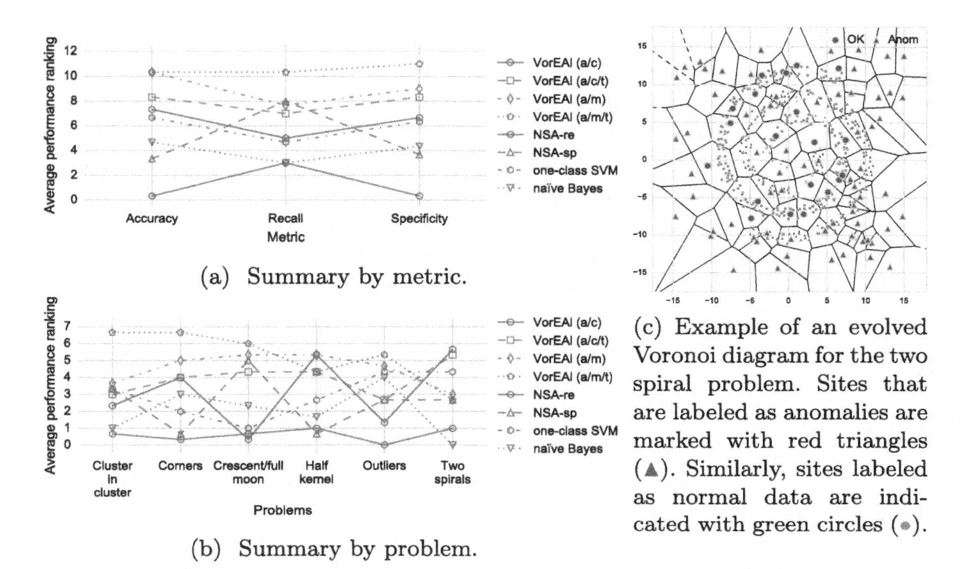

(a) Summary by metric.

(b) Summary by problem.

(c) Example of an evolved Voronoi diagram for the two spiral problem. Sites that are labeled as anomalies are marked with red triangles (▲). Similarly, sites labeled as normal data are indicated with green circles (●).

Fig. 5. Summaries of the statistical tests and an illustrative example. (Color figure online)

When many tests are carried out, a comprehensive analysis of the results is rather difficult as it implies cross-examining and comparing the results presented separately. Consequently, we present them in summarized form in Table 1. It should be noted that experiment parameters and results are available online at http://lmarti.com/VorEAl. To further simplify the understanding of results, why we decided to adopt a more integrative representation like the one proposed in [1]. Figure 5 summarizes the outcome of the hypothesis tests by grouping them by metric and problem, as explained in the previous section. Here it is clearly visible how VorEAl with multiplicative compactness and total empty volume objectives is generally able to yield better results. Finally, as an illustrative example, we show in Fig. 5c an example of an evolved Voronoi diagram.

5 Discussion and Conclusion

In this paper we have presented VorEAl, a multi-objective evolutionary algorithm that relies on Voronoi diagrams for representation. VorEAl has been devised with the problem of anomaly detection in mind. The experimental results obtained as part of this work point out that this is a promising direction of work. However, there are many areas that should be further studied and explored. From an algorithmic point of view, we should explore other classification objectives (metrics).

It is important to try other multi-objective fitness assignments, like those based on multi-objective performance indicators or reference points. This last approach is of particular interest as, as we already mentioned, in our case we have an *a priori* known ideal solution that can be used to guide the search. In parallel, work should the done in understanding and reducing the computational complexity of the algorithm. In this direction, we are already working on creating approximative versions of the volume meant to decrease the computational cost of the computation of the objective functions.

Acknowledgements. This work has been funded by the project PIA-FSN-P3344-146479. Authors wish to thank the reviewers for their fruitful comments.

References

1. Bader, J.: Hypervolume-Based Search for Multiobjective Optimization: Theory and Methods. Ph.D. thesis, ETH Zurich, Switzerland (2010)
2. Chandola, V., Banerjee, A., Kumar, V.: Anomaly detection: a survey. ACM Comput. Surv. (CSUR) **41**(3), Article No. 15 (2009)
3. Deb, K., Pratap, A., Agarwal, S., Meyarivan, T.: A fast and elitist multiobjective genetic algorithm: NSGA-II. IEEE Trans. Evolut. Comput. **6**(2), 182–197 (2002)
4. García, S., Herrera, F.: An extension on "statistical comparisons of classifiers over multiple data sets" for all pairwise comparisons. J. Mach. Learn. Res. **9**, 2677–2694 (2008)
5. Hamda, H., Jouve, F., Lutton, E., Schoenauer, M., Sebag, M.: Compact unstructured representations for evolutionary design. Appl. Intell. **16**, 139–155 (2002)
6. Ji, Z., Dasgupta, D.: Real-valued negative selection algorithm with variable-sized detectors. In: Deb, K., Tari, Z. (eds.) GECCO 2004. LNCS, vol. 3102, pp. 287–298. Springer, Heidelberg (2004)
7. Kim, J., Bentley, P.J., Aickelin, U., Greensmith, J., Tedesco, G., Twycross, J.P.: Immune system approaches to intrusion detection a review. Nat. Comput. **6**(4), 413–466 (2007)
8. Martí, L., Fansi-Tchango, A., Navarro, L., Schoenauer, M.: VorAIS: a multi-objective Voronoi diagram-based artificial immune system. In: Proceedings of the 11th Annual Conference on Genetic and Evolutionary Computation (GECCO 2009), ACM (2016)
9. Northcutt, S., Novak, J.: Network Intrusion Detection. Sams Publishing, Indianapolis (2002)
10. Schoenauer, M.: Shape representation for evolutionary optimization and identification in structural mechanics. In: Winter, G., Périaux, J., Galán, M., Cuesta, P. (eds.) Genetic Algorithms in Engineering and Computer Science (EUROGEN 1995), pp. 443–464 (1995)
11. Schwefel, H.P.: Numerical Optimization of Computer Models. Wiley, New York (1981). (1995–2nd edn.)
12. Tax, D.M.J., Duin, R.P.W.: Support vector data description. Mach. Learn. **54**(1), 45–66 (2004)

Evolving Spatially Aggregated Features from Satellite Imagery for Regional Modeling

Sam Kriegman(✉), Marcin Szubert, Josh C. Bongard, and Christian Skalka

University of Vermont, Burlington, VT 05405, USA
sam.kriegman@uvm.edu

Abstract. Satellite imagery and remote sensing provide explanatory variables at relatively high resolutions for modeling geospatial phenomena, yet regional summaries are often desirable for analysis and actionable insight. In this paper, we propose a novel method of inducing spatial aggregations as a component of the machine learning process, yielding regional model features whose construction is driven by model prediction performance rather than prior assumptions. Our results demonstrate that Genetic Programming is particularly well suited to this type of feature construction because it can automatically synthesize appropriate aggregations, as well as better incorporate them into predictive models compared to other regression methods we tested. In our experiments we consider a specific problem instance and real-world dataset relevant to predicting snow properties in high-mountain Asia.

Keywords: Spatial aggregation · Feature construction · Genetic programming · Symbolic regression

1 Introduction

Regional modeling focuses on explaining phenomena occurring at a regional, as opposed to site-specific or global scales [11]. Regional models are of interest in many remote sensing applications, as they provide meaningful units for analysis and actionable insight to policymakers. Yet satellite imagery and remote sensing provide variables at relatively high resolutions. Consequently, studies often involve decisions concerning how to integrate this information in order to model regional processes. Considering measurements at each individual spatial unit as a separate model feature can result in a high dimensional problem in which high variance and overfitting are major concerns. For this reason, spatial aggregation is often applied in this setting to uniformly up-sample variables to be consistent with the response. Although in averaging variables across all spatial units in the region, we discard information which could in turn diminish prediction accuracy and our understanding of underlying phenomena.

Rather than strictly incorporating individual spatial units or uniformly up-sampling, it might instead be beneficial to construct features of a regional model using particularly important subsets of geographical space. In this paper, we

J. Handl et al. (Eds.): PPSN XIV 2016, LNCS 9921, pp. 707–716, 2016.
DOI: 10.1007/978-3-319-45823-6_66

move away from uniform up-sampling aggregations towards more flexible and interesting aggregation operations predicated on their subsequent use as features of a regional model. We propose a novel method of inducing spatial aggregations as a component of the machine learning process, yielding features whose construction is driven by model performance rather than prior assumptions.

In experiments designed to explore these techniques, we consider a specific problem and real dataset: estimating regional Snow Water Equivalent (SWE) in high-mountain Asia with satellite imagery. Improved estimation of SWE in mountainous regions is critical [3] but is difficult due in part to complex characteristics of snow distribution [2].

2 Methods

We take a comparative approach to the SWE problem, considering ridge regression, lasso, and GP-based symbolic regression[1]. For each regression model, we consider a filter-based method of feature construction in addition to a second, more dynamic method. For linear regression, we incorporate a wrapper approach in which constructed features and the regression model are induced in separate learning processes, with feedback between the two. For symbolic regression, we use an embedded approach where constructed features and the regression model are induced simultaneously over the course of an evolutionary run.

The Dataset. The SWE dataset[2] is derived from data collected by NASA's Advanced Microwave Scanning Radiometer (AMSR2/E) and Moderate Resolution Imaging Spectroradiometer (MODIS) for March 1 - September 30, in 2003 - 2011, over an area that spans most of the high mountain Asia. We have three explanatory variables measured daily across a 113 $\times$ 113 regular grid for 1935 days: (1) mean and (2) standard deviation of sub-pixel Snow Covered Area [4,10], as well as (3) an estimate of SWE derived from passive microwaves [15]. Our response variable is regional SWE, an attribute of the entire study region, represented as a single value for each of the 1935 days. The response was "reconstructed" by combining snow cover depletion record with a calculation of the melt rate to retroactively estimate how much snow had existed in the region [9].

2.1 Regression Models

Ridge regression [5] is similar to ordinary least squares (OLS) but subject to a bound on the L_2-norm of the coefficients. Because of the nature of its quadratic constraint, ridge regression cannot produce coefficients exactly equal to zero and keeps all of the features in its model. Lasso (Least Absolute Shrinkage and Selection Operator, [16]) modifies the ridge penalty and is subject to a bound

[1] The source code necessary for reproducing our results is available at https://github.com/skriegman/ppsn_2016.

[2] Raw satellite data was pre-processed by Dr. Jeff Dozier (UCSB) using previously reported techniques and is available upon request.

on the L_1-norm of the coefficients. The geometry of this L_1-penalty has a strong tendency to produce sparse solutions with coefficients exactly equal to zero. In many high dimensional settings, lasso is the state-of-the-art regression method given its ability to produce parsimonious models with excellent generalization performance. For both lasso and ridge regression, the parameter constraining the coefficients is set through cross-validation.

Genetic Programming (GP, [7]) is a very flexible heuristic technique which can conveniently represent free-form mathematical equations (candidate regression models) as parse trees. GP's inherent flexibility is well-suited for our particular problem because it can efficiently express spatial aggregations and seamlessly combine them into the learning process with minimal assumptions. Furthermore, the "white box" nature of GP may provide physical insights about this complex problem that is currently lacking, as in other domains [1,13].

To search the space of possible GP trees we use a variant of Age-Fitness Pareto Optimization (AFPO, [12]). AFPO is a multiobjective method that relies on the concept of genotypic age, an attribute intended to preserve diversity. We extend AFPO to include an additional objective of model size, defined as the syntactic length of an individual tree. The size attribute protects parsimonious models which are less prone to overfitting the training data. The GP algorithm therefore identifies the Pareto front using three objectives (all minimized): age, error (fitness), and size. For the fitness objective, we use a correlation-based function rather than pure error, and define $f_{COR} = 1 - |\phi(\hat{s}, s)|$, where $\phi(\hat{s} - s)$ denotes Pearson correlation between model predictions ($\hat{s}$) and actual values of our response (s), regional SWE. Correlation has recently been shown to outperform error-based search drivers given that if a model makes a systematic error it could be easily eliminated by linearly scaling the output and therefore should be protected [14]. Accordingly, for all GP implementations, we apply a linear transformation after f_{COR} -driven evolution has concluded, by using an individual program (model) output as the single input of OLS on the training data.

Our implemented GP experiments used ramped half-and-half initialization with a height range of 2–6 and an instruction set including unary ($\{\sin, \cos, \log, \exp\}$) and binary functions ($\{\times, +, -, /\}$). One thousand individuals in the population are subject to crossover (with probability 0.75) and mutation (with probability 0.01) over the course of 1000 generations. There is a static limit on the tree height (17) as well as the tree size (300 nodes). Each experiment consists of 30 evolutionary runs, from which the best model (lowest training f_{COR}) is selected. The selected model is then transformed using OLS, and subsequently validated using unseen test data.

Standard Methods. Ridge regression, lasso, and GP may be performed on the raw data using each variable at each individual spatial unit as a separate feature. We denote these methods as Standard Ridge (SR), Standard Lasso (SL) and Standard GP (SGP). SR, SL and SGP each have access to $113 \times 113 \times 3 = 38307$ features, but only 1720 observations in each fold of data.

2.2 Feature Construction Methods

Feature construction is a well studied problem and the utility of genetic programming for feature construction has been recognized in many previous studies [8]. The key difference in our work from this past work is the nature of the data being modeled. We presume that there exist spatial autocorrelations of varying size and shape that, if aggregated to improve the signal to noise ratio, yield features supporting more accurate predictions.

In a regional model, we can construct features by aggregating higher dimensional variables across space. However, it is not entirely clear what kind of aggregations are useful as features of a predictive model. Grouping variables based on similarity or dissimilarity does not necessarily produce useful regional features. In this paper, we make an assumption about the importance of distance and continuity in effective spatial aggregations, based on Tobler's first law of geography [17] which states that "everything is related to everything else, but near things are more related than distant things." Accordingly, we limit the space of possible spatial aggregations to be an average of values within a circular spatial area defined by its centerpoint and radius. However, where to aggregate, how many aggregations to perform, and how to combine the aggregates must still be determined manually or decided during model optimization. We view filters and wrappers as intermediary steps in relaxing assumptions towards our embedded approach, which automates all three of these aspects.

The Filter Method. Filter-based feature construction methods transform or "filter" the original variables as a preprocessing step, prior to modeling. Our filter for the SWE problem represents a static up-sampling transformation of the original variables. Each variable is decomposed in space by a grid of overlapping circles[3] of equal radii centered on a square lattice pattern of points (see Fig. 1a, c and e for example). Each constructed feature corresponds to the average (arithmetic mean) of a particular variable sampled within a particular circle of space. Units that reside in an overlapping region of two separate circles are included in the calculation of both features. Since there are three explanatory variables in the SWE dataset, an $R \times R$ grid corresponds to $p = 3R^2$ constructed features. The constructed features are then used as inputs for ridge regression, lasso, and GP, which we will refer to as Filtered Ridge (FR), Filtered Lasso (FL), and Filtered GP (FGP). We will also specify the value of R used in a particular model instance as a subscript, e.g. FR_{15} denotes Filtered Ridge with $R = 15$. We consider filters with $R \in \{1, 2, \ldots, 20\}$, however note that the standard methods are essentially filters with $R = 113$, albeit with the non-overlapping square pixels.

The Wrapper Method. Wrapper-based feature construction methods incorporate feedback from the fit of the model. We implement wrappers around both ridge regression and lasso in order to enable the circular sampling regions to

[3] The shape of circles are in reality so-called "small circles," as they lie on the surface of earth.

define their own center and radius. The circles are no longer fixed on a grid with a predetermined size. Instead, each constructed feature is uniquely parameterized by the coordinates of a center unit (x, y), as a latitude and longitude tuple, and a radius r, as a single value floating point number in km. The center can be any spatial unit in the region, including one at the edge of the raster. The radius is restricted to be within 0 and 1000 km, which is flexible enough to contain only a single unit or span the entire region (see Fig. 1b and d for example).

Wrapped Ridge (WR) and Wrapped Lasso (WL) separately use a ridge/lasso-driven hill climbing algorithm to construct features that minimize Mean Absolute Error (MAE), i.e. $\frac{1}{n}\sum_{i=1}^{n}|\hat{s_i} - s_i|$, where s_i is the actual value of our response (regional SWE) and $\hat{s_i}$ is output predicted by the model over n observations. The algorithm uses the same number of circles for each of the three variables, initializing their parameters (x, y, r) randomly. For 1000 iterations, a single constructed feature (circle) is randomly selected and subject to a Gaussian mutation on one of its parameters with standard deviation equal to 25 % of the radius and centered at zero. A new ridge/lasso model is then refit on the mutated set of features using a random subset of data sampled without replacement. If the mutation lowered model error on the complementing set of training data left out, then the change is accepted. Otherwise, the mutation is undone. If a proposed mutation to the radius would take it outside the restricted range of 0 – 1000 km, then it is "bounced-back" the distance it would have exceeded the boundary. For example, a random mutation that would result in a radius of 1200 km, becomes 1000 – (1200 – 1000) = 800 km. Thirty restarts are used from which the best model based on training data is selected. We consider $R \in \{1, 2, 3, 4\}$ for wrappers corresponding to $3 \times R^2$ features which really means $3 \times 3 \times R^2$ modifiable parameters.

The Embedded Method. By using GP, we can allow for flexibility with respect to the placement and number of aggregations as well as the way in which they are combined to form a model. However, stochastic optimization methods like GP cannot be easily "refit" in the same manner as deterministic algorithms like ridge regression or lasso. Therefore using wrapper approach for GP is computationally infeasible. Instead, modifications to aggregated features are implemented through mutation-based operators.

In Genetic Programming with Embedded Spatial Aggregation (GPESA) introduced here, our constructed features are represented as parameterized tree terminals, with parameters (x, y, r). Constructed features are randomly initialized in the same manner as the wrapper method, but separately for each terminal of each individual in the population. Greedy Gaussian mutations to the parameters (x, y, r) of a randomly selected constructed feature occur in the population with 20 % probability, each generation. Mutations to r have mean zero and a standard deviation of 25 %, subject to the bounce-back rule. Similarly, mutations to (x, y) have mean distance zero and a standard deviation of $0.25r$. For 25 iterations, greedy mutations modify the parameterized terminals within a particular GP tree. A modification is accepted if it successfully reduces average error (f_{COR}) on random subsets of training data sampled with replacement. Aside from the stochastic application, another key difference between the wrapper method's hill climbing algorithm and the GPESA's greedy mutations is that the overall regression model stays the same between mutations rather than being "refit" after each mutation.

Validation. In order to validate the generalization of models we partition the dataset into nine overlapping folds. Each fold corresponds to leaving out one year for testing and training on the remaining eight (using years 2003–2011). We use MAE on the unseen test data as a metric to assess model performance. To account for a difference in scale across any set of features, all input model features are standardized over time by removing the mean and scaling to unit variance. This means that as wrapper and embedded methods construct new aggregations, the sampled data is scaled over time prior to being averaged over space. Since our goal is near-real-time estimation for a future day, the training values of a feature's mean and variance are reapplied when scaling the same feature in validation.

3 Results

Table 1 displays the test error of each valid regression and feature construction method combination. For filters and wrappers, only the best performing model is displayed and we indicate the particular value of parameter R as a subscript. Since the ultimate goal of our paper is to synthesize a method better than existing approaches, we must statistically compare GPESA to SL, the state-of-the-art linear regression/variable selection algorithm. The null hypothesis of interest here is that of no difference between GPESA and a SL. Therefore we perform yearly Wilcoxon signed rank tests [6] comparing GPESA to SL with Bonferroni correction across the nine years. For five out of the nine test years, GPESA is significantly better than SL, while for the other four years there is no significant difference with SL.

Through displaying only the best testing filters and wrappers, we aim to focus speculation about GPESA performance through a conservative lens. Yet we ultimately view filters and wrappers as intermediary steps "working up" to GPESA. Accordingly, the best test error better represents a bound on the potential performance of a particular intermediary method even though it may not be possible to achieve such performance through a parameter sweep based on the training data. And indeed, across all methods tested, GPESA reported the lowest recorded median mean-absolute error within all but two years (7 of 9) where it has the second lowest.

Table 1. Median mean-absolute error with corresponding standard errors in parentheses. Only the best testing filter- and wrapper-based results (choice of R) are displayed. We explicitly compare GPESA with the state-of-art, SL. Bold values indicate significance (at 0.05 level with Bonferroni correction) under a Wilcoxon singed rank test in which the null hypothesis asserts that distribution of the differences between GPESA and SL is symmetrically distributed about 0.

Year	SR	SL	SGP	FR_4	FL_{19}	FGP_{19}	WR_2	WL_3	GPESA
2003	0.86	0.51	0.35 (0.14)	0.50	0.46	0.44 (0.08)	0.43 (0.10)	0.49 (0.09)	**0.29 (0.09)**
2004	0.47	0.30	0.32 (0.10)	0.34	0.29	0.26 (0.05)	0.37 (0.16)	0.35 (0.16)	**0.17 (0.05)**
2005	0.95	0.44	0.50 (0.13)	0.61	0.40	0.52 (0.06)	0.58 (0.11)	0.63 (0.09)	**0.32 (0.07)**
2006	0.66	0.27	0.41 (0.29)	0.57	0.52	0.36 (0.06)	0.53 (0.11)	0.54 (0.11)	0.27 (0.05)
2007	0.72	0.33	0.44 (0.10)	0.42	0.38	0.34 (0.05)	0.52 (0.13)	0.50 (0.11)	**0.24 (0.06)**
2008	1.46	0.46	0.60 (0.13)	0.71	0.64	0.58 (0.11)	0.70 (0.31)	0.54 (0.26)	0.52 (0.18)
2009	0.81	0.41	0.65 (0.08)	0.90	0.61	0.56 (0.08)	0.98 (0.10)	1.03 (0.09)	0.41 (0.10)
2010	0.62	0.48	0.44 (0.12)	0.43	0.47	0.41 (0.06)	0.43 (0.11)	0.52 (0.11)	**0.32 (0.07)**
2011	0.87	0.48	0.61 (0.17)	0.77	0.60	0.53 (0.10)	0.82 (0.20)	0.93 (0.16)	0.45 (0.12)
Mean	0.82	0.41	0.48	0.58	0.49	0.44	0.58	0.61	0.33

4 Discussion

Our results show that incorporating dynamic aggregations of higher resolution variables into a regional model is beneficial in our particular problem setting, as compared to both uniform up-sampling of variables and a state-of-the-art linear regression technique (SL) that incorporates individual spatial units. SL achieves competitive prediction performance through a sparse linear combination of the individual spatial units, on par with SGP which is not linearly constrained. Ultimately, GPESA performed significantly better (lower median test error) than SL on a majority (5 of 9) of cross validation folds. Moreover, whenever GPESA was not significantly better than SL it was not significantly worse.

A main reason why GPESA has an advantage in this application is the difficulty of knowing a priori what the most important spatial datapoints are, and how to best aggregate them. Additionally, the structure of the model itself is unknown and it depends on the resulting aggregations. Therefore this is not a fixed length optimization problem, which makes it well-suited for GPESA, which can search over different numbers and non-linear combinations of spatial aggregations. While SL can theoretically perform the same aggregation as a GPESA terminal (mean within a radius of a geographical point), SL is restricted to a single linear solution while GPESA is not.

However, it's important to emphasize that the computational cost of GPESA is higher than that of traditional GP and much higher than that of linear regression. In particular, the most expensive operation is the "on the fly" aggregation component of GPESA which makes the fitness evaluation require 500 % more time than in SGP. Part of the incurred cost is due to inefficiencies of our implementation that necessitated a copy with all spatial aggregation operations. In future

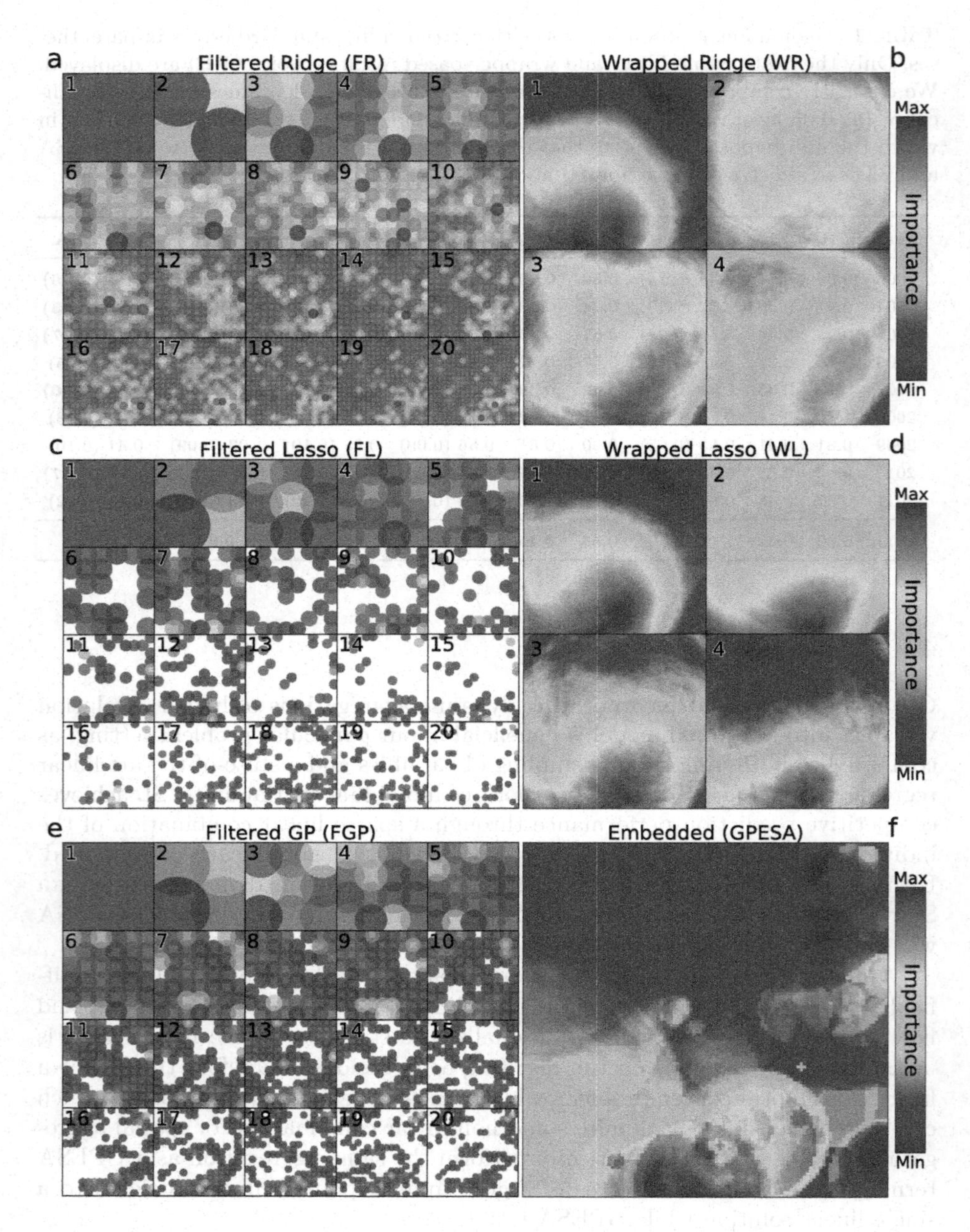

Fig. 1. Importance (defined in Sect. 4) of spatial units. For filters (a.) FR, (c.) FL, and (e.) FGP, importance is displayed at each resolution $R \in \{1, 2, \ldots 20\}$ and each individual filter subplot is annotated with the corresponding R. For wrappers (b.) WR and (d.) WL, $R \in \{1, 2, 3, 4\}$. Finally, (f.) GPESA, which has no R parameter. White areas indicate spatial units unused in feature construction across all three exploratory variables.

work we will look at reducing this overhead through more efficient data structures (e.g. k-d trees).

Importance of Spatial Data. To better understand the relevance of particular spatial locations, we define the *importance* of a spatial unit for both linear and symbolic methods, separately. For ridge regression and lasso, we can define importance by exploiting the disposition of coefficients to be larger for variables with a stronger correlation to the response, relative to a particular feature set. We define linear regression importance of a particular spatial unit as the average absolute coefficient of features that incorporate the unit into a regression model. While we cannot as easily determine relative importance within nonlinear models, we can instead define importance by exploiting the multiple candidate solutions provided from stochastic multiobjective optimization. We define GP importance of a particular spatial unit as the average absolute correlation $(1 - f_{COR})$ of nondominated solutions that incorporate the unit.

To visualize the importance of spatial information, we generated a series of heatmaps (Fig. 1). In Figs. 1a, c and e we show regional importance values of filter methods for each $R \in \{1, ..., 20\}$, with the relevant value of R annotated in the upper left corner of each box. Note that in lasso- and GP-based approaches, some variables are unused (white), while ridge cannot perform variable selection and uses all. Figures 1b and d plot WR and WL for $R \in \{1, 2, 3, 4\}$. Finally, Figs. 1e and f plot the importance of spatial information in the GP sense, for FGP and GPESA, respectively. Overall, this visualization indicates an agreement among all methods on the relatively higher importance of information in the lower center/right region of the image.

5 Conclusion

In this work we developed a novel method to address the problem of modeling a regional response with high resolution satellite imagery. We moved away from uniform up-sampling aggregations towards more flexible and interesting aggregation operations predicated on their subsequent use as features of a regional model. Our proposed technique, GPESA, is general and intended to apply to a variety of modeling problems on spatially organized data. But as an application example, and as a setting in which to evaluate our techniques, we considered the problem of estimating snow water equivalent in high mountain Asia using satellite imagery. Our results showed that using GP to evolve spatial aggregations outperforms lasso, the state-of-the-art method for directly incorporating individual spatial units into a sparse linear model.

In future work we plan to explore more flexible spatial and temporal aggregations for more predictive modeling in real earth science applications.

Acknowledgements. Thanks to Dr. Jeff Dozier (UCSB) for posing the high-mountain Asia SWE problem and providing associated datasets.

References

1. Bongard, J., Lipson, H.: Automated reverse engineering of nonlinear dynamical systems. Proc. Nat. Acad. Sci. **104**(24), 9943–9948 (2007)
2. Buckingham, D., Skalka, C., Bongard, J.: Inductive learning of snowpack distribution models for improved estimation of areal snow water equivalent. J. Hydrol. **524**, 311–325 (2015)
3. Dong, J., Walker, J.P., Houser, P.R.: Factors affecting remotely sensed snow water equivalent uncertainty. Remote Sens. Environ. **97**(1), 68–82 (2005)
4. Dozier, J., Painter, T.H., Rittger, K., Frew, J.: Time-space continuity of daily maps of fractional snow cover and albedo from MODIS. Adv. Water Resour. **31**(11), 1515–1526 (2008)
5. Hoerl, A.E., Kennard, R.W.: Ridge regression: biased estimation for nonorthogonal problems. Technometrics **12**(1), 55–67 (1970)
6. Hollander, M., Wolfe, D.A., Chicken, E.: Nonparametric Statistical Methods. Wiley, Hoboken (2013)
7. Koza, J.R.: Genetic Programming: On the Programming of Computers by Means of Natural Selection. MIT Press, Cambridge (1992)
8. Krawiec, K.: Genetic programming-based construction of features for machine learning and knowledge discovery tasks. Genet. Program Evolvable Mach. **3**(4), 329–343 (2002)
9. Martinec, J., Rango, A.: Areal distribution of snow water equivalent evaluated by snow cover monitoring. Water Resour. Res **17**(5), 1480–1488 (1981)
10. Painter, T.H., Rittger, K., McKenzie, C., Slaughter, P., Davis, R.E., Dozier, J.: Retrieval of subpixel snow-covered area, grain size, and albedo from MODIS. Remote Sens. Environ. **113**, 868–879 (2009)
11. Rees, J., Gibson, A., Harrison, M., Hughes, A., Walsby, J.: Regional modelling of geohazard change. Geol. Soc. Lond. Eng. Geol. Spec. Publ. **22**(1), 49–63 (2009)
12. Schmidt, M., Lipson, H.: Age-fitness pareto optimization. In: Riolo, R., McConaghy, T., Vladislavleva, E. (eds.) Genetic Programming Theory and Practice VIII. Genetic and Evolutionary Computation, vol. 8, pp. 129–146. Springer, Heidelberg (2011)
13. Schmidt, M.D., Vallabhajosyula, R.R., Jenkins, J.W., Hood, J.E., Soni, A.S., Wikswo, J.P., Lipson, H.: Automated refinement and inference of analytical models for metabolic networks. Phys. Biol. **8**(5), 055011 (2011)
14. Stanislawska, K., Krawiec, K., Vihma, T.: Genetic programming for estimation of heat flux between the atmosphere and sea ice in polar regions. In: Proceedings of the 2015 on Genetic and Evolutionary Computation Conference, pp. 1279–1286. ACM (2015)
15. Tedesco, M., Narvekar, P.S.: Assessment of the NASA AMSR-E SWE product. IEEE J. Sel. Top. Appl. Earth Obs. Remote Sens. **3**(1), 141–159 (2010)
16. Tibshirani, R.: Regression shrinkage and selection via the lasso. J. Roy. Stat. Soc. Ser. B (Methodol.) **58**(1), 267–288 (1996)
17. Tobler, W.R.: A computer movie simulating urban growth in the detroit region. Econ. Geogr. **46**, 234–240 (1970)

A Hybrid Autoencoder and Density Estimation Model for Anomaly Detection

Van Loi Cao(✉), Miguel Nicolau, and James McDermott

NCRA Group, University College Dublin, Dublin, Ireland
loi.cao@ucdconnect.ie, {miguel.nicolau,james.mcdermott2}@ucd.ie
http://ncra.ucd.ie

Abstract. A novel one-class learning approach is proposed for network anomaly detection based on combining autoencoders and density estimation. An autoencoder attempts to reproduce the input data in the output layer. The smaller hidden layer becomes a bottleneck, forming a compressed representation of the data. It is now proposed to take low density in the hidden layer as indicating an anomaly. We study two possibilities for modelling density: a single Gaussian, and a full kernel density estimation. The methods are tested on the NSL-KDD dataset, and experiments show that the proposed methods out-perform best-known results on three out of four sub-datasets.

Keywords: Anomaly detection · Autoencoder · Density estimation

1 Introduction

Anomaly detection plays an important role in a variety of application domains ranging from intrusion detection in network security, credit card fraud detection, health care and insurance to fault detection in safety critical systems [1,3]. This is due to the fact that anomalies often translate to critical, actionable information or potentially dangerous situations and events. In network security, anomaly detection is the task of distinguishing illegal, malicious activities from normal traffic or behavior of systems [3,13]. This has become increasingly important due to valuable resources and the widespread use of computer networks in recent years.

Network anomaly detection models must be sufficiently flexible to keep up with the continuous evolution of attacks or malicious activities over time, and the occurrence of new, unknown anomalies [8]. Moreover, labeled anomaly data may not be available, due to the rarity of intrusions, difficulty of labeling, and the privacy and security concerns of computer networks [8,19]. For these reasons, one-class learning or novelty detection is a common approach for network anomaly detection. A one-class classifier constructed from only normal (target) data is employed to classify whether an unseen instance belongs to the normal class or anomaly (non-target) class [15].

J. Handl et al. (Eds.): PPSN XIV 2016, LNCS 9921, pp. 717–726, 2016.
DOI: 10.1007/978-3-319-45823-6_67

We are continuing previous research on one-class classification (OCC) with Kernel Density Estimation (KDE) [2]. It works by defining a threshold on *density* of the normal data: query points below the threshold are classed as anomalies. Several approaches to modeling density are possible, e.g. a single Gaussian, multiple independent Gaussians, and negative mean distance [21], but KDE is the most flexible of all. We found that KDE performed very well (better than One-class SVM [18]), but was slow at query time, so we provided a method to speed it up using Genetic Programming (GP).

Another method commonly used for anomaly detection is autoencoders (AEs). This design was named an "autoencoder" by Japkowicz et al. [11], who applied it for novelty detection in 1995. An autoencoder is a neural network which learns to reconstruct its input at the output layer. A narrow middle layer compresses redundancies in the input data while non-redundant information remains [11]. The effect is rather like a non-linear PCA. AEs are commonly used as building blocks in deep neural networks [10], and a key idea is that after training, the output layer is discarded, and the hidden layer is used as a new feature representation. In the one-class learning context, the reconstruction error (RE) of trained AEs is commonly used as a measure of "anomalyness".

In this paper, we investigate the distribution of data in the AE hidden layer. Based on this, we will propose a novel one-class learning method which models density of the compressed data from hidden layer on a trained AE. Two well-known density estimators are employed to model the density from hidden layer, a single Gaussian and a full KDE. An autoencoder is first trained on the normal class to minimize RE. The normal data is then passed through the trained AE again, and its density in the hidden layer is estimated. At the testing stage, a query point is first passed through the trained AE, and its value at the hidden layer is classified into normal or anomaly class by the density models.

The rest of this paper is organized as follows. We briefly review some work related to OCC based on AEs. In Sect. 3, we give a short introduction to AEs and density estimation. This is followed by a section proposing OCC using AEs and density estimation together. Experiments, Results and Discussion are presented in Sects. 5 and 6 respectively. The paper concludes with highlights and future directions.

2 Related Work

Recently, autoencoders or bottleneck neural networks became popular for anomaly detection as one-class learning techniques [17,22]. Hawkins et al. [9] trained a replicator neural network with narrow middle layers on normal data to construct a one-class classifier using reconstruction error as an indicator of anomalies. They used a step-wise activation function for the hidden layer to divide the continuously distributed data into clusters. Similarly, Sakurada and Yairi [17] compared classifiers based on AE, denoising AE, linear PCA, and kernel PCA. The classifiers were evaluated on spacecraft telemetry data. The learned features in the hidden layer were also examined.

Veeramachaneni et al. [22] proposed an ensemble learner to combine three single classifiers: AE, density-based, and matrix decomposition-based. They also used a human expert to provide ongoing correct labels for the algorithms to learn from. They tested their model on a large network log file dataset, with good results.

Erfani et al. [7] proposed a hybrid of a Deep Belief Network (DBN) and a linear one-class SVM for high-dimensional anomaly detection. A one-class SVM was built on the top of the trained DBN. This structure takes advantage of high decision classification accuracy from one-class SVMs and non-linear feature reduction from DBNs. The model was tested on eight UCI datasets, with comparable results to AE, and a significant improvement at query time.

In our work, we present a new approach for anomaly detection. We apply density estimation on the compressed data in the hidden layer. This method is distinct from those discussed above.

3 Preliminaries

3.1 Autoencoder

An autoencoder is a neural network with a (typically) narrow middle layer ("bottleneck"). It attempts to reproduce the input at the output, as illustrated in Fig. 1(a). It is commonly used for novelty detection and deep learning [9,11].

Let $x \in \mathbb{R}^n$ be an input example. The hidden representation $z(x) \in \mathbb{R}^m$ is represented in Eq. 1,

$$z\left(x\right) = f_1\left(W_1 x + b_1\right) \tag{1}$$

where f_1 is a non-linear activation function, $W_1 \in \mathbb{R}^{n\times m}$ is a weight matrix, $b_1 \in \mathbb{R}^m$ is a bias vector. The latent representation z is then mapped back into a reconstruction $\hat{x} \in \mathbb{R}^n$ in the output layer:

$$\hat{x} = f_2\left(W_2 z(x) + b_2\right) \tag{2}$$

where $W_2 \in m \times n$ and $b_2 \in \mathbb{R}^n$ are the weight matrix and bias vector of the output layer. f_2 is the output function. In this work, the logistic function (Eq. 3) and the identity function are used for hidden and output layers respectively. In Eq. 3, k is a steepness parameter.

$$f_1\left(z\right) = \frac{1}{1 + e^{(-kz)}} \tag{3}$$

The parameters of the network, $\theta = \{W_1, W_2, b_1, b_2\}$, are optimized such that the average reconstruction error (RE) is minimized. RE can be measured in many ways, and mean square error (MSE) is commonly used in training neural networks. In order to minimise the RE, stochastic gradient descent (SGD) is commonly used to train the network.

For anomaly detection, a model trained on normal data tends to fail to reproduce anomaly data, and produces high RE. Therefore, the reconstruction error is used as anomaly score. A test instance will be regarded as an anomaly if its RE is higher than a pre-determined error threshold.

3.2 Density Estimation

In this section, we briefly describe two methods of estimating density, *Centroid* and KDE. The Centroid method uses a single Gaussian, whose mean is placed at the centroid of the training data. The standard deviation is chosen to equal the standard deviation of the data, but in fact is unimportant: when we impose a threshold on density, the method becomes equivalent to imposing a threshold on distance (i.e. radius) from the centroid.

KDE is a non-parametric method of estimating probability density given a sample. Let $x_1, x_2, ..., x_n$ be a set of d-dimensional samples in $\mathbb{R}^d$ drawn from an unknown distribution with density function $p(x)$. An estimate $\hat{p}(x)$ of the density at x can be calculated using

$$\hat{p}(x) = \frac{1}{n} \sum_{i=1}^{n} K_h (x - x_i) \tag{4}$$

where $K_h : \mathbb{R}^d \to \mathbb{R}$ is a kernel function with a parameter h called the *bandwidth*. The Gaussian kernel (Eq. 5) is common in applications and is the only one used in this paper. As illustrated in Fig. 1(b) in KDE each point contributes a small "bump" to the overall density, with its shape controlled by the kernel and bandwidth. The bandwidth parameter h controls the trade-off between bias of the estimator and its variance.

$$K_h (x) = \exp{(-\frac{x^2}{2h^2})} \tag{5}$$

4 Proposed Approach

Our proposed approach is to use density estimation on the hidden layer of an autoencoder. Our motivation for this is the same as that for RE-based OCC: anomaly data is poorly reconstructed by an AE trained on normal data, and part of this must be due to anomaly data occupying an unusual position in the hidden layer. We demonstrate this in Fig. 3. There are two phases in our method, training and testing, as illustrated in Fig. 2. In the training phase, an AE is first trained on a normal training set, and the training set is then passed through the trained AE again. The training data, compressed in the hidden layer, is used to build a density model. Based on the training stage, a density threshold is set, for example keeping 95 % of the training set. The compressed data will be classified as normal or anomaly by a threshold on the density model. Two density estimation methods are employed: Centroid and KDE.

The combination of an AE and density estimation takes advantage of their different strengths. AEs can compress input data to fewer dimensions while retaining non-redundant information, while density estimation works best in lower-dimensional spaces.

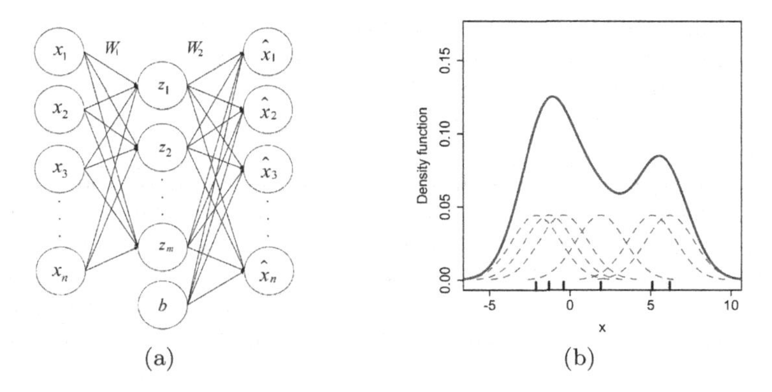

Fig. 1. (a) An autoencoder. (b) Density estimated by KDE (Figure from https://en.wikipedia.org/w/index.php?title=Kernel_density_estimation)

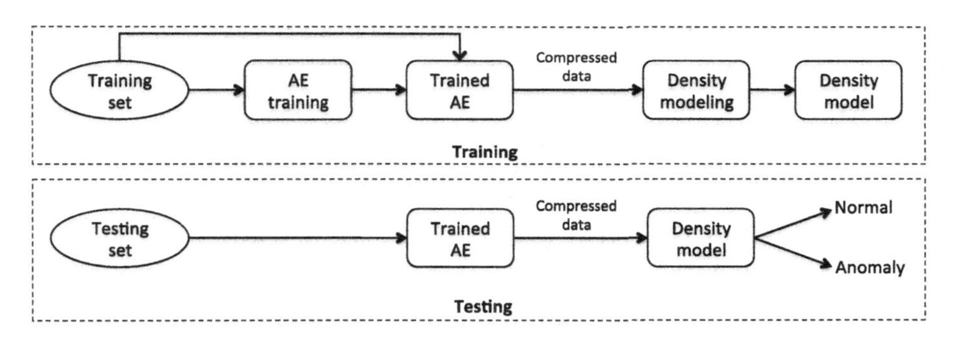

Fig. 2. The proposed anomaly detection model

5 Experiments

5.1 Datasets

The approach of simulating a one-class dataset by throwing away data from a binary dataset is a common approach in previous work [2,5,21]. In this work, we choose datasets that have one class considered as normal class and other classes treated as an anomaly class [4,21]. Four UCI datasets [14], namely Wisconsin Breast Cancer Database (WBC), Wisconsin Diagnostic Breast Cancer (WDBC), Cleveland heart disease (C-heart) and Australian Credit Approval (ACA), and NS-KDD dataset [20] are employed for our experiments. For the UCI datasets, we randomly sample 70 % for training and 30 % for testing. The normal training set is formed by removing all anomaly examples.

NSL-KDD dataset is a filtered version of the KDD Cup 1999 dataset [12] after removing all redundant instances and making the task more difficult. Each record in the dataset is labeled as either normal or as a specific kind of attack belonging to one of the four main categories: Denial of Service (DoS), Remote to Local (R2L), User to Local (U2R) and Probe. NSL-KDD consists of two datasets: $KDDTrain^{+}$ and $KDDTest^{+}$ which are drawn from different distributions. Several of the variables in the dataset are categorical or discrete. We simply treat them as real-valued. As shown in Sect. 6 this gives good results, but better encodings are possible.

In this work, we plan to conduct our experiments on four groups of attacks separately. The aim is to see how efficiently our method performs on each group of attacks. In order to transfer hyperparameter values from the UCI to the NSL-KDD datasets, we wish to have a similar-sized dataset. We randomly subsample 350 normal instances from $KDDTrain^{+}$. We use all normal and anomaly instances from $KDDTest^{+}$ as our labelled test set. The details are shown in Table 1.

Table 1. One-class classification datasets

Dataset	Features	Training set	Testing set	
		Normal	Normal	Anomaly
C-heart	13	112	48	42
ACA	14	268	115	93
WBC	9	310	134	72
WDBC	30	249	108	64
DoS	41	350	9711	7458
R2L	41	350	9711	2887
U2R	41	350	9711	67
Probe	41	350	9711	2421

5.2 Experimental Settings

In this work, the classifiers will be constructed from the normal class only. There is no validation set for doing cross-validation. Therefore, we plan to conduct two experiments, one preliminary experiment for tuning the hyperparameters of the proposed models and one main experiment for evaluating the models. We use the terms OCCEN, OCKDE, and OCAE to refer to one-class classifiers based on the hybrid of AE and Centroid, the hybrid of AE and KDE, and AE itself respectively. The choice of a threshold for classifiers in practice varies from domain to domain, but in this work we try many different thresholds, and evaluate the area under the resulting ROC curve (AUC).

The parameters that will not be estimated from the preliminary experiment are set to common values. The Adaptive Gradient Algorithm (Adagrad) [6] with a common value for the learning rate, $\alpha = 0.01$, and smoothing term $\varepsilon = 10^{-8}$ will be used to train AEs. Hawkins et al. [9] chose different values for *epochs* (from 1000 to 40000), but in this work we choose a single value for *epochs* = 5000. The Gaussian kernel is used for KDE and its bandwidth, $h = \sqrt{\frac{hidden\ size}{2}}$ as in [16].

The preliminary experiment is done on the four UCI datasets to investigate steepness, k, and estimate the size of hidden layer, m, of the models for later testing on the NS-KDD dataset. Firstly, we visualise the data distribution in the hidden layer. In Fig. 3, normal and anomaly data are plotted with different values of k (0.1, 0.5, 1.0) and $m = 2$. We see normal data is approximately Gaussian for $k = 0.1$ whereas for $k = 1.0$ it seems to be distributed along the borders of a hyperbox. Even in this 2D example, for $k = 1.0$, the anomaly data is strongly concentrated in a single area, allowing good separation. We choose a common value $k = 1.0$ for our main experiment.

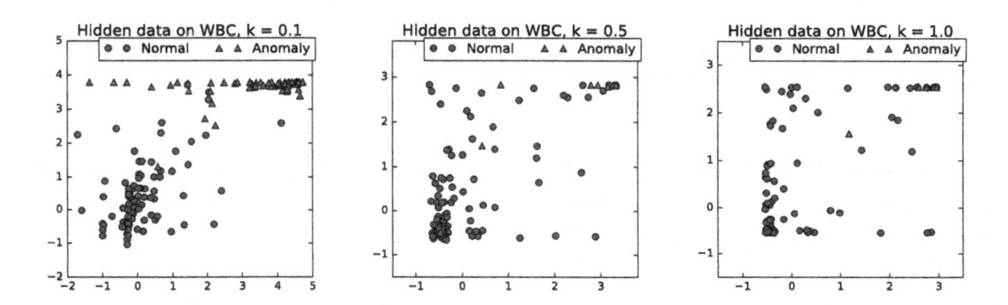

Fig. 3. Data on hidden layer with respect to hidden size, $m = 2$

Secondly, we run the models with different sizes of hidden layer on the four UCI datasets. In Fig. 4, the AUC from OCCEN, OCKDE and OCAE are plotted against hidden size. The figures illustrate that both the three classifiers produce very high AUC values at hidden size, $m = 4$ and 8 on WBC, and $m = 4$ on ACA. However, on C-heart, OCCEN and OCKDE perform very well at $m = 3$ and 6 whereas OCAE produces the highest AUC at $m = 4$. The highest and the second highest AUC values from the three classifiers on WBCD are obtained at $m = 8$ and 6 respectively. Overall, these classifiers produce good accuracy at $m = 3$ or 4 on WBC, ACA and C-heart, and at $m = 6$ or 8 on WBCD.

Therefore, we propose a rule of thumb for choosing hidden size, $m = [1 + \sqrt{n}]$, where n is the number of features. Based on this rule, we calculate parameter m for the models on these datasets, $m = 4$ for WBC, ACA and C-heart, and $m = 6$ for WBCD. For the NSL-KDD dataset, m will be equal to 7.

The main experiment is to investigate the performance of the methods OCAE, OCCEN, OCKDE on the four groups of attacks in NSL-KDD dataset. The classifiers are set up with the set of parameters presented above, and the results are shown in Table 2 and Fig. 5. The code of the experiments is available on github[1].

6 Results and Discussion

This section presents the experimental results of evaluating the proposed one-class classifiers on the four groups of attacks in NSL-KDD dataset. The performance of

[1] https://github.com/caovanloi/AEDensityEstimation.

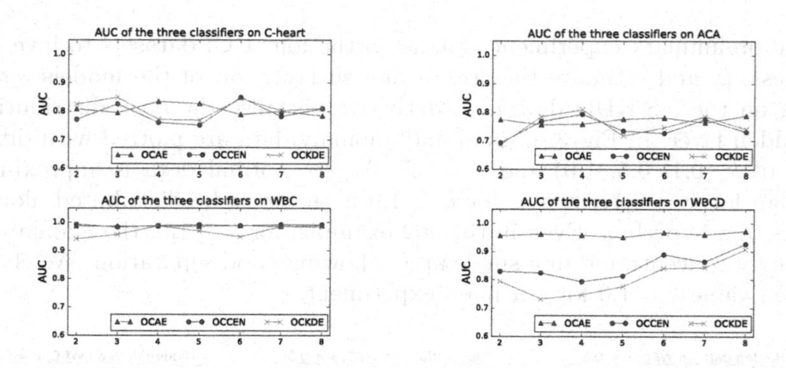

Fig. 4. Plotting AUC values against different hidden sizes on the UCI datasets

the three one-class classifiers, OCAE, OCCEN, OCKDE is evaluated using AUC. The results are summarized in Table 2. The ROC curves of the density-based classifiers are shown against those of OCAE in Fig. 5.

Table 2 illustrates the AUC values from the three one-class classifiers. It can be seen from the table that OCKDE performs very well in terms of accuracy, and better than OCAE on DoS, U2R and Probe. The AUC values from OCCEN are also higher than those from OCAE on Probe. However, the performance of OCCEN is similar to or worse than that of OCAE on three other groups.

The ROC curves are displayed in Fig. 5. The ROC curves of OCCEN, OCKDE are plotted against the ROC curve of OCAE. It can be seen that the curves of OCKDE is usually higher than the curves of the two other classifiers on the NSL-KDD dataset.

Table 2. The AUC results from the three classifiers on NSL-KDD dataset

Dataset	RE	AUC		
		OCAE	OCCEN	OCKDE
DoS	0.459	0.960	0.956	0.974
R2L	0.459	0.909	0.839	0.891
U2R	0.459	0.928	0.888	0.945
Probe	0.459	0.971	0.986	0.987

Overall, these results suggest that the proposed density-based classifiers, OCCEN and OCKDE, tend to perform well in terms of accuracy on datasets in which the normal and anomaly classes are highly separated (e.g. Probe, DoS or U2R). The KDE-based one-class classifier, OCKDE, is more powerful than OCCEN and OCAE in detecting anomalies from NSL-KDD dataset.

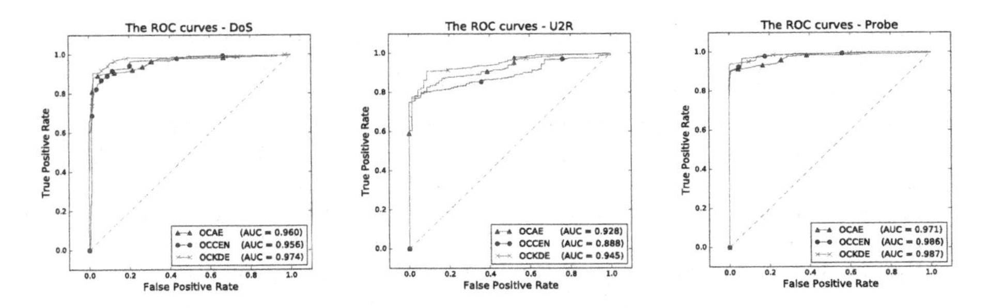

Fig. 5. The ROC curves of the three classifiers on NSL-KDD dataset

7 Conclusion and Further Work

In this paper, we have proposed a novel method for anomaly detection based on estimating density in the compressed hidden-layer representation of autoencoders. We have motivated this method through visualization of density in the hidden layer. We investigated the hyperparameters of AE based on the UCI datasets, and proposed an equation for estimating the size of hidden layer for later evaluating the models on NSL-KDD dataset.

The experimental results suggest that our proposed model performs well, and often out-performs a typical autoencoder approach based on reconstruction error on the dataset from security domain. The model also tend to work efficiently on the datasets in which the normal and anomaly classes are highly separated. This will help our model become an universal method for anomaly detection. Further work will focus on how to speed up the query stage of the models.

Acknowledgements. This work is funded by Vietnam International Education Development (VIED) and by agreement with the Irish Universities Association.

References

1. Aggarwal, C.C.: Outlier Analysis. Springer Science & Business Media, Berlin (2013)
2. Cao, V.L., Nicolau, M., McDermott, J.: One-class classification for anomaly detectionwith kernel density estimation and genetic programming. In: Heywood, M.I., McDermott, J., Castelli, M., Costa, E., Sim, K. (eds.) EuroGP 2016. LNCS, vol. 9594, pp. 3–18. Springer, Berlin (2016)
3. Chandola, V., Banerjee, A., Kumar, V.: Anomaly detection: a survey. ACM Comput. Surv. (CSUR) **41**(3), 15 (2009)
4. Curry, R., Heywood, M.: One-class learning with multi-objective genetic programming. In: IEEE International Conference on Systems, Man and Cybernetics, ISIC, pp. 1938–1945. IEEE (2007)
5. Curry, R., Heywood, M.I.: One-class genetic programming. In: Vanneschi, L., Gustafson, S., Moraglio, A., De Falco, I., Ebner, M. (eds.) EuroGP 2009. LNCS, vol. 5481, pp. 1–12. Springer, Heidelberg (2009)
6. Duchi, J., Hazan, E., Singer, Y.: Adaptive subgradient methods for online learning and stochastic optimization. J. Mach. Learn. Res. **12**, 2121–2159 (2011)

7. Erfani, S.M., Rajasegarar, S., Karunasekera, S., Leckie, C.: High-dimensional and large-scale anomaly detection using a linear one-class SVM with deep learning. Pattern Recogn. **58**, 121–134 (2016)
8. Fiore, U., Palmieri, F., Castiglione, A., De Santis, A.: Network anomaly detection with the restricted Boltzmann machine. Neurocomputing **122**, 13–23 (2013)
9. Hawkins, S., He, H., Williams, G.J., Baxter, R.A.: Outlier detection using replicator neural networks. In: Kambayashi, Y., Winiwarter, W., Arikawa, M. (eds.) DaWaK 2002. LNCS, vol. 2454, pp. 170–180. Springer, Heidelberg (2002)
10. Hinton, G.E., Salakhutdinov, R.R.: Reducing the dimensionality of data with neural networks. Science **313**(5786), 504–507 (2006)
11. Japkowicz, N., Myers, C., Gluck, M., et al.: A novelty detection approach to classification. In: IJCAI, pp. 518–523 (1995)
12. KDD Cup Dataset (1999). http://kdd.ics.uci.edu/databases/kddcup99/kddcup99.html
13. Lee, W., Stolfo, S.J., Mok, K.W.: A data mining framework for building intrusion detection models. In: Proceedings of the 1999 IEEE Symposium on Security and Privacy, pp. 120–132. IEEE (1999)
14. Lichman, M.: UCI Machine Learning Repository (2013). http://archive.ics.uci.edu/ml
15. Moya, M.M., Koch, M.W., Hostetler, L.D.: One-class classifier networks for target recognition applications. Technical report, Sandia National Labs., Albuquerque, NM (United States) (1993)
16. Pedregosa, F., Varoquaux, G., Gramfort, A., Michel, V., Thirion, B., Grisel, O., Blondel, M., Prettenhofer, P., Weiss, R., Dubourg, V., Vanderplas, J., Passos, A., Cournapeau, D., Brucher, M., Perrot, M., Duchesnay, E.: Scikit-learn: machine learning in Python. J. Mach. Learn. Res. **12**, 2825–2830 (2011)
17. Sakurada, M., Yairi, T.: Anomaly detection using autoencoders with nonlinear dimensionality reduction. In: Proceedings of MLSDA 2014 2nd Workshop on Machine Learning for Sensory Data Analysis, p. 4. ACM (2014)
18. Schölkopf, B., Platt, J.C., Shawe-Taylor, J., Smola, A.J., Williamson, R.C.: Estimating the support of a high-dimensional distribution. Neural Comput. **13**(7), 1443–1471 (2001)
19. Shafi, K., Abbass, H.A.: Evaluation of an adaptive genetic-based signature extraction system for network intrusion detection. Pattern Anal. Appl. **16**(4), 549–566 (2013)
20. Tavallaee, M., Bagheri, E., Lu, W., Ghorbani, A.A.: NSL-KDD Dataset (2009). http://www.unb.ca/research/iscx/dataset/iscx-NSL-KDD-dataset.html
21. To, C., Elati, M.: A parallel genetic programming for single class classification. In: Proceedings of 15th Annual Conference Companion on Genetic and Evolutionary Computation, pp. 1579–1586. ACM (2013)
22. Veeramachaneni, K., Arnaldo, I., Cuesta-Infante, A., Korrapati, V., Bassias, C., Li, K.: AI^2: training a big data machine to defend. In: International Conference on Big Data Security. IEEE, New York (2016)

Theory

Parameterized Analysis of Multi-objective Evolutionary Algorithms and the Weighted Vertex Cover Problem

Mojgan Pourhassan[1], Feng Shi[1,2], and Frank Neumann[1](✉)

[1] Optimisation and Logistics, School of Computer Science, The University of Adelaide, Adelaide, Australia
frank@cs.adelaide.edu.au
[2] School of Information Science and Engineering, Central South University, Changsha, People's Republic of China

Abstract. A rigorous runtime analysis of evolutionary multi-objective optimization for the classical vertex cover problem in the context of parameterized complexity analysis has been presented by Kratsch and Neumann [1]. In this paper, we extend the analysis to the weighted vertex cover problem and provide a fixed parameter evolutionary algorithm with respect to OPT, the cost of the optimal solution for the problem. Moreover, using a diversity mechanism, we present a multi-objective evolutionary algorithm that finds a 2−approximation in expected polynomial time.

1 Introduction

The area of runtime analysis has provided many rigorous new insights into the working behaviour of bio-inspired computing methods such as evolutionary algorithms and ant colony optimization [2–4]. In recent years, the parameterized analysis of bio-inspired computing has gained additional interest [1,5,6]. Here the runtime of bio-inspired computing is studied in dependence of the input size and additional parameters such as the solution size and/or other structural parameters of the given input.

One of the classical problems that has been studied extensively in the area of runtime analysis is the classical NP-hard vertex cover problem. Here, an undirected graph is given and the goal is to find a minimum set of nodes V' such that each edge has at least one endpoint in V'. Friedrich et al. [7] have shown that the single-objective evolutionary algorithm (1+1) EA can not achieve a better than trivial approximation ratio in expected polynomial time. Furthermore, they have shown that a multi-objective approach using Global SEMO gives a factor $O(\log n)$ approximation for the wider classes of set cover problems in expected polynomial time. Further investigations regarding the approximation behaviour of evolutionary algorithms for the vertex cover problem have been carried out in [8,9]. Edge-based representations in connection with different fitness functions have been investigated in [10,11] according to their approximation behaviour in

J. Handl et al. (Eds.): PPSN XIV 2016, LNCS 9921, pp. 729–739, 2016.
DOI: 10.1007/978-3-319-45823-6_68

the static and dynamic setting. Kratsch and Neumann [1] have studied evolutionary algorithms and the vertex cover problem in the context of parameterized complexity. They have shown that Global SEMO, with a problem specific mutation operator is a fixed parameter evolutionary algorithm for this problem and finds 2−approximations in expected polynomial time. Kratsch and Neumann [1] have also introduced an alternative mutation operator and have proved that Global SEMO using this mutation operator finds a $(1+\varepsilon)$−approximation in expected time $O(n^2 \log n + OPT \cdot n^2 + n \cdot 4^{(1-\varepsilon)OPT})$. Jansen et al. [10] have shown that a 2-approximation can also be obtained by using an edge-based representation in the (1+1) EA combined with a fitness function formulation based on matchings.

To our knowledge all investigations so far in the area of runtime analysis consider the (unweighted) vertex cover problem. In this paper, we consider the weighted vertex cover problem where in addition weights on the nodes are given and the goal is to find a vertex cover of minimum weight. We extend the investigations carried out in [1] to the weighted minimum vertex cover problem. In [1], multi-objective models in combination with a simple multi-objective evolutionary algorithm called Global SEMO are investigated. One key argument for the results presented for the (unweighted) vertex cover problem is that the population size is always upper bounded by $n+1$. This argument does not hold in the weighted case. Therefore, we study how a variant of Global SEMO using an appropriate diversity mechanism is able to deal with the weighted case.

Our focus is on finding good approximations of an optimal solution. We analyse the time complexity with respect to n, W_{max}, and OPT, which denote the number of vertices, the maximum weight in the input graph, and the cost of the optimal solution respectively. We first study the expected time of finding a solution with expected approximation ratio $(1+\varepsilon)$ for this problem by Global SEMO with alternative mutation operator. Afterwards, we consider DEMO, a variant of Global SEMO, which incorporates ε-dominance [12] as diversity mechanism. We show that DEMO using standard mutation finds a 2-approximation in expected polynomial time.

The outline of the paper is as follows. In Sect. 2, the problem definition is presented as well as the classical Global SEMO algorithm and DEMO algorithm. Runtime analysis for finding a $(1+\varepsilon)$−approximation by Global SEMO is presented in Sect. 3. Section 4 includes the analysis that shows DEMO can find 2−approximations of the optimum in expected polynomial time. At the end, in Sect. 5 we summarize and conclude.

2 Preliminaries

We consider the weighted vertex cover problem defined as follows. Given a graph $G=(V,E)$ with vertex set $V=\{v_1,\dots,v_n\}$ and edge set $E=\{e_1,\dots,e_m\}$, and a positive weight function $w: V \rightarrow \mathbb{N}^+$ on the vertices, the goal is to find a subset of nodes, $V_C \subseteq V$, that covers all edges and has minimum weight, i.e. $\forall e \in E, e \cap V_C \neq \emptyset$ and $\sum_{v \in V_C} w(v)$ is minimized. We consider the standard node-based approach, i.e. the search space is $\{0,1\}^n$ and for a solution $x =$

1. Choose $x \in \{0,1\}^n$ uniformly at random and set $P = \{x\}$;
2. while (not termination condition)
   - Choose $x \in P$ uniformly at random and set $x' = x$;
   - Let $E(x) \subseteq E$ denote the set of edges that are not covered by x and $S(x) \subseteq \{1, \dots, n\}$ the vertices being incident on the edges in $E(x)$.
   - Choose $b \in \{0,1\}$ uniform at random.
   - If $b = 0$ flip each bit of x' independently with probability $1/n$.
   - Otherwise flip each bit of $S(x')$ independently with probability $1/2$ and each other bit independently with probability $1/n$.
   - If there is no $y \in P$ with $f(y) \leq f(x')$ then delete all $z \in P$ with $f(x') \leq f(z)$ from P and add x' to P.

Algorithm 1. Global SEMO

$(x_1, \dots, x_n)$ the node v_i is chosen iff $x_i = 1$. The Integer Linear Programming (ILP) formulation of this problem is:

$$\begin{aligned} \min \sum_{i=1}^{n} w(v_i) \cdot x_i \\ s.t.\ \ x_i + x_j \geq 1 \quad \forall (i,j) \in E \\ x_i \in \{0,1\} \end{aligned}$$

By relaxing the constraint $x_i \in \{0,1\}$ to $x_i \in [0,1]$, the linear program formulation of Fractional Weighted Vertex Cover is obtained.

We consider primarily multi-objective approaches for the weighted vertex cover problem. Given a multi-objective fitness function $f = (f_1, \dots, f_d)\colon S \to \mathbb{R}$ where all d objectives should be minimized, we have $f(x) \leq f(y)$ iff $f_i(x) \leq f_i(y)$, $1 \leq i \leq d$. We say that x (weakly) dominates y iff $f(x) \leq f(y)$.

Let $G(x)$ be the graph obtained from G by removing all nodes chosen by x and the corresponding covered edges. Formally, we have $G(x) = (V(x), E(x))$ where $V(x) = V \setminus \{v_i \mid x_i = 1\}$ and $E(x) = E \setminus \{e \mid e \cap (V \setminus V(x)) \neq \emptyset\}$. Kratsch and Neumann [1] investigated a multi-objective baseline algorithm called Global SEMO using the LP-value for $G(x)$ as one of the fitness values for the (unweighted) minimum vertex cover problem. In order to expand the analysis on behaviour of multi-objective evolutionary algorithms to the Weighted Vertex Cover problem, we modify the fitness function that was used in Global SEMO in [1], to match the weighted version of the problem. We investigate the multi-objective fitness function $f(x) = (Cost(x), LP(x))$, where

- $Cost(x) = \sum_{i=1}^{n} w(v_i)x_i$ is the sum of weights of selected vertices
- $LP(x)$ is the value of an optimal solution of the LP for $G(x)$.

We investigate Global SEMO with alternative mutation operator (Algorithm 1) introduced in [1]. Here, the nodes that are adjacent to uncovered edges are mutated with probability $1/2$ in some steps. In the fitness function used in Global SEMO, both $Cost(x)$ and $LP(x)$ can be exponential with respect to the input size; therefore, we need to deal with exponentially large number of solutions, even if we only keep the Pareto front.

1. Choose $x \in \{0,1\}^n$ uniformly at random and set $P = \{x\}$;
2. while (not termination condition)
   - Choose $x \in P$ uniformly at random and set $x' = x$;
   - Flip each bit of x' independently with probability $1/n$.
   - If there is a $y \in P$ where $(f(y) \leq f(x') \wedge f(y) \neq f(x'))$ or $(b(y) = b(x') \wedge Cost(y) + 2 \cdot LP(y) \leq Cost(x') + 2 \cdot LP(x'))$ then keep P unchanged and go to 4;
   - Otherwise delete all $z \in P$ with $f(x') \leq f(z) \vee b(z) = b(x')$ from P and add x' to P.

Algorithm 2. DEMO

One approach for dealing with this problem is using the concept of $\varepsilon-$dominance [12]. The concept of $\varepsilon-$dominance has previously been proved to be useful for coping with exponentially large Pareto fronts in some problems [13, 14]. Having two objective vectors $u = (u_1, \cdots, u_m)$ and $v = (v_1, \cdots, v_m)$, u $\varepsilon-$dominates v, denoted by $u \preceq_\varepsilon v$, if for all $i \in \{1, \cdots, m\}$ we have $(1+\varepsilon)u_i \leq v_i$. In this approach, the objective space is partitioned into a polynomial number of boxes in which all solutions $\varepsilon-$dominate each other, and at most one solution from each box is kept in the population.

Motivated by this approach, DEMO (Diversity Evolutionary Multi-objective Optimizer) has been investigated in [14,15]. In Sect. 4, we analyze DEMO (see Algorithm 2) in which only one non-dominated solution can be kept in the population for each box based on a predefined criteria. In our setting, among two solutions x and y from one box, y is kept in P and x is discarded if $Cost(y) + 2 \cdot LP(y) \leq Cost(x) + 2 \cdot LP(x)$. To implement the concept of $\varepsilon-$dominance in DEMO, we use the parameter $\delta = \frac{1}{2n}$ and define the boxing function $b : \{0,1\}^n \rightarrow \mathbb{N}^2$ as $b_1(x) = \lceil \log_{1+\delta}(1 + Cost(x)) \rceil$ and $b_2(x) = \lceil \log_{1+\delta}(1 + LP(x)) \rceil$.

Analysing the runtime of our evolutionary algorithms, we are interested in the expected number of rounds of the while loop until a solution of desired quality has been obtained. We call this the expected time until the considered algorithm has achieved its desired goal.

3 Analysis of Global SEMO

In this section, we analyse the expected time of Global SEMO (Algorithm 1) to find a (1+ε)-approximation. Before we present our analysis for Global SEMO, we state some basic properties of the solutions in our multi-objective model. The following theorem shown by Balinski [16] states that all basic feasible solutions of the fractional vertex cover, which are the extremal points or the corner solutions of the polyhedron that forms the feasible space, are half-integral.

Theorem 1. *Each basic feasible solution x of the relaxed Vertex Cover ILP is half-integral, i.e., $x \in \{0, 1/2, 1\}^n$* [16].

As a result, there always exists a half integral optimal LP solution for a vertex cover problem. This result and the following lemmata are used in the analysis of Theorem 2 which presents the main approximation result for Global SEMO. The proof of Lemma 3 can be found in [17].

Lemma 1. *For any* $x \in \{0,1\}^n$*,* $LP(x) \leq LP(0^n) \leq OPT$*.*

Proof. Let y be the LP solution of $LP(0^n)$. Also, for any solution x, let $G(x)$ be the graph obtained from G by removing all vertices chosen by x and their edges. The solution 0^n contains no vertices; therefore, y is the optimal fractional vertex cover for all edges of the input graph. Thus, for any solution x, y is a (possibly non-optimal) fractional cover for $G(x)$; therefore, $LP(x) \leq LP(0^n)$. Moreover, we have $LP(0^n) \leq OPT$ as $LP(0^n)$ is the optimal value of the LP relaxation. □

Lemma 2. *Let* $x = \{x_1, \cdots, x_n\}, x_i \in \{0,1\}$ *be a solution and* $y = \{y_1, \cdots, y_n\}$*,* $y_i \in [0,1]$ *be a fractional solution for* $G(x)$*. If there is a vertex* v_i *where* $y_i \geq \frac{1}{2}$*, mutating* x_i *from 0 to 1 results in a solution* x' *for which* $LP(x') \leq LP(x) - y_i \cdot w(v_i) \leq LP(x) - \frac{1}{2}w(v_i)$*.*

Proof. The graph $G(x')$ is the same as $G(x)$ excluding the edges connected to v_i. Therefore, the solution $y' = \{y_1, \cdots, y_{i-1}, 0, y_{i+1}, y_n\}$ is a fractional vertex cover for $G(x')$ and has a cost of $LP(x) - y_i w(v_i)$. The cost of the optimal fractional vertex cover of $G(x')$ is at most as great as the cost of y'; thus $LP(x') \leq LP(x) - y_i \cdot w(v_i) \leq LP(x) - \frac{1}{2}w(v_i)$. □

Lemma 3. *The population size of Global SEMO (Algorithm 1) is upper bounded by* $2 \cdot OPT + 1$ *and the search point* 0^n *is included in the population of Global SEMO, in expected time* $O(OPT \cdot n(\log W_{max} + \log n))$*.*

Lemma 4. *A solution* x *fulfilling the two properties*

1. $LP(x) = LP(0^n) - Cost(x)$ *and*
2. *there is an optimal solution of the LP for* $G(x)$ *which assigns 1/2 to each non-isolated vertex of* $G(x)$

is included in the population of Global SEMO in expected time $O(OPT \cdot n(\log W_{max} + \log n + OPT))$*.*

Proof. The search point 0^n which satisfies property 1 is included in the population in expected time of $O(OPT \cdot n(\log W_{max} + \log n))$, due to Lemma 3. Let $P' \subseteq P$ be a set of solutions such that for each solution $x \in P'$, $LP(x) + Cost(x) = LP(0^n)$. Let $x_{min} \in P'$ be a solution such that $LP(x_{min}) = min_{x \in P'} LP(x)$.

If the optimal fractional vertex cover for $G(x_{min})$ assigns 1/2 to each non-isolated vertex of $G(x_{min})$, then the conditions of the lemma hold. Otherwise, it assigns 1 to some non-isolated vertex, say v. The probability that the algorithm selects x_{min} and flips the bit corresponding to v, is $\Omega(\frac{1}{OPT \cdot n})$, because the population size is $O(OPT)$ (Lemma 3). Let x_{new} be the new solution. We have

$Cost(x_{new}) = Cost(x_{min}) + w(v)$, and by Lemma 2, $LP_w(x_{new}) \leq LP_w(x_{min}) - w(v)$. This implies that $LP(x_{new}) + Cost(x_{new}) = LP(0^n)$; hence, x_{new} is a Pareto Optimal solution and is added to the population P.

Since $LP_w(x_{min}) \leq OPT$ (Lemma 1) and the weights are at least 1, assuming that we already have the solution 0^n in the population, by means of the method of fitness based partitions, we find the expected time of finding a solution that fulfils the properties given above as $O(OPT^2 \cdot n)$. Since the search point 0^n is included in expected time $O(OPT \cdot n(\log W_{max} + \log n))$, the expected time that a solution fulfilling the properties given above is included in P is $O(OPT \cdot n(\log W_{max} + \log n + OPT))$. □

Theorem 2. *The expected time until Global SEMO has obtained a solution that has expected approximation ratio* $(1+\varepsilon)$ *is* $O(OPT \cdot 2^{\min\{n, 2(1-\varepsilon)OPT\}} + OPT \cdot n(\log W_{max} + \log n + OPT))$.

Proof. By Lemma 4, a solution x that satisfies the two properties given in Lemma 4 is included in the population in expected time of $O(OPT \cdot n(\log W_{max} + \log n + OPT))$. For a set of nodes, X', we define $Cost(X') = \sum_{v \in X'} w(v)$. Let X be the vertex set of graph $G(x)$. Also, let $S \subseteq X$ be a vertex cover of $G(x)$ with the minimum weight over all vertex covers of $G(x)$, and T be the set containing all non-isolated vertices in $X \setminus S$. Note that all vertices in $X \setminus (S \cup T)$ are isolated vertices in $G(x)$. Due to property 2 of Lemma 4, $\frac{1}{2}Cost(S) + \frac{1}{2}Cost(T) = LP(x) \leq Cost(S)$; therefore, $Cost(T) \leq Cost(S)$. Let $OPT' = OPT - Cost(x)$. Observe that $OPT' = Cost(S)$.

Let $s_1, \ldots, s_{|S|}$ be a numbering of the vertices in S such that $w(s_i) \leq w(s_{i+1})$, for all $1 \leq i \leq |S| - 1$. And let $t_1, \ldots, t_{|T|}$ be a numbering of the vertices in T such that $w(t_i) \geq w(t_{i+1})$, for all $1 \leq i \leq |T| - 1$. Let $S_1 = \{s_1, s_2, \ldots, s_\rho\}$, where $\rho = min\{|S|, (1-\varepsilon) \cdot OPT'\}$, and $T_1 = \{t_1, t_2, \ldots, t_\eta\}$, where $\eta = min\{|T|, (1-\varepsilon) \cdot OPT'\}$.

With probability $\Omega(\frac{1}{OPT})$, the algorithm Global SEMO selects the solution x, and sets $b = 1$. With $b = 1$, the probability that the bits corresponding to all vertices of S_1 are flipped, is $\Omega((\frac{1}{2})^\rho)$, and the probability that none of the bits corresponding to the vertices of T_1 are flipped is $\Omega((\frac{1}{2})^\eta)$. Also, the bits corresponding to the isolated vertices of $G(x)$ are flipped with probability $\frac{1}{n}$; hence, the probability that none of them flips is $\Omega(1)$. As a result, with probability $\Omega(\frac{1}{OPT} \cdot (\frac{1}{2})^{\rho+\eta})$, solution x is selected, the vertices of S_1 are included, and the vertices of T_1 and isolated vertices are not included in the new solution x'. Since $\rho + \eta \leq 2(1-\varepsilon) \cdot OPT' \leq 2(1-\varepsilon) \cdot OPT$, and also $\rho + \eta \leq n$; the expected time until solution x' is found after reaching solution x, is $O(OPT \cdot 2^{\min\{n, 2(1-\varepsilon)OPT\}})$.

Note that the bits corresponding to vertices of $S_2 = S \setminus S_1$ and $T_2 = T \setminus T_1$, are arbitrarily flipped in solution x' with probability $1/2$ by the Alternative Mutation Operator. Here we show that for the expected cost and the LP value of x', the following constraint holds: $E[Cost(x')] + 2 \cdot LP(x') \leq (1+\varepsilon) \cdot OPT$.

Let $S' \subseteq S$ and $T' \subseteq T$ denote the subset of vertices of S and T that are actually included in the new solution x' respectively. In the following, we show that for the expected values of $Cost(S')$ and $Cost(T')$, we have:

$$E\left[Cost(S')\right] \geq (1-\varepsilon) \cdot OPT' + E\left[Cost(T')\right] \tag{1}$$

Since the bits corresponding to the vertices of S_2 and T_2 are flipped with probability 1/2, for the expected values of $Cost(S')$ and $Cost(T')$ we have:

$$\begin{aligned} E\left[Cost(S')\right] &= Cost(S_1) + \frac{Cost(S_2)}{2} = Cost(S_1) + \frac{Cost(S) - Cost(S_1)}{2} \\ &= 1/2Cost(S) + 1/2Cost(S_1) \end{aligned}$$

and $E\left[Cost(T')\right] = 1/2Cost(T_2)$.

If $\rho = |S|$, then $S_1 = S$ and $Cost(S_1) = Cost(S) = OPT'$. If $\rho = (1-\varepsilon) \cdot OPT'$, we have $Cost(S_1) \geq (1-\varepsilon) \cdot OPT'$, since each vertex has a weight of at least 1. Using $Cost(S) = OPT'$ and the inequality above, we have

$$E\left[Cost(S')\right] \geq (1-\varepsilon) \cdot OPT' + \frac{\varepsilon \cdot OPT'}{2}$$

We divide the analysis into two cases based on the relation between η and $|T|$.

Case (I). $\eta = |T|$. Then $T_2 = T' = \emptyset$. Thus, $E\left[Cost(T')\right] = 0$ and Inequality (1) holds true.

Case (II). $\eta = (1-\varepsilon) \cdot OPT' < |T|$. Since $w(t_i) \geq w(t_{i+1})$ for $1 \leq i \leq |T| - 1$ and $Cost(T) \leq Cost(S) = OPT'$, we have

$$\begin{aligned} Cost(T_2) &\leq \frac{|T| - \eta}{|T|} Cost(T) \leq \frac{OPT' - (1-\varepsilon) \cdot OPT'}{OPT'} Cost(T) \\ &\leq \varepsilon Cost(S) = \varepsilon \cdot OPT' \end{aligned}$$

Thus for the expected value of $Cost(T')$, we have $E\left[Cost(T')\right] = \frac{1}{2}Cost(T_2) \leq \frac{\varepsilon \cdot OPT'}{2}$.

Summarizing above analysis, we can get that the Inequality (1) holds. Using this inequality, we prove that in expectation, the new solution x' satisfies the inequality $Cost(x') + 2 \cdot LP(x') \leq (1+\varepsilon) \cdot OPT$:

$$\begin{aligned} & E\left[Cost(x')\right] + 2 \cdot LP(x') = Cost(x) + E\left[Cost(S')\right] + E\left[Cost(T')\right] + 2 \cdot LP(x') \\ & \leq Cost(x) + E\left[Cost(S')\right] + E\left[Cost(S')\right] - (1-\varepsilon) \cdot OPT' + 2 \cdot LP(x') \\ & \leq Cost(x) + 2E\left[Cost(S')\right] - (1-\varepsilon) \cdot OPT' + 2 \cdot (OPT' - E\left[Cost(S')\right]) \\ & = Cost(x) + (1+\varepsilon) \cdot OPT' = Cost(x) + (1+\varepsilon) \cdot (OPT - Cost(x)) \\ & \leq (1+\varepsilon) \cdot OPT. \end{aligned}$$

Now we analyze whether the new solution x' could be included in the population P. If x' could not be included in P, then there is a solution x''

dominating x, i.e., $LP(x'') \leq LP(x')$ and $Cost(x'') \leq Cost(x')$. This implies $Cost(x'') + 2 \cdot LP(x'') < Cost(x') + 2 \cdot LP(x') \leq (1+\varepsilon) \cdot OPT$. Therefore, after having a solution that fulfils the properties of Lemma 4 in P, in expected time $O(OPT \cdot 2^{\min\{n, 2(1-\varepsilon)OPT\}})$, the population would contain a solution y such that $Cost(y) + 2 \cdot LP(y) \leq (1+\varepsilon) \cdot OPT$.

Let P' contain all solutions $x \in P$ such that $Cost(x)+2 \cdot LP(x) \leq (1+\varepsilon) \cdot OPT$, and let $x_{\min}$ be the one that minimizes LP. Let $y = \{y_1, \cdots, y_n\}$ be a basic LP solution for $G(x_{\min})$. According to Theorem 1, y is a half-integral solution.

Let Δ^t be the improvement that happens on the minimum LP value in p' at time step t. Also let k be the number of nodes that are assigned at least $\frac{1}{2}$ by y. Flipping only one of these nodes by the algorithm happens with probability at least $\frac{k}{e \cdot n}$. According to Lemma 2, flipping one of these nodes, v_i, results in a solution x' with $LP(x') \leq LP(x_{\min}) - y_i \cdot w(v_i) \leq LP(x_{\min}) - \frac{1}{2} \cdot w(v_i)$. Observe that the constraint of $Cost(x') + 2 \cdot LP(x') \leq 2 \cdot OPT$ holds for solution x'. Therefore, $\Delta^t \geq y_i \cdot w(v_i)$, which is in expectation at least $\frac{LP(x_{\min})}{k}$ due to definition of $LP(x_{\min})$.

Moreover, at each step, the probability that $x_{\min}$ is selected and only one of the k bits defined above flips is $\frac{k}{(2 \cdot OPT+1) \cdot e \cdot n}$, As a result we have:

$$E[\Delta^t \mid x_{\min}] \geq \frac{k}{(2 \cdot OPT + 1) \cdot e \cdot n} \cdot \frac{LP(x_{\min})}{k} = \frac{LP(x_{\min})}{e \cdot n \cdot (2 \cdot OPT + 1)}$$

According to Lemma 1 for any solution x, we have $LP(x) \leq OPT$. We also know that for any solution x which is not a complete cover, $LP(x) \geq 1$, because the weights are positive integers. Using the method of Multiplicative Drift Analysis [18] with $s_0 \leq OPT$ and $s_{min} \geq 1$, we get the expected time $O(OPT \cdot n \log OPT)$ to find a solution z for which $LP(z) = 0$ and $Cost(z) + 2 \cdot LP(z) \leq (1+\varepsilon) \cdot OPT$.

Overall, the expected number of iterations of Global SEMO, for finding a $(1+\varepsilon)$-approximate weighted vertex cover, is bounded by $O(OPT \cdot 2^{\min\{n, 2(1-\varepsilon)OPT\}} + OPT \cdot n(\log W_{max} + \log n + OPT))$. □

4 Analysis of DEMO

In this section, we analyse the other evolutionary algorithm, DEMO (Algorithm 2), that uses some diversity handling mechanisms for dealing with exponentially large population sizes. We are making use of the following lemma whose proof can be found in [17].

Lemma 5. *The population size of DEMO is upper bounded by $O(n \cdot (\log n + \log W_{max}))$ and the search point 0^n is included in the population in expected time of $O(n^3(\log n + \log W_{max})^2)$.*

Lemma 6. *Let $x \in P$ be a search point such that $Cost(x) + 2 \cdot LP(x) \leq 2 \cdot OPT$ and $b_2(x) > 0$. There exists a 1-bit flip leading to a search point x' with $Cost(x') + 2 \cdot LP(x') \leq 2 \cdot OPT$ and $b_2(x') < b_2(x)$.*

Proof. Let $y = \{y_1 \cdots y_n\}$ be a basic half integral LP solution for $G(x)$. Since $b_2(x) = LP(x) \neq 0$, there must be at least one uncovered edge; hence, at least one vertex v_i has a $y_i \geq \frac{1}{2}$ in LP solution y. Consider v_j the vertex that maximizes $y_i w(v_i)$ among vertices v_i, $1 \leq i \leq n$. Also, let x' be a solution obtained by adding v_j to x. Since solutions x and x' are only different in one vertex, v_j, we have $Cost(x') = Cost(x) + w(v_j)$. Moreover, according to Lemma 2, $LP(x') \leq LP(x) - \frac{1}{2} \cdot w(v_j)$. Therefore,

$$\begin{aligned} Cost(x') + 2 \cdot LP(x') &\leq Cost(x) + w(v_j) + 2\left(LP(x) - \frac{w(v_j)}{2}\right) \\ &\leq Cost(x) + 2 \cdot LP(x) \leq 2 \cdot OPT \end{aligned}$$

which means solution x' fulfils the mentioned constraint. If $LP(x) = W$, then $y_j w(v_j) \geq \frac{W}{n}$, because n is an upper bound on the number of vertices selected by the LP solution. As a result, using Lemma 2, we get $LP(x') \leq W \cdot (1 - \frac{1}{n})$. Therefore, we have:

$$\begin{aligned} (1+\delta)(1+LP(x')) &\leq 1 + \delta + W\left(1 - \frac{1}{n}\right)(1+\delta) \\ &\leq 1 + \delta + W + W(\delta - \frac{1}{n} - \frac{\delta}{n}) \\ &\leq 1 + W + W(2\delta - \frac{1}{n} - \frac{\delta}{n}) \leq 1 + W \end{aligned}$$

which implies $1 + \log_{1+\delta}(1 + LP(x')) \leq \log_{1+\delta}(1+W)$. As a result, $b_2(x') < b_2(x)$ holds for x', which is obtained by performing a 1-bit flip on x, and the lemma is proved. □

Theorem 3. *The expected time until DEMO constructs a 2-approximate vertex cover is* $O\left(n^3 \cdot (\log n + \log W_{max})^2\right)$.

Proof. Consider solution $x \in P$ that minimizes $b_2(x)$ under the constraint that $Cost(x) + 2 \cdot LP(x) \leq 2 \cdot OPT$. Note that 0^n fulfils this constraint and according to Lemma 5, the solution 0^n will be included in P in time $O\left(n^3(\log n + \log W_{max})^2\right)$.

If $b_2(x) = 0$ then x covers all edges and by selection of x we have $Cost(x) \leq 2 \cdot OPT$, which means that x is a 2−approximation.

In case $b_2(x) \neq 0$, according to Lemma 6 there is a one-bit flip on x that results in a new solution x' for which $b_2(x') < b_2(x)$, while the mentioned constraint also holds for it. Since the population size is $O\left(n \cdot (\log n + \log W_{max})\right)$ (Lemma 5), this 1-bit flip happens with a probability of $\Omega\left(n^{-2} \cdot (\log n + \log W_{max})^{-1}\right)$ and x' is obtained in expected time of $O(n^3 \cdot (\log n + \log W_{max})^2)$. This new solution will be added to P because a solution y with $Cost(y) + 2 \cdot LP(y) > 2 \cdot OPT$ can not dominate x' with $Cost(x') + 2 \cdot LP(x') \leq 2 \cdot OPT$, and x' has the minimum value of b_2 among solution that fulfil the constraint. Moreover, if there already is a solution, x_{prev}, in the same box as x', it will be replaced by x' because $Cost(x_{prev})+2\cdot LP(x_{prev}) > 2\cdot OPT$; otherwise, it would have been selected as x.

There are at most $A = 1 + \lceil\frac{\log n+\log W_{max}}{\log(1+\delta)}\rceil$ different values for b_2 in the objective space, and since $\delta = \frac{1}{2n}$, $A = O(n\cdot(\log n + \log W_{max}))$. Therefore, the expected time until a solution x'' is found so that $b_2(x'') = 0$ and $Cost(x'') + 2\cdot LP(x'') \leq 2\cdot OPT$, is at most $O(n^3\cdot(\log n + \log W_{max})^2)$. □

5 Conclusion

The minimum vertex cover problem is one of the classical NP-hard combinatorial optimization problems. In this paper, we have generalized previous results of Kratsch and Neumann [1] for the unweighted minimum vertex cover problem to the weighted case where in addition weights on the nodes are given. We have studied the expected time required by Global SEMO to find a $(1+\varepsilon)$-approximation. Furthermore, our investigations show that the algorithm DEMO using the ε-dominance approach reaches a 2-approximation in expected polynomial time.

Acknowledgements. This research has been supported by Australian Research Council grants DP140103400 and DP160102401.

References

1. Kratsch, S., Neumann, F.: Fixed-parameter evolutionary algorithms and the vertex cover problem. Algorithmica **65**(4), 754–771 (2013)
2. Neumann, F., Witt, C.: Bioinspired Computation in Combinatorial Optimization: Algorithms and Their Computational Complexity, 1st edn. Springer, New York (2010)
3. Auger, A., Doerr, B.: Theory of Randomized Search Heuristics: Foundations and Recent Developments. World Scientific Publishing Co., Inc., River Edge (2011)
4. Jansen, T.: Analyzing Evolutionary Algorithms - The Computer Science Perspective. NCS. Springer, Berlin (2013)
5. Sutton, A.M., Neumann, F.: A parameterized runtime analysis of simple evolutionary algorithms for makespan scheduling. In: Coello, C.A.C., Cutello, V., Deb, K., Forrest, S., Nicosia, G., Pavone, M. (eds.) PPSN 2012, Part I. LNCS, vol. 7491, pp. 52–61. Springer, Heidelberg (2012)
6. Sutton, A.M., Neumann, F., Nallaperuma, S.: Parameterized runtime analyses of evolutionary algorithms for the planar euclidean traveling salesperson problem. Evol. Comput. **22**(4), 595–628 (2014)
7. Friedrich, T., Hebbinghaus, N., Neumann, F., He, J., Witt, C.: Approximating covering problems by randomized search heuristics using multi-objective models. In: Proceedings of 9th Annual Conference on Genetic and Evolutionary Computation, GECCO 2007, pp. 797–804. ACM, New York (2007)
8. Friedrich, T., He, J., Hebbinghaus, N., Neumann, F., Witt, C.: Analyses of simple hybrid algorithms for the vertex cover problem. Evol. Comput. **17**(1), 3–19 (2009)

9. Oliveto, P.S., He, J., Yao, X.: Analysis of the (1+1)-EA for finding approximate solutions to vertex cover problems. IEEE Trans. Evol. Comput. **13**(5), 1006–1029 (2009)
10. Jansen, T., Oliveto, P.S., Zarges, C.: Approximating vertex cover using edge-based representations. In: Neumann, F., Jong, K.A.D. (eds.) Foundations of Genetic Algorithms XII, FOGA 2013, Adelaide, SA, Australia, 16–20 January 2013, pp. 87–96. ACM (2013)
11. Pourhassan, M., Gao, W., Neumann, F.: Maintaining 2-approximations for the dynamic vertex cover problem using evolutionary algorithms. In: Proceedings of Genetic and Evolutionary Computation Conference, GECCO 2015, Madrid, Spain, pp. 903–910. ACM (2015)
12. Laumanns, M., Thiele, L., Deb, K., Zitzler, E.: Combining convergence and diversity in evolutionary multiobjective optimization. Evol. Comput. **10**(3), 263–282 (2002)
13. Horoba, C., Neumann, F.: Benefits and drawbacks for the use of ϵ-dominance in evolutionary multi-objective optimization. In: Proceedings of GECCO 2008 (2008)
14. Neumann, F., Reichel, J., Skutella, M.: Computing minimum cuts by randomized search heuristics. Algorithmica **59**(3), 323–342 (2011)
15. Neumann, F., Reichel, J.: Approximating minimum multicuts by evolutionary multi-objective algorithms. In: Rudolph, G., Jansen, T., Lucas, S., Poloni, C., Beume, N. (eds.) PPSN 2008. LNCS, vol. 5199, pp. 72–81. Springer, Heidelberg (2008)
16. Balinski, M.: On the maximum matching, minimum covering. In: Proceedings of Symposium on Mathematical Programming, pp. 434–445. Princeton University Press (1970)
17. Pourhassan, M., Shi, F., Neumann, F.: Parameterized analysis of multi-objective evolutionary algorithms and the weighted vertex cover problem (2016). CoRR http://arXiv.org/abs/1604.01495
18. Doerr, B., Johannsen, D., Winzen, C.: Multiplicative drift analysis. Algorithmica **64**(4), 673–697 (2012)

Fixed-Parameter Single Objective Search Heuristics for Minimum Vertex Cover

Wanru Gao[1], Tobias Friedrich[1,2], and Frank Neumann[1(✉)]

[1] School of Computer Science, The University of Adelaide, Adelaide, Australia
frank@cs.adelaide.edu.au
[2] Hasso Plattner Institute, Potsdam, Germany

Abstract. We consider how well-known branching approaches for the classical minimum vertex cover problem can be turned into randomized initialization strategies with provable performance guarantees and investigate them by experimental investigations. Furthermore, we show how these techniques can be built into local search components and analyze a basic local search variant that is similar to a state-of-the-art approach called NuMVC. Our experimental results for the two local search approaches show that making use of more complex branching strategies in the local search component can lead to better results on various benchmark graphs.

1 Introduction

The parameterized analysis of heuristic search methods has gained a lot of attention during the last few years [1,3,6–8,10]. It provides a mechanism for understanding how and why heuristic methods work for prominent combinatorial optimization problems. There are different methods closely related to the notion of fixed parameter algorithms. One popular paradigm to design parameterized algorithms are *bounded search tree algorithms* which search for a good solution by branching according to different rules that may be applied to solve the underlying problem.

For the classical vertex cover problem, different branching algorithms are available to answer the question whether a given graph has a vertex cover of size at most k. We investigate two common strategies resulting in fixed parameter algorithms running in time $\mathcal{O}^*(2^k)$ and $\mathcal{O}^*(\alpha^k)$[1], where $\alpha = 1.4656$, to solve this problem.

We show how these search tree algorithms can be turned into initialization approaches that produce initial solutions in linear time. We start by presenting an edge-based initialization approach which obtains a vertex cover having at most $k = 2\,OPT - r$, $0 \leq r \leq OPT$, nodes with probability at least $\binom{k}{OPT} \cdot 2^{-k}$.

[1] We use $\mathcal{O}^*(\cdot)$ to describe the essential functional behavior, ignoring all terms of lower order. For exponential expressions all polynomials are omitted: $\mathcal{O}^*(g(n)) = \mathcal{O}(g(n) \operatorname{poly} g(n))$.

J. Handl et al. (Eds.): PPSN XIV 2016, LNCS 9921, pp. 740–750, 2016.
DOI: 10.1007/978-3-319-45823-6_69

Furthermore, we present a node-based initialization approach which obtains an optimal solution with probability at least α^{-OPT}.

After having considered initialization approaches, we turn the branching rules into local search approaches and investigate their behaviour on different types of graphs. Both local search approaches start with a given vertex cover and try to find a smaller vertex cover by searching in the infeasible region of the search space. Our edge-based approach captures the essential ideas of a state-of-the-art local search algorithm for minimum vertex cover called NuMVC [2]. Having a vertex cover of size k, one node is removed to obtain a set of $k-1$. If this set is still a vertex cover, the algorithm searches for a vertex cover of size $k-2$ and so on. If the set is not yet a vertex cover an additional node is taken out and a node covering an uncovered edge is chosen. We turn these ideas in combination with our theoretical insights into an edge-based local search approach which obtains a vertex cover of size at most $k = 2\,OPT - r$ in an expected number of 2^{r+1} phases where each phase consists of a sequence of k local search steps.

Furthermore, we turn the node-based initialization approach into a similar local search approach and compare both local search strategies on different benchmark graphs. Our experimental results show that the node-based approach usually leads to a local search approach that obtains better solutions than the edge-based local search approach.

The paper is structured as follows. In Sect. 2, we provide some background material on parameterized algorithms and the minimum vertex cover problem. Section 3 introduces our two initialization heuristics and examines them from a theoretical and experimental perspective. Section 4 presents our two local search approaches and studies them on different types of benchmark instances. Finally, we provide some concluding remarks.

2 Preliminaries

The vertex cover problem is one of the best-known combinatorial optimization problems. Given an undirected graph $G = (V, E)$, the goal is to find a minimum set of vertices V' such that edge has at least one end vertex in V'. The problem is NP-hard and several 2-approximation algorithms are known. Furthermore, the problem has been studied extensively in the area of parameterized complexity. In fact, it is the archetypical problem in this area. Various kernelization approaches leading to fixed parameter algorithms of different runtime quality are known.

We make use of two branching approaches from the area of parameterized complexity [4]. Both have been introduced to determine whether a given graph $G = (V, E)$ contains a vertex cover of at most k nodes. The first approach builds on the fact that a vertex cover has to contain for each edge at least 1 node. It starts with G, picks an edge $e = \{u, v\}$ currently not covered, and branches according to the two options of including u or v. This allows to answer the question of whether G contains a vertex cover of size at most k in time $O^*(2^k)$.

The second approach makes more sophisticated decisions according to the degree of a node with respect to the uncovered edges. Considering a degree 1 node, it's always safe to take its neighbor. In the case of dealing with a degree 2

Algorithm 1. Edge-based Initialization Heuristic

```
1 C := ∅;
2 repeat
3     Let e = {u, v} be a random uncovered edge, i.e., e ∈ G[C];
4     with probability 1/2 do
5         C := C ∪ {u}
6     else
7         C := C ∪ {v}
8 until C is a vertex cover of G;
9 Return C;
```

node u, one has to choose either the two neighbors v and w of u or all neighbors (including u) of v and w. Finally, for a node u of degree at least 3, one has to choose u or all its neighbours. This approach allows to answer the question of whether G contains a vertex cover of size at most k in time $\mathcal{O}^*(\alpha^k)$, where $\alpha = 1.4656$.

We build on these two fixed parameter algorithms for the decision version of the vertex cover problem and study how to turn them into randomized initialization strategies with provable guarantees on their probability of achieving a solution of certain quality. In addition, we explore how they can be turned into local search approaches and study the performance of these approaches on benchmark instances.

For describing our algorithms we need one more piece of notation for each vertex cover $C \subseteq V$ of a graph $G = (V, E)$. We denote the subgraph of G consisting of the edges not covered by C and the corresponding non-isolated vertices by $G[C] := (V_C, E_C)$ with

$$E_C := E \setminus \{e \in E \mid e \cap C \neq \emptyset\} \text{ and}$$
$$V_C := \{v \in V \mid v \cap E_C \neq \emptyset\}.$$

Furthermore, we denote by $\deg_{G[C]}(u)$ the degree of a node u in $G[C]$ and by $N_{G[C]}[u]$ the set of neighbours of u in $G[C]$.

3 Initialization Strategies

We now describe two randomized initialization strategies based on the branching approaches described in the previous section. Both start with an empty set of nodes and add vertices until a vertex cover has been obtained. The edge-based initialization outlined in Algorithm 1 randomly selects in each step an uncovered edge and adds one of its endpoints chosen uniformly at random to the vertex cover.

For the edge-based initialization we can give a tradeoff between size of the obtained vertex cover and success probability.

Algorithm 2. Vertex-based Initialization Heuristic

```
1  C := ∅;
2  repeat
3      if mindeg(G[C]) = 1 then
4          Let u be a random node with deg_{G[C]}(u) = 1;
5          C := C ∪ N_{G[C]}[u] ;                              /* degree 1 rule */
6      else
7          Let u be a node chosen uniformly at random from G[C];
8          if deg_{G[C]}(u) = 2 then
9              Let v, w ∈ V such that N_{G[C]}[u] = {v, w};
10             with probability α^{-|N_{G[C]}[v] ∪ N_{G[C]}[w]|} do
11                 C := C ∪ N_{G[C]}[v] ∪ N_{G[C]}[w]
12             else
13                 C := C ∪ N_{G[C]}[u] ;                      /* degree 2 rule */
14         else
15             with probability α^{- deg_{G[C]}(u)} do
16                 C := C ∪ N_{G[C]}[u]
17             else
18                 C := C ∪ {u} ;                              /* degree ≥ 3 rule */
19 until C is a vertex cover of G;
20 Return C;
```

Theorem 1. *For all r with $0 \leq r \leq OPT$, the edge-based initialization heuristic obtains a vertex cover of size at most $k := 2 \cdot OPT - r$ with probability at least $\binom{k}{OPT} \cdot 2^{-k}$.*

Proof. Let C^* be an optimal solution of value OPT. For each edge e at least one of its endpoints is contained in C^*. Hence, each step in the initialization process increases the number of nodes chosen from C^* by 1 with probability at least 1/2. We call a step increasing the number of nodes already chosen from C^* a *success*. OPT successes are sufficient to obtain a vertex cover. The probability to have OPT successes during k steps is at least $\binom{k}{OPT} \cdot 2^{-k}$. □

Observe that for $r := 0$ (and $k = 2\,OPT$), the edge-based initialization heuristic therefore obtains a 2-approximation of the minimum vertex cover with probability at least $\binom{2\,OPT}{OPT} \cdot 2^{-2\,OPT} = \Theta(1/\sqrt{OPT})$. On the other hand, for $r := OPT$ (and $k = OPT$), the edge-based initialization heuristic obtains a minimum vertex cover with probability at least 2^{-OPT}.

We now introduce an initialization heuristic based on more complex vertex-based branching. The vertex-based initialization given in Algorithm 2 first handles degree 1 nodes in the graph $G[C]$. If there is no degree 1 node in $G[C]$ then a node u in $G[C]$ is chosen uniformly at random and the degree rule for u is applied in a probabilistic way. To be more precise, if u is of degree 2 and v, w are its neighbours in $G[C]$ then all neighbours of v and w are added with

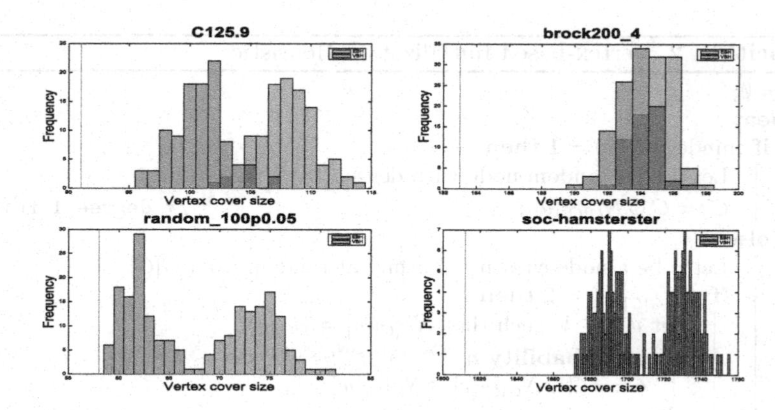

Fig. 1. The histograms show the frequency that each algorithm gets the initial vertex cover of certain size. The optimal vertex cover size of each instance is indicated with red vertical line in each figure.

probability $\alpha^{-|N_{G[C]}[v] \cup N_{G[C]}[w]|}$ while v and w are added otherwise. Similarly, if u is of degree at least 2 in $G[C]$ then all neighbours of u in $G[C]$ are added with probability at least $\alpha^{-\deg_{G[C]}(u)}$ while u is added otherwise.

We provide a lower bound on the probability that the vertex-based initialization obtains an optimal solution.

Theorem 2. *The vertex-based initialization heuristic obtains a vertex cover of size OPT with probability at least α^{-OPT}, where $\alpha = 1.4656$.*

Proof. The vertex-based initialization heuristics carries out a randomized branching according to the different rules. We distinguish the different cases regarding the degree of a node. For any graph, there is an optimal vertex cover that does not contain the node u if u is a degree one node. We investigate the degree 2 and 3 rules and show that each step i which requires selecting OPT_i nodes corresponding to an optimal solution occurs with probability at least α^{-OPT_i}. For a degree 2 node, there is an optimal vertex cover that contains either the neighbors v and w of u or all the neighbors of v and w. Note that a degree 2 rule is only applied if there is no node of degree 1 in $G[C]$. This implies that both v and w have to be connected to a node different from u. The probability of selecting v and w is $1 - \alpha^{-|N_{G[C]}[v] \cup N_{G[C]}[w]|}$ which is at least α^{-2} if $|N_{G[C]}[v] \cup N_{G[C]}[w]| \geq 2$. If $|N_{G[C]}[v] \cup N_{G[C]}[w]| = 1$, then v and w are connected and we have a cycle of length 3 $(u - v - w - u)$ for which selecting any subset of 2 nodes is optimal. Selecting u leads to an isolated edge $\{v, w\}$ for which the degree 1 rule selects a single vertex and therefore situations where $|N_{G[C]}[v] \cup N_{G[C]}[w]| = 1$ always lead to an optimal solution for the cycle of length 3. Finally, if u is of degree at least 3 there is an optimal vertex cover which either contains u or all the neighbors of u. The probability of selection u is $1 - \alpha^{-\deg_{G[C]}(u)} > \alpha^{-1}$.

Hence, the probability of selecting, in each step, a set of nodes leading to an optimal solution is at least

$$\prod_{i=1}^{\ell} \alpha^{-OPT_i} = \alpha^{-OPT}$$

where are ℓ is the number of iterations of the algorithm to produce the vertex cover. □

3.1 Experimental Investigations

In this section, we discuss about the experiments aiming at comparing the performance of Algorithms 1 and 2. Both algorithms are evaluated on sample Vertex Cover instances chosen from different benchmarks categories, which are *DIMACS* benchmarks, random generated undirected graphs and real world graphs.

There are some vertex cover benchmarks that are widely used to evaluated the performance of minimum vertex cover solver. One of these benchmarks is the *DIMACS* benchmark which is a set of challenge problems coming from the Second *DIMACS* Implementation Challenge for Maximum Clique, Graph Coloring and Satisfiability [5]. The original Max Clique problems from the challenge are converted to complement graphs and used as vertex cover problems. The random undirected graphs are generated with a pre-defined instance size and selection rate of edges. An edge between any two nodes is added to the graph with a certain pre-defined probability. In [9], there are a number of real world graphs with various number of vertices and edges. The sample graphs are selected from the undirected unweighted graphs.

Both of the algorithms are implemented in JAVA and the programs are executed for 101 independent repeated runs on each instance to obtain the statistics. The histograms in Fig. 1 are achieved by comparing the vertex cover sizes that the two algorithms get from running on four instances from different categories. The distribution of the solutions obtained in 101 independent runs is visualized with the histograms. In the first histogram and those lying in the second row, it is clear that vertex-based initialization generated smaller solutions for these two instances. For the instance brock200_4 from *DIMACS* benchmarks, the vertex-based approach has higher probability to generate better initial solutions than its edge-based counterpart.

Table 1 shows the five-number summary of each ranked set of 101 results testing on specific instance. From Table 1, the initial solutions of real world graphs generated by Algorithm 2 are all smaller than those from Algorithm 1. For the graphs from random and *DIMACS* benchmarks, the vertex-based approach can give better initial solutions for most times. Moreover, Algorithm 2 is able to generate solutions that are already global optimum for some of the instances in random and real world category.

Table 1. Experimental results on instances comparing the statistics between Algorithms 1 and 2.

Instance				EBH					VBH				
Name	$\|V\|$	$\|E\|$	OPT	min	Q1	Median	Q3	Max	Min	Q1	Median	Q3	Max
random_50p0.1	50	117	28	31	35	36	37	40	28	29	30	31	33
random_50p0.1-2	50	139	31	34	37	38	39	43	31	32	33	34	36
random_100p0.05	100	288	58	68	72	74	75	81	59	61	62	63	67
random_100p0.05-2	100	261	58	67	71	73	75	79	58	60	61	62	66
random_500p0.01	500	1 206	284	344	353	357	362	371	292	296	298	301	308
random_500p0.01-2	500	1 282	284	344	358	362	365	372	290	298	300	302	308
soc-hamsterster	2 426	16 630	1 612	1 709	1 726	1 731	1 737	1 755	1 672	1 684	1 690	1 695	1 716
soc-wiki-Vote	889	2 914	406	486	501	508	513	532	406	406	407	409	412
web-edu	3 031	6 474	1 451	1 742	1 765	1 771	1 780	1 793	1 451	1 452	1 453	1 454	1 457
web-google	1 299	2 773	498	582	596	604	611	632	501	506	508	509	517
bio-celegans	453	2 025	249	286	293	298	300	306	254	260	263	266	277
bio-yeast	1 458	1 948	456	583	608	618	626	656	456	459	460	462	468
brock200_4	200	6 811	183	192	194	195	196	198	190	193	194	194	197
brock400_4	400	20 035	367	390	392	393	394	396	387	390	391	392	395
brock800_4	800	111 957	774	792	794	795	796	798	792	793	794	794	797
C125.9	125	787	91	102	107	108	110	114	96	100	101	102	107
C250.9	250	3 141	206	227	231	232	234	238	222	225	226	228	232
C500.9	500	12 418	443	474	479	481	483	487	467	474	476	477	480

4 Local Search

We now introduce local search algorithms that make use of the aforementioned branching ideas. Both local search algorithms work with a list C representing a set of nodes and adding nodes to C in both algorithms always means adding them to the end of the list.

The edge-based local algorithm (see Algorithm 3) is a simplified version of one of the most successful approaches for solving the vertex cover problem, namely NuMVC [2]. It starts with a vertex cover of size $k+1$ and tries to find a smaller vertex cover of size k by removing one node. If this step violates the property of a vertex cover, it removes an additional node, picks an uncovered edge and adds one of its nodes uniformly at random. After a vertex cover of size k is obtained, it continues the process to search for a vertex cover of size $k-1$ and so on.

In the following, we give an upper bound on the number of steps of edge-based local search to find a vertex cover of size k. For our analysis, we partition the run of edge-based local search into distinct phases of length k which consist of k iterations of the while-loop.

Theorem 3. *For all r with $0 \leq r \leq OPT$, the edge-based local search finds a vertex cover of size $k := 2\,OPT - r$ after (expected) at most 2^{r+1} phases of length k.*

Proof. We investigate the probability that during k steps of the while-loop a vertex cover has been found at least once. We call this a *success* during a phase of k steps. Let C^* be a vertex cover of size OPT. As C^* is a vertex cover, it

Algorithm 3. Edge-based Local Search

```
1  Let C be an initial vertex cover represented as a list;
2  repeat
3      Choose a node v ∈ C uniformly at random and set C := C \ v;
4      while ((C is not a vertex cover of G) and (not termination condition)) do
5          Choose the first node v of C and set C := C \ v;
6          Let e = {u, v} be a random uncovered edge, i.e., e ∈ G[C];
7          with probability 1/2 do
8              C := C ∪ {u}
9          else
10             C := C ∪ {v}
11 until termination condition;
12 Return C;
```

contains for each edge $e \in E$ at least one vertex. Consider an edge $e = \{u, v\}$. At each iteration, a vertex $z \in C^*$ is picked with probability at least $1/2$ and each node of C^* is picked at most once as only uncovered edges are chosen. The expected number of distinct vertices contained in C^* during a phase of k steps is therefore at least $k/2 = (2\,OPT - r)/2$. The probability that during the first r steps only nodes of C^* are picked is at least 2^{-r}. The expected number of nodes of C^* picked in the remaining $2\,OPT - 2r$ steps (before a vertex cover is reached) is at least $OPT - r$. Furthermore, it is at least $OPT - r$ with probability $1/2$. Hence, the algorithm picks all OPT nodes during a phase of $k = 2\,OPT - r$ steps with probability at least $2^{-(r+1)}$. The expected number of phases of length k needed to find a vertex cover is therefore at most 2^{r+1}. □

We also turn the vertex-based branching approach into a vertex-based local search algorithm (see Algorithm 4). This approach searches for a vertex cover after removing a node together with all its neighbors. Afterwards, it tries to obtain a new vertex cover by picking a random node of minimum degree in the graph consisting of currently all uncovered edges. Based on the degree of this node the degree rules are applied with the already introduced biased probabilities. The last step is iterated until a vertex cover is found again.

4.1 Experimental Investigations

We test Algorithms 3 and 4 on some sample instances to evaluate their performance. Both algorithms are given an initial vertex cover produced by Algorithm 1 and the cut off generation is set to 100 000. Both algorithms are implemented in JAVA and their performance is measured by the number of iterations it takes to make improvement.

Algorithm 4. Vertex-based Local Search

```
1  Set α := 1.4656;
2  Let C be an initial vertex cover represented as a list;
3  repeat
4      Choose the first node v of C and set C := C \ N²_G[v];
5      repeat
6          Let u be a random node with deg_G[C](u) = mindeg(G[C]);
7          if deg_G[C](u) = 1 then
8              C := C ∪ N_G[C][u] ;                          /* degree 1 rule */
9          else if deg_G[C](u) = 2 then
10             Let v, w ∈ V such that N_G[C][u] = {v, w};
11             with probability α^(−|N_G[C][v] ∪ N_G[C][w]|) do
12                 C := C ∪ N_G[C][v] ∪ N_G[C][w]
13             else
14                 C := C ∪ N_G[C][u] ;                      /* degree 2 rule */
15         else
16             with probability α^(− deg_G[C](u)) do
17                 C := C ∪ N_G[C][u]
18             else
19                 C := C ∪ {u} ;                            /* degree ≥ 3 rule */
20     until C is a vertex cover of G (or termination condition);
21 until termination condition;
22 Return C;
```

Figure 2 shows the improvement of the two algorithms on example instances over iterations. $|C| - OPT$ denotes the size difference between the best solution so far and the globally optimal solution. The stairstep lines are drawn for three independent runs for each instance and algorithm. The vertex-based heuristic makes significant improvement before 2 000 generations for these three instances from the observation of the solid lines while the solution of edge-based heuristic does not improve much until 100 000 which is the cutoff bound. For the random graphs, the vertex-based approach is able to find a global optimum before 10 000 iterations whereas the edge-based heuristic does not reach the optimal solution before 100 000 iterations.

More results are shown in Table 2. The average best vertex cover sizes at certain number of iterations from 10 independent runs of these two algorithms on a certain vertex cover problem are listed in the table. From the statistics in Table 2, vertex-based approach produces better results for 15, 15, 16 and 16 out of the 17 instances after 10 000, 50 000, 100 000 and 200 000 iterations, respectively. Moreover, Algorithm 4 has a success rate of 100 % in solving 8 instances from different categories.

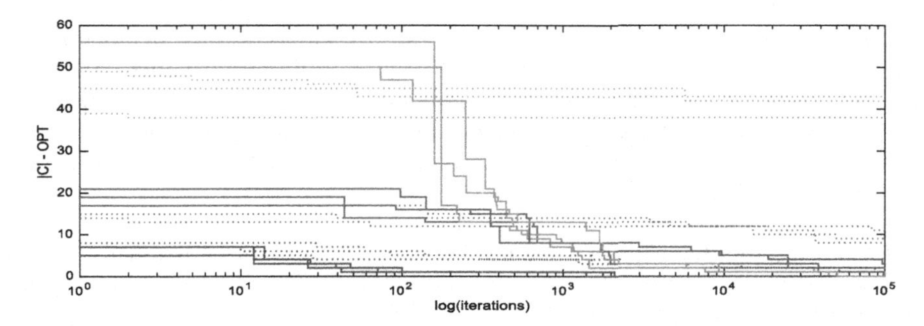

Fig. 2. The improvement of both algorithms in three example instances over iterations. The lines in blue, red and green color represent an independent run on the instance random-50prob10, C125.9 and bio-celegans, respectively. The dotted lines and solid lines denote the results from Algorithms 3 and 4.

Table 2. Performance comparison between Algorithms 3 and 4 on some sample instances. The average vertex cover size is listed after running each algorithm for certain number of iterations.

Instance				EBH				VBH			
Name	$\|V\|$	$\|E\|$	OPT	10 000	50 000	100 000	200 000	10 000	50 000	100 000	200 000
random_50p0.1	50	117	28	29.8	29.4	28.9	28.8	28.0	28.0	28.0	28.0
random_50p0.1-2	50	139	31	33.0	32.6	32.1	32.0	31.0	31.0	31.0	31.0
random_100p0.05	100	288	58	66.7	65.4	65.1	64.8	58.0	58.0	58.0	58.0
random_100p0.05-2	100	261	58	66.4	64.8	64.3	64.0	58.0	58.0	58.0	58.0
random_500p0.01	500	1 206	284	351.3	348.8	348.2	346.6	286.4	284.9	284.4	284.4
random_500p0.01-2	500	1 282	284	357.0	354.9	353.1	352.3	286.3	284.2	284.2	284.0
bio-celegans	453	2 025	249	291.4	290.7	290.0	289.8	250.7	249.7	249.3	249.3
bio-diseasome	516	1 188	285	316.2	314.5	314.5	313.3	288.9	287.3	287.0	286.6
soc-dolphins	62	159	34	36.3	35.7	35.4	34.9	34.0	34.0	34.0	34.0
soc-wiki-Vote	889	2 914	406	502.2	502.2	502.2	502.2	406.2	406.0	406.0	406.0
ca-netscience	379	914	214	243.9	241.1	240.3	238.7	216.7	215.7	215.1	214.7
ca-Erdos992	6 100	7 515	461	819.1	808.3	801.2	794.9	461.0	461.0	461.0	461.0
C125.9	125	787	91	102.7	101.1	100.5	100.4	95.3	93.3	92.8	92.8
C250.9	250	3 141	206	228.2	226.8	226.3	225.6	232.5	231.5	231.2	230.6
MANN_a27	378	702	252	261.0	260.8	260.4	260.1	252.9	252.6	252.3	252.1
MANN_a45	1 035	1 980	690	705.0	705.0	705.0	705.0	701.6	694.2	693.3	692.7
MANN_a81	3 321	6 480	2 221	2 241.0	2 241.0	2 241.0	2 241.0	2 241.4	2 241.1	2 239.0	2 235.1

5 Conclusions

We have shown how well-known fixed parameter branching algorithms for the minimum vertex cover problem can be turned into randomized initialization strategies and guarantee the probabilities of obtaining good solutions. Furthermore, we have incorporated the branching rules into local search algorithms and observed that the edge-based local search algorithm is equivalent to the core component of the state-of-the-art local search algorithm called NuMVC. Considering the edge-based local search algorithm from a theoretical perspective we have shown fixed parameter and trade-off results on its performance. Additionally, we have demonstrated how the more complex vertex-based branching rules can be incorporated into the vertex-based local search algorithm and shown that this usually leads to better results on random graphs and social networks than edge-based local search.

Acknowledgements. This research has been supported by the Australian Research Council (ARC) under grant agreement DP140103400 and the German Science Foundation (DFG) under grant agreement no. FR 2988.

References

1. Bringmann, K., Friedrich, T.: Parameterized average-case complexity of the hypervolume indicator. In: Genetic and Evolutionary Computation Conference (GECCO), pp. 575–582 (2013)
2. Cai, S., Su, K., Sattar, A.: Two new local search strategies for minimum vertex cover. In: Twenty-Sixth AAAI Conference on Artificial Intelligence (2012)
3. Corus, D., Lehre, P.K., Neumann, F., Pourhassan, M.: A parameterised complexity analysis of bi-level optimisation with evolutionary algorithms. Evol. Comput. **24**, 183–203 (2015)
4. Downey, R.G., Fellows, M.R.: Fundamentals of Parameterized Complexity. Texts in Computer Science. Springer, Heidelberg (2013)
5. Johnson, D.J., Trick, M.A. (eds.): Cliques, Coloring, and Satisfiability: Second DIMACS Implementation Challenge, Workshop, October 11–13, 1993. American Mathematical Society, Boston (1996)
6. Kratsch, S., Neumann, F.: Fixed-parameter evolutionary algorithms and the vertex cover problem. Algorithmica **65**, 754–771 (2013)
7. Kratsch, S., Lehre, P.K., Neumann, F., Oliveto, P.S.: Fixed parameter evolutionary algorithms and maximum leaf spanning trees: a matter of mutation. In: Schaefer, R., Cotta, C., Kołodziej, J., Rudolph, G. (eds.) PPSN XI. LNCS, vol. 6238, pp. 204–213. Springer, Heidelberg (2010)
8. Nallaperuma, S., Sutton, A.M., Neumann, F.: Parameterized complexity analysis and more effective construction methods for ACO algorithms and the euclidean traveling salesperson problem. In: Proceedings of the IEEE Congress on Evolutionary Computation, CEC, pp. 2045–2052. IEEE (2013)
9. Rossi, R.A., Ahmed, N.K.: The network data repository with interactive graph analytics and visualization. In: AAAI, pp. 4292–4293 (2015)
10. Sutton, A.M., Neumann, F., Nallaperuma, S.: Parameterized runtime analyses of evolutionary algorithms for the planar euclidean traveling salesperson problem. Evol. Comput. **22**, 595–628 (2014)

What Does the Evolution Path Learn in CMA-ES?

Zhenhua Li(✉) and Qingfu Zhang

City University of Hong Kong, Tat Chee Avenue, Kowloon Tong, Hong Kong
ilizhenhua@gmail.com, qingfu.zhang@cityu.edu.hk

Abstract. The Covariance matrix adaptation evolution strategy (CMA-ES) evolves a multivariate Gaussian distribution for continuous optimization. The evolution path, which accumulates historical search directions in successive generations, plays a crucial role in the adaptation of covariance matrix. In this paper, we investigate what the evolution path learns in the optimization procedure. We show that the evolution path accumulates natural gradient with respect to the distribution mean, and acts as a momentum under stationary condition. The experimental results suggest that the evolution path learns relative scales of the eigenvectors, expanded by singular values along corresponding eigenvectors of the inverse Hessian. Further, we show that the outer product of evolution path serves as a rank-1 momentum term for the covariance matrix.

1 Introduction

The covariance matrix adaptation evolution strategy (CMA-ES) is a popular evolutionary algorithm for continuous optimization. It samples a set of candidate solutions from the multivariate Gaussian distribution, then selects the best ones for adapting the distribution mean, covariance matrix, and global step-size [1,2].

CMA-ES is invariant to linear transformations of the search space [3,4] due to its covariance matrix. It approximates the inverse Hessian matrix, learning the objective condition and pair-wise parameter dependencies. The covariance matrix is adapted by a *rank-1* update with a evolution path and a *rank-μ* update with a weighted maximum likelihood estimation using selected solutions. By averaging the search directions over a few generations, the evolution path cancels out opposite search directions and accumulates consistent directions in successive generations [1,5]. It is constructed into rank-1 update for the covariance matrix for increasing the likelihood of producing solutions along this direction. Experimental results show that the evolution path dramatically affects the algorithm performance [6].

Yet, it is still not fully understood what the evolution path approximates in CMA-ES. This paper is dedicated to experimentally investigate what the evolution path approaches in the adaptation for covariance matrix. The main conclusions of this paper are as follows.

J. Handl et al. (Eds.): PPSN XIV 2016, LNCS 9921, pp. 751–760, 2016.
DOI: 10.1007/978-3-319-45823-6_70

- The evolution path accumulates the natural gradient with respect to the distribution mean, and acts as a momentum term under a stationary condition for guiding the search.
- The evolution path learns the relative scales of the eigenvectors of the inverse Hessian.
- The update term for the evolution path provides an efficient approximation to the first principal component of the rank-μ weighed maximum likelihood estimation of the covariance.
- The rank-1 update with the evolution path serves as a rank-1 momentum term of for covariance matrix.

The remainder of this paper is organized as follows. In Sect. 2, we present a brief description of CMA-ES. Section 3 shows that the evolution path accumulates the natural gradient with respect to the distribution mean, and serves as a momentum term. In Sect. 4, we conduct experiments to investigate geometric properties of the evolution path. Section 5 concludes the paper and discusses the future research.

2 The CMA-ES

For a given optimization problem $\min_{\mathbf{x}\in\mathbb{R}^n} f(\mathbf{x})$, the CMA-ES algorithm is outlined in Algorithm 1. At iteration t, it samples λ solutions from the current distribution by adding a Gaussian mutation to the distribution mean (line 5 to line 7).

The sampled solutions are evaluated with the objective function $f(\mathbf{x})$. Then they are sorted according to the objective values for updating the parameters. The best μ solutions are selected to recombined into the distribution mean for the next generation with specific weights (line 9).

The step-size is adapted based on cumulative step-size adaptation method using a cumulative evolution path $\mathbf{p}_t^\sigma$. Consecutive search directions $\mathbf{C}_t^{-\frac{1}{2}}(\mathbf{m}_{t+1} - \mathbf{m}_t)/\sigma_t$ are accumulated into the evolution path $\mathbf{p}_t^\sigma$ with specific learning rate (line 10), with $\mathbf{C}_t^{-\frac{1}{2}} = \mathbf{B}^T\mathbf{D}^{-1}\mathbf{B}$. It exploits the correlations between successive search directions. The step-size tends to increase, if they are correlated and the length of $\mathbf{p}_\sigma^{t+1}$ is larger than the expected length under random selection. Otherwise, the step-size tends to decrease if consecutive search directions are anti-related (line 14).

The covariance matrix is adapted with two terms, the rank-1 update with evolution path and rank-μ update with weighted maximum likelihood estimation of the covariance matrix (line 13) [1,2]. The evolution path accumulates the mean difference between consecutive generations (line 12). Historical search directions damp with a factor $(1 - c_c)$ for reducing the importance. A trigger h_σ is used for preventing the evolution path from growing too large (line 11). A weighted maximum likelihood estimation of the covariance matrix based on selected solutions contributes rank-μ update for the covariance matrix (line 13).

Algorithm 1. $(\mu/\mu_w, \lambda)$-CMA-ES

1: **Given** $\lambda = 4 + \lfloor 3 \ln n \rfloor$, $\mu = \lfloor \frac{\lambda}{2} \rfloor$, $w_i = \frac{\ln(\mu+1)-\ln i}{\mu \ln(\mu+1) - \sum \ln(j)}$ for $i = 1, \cdots, \mu$, $\mu_{\text{eff}} = \frac{1}{\sum_{i=1}^{\mu} w_i^2}$, $c_\sigma = \frac{\mu_{\text{eff}}+2}{n+2\mu_{\text{eff}}+3}$, $d_\sigma = 1 + c_\sigma + 2\max(0, \sqrt{\frac{\mu_{\text{eff}}-1}{n+1}} - 1)$, $c_c = \frac{4}{n+4}$, $c_1 = \frac{2\min(1,\lambda/6)}{(n+1.3)^2+\mu_{\text{eff}}}$, $c_\mu = \frac{2(\mu_{\text{eff}}-2+1/\mu_{\text{eff}})}{(n+2)^2+\mu_{\text{eff}}}$
2: **Initialize** $\mathbf{m}_0, \sigma_0, \mathbf{C}_0 = \mathbf{I}, \mathbf{p}_0 = \mathbf{0}, \mathbf{p}_\sigma^0 = \mathbf{0}, t = 0$
3: **repeat**
4: **for** 1 to λ **do**
5: $\mathbf{z}_i \sim \mathcal{N}(\mathbf{0}, \mathbf{I})$
6: $\mathbf{y}_i = \mathbf{B}\mathbf{D}\mathbf{z}_i \sim \mathcal{N}(\mathbf{0}, \mathbf{C}_t)$
7: $\mathbf{x}_i = \mathbf{m}_t + \sigma_t \mathbf{y}_i \sim \mathcal{N}(\mathbf{m}_t, \sigma_t^2 \mathbf{C}_t)$
8: **end for**
9: $\mathbf{m}_{t+1} = \sum_{i=1}^{\mu} w_i \mathbf{x}_{i:\lambda}$, where $f(\mathbf{x}_{1:\lambda}) \leq f(\mathbf{x}_{2:\lambda}) \leq \cdots \leq f(\mathbf{x}_{\mu:\lambda})$
10: $\mathbf{p}_{t+1}^\sigma = (1 - c_\sigma)\mathbf{p}_t^\sigma + \sqrt{c_\sigma(2 - c_\sigma)\mu_{\text{eff}}} \mathbf{C}_t^{-\frac{1}{2}} (\mathbf{m}_{t+1} - \mathbf{m}_t)/\sigma_t$
11: $h_\sigma = \mathbb{1}\left(\|\mathbf{p}_{t+1}^\sigma\| < \sqrt{1 - (1 - c_\sigma)^{2(t+1)}} (1.4 + \frac{2}{n+1}) \mathbb{E}\|\mathcal{N}(\mathbf{0}, \mathbf{I})\| \right)$
12: $\mathbf{p}_{t+1} = (1 - c_c)\mathbf{p}_t + h_\sigma \sqrt{c_c(2 - c_c)\mu_{\text{eff}}} (\mathbf{m}_{t+1} - \mathbf{m}_t)/\sigma_t$
13: $\mathbf{C}_{t+1} = (1 - c_1 - c_\mu)\mathbf{C}_t + c_1 \mathbf{p}_{t+1} \mathbf{p}_{t+1}^T + c_\mu \sum_{i=1}^{\mu} w_i \mathbf{y}_{i:\lambda} \mathbf{y}_{i:\lambda}^T$
14: $\sigma_{t+1} = \sigma_t \cdot \exp\left(\frac{c_\sigma}{d_\sigma} \left(\frac{\|\mathbf{p}_{t+1}^\sigma\|}{E\|\mathcal{N}(\mathbf{0},\mathbf{I})\|} - 1 \right) \right)$
15: $t = t + 1$
16: **until** stopping criterion is met
17: **return**

3 The Evolution Path Acts as Momentum Term

We illustrate in the following that the evolution path acts as a momentum term by accumulating natural gradients.

Typically, the evolution path exploits sign information by accumulating historical search directions. For a given mutation direction $\mathbf{y} \in \mathcal{N}(0, \mathbf{C})$, the outer products $\mathbf{y}\mathbf{y}^T$ and $(-\mathbf{y})(-\mathbf{y})^T$ give the same result. Hence, if updating the covariance with only current mutation directions, the sign information gets lost. The evolution path is designed for averaging the search directions over successive steps. It is updated by the direction

$$\mathbf{m}_{t+1} - \mathbf{m}_t = \sum_{i=1}^{\mu} w_i (\mathbf{x} - \mathbf{m}_t). \tag{1}$$

with an coefficient $\sqrt{\mu_{\text{eff}}}/\sigma_t$ such that the update direction $\sqrt{\mu_{\text{eff}}}(\mathbf{m}_{t+1} - \mathbf{m}_t)/\sigma_t \sim \mathcal{N}(0, \mathbf{C}_t)$ under random selection. Hence, with the learning rate designed according to $(1 - c_c)^2 + (\sqrt{c_c(2 - c_c)})^2 = 1$, the evolution path is also distributed according to $\mathcal{N}(0, \mathbf{C}_t)$ under random selection. This is known as stationary condition [4].

The information geometric optimization framework (IGO) [7] provides us another way to investigate the evolution path. The IGO framework considers to optimize the expected fitness

$$J(\theta) = \int W^f_{\theta_t}(\mathbf{x}) p_\theta(\mathbf{x}) \mathrm{d}\mathbf{x}, \tag{2}$$

where θ is the parameters of the distribution family, and $W^f_{\theta_t}(\mathbf{x})$ is the transformed objective function of f and the current distribution θ_t. It determines the selection scheme [7].

Given $\theta = (\mathbf{m}, \mathbf{C})$ for Gaussian distributions, the parameters are iteratively updated along the natural gradient direction, which presents the steepest descent direction on the statistic manifold. The natural gradient is generally estimated based on samples. Specifically, the estimation of natural gradient with respect to the distribution mean is given by (Eqs. 41–42 in [7])

$$\tilde{\nabla}_m J(\theta) = \sum_{i=1}^{\lambda} \hat{w}_i (\mathbf{x}_i - \mathbf{m}_t), \tag{3}$$

where $\hat{w}_i$ are empirical selection scheme as $\hat{w}_i = \frac{1}{\lambda} \mathbb{1}\left(\frac{rk(\mathbf{x}_i)+0.5}{\lambda} < 0.5\right)$ (see Eq. 14 in [7]), where $rk(\mathbf{x}_i) = |\{j | f(\mathbf{x}_j) < f(\mathbf{x}_i)\}|$ is the number of solutions which are superior to $\mathbf{x}_i$.

As we only care about the direction, the similarity between the Eqs. (1) and (3) suggests that the update term for the evolution path is actually an estimation of the natural gradient with respect to the distribution mean. Consequently, the evolution path accumulates the natural gradients with respect to the distribution mean. Thus, the evolution path acts as a momentum term under the stationary condition.

The momentum technique is commonly used in optimization of machine learning [8–10]. Further research pointed out that the momentum method is actually a stationary version of the conjugate gradient method [11]. Hence, searching along the evolution path can be considered as a stochastic approximation to the conjugate gradient method.

4 Empirical Results on Evolution Path

In this section, we conduct some experiments on commonly used test problems to investigate what the evolution path learns. Although the experimental results may differ from run to run due to the sample randomness, we can still obtain some common results. In these experiments, all the parameters of CMA-ES are set as [3]. The test problems are presented in Table 1. The experiments are conducted on dimension 10 unless exceptionally and clearly specified.

4.1 Evolution Path Learns the Relative Scale

Consider a quadratic function

$$f(\mathbf{x}) = \frac{1}{2} \mathbf{x}^T \mathbf{H} \mathbf{x}, \tag{4}$$

Table 1. Test problems

$f_{\text{cigar}}(\mathbf{x}) = x_1^2 + 10^6 \cdot \sum_{i=2}^{n} x_i^2$	$f_{\text{cigtab}}(\mathbf{x}) = x_1^2 + 10^4 \cdot \sum_{i=2}^{n-1} x_i^2 + 10^6 \cdot x_n^2$
$f_{\text{elli100}}(\mathbf{x}) = \sum_{i=1}^{n} 10^{2 \cdot \frac{i-1}{n-1}} \cdot x_i^2$	$f_{\text{elli}}(\mathbf{x}) = \sum_{i=1}^{n} 10^{6 \cdot \frac{i-1}{n-1}} \cdot x_i^2$
$f_{\text{tablet}}(\mathbf{x}) = 10^6 \cdot x_1^2 + \sum_{i=2}^{n} x_i^2$	$f_{\text{twoaxes}}(\mathbf{x}) = x_1^2 + 10^6 \cdot \sum_{i=2}^{n} x_i^2$
$f_{\text{diffpow}}(\mathbf{x}) = \sum_{i=1}^{n} x_i^{2+10*\frac{i-1}{n-1}}$	$f_{\text{rosen}}(\mathbf{x}) = \sum_{i=1}^{n-1}(100(x_{i+1} - x_i^2)^2 + (x_i - 1)^2)$

where $\mathbf{H}$ is the positive-definite Hessian matrix. Its inverse $\mathbf{H}^{-1}$ can be decomposed as $\mathbf{H}^{-1} = \mathbf{U}\mathbf{\Lambda}^2\mathbf{U}^T$, with $\mathbf{U}$ as orthogonal matrix and $\mathbf{\Lambda} = diag(\lambda_1, \cdots, \lambda_n)$ as diagonal matrix. Let $\mathbf{s}$ be the direction expanded as

$$\mathbf{s} = \lambda_1 \mathbf{u}_1 + \lambda_2 \mathbf{u}_3 + \cdots + \lambda_n \mathbf{u}_n,$$

As the eigenvectors are restricted to be unit, the coefficients $\lambda_i, i = 1, \cdots, n$, actually present the relative scales along each eigenvector. Figure 1 shows a scale vector on quadratic functions.

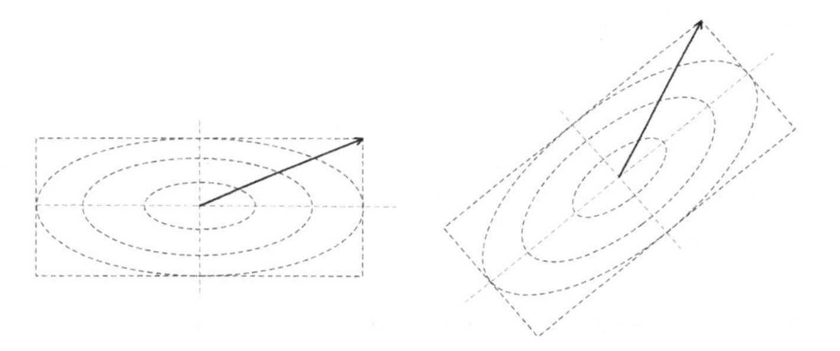

Fig. 1. The direction $\mathbf{s}$ on quadratic functions. Any mirror direction with respect to any eigenvector or origin is equivalent with $\mathbf{s}$.

A Case Study. Figure 2 presents a typical run of CMA-ES on the $f_{\text{cigar}}(\mathbf{x})$ function, which is characterized by a predominant long search direction. Clearly, the optimization procedure can be divided into three phases according to the step-size. First, the objective value and the step-size descend, as the distribution mean approaches the long search direction. Then, the step-size increases as the evolution path gradually learns scale direction. At the third phase, the step-size and the objective value descend, while the evolution path oscillates around the scale direction.

Experimental Results Analysis. Figure 3 presents experimental results of a typical run on the test problems with dimension 50. Depicted are the true scale direction on each component, the evolution path, and an average evolution path over $n = 50$ generations. We consider only the relative scale of the

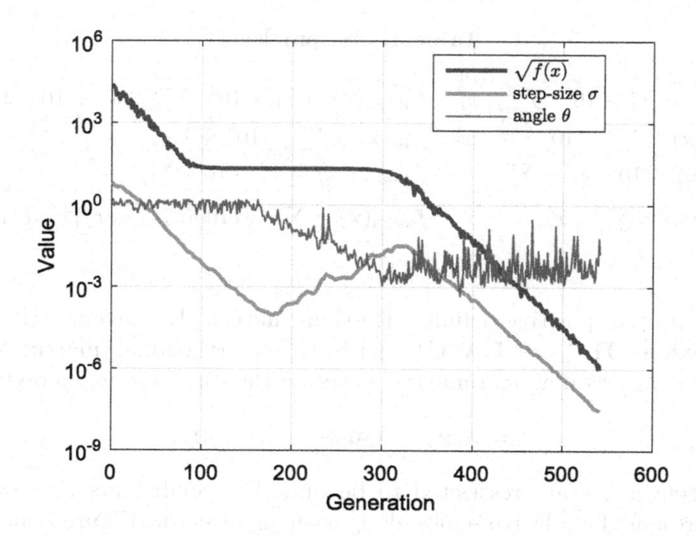

Fig. 2. The optimization procedure of a typical run $f_{\text{cigar}}(\mathbf{x})$ function. Depicted in the figure include the square root of the objective function, the step-size, and the angle between evolution path and the predominant search direction $\mathbf{e} = (1, 0, \cdots, 0)$. (Color figure online)

evolution path among the components. These experimental results suggest that, although affected dramatically by the randomness, the evolution path approximately learns a scale direction.

4.2 Rank-1 Approximation to the Rank-μ Update

Let $\mathbf{y}_w$ be the update term for the evolution path $\mathbf{y}_w = \frac{1}{\sigma_t}(\mathbf{m}_{t+1} - \mathbf{m}_t) = \sum_{i=1}^{\mu} w_i \mathbf{y}_{i:\lambda}$. The outer product $\mathbf{y}_w \mathbf{y}_w^T$ provides an efficient approximation to the rank-μ update term (5).

In addition to rank-1 update with evolution path, the covariance matrix in CMA-ES is updated by the rank-μ update with weighted maximum likelihood estimation using current population. The rank-μ update term gives

$$C_\mu = \sum_{i=1}^{\mu} w_i \mathbf{y}_{i:\lambda} \mathbf{y}_{i:\lambda}^T. \tag{5}$$

This can be obtained from the IGO perspective [7], as an estimation to the natural gradient direction of the expected fitness with respect to the covariance.

Typically, the outer product of the first principal component of covariance matrix C_μ serves as the optimal rank-1 approximation to C_μ. Thus, we test whether the direction $\mathbf{y}_w$ approximates the first principal component of C_μ, denoted as $\mathbf{e}_1$. We calculate the angle between the evolution path and the first principal component $\cos\theta = \mathbf{y}_w^T \mathbf{e}_1 / \|\mathbf{y}_w\|$.

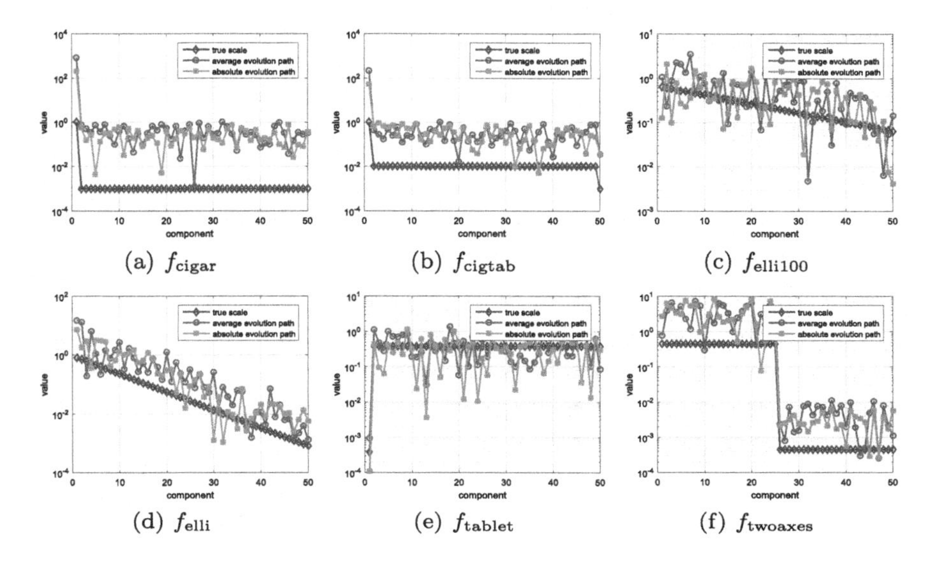

(a) f_{cigar} (b) f_{cigtab} (c) f_{elli100}

(d) f_{elli} (e) f_{tablet} (f) f_{twoaxes}

Fig. 3. The absolute evolution path learns a scale direction of the objective function, which is expanded by singular values on the eigenvectors. The scale direction $\mathbf{s}$ is normalized to be unit.

Figure 4 presents the frequency of $|\cos\theta|$ in the optimization procedure of a typical single run. It shows that the angle $|\cos\theta| > 0.95$ presents a predominant frequency on all test problems. Definitely, this indicates that the evolution path approximately learns the first principal component of the rank-μ covariance matrix. Consequently, the outer product of $\mathbf{y}_w$ provides an efficient rank-1 approximation to the rank-μ update term C_μ.

4.3 Rank-1 Approximation to the Covariance Matrix

In this subsection, we investigate how the evolution path approximates the covariance matrix with both rank-1 and rank-μ updates. We consider how the evolution path approximates the first principal component of the covariance matrix $\mathbf{C}$, which represents the longest search direction of the fitness local landscape of the objective function.

We calculate the frequencies which eigenvector is the closest to the evolution path. At each generation, the covariance matrix is decomposed into eigenvectors, and we calculate the angle between any eigenvector to the evolution path $\cos\phi_i = \mathbf{p}_c^T\mathbf{b}_i, i = 1, \ldots, n$, with $\mathbf{b}_i$ denoting the eigenvector corresponding to the i-th largest eigenvalue of $\mathbf{C}_{t+1}$. The one with largest absolute value corresponds to the eigenvector closest to the evolution path.

Figure 5 presents the experimental results. On f_{cigar} and f_{cigtab} functions which have a long search direction, the first principal component possess the dominant frequency as the closest eigenvector among all eigenvectors. On all test functions except for f_{diffpow}, the first principal component corresponds to

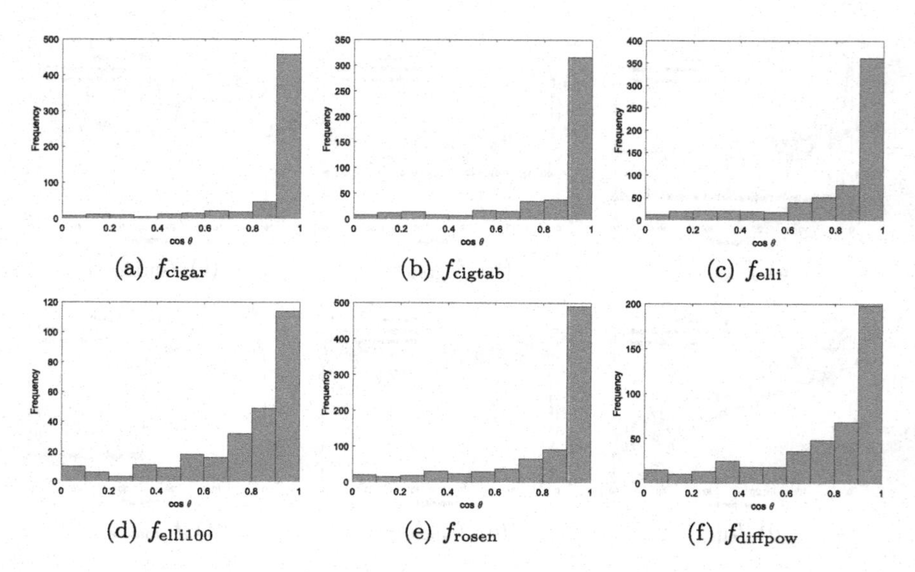

(a) f_{cigar} (b) f_{cigtab} (c) f_{elli}

(d) f_{elli100} (e) f_{rosen} (f) f_{diffpow}

Fig. 4. The frequencies of $\cos\theta$ values between $\mathbf{y}_w$ and the first principal component of C_μ in a typical single run.

the largest frequency among all eigenvectors, including the non-quadratic f_{rosen} function. Special attention should be payed to the f_{rosen} function which has a dramatically changed fitness landscape. The frequency is even higher than that of f_{elli100} function. This indicates that the evolution path roughly learns the first principal component of the covariance matrix. Consequently, the outer product of evolution path provides an efficient rank-1 approximation to the covariance matrix.

4.4 Momentum for the Covariance Matrix

The rank-μ update with weighted maximum likelihood estimation for the covariance can be obtained from the IGO perspective [7,12]. In the training for neural networks, the momentum term can effectively cancel out opposite search directions and accumulate consistent search directions [9,10]. However, as the zeroth-order update for the covariance matrix can only increase the likelihood along the selected directions, directly accumulating the rank-μ as the momentum for the covariance cannot cancel out opposite search directions, neither reduces the likelihood along these directions. This means that the pure rank-μ update for the covariance matrix cannot exploit the sign information [4].

It is the evolution path that accumulates historical search directions in successive generations and cancels out opposite search directions. Updating the covariance matrix with the evolution path can effectively increase only the likelihood along consistent search directions. Consequently, the rank-1 update with evolution path acts as a rank-1 momentum term for covariance matrix.

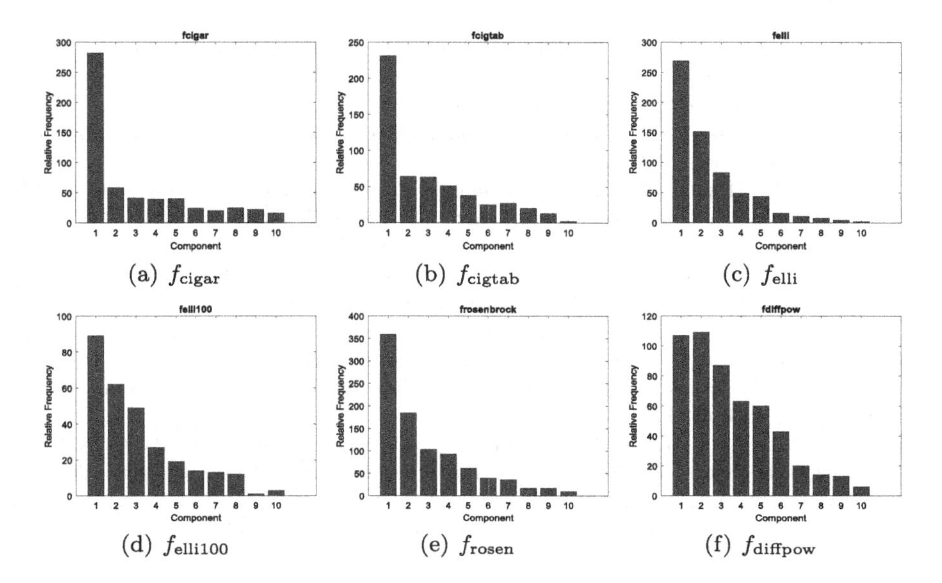

(a) f_{cigar} (b) f_{cigtab} (c) f_{elli}

(d) f_{elli100} (e) f_{rosen} (f) f_{diffpow}

Fig. 5. The frequencies of eigenvectors presenting closest one to evolution path in a typical single run.

5 Conclusion and Future Research

In this paper, we investigate what the evolution path approximates in the optimization procedure of CMA-ES. By accumulating successive natural gradient on the distribution mean, it can be viewed as a momentum under stationary condition. Further, experimental results suggest that the evolution path actually approximates a scale direction expanded by singular values of the inverse Hessian on quadratic functions. As the evolution path cancels out opposite search directions, it increases the likelihood along consistent search directions. Consequently, the rank-1 update term with evolution path serves as rank-1 momentum term for the covariance matrix.

Researches on the momentum methods pointed out that it can be viewed as discretization of a non-zero mass Newtonian equation with the influence of a conservative force field [13]. In the future, we will investigate how to extend the information geometric optimization framework to second order methods to include the evolution path. This can provides us a class of randomized optimization algorithms.

References

1. Hansen, N., Ostermeier, A.: Completely derandomized self-adaptation in evolution strategies. Evol. Comput. **9**(2), 159–195 (2001)
2. Hansen, N., Müller, S.D., Koumoutsakos, P.: Reducing the time complexity of the derandomized evolution strategy with covariance matrix adaptation (CMA-ES). Evol. Comput. **11**(1), 1–18 (2003)

3. Hansen, N.: The CMA evolution strategy: a comparing review. In: Lozano, J.A., Larrañaga, P., Inza, I., Bengoetxea, E. (eds.) Towards a New Evolutionary Computation, pp. 75–102. Springer, Heidelberg (2006)
4. Hansen, N., Auger, A.: Principled design of continuous stochastic search: from theory to practice (2013)
5. Suttorp, T., Hansen, N., Igel, C.: Efficient covariance matrix update for variable metric evolution strategies. Mach. Learn. **75**, 167–197 (2009)
6. Stich, S.U.: On low complexity acceleration techniques for randomized optimization. In: Bartz-Beielstein, T., Branke, J., Filipič, B., Smith, J. (eds.) PPSN 2014. LNCS, vol. 8672, pp. 130–140. Springer, Heidelberg (2014)
7. Ollivier, Y., Arnold, L., Auger, A., Hansen, N.: Information-geometric optimization algorithms: a unifying picturevia invariance principles. arXiv (2013)
8. Bishop, C.M.: Neural Networks for Pattern Recognition. Oxford University Press Inc., New York (1995)
9. Rumelhart, D.E., Hinton, G.E., Williams, R.J.: Learning representations by back-propagating errors. In: Neurocomputing: Foundations of Research, pp. 696–699. MIT Press, Cambridge (1988)
10. Moreira, M., Fiesler, E.: Neural networks with adaptive learning rate and momentum terms. Technical report, Idiap-RR-04-1995. IDIAP, Martigny, Switzerland, October 1995
11. Bhaya, A., Kaszkurewicz, E.: Steepest descent with momentum for quadratic functions is a version of the conjugate gradient method. Neural Netw. **17**(1), 65–71 (2004)
12. Akimoto, Y., Nagata, Y., Ono, I., Kobayashi, S.: Theoretical foundation for CMA-ES from information geometry perspective. Algorithmica **64**, 698–716 (2012)
13. Qian, N.: On the momentum term in gradient descent learning algorithms. Neural Netw. **12**(1), 145–151 (1999)

Graceful Scaling on Uniform Versus Steep-Tailed Noise

Tobias Friedrich, Timo Kötzing, Martin S. Krejca(✉), and Andrew M. Sutton

Hasso Plattner Institute, Potsdam, Germany
martin.krejca@hpi.de

Abstract. Recently, different evolutionary algorithms (EAs) have been analyzed in noisy environments. The most frequently used noise model for this was additive posterior noise (noise added after the fitness evaluation) taken from a Gaussian distribution. In particular, for this setting it was shown that the $(\mu + 1)$-EA on OneMax does not scale gracefully (higher noise cannot efficiently be compensated by higher μ).

In this paper we want to understand whether there is anything special about the Gaussian distribution which makes the $(\mu + 1)$-EA not scale gracefully. We keep the setting of posterior noise, but we look at other distributions. We see that for exponential tails the $(\mu + 1)$-EA on OneMax does also not scale gracefully, for similar reasons as in the case of Gaussian noise. On the other hand, for uniform distributions (as well as other, similar distributions) we see that the $(\mu + 1)$-EA on OneMax does scale gracefully, indicating the importance of the noise model.

Keywords: Evolutionary algorithm · Noisy fitness · Theory

1 Introduction

A major challenge to the theoretical analysis of randomized search heuristics is developing a rigorous understanding of how they behave in the presence of uncertainty. Uncertain problems are pervasive in practice, and practitioners often rely on heuristic techniques in these settings because classical tailored approaches often cannot cope with uncertain environments such as noisy objective functions and dynamically changing problems [1,10].

It is therefore very important to understand the effect that different properties of uncertainty have on algorithm behavior. In *stochastic* optimization, the fitness of a candidate solution does not have a deterministic value, but instead follows some given (but fixed) *noise distribution*. We are interested in understanding what properties of the noise distribution pose a direct challenge to optimization, and what problems might be overcome by different features of the algorithm. Prior work on stochastic optimization is mostly concerned with the *magnitude* of noise (usually measured by the variance). Our goal in this paper is to also understand how different *kinds* of distributions might affect optimization.

This is in contrast to most recent work on the theoretical analysis of randomized search heuristics in stochastic environments. For ant colony optimization,

J. Handl et al. (Eds.): PPSN XIV 2016, LNCS 9921, pp. 761–770, 2016.
DOI: 10.1007/978-3-319-45823-6_71

a series of papers studied the performance of ACO on stochastic path finding problems [3,5,13], see also [8,9] for early work in this area. For evolutionary algorithms, Gießen and Kötzing [7] analyzed the $(\mu + \lambda)$-EA on noisy OneMax and LeadingOnes and found that populations make the EA robust to specific distributions of prior and posterior noise, while [2] considers non-elitist EAs and gives run time bounds in settings of partial information. None of these works aimed at showing difference between noise settings. Posterior noise from a Gaussian was considered in [6,12] for various algorithms.

We follow [6] in their definition of what counts as a desirable property of an algorithm with respect to noisy optimization: graceful scaling. An algorithm is said to *scale gracefully* if, for any noise strength v, there is a suitable parameter for this algorithm such that optimization is possible in a time polynomial in v (a typical measure of noise strength v is the variance of the noise). We give the formal details in Definition 1.

In this paper we consider the $(\mu + 1)$-EA optimizing the classical OneMax fitness function with additive posterior noise coming from some random variable D. The case of D a Gaussian distribution was considered in [6], where the authors found that the $(\mu + 1)$-EA does not scale gracefully. In this paper we investigate what properties of D lead to graceful scaling, and what properties do not.

In Sect. 3 we consider the case of exponentially decaying tails in the distribution of the noise. This is similar to the case of Gaussian noise, which decays even faster. In fact, we use a similar proof to show that also in this case the $(\mu + 1)$-EA does not scale gracefully with noise.

After this we turn to another extreme case, uniform distributions. In Sect. 4 we show that, for noise taken from a uniform distribution, the $(\mu + 1)$-EA scales gracefully. Our proof makes use of the fact that the uniform distribution is *truncated* at its lower end: there is a value k such that the noise never takes values below k, but values between k and $k + 1$ are still fairly frequent. Thus, our results generalize to all noise distributions with this property.

Our results have some interesting implications. First, if it is possible to truncate the noise distribution artificially, then this can potentially improve the run time of an EA. This is an attractive option since no noise-specific modifications need to be made to the algorithm (such as performing re-evaluations to sample the distribution and thereby reduce the variance). Second, there are settings where even very large populations do not sufficiently reduce the effect of the noise, so that other techniques are required. This serves as a cautionary tale to practitioners that increasing the population size does not always improve an EA's robustness to noise.

Before we discuss our results on exponential tails and uniform distributions in Sects. 3 and 4, respectively, we introduce the algorithm and noise model in Sect. 2. Finally in Sect. 5 we summarize our findings and conclude the paper.

2 Preliminaries

In this paper we study the optimization of pseudo-Boolean functions (mapping $\{0,1\}^n$ for fixed n to real numbers). As our main test function we use OneMax, where

$$\text{OneMax}: \{0,1\}^n \to \mathbb{R}, \quad x \mapsto \|x\|_1 := |\{i : x_i = 1\}|.$$

As algorithm for the optimization, we consider the $(\mu+1)$-EA, defined in Algorithm 1. The $(\mu+1)$-EA is a simple mutation-only evolutionary algorithm that maintains a population of μ solutions and uses elitist survival selection.

Algorithm 1. The $(\mu+1)$-EA.

```
1 t ← 1;
2 P_t ← μ elements of {0,1}^n uniformly at random;
3 while termination criterion not met do
4     Select x ∈ P_t uniformly at random;
5     Create y by flipping each bit of x with probability 1/n;
6     P_{t+1} ← P_t ∪ {y} \ {z} where ∀v ∈ P_t: f(z) ≤ f(v);
7     t ← t + 1;
```

2.1 Noise Model

We consider *additive posterior noise*, meaning that the noisy fitness value is given by the actual fitness value plus some term sampled (independently for each sample) from some fixed random variable D. For OneMax and a fixed distribution, we call this noisy fitness function OneMax_D.

Let F be a family of pseudo-Boolean functions $(F_n)_{n \in \mathbb{N}}$ where each F_n is a set of functions $f: \{0,1\}^n \to \mathbb{R}$. Let D be a family of distributions $(D_v)_v$ such that for all $D_v \in D$, $\mathrm{E}(D_v) = 0$. We define F with additive posterior D-noise as the set $F[D] := \{f_n + D_v : f_n \in F_n, D_v \in D\}$.

Definition 1. *An algorithm A* scales gracefully with noise *on $F[D]$ if there is a polynomial q such that, for all $g_{n,v} = f_n + D_v \in F[D]$, there exists a parameter setting p such that $A(p)$ finds the optimum of f_n using at most $q(n,v)$ calls to $g_{n,v}$.*

We will need the following result regarding noisy optimization from [6] for our negative results.

Theorem 2 [6]. *Let $\mu \geq 1$ and D a distribution on $\mathbb{R}$. Let Y be the random variable describing the minimum over μ independent copies of D. Suppose*

$$\Pr(Y > D + n) \geq \frac{1}{2(\mu+1)}.$$

Consider optimization of OneMax$_D$ *by* $(\mu+1)$*-EA. Then, for* μ *bounded from above by a polynomial, the optimum will* not *be evaluated after polynomially many iterations w.h.p.*

Intuitively, whenever the selection pressure is so weak that selection is almost uniform (which would mean a probability of $1/(\mu+1)$ for choosing any particular individual), optimization will not succeed.

2.2 Drift Analysis

For the theoretical analysis we will use the following *drift theorem.*

Theorem 3 (Multiplicative Drift [4]). *Let* $(X_t)_{t\geq 0}$ *be a sequence of random variables over* $\mathbb{R}_{\geq 0}$*. Let* T *be the random variable that denotes the earliest point in time* $t \leq 0$ *such that* $X_t < 1$*. If there exist* $c > 0$ *such that, for all* a*,*

$$E[X_t - X_{t+1} \mid T > t,\, X_t = a] \geq c\, a,$$

then, for all a*,*

$$E[T \mid X_0 = a] \leq \frac{1 + \ln(a)}{c}.$$

3 Exponential Tails

In this section we consider noise taken from a random variable D that decays exponentially fast, i.e., we assume

$$F(t) := \Pr(D < k) = \tfrac{1}{2}\mathrm{e}^{ct} \quad \text{if } t \leq 0 \text{ and}$$
$$F(t) := 1 - \tfrac{1}{2}\mathrm{e}^{-ct} \quad \text{if } t > 0,$$

for some constant c. By taking the derivative of F, we get the probability mass function p of D, i.e.,

$$p(t) = F'(t) = \tfrac{c}{2}\mathrm{e}^{ct} \quad \text{if } t \leq 0 \text{ and}$$
$$p(t) = \tfrac{c}{2}\mathrm{e}^{-ct} \quad \text{if } t > 0.$$

This is basically a symmetric variant of the exponential distribution. Note that D is a distribution, since F is non-negative and monotonically increasing, $\lim_{t\to-\infty} F(t) = 0$, and $\lim_{t\to\infty} F(t) = 1$. Because p is symmetric around 0, it follows that D has mean 0.

We calculate the variance of D:

$$\begin{aligned}\mathrm{Var}(D) &= \textstyle\int_{-\infty}^{\infty} t^2 p(t)\mathrm{d}t = \frac{c}{2}\left(\int_{-\infty}^{0} t^2\mathrm{e}^{ct}\mathrm{d}t + \int_{0}^{\infty} t^2\mathrm{e}^{-ct}\mathrm{d}t\right)\\ &= c\textstyle\int_{-\infty}^{0} t^2\mathrm{e}^{ct}\mathrm{d}t = c\left[\frac{(2-2ct+t^2c^2)\mathrm{e}^{ct}}{c^3}\right]_{-\infty}^{0}\\ &= \tfrac{2}{c^2} =: \sigma^2,\end{aligned}$$

where the integral can be computed by integrating by parts twice. This leads to

$$F(t) = \tfrac{1}{2}e^{\sqrt{2}\frac{t}{\sigma}} \qquad \text{if } t \leq 0\text{; and}$$
$$F(t) = 1 - \tfrac{1}{2}e^{-\sqrt{2}\frac{t}{\sigma}} \qquad \text{if } t > 0.$$

We now want to show that, in this setting and for sufficiently large variance, the $(\mu+1)$-EA is not successful. We will start with the case of $\mu \in \{1,2\}$, as this case is not covered by our main theorem of this section (Theorem 5 below). The proof is instructive, since its structure is similar to the proof of Theorem 5, while it is a bit simpler in the details.

Proposition 4. *Consider optimization of* OneMax_D *by the* $(\mu+1)$*-EA with* $\mu \in \{1,2\}$. *Suppose* $\sigma^2 = \omega(n^2)$. *Then the optimum will* not *be evaluated after polynomially many iterations w.h.p.*

Proof. We set up to use Theorem 2. Thus, let $t^- < 0$ and $t^+ > 0$ be such that $\Pr(D < t^-) = \Pr(D \geq t^+) = 1/4$. Hence, $\Pr(t^- \leq D < 0) = \Pr(0 \leq D < t^+) = 1/4$ because D is symmetric around 0.

We consider D and μ copies of it: D_i^*, for $i = 1, \ldots, \mu$. We want to bound

$$\Pr\left(\min_{i=1,\ldots,\mu}\{D_i^*\} > D + n\right) = \Pr\left(\bigwedge_{i=1}^{\mu} D_i^* > D + n\right).$$

We lower-bound the above probability as follows:

$$\begin{aligned}
\Pr\left(\bigwedge_{i=1}^{\mu} D_i^* > D + n\right) &\geq \Pr(D < t^-)\prod_{i=1}^{\mu}\Pr(D_i^* \geq t^- + n) + \\
&\quad \Pr(t^- < D < 0)\prod_{i=1}^{\mu}\Pr(D_i^* \geq n) + \\
&\quad \Pr(0 < D < t^+)\prod_{i=1}^{\mu}\Pr(D_i^* \geq t^+ + n) \\
&= \tfrac{1}{4}\left(\prod_{i=1}^{\mu}\Pr(D_i^* \geq t^- + n) + \prod_{i=1}^{\mu}\Pr(D_i^* \geq n) + \right. \\
&\quad \left. \prod_{i=1}^{\mu}\Pr(D_i^* \geq t^+ + n)\right).
\end{aligned}$$

Thus, we have to bound the probabilities of the form $\Pr(D_i^* \geq a + n)$. We do so by showing $\Pr(D_i^* \geq a+n) \geq \big(1 - o(1)\big)\Pr(D_i^* \geq a)$. First, consider $a \geq 0$.

$$\begin{aligned}
\Pr(D_i^* \geq a + n) &= \tfrac{1}{2}e^{-\sqrt{2}\frac{a+n}{\sigma}} = \tfrac{1}{2}e^{-\sqrt{2}\left(\frac{a}{\sigma}+\frac{n}{\sigma}\right)} = \tfrac{1}{2}e^{-\sqrt{2}\left(\frac{a}{\sigma}+o(1)\right)} \\
&= \big(1 - o(1)\big)\Pr(D_i^* \geq a).
\end{aligned}$$

Note that if $a = t^-$, we get $t^- + n < 0$, because we assume that $\Pr(D < t^-) = 1/4$, which means that $t^- = -\Theta(\sigma) = -\omega(n^2)$.

$$\begin{aligned}
\Pr(D_i^* \geq t^- + n) &= 1 - \Pr(D_i^* < t^- + n) = 1 - \tfrac{1}{2}e^{\sqrt{2}\frac{t^-+n}{\sigma}} \\
&= 1 - \big(1 + o(1)\big)e^{\sqrt{2}\frac{t^-}{\sigma}} = \big(1 - o(1)\big)\Pr(D_i^* \geq t^-).
\end{aligned}$$

This results in

$$\Pr\left(\bigwedge_{i=1}^{\mu} D_i^* > D + n\right) \geq (1 - o(1))\frac{1}{4}\left(\left(\frac{3}{4}\right)^{\mu} + \left(\frac{2}{4}\right)^{\mu} + \left(\frac{1}{4}\right)^{\mu}\right),$$

which is at least $1/(2(\mu+1))$ for $\mu \in \{1, 2\}$. Applying Theorem 2 completes the proof. □

We now turn to the more general case.

Theorem 5. *Consider optimization of* OneMax$_D$ *by the* $(\mu+1)$*-EA with* $\mu \geq 3$ *and* μ *bounded from above by a polynomial in* n. *Suppose* $\sigma^2 = \omega(n^2)$. *Then the optimum will* not *be evaluated after polynomially many iterations w.h.p.*

Proof. This proof follows the ideas of the one of Corollary 6 from [6]. Let $a = \omega(1)$ be such that $\sigma^2 \geq (na)^2$.

Again, we want to use Theorem 2, thus, let Y be the minimum of μ independent copies of D, whereas D is a distribution as defined above. We want to bound $\Pr(D + n < Y)$. Hence, we choose two points $t_0 < t_1 < 0$ such that $\Pr(D < t_0) = 0.7/\mu$ and $\Pr(D < t_1) = 1.4/\mu$. Note that $\Pr(D < t_0) < \Pr(D < t_1) < 1/2$, since $\mu \geq 3$. Thus, t_0 and t_1 actually exist.

We define the following two disjoint events that are a subset of the event $D + n < Y$:

A: The event that $D < t_0 - n$ and $t_0 < Y$.
B: The event that $t_0 - n < D < t_1 - n$ and $t_1 < Y$.

We first focus on bounding $\Pr(D < t_0 - n)$ and do so showing that $t_0 \leq -na/32$ holds via contraposition.

Assume that $t_0 > -na/32$ (still $t_0 < 0$). Because we assume $\sigma \geq na$ as well, we get that $s := -t_0/\sigma < 1/32$. Due to the definition of D we get

$$\Pr(D < t_0) = F(t_0) = \frac{1}{2}\mathrm{e}^{-\sqrt{2}s} > \frac{1}{2}\mathrm{e}^{-\frac{\sqrt{2}}{32}} > \frac{0.7}{3}.$$

Since we assume $\mu \geq 3$, this contradicts the definition of t_0. Hence, the bound $t_0 \leq -na/32$ holds which is equivalent to $t_0(1+32/a) \leq t_0 - n$. Thus, we estimate

$$\Pr(D < t_0 - n) \geq \Pr\left(t_0\left(1 + \tfrac{32}{a}\right)\right) = \tfrac{1}{2}\mathrm{e}^{\sqrt{2}\frac{t_0\left(1+\frac{32}{a}\right)}{\sigma}} = \tfrac{\mathrm{e}^{-o(1)}}{2}\mathrm{e}^{\sqrt{2}\frac{t_0}{\sigma}}$$
$$= (1 - o(1))\tfrac{1}{2}\mathrm{e}^{\sqrt{2}\frac{t_0}{\sigma}} = (1 - o(1))\Pr(D < t_0).$$

Because $\Pr(D < t_0 - n) \leq \Pr(D < t_0)$ holds trivially, we have $\Pr(D < t_0 - n) = (1 - o(1))\Pr(D < t_0)$.

We now bound $\Pr(t_0 - n < D < t_1 - n) = \Pr(D < t_1 - n) - \Pr(t_0 - n < D)$, where we are left with bounding $\Pr(D < t_1 - n)$. We do so analogously to the calculations before. This time, we bound $t_1 \leq -na/64$, since assuming $t_1 > -na/64$ leads to

$$\Pr(D < t_1) > \frac{1}{2}\mathrm{e}^{-\frac{\sqrt{2}}{64}} > \frac{1.4}{3}.$$

All the remaining calculations can be done as before, and we get $\Pr(t_0 - n < D < t_1 - n) = \big(1 - \mathrm{o}(1)\big)1.4/\mu - \big(1 - \mathrm{o}(1)\big)0.7/\mu = \big(1 - \mathrm{o}(1)\big)0.7/\mu$. Overall, we have

$$\begin{aligned}\Pr(Y > D + n) &\geq \Pr(A) + \Pr(B) = \big(1 - \mathrm{o}(1)\big)\tfrac{0.7}{\mu}\left(\left(1 - \tfrac{0.7}{\mu}\right)^{\mu} + \left(1 - \tfrac{1.4}{\mu}\right)^{\mu}\right) \\ &\geq \big(1 - \mathrm{o}(1)\big)^2 \tfrac{0.7}{\mu}\left(\mathrm{e}^{-0.7} + \mathrm{e}^{-1.4}\right) \geq \tfrac{1}{2\mu} \geq \tfrac{1}{2(\mu+1)}.\end{aligned}$$

Applying Theorem 2 completes the proof. □

The statement of Theorem 5 is basically that if the standard deviation of the noise is asymptotically larger than the greatest OneMax value (n), the noise will dominate and optimization will fail. The proof idea can be expanded to the case when $\Pr(D < t_0) = \Omega(1)$ if $t_0 = -\Omega(\sigma)$, i.e., there is at least a constant probability to deviate by at least one standard deviation from the mean. In such a case the OneMax value is irrelevant, since it will easily be dominated by the noise.

Overall, we get the following statement regarding graceful scaling.

Corollary 6. *The* $(\mu{+}1)$*-EA does* not *scale gracefully on OneMax with additive posterior noise from a distribution with exponential tails as given above (parametrized in the variance).*

4 Truncated Distributions

In this section we consider *truncated* distributions; these distributions are a generalization of uniform distributions, which capture the essence of what our proofs need to show that the $(\mu + 1)$-EA can scale gracefully. Truncated distributions are distributions whose density functions vanish above (respectively, below) some point k and whose mass near that point is bounded from below by some value q.

Definition 7. *Let* D *be a random variable. If there are* $k, q \in \mathbb{R}$ *such that*

$$\Pr(D > k) = 0 \wedge \Pr\big(D \in (k - 1, k]\big) \geq q,$$

then we call D upper q-truncated. *Analogously, we call* D lower q-truncated *if there is a* $k \in \mathbb{R}$ *with*

$$\Pr(D < k) = 0 \wedge \Pr\big(D \in [k, k + 1)\big) \geq q.$$

From [14] we know that the run time of the $(\mu+1)$-EA on OneMax is $\mathcal{O}(\mu n + n \log n)$ when no noise is present. The following theorem looks at the optimization behavior for truncated noise and gives a slightly weaker run time bound in the presence of noise which is suitably truncated.

Theorem 8. *Let $\mu \geq 1$ and let D be lower $2\log(n\mu)/\mu$-truncated. Consider optimization of* OneMax_D *by the $(\mu+1)$-EA. Then the optimum will be evaluated, in expectation, after $\mathcal{O}(\mu n \log n)$ iterations.*

Proof. We argue with drift on the number of 1s in the search point with the most number of 1s. If the search point with the most number of 1s is never removed within the first $\mathcal{O}(\mu n \log n)$ iterations, then multiplicative drift (Theorem 3) will give us the result. If any other search point evaluates in the minimal bracket $[k, k+1)$, then the best search point is safe (if there are multiple best, then it is safe anyway). The probability that none evaluates in the minimal bracket is at most

$$\begin{aligned}\Pr\left(D \notin [k, k+1)\right)^{\mu} &\leq \left(1 - \frac{2\log(n\mu)}{\mu}\right)^{\mu} \\ &\leq \exp(-2\log(n\mu)) \\ &\leq \mathcal{O}\left(\frac{1}{(n\mu)^2}\right).\end{aligned}$$

Thus, the expected number of iterations until the best search point decreases in number of 1s is $\omega(\mu n \log n)$ iterations. Since this holds from any starting configuration, the result follows from the fact that the optimum can be found by iteratively increasing the best individual in $\mathcal{O}(\mu n \log n)$ iterations. □

As a corollary to Theorem 8, we turn the statement of the previous theorem around and show how large a population is required for efficient optimization in the presence of truncated noise.

Corollary 9. *Let D be lower q-truncated. Then, for all $\mu \geq 3q^{-1}\log(nq^{-1})$, in the optimization of* OneMax_D *by $(\mu+1)$-EA, the optimum will be evaluated, in expectation, after $\mathcal{O}(\mu n \log n)$ iterations.*

Corollary 10. *Let $\mu, r \geq 1$. Consider optimization of OneMax with reevaluated additive posterior noise uniformly from $[-r, r]$ by $(\mu+1)$-EA without crossover. Then the optimum*

1. *will be evaluated within $\mathcal{O}(\mu n \log n)$ iterations in expectation if $r \leq \mu/(4\log(n\mu))$;*
2. *will* not *be evaluated within polynomially many iterations w.h.p. if $r \geq n(\mu+1)$.*

Proof. Let D be the uniform distribution on $[-r, r]$.

Regarding the first claim, we note that D is lower $1/(2r)$-truncated, so the result follows from Theorem 8.

For the second claim we want to use Theorem 2. Define

$$f\colon [-r,r] \to [-r,r]+n, \quad x \mapsto \begin{cases} x & \text{if } x \geq -r+n\,; \\ r+n-x & \text{otherwise.} \end{cases}$$

Then we have that $D+n$ and $f(D)$ have the same distribution. Let Y be the minimum over μ independent copies of D. Due to symmetry, we have $\Pr(Y > D) = 1/(\mu+1)$. Thus, we have

$$\begin{aligned} \Pr(Y > D+n) &= \Pr\left(Y > f(D)\right) \\ &\geq \Pr(Y > f(D) \wedge D > -r+n) \\ &= \Pr(Y > D \wedge D > -r+n) \\ &\geq \Pr(Y > D) - \Pr(D \leq -r+n) \\ &= \frac{1}{\mu+1} - \frac{n}{2r} \\ &\geq \frac{1}{2(\mu+1)}\,. \end{aligned}$$

The result now follows with Theorem 2. □

From this we get the result regarding graceful scaling on uniform noise.

Corollary 11. *The $(\mu+1)$-EA scales gracefully on OneMax with additive posterior noise from the uniform distribution on $[-r,r]$.*

5 Summary

In this work we saw indications that the shape of the distributions plays an important role in settings with noisy fitness functions. For the case of the uniform distributions, we can give bounds for when optimization is successful and for when it is not. The analysis is significantly more complicated for other distributions, but our results still suggest that more even distributions make optimization easier.

It seems that further results are hard to come by and probably require a new way of dealing with diversity of populations (a long standing open problem). In particular, the only theorem for lower bounds we have is Theorem 2, which makes significant worst-case assumptions about the diversity of the population. Similarly, all upper bounds usually make worst-case assumptions on the diversity, but this time in the other direction (namely that the population is clustered, while Theorem 2 is based on the assumption that a single good individual works against many bad individuals). This also explains the gap in the bounds for the uniform distribution (Corollary 10). An alternative route could be in adapting drift theorems specifically for populations [11].

It is open whether we need better tools for showing lower or upper bounds; a useful first step could thus be to conjecture run time bounds based on empirical evidence and analyzing also the spread of the population carefully in dependence on the distribution.

Acknowledgments. The research leading to these results has received funding from the European Union Seventh Framework Programme (FP7/2007-2013) under grant agreement no. 618091 (SAGE) and the German Research Foundation (DFG) under grant agreement no. FR 2988 (TOSU).

References

1. Bianchi, L., Dorigo, M., Gambardella, L., Gutjahr, W.: A survey on metaheuristics for stochastic combinatorial optimization. Nat. Comput. **8**, 239–287 (2009)
2. Dang, D.-C., Lehre, P.K.: Evolution under partial information. In: Proceedings of GECCO 2014, pp. 1359–1366 (2014)
3. Doerr, B., Hota, A., Kötzing, T.: Ants easily solve stochastic shortest path problems. In: Proceedings of GECCO 2012, pp. 17–24 (2012)
4. Doerr, B., Johannsen, D., Winzen, C.: Multiplicative drift analysis. Algorithmica **64**(4), 673–697 (2012)
5. Feldmann, M., Kötzing, T.: Optimizing expected path lengths with ant colony optimization using fitness proportional update. In: Proceedings of FOGA 2013, pp. 65–74 (2013)
6. Friedrich, T., Kötzing, T., Krejca, M.S., Sutton, A.M.: The benefit of recombination in noisy evolutionary search. In: Elbassioni, K., Makino, K. (eds.) ISAAC 2015. LNCS, vol. 9472, pp. 140–150. Springer, Heidelberg (2015). doi:10.1007/978-3-662-48971-0_13
7. Gießen, C., Kötzing, T.: Robustness of populations in stochastic environments. In: Proceedings of GECCO 2014, pp. 1383–1390 (2014)
8. Gutjahr, W.J.: A converging ACO algorithm for stochastic combinatorial optimization. In: Albrecht, A.A., Steinhöfel, K. (eds.) SAGA 2003. LNCS, vol. 2827, pp. 10–25. Springer, Heidelberg (2003)
9. Gutjahr, W., Pflug, G.: Simulated annealing for noisy cost functions. J. Glob. Optim. **8**, 1–13 (1996)
10. Jin, Y., Branke, J.: Evolutionary optimization in uncertain environments–a survey. IEEE Trans. Evol. Comp. **9**, 303–317 (2005)
11. Lehre, P.K.: Negative drift in populations. In: Schaefer, R., Cotta, C., Kołodziej, J., Rudolph, G. (eds.) PPSN XI. LNCS, vol. 6238, pp. 244–253. Springer, Heidelberg (2010)
12. Prügel-Bennett, A., Rowe, J.E., Shapiro, J.: Run-time analysis of population-based evolutionary algorithm in noisy environments. In: Proceedings of FOGA 2015, pp. 69–75 (2015)
13. Sudholt, D., Thyssen, C.: A simple ant colony optimizer for stochastic shortest path problems. Algorithmica **64**, 643–672 (2012)
14. Witt, C.: Runtime analysis of the $(\mu + 1)$ EA on simple pseudo-boolean functions. Evol. Comput. **14**, 65–86 (2006)

On the Robustness of Evolving Populations

Tobias Friedrich, Timo Kötzing, and Andrew M. Sutton(✉)

Hasso Plattner Institute, Potsdam, Germany
andrew.sutton@hpi.de

Abstract. Most theoretical work that studies the benefit of recombination focuses on the ability of crossover to speed up optimization time on specific search problems. In this paper, we take a slightly different perspective and investigate recombination in the context of evolving solutions that exhibit *mutational robustness*, i.e., they display insensitivity to small perturbations. Various models in population genetics have demonstrated that increasing the effective recombination rate promotes the evolution of robustness. We show this result also holds in the context of evolutionary computation by rigorously proving crossover promotes the evolution of robust solutions in the standard $(\mu + 1)$ GA. Surprisingly, our results show that this effect is still present even when robust solutions are at a selective disadvantage due to lower fitness values.

1 Introduction

The role of crossover in evolutionary computation is still a major open problem in the theory of evolutionary algorithms. In some cases, it can be provably helpful for optimization obtaining quantifiable speed-ups on functions like JUMP and ONEMAX, and particular combinatorial optimization problems on graphs [2,5,7,10,11]. In other cases, recombination can actually be seen as a destructive operator that is detrimental to optimization [13]. The goal of this work is to contribute to our understanding of environments in which crossover can be helpful.

In population genetics, an increased recombination rate has been shown to increase *mutational robustness*: the resistance of fitness to mutational perturbations [3]. In this paper, we want to examine this effect in the context of runtime analysis for evolutionary algorithms. In particular, we introduce a model landscape for which we prove crossover favors regions of higher neutrality. This effect can be seen even when robust solutions have a much weaker fitness gradient than non-robust solutions. On the other hand, as the recombination rate is tuned to zero, greedy hill-climbing behavior takes over and favors regions with sharper fitness gradients: even when these regions contain solutions that are not as robust to perturbations.

Our model landscape is motivated by the fact that in some optimization problems, there could be *sensitive* decision variables that correspond to non-robust solutions and *non-sensitive* decision variables that induce more robust solutions because they correspond to a large plateau. In practice, we will not

J. Handl et al. (Eds.): PPSN XIV 2016, LNCS 9921, pp. 771–781, 2016.
DOI: 10.1007/978-3-319-45823-6_72

know where these sensitive and non-sensitive variables are. However, our results suggest that if they exist, then using crossover tends to favor these more robust solutions even when they are at a fitness disadvantage.

1.1 Visualizing the Evolution of Robustness

We begin with a simple visualization that gives an intuition for why high crossover rates favor robust solutions. The purpose of this short section is to gain some geometric insight into the proofs contained in Sects. 2 and 3. We consider an extension of the cycle $\mathbb{Z}_n$ to a cylinder (formally $\mathbb{Z}_n^2$, but without the added wrap-around in the second dimension). We consider all individuals at the "bottom" of the cylinder to have a small fitness of, say, 5. Fitness grows "upward", but only on certain paths (i.e., for certain x-values). We consider the case of one wide path and many narrow paths. Figure 1 depicts this fitness landscape (darker colors indicate higher fitness). Maximal fitness can be achieved for all maximal y values. We assume that fitness grows more quickly along narrow paths, by a factor of 1.1. Individuals not on any paths with x-value larger than 0 are dead.

To formulate a simple illustrative example of how recombination can favor robust optimization we present the following short experiment (details omitted due to space constraints) to motivate the rest of the paper. We construct a cylinder of size $10^2 \times 10^4$; the wide path has a width of 10 and there are 14 evenly-spaced narrow paths. We consider a $(\mu + \lambda)$ GA with $\mu = 50$ and $\lambda = 250$ evolving on this fitness landscape. During a run, populations with high crossover rates favor the wide path, focusing on the middle of the wide path. On the other hand, mainly asexually reproducing populations favor narrow paths.

Now consider the case where the fitness landscape changes dynamically, but rarely. More precisely, we consider a random shift of the complete fitness landscape in x-direction by ± 2 after 6000 iterations, simulating a small but

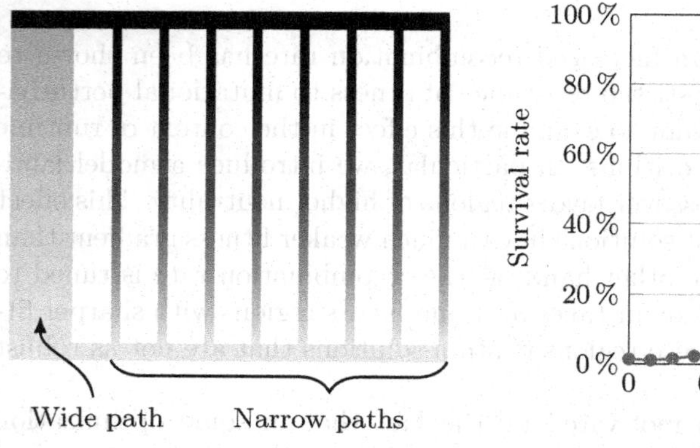
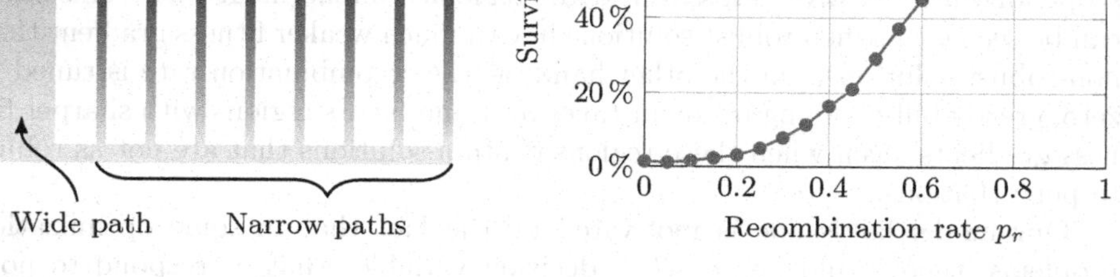

Fig. 1. Left: landscape with many paths leading to the global optimum on the cylinder. Right: prob. GA w/recombination rate p_r does not get extinct during optimization.

significant change in the environment. Populations that exclusively focus on climbing narrow paths will go extinct when the dynamic change of the fitness landscape occurs; populations not too close to the edge of the wide path will survive.

We plot the ratio of populations going extinct before reaching the top y-value, in dependence on the crossover rate p_r (Fig. 1, right). As we can see, with growing crossover rate, it is more and more likely that a population survives and reaches the optimum, due to choosing the wide path to walk up instead of the narrow paths. In particular, an asexually reproducing population will go extinct with high probability, while a population employing crossover in every iteration will go extinct with only very low probability.

2 Preliminaries

We now turn to a more formal analysis in order to prove rigorous statements about how robustness can evolve using recombination. Our aim is to construct a pseudo-Boolean function $f\colon \{0,1\}^n \to \mathbb{R}$ that is structurally similar to the landscape in Fig. 1, but is somewhat easier to work with mathematically. In particular, we want to define a function that has a set of solutions corresponding to a "wide path", on which each level has an exponential number of solutions, and a collection of sets corresponding to "narrow paths" where there is only a small plateau (each fitness level has only a polynomial number of solutions). We also require the fitness values of the wide path to have a gradual slope, whereas the fitness values of narrow paths have a sharper slope.

Let $n = 2^k$ for some $k \in \mathbb{N}$. We partition a bitstring $x \in \{0,1\}^n$ into three consecutive segments of length k, length $n/2$, and length $(n/2 - k)$.

$$\underbrace{x_1 \cdots x_k}_{\text{first segment}}\ \underbrace{x_{k+1} \cdots x_{k+n/2}}_{\text{second segment}}\underbrace{x_{k+n/2+1} \cdots x_n}_{\text{third segment}}.$$

Denote $[x]_1 = (x_1 x_2 \cdots x_k)$ as the length-k string corresponding to the first segment substring of x. Similarly, $[x]_2$ and $[x]_3$ are the length-$n/2$ and length-$(n/2 - k)$ strings formed from the second and third segments of x.

We say that a bit string is *on the wide path* iff the first segment is all 0s. We fix a set $H = \{h_1, h_2, \ldots, h_{n/2}\}$ of $n/2$ unique bit strings of length $n/2$. We say a bitstring is *on a narrow path* if (1) it contains at least $\log k$ 1s in the first segment, and (2) the substring in its second segment belongs to H. The first condition ensures adequate separation from the wide path; the second condition defines a collection of subspaces of $\{0,1\}^n$ that contain each narrow path. Formally, we define the set $\mathcal{W}$ of wide-path solutions as $\mathcal{W} = \{x \in \{0,1\}^n : x_1 = \cdots = x_k = 0\}$ and the set $\mathcal{N}$ of narrow-path solutions as $\mathcal{N} = \{x \in \{0,1\}^n : x_1 + \cdots + x_k \geq \log k \wedge [x]_2 \in H\}$. Each narrow path is associated with a unique $h_i \in H$.

To provide a concrete definition of the set H of narrow path keys, we employ the concept of Hadamard codes [1]. Our motivation is that Hadamard codes provide a clean way of ensuring the narrow paths are sufficiently distant from one another, while simplifying many of the proofs.

A Hadamard code is an error-correcting linear code over binary strings. For our setting, we consider the set of dimension $k-1$ Hadamard codes of length $2^{k-1} = n/2$. To construct this code, we define the inner product between two length $k-1$ bitstrings x and y as $\langle x, y\rangle = \sum_{i=1}^{k-1} x_i y_i \pmod 2$. Let $\sigma\colon \{1,\ldots,2^{k-1}\} \to \{0,1\}^{k-1}$ be a bijection. Then, for each $i \in \{1,\ldots,2^{k-1}\}$, the i-th codeword is the string $h_i \in \{0,1\}^{2^{k-1}}$ such that $h_{ij} = \langle\sigma(i), \sigma(j)\rangle$ for all $j \in \{1,\ldots,2^{k-1}\}$. There are hence $2^{k-1} = n/2$ unique codewords of length $n/2$ and we set $H = \{h_i \mid i \in \{1,\ldots,2^{k-1}\}\}$. We explicitly rely on the *minimum distance* property of a dimension $k-1$ Hadamard code: each pair of distinct codewords is separated by Hamming distance at least $2^{k-1}/2$.

We define a pseudo-Boolean fitness function f in such a way so that there is a steeper fitness gradient on the narrow paths. Let $c > 1$ be a constant.

$$f(x) = \begin{cases} \text{LEADINGONES}([x]_3), & x \in \mathcal{W}; \\ c\text{LEADINGONES}([x]_3), & x \in \mathcal{N}; \\ -\infty, & \text{otherwise}. \end{cases} \tag{1}$$

Here $\text{LEADINGONES}(x) := \sum_{i=1}^{n} \prod_{j=1}^{i} x_j$ counts the number of leading ones of its argument. We say an individual is *non-viable* if its fitness is negative infinity. Such an individual corresponds to an *infeasible* solution.

The uniform crossover of two individuals on the wide path always results in an individual on the wide path (with fitness at least the minimal of the two parent fitnesses). On the other hand, uniform crossover of two individuals on separate narrow paths will very likely be non-viable, as we now see.

Lemma 1. *Let $x, y \in H \subseteq \{0,1\}^{n/2}$ such that $x \neq y$. Define $0 < \epsilon < 1$ to be an arbitrary constant. Then with probability $1 - 2^{-\Omega(n)}$, the offspring produced by uniform crossover of x and y is at distance at least n^ϵ from any string in H.*

Proof. Let $z \in H \subseteq \{0,1\}^{n/2}$ be an arbitrary length $n/2$ Hadamard code. Let $B_r(z) \subseteq \{0,1\}^{n/2}$ denote the ball of radius $r < n/4$ around z. By the properties of the dimension $k-1$ Hadamard code, each codeword has minimum distance $2^{k-1}/2$ to any other codeword and thus $d(x, y) \geq 2^{k-1}/2 = n/4$. Therefore, every element of $B_r(z)$ must lie at distance at least $\max\{d(x, z), d(y, z)\} - r \geq n/4 - r$ from at least one of x or y. Therefore, the probability that crossover produces an offspring $w \in B_r(z)$ is at most $(1/2)^{\max\{d(x,w),d(y,w)\}} \leq 2^{-n/4+r}$.

We now bound the probability of the offspring of x and y belonging to a set of solutions that lies within some radius-r ball of any narrow-path solution in $\mathcal{N}$. Let $w \in \{0,1\}^{n/2}$ be the offspring produced by uniform crossover of x and y. There are $|H| = n/2$ distinct Hadamard codes, and $|B_r(z)| \leq (n/2+1)^r$. Applying a union bound, the probability that w lies within a ball of radius r around any Hadamard code $z \in H$ is at most

$$\Pr\left(w \in \bigcup_{z \in H} B_r(z)\right) \leq |H|(n/2+1)^r 2^{-n/4+r} = 2^{-n/4+O(r \log n)}.$$

Setting $r := n^\epsilon$ completes the proof. $\square$

Lemma 1 implies that crossing over two solutions on distinct narrow paths is usually fatal: with overwhelming probability, the offspring is non-viable (or it lies on the wide path) because its narrow-path segment lies sufficiently far from any Hadamard code. Moreover, it is exponentially unlikely that a subsequent mutation operation applied to the offspring could repair the damage, since it would need to flip at least n^{ϵ} bits to get to the nearest narrow path solution.

3 Formal Analysis

We now prove that a high recombination rate favors wide-path solutions during evolution, whereas a low recombination rate favors narrow-path solutions. Individuals that lie on the wide path are *robust* in the following sense. Let x be on the wide path and let y be on a narrow path. Consider any perturbation process that that changes some bits in a string, subject to the constraint that a constant number of bits change in expectation, and that number is concentrated around its expectation (e.g., uniform mutation with a $\Theta(1/n)$ mutation rate, or changing $\Theta(1)$ bits uniformly at random). If x undergoes this perturbation, it is non-viable only with probability $\Theta(k/n) = o(1)$. On the other hand, such a mutation on y results in a non-viable solution already with constant probability. It is therefore easy to see that in a dynamic environment where perturbations occur during evolution (as with our example in Sect. 1.1), a process following the wide path will in general be more successful. The result is also interesting in a static context where the algorithm produces a string robust to changes *after* evolution, or in homologous landscapes in which the optimal solution lies only at the end of the wide path.

3.1 Algorithm

We study a simple population-based evolutionary algorithm equipped with a recombination rate parameter $p_r \in [0, 1]$ that dictates the frequency with which recombination is employed to generate offspring. The $(\mu + 1)$ GA (see Algorithm 1) is a steady-state genetic algorithm that maintains a population of μ elements of $\{0, 1\}^n$ and uses uniform parent selection and truncation survival selection. In each iteration, with probability p_r two parents are chosen uniformly at random without replacement (this condition is not necessary for the result, but necessary for a simpler proof). An offspring is then produced by uniform crossover followed by mutation. Otherwise, with probability $1-p_r$ a single individual is chosen uniformly at random and an offspring is produced by mutation only. We examine its behavior at extremal recombination rates $p_r \in \{0, 1\}$.

We construct the initial population P_0 by selecting exactly one element uniformly at random from each path, hence $\mu = n/2+1$. For each length-$n/2$ string $z' \in H$, we construct a $x \in \{0, 1\}^n$ where the $[x]_1$ is drawn uniformly at random from the set of length-k binary strings with at least $\log k$ ones, $[x]_2 := z'$, and the remaining positions are initialized uniformly at random. For the wide path, we choose a solution uniformly at random from $\mathcal{W}$ by creating a string with the

Algorithm 1. The $(\mu+1)$ GA with recombination rate p_r

```
for t ← 0 to ∞ do
    Select r uniformly at random from [0, 1];
    Select {x, y} ⊆ P_t uniformly at random;
    if r < p_r then x ← UniformCrossover(x,y);
    Create z by flipping each bit of x independently with probability 1/n;
    Let u ∈ P_t ∪ {z} chosen s.t. ∀v ∈ P_t ∪ {z}: f(u) ≤ f(v);
    P_{t+1} ← (P_t ∪ {z}) \ {u};
```

first k bits set to zero, and the remaining string initialized uniformly at random. Our use of dimension-$\Omega(\log n)$ Hadamard codes requires a linear population size. In this paper we leave the effect of different orders of μ as an open question.

Lemma 2. *Let P_0 be the initial population described above. With probability $1-e^{-\Omega(n)}$, at least $(1-\epsilon)n/4$ narrow-path solutions have zero fitness.*

Proof. The third segment of each string in P_0 is drawn uniformly at random and so the number of leading ones is geometrically distributed. The event that a string has zero fitness occurs independently with probability $1/2$. The count of zero-fitness narrow-path strings in P_0 is binomially distributed and a simple application of Chernoff bounds completes the proof. □

Lemma 3. *Let P_0 be the initial population described above. Let $c>1$ be the multiplicative constant defined in Eq.* (1) *and let $a>1$ be an arbitrary constant. With probability* $1-O(n^{-(a-1)})$, $\max\{f(x)\colon x\in P_0\}\le ac\log n$.

Proof. Since the initial fitnesses are geometrically distributed, the probability that a given leading-ones segment has ℓ leading ones is $(1/2)^{\ell+1}$. Taking a union bound over all $\mu=n/2+1$ solutions, the probability that no string has more than ℓ leading ones is at least $1-(n/2+1)(1/2)^{\ell+1}$.

The claim is proved by setting $\ell=a\log n$, since the fitness can be no higher than c multiplied by the number of leading ones in the third segment. □

3.2 No Recombination

We show that mutation-only strategies favor the non-robust narrow-path solutions. We begin by proving it is unlikely that the initial fitness of any wide-path solution is improved within $O(n\log^{1/c} n)$ generations.

Lemma 4. *Let P_0 be the initial population described above and $c>1$ be the multiplicative constant defined in Eq.* (1). *Let f_0 be the initial fitness of the wide-path solution in P_0. Then with probability $1-o(1)$, after $an\log^{1/c} n$ steps of the $(\mu+1)$ GA with $p_r=0$, for any constant $a>1$, every wide path solution has fitness at most f_0.*

Proof. Let $\mathcal{E}$ be the event that the number of wide-path solutions after $an\log^{1/c} n$ iterations is strictly less than $n/\log^2 n$. We first bound the probability of $\mathcal{E}$.

Denote as T_i the waiting time until the number of wide-path solutions increases, measured from the first generation in which there are i wide-path solutions. To jump to the wide-path from a narrow-path, mutation must flip $\log\log n$ bits. This occurs in any generation with probability that vanishes superpolynomially fast, so we assume it does not happen during $an\log^{1/c} n$ steps.

Each T_i is geometrically distributed with success probability at most i/μ since at the very least we must select a wide-path solution for mutation. Let T be first time there are $n/\log^2 n$ wide-path solutions in the population. Thus, $\mathrm{E}(T) = \sum_{i=1}^{n/\log^2 n} \mathrm{E}(T_i) \geq \mu \sum_{i=1}^{n/\log^2 n} 1/i = \Theta(n\log n)$. The probability that the number of wide-path solutions exceeds $n/log^2 n$ after $an\log^{1/c} n$ generations is

$$\Pr(\mathcal{E}) = 1 - \Pr(T \leq an\log^{1/c} n) \geq 1 - \Pr\left(T \leq \frac{a}{\log^{1-1/c} n} E(T)\right) = 1 - o(1).$$

Here we have applied tail bounds on the sum of independent geometric random variables [4]. Assume there are i wide-path solutions in iteration $t < an\log^{1/c} n$. Then, under condition $\mathcal{E}$, the probability that the fitness of *any* wide-path solution is increased is at most $i/(\mu n) \leq 2/(n\log^2 n)$.

Let $\mathcal{F}$ be the event that no wide-path fitness ever increases during $an\log^{1/c} n$ generations. By the law of total probability, $\Pr(\mathcal{F}) \geq \Pr(\mathcal{E})\Pr(\mathcal{F} \mid \mathcal{E})$, so

$$\Pr(\mathcal{F}) \geq \Pr(\mathcal{E})\left(1 - \frac{2}{n\log^2 n}\right)^{an\log^{1/c} n} \geq \Pr(\mathcal{E})\left(1 - \frac{2a}{\log^{2-1/c} n}\right) = 1 - o(1),$$

where we have applied Bernoulli's inequality. □

Theorem 5. *Consider a run of the ($\mu + 1$) GA with $p_r = 0$ initialized with P_0. With probability $1 - o(1)$, there exists a polynomial $\mathrm{poly}(n)$ such that for all $t > \mathrm{poly}(n)$ all elements of P_t are on some narrow path.*

Proof. We first argue that the fitness of the initial wide-path solution is $f_0 \leq \log\log n$ with probability $1 - o(1)$. The fitness of this individual depends only on what position in the leading-ones segment the first zero appears. This value is distributed geometrically with success probability $1/2$. So the probability that the first zero appears beyond the ($\log\log n$)-th position is $1 - 2^{-\log\log n} = 1 - o(1)$.

We say a solution $x \in \{0,1\}^n$ is *high-fitness* if $f(x) > f_0$. We now argue that there are many high-fitness narrow-path solutions in P_0. Each narrow-path solution is high-fitness if it has more than $(1/c)\log\log n$ leading ones, because its fitness is then strictly greater than $\log\log n = f_0$. Hence, the probability that an individual narrow-path solution is high-fitness is $2^{-(1/c)\log\log n}$. The initial $n/2$ fitness-values are independent, so by Chernoff bounds, for some positive constant $\gamma > 0$, there are at least $n/(\gamma\log^{1/c} n)$ high-fitness solutions with high probability. For the remainder of the proof, we assume this property holds.

Since an individual in the population can only be replaced by an offspring with a larger or equal fitness value, the count of high-fitness solutions never decreases. Let X_t denote the 0–1 random variable such that $X_t = 1$ if and only if a high-fitness solution is cloned in generation t. We have

$$\Pr(X_t = 1) \geq \frac{1}{\mu} \frac{n}{\gamma \log^{1/c} n} \left(1 - \frac{1}{n}\right)^n \geq \frac{1}{a \log^{1/c} n},$$

for $a > 0$ a positive constant. Let $S = \sum_{t=0}^{an \log^{1/c} n} X_t$. Obviously, S is a lower bound on the number of high-fitness solutions in the population after $an \log^{1/c} n$ generations. Thus we have $\mathrm{E}(S) \geq n = 2(\mu - 1)$ and by Chernoff bounds, $\Pr(S < \mu) \leq e^{-\Omega(n)}$. Hence, with high probability, there are only high-fitness solutions in the population by generation $an \log^{1/c} n$.

Finally, we can apply Lemma 4 to conclude that with probability $1 - o(1)$, no wide-path solution was ever improved during the take-over period of high-fitness solutions. Under these events, no wide-path solution remains in the population.

After this point, a new wide-path solution appears only if a narrow-path solution is mutated onto the wide path. This requires changing at least $\log \log n$ bits in the first segment for which we derive a superpolynomial waiting time w.h.p. When $p_r = 0$, the algorithm is identical to the (μ+1) EA, which solves leading ones in polynomial time [12]. Iterations with no viable offspring only slow the process by a constant factor. After poly(n) steps, all individuals have fitness at least $(n/2 - k) + 1$ and so no wide-path solution will ever be accepted. □

3.3 Full Recombination

We now prove that if the recombination rate is one, the $(\mu + 1)$ GA favors robust wide-path solutions. The following lemma states narrow-path solutions are difficult to create by the crossover operation.

Lemma 6. *Consider a run of the $(\mu + 1)$ GA with $p_r = 1$ initialized as above. With probability $1 - o(1)$, no new narrow-path offspring are accepted within n^3 generations.*

Proof. Let $\mathcal{E}_t$ denote the event that the first narrow-path offspring is generated in generation t. We argue that $\Pr(\mathcal{E}_t)$ is sufficiently close to zero for all $t \leq n^3$. Consider a generation in which no new narrow-path offspring have been created yet. There are three possibilities for parent selection: (1) two wide-path solutions are selected as parents, (2) two narrow-path solutions are selected as parents, and (3) wide-path solution and a narrow path solution are selected as parents.

In the first case, the result of uniform crossover must lie on the wide-path since the offspring inherits the entire first segment from both parents. In this case it is up to mutation alone to move the offspring to a narrow path. However, since each narrow-path solution must have $\log \log n$ ones in the first segment, mutation must flip $\log \log n$ bits, which only happens with probability $o(1)$.

In the second case, since we assume no new narrow-path solutions have been produced by generation t, each pair of narrow-path solutions lie on distinct paths. By Lemma 1, crossover between any two narrow-path solutions results in an offspring whose second segment is at least n^ϵ-far from any Hadamard code.

For the third case, we can focus on a single Hadamard code, namely $h_0 = 0^{n/2}$, and then take a union bound over all $n/2$ codes. We define the *Hamming weight* of a binary string to be the number of ones it contains. Let X_t be the minimum Hamming weight in the second segment over all wide-path solutions at time t. If a wide-path solution is recombined with the unique narrow-path solution $y \in \mathcal{N}$ with $[y]_2 = h_0$, then the expected Hamming weight of the offspring is $X_t/2$. Otherwise, if we cross a wide-path solution with some string z having $m > 0$ ones in $[z]_2$, the expected Hamming weight of the offspring is $X_t/2+m/2$. The probability of selecting y is $1/\mu$. It is straight-forward to apply a negative drift argument [8,9] on the potential $\log(X_t)$ to show X_t does not hit zero within n^3 generations with probability at most $1/n^{4+\delta}$ for a constant $\delta > 0$. In total, the probability for the third case is at most $2/n^{3+\delta}$.

Therefore, letting $p = \max_{0 \le t \le n^3}\{\Pr(\mathcal{E}_t)\}$, the probability that the count of narrow-path solutions does not increase in the first n^3 steps is bounded by $\prod_{t=1}^{n^3} (1 - \Pr(\mathcal{E}_t)) \ge (1-p)^{n^3} \ge 1 - pn^3 = 1 - o(1)$. □

Theorem 7. *Starting from the initial population described above, with high probability, after $T = O(n^2 \log n)$ iterations of the (μ + 1) GA, all elements of P_T are on the wide path and remain there for any polynomial number of steps.*

Proof. We show that with probability $1 - o(1)$ the entire population converges to the wide path in $O(n^2 \log n)$ steps. Subsequently, only wide-path solutions reproduce so we get superpolynomial waiting time for new narrow-path solutions.

Define W_t and Z_t to be the count of wide-path solutions and zero-fitness narrow path solutions in P_t, respectively. We condition on the following set of events, each holding with high probability: (1) a narrow-path offspring does not appear within n^3 steps (Lemma 6), (2) $Z_0 \ge (1-\epsilon)n/4$ (Lemma 2), and (3) the fitness of any solution in P_0 is at most $ac \log n$ for constants $a, c > 1$ (Lemma 3).

We divide a run of the $(\mu + 1)$ GA into two phases. The first phase begins at $t = 0$ and lasts until there are $\Omega(n)$ wide-path solutions in the population. Let $T_1 = \inf\{t \in \mathbb{N} \colon W_t > \mu/8\}$. The first phase begins at $t = 0$ and ends at $t = T_1$. During this phase, since we assume no narrow-path solutions are spontaneously created, $W_t \le \mu/8$ and $Z_t \ge Z_0 - \mu/8$. Moreover, $W_{t+1} - W_t \ge 0$ for all $t \le T_1$.

During this phase, a wide-path solution is chosen as a parent with probability W_t/μ and the resulting offspring is on the wide path with probability at least $1/(2en)$. Under this event, $W_{t+1} = W_t + 1$ only if a narrow-path solution is replaced in the selection phase. Since the fitness of the offspring is at least zero, the probability that a narrow path solution is selected for deletion is at least $Z_t/\mu \ge (3-4\epsilon)/8$. Thus at each iteration $t \le T_1$ in the first phase, we have $\mathrm{E}(W_{t+1} - W_t \mid W_t) \ge W_t(3-4\epsilon)/(16\mu en) = \Omega(W_t/n^2)$. We can bound the hitting time of W_t to $\mu/8 = n/16$ using the General Drift Theorem of Lehre and Witt [6] to get $T_1 = O(n^2 \log n)$ with probability $1 - o(1)$.

The second phase begins at time $T_1 + 1$. In this phase, $W_t \ge \mu/8$ so two wide-path solutions are selected as parents for crossover with probability $\Omega(1)$.

If two wide-path solutions are selected as parents, the result of crossover (before mutation) must be on the wide path, and must have a fitness at least as high as the lowest fitness of the parents. With probability at least $1/(en)$, mutation can then flip the first zero in the leading-ones segment to improve the fitness of the offspring by at least 1. This offspring is accepted since it is strictly more fit than at least one element of P_t (its least fit parent). We call such a result a *success.*

The probability of a success in each iteration is $\Omega(1/n)$ and independent. After $c'\mu n$ iterations for an appropriate constant $c' > 0$, we have had at least μ successes in expectation. We call a sequence of μ successes a *round.* Let $m = \min\{f(x) : x \in P_{T_1+1}\} \geq 0$ be the minimum fitness in the population at the start of phase two. After one round, μ offspring have been accepted with fitness at least $m+1$ so the minimum fitness in the population after the first round is at least 1. After $1 + ac\log n$ rounds, the minimum fitness is at least $(m+1) + ac\log n > ac\log n$. Since no new narrow-path solutions spontaneously appear during this time, it follows by Lemma 3 that all narrow-path solutions present in the initial population must have been replaced by wide-path solutions during this phase.

Applying Chernoff bounds to the success count, with probability $1 - e^{-\Omega(n)}$, each round takes at most $(1+\epsilon)c'\mu n = O(n^2)$ steps and we conclude all narrow-path solutions are replaced after $O(n^2 \log n)$ rounds during the second phase. □

Acknowledgements. The research leading to these results has received funding from the European Union Seventh Framework Programme (FP7/2007-2013) under grant agreement no. 618091 (SAGE) and the German Research Foundation (DFG) under grant agreement no. FR 2988 (TOSU).

References

1. Arora, S., Barak, B.: Computational Complexity: A Modern Approach, 1st edn. Cambridge University Press, New York (2009)
2. Doerr, B., Happ, E., Klein, C.: Crossover can provably be useful in evolutionary computation. Theor. Comput. Sci. **425**, 17–33 (2012)
3. Gardner, A., Kalinka, A.T.: Recombination and the evolution of mutational robustness. J. Theor. Biol. **241**(4), 707–715 (2006)
4. Janson, S.: Tail bounds for sums of geometric and exponential variables. (2014). http://www2.math.uu.se/svante/papers/sjN14.pdf
5. Kötzing, T., Sudholt, D., Theile, M.: How crossover helps in pseudo-Boolean optimization. In: GECCO, pp. 989–996 (2011)
6. Lehre, P.K., Witt, C.: General drift analysis with tail bounds. arXiv:1307.2559 [cs.NE] (2013)
7. Lehre, P.K., Yao, X.: Crossover can be constructive when computing unique input-output sequences. Soft Comput. **15**(9), 1675–1687 (2011)
8. Oliveto, P.S., Witt, C.: Simplified drift analysis for proving lower bounds in evolutionary computation. Algorithmica **59**(3), 369–386 (2011)
9. Oliveto, P.S., Witt, C.: Erratum: simplified drift analysis for proving lower bounds in evolutionary computation. arXiv:1211.7184 [cs.NE] (2012)

10. Sudholt, D.: Crossover is provably essential for the Ising model on trees. In: GECCO, pp. 1161–1167 (2005)
11. Sudholt, D.: Crossover speeds up building-block assembly. In: GECCO, pp. 689–696 (2012)
12. Witt, C.: Runtime analysis of the $(\mu + 1)$ EA on simple Pseudo-Boolean functions. Evol. Comput. **14**, 65–86 (2006)
13. Yao, X.: Evolving artificial neural networks. Proc. IEEE **87**(9), 1423–1447 (1999)

Provably Optimal Self-adjusting Step Sizes for Multi-valued Decision Variables

Benjamin Doerr[1], Carola Doerr[2(✉)], and Timo Kötzing[3]

[1] École Polytechnique, Palaiseau, France
[2] CNRS and Sorbonne Universités, UPMC Univ Paris 06, LIP6, Paris, France
carola.doerr@lip6.fr
[3] Hasso-Plattner-Institut, Potsdam, Germany

Abstract. We regard the problem of maximizing a OneMax-like function defined over an alphabet of size r. In previous work [GECCO 2016] we have investigated how three different mutation operators influence the performance of Randomized Local Search (RLS) and the (1+1) Evolutionary Algorithm. This work revealed that among these natural mutation operators none is superior to the other two for any choice of r. We have also given in [GECCO 2016] some indication that the best achievable run time for large r is $\Theta(n \log r(\log n + \log r))$, regardless of how the mutation operator is chosen, as long as it is a *static choice* (i.e., the distribution used for variation of the current individual does not change over time).

Here in this work we show that we can achieve a better performance if we allow for *adaptive* mutation operators. More precisely, we analyze the performance of RLS using a *self-adjusting mutation strength*. In this algorithm the size of the steps taken in each iteration depends on the success of previous iterations. That is, the mutation strength is increased after a successful iteration and it is decreased otherwise. We show that this idea yields an expected optimization time of $\Theta(n(\log n + \log r))$, which is optimal among all comparison-based search heuristics. This is the first time that self-adjusting parameter choices are shown to outperform static choices on a discrete multi-valued optimization problem.

Keywords: Run time analysis · Adaptive parameter choices · Mutation · Theory

1 Introduction

We combine in this work two ideas that came up quite recently in the theory of randomized search heuristics for the optimization of discrete problems: the study of multi-valued functions $f : \{0, 1, \ldots, r-1\}^n \to \mathbb{R}$ and an adaptive choice of the parameters. For the multi-valued generalization of OneMax-type functions we present a variant of Randomized Local Search (RLS) that chooses its step sizes in a self-adjusting manner. We prove that this algorithm is optimal among all comparison-based black-box optimizers. Even more, its expected optimization

J. Handl et al. (Eds.): PPSN XIV 2016, LNCS 9921, pp. 782–791, 2016.
DOI: 10.1007/978-3-319-45823-6_73

time is strictly smaller than that of any comparison-based search heuristic using static parameter choices. After the work presented in [5] this is only the second time that a self-adjusting parameter setting is proven to outperform any static choice for a discrete optimization problem, and it is the first time that this is shown for a problem over multiple decision variables.

Some background information and references for the concepts used in this work follow.

1.1 Optimization of Multi-valued OneMax Functions

Most research in discrete evolutionary computation theory regards problems that are defined over the n-dimensional Hamming cube $\{0,1\}^n$, while many experimental results exist also for other discrete search spaces (see, for example, [20] and the references therein for early examples). Only few theoretical works exist that study the extension of evolutionary algorithms and other randomized search heuristics to more general domains Ω, cf. [7] for a discussion. In [7] we considered the minimization of r-valued OneMax-type functions f_z assigning to each string $x \in \{0,1,\ldots,r-1\}^n$ the sum $\sum_{i=1}^{n} d(x_i, z_i)$ of the component-wise distances to a fixed unknown string $z \in \{0,1,\ldots,r-1\}^n$ (cf. Sect. 2 for detailed definitions).

We have analyzed in [7] three different ways to extend RLS and the $(1+1)$ Evolutionary Algorithm (EA) to black-box optimizers for r-valued functions. All three versions maintain the property that for RLS in each iteration the entry of exactly one position $i \in \{1,\ldots,n\}$ is changed, while for the (1+1) EA an independent coin flip with success probability $1/n$ decides whether or not the entry of the i-th position is subject to change. The three variants thus differ in how entries selected for modification are updated. The uniform step operator replaces an entry by different one chosen uniform at random, while the ± 1 step operator adds or subtracts 1 from the current entry. For large r the operator with the best performance on the r-valued generalizations of OneMax is the Harmonic one which adds or subtracts to the current entry a number $j \leq r$ that is chosen with probability proportional to $1/j$. Its expected optimization time on r-valued OneMax is $\Theta(n \log r(\log n + \log r))$.

A natural question to ask is whether a better performance with respect to r can be achieved. However, from [4] we know that no *static* distribution of step sizes can achieve a better run time than $\Omega((\log r)^2)$ for $n = 1$ (see [7] for a discussion).

1.2 RLS with Self-adjusting Step Sizes

In this work we show that a better dependence on r can be achieved if we allow the step operator to change over time. More precisely, we regard the algorithm $\text{RLS}_{a,b}$ which works as follows. A current search point $x \in \{0,1,\ldots,r-1\}^n$ is maintained, along with a real-valued velocity vector $v \in [1, r/4]^n$ denoting the *step size* in each dimension. In each iteration, one dimension $i \leq n$ is chosen uniformly at random for variation; with probability $1/2$ the value x_i is increased by $\lfloor v_i \rfloor$, otherwise decreased by $\lfloor v_i \rfloor$, all other dimension remain as in x. If this new search point has better fitness, the old search point is discarded and the step size v_i is increased to av_i (for some constant $a > 1$). If the new search point has worse fitness, it is discarded and the velocity v_i is decreased to bv_i (for a positive constant $b < 1$). See Algorithm 1 in Sect. 2.2 for details.

We show that, for suitable constants a and b, the expected optimization time of $\text{RLS}_{a,b}$ on the set of r-valued OneMax functions is $O(n(\log n + \log r))$, thus gaining a factor of at least $\log r$ over any RLS variant using static step sizes. This bound is provably optimal among all comparison-based algorithms. That is, no black-box algorithm can achieve a better performance on r-valued OneMax functions unless it explicitly exploits absolute fitness-values.

1.3 Self-adjusting Parameter Choices

One easily observes that in continuous optimization static parameter choices are not very meaningful. This is why for such problems several examples exist where adaptive parameter choices are well understood also from a theoretical perspective (for example, the works [2,14,15] analyze the convergence rates of different evolution strategies). As has been noted in [7], however, such results are difficult to compare to performance guarantees in discrete optimization let alone being transferable to such problems. This is mostly due to the fact that in discrete optimization we do not study the speed of convergence but the time needed to hit an optimal solution. But even if one studies continuous optimization with an a-priori fixed target precision (see [16] and the references therein), then typically the norms used to evaluate a solution differ from the typically regarded 1-norm used in discrete optimization.

For the discrete domain, several empirical works exist that suggest an advantage of adaptive parameter updates (cf. [12], [13, Chap. 8], and [17] for surveys). However, the first work formally showing an asymptotic gain over static parameter selection is the self-adjusting choice of the population size of the $(1+(\lambda,\lambda))$ GA proposed and analyzed in [5]. In that work the advantage is shown for the classic OneMax functions $f_z : \{0,1\}^n \to \mathbb{R}, x \mapsto |\{1 \leq i \leq n \mid x_i = z_i\}|$. Our result is hence the first of its type for a multi-valued search problem in the discrete domain.

Also when we include in our consideration other adaptive parameter choices[1] only few situations exist for which an advantage over static parameter choices could be proven. All these works study the optimization of pseudo-Boolean functions $f : \{0,1\}^n \to \mathbb{R}$. To be more precise, the only theoretical investigations of adaptive parameter choices in discrete optimization that we are aware of

[1] Following the terminology introduced in [5, Sect. 3.1] we distinguish between functionally-dependent and self-adjusting parameter choices. While *functionally-dependent* parameter choices depend only on the current state of the algorithm, they may explicitly use absolute fitness values. Fitness-dependent mutation rates are a typical example for such *functionally-dependent* parameter choices. *Self-adjusting* parameter choices, in contrast, do not depend on absolute fitness information but rather on the success of previous iterations. This is the case of the parameter updates of the $\text{RLS}_{a,b}$ considered in this work.

analyze advantages of a fitness-dependent mutation rate for the (1+1) EA optimizing LeadingOnes [3] and for RLS optimizing OneMax [8], a self-adjusting choice of the number of parallel evaluations in a parallel EA [19] as well as a fitness-dependent [6] and the above-mentioned self-adjusting [5] choice of the population size for the $(1 + (\lambda, \lambda))$ GA.

We believe that self-adjusting parameter choices provide a possibility for significant improvement of many search heuristics, and theoretical analyses can offer guidance for how to design such self-adjustment mechanisms. Our work shows that our mathematical toolbox, in particular drift analysis, is well-suited to analyze such systems.

2 Preliminaries

For any positive integer n we set $[n] := \{1, 2, \ldots, n\}$ and $[0..n] := \{0\} \cup [n]$. We regard in this work r-valued functions over strings of length n, i.e., functions $f : [0..r-1]^n \to \mathbb{R}$. The value of r may or may not depend on n, and it may or may not be smaller or larger than n.

We briefly define below the problem setting and the self-adjusting version of RLS that we aim at analyzing.

2.1 Multi-valued OneMax Problems

As in [7] we regard two classes of r-valued OneMax functions. These classes are the collection of functions $f_z : [0..r-1]^n \to \mathbb{R}; x \mapsto \sum_{i=1}^{n} d(x_i, z_i)$, $z \in [0..r-1]^n$. They differ in the metric d used to evaluate the *distance* of x_i to z_i. The first metric, which we call the *interval-metric* d_{int}, is the usual metric on the integers, i.e.,

$$d_{\text{int}}(a, b) := |b - a|.$$

Note that in the interval-metric the fitness landscapes of the r-valued OneMax functions are not isomorphic to each other. This can be easily seen, for example, by comparing $f_{(0,\ldots,0)}$ with $f_{(r/2,\ldots,r/2)}$ which has a much more symmetric fitness landscape. Note that, for the boundary handling we employ in this paper (see Sect. 2.2), our results are unaffected by the exact choice of r-valued OneMax function. This is why we also consider a second metric, which we call the *ring-metric* d_{ring}. This metric connects the two endpoints of the interval $[0..r-1]$ such that it forms a ring, i.e.,

$$d_{\text{ring}}(a, b) := \min\{|b - a|, |b - a + r|, |b - a - r|\}.$$

Unlike the name OneMax suggests, we regard in this work the *minimization* of the r-valued OneMax functions. It is easily seen that, regardless of the metric in place, the unique global optimum of f_z is thus the string z. We call z the *target vector* of f_z.

2.2 RLS with Self-adjusting Mutation Strength

We investigate the following natural generalization of RLS to a multi-valued algorithm $\text{RLS}_{a,b}$ with a self-adjusting mutation strength whose update rules are parametrized by the constants $1 < a \leq 2$ and $1/2 < b < 1$. The algorithm is summarized in Algorithm 1.

Algorithm 1. $\text{RLS}_{a,b}$ with self-adjusting step sizes maximizing a function $f : [0..r-1]^n \to \mathbb{R}$

1 **Initialization:** Let $v \in [1, \lfloor r/4 \rfloor]^n$ uniformly at random;
2 Sample $x \in [0..r-1]^n$ uniformly at random and query $f(x)$;
3 **Optimization: for** $t = 1, 2, 3, \ldots$ **do**
4 Choose $i \in [n]$ uniformly at random;
5 **for** $j = 1, \ldots, n$ **do**
6 **if** $j = i$ **then** with probability 1/2 let $y_j \leftarrow x_j - \lfloor v_j \rfloor$ and let $y_j \leftarrow x_j + \lfloor v_j \rfloor$ otherwise
7 **else** $y_j \leftarrow x_j$
8 Query $f(y)$;
9 **if** $f(y) < f(x)$ **then** $v_i \leftarrow \min\{av_i, \lfloor r/4 \rfloor\}$ **else** $v_i \leftarrow \max\{1, bv_i\}$
10 **if** $f(y) \leq f(x)$ **then** $x \leftarrow y$

$\text{RLS}_{a,b}$ maintains a search point $x \in [0..r-1]^n$ as well as a real-valued *velocity vector* $v \in [1, \lfloor r/4 \rfloor]^n$; we use real values for the velocity to circumvent rounding problems. Both these strings are initialized uniformly at random, but it is not difficult to verify that all results shown in this paper apply to any arbitrary initialization of x and v. In one iteration of the algorithm a position $i \in [n]$ is chosen uniformly at random. The entry x_i is replaced by $x_i - \lfloor v_i \rfloor$ with probability 1/2 and by $x_i + \lfloor v_i \rfloor$ otherwise (see below for how to deal with overstepping the endpoints of the interval $[0, r-1]$). The entries in positions $j \neq i$ are not subject to mutation. The resulting string y replaces x if its fitness is at least as good as the one of x, i.e., if $f(y) \leq f(x)$ holds (recall that we regard the minimization of f). If the offspring y is strictly better than its parent x, i.e., if $f(y) < f(x)$, we increase the velocity v_i in the i-th component by multiplying it with the constant a and we decrease v_i to bv_i otherwise. The algorithm proceeds this way until we decide to stop it. Since we regard in this work the time needed until $\text{RLS}_{a,b}$ evaluates for the first time an optimal solution (this random variable is called the *run time* of Algorithm 1), we do not specify any stopping criterion here.

We will now discuss some technical details.

It may happen that $x_i - \lfloor v_i \rfloor < 0$ or $x_i + \lfloor v_i \rfloor > r - 1$. If we are working with the interval-metric then we assume that the algorithm does not change its current position, that is, the offspring is discarded and the velocity is not adjusted (decreasing the velocity in this case would lead to the same results). In the ring-metric we identify all values *modulo* r, i.e., we identify values $p < 0$ with $p + r$ and values $p > r - 1$ with $p - r$. Note that in the ring-metric it can happen that we decrease the fitness regardless of whether we add or subtract from x_i the value $\lfloor v_i \rfloor$. This in particular applies when x_i is close to $r/2$.

Furthermore, we emphasize that the velocity vector is an element in the real interval $[1, \lfloor r/4 \rfloor]$, that is, it does not necessarily take integer values. This technicality avoids that rounding inaccuracies accumulate over several velocity adaptations. The velocity is capped at 1 (to avoid situations in which we do not move at all) and at $\lfloor r/4 \rfloor$ (to avoid too large jumps).

To further lighten the notation, we say that the algorithm *"moves in the right direction"* or *"towards the target value"* if the distance to the target is actually decreased by $\lfloor v_i \rfloor$. Analogously, we speak otherwise of a step *"away from the target"* or *"in the wrong direction"*.

2.3 Drift Analysis

The idea of drift analysis is to map the optimization process to a series of real-valued random variables that measure, in a suitable way, the expected progress that the algorithm achieves in one iteration. The hope is to show that this expected progress systematically depends on the current state of the algorithm, for example, in an additive or a multiplicative way. Drift theorems then help to convert the expected progress made in one iteration to bounds on the time needed to hit a certain goal such as identifying an optimal search point; cf. [10, 18] for a more detailed discussion of drift theory.

In the context of $\text{RLS}_{a,b}$ the state of the algorithm can be described by the pair (x, v) consisting of the current search point $x \in [0..r-1]^n$ and the current velocity vector $v \in [1, \lfloor r/4 \rfloor]^n$. We will design in Sect. 3 a potential function g that maps these states to real numbers in a way that the expected progress of one iteration of $\text{RLS}_{a,b}$ depends on the current potential $g(x, v)$ in a multiplicative way. That is, for y and v' denoting the resulting search point and velocity vector after one iteration of $\text{RLS}_{a,b}$, we will show that $E(g(x,v) - g(y,v')) \geq \delta g(x,v)$ for some positive constant δ. The following drift theorem will then allow us to derive bounds on the expected run time of $\text{RLS}_{a,b}$ on any r-valued OneMax function. This *multiplicative drift theorem* had first been introduced to the theory of randomized search heuristics in [10]. A more direct proof of this results, that also gives large deviation bounds, can be found in [9]. The variables $X^{(t)}$ in the statement correspond to the state $g(x, v)$ of the algorithm after t iterations.

Theorem 1 (from [10]). *Let $X^{(0)}, X^{(1)}, \ldots$ be a random process taking values in $S := \{0\} \cup [s_{\min}, \infty) \subseteq \mathbb{R}$. Assume that $X^{(0)} = s_0$ with probability one. Assume that there is a $\delta > 0$ such that for all $t \geq 0$ and all $s \in S$ with $\Pr[X^{(t)} = s] > 0$ we have $E[X^{(t+1)} | X^{(t)} = s] \leq (1-\delta)s$. Then $T := \min\{t \geq 0 \mid X^{(t)} = 0\}$ satisfies $E[T] \leq \frac{\ln(s_0/s_{\min})+1}{\delta}$.*

3 Main Result

In this section we sketch the proof of the following statement (the full proof does not fit the available space).

Theorem 2. *For constants a, b satisfying $1 < a \leq 2$, $1/2 < b \leq 0.9$, $2ab - b - a > 0$, $a + b > 2$, and $a^2 b > 1$ (one can choose, for example, $a = 1.7$ and $b = 0.9$) the expected run time of $RLS_{a,b}$ (Algorithm 1) on any generalized r-valued* OneMax *function is $\Theta(n(\log n + \log r))$ and this is optimal among all comparison-based algorithms.*

The lower bound as well as the statement that no comparison-based algorithm can have an expected run time of smaller order easily follows from a coupon collector argument and the information-theoretic lower bound. In a bit more detail, we note that in the initial solution there are, with high probability, $\Theta(n)$ positions i in which the value x_i does not agree with that of the target string. The algorithm has to touch each of these positions at least once, which by the well-known coupon collector theorem (cf. [1, Sect. 1] for an introduction to this problem) requires $\Theta(n \log n)$ iterations on average and with high probability. The $\Omega(n \log r)$ follows from the observation that there are r^n possible target strings in total. Since $\text{RLS}_{a,b}$ exploits only the information whether or not the offspring has a fitness value that is at least as good as that of its parent (in the decision of whether or not to replace the parent) and whether or not its fitness is strictly better (in the decision how to update the velocity), it is a *comparison-based algorithm* that uses only $\log_2(3)$ bits of information per iteration. As such it therefore needs $\Omega(\log(r^n)) = \Omega(n \log r)$ iterations in expectation to optimize any unknown r-valued OneMax function. See [11] for how to turn the latter information-theoretic consideration into a formal proof.

To prove the upper bound we use *drift analysis*; multiplicative drift analysis to be more precise. To this end, as explained in Sect. 2.3, we need to find a mapping of the state (x, v) of the algorithm to a real value. This *potential function* should measure some sort of distance to the target state. We briefly discuss this potential function below. Proving that it yields the required multiplicative drift is the purpose of Lemma 3.

To simplify the notation below, for a given search point x and the target bit string z and the chosen metric d, we let $d_i = d(x_i, z_i)$ (for all $i \leq n$) be the distance vector of x to z. Thus, the goal is to reach a state in which the distance vector is $(0, \ldots, 0)$. We now want to define a potential function in dependence on (d, v) (where of course d is dependent on x) such that it is 0 when d is $(0, \ldots, 0)$ and strictly positive for any $x \neq (0, \ldots, 0)$. Furthermore, we easily see that there are two important ways to make progress, either by advancing in terms of fitness or by adjusting the velocity to a value that is more suitable to make progress in future iterations. This has to be reflected in the potential function. Our ultimate goal being the minimization of fitness, it is not difficult to see that some preference should be given to a progress in fitness. This can be achieved by multiplying the term accounting for the appropriateness of the velocity with some constant $c < 1$. We measure the appropriateness of the velocity as the maximum of the ratios $d_i/(2v_i)$ and $2v_i/d_i$, reflecting the fact that a velocity of $d_i/2$ is very well-suited for progress; smaller values give less progress, while larger values lead to a badly adjusted velocity in the next iteration (and very large values make progress in fitness impossible).

One problem in getting good drift is that velocities v_i just below $2d_i$ allow for jumping over the target while increasing the (already too large) velocity. We get around this problem by observing that it is equally likely that the large velocity is reduced because of a jump in the wrong direction, and then, while still larger than d_i, will still give a good improvement when overstepping the goal. We reflect this in the potential function by giving a penalty term of pd_i (for some suitable constant p) on any state (d, v) having a too large velocity.

To sum up this discussion we use as potential function the following map $g : [0..r-1]^n \times [1, \lfloor r/4 \rfloor]^n \to \mathbb{R}, (x, v) \mapsto \sum_{i=1}^{n} g_i(d_i, v_i)$ where $g_i(d_i, v_i) := 0$ for $d_i = 0$ and for $d_i \geq 1$

$$g_i(d_i, v_i) := d_i + \begin{cases} cd_i \max\{2v_i/d_i, d_i/(2v_i)\}, & \text{if } v_i \leq 2bd_i; \\ cd_i \max\{2v_i/d_i, d_i/(2v_i)\} + pd_i, & \text{otherwise} \end{cases} \tag{1}$$

and c, p are (small) constants specified below.

Summarizing all the conditions needed below, we require that the constants a, b, c, p satisfy $1 < a \leq 2$, $1/2 < b \leq 0.9$, $2ab - b - a > 0$, $a + b > 2$, $a^2 b > 1$, $8abc + 2p + 4c/b \leq 1/16$, $p > 8c\left(\frac{a+b}{2} - 1\right)$, and $p > 4(a-1)c > 0$.

We can thus choose, for example, $a = 1.7$, $b = 0.9$, $p = 0.01$, and $c = 0.001$.

The following lemma, together with the observation that the initial potential is of order at most nr^2 plugged into the multiplicative drift theorem (Theorem 1) proves the desired overall expected run time of $O(n \log(nr))$.

Lemma 3. *Let $d \neq (0, \ldots, 0)$ and $v \in [1, \lfloor r/4 \rfloor]^n$. Let (d', v') be the state of Algorithm 1 started in (d, v) after one iteration (i.e., after a possible update of x and v). The expected difference in potential satisfies*

$$E\left(g(d, v) - g(d', v') \mid d, v\right) \geq \frac{\delta}{n} g(d, v)$$

for some positive constant δ.

4 Conclusions

While in [7] we analyzed static mutation operators for optimizing multi-valued functions $f : [0..r-1]^n \to \mathbb{R}$, in this paper we gave an operator based on self-adjusting step sizes. We proved that, in the case of RLS, this leads to a provably optimal run time for r-valued OneMax functions.

Already for the analysis of RLS we gave an intricate drift-argument, with many different cases to consider and penalty terms for resolving situations which would otherwise allow for search points with negative drift. Extending our results to the case of the (1+1) EA might thus be a very challenging task, pushing the limits of drift theory.

Note that we chose a specific step size adaptation scheme which guarantees optimal run time. It would also be interesting to investigate other adaptation schemes. For example, the step size, in each iteration, could be drawn from a distribution (just as in one of the operators presented in [7]), and the parameters of this distribution are adapted.

Another issue with step sizes is that infeasible areas of the search space might be reached (in our setting this can happen if we use the interval metric). The issue of boundary handling is a known problem, and our boundary handling technique is by no means the only way for dealing with it. We believe that our choice is natural and leads to a "fair" treatment of all parts of the search space, and it leads to an optimal run time for our setting. It might be interesting to see whether there are other settings where a different boundary handling is more natural, or gives better run time.

Acknowledgments. This research benefited from the support of the "FMJH Program Gaspard Monge in optimization and operation research", and from the support to this program from EDF (Électricité de France).

References

1. Auger, A., Doerr, B.: Theory of Randomized Search Heuristics. World Scientific, Singapore (2011)
2. Auger, A., Hansen, N.: Linear convergence on positively homogeneous functions of a comparison based step-size adaptive randomized search: the (1+1) ES with generalized one-fifth success rule. CoRR, abs/1310.8397 (2013). http://arxiv.org/abs/1310.8397
3. Böttcher, S., Doerr, B., Neumann, F.: Optimal fixed and adaptive mutation rates for the leadingones problem. In: Schaefer, R., Cotta, C., Kołodziej, J., Rudolph, G. (eds.) PPSN XI. LNCS, vol. 6238, pp. 1–10. Springer, Heidelberg (2010)
4. Dietzfelbinger, M., Rowe, J.E., Wegener, I., Woelfel, P.: Tight bounds for blind search on the integers and the reals. Comb. Probab. Comput. **19**, 711–728 (2010)
5. Doerr, B., Doerr, C.: Optimal parameter choices through self-adjustment: applying the 1/5-th rule in discrete settings. In: Proceedings of the ACM Genetic and Evolutionary Computation Conference (GECCO 2015), pp. 1335–1342. ACM (2015)
6. Doerr, B., Doerr, C., Ebel, F.: From black-box complexity to designing newgenetic algorithms. Theor. Comput. Sci. **567**, 87–104 (2015)
7. Doerr, B., Doerr, C., Kötzing, T.: The right mutation strength for multi-valued decision variables. In: Proceedings of the ACM Genetic and Evolutionary Computation Conference (GECCO 2016). ACM (2016, to appear). http://arxiv.org/abs/1604.03277
8. Doerr, B., Doerr, C., Yang, J.: Optimal parameter choices via precise black-box analysis. In: Proceedings of the ACM Genetic and Evolutionary Computation Conference (GECCO 2016). ACM (2016, to appear)
9. Doerr, B., Goldberg, L.A.: Adaptive drift analysis. Algorithmica **65**, 224–250 (2013)
10. Doerr, B., Johannsen, D., Winzen, C.: Multiplicative drift analysis. Algorithmica **64**, 673–697 (2012)
11. Droste, S., Jansen, T., Wegener, I.: Upper and lower bounds for randomized search heuristics in black-box optimization. Theor. Comput. Syst. **39**, 525–544 (2006)
12. Eiben, A.E., Hinterding, R., Michalewicz, Z.: Parameter control in evolutionary. IEEE Trans. Evol. Comput. **3**, 124–141 (1999)

13. Eiben, A.E., Smith, J.E.: Introduction to Evolutionary Computing. Springer, Heidelberg (2003)
14. Hansen, N., Gawelczyk, A., Ostermeier, A.: Sizing the population with respect to the local progress in (1,λ)-evolution strategies - a theoretical analysis. In: Proceedings of the IEEE Congress on Evolutionary Computation (CEC 1995), pp. 80–85. IEEE (1995)
15. Jägersküpper, J.: Rigorous runtime analysis of the (1+1) ES: 1/5-rule and ellipsoidal fitness landscapes. In: Wright, A.H., Vose, M.D., De Jong, K.A., Schmitt, L.M. (eds.) FOGA 2005. LNCS, vol. 3469, pp. 260–281. Springer, Heidelberg (2005)
16. Jägersküpper, J.: Oblivious randomized direct search for real-parameter optimization. In: Halperin, D., Mehlhorn, K. (eds.) ESA 2008. LNCS, vol. 5193, pp. 553–564. Springer, Heidelberg (2008)
17. Karafotias, G., Hoogendoorn, M., Eiben, A.: Parameter control in evolutionary algorithms: trends and challenges. IEEE Trans. Evol. Comput. **19**, 167–187 (2015)
18. Kötzing, T.: Concentration of first hitting times under additive drift. Algorithmica **75**, 490–506 (2016)
19. Lässig, J., Sudholt, D.: Adaptive population models for offspring populations and parallel evolutionary algorithms. In: Proceedings of the ACM Workshop on Foundations of Genetic Algorithms (FOGA 2011), pp. 181–192. ACM (2011)
20. Rudolph, G.: An evolutionary algorithm for integer programming. In: Davidor, Y., Schwefel, H.-P., Mönner, R. (eds.) (PPSN 1994). LNCS, pp. 139–148. Springer, Heidelberg (1994)

Example Landscapes to Support Analysis of Multimodal Optimisation

Thomas Jansen[1] and Christine Zarges[2(✉)]

[1] Department of Computer Science,
Aberystwyth University, Aberystwyth SY23 3DB, UK
t.jansen@aber.ac.uk

[2] School of Computer Science, University of Birmingham,
Birmingham B15 2TT, UK
c.zarges@cs.bham.ac.uk

Abstract. Theoretical analysis of all kinds of randomised search heuristics has been and keeps being supported and facilitated by the use of simple example functions. Such functions help us understand the working principles of complicated heuristics. If the function represents some properties of practical problem landscapes these results become practically relevant. While this has been very successful in the past for optimisation in unimodal landscapes there is a need for generally accepted useful simple example functions for situations where unimodal objective functions are insufficient: multimodal optimisation and investigation of diversity preserving mechanisms are examples. A family of example landscapes is defined that comes with a limited number of parameters that allow to control important features of the landscape while all being still simple in some sense. Different expressions of these landscapes are presented and fundamental properties are explored.

1 Introduction

Most real-world optimisation problems do not have a single best solution but many locally or globally optimal ones. The field of multimodal optimisation deals with tackling such problems and nature-inspired techniques have proven to be very popular and powerful to tackle these types of problems [17].

Over the last decade a rich set of benchmarks for the systematic and sound comparison of different optimisation methods has been developed[1]. Many problems in these benchmarks are multimodal. However, they are usually restricted to real-parameter optimisation problems and not accessible to theoretical analysis.

The authors want to thank the organisers of the Dagstuhl Seminar 15211 'Theory of Evolutionary Algorithms' for encouraging discussions that motivated this work. This article is based upon work from COST Action CA15140 'Improving Applicability of Nature-Inspired Optimisation by Joining Theory and Practice (ImAppNIO)' supported by COST (European Cooperation in Science and Technology).

[1] see, e.g., www.epitropakis.co.uk/cec16-niching/competition and coco.gforge.inria.fr.

J. Handl et al. (Eds.): PPSN XIV 2016, LNCS 9921, pp. 792–802, 2016.
DOI: 10.1007/978-3-319-45823-6_74

There has been some debate on appropriate optimisation goals in multimodal optimisation [2]. On one hand, one could be interested in the global perspective of locating a single (local or global) optimum. On the other hand, practitioners are often aiming at a multi-local perspective, i.e., they want to identify a multitude of different optima, either in a simultaneous or sequential fashion[2]. When considering such a multi-local perspective, *niching* techniques [19] are very common, i.e., techniques that prevent the algorithm from converging to a single solution and thus, enable it to explore multiple peaks of the search space in parallel. Some previous theoretical work consider Ising model problems [5,21] or simple bi-modal example function [6,16] exist. However, no common set of benchmarks suitable for theoretical analysis is available to date.

This lack of suitable benchmark functions is a serious impediment for the development of a theory of multimodal optimisation. In the area of classical optimisation where one is 'only' interested in finding an optimal search point simple example functions have been at the heart of the development of a powerful and useful theoretical framework and a multitude of strong theoretical results. Consider for example the well-known example function OneMax, used as early as 1992 to derive run time results for a simple evolutionary algorithm [15]. It has given rise to a natural generalisation, the class of linear functions [4] which in turn has motivated the introduction of a powerful proof technique: drift analysis [7]. And still today it is the function to consider when introducing novel perspectives [9] or expanding the horizon of theoretical analysis [12]. Clearly, OneMax is not the only useful and important example function but it is one of a relatively small number of example functions, most of which are unimodal (see [8] for a broad overview). Multimodal example functions are rarely considered–one noteworthy exception being TwoMax, a simple bi-model problem that can be seen as the maximum of OneMax and ZeroMax [6].

We address this need by introducing a family of landscapes with a limited number of parameters. We want to allow for the control of important features of problems that are simple enough for theoretical analysis. We explore properties of these 'theoretical' landscapes in the spirit of fitness landscape analysis that usually considers landscapes underlying real-world problems such as satisfiability [18] or are inspired by biology [20].

It is important to note that there has been some debate on appropriate example functions and optimisation goals in multimodal optimisation [2]. While the research in this paper is inspired by this discussion it goes beyond the initial ideas presented in [2] by introducing three different ways of implementation. One might want to argue that our example functions are inspired by and a generalisation of TwoMax, similar to linear functions being inspired by and a generalisation of OneMax. We think that the set of example functions presented here is a richer and more interesting generalisation. It bears resemblance with 'older' problem classes (e. g., [10,11]) but allows for more control. It is similar to the moving peaks benchmark [1] but it is static, of course.

[2] see, e.g., www.epitropakis.co.uk/ppsn2016-niching.

In the next section we present our main ideas behind our example functions. We describe the properties of some interesting landscapes in Sect. 3. We hint at the richness of the different example functions in Sect. 4.

2 Defining Landscapes and Objective Functions

We define our example functions based on an abstract idea of a landscape. It is important to note that we use landscape in a general, colloquial sense that does not coincide with the technical meaning of a fitness landscape as something that is defined by a neighbourhood graph and function values. We will be considering this latter kind of landscape (calling them *fitness landscape* to emphasise the difference) when we have defined objective functions.

We fix the set of bit strings of length n (equivalently, the Boolean hypercube of dimension n) as our search space. This is a complex, high-dimensional search space. Nevertheless, we *think* of it as a flat landscape where we introduce *peaks* that are defined by their *position*, their *slope* and their *height* (where we will give the height in an indirect way). The objective of an optimisation algorithm operating in this landscape is to identify peaks: a highest peak in exact optimisation, a collection of peaks in multimodal optimisation.

The kind of search heuristics we consider usually conduct search by modifying one or several bit strings they have explored already to get to another, yet unexplored bit string. The modifications tend to change only a limited number of bits and, therefore, it makes sense to use the Hamming distance between two bit strings as metric in our search space. The Hamming distance of x and y, $\mathrm{H}(x, y)$, equals the number of bits that have different values in x and y. Clearly, it is a value between 0 and n. If x is a point in our landscape we currently have and y is a point we want to reach then $\mathrm{H}(x, y) = 0$ indicates that we have reached the target point y. Since we will be considering maximisation it is more convenient to consider $n - \mathrm{H}(x, y)$ instead.

Definition 1. *For* $x, y \in \{0,1\}^n$ *let* $\mathrm{H}(x,y) := \sum_{i=0}^{n-1} |x[i] - y[i]|$ *denote the Hamming distance of* x *and* y. *We also define* $\mathrm{G}(x,y) := n - \mathrm{H}(x,y)$.

We now introduce our notion of a landscape that is defined by some number of peaks with their parameters that are introduced to the search space. We want to find these peaks and therefore consider the distance to a nearest peak.

Definition 2. *A landscape is defined by the number of peaks* $k \in \mathbb{N}$ *and the definition of the* k *peaks (numbered* $1, 2, \ldots, k$*) where the* i*-th peak is defined by its position* $p_i \in \{0,1\}^n$*, its slope* $a_i \in \mathbb{R}^+$*, and its offset* $b_i \in \mathbb{R}_0^+$.

For a search point $x \in \{0,1\}^n$ *we define its closest peak (given by its index* i*) as* $cp(x) := \operatorname*{arg\,min}_{i \in \{1,2,\ldots,k\}} \mathrm{H}(x, p_i)$. *In cases where there are multiple* i *that minimise* $\mathrm{H}(x, p_i)$ *we define as tie breaking rule that* i *should additionally maximise* $a_i \cdot \mathrm{G}(x, p_i) + b_i$. *If this is still not unique an arbitrary* i *that minimises* $\mathrm{H}(x, p_i)$ *and among those maximises* $a_i \cdot \mathrm{G}(x, p_i) + b_i$ *can be selected.*

The tie breaking rule we introduce is tailored towards the way we calculate fitness (which we define in Definition 3). Since we are interested in finding peaks it makes sense to concentrate on a higher one if there are multiple nearest peaks. Since we only care about distance and height we do not care about any tertiary criterion.

The general idea of our landscape is that the fitness value of a search point depends on peaks in its vicinity. For the sake of clarification, let us consider the situation for a landscape with only a single peak, i. e., $k = 1$ and the parameters of the peak are p_1, a_1, b_1. The fitness of $x \in \{0,1\}^n$ is given as $a_1 \cdot \mathrm{G}(x, p_1) + b_1$. We see that the peak itself has fitness $a_1 \cdot \mathrm{G}(p_1, p_1) + b_1 = a_1 \cdot n + b_1$. We call $a_1 n + b_1$ the height of the peak p_1.

It remains to be determined how we deal with multiple peaks. There are different ways this can be handled and there is no correct or incorrect way of doing it. It depends on what you want to achieve. We consider three different options and briefly discuss what we have in mind for the different versions.

Definition 3. *Let $k \in \mathbb{N}$ and k peaks $(p_1, a_1, b_1), (p_2, a_2, b_2), \ldots, (p_k, a_k, b_k)$ be given. We define the following three objective functions (also called fitness functions).*

- $f_1(x) := a_{cp(x)} \cdot \mathrm{G}\big(x, p_{cp(x)}\big) + b_{cp(x)}$, *called the nearest peak function*
- $f_2(x) := \max\limits_{i \in \{1,2,\ldots,k\}} a_i \cdot \mathrm{G}(x, p_i) + b_i$, *called the weighted nearest peak function*
- $f_3(x) := \sum\limits_{i \in \{1,2,\ldots,k\}} a_i \cdot \mathrm{G}(x, p_i) + b_i$, *called the all peaks function*

The nearest peak function, f_1, has the fitness of a search point x determined by the closest peak. The fitness is given as discussed above, $a_i \cdot \mathrm{G}(x, p_i) + b_i$, and the peak i that determines the slope a_i and offset b_i is the closest peak, $i = \mathrm{cp}(x)$. It implements a very local point of view where the height of other peaks is ignored even if their height is very much higher and they are only a little farther.

The weighted nearest peak function, f_2, takes the height of peaks into account. It considers $a_i \cdot \mathrm{G}(x, p_i) + b_i$ for all k peaks and uses the peak that yields the largest value to determine the function value. This implies that peaks with bigger height determine the function value in a larger area of the search space in comparison to smaller peaks.

The all peaks function, f_3, takes into account $a_i \cdot \mathrm{G}(x, p_i) + b_i$ for all peaks simultaneously and simply adds them up. Note that $f_3(x)/k$ yields the average influence of all peaks and in this sense we can view f_3 as an 'averaged' fitness landscape. Since many randomised search heuristics are rank-based [3] the difference between $f_3(x)$ and $f_3(k)/k$ is inconsequential.

We use the following visualisation of fitness landscapes resulting from the above definitions: We project the n-dimensional Boolean hypercube onto a 2-dimensional plane and connect direct Hamming neighbours by edges. We use a third dimension for the resulting fitness values, indicated by both height and colour (where blue indicates low fitness and red high fitness). An example for f_1

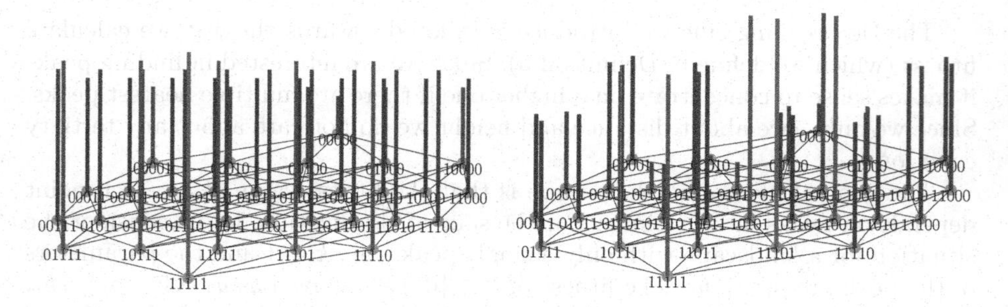

Fig. 1. Visualisation of fitness landscapes: f_1 with $n = 5$, $k = 1$, $p_1 = 1^n$, $a_1 = 1$ and $b_1 = 0$ (left) and f_3 with $p_1 = 11111$, $p_2 = 11001$, $p_3 = 10101$, $a_1 = a_2 = a_3 = 5$ and $b_1 = b_2 = b_3 = 0$ (right) (Color figure online)

with $n = 5$, $k = 1$, $p_1 = 1^n$ (the all-ones bit string), $a_1 = 1$ and $b_1 = 0$, which is identical to the well-known OneMax function, is shown in Fig. 1 (left).

We will discuss the differences between the three different fitness functions in more detail in the next two sections. We will also discuss which properties of the different fitness functions are of particular interest.

3 Properties

All three objective functions yield the same fitness landscapes for $k = 1$. They are all OneMax-like, i.e., p_1 is the single local and global optimum, fitness strictly decreases with increasing Hamming distance to p_1 and all points with equal Hamming distance to p_1 have the same fitness value. Consequently, we restrict ourselves to the more interesting case of $k > 1$.

When analysing fitness landscapes a variety of criteria can be considered (see, e.g., [20] for an overview). In this paper, we are particularly interested in the number of local and global optima and their locations in the search space. We additionally consider the so-called basin of attraction of a local optimum, i.e., the set of search points that are guaranteed to lead to it when using a simple hill-climber such as Random Local Search (RLS, Algorithm 1), and use as a measure for its size the probability that this happens when starting from a search point selected uniformly at random (u.a.r.).

Algorithm 1. Random Local Search (RLS)

1 Choose $x \in \{0,1\}^n$ u.a.r.
2 **repeat**
3 | Create offspring $y := x$. Select $i \in \{0, \ldots, n-1\}$ u.a.r. and flip bit $y[i]$.
4 | **if** $f(y) \geq f(x)$ **then** $x := y$
5 **until** *forever*

Additionally, we are interested in the influence of different parameters of landscapes (see Definition 2). This includes particularly the number of peaks k, their positions and heights (as defined by their slope and offset). Note, the peaks that we use to define a landscape do not necessarily correspond to a local optimum of the resulting fitness landscape (see Sect. 4.2).

4 Results

We provide some first insights into properties of our proposed set of example functions by considering a number of properties that are similar to properties of known example functions and are, we hope, of some general interest. The results in these sections hint at properties of different instantiations of our families of example functions that could be starting point for useful analysis of different randomised search heuristics. We examine a generalisation of the well-known TwoMax function [6] for all three fitness functions from Definition 3 in Sect. 4.1. Section 4.2 is dedicated to the comparison of f_1 and f_2. Looking at randomly distributed peaks we compare the influence of the slope as it manifests itself in f_1 and f_2. Looking at f_3 in Sect. 4.3 we consider an important property by means of a specific configuration of peaks.

4.1 Generalisation of TwoMax

As a starting point, we consider a landscape with two peaks $p_1 = 0^n$ and $p_2 = 1^n$ and see that for f_1 with offsets $b_1 = b_2 = 0$ and slopes $a_1 = a_2 = 1$ this is identical to the well-known bi-modal example function $\text{TwoMax}(x) := \max\{\sum_{i=1}^{n} x[i],\ n - \sum_{i=1}^{n} x[i]\}$. We examine all three fitness functions and different settings for the two offsets and slopes. In the following, let $|x|_1$ denote the number of 1-bits in x and $|x|_0$ the number of 0-bits.

It is easy to see that f_1 has exactly the two local maxima p_1 and p_2. Offsets and slopes influence only the fitness values but not the basins of attractions.

Theorem 1. *Let $p_1 = 0^n$ and $p_2 = 1^n$ with arbitrary $a_1, a_2 \in \mathbb{R}^+$, $b_1, b_2 \in \mathbb{R}_0^+$. The fitness landscape defined by f_1 has exactly two local maxima, p_1 and p_2, with fitness $a_1 \cdot |x|_0 + b_1$ and $a_2 \cdot |x|_1 + b_2$, respectively. RLS reaches p_1 with probability $1/2$ and p_2 otherwise.*

Proof. As discussed in Sect. 2, the fitness is only determined by the closest peak. It follows immediately, that the two peaks are both locally optimal and that each search point is in the basin of attraction of its closest peak. Plugging all parameters into Definition 3 yields the first statement. Let $\mathcal{B}_i$ denote the basin of attraction of p_i. For the second statement we need to prove that the RLS starts in $\mathcal{B}_1$ or $\mathcal{B}_2$ with equal probability. From the above, we see that all x with $|x|_0 > n/2$ are in $\mathcal{B}_1$ while all x with $|x|_1 > n/2$ are in $\mathcal{B}_2$. As both sets of points are of equal size RLS starts in either of them with equal probability. Points with $|x|_1 = |x|_0 = n/2$ have equal distance to p_1 and p_2 and belong to neither basis of attraction. Given such a point x, we know that RLS flips a 1-bit with probability $1/2$ and a 0-bit otherwise. Thus, after one step, we are in one of the two previous cases. □

Things are different for f_2 as larger peaks have influence in a larger area of the search space in comparison to smaller peaks and thus will have a larger basin of attraction. We remark that our choice of p_1 and p_2 implies that two search points with the same number of 0-bits have equal fitness value. Thus, we can derive a bound on $|x|_0$ that determines the boundary of the basins of attractions of p_1 and p_2.

Theorem 2. *Let $p_1 = 0^n$ and $p_2 = 1^n$ with arbitrary $a_1, a_2 \in \mathbb{R}^+$, $b_1, b_2 \in \mathbb{R}_0^+$ and consider the fitness landscape defined by f_2. The basin of attraction of p_1 contains all search points x with $|x|_0 > a_2/(a_1 + a_2) \cdot n + (b_2 - b_1)/(a_1 + a_2)$.*

Proof. According to Definition 3, the fitness of a search point x is determined by p_1 if $a_1 \cdot (n - \mathrm{H}(x, 0^n)) + b_1 > a_2 \cdot (n - \mathrm{H}(x, 1^n)) + b_2$. We see that $|x|_1 = \mathrm{H}(x, 0^n)$ and thus, $|x|_0 = n - \mathrm{H}(x, 0^n)$. Similarly, we have $|x|_1 = n - \mathrm{H}(x, 1^n)$. We get $a_1 \cdot |x|_0 + b_1 > a_2 \cdot (n - |x|_0) + b_2$ which is equivalent to $|x|_0 > a_2/(a_1 + a_2) \cdot n + (b_2 - b_1)/(a_1 + a_2)$ and see that all x with this property are in the basin of attraction of p_1. □

We see that RLS is initialised in the basin of attraction of p_1 with probability $1 - o(1)$ if $(a_2 n + b_2 - b_1) / (a_1 + a_2) = n/2 - \omega(\sqrt{n})$.

For f_3 all peaks have an influence on a search point's fitness. This leads to a very different structure of the fitness landscape.

Theorem 3. *Let $p_1 = 0^n$ and $p_2 = 1^n$ with arbitrary $a_1, a_2 \in \mathbb{R}^+$, $b_1, b_2 \in \mathbb{R}_0^+$. If $a_1 \neq a_2$, the fitness landscape defined by f_3 has a unique global optimum. If $a_1 > a_2$, this global optimum is p_1. Otherwise it is p_2.*

If $a_1 = a_2$, all search points have the same fitness $a_2 \cdot n + b_1 + b_2$.

Proof. According to Definition 3, the fitness of a search point x is

$$f_3(x) = (a_1 G(x, 0^n) + b_1) + (a_2 G(x, 1^n) + b_2) = (a_1 - a_2) \cdot |x|_0 + a_2 \cdot n + b_1 + b_2.$$

We see that $a_1 = a_2$ implies $f_3(x) = a_2 \cdot n + b_1 + b_2$, which is independent of x, proving the second statement. For $a_1 > a_2$ the fitness increases with increasing number of zeros and thus, p_1 is the unique global optimum. Similarly, it decreases with increasing number of zeros if $a_1 < a_2$. □

4.2 Comparing f_1 and f_2

The fitness landscapes defined as f_1 and f_2 are similar in nature. For both fitness landscapes the fitness is defined by only one of the peaks: for f_1 it is always the nearest peak; for f_2 the slope and offset of the peaks are taken into account so that 'higher' peaks can 'overrule' closer but smaller peaks. We formalise this by considering the set of local optima.

Theorem 4. *For f_1 and f_2 the set of local maxima is a subset of the peak locations $\{p_1, p_2, \ldots, p_k\}$. If the minimum Hamming distance between two peaks is at least 3 then the set of local maxima for f_1 is the set of peaks and, for f_2, the set of local maxima is a subset of the set of local maxima of f_1.*

Proof. If a point x is not a peak it has a Hamming neighbour with smaller Hamming distance to the peak that defines the function value of x. This proves that x cannot be a local maximum. Now, consider f_1 for a set of peaks that have minimum Hamming distance 3. Each Hamming neighbour y of a peak p has p as its nearest neighbour because the other peaks have Hamming distance at least 2 from y. This implies that $f_1(y) < f_1(p)$ and since this holds for each Hamming neighbour y we have that p is a local optimum. Finally, consider a peak p_i that is local maximum for f_2. We want to prove that p_i is also a local maximum for f_1. If the nearest other peak has Hamming distance at least 3 we are done. Consider a peak p_j with Hamming distance 1. We have that p_i is not a local optimum for f_1 if $f_1(p_j) > f_1(p_i)$ holds. But in this case $f_2(p_j) > f_2(p_i)$, too, so p_i is not a local maximum for f_2, either. Finally, consider a peak p_j with Hamming distance 2. Again, we have that p_i is not a local optimum if $f_1(p_j) > f_1(p_i)$ holds. But in the same way this implies $f_2(p_j) > f_2(p_i)$ and p_i is a local maximum for f_1. □

Clearly, the question if the set of local optima for f_1 and f_2 differ for a given set of peaks depends on the parameters of the peaks. We consider the case of peaks with random positions to show a remarkable phase transition with respect to the other parameters, slope and offset. While the relative slope difference a_i/a_j can be arbitrarily large (measured in n) it turns out that constant bounds on the smallest and largest relative difference determine if f_1 and f_2 have completely equal or almost completely different local optima.

Theorem 5. *Let an at most polynomial number $k = n^{O(1)}$ of peaks (p_1, a_1, a_2), (p_2, a_2, b_2), ..., (p_k, a_k, b_k) with $a_1, a_2, \ldots, a_k \in \mathbb{R}^+$, $b_1, b_2, \ldots, b_k \in \mathbb{R}_0^+$ and $b_i \leq a_i$ for all $i \in \{1, 2, \ldots, k\}$ be given where the peak positions $p_1, p_2, \ldots, p_k$ are chosen independently, uniformly at random from $\{0,1\}^n$. Let the minimum and maximum relative slope differences be $m := \min\limits_{i \neq j \in \{1,2,\ldots,k\}} a_i/a_j$ and $M := \max\limits_{i \neq j \in \{1,2,\ldots,k\}} a_i/a_j$. There exist constants $0 < c_1 < c_2 < 1$ such that if $m > c_2$ the set of local optima of f_1 and f_2 are equal to $\{p_1, p_2, \ldots, p_k\}$ with probability $1 - o(1)$ and if $M < c_1$ there are peak parameters with this value of M such that the set of local optima of f_1 and f_2 have only one element in common with probability $1 - o(1)$.*

Proof. We first show that the peaks are all in linear Hamming distance of each other with overwhelming probability. Consider two arbitrary peaks p_i and p_j. Considering p_i fixed, the expected number of bits equal in p_i and p_j when choosing $p_j \in \{0,1\}^n$ uniformly at random equals $n/2$. Application of Chernoff bounds [14] and application of a simply union bound yields that for all pairs of peak positions p_i, p_j with $i \neq j$ we have $\Pr(\mathrm{H}(p_i, p_j) \in [(1-\varepsilon)n/2, (1+\varepsilon)n/2]) = 1 - e^{-\Omega(n)}$. We consider only the situation where this is the case.

We have $f_1(p_i) = a_i \cdot n + b_i$ and $f_2(y) = a_i \cdot (n-1) + b_i$ for any Hamming neighbour y of p_i. We want to show that $f_2(p_i) = f_1(p_i)$ and $f_2(y) = f_1(y)$ holds which implies that p_i is a local optimum of f_2. We consider only p_i since the case y is very similar. We have $f_2(p_i) = \max\limits_{j \in \{1,\ldots,k\} \setminus \{i\}} \{a_i \cdot n + b_i, a_j \cdot (n - \mathrm{H}(p_i, p_j)) + b_j\}$.

Thus, we want to prove that $a_i \cdot n + b_i > a_j \cdot (n - \mathrm{H}(p_i, p_j)) + b_j$ holds. Remember that we have $b_j \leq a_j$ and $\mathrm{H}(p_i, p_j) \geq ((1-\varepsilon)/2)n$. Thus, it suffices if $a_i \cdot n > a_j \cdot n \cdot (((1+\varepsilon)/2) + (1/n))$ holds. With $a_i/a_j > ((1+\varepsilon)/2) + (1/n)$ this is the case so that choosing any $c_2 > (1+\varepsilon)/2$ suffices (because $a_i/a_j \geq m$).

On the other hand, we are also in the situation where $\mathrm{H}(p_i, p_j) < (1+\varepsilon)n/2$. We have $a_i n + b_i \leq (1+1/n)a_i n$ and $a_j \cdot (n - \mathrm{H}(p_i, p_j)) + b_j \geq a_j \cdot n \cdot ((1-\varepsilon)/2)$. Thus, if $a_i/a_j < ((1-\varepsilon)/2)/(1+1/n)$ we have that $f_2(p_i)$ is determined by the peak (p_j, a_j, b_j). Clearly, any constant $c_1 < (1-\varepsilon)/2$ suffices (because $a_i/a_j < M$). It is not hard to see that we can set the peak slopes in a way that f_2 is defined by the same peak (p_j, a_j, b_j) making p_j the only local (and thus also global) optimum. □

4.3 Considering Properties of f_3

As a third example we consider an important property of f_3. For this we look at a landscape on $n = 5d$ bits with three clustered peaks $p_1 = 1^n$, $p_2 = 1^{2d}0^{2d}1^d$ and $p_3 = 1^d0^d1^d0^d1^d$, $a_1 = a_2 = a_3$ and arbitrary b_1, b_2 and b_3. Note, that the three peaks have pairwise equal Hamming distance $\mathrm{H}(p_i, p_j) = 2d$. We first observe that the fitness landscape based on f_3 has a unique global optimum that coincides with the centre of mass of the three peaks. An example for $d = 1$ is shown in Fig. 1 (right).

Theorem 6. *Let $p_1 = 1^n$, $p_2 = 1^{2d}0^{2d}1^d$ and $p_3 = 1^d0^d1^d0^d1^d$, $a_1, a_2, a_3 \in \mathbb{R}^+$ with $a_1 = a_2 = a_3$ and arbitrary $b_1, b_2, b_3 \in \mathbb{R}_0^+$. The centre of mass of the three peaks, i.e., $1^{3d}0^d1^d$, is the unique global optimum of the fitness landscape defined by f_3.*

Proof. Recall that $f_3(x) := \sum_{i \in \{1,2,\ldots,k\}} a_i \cdot (n - \mathrm{H}(x, p_i)) + b_i$. We first observe that the offsets b_i do not have an influence on the ranking of search points as $b_1 + b_2 + b_3$ is added to the fitness of all search points. Thus, we can ignore the b_i in the following. As $a_1 = a_2 = a_3$, search points maximising $\sum_{i \in \{1,2,\ldots,k\}} n - \mathrm{H}(x, p_i)$ will be assigned the maximal fitness value. It is easy to see that these are exactly the points that minimise the average Hamming distance to the given peaks. Using this, the first statement follows directly from the proof of Theorem 1 in [13] and we obtain the centre of mass by performing a simple majority vote for each bit position. □

We remark that the above approach can be used to determine the set of global maxima for arbitrary sets of peaks. If $a_1 = a_2 = a_3$, we first obtain the set of search points with maximal fitness value by performing a simple majority vote for each bit position. Note, that in case of ties, search points with both bit values are assigned maximal fitness. For example, let us consider the above peaks with $d = 1$, i.e., $p_1 = 11111$, $p_2 = 11001$ and $p_3 = 10101$, and $p_4 = 00000$. We see that we have a tie for the 2nd and 3rd bits. Thus, we have four search points with maximal fitness value: 11101, 11001, 10101 and 10001. Given the set of search points with maximal fitness values we can then easily determine the set of global maxima.

The approach can also be generalised to peaks with different slopes by using a weighted majority vote where each a bit in p_i is assigned weight a_i. Let $W_0 = \sum_{i \text{ with } p_i[j]=0} a_i$ and $W_1 = \sum_{i \text{ with } p_i[j]=1} a_i$. We set the j-th bit to 0 if $W_0 > W_1$ and to 1 if $W_1 > W_0$. Ties are handled as discussed above.

References

1. Branke, J.: Memory enhanced evolutionary algorithms for changing optimization problems. In: Proceedings of CEC, pp. 1875–1882. IEEE Press (1999)
2. Doerr, B., Hansen, N., Igel, C., Thiele, L.: Theory of evolutionary algorithms (Dagstuhl seminar 15211). Dagstuhl Rep. **5**(5), 57–91 (2016)
3. Doerr, B., Winzen, C.: Ranking-based black-box complexity. Algorithmica **68**(3), 571–609 (2014)
4. Droste, S., Jansen, T., Wegener, I.: A rigorous complexity analysis of the (1 + 1) evolutionary algorithm for linear functions with Boolean inputs. In: Proceedings of ICEC, pp. 499–504. IEEE Press (1998)
5. Fischer, S., Wegener, I.: The one-dimensional Ising model: mutation versus recombination. Theor. Comput. Sci. **344**(2–3), 208–225 (2005)
6. Friedrich, T., Oliveto, P.S., Sudholt, D., Witt, C.: Analysis of diversity-preserving mechanisms for global exploration. Evol. Comput. **17**(4), 455–476 (2009)
7. He, J., Yao, X.: Drift analysis and average time complexity of evolutionary algorithms. Artif. Intell. **127**, 57–85 (2001)
8. Jansen, T.: Analyzing Evolutionary Algorithms. The Computer Science Perspective. Springer, Heidelberg (2013)
9. Jansen, T., Zarges, C.: Performance analysis of randomised search heuristics operating with a fixed budget. Theor. Comput. Sci. **545**, 39–58 (2014)
10. Jong, K.D., Spears, W.M.: An analysis of the interacting roles of population size and crossover in genetic algorithms. In: Schwefel, H.-P., Manner, R. (eds.) Proceedings of PPSN, pp. 38–47. Springer, Heidelberg (1990)
11. Kennedy, J., Spears, W.M.: Matching algorithms to problems: an experimental test of the particle swarm and some genetic algorithms on the multimodal problem generator. In: Proceedings of WCCI, pp. 78–83. IEEE Press (1998)
12. Kötzing, T., Lissovoi, A., Witt, C.: (1+1) EA on generalized dynamic onemax. In: Proceedings of FOGA, pp. 40–51. ACM Press (2015)
13. Moraglio, A., Johnson, C.G.: Geometric generalization of the Nelder-Mead algorithm. In: Cowling, P., Merz, P. (eds.) EvoCOP 2010. LNCS, vol. 6022, pp. 190–201. Springer, Heidelberg (2010)
14. Motwani, R., Raghavan, P.: Randomized Algorithms. Cambridge University Press, Cambridge (1995)
15. Mühlenbein, H.: How genetic algorithms really work: mutation and hillclimbing. In: Proceedings of PPSN, pp. 15–26. Elsevier (1992)
16. Oliveto, P.S., Sudholt, D., Zarges, C.: On the runtime analysis of fitness sharing mechanisms. In: Bartz-Beielstein, T., Branke, J., Filipič, B., Smith, J. (eds.) PPSN 2014. LNCS, vol. 8672, pp. 932–941. Springer, Heidelberg (2014)
17. Preuss, M.: Multimodal Optimization by Means of Evolutionary Algorithms. Springer, Heidelberg (2015)
18. Prügel-Bennett, A., Tayarani-Najaran, M.: Maximum satisfiability: anatomy of the fitness landscape for a hard combinatorial optimization problem. IEEE Trans. Evol. Comput. **16**(3), 319–338 (2012)

19. Shir, O.M.: Niching in evolutionary algorithms. In: Rozenberg, G., Bäck, T., Kok, J.N. (eds.) Handbook of Natural Computing, pp. 1035–1070. Springer, Heidelberg (2012)
20. Stadler, P.: Fitness landscapes. Biol. Evol. Stat. Phys. **585**, 183–204 (2002)
21. Sudholt, D.: Crossover is provably essential for the Ising model on trees. In: Proceedings of GECCO, pp. 1161–1167. ACM Press (2005)

Self-adaptation of Mutation Rates in Non-elitist Populations

Duc-Cuong Dang and Per Kristian Lehre(✉)

School of Computer Science, University of Nottingham, Nottingham, UK
{duc-cuong.dang,PerKristian.Lehre}@nottingham.ac.uk

Abstract. The runtime of evolutionary algorithms (EAs) depends critically on their parameter settings, which are often problem-specific. Automated schemes for parameter tuning have been developed to alleviate the high costs of manual parameter tuning. Experimental results indicate that self-adaptation, where parameter settings are encoded in the genomes of individuals, can be effective in continuous optimisation. However, results in discrete optimisation have been less conclusive. Furthermore, a rigorous runtime analysis that explains how self-adaptation can lead to asymptotic speedups has been missing. This paper provides the first such analysis for discrete, population-based EAs. We apply level-based analysis to show how a self-adaptive EA is capable of fine-tuning its mutation rate, leading to exponential speedups over EAs using fixed mutation rates.

1 Introduction

An obstacle when applying Evolutionary Algorithms (EAs) is that their efficiency depends crucially, and sometimes unpredictably, on their parameter settings, such as selective pressure and mutation rates [12]. *Parameter tuning* [7], where the parameters are fixed before running the algorithm, is the most common way of choosing the parameters. A weakness with parameter tuning is that optimal parameter settings may depend on the current state of the search process. In contrast, *parameter control* allows the parameters to change during the execution of the algorithm, e. g. according to a fixed schedule as in simulated annealing, through feedback from the search, or via self-adaptation [7]. Adaptive parameters can be essential and advantageous (e. g. covariance-matrix adaptation [9]) in continuous search spaces. In discrete spaces, it has been shown that changing the mutation rate as a function of the current fitness [2] can improve the runtime, and the 1/5-rule has been used to adapt the population size [5].

While previous studies have shown the benefit of adaptive parameters, only global parameters were analysed. Instead, we look at so-called "evolution of evolution" or true self-adaptation [7], in which the parameter is encoded in the genome of individual solutions. The existing studies on this topic from the EC literature is mostly experimental [1,7,13], or about proving the convergence of the population model at their limit [1], i.e. infinite population.

We study evolution of mutation rates in *non-elitist* populations, where the mutation rates of individuals are encoded in their own genomes. The mutation

J. Handl et al. (Eds.): PPSN XIV 2016, LNCS 9921, pp. 803–813, 2016.
DOI: 10.1007/978-3-319-45823-6_75

rate of the mutation rate is a *strategy parameter* p, which in *endogenous* control is itself evolved [1,14]. We consider *exogenous* control, where the parameter p is fixed globally. Our contribution is twofold: using a benchmark function, we provide necessary and sufficient conditions for self-adaptation to be effective; we show that self-adaptation is necessary in optimising a variant of this function. More precisely, an EA with a fixed or uniform mixing of mutation rates requires exponential time, while self-adaptation is efficient. As a by-product, we also prove that a non-elitist EA can outperform the elitist $(\mu + \lambda)$ EA.

2 Preliminaries

For any $n \in \mathbb{N}$, define $[n] := \{1, \ldots, n\}$. The natural logarithm is denoted by $\ln(\cdot)$, and the logarithm to the base 2 is denoted by $\log(\cdot)$. For $x \in \{0,1\}^n$, we write $x(i)$ for the i-th bit value. The Hamming distance is denoted by $\mathrm{H}(\cdot, \cdot)$ and the Iverson bracket by $[\cdot]$. Given a partition of a search space $\mathcal{X}$ into m ordered "levels" $(A_1, \ldots, A_m)$, we define $A_{\geq j} := \cup_{i=j}^{m} A_i$. A *population* is a vector $P \in \mathcal{X}^\lambda$, where the i-th element $P(i)$ is called the i-th *individual*. Given $A \subseteq \mathcal{X}$, we let $|P \cap A| := |\{i \mid P(i) \in A\}|$ be the number of individuals in population P that belong to the subset A.

All algorithms considered here are of the form of Algorithm 1 [4]. A new population P_{t+1} is generated by independently sampling λ individuals from an existing population P_t according to a selection mechanism p_{sel}, and perturbing each of the selected individuals by a variation operator p_{mut}. A fitness function $g : \mathcal{Y} \to \mathbb{R}$ is implicitly embedded in the selection mechanism p_{sel}.

Algorithm 1. [4]

Require: Finite search space $\mathcal{Y}$ with an initial population $P_0 \in \mathcal{Y}^\lambda$.
1: **for** $t = 0, 1, 2, \ldots$ until a termination condition is met **do**
2:    **for** $i = 1$ to λ **do**
3:       Sample $I_t(i) \in [\lambda]$ according to $p_{\text{sel}}(P_t)$, and set $x := P_t(I_t(i))$.
4:       Sample $x' \in \mathcal{Y}$ according to $p_{\text{mut}}(x)$, and set $P_{t+1}(i) := x'$.

We consider the standard bitwise mutation operator, where for any pair of bitstrings $x, x' \in \{0,1\}^n$ and any *mutation rate* $\chi \in (0, n]$, the probability of obtaining x' from x is $\Pr(x' = \text{mut}(x, \chi)) = (\chi/n)^{H(x,x')} (1 - \chi/n)^{n - H(x,x')}$. To model the parameter control problem, we assume that Algorithm 1 must choose the mutation rate parameter χ from a predefined set $\mathcal{M}$.

Uniform mixing, denoted $p_{\text{mut}}^{\text{mix}}$, chooses the mutation rate χ uniformly at random from the set $\mathcal{M}$ every time an individual is mutated, $p_{\text{mut}}^{\text{mix}}(x) := \text{mut}(x, \chi)$, where $\chi \sim \text{Unif}(\mathcal{M})$. The special case of $|\mathcal{M}| = 1$, i.e. a fixed mutation rate, has been studied extensively [4,12]. Here, we focus on $|\mathcal{M}| > 1$. It is known that such mixing of mutation operators can be beneficial [6,11].

Self-adaptation uses an extended search space $\mathcal{Y} = \mathcal{X} \times \mathcal{M}$, where each element (x, χ) consists both of a search point $x \in \mathcal{X}$ and a mutation rate $\chi \in \mathcal{M}$. A fitness function $g : \mathcal{Y} \to \mathbb{R}$ is defined by $g((x, \chi)) := f(x)$ for all $(x, \chi) \in \mathcal{Y}$. The mutation operator p_{mut} is written as $p_{\text{mut}}^{\text{adapt}}$ and it is parameterised by a globally fixed parameter $p \in (0, 1/2]$ such that $p_{\text{mut}}^{\text{adapt}}((x, \chi)) := (x', \chi')$ where $\chi' = \chi$ with probability $1 - p$, and $\chi' \sim \text{Unif}(\mathcal{M} \setminus \{\chi\})$ otherwise, and $x' = \text{mut}(x, \chi')$.

We analyse the runtime of Algorithm 1 using the level-based theorem [3]. This theorem applies to any population-based process where the individuals in P_{t+1} are sampled independently from the same distribution $D(P_t)$, where D maps populations to distributions over the search space $\mathcal{X}$. In Algorithm 1, the map is $D = p_{\text{mut}} \circ p_{\text{sel}}$, i.e., composition of selection and mutation.

Theorem 1 ([3]). *Given a partition $(A_1, \ldots, A_{m+1})$ of $\mathcal{X}$, define $T := \min\{t\lambda \mid |P_t \cap A_{m+1}| > 0\}$ to be the first point in time that elements of A_{m+1} appear in P_t of Algorithm 1. If there exist parameters $z_1, \ldots, z_m, z_* \in (0, 1]$, $\delta > 0$, a constant $\gamma_0 \in (0, 1)$ and a function $z_0 : (0, \gamma_0) \to \mathbb{R}$ such that for all $j \in [m]$, $P \in \mathcal{X}^\lambda$, $y \sim D(P)$ and $\gamma \in (0, \gamma_0]$ we have*

(G1) $\Pr(y \in A_{\geq j} \mid |P \cap A_{\geq j-1}| \geq \gamma_0 \lambda) \geq z_j \geq z_*$
(G2) $\Pr(y \in A_{\geq j} \mid |P \cap A_{\geq j-1}| \geq \gamma_0 \lambda, |P \cap A_{\geq j}| \geq \gamma \lambda) \geq z_0(\gamma) \geq (1 + \delta)\gamma$
(G3) $\lambda \geq \frac{2}{a} \ln\left(\frac{16m}{ac\varepsilon z_*}\right)$ *with* $a = \frac{\delta^2 \gamma_0}{2(1+\delta)}$, $\varepsilon = \min\{\delta/2, 1/2\}$ *and* $c = \varepsilon^4/24$

then $\mathbf{E}[T] \leq (2/c\varepsilon)(m\lambda(1 + \ln(1 + c\lambda)) + \sum_{j=1}^{m} 1/z_j)$.

We apply the *negative drift theorem for populations* [10] to obtain tail bounds on the runtime of Algorithm 1. For any individual $P_t(i)$, where $t \in \mathbb{N}$ and $i \in [\lambda]$, define $R_t(i) := |\{j \in [\lambda] \mid I_t(j) = i\}|$, i.e., the number of times the individual was selected. We define the *reproductive rate* of the individual $P_t(i)$ to be $\mathbf{E}[R_t(i) \mid P_t]$, i.e., the expected number of offspring from individual $P_t(i)$. Informally, the theorem states that if all individuals close to a given search point $x^* \in \mathcal{X}$ have reproductive rate below a certain threshold α_0, then the algorithm needs exponential time to reach x^*. The threshold depends on the mutation rate. Here, we derive a variant of this theorem for algorithms that use multiple mutation rates. In particular, we assume that the algorithm uses m mutation rates, where mutation rate χ_i/n for $i \in [m]$ is chosen with probability q_i. The proof of this theorem is similar to that of Theorem 4 in [10], and thus omitted.

Theorem 2. *For any $x^* \in \{0, 1\}^n$, define $T := \min\{t \mid x^* \in P_t\}$, where P_t is the population of Algorithm 1 at time $t \in \mathbb{N}$. If there exist constants $\alpha_0, c, c', \delta > 0$ such that with probability $1 - e^{-\Omega(n)}$*

- *the initial population satisfies $H(P_0, x^*) \geq c'n$*
- *for all $t \leq e^{cn}$ and $i \in [\lambda]$, if $H(P_t(i), x^*) \leq c'n$, then the reproductive rate of individual $P_t(i)$ is no more than α_0,*
- *$\sum_{j=1}^{m} q_j e^{-\chi_j} \leq (1 - \delta)/\alpha_0$, and $\max_j \chi_j \leq \chi_{\max}$ for a constant $\chi_{\max}$,*

then $\Pr\left(T \leq e^{c''n}\right) = e^{-\Omega(n)}$ *for a constant $c'' > 0$.*

3 General Negative Results

Using Theorem 2, we can now show general negative results for uniform mixing and self-adaptation of two mutation rates for any function with a unique global optimum x^*, assuming that the initial population is positioned sufficiently far away from x^*. The following theorem is a special case of Theorem 2 for $|\mathcal{M}| = 1$.

Theorem 3. *The runtime of Algorithm 1 with reproductive rate* α_0 *and mutation rate* $\chi_{high}/n \geq (\ln(\alpha_0) + \delta)/n$ *for some constant* $\delta > 0$ *satisfies* $\Pr(T \leq e^{cn}) = e^{-\Omega(n)}$ *on any function with a unique global optimum* x^* *assuming that* $\mathrm{H}(P_0, x^*) \geq c'n$ *for two constants* $c > 0$ *and* $c' \in (0, 1)$.

For binary tournament and (μ, λ)-selection, α_0 is bounded from above by 2 and λ/μ respectively. Hence, any mutation rate above $\ln(2)$ for 2-tournament selection and $\ln(\lambda/\mu)$ for (μ, λ)-selection by a constant renders the EA inefficient.

For $|\mathcal{M}| = 2$, we have the following general result, again due to Theorem 2.

Theorem 4. *Consider Algorithm 1 with reproductive rate* α_0 *and mutation rates* χ_{low}/n *and* χ_{high}/n. *If there exist constants* $\delta_1, \delta_2, \varepsilon > 0$ *such that*

- $\chi_{low} \geq \ln(\alpha_0) - \ln(1 + \delta_1)$ *and* $\chi_{high} \geq \ln(\alpha_0) - \ln(1 - \delta_2)$,
- *the EA chooses mutation rate* χ_{high} *with probability at least* $\frac{\delta_1(1+\varepsilon)}{\delta_1+\delta_2}$,

then $\Pr(T \leq e^{cn}) = e^{-\Omega(n)}$ *on any function with a unique optimum* x^* *given that* $\mathrm{H}(P_0, x^*) \geq c'n$ *for some constants* $c', c > 0$.

Uniform mixing selects the mutation rate χ_{high}/n with probability 1/2. Thus, if $\delta_1/(\delta_1 + \delta_2)$ is below 1/2 by a constant then the EA is inefficient. For example, in binary tournament, the setting $\chi_{\text{low}} \geq \ln(3/2) - \ln(100/99)$ and $\chi_{\text{high}} \geq \ln 3 + \ln(33/32)$ satisfies the conditions for $\delta_1 = 103/297$, $\delta_2 = 105/297$ and $\delta_1/(\delta_1 + \delta_2) = 103/208 < 1/2$. In contrast, Theorem 8 shows that self-adaptation is efficient in this setting. In self-adaptation, χ_{high}/n is selected with at least probability p, thus self-adaptation becomes inefficient if $p > \delta_1/(\delta_1 + \delta_2)$.

4 Robustness of Self-adaptation

The previous section showed how critically non-elitist EAs depend on having appropriate mutation rates. A slightly too high mutation rate χ_{high} can lead to an exponential increase in runtime. Uniform mixing of mutation rates can fail if the set of allowed mutation rates $\mathcal{M}$ contains one mutation rate which is too high, even though the set also contains an appropriate mutation rate χ_{low}.

Self-adaptation has a similar problem if the strategy parameter p is chosen too high. However, we will prove for a simple, unimodal fitness function that for a sufficiently small strategy parameter p, self-adaptation becomes highly robust, and is capable of fine-tuning the mutation rate. For the rest of this section, we consider a set of two mutation rates $\mathcal{M} = \{\chi_{\text{low}}, \chi_{\text{high}}\}$ which for arbitrary parameters $\ell \in [n]$ and $\varepsilon > 0$ are defined by $\left(1 - \frac{\chi_{\text{high}}}{n}\right)^{\ell} < \frac{\mu}{\lambda} \leq$

$\left(1-\frac{\chi_{\text{high}}}{n}\right)^{\ell-1}$ and $\frac{\mu}{\lambda}(1+\varepsilon) \leq \left(1-\frac{\chi_{\text{low}}}{n}\right)^{n}$. By the previous section, if ℓ is chosen sufficiently small, and hence χ_{high} sufficiently high, then uniform mixing will fail on any problem with a unique optimum. In contrast, using a Chernoff and a union bound, the following lemma shows that individuals that have chosen χ_{high} will quickly vanish from a self-adapting population, and the population will be dominated by individuals choosing the appropriate mutation parameter χ_{low}.

Lemma 1. *Let $Y_t := |P_t \cap A_{-1}|$ where P_t is the population of Algorithm 1 at time $t \in \mathbb{N}$ with (μ, λ)-selection on* LEADINGONES *and the set A_{-1} is as defined in Eq. (1). Then* $\Pr\left(Y_t \geq \max((3/4)\mu, (1-p/3)^t Y_0)\right) \leq t \cdot e^{-\Omega(\lambda)}$ *for all $t \in \mathbb{N}$.*

Theorem 5. *Algorithm 1 with (μ, λ)-selection where $\lambda \geq c\ln(n)$ for a sufficiently large constant $c > 0$, and self-adaptation from the set $\mathcal{M} = \{\chi_{low}, \chi_{high}\}$ using a sufficiently small constant strategy parameter p satisfying $(1+\varepsilon)(1-p) \geq 1+p\varepsilon$ has expected runtime $O(n\lambda\log(\lambda)+n^2)$ on* LEADINGONES.

Proof. We partition the search space into the following $n+2$ levels

$$A_j := \begin{cases} \{(x, \chi_{\text{high}}) \mid \text{LO}(y) \geq \ell\} & \text{if } j = -1 \\ \{(x, \chi_{\text{low}}), (x, \chi_{\text{high}}) \mid \text{LO}(x) = j\} & \text{if } 0 \leq j \leq \ell - 1 \\ \{(x, \chi_{\text{low}}) \mid \text{LO}(x) = j\} & \text{if } \ell \leq j \leq n. \end{cases} \tag{1}$$

The special level A_{-1} contains search points with too high mutation rate. We first estimate the expected runtime assuming that there are never more than $(3/4)\mu$ individuals in level A_{-1}. In the end, we will account for the generations where this assumption does not hold.

We now show that conditions (G1) and (G2) of the level-based theorem hold for the parameters $\gamma_0 := (1/8)(\mu/\lambda)$, $\delta := p\varepsilon$, and $z_j = \Omega(1/n)$. Assume that the current population has at least $\gamma_0\lambda = \mu/8$ individuals in $A_{\geq j-1}$ and $\gamma\lambda < \gamma_0\lambda$ individuals in $A_{\geq j}$, for $0 \leq j \leq n$ and $\gamma \in [0, \gamma_0)$. If $0 \leq j \leq \ell - 1$, then an individual can be produced in levels $A_{\geq j}$ if one of the $\gamma\lambda$ individuals in these levels is selected, and none of the first j bits are mutated. Assuming in the worst case that the selected individual has chosen the high mutation rate, the probability of this event is at least $\left(\frac{\gamma\lambda}{\mu}\right)\left(\left(1-\frac{\chi_{\text{high}}}{n}\right)^{j}(1-p)+\left(1-\frac{\chi_{\text{low}}}{n}\right)^{j}p\right) > \left(\frac{\gamma\lambda}{\mu}\right)\left(\left(1-\frac{\chi_{\text{high}}}{n}\right)^{\ell-1}(1-p)+\left(1-\frac{\chi_{\text{low}}}{n}\right)^{n}p\right) \geq \gamma(1+p\varepsilon)$. All individuals in levels $j \geq \ell$ use the low mutation rate. Hence, an individual in levels $A_{\geq j}$ can be produced by selecting one of the $\gamma\lambda$ individuals in this level, not change the mutation rate, and not flip any of the first $j \leq n$ leading 1-bits. The probability of this event is at least $\frac{\gamma\lambda}{\mu}\left(1-\frac{\chi_{\text{low}}}{n}\right)^{j}(1-p) > \frac{\gamma\lambda}{\mu}\left(\frac{\mu}{\lambda}(1+\varepsilon)(1-p)\right) \geq \gamma(1+\delta)$. Condition (G2) is therefore satisfied for all levels. For condition (G1), assume that the population does not contain any individuals in $A_{\geq j}$. Then in the worst case, it suffices to select one of the at least $\gamma_0\lambda$ individuals in level A_j, switch the mutation rate, and only flip the first 0-bit and no other bits. The probability of this event is higher than $\frac{\gamma_0\lambda}{\mu}\left(\frac{\chi_{\text{low}}}{n}\right)\left(1-\frac{\chi_{\text{high}}}{n}\right)^{n-1}p = \Omega(1/n)$.

Condition (G3) holds for any population size $\lambda \geq c\ln(n)$ and a sufficiently large constant c, because γ_0 and δ are constants. It follows that the expected number of generations until the optimum is found is $t_1(n) = O(n \log(\lambda) + n^2/\lambda)$. By Markov's inequality, the probability that the algorithm has not found the optimum after $2t_1(n)$ generations is less than 1/2.

Finally, we account for the generations with more than $(3/4)\mu$ individuals in level A_{-1}. We call a phase *good* if after $t_0(n) = O(\log(\lambda))$ generations and for the next $2t_1(n)$ generations, there are fewer than $(3/4)\mu$ individuals in level A_{-1}. By Lemma 1, a phase is good with probability $1 - (t_0(n) + 2t_1(n)) \cdot e^{-\Omega(\lambda)} = \Omega(1)$, for $\lambda \geq c\ln(n)$ and c a sufficiently large constant. By the level-based analysis, the optimum is found with probability at least 1/2 during a good phase. Hence, the expected number of phases required to find the optimum is $O(1)$. The theorem now follows by keeping in mind that each generation costs λ evaluations. □

We have shown that the EA can self-adapt to choose the low mutation parameter χ_{low} when required. Nevertheless, uniform mixing of mutation rates with a sufficiently small χ_{low} could achieve the same asymptotic performance. Furthermore, naively picking a mutation rate from the beginning also has a constant probability of optimising the function in polynomial time. Our aim is therefore to show that there exists a setting for which all the above approaches, except self-adaptation, fail. To prove this, we have identified a problem f_m where a high mutation rate is required in one part of the search space, and a low mutation rate is required in another part.

$$f_m(x) := \begin{cases} m & \text{if } x = 0^n, \text{ and} \\ \textsc{LeadingOnes}(x) & \text{otherwise.} \end{cases}$$

We call the local optimum 0^n the *peak*, and assume that all individuals in the initial population are peak individuals. The elitist $(\mu+\lambda)$ EA without any diversity mechanism will only accept a search point if it has at least m leading 1-bits.

Theorem 6. *Starting at 0^n, the $(\mu+\lambda)$ EA has expected runtime $n^{\Omega(m)}$ on f_m.*

To reach the optimal search point more efficiently, it is necessary to accept worse individuals into the population, e.g. a non-elitist selection scheme should be investigated. Since f_m has a unique global optimum, either using only a too high mutation rate or uniformly mixing a correct mutation rate with a too high one can lead to exponential runtime as discussed above. Analogously to the $(\mu + \lambda)$ EA, we also prove that using a too low mutation rate fails because the population is trapped on the peak (e. g. due to Theorem 2, individuals fell off the peak have too low reproductive rate to optimise m leading 1-bits). Subsequent proofs use the two functions $q(i) := (1 - \chi_{\text{low}}/n)^i$ and $r(i) := (1 - \chi_{\text{high}}/n)^i$, which are the probabilities of not flipping the first $i \in [n]$ bits using mutation rate χ_{low}/n and χ_{high}/n respectively. Clearly, $q(i)$ and $r(i)$ are monotonically decreasing in i. We also use the function $\beta(\gamma) := 2\gamma(1-\gamma/2)$, which is the probability that binary tournament selection chooses one of the $\gamma\lambda$ fittest individuals.

Theorem 7. *The runtime of Algorithm 1 on f_m with tournament size 2, initialised with the population at 0^n and fixed mutation rate $\chi \leq \ln(3/2) - \varepsilon$ for any constant $\varepsilon \in (0, \ln(3/2))$ satisfies $\Pr(T \leq e^{cn}) = e^{-\Omega(\lambda)}$ for a constant $c > 0$.*

Theorem 8. *If $\mathcal{M} = \{\chi_{low}, \chi_{high}\}$ where $\chi_{low} := \ln(\frac{3}{2}) - \varepsilon$ for any constant $\varepsilon \in (0, \ln(\frac{100}{99}))$, and $\ln(3) \leq \chi_{high} = O(1)$, then there exists an $m \in \Theta(n)$ such that Algorithm 1 starting with the population at 0^n, with tournament size 2, population size $\lambda \geq c \ln n$ for some constant $c > 0$ and self-adaptation of $\mathcal{M}$ with $p = 1/20$ has expected runtime $O(n\lambda \log(\lambda) + n^2)$ on f_m.*

Recall that uniform mixing is inefficient in this setting. Our intuition is that with sufficiently high mutation rate, some individuals fall off the peak and form a sub-population which starts optimising the LEADINGONES part of the problem. If the selective pressure is not too high, the sub-population should escape the local optimum, adapt the mutation rate, and reach the optimal search point 1^n. We used the level-based technique to infer constraints on the mutation rates and the strategy parameter p that allow this to happen. We use Lemma 2 to show that there are few individuals on the peak, or with "incorrect" mutation rates.

Lemma 2. *Given any subset $A \subset \mathcal{X}$, let $Y_t := |P_t \cap A|$ be the number of individuals in generation $t \in \mathbb{N}$ of Algorithm 1 with tournament size 2, that belong to subset A. If there exist three parameters $\rho, \sigma, \varepsilon \in (0, 1)$ such that $\Pr(p_{\text{mut}}(y) \in A) \leq \rho$ for all $y \in A$ and $\Pr(p_{\text{mut}}(y) \in A) \leq \sigma\gamma_* - \varepsilon$ for all $y \notin A$, where $\gamma_* := 2 - (1 - \sigma)/\rho$, then $\Pr(Y_t \geq \max(\gamma_*\lambda, (1 - \varepsilon/2)^t Y_0)) \leq t \cdot e^{-\Omega(\lambda)}$.*

Proof (of Theorem 8). We apply the level-based theorem with respect to a partitioning of the search space $\mathcal{X} = \{0, 1\}^n \times \mathcal{M}$ into the following $n + 2$ levels

$$A_j := \begin{cases} \{(0^n, \chi_{\text{low}}), (0^n, \chi_{\text{high}})\} & \text{if } j = -1, \\ \{(x, \chi_{\text{low}}), (x, \chi_{\text{high}}) \mid \text{LO}(x) = 0 \wedge x \neq 0^n\} & \text{if } j = 0, \\ \{(x, \chi_{\text{low}}), (x, \chi_{\text{high}}) \mid \text{LO}(x) = j\} & \text{if } 1 \leq j \leq \ell - 2, \\ \{(x, \chi_{\text{low}}), (y, \chi_{\text{high}}) \mid \text{LO}(x) = \ell - 1, \text{LO}(y) \geq \ell - 1\} & \text{if } j = \ell - 1, \\ \{(x, \chi_{\text{low}}) \mid \text{LO}(x) = j\} & \text{if } \ell \leq j \leq n. \end{cases}$$

where $\ell \in [n]$ is the unique integer such that $\left(1 - \frac{\chi_{\text{high}}}{n}\right)^{\ell} < \frac{85}{171} \leq \left(1 - \frac{\chi_{\text{high}}}{n}\right)^{\ell - 1}$. Note that as long as $m \leq \ln(171/85)(n - 1)/\chi_{\text{high}}$, we have $\left(1 - \frac{\chi_{\text{high}}}{n}\right)^m \geq \left(e^{-\chi_{\text{high}}}\right)^{\frac{m}{n-1}} \geq \frac{85}{171} > \left(1 - \frac{\chi_{\text{high}}}{n}\right)^{\ell}$, hence $\ell > m$.

We first estimate the expected runtime assuming that every population contains less than $\psi\lambda$ individuals in A_{-1}, and less than $\xi\lambda$ individuals in the set $B := \{(y, \chi_{\text{high}}) \mid \text{LO}(y) \geq \ell\}$, where $\psi := 123/250$ and $\xi := 1/5$. In the end, we will account for the generations where these assumptions do not hold. We begin by showing that condition (G2) of the level-based theorem hold for all levels.

Levels $0 \leq j \leq m$: Assume that the population contains $\gamma\lambda$ individuals in $A_{\geq j}$ for any $\gamma \in (0, \gamma_0)$. An individual in $A_{\geq j}$ will be selected if the tournament contains at least one individual in $A_{\geq j}$, and no individuals in A_{-1}. The probability of this event is $\beta(\gamma) \geq 2\gamma(1 - \gamma_0/2 - \psi)$. The mutated offspring of the selected individual will belong to levels $A_{\geq j}$ if none of the first $j \leq m$ bits are flipped, which occurs with probability at least $r(m)$. Hence, condition (G2) is satisfied if there exists a $\gamma_0 \in (0, 1)$ and a constant $\delta > 0$ such that for all

$\gamma \in (0, \gamma_0]$, it holds $\beta(\gamma)r(m) \geq \gamma(1+\delta)$, i.e., it is sufficient to choose $m \in \mathbb{N}$ sufficiently small such that $r(m) = \left(1 - \frac{\chi_{\text{high}}}{n}\right)^m \geq \frac{1+\delta}{2(1-\gamma_0/2-\psi)}$. Note that such an $m = \Theta(n)$ exists, because $2(1-\gamma_0/2-\psi) = \frac{127}{125} - \gamma_0 > 1+\delta$ when γ_0 and δ are sufficiently small.

Levels $m+1 \leq j < \ell$: The probability of mutating an individual from $A_{\geq j}$ into $A_{\geq j}$, pessimistically assuming that the selected individual uses the high mutation rate χ_{high}, is at least $r(\ell-1)(1-p)+q(\ell-1)p > r(\ell-1)(1-p)+q(n)p > (85/171)(1-p)+(2/3)p = 1/2+1/180$. Hence, assuming that the current population has $\gamma\lambda$ individuals in $A_{\geq j}$ where $\gamma \in (0, \gamma_0)$, the probability of selecting one of these individuals and mutating them into $A_{\geq j}$ is at least $\beta(\gamma)(r(\ell-1)(1-p)+q(\ell-1)p) > 2\gamma(1-\gamma_0/2)(1/2+1/180) = \gamma(1-\gamma_0/2)(1+1/90) > \gamma(1+\delta')$ for some $\delta' > 0$ given that γ_0 is a sufficiently small constant. Note that the lower bound on $\beta(\gamma)$ here does not depend on ψ, and nor on ξ because in this setting the peak individuals have lower fitness than the individuals in A_j, and $B \subset A_{\geq j}$.

Levels $\ell \leq j \leq n$: By the level-partitioning, any individual in these levels uses the low mutation rate χ_{low}, and other individuals with at least ℓ leading 1-bits belong to the set B. Assume that the current population contains $\gamma \in (0, \gamma_0)$ individuals in $A_{\geq j}$. An individual in $A_{\geq j}$ can be produced by having a binary tournament with at least one individual from $A_{\geq j}$ and none of the at most $\xi\lambda$ individuals in B, not mutating any of the bits, and not changing the mutation rate. The probability of this event is at least $2\gamma(1-\gamma_0/2-\xi)q(n)(1-p) \geq \gamma(4/5-\gamma_0/2)(19/15) = \gamma(1+1/75-(19/30)\gamma_0) > \gamma(1+\delta')$ for some constant $\delta' > 0$, assuming that γ_0 is sufficiently small.

We now show that condition (G1) is satisfied for a parameter $z = \Omega(1/n)$ in any level j. Assume that there are at least $\gamma_0\lambda$ individuals in $A_{\geq j}$. Then, to create an individual in $A_{\geq j+1}$, it is sufficient to create a tournament of two individuals from $A_{\geq j}$, flip at most one bit, and either keep or switch the mutation rate. The probability of such an event is at least $\gamma_0^2(\chi_{\text{low}}/n)(1-\chi_{\text{high}}/n)^{n-1}p = \Omega(1/n)$.

To complete the application of the level-based theorem, we note that since δ and γ_0 are constants, condition (G3) is satisfied when $\lambda \geq c\ln n$ for some constant c. Hence, under the assumptions on the number of individuals in level A_{-1} and B described above, the level-based theorem implies that the algorithm obtains the optimum in expected $t_1(n) = O(n\log(\lambda)+n^2/\lambda)$ generations. Furthermore, by Markov's inequality, the probability that the optimum has not been found within $2t_1(n)$ generations is less than $1/2$.

To complete the proof, we justify the assumption that less than $\psi\lambda$ individuals belong to level A_{-1}, and less than $\xi\lambda$ individuals belong to B. We will show using Lemma 2 that starting with any population, these assumptions hold after an initial phase of $t_0(n) = O(\log(\lambda))$ generations. We call a phase *good* if the assumptions hold for the next $t_1(n) < e^{c\lambda}$ generations.

To apply Lemma 2 with respect to level A_{-1}, we note that the probability of obtaining an individual in A_{-1} by mutating an individual in A_{-1} is bounded from above by $q(n)(1-p)+r(n)p \leq (2/3)e^{\varepsilon}(1-p)+p/3 \leq 65/99$. Furthermore, to mutate an individual from $\mathcal{X} \setminus A_{-1}$ into A_{-1}, it is necessary to flip at least one specific bit-position, i.e., with probability $O(1/n)$. Therefore, by Lemma 2 with $\sigma = 49/4950$ and $\rho = 65/99$, it holds for all t where $t_0(n) < t < e^{cn}$ and $t_0(n) = O(\log(\lambda))$ that $\Pr(|P_t \cap A_{-1}| \geq \psi\lambda) = e^{-\Omega(\lambda)}$ where $\psi := 123/250$.

Similarly, the probability of not destroying a B-individual with mutation is by definition of ℓ at most $\left(1 - \frac{\chi_{\text{high}}}{n}\right)^{\ell}(1-p) \leq \left(\frac{85}{171}\right)\left(\frac{19}{20}\right) = \frac{17}{36} =: \rho$. To create a B-individual from $\mathcal{X} \setminus B$, it is in the best case necessary to change the mutation rate from χ_{low} to χ_{high} and not mutate the first ℓ bit-positions. The probability of this event is $\left(1 - \frac{\chi_{\text{high}}}{n}\right)^{\ell} p \leq \left(\frac{85}{171}\right)\left(\frac{1}{20}\right) = \frac{17}{684}$. Therefore, by Lemma 2 wrt $\sigma := 3/20$ and the above value of ρ, for every generation t where $t_0(n) < t < e^{c\lambda}$ and $t_0(n) = O(\log(\lambda))$ it holds $\Pr(|P_t \cap B| \geq \xi\lambda) = e^{-\Omega(\lambda)}$, where $\xi := 1/5$.

To summarise, starting from any configuration of the population, a phase of length $t_0(n) + 2t_1(n) = O(n\log(\lambda) + n^2/\lambda)$ generations is *good* with probability $1 - e^{-\Omega(\lambda)}$. If a phase is good, then the optimum will be found by the end of that phase with probability at least 1/2. Hence, the expected number of phases required to find the optimum is $O(1)$, and the theorem follows, keeping in mind that each generation costs λ function evaluations. □

5 Experiments

Below are results from 1000 experiments with the self-adaptive EA on the LeadingOnes function for $n = 200$, $p = 1/1000$ using (μ, λ)-selection for $\mu = 500$, $\lambda = 4\mu$, and mutation parameters $\mathcal{M} = \{2/5, 2\}$. For each $j \in [n]$, the figure contains a box-plot describing the distribution of the fraction of the population choosing χ_{low} over all generations where the (1/10)-ranked individual in the population has j leading one-bits.

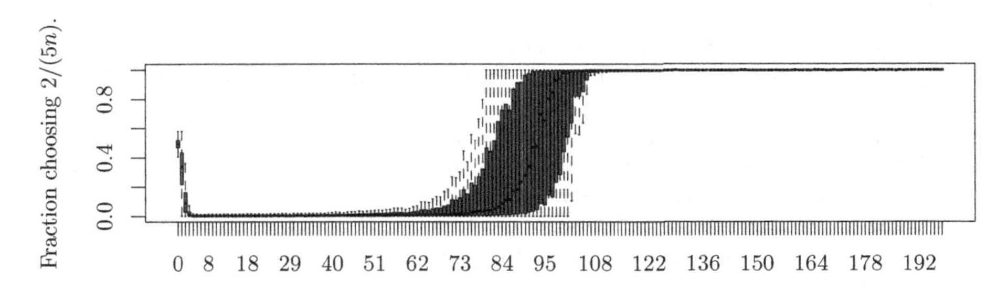

The initial population, including mutation rates, are sampled uniformly at random. Hence the (1/10)-ranked individual will have fitness close to 1 in the first generations. For $j \leq 5$, i.e. early in the run, approximately half of the population chooses the low mutation. However, the population quickly switches to the higher mutation χ_{high} until the (1/10)-ranked individual in the population reaches a value approximately $j \geq 60$ where the population switches to the lower mutation χ_{low}. Almost all individuals choose χ_{low} for $j \geq 108$. These experimental results confirm that the population adapts the mutation rate according to the region of the fitness landscape currently searched.

6 Conclusion

In this first runtime analysis of self-adaptation, we have shown that self-adaptation with a sufficiently low strategy parameter can robustly control mutation-rates in non-elitist EAs, and that this automated control can lead to exponential speedups compared to EAs that use fixed mutation rates, or uniform mixing of mutation rates. The results were obtained via level-based analysis, further demonstrating the strength of this technique in handling complex population dynamics.

Acknowledgements. This work received funding from the European Union Seventh Framework Programme (FP7/2007-2013) under grant agreement no. 618091 (SAGE).

References

1. Bäck, T.: Self-adaptation in genetic algorithms. In: Proceedings of ECAL 1992, pp. 263–271 (1992)
2. Böttcher, S., Doerr, B., Neumann, F.: Optimal fixed and adaptive mutation rates for the leadingones problem. In: Schaefer, R., Cotta, C., Kołodziej, J., Rudolph, G. (eds.) PPSN XI. LNCS, vol. 6238, pp. 1–10. Springer, Heidelberg (2010)
3. Corus, D., Dang, D.-C., Eremeev, A.V., Lehre, P.K.: Level-based analysis of genetic algorithms and other search processes. In: Bartz-Beielstein, T., Branke, J., Filipič, B., Smith, J. (eds.) PPSN 2014. LNCS, vol. 8672, pp. 912–921. Springer, Heidelberg (2014)
4. Dang, D.-C., Lehre, P.K.: Refined upper bounds on the expected runtime of non-elitist populations from fitness-levels. In: Proceedings of GECCO 2014, pp. 1367–1374 (2014)
5. Doerr, B., Doerr, C.: Optimal parameter choices through self-adjustment: applying the 1/5-th rule in discrete settings. In: Proceedings of GECCO 2015, pp. 1335–1342 (2015)
6. Doerr, B., Doerr, C., Kötzing, T.: Solving problems with unknown solution length at (almost) no extra cost. In: Proceedings of GECCO 2015, pp. 831–838 (2015)
7. Eiben, A.E., Michalewicz, Z., Schoenauer, M., Smith, J.E.: Parameter control in evolutionary algorithms. In: Lobo, F.G., Lima, C.F., Michalewicz, Z. (eds.) Parameter Setting in Evolutionary Algorithms. SCI, vol. 54, pp. 19–46. Springer, Heidelberg (2007)
8. Gerrish, P.J., Colato, A., Perelson, A.S., Sniegowski, P.D.: Complete genetic linkage can subvert natural selection. PNAS **104**(15), 6266–6271 (2007)
9. Hansen, N., Ostermeier, A.: Completely derandomized self-adaptation in evolution strategies. Evol. Comput. **9**(2), 159–195 (2001)
10. Lehre, P.K.: Negative drift in populations. In: Schaefer, R., Cotta, C., Kołodziej, J., Rudolph, G. (eds.) PPSN XI. LNCS, vol. 6238, pp. 244–253. Springer, Heidelberg (2010)
11. Lehre, P.K., Özcan, E.: A runtime analysis of simple hyper-heuristics: to mix or not to mixoperators. In: Proceedings of FOGA 2013, pp. 97–104 (2013)
12. Lehre, P.K., Yao, X.: On the impact of mutation-selection balance on the runtime of evolutionary algorithms. IEEE Trans. Evol. Comput. **16**(2), 225–241 (2012)

13. van Rijn, S., Emmerich, M.T.M., Reehuis, E., Bäck, T.: Optimizing highly constrained truck loadings using a self-adaptive genetic algorithm. In: Proceedings of CEC 2015, pp. 227–234 (2015)
14. Xue, J.Z., Kaznatcheev, A., Costopoulos, A., Guichard, F.: Fidelity drive: a mechanism for chaperone proteins to maintain stable mutation rates in prokaryotes over evolutionary time. J. Theor. Biol. **364**, 162–167 (2015)

Hypervolume Sharpe-Ratio Indicator: Formalization and First Theoretical Results

Andreia P. Guerreiro(✉) and Carlos M. Fonseca

CISUC, Department of Informatics Engineering, University of Coimbra,
Pólo II, Pinhal de Marrocos, 3030-290 Coimbra, Portugal
{apg,cmfonsec}@dei.uc.pt

Abstract. Set-quality indicators have been used in Evolutionary Multi-objective Optimization Algorithms (EMOAs) to guide the search process. A new class of set-quality indicators, the Sharpe-Ratio Indicator, combining the selection of solutions with fitness assignment has been recently proposed. This class is based on a formulation of fitness assignment as a Portfolio Selection Problem which sees solutions as assets whose returns are random variables, and fitness as the investment in such assets/solutions. An instance of this class based on the Hypervolume Indicator has shown promising results when integrated in an EMOA called POSEA. The aim of this paper is to formalize the class of Sharpe-Ratio Indicators and to demonstrate some of the properties of that particular Sharpe-Ratio Indicator instance concerning monotonicity, sensitivity to scaling and parameter independence.

Keywords: Sharpe Ratio · Portfolio selection · Evolutionary algorithms · Multiobjective optimization

1 Introduction

Indicator-based Evolutionary Multiobjective Optimization Algorithms (EMOAs) are currently among the state-of-the-art in Evolutionary Multiobjective Optimization. These EMOAs rely on quality indicators to guide the search, which map a point set into a scalar value, such as the Hypervolume Indicator [5,9]. Good quality indicators capture in a single value the proximity to the Pareto front and the sparsity/diversity of the set, which tends to enhance the capability of indicator-based EMOAs to find well-spread sets of good solutions.

Studies of quality-indicator properties have shown the abilities and limitations of indicator-based EMOAs. Such properties allow one to better understand, for example, whether an indicator-based EMOA aiming at the maximization of the indicator, is able to converge to the Pareto Front (monotonicity [10]) or understand which distribution each indicator favors (optimal μ-distributions [1]).

Yevseyeva *et al.* [8] established a link between the theory of Portfolio Selection and selection in Evolutionary Algorithms (EAs) by making an analogy between assets and individuals, expected return and individual quality, and return covariance and lack of diversity. They proposed that individuals be assessed through

J. Handl et al. (Eds.): PPSN XIV 2016, LNCS 9921, pp. 814–823, 2016.
DOI: 10.1007/978-3-319-45823-6_76

the optimization of a Portfolio Selection Problem (PSP), formalized as the bi-objective problem of assigning investment to a set of assets so as to maximize expected return while minimizing return variance (associated to risk). This translates into the problem of assigning fitness to an EA population so as to maximize overall population quality while minimizing lack of diversity. Due to the bi-objective nature of the PSP, different optimal investment strategies balancing risk and expected return may be defined, such as the Sharpe Ratio, a risk-adjusted performance index well known in Finance [3]. A new indicator related to the Hypervolume Indicator, but based on the maximization of the Sharpe Ratio, was proposed and integrated in an EMOA with promising results. However, its theoretical properties have not been considered so far.

The goal of this paper is to formalize the class of Sharpe-Ratio Indicators and to study some of the properties of the indicator proposed by Yevseyeva *et al.* [8]. Section 2 provides the background. Section 3 details and formalizes the class of indicators based on the Sharpe Ratio and reintroduces the indicator proposed by Yevseyeva *et al.*, which will be called Hypervolume Sharpe Ratio (HSR) Indicator, as an instance of this class. Then, some properties of the HSR Indicator regarding monotonicity, reference points, and scaling independence, will be demonstrated in Sect. 4. Some conclusions are drawn in Sect. 5.

2 Background

2.1 Definitions

In multiobjective optimization, each solution is mapped according to d objective functions onto a point in the objective space, $\mathbb{R}^d$. For simplicity, only those points in objective space will be considered throughout this paper. Note that a number in parentheses in superscript is used for enumeration (e.g. $a^{(1)}, a^{(2)}, a^{(3)} \in \mathbb{R}^d$) while a number in subscript is used to refer to a coordinate of a point/vector (e.g. v_i is the ith coordinate of $v \in \mathbb{R}^d$). As the objective space is a partially ordered set, the Pareto dominance relation is introduced [4,11]:

Definition 1 ***(Dominance).*** *A point $u \in \mathbb{R}^d$ is said to weakly dominate a point $v \in \mathbb{R}^d$, iff $u_i \leq v_i$ for all $1 \leq i \leq d$, and this is represented as $u \leq v$. If, in addition $u \neq v$, then u is said to dominate v and is represented as $u < v$. If $u_i < v_i$ for all $1 \leq i \leq d$, then u is said to strongly dominate v, and this is represented as $u \ll v$.*

Definition 2 ***(Set dominance).*** *A set $\mathrm{A} \subset \mathbb{R}^d$ is said to weakly dominate a set $\mathrm{B} \subset \mathbb{R}^d$ iff $\forall_{b \in \mathrm{B}}, \exists_{a \in \mathrm{A}} : a \leq b$. This is represented as $\mathrm{A} \preceq \mathrm{B}$. A is said to dominate a set B iff $\mathrm{A} \preceq \mathrm{B}$ and $\mathrm{B} \npreceq \mathrm{A}$, and this is represented as $\mathrm{A} \prec \mathrm{B}$.*

2.2 Properties

A set-indicator is a function I that assigns a real value to a non-empty set of points in $\mathbb{R}^d$ [10]. Among the properties a set-indicator may possess [10], this paper will cover parameter independence, sensitivity to scaling and monotonicity.

Typically, an indicator is easier to use the lower is the number of parameters that must be set. A *scaling invariant* indicator (e.g. the cardinality indicator [10]) guarantees that the indicator value for any subset of the objective space remains unchanged when the objective space is scaled. A weaker form of invariance, called *scaling independence*, ensures that the order defined by an indicator among all subsets of the objective space is kept when the objective space is scaled.

Monotonicity is an important property as it formalizes the empirical notion of agreement between indicator values and set dominance. A monotonic indicator guarantees that a set of nondominated solutions is never considered to be worse than another set which it dominates. A definition of (weak) monotonicity of a set-quality indicator with respect to set dominance is given in [10]:

Definition 3 ***(Monotonicity).*** *A set-indicator I is weakly monotonic w.r.t set dominance iff, given two point sets* $\mathrm{A}, \mathrm{B} \subset \mathbb{R}^d$, $\mathrm{A} \prec \mathrm{B}$ *implies* $I(\mathrm{A}) \geq I(\mathrm{B})$.

The above properties have been studied for indicators such as the hypervolume indicator (strictly monotonic [10] for sets of points that strongly dominate the reference point, parameter-dependent [1], scaling independent [5,9]) and the additive ϵ-indicator (weakly monotonic [10], dependent on multiple parameters [10]), thereby motivating their use in EMOAs as well as in performance assessment. Not holding such properties may discourage the use of an indicator in EMOAs. For example, a non-monotonic indicator may prefer non-Pareto Front solutions over Pareto front solutions dominating them, as is the case with the Average Hausdorff distance [7] and cardinality [10].

2.3 Sharpe Ratio

A portfolio balancing return and risk, is obtained by optimizing Problem 1:

Problem 1 (**Sharpe-Ratio Maximization**). Let $\mathrm{A} = \{a^{(1)}, \ldots, a^{(n)}\}$ be a non-empty set of assets, let vector $r \in \mathbb{R}^n$ denote the expected return of these assets and matrix $Q \in \mathbb{R}^{n \times n}$ denote the return covariance between pairs of assets. Let $x \in [0,1]^n$ be the investment vector where x_i denotes the investment in asset $a^{(i)}$. The Sharpe-Ratio maximization problem is defined as:

$$\max_{x \in [0,1]^n} \quad h(x) = \frac{r^T x - r_f}{\sqrt{x^T Q x}} \quad \text{s.t.} \quad \sum_{i=1}^{n} x_i = 1 \tag{1}$$

where r_f represents the return of a riskless asset and $h(x)$ is the Sharpe Ratio [3].

Although Problem 1 is non-linear, $h(x)$ may be homogenized and thus, it may be restated as an equivalent convex quadratic programming (QP) problem [3]:

Problem 2 (**Sharpe-Ratio Maximization - QP Formulation**).

$$\min_{y \in \mathbb{R}^n} \quad g(y) = y^T Q y \tag{2a}$$

$$\text{s.t.} \quad \sum_{i=1}^{n} (r_i - r_f) y_i = 1 \tag{2b}$$

$$y_i \geq 0, \quad i = 1, \ldots, n \tag{2c}$$

The optimal investment x^* for Problem 1, i.e., the optimal risky portfolio, is given by $x^* = y^*/k$, where y^* is the optimal solution of Problem 2 and $k = \sum_{i=1}^{n} y_i^*$.

So far, the set of assets A has been considered to be fixed and so have r and Q. However, in this paper, r and Q are computed as function of a set of assets A that is not fixed and thus, $h^{\mathrm{A}}(x)$ and $g^{\mathrm{A}}(y)$ will be used instead of $h(x)$ and $g(y)$, respectively, to highlight this dependence where needed. Moreover, with a slight abuse of language, a solution y to Problem 2 will also be called an investment vector, as for a solution x for Problem 1.

3 Sharpe-Ratio Indicator

In this section, the class of Sharpe-Ratio Indicators is formalized, and the Hypervolume Sharpe-Ratio Indicator proposed by Yevseyeva *et al.* [8] is instantiated.

The return of each individual is related to the preferences of a Decision Maker (DM) and different methods can be used to model the uncertainty surrounding DM preferences. Yevseyeva *et al.*'s [8] interpretation of selection in EAs as a portfolio selection problem sees the return of each individual asset as a random variable whose expected values can be computed.

Problem 1 does not state what the expected return and covariance of assets/individuals are. Different preferences lead to different ways of modeling return (and vice-versa) which may lead to different investment strategies in EAs. Therefore, a broad class of indicators based on the Sharpe Ratio can be defined:

Definition 4 ***(Sharpe-Ratio Indicator).*** *Given a non-empty set of assets* $\mathrm{A} = \{a^{(1)}, \ldots, a^{(n)}\}$, *the corresponding expected return,* r, *and covariance matrix,* Q, *the Sharpe-Ratio Indicator,* $I_{SR}(\mathrm{A})$, *is defined as follows:*

$$I_{SR}(\mathrm{A}) = \max_{x \in \Omega} \; h^{\mathrm{A}}(x) \tag{3}$$

where $\Omega \subset [0,1]^n$ *is the set of solutions that satisfy the constraints of Problem 1.*

Note that the Sharpe-Ratio Indicator simultaneously evaluates the quality of the set A through a scalar, $I_{SR}(\cdot)$, and also the importance of each solution in that set through the optimal investment vector x^*.

The Hypervolume Sharpe-Ratio Indicator (HSR Indicator) is an instance of the Sharpe-Ratio Indicator where the expected return vector and the return covariance matrix are computed based on the Hypervolume Indicator as proposed by Yevseyeva *et al.* [8]. The expected return of a solution is the probability of that solution being satisfactory to the DM, assuming a uniform distribution of the DM's goal vector in an orthogonal range $[l, u]$, $l, u \in \mathbb{R}^d$. For the ith individual in the population, this is represented by component p_i of a vector p, whereas the return covariance between the ith and j^{th} individuals is represented by element q_{ij} of a matrix Q $(i, j = 1, \ldots, n)$. Let:

$$p_{ij}(l,u) = \frac{\Lambda([l,u] \cap [a^{(i)}, \infty[\cap [a^{(j)}, \infty[)}{\Lambda([l,u])} = \frac{\prod_{k=1}^{d}(u_k - \max(a_k^{(i)}, a_k^{(j)}))}{\prod_{k=1}^{d}(u_k - l_k)} \tag{4}$$

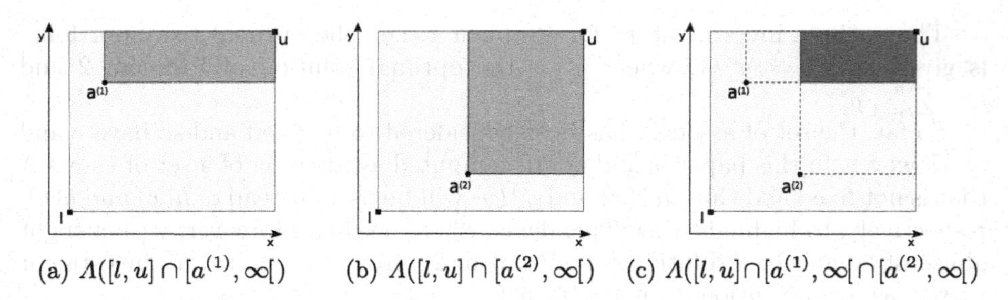

(a) $\Lambda([l, u] \cap [a^{(1)}, \infty[)$ (b) $\Lambda([l, u] \cap [a^{(2)}, \infty[)$ (c) $\Lambda([l, u] \cap [a^{(1)}, \infty[\cap [a^{(2)}, \infty[)$

Fig. 1. An example of the region measured to compute p_{ij}, given a point set $\mathrm{A} = \{a^{(1)}, a^{(2)}\} \subset \mathbb{R}^2$. The region measured to compute p_1 and p_2 and p_{12} is depicted in darker gray in Figures (a), (b) and (c), respectively.

where $l, u \in \mathbb{R}^d$ are two reference points and $\Lambda(\cdot)$ denotes the Lebesgue measure [2]. Note that p_{ij} is, therefore, the normalized hypervolume indicator of the region jointly dominated by $a^{(i)}$ and $a^{(j)}$ inside the region of interest, $[l, u]$. Moreover, from the formulation [8], $r_i(l, u) = p_i(l, u) = p_{ii}(l, u)$ and $q_{ij}(l, u) = p_{ij}(l, u) - p_i(l, u)p_j(l, u)$. For the sake of readability, $P = [p_{ij}]_{n \times n}$ and $Q = [q_{ij}]_{n \times n}$ will be assumed to have been previously calculated and, therefore, parameters l and u from expression (4) will be omitted as long as no ambiguity arises. Note that, from the definition of q_{ij}, $Q = P - pp^T$.

In Fig. 1, assuming w.l.o.g. that $l = (0, 0)$ and $u = (1, 1)$, and thus, $\Lambda([l, u]) = 1$, the area of the darker regions in Figs. 1(a) to (c) are, exactly, p_1, p_2 and p_{12}, respectively. Note that p_{ii} is related to the area dominated by $a^{(i)}$ inside the region $[l, u]$, while p_{ij} is related to the area simultaneously dominated by $a^{(i)}$ and $a^{(j)}$ inside the region $[l, u]$.

The Sharpe Ratio $h^{\mathrm{A}}(x)$ for the set of solutions A where r and Q are defined as in (4) will be represented by $h^{\mathrm{A}}_{\mathrm{HSR}}(x, l, u)$. Analogously to the Sharpe-Ratio Indicator, the HSR Indicator is formally defined as follows:

Definition 5 ***(Hypervolume Sharpe-Ratio Indicator).*** *Given a non-empty point set* $\mathrm{A} = \{a^{(1)}, \ldots, a^{(n)}\} \subset \mathbb{R}^d$*, the points* $l, u \in \mathbb{R}^d$*, the expected return* p *and the covariance* Q *computed as expressed in (4), the Hypervolume Sharpe-Ratio Indicator* $I_{\mathrm{HSR}}(\mathrm{A}, l, u)$ *is given by:*

$$I_{\mathrm{HSR}}(\mathrm{A}, l, u) = \max_{x \in \Omega} \; h^{\mathrm{A}}_{\mathrm{HSR}}(x, l, u) \tag{5}$$

where $\Omega \subset [0, 1]^n$ *is the set of solutions that satisfy the constraints of Problem 1.*

As Yevseyeva *et al.* [8] pointed out, it follows from the definition of q_{ij} that the riskless asset is such that $r_f = 0$. Consequently, Problem 2 may be simplified by noting that the constraint (2b) must always be satisfied. Therefore, the following is true for any solution y in the feasible space Ω:

$$y^T Q y = \sum_{i=1}^{n}\sum_{j=1}^{n} p_{ij} y_i y_j - \sum_{i=1}^{n} p_i y_i \sum_{j=1}^{n} p_j y_j = \sum_{i=1}^{n}\sum_{j=1}^{n} p_{ij} y_i y_j - 1 = y^T P y - 1 \tag{6}$$

Note that this simplification of Problem 2 is applicable to any DM preference model where r_f is zero.

4 Properties of the HSR Indicator

In the following, the optimal investment is shown to be invariant to the setting of l under certain conditions. Varying l can also be interpreted as applying linear transformations to the objective space, under which the indicator is scaling independent. Finally, the HSR Indicator is shown to be weakly monotonic.

4.1 Reference Points and Linear Scaling

Given a non-empty point set $\mathrm{A} \subset \mathbb{R}^d$ and the reference points $l, u \in \mathbb{R}^d$, such that for all $a \in \mathrm{A}$, $l \leq a \ll u$ holds, the location of l can be shown to have no effect on the optimal investment in A as long as $\{l\} \leq \mathrm{A}$ and u remains fixed. This is equivalent to applying a linear transformation to the objective space, with u as the center of the transformation. Thus, in practice, only one parameter of the HSR Indicator needs to be set (the upper reference point, u). Formally:

Theorem 1. *Let $\mathrm{A} \subset \mathbb{R}^d$ be a non-empty point set, let $l, u \in \mathbb{R}^d$ be two reference points such that $\forall_{a \in \mathrm{A}}$, $l \leq a \ll u$, and let $x^* \in [0,1]^n$ be such that $I_{\mathrm{HSR}}(\mathrm{A}, l, u) = h_{\mathrm{HSR}}^{\mathrm{A}}(x^*, l, u)$. If $l' \in \mathbb{R}^d$ is such that $\{l'\} \leq \mathrm{A}$, then x^* also satisfies $I_{\mathrm{HSR}}(\mathrm{A}, l', u) = h_{\mathrm{HSR}}^{\mathrm{A}}(x^*, l', u)$.*

Proof. Recall expression (4), of p_{ij}, for a given point set $\mathrm{A} = \{a^{(1)}, \ldots, a^{(n)}\} \subset \mathbb{R}^d$, where $p = [p_{ii}]_{n \times 1}$ and $P = [p_{ij}]_{n \times n}$ $(i, j = 1, \ldots, n)$. $P(l', u)$ and $p(l', u)$ may be defined as functions of $P(l, u)$ and $p(l, u)$, respectively, in the following way:

$$P(l', u) = \frac{v}{v'} P(l, u) \tag{7a}$$

$$p(l', u) = \frac{v}{v'} p(l, u) \tag{7b}$$

where $v = \Lambda([l, u])$ and $v' = \Lambda([l', u])$.

Assume that $y \in \mathbb{R}^n$ is the vector of variables of Problem 2 (minimizing $g_{\mathrm{HSR}}^{\mathrm{A}}(y, l, u) = y^T P y - 1$), when l is set as the lower reference point and that, analogously, $y' \in \mathbb{R}^n$ is the corresponding vector of variables when l' is used instead. Taking into account expressions (7b) and the equality constraint of Problem 2, the following is derived:

$$p(l, u)^T y = p(l', u)^T y' \quad \Leftrightarrow \quad p(l, u)^T y = \frac{v}{v'} p(l, u)^T y' \Leftrightarrow \quad y = \frac{v}{v'} y' \tag{8}$$

which implies that when y is such that $y = \frac{v}{v'} y'$, if $y' > 0$ then $y > 0$ and therefore, if y' is feasible so is y and vice-versa. Hence, the following holds:

$$g_{\mathrm{HSR}}^{\mathrm{A}}(y', l', u) = y'^T P(l', u) y' - 1 = \frac{v'}{v} y^T P(l, u) y - 1 = \frac{v'}{v} g_{\mathrm{HSR}}^{\mathrm{A}}(y, l, u) - 1 + \frac{v'}{v}$$

Therefore, the optimal solution y'^* for Problem 2, given l', can be obtained from the optimal solution y^*, given l, i.e., $y'^* = \frac{v'}{v}y^*$. Consequently, the optimal solution x^* for Problem 1:

$$x^* = \frac{y^*}{\sum_{i=1}^n y_i^*} = \frac{\frac{v}{v'}y'^*}{\frac{v}{v'}\sum_{i=1}^n y_i'^*} = \frac{y'^*}{\sum_{i=1}^n y_i'^*}. \tag{9}$$

Hence, $I_{\text{HSR}}(\text{A}, l, u) = h_{\text{HSR}}(x^*, l, u)$ implies that $I_{\text{HSR}}(\text{A}, l', u) = h_{\text{HSR}}(x^*, l', u)$ thus, Theorem 1 is proved.

Note that moving the lower reference point, l, for example, to a lower value of one of the objectives while the others are kept the same, is equivalent to scaling down that objective with respect to the other objectives. Thus, the placement of l can also be seen as a way of linearly scaling the objective functions (as long as this reference point continues to dominate A). Therefore, by Theorem 1, scaling the objective space under such conditions does not affect the optimal investment.

Scaling through l comes down to multiplying p_i and p_{ij} by a positive constant as in the proof of Theorem 1. Observing the Sharpe Ratio expression $h(x)$ in Problem 1, the HSR-indicator is not scaling invariant, i.e., scaling the objective space will affect the indicator value. However, the HSR-indicator is scaling independent under these linear transformations, as shown next.

Theorem 2 ***(Linear-Scaling Independence of*** I_{HSR}***).*** *Consider two point sets* $\text{A}, \text{B} \subset \mathbb{R}^d$ *and two reference points* $l, u \in \mathbb{R}^d$ *such that* $\forall_{a \in \text{A}, b \in \text{B}}, l \leq a, b \ll u$. *Assume w.l.o.g. that* A *and* B *are such that* $I_{\text{HSR}}(\text{A}, l, u) \leq I_{\text{HSR}}(\text{B}, l, u)$. *Then,* $I_{\text{HSR}}(\text{A}, l', u) \leq I_{\text{HSR}}(\text{B}, l', u)$ *holds for any* $l' \in \mathbb{R}^d$ *such that* $\{l'\} \leq \text{A}, \text{B}$.

Proof. Let p_A, P_A and Q_A denote, respectively, the expected return vector, the matrix of expected return and the return covariance matrix with respect to point set A. Scaling is applied to A and B in expression $h(x)$ in Problem 1 by multiplying a constant $t > 0$ by each p_i and p_{ij} and, therefore, $p'_\text{A} = tp_\text{A}$ and $P'_\text{A} = tP_\text{A}$, where $t = \frac{\Lambda([l,u])}{\Lambda([l',u])}$. Consequently,

$$\begin{aligned} I_{\text{HSR}}(\text{A}, l', u) \leq I_{\text{HSR}}(\text{B}, l', u) \quad &\Leftrightarrow \\ \frac{tp_\text{A}^T x_\text{A}}{\sqrt{tx_\text{A}^T P_\text{A} x_\text{A} - t^2 x_\text{A}^T p_\text{A} p_\text{A}^T x_\text{A}}} \leq \frac{tp_\text{B}^T x_\text{B}}{\sqrt{tx_\text{B}^T P_\text{B} x_\text{B} - t^2 x_\text{B}^T p_\text{B} p_\text{B}^T x_\text{B}}} \quad &\Leftrightarrow \\ \frac{1}{t}(x_\text{A}^T p_\text{A} p_\text{A}^T x_\text{A})(x_\text{B}^T P_\text{B} x_\text{B}) \leq \frac{1}{t}(x_\text{B}^T p_\text{B} p_\text{B}^T x_\text{B})(x_\text{A}^T P_\text{A} x_\text{A}) \end{aligned} \tag{10}$$

Since the constant t vanishes from the inequality, which includes the case where the lower reference point is not changed ($t = 1$), Theorem 2 is proved.

4.2 Monotonicity

The property of monotonicity may now be stated for the HSR Indicator:

Theorem 3 *(Weak Monotonicity of the Hypervolume Sharpe-Ratio Indicator). Consider two reference points $l, u \in \mathbb{R}^d$ and two point sets $\mathrm{A}, \mathrm{B} \subset [l, u[$ such that $\mathrm{A} \prec \mathrm{B}$. Then $I_{\mathrm{HSR}}(\mathrm{A}, l, u) \geq I_{\mathrm{HSR}}(\mathrm{B}, l, u)$.*

In order to prove this theorem, two auxiliary results are stated first. Lemma 1 is used to prove Lemma 2, which is then used in the proof of the theorem. Similarly to expression (4), for any two points $a, b \in [l, u[$, let p_{ab} denote the measure of the region bounded above by $u \in \mathbb{R}^d$ that is dominated simultaneously by a and b, and let $p_a = p_{aa}$. Note that $p_c > 0$ for any point $c \in [l, u[$.

Lemma 1. *Consider two points $a, b \in [l, u[$ such that $a < b$. Then, for all $c \in [l, u[\subset \mathbb{R}^d$, $p_b p_{ac} \leq p_{bc} p_a$ holds.*

Proof. Consider w.l.o.g. that $l = (0, ..., 0)$ and $u = (1, \ldots, 1)$ and therefore, $\Lambda([l, u]) = 1$. Lemma 1 will be proved by contradiction. Hence, suppose that, for some choice of $c \in [l, u[$:

$$\begin{gathered} p_b p_{ac} > p_{bc} p_a \Leftrightarrow \\ \prod_{i=1}^{d} (1 - b_i)(1 - \max(a_i, c_i)) > \prod_{i=1}^{d} (1 - \max(b_i, c_i))(1 - a_i) \end{gathered} \tag{11}$$

Thus, there should be, at least, a dimension i for which the following holds:

$$(1 - b_i)(1 - \max(a_i, c_i)) > (1 - \max(b_i, c_i))(1 - a_i) \tag{12}$$

However, by manipulating expression (12), it is possible to verify that $b_i \geq c_i$ implies $a_i > \max(a_i, c_i)$, and that $b_i < c_i$ implies $a_i > b_i$, which are both untrue. Consequently, expression (11) does not hold either, and Lemma 1 is proved.

Lemma 2. *Consider a point set $\mathrm{A} = \{a^{(1)}, \ldots, a^{(n)}\} \subset [l, u[$, where $n \geq 2$, and, without loss of generality, assume that $a^{(2)} < a^{(1)}$. Then, the investment vector $x^* \in [0, 1]^n$ that maximizes the Sharpe Ratio for the set A is such that the investment in $a^{(1)}$, denoted by x_1^*, is zero.*

Proof. Note that, for constraint (2b) to be satisfied, there has to be a strictly positive investment in, at least one asset and thus, all constraints are linearly independent for any feasible solution to Problem 2. Thus, the prerequisites of the first-order necessary optimality conditions (KKT conditions) [6] are satisfied.

Following the notation and definitions in Nocedal and Wright [6], the KKT conditions state that if a feasible solution y^* is optimal, then there is a Lagrange multiplier vector λ^* for which all components associated to an inequality constraint are nonnegative and the product of each component of λ^* and the corresponding constraint at y^* is zero. Moreover, the gradient of the Lagrangian function w.r.t y^* is zero ($\nabla_y \mathcal{L}(y^*, \lambda^*) = 0$). The Lagrangian function, for the HSR Indicator (in Problem 2) is:

$$\mathcal{L}(y, \lambda) = y^T P y - 1 - \lambda_1 p^T y - \sum_{i=2}^{n+1} \lambda_i y_{i-1} \tag{13}$$

and the corresponding partial derivative w.r.t. y_k at (y^*, λ^*) for $k = 1, ..., n$ is:

$$\frac{\partial \mathcal{L}(y^*, \lambda^*)}{\partial y_k} = 2\sum_{i=1}^{n} p_{ik} y_i^* - p_k \lambda_1^* - \lambda_{k+1}^* = 0 \tag{14}$$

Lemma 2 is proved by contradiction. Let y_1^* and y_2^* represent the investments in $a^{(1)}$ and $a^{(2)}$, respectively. Since $a^{(2)}$ dominates $a^{(1)}$, the following holds:

$$p_1 = p_{12}, \quad p_1 < p_2 \quad \text{and} \quad p_{1i} \leq p_{2i}, \quad i = 3, \ldots, n \tag{15}$$

Suppose that the optimal investment y^* is such that $y_1^* > 0$. Then, the KKT conditions imply that $\lambda_2^* = 0$. By manipulating Eq. (14) for $k = 1, 2$ using the conditions in (15), the following condition on λ_3^* is obtained:

$$p_1(p_{12} - p_2)y_1^* + \sum_{i=3}^{n}(p_1 p_{2i} - p_{1i} p_2)y_i^* = \frac{p_1 \lambda_3^*}{2} \geq 0 \tag{16}$$

$\lambda_3^* \geq 0$ must be true so that it is a valid Lagrange multiplier. Therefore, since $p_1 > 0$, the left-hand side of expression (16) must be zero or positive. However, the first term is clearly negative since $p_{12} = p_1 < p_2$, and the sum is non-positive by Lemma 1.

Therefore, no optimal Lagrange multiplier vector λ^* exists for which the KKT conditions hold true when y_1^* is strictly positive, and consequently, y^* cannot be optimal. Therefore, $y_1^* = 0$ which implies that $x_1^* = 0$ and proves Lemma 2.

Proof (Theorem 3). Consider two point sets $\mathrm{A}, \mathrm{B} \subset [l, u[\subset \mathbb{R}^d$, such that $|\mathrm{A}|, |\mathrm{B}| \geq 1$ and $\mathrm{A} \prec \mathrm{B}$. Since any points in $\mathrm{B} - \mathrm{A}$ are dominated points in $\mathrm{A} \cup \mathrm{B}$, by Lemma 2 they are assigned zero investment, and $I_{\mathrm{HSR}}(\mathrm{A} \cup \mathrm{B}) = I_{\mathrm{HSR}}(\mathrm{A})$ must hold true. Suppose that $I_{\mathrm{HSR}}(\mathrm{B}) > I_{\mathrm{HSR}}(\mathrm{A})$. Then, an investment strategy in $\mathrm{A} \cup \mathrm{B}$ with Sharpe Ratio greater than $I_{\mathrm{HSR}}(\mathrm{A} \cup \mathrm{B})$ where zero investment is given to the points in $\mathrm{A} - \mathrm{B}$ would exist, which leads to a contradiction and proves the theorem.

5 Concluding Remarks

The Sharpe-Ratio Indicator class has been formalized, and theoretical results on the particular HSR Indicator have been presented regarding the independence of one of the reference points, scaling independence and the monotonicity property. Although the formulation of the HSR Indicator involves two reference points, only one needs to be set in practice. The second reference point is just a technical parameter that is required by the formulation. Indeed, the optimal investment is not affected by the linear objective rescaling implied by changes to this second reference point, and the indicator is scaling independent under such transformations. Thus, the HSR Indicator does not require more parameters to be set than, for example, the Hypervolume Indicator. The HSR Indicator is also weakly monotonic w.r.t. set dominance.

The study of other properties of interest, including optimal μ-distributions for the HSR Indicator, will be the subject of future work.

Acknowledgments. This work was supported by national funds through the Portuguese Foundation for Science and Technology (FCT), by the European Regional Development Fund (FEDER) through COMPETE 2020 – Operational Program for Competitiveness and Internationalization (POCI). A. P. Guerreiro acknowledges FCT for Ph.D. studentship SFHR/BD/77725/2011, co-funded by the European Social Fund and by the State Budget of the Portuguese Ministry of Education and Science in the scope of NSRF–HPOP–Type 4.1–Advanced Training.

References

1. Auger, A., Bader, J., Brockhoff, D., Zitzler, E.: Theory of the hypervolume indicator: optimal μ-distributions and the choice of the reference point. In: Foundations of Genetic Algorithms (FOGA 2009), pp. 87–102. ACM (2009)
2. Beume, N., Naujoks, B., Emmerich, M.: SMS-EMOA: multiobjective selection based on dominated hypervolume. EJOR **181**, 1653–1669 (2007)
3. Cornuejols, G., Tuntuncu, R.: Optimization Methods in Finance. Cambridge University Press, Cambridge (2007)
4. Ehrgott, M.: Multicriteria Optimization, 2nd edn. Springer, Heidelberg (2005)
5. Knowles, J.D.: Local-search and hybrid evolutionary algorithms for Pareto optimization. Ph.D. thesis, Department of Computer Science, University of Reading (2002)
6. Nocedal, J., Wright, S.J.: Numerical Optimization. Springer Series in Operations Research and Financial Engineering, 2nd edn. Springer, New York (2006)
7. Rudolph, G., Schütze, O., Trautmann, H.: On the closest averaged Hausdorff archive for a circularly convex Pareto front. In: Squillero, G., Burelli, P. (eds.) EvoApplications 2016. LNCS, vol. 9598, pp. 42–55. Springer, Heidelberg (2016). doi:10.1007/978-3-319-31153-1_4
8. Yevseyeva, I., Guerreiro, A.P., Emmerich, M.T.M., Fonseca, C.M.: A portfolio optimization approach to selection in multiobjective evolutionary algorithms. In: Bartz-Beielstein, T., Branke, J., Filipič, B., Smith, J. (eds.) PPSN 2014. LNCS, vol. 8672, pp. 672–681. Springer, Heidelberg (2014)
9. Zitzler, E.: Evolutionary algorithms for multiobjective optimization: methods and applications. Ph.D. thesis, ETH Zurich, Switzerland (1999)
10. Zitzler, E., Knowles, J.D., Thiele, L.: Quality assessment of pareto set approximations. In: Branke, J., Deb, K., Miettinen, K., Słowiński, R. (eds.) Multiobjective Optimization. LNCS, vol. 5252, pp. 373–404. Springer, Heidelberg (2008)
11. Zitzler, E., Thiele, L., Laumanns, M., Fonseca, C.M., da Fonseca, V.G.: Performance assessment of multiobjective optimizers: an analysis and review. IEEE Trans. Evol. Comput. **7**(2), 117–132 (2003)

k-Bit Mutation with Self-Adjusting k Outperforms Standard Bit Mutation

Benjamin Doerr[1], Carola Doerr[2(✉)], and Jing Yang[1]

[1] École Polytechnique, Palaiseau, France
[2] CNRS and Sorbonne Universités, UPMC University Paris 06, LIP6, Paris, France
doerr@lip6.fr

Abstract. When using the classic standard bit mutation operator, parent and offspring differ in a random number of bits, distributed according to a binomial law. This has the advantage that all Hamming distances occur with some positive probability, hence this operator can be used, in principle, for all fitness landscapes. The downside of this "one-size-fits-all" approach, naturally, is a performance loss caused by the fact that often not the ideal number of bits is flipped. Still, the fear of getting stuck in local optima has made standard bit mutation become the preferred mutation operator.

In this work we show that a self-adjusting choice of the number of bits to be flipped can both avoid the performance loss of standard bit mutation and avoid the risk of getting stuck in local optima. We propose a simple mechanism to adaptively learn the currently optimal mutation strength from previous iterations. This aims both at exploiting that generally different problems may need different mutation strengths and that for a fixed problem different strengths may become optimal in different stages of the optimization process.

We experimentally show that our simple hill climber with this adaptive mutation strength outperforms both the randomized local search heuristic and the (1+1) evolutionary algorithm on the LeadingOnes function and on the minimum spanning tree problem. We show via mathematical means that our algorithm is able to detect precisely (apart from lower order terms) the complicated optimal fitness-dependent mutation strength recently discovered for the OneMax function. With its self-adjusting mutation strength it thus attains the same runtime (apart from $o(n)$ lower-order terms) and the same (asymptotic) 13 % fitness-distance improvement over RLS that was recently obtained by manually computing the optimal fitness-dependent mutation strength.

1 Introduction

When using a bit-string representation in evolutionary computation, that is, when the search space is $\Omega = \{0,1\}^n$, then *standard bit mutation* is the by far most-employed mutation operator. It creates a new individual (*offspring*) from an existing one (*parent*) by flipping each bit of the parent independently with some probability p, often with $p = 1/n$. By this, the Hamming distance of parent

J. Handl et al. (Eds.): PPSN XIV 2016, LNCS 9921, pp. 824–834, 2016.
DOI: 10.1007/978-3-319-45823-6_77

and offspring, that is, the number of positions in which the strings differ, follows a binomial distribution with parameters n and p. If $p = 1/n$, then the expected distance is one (following the idea that mutation should be a minimalistic change of the individual), but all distances in $[0..n] := \{0, 1, \ldots, n\}$ occur with positive probability. Consequently, a hill climber using this mutation operator (e.g., the $(1 + 1)$ evolutionary algorithm (EA)) cannot get permanently stuck in a local optimum, no matter what the fitness landscape of the underlying optimization problem looks like.

The downside of this "one-size-fits-all" approach, naturally, is that it does not exploit particular properties of the landscape. It is well-known (though not in all cases explicitly proven) that for simple fitness landscapes like those of OneMax, linear functions, LeadingOnes, and royal-road functions, flipping only single bits gives smaller optimization times than using standard bit mutation (often, the improvement is by a factor of $e \approx 2.718$). For the minimum spanning tree (MST) problem, a mutation operator randomly choosing between flipping a single random bit or two random bits gives again better results than standard bit mutation, whereas flipping always only one bit or always exactly two bits in most cases lets the algorithm get stuck in a local optimum.

Our Results: The examples above show that using a problem-specific optimal mutation strength can lead to a fair speed-up over standard bit mutation, however, with the risk of making the algorithm fail badly when choosing a wrong mutation strength. For this reason, we design a simple hill climber that autonomously tries to choose the optimal mutation rate by analyzing the past performance of the different mutation strengths. This aims both at exploiting that different problems ask for different mutation strengths and at exploiting that for a fixed problem the optimal mutation strength may change during the optimization process; a problem even less understood than the right problem-specific static mutation strength.

We experimentally analyze our new algorithm on the LeadingOnes and the MST problem. We observe that, for suitable parameter settings, it clearly outperforms the $(1 + 1)$ EA. Interestingly, it even beats the randomized local search (RLS) heuristic (flipping always one bit) for the LeadingOnes problem and the variant of RLS flipping one or two bits for the MST problem. This shows that for these problems a better performance can be obtained from a mutation strength that changes over time, and that our algorithm is able to find such superior fitness-dependent mutation strengths.

The heart of our work is making this effect mathematically precise for OneMax. For this function, an optimal fitness-dependent mutation strength was recently found in [4]. This optimal mutation strength is quite particular. It uses, for all but a lower order fraction of the runtime, the mutation strength one (that is, flips a random bit). In a short initial segment of the optimization process, flipping a larger number of bits is superior. The optimal number of bits is decreasing with increasing fitness, but is always an odd number. Despite differing from RLS only in a short period, the simple hill climber using this fitness-dependent mutation strength with a fixed budget of iterations computes solutions that have an expected fitness distance that is 13 % smaller than those computed by RLS, making it the current best unbiased mutation-based optimizer for OneMax (cee [8] for a discussion on the fixed-budget performance measure). However, due to its complicated nature, it is not clear how a non-expert should find such fitness-dependent mutation strengths.

For our new algorithm with self-adjusting mutation strength, we show that it essentially is able to find this optimal mutation schedule on the fly. More precisely, with high probability our algorithm always (apart from a lower-order fraction of the iterations) uses a mutation strength which gives an expected progress equal to the best possible progress (again, apart from lower order terms). Consequently, our algorithm has the same optimization time (apart from an $o(n)$ additive lower order term) and the same asymptotic 13 % superiority in the fixed budget perspective as the algorithm with the hand-crafted mutation strength schedule from [4].

These first results indicate that a self-adjusting mutation strength both works well for problems with different optimal mutation strengths (in an even better way than the "one-size-fits-all" approach of standard bit mutation) and, beyond this, can also find good fitness-dependent mutation schedules. We defer the details to the following sections, where we propose our new algorithm (Sect. 2), give some experimental evidence for its superiority (Sect. 3), conduct a rigorous runtime analysis for OneMax (including the proof that the optimal mutation strength essentially is always employed) in Sect. 4, and discuss how to choose parameters and take other design choices (Sect. 5).

Discussion of Previous Works on Adaptive Mutation Operators: Given the importance of mutation, not surprisingly, there is a plethora of works on adaptive uses of mutation. With very few exceptions, these works are experimental in nature. They mostly indicate that an adaptive change of how mutation is performed can be beneficial. However, it seems hard to derive generally accepted design rules from these works. For reasons of space, we cannot avoid referring the reader to some of the central works [1,5,9,12] and the extensive follow-up work.

On the theoretical side, a first dynamic setting of the mutation rate was proposed and analyzed in [7]. They propose to use the $(1+1)$ EA with a mutation rate that, depending on the iteration counter, takes a value in $\{2^k/n \mid k = 0, 1, 2, \ldots, \lceil \log_2(n) \rceil - 2\}$. They construct an example function where the $(1+1)$ EA with this dynamic mutation rate greatly outperforms the $(1+1)$ EA with any fixed mutation rate. However, they also show that their EA has an asymptotically larger runtime on most classic test functions. In [2], a fitness-dependent choice of the mutation rate was proposed that improves the runtime of the $(1 + 1)$ EA on the LeadingOnes function from approximately $0.86n^2$ for the fixed mutation probability $1/n$ to approximately $0.68n^2$. For population-based EAs a rank-based mutation rate has been investigated in [11].

All of the works discussed above use standard bit mutation, that is, flip each bit independently with a certain, adaptively chosen, probability p. Our main point in this work is that flipping a fixed number of r bits, where the mutation

Algorithm 1. RLS with fitness-dependent mutation strength. When maximizing functions $f : \{0,1\}^n \to D$, the algorithm takes as parameter a mutation strength function $r : D \to [1..n]$ describing how many bits to flip given a certain fitness of the current search point. The operator flip(x, r) generates from x a new search point by flipping exactly r random bit positions.

1 **Initialization:** Choose $x \in \{0,1\}^n$ uniformly at random;
2 **Optimization: for** $t = 1, 2, 3, \ldots$ **do**
3 $\quad y \leftarrow \text{flip}(x, r(f(x)))$;
4 $\quad$ **if** $f(y) \geq f(x)$ **then** $x \leftarrow y$;

strength r is chosen in a self-adjusting manner, is more profitable because it greatly reduces the use of r-bit flips with a sub-optimal mutation strength r. We are not aware of any work on this type of self-adjusting mutation. An optimal fitness-dependent choice of r for OneMax was determined in [4] recently.

2 Randomized Local Search with Fitness-Dependent and Self-adjusting Mutation Strength

In [4], a variant of the classic *randomized local search* (RLS) heuristic with fitness-dependent mutation strength was proposed (Algorithm 1). Whereas the classic version of RLS creates a new search point always by flipping a single random bit, *RLS with fitness-dependent mutation strength* flips a number of bits ("mutation strength") functionally depending on the current fitness.

While it is clear that choosing the best mutation strength for each fitness level can improve the performance, it is not so clear how to find a good mutation strength function. The example of OneMax studied in [4] indicates that a substantial understanding of the underlying optimization problem is necessary to profit from varying the mutation strength depending on the fitness.

To overcome this difficulty, in this work we propose to choose the mutation strength in each iteration based on the experience in the optimization process so far. We enforce gaining a certain experience by designating each iteration with probability δ as a *learning iteration.* In a learning iteration we flip a random number of bits (chosen uniformly at random from a domain $[1..r_{\max}]$) and store (in an efficient manner) the progress made in these iterations. In all regular iterations, we use the experience made in these learning iterations to determine the most promising mutation strength and create the offspring with this mutation strength.

More precisely, let us denote by x_t the search point after the t-th iteration, that is, after the mutation and selection step of iteration t. Denote by x_0 the random initial search point. If t is a learning iteration, denote by r_t the random mutation strength r used in this iteration. Otherwise set $r_t = 0$.

The main idea of our algorithm is to learn the efficiency of the mutation strengths, that is, the expected progress made when flipping r bits, for all

$r \in [1..r_{\max}]$. We do so via a time-discounted average of the progresses observed in the learning iterations: We define an estimate for the future progress, called *velocity* in the absence of a better name, after the t-th iteration by

$$v_t[r] := \frac{\sum_{s=1}^{t} \mathbf{1}_{r_s=r}(1-\varepsilon)^{t-s}(f(x_s) - f(x_{s-1}))}{\sum_{s=1}^{t} \mathbf{1}_{r_s=r}(1-\varepsilon)^{t-s}}. \tag{1}$$

In this expression, the parameter ε, called *forgetting rate*, determines the decrease of the importance of older information. Since $(1-\varepsilon)^{1/\varepsilon} = (1/e) + o(1)$ for all $\varepsilon = o(1)$, the reciprocal $1/\varepsilon$ of the forgetting rate is (apart from constant factors) the information half-life.

We first observe that we can compute the velocities iteratively and thus, unlike equation (1) might suggest, do not need to store the full history of the learning iterations. To this aim, we need to store one additional value for each r, namely the sum of the $(1-\varepsilon)^{t-s}$ terms used in the weighted average, that is,

$$w_t[r] := \sum_{s=1}^{t} \mathbf{1}_{r_s=r}(1-\varepsilon)^{t-s}.$$

Then the following recursive description of the velocities and weight sums is easily seen: If in iteration $t+1$ we have not done a learning step with mutation strength r, that is, $r_{t+1} \neq r$, then $v_{t+1}[r] = v_t[r]$ and $w_{t+1}[r] = (1-\varepsilon)w_t[r]$. If $r_{t+1} = r$, then

$$v_{t+1}[r] = \frac{(1-\varepsilon)w_t[r]v_t[r] + f(x_{t+1}) - f(x_t)}{(1-\varepsilon)w_t[r] + 1},$$
$$w_{t+1}[r] = (1-\varepsilon)w_t[r] + 1.$$

For exploiting the experience gained in the learning iterations, we adopt a greedy strategy and always choose the mutation strength with highest velocity (breaking ties randomly, but giving preference to the previous-best mutation strength). While we generally postpone a discussion on parameter settings and other design choices to Sect. 5, let us remark already here that our greedy choice of the mutation strength might be detrimental for fitness landscapes in which the optimal mutation strength changes very frequently. There a velocity-weighted random choice might be more fruitful.

From this discussion, we derive the algorithm *RLS with self-adjusting mutation strength* (Algorithm 2).

3 Experimental Results

In this section we describe some experimental results for our algorithm. These are by no means intended to account for a thorough scientific investigation, both for reasons of space and because we feel that the mathematical investigation in the subsequent section is more insightful, also with respect to *why* the proposed ideas

Algorithm 2. RLS with self-adjusting mutation strength. The parameters of the algorithm are the maximum mutation strength $r_{\max}$, the learning rate δ, and the forgetting rate ε.

```
1  Initialization:
2      Choose x ∈ {0,1}^n uniformly at random;
3      for r = 1 to r_max do  v[r] := 0 and w[r] := 0 ;
4      r* ← 1;
5  Optimization: for t = 1,2,3,... do
6      z ← random([0,1]);
7      if z ≤ δ then    % learning iteration
8          r ← random({1,...,r_max});
9          y ← flip(x, r);
10         v[r] ← ((1−ε)w[r]v[r] + max{0, f(y) − f(x)}) / ((1−ε)w[r] + 1);
11         w[r] ← (1 − ε)w[r] + 1;
12         if f(y) ≥ f(x) then x ← y;
13         for r' ∈ {1,...,r_max} \ {r} do w[r'] ← (1 − ε)w[r'];
14     else
15         r+ ← random(argmax_r(v[r]));
16         if v[r+] > v[r*] then r* ← r+;
17         y ← flip(x, r*);
18         if f(y) ≥ f(x) then x ← y;
19         for r ∈ {1,...,r_max} do w[r] ← (1 − ε)w[r];
```

work well. Nevertheless, the experimental results indicate that our new algorithm gives good results also for problems other than OneMax, they give some hints on how to choose the parameters $r_{\max}$, δ and ε (more on this in Sect. 5), and they taught us that finding suitable parameters was not very difficult—we were immediately faster than the $(1+1)$ EA and with at most a few trials were able to beat RLS. All experiments were repeated 100 times; all numbers given below are the averages of these 100 runs.

LeadingOnes Function: The LeadingOnes function is defined by $\text{Lo}(x) := \max\{i \in [0..n] \mid \forall j \leq i : x_j = 1\}$, that is, it counts, starting from the left end, how many consecutive ones the bit-string x contains. The expected optimization time (number of iterations until the optimum is found) for RLS is $0.5n^2 \pm O(n)$, that of the $(1+1)$ EA with mutation rate $p = 1/n$ is $0.5n^2(1-1/n)((1-1/n)^{-n}-1) = 0.5(e-1)n^2 \pm O(n) \approx 0.8591n^2$. When taking the asymptotically optimal mutation rate of approximately $1.59/n$, the optimization time drops to approximately $0.7720n^2$. When taking a (best-possible) fitness-dependent mutation rate of $p_i = 1/(i+1)$ at fitness i, then the optimization time drops to $(e/4)n^2 \pm O(n) \approx 0.6796n^2$ [2].

Experimentally, for $n = 10,000$ and taking the parameters $r_{\max} = 5$, $\delta = 0.1$ and $\varepsilon = 1/(5,000,000)$, we observed an average optimization time of 45.0 million iterations, that is, $0.450n^2$, which clearly beats RLS and all $(1+1)$ EA results described above. The relative standard deviation is low, 4.36 % to be precise.[1]

Minimum Spanning Trees: Given a connected undirected graph $G = (V, E)$ with edge weights $w : E \to \mathbb{R}_{>0}$, the *minimum spanning tree problem* asks for finding a tree in G that connects all vertices and that has minimal total weight. This problem can be solved via evolutionary methods by taking a bit-string representation (each bit describes whether some edge is part of the tree or not) and taking as fitness function (to be minimized) the sum of the weights of the edges in the string representation plus a punishment term for each connected component (except the first one). For this representation of the problem, both RLS (flipping one or two bits with equal probability) and the $(1+1)$ EA find an optimal solution in any input in expected time $O(|E|^2 \log(|E| w_{\max})$, where $w_{\max}$ is the maximum weight of an edge (see [10]).

We ran the following experiments. We took as graph G the complete graph on 50 vertices (hence $|E| = 1225$) with edge weights chosen independently at random in $[0, 1]$, thus having a unique minimum spanning tree. On this instance, RLS in the variant that flips either one or two bits (random choice between these two alternatives) took $5.08 \cdot 10^6 \pm 37.75\,\%$ iterations. Our algorithm with $r_{\max} = 5$, $\delta = 0.1$, and $\varepsilon = 1/(20,000)$ took $2.70 \cdot 10^6 \pm 36.34\,\%$ iterations. Analyzing these runs in more detail, we observe that the preferred mutation strength r after a short initial phase takes the maximum value 5, then decreases to one, and finally goes back to two, which is then used for the large remainder of the optimization process. For reasons of computation time, we could not evaluate the $(1+1)$ EA on graphs on 50 vertices. For graphs on 20 vertices, the $(1+1)$ EA was roughly 2.7 times slower than RLS.

4 Mathematical Runtime Analysis on OneMax

In this section, we analyze via mathematical means how our algorithm optimizes OneMax. This is an asymptotic analysis in terms of the problem size n. We refer to the previous well-established runtime analysis literature for more details on the motivations of mathematical runtime analysis and on the meaning of asymptotic results, cf. [6].

The main result of this section is a proof that our algorithm with reasonable parameter settings very precisely detects the optimal mutation strength. It thus, apart from the learning iterations, has the same performance as the recently proposed randomized local search algorithm with fitness-dependent mutation strength [4]. The main technical challenge in this analysis are the dependencies between the progress of the algorithm and the learning system trying to estimate the velocities. We overcome these, among others, via a domination argument developed in [3, Lemma 1.20].

Throughout this section, we assume that $r_{\max}$ is a constant independent of n. For simplicity, we only regard the parameters $\varepsilon = n^{-0.99}$ and $\delta = n^{-0.01}$ but

[1] We report in the following the mean and relative standard deviations of our experiments by expressing, for example, the previous numbers as $45.0 \cdot 10^6 \pm 4.36\,\%$.

remark that broader ranges of these parameters would work as well. In addition to the notation introduced in Sect. 2, we write r_t^* for the number of bits flipped in a non-learning iteration. We also define the fitness distance $d(x) = n - f(x)$ for all $x \in \{0,1\}^n$.

For reasons of space, all proofs had to be removed from this extended abstract. They will be made available in a full journal version.

The following lemma states that, apart from an initial segment of the optimization process, the values of $w_t[r]$ can essentially assumed to be constant over time.

Lemma 1. *Let $r \in [1..r_{\max}]$, $t \geq H := (1/\varepsilon)\ln(n)$ and $w^* := \delta/r_{\max}\varepsilon$. Then with probability $1 - \exp(-n^{\Omega(1)})$, $|w_t[r] - w^*| \leq w^* O(n^{-0.002})$.*

The following two lemmas show how well our learning mechanism is able to detect the currently most profitable mutation strength. We denote in the following the progress from flipping r bits when in distance d from the optimum by $X_d^r := \max\{d(x) - d(\text{flip}(x,r)), 0\}$, where x is any search point with $d(x) = d$.

Lemma 2. *Let $r \in [1..r_{\max}]$, $t \geq 1$ and $H = (1/\varepsilon)2\ln(n)$. Then with probability at least $1 - \exp(-n^{\Omega(1)})$, we have $v_{t+H}[r] \leq (1 + O(n^{-0.002}))\max\{E[X^r_{d(x_{t+H})}], (\varepsilon/\delta)n^{0.01}\}$.*

Lemma 3. *Let $r \in [1..r_{\max}]$, $t \geq 1$ and $H = (1/\varepsilon)\ln(n)$. Assume that $E[X^r_{d(x_{t+H})}] = \Omega(n^{0.01}\varepsilon/\delta)$. Then with probability at least $1 - \exp(-n^{\Omega(1)})$, we have $v_{t+H}[r] \geq (1 - O(n^{-0.002}))E[X^r_{d(x_{t+H})}]$.*

The fact that our algorithm very precisely detects the optimal mutation strength implies that its fitness progress in each iteration is very close to the maximum possible (Theorem 1) and that it has a performance very close to the algorithm developed in [4] (Theorem 2).

Theorem 1. *Let T be the optimization time of our algorithm with parameters $\delta = n^{-0.01}$ and $\varepsilon = n^{-0.99}$ on* OneMax*. Let $T' = \min\{T, 2n\ln(n)\}$. Then with probability at least $1 - O(n^{0.19})$, for each non-learning iteration $t \in [2\ln(n)/\varepsilon, T']$, we have $E[X^{r_t^*}_{d(x_{t-1})}] \geq (1 - O(n^{-0.002}))\max\{E[X^r_{d(x_{t-1})}] \mid r \in [1..r_{\max}]\}$.*

Theorem 2. *Let $T_{r_{\max}}$ be the minimal expected runtime on the* OneMax *problem among all randomized local search algorithms with fitness-dependent mutation strength flipping at most $r_{\max}$ bits. Then the expected runtime T of our algorithm A is at most $T_{r_{\max}} + o(n)$. Consequently, by taking $r_{\max}$ large enough, our algorithm has the same expected runtime (apart from $o(n)$ terms) as the algorithm using the optimal fitness-dependent mutation strength of [4]. Also, let x_A be the current solution of our algorithm and x_{RLS} be the current solution of randomized local search after a fixed budget of $B \geq 0.2675n$ iterations. Then the expected Hamming distances to the optimum x^* satisfy $E[H(x_A, x^*)] \leq (1 + o(1))0.872E[H(x_{RLS}, x^*)]$.*

5 Parameter Choice and Design Alternatives

In this work we have proposed and analyzed a first hill climber that adaptively—based on previous fitness improvements—decides how many bits to flip in the mutation step. The required design choices were influenced by the positive experimental results and by our desire to prove with mathematical means that this algorithm tracks well the optimal mutation strength recently exhibited for OneMax. We now discuss two design variants that might prove useful for other optimization problems and give some general hints on how to choose the parameters of the algorithm.

A first observation is that our algorithm does not necessarily converge to an optimal solution. If there are local optima that can only be left by flipping more than $r_{\max}$ bits, then our algorithm has a positive probability of being stuck in such an optimum indefinitely. A simple way to overcome this is to use the mechanism proposed in this work only to redistribute the probability mass on r-bit-flips with $r \in [1..r_{\max}]$ and keep the probability distribution from standard bit mutation for the rest. In other words, in the main optimization loop in a non-learning iteration with probability $\Pr[\mathcal{B}(n, 1/n) > r_{\max}]$, the algorithm determines r according to the binomial distribution $\mathcal{B}(n, 1/n)$ conditional on being greater than $r_{\max}$; and it determines r from the learned velocities otherwise. This obviously ensures that the algorithm converges.

A second observation is that the highly greedy choice of r as the maximizer of the learned velocities might be too greedy for less well-behaved optimization problems in which the ideal mutation strength changes very frequently. In such situations it might be preferable to use the learned velocities only to give a mild preference to seemingly more profitable strengths. For example, in line 15 of Algorithm 2 one could choose r^+ with probability proportional to $v[r]$ and then flip r^+ bits.

A final modification that we want to propose is to use the progress experienced in any iteration (and not only the learning iterations) to update the velocities. Our update rule is designed in a way that different frequencies of the r values impose no problems. Hence in a sense, we are currently wasting the information available from the non-learning iterations. Our main motivation for doing so is that the mathematical analysis would have been more difficult, in particular, Lemma 1 would not be true with updates in each iteration. In experiments, the version with updates in each iteration usually, but not consistently, performed better.

A word on the parameters: it seems advisable to choose $r_{\max}$ small, because the learning effort is proportional to $r_{\max}$ and because in the vast majority of the iterations a small r was optimal. In our experiments, we always obtained good results with $r_{\max} = 5$, but we admit that in this first study we have not conducted an exhaustive series of experiments (also not for the other parameters). For the learning rate δ, we obtained good results with $\delta = 0.1$. It is immediately clear that $\delta(1-1/r_{\max})$ is the rate of iterations using a non-optimal mutation strength (unless two strengths are equally good), which gives some motivation to keep δ small. Of course, often a non-optimal mutation strength still has a reasonable chance of giving progress, so these iterations often are not completely wasted (that is, only spent on learning and not on optimization). The parameter hardest to set is ε. A large value implies that we quickly forget the outcomes of previous iterations. This may allow a quick adaption to a changed environment, but also carries the risk that a rare exceptional success with a non-ideal r-value has a too large influence. In our experiments, the latter aspect was seemingly more dominant and we obtained the best results with relatively small ε-values like 0.01 or even the reciprocal of 0.1 times the expected total number of iterations.

6 Conclusion

We proposed and analyzed a simple hill climber using k-bit flips with a self-adjusting choice of the mutation strength k. This use of k-bit flips instead of the usually preferred standard bit mutation with its random mutation strength allowed to much better exploit the most effective mutation strength. At the same time, the self-adjusting choice allowed to find the optimal mutation strength automatically and on-the-fly. By this, also the risk of getting stuck in local optima, the known draw-back of k-bit flips, was overcome. We are confident that replacing standard bit mutation by k-bit flips with a self-adjusting choice of k will lead to performance gains for many optimization problem beyond the ones regarded in this work.

References

1. Bäck, T.: An overview of parameter control methods by self-adaption in evolutionary algorithms. Fundam. Inform. **35**(1–4), 51–66 (1998)
2. Böttcher, S., Doerr, B., Neumann, F.: Optimal fixed and adaptive mutation rates for the LeadingOnes problem. In: Schaefer, R., Cotta, C., Kołodziej, J., Rudolph, G. (eds.) PPSN XI. LNCS, vol. 6238, pp. 1–10. Springer, Heidelberg (2010)
3. Doerr, B.: Analyzing randomized search heuristics: tools from probability theory. In: Auger, A., Doerr, B. (eds.) Theory of Randomized Search Heuristics, pp. 1–20. World Scientific Publishing, Singapore (2011)
4. Doerr, B., Doerr, C., Yang, J.: Optimal parameter choices via precise black-box analysis. In: GECCO 2016. ACM (2016, to appear)
5. Eiben, A.E., Hinterding, R., Michalewicz, Z.: Parameter control in evolutionary algorithms. IEEE Trans. Evol. Comput. **3**, 124–141 (1999)
6. Jansen, T.: Analyzing Evolutionary Algorithms–The Computer Science Perspective. Natural Computing Series. Springer, Heidelberg (2013)
7. Jansen, T., Wegener, I.: On the analysis of a dynamic evolutionary algorithm. J. Discrete Algorithms **4**, 181–199 (2006)
8. Jansen, T., Zarges, C.: Performance analysis of randomised search heuristics operating with a fixed budget. Theor. Comput. Sci. **545**, 39–58 (2014)
9. Karafotias, G., Hoogendoorn, M., Eiben, A.: Parameter control in evolutionary algorithms: trends and challenges. IEEE Trans. Evol. Comput. **19**, 167–187 (2015)

10. Neumann, F., Wegener, I.: Randomized local search, evolutionary algorithms, and the minimum spanning tree problem. Theor. Comput. Sci. **378**, 32–40 (2007)
11. Oliveto, P.S., Lehre, P.K., Neumann, F.: Theoretical analysis of rank-based mutation - combining exploration and exploitation. In: CEC 2009, pp. 1455–1462. IEEE (2009)
12. Thierens, D.: Adaptive mutation rate control schemes in genetic algorithms. In: CEC 2002, pp. 980–985. IEEE (2002)

Selection Hyper-heuristics Can Provably Be Helpful in Evolutionary Multi-objective Optimization

Chao Qian[1,2](✉), Ke Tang[1], and Zhi-Hua Zhou[2]

[1] UBRI, School of Computer Science and Technology, University of Science and Technology of China, Hefei 230027, China
{chaoqian,ketang}@ustc.edu.cn

[2] National Key Laboratory for Novel Software Technology, Nanjing University, Nanjing 210023, China
zhouzh@nju.edu.cn

Abstract. Selection hyper-heuristics are automated methodologies for selecting existing low-level heuristics to solve hard computational problems. They have been found very useful for evolutionary algorithms when solving both single and multi-objective real-world optimization problems. Previous work mainly focuses on empirical study, while theoretical study, particularly in multi-objective optimization, is largely insufficient. In this paper, we use three main components of multi-objective evolutionary algorithms (selection mechanisms, mutation operators, acceptance strategies) as low-level heuristics, respectively, and prove that using heuristic selection (i.e., mixing low-level heuristics) can be exponentially faster than using only one low-level heuristic. Our result provides theoretical support for multi-objective selection hyper-heuristics, and might be helpful for designing efficient heuristic selection methods in practice.

1 Introduction

Hyper-heuristics are automated methodologies for selecting or generating heuristics to solve hard computational problems [4]. There are two main hyper-heuristic categories: heuristic selection and heuristic generation. This paper focuses on the former type. Given a set of low-level heuristics, a heuristic selection method chooses an appropriate one to be applied at each decision point.

Selection hyper-heuristics have been widely and successfully applied for evolutionary algorithms (EAs) solving single-objective optimization problems such as personnel scheduling, packing, vehicle routing, etc [3]. After that, they start to emerge in evolutionary multi-objective optimization. Burke et al. [5] first proposed a multi-objective hyper-heuristic approach based on tabu search for

This work was supported by the NSFC (61333014, 61329302), the Fundamental Research Funds for the Central Universities (WK2150110002), and the Collaborative Innovation Center of Novel Software Technology and Industrialization.

J. Handl et al. (Eds.): PPSN XIV 2016, LNCS 9921, pp. 835–846, 2016.
DOI: 10.1007/978-3-319-45823-6_78

the space allocation and timetabling problems. McClymont and Keedwell [17] developed a Markov chain based hyper-heuristic method for designing water distribution network. By using NSGAII, SPEA2 and MOGA as low-level heuristics, Maashi et al. [16] designed a choice function based hyper-heuristic to solve the vehicle crashworthiness design problem. Selection hyper-heuristics also achieved successes in other real-world multi-objective optimization problems, e.g., the 2D guillotine strip packing problem [18] and the integration and test order problem in software engineering [10].

Most previous work focuses on empirical study. Meanwhile, theoretical analysis, particularly running time analysis, is important for enhancing our understanding and designing efficient hyper-heuristics, as Burke et al. stated in [3]. However, the running time analysis on selection hyper-heuristics is difficult due to their complexity and randomness, and few results have been reported. By using mutation operators as low-level heuristics, He et al. [11] first gave some conditions under which the asymptotic hitting time of the (1+1)-EA (a simple EA) with a mixed strategy is not larger than that with any pure strategy. Note that a mixed strategy (which corresponds to heuristic selection) chooses one low-level heuristic according to some distribution each time, while a pure strategy uses only one fixed low-level heuristic. Their result was then extended to the expected running time measure and to population-based EAs in [12]. Lehre and Özcan [15] later gave concrete evidence that mixing two specific mutation operators is more efficient than using only one operator for the (1+1)-EA solving the GapPath function. They also proved the benefit of mixing acceptance strategies for the (1+1)-EA solving the RR_k function. In [19], by mixing global and local mutation operators, the (1+1)-EA was proved to be a polynomial time approximation algorithm for the NP-hard single machine scheduling problem. In [6], the (1+1)-EA mixing two specific mutation operators was shown to be able to solve the easiest function for each mutation operator efficiently. The above studies investigate whether selection hyper-heuristics can bring an improvement on the performance. Alanazi and Lehre [1] also compared different heuristic selection methods (i.e., mixed strategies with different distributions), and proved their similar performance for the (1+1)-EA solving the LeadingOnes function.

All of the above-mentioned studies consider single-objective optimization. To the best of our knowledge, there has been no theoretical work supporting the effectiveness of selection hyper-heuristics in multi-objective optimization. In this paper, we prove that using heuristic selection can speed up evolutionary multi-objective optimization exponentially via rigorous running time analysis. The widely used multi-objective EA GSEMO in previous theoretical analyses [8,14,21] is employed. It repeats three steps: choosing a solution by some selection mechanism, reproducing a new solution by mutation, and updating the population by some acceptance strategy. This paper considers the three main components of GSEMO, i.e., selection mechanism, mutation operator and acceptance strategy, as the low-level heuristic, respectively. For each kind of low-level heuristic, we give a bi-objective pseudo-Boolean function, and prove that the expected running time of GSEMO with a mixed strategy is polynomial while GSEMO with a pure strategy needs at least exponential running time. The analysis also shows that the helpfulness of selection hyper-heuristics is because the strengths of one heuristic can compensate for the weaknesses of another. For mixing acceptance strategies, we also empirically compare the running time of GSEMO with different mixed strategies, and the results imply the importance of choosing a proper heuristic selection method.

The rest of this paper is organized as follows. Section 2 introduces some preliminaries. The helpfulness of mixing selection mechanisms, mutation operators and acceptance strategies is then theoretically analyzed. Finally, we conclude the paper.

2 Preliminaries

Multi-objective optimization requires to simultaneously optimize two or more objective functions, as shown in Definition 1. Note that maximization is considered in this paper. The objectives are usually conflicted, and thus there is no canonical complete order on the solution space $\mathcal{X}$. The comparison between solutions relies on the *domination* relationship, as presented in Definition 2. A solution is *Pareto optimal* if there is no other solution in $\mathcal{X}$ that dominates it. The set of objective vectors of all the Pareto optimal solutions constitutes the *Pareto front*. The goal of multi-objective optimization is to find the Pareto front, that is, to find at least one corresponding solution for each element in the Pareto front. In this paper, we consider the Boolean space, i.e., $\mathcal{X} = \{0,1\}^n$.

Definition 1 (Multi-objective Optimization). *Given a feasible solution space $\mathcal{X}$ and objective functions $f_1, \ldots, f_m$, multi-objective optimization can be formulated as*

$$\max_{\boldsymbol{x} \in \mathcal{X}} \ \big(f_1(\boldsymbol{x}), f_2(\boldsymbol{x}), ..., f_m(\boldsymbol{x})\big).$$

Definition 2 (Domination). *Let $\boldsymbol{f} = (f_1, f_2, \ldots, f_m) : \mathcal{X} \to \mathbb{R}^m$ be the objective vector. For two solutions $\boldsymbol{x}$ and $\boldsymbol{x}' \in \mathcal{X}$:*

1. $\boldsymbol{x}$ weakly dominates $\boldsymbol{x}'$ *if,* $\forall 1 \leq i \leq m, f_i(\boldsymbol{x}) \geq f_i(\boldsymbol{x}')$, *denoted as* $\boldsymbol{x} \succeq \boldsymbol{x}'$*;*
2. $\boldsymbol{x}$ dominates $\boldsymbol{x}'$ *if,* $\boldsymbol{x} \succeq \boldsymbol{x}'$ *and* $f_i(\boldsymbol{x}) > f_i(\boldsymbol{x}')$ *for some* i*, denoted as* $\boldsymbol{x} \succ \boldsymbol{x}'$*.*

Evolutionary algorithms (EAs) have become a popular tool for multi-objective optimization, due to their population-based nature. In previous theoretical studies, GSEMO is the most widely used multi-objective EA (MOEA) [8,14,21]. As described in Algorithm 1, it first randomly selects an initial solution, then repeats the three steps (selection, mutation, acceptance) to improve the quality of the population. In selection, a solution is uniformly selected from the current population; in mutation, a new solution is generated by flipping each bit of the selected solution with probability $\frac{1}{n}$; in acceptance, the new solution is compared with the solutions in the population, and then only non-dominated solutions are kept. Although simple, GSEMO explains the common structure of various MOEAs, and hence will be used in this paper as well.

Algorithm 1. GSEMO

Given the solution space $\mathcal{X} = \{0,1\}^n$ and the objective function vector $\boldsymbol{f}$, GSEMO consists of the following steps:

```
1: Choose x ∈ X uniformly at random
2: P ← {x}
3: repeat
4:    [Selection] Choose x from P uniformly at random
5:    [Mutation] Create x' by flipping each bit of x with probability 1/n
6:    [Acceptance] if ∄z ∈ P such that z ≻ x'
7:                    P ← (P − {z ∈ P | x' ⪰ z}) ∪ {x'}
8:                 end if
9: until some criterion is met
```

Selection hyper-heuristics manage a set of low-level heuristics, and select an appropriate one to be applied at each decision point. Despite their practical successes, the theoretical analysis is still in its infancy, particularly for multi-objective optimization. In this paper, we take the three components of GSEMO, i.e., selection, mutation and acceptance, as the low-level heuristic, respectively, and compare the performance of GSEMO with a mixed strategy and a pure strategy. For each component of GSEMO, we will use two concrete low-level heuristics. A typical mixed strategy employed in our analysis (denoted by GSEMO_p) is to use the first low-level heuristic with probability $p \in [0,1]$ in each iteration of GSEMO, and use the second one otherwise. Note that a mixed strategy corresponds to using heuristic selection, while a pure strategy only uses one specific low-level heuristic and thus implies that heuristic selection is not employed.

The performance of the comparison algorithms is measured by their running time complexity. Note that running time analysis has been a leading theoretical aspect for randomized search heuristics [2,20]. The running time of a MOEA is usually counted by the number of fitness evaluations (the most costly computational process) until finding the Pareto front [8,14,21].

3 Mixing Selection Mechanisms

In this section, we use two fair selection mechanisms [9,14] as low-level heuristics:

- **fair selection w.r.t. the decision space:** Each solution in the current population has a counter c_1, which records the number of its offsprings. The solution with the smallest c_1 value will be selected for reproduction in each iteration. That is, line 4 of Algorithm 1 changes to be "Choose $\boldsymbol{x} \in \{\boldsymbol{y} \in P \mid c_1(\boldsymbol{z}) \geq c_1(\boldsymbol{y}), \forall \boldsymbol{z} \in P\}$ uniformly at random".
- **fair selection w.r.t. the objective space:** Each counter (denoted as c_2) is associated with an objective vector rather than a decision vector. Line 4 thus changes to be "Choose $\boldsymbol{x} \in \{\boldsymbol{y} \in P \mid c_2(f(\boldsymbol{z})) \geq c_2(f(\boldsymbol{y})), \forall \boldsymbol{z} \in P\}$ uniformly at random".

Fairness is employed to balance the number of offsprings of all solutions in the current population, and thus to achieve a good spread over the Pareto front. GSEMO with these two mechanisms are denoted by GSEMO_{ds} and GSEMO_{os}, respectively. For GSEMO with the mixed strategy (denoted by GSEMO_p), it uses the fairness w.r.t the decision space with probability $p \in [0,1]$ in each iteration; otherwise, it uses the fairness w.r.t the objective space. Note that GSEMO_{ds} and GSEMO_{os} are GSEMO_p with $p = 1$ and $p = 0$, respectively.

We then compare their running time on the ZPLG function. As shown in Definition 3, ZPLG can be divided into three parts: ZeroMax, a plateau, and a path with little gaps. It is obtained from the PLG function in [9] by replacing the second objective value 1 in the ZeroMax part with 2. The Pareto front is $\{(n,2),(n+1,1),(\frac{9n}{8}+2,0)\}$, and the corresponding Pareto optimal solutions are 0^n, SP_1 and 1^n, respectively.

Definition 3 (ZPLG).

$$ZPLG(\boldsymbol{x}) = \begin{cases} (|\boldsymbol{x}|_0, 2) & \boldsymbol{x} \notin SP_1 \cup SP_2 \\ (n+1, 1) & \boldsymbol{x} \in SP_1 \\ (n+2+i, 0) & \boldsymbol{x} = 1^{3n/4+2i}0^{n/4-2i} \in SP_2, \end{cases}$$

where $|\boldsymbol{x}|_0 = \sum_{j=1}^{n}(1-x_j)$ *denotes the number of 0-bits,* $SP_1 = \{1^i0^{n-i} \mid 1 \leq i < 3n/4\}$, $SP_2 = \{1^{3n/4+2i}0^{n/4-2i} \mid 0 \leq i \leq n/8\}$ *and* $n = 8m, m \in \mathbb{N}$.

Theorem 1 shows that GSEMO with a pure strategy needs exponential running time with a high probability. The result of GSEMO_{os} on ZPLG is directly from that on the PL function (i.e., Theorem 1) in [9], since ZPLG has the same structure as PL by treating its SP_2 part as a whole. The inefficiency is because GSEMO_{os} allows the Pareto optimal solution 0^n to generate new solutions in SP_1, which stop the random walk on the plateau SP_1 and thus prevent from reaching SP_2. The result of GSEMO_{ds} on ZPLG can be directly from that on the PLG function (i.e., Theorem 4) in [9], since their proof relies on SP_1 and SP_2, which are the same for ZPLG and PLG. The inefficiency is because GSEMO_{ds} easily gets trapped in the random walk on SP_1, which prevents from following the path SP_2 to find the Pareto optimal solution 1^n. We then prove in Theorem 2 that by using the mixed strategy, GSEMO_p can solve ZPLG in polynomial running time. The idea is that first employing GSEMO_{ds} allows the random walk on SP_1 to reach SP_2, and then employing GSEMO_{os} allows following the path SP_2 to find 1^n. Thus, we can see that the advantage of using heuristic selection is that the strengths of one heuristic can compensate for the weaknesses of another.

Theorem 1. *On ZPLG, the running time of* GSEMO_{ds} *is* $2^{\Omega(n^{1/2})}$ *with probability* $1 - 2^{-\Omega(n^{1/2})}$, *and that of* GSEMO_{os} *is* $2^{\Omega(n^{1/4})}$ *with probability* $1 - e^{-\Omega(n^{1/3})}$.

Theorem 2. *The expected running time of* GSEMO_p *with* $p = 1 - \frac{1}{n^3}$ *on ZPLG is* $O(n^6)$.

Proof. We divide the optimization process into two phases: (1) starts after initialization and finishes until the population P contains 0^n, a solution from SP_1 and a solution from SP_2; (2) starts after phase (1) and finishes until P contains 0^n, a solution from SP_1 and 1^n, i.e., the Pareto front is found.

For the first phase, we can follow the analysis of GSEMO$_{ds}$ on PL (i.e., Theorem 2) in [9]. In their proof, the only part relying on the fair selection w.r.t. the decision space is to allow a consecutive random walk of length δn^3 (δ is a constant) on the plateau SP_1, under the condition that the c_1 value of the maintained solution from SP_1 is always smaller than that of the Pareto optimal solution 0^n. Note that the fair selection w.r.t. the decision space is used with probability $1-1/n^3$ in each iteration of GSEMO$_p$. Such a random walk happens with probability $(1-\frac{1}{n^3})^{\delta n^3} \geq (2e)^{-\delta} \in \Omega(1)$. Thus, the asymptotic running time is not affected, and the expected running time of this phase is the same as that of GSEMO$_{ds}$ on PL, i.e., $O(n^3 \log n)$.

For the second phase, the population P always contains three solutions, 0^n, a solution from SP_1 and a solution from SP_2. The probability that a better solution from SP_2 is found under the condition that a solution from SP_2 has been selected for mutation is at least $\frac{1}{n^2}(1-\frac{1}{n})^{n-2} \geq \frac{1}{en^2}$, since it is sufficient to flip the leftmost two 0-bits. It is easy to see that at most $\frac{n}{8}$ such improvements are sufficient to find the Pareto optimal solution 1^n. The worst case is reached when the first found solution from SP_2 is $1^{3n/4}0^{n/4}$. We consider that the fair selection w.r.t. the objective space is used, which happens with probability $\frac{1}{n^3}$ in each iteration of GSEMO$_p$. Because the c_2 values of $(n, 2)$ and $(n+1, 1)$ (i.e., the objective vectors of 0^n and the solution from SP_1) are never decreased, the solution from SP_2 is selected for reproduction at least once in three consecutive iterations. Thus, the expected running time of this phase is at most $n^3 \cdot 3 \cdot \frac{n}{8} \cdot en^2 \in O(n^6)$. □

4 Mixing Mutation Operators

In this section, we use two mutation operators [15] as low-level heuristics:

- **one-bit mutation:** Line 5 of Algorithm 1 changes to be "Create $\boldsymbol{x}'$ by flipping one randomly chosen bit of $\boldsymbol{x}$". Note that one specific bit is chosen with probability $\frac{1}{n}$.
- **two-bit mutation:** Line 5 of Algorithm 1 changes to be "Create $\boldsymbol{x}'$ by flipping two different and randomly chosen bits of $\boldsymbol{x}$". Note that two specific bits are chosen with probability $1/\binom{n}{2} = \frac{2}{n(n-1)}$.

GSEMO with these two operators are denoted by GSEMO$_{1b}$ and GSEMO$_{2b}$, respectively. GSEMO with the mixed strategy (denoted by GSEMO$_p$) uses one-bit mutation with probability $p \in [0, 1]$ in each iteration; otherwise, it uses two-bit mutation.

We then compare their running time on the SPG function. As shown in Definition 4, SPG has a short path SP with increasing fitness except the solutions 1^i0^{n-i} with $i \bmod 3 = 1$. The construction of SPG is inspired from the GapPath function in [15]. The Pareto front is $\{(n,1),(n^2,0)\}$, and the corresponding Pareto optimal solutions are 0^n and 1^n, respectively.

Definition 4 (SPG).

$$SPG(\boldsymbol{x}) = \begin{cases} (|\boldsymbol{x}|_0, 1) & \boldsymbol{x} \notin SP \\ (-1, 0) & \boldsymbol{x} = 1^i 0^{n-i} \in SP,\ i \bmod 3 = 1 \\ (in, 0) & \boldsymbol{x} = 1^i 0^{n-i} \in SP,\ i \bmod 3 = 0 \text{ or } 2, \end{cases}$$

where $SP = \{1^i 0^{n-i} \mid 1 \leq i \leq n\}$ *and* $n = 3m, m \in \mathbb{N}$.

The following two theorems show that the expected running time of GSEMO with a pure strategy is infinite while that of GSEMO with the mixed strategy is polynomial. The proof idea is straightforward. In every three adjacent solutions on the path SP, there is a bad one $1^i 0^{n-i}$ with i mod $3 = 1$. Using one-bit and two-bit mutation alternatively can jump over those bad solutions on SP and finally reach the Pareto optimal solution 1^n, while using only one-bit or two-bit mutation obviously will get stuck in some solution $1^i 0^{n-i}$ with i mod $3 = 0$ or 2.

Theorem 3. *The expected running time of GSEMO$_{1b}$ and GSEMO$_{2b}$ on SPG is infinite.*

Proof. We consider that the initial solution is the Pareto optimal solution 0^n, which has the objective vector $(n,1)$. This happens with probability $\frac{1}{2^n}$ due to the uniform sampling. For GSEMO$_{1b}$, one-bit mutation on 0^n can only generate solutions with the objective vector $(n-1,1)$ or $(-1,0)$, which are dominated by 0^n. Thus, the population P will always contain only 0^n. For GSEMO$_{2b}$, two-bit mutation on 0^n generates solutions with the objective vector $(n-2,1)$ or $(2n,0)$. Thus, P contains 0^n and $1^2 0^{n-2}$ after a while. Since two-bit mutation on $1^2 0^{n-2}$ cannot generate better solutions on SP, P will always keep in this state. Thus, starting from 0^n, either GSEMO$_{1b}$ or GSEMO$_{2b}$ cannot find the Pareto front, which implies that the expected running time is infinite. □

Theorem 4. *The expected running time of GSEMO$_p$ with $p \in [0,1]$ being a constant on SPG is $O(n^3)$.*

Proof. The population P contains at most two solutions, because the second objective of SPG has only two values 0 and 1. We first analyze the expected running time until the Pareto optimal solution 0^n is found. Let $\boldsymbol{x}$ denote the solution with the second objective value 1 in P. Such a solution will exist in P after at most n expected running time. This is because a solution from SP can generate an offspring solution not from SP by flipping the first 1-bit, which happens with probability at least $\frac{1}{n}$ by either one-bit or two-bit mutation. Assume that the number of 0-bits of $\boldsymbol{x}$ is j $(j \geq 1)$. It is easy to see that j cannot decrease, and it can increase by flipping one 1-bit (but not the last) using one-bit mutation. Because the probability of selecting $\boldsymbol{x}$ for mutation is at least $\frac{1}{2}$ and one-bit mutation is used with probability p, the probability of increasing j by 1 in one iteration is at least $\frac{1}{2} \cdot p \cdot \frac{n-j-1}{n}$ for $j \leq n-2$ and $\frac{1}{2} \cdot p \cdot \frac{1}{n}$ for $j = n-1$. Thus, the expected running time to find 0^n is at most $\sum_{j=1}^{n-2} \frac{2n}{p(n-j-1)} + \frac{2n}{p} \in O(n \log n)$.

When finding 0^n, we pessimistically assume that the solution from SP has not been found. Starting from 0^n, the solution $1^2 0^{n-2}$ can be found by flipping the first two 0-bits using two-bit mutation. This happens with probability $(1-p) \cdot \frac{2}{n(n-1)}$, and thus the expected running time is $\frac{n(n-1)}{2(1-p)} \in O(n^2)$. Once a solution from SP with $i \bmod 3 \neq 1$ has been found, using one-bit and two-bit mutation alternatively can follow the path SP to find the Pareto optimal solution 1^n. If $i \bmod 3 = 2$, flipping its first 0-bit by one-bit mutation can generate a better solution. This happens with probability $\frac{1}{2} \cdot p \cdot \frac{1}{n}$. If $i \bmod 3 = 0$, flipping its first two 0-bits by two-bit mutation can generate a better solution. This happens with probability $\frac{1}{2} \cdot (1-p) \cdot \frac{2}{n(n-1)}$. Since $\frac{n}{3}$ such two improvements are sufficient to find 1^n, the expected running time is at most $\frac{n}{3} \cdot (\frac{2n}{p} + \frac{n(n-1)}{(1-p)}) \in O(n^3)$. Thus, the theorem holds. □

5 Mixing Acceptance Strategies

In this section, we use two acceptance strategies as low-level heuristics:

- **elitist acceptance:** As lines 6–8 of Algorithm 1, only non-dominated solutions are kept in the population, and the existing solution in P with the same objective vector as the newly generated solution will be replaced.
- **strict elitist acceptance:** It is the same as elitist acceptance, except that the old solution with the same objective vector as the newly generated solution will not be replaced. That is, line 6 of Algorithm 1 changes to be "if $\nexists \boldsymbol{z} \in P$ such that $\boldsymbol{z} \succeq \boldsymbol{x}'$".

The difference between these two strategies is to accept or reject the solution with the same fitness. This has been theoretically shown to have a significant effect on the performance of EAs in single-objective optimization [13]. Note that GSEMO with elitist acceptance is just GSEMO. GSEMO with the strict strategy is denoted by GSEMO_{strict}. In each iteration of GSEMO with the mixed strategy (denoted by GSEMO_{mixed}), elitist acceptance is used if the newly generated solution $\boldsymbol{x}'$ and the parent solution $\boldsymbol{x}$ have the same objective vector; otherwise, strict elitist acceptance is used. Note that the mixed strategy employed here is different from that of GSEMO_p.

We then compare their running time on the PL function. As shown in Definition 5, PL has a short path $SP - \{1^n\}$ with constant fitness. The Pareto front is $\{(n,1),(n+2,0)\}$, and the corresponding Pareto optimal solutions are 0^n and 1^n, respectively.

Definition 5 (PL) [8].

$$PL(\boldsymbol{x}) = \begin{cases} (|\boldsymbol{x}|_0, 1) & \boldsymbol{x} \notin SP = \{1^i 0^{n-i} \mid 1 \leq i \leq n\} \\ (n+1, 0) & \boldsymbol{x} \in \{1^i 0^{n-i} \mid 1 \leq i < n\} \\ (n+2, 0) & \boldsymbol{x} = 1^n. \end{cases}$$

Theorem 5 shows that GSEMO with a pure strategy on PL needs exponential running time. The result of GSEMO was proved in [8], and its inefficiency is because a solution not from SP can generate a new solution from SP, which stops the ongoing random walk on SP. The inefficiency of GSEMO$_{strict}$ is because the first found solution from SP is far from the Pareto optimal solution 1^n, and strict elitist acceptance does not allow the random walk on SP. We then prove in Theorem 6 that GSEMO with the mixed strategy can solve PL in polynomial running time. It works by allowing accepting the solution with the same fitness only in the random walk procedure.

Theorem 5. *On PL, the running time of GSEMO is* $2^{\Omega(n^{1/24})}$ *with probability* $1-e^{-\Omega(n^{1/24})}$ [8], *and that of GSEMO*$_{strict}$ *is* $n^{\Omega(\frac{n}{5})}$ *with probability* $1-2^{-\Omega(n)}$.

Proof. The initial solution is not in SP with probability $1-\frac{n}{2^n}$ due to uniform selection, and it has at most $\frac{2n}{3}$ 1-bits with probability $1-e^{-\Omega(n)}$ by Chernoff bounds. The population P contains at most two solutions, since the second objective of PL has only two different values. Note that the number of 1-bits of the solution not from SP will never increase, since the first objective is to maximize the number of 0-bits. Because the probability of flipping at least $\frac{n}{12}$ bits simultaneously in one step is less than $n^{-\frac{n}{12}}$, the first found solution from SP has at most $\frac{3n}{4}$ 1-bits with probability at least $1-n^{-\frac{n}{12}}$. Once a solution from $SP-\{1^n\}$ has been found, it will never change because $SP-\{1^n\}$ is a plateau and GSEMO$_{strict}$ will not replace the solution with the same fitness. Thus, P will always contain two solutions, a solution $\boldsymbol{x}$ from $SP-\{1^n\}$ with $|\boldsymbol{x}|_1 \leq \frac{3n}{4}$ and a solution $\boldsymbol{y}$ not from SP with $|\boldsymbol{y}|_1 \leq \frac{2n}{3}$. The probabilities of mutating $\boldsymbol{x}$ and $\boldsymbol{y}$ to 1^n in one step are at most $n^{-\frac{n}{4}}$ and $n^{-\frac{n}{3}}$, respectively. Thus, after $n^{\frac{n}{5}}$ steps, the Pareto optimal solution 1^n is generated with probability at most $n^{\frac{n}{5}} \cdot n^{-\frac{n}{4}} = n^{-\frac{n}{20}}$ by the union bound. By combining all the above probabilities, we get that the running time is $n^{\Omega(\frac{n}{5})}$ with probability $1-2^{-\Omega(n)}$. □

Theorem 6. *The expected running time of GSEMO*$_{mixed}$ *on PL is* $O(n^3)$.

Proof. Since the function PL outside SP has the same structure as OneMax, the expected steps to find 0^n is $O(n \log n)$ by using the analysis result of the (1+1)-EA on OneMax [7]. Then, it needs $O(n)$ expected steps to find a solution from SP, as it suffices to flip the leftmost 0-bit of 0^n. For GSEMO$_{mixed}$, if an offspring solution from SP is generated by mutation on the solution not from SP, it will not replace the solution from SP in the current population; but if it is generated by mutation on the current solution from SP, the replacement will be implemented. Thus, the algorithm will perform the random walk on the plateau SP and the solution not from SP will not influence it. Note that the solution from SP is selected for mutation with probability $\frac{1}{2}$. Using the analysis result of the (1+1)-EA on SPC (i.e., Theorem 7) in [13], we get that the random walk needs $O(n^3)$ expected running time to find 1^n. □

Note that the mixed strategy employed by GSEMO$_{mixed}$ here is different from that by GSEMO$_p$ for mixing selection mechanisms or mutation operators.

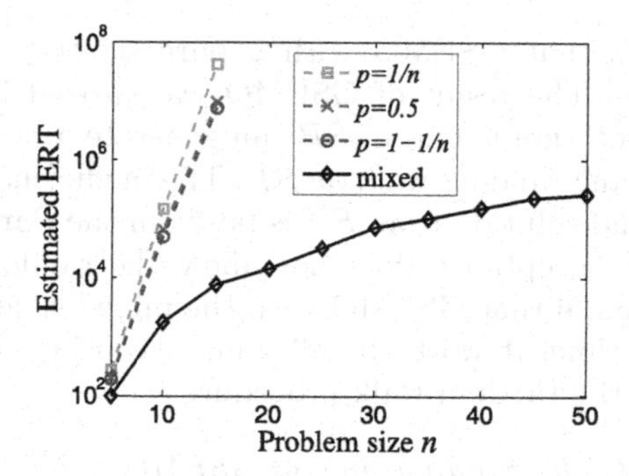

Fig. 1. Estimated ERT of GSEMO_{mixed} and GSEMO_p for solving the PL problem, where a base 10 logarithmic scale is used for the y-axis.

GSEMO_p uses the first low-level heuristic with probability $p \in [0,1]$ in each iteration and uses the second one otherwise. To investigate the influence of different mixed strategies, we conduct experiments to compare GSEMO_{mixed} with GSEMO_p for mixing elitist and strict elitist acceptance. The parameter p is set as $\frac{1}{n}$, 0.5 and $1-\frac{1}{n}$, respectively. For each comparison algorithm on each problem size $n \in \{5, 10, \ldots, 50\}$, we run the algorithm 100 times independently, where each run stops when the Pareto front of the PL problem is found. The average number of fitness evaluations is used as the estimation of the expected running time (ERT). The result is plotted in Fig. 1. Note that the ERT of GSEMO_p for $n \geq 20$ is too large to estimate. We can observe that GSEMO_{mixed} is much more efficient than GSEMO_p. The curves of GSEMO_{mixed} and GSEMO_p grow in a closely logarithmic and linear trend, respectively, which implies that their ERT is approximately polynomial and exponential, respectively. Thus, these empirical results suggest that choosing a proper threshold selection method is important.

6 Conclusion

This paper presents a theoretical study on the effectiveness of selection hyper-heuristics for multi-objective optimization. Rigorous running time analysis showed that applying selection hyper-heuristics to any of the three major components of a MOEA, i.e., selection, mutation and acceptance, can exponentially speed up the optimization. From the analysis, we find that selection hyper-heuristics work by allowing the strengths of one heuristic to compensate for the weaknesses of another. Our result provides theoretical support for multi-objective selection hyper-heuristics. The empirical comparison on different mixed strategies also implies the importance of choosing a proper heuristic selection method.

References

1. Alanazi, F., Lehre, P.K.: Runtime analysis of selection hyper-heuristics with classical learning mechanisms. In: Proceedings of CEC 2014, pp. 2515–2523, Beijing, China (2014)
2. Auger, A., Doerr, B.: Theory of Randomized Search Heuristics: Foundations and Recent Developments. World Scientific, Singapore (2011)
3. Burke, E.K., Gendreau, M., Hyde, M., Kendall, G., Ochoa, G., Özcan, E., Qu, R.: Hyper-heuristics: a survey of the state of the art. J. Oper. Res. Soc. **64**(12), 1695–1724 (2013)
4. Burke, E.K., Hyde, M., Kendall, G., Ochoa, G., Özcan, E., Woodward, J.R.: A classification of hyper-heuristic approaches. In: Gendreau, M., Potvin, J.-Y. (eds.) Handbook of Metaheuristics. International Series in Operations Research & Management Science, pp. 449–468. Springer, Heidelberg (2010)
5. Burke, E.K., Silva, J.D.L., Soubeiga, E.: Multi-objective hyper-heuristic approaches for space allocation and timetabling. In: Ibaraki, T., Nonobe, K., Yagiura, M. (eds.) Metaheuristics: Progress as Real Problem Solvers. Operations Research/Computer Science Interfaces Series, vol. 32, pp. 129–158. Springer, Heidelberg (2005)
6. Corus, D., He, J., Jansen, T., Oliveto, P.S., Sudholt, D., Zarges, C.: On easiest functions for somatic contiguous hypermutations and standard bit mutations. In: Proceedings of GECCO 2015, pp. 1399–1406, Madrid, Spain (2015)
7. Droste, S., Jansen, T., Wegener, I.: On the analysis of the (1+1) evolutionary algorithm. Theoret. Comput. Sci. **276**(1–2), 51–81 (2002)
8. Friedrich, T., Hebbinghaus, N., Neumann, F.: Plateaus can be harder in multi-objective optimization. Theoret. Comput. Sci. **411**(6), 854–864 (2010)
9. Friedrich, T., Horoba, C., Neumann, F.: Illustration of fairness in evolutionary multi-objective optimization. Theoret. Comput. Sci. **412**(17), 1546–1556 (2011)
10. Guizzo, G., Fritsche, G.M., Vergilio, S.R., Pozo, A.T.R.: A hyper-heuristic for the multi-objective integration and test order problem. In: Proceedings of GECCO 2015, pp. 1343–1350, Madrid, Spain (2015)
11. He, J., He, F., Dong, H.: Pure strategy or mixed strategy? In: Hao, J.-K., Middendorf, M. (eds.) EvoCOP 2012. LNCS, vol. 7245, pp. 218–229. Springer, Heidelberg (2012)
12. He, J., Hou, W., Dong, H., He, F.: Mixed strategy may outperform pure strategy: an initial study. In: Proceedings of CEC 2013, pp. 562–569, Cancun, Mexico (2013)
13. Jansen, T., Wegener, I.: Evolutionary algorithms-how to cope with plateaus of constant fitness and when to reject strings of the same fitness. IEEE Trans. Evol. Comput. **5**(6), 589–599 (2001)
14. Laumanns, M., Thiele, L., Zitzler, E.: Running time analysis of multiobjective evolutionary algorithms on pseudo-Boolean functions. IEEE Trans. Evol. Comput. **8**(2), 170–182 (2004)
15. Lehre, P.K., Özcan, E.: A runtime analysis of simple hyper-heuristics: to mix or not to mix operators. In: Proceedings of FOGA 2013, pp. 97–104, Adelaide, Australia (2013)
16. Maashi, M., Özcan, E., Kendall, G.: A multi-objective hyper-heuristic based on choice function. Expert Syst. Appl. **41**(9), 4475–4493 (2014)
17. McClymont, K., Keedwell, E.C.: Markov chain hyper-heuristic (MCHH): an online selective hyper-heuristic for multi-objective continuous problems. In: Proceedings of GECCO 2011, pp. 2003–2010, Dublin, Ireland (2011)

18. Miranda, G., De Armas, J., Segura, C., León, C.: Hyperheuristic codification for the multi-objective 2D guillotine strip packing problem. In: Proceedings of CEC 2010, pp. 1–8, Barcelona, Spain (2010)
19. Mitavskiy, B., He, J.: A polynomial time approximation scheme for a single machine scheduling problem using a hybrid evolutionary algorithm. In: Proceedings of CEC 2012, pp. 1–8, Brisbane, Australia (2012)
20. Neumann, F., Witt, C.: Bioinspired Computation in Combinatorial Optimization: Algorithms and Their Computational Complexity. Springer, Berlin (2010)
21. Qian, C., Yu, Y., Zhou, Z.H.: An analysis on recombination in multi-objective evolutionary optimization. Artif. Intell. **204**, 99–119 (2013)

Diversity and Landscape Analysis

RK-EDA: A Novel Random Key Based Estimation of Distribution Algorithm

Mayowa Ayodele(✉), John McCall, and Olivier Regnier-Coudert

Robert Gordon University, Aberdeen, UK
{m.m.ayodele,j.mccall,o.regnier-coudert}@rgu.ac.uk

Abstract. The challenges of solving problems naturally represented as permutations by Estimation of Distribution Algorithms (EDAs) have been a recent focus of interest in the evolutionary computation community. One of the most common alternative representations for permutation based problems is the Random Key (RK), which enables the use of continuous approaches for this problem domain. However, the use of RK in EDAs have not produced competitive results to date and more recent research on permutation based EDAs have focused on creating superior algorithms with specially adapted representations. In this paper, we present RK-EDA; a novel RK based EDA that uses a cooling scheme to balance the exploration and exploitation of a search space by controlling the variance in its probabilistic model. Unlike the general performance of RK based EDAs, RK-EDA is actually competitive with the best EDAs on common permutation test problems: Flow Shop Scheduling, Linear Ordering, Quadratic Assignment, and Travelling Salesman Problems.

Keywords: Estimation of distribution algorithm · Random key · Permutation problems · Cooling scheme · Univariate model

1 Introduction

Estimation of Distribution Algorithms (EDAs) are Evolutionary Algorithms (EAs) that generate solutions by sampling a Probabilistic Model (PM) of promising solutions. The ability to model the features of more promising solutions is a major attribute that differentiates them from most other EAs [7]. They benefit from the use of machine learning techniques, which makes them better at solving certain categories of larger and more difficult problems [12]. Problems naturally represented as permutations have however been identified as challenging for EDAs. This is attributed to the fact that EDAs have not been extensively explored to solve this class of problems [3]. EDAs for permutation spaces have therefore been a focus of research in recent years.

EDAs applied to permutations have been categorised into ad hoc approaches with varying strategies, integer space based and continuous space based [3]. One of the common continuous representations for solving permutations in EAs is the well-known Random Key (RK). RKs have an advantage over most other permutation representations as they always produce permutation feasible solutions.

J. Handl et al. (Eds.): PPSN XIV 2016, LNCS 9921, pp. 849–858, 2016.
DOI: 10.1007/978-3-319-45823-6_79

This is particularly not the case for integer based EDAs as they often require a procedure to handle the mutual exclusivity constraint.

RK based EDAs have however been considered the poorest [3] of the EDAs designed for permutation problems. RK representation has not been sufficiently adapted to benefit from the operation of EDA. It contains some inherent redundancy as a result of several RKs producing the same ordering thereby introducing plateaux to the search space [2,3,13]. Also, variability in the values that capture the same priority across solutions of a population limits the information captured by the probabilistic model. They therefore struggle to produce competitive results [7]. Models that are more specific to permutations such as histogram models [16,17], permutation distribution models [4–6] and factoradics [14] have shown better performances.

Some classical examples of RK based EDAs are REDA [15], $EGNA_{ee}$ &$UMDA_c$ [10]. REDA uses the triangulation of Bayesian network approach and focuses on model efficiency by modelling subset nodes of a problem. $EGNA_{ee}$ builds a Gaussian network where the structure of a problem is learnt using edge exclusion tests [10]. The $UMDA_c$ which is also a structure identification algorithm based on Gaussian network performs hypothesis tests to identify the density of its model's components. In addition, IDEA-ICE [2] can also be classified as a RK based EDA, although it uses a crossover operator to preserve building blocks in addition to its probabilistic model. Also, RKs associated with the building blocks are rescaled to improve the likelihood of them being properly combined. The IDEA-ICE shows better performance compared to the classical RK based EDAs.

The proposed Random Key Estimation of Distribution Algorithm (RK-EDA) attempts to capture some of the identified limitations of RKs as well as exploit their advantages.

The rest of this paper is described as follows. Section 2 motivates and describes the novel algorithm, RK-EDA. A discussion of problem sets and experimental design is presented in Sect. 3. Section 4 presents and discusses results while conclusions are presented in Sect. 5.

2 RK-EDA

The proposed RK-EDA is a univariate EDA whose probabilistic model, similar to $UMDA_c$, is based on mean values of genes in more promising solutions of a population. It exploits already found good genes by sampling a Gaussian distribution based on mean and variance values. Unlike $UMDA_c$, RK-EDA imposes a user defined variance parameter rather than a population generated one. This is because we achieved better performance using a controlled variance value. Furthermore, we propose to use a cooling rate parameter to control exploration and exploitation. This controls the level of variance in solutions of a population such that there is more exploration at the start of the algorithm, which automatically cools as the search progresses.

In this section, we present the algorithmic details of RK-EDA.

Algorithm 1. RK-EDA

```
1: Initialise σ, t_s and p_s
2: Generate initial population P of size p_s
3: for g = 1 to MaxGen do
4:    Evaluate and rescale individuals in P
5:    Select best t_s < p_s solutions to form S
6:    Calculate μ_S
7:    c = 1 − g/MaxGen
8:    σ_g = σ * c
9:    M = N(μ_S, σ_g)
10:   P_new = ∅
11:   repeat
12:      Sample M to generate offspring off
13:      Add off to P_new
14:   until |P_new| = p_s
15:   P = P_new
16: end for
```

As shown in Algorithm 1, RK-EDA requires the initialisation of three parameters which are initial variance σ, truncation size t_s and population size p_s. Since the stopping criteria is based on the number of fitness evaluations allowed (FEs), the maximum number of generations $MaxGen$ is estimated by dividing FEs by p_s.

A population P of RKs is randomly generated, evaluated and rescaled. The rescaling procedure requires the conversion of RKs to ranks e.g. [0.12, 0.57, 0.23, 0.25, 0.99] becomes [1, 4, 2, 3, 5]. The ranks are then rescaled to values between 0 and 1. This is done by setting $rescaledRK_i = \frac{rank_i - 1}{n-1}$ where $rescaledRK_i$ and $rank_i$ are respectively the rescaled RK and rank of gene i, and n is the problem size. The RK in the previous example therefore becomes [0.00, 0.75, 0.25, 0.50, 1.00]. With this approach, another set of RKs [0.01, 0.06, 0.03, 0.04, 0.2] which is the same solution as the previous example will have the same rescaled RK value [0.00, 0.75, 0.25, 0.50, 1.00]. With this approach, we are able to minimise redundancy and improve the information captured by the probabilistic model.

The best t_s solutions of the population are selected to generate a population S. Also, μ_S in ln. (6) is an array $\mu_{S_1}, ..., \mu_{S_n}$ that saves the mean of all RKs at indexes $\{1 \cdots n\}$ in the selected population S. Note that μ_{S_n} refers to the mean of all RKs in the nth index of each solution of S.

Cooling Rate c is calculated with respect to the particular generation such that its value is higher for the first few iterations and lower at the last set of iterations. As shown in ln. (8), c is used to generate generational variance σ_g. Multiplying c with σ to form σ_g makes it possible to achieve higher exploration at the start of the algorithm and more exploitation as g increases.

Furthermore, M is defined as a normal distribution $N(\mu_S, \sigma_g)$ and is updated for each generation g. Unlike μ_S which is an array of values, σ_g is not an array but a single value. An offspring solution *off* is generated by sampling M. Each gene i ($1 \leq i \leq n$) of *off* is generated based on σ_g and μ_{S_i}, *off* is repeatedly added to the offspring population P_{new} until its size equals p_s. At the end of each generation, P_{new} completely replaces the parent population P.

3 Experimental Settings

In this section we present the permutation problem instances as well as the parameter settings.

3.1 Permutation Problems

To assess the performance of RK-EDA, we apply it to a range of permutation benchmark problems. These problems include Flow Shop Scheduling Problem (FSSP), Linear Ordering Problem (LOP), Quadratic Assignment Problem (QAP) and Travelling Salesman Problem (TSP). These are formerly defined in [3], we have used the same objective functions as presented in the review paper and is summarised in Table 1. Note that we also consider the more recently used Total Flow Time (TFT) criteria for further experiments on the FSSP.

Table 1. Definition of the permutation problems

PPs	Objective functions	Definition of symbols
TSP	$min\left\{\sum_{i=2}^{n} d_{c_{i-1},c_i} + d_{c_n,c_1}\right\}$	c_i - i_{th} city d_{c_{i-1},c_i} - distance between c_{i-1} and c_i
FSSP	$min\{c_{j_n,m}\}\ c_{j_i,m} = max(c_{j_i,m-1}, c_{j_{i-1},m}) + p_{j_i,m}$	j_i - i_{th} job m - machine m $c_{j_i,m}$ - completion time for j_i on m $p_{j_i,m}$ - processing time for j_i on m
QAP	$min\left\{\sum_{i=1}^{n}\sum_{j=1}^{n} h_{a,b} \times d_{l_a,l_b}\right\}$	l_i - i_{th} location $h_{a,b}$ - flow between facilities a and b d_{l_a,l_b} - distance between l_a and l_b
LOP	$max\left\{\sum_{i=1}^{n}\sum_{j=1}^{n} d_{\omega_i\omega_j}\right\}$	ω_i - index of row and column at position i Matrix $D = [d_{ij}]$

3.2 Problem Sets

We evaluate RK-EDA using the selected permutation problems in [14]. We acknowledge that many of the problems are small instances especially the FSSP. Also, results from running RK-EDA on the FSSP problem instances gives an intuition that the algorithm is more competitive on the FSSP. We therefore added four larger FSSP problems.

The problem sets used in this paper are listed below.

1. TSP: *bays29, berlin52, dantzig42* and *fri26*
2. FSSP: *tai20-5-0, tai20-5-1, tai20-10-0* and *tai20-10-1* (smaller instances) *tai50-10-0, tai50-10-1, tai100-20-0* and *tai100-20-1* (larger instances)
3. QAP: *tai15a, tai15b, tai40a* and *tai40b*
4. LOP: *t65b11, be75np* and *be75oi*

These are commonly used problems and we consider them useful for comparing with other EDAs for permutation problems.

3.3 Parameter Setting

To be able to understand the parameter settings that suit RK-EDA, we explored a range of values and found different parameters suitable for different problem classes and sizes. To be able to make a fair comparison between RK-EDA and the considered algorithms, we use a set of parameters across all problems as done in the review [3]. The set of parameters used for RK-EDA is shown in Table 2. Based on preliminary tests, these parameters produce relatively good quality solutions across all problem classes and instances.

Table 2. Parameter values for RK-EDA

Parameter	Values
Population Size (p_s)	50
Truncation Size (t_s)	$0.1^{*}p_s$
Variance (σ)	$1/(3.14 * log_{10}n)$
Stopping Criteria	$1000n^2$ FEs
Maximum Number of Generations ($MaxGen$)	$20n^2$
Number of Runs	10

4 Results and Discussion

In this section, we present the results of running RK-EDA on the aforementioned permutation problem sets. Table 3 shows the minimum, maximum, average and standard deviation based on 10 runs of RK-EDA using the parameters presented in Table 2. Results are compared based on averages and standard deviations. We have highlighted results where optimal solution was found (appended *). We also highlight results that are significantly better (appended ✓) or not significantly different (appended **) from the best of the reviewed algorithms. We used the student t-test to measure statistical significance with a 0.05 significance level.

The results in Table 3 are presented according to problem classes. Note that $FSSP_s$ and $FSSP_l$ respectively denote the smaller and larger instances of the FSSP.

Table 3. Average performance of RK-EDA on benchmark problems

Groups	Problems	Minimum	Maximum	Mean	Stdev
TSP	**bays29**	**2020.0**	**2091.0**	**2041.5**	**21.3***
	berlin52	8207.0	8742.0	8404.6	164.0
	dantzig42	729.0	824.0	771.2	35.6
	fri26	**937.0**	**968.0**	**949.5**	**11.9***
$FSSP_s$	**tai20-5-0**	**1278.0**	**1279.0**	**1278.1**	**0.3*** ✓
	tai20-5-1	**1359.0**	**1360.0**	**1359.5**	**0.5****
	tai20-10-0	1586.0	1618.0	1602.9	11.1
	tai20-10-1	1680.0	1691.0	1685.2	3.2
$FSSP_l$	**tai50-10-0**	**3046.0**	**3119.0**	**3090.7**	**24.2****
	tai50-10-1	**2923.0**	**2964.0**	**2937.6**	**14.9** ✓
	tai100-20-0	**6344.0**	**6424.0**	**6386.4**	**21.0** ✓
	tai100-10-1	**6291.0**	**6381.0**	**6338.6**	**27.2** ✓
QAP	tai15a	393496.0	412072.0	404616.6	5350.2
	tai15b	51968294.0	52238818.0	52088443.6	72876.7
	tai40a	3353650.0	3418792.0	3391139.0	20951.9
	tai40b	642257062.0	659424886.0	652079961.9	4690584.3
LOP	t65b11	355180.0	356311.0	356028.2	295.6
	be75np	**716221.0**	**716930.0**	**716644.3**	**249.8****
	be75oi	110928.0	111156.0	111012.3	77.8

Table 4 shows the performance of each algorithm on the considered problems. The table is ordered according to the overall ranks shown in column "ALL". Columns TSP, $FSSP_s$, QAP, LOP and $FSSP_l$ show the average ranks of algorithms on instances of their respective problem classes. Column ALL is the average rank of algorithms on all instances of TSP, $FSSP_s$, QAP and LOP. Since one of the motivations for selecting the additional problems ($FSSP_l$) is that we ranked relatively high on $FSSP_s$, $FSSP_l$ was not used to create the overall rank so as to eliminate bias towards performance on FSSP. Also, since one of the reviewed algorithms was not applied to instances of $FSSP_l$, it will be impossible to generate an overall rank for the algorithm. To generate the ranks shown in the table, we use the average fitness recorded by each algorithm as reported in [3] and [14] as well as that of RK-EDA shown in Table 3. All algorithms are ranked from best to worst for each problem.

We used "-" to denote missing results where authors have not applied their algorithm to the given problem class.

According to the review presented in [3], $EHBSA_{WT}$ and $NHBSA_{WT}$ were recognised as the best performing algorithms. A similar result is depicted by the overall rank of these algorithms in Table 4. $EHBSA_{WT}$ ranks 1st while RK-EDA ranks 2nd with $NHBSA_{WT}$.

Table 4. Average ranks of algorithms

Algorithms	TSP	$FSSP_s$	QAP	LOP	ALL	$FSSP_l$
$EHBSA_{WT}$ [16]	1.00	1.75	4.00	2.00	2.13	3.25
RK-EDA	3.75	2.50	7.00	2.25	4.00	1.00
$NHBSA_{WT}$ [17]	8.50	3.00	2.00	1.75	4.00	3.00
$NHBSA_{WO}$ [17]	6.00	4.50	2.50	4.25	4.27	4.75
Factoradics [14]	6.50	6.25	6.75	7.00	6.47	-
UMDA [9]	8.25	6.75	4.75	6.25	6.53	7.00
$EBNA_{BIC}$ [1]	8.25	7.50	3.75	6.50	6.67	7.00
$EHBSA_{WO}$ [16]	2.25	6.00	10.00	10.75	7.27	9.75
MIMIC [1]	10.50	8.00	6.25	6.50	7.80	3.50
TREE [13]	12.25	10.50	8.75	9.75	10.33	7.00
IDEA-ICE [2]	11.25	10.75	10.50	9.75	10.53	8.75
$REDA_{UMDA}$ [15]	14.50	11.00	12.00	11.50	12.27	12.25
$REDA_{MIMIC}$ [15]	8.50	14.25	14.00	13.25	12.47	12.25
$EGNA_{ee}$ [11]	9.00	14.75	13.25	15.00	12.93	12.25
omeGA [8]	14.25	12.00	14.75	14.50	13.80	14.75
$UMDA_c$ [11]	10.25	16.00	15.75	15.00	14.13	13.50

We observed that the RK based EDAs such as $REDA_{UMDA}$, $REDA_{MIMIC}$, $EGNA_{ee}$, $UMDA_c$ as well as the RK based GA (OmeGA) are ranked least in Table 4 which is similar to the conclusion in the review Ceberio et al. [3]. OmeGA had been introduced in the review to compare with the performance of the EDAs in general. RK-EDA however shows a different trait outperforming all other RK based algorithms.

Furthermore, the performance of RK-EDA varies with different classes of problems. It produced competitive results on the FSSP, ranking 2nd on $FSSP_s$ and 1st on $FSSP_l$. RK-EDA produced statistically better results than the best of the reviewed algorithms on three $FSSP_l$ instances. It also produced competitive results for the TSP and LOP but much less competitive performance on the QAP. This may be attributed to the fact that parameters that suit other problem classes are not particularly suitable for the search space presented by the QAP.

In addition to the reviewed algorithms, other permutation based EDAs exist but were not included in the previous comparison because their results are not reported on the selected problems. GM-EDA [4] exhibits the best results on FSSP when hybridised with local search procedures such as variable neighbourhood search (VNS). We therefore compare RK-EDA with GM-EDA on a selected set of FSSP instances. In order to compare the two EDAs in a fair way, we use the reported results of GM-EDA without VNS.

We use the same set of parameters presented by the authors in [4] except that we do not consider elitism. This is because preliminary experiments show that

Table 5. Parameter values and stopping criteria for experiments on FSSP based on TFT

Parameter settings:	Parameter	Values
	Population size (p_s)	$10n$
	Truncation size (t_s)	$0.1{*}p_s$
	Variance (σ)	0.15
	$MaxGen$	FEs/p_s
	Number of runs	20
Stopping criteria:	Problem sizes	FEs
	20 × 05	182224100
	20 × 10	224784800
	50 × 10	256208100
	100 × 20	283040000

Table 6. Average TFT for FSSP

Problems	Algorithm	Average	Stdev
tai20-5-0	RK-EDA	14085	14
	GM-EDA	**14058**	**13**
tai20-5-1	RK-EDA	15223	20
	GM-EDA	15224	46
tai20-10-0	RK-EDA	21003	14
	GM-EDA	21006	46
tai20-10-1	RK-EDA	22660	81
	GM-EDA	**22561**	**135**
tai50-10-0	RK-EDA	89233	292
	GM-EDA	89041	400
tai50-10-1	RK-EDA	84858	138
	GM-EDA	84849	326
tai100-20-0	**RK-EDA**	**373607**	**523**
	GM-EDA	374708	1388
tai100-20-1	**RK-EDA**	**379947**	**501**
	GM-EDA	380750	868

elitism does not improve the performance of RK-EDA. In addition, 0.15 initial variance value particularly produced competitive results for FSSP instances. Table 5 shows the parameters of RK-EDA, which are adapted for solving the FSSP.

In Table 6, we present the average fitness over 20 runs for RK-EDA as well as GM-EDA. The results are based on the Total Flow Time (TFT) objective function and we compare using instances of $FSSP_s$ and $FSSP_l$. The results for GM-EDA have been extracted from [4]. Values that are significantly better are presented in bold. The results show that the GM-EDA is significantly better on two of the smallest problems (*tai20-5-0* and *tai20-10-1*) while RK-EDA shows significant improvement on the largest problems (*tai100-20-0* and *tai100-20-1*). There are however no significant difference between the performance of the algorithms on other instances.

Results from comparing RK-EDA with GM-EDA as shown in Table 6 also indicate that RK-EDA is competitive and should be further explored to solve bigger and more complex problems.

5 Conclusions

EDAs based on RKs have previously been considered the poorest of permutation based EDAs [3]. One of the problems posed by RKs is attributed to the variety of ways of representing an ordering [13]. In this paper, we introduce a novel RK based EDA (RK-EDA) which addresses this by rescaling the RKs uniformly. This approach improves the information captured by the probabilistic model. Furthermore, RK-EDA uses a cooling scheme to manage the rate of exploration/exploitation of the search space such that there is better exploration at the start of the algorithm and better exploitation of already found good pattern as the search progresses.

Furthermore, learning a probability structure is considered the most expensive operation in EDAs [2], we present a simple model, which only saves the mean of solutions in a selected population. This is relatively computationally efficient. RK-EDA whose procedure is comparatively simple produces very competitive results. It outperforms other reviewed continuous EDAs. It is also competitive with the best permutation EDAs in general.

RK-EDA's most competitive performance is seen on FSSP and the least on QAP. It's performance on FSSP gets more competitive as the problem size increases presenting the best results on the largest of the considered FSSP instances. The performance of RK-EDA on larger problems is therefore recommended for further investigation.

In addition, the use of local search has been reported to improve the performance of the GM-EDA, hybridisation of the RK-EDA may also improve its performance.

References

1. Bengoetxea, E., Larrañaga, P., Bloch, I., Perchant, A., Boeres, C.: Inexact graph matching by means of estimation of distribution algorithms. Pattern Recogn. **35**(12), 2867–2880 (2002)
2. Bosman, P.A., Thierens, D.: Crossing the road to efficient IDEAs for permutation problems. In: Proceedings of the 6th annual conference on Genetic and evolutionary computation, pp. 219–226. ACM (2001)

3. Ceberio, J., Irurozki, E., Mendiburu, A., Lozano, J.A.: A review on estimation of distribution algorithms in permutation-based combinatorial optimization problems. Prog. Artif. Intell. **1**(1), 103–117 (2012)
4. Ceberio, J., Irurozki, E., Mendiburu, A., Lozano, J.A.: A distance-based ranking model estimation of distribution algorithm for the flowshop scheduling problem. IEEE Trans. Evolut. Comput. **18**(2), 286–300 (2014)
5. Ceberio, J., Mendiburu, A., Lozano, J.A.: The plackett-luce ranking model on permutation-based optimization problems. In: 2013 IEEE Congress on Evolutionary Computation (CEC), pp. 494–501. IEEE (2013)
6. Ceberio, J., Mendiburu, A., Lozano, J.A.: Kernels of mallows models for solving permutation-based problems. In: Proceedings of the 2015 on Genetic and Evolutionary Computation Conference, pp. 505–512. ACM (2015)
7. Hauschild, M., Pelikan, M.: An introduction and survey of estimation of distribution algorithms. Swarm Evolut. Comput. **1**(3), 111–128 (2011)
8. Knjazew, D., Goldberg, D.E.: Omega-ordering messy GA: solving permutation problems with the fast messy genetic algorithm and random keys. In: Proceedings of Genetic and Evolutionary Computation Conference, pp. 181–188 (2000)
9. Larrañaga, P., Etxeberria, R., Lozano, J.A., Peña, J.M.: Optimization in continuous domains by learning and simulation of Gaussian networks. In: Workshop in Optimization by Building and using Probabilistic Models, pp. 201–204. A Workshop withing the 2000 Genetic and Evolutionary Computation Conference, GECCO 2000, Las Vegas, Nevada, USA (2000)
10. Larrañaga, P., Lozano, J.A.: Estimation of Distribution Algorithms: A New Tool for Evolutionary Computation, vol. 2. Springer, New York (2002)
11. Lozano, J., Mendiburu, A.: Estimation of distribution algorithms applied to the job shop scheduling problem: Some preliminary research. In: Larrañaga, P., Lozano, J.A. (eds.) Estimation of Distribution Algorithms, pp. 231–242. Springer, Heidelberg (2002)
12. Pelikan, M., Sastry, K., Cantú-Paz, E.: Scalable Optimization via Probabilistic Modeling: From Algorithms to Applications. Springer, Heidelberg (2006)
13. Pelikan, M., Tsutsui, S., Kalapala, R.: Dependency trees, permutations, and quadratic assignment problem. In: Genetic and Evolutionary Computation Conference: Proceedings of the 9th Annual Conference on Genetic and Evolutionary Computation, vol. 7, pp. 629–629 (2007)
14. Regnier-Coudert, O., McCall, J.: Factoradic representation for permutation optimisation. In: Bartz-Beielstein, T., Branke, J., Filipič, B., Smith, J. (eds.) PPSN 2014. LNCS, vol. 8672, pp. 332–341. Springer, Heidelberg (2014)
15. Romero, T., Larrañaga, P.: Triangulation of Bayesian networks with recursive estimation of distribution algorithms. Int. J. Approx. Reason. **50**(3), 472–484 (2009)
16. Tsutsui, S.: Probabilistic model-building genetic algorithms in permutation representation domain using edge histogram. In: Guervós, J.J.M., Adamidis, P.A., Beyer, H.-G., Fernández-Villacañas, J.-L., Schwefel, H.-P. (eds.) PPSN 2002. LNCS, vol. 2439, pp. 224–233. Springer, Heidelberg (2002)
17. Tsutsui, S., Pelikan, M., Goldberg, D.E.: Node histogram vs. edge histogram: a comparison of PMBGAs in permutation domains. MEDAL Report (2006009) (2006)

REMEDA: Random Embedding EDA for Optimising Functions with Intrinsic Dimension

Momodou L. Sanyang[1,2](✉) and Ata Kabán[1]

[1] School of Computer Science, University of Birmingham, Edgbaston B15 2TT, UK
{M.L.Sanyang,A.Kaban}@cs.bham.ac.uk
[2] School of Information Technolgy and Communication, University of the Gambia, Brikama Campus, P.O. Box 3530, Serekunda, The Gambia
MLSanyang@utg.edu.gm

Abstract. It has been observed that in many real-world large scale problems only few variables have a major impact on the function value: While there are many inputs to the function, there are just few degrees of freedom. We refer to such functions as having a low intrinsic dimension. In this paper we devise an Estimation of Distribution Algorithm (EDA) for continuous optimisation that exploits intrinsic dimension without knowing the influential subspace of the input space, or its dimension, by employing the idea of random embedding. While the idea is applicable to any optimiser, EDA is known to be remarkably successful in low dimensional problems but prone to the curse of dimensionality in larger problems because its model building step requires large population sizes. Our method, Random Embedding in Estimation of Distribution Algorithm (REMEDA) remedies this weakness and is able to optimise very large dimensional problems as long as their intrinsic dimension is low.

Keywords: Estimation of distribution algorithm · Black-box optimization · Intrinsic dimension

1 Introduction

Optimisation over a high dimensional search space is challenging. However, it has been noted that in certain classes of functions most decision variables have a limited impact on the objective function. Examples include hyperparameter optimisation for neural and deep belief networks [1], automatic configuration of state-of-the algorithms for solving NP-hard problems [8], optimisation problems in robotics [14], and others [3]. In other words, these problems have low intrinsic dimensionality. In the numerical analysis literature [3] the influential parameter subspace has been termed as the 'active subspace', and methods have been developed to estimate this subspace. Fortunately, for optimisation, estimating the influential subspace is not required: In [14] it was shown that a sufficiently large *random* subspace contains an optimum with probability 1, and this was

J. Handl et al. (Eds.): PPSN XIV 2016, LNCS 9921, pp. 859–868, 2016.
DOI: 10.1007/978-3-319-45823-6_80

used to dramatically improve the efficiency of Bayesian optimisation by exploiting the low intrinsic dimensionality of problems.

In this paper we further develop the random embedding technique, and introduce it to evolutionary search, by employing it to scale up Estimation of Distribution Algorithms (EDA) for problems with low intrinsic dimension. Although the underlying theoretical considerations are applicable to any optimisation method, our focus on EDA is due to it being one of the most successful methods in low dimensional problems [11] and most unsuccessful or expensive in high dimensions [5,9,12].

Definition. A function $f : \mathcal{R}^D \to R$ has **intrinsic dimension** d_i, with $d_i < D$, if there exists a d_i dimensional subspace Υ such that $\forall x \in \mathcal{R}^D$, $f(x) = f(\text{Proj}_\Upsilon(x))$.
In the above, $\text{Proj}_\Upsilon(x)$ denotes the orthogonal projection, i.e. $\text{Proj}_\Upsilon(x) = \Phi\Phi^T x$, where $\Phi \in R^{D \times d_i}$ has columns holding a linear basis of Υ.

The following result in [14] shows that, for such functions, a global optimum exists in a randomly chosen linear subspace – hence a low dimensional search is sufficient.

Theorem 1 [14]. *Assume we are given a function $f : \mathcal{R}^D \to \mathcal{R}$ with intrinsic dimension $d_i < d$ and a random matrix $R \in \mathcal{R}^{D \times d}$ with independent entries sampled from a standard Gaussian. Then, with probability 1, for any $x \in \mathcal{R}^D$, there exists a $y \in \mathcal{R}^d$ such that $f(x) = f(Ry)$.*

Given some box constraints on the original problem, the authors [14] develop an upper bound on the search box required for the low dimensional search. However, their proof only applies to the case when $d = d_i$, and in practice they recommend a smaller search box and use a slightly larger d. Recall, in practice we have no knowledge of the value of d_i. However, on synthetic problems the experimental results do appear to be better when d is slightly larger than d_i.

In the next section we derive a bound on the search box that holds true for $d > d_i$, and show that the required box size that guarantees to contain a global optimum is indeed smaller when d is larger. Secondly, we devise an EDA optimisation algorithm that implements these ideas employing a random Gaussian embedding.

2 REMEDA: Random Embedding EDA

In this section we present our REMEDA algorithm and explain how it exploits the intrinsic dimensionality of problems. Instead of optimising in the high dimensional space, REMEDA will do a random embedding, using the random matrix $R \in \mathcal{R}^{D \times d}$, $d \ll D$ with i.i.d. entries drawn from a standard Gaussian, and then optimises the function $g(y) = f(Ry)$, $y \in \mathcal{R}^d$ in the lower dimensional space.

The psuedo-code of REMEDA is given in Algorithm 1. It takes the population size N, box constraints for a D-dimensional problem, and the internal working dimension $d \ll D$. As in basic EDA, the REMEDA algorithm then proceeds by initially generating a population of individuals uniformly randomly. However, these individuals are generated in the d-dimensional space, within some suitable box constraints in this space that are determined from the given D-dimensional box. The details of how this is done will follow shortly. The algorithm then evaluates the fitness of these individuals with the use of a random embedding matrix $R \in \mathcal{R}^{D\times d}$ that transforms the d-dimensional points into the original D-dimensional space of decision variables. The matrix R has entries drawn i.i.d. from a standard Gaussian distribution, as in Theorem 1. Based on the fitness values obtained, the fittest individuals are selected using a selection method, such as truncation selection. The maximum likelihood estimates (MLE) of the mean $\mu \in \mathcal{R}^d$ and the covariance $\Sigma \in \mathcal{R}^{d\times d}$ of the promising solutions are computed from the set of selected fittest individuals, and these are used to generate the new generation by sampling from a multivariate Gaussian distribution. The new population is formed by replacing the old individuals by the new ones. We also use elitism, whereby the best individual of the previous generation is kept.

Algorithm 1. The Pseudocode of REMEDA with Population size N and intrinsic dimensionality of the problems, d_i

Inputs: N, D, d, Box
(1) Set the search box boundaries in the low-dimensional space $\mathcal{R}^d$ (cf. Theorem 2 &text)
(2) Set $P \leftarrow$ Generate N points uniformly randomly within the box in $\mathcal{R}^d$ to give an initial population
(3) Set $R \leftarrow$ Generate a random embedding matrix, $R \in \mathcal{R}^{D\times d}$.
Do
 (4) Evaluate the fitness of y_i as $f(Ry_i)$, $i = 1...N$
 (5) Select best individuals P^{sel} from P based on their fitness values
 (6) Calculate the mean μ and covariance Σ of P^{sel}
 (7) Use the μ and Σ to sample new population, P^{new}
 (8) $P \leftarrow P^{new}$
Until Termination criteria are met
Output: P

We have not yet specified how to determine the d-dimensional box constraints that correspond to the given D-dimensional ones. Given some box constraints in $\mathcal{R}^D$, the following theorem gives the required box constraints for the search in $\mathcal{R}^d$.

Theorem 2. *Let $f : \mathcal{R}^D \rightarrow \mathcal{R}$ be a function with intrinsic dimension $d_i < d < D$ that we want to optimise subject to the box constraint $\mathcal{X} \subset \mathcal{R}^D$, where $\mathcal{X}$ is centered around 0. Denote the intrinsic subspace by Υ, and let Φ be a $D \times d_i$ matrix whose columns form an orthonormal basis for Υ. Denote by $x_t^* \in \Upsilon \cap \mathcal{X}$ an optimiser of f inside Υ. Let R be a $D \times d$ random matrix with independent standard Gaussian entries. Then there exists an optimiser $y^* \in \mathcal{R}^d$ such that $f(Ry^*) = f(x_t^*)$ w.p. 1, and for any choice of $\epsilon \in (0,1)$,* *if* $d > (\sqrt{d_i} + \sqrt{2\ln(1/\epsilon)})^2$, *then* $||y^*||_2 \leq \frac{||x_t^*||_2}{\sqrt{d}-\sqrt{d_i}-\sqrt{2\ln(\frac{1}{\epsilon})}}$ *with probability at least* $1 - \epsilon$.

Proof. The existence of y^* is guaranteed by Theorem 1, and global optimisers outside the subspace Υ are irrelevant since the function takes all its range of values in Υ. So our focus is to upper bound the length of y^*.

From the proof of Theorem 1 in [14] we know that $\exists y^* \in \mathcal{R}^d$ s.t.

$$\Phi\Phi^T R y^* = x_t^* \tag{1}$$

Hence,

$$||x_t^*|| = ||\Phi\Phi^T R y^*|| \geq s_{\min}(\Phi\Phi^T R)||y^*|| \tag{2}$$

where we use the Rayleigh quotient inequality, and $s_{\min}(\cdot)$ denotes the smallest singular value.

Note that $\Phi\Phi^T R$ is a $d_i \times d$ random matrix with i.i.d. Gaussian entries. When $d = d_i$ it is a square matrix, and a bound on its smallest singular value was applied in [14]. Instead, for the case $d > d_i$ we employ the bound of Davidson and Szarek that applies to rectangular Gaussian matrices [4]. We have for any $\epsilon \in (0, 1)$ for which $\sqrt{d} - \sqrt{d_i} - \epsilon > 0$, that:

$$||y^*|| \leq \frac{||x_t^*||}{\sqrt{d} - \sqrt{d_i} - \epsilon} \tag{3}$$

with probability $1 - \exp(-\frac{\epsilon^2}{2})$. Now setting $\exp(-\epsilon^2/2) = \tau$ and solving for ϵ we get $\epsilon = \sqrt{2\ln(\frac{1}{\tau})}$. Plugging this back and renaming τ to ϵ completes the proof. □

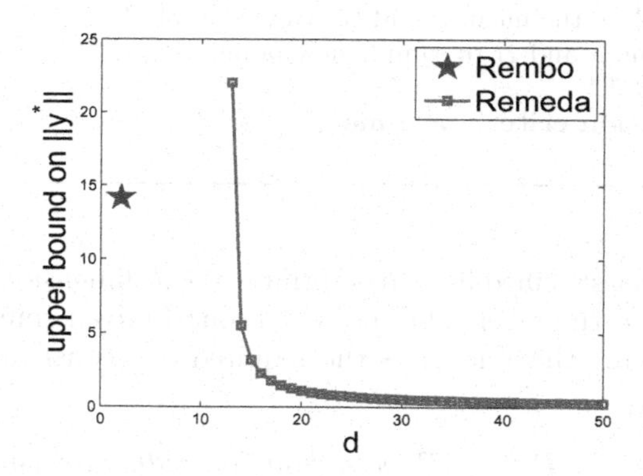

Fig. 1. Comparison of our theoretical bound (REMEDA), with various values of $d > d_i$ versus the bound of [14] (REMBO), which holds when $d = d_i$.

In Fig. 1 we plotted the bound on the search box from our Theorem 2 for various values of $d > d_i$ in comparison with the bound in [14] for $d = d_i$. We see that our result is tighter for nearly all values of d and it explains why a smaller search box is sufficient when $d > d_i$. The single point for Rembo in Fig. 1 is for $d = d_i$ where as the curve for Remeda is for values of $d > d_i$.

In practice, of course, we typically have no knowledge of the value of d_i, in which case we cannot use theoretical bounds directly to set our search box. However, we can fix the search box – for instance to $\sqrt{d}$-times the coordinates if the original box, as suggested in [14], and our Theorem 2 then suggests that increasing d can eventually make this fixed-size box sufficiently large to contain the optimiser y^*. This is what we used in the experiments reported.

3 Related Work

Model building in high dimensions is the subject of many recent research efforts, as high dimensionality limits the usefulness of optimisers in practice. Many approaches were proposed, here we will limit ourselves to a few of the most relevant ones.

Among these methods, the Eigendecomposition EDA ($ED-EDA$) [6] proposes to utilise a repaired version of the full covariance matrix estimate, with the aim to capture interactions among all decision variables and guide exploration of the search space. Other methods use limited dependencies. For example, Cooperative Co-evolution with Variable Interaction Learning ($CCVIL$) proposed by *Weicker et al.* in [15] is a deterministic method to uncover dependencies between decision variables, which has later been extended to the CCVIL framework by *Chen et al.* in [13]. EDA with Model Complexity Control ($EDA-MCC$) [5] also employs a deterministic algorithm to split the decision variables into two disjoint subsets, of which one set contains decision variables with only minor interaction and the other set contains the strongly dependent variables that are further grouped randomly, and inter-group dependencies are neglected. Other methods include Covariance Matrix Adaptation *(CMA-ES)* [7], separable CMA-ES (*sep-CMA-ES*) [10] and Multilevel Cooperate Co-evolution ($MLCC$)[16].

There are also methods that apply dimensionality reduction techniques to reduce the dimension of the problems in order to avail EDA the opportunity to demonstrate its capabilities. An example of this type of techniques are random projections [9,12,14]. However, none of these methods have been designed to take advantage of the intrinsic structure of the problems as our REMEDA approach does.

4 Experiments

4.1 Test Functions and Performance Measures

We created test functions with intrinsic dimension 5 from existing benchmark functions, by embedding the d_i-dimensional versions of these problems into higher D-dimensions. That is, we add $D-5$ additional dimensions which do not impact on the function value, and (optionally) rotate the search space around the origin in a random direction. Hence, the functions will take D-dimensional inputs, but only 5 linear combinations of these input variables determine the function value. The algorithm will have no knowledge of which these directions are, not even that there are 5, but it has knowledge that the number of important directions is much less that D. The functions we employed in this way here are the following: Shifted Ellipse, Shifted Schwefel's problem 1.2, Shifted Rotated High Conditional Elliptic function, and Shifted Rosenbrock function. We also took the Branin function from [14] which has intrinsic dimension 2. The functions are listed in Table 1.

Table 1. Test functions of low intrinsic dimension of 2 or 5. o is the shift vector.

PN	Name	Expression
1	Sphere	$\sum_{j=1}^{d_i}(x_j - o_j)^2$
2	Ackley's	$20 - 20\exp(-0.2\sqrt{\frac{1}{d_i}\sum_{j=1}^{d_i}((x_j - o_j) * M)^2})$- $\exp(\frac{1}{d_i}\sum_{j=1}^{d_i}(\cos(2\pi(x_j - o_j) * M))$+e
3	Elliptic	$\sum_{j=1}^{d_i}(10^6)^{\frac{j-1}{d_i-1}} * (x_j - o_j) * M)$
4	Rosenbrock	$\sum_{j=1}^{d_i-1}(100(z_j^2 - z_{j+1})^2 + (z_j - 1)^2)$ $z = x - o + 1$
5	Branin	$(-1.275\frac{x_1^2}{\pi^2} + 5\frac{x_1}{\pi} + x_2 - 6)^2$ $+(10 - \frac{5}{4\pi})\cos(x_1) + 10$

We employ two common performance indicators: (i) The fitness gap achieved under a fixed budget is the difference between the best fitness achieved and the true optimum; (ii) The scalability is the budget of function evaluations needed to reach a pre-defined value to reach.

4.2 Results and Discussion

Experiments on a d_i= 2 Problem. In the first set of experiments we consider the $D = 25$ dimensional Branin function that has intrinsic dimension $d_i = 2$. Though, we should note that D can be as large in principle, as we like since the working strategy and the budget usage of REMEDA are independent of D. In this experiment, we vary the internal working dimension d, and the population size N, under a fixed budget of 500 function evaluations.

The results are shown in Table 2, as obtained from 50 independent repetitions of each experiment. We can see from Table 2 that $d = d_i = 2$ is not the best choice, as the size of the search box is not sufficient at $d = d_i$. Also observe that increasing d beyond 4 drops the performance – this is because searching in a larger dimensional space is not necessary and is less resource-effective. Furthermore, we see for all d tested, the higher the population sizes, the worse the

Table 2. Fitness gap achieved by REMEDA on the Branin function ($d_i = 2$ embedded in $D = 25$), with a total budget of 500 function evaluations.

Pop. size	d = 2		d = 4		d = 6	
	Mean	std	Mean	std	Mean	std
300	1.4297	2.601	2.4908	3.1013	3.9007	3.0322
150	0.4128	0.6607	1.1701	1.313	2.1368	2.1046
80	0.826	2.9973	0.4193	0.4459	0.8303	0.9331
40	0.6375	2.9073	0.04	0.0969	0.1927	0.3865
30	0.6737	2.4939	0.0336	0.0853	0.1038	0.2615

performance. This is because a large population unnecessarily uses up the budget when the search only happens in a small dimensional subspace. With these insights in place, next we carry out a more comprehensive study.

Results and Comparisons on Problems with $d_i = 5$. In this section, we compare our method with state of the art approaches in heuristic optimisation, on problems with intrinsic dimension $d_i = 5$. The ambient dimension was $D = 1000$ in these experiments, but as already mentioned this can be much higher without causing problems as long as d_i stays low.

We expect that REMEDA should gain advantage from its ability to exploit intrinsic dimensional property while other methods have not been designed to make use of such structure. On the other hand, REMEDA needs to find a good value for its internal dimension d without knowing d_i (as this information is normally not available in practice). This will use up part of the budget, but the hope is that it will pay off by a speedy progress in the search.

We start with $d = 1$, using convergence as a stopping criterion, and move up progressively to higher values of d until the fitness reached upon convergence is no longer improved by the increase of d. Within each value of d tried, we run REMEDA to convergence, until the relative change in fitness is below a threshold: $\frac{f(t)-f(t+1)}{f(t)} < 10^{-8}$, where t is the generation count and f is the fitness value. When this condition is satisfied, we move on to the next value of d, and re-initialise the population randomly (although other schemes could also be investigated). The total number of fitness evaluations used throughout this process is the total budget that we then provide to the competing algorithms.

The bar chart in the leftmost plot of Fig. 2 shows an example of the fitness gaps achieved at convergence with consecutive values of d. The error bars show one standard error from 25 independent repetitions. In the rightmost plot we show the evolution of the best fitness. Superimposed, we also show the trajectories of competing state of the art methods: EDA-MCC [5], RP-EDA [9], and tRP-EDA [12]. All use the same budget and same population size (Table 3). For all the d tried, we plot their concatenated trajectories in such a way that the next starts from the end of the current one.

From Fig. 2 we can see that REMEDA attains a fitness value close to the optimum efficiently, while the other methods are not able to achieve the same within the same budget. We also superimposed an idealised version of plain EDA – that is a plain EDA that receives the d_i-dimensional version of the problem – and we see that REMEDA is nearly as good.

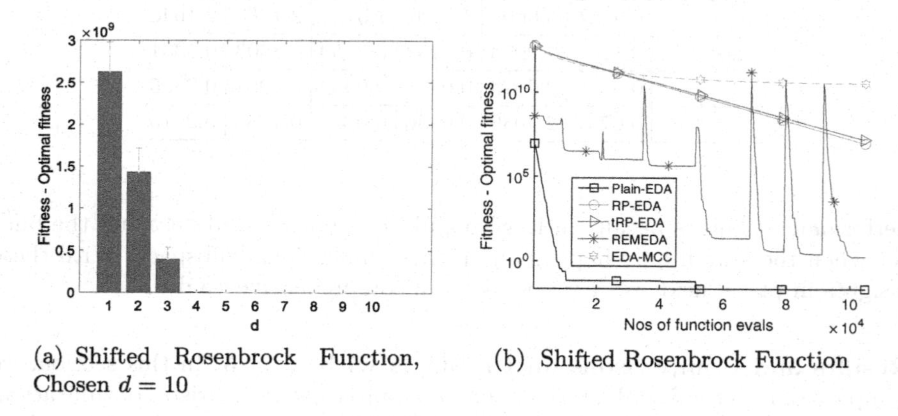

(a) Shifted Rosenbrock Function, Chosen $d = 10$

(b) Shifted Rosenbrock Function

Fig. 2. Finding d (left) and evolution of best fitness (right) for REMEDA, 3 competing methods, and a d_i-dimensional EDA on the idealised problem. Results are averaged over 25 independent runs. All methods use the same population size.

Table 3. Comparing REMEDA with other state of the art methods.

Fn	REMEDA		sep-CMA-ES		EDA-MCC		tRP-ENS-EDA		RP-ENS-EDA	
	Mean	std	Mean	std	Mean	std	Mean	std	Mean	std
F1	0	0	3.81E+04	2.16E+04	6.41E+04	8.21E+03	37.80	3.17	25.79	2.05
F2	0	0	1.00E+07	7.92E+05	1.99E+06	1.42E+06	1.40E+08	7.99E+06	1.38E+08	6.59E+06
F3	**0.18**	**0.91**	4.75E+07	3.47E+06	2.90E+09	2.29E+08	6.81E+08	5.60E+07	2.84E+08	2.62E+07
F6	**45.92**	**216.11**	5.44E+06	2.09E+06	3.56E+10	4.93E+09	1.60E+07	1.88E+06	9.64E+06	1.44E+06
F8	**14.19**	**7.89**	21.67	0.01	21.34	0.06	21.66	0.01	21.43	0.06

4.3 Scalability Experiments

Function evaluation is costly in most practical problems. Here we study what is the required budget of function evaluations to reach a specified value of the fitness gap.

Before running these scalability experiments, we carried out some experiments to determine the required population size as a function of the intrinsic dimension of the problem, so that we can vary the latter and set the population size automatically. For this we use a bisection method as in [2]. To find the population size that results in the lowest number of evaluations required to reach a pre-determined fitness gap value to reach (VTR), we start from a very large population size that can solve the problem within a large predetermined budget and then search for a small population size that cannot solve the problem anymore. In between these limits we use binary search to find the optimal population size. We repeated this 25 times and took the average.

We fix the value to reach (VTR) to 10^{-5}, and vary the intrinsic dimensionality of the problem $d_i \in [2, 50]$. We count the number of fitness evaluations needed for our proposed REMEDA to reach the VTR. The same experiment was repeated for three other choices of VTR: 10^{-3}, 10^2 and 10^3 in order to make sure that the conclusions will not be specific to a particular choice of the VTR. In all these experiments the maximum fitness evaluations was fixed to 6×10^3, so the algorithm stops when the budget is exhausted.

Figure 3 shows the average number of function evaluations as computed from the successful runs out of 25 independent repetitions for each problem, with each intrinsic dimension tested. From the figure, we observe a linear fit on the scalability measurements.

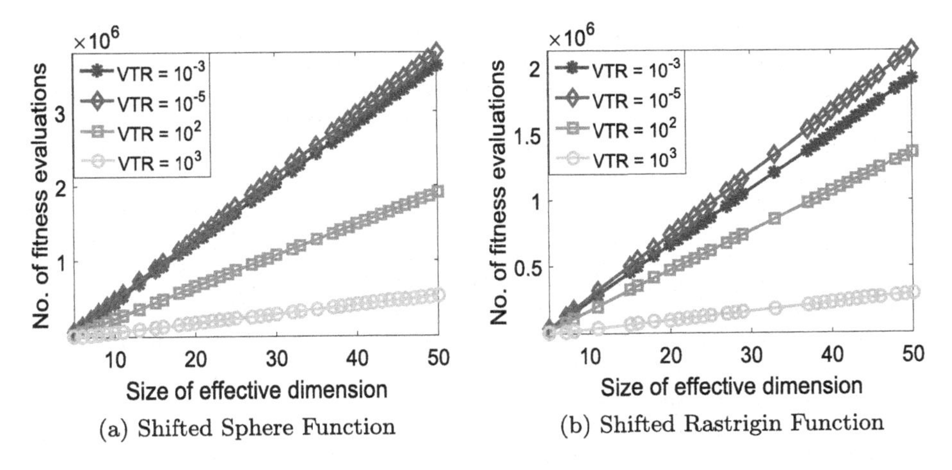

(a) Shifted Sphere Function (b) Shifted Rastrigin Function

Fig. 3. Number of function evaluations taken by successful runs of REMEDA to reach a pre-specified value to reach (VTR) as the problem intrinsic dimensionality is varied in $d_i \in [2, 50]$. The markers represent averages computed from 25 independent repetitions.

5 Conclusions and Future Work

We proposed random embedding in Estimation of Distribution Algorithm to scale up EDA by exploiting the intrinsic dimension of problems whereby the search takes place in a much lower dimensional space than that of the original problem. Our method is suited for large scale problems that take a large number of inputs but only depend on a few linear combinations of them. On such problems we have demonstrated that our method outperforms the best state of the art algorithms in evolutionary computation. Our technique and its theoretical basis are applicable in principle to any optimisation method, and in the light that problems with intrinsic dimension are quite prevalent in real-world applications, it seems a worthwhile avenue for future work to make use of it more widely.

References

1. Bergstra, J., Bengio, Y.: Random search for hyper-parameter optimization. Mach. Learn. Res. **13**, 281–305 (2012)
2. Bosman, P.: On empirical memory design, faster selection of Bayesian factorizations and parameter-free Gaussian EDAs. In: Proceedings of Genetic and Evolutionary Computation Conference (GECCO), pp. 389–396. ACM (2009)
3. Constantine, P.: Active subspace methods in theory and practice: applications to kriging surfaces. SIAM J. Sci. Comput. **36**(4), 1500–1524 (2013)
4. Davidson, K.R., Szarek, S.J.: Local operator theory, random matrices and banach spaces. In: Handbook of the Geometry of Banach Spaces, vol. 1. pp. 317–366 (2001)
5. Dong, W., Chen, T., Tino, P., Yao, X.: Scaling up estimation of distribution algorithm for continuous optimisation. IEEE Trans. Evol. Comput. **17**(6), 797–822 (2013)
6. Dong, W., Yao, X.: Unified eigen analysis on multivariate Gaussian based estimation of distribution algorithms. Inf. Sci. **178**, 3000–3023 (2008)
7. Hansen, N.: The CMA evolution strategy: a comparing review. In: Lozano, J.A., Larrañaga, P., Inza, I., Bengoetxea, E. (eds.) Towards a New Evolutionary Computation: Advances in the Estimation of Distribution Algorithms. Studies in Fuzziness and Soft Computing, vol. 192. Springer, Heidelberg (2006)
8. Hutter, F.: Automated configuration of algorithms for solving hard computational problems. Ph.D. thesis (2009)
9. Kabán, A., Bootkrajang, J., Durrant, R.J.: Towards large scale continuous EDA: a random matrix theory perspective. Evol. Comput. (2015). MIT Press
10. Ros, R., Hansen, N.: A simple modification in CMA-ES achieving linear time and space complexity. In: Rudolph, G., Jansen, T., Lucas, S., Poloni, C., Beume, N. (eds.) PPSN 2008. LNCS, vol. 5199, pp. 296–305. Springer, Heidelberg (2008)
11. Sanyang, M.L., Kaban, A.: Multivariate cauchy EDA optimisation. In: Corchado, E., Lozano, J.A., Quintián, H., Yin, H. (eds.) IDEAL 2014. LNCS, vol. 8669, pp. 449–456. Springer, Heidelberg (2014)
12. Sanyang, M.L., Kabán, A.: Heavy tails with parameter adaptation in random projection based continuous EDA. In: IEEE Congress on Evolutionary Computation (CEC), pp. 2074–2081 (2015)
13. Yang, Z., Chen, W., Weise, T., Tang, K.: Large-scale global optimization using cooperative co-evolution with variable interaction learning. In: Parallel Problem Solving from Nature (PPSN), vol. 11, pp. 300–309 (2010)
14. Wang, Z., Zoghi, M., Hutter, F., Matheson, D., de Freitas, N.: Bayesian optimization in a billion dimensions via random embeddings. In: IJCAI (2013)
15. Weicker, K., Weicker, N.: On the improvement of co-evolutionary optimizers by learning variable inter-dependencies. In: Proceedings of Congress on Evolutionary Computation, vol. 3 (1999)
16. Tang, K., Yang, Z., Yao, X.: Multilevel cooperative co-evolution for large scale optimization. In: IEEE World Congress on Computational Intelligence 2008 (CEC 2008) (2008)

Feature-Based Diversity Optimization for Problem Instance Classification

Wanru Gao[1], Samadhi Nallaperuma[2], and Frank Neumann[1(✉)]

[1] School of Computer Science, The University of Adelaide, Adelaide, Australia
frank@cs.adelaide.edu.au

[2] Department of Computer Science, The University of Sheffield, Sheffield, UK

Abstract. Understanding the behaviour of heuristic search methods is a challenge. This even holds for simple local search methods such as 2-OPT for the Traveling Salesperson problem. In this paper, we present a general framework that is able to construct a diverse set of instances that are hard or easy for a given search heuristic. Such a diverse set is obtained by using an evolutionary algorithm for constructing hard or easy instances that are diverse with respect to different features of the underlying problem. Examining the constructed instance sets, we show that many combinations of two or three features give a good classification of the TSP instances in terms of whether they are hard to be solved by 2-OPT.

1 Introduction

Heuristic search methods such as local search, simulated annealing, evolutionary algorithms and ant colony optimization have been shown to be very successful for various combinatorial optimization problems. Although they usually don't come with performance guarantees on their runtime and/or approximation behaviour, they often perform very well in several situations. Understanding the conditions under which optimization algorithms perform well is essential for automatic algorithm selection, configuration and effective algorithm design. In both the artificial intelligence (AI) [1–4] and operational research communities [5,6], this topic has become a major point of interest.

The feature-based analysis of heuristic search algorithms has become an important part in understanding such type of algorithms [7,8]. This approach characterizes algorithms and their performance for a given problem based on features of problem instances. Thereby, it provides an important tool for bridging the gap between pure experimental investigations and mathematical methods for analysing the performance of search algorithms [9–11]. Current methods for the feature-based analysis are based on constructing hard and easy instances for an investigated search heuristic and a given optimization problem by evolving instances using an evolutionary algorithm [7,12,13]. This evolutionary algorithm constructs problem instances where the examined algorithm either shows a bad (good) approximation behaviour and/or requires a large (small) computational

J. Handl et al. (Eds.): PPSN XIV 2016, LNCS 9921, pp. 869–879, 2016.
DOI: 10.1007/978-3-319-45823-6_81

effort to come up with good or optimal solutions. Although the evolutionary algorithm for constructing such instances is usually run several times to obtain a large set of hard (easy) instances, the question arises whether the results in terms of features of the instances obtained give a good characterization of problem difficulty.

In the paper, we propose a new approach of constructing hard and easy instances. Following some recent work on using evolutionary algorithms for generating diverse sets of instances that are all of high quality [14,15], we introduce an evolutionary algorithm which maximizes diversity of the obtained instances in terms of a given set of features. Our approach allows to generate a set of instances that is guaranteed to be diverse with respect to the problem features at hand. Carrying out this process for several combinations of features of the considered problem and algorithm gives a much better classification of instances according to their difficulty of being solved by the considered algorithm.

To show the benefit of our approach compared to previous methods, we consider the classical 2-OPT algorithm for the TSP. Previous feature-based analyses have already considered hard and easy instances in terms of approximation ratio and analyzed the features of such hard (easy) instances obtained by an evolutionary algorithm. The experimental results of our new approach show that diversity optimization of the features results in an improved coverage of the feature space over classical instance generation methods. In particular, the results show that for some combinations of two features it is possible to classify hard and easy instances into two clusters with a wider coverage of the feature space compared to the classical methods. Moreover, the three-feature combinations further improve the classification of hard and easy instances for most of the feature combinations. Furthermore, a classification model is built using these diverse instances that can classify TSP instances based on hardness for 2-OPT.

The remainder of this paper is organized as follows. Firstly, we introduce the Euclidean TSP and the background on feature based analysis. Afterwards, we state our diversity optimization approach for evolving instances according to feature values and report on the impact of diversity optimization in terms of the range of feature values. As feature values can be very diverse both for easy and hard instances, we consider the combinations of several features for instance classification afterwards. We then build a classification model that can classify instances based on hardness and finally finish with some conclusions.

2 Background

We consider the classical NP-hard Euclidean Traveling Salesperson problem (TSP) as the example problem for evolving hard and easy instances which have a diverse set of features. Our methodology can be applied to any optimization problem, but using the TSP in our study has the advantage that it has already been investigated extensively from different perspectives including the area of feature-based analysis.

The input of the problem is given by a set $V = \{v_1, \ldots, v_n\}$ of n cities in the Euclidean plane and Euclidean distances $d : V \times V \rightarrow \mathbb{R}_{\geq 0}$ between

Algorithm 1. $(\mu + \lambda)$-EA_D

1. Initialize the population P with μ TSP instances of approximation ratio at least α_h.
2. Let $C \subseteq P$ where $|C| = \lambda$.
3. For each $I \in C$, produce an offspring I' of I by mutation. If $\alpha_A(I') \geqslant \alpha_h$, add I' to P.
4. While $|P| > \mu$, remove an individual $I = \arg\min_{J \in P} d(J, P)$ uniformly at random.
5. Repeat step 2 to 4 until termination criterion is reached.

the cities. The goal is to find a Hamiltonian cycle whose sum of distances is minimal. A candidate solution for the TSP is often represented by a permutation $\pi = (\pi_1, \ldots, \pi_n)$ of the n cities and the goal is to find a permutation π^* which minimizes the tour length given by $c(\pi) = d(\pi_n, \pi_1) + \sum_{i=1}^{n-1} d(\pi_i, \pi_{i+1})$.

For our investigations cities are always in the normalized plane $[0, 1]^2$, i. e. each city has an x- and y-coordinate in the interval $[0, 1]$. In following, a TSP instance always consists of a set of n points in $[0, 1]^2$ and the Euclidean distances between them.

Local search heuristics have been shown to be very successful when dealing with the TSP and the most prominent local search operator is the 2-OPT operator [16]. The resulting local search algorithm starts with a random permutation of the cities and repeatedly checks whether removing two edges and reconnecting the two resulting paths by two other edges leads to a shorter tour. If no improvement can be found by carrying out any 2-OPT operation, the tour is called locally optimal and the algorithm terminates.

The key factor in the area of feature-based analysis is to identify the problem features and their contribution to the problem hardness for a particular algorithm and problem combination. This can be achieved through investigating hard and easy instances of the problem. Using an evolutionary algorithm, it is possible to evolve sets of hard and easy instances by maximizing or minimizing the fitness (tour length in the case of the TSP) of each instance [5–8]. However, none of these approaches have considered the diversity of the instances explicitly. Within this study we expect to improve the evolutionary algorithm based instance generation approach by introducing diversity optimization.

The structural features are dependent on the underlying problem. In [7], there are 47 features in 8 groups used to provide an understanding of algorithm performance for the TSP. The different feature classes established are distance features, mode features, cluster features, centroid features, MST features, angle features and convex hull features. The feature values are regarded as indicators which allow to predict the performance of a given algorithm on a given instance.

3 Feature-Based Diversity Optimization

In this section, we introduce our approach of evolving a diverse set of easy or hard instances which are diverse with respect to important problem features.

As in previous studies, we measure hardness of a given instance by the ratio of the solution quality obtained by the considered algorithm and the value of an optimal solution.

The approximation ratio of an algorithm A for a given instance I is defined as

$$\alpha_A(I) = A(I)/OPT(I)$$

where $A(I)$ is value of the solution produced by algorithm A for the given instance I, and $OPT(I)$ is value of an optimal solution for instance I. Within this study, $A(I)$ is the tour length obtained by 2-OPT for a given TSP instance I and $OPT(I)$ is the optimal tour length which we obtain in our experiments by using the exact TSP solver Concorde [17].

We propose to use an evolutionary algorithm to construct sets of instances of the TSP that are quantified as either easy or hard in terms of approximation and are diverse with respect to underlying features of the produced problem instances. Our evolutionary algorithm (shown in Algorithm 1) evolves instances which are diverse with respect to given features and meet given approximation ratio thresholds.

The algorithm is initialized with a population P consisting of μ TSP instances which have an approximation ratio at least α_h in the case of generating a diverse set of hard instances. In the case of easy instances, we start with a population where all instances have an approximation ratio of at most α_e and only instances of approximation ratio at most α_e can be accepted for the next iteration. In each iteration, $\lambda \leq \mu$ offspring are produced by selecting λ parents and applying mutation to the selected individuals. Offsprings that don't meet the approximation threshold are rejected immediately.

The new parent population is formed by reducing the set consisting of parents and offsprings satisfying the approximation threshold until a set of μ solutions is achieved. This is done by removing instances one by one based on their contribution to the diversity according to the considered feature.

The core of our algorithm is the selection among individuals meeting the threshold values for the approximation quality according to feature values. Let $I_1, \ldots, I_k$ be the elements of P and $f(I_i)$ be their features values. Furthermore, assume that $f(I_i) \in [0, R]$, i.e. feature values are non-negative and bounded above by R.

We assume that $f(I_1) \leq f(I_2) \leq \ldots \leq f(I_k)$ holds. The diversity contribution of an instance I to a population of instances P is defined as

$$d(I, P) = c(I, P),$$

where $c(I, P)$ is a contribution based on other individuals in the population

Let I_i be an individual for which $f(I_i) \neq f(I_1)$ and $f(I_i) \neq f(I_k)$. We set

$$c(I_i, P) = (f(I_i) - f(I_{i-1})) \cdot (f(I_{i+1}) - f(I_i)),$$

which assigns the diversity contribution of an individual based on the next smaller and next larger feature values. If $f(I_i) = f(I_1)$ or $f(I_i) = f(I_k)$, we set $c(I_i, P) = R^2$ if there is no other individual $I \neq I_i$ in P with $f(I) = f(I_i)$ and $c(I_i, P) = 0$ otherwise. This implies an individual I_i with feature value equal to any other instances in the population gains $c(I_i, P) = 0$. Furthermore, an individual with the unique smallest and largest feature value always stays in the population when working with $\mu \geq 2$.

In [7], 47 features of TSP instances for characterizing easy and hard TSP instances have been studied. We consider 7 features coming from different feature classes which have shown to be well suited for classification and prediction. These features are: *angle_mean*, *centroid_mean_distance_to_centroid*, *chull_area*, *cluster_10pct_mean_distance_to_centroid*, *mst_depth_mean*, *nnds_mean* and *mst_dists_mean*.

We refer the reader to [7] for a detailed explanation for each feature. We carry out our diversity optimization approach for these features and use the evolutionary algorithm to evolve for each feature a diverse population of instances that meets the approximation criteria for hard/easy instances given by the approximation ratio thresholds.

All programs in our experiments are written in R and run in R environment [18]. We use the functions in tspmeta package to compute the feature values [7].

The setting of the evolutionary algorithm for diversity optimization used in our experiments is as follows. We use $\mu = 30$ and $\lambda = 5$ for the parent and offspring population size, respectively. The 2-OPT algorithm is executed on each instance I five times with different initial solutions and we set $A(I)$ to the average tour length obtained. The examined instance sizes n are 25, 50 and 100, which are denoted by the number of cities in one instance. Based on previous investigations in [7] and initial experimental investigations, we set $\alpha_e = 1$ for instances of size 25 and 50, and $\alpha_e = 1.03$ for instances of size 100. Evolving hard instances, we use $\alpha_h = 1.15, 1.18, 1.2$ for instances of size $n = 25, 50, 100$, respectively. The mutation operator picks in each step one city for the given parent uniformly at random and changes its x- and y-coordinator by choosing an offset according to the Normal-distribution with standard deviation σ. Coordinates that are out of the interval are reset to the value of the parent. Based on initial experiments we use two mutation operators with different values of σ. We use $\sigma = 0.025$ with probability 0.9 and $\sigma = 0.05$ with probability 0.1 in a mutation step. The evolutionary algorithm terminates after 10,000 generations which allows to obtain a good diversity for the considered features. For each $n = 25, 50, 100$ and each of the 7 features, a set of easy and hard instances are generated, which results in 42 independent runs of the $(\mu+\lambda)$-EA$_D$.

4 Range of Feature Values

We first evaluate our diversity optimization approach in terms of the diversity that is obtained with respect to a single feature. Focusing on a single feature in each run provides the insight of the possible range of a certain feature value for hard or easy instances. The previous study [7], suggests that there are some

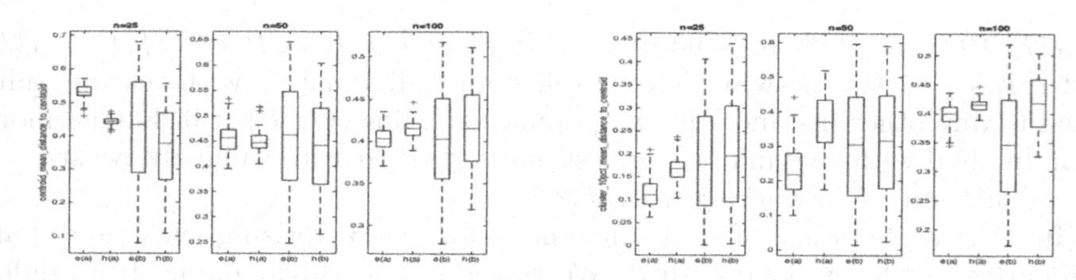

Fig. 1. (left) The boxplots for centroid mean distance to centroid feature values of a population consisting of 100 different hard or easy TSP instances of different number of cities without or with diversity mechnism. (right) The boxplots for cluster 10 % distance distance to centroid feature values of a population consisting of 100 different hard or easy TSP instances of different number of cities without or with diversity mechnism. Easy and hard instances from conventional approach and diversity optimization are indicated by e(a), h(a) and e(b), h(b) respectively.

differences in the possible range of feature values for easy and hard instances. We study the effect of the diversity optimization on the range of features by comparing the instances generated by diversity optimization to the instances generated by the conventional approach in [7]. Evolving hard instances based on the conventional evolutionary algorithm, the obtained instances have mean approximation ratios of 1.12 for $n = 25$, 1.16 for $n = 50$, and 1.18 for $n = 100$. For easy instances, the mean approximation ratios are 1 for $n = 25, 50$ and 1.03 for $n = 100$.

Figure 1 (left) presents the variation of the mean distance of the distances between points and the centroid feature (*centroid_mean_distance_to_centroid*) for hard and easy instances of the three considered sizes 25, 50 and 100. Each set consists of 100 instances generated by independent runs [7]. As shown in Fig. 1 (left) the hard instances have higher feature values than for easy instances for all instance sizes. For example, for instance size 100 and for the hard instances the median value (indicated by the red line) is 0.4157 while its only 0.0.4032 for the easy instances. The respective range of the feature value is 0.0577 for the hard instances and 0.0645 for the easy instances. For the instances generated by diversity optimization (easy and hard instances are indicated by e(b) and h(b) respectively), there is a difference in the median feature values for the hard and easy instances similar to the instances generated by the conventional approach. Additionally, the range of the feature values for both the hard and easy instances has significantly increased. For example, for the instance size 100, the median value for easy instances is 0.4028 and the range is 0.2382. For the hard instances of the same size, the median is 0.04157 while the range is 0.1917 (see Fig. 1 (left)).

Similarly, Fig. 1 (right) presents the variation of cluster 10 % distance to centroid (*cluster_10pct_distance_to_centroid*) feature for the hard and easy instances generated by the conventional approach (indicated by (e(a) and h(a)) and for the hard and easy instances generated by diversity optimization (indicated by (e(b) and h(b))). The general observations from these box plots are quite similar to the observations from the *mst_dist_mean* shown in Fig. 1 (left).

The above results suggest that the diversity optimization approach has resulted in a significant increase in the coverage over the feature space. Having the threshold for approximation ratios (α_e and α_h) our method guarantees the hardness of the instances. These approximation thresholds are more extreme than the mean approximation values obtained by the conventional method. Being able to discover all these instances spread in the whole feature space, our approach provides a strong basis for more effective feature based prediction.

As a result of the increased ranges and the similar gap in median feature values for hard and easy instances compared to the conventional instances, there is a strong overlap in the ranges of the features for easy and hard instances generated by the diversity optimization. This is observed in the results for *mst_dist_mean* and *cluster_10pct_distance_to_centroid* shown in Fig. 1. Similar pattern holds for the other features as well. This prevents a good classification of problem instances based on single feature value.

5 Classification Based on Multiple Features

As a single feature is not capable in clearly classifying the hard/easy instances, combinations of two or three different features are examined in the following. Our analysis mainly focuses on combinations of the 7 previously introduced features.

According to the observation and discussion in [7], the two features *distance_max* and *angle_mean* can be considered together to provide an accurate classification of the hard and easy instances. Whereas after increasing the diversity over the seven different feature values and a wider coverage of the 2D space is achieved, the separation of easy and hard instances is not so obvious, as shown in Fig. 2. There are large overlapping areas lying between the two groups of instances. As the number of cities in an instance increases, the overlapping area becomes larger. It is hard to do classification based on this. Therefore the idea of combining three different feature is put forward.

Support vector machines (SVMs) are well-known supervised learning models in machine learning which can be used for classification, regression and outliers

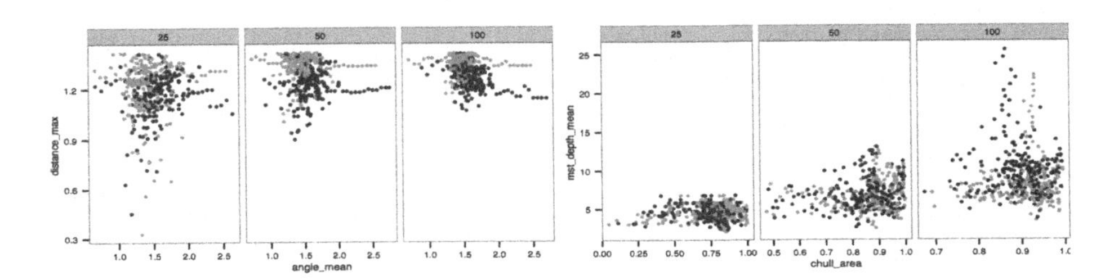

Fig. 2. 2D Plots of feature combinations which provide a separation between easy and hard instances. The blue dots and orange dots represent hard and easy instances respectively. (Color figure online)

detection [19]. In order to quantify the separation between instances of different hardness based on the feature values, SVM models are constructed for each combination of features.

Let ACC_n be the training accuracy of a feature combination in separating the hard and easy instances of size n. We define ACC_n as the ratio of number of instances which are correctly classified by the model to the total number of instances in the dataset. All classification experiments are done in R with library{$e1071$} [20].

The training data of the SVM models are the population of 420 instances generated as in Sect. 3 and the training accuracy is regarded as a quantified measurement of the separation between hard and easy instances. The feature combinations used for classification are the 21 two-feature combinations and 35 three-feature combinations discussed in Sect. 5.

The linear classifier is the first model tried in classifying the dataset. Since the dataset is not linearly separable, taken the trade-off between maximizing the margin and minimizing the number of misclassified data points into consideration, the soft-margin SVM is used for classification. From experiment results, most of the accuracies of different feature combinations lie in the range of 0.6 to 0.7, which implies the high possibility that the linear models are not suitable for separating the hard and easy instances based on most of the feature combinations. Therefore we move to applying kernel functions for non-linear mapping of the feature combination. The Radial Basis Function (RBF) kernel is one of the well-known kernel functions used in SVM classification.

There are two parameters needed when applying RBF kernel, which are C(cost) and γ. The parameter setting for RBF is crucial, since increasing C and γ leads to accurate separation of the training data but at the same time causes

Table 1. The accuracy of SVM with RBF kernel separating the hard and easy instances in different two-feature space.

Feature 1	Feature 2	ACC_{25}	ACC_{50}	ACC_{100}
angle_mean	centroid_mean_distance_to_centroid	0.8476	0.8071	0.8071
angle_mean	chull_area	0.7857	0.7810	0.7929
angle_mean	cluster_10pct_mean_distance_to_centroid	0.7810	0.7786	0.8000
angle_mean	mst_depth_mean	0.7524	0.7381	0.8000
angle_mean	nnds_mean	0.8167	0.8833	0.8452
angle_mean	mst_dists_mean	0.8119	0.8024	0.8405
centroid_mean_distance_to_centroid	chull_area	0.8619	0.7667	0.8381
centroid_mean_distance_to_centroid	cluster_10pct_mean_distance_to_centroid	0.8524	0.8357	0.7548
centroid_mean_distance_to_centroid	mst_depth_mean	0.8381	0.7643	0.8095
centroid_mean_distance_to_centroid	nnds_mean	0.8786	0.9524	0.8476
centroid_mean_distance_to_centroid	mst_dists_mean	0.8905	0.8571	0.8762
chull_area	cluster_10pct_mean_distance_to_centroid	0.8000	0.7881	0.8548
chull_area	mst_depth_mean	0.7429	0.7429	0.7571
chull_area	nnds_mean	0.8071	0.8905	0.8452
chull_area	mst_dists_mean	0.8619	0.8643	0.9024
cluster_10pct_mean_distance_to_centroid	mst_depth_mean	0.7619	0.7714	0.7929
cluster_10pct_mean_distance_to_centroid	nnds_mean	0.8190	0.8833	0.8643
cluster_10pct_mean_distance_to_centroid	mst_dists_mean	0.8095	0.8095	0.8738
mst_depth_mean	nnds_mean	0.7786	0.8595	0.8405
mst_depth_mean	mst_dists_mean	0.8095	0.8214	0.8810
nnds_mean	mst_dists_mean	0.8500	0.9143	0.9024

Table 2. The accuracy of SVM with RBF kernel separating the hard and easy instances in different three-feature space.

Feature 1	Feature 2	Feature 3	ACC_{25}	ACC_{50}	ACC_{100}
angle_mean	centroid_mean_distance_to_centroid	chull_area	0.9500	0.9190	0.9452
angle_mean	centroid_mean_distance_to_centroid	cluster_10pct_mean_distance_to_centroid	0.9405	0.9357	0.8214
angle_mean	centroid_mean_distance_to_centroid	mst_depth_mean	0.9548	0.9548	0.9214
angle_mean	centroid_mean_distance_to_centroid	nnds_mean	0.9452	0.9952	0.9833
angle_mean	centroid_mean_distance_to_centroid	mst_dists_mean	0.9571	0.9500	0.9524
angle_mean	chull_area	cluster_10pct_mean_distance_to_centroid	0.9524	0.9310	0.8881
angle_mean	chull_area	mst_depth_mean	0.9357	0.9238	0.9500
angle_mean	chull_area	nnds_mean	0.9405	0.9714	0.9571
angle_mean	chull_area	mst_dists_mean	0.9667	0.9619	0.9143
angle_mean	cluster_10pct_mean_distance_to_centroid	mst_depth_mean	0.9214	0.9143	0.9810
angle_mean	cluster_10pct_mean_distance_to_centroid	nnds_mean	0.9476	0.9690	0.9333
angle_mean	cluster_10pct_mean_distance_to_centroid	mst_dists_mean	0.9571	0.9143	0.9405
angle_mean	mst_depth_mean	nnds_mean	0.9310	0.9762	0.9238
angle_mean	mst_depth_mean	mst_dists_mean	0.9476	0.9262	0.9476
angle_mean	nnds_mean	mst_dists_mean	0.9429	0.9762	0.8833
centroid_mean_distance_to_centroid	chull_area	cluster_10pct_mean_distance_to_centroid	0.9476	0.9333	0.9310
centroid_mean_distance_to_centroid	chull_area	mst_depth_mean	0.9595	0.8762	0.9762
centroid_mean_distance_to_centroid	chull_area	nnds_mean	0.9667	0.9881	0.9929
centroid_mean_distance_to_centroid	chull_area	mst_dists_mean	0.9714	0.9714	0.8381
centroid_mean_distance_to_centroid	cluster_10pct_mean_distance_to_centroid	mst_depth_mean	0.9476	0.9286	0.8571
centroid_mean_distance_to_centroid	cluster_10pct_mean_distance_to_centroid	nnds_mean	0.9643	0.9905	0.8810
centroid_mean_distance_to_centroid	cluster_10pct_mean_distance_to_centroid	mst_dists_mean	0.9500	0.9595	0.9190
centroid_mean_distance_to_centroid	mst_depth_mean	nnds_mean	0.9500	0.9881	0.9595
centroid_mean_distance_to_centroid	mst_depth_mean	mst_dists_mean	0.9548	0.9548	0.9595
centroid_mean_distance_to_centroid	nnds_mean	mst_dists_mean	0.9667	1.0000	0.9952
chull_area	cluster_10pct_mean_distance_to_centroid	mst_depth_mean	0.9286	0.9524	0.9333
chull_area	cluster_10pct_mean_distance_to_centroid	nnds_mean	0.9524	0.9667	0.9667
chull_area	cluster_10pct_mean_distance_to_centroid	mst_dists_mean	0.9595	0.9595	0.9929
chull_area	mst_depth_mean	nnds_mean	0.9381	0.9857	0.9476
chull_area	mst_depth_mean	mst_dists_mean	0.9476	0.9738	0.9833
chull_area	nnds_mean	mst_dists_mean	0.9714	0.9857	0.9667
cluster_10pct_mean_distance_to_centroid	mst_depth_mean	nnds_mean	0.9214	0.9857	0.9738
cluster_10pct_mean_distance_to_centroid	mst_depth_mean	mst_dists_mean	0.9500	0.9476	0.9643
cluster_10pct_mean_distance_to_centroid	nnds_mean	mst_dists_mean	0.9643	0.9833	0.9976
mst_depth_mean	nnds_mean	mst_dists_mean	0.9429	0.9929	0.9929

over-fitting. The SVMs here are generated for quantifying the separation rate between hard and easy instances rather than classifying other instances. After some initial trials, (C, γ) is set to $(100, 2)$ in all the tests to avoid over-fitting. This parameter setting may not be the best for the certain feature combination in SVM classifying, but it helps us to gain some understanding of the separation of hard and easy instances generated from previous experiments based on the same condition.

Tables 1 and 2 show the accuracy of different two-feature or three-feature combination in hard and easy instances separation. With RBF kernel, SVM with certain parameter setting can generate a model separating the dataset with average accuracy of 0.8170, 0.8244 and 0.8346 in 2D feature space for instance size 25, 50 and 100 respectively. Whereas with three features, SVM with the same parameter setting provides a separation with average accuracy of 0.9503, 0.9584 and 0.9422 for instance size 25, 50 and 100 respectively.

From the results, it can be concluded that there are better separations between hard and easy instances in the 3D feature space.

6 Conclusions

With this paper, we have introduced a new methodology of evolving easy/hard instances which are diverse with respect to feature sets of the optimization problem at hand. Using our diversity optimization approach we have shown that the easy and hard instances obtained by our approach covers a much wider range in the feature space than previous methods. The diversity optimization approach provides instances which are diverse with respect to the investigated features. The proposed population diversity measurements provide good evaluation of the diverse over single or multiple feature values. Our experimental investigations for 2-OPT and TSP have shown that our large set of diverse instances can be classified quite well into easy and hard instances when considering a suitable combination of multiple features which provide some guidance for predication as the next step. In particular, the SVM classification model built with the diverse instances that can classify TSP instances based on problem hardness provides a strong basis for future performance prediction models that lead to automatic algorithm selection and configuration. Building such models would require further experimentation to determine the minimal set of strong features that can predict performance accurately.

Acknowledgement. This research has been supported by the European Union Seventh Framework Programme (FP7/2007–2013) under grant agreement no. 618091 (SAGE) and by the Australian Research Council under grant agreement DP140103400.

References

1. Hutter, F., Xu, L., Hoos, H.H., Leyton-Brown, K.: Algorithm runtime prediction: methods & evaluation. Artif. Intell. **206**, 79–111 (2014)

2. Vilalta, R., Drissi, Y.: A perspective view and survey of meta-learning. Artif. Intell. Rev. **18**(2), 77–95 (2002)
3. Eggensperger, K., Hutter, F., Hoos, H.H., Leyton-Brown, K.: Efficient benchmarking of hyperparameter optimizers via surrogates. In: Proceedings of the Twenty-Ninth AAAI Conference on Artificial Intelligence, 25–30 January 2015, Austin, Texas, USA, pp. 1114–1120 (2015)
4. Feurer, M., Springenberg, J.T., Hutter, F.: Initializing bayesian hyperparameter optimization via meta-learning. In: Proceedings of the Twenty-Ninth AAAI Conference on Artificial Intelligence, 25–30 January 2015, Austin, Texas, USA, pp. 1128–1135 (2015)
5. Smith-Miles, K., Lopes, L.: Measuring instance difficulty for combinatorial optimization problems. Comput. Oper. Res. **39**(5), 875–889 (2012)
6. van Hemert, J.I.: Evolving combinatorial problem instances that are difficult to solve. Evol. Comput. **14**(4), 433–462 (2006)
7. Mersmann, O., Bischl, B., Trautmann, H., Wagner, M., Bossek, J., Neumann, F.: A novel feature-based approach to characterize algorithm performance for the traveling salesperson problem. Ann. Math. Artif. Intell. **69**(2), 151–182 (2013)
8. Smith-Miles, K., van Hemert, J., Lim, X.Y.: Understanding TSP difficulty by learning from evolved instances. In: Blum, C., Battiti, R. (eds.) LION 4. LNCS, vol. 6073, pp. 266–280. Springer, Heidelberg (2010)
9. Neumann, F., Witt, C.: Bioinspired Computation in Combinatorial Optimization: Algorithms and Their Computational Complexity, 1st edn. Springer, New York (2010)
10. Kötzing, T., Neumann, F., Röglin, H., Witt, C.: Theoretical analysis of two ACO approaches for the traveling salesman problem. Swarm Intell. **6**(1), 1–21 (2012)
11. Englert, M., Röglin, H., Vöcking, B.: Worst case and probabilistic analysis of the 2-Opt algorithm for the TSP. Algorithmica **68**(1), 190–264 (2014)
12. Nallaperuma, S., Wagner, M., Neumann, F., Bischl, B., Mersmann, O., Trautmann, H.: A feature-based comparison of local search and the Christofides algorithm for the travelling salesperson problem. In: FOGA 2013, pp. 147–160 (2013)
13. Nallaperuma, S., Wagner, M., Neumann, F.: Parameter prediction based on features of evolved instances for ant colony optimization and the traveling salesperson problem. In: Bartz-Beielstein, T., Branke, J., Filipič, B., Smith, J. (eds.) PPSN 2014. LNCS, vol. 8672, pp. 100–109. Springer, Heidelberg (2014)
14. Ulrich, T., Bader, J., Thiele, L.: Defining and optimizing indicator-based diversity measures in multiobjective search. In: Schaefer, R., Cotta, C., Kołodziej, J., Rudolph, G. (eds.) PPSN XI. LNCS, vol. 6238, pp. 707–717. Springer, Heidelberg (2010)
15. Ulrich, T., Bader, J., Zitzler, E.: Integrating decision space diversity into hypervolume-based multiobjective search. In: GECCO, pp. 455–462 (2010)
16. Croes, G.A.: A method for solving traveling-salesman problems. Oper. Res. **6**(6), 791–812 (1958)
17. Applegate, D., Cook, W., Dash, S., Rohe, A.: Solution of a min-max vehicle routing problem. INFORMS J. Comput. **14**(2), 132–143 (2002)
18. Core Team: R: A Language and Environment for Statistical Computing. R Foundation for Statistical Computing, Vienna, Austria (2015)
19. Cortes, C., Vapnik, V.: Support-vector networks. Mach. Learn. **20**(3), 273–297 (1995)
20. Meyer, D., Dimitriadou, E., Hornik, K., Weingessel, A., Leisch, F.: e1071: Misc Functions of the Department of Statistics, Probability Theory Group (Formerly: E1071), TU Wien. R package version 1.6-7 (2015)

Searching for Quality Diversity When Diversity is Unaligned with Quality

Justin K. Pugh(✉), L.B. Soros, and Kenneth O. Stanley

Department of Computer Science, University of Central Florida,
4328 Scorpius Street, Orlando, FL 32816-2362, USA
{jpugh,lsoros,kstanley}@cs.ucf.edu

Abstract. Inspired by natural evolution's affinity for discovering a wide variety of successful organisms, a new evolutionary search paradigm has emerged wherein the goal is not to find the single best solution but rather to collect a diversity of unique phenotypes where each variant is as good as it can be. These *quality diversity* (QD) algorithms therefore must explore multiple promising niches simultaneously. A QD algorithm's diversity component, formalized by specifying a *behavior characterization* (BC), not only generates diversity but also promotes quality by helping to overcome deception in the fitness landscape. However, some BCs (particularly those that are *unaligned* with the notion of quality) do not adequately mitigate deception, rendering QD algorithms unable to discover the best-performing solutions on difficult problems. This paper introduces a solution that enables QD algorithms to pursue arbitrary notions of diversity without compromising their ability to solve hard problems: driving search with multiple BCs simultaneously.

Keywords: Novelty search · Non-objective search · Quality diversity · Behavioral diversity · Neuroevolution

1 Introduction

Evolutionary computation (EC) has developed increasingly sophisticated search algorithms around the idea that increasing fitness is a powerful mechanism for optimization [2]. However, natural evolution is more than an optimizer. Unlike conventional optimization, nature has no single unifying target and often rewards being different in addition to being better. Indeed, natural evolution has discovered a vast diversity of organisms and ways of being, simultaneously solving an uncountable and ever-changing array of problems from sight to ambulation to cognition, not by finding a single "best" solution to each but instead by collecting a breadth of viable alternatives.

In a step away from EC's longstanding fixation on fitness for the purpose of optimization, a new algorithm called novelty search (NS) [8] was introduced, which searches only for diversity and is notably free from objective pressure. Ironically, novelty search and its variants [4,13] were initially heralded themselves

J. Handl et al. (Eds.): PPSN XIV 2016, LNCS 9921, pp. 880–889, 2016.
DOI: 10.1007/978-3-319-45823-6_82

as powerful tools for optimization because their agnosticism to the objective sometimes allows them to bypass the problem of *deception* and thus succeed on tasks where traditional objective-based approaches fail. However, NS's ability to collect a wide breadth of phenotypes is largely unappreciated when applied as an optimization algorithm: any accumulated diversity in such an application is eventually discarded in favor of saving only the best-performing individual.

Taking NS in a different direction, a unique search paradigm has begun to emerge within EC wherein diversity itself is a desirable end product. New algorithms such as Novelty Search with Local Competition (NSLC) [9] and MAP-Elites [12] stand apart from the usual focus on optimization in that rather than simply trying to find the single best individual (or tradeoffs among a set of objective targets [3]), these algorithms are instead designed to find *quality diversity* – a maximally diverse collection of individuals in which each member is as high-performing as possible. For example, one classic application of QD is to collect as many successful ambulating virtual creature morphologies as possible [9,18]. QD is distinct from other approaches designed to return multiple results (such as those that seek to return a handful of local optima) in that all parts of the diversity space are considered equally important and the goal is to sample the entire space, returning the best possible performance in each region (even lower-performing regions). Compared to simple optimization, QD represents a new style of search that more closely embodies the spirit of natural evolution and for which evolutionary treatments are uniquely well-suited due to their natural inclination for exploring many promising directions at the same time.

Applying QD algorithms such as NSLC or MAP-Elites requires both a notion of quality (a fitness function) and a notion of diversity, called the *behavior characterization* (BC), which defines the degree of difference between two individuals. So far, applications of QD have largely featured characterizations that are *unaligned* with quality, which means that where an individual is located in the diversity space has little bearing on its potential performance; examples include the time individual legs of a hexapod robot spend on the ground [1], the specific image class targeted by a generated image [14,15], and the size and shape of ambulating stick-figure creatures [18]. We can see that this focus on characterizations that are orthogonal to quality is natural by examining our intuitive sense of QD in nature. Indeed, Earth has accumulated a diverse repertoire of organisms with respect to intuitive characterizations such as size, appearance, or locomotion strategy, but those characterizations themselves are not good predictors of a particular organism's reproductive capacity or cognitive function (intuitive measures of quality). In effect, the types of diversity that we consider to be interesting or salient are often unaligned with our notions of quality.

This observation is interesting because a recent study comparing state-of-the-art QD algorithms in a relatively easy maze domain called the "QD-Maze" (inspired by the "HardMaze" domain [8] that has become ubiquitous in studies involving NS) indicates that the degree of characterization-quality alignment has a significant impact on performance and furthermore suggests that unaligned characterizations may be sub-optimal for driving search [16]. If this hypothesis is true, then typical approaches to QD may break down on harder problems.

The goal of this paper is to specifically address the challenge of finding "unaligned QD" in the context of a difficult maze domain (such that finding solutions is non-trivial even for the most sophisticated approaches). Experimental evidence in this domain confirms that driving search with an unaligned BC indeed has catastrophic effects on the ability of QD algorithms to successfully collect QD. As a solution, this paper introduces the idea of driving search with multiple BCs *simultaneously*. The success of this new approach offers a promising strategy for applying QD algorithms even when there is an incongruity between the desired notion of diversity and the ideal characterization for driving search, thus opening the door to a wider breadth of potential domains in the future.

2 Domain: QD-Gauntlet

Because the QD-Maze domain of Pugh et al. [16] is relatively easy, its results in effect speak to search spaces with a variety of relatively simple solutions. Yet many spaces of interest in the future will likely require a significant degree of complexity to find the interesting needles in the haystack. Earth itself has this quality, where there are innumerable different species, yet each is highly complex in its own right. Thus the new *QD-Gauntlet* domain in this paper builds on the old QD-Maze by greatly increasing the complexity of the possible paths to solutions, but still providing the opportunity for variety among those paths. In particular, like the QD-Maze [16], the QD-Gauntlet is a maze domain featuring an egocentric robot with multiple viable paths to the goal. An egocentric maze is an appealing platform for studying QD algorithms because the results are easily visualized and egocentric mazes are well-studied in the context of novelty search, where results originally obtained on HardMaze [8] have been shown to generalize to a variety of other domains such as quadruped locomotion [11], game content generation [10], swarm robotics [4], and image classification [19]. Furthermore, while the *Euclidean distance to the goal* heuristic that traditionally drives search in maze domains is known to be deceptive (which is the primary source of difficulty), in this domain we can also compute the perfect solution paths, enabling concrete measurements of the progress towards solving the maze.

The new maze, QD-Gauntlet (Fig. 1), is significantly more complex than its predecessors. In QD-Gauntlet, there are four distinct corridors leading horizontally to a goal point on the right side of the maze. Each corridor (i.e. leg) of the maze is composed of four successive segments, where each segment is designed to be approximately equivalent to the size and complexity of the canonical HardMaze [8]. To ensure that the four legs are similarly difficult, the legs are near mirror images of each other with slight variations. Importantly, QD-Gauntlet contains several long, straight corridors that terminate in a dead-end close to the goal. These dead-end corridors serve to increase maze difficulty by deceiving quality-seeking (i.e. fitness-based) search mechanisms. To further increase maze difficulty, agents are given strict time constraints such that deviating too far from one of the optimal paths will cause the agent to run out of time before reaching the goal.

Fig. 1. QD-Gauntlet. An egocentric agent begins on the start point (left) and is presented with four viable paths to reach the goal point (right). The maze is riddled with dead-ends that serve to deceive objective-oriented search algorithms.

As in Lehman and Stanley [8], agents are driven by evolved neural networks and are equipped with a set of six wall-sensing rangefinders (five spanning the frontal 180 degrees and one facing the rear), four pie-slice sensors to sense the distance and relative direction of the goal point, and a single output to specify left-right turns. The challenge then is not to evolve a path, but rather to evolve a neural network that can correctly guide the agent through one of the long and deceptive corridors based on its sensory inputs. Furthermore, the hope is that QD algorithms can find *multiple* such solutions in the same run, corresponding to a variety of different driving strategies. Doing so would be a proxy for finding QD in any domain whose solutions are challenging and deceptive to reach.

3 Algorithms

This section describes the algorithms considered in this study, focusing first on variants of novelty search from the literature. Then, two additional algorithms are introduced to address the problem of finding QD when the desired notion of diversity is not aligned with quality. The hope is that the results from this study will also provide guidance for other QD algorithms in the future, such as MAP-Elites [1,12].

Fitness. Included as a baseline to establish domain difficulty, a purely fitness-based search is implemented as standard generational NEAT [17] with a population size of 500, where fitness is the deceptive *Euclidean distance to the goal* heuristic (which also drives the quality portion of the QD algorithms that follow).

With the exception of Fitness, all of the other algorithms in this paper are implemented as steady state (only a small portion of the population is replaced at a time to avoid radical shifts in what is considered "novel" from one tick to the next: genomes are replaced in batches of 32 to allow moderate parallelism).

NS. While not technically a "quality diversity" algorithm because there is no quality component, the *novelty search* (NS) [8] algorithm forms the foundation of a number of other algorithms in this study and serves as another baseline for comparison, establishing what is possible without a drive towards quality. Novelty search works by rewarding *novelty* instead of fitness, where novelty measures how different an individual's behavior is from those that have been seen before. More formally, novelty is calculated by summing the distance to the k-nearest behaviors (in this paper, $k = 20$) from a set composed of the current population and an *archive* of past behaviors. The distance between two behaviors is simply the Euclidean distance between those behaviors when represented as a vector of numbers (called a *behavior characterization*). While there exist several different strategies for managing the archive [5], preliminary experiments indicated that a powerful strategy is to add all individuals to an archive with a maximum size that is enforced by deleting those with the lowest novelty (novelty is recomputed against the archive before each deletion). In all cases, NS is run with a population size of 500 and a maximum archive size of 2,500.

NSLC. *Novelty search with local competition* (NSLC) [9] combines the diversifying pressure of NS with a localized drive towards quality called *local competition* (LC), calculated as the proportion of 20 nearest behavioral neighbors with a lower fitness score. LC encourages increasing performance within local behavioral neighborhoods without suffering the deleterious effects of a global objective pressure. Novelty and LC are combined by Pareto ranking as in the NSGA-II multi-objective optimization algorithm [3].

3.1 Multi-BC QD Algorithms

In each of NS and NSLC, search is driven by some notion of behavioral diversity (i.e. a BC). Traditionally, the BC that drives search corresponds to the type of diversity that the researcher is interested in collecting (e.g. different types of robot morphologies or different walking gaits) and thus is typically unaligned with the notion of quality. Unaligned BCs are less capable of overcoming deception [7,16] and on difficult tasks (such as QD-Gauntlet) may altogether fail to obtain high-performing solutions, creating a problem for researchers interested in finding unaligned QD. As a solution, we introduce the idea here of driving search with *multiple* BCs simultaneously and propose two possible methods for doing so. While each method can conceivably support three or more BCs, for simplicity the experiments that follow are restricted to only two BCs.

NS-NS. The basic NS algorithm can be extended to support multiple BCs simultaneously by combining their respective novelty scores in a multi-objective formulation (with NSGA-II [3]). In this algorithm, dubbed NS-NS, each BC maintains its own independent archive and individuals are evaluated against each archive in turn to calculate one novelty score per BC. There is only a single breeding population where the breeding potential for each member is decided by Pareto ranking according to novelty scores. The key idea is that in a two-BC formulation, one BC may be ideal for driving search while the other corresponds to the type of diversity the user is interested in collecting.

NS-NSLC. While NS-NS facilitates searching with multiple concepts of diversity, it lacks the explicit drive towards quality that is essential to QD algorithms. This omission is remedied in NS-NSLC by adding a local competition objective (in the same way as in NSLC) where behavioral neighbors are decided by the (unaligned) BC that corresponds to the user's desired notion of diversity.

4 Experiment

As discussed in Sect. 3, a common component of all QD algorithms is the BC, which formalizes the notion of diversity so that it can drive the search explicitly. In domains where there is really only one desired objective behavior, the BC serves only to drive search towards better solutions and thus a strongly-aligned BC is most appropriate (e.g. NS quickly solves the difficult HardMaze domain [8] because it circumvents the problem of deception by pursuing novel *endpoints*). However, when diversity itself is a desirable product of search, researchers must choose a BC that expresses the type of diversity they want to collect; often this choice results in a BC that is not well-aligned with the notion of quality, which recent research suggests may not be optimal for driving search towards better solutions [7,16]. To investigate how the performance gap between aligned and unaligned BCs extends to hard problems, this paper compares the strongly-aligned EndpointBC from Lehman and Stanley [8] and the unaligned DirectionBC from Pugh et al. [16] on the much more challenging QD-Gauntlet (Fig. 1). However, congruent with the common practice of searching for unaligned QD, in this paper QD is always *collected* with respect to DirectionBC.

EndpointBC simply characterizes agent behavior by its (x, y) location at the end of its trial. This strongly-aligned BC is a powerful way to drive search on maze domains because it explores progressively more remote locations until the goal point is found. On the other end of the alignment spectrum, **DirectionBC** characterizes *how* the agent drives instead of where. DirectionBC consists of five values indicating whether the agent was most frequently facing north (0.125), east (0.375), south (0.625), or west (0.875) during each fifth of its trial. When driving search with this unaligned BC, it is possible to exhaust the entire behavior space without ever discovering high-performing solutions.

Each of the algorithms from Sect. 3 is implemented with each of DirectionBC and EndpointBC for a total of seven treatments: Fitness, NS_d, NS_e,

$NSLC_d$, $NSLC_e$, NS_eNS_d, and NS_eNSLC_d. Each treatment is run 20 times on QD-Gauntlet, each for 1,000,000 evaluations (by which time all treatments reach a performance plateau). Networks are evolved with a modified version of SharpNEAT 1.0 [6] with mutation parameters validated by Pugh et al. [16]: 60 % mutate connection, 10 % add connection, 0.5 % add neuron. Networks are feed-forward and restricted to asexual reproduction; other settings follow SharpNEAT 1.0 defaults.

The performance of each treatment is evaluated according to the *QD-score* metric introduced by Pugh et al. [16] (and similar to the "global reliability" metric in Mouret and Clune [12]). Over the course of a run, a collection of individuals called the "QD grid" is gathered: the QD grid is managed such that each behavioral bin remembers the highest quality individual seen so far. While these bins are reminiscent of the bins in a MAP-Elites grid, the QD grid is completely external to the breeding population and thus does not interfere with or influence evolution. The QD-score is calculated as the total quality across all filled bins within the QD grid and reflects both how many distinct behaviors have been discovered and how good those behaviors are. Regardless of the BC driving search, QD-score for this study is always calculated with respect to DirectionBC (i.e. the QD grid represents a collection of different ways of driving). *Consequently, the EndpointBC-driven treatments in effect test whether QD with respect to one BC can be achieved passively by driving search with another BC altogether.*

While fitness for algorithms in this paper is the Euclidean distance to the goal, this heuristic is deceptive and does not accurately characterize how close collected behaviors are to actually solving the maze. Thus, for the purposes of evaluation, quality for individuals within the QD grid is instead represented by a *progress score* that respects that agents cannot drive through walls. Progress is defined as inversely proportional to the length of the shortest valid path between the agent's final location and the goal point of the maze. Importantly, this measure of quality (which draws a perfect, non-deceptive gradient over the drivable area of the maze) is not available to drive search and is only used for assessment.

5 Results

Figure 2 depicts the final QD-score (averaged over 20 runs) for each treatment after 1,000,000 evaluations. Unsurprisingly, Fitness performs significantly worse[1] than all other treatments ($p < 0.001$), underscoring the importance of specialized approaches to QD from outside the realm of conventional optimization. Notably, treatments driven by DirectionBC perform significantly worse than the EndpointBC and multi-BC treatments ($p < 0.001$), even though EndpointBC has nothing to do with the type of diversity being collected. Explaining this disparity, DirectionBC treatments fail to find even a single solution across all 40 runs, while each of the EndpointBC and multi-BC treatments consistently find multiple maze solutions per run – thus DirectionBC's low performance represents an inability to overcome deception in the QD-Gauntlet.

[1] Statistical significance is determined by an unpaired two-tailed Student's t-test.

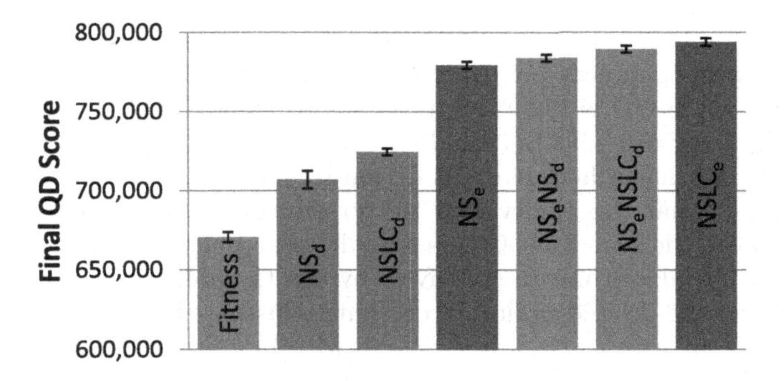

Fig. 2. Final QD-score. The final QD-score achieved by each treatment is shown (averaged over 20 runs). Error bars represent standard error. Bars are color-coded according to which BC drives search: DirectionBC-driven treatments (subscript d) are drawn in green, EndpointBC-driven (subscript e) in red, and multi-BC in blue. (Color figure online)

Among the best surveyed approaches, $\mathrm{NS}_e\mathrm{NSLC}_d$ and NSLC_e are not significantly different after the full 1,000,000 evaluations ($p = 0.163$). However, EndpointBC and multi-BC approaches demonstrate fundamentally different trends of QD-score over time (not shown): while multi-BC increases rapidly to an early plateau, EndpointBC takes much longer to reach similarly high scores. Correspondingly, *NS_eNSLC_d scores significantly better than $NSLC_e$ ($p < 0.05$) until 224,000 evaluations*, after which point NSLC_e catches up.

6 Discussion and Conclusion

The QD-Gauntlet task, like the HardMaze [8] before it, contains a pronounced level of deception wherein the natural "Euclidean distance to the goal" fitness function does not always point the way to the goal but rather often leads search into an evolutionary dead end. In this task, as in any domain where deception is sufficiently present, strictly following the compass of increasing fitness (as in the Fitness treatment in this paper) is doomed to fail. While the idea of pursuing behavioral diversity instead of objective fitness (as in novelty search) is often recognized as a powerful way to confront such problems, it is clear that to be effective behavioral diversity cannot be applied naively.

Such a naive application of behavioral diversity lies in DirectionBC. Because it is unaligned with the concept of quality (driving closer to the goal), diversity with respect to DirectionBC can be achieved without making any progress towards conquering the problem of deception. Indeed, in the QD-Gauntlet, DirectionBC alone is incapable of unlocking the best-performing parts of the search space. Even if QD algorithms driven by such a BC succeed in generating a collection of locally optimal niches, if the best-performing behaviors cannot be found then the system is only operating at a fraction of its potential.

Experimental evidence on the QD-Gauntlet supports the idea that achieving QD with respect to quality-unaligned BCs in the traditional way is problematic: surprisingly, more QD is found with respect to DirectionBC by driving search with a completely *different* BC (EndpointBC – which is known to excel at solving maze tasks) than when search is driven by DirectionBC itself (Fig. 2). This result suggests that the ability of a BC to overcome deception can be just as instrumental in the search for QD as actually searching for diversity.

To allow QD algorithms to effectively bypass the problem of deception without sacrificing the desired notion of diversity, this paper introduces the idea of multi-BC QD algorithms that drive search with more than one BC at the same time: one that targets diversity and another that is better suited to driving search. Of the multi-BC variants surveyed, the best performance is achieved by NS_eNSLC_d, which is on par with $NSLC_e$ for achieving the highest QD-score on the QD-Gauntlet (Fig. 2). Although EndpointBC alone in $NSLC_e$ succeeds in collecting diversity with respect to DirectionBC in this study, it would be naive to assume that collecting diversity with respect to one BC can *always* succeed at covering diversity in arbitrary unrelated characterization spaces. Therefore, while NSLC with an aligned BC emerges highly successful in this study, for researchers aiming for long-term collection of unaligned QD, the multi-BC formulations offer an attractive alternative that allows explicitly searching for *both* unaligned and aligned diversity while potentially losing no significant performance.

While this paper has focused on QD variants of NS, in particular centered on NSLC, an interesting future direction will be to validate the advantage of multiple BCs on other QD algorithms such as MAP-Elites. In MAP-Elites, multiple BCs are possible because each BC can in principle control a separate grid, and all the grids can be run at the same time. In this way, it is possible to apply lessons from experiments here to a broad range of future QD algorithms.

Overall, this study reveals that the currently-accepted practice within QD of exclusively driving search with the desired notion of diversity (BC) can break down on hard problems and that an effective solution is to drive search with multiple BCs simultaneously. The insights gained here therefore expand the reach of quality diversity to more difficult problems and to arbitrary notions of diversity, bringing evolutionary computation one step closer to emulating the inventive power of natural evolution.

Acknowledgments. This work was supported by the National Science Foundation under grant no. IIS-1421925. Any opinions, findings, and conclusions or recommendations expressed in this material are those of the authors and do not necessarily reflect the views of the National Science Foundation.

References

1. Cully, A., Clune, J., Tarapore, D., Mouret, J.-B.: Robots that can adapt like animals. Nature **521**(7553), 503–507 (2015)
2. De Jong, K.A.: Evolutionary Computation: A Unified Perspective. MIT Press, Cambridge (2002)

3. Deb, K., Pratap, A., Agarwal, S., Meyarivan, T.: A fast and elitist multiobjective genetic algorithm: NSGA-II. IEEE Trans. Evolut. Comput. **6**(2), 182–197 (2002)
4. Gomes, J., Urbano, P., Christensen, A.L.: Evolution of swarm robotics systems with novelty search. Swarm Intell. **7**(2–3), 115–144 (2013)
5. Gomes, J., Mariano, P., Christensen, A.L.: Devising effective novelty search algorithms.: a comprehensive empirical study. In: Proceedings of the 17th Annual Conference on Genetic and Evolutionary Computation (GECCO 2015), pp. 943–950. ACM (2015)
6. Green, C.: SharpNEAT homepage (2003–2006). http://sharpneat.sourceforge.net/
7. Kistemaker, S., Whiteson, S.: Critical factors in the performance of novelty search. In: Proceedings of the 13th Annual Conference on Genetic and Evolutionary Computation (GECCO 2011), pp. 965–972. ACM (2011)
8. Lehman, J., Stanley, K.O.: Abandoning objectives: evolution through the search for novelty alone. Evolut. Comput. **19**(2), 189–223 (2011)
9. Lehman, J., Stanley, K.O.: Evolving a diversity of virtual creatures through novelty search and local competition. In: Proceedings of the 13th Annual Conference on Genetic and Evolutionary Computation (GECCO 2011), pp. 211–218. ACM (2011)
10. Liapis, A., Yannakakis, G.N., Togelius, J.: Sentient sketchbook: Computer-aided game level authoring. In: FDG, pp. 213–220 (2013)
11. Morse, G., Risi, S., Snyder, C.R., Stanley, K.O.: Single-unit pattern generators for quadruped locomotion. In: Proceedings of the 15th Annual Conference on Genetic and Evolutionary Computation (GECCO 2013), pp. 719–726. ACM (2013)
12. Mouret, J.-B., Clune, J.: Illuminating search spaces by mapping elites (2015). arXiv preprint arXiv:1504.04909
13. Mouret, J.-B., Doncieux, S.: Encouraging behavioral diversity in evolutionary robotics: an empirical study. Evolut. Comput. **20**(1), 91–133 (2012)
14. Nguyen, A., Yosinski, J., Clune, J.: Deep neural networks are easily fooled: high confidence predictions for unrecognizable images. In: Proceedings of the 2015 IEEE Conference on Computer Vision and Pattern Recognition, IEEE (2015)
15. Nguyen, A., Yosinski, J., Clune, J.: Innovation engines: automated creativity and improved stochastic optimization via deep learning. In: Proceedings of the 17th Annual Conference on Genetic and Evolutionary Computation (GECCO 2015), New York, NY, USA, ACM (2015)
16. Pugh, J.K., Soros, L.B., Szerlip, P.A., Stanley, K.O.: Confronting the challenge of quality diversity. In: Proceedings of the 17th Annual Conference on Genetic and Evolutionary Computation, New York, NY, USA, ACM (2015)
17. Stanley, K.O., Miikkulainen, R.: Evolving neural networks through augmenting topologies. Evolut. Comput. **10**, 99–127 (2002)
18. Szerlip, P., Stanley, K.O.: Indirectly encoded sodarace for artificial life. Adv. Artif. Life **12**, 218–225 (2013). Proceedings of the European Conference on Artificial Life (ECAL 2013)
19. Szerlip, P.A., Morse, G., Pugh, J.K., Stanley, K.O.: Unsupervised feature learning through divergent discriminative feature accumulation. In: Proceedings of the 29th AAAI Conference on Artificial Intelligence (AAAI 2015), Menlo Park, CA, AAAI Press (2015)

Emergence of Diversity and Its Benefits for Crossover in Genetic Algorithms

Duc-Cuong Dang[1], Tobias Friedrich[2], Timo Kötzing[2], Martin S. Krejca[2], Per Kristian Lehre[1], Pietro S. Oliveto[3(✉)], Dirk Sudholt[3], and Andrew M. Sutton[2]

[1] University of Nottingham, Nottingham, UK
[2] Hasso Plattner Institute, Potsdam, Germany
[3] University of Sheffield, Sheffield, UK
p.oliveto@sheffield.ac.uk

Abstract. Population diversity is essential for avoiding premature convergence in Genetic Algorithms (GAs) and for the effective use of crossover. Yet the dynamics of how diversity emerges in populations are not well understood. We use rigorous runtime analysis to gain insight into population dynamics and GA performance for a standard (μ+1) GA and the Jump_k test function. By studying the stochastic process underlying the size of the largest collection of identical genotypes we show that the interplay of crossover followed by mutation may serve as a catalyst leading to a sudden burst of diversity. This leads to improvements of the expected optimisation time of order $\Omega(n/\log n)$ compared to mutation-only algorithms like the (1+1) EA.

Keywords: Genetic algorithms · Crossover · Diversity · Runtime analysis · Theory

1 Introduction

Genetic Algorithms (GAs) are powerful general-purpose optimisers that perform surprisingly well in many applications. Their wide-spread success is based on a number of factors: using populations to diversify search, using mutation to generate novel solutions, and using crossover to combine features of good solutions. Crossover can combine building blocks of good solutions, and help to focus search on bits where parents disagree. For both tasks the population needs to be diverse enough for crossover to be effective. A common problem in the application of GAs is the loss of diversity when the population converges to copies of the same search point, often called "premature convergence".

Understanding population diversity and crossover has proved elusive. The first example function where crossover was proven to be beneficial is called Jump_k. In this problem, GAs have to overcome a fitness valley such that all local optima have Hamming distance k to the global optimum. Jansen and Wegener [7] showed that, while mutation-only algorithms such as the (1+1) EA

J. Handl et al. (Eds.): PPSN XIV 2016, LNCS 9921, pp. 890–900, 2016.
DOI: 10.1007/978-3-319-45823-6_83

require expected time $\Theta(n^k)$, a simple (μ+1) GA with crossover only needs time $O(\mu n^2 k^3 + 4^k/p_c)$ where p_c is the crossover probability. This time is $O(4^k/p_c)$ for large k, and hence significantly faster than mutation-only GAs. However, their analysis requires an unrealistically small crossover probability $p_c \le 1/(ckn)$ for a large constant $c > 0$. Hence the analysis does not reflect the typical behaviour in GA populations with constant crossover probabilities $p_c = \Theta(1)$ as used in practice. Kötzing et al. [8] later refined these results towards a crossover probability $p_c \le k/n$, which is still unrealistically small. Both approaches focus on creating diversity through a sequence of lucky mutations, relying on crossover to create the optimum once sufficient diversity has been created. Their arguments break down if crossover is applied frequently.

Here we provide a novel approach to show that diversity can also be created by frequent applications of crossover followed by mutation. For the maximum crossover probability $p_c = 1$ we prove that on Jump_k diversity emerges naturally in a population: the interplay of crossover, followed by mutation, can serve as a catalyst for creating a diverse range of search points out of few different individuals. This allows to prove a speedup of order $n/\log n$ for $k \ge 3$ compared to mutation-only algorithms such as the (1+1) EA. Both operators are vital: mutation alone requires $\Theta(n^k)$ expected iterations to hit the optimum from a local optimum. As shown in [8, Theorem 8] using only crossover with $p_c = \Omega(1)$ but no mutation, diversity reduces quickly, leading to inefficient runtimes for small population sizes ($\mu = O(\log n)$).

After defining the algorithm, the (μ+1) GA from [7], and the Jump_k function in Sect. 2, we elaborate on the population dynamics in Sect. 3, preparing the ground for the following runtime result. For the standard mutation rate, we show in Sect. 4 that the (μ+1) GA with $p_c = 1$, $\mu = O(n)$, $k = O(1)$ optimises Jump_k in expected time $O\big(\mu n \log(\mu) + n^k/\mu + n^{k-1}\log(\mu)\big)$. Compared to the expected time $\Theta(n^k)$ for the (1+1) EA this corresponds to a speedup of order $n/\log n$ for $k \ge 3$ and $\sqrt{n/\log n}$ for $k = 2$, for the best possible choice of μ.

2 Preliminaries

The $\text{Jump}_k\colon \{0,1\}^n \to \mathbb{N}$ class of pseudo-Boolean fitness functions was originally introduced by Jansen and Wegener [7]. The function value increases with the number of 1-bits in the bitstring until a *plateau* of local optima is reached consisting of all points with $n-k$ 1-bits. However, its only global optimum is the all-ones string 1^n. Between the plateau and the global optimum there is a gap of Hamming distance k which has to be "jumped over" for the function to be optimised. The function is formally defined as

$$\text{Jump}_k(x) = \begin{cases} k + |x|_1 & \text{if } |x|_1 = n \text{ or } |x|_1 \le n-k, \\ n - |x|_1 & \text{otherwise,} \end{cases}$$

where $|x|_1 = \sum_{i=1}^{n} x_i$ is the number of 1-bits in x.

We will analyse the performance of a standard steady-state (μ+1) GA [7] with $p_c = 1$ using at each step uniform crossover (i.e., each bit of the offspring is

Algorithm 1. (μ+1) GA with $p_c = 1$

```
1 P ← μ individuals, uniformly at random from {0,1}^n;
2 while 1^n ∉ P do
3     Choose x, y ∈ P uniformly at random;
4     z ← mutate(crossover(x, y));
5     P ← P ∪ {z};
6     Remove one element from P with lowest fitness, breaking ties u.a.r.;
```

chosen uniformly at random from one of the parents) and standard bit mutation (i.e., each bit is flipped with probability $p_m = \chi/n = \Theta(1/n)$). Algorithm 1 shows the pseudo code for the (μ+1) GA, tailored to $p_c = 1$.

3 Population Dynamics

The following lemma gives a on bound the expected time for the whole population to reach the plateau. It is proved using level-based arguments, similar to those in [3]. The proof is omitted due to space restrictions.

Lemma 1. *The expected time until the entire population of (μ+1) GA with* $p_m = \Theta(1/n)$ *reaches the plateau of* Jump_k *for* $k = O(1)$ *(or the optimum has been found) is* $O(n\mu \log \mu + n \log n)$.

In the remainder of the analysis we study the algorithm's behaviour once all individuals are on the plateau. Previous observations of simulations have revealed the following behaviour. Assume the algorithm has reached a population where all individuals are identical. We refer to identical individuals as a *species*, hence in this case there is only one species. Eventually, a mutation will create a different search point on the plateau, leading to the creation of a new species. Both species may shrink or grow in size, and there is a chance that the new species disappears and we go back to one species only.

However the existence of two species also serves as a catalyst for creating further species in the following sense. Say two parents 0001111111 and 0010111111 are recombined, then crossover has a good chance of creating an individual with $n - k + 1$ 1s, e.g. 0011111111. Then mutation has a constant probability of flipping any of the $n - k - 1$ unrelated 1-bits to 0, leading to a new species, e.g. 0011111011. This may lead to a sudden burst of diversity in the population.

Due to the ability to create new species, the size of the largest cluster performs an almost fair random walk. Once its size has decreased significantly from its maximum μ, there is a good chance for recombining two parents from different species. This helps in finding the global optimum as crossover can increase the number of 1s in the offspring, compared to its parents, such that fewer bits need to be flipped by mutation to reach the optimum. This is formalised in the following lemma.

Lemma 2. *The probability that the global optimum is constructed by a uniform crossover of two parents on the plateau with Hamming distance $2d$, followed by mutation ($p_m = \chi/n$), is*

$$\sum_{i=0}^{2d} \binom{2d}{i} \frac{1}{2^{2d}} \left(\frac{\chi}{n}\right)^{k+d-i} \left(1 - \frac{\chi}{n}\right)^{n-k-d+i} \geq \frac{1}{2^{2d}} \left(\frac{\chi}{n}\right)^{k-d} \left(1 - \frac{\chi}{n}\right)^{n-k+d} \quad (1)$$

Proof. For a pair of search points on the plateau with Hamming distance $2d$, both parents have d ones among the $2d$ bits that differ between parents, and $n-k-d$ 1s outside this area. Assume that crossover sets i out of these $2d$ bits to 1, which happens with probability $\binom{2d}{i} \cdot 2^{-2d}$. Then mutation needs to flip the remaining $k+d-i$ 0s to 1. The probability that such a pair creates the optimum is hence

$$\sum_{i=0}^{2d} \binom{2d}{i} \frac{1}{2^{2d}} \left(\frac{\chi}{n}\right)^{k+d-i} \left(1 - \frac{\chi}{n}\right)^{n-k-d+i}.$$

The second bound is obtained by ignoring summands $i < 2d$ for the sum. □

Note that even a Hamming distance of 2, i.e. $d = 1$, leads to a probability of $\Omega(n^{-k+1})$, provided that such parents are selected for reproduction. The probability is by a factor of n larger than the probability $\Theta(n^{-k})$ of mutation without crossover reaching the optimum from the plateau. We will show that this effect leads to a speedup of nearly n for the (μ+1) GA, compared to the expected time of $\Theta(n^k)$ for the (1+1) EA [5] and other EAs only using mutation.

The idea behind the analysis is to investigate the random walk underlying the size of the largest species. We bound the expected time for this size to decrease to $\mu/2$, and then argue that the (μ+1) GA is likely to spend a good amount of time with a population of good diversity, where the probability of creating the optimum in every generation is $\Omega(n^{-k+1})$ due to the chance of recombining parents of Hamming distance at least 2.

In the following we refer to $Y(t)$ as the size of the largest species in the population at time t. Define

$$p_+(y) := \Pr\left(Y(t+1) - Y(t) = 1 \mid Y(t) = y\right),$$
$$p_-(y) := \Pr\left(Y(t+1) - Y(t) = -1 \mid Y(t) = y\right),$$

i.e., $p_+(y)$ is the probability that the size of the largest species increases from y to $y+1$, and $p_-(y)$ is the probability that it decreases from y to $y-1$.

The following lemma gives bounds on these transition probabilities, unless two parents of Hamming distance larger than 2 are selected for recombination (this case will be treated later in Lemma 4). We formulate the lemma for arbitrary mutation rates $\chi/n = \Theta(1/n)$ and restrict our attention to sizes $Y(t) \geq \mu/2$ as we are only interested in the expected time for the size to decrease to $\mu/2$.

Lemma 3. *For every population on the plateau of* Jump_k *for* $k = O(1)$ *the following holds. Either the* $(\mu+1)$ *GA with mutation rate* $\chi/n = \Theta(1/n)$ *performs a crossover of two parents whose Hamming distance is larger than 2, or the size* $Y(t)$ *of the largest species changes according to transition probabilities* $p_-(\mu) = \Omega(1/n)$ *and, for* $\mu/2 \leq y < \mu$,

$$p_+(y) \leq \frac{y(\mu-y)(\mu+y)}{2\mu^2(\mu+1)}\left(1-\frac{\chi}{n}\right)^n + \mathrm{O}\left(\frac{(\mu-y)^2}{\mu^2 n}\right),$$

$$p_-(y) \geq \frac{y(\mu-y)(\mu+\chi y)}{2\mu^2(\mu+1)}\left(1-\frac{\chi}{n}\right)^n - \mathrm{O}\left(\frac{\mu-y}{\mu n}\right).$$

Proof. We call an individual belonging to the current largest species a y-individual and all the others non-y individuals. In each generation, there is either no change, or one individual is added to the population and one individual chosen uniformly at random is removed from the population. In order to increase the number of y-individuals, it is necessary that a y-individual is added to the population, and a non-y individual is removed from the population. Analogously, in order to decrease the number of y-individuals, it is necessary that a non-y individual is added to the population, and a y-individual is removed from the population.

Given that $Y(t) = y$, let $p(y)$ be the probability that a y-individual is created at time $t+1$, and $q(y)$ the probability that a non-y individual is created.

We now estimate an upper bound on $p(y)$. We may assume that the Hamming distance between parents is at most 2 as otherwise there is nothing to prove. A y-individual can be created in the following three ways:

- Two y-individuals are selected. Crossing over two y-individuals produces another y-individual, which survives mutation if no bits are flipped, i.e., with probability $(1-\chi/n)^n$.
- One y-individual and one non-y individual are selected. The crossover operator produces a y-individual with probability $1/4$ and mutation does not flip any bits with probability $(1-\chi/n)^n$. If the crossover operator does not produce a y-individual, then to produce a y-individual at least one specific bit-position must be mutated, which occurs with probability $O(1/n)$. The overall probability is hence $(1/4)(1-\chi/n)^n + O(1/n)$.
- Two non-y individuals are selected. These two individuals are either identical or have Hamming distance 2 (i.e., by assumption). In the first case they both have one of the k 0-bit positions of a y-individual set to 1. In the second case they either both have one of the k 0-bit positions of a y-individual set to 1 or they both have one of the $n-k$ 1-bit positions set to 0. In both cases, crossover cannot change the value of such bit. Thus, at least one specific bit-position must be flipped, which occurs with probability $O(1/n)$.

Taking into account the probabilities of the three selection events above, the probability of producing a y-individual is

$$
\begin{aligned}
p(y) &= \left(\frac{y}{\mu}\right)^2 \left(1-\frac{\chi}{n}\right)^n + 2\left(\frac{y}{\mu}\right)\left(1-\frac{y}{\mu}\right)\left[\left(\frac{1}{4}\right)\left(1-\frac{\chi}{n}\right)^n + O\left(\frac{1}{n}\right)\right] \\
&\quad + \frac{(\mu-y)^2}{\mu^2} O\left(\frac{1}{n}\right) \\
&= \left(1-\frac{\chi}{n}\right)^n \left(\frac{y}{\mu}\right)\left(\frac{y}{\mu}+\frac{\mu-y}{2\mu}\right) + O\left(\frac{y(\mu-y)}{\mu^2}\cdot\frac{1}{n}\right) + O\left(\frac{(\mu-y)^2}{\mu^2}\cdot\frac{1}{n}\right) \\
&= \frac{y(\mu+y)}{2\mu^2}\left(1-\frac{\chi}{n}\right)^n + O\left(\frac{\mu(\mu-y)}{\mu^2}\cdot\frac{1}{n}\right)
\end{aligned}
$$

We then estimate a lower bound on $q(y)$. In the case where $y=\mu$, a non-y individual can be added to the population if:

- two y-individuals are selected, and the mutation operator flips one of the k 0-bits and one of the $n-k$ 1-bits. This event occurs with probability $q(\mu) = k(n-k)\left(\frac{\chi}{n}\right)^2\left(1-\frac{\chi}{n}\right)^{n-2} = \Omega(1/n)$, where we used that $k=O(1)$.

In the other case where $y<\mu$, then a non-y individual can be added to the population in the following two ways:

- A y-individual and a non-y individual are selected. Crossover produces a copy of the non-y individual with probability $1/4$, which is unchanged by mutation with probability $(1-\chi/n)^n$. Or with probability $1/4$, crossover produces an individual with $k-1$ 0-bits. Mutation then creates a non y-individual by flipping a single of the $n-k$ 1-bit positions. This event occurs with probability $(1/4)(n-k)\left(\frac{\chi}{n}\right)\left(1-\frac{\chi}{n}\right)^{n-1} \geq (\chi/4)\left(1-\frac{\chi}{n}\right)^n - O(1/n)$ using again that $k=O(1)$.
- Two non y-individuals are selected. In the worst case, the selected individuals are different, hence crossover produces an individual on the plateau with probability at least $1/2$, which mutation does not destroy with probability $(1-\chi/n)^n$.

Assuming that $\mu/2 \leq y < \mu$ and n is sufficiently large, the probability of adding a non-y individual is

$$
\begin{aligned}
q(y) &\geq 2\left(\frac{y}{\mu}\right)\left(1-\frac{y}{\mu}\right)\left[\left(\frac{\chi+1}{4}\right)\left(1-\frac{\chi}{n}\right)^n - O\left(\frac{1}{n}\right)\right] + \frac{1}{2}\left(1-\frac{y}{\mu}\right)^2\left(1-\frac{\chi}{n}\right)^n \\
&= \frac{(\mu-y)(\mu+\chi y)}{2\mu^2}\left(1-\frac{\chi}{n}\right)^n - O\left(\frac{\mu-y}{\mu}\cdot\frac{1}{n}\right).
\end{aligned}
$$

Multiplying $p(y)$ and $q(y)$ by the respective survival probabilities, we get

$$p_-(y) \geq \left[\frac{(\mu-y)(\mu+\chi y)}{2\mu^2}\left(1-\frac{\chi}{n}\right)^n - O\left(\frac{\mu-y}{\mu}\cdot\frac{1}{n}\right)\right]\left(\frac{y}{\mu+1}\right)$$
$$= \frac{(\mu-y)(\mu+\chi y)y}{2\mu^2(\mu+1)}\left(1-\frac{\chi}{n}\right)^n - O\left(\frac{(\mu-y)}{\mu}\cdot\frac{1}{n}\right).$$

$$p_+(y) = \left[\frac{y(\mu+y)}{2\mu^2}\left(1-\frac{\chi}{n}\right)^n + O\left(\frac{y(\mu-y)}{\mu^2}\cdot\frac{1}{n}\right)\right]\left(\frac{\mu-y}{\mu+1}\right)$$
$$= \frac{(\mu^2-y^2)y}{2\mu^2(\mu+1)}\left(1-\frac{\chi}{n}\right)^n + O\left(\frac{(\mu-y)^2}{\mu^2}\cdot\frac{1}{n}\right).$$

Both equalities hold for values of y between $\mu/2$ and μ. □

Steps where crossover recombines two parents with larger Hamming distance were excluded from Lemma 3 as they require different arguments. The following lemma shows that conditional transition probabilities in this case are favourable in that the size of the largest species is more likely to decrease than to increase.

Lemma 4. *Assume that $y \geq \mu/2$ and the (μ+1) GA on* Jump$_k$ *with $k = O(1)$ and mutation rate $\chi/n = \Theta(1/n)$ selects two individuals on the plateau with Hamming distance larger than 2, then for conditional transition probabilities $p_-^*(y)$ and $p_+^*(y)$ for decreasing or increasing the size of the largest species, $p_-^*(y) \geq 2p_+^*(y)$.*

Proof. Assume that the population contains two individuals x and z with Hamming distance $2\ell \leq 2k$, where $\ell \geq 2$. Without loss of generality, let us assume that they differ in the first 2ℓ bit positions.

In the case that the majority individual y has ℓ 0-bits in the first 2ℓ positions, then a y-individual may be produced by creating the ℓ 0-bits and ℓ 1-bits in the exact positions by crossover and no mutation should occur. Alternatively, at least one exact bit has to be flipped by mutation. Then, the probability of producing a y-individual from x and z, and replacing a non y-individual with y is less than

$$p_+^*(y) \leq \left[\left(\frac{1}{2}\right)^{2\ell}\left(1-\frac{\chi}{n}\right)^n + O\left(\frac{1}{n}\right)\right]\left(\frac{\mu-y}{\mu+1}\right)$$

On the other hand, the probability of producing an individual on the plateau different from y, and replacing a y-individual is at least (for sufficiently large n)

$$p_-^*(y) \geq \left(\binom{2\ell}{\ell}-1\right)\left(\frac{1}{2}\right)^{2\ell}\left(1-\frac{\chi}{n}\right)^n\left(\frac{y}{\mu+1}\right) > 2p_+^*(y).$$

In the other case, assume that the majority individual y does not have ℓ 0-bits in the first 2ℓ bit-positions. Then the mutation operator must flip at least one specific bit among the last $n-2\ell$ positions to produce y, which occurs with probability $O(1/n)$, while the probability to produce a non y-individual on the plateau is still $\Omega(1)$. □

4 Standard Mutation Rate

In this section we state the main result. Herein we consider $p_m = 1/n$.

Theorem 1. *The expected optimisation time of the (μ+1) GA with $p_c = 1$, $p_m = 1/n$ and $\mu \leq \kappa n$, for some constant $\kappa > 0$, on* Jump_k*, $k = O(1)$, is $O(\mu n \log(\mu) + n^k/\mu + n^{k-1}\log(\mu))$.*

For $k \geq 3$ the best speedup compared to the expected time of $\Theta(n^k)$ for the (1+1) EA [5] and other EAs only using mutation is of order $\Omega(n/\log n)$ for $\mu = \kappa n$. For $k = 2$ the best speedup is of order $\Omega(\sqrt{n/\log n})$ for $\mu = \Theta(\sqrt{n/\log n})$.

Note that for mutation rate $1/n$, the dominant terms in Lemma 3 are equal, hence the size of the largest species performs a fair random walk, up to a bias resulting from small-order terms. This confirms our intuition from observing simulations. The following lemma formalises this fact: in steps where the size $Y(t)$ of the largest species changes, it performs an almost fair random walk.

Lemma 5. *For the random walk induced by the size of the largest species, conditional on the current size y changing, for $\mu/2 < y < \mu$, the probability of increasing y is at most $1/2 + O(1/n)$ and the probability of decreasing it is at least $1/2 - O(1/n)$.*

We use these transition probabilities to bound the expected time for the random walk to hit $\mu/2$.

Lemma 6. *Consider the random walk of $Y(t)$, starting in state $X_0 \geq \mu/2$. Let T be the first hitting time of state $\mu/2$. If $\mu = O(n)$, then $\mathrm{E}(T \mid X_0) = O(\mu n + \mu^2 \log \mu)$ regardless of X_0.*

Proof. Let E_i abbreviate $\mathrm{E}(T \mid X_0 = i)$, then $E_{\mu/2} = 0$ and $E_\mu = O(n) + E_{\mu-1}$ as $p_-(\mu) = \Omega(1/n)$ by Lemma 3.

For $\mu/2 < y < \mu$ the probability of leaving state y is always (regardless of Hamming distances between species) bounded from below by the probability of selecting two y-individuals as parents, not flipping any bits during mutation, and choosing a non-y individual for replacement (cf. Lemma 3, Lemma 4):

$$p_+(y) + p_-(y) \geq \frac{y^2}{\mu^2} \cdot \left(1 - \frac{1}{n}\right)^n \cdot \frac{\mu - y}{\mu + 1} \geq \frac{\mu - y}{24\mu}$$

as $y \geq \mu/2$, $\mu + 1 \leq 3\mu/2$ (since $\mu \geq 2$), and $(1 - 1/n)^n \geq 1/4$ for $n \geq 2$. Using conditional transition probabilities $1/2 \pm \delta$ for $\delta = O(1/n)$ according to Lemma 5, E_i is bounded as $E_i \leq \frac{24\mu}{\mu - i} + \left(\frac{1}{2} - \delta\right) E_{i-1} + \left(\frac{1}{2} + \delta\right) E_{i+1}$.

This is equivalent to $\left(\frac{1}{2} - \delta\right) \cdot (E_i - E_{i-1}) \leq \frac{24\mu}{\mu - i} + \left(\frac{1}{2} + \delta\right) \cdot (E_{i+1} - E_i)$. Introducing $D_i := E_i - E_{i-1}$, this is equivalent to

$$D_i \leq \frac{\frac{24\mu}{\mu - i} + \left(\frac{1}{2} + \delta\right) \cdot D_{i+1}}{\frac{1}{2} - \delta} \leq \frac{50\mu}{\mu - i} + \alpha \cdot D_{i+1}$$

for $\alpha := \frac{1+2\delta}{1-2\delta} = 1+O(1/n)$, assuming n is large enough. From $E_\mu = O(n)+E_{\mu-1}$ we get $D_\mu = O(n)$, hence an induction yields $D_i \leq \sum_{j=i}^{\mu-1} \frac{50\mu}{\mu-j} \cdot \alpha^{j-i} + \alpha^{\mu-i} \cdot O(n)$.

Combining $\alpha = 1 + O(1/n)$ and $1 + x \leq e^x$ for all $x \in \mathbb{R}$, we have $\alpha^\mu \leq e^{O(\mu/n)} \leq e^{O(1)} = O(1)$. Bounding both α^{j-i} and $\alpha^{\mu-i}$ in this way, we get

$$D_i \leq O(n) + O(\mu) \cdot \sum_{j=i}^{\mu-1} \frac{1}{\mu - j} = O(n + \mu \log \mu)$$

as the sum is equal to $\sum_{j=1}^{\mu-i} 1/j = O(\log \mu)$.

Now, $D_{\mu/2+1} + D_{\mu/2+2} + \cdots + D_i = (E_{\mu/2+1} - E_{\mu/2}) + (E_{\mu/2+2} - E_{\mu/2+1}) + \cdots + (E_i - E_{i-1}) = E_i - E_{\mu/2} = E_i$. Hence we get $E_i = \sum_{k=(\mu/2)+1}^{i} D_k \leq O(\mu n + \mu^2 \log \mu)$. □

Now we show that, when the largest species has decreased its size to $\mu/2$, there is a good chance that the optimum will be found within the following $\Theta(\mu^2)$ generations.

Lemma 7. *Consider the (μ+1) GA with $p_c = 1$ on* Jump_k*. If the largest species has size at most $\mu/2$ and $\mu \leq \kappa n$ for a sufficiently small constant $\kappa > 0$, the probability that during the next $c\mu^2$ generations, for some constant $c > 0$, the global optimum is found is $\Omega\left(1/(1 + n^{k-1}/\mu^2)\right)$.*

Proof. We show that during the $c\mu^2$ generations the size of the largest species never rises above $(3/4)\mu$ with at least constant probability. Then we calculate the probability of jumping to the optimum during the phase, given this happens.

Let X_i, $1 \leq i \leq c\mu^2$ be random variables indicating the increase in number of individuals of the largest species at generation i. We pessimistically ignore self-loops, thus the size of the species either increases or decreases in each generation. Using the conditional probabilities from Lemma 5, we get that the expected increase in each step is $1 \cdot (1/2 + O(1/n)) - 1 \cdot (1/2 - O(1/n)) = O(1/n)$. Then the expected increase in size of the largest species at the end of the phase is

$$\mathrm{E}(X) = \sum_{i=1}^{c\mu^2} X_i = \sum_{i=1}^{c\mu^2} O(1/n) = (c'\mu^2)/n \leq c'\kappa\mu \leq (1/8)\mu,$$

where we use that $\mu \leq \kappa n$ and κ is chosen small enough.

By an application of Hoeffding bounds $\Pr(X \geq \mathrm{E}(X) + \lambda) \leq \exp(-2\lambda^2/\sum_i c_i^2)$ with $\lambda = \mu/8$ and $c_i = 2$, we get that $\Pr(X \geq (2/8)\mu) \leq \exp(-c') = 1 - \Omega(1)$. We remark that the bounds also hold for any partial sum of the sequence X_i ([1], Chap. 1, Theorem 1.13), i.e. with probability $\Omega(1)$ the size *never* exceeds $(3/4)\mu$ in the considered phase of length $c\mu^2$ generations.

While the size does not exceed $(3/4)\mu$, in every step, there is a probability of at least $1/4 \cdot 3/4 = \Omega(1)$ of selecting parents from two different species, and by Lemma 2 the probability of creating the optimum is $\Omega(n^{-k+1})$.

Finally, the probability that at least one successful generation occurs in a phase of $c\mu^2$ is, using $(1-(1-p))^\lambda \geq (\lambda p/(1+\lambda p))$ for $\lambda \in \mathbb{N}, p \in [0,1]$ [2, Lemma 10], the probability that the optimum is found in one of these steps is

$$1-\left(1-\frac{1}{\Omega(n^{-k+1})}\right)^{c\mu^2} \geq \Omega\left(\frac{\mu^2 \cdot n^{-k+1}}{1+\mu^2 \cdot n^{-k+1}}\right).$$

Finally, we assemble all lemmas to prove our main result.

Proof (of Theorem 1). The expected time for the whole population to reach the plateau is $O(\mu n \log(\mu) + n \log n)$ by Lemma 1. Once the population is on the plateau, we wait till the largest species has decreased its size to at most $\mu/2$. According to Lemma 6, the time for the largest species to reach size $\mu/2$ is $O(\mu n + \mu^2 \log \mu)$. By Lemma 7, the probability that in the next $c\mu^2$ steps the optimum is found is $\Omega\left(1/(1+n^{k-1}/\mu^2)\right)$. If not, we repeat the argument. The expected number of such trials is $O(1+n^{k-1}/\mu^2)$ and the expected length of one trial is $O(\mu n + \mu^2 \log \mu) + c\mu^2 = O(\mu n + \mu^2 \log \mu)$. The expected time for reaching the optimum from the plateau is hence at most $O(\mu n + \mu^2 \log(\mu) + n^k/\mu + n^{k-1} \log(\mu))$.

Adding up all times and subsuming terms $O(\mu^2 \log(\mu)) = O(\mu n \log \mu)$ and $O(n \log n) = O(n^k/\mu + n^{k-1} \log \mu)$ completes the proof. □

5 Conclusion

A rigorous analysis of the $(\mu+1)$ GA has been presented showing how the use of both crossover and mutation considerably speeds up the runtime for Jump_k compared to algorithms using mutation only. Traditionally it has been believed that crossover may be useful only if sufficient diversity is readily available and that the emergence of diversity in the population is due to either mutation alone or should be enforced by the introduction of diversity mechanisms [4,6,9]. Indeed, previous work highlighting that crossover may be beneficial for Jump_k used unrealistically low crossover probabilities to allow mutation alone to create sufficient diversity. Conversely, our analysis shows that the interplay between crossover and mutation on the plateau of local optima of the Jump_k function quickly leads to a burst of diversity that is then exploited by both operators to reach the global optimum.

Acknowledgements. The research leading to these results has received funding from the European Union Seventh Framework Programme (FP7/2007–2013) under grant agreement no. 618091 (SAGE) and from the EPSRC under grant no. EP/M004252/1 and is based upon work from COST Action CA15140 ‘Improving Applicability of Nature-Inspired Optimisation by Joining Theory and Practice (ImAppNIO)’.

References

1. Auger, A., Doerr, B.: Theory of Randomized Search Heuristics. World Scientific, Singapore (2011)
2. Badkobeh, G., Lehre, P.K., Sudholt, D.: Black-box complexity of parallel search with distributed populations. In: Proceedings of FOGA 2015, pp. 3–15 (2015)
3. Corus, D., Dang, D.-C., Eremeev, A.V., Lehre, P.K.: Level-based analysis of genetic algorithms and other search processes. In: Bartz-Beielstein, T., Branke, J., Filipič, B., Smith, J. (eds.) PPSN 2014. LNCS, vol. 8672, pp. 912–921. Springer, Heidelberg (2014)
4. Dang, D.-D., Friedrich, T., Kötzing, T., Krejca, M.S., Lehre, P.K., Oliveto, P.S., Sudholt, D., Sutton, A.M.: Escaping local optima with diversity mechanisms and crossover. In: GECCO 2016 (2016, to appear)
5. Droste, S., Jansen, T., Wegener, I.: On the analysis of the (1+1) evolutionary algorithm. Theoret. Comput. Sci. **276**, 51–81 (2002)
6. Friedrich, T., Oliveto, P.S., Sudholt, D., Witt, C.: Analysis of diversity-preserving mechanisms for global exploration. Evolut. Comput. **17**(4), 455–476 (2009). doi:10.1162/evco.2009.17.4.17401. ISSN 1063-6560
7. Jansen, T., Wegener, I.: The analysis of evolutionary algorithms - a proof that crossover really can help. Algorithmica **34**(1), 47–66 (2002)
8. Kötzing, T., Sudholt, D., Theile, M.: How crossover helps in pseudo-boolean optimization. In: Proceedings of GECCO 2011, pp. 989–996 (2011)
9. Oliveto, P.S., Zarges, C.: Analysis of diversity mechanisms for optimisation in dynamic environments with low frequencies of change. Theoret. Comput. Sci. **561**, 37–56 (2015)

Coarse-Grained Barrier Trees of Fitness Landscapes

Sebastian Herrmann[1(✉)], Gabriela Ochoa[2], and Franz Rothlauf[1]

[1] Department of Information Systems and Business Administration, Johannes Gutenberg-Universität, Jakob-Welder-Weg 9, 55128 Mainz, Germany
{s.herrmann,rothlauf}@uni-mainz.de

[2] Department of Computing Science and Mathematics, University of Stirling, Stirling FK9 4LA, Scotland
goc@cs.stir.ac.uk

Abstract. Recent literature suggests that local optima in fitness landscapes are clustered, which offers an explanation of why perturbation-based metaheuristics often fail to find the global optimum: they become trapped in a sub-optimal cluster. We introduce a method to extract and visualize the global organization of these clusters in form of a barrier tree. Barrier trees have been used to visualize the barriers between local optima basins in fitness landscapes. Our method computes a more coarsely grained tree to reveal the barriers between clusters of local optima. The core element is a new variant of the flooding algorithm, applicable to local optima networks, a compressed representation of fitness landscapes. To identify the clusters, we apply a community detection algorithm. A sample of 200 NK fitness landscapes suggests that the depth of their coarse-grained barrier tree is related to their search difficulty.

Keywords: Fitness landscape analysis · Barrier tree · Disconnectivity graph · Local optima networks · Big valley · Search difficulty · NK-landscapes

1 Introduction

To overcome the problem of getting stuck in a local optimum, many metaheuristics based on local search apply a perturbation operator. The perturbation is supposed to "kick" an algorithm away from the current region of the search space. This principle is known as iterated local search (ILS) [1], e.g. as implemented in the Lin and Kernighan Heuristic [2,3]. The "big valley" hypothesis [4] states that the local optima in many fitness landscapes are not randomly distributed, but clustered and surrounding the global optimum. Consequently, one might assume that once a local optimum has been reached, ILS-based algorithms should easily find the global optimum after a limited number of perturbations. However, we know that this is by no means the case in practice. An approach

G. Ochoa—Acknowledges funding from the Leverhulme Trust, UK [award number RPG-2015-395].

J. Handl et al. (Eds.): PPSN XIV 2016, LNCS 9921, pp. 901–910, 2016.
DOI: 10.1007/978-3-319-45823-6_84

to explain this observation is given in the most recent literature [4–7]: instead of one big valley, fitness landscapes consist of multiple clusters (or funnels). The existence of such a structure offers a new explanation for the search difficulty of landscapes: since the connections between clusters are sparse, perturbation steps fail to escape from sub-optimal clusters to the cluster of the global optimum.

The objective of this paper is to complement the recent literature on the multi-cluster structure of landscapes with a new approach to study this structure, and to draw conclusions on search difficulty. A method that has been used to characterize the structure of fitness landscapes are barrier trees [8]. A barrier tree shows in a hierarchical structure how the local optima basins are connected in the landscape. The leaf nodes are the local optima and the branching nodes are the saddle points connecting the basins [9]. Due to the ability of ILS to easily move from local optimum to local optimum, we are primarily not interested in the barriers between their basins. The core issue for ILS is that local optima are clustered. Thus, we need to study which barriers exist between these clusters. The method we introduce here addresses this purpose. It allows us to compute a coarse-grained barrier tree and to characterize the landscape on the level of clusters. To reveal the clustering structure of landscapes, local optima networks (LONs) [10] have been used. A LON is a compressed representation of a fitness landscape. In a LON, each node is a local optimum, and the edges represent the transitions of an algorithm between the basins around the local optima. A problem with LONs is that it can be difficult to visualize their structure when they consist of a large number of nodes and edges. To identify clusters in fitness landscapes, statistical measures have been applied to LONs, e.g. counting the network graph's connected components [5] or community detection [7].

Our contribution is a modified version of the "flooding algorithm", which accepts as an input (i) a LON of a fitness landscape and (ii) a pre-computed clustering structure of the LON. The output is a coarse-grained picture of the landscape which retains the global structure and allows the eventual visualization of larger landscapes. We demonstrate our method with instances of the Kauffman NK model. For each instance, we computed the LON and the clusters. We obtained the clusters by community detection with the Markov cluster algorithm [11], as proposed in an earlier study [7]. We analyze the resulting barrier trees by visual inspection and a statistical approach. We provide an indication how the structure of the barrier tree is related to the search difficulty of a landscape.

The article is structured as follows: Sect. 2 introduces the concept of fitness landscapes for the study of problems and heuristic search. In Sect. 3, we explain how to construct a standard barrier tree for fitness landscape analysis. In order to construct a coarse-grained barrier tree (based on the local optima clusters), we need a method to identify the clustering structure. In Sect. 4, we introduce local optima networks as a compressed representation of fitness landscapes, and the Markov cluster algorithm to reveal the clustering structure of a fitness landscape. In Sect. 5, we present the algorithm to calculate the coarse-grained barrier tree of a fitness landscape. We visualize instances and examine the search difficulty. A brief summary and conclusions are in Sect. 6.

2 Fitness Landscapes

The concept Fitness Landscapes was introduced to study the reproductive success of genotypes in theoretical biology [12]. Fitness landscapes have been adopted in combinatorial optimization to study the structure of problems and the dynamics of heuristic search. A fitness landscape is defined as a triplet of the search space S, the fitness function f, and the neighborhood structure $N(S)$. The search space S contains all valid solutions. The fitness function $f : S \rightarrow \mathbb{R}_{\geq 0}$ assigns a fitness value to each $s \in S$ (we assume non-negative values and a maximization problem). The neighborhood function $N : S \rightarrow \mathcal{P}(S)$ assigns a set of neighbors $N(s)$ to every $s \in S$. Two solutions are neighbors if they are mutually reachable by one step of local search.

A local optimum is a solution that has a higher fitness than its neighbors [13]. A higher number of local optima (modality) leads to a landscape that is more "rugged", which increases the search difficulty for local search-based algorithms [14]. A local optimum is surrounded by a *basin of attraction*. The basin around an optimum is the set of solutions from which the optimum attracts a local search algorithm. We define a function for the basin around a local optimum lo as $B : lo \rightarrow \mathcal{P}(S \backslash LO)$. B assigns an element from the set of all subsets (power set $\mathcal{P}$) of solutions over the search space to each local optimum $lo \in LO$ (the set of all local optima).

The Kauffman NK model of landscapes [15] is frequently used for the study of fitness landscapes. The NK model is a combinatorial optimization problem from the class of pseudo-Boolean functions. An instance is defined by the two parameters N and K, where N is the number of binary variables. The size of the search space S is $|S| = 2^N$. K is the number of variables interacting with each other (epistasis). To instantiate the model, the co-variables are randomly selected. A higher value of K leads to a higher search difficulty [14]. The distance between two solutions $x, y \in S$ is the number of differing bits (Hamming-distance).

3 Barrier Trees of Fitness Landscapes

Barrier trees were introduced in computational chemistry to study the structure of potential energy landscapes [16,17], i.e. to examine the barriers that exist between the optima basins. Barrier trees are sometimes referred to as *disconnectivity graphs* [18,19]. Even though Barrier trees have been used to study heuristic search [8,9], the literature on this topic is rather sparse. To construct the barrier tree of a fitness landscape, a database of the local optima (we assume local maxima in this paper), and the transition states connecting at least two basins around different local optima, is required. The transition states are also called saddle points. In a 2-dimensional landscape, a saddle point is a local minimum. In a higher dimensional landscape, multiple of local minima, connecting two basins, may exist. In such a case, the saddle point is the local minimum with maximal fitness. Since the fitness of the saddle point is lower than the fitness of the two connected local optima, it can be interpreted as a barrier between them: to move from one of the local optima to the other, an algorithm has to accept a fitness deterioration down to the level of the local minimum. To visualize the barrier tree, local optima are identified with leaves, while the branching nodes represent saddle points separating groups of local optima.

A method to compute the barrier tree of a fitness landscape is the so-called "flooding algorithm" [9]. We think that a comprehensive understanding of this method is essential; hence we depict the mechanism in Fig. 1. For a maximization problem, the algorithm iterates over all solutions in the search space in a descending order (in terms of fitness): the landscape is "flooded". When a local optimum is found, a node is added to the barrier tree (steps 1 and 2). When a saddle point is found, a branching node is added to the tree, and edges are added to connect the saddle point to the adjacent local optima. From here, the saddle point now represents the basins of all adjacent local optima (step 3, the basins are merged by the flooding). This procedure is repeated until the last local optimum or saddle point has been found (step 4).

Since we are interested in the barriers that exist between the clusters of local optima in a landscape, we present a variant of the flooding algorithm suitable for this purpose in Sect. 5. Before, we need to explain how to characterize funnels in fitness landscapes. For this purpose, we introduce a special representation of fitness landscapes known as local optima networks (LONs) and a method using this representation to characterize funnels in the next Sect. 4.

4 Clusters of Local Optima in Fitness Landscapes

Local Optima Networks (LONs) are a novel approach to study the structure of fitness landscapes [10] and have recently been used to reveal the structure of multiple clusters [5–7,20]. LONs were originally inspired by the study of energy landscapes [21]. A LON is a complex network in which the nodes represent the local optima in a landscape (and their basins, resp.). The edges reflect an algorithm's transition between the basins. The concept of LONs allows the study of fitness landscapes from a network perspective and has the potential to deepen our understanding of metaheuristics and problems.

A network is a graph $G = (V, E)$ with the set of vertices V and the set of edges E. In a LON, the vertex set V contains the local optima of the fitness landscape. There exists an edge between two local optima if their basins are in some way connected, leading to a potential transition between the two local optima. An escape edge [22] is defined by the distance function of the fitness landscape d (minimal number of moves between two solutions): there exists a directed edge e_{xy} from local optimum lo_x to lo_y if there is a solution s such that $d(s, lo_x) \leq D \wedge s \in B(lo_y)$. The weight w_{xy} of edge e_{xy} is the probability that a search algorithm can escape from the local optimum lo_x into the basin around lo_y. The constant $D > 0$ determines the maximum distance that an algorithm uses during a perturbation step.

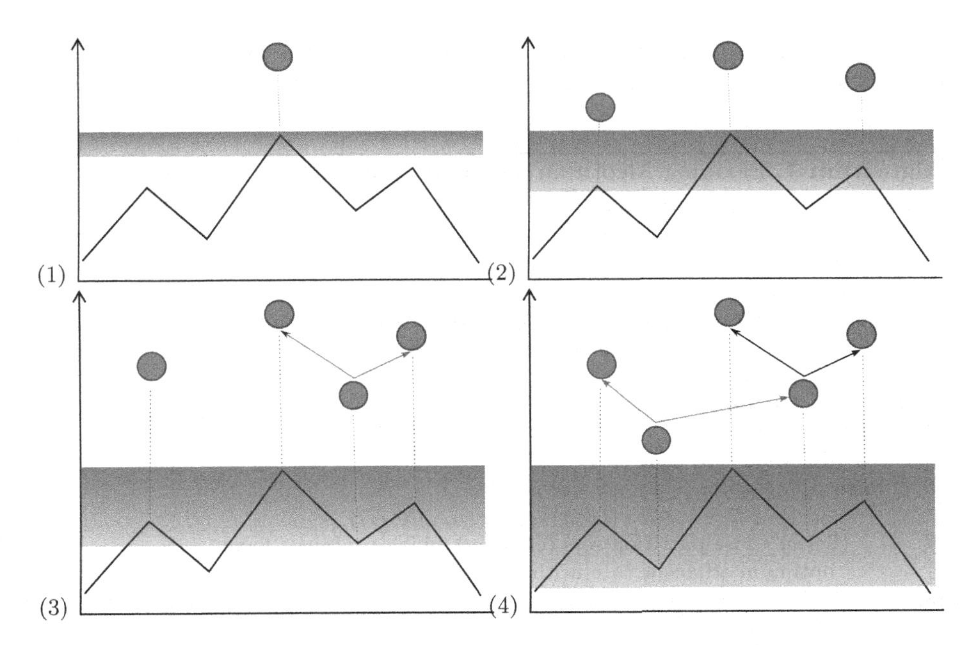

Fig. 1. Four steps of the flooding algorithm, creating the barrier tree of a fitness landscape. The vertical axis is the fitness, the horizontal axis is the landscape. Since we use a maximization problem, the space is "flooded" from the top to the bottom.

To reveal the clustering structure of fitness landscapes, we proposed to apply "community detection" to local optima networks [7]. Community detection is an exploratory variant of graph partitioning [23]. The objective of this method is to partition the network graph in a discipline-related, meaningful way. A very general definition of a community is a group of nodes that have more links among each other than to nodes in other communities. However, the definition of a community depends on the discipline applied and there exists a variety of algorithms that have been validated for different purposes [24,25].

Community detection in LONs has been done in earlier studies [26,27]. However, we [7] found that in particular, the Markov Cluster Algorithm (MCL, [11]) is an appropriate method of community detection to detect clusters in LONs and characterize the clustering structure of fitness landscapes. An explanation for this is that MCL is based on stochastic flows. LONs model the stochastic process of an algorithm in a fitness landscape. For this reason, the application of MCL matches the network model and produces meaningful results.

5 Coarse-Grained Barrier Trees of Fitness Landscapes

In order to escape from a cluster of local optima to another cluster, ILS needs to pass a barrier by a deterioration of the fitness. To visualize the structure of the barriers between the clusters in the landscape, we present a variant of the flooding algorithm [9] as introduced in Sect. 3 and Fig. 1. The pseudo code can be obtained from Algorithm 1. As an input, the algorithm accepts a LON and a

Algorithm 1. Flooding Algorithm for LONs (Maximization Problem)

Require: Local Optima Network $G = (V, E)$, Partition P over V (the cluster sets)
1: Let R be an empty set
2: **for all** $p \in P$ **do**
3: Add the local optimum of p with max. fitness to R
4: **end for**{R contains one representing local optimum per cluster in P}
5: Let $T = (V_{Tree}, E_{Tree})$ be the empty Barrier Tree
6: Order V by f in descending order
7: **for all** $v \in V$ **do**
8: **if** $v \in R$ **then**
9: Add Node v to Barrier Tree V_{Tree}
10: **else**
11: $C = \{p \in P \,|\, \exists n \in p \,|\, ((v, n) \in E \vee (n, v) \in E)\}$
12: {Select those partition sets (clusters) which contain a local optimum adjacent to v in the LON graph}
13: **if** $|B| > 1$ **then** {v connects at least two clusters, i.e. v is a saddle point}
14: Add Node v to Barrier Tree V_{Tree}
15: **for all** $c \in C$ **do** {For each cluster set c connected to saddle point v}
16: $r = c \wedge R$ {Choose node r representing connected cluster set c}
17: Add Edge (v, r) to E_{Tree}
18: Update P: Merge Partition set containing v and c
19: Remove r from R {Flood the connected cluster}
20: **end for**
21: **end if**
22: **end if**
23: **end for**
24: **return** T

partition of the LON's vertex set, i.e. a set with the clustering structure of the landscape. To obtain the clusters, we propose to apply the Markov cluster algorithm to the LON. As a first step, the algorithm selects the best local optimum for each cluster (set R). Then, the set of local optima nodes V is ordered by fitness in descending order. The algorithm iterates over each node. If the node is a representing node (in R), it is added to the barrier tree. Else, the algorithm determines the number of clusters adjacent to the current node in the LON. If the number is higher than one, the node is a saddle point and is also added to the tree. Then, the algorithm connects the saddle point to the nodes representing the adjacent clusters in the tree. From here, the saddle point represents all adjacent clusters ("flooding"): the clusters of the current and all the adjacent nodes are merged in the partition set, and the representers of the adjacent clusters are removed from R. This process is repeated until the whole LON is flooded (merged into one partition).

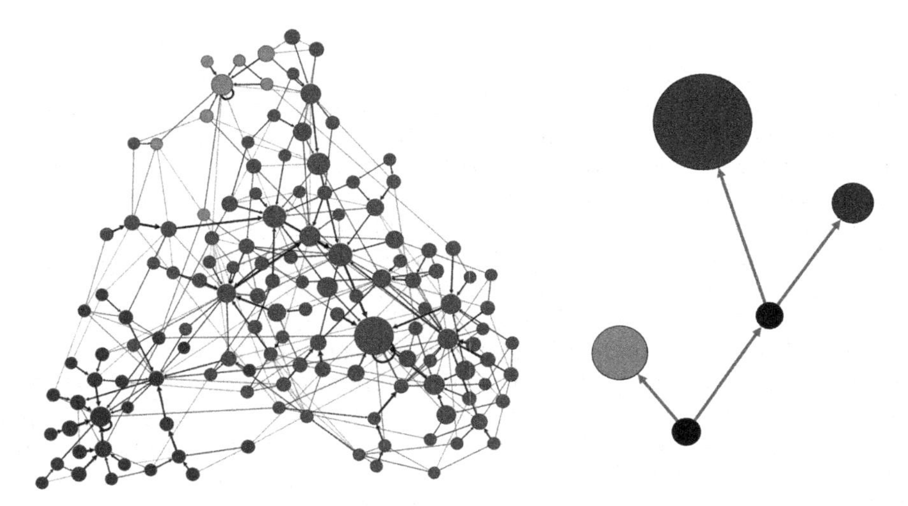

Fig. 2. Local optima network (left) and the coarse-grained barrier tree (right) of an NK landscape ($N = 20$, $K = 5$) with low search difficulty (success rate of ILS: 0.76). The color of the nodes represents the cluster (global optimum cluster is red in both graph types). In the tree, the branching nodes are black. In the local optima network, the size of the node represents the fitness, whereas the node size in the tree is the size of the cluster by the number of local optima. In the tree, the fitness is visualized by the node height (higher distance to the root means higher fitness). The layout of the local optima network is based on the ForceAtlas2 algorithm [28]. The local optima network shows only the best 20 % of nodes (all clusters still visible). (Color figure online)

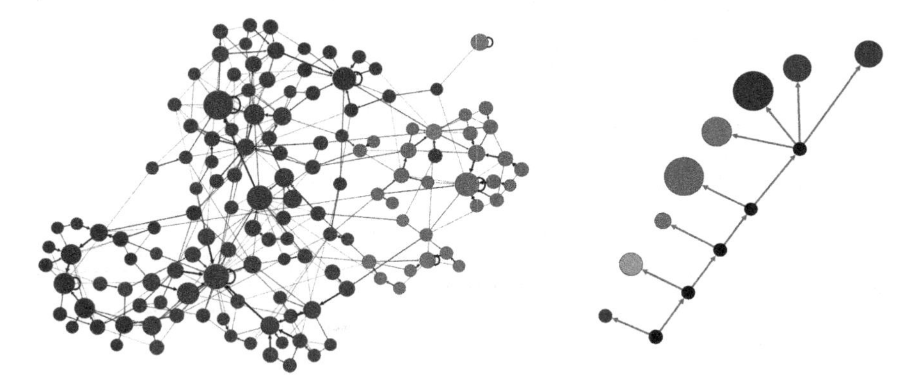

Fig. 3. LON and Coarse-Grained Barrier Tree of an NK landscape ($N = 20$, $K = 5$) with high search difficulty (success rate of ILS: 0.22). Please cf. Fig. 2 for further explanations.

To demonstrate our method, we selected an easy and a hard instance of the Kauffman NK model ($N = 20$, $K = 5$). To determine their difficulty, we performed 1000 independent runs of ILS per instance and measured the success rates (0.76 and 0.22). The ILS stopped after a limited number of fitness function evaluations ($^1/_{5th}$ of the search space), or when the global optimum was found. We extracted the LONs and computed the clusters in both LONs with MCL. We used the LONs and the clusters to construct the coarse-grained barrier trees with our variant of the flooding algorithm. Figures 2 and 3 plot the LON and the corresponding tree. Visual inspection of the LONs (left) confirms that the clustering as obtained by MCL is meaningful: nodes of the same color have a higher proximity to their own cluster than to those of a different cluster. Comparing both barrier trees (right), we observe a much deeper tree and thus a higher number of barriers in the case of the hard instance.

Even though a deeper study on the search difficulty is out of the scope of this paper, we conducted a first systematic approach towards this observation. We generated 200 instances of NK landscapes ($N = 20$, $K = 5$). We grouped the landscapes by the depth of the coarse-grained barrier tree and compared their difficulties for ILS. The results can be obtained from Fig. 4. For landscapes with a very short tree, we observe that the difficulty has a high variety, even though the median indicates a low difficulty (≈ 0.6). The median success rates get lower with a deeper tree, which means that their difficulty increases. This is not surprising: a deeper tree means that a traversal to the global optimum has—by average—a longer path. A search algorithm needs to pass more barriers then, and the difficulty is higher. This finding is consistent with the previous literature on regular barrier trees [9], however the observation that many landscapes with a low number of barriers can be difficult is counter-intuitive. We suggest that in these cases, additional factors, like the cluster size of the global optimum [7] need to be considered. We plan to conduct more research towards this direction.

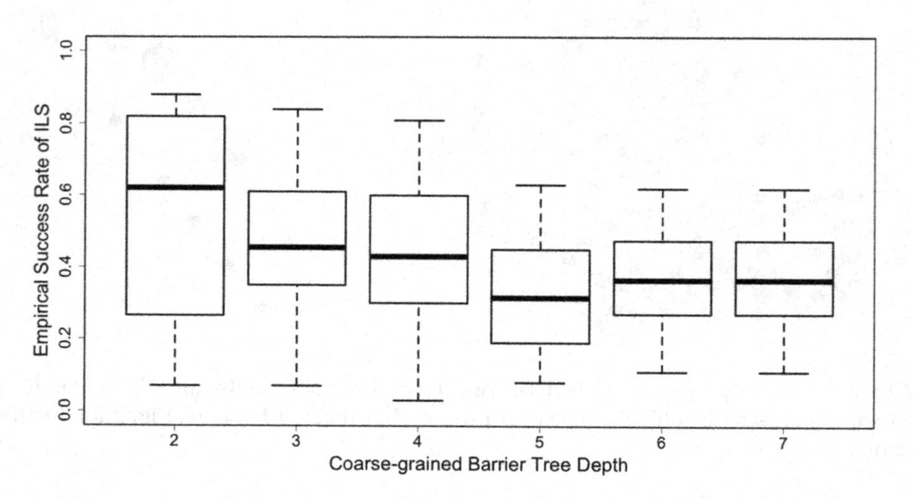

Fig. 4. Success Rate of ILS (difficulty) for different values of tree depth. The median success rate declines (search difficulty increases) with a higher depth of the tree.

6 Summary and Conclusion

As our main contribution, we presented a new method to visualize fitness landscapes and characterize them by the barriers between clusters of local optima. The existence of a multiple-cluster structure has recently emerged [6,7] as a refinement of the big valley hypothesis. We applied our method to a limited set of instances of the Kauffman NK model. Our results suggest that the tree depth might be related to the search difficulty of the landscapes for iterated local search. This is consistent with previous findings on difficulty in the literature [9]. A possible explanation is that the existence of barriers prevents iterated local search from escaping local optima clusters. This finding is rather preliminary and needs further investigation. Other structural properties of the landscapes must be taken into consideration, too. For further research, it would be interesting to see how the coarse-grained trees look for NK landscapes with higher levels of epistasis. It is also unclear whether or not there are differences between the NK model with random and adjacent co-variables. The adjacent NK model is often considered to be solvable with less effort. It would be worthwhile to examine if the tree depths are different between both models. We think that the method introduced here points to a new direction in studies of fitness landscapes.

References

1. Lourenço, H.R., Martin, O.C., Stützle, T.: Iterated local search. In: Handbook of Metaheuristics, pp. 320–353. Kluwer Academic Publishers, Boston (2003)
2. Applegate, D., Cook, W., Rohe, A.: Chained Lin-Kernighan for large traveling salesman problems. INFORMS J. Comput. **15**, 82–92 (2003)
3. Lin, S., Kernighan, B.W.: An effective heuristic algorithm for the traveling-salesman problem. Oper. Res. **21**, 498–516 (1973)
4. Hains, D.R., Whitley, D.L., Howe, A.E.: Revisiting the big valley search space structure in the TSP. J. Oper. Res. Soc. **62**, 305–312 (2011)
5. Ochoa, G., Veerapen, N., Whitley, D., Burke, E.K.: The multi-funnel structure of TSP fitness landscapes: a visual exploration. In: Bonnevay, S., Legrand, P., Monmarché, N., Lutton, E., Schoenauer, M. (eds.) EA 2015. LNCS, vol. 9554, pp. 1–13. Springer, Heidelberg (2016). doi:10.1007/978-3-319-31471-6_1
6. Ochoa, G., Veerapen, N.: Deconstructing the big valley search space hypothesis. In: Chicano, F., Hu, B., García-Sánchez, P. (eds.) EvoCOP 2016. LNCS, vol. 9595, pp. 58–73. Springer, Heidelberg (2016). doi:10.1007/978-3-319-30698-8_5
7. Herrmann, S., Ochoa, G., Rothlauf, F.: Communities of local optima as funnels in fitness landscapes. In: Proceedings of 2016 Genetic and Evolutionary Computation Conference - GECCO 2016 (2016)
8. Hallam, J., Prügel-Bennett, A.: Large barrier trees for studying search. IEEE Trans. Evol. Comput. **9**, 385–397 (2005)
9. van Stein, B., Emmerich, M., Yang, Z.: Fitness landscape analysis of nk landscapes and vehicle routing problems by expanded barrier trees. In: Emmerich, M., et al. (eds.) Evolutionary Computation IV. AISC, vol. 227, pp. 75–89. Springer, Heidelberg (2013)

10. Ochoa, G., Tomassini, M., Vérel, S., Darabos, C.: A study of NK landscapes' basins and local optima networks. In: Proceedings of 10th Annual Conference on Genetic and Evolutionary Computation - GECCO 2008, p. 555. ACM Press, New York (2008)
11. van Dongen, S.: Graph clustering by flow simulation. Ph.D. thesis, Utrecht University (2001)
12. Wright, S.: The roles of mutation, inbreeding, crossbreeding, and selection in evolution. In: Proceedings of 6th International Congress of Genetics, pp. 356–366 (1932)
13. Glover, F.: Future paths for integer programming and links to artificial intelligence. Comput. Oper. Res. **13**, 533–549 (1986)
14. Weinberger, E.: Correlated and uncorrelated fitness landscapes and how to tell the difference. Biol. Cybern. **336**, 325–336 (1990)
15. Kauffman, S.A., Weinberger, E.D.: The NK model of rugged fitness landscapes and its application to maturation of the immune response. J. Theor. Biol. **141**, 211–245 (1989)
16. Becker, O.M., Karplus, M.: The topology of multidimensional potential energy surfaces: theory and application to peptide structure and kinetics. J. Chem. Phys. **106**, 1495 (1997)
17. Flamm, C., Hofacker, I.L., Stadler, P.F., Wolfinger, M.T.: Barrier trees of degenerate landscapes. Zeitschrift füer Physikalische Chemie **216**, 155 (2002)
18. Doye, J.P.K., Miller, M.A., Wales, D.J.: Evolution of the potential energy surface with size for Lennard-Jones clusters. J. Chem. Phys. **111**, 8417 (1999)
19. Doye, J.P.K., Miller, M.A., Wales, D.J.: The double-funnel energy landscape of the 38-atom Lennard-Jones cluster. J. Chem. Phys. **110**, 6896 (1999)
20. Ochoa, G., Veerapen, N.: Additional dimensions to the study of funnels in combinatorial landscapes. In: Proceedings of 2016 Genetic and Evolutionary Computation Conference - GECCO 2016 (2016)
21. Stillinger, F.H.: A topographic view of supercooled liquids and glass formation. Science **267**, 1935–1939 (1995)
22. Vérel, S., Daolio, F., Ochoa, G., Tomassini, M.: Local optima networks with escape edges. In: Hao, J.-K., Legrand, P., Collet, P., Monmarché, N., Lutton, E., Schoenauer, M. (eds.) EA 2011. LNCS, vol. 7401, pp. 49–60. Springer, Heidelberg (2012)
23. Talbi, E., Bessière, P.: A parallel genetic algorithm for the graph partitioning problem. In: Proceedings of 5th International Conference on Supercomputing - ICS 1991, pp. 312–320. ACM Press, New York (1991)
24. Fortunato, S.: Community detection in graphs. Phys. Rep. **486**, 75–174 (2010)
25. Porter, M.A., Onnela, J.P., Mucha, P.J.: Communities in networks. Not. AMS **486**, 1082–1097 (2009)
26. Daolio, F., Tomassini, M., Vérel, S., Ochoa, G.: Communities of minima in local optima networks of combinatorial spaces. Phys. A: Stat. Mech. Appl. **390**, 1684–1694 (2011)
27. Iclanzan, D., Daolio, F., Tomassini, M.: Data-driven local optima network characterization of QAPLIB instances. In: Proceedings of 2014 Conference on Genetic and Evolutionary Computation - GECCO 2014, pp. 453–460. ACM Press, New York (2014)
28. Jacomy, M., Venturini, T., Heymann, S., Bastian, M.: ForceAtlas2, a continuous graph layout algorithm for handy network visualization designed for the Gephi software. PLoS ONE **9**, e98679 (2014)

Rapid Phenotypic Landscape Exploration Through Hierarchical Spatial Partitioning

Davy Smith(✉), Laurissa Tokarchuk, and Geraint Wiggins

School of Electronic Engineering and Computer Science,
Queen Mary University of London, London E1 4NS, UK
{david.smith,laurissa.tokarchuk,geraint.wiggins}@qmul.ac.uk

Abstract. Exploration of the search space through the optimisation of phenotypic diversity is of increasing interest within the field of evolutionary robotics. Novelty search and the more recent MAP-Elites are two state of the art evolutionary algorithms which diversify low dimensional phenotypic traits for divergent exploration. In this paper we introduce a novel alternative for rapid divergent search of the feature space. Unlike previous phenotypic search procedures, our proposed Spatial, Hierarchical, Illuminated Neuro-Evolution (SHINE) algorithm utilises a tree structure for the maintenance and selection of potential candidates. SHINE penalises previous solutions in more crowded areas of the landscape. Our experimental results show that SHINE significantly outperforms novelty search and MAP-Elites in both performance and exploration. We conclude that the SHINE algorithm is a viable method for rapid divergent search of low dimensional, phenotypic landscapes.

Keywords: Algorithm design · Phenotypic diversity · Neuroevolution · Evolutionary robotics

1 Introduction

Divergent evolutionary search methods are receiving increasing interest in the evolutionary robotics community. Optimising phenotypic diversity within a population has been shown to avoid convergence towards local optima [5], to provide diverse ranges of solutions in a given domain, [4,7,8] and to assist with the adaptability of robot controllers [2]. Novelty search, introduced in [5] and the more recent multi-dimensional archive of phenotypic elites (MAP-Elites) [10], are two algorithms which utilise divergent phenotypic search. In this paper we introduce the Spatial, Hierarchical, Illuminated Neuro-Evolution (SHINE) algorithm, a novel method which the authors show explores low dimensional phenotypic landscapes more thoroughly and rapidly than the current state of the art. Similarly to MAP-Elites, our proposed SHINE algorithm selects future populations from an archive of previous solutions. However, the archive in the SHINE algorithm is maintained within an hierarchical, spatially partitioned tree structure. Both the weighting of offspring selection and the number of representatives assigned

J. Handl et al. (Eds.): PPSN XIV 2016, LNCS 9921, pp. 911–920, 2016.
DOI: 10.1007/978-3-319-45823-6_85

to the archive are calculated from the depth of the vertices within which the solutions reside. Candidate solutions which exhibit phenotypic traits in more crowded areas of the landscape are assigned to vertices deeper within the tree, and are penalised accordingly. This allows the evolutionary trajectory to focus on larger, shallower areas of the landscape, producing a divergent, and iteratively more focused search procedure.

This paper is organised as follows. In Sect. 2 we give a brief overview of the use of divergent phenotypic search within evolutionary robotics. In Sect. 3, we introduce our proposed SHINE algorithm, highlighting the methods for archive management, spatial partitioning and selection of offspring in a 2-dimensional, quadtree implementation. An initial experimental domain, selected to assess the ability of the SHINE algorithm to explore the phenotypic landscape, is presented in Sect. 4. Our results, which are presented in Sect. 5, highlight that SHINE significantly outperforms both novelty search and MAP-Elites. In Sect. 6 we conclude that the hierarchical procedure adopted by the SHINE algorithm is a promising method for rapid divergent phenotypic search.

2 Related Work

Novelty Search. Novelty search, as proposed by Lehman and Stanley [5], is an algorithm which removes the need for a traditional objective function through the assignment of high fitness values to novel behaviours in a population. The objective fitness function is replaced by a behavioural distance metric, which is used to determine the novelty of an individual in a population. High novelty is assigned to individuals which exhibit features with a large distance to both the rest of the population and an archive of previously encountered, highly novel phenotypic traits.

Although novelty search has been shown to outperform objective fitness search, especially in deceptive domains, it has been shown that the assessment of behavioural novelty alone is insufficient as a generalisable evolutionary technique in many tasks, especially in domains with large feature spaces [1,9].

MAP-Elites. More recently, the MAP-Elites algorithm, as introduced in [2,10] is an evolutionary procedure that aims to find the highest performing solution at each point in a low dimensional behaviour space. It is a hybridization of objective driven and divergent search. In MAP-Elites, evolution proceeds through the maintenance of an archive of previously high performing individuals, with each individual being assigned to bin within a discrete, low dimensional representation of the feature space. Offspring for subsequent generations are randomly selected from the archive of high performing, yet phenotypically diverse individuals.

Due to the ability of MAP-Elites to highlight the highest performing solutions in a phenotypic landscape, Mouret and Clune introduce the term *illumination algorithm* to separate it from traditional optimisation algorithms [10].

3 Spatial, Hierarchical, Illuminated Neuro-Evolution

SHINE is an illumination algorithm designed for rapid exploration of low-dimensional feature spaces. SHINE promotes divergent search through penalising solutions which are in more crowded areas of a predefined, low dimensional phenotypic landscape. The algorithm utilises a spatially partitioned tree for the maintenance of an archive of phenotypic representatives. The mechanisms applied to both the storage and selection of the representatives are designed specifically to weight subsequent generations towards more offspring in sparse areas of the landscape.

The SHINE algorithm shares similarities to both novelty search and MAP-Elites. As in MAP-Elites, SHINE maintains an archive of previous solutions which are selected for inclusion by low-dimensional discrete phenotypic traits. However, SHINE utilises an hierarchical, spatially partitioned tree structure for archive maintenance. MAP-Elites stores a single elite within each area of the feature space; the current best performing individual at an objective function. SHINE maintains multiple individuals within each vertex of the archive tree which are chosen by their distance to the boundaries of their particular phenotypic trait, in a manner more aligned with novelty search. Therefore, the SHINE algorithm also differs from MAP-Elites in that it directly aims to optimise sparse areas of the feature space. Here we introduce the main SHINE procedure, outlining a 2-dimensional implementation which utilises a quadtree structure [11].

3.1 The Algorithm

The main procedure of the SHINE algorithm, (Algorithm 1) begins by initializing a random population P with n random individuals (Lines 1–5). In each generation, every individual ρ is assessed in the domain and a phenotypic descriptor is measured and assigned to μ (lines 7–9). The tree, $\mathcal{T}$, is queried with the descriptor μ (line 9). After all individuals in the current population have been assessed and the tree structure updated, P is added to the archive (line 11). A new archive is calculated and assigned to $\mathcal{X}$ (line 12). All individuals are removed from the population, which is then repopulated with mutated offspring from the updated archive $\mathcal{X}$ via weighted roulette selection (lines 14–18). This procedure is repeated until a terminating condition is met, or alternatively after a predefined number of generations (line 19).

Phenotypic Tree. In a similar manner to MAP-Elites, the SHINE algorithm progresses through the maintenance of an archive of genomes which are selected for inclusion by a measured phenotypic trait. However, SHINE maintains an archive of potential genomes in an hierarchical, spatially partitioned tree.

The number of dimensions and the bounding volume of the phenotypic descriptor are required to initialise the root vertex of the phenotypic tree. In this paper, we focus upon the 2-dimensional implementation of the algorithm, resulting in a quadtree structure [11]. We define a phenotypic descriptor as an

Algorithm 1. Main SHINE procedure

Require: α: max tree depth, β: vertex division level, $\mathcal{V}$: phenotypic tree

```
 1: procedure SHINE
 2:     P ← ∅
 3:     while |P| < n do
 4:         P ← RandomIndividual()
 5:     end while
 6:     do
 7:         for ρ ∈ P do
 8:             μ ← PerformTrial(ρ)
 9:             QueryTree(μ, V)
10:         end for
11:         UpdateArchive(P, V)
12:         X ← CurrentRepresentatives(V)
13:         P ← ∅
14:         while |P| < n do
15:             x ← RouletteSelection(X)
16:             x' ← Mutate(x)
17:             P ← P ⋃ x'
18:         end while
19:     while Terminate() is false
20: end procedure
```

ordered pair $\mu = (x, y)$. However, the algorithm may be extended to phenotypic descriptors with higher numbers of dimensions. Let $|\mu|$ represent the number of dimensions of a phenotypic descriptor and let $c = 2^{|\mu|}$. Each vertex will be subdivided into c child vertices (each dimension being split into 2 equal regions). Therefore, 3-dimensional traits ($|\mu| = 3$) would require an octree ($c = 2^3$) structure.

The SHINE algorithm requires 2 pre-defined constants to control the subdivision of the tree. We define constant α to be the maximium depth of the tree and β as the maximum number of points which may fall within a leaf vertex before it is divided. These constants are used to determine both the underlying phenotypic tree structure and the archive of representatives.

A series of trial runs in our experimental domain were performed with a range of α and β values: $\alpha = (3, 4, 5, \ldots, 12, 13, 14)$, $\beta = (20, 40, 60, \ldots, 120, 140, 160)$. The values $\alpha = 7$ and $\beta = 80$ produced the most reliable and optimal results and are therefore used in our experimental setup. Testing in further domains and with differing population sizes would be required to ascertain whether these values are universally optimal.

The QueryTree$(\mu, \mathcal{V})$ method (line 9, Algorithm 1) determines the development of the tree structure. Figure 1 illustrates an example quadtree structure with parameters $\alpha = 4$ and $\beta = 2$. During each generation, all individuals are assessed and the tree is queried with their phenotypic descriptor, μ. Let v represent the relevant vertex of $\mathcal{V}$. Let the bounding area of $v = [v_{x1} : v_{x2}] \times [v_{y1} : v_{y2}]$, where $v_{x1} < \mu_x \leq v_{x2} \wedge v_{y1} < \mu_y \leq v_{y2}$. Let v_d be the depth within the tree

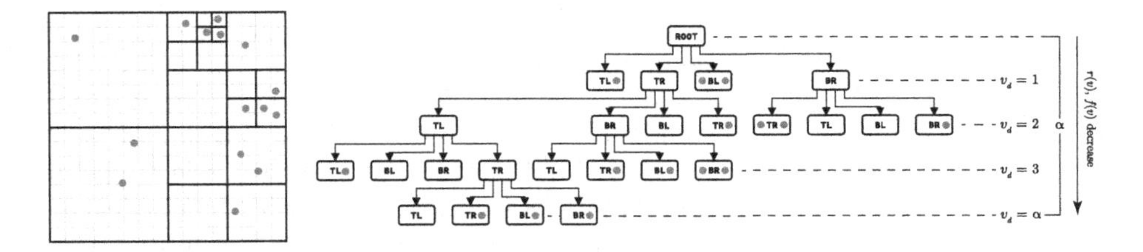

Fig. 1. An overview of spatial partitioning in the SHINE archive. ($\alpha = 4, \beta = 2$).

and $|v|$ be the number of descriptors currently assigned to v. If the capacity of v has been exceeded and the maximum depth has not been reached, such that $|v| > \beta \wedge v_d < \alpha$, then v is subdivided into 4 equal sized regions, i. e., top-left, top-right, bottom-left and bottom-right (TL, TR, BL, BR, Fig. 1). All descriptors within v are then assigned to their relevant child vertices.

Archive Management. After the tree has been queried by the population, the resulting structure is utilised to determine the distribution of the archive of representatives from which subsequent populations are selected. Membership of the archive is weighted dependant upon the depth of the representatives' containing vertex. Shallower vertices in the tree structure are assigned more representatives. Representatives do not alter the structure of the tree, rather the relevant vertex for a potential representative's phenotypic descriptor determines whether it is added to the archive. Let $|\mu|$ represent the dimensions of a phenotypic descriptor and let $c = 2^{|\mu|}$. Equation (1) defines the maximum number of representatives $r(v)$ which may be assigned to a particular vertex.

$$r(v) = (v_d - \alpha + 1)^c \tag{1}$$

The number of representatives within a single vertex will therefore fall within the range $1 \leq r(v) \leq (\alpha + 1)^c$. Let $\mathcal{X}_v$ be the set of all representative within a vertex, v. If the capacity of v is reached, such that $|\mathcal{X}_v| = r(v)$, representatives from $\mathcal{X}_v$ are selected for addition or removal based upon a distance function $d(x)$. This distance function determines the distribution of representatives *within a single leaf vertex*. In alignment with this, let x be a potential representative for inclusion within the archive, where $x \notin \mathcal{X}_v$. Let $w \in \mathcal{X}$ be the weakest current representative $w = \arg\max_{\forall i \in \mathcal{X}_v} d(i)$. The updated archive of representatives, which we define as $\mathcal{X}_v^{'}$, is determined as in Eq. (2).

$$\mathcal{X}_v^{'} = \begin{cases} \mathcal{X}_v \bigcup x & \text{if } |\mathcal{X}_v| < r(v) \\ \mathcal{X}_v & \text{if } |\mathcal{X}_v| = r(v) \text{ and } d(x) > d(w) \\ \{\mathcal{X}_v \setminus w\} \bigcup x & \text{if } |\mathcal{X}_v| = r(v) \text{ and } d(x) \leq d(w) \end{cases} \tag{2}$$

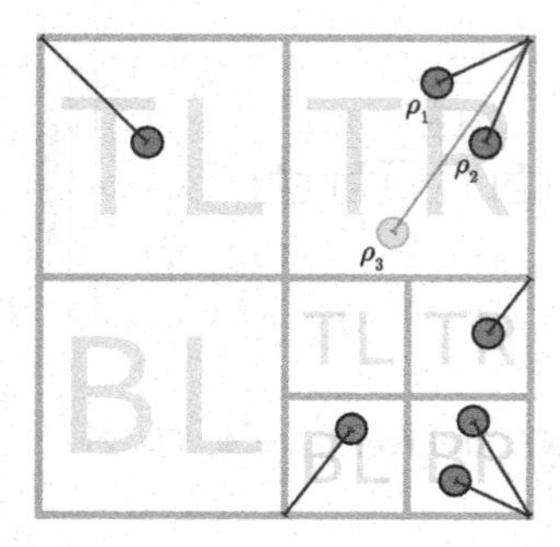

Fig. 2. An illustration of the *corner* sorting method for representative selection.

Dependant upon the particular type of search required, various metrics may be proposed. For example, defining $d(x)$ as an objective function would allow the archive to behave in a similar manner to the MAP-Elites algorithm [10], selecting elite representatives for inclusion within the phenotypic tree. We also suggest that metrics based upon novelty search [5] or hybrid novelty-objective measures [12] may be of particular interest for further testing of the algorithm in different domains.

In our experiment, presented in Sect. 4, we utilise the *corner* distance metric, a function which favours representatives in the outer corners of the containing vertex, encouraging representatives to focus on the areas closest to neighbouring vertices and increasing the chance of mutated offspring to acquire phenotypic traits in neighbouring cells. Figure 2 illustrates our *corner* method for representative selection. Representatives are sorted by distance from the outer corner of their assigned vertex's position in the quad tree structure (i.e. representatives in top-left vertices are sorted by their distance from top left corner of the vertex). Once the number of representatives exceeds the maximal threshold, as defined in Eq. (1), the representative with the largest distance is removed.

Proportional Selection. SHINE utilises a traditional roulette wheel method for the selection of offspring. Potential solutions are selected from the complete set of current representatives within the tree $\mathcal{X} = \{\mathcal{X}_{v_1} \bigcup, ..., \bigcup \mathcal{X}_{v_{|V|}}\}$. The fitness $f(x)$ of a representative x in vertex v is obtained by calculating the reciprocal of the sum of the vertices' depth v_d and its normalised population $\frac{v_p}{\beta}$. Defined as $1/(v_d + \frac{v_p}{\beta})$ and simplified in Eq. (3).

$$f(x) = \frac{\beta}{\beta v_d + v_p} \tag{3}$$

This fitness assignment results in a lower probability of selection of representatives within smaller (deeper within the tree) and more crowded (higher population) areas of the phenotypic landscape, allowing the search procedure to concentrate on larger and sparser vertices within the tree.

4 Experimental Evaluation

Domain. The aim of our experiment is to assess the diversity and thoroughness of phenotypic exploration in an evolutionary trajectory optimised with the SHINE algorithm in comparison to novelty search and MAP-Elites. Therefore, we select a domain with a deceptive objective function and which requires a high level of exploration to produce a successful solution.

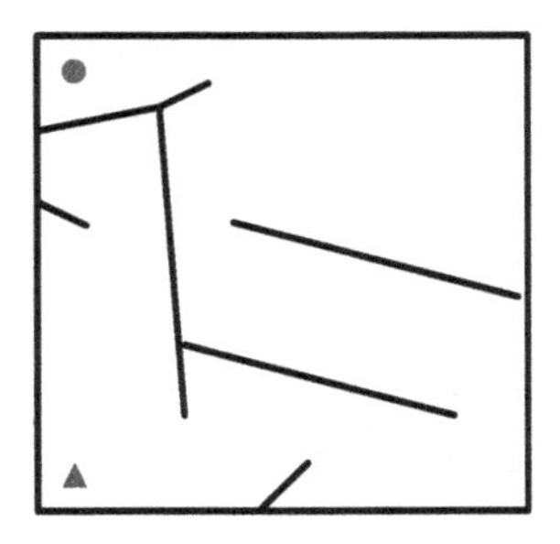

Fig. 3. The HARD maze domain. Triangle indicates agent start position, circle indicates exit.

Our experimental domain is taken directly from previous studies which have assessed novelty search and variants of the algorithm [3,5,6]. The maze used in our experiment, the HARD maze, is classified as a deceptive domain, particularly difficult for objective algorithms to reliably find solutions (Fig. 3). The maze is of the size 1000×1000 units, the agent has a size of 20 units and successfully reaching the exit requires the agent to be within 20 units. Each agent is given 4000 time steps to complete the maze. Populations of 200 controllers were optimised for 1000 generations. The agent controllers are neural networks which are evolved using the NEAT algorithm [13], with the speciation mechanism deactivated. As in [3,5,6], the objective fitness of a solution ρ is calculated as $f(\rho) = l - dist(\rho, e)$, where l is the diagonal length of the maze and e is the exit to the maze. The phenotypic descriptor is calculated from the ending position of the agent, $\mu = (\rho_x, \rho_y)$.

We assess 4 algorithms in our experiment — traditional objective based search (OBJECTIVE), novelty search (NOVELTY), MAP-Elites (MAP-ELITES), and our proposed SHINE algorithm (SHINE). The algorithms were repeated in each domain 50 times with a different random seed in each trial. In order to ensure consistency between algorithms, identical random seed values were given to each of the algorithms in each trial. The performance of each algorithm was determined by the number of generations taken to locate the exit in the domain.

The simulation was performed using a bespoke domain written in the C++ programming language, developed to be similar to the original maze domain experiments in [5,6]. The implementation of the NEAT algorithm used was developed as an extension to the MultiNEAT software in the C++ language[1].

Domain Coverage. The cumulative coverage of the domain is calculated at each generation in the trial over 1000 generations. The domain is divided into a 2-dimensional matrix M, where $|M| = n \times n$. In our presented results, $n = 30$. The final position of an individual (ρ_x, ρ_y) is mapped to the corresponding region of M. Let M' be the set of the regions of M which contain individuals: $M' = \{x : x \in M \land |x| > 0\}$. Domain coverage is then calculated as $\frac{|M'|}{|M|}$.

Exploration Uniformity. The spread of the population is measured through the calculation of *exploration uniformity* in a similar manner to [3]. To ascertain the speed at which exploration occurs for each algorithm, values are calculated at each generation in the trial rather than cumulatively over the whole trial as in [3]. Again, the population is mapped to the discrete matrix M. Let P_t be the set of individuals in the population at generation t and let Ψ_t be the distribution of P_t over M. The exploration uniformity of the population, $D(P_t)$, is calculated as

[1] ©2012 Peter Chervenski. http://multineat.com/index.html.

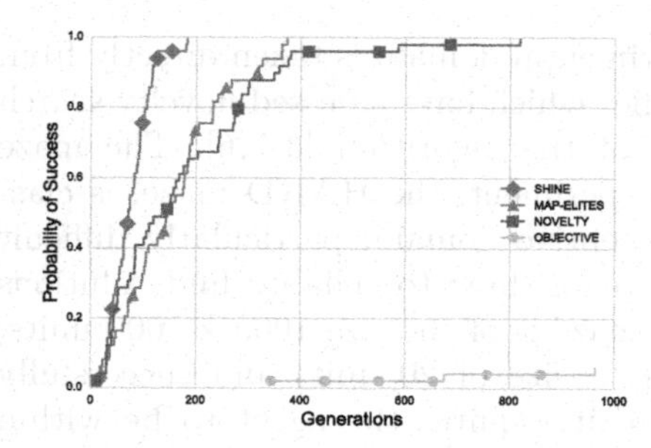

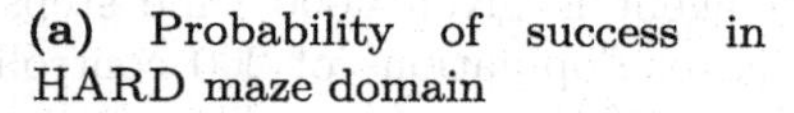
(a) Probability of success in HARD maze domain

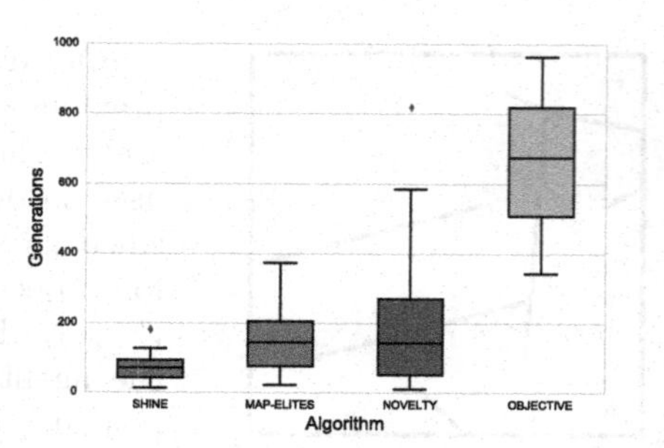

(b) Number of Generations taken to locate the exit in the HARD maze domain in successful trials.

Fig. 4. Performance results from HARD maze domain (SHINE, NOVELTY and MAP-ELITES were successful in all trials.)

the similarity between Ψ_t and the uniform distribution U. As in [3] the distance metric used is the *Jensen-Shannon* distance (JSD). The exploration uniformity at generation t is thus defined as:

$$D(P_t) = 1 - JSD(\Psi_t, U),\ where:$$
$$\Psi_t = \left(\frac{|I_1|}{|P_t|}, ..., \frac{|I_{|P_t|}|}{|P_t|}\right), I_r = \{i \in P_t : region(i) = r\}$$
$$U = \left(\overbrace{\frac{1}{|M|} \times \cdots \times \frac{1}{|M|}}^{n^2 \text{ times}}\right) \tag{4}$$

5 Results

Performance. As illustrated in Fig. 4a, all 3 algorithms located solutions to the maze in all 50 trials, resulting in a probability of success of 1.0. Maximum probability of success is reached significantly faster ($p < 0.001$) by the SHINE algorithm, after 182 generations, compared with 374 generations for MAP-ELITES and 819 generations for NOVELTY. Both NOVELTY and MAP-ELITES follow a similar gradient of ascent, however NOVELTY requires a higher number of generations to locate a solution in 3 of the trials.

Figure 4b shows the number of generations taken to find a successful solution. The SHINE algorithm requires a significantly fewer number of generations, with a median value of 71. MAP-ELITES and NOVELTY achieve similar results, with median values of 146 and 141 generations respectively.

Diversity. Figure 5a shows the exploration uniformity for each of the algorithms over 1000 generations. The maximum mean level of exploration uniformity is

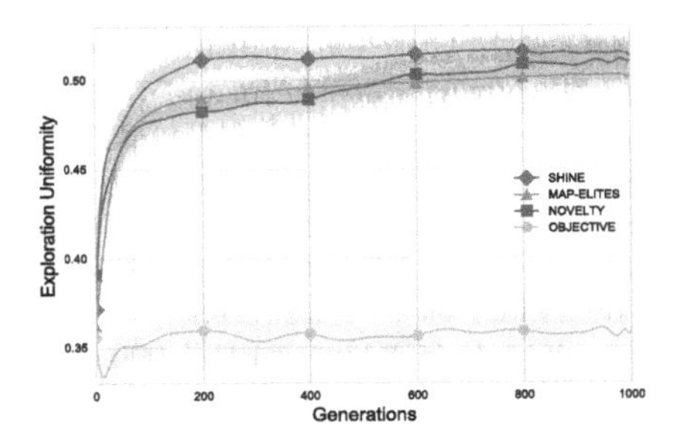

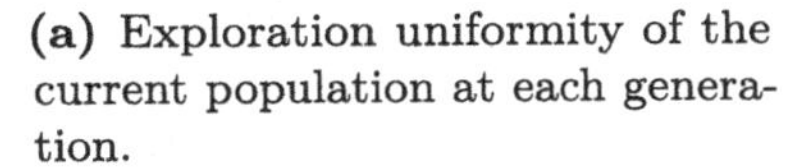
(a) Exploration uniformity of the current population at each generation.

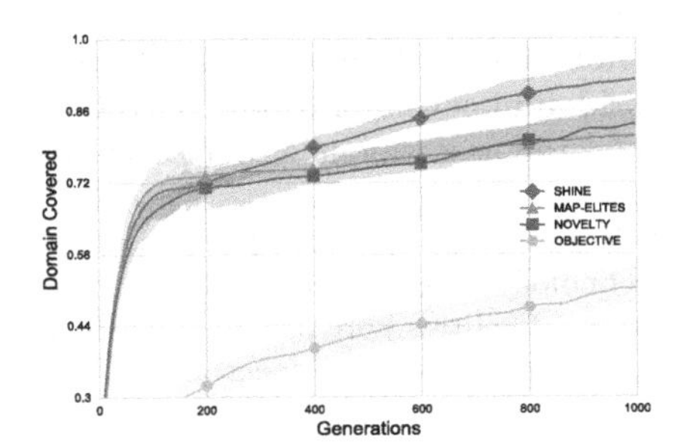

(b) Cumulative proportion of domain coverage by the population after each generation.

Fig. 5. Diversity of the algorithms within the HARD domain. (Shaded area indicates 25th to 75th percentiles.)

achieved by the SHINE algorithm, 0.51912 after 772 generations. However, it achieves comparably high levels after 232 generations, remaining relatively stable throughout the evolution. Both MAP-ELITES and NOVELTY fail to achieve this maximal level within 1000 generations, however the exploration uniformity is still increasing for both algorithms at the end of the trial. The maximum mean level achieved by MAP-ELITES is 0.50584 after 984 generations. NOVELTY achieves a maximal value of 0.51408 after 988 generations. Therefore an evolutionary run with a higher number of generations may allow MAP-ELITES and NOVELTY to achieve a level of exploration uniformity similar to SHINE. Figure 5b shows the proportion of the domain covered by the population. All three algorithms produce similar levels of domain coverage for the initial 400 generations. However, beyond this SHINE covers significantly more of the domain than both NOVELTY and MAP-ELITES.

6 Conclusion

In this paper we have introduced a novel method for rapid exploration of low dimensional feature spaces. Our experimental evaluation in a deceptive simulated maze domain shows that the SHINE algorithm outperforms both novelty search and MAP-Elites, two state of the art algorithms for divergent phenotypic search. We have shown that the hierarchical tree structure and approach taken for archive maintenance and offspring selection in the SHINE algorithm are viable methods for rapid phenotypic exploration.

Further experimental validation is required in order to establish the performance of the SHINE algorithm in domains with a less direct mapping between the feature space and the objective landscape. The authors suggest that a replacement of the *corner* method presented in this paper to an objective function would allow SHINE to be compared more directly with MAP-Elites in objective focussed domains. The authors are aware of the limitations in testing within a simulated environment. MAP-Elites has been shown to be extendible to the real world application of robot controllers [2]. Therefore we suggest a future direction to be the assessment of SHINE beyond simulation, in real world domains.

Acknowledgements. This work was funded by EPSRC through the Media and Arts Technology Programme, an RCUK Doctoral Training Centre EP/G03723X/1. Computational facilities were provided by the MidPlus Regional Centre of Excellence for Computational Science, Engineering and Mathematics, under EPSRC grant EP/K000128/1.

References

1. Cuccu, G., Gomez, F.: When novelty is not enough. In: Di Chio, C., et al. (eds.) EvoApplications 2011, Part I. LNCS, vol. 6624, pp. 234–243. Springer, Heidelberg (2011)
2. Cully, A., Clune, J., Tarapore, D., Mouret, J.-B.: Robots that can adapt like animals. Nature **521**(7553), 503–507 (2015)
3. Gomes, J., Mariano, P., Christensen, A.L., Devising effective novelty search algorithms: a comprehensive empirical study. In: Proceedings of the 2015 Genetic and Evolutionary Computation Conference, pp. 943–950. ACM (2015)
4. Gomez, F.J.: Sustaining diversity using behavioral information distance. In: Proceedings of the 11th Annual Conference on Genetic and Evolutionary Computation, pp. 113–120. ACM (2009)
5. Lehman, J., Stanley, K.O.: Exploiting open-endedness to solve problems through the search for novelty. In: ALIFE, pp. 329–336 (2008)
6. Lehman, J., Stanley, K.O.: Abandoning objectives: evolution through the search for novelty alone. Evol. Comput. **19**(2), 189–223 (2011)
7. Lehman, J., Stanley, K.O.: Evolving a diversity of virtual creatures through novelty search and local competition. In: Proceedings of the 13th Annual Conference on Genetic and Evolutionary Computation, pp. 211–218. ACM (2011)
8. Lehman, J., Stanley, K.O., Miikkulainen, R.: Effective diversity maintenance in deceptive domains. In: Proceedings of the 15th Annual Conference on Genetic and Evolutionary Computation, pp. 215–222. ACM (2013)
9. Mouret, J.-B.: Novelty-based multiobjectivization. In: Doncieux, S., Bredèche, N., Mouret, J.-B. (eds.) New Horizons in Evolutionary Robotics. Studies in Computational Intelligence, vol. 341, pp. 139–154. Springer, Heidelberg (2011)
10. Mouret, J.-B., Clune, J.: Illuminating search spaces by mapping elites (2015). arXiv preprint arXiv:1504.04909
11. Samet, H.: The quadtree and related hierarchical data structures. ACM Comput. Surv. (CSUR) **16**(2), 187–260 (1984)
12. Smith, D., Tokarchuk, L., Wiggins, G., Exploring conflicting objectives with madns: multiple assessment directed novelty search. In: Companion Proceedings of the 2016 Genetic and Evolutionary Computation Conference. ACM (2016)
13. Stanley, K.O., Miikkulainen, R.: Evolving neural networks through augmenting topologies. Evol. Comput. **10**(2), 99–127 (2002)

Understanding Environmental Influence in an Open-Ended Evolutionary Algorithm

Andreas Steyven(✉), Emma Hart, and Ben Paechter

School of Computing, Edinburgh Napier University,
10 Colinton Road, Edinburgh, Scotland, UK
{a.steyven,e.hart,b.paechter}@napier.ac.uk

Abstract. It is well known that in open-ended evolution, the nature of the environment plays in key role in directing evolution. However, in Evolutionary Robotics, it is often unclear exactly how parameterisation of a given environment might influence the emergence of particular behaviours. We consider environments in which the total amount of energy is parameterised by availability and value, and use surface plots to explore the relationship between those environment parameters and emergent behaviour using a variant of a well-known distributed evolutionary algorithm (mEDEA). Analysis of the resulting landscape show that it is crucial for a researcher to select appropriate parameterisations in order that the environment provides the right balance between facilitating survival and exerting sufficient pressure for new behaviours to emerge. To the best of our knowledge, this is the first time such an analysis has been undertaken.

Keywords: Evolutionary robotics · Parameter selection · Environment-driven evolution · Distributed online adaptation

1 Introduction

Due to technological advances in both hardware and software, the vision of sending swarms of robots into unchartered terrains to monitor and map environments is becoming much closer to being realised. This brings significant new challenges for evolutionary robotics, with the need for completely distributed evolutionary algorithms to evolve controllers that enable robots to survive for long-periods of time. The issue of survival is key if robots are to effectively accomplish any kind of task: user-driven tasks cannot even be achieved if the integrity of the swarm is compromised through lack of ability to survive.

A number of recent algorithms tackle this issue, notably mEDEA [1] and its variations e.g. mEDEA$_{rf}$ [7] and MONEE [4,6]. However, the emerging behaviours arising from the interactions of an open-ended evolutionary algorithm with its environment are not well understood, perhaps in part due to the time-consuming experimentation that needs to be done to conduct sweeps of the parameters that define the environment. It is common in optimisation to

J. Handl et al. (Eds.): PPSN XIV 2016, LNCS 9921, pp. 921–931, 2016.
DOI: 10.1007/978-3-319-45823-6_86

explore the relationship between algorithmic parameters and fitness. However, evolutionary robotics adds an additional dimension in that it is not only the algorithms parameters that change but also the *environmental* parameters.

Given that it is the environment that provides the pressure to adapt in a purely open-ended scenario, it is crucial to gain some understanding of these landscapes. Particularly in simulation, it is easy to arbitrarily select environmental parameters such as the number of available energy sources or their corresponding energy-values. However, arbitrary choices can inadvertently create landscapes which have a major influence on the evolution of behaviour. For example, assume a researcher wishes to investigate whether individual learning speeds up environment-driven evolution: if an environment is created that has too much energy available then it is unlikely to exert sufficient pressure for individual learning to be beneficial or even emerge. Quantifying 'too much' (or 'too little') is of course difficult. In order to address this, we conduct an analysis of an open-ended evolution algorithm operating in a variable environment. To the best of our knowledge, this is the first time this has been attempted.

Using an open-ended evolutionary algorithm, mEDEA$_{rf}$ [7], we consider evolved behaviours in environments in which the total energy available is parameterised by two variables that determine the availability and value of energy pellets within in the environment. Using a 3-dimensional visualisation of the energy landscape for mEDEA$_{rf}$ we show:

- the energy landscape contains three distinct regions: energy-poor, energy-neutral and energy-rich, as well as a 'dead-zone' in which robots cannot survive
- the energy-rich region is relatively large compared to other regions but is very rugged
- that on the energy-neutral line, distinct behaviours evolve at different places along the line

We propose that the energy-neutral region provides the most obvious settings for conducting experimentation that aims to extend a robots ability to survive or accomplish tasks.

2 Related Work

The completely distributed evolutionary algorithm for open-ended evolution *mEDEA* was first proposed in [1]. It was tested using a scenario in which environmental pressure forces robots to compete for limited resources in order to gain energy. The algorithm was demonstrated to be both efficient with regard to providing distributed evolutionary adaptation in unknown environments, and robust to unpredicted changes in the environment. The basic algorithm has been extended in a number of ways.

Haasdijk *et al.* [6] extended mEDEA so that in addition to surviving and operating reliably in an environment, a robot could also perform user-defined tasks. Their new framework MONEE (Multi-Objective aNd open-Ended Evolution algorithm) showed initially that task-driven behaviour can be promoted without compromising environmental adaptation. More recently, they investigated the trade-off between the survival and task-accomplishment that evolution must establish when the task is detrimental to survival, finding that task-based selection exerts a higher pressure than the environment. Fernández Pérez *et al.* [3] study the impact of adding explicit selection methods to the mEDEA algorithm in a task-driven scenario. They evaluate four selection methods that induce different intensities of selection pressure, using tasks that include obstacle avoidance and foraging, finding that higher selection pressure results in improved performances, especially in more challenging tasks. Hart [7] also extended mEDEA by including selection based on a fitness value that was calculated relative to those robots in the immediate vicinity, thus maintaining the decentralised nature of the algorithm, and additionally using this relative fitness value to control the frequency and range of broadcasting. Parameter tuning of *algorithmic* parameters to optimise algorithmic task-performance was investigated by [5]. However, to the best of our knowledge, no methodical investigation of *environment* parameter settings has been conducted: researchers tend to select arbitrary values or simply use those defined in previous papers.

3 Algorithm Description

Evolution of robot controllers is performed by the mEDEA$_{rf}$, first introduced in [7]. The algorithm is an extension of the original mEDEA algorithm of Bredeche and Montanier [1] with the addition of an explicit fitness measure. This influences the spread of genomes through the population in order to increase survivability, thus ensuring the integrity of the swarm.

mEDEA$_{rf}$ utilises an agent driven by a control architecture whose parameters are defined by the currently active genome. The genome defines the weights of an Elman recurrent neural network (RNN) consisting of 16 sensory inputs, one bias node (feeding into the hidden layer) and 2 motor outputs (translational and rotational speeds). 8 ray-sensors are distributed around the robot's body. They detect the proximity to the nearest object and its type. The RNN has 1 hidden layer with 16 nodes, thus 322 weights are defined by the genome. This setup is adapted from [1]. An overview of the algorithm is given in Algorithm 1 and reader is referred to [7] for more detail. In brief, for a fixed period, robots move according to their control algorithm, broadcasting their genome that is received and stored by any robot within range. At the end of this period, a robot uses roulette-wheel selection to choose a genome from its list of collected genomes according to a relative fitness value, and applies a variation operator. This takes the form of a Gaussian random mutation operator, inspired from Evolution Strategies. Robots that have not collected any genomes temporarily become inactive, thus reducing the population size.

Each robot estimates its fitness in terms of its ability to survive based on the balance between energy lost and energy gained, delta Energy (δ_E): this term is initialised to 0 at $t = 0$ (when the current genome was activated) and is decreased by 1 at each time-step, and increased by E_{token} if it crosses an energy token.

Given δ_E, a robot calculates a fitness value which is relative to those robots in a range r according to Eq. 1, where f_i' is the relative fitness of robot i at time t, $mean_{sub_i}$ is the mean δ_E of the robots within the subpopulation defined by all robots in range r of robot i, and sd_{sub_i} is the standard deviation of the δ_E of the subpopulation.

$$f_i'(t) = \frac{\delta_i(t) - mean_{sub_i}(t)}{sd_{sub_i}(t)} \tag{1}$$

Note that evolution is asynchronous, in keeping with the paradigm of a distributed algorithm without central control. If a robot runs out of energy and has an empty genome list, it remains stationary until it receives a new genome from a passing robot at which point it starts a new lifetime. Thus at any time-step, each robot potentially has a different 'age'.

```
genome.randomInitialise();
agent.load(genome);
while forever do
    if genome.isNotEmpty() then
        while lifetime < maxLifetime and energy > 0 do
            agent.move();
            if neighbourhood.isNotEmpty() then
                rf = agent.calculateRelativeFitness(neighbourhood);   // eq. 1
                broadcast(genome,rf);
            end
        end
        genome.empty();
    end
    if genomeList.size() > 0 then
        genome = applyVariation(select_rhoulette-wheel(genomeList));
        agent.load(genome);
        genomeList.empty();
    end
end
```

Algorithm 1. Pseudo code of our adapted version of the mEDEA algorithm based on vanilla mEDEA by Bredeche and Montanier [1]

4 Method

All experiments are conducted in simulation using Roborobo! by Bredeche et al. from [2]. A static environment is created, using an arena previously described in [3,4,6]. The robot cannot pass through the outer and inner walls, however, it is possible to broadcast through an obstacle. Energy *tokens* are randomly scattered in the environment. If a robot moves over a token, its energy is increased by an amount E_{token}. The energy token disappears when consumed and reappears after a fixed amount of time later at a different random location. Fixed parameters describing the simulation are given in Table 1.

Energy is consumed in three ways. There is a fixed cost to 'living' of 0.5 units per timestep, regardless of whether the robot moves or not. A robot moving consumes an amount of energy E_m that is related to its rotational speed v_{rot}, translational speed v_{trans}, and their respective maximum values $v_{\text{rot}_{\text{MAX}}}$ and $v_{\text{trans}_{\text{MAX}}}$, and is given by

$$E_m = (v_{\text{rot}}/v_{\text{rot}_{\text{MAX}}} + v_{\text{trans}}/v_{\text{trans}_{\text{MAX}}})/4 \quad (2)$$

Finally, a robot consumes energy when communicating. This is an important factor in the real-word but one that it is often overlooked in simulation models. The model used is exactly as described in [10], with an energy cost of $E_{RX} = 0.082$ units for receiving and a cost of $E_{TX}(r) = 0.075$ units for transmitting.

The goal of the experiments is to understand the energy landscape in terms of the median δEnergy of a robot in the population as a function of the two environmental parameters: *count*, the number of energy tokens available, and *value*, the energy value of each token. Table 1 shows the ranges of values considered for each parameter. Parameters are set before the beginning of the experiment and remain fixed throughout. Each experiments was repeated for 5 independent runs. This number is rather low for a noisy application of this type but was chosen to speed up computation due to the high number of experiments that had to be run in total.

Table 1. Simulation and experimental parameters for all experiments

Simulation parameters	
Arena size	1024 pixel by 1024 pixel
Max. robot lifetime	2500 iterations
Token re-spawn time	500 iterations
Sensor range	196 pixel
Variable parameters	
Number of robots	50, 75, 100
Number of tokens (*count*)	0–1300 (in steps of 50)
Energy value per token (*value*)	0–1400 (in steps of 50)
Experimental parameters	
Number of runs	5
Maximum iterations	375000 (=150 × 2500)
Start energy	500
Maximum range $r_{\max}$	128

Data is gathered from the robots every 2500 iterations. Recall from Sect. 3 that each robot chooses a new genome once it has depleted all its energy or reached the maximum lifetime, leading to asynchronous generation changes throughout the population. Hence, the data gathered at each interval represents a snapshot across robots of multiple ages and therefore does not necessarily capture the peak performance of each robot (i.e. it may include very 'young' robots). However, given that the goal of the experiment is to understand the interplay of the specific algorithm and environment under consideration, this is not a relevant factor.

5 Analysis

Figure 1 shows three rotated 3-dimensional plots of surface obtained using 100 robots after 375,000 iterations. The x and y axes represent the *count* and *value* variables, while the z axis represent the median δ_E of the robot population over the last 2500 iterations. The grey plane marks a value for δ_E of zero, at which point robots have an energy balance of zero, i.e. the same amount of energy as they started the experiment with. Three broad regions are noticeable: a large region in which the robots have positive δ_E (green and blue value above the grey plane), a region lying on the plane itself, and finally a region below the plane in which robots are spending more energy than they are collecting, i.e. $\delta_E < 0$. In order to explore this in more detail, a 2-dimensional top-down projection is shown in Fig. 2 obtained from populations of 50, 75 and 100 robots, and is discussed in detail below.

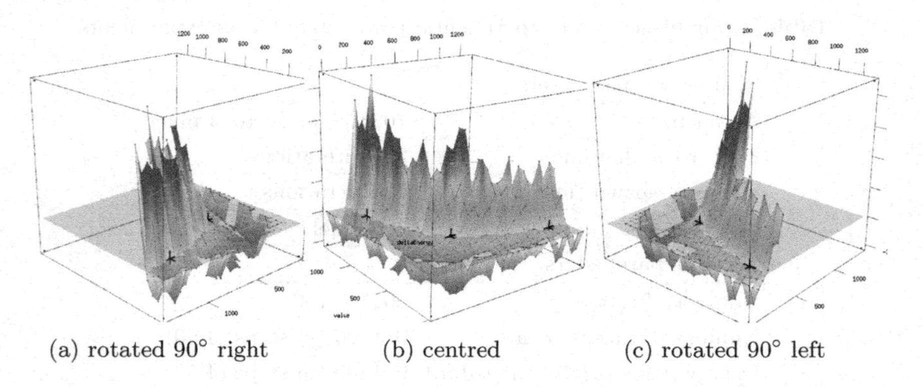

(a) rotated 90° right (b) centred (c) rotated 90° left

Fig. 1. View on the resulting surface from different angles. The figure was created by plotting the median δ_E of the last 2500 iterations of the experiment. The grey plane marks a value for δ_E of zero, at which point robots in an experiment have an energy balance of zero. In other words, the same amount of energy as they started the experiment with. A 3D model can be found at [9] (Color figure online)

5.1 Different Performance Regions

Figure 2 shows clearly that the landscape is defined by four different regions:

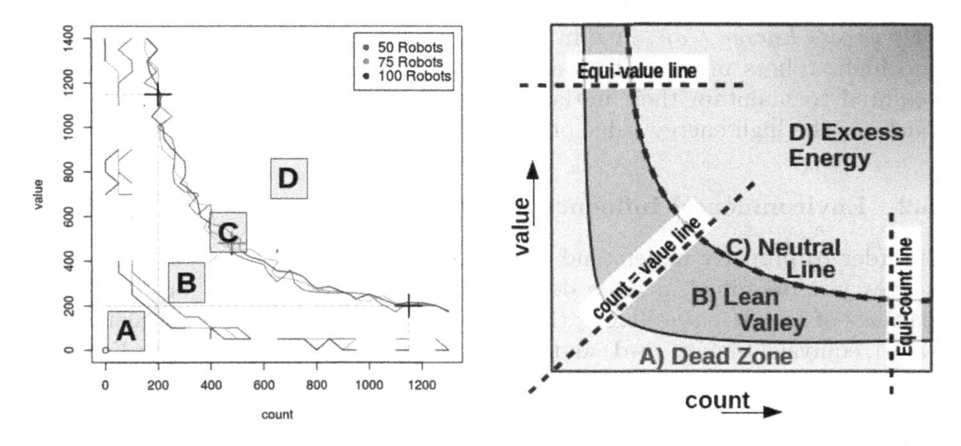

Fig. 2. Overview of andscape, as plot of the real data on the left and as a cartoon version on the right. 4 different regions are shown: (A) Dead Zone, (B) Lean Valley, (C) Neutral Line, (D) Excess Energy. (Color figure online)

(A) Dead Zone: In this region, the environment does not provide enough energy for the algorithm to evolve controller that can survive a full run. Low values for both parameters, *count* and *value* result in the extinction of the whole robot population within a few generations. The random genomes that the controllers are initialised with generally result in a random spinning behaviour, rather than movement. This random behaviour, combined with the lack of energy tokens in the immediate vicinity in which the robot is born, mean that robots cannot survive given its inability to move.

(B) Lean Valley (negative δ_E): This region starts at the edge of the dead zone that marks the point where there is just enough energy available that some robots survive until the end of the experiment, i.e. it marks the point where a robot has spent all its initial energy and started picking up tokens from the environment. Moving down towards the bottom of the valley, an increasing number of robots survive as there is more energy in environment, with the corollary that each robot has less total energy — the energy available is shared between more robots. The bottom of the valley marks the minimum δ_E that still enables survival. Moving upwards out of the valley on the other side, robots gradually get better in both harvesting energy from the environment and managing their residual energy as a result of evolving better strategies. For example, good strategies optimise movement, or avoid moving towards tokens in which there are other robots close by.

(C) Neutral Line ($\delta_E = 0$): This line marks the points in the environment where the environment provides exactly enough energy to enable a robot to maintain an energy balance of zero, i.e. the costs of moving and communicating are just balanced by energy harvested.

(D) Excess Energy ($\delta_E > 0$): In the final region, in which both *cost* and *value* are high, robots are able to locate more energy in the environment than is required to maintain their initial energy E_0, either due to the abundance of pucks or the high energy value of pucks.

5.2 Environmental Influence on Behaviour

In order to properly understand the evolved behaviours that lead to the landscapes just described, a more detailed analysis is required. Figure 3 examines pairings of $(count, value)$ along the three dashed lines in 2, i.e. equivalent-*value* (**a-b**), equivalent-*count* (**c-d**) and the diagonal in which $count = value$ line (**e-f**). The figure shows boxplots of the δ_E values at specific pairings of $(count, value)$ and the ratio of genome broadcasts made to unique genomes received over a lifetime. The latter quantity leads to insights into behaviour as it relates to the number of *unique* robots encountered by an individual robot: a robot will broadcast indiscriminately to any robot in its range but will only collect unique genomes. At the equivalent-count and equivalent-value lines, we fix the parameter *count* and *value* respectively, and successively increase the other parameter in steps of 50.

5 points are shown. The first point on (a) corresponds to a total energy E_{tot} that is the same as the first points on graphs (c) and (e) below it etc.[1]. For a specific value of E_{tot}, then is clear that high *value* combined with low *count* leads to robots that have increased δ_E when compared to robots with high *count* but low *value* (graph (a) compared to graph (e)). Robots must therefore evolve behaviours that enable them to seek out the rare but high-value pucks. These robots also have high broadcast:genome ratios, suggesting the robots are frequently coming into contact with the *same* robots. A possible explanation lies in the fact that the robots appear travel in small groups, thus broadcasting continually to the same robots; the rare occurrence of pucks leads to many robots having to travel towards the same regions of the space. On the other hand, a high *count* leads to robots that receive more unique genomes than in the high *value* case: this is suggestive of a more random movement pattern that enables each robot to encounter many unique robots during its lifetime. In this case there is low selection pressure to evolve focused movement due to the abundance of pucks.

5.3 Behaviours in the Neutral Region

We propose that the energy neutral region is of greatest interest for researchers wishing to conduct research moving beyond genetic evolution of survival, for example using individual or social learning [8] or task-driven research [4]. In this region, on the one hand, robots are able to survive, while on the other, the environment does not *over*-provide, thus ensuring that there is scope for robots

[1] While this is exactly true for the first and third rows, in the middle row which represents equal count/value it is necessary to approximate.

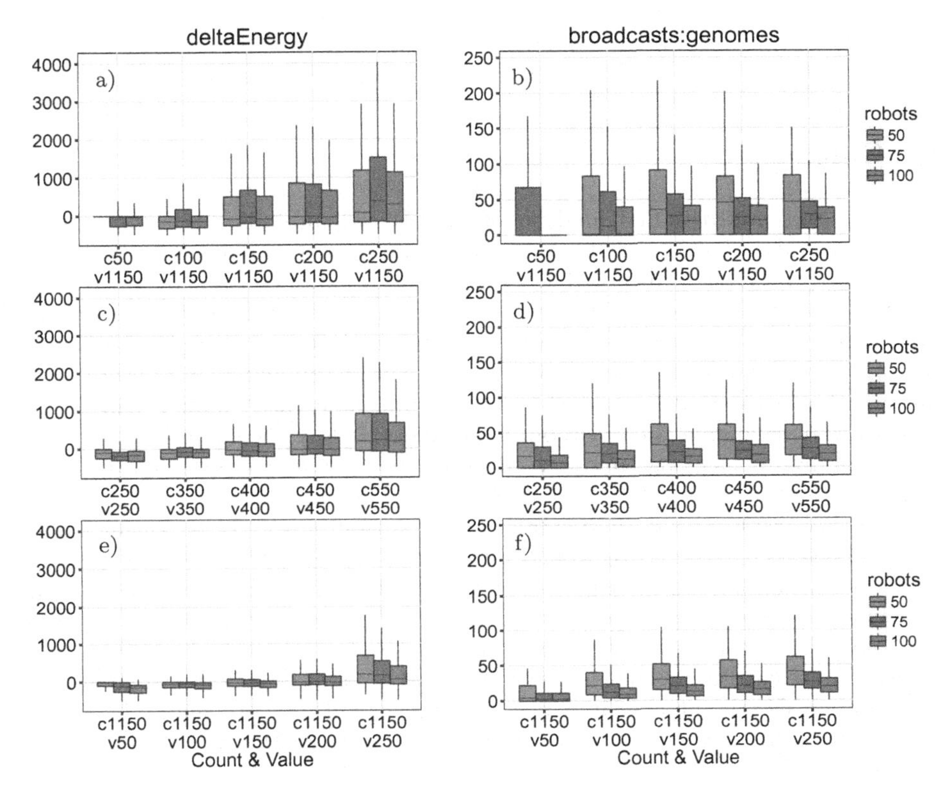

Fig. 3. Cuts through different parts of the landscape. Points towards different behaviours in terms of exploration. (a-b) *value* = 1150, vary *count*; (c-d) *count* = *value*; (e-f) *count* =1150, vary *value*. (Color figure online)

Table 2. Results obtained at three configurations within the neutral region

Count	Value	Robots								
		50			75			100		
		Age	Age/Genome	Brodcasts/Genome	Age	Age/Genome	Brodcasts/Genome	Age	Age/Genome	Brodcasts/Genome
200	1150	*770*	*107.94*	*46.61*	*767.5*	*63.86*	*28.99*	*667*	*49.97*	*21.18*
500	500	*1026.5*	*86.12*	*39.11*	*1038.5*	*52.39*	*25.93*	*928.5*	*41.26*	*19.67*
1150	200	*1173.5*	*79.83*	*33.43*	*1093*	*47.39*	*21.95*	*1059*	*36.52*	*15.49*

to learn novel behaviours. We further investigate three specific points within this region there is approximately the *same* amount of energy available in the environment (Table 2). The table shows the median age increases with increasing *count* — it is easier to maintain sufficient energy to survive as availability increases. The lower median observed at low *count* reflects the fact that many robots do not survive long. The time to find a new unique genome (age:genome) is shortest at high *count*, reflecting frequent encounters with novel robots. Broadcast:genomes is highest at low *count* as observed in the previous section. All three configurations lead to the same energy balance of 0, but diverse behaviours result in the gain in energy being offset by movement and broadcasting in each case.

6 Conclusion

We have presented the first analysis of the fitness landscape (as a function of environmental parameter) that results from running an open-ended evolutionary algorithm (mEDEA$_r f$) in an environment that is parameterised by two values that control the distribution of energy in the environment. Adjusting the availability and value of energy pucks results in the evolution of a range of different behaviours. Rather than arbitrarily selecting parameters in which to study evolution, we suggest that it is vital to understand *how* these choices will direct evolution, by changing the selection pressure exerted by the environment.

Three distinct regions are observed in which the final energy balance can be negative, neutral, or positive. A fourth region is found in which robots cannot survive. We propose that the energy neutral region is a good region in which to undertake experiments. It provides an environment in which robots are able to survive, enabling experimentation, while at the same time, will reward new behaviours which are able to more efficiently harness energy from the environment. It is clear that the environment plays a key role in influencing what kind of behaviours emerge, in that it is not the total amount of energy available that matters but also the manner in which it is spread. Future work should be aimed at understanding the landscape in more detail, and in particular, explaining the ruggedness of some regions.

References

1. Bredeche, N., Montanier, J.-M.: Environment-driven embodied evolution in a population of autonomous agents. In: Schaefer, R., Cotta, C., Kołodziej, J., Rudolph, G. (eds.) PPSN XI. LNCS, vol. 6239, pp. 290–299. Springer, Heidelberg (2010)
2. Bredeche, N., Montanier, J.M., Weel, B., Haasdijk, E.: Roborobo! a fast robot simulator for swarm and collective robotics. CoRR abs/1304.2, April 2013
3. Fernández Pérez, I., Boumaza, A., Charpillet, F.: Comparison of selection methods in on-line distributed evolutionary robotics. In: ALIFE 2014, pp. 282–289. MIT Press (2014)
4. Haasdijk, E.: Combining conflicting environmental and task requirements in evolutionary robotics. In: 2015 IEEE 9th International Conference on Self-adaptive and Self-organizing Systems, pp. 131–137. IEEE, September 2015
5. Haasdijk, E., Smit, S.K., Eiben, A.E.: Exploratory analysis of an on-line evolutionary algorithm in simulated robots. Evol. Intell. **5**(4), 213–230 (2012)
6. Haasdijk, E., Weel, B., Eiben, A.E.: Right on the MONEE. In: Blum, C. (ed.) Proceedings of GECCO 2013, pp. 207–214. ACM Press (2013)
7. Hart, E., Steyven, A., Paechter, B.: Improving survivability in environment-driven distributed evolutionary algorithms through explicit relative fitness and fitness proportionate communication. In: Silva, S. (ed.) Proceedings of GECCO 2015, pp. 169–176. ACM Press (2015)

8. Heinerman, J., Rango, M., Eiben, A.E.: Evolution, individual learning, and social learning in a swarm of real robots. In: 2015 IEEE Symposium Series on Computational Intelligence, pp. 1055–1062. IEEE (2015)
9. Steyven, A.: Interactive 3D model of mEDEA_rf parameter sweep fitness landscape (2016). http://research.steyven.de/conf/ppsn2016/
10. Steyven, A., Hart, E., Paechter, B.: The cost of communication. In: Silva, S. (ed.) GECCO Companion 2015, pp. 1239–1240. ACM Press (2015)

Simple Random Sampling Estimation of the Number of Local Optima

Khulood Alyahya(✉) and Jonathan E. Rowe

School of Computer Science, University of Birmingham, Birmingham B15 2TT, UK
{kya020,J.E.Rowe}@cs.bham.ac.uk
http://www.cs.bham.ac.uk

Abstract. We evaluate the performance of estimating the number of local optima by estimating their proportion in the search space using simple random sampling (SRS). The performance of this method is compared against that of the jackknife method. The methods are used to estimate the number of optima in two landscapes of random instances of some combinatorial optimisation problems. SRS provides a cheap, unbiased and accurate estimate when the proportion is not exceedingly small. We discuss choices of confidence interval in the case of extremely small proportion. In such cases, the method more likely provides an upper bound to the number of optima and can be combined with other methods to obtain a better lower bound. We suggest that SRS should be the first choice for estimating the number of optima when no prior information is available about the landscape under study.

1 Introduction

Local search algorithms are widely used to find solutions to many optimisation problems either on their own or as a part of other metaheuristics. The neighbourhood operator they employ defines a structure over the search space; the properties of that structure can strongly influence their performance. One of these properties is the number of local optima, which combined with the additional knowledge of other properties such as the quality of the optima and the correlation between the basin size and fitness can give an indication of the structure difficulty. Nonetheless, knowing only the number of local optima can still provide some guidance in informing the choice of the neighbourhood operator. The knowledge of the number of local optima can also be used to study its growth behaviour, as the dimensionality increases, or across different values of problem parameters (e.g. phase transition control parameter). However, the number of local optima in a given instance is not known in advance and counting them is infeasible in most cases, apart from very small problem sizes. Therefore, the need for obtaining a statistical estimate of the number of local optima arises. Having an estimate of the total number of optima can also be helpful in commenting on the quality of the found local optima or the confidence that the global has been seen [17]. In the last two decades, a number of approaches have

J. Handl et al. (Eds.): PPSN XIV 2016, LNCS 9921, pp. 932–941, 2016.
DOI: 10.1007/978-3-319-45823-6_87

been proposed for estimating the number of local optima in combinatorial optimisation problems (COPs) [4,6,8,9,15–17]. Most of these methods start from a random sample of different configurations and apply local search to them until a local optimum is reached. Some of the methods are non-parametric estimators such as jackknife and bootstrap [6], while others assume some parametric distribution of the basin sizes (e.g. gamma distributions) [8,9]. However, each of these methods has its particular limitations and none of them provide a good estimate in all scenarios (e.g. when the basin sizes are different or when the number of optima is small). For example, the jackknife method [6] requires the sample size to increase as the number of optima increases, which is impractical since the number of optima grows exponentially or sub-exponentially with the problem size in most problems [13,15]. One drawback of the bootstrap method is its computational demands to carry out the re-samplings [6]. The approach proposed by [8] models the basin sizes using gamma distribution and requires an estimate of the parameter value of the distribution, which may not be practical. Another possible limitation of all the methods that apply local search to an initial random sample is the time needed to converge to a local optimum. In many cases, this time is linear or superlinear in problem size [15,21], but it can be exponential in other cases [5]. A review and an evaluation for several of these methods and others from the statistical literature can be found in [12].

The problem of estimating the number of local optima in COPs can be considered as the classical problem of estimating a population proportion in statistics. However, the use of this method to estimate the number of local optima is seldom found in the literature. It has been used to estimate number of optima in the multidimensional assignment problem [11], and in the quadratic assignment problem [19,20]. [4] mentioned the attractiveness of the simplicity and the unbiased estimate provided by this method, but they argued against it as the required sample size can be very large when the proportion is exceedingly small. They also criticised that in such a case, the method is more likely to provide an upper bound estimate rather than a lower bound one. [12] recommends using it only when all or most of the sampled optima have been seen once, after applying local search to an initial sample of points. This method is problem-independent and we argue that it is the best for estimating the number of local optima in terms of simplicity, accuracy and computational requirement when their proportion is large. As mentioned before, the required sample size for an accurate estimate increases as the proportion decreases, which makes obtaining an accurate estimate very expensive. However, an upper bound on the number of optima in such cases can still be obtained with reasonable sample sizes, giving some useful information about the studied landscapes. In the rest of this paper, we refer to estimating the number of local optima by estimating their proportion as simple random sampling (SRS). To provide a baseline, we compare the performance of SRS with the performance of the jackknife method. In Sect. 2, we introduce some preliminaries. In Sect. 3, we describe SRS and jackknife, and discuss different choices of confidence intervals for SRS. In Sect. 4 we describe the experimental settings and discuss the results.

2 Preliminaries

Search Space: The search space X is the finite set of all the candidate solutions. The fitness functions of all the studied problems in this paper are pseudo-Boolean functions, hence the search space size is $|X| = 2^n$.

Neighbourhood: A neighbourhood is a mapping $N : X \to P(X)$, that associates each solution with a set of candidate solutions, called neighbours, which can be reached by applying the neighbourhood operator once. The set of neighbours of x is called $N(x)$, and $x \notin N(x)$. We consider two different neighbourhood operators: the Hamming 1 operator (*H1*) and the 1+2 Hamming operator (*H1+2*). The neighbourhood of the *H1* operator is the set of points that are reached by a 1-bit flip mutation of the current solution x, hence the neighbourhood size is $|N(x)| = n$. The neighbourhood of the *H1+2* operator includes the Hamming one neighbours in addition to the Hamming two neighbours of the current solution x, which can be reached by a 2-bits flip mutation. The neighbourhood size for this operator is $|N(x)| = (n^2 + n)/2$.

Fitness Landscape: The fitness landscape of a combinatorial optimisation problem is a triple (X, N, f), where f is the objective function $f : X \to R$, X is the search space and N is the neighbourhood operator function [18].

Local Optima: We define a local minimum $x^* \in X$ as $f(y) > f(x^*)$ for all $y \in N(x^*)$. A local maximum is defined analogously. We use the term local optimum to denote either a local maximum or a local minimum. We refer to the actual number of optima in a given landscape as v.

Local Search: The local search strategy we use is the best improving move, stopping when a local optimum is reached.

Basin of Attraction: The basin of attraction $B(x^*)$ for an optimum $x^* \in X$ is the set of points that leads to it after applying local search to them, $B(x^*) = \{x \in X \mid \text{local-search}(x) = x^*\}$.

3 Estimation Methods

3.1 Simple Random Sampling

Suppose that a random sample of size s is taken from the search space, and that Y optima has been observed in the sample ($0 \leq Y \leq s$), p is the unknown proportion of the optima in the search space. Since the sample size is fixed, and the sampled configurations are independent and have a constant probability of being an optimum given by p, then Y has a Binomial distribution, $\mathrm{B}(s, p)$, with s trials and p success probability. The unbiased point estimate of the population proportion is given by $\hat{p} = Y/s$ and the estimated number of local optima can then be directly calculated by multiplying $\hat{p}$ by the search space size $S = |X|$. There are several methods for computing confidence interval estimates for p; the most referred ones are based on the approximation of the binomial distribution by the normal distribution [14]. A rule of thumb, that is frequently mentioned, is that the binomial distribution is suitable for approximation by the normal distribution as long as $sp \geq 5$ and $s(1-p) \geq 5$ [2,22]. The most widely used confidence interval for p is the standard Wald confidence interval ($\mathrm{CI_s}$) [2,14,22]:

$$\mathrm{CI_s} = \hat{p} \pm z_{\alpha/2}\sqrt{\frac{\hat{p}(1-\hat{p})}{s}} \tag{1}$$

where $z_{\alpha/2}$ is the z-score for $(1-\alpha)100\,\%$ confidence level and $z_{\alpha/2}\sqrt{\frac{\hat{p}(1-\hat{p})}{s}}$ is the error margin e. The error margin can be corrected for a finite population of size S to be equal to $e = z_{\alpha/2}\sqrt{\frac{\hat{p}(1-\hat{p})}{s}}\sqrt{\frac{S-s}{S-1}}$, where the value $\sqrt{\frac{S-s}{S-1}}$ is the finite population correction (fpc) factor [22]. The value of fpc is approximately one when S is large compared to s, and is obviously equal to zero when $s = S$. The sample size for a desired confidence level and a desired margin of error can be determined for an infinite population by:

$$s_0 = \frac{z^2_{\alpha/2}\hat{p}(1-\hat{p})}{e^2} \tag{2}$$

If no prior information about p or no initial estimate of $\hat{p}$ is available, then $\hat{p}$ can conservatively be set to 0.5 where the expression $\hat{p}(1-\hat{p})$ is maximised. This will ensure that the sample size is at its maximum for the desired e. However, the proportion of optima is typically much smaller than that, thus it might be more wise to set p to a smaller value and set e to a much smaller value. The sample size can be corrected for a finite population by the following formula:

$$s_1 = \frac{s_0 S}{s_0 + (S-1)} \tag{3}$$

From Eq. (2) we can see that the sample size does not depend on the population size but only on the desired confidence level, the desired margin of error, and the estimate of p. The behaviour of Wald interval is poor when p is close to 0 or 1, and when $Y = 0$ or $Y = s$, the length of the Wald interval is zero [1,2,14]. The exact Clopper-Pearson interval (exact in the sense of using the binomial distribution rather than the approximation by the normal distribution) is an alternative method to consider in such cases. However, and because of the inherent conservativeness of exact methods, other approximate methods are more useful [1]. The *Agresti-Coull* confidence interval ($\mathrm{CI_{AC}}$) is recommended for correcting the Wald interval. It recentres the Wald interval by adding the value $z^2_{\alpha/2}/2$ to Y so it becomes $\tilde{Y} = Y + z^2_{\alpha/2}/2$ and adding the value $z^2_{\alpha/2}$ to s to become $\tilde{s} = s + z^2_{\alpha/2}$. When the z-score for the 95 % confidence level ($z^2_{0.05/2} = 1.96$) is approximated to 2, the *Agresti-Coull* interval is equivalent to adding two successes and two failures to the sample [1,2]. The corrected point estimate is $\tilde{p} = \tilde{Y}/\tilde{s}$ and the confidence interval is given by:

$$\mathrm{CI_{AC}} = \tilde{p} \pm z_{\alpha/2}\sqrt{\frac{\tilde{p}(1-\tilde{p})}{\tilde{s}}} \tag{4}$$

Using *Agresti-Coull* confidence interval, the SRS estimation of the number of local optima is given by:

$$\hat{v}^{SRS} = \tilde{p}S \tag{5}$$

3.2 Jackknife

Jackknife is a non-parametric method based on the idea of re-sampling to reduce the bias of the estimate. The use of jackknife to estimate the number of local optima was first proposed by [6]. We selected the jackknife method as a comparison baseline for two reasons: jackknife has an attractive simple and fast closed-form computation, and it is recommend to be used when the size of the sample is adequate with respect to v [6,12]. Starting from s different randomly sampled configurations and after applying local search to each one of them, the jackknife estimate of the number of local optima is given by:

$$\hat{v}^{JK} = \beta + \frac{s-1}{s}\beta_1 \tag{6}$$

where β_1 is the number of optima that have been seen once and $\beta = \sum_{i=1}^{r} \beta_i$ is the number of distinct optima seen. Note that this is a special case of the jackknife estimator where one point is left out of the original sample s at a time. A generalised estimator that considers leaving out $1, \ldots, 5$ points at a time can be found in [3]. As pointed out by [17], the choice of the most suitable number of points to leave out in order to achieve a better estimate is problem-dependent.

4 Experiments

We obtain statistical estimates of the number of optima in randomly generated instances of the number partitioning problem and the 0–1 knapsack problem. The aim of the experiments is twofold: compare the estimates of SRS with that of jackknife, and examine the effect of the sample size on the accuracy of the SRS estimate. We compare the performance of the two methods using two sample sizes to allow for a fair comparison, since SRS uses at most $s(|N(x)|+1)$ number of fitness evaluations compared to $s(|N(x)|+1)+t|N(x)|$ fitness evaluations used by jackknife, where t is the total number of steps taken when descending(ascending) from each initial configuration. We describe the settings of the two sample sizes in more details in the results subsection.

4.1 Combinatorial Optimization Problems

Number Partitioning Problem (NPP). Given a set $W = \{w_1, \ldots, w_n\}$ of m-bit positive integers (weights) drawn at random from the set $\{1, 2, \ldots, M\}$ with $M = 2^m$, the goal is to partition W into two disjoint subsets S, S' such that the discrepancy between them $|\sum_{w_i \in S} w_i - \sum_{w_i \in S'} w_i|$ is minimised. The instances we study have weights drawn from a uniform distribution and $m = n$.

When the weights are drawn from a uniform distribution, the theoretical average proportion of the local optima in the *H1* landscape is given by the following formula that was obtained using statistical mechanics analysis [7]:

$$\langle p \rangle^{\mathrm{NPP}} = \sqrt{\frac{24}{\pi}} n^{-3/2} \tag{7}$$

0-1 Knapsack Problem (0-1KP) is defined as follows: given a knapsack of capacity C and a set of n items each with associated weight w_i and profit p_i, the aim is to find a subset of items that maximises $f(x) = \sum_{i=1}^{n} x_i p_i$, subject to $\sum_{i=1}^{n} x_i w_i \leq C, where\ x \in \{0,1\}^n,\ C = \lambda \sum_{i=1}^{n} w_i,\ and\ 0 \leq \lambda \leq 1$. Infeasible solutions that violate the given constraint are penalised by subtracting this value from the fitness function: $\text{Pen}(x) = \rho \left(\sum_{i=1}^{n} x_i w_i - C\right) + \sum_{i=1}^{n} p_i$, where $\rho = \max_{i=1,\ldots,n} \{p_i\} / \min_{i=1,\ldots,n} \{w_i\}$. The weights of the instances studied in this paper are drawn from a discretised normal distribution $\mathcal{N}(2^{n-1}, \frac{2^n}{10})$.

4.2 Results

The mean estimates of v in the two landscape of the 0-1KP is shown as n grows in Fig. 1 (note that some data points lie on top of each other). The estimates were obtained by the jackknife and SRS, and were averaged over 10 samples for each sample size. The sample sizes are set as follows: first we obtained the sample size s for each n from Eqs. (2) and (3) by setting $e = 0.005$, $\hat{p} = 0.3$ and $z_{\alpha/2} = 2.576$. Note that the sample size, only changes slightly as n increases, starting from $s = 45,701$ when $n = 18$, until it reaches $s = 55,351$ when $n = 100$. After obtaining s, we then set the small sample size of SRS to s and the small sample size of jackknife to $s - t + t/(|N(x)| + 1)$ (i.e. we subtract the fitness evaluations used when ascending from the sample budget). We set the large sample size of jackknife to s and the large sample size of SRS to $s+t-t/(|N(x)|+1)$, where t is the total number of steps taken by jackknife with the large sample. The samples are drawn without replacement for $n \leq 24$. The figure shows that SRS using both small and large sample sizes accurately estimates the real proportions in both landscapes, apart from $n = 100$ in the *H1+2* landscape. The discrepancy between estimates of the large and small samples in this case, in addition to the larger standard deviations, indicate that the proportion is small and that the sample size, in particular the small one is probably inadequate. As for the jackknife, both sample sizes quickly become inadequate as the number of optima

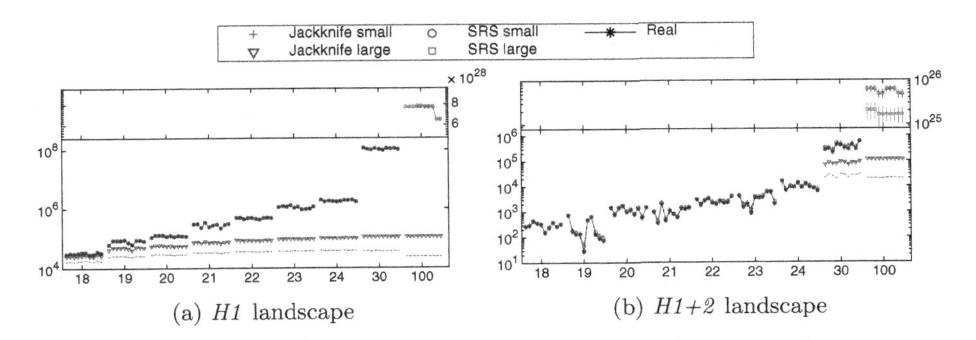

(a) *H1* landscape

(b) *H1+2* landscape

Fig. 1. SRS and Jackknife estimates of the optima number (in log scale) as the problem size grows. Each data point represents the average estimate of 10 samples from a single instance of 0-1KP. The error bars show the standard deviations.

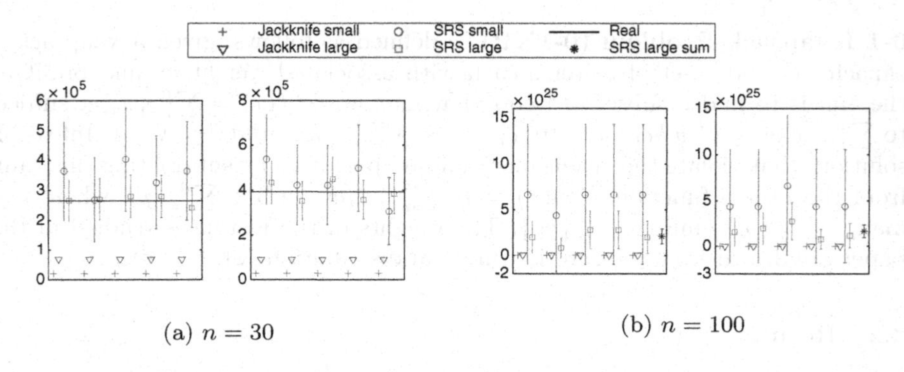

(a) $n = 30$ (b) $n = 100$

Fig. 2. Each figure shows the estimates of the number of optima in a single instance of 0-1KP, and each data point shows the estimate of a single sample. The error bars around SRS estimates are the 95 % $\mathrm{CI_{AC}}$.

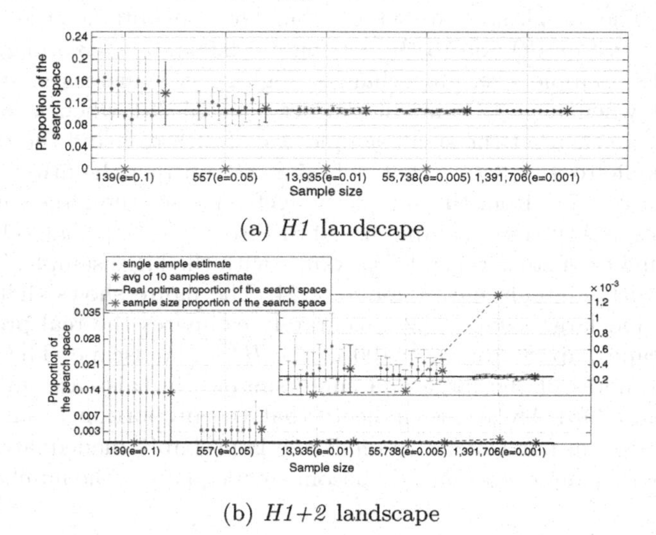

(a) *H1* landscape

(b) *H1+2* landscape

Fig. 3. SRS estimates of the optima proportion versus s. The sample sizes are obtained from Eqs. (2) and (3) by setting $\hat{p} = 0.3$ and $z_{\alpha/2} = 2.576$ (corresponding to 99 % confidence level). The results are for a single instance of 0-1KP of size $n = 30$. The error bars are the 95 % $\mathrm{CI_{AC}}$.

seen once quickly grows with n until all the optima that have been seen were only seen once. Thus, the method fails to provide accurate estimates and grossly underestimates v. This is more noticeable in the *H1* landscape where v is large. The $\mathrm{CI_{AC}}$ of SRS estimates are very narrow in *H1* landscape across all n, but they get wider as n increases in the *H1+2* landscape. In Fig. 2, we look closely at the results of four instances of size $n = 30, 100$ from Fig. 1. The figure shows the confidence interval around 5 estimates of each method with each sample size. The width of the $\mathrm{CI_{AC}}$ decreased with the large sample size as expected.

Table 1. NPP sample sizes

	n	24	30	100	1000
s	$e = \langle p \rangle^{\text{NPP}}$	276	388	2,395	75,915
	$e = \frac{\langle p \rangle^{\text{NPP}}}{5}$	6,889	9,697	59,855	1,897,856
	$e = \frac{\langle p \rangle^{\text{NPP}}}{10}$	27,520	38,785	239,420	7,591,421

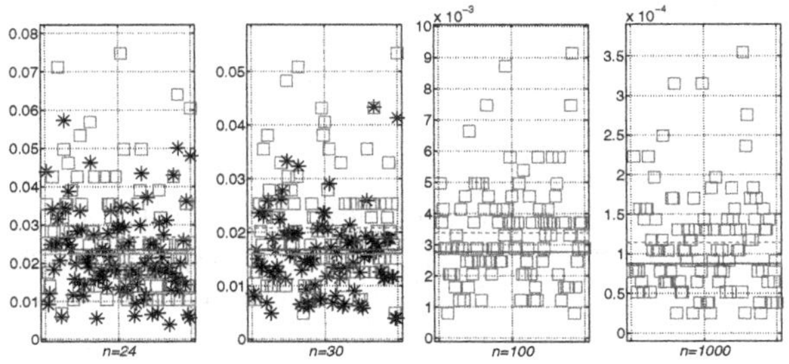

(a) $e = \langle p \rangle^{\text{NPP}}$

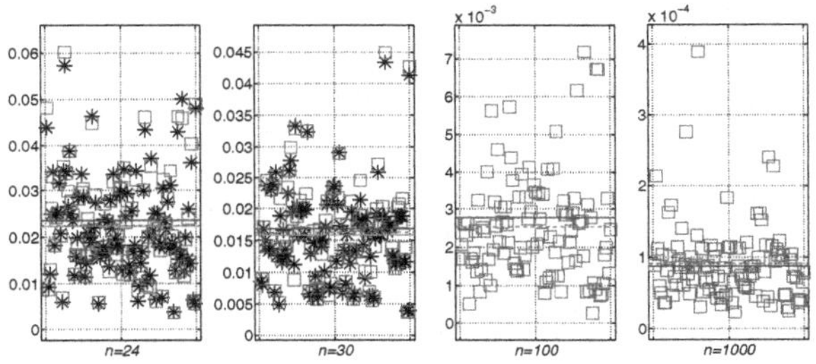

(b) $e = \frac{\langle p \rangle^{\text{NPP}}}{5}$

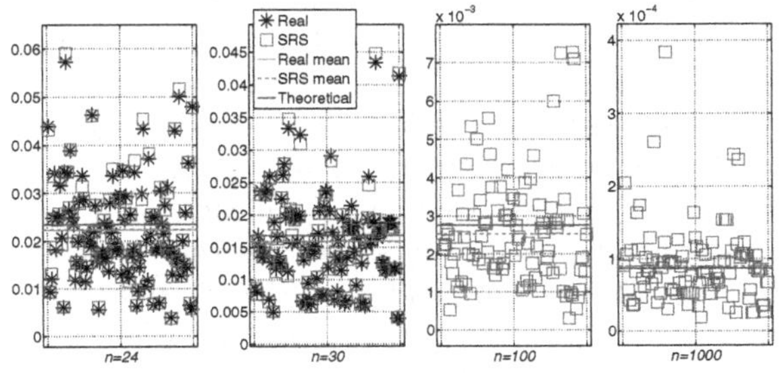

(c) $e = \frac{\langle p \rangle^{\text{NPP}}}{10}$

Fig. 4. Optima proportion in the *H1* landscape of NPP for different vales of n. SRS estimates are shown when the sample size is obtained with 3 different desired error margins e (shown in Table 1). The results are for 100 random instances for each n. Obtaining the real proportion was only computationally feasible for $n = 24, 30$. The theoretical mean proportions are obtained from Eq. (7).

The SRS large sample size for $n = 30$ is around 2×10^5 and around 3×10^5 for $n = 100$. Obtaining the real number of optima was infeasible for $n = 100$ (note that methods that exploit some knowledge of f can obtain v of larger n than that feasible by exhaustive search of X [10]), therefore we show the estimate of SRS with a larger sample size by setting Y to the sum of the number of optima found in all the large samples and s to the sum of the large sample sizes. The outcome $\hat{v}^{SRS}$ of both instances are around 10^{-5}. The very wide $\mathrm{CI_{AC}}$ with negative lower bounds around the small sample size estimates of SRS in $n = 100$ indicate that the proportion is much smaller than what SRS can precisely estimate with this sample size. In such a case, the $\hat{v}^{SRS}$ more likely provides an upper bound to v. However, we suggest combining the results of the two methods in such cases by using the result of the jackknife method for a better lower bound than just zero.

Figures 3 and 4 show how the accuracy of SRS estimates increases as the desired error margin e decreases. Decreasing e consequently increases s. The figures also show how SRS is able to accurately estimate the fraction of v with relatively small s. As we mentioned before, the required s does not directly depend on n, but since the fraction of v usually declines as n grows [7], the required s will increase with n as shown in Table 1. The values of s in Table 1 are obtained from Eqs. (2) and (3) by setting $\hat{p} = \langle p \rangle^{\mathrm{NPP}}$ (obtained from Eq. (7)), $z_{\alpha/2} = 2.576$ and e as shown in the table. In both problems and in both landscapes, most of the basin sizes are small and only very few ones are large.

5 Conclusions

Simple random sampling with the $\mathrm{CI_{AC}}$ provides a simple way to obtain an unbiased statistical estimate of the number of local optima. The accuracy of the obtained estimate depends on the sample size s, which can be determined for a desired margin of error e. A negative lower bound of the $\mathrm{CI_{AC}}$ usually indicates that the proportion is smaller than the desired e. In such a case, s can be increased considering that it only costs at most $|N(x)| + 1$ fitness evaluations per configuration. This is practical as long as the proportion is not exceedingly small. Alternatively, the estimate of SRS can be used as an upper bound as it is more likely to provide an overestimate in such cases. It can be combined with the estimate of another method that applies local search to an initial sample for a lower bound other than zero (since these methods usually tend to provide an underestimate [12]). We recommend that SRS should be the first method to use for estimating the number of optima, especially when no prior information is available about the problem being studied.

References

1. Agresti, A., Coull, B.A.: Approximate is better than "exact" for interval estimation of binomial proportions. Am. Statistician **52**(2), 119–126 (1998)
2. Brown, L.D., Cai, T.T., DasGupta, A.: Interval estimation for a binomial proportion. Stat. Sci. **16**(2), 101–117 (2001)

3. Burnham, K.P., Overton, W.S.: Estimation of the size of a closed population when capture probabilities vary among animals. Biometrika **65**(3), 625–633 (1978)
4. Caruana, R., Mullin, M.: Estimating the number of local minima in big, nasty search spaces. In: Proceedings of IJCAI-1999 Workshop on Statistical Machine Learning for Large-Scale Optimization (1999)
5. Englert, M., Röglin, H., Vöcking, B.: Worst case and probabilistic analysis of the 2-Opt algorithm for the TSP. Algorithmica **68**(1), 190–264 (2013)
6. Eremeev, A.V., Reeves, C.R.: Non-parametric estimation of properties of combinatorial landscapes. In: Cagnoni, S., Gottlieb, J., Hart, E., Middendorf, M., Raidl, G.R. (eds.) EvoIASP 2002, EvoWorkshops 2002, EvoSTIM 2002, EvoCOP 2002, and EvoPlan 2002. LNCS, vol. 2279, pp. 31–40. Springer, Heidelberg (2002)
7. Ferreira, F.F., Fontanari, J.F.: Probabilistic analysis of the number partitioning problem. J. Phys. A: Math. Gen. **31**(15), 3417 (1998)
8. Garnier, J., Kallel, L.: How to detect all maxima of a function. In: Kallel, L., Naudts, B., Rogers, A. (eds.) Theoretical Aspects of Evolutionary Computing, pp. 343–370. Springer, Heidelberg (2001)
9. Garnier, J., Kallel, L.: Efficiency of local search with multiple local optima. SIAM J. Discret. Math. **15**(1), 122–141 (2002)
10. Goldman, B.W., Punch, W.F.: Hyperplane elimination for quickly enumerating local optima. In: Chicano, F., et al. (eds.) EvoCOP 2016. LNCS, vol. 9595, pp. 154–169. Springer, Heidelberg (2016)
11. Grundel, D.A., Krokhmal, P.A., Oliveira, C.A.S., Pardalos, P.M.: On the number of local minima for the multidimensional assignment problem. J. Comb. Optim. **13**(1), 1–18 (2007)
12. Hernando, L., Mendiburu, A., Lozano, J.A.: An evaluation of methods for estimating the number of local optima in combinatorial optimization problems. Evol. Comput. **21**(4), 625–658 (2013)
13. Mathias, K.E., Whitley, L.D.: Transforming the search space with gray coding. In: IEEE WCCI, pp. 513–518, vol. 1 (1994)
14. Pires, A.M., Amado, C.: Interval estimators for a binomial proportion: comparison of twenty methods. REVSTAT-Stat. J. **6**(2), 165–197 (2008)
15. Prügel-Bennett, A., Tayarani-N, M.-H.: Maximum satisfiability: anatomy of the fitness landscape for a hard combinatorial optimization problem. IEEE Trans. Evol. Comput. **16**(3), 319–338 (2012)
16. Reeves, C.R.: Direct statistical estimation of GA landscape properties. Found. Genet. Algorithms **6**, 91–107 (2001)
17. Reeves, C.R., Eremeev, A.V.: Statistical analysis of local search landscapes. J. Oper. Res. Soc. **55**(7), 687–693 (2004)
18. Stadler, P.F., Stephens, C.R.: Landscapes and effective fitness. Comments Theor. Biol. **8**(4–5), 389–431 (2002)
19. Tayarani-N, M.-H., Prügel-Bennett, A.: On the landscape of combinatorial optimization problems. IEEE Trans. Evol. Comput. **18**(3), 420–434 (2014)
20. Tayarani-N, M.-H., Prügel-Bennett, A.: Quadratic assignment problem: a landscape analysis. Evol. Intell. **8**(4), 165–184 (2015)
21. Tovey, C.A.: Hill climbing with multiple local optima. SIAM J. Algebraic Discrete Methods **6**(3), 384–393 (1985)
22. Triola, M.F.: Elementary Statistics, 12th edn. Pearson, Upper Saddle River (2012)

evoVision3D: A Multiscale Visualization of Evolutionary Histories

Justin J. Kelly(✉) and Christian Jacob(✉)

Department of Computer Science, University of Calgary,
2500 University Dr NW, Calgary, AB T2N 1N4, Canada
{kellyjj,cjacob}@ucalgary.ca
http://www.ucalgary.ca

Abstract. Evolutionary computation is a field defined by large data sets and complex relationships. Because of this complexity it can be difficult to identify trends and patterns that can help improve future projects and drive experimentation. To address this we present *evoVision3D*, a multiscale 3D system designed to take data sets from evolutionary design experiments and visualize them in order to assist in their inspection and analysis. Our system is implemented in the Unity 3D game development environment, for which we show that it lends itself to immersive navigation through large data sets, going even beyond evolution-based search and interactive data exploration.

Keywords: Evolutionary computation · Multiscale · Visualization · Game engine

1 Introduction

It is said that history is the greatest teacher. Sometimes in order to move forward one must review past decisions and choices in order to identify common trends and patterns to predict future outcomes. This historical evaluation is especially valuable in interactive evolutionary algorithms [6] and genetic programming [12], where users review past experiments and trends in order to improve and refine their selection algorithms and fitness evaluations. However, evolutionary systems often produce very large data sets filled with complex relationships, making it difficult for a human to effectively process. Additionally, there are times when a system's requirements can suddenly change, rendering previous evaluations insufficient and forcing the user to begin their review from scratch. To address these issues we present *evoVision3D*, a multi-level visualization environment, displaying complex evolutionary data in a 3-dimensional, immersive scene (Fig. 1). In this paper we will explore *evoVision3D*'s features and how we have expanded upon *evoVersion*, an evolutionary data tracking and synchronization tool, we have developed earlier [10].

J. Handl et al. (Eds.): PPSN XIV 2016, LNCS 9921, pp. 942–951, 2016.
DOI: 10.1007/978-3-319-45823-6_88

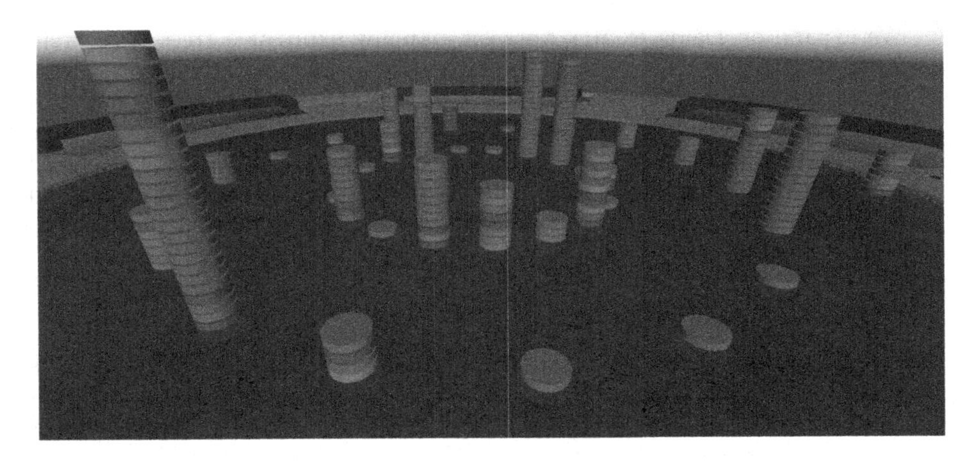

Fig. 1. Example of an evolutionary design workspace in 3D virtual reality using *evoVision3D*: columns represent experiments, disks denote populations, whose colours depict average fitness.

2 Related Work

With the advent of highly capable video game development environments, such as Unity 3D [4], it has become more and more common to use game engines for scientific research and visualization [15]. Taking advantage of advanced visualisation libraries and built-in physics engines, there is substantial opportunity for their integration into professional research. In this paper, we present related work in the areas of (1) evolutionary visualization, (2) VR technology and (3) *evoVersion*, one of our previous systems implemented in the Unity 3D Game Engine.

2.1 Evolutionary Visualization

Building upon the foundation laid by *evoVersion*, we draw inspiration from previous works. We combine a node-ring graph visualization [9] with a multiscale visualization model [14] to display data efficiently and with a dense arrangement of visual information without becoming overwhelming. We organize each session into a set of discrete generations represented by a series of stacked rings (Fig. 1).

We have drawn inspiration from the *EvoShelf* system, which applies techniques normally found in photo management software, organizing evolutionary data in a manner reminiscent to programs such as iTunes^{TM} [8]. We use a similar modular design, providing a flexible and plug-in friendly environment. As we will demonstrate, the straightforward presentation of data makes searching through larger populations smoother and less cumbersome. *evoVision3D* differs from *EvoShelf* due to our use of 3D visualization rather than 2D with respect to result presentation and navigation (see Sect. 3.1). With *evoVision3D* we provide a tool to coordinate collaborative development among multiple users, rather than just one unsynchronized account.

We expand on Daida et al.'s work on mapping expression trees to a circular 2D grid, which provides a simple visualization, facilitating the identification of trends and patterns across a genetic programming session [7]. In comparison, *evoVision3D* enables data inspection, filtering, and analysis across multiple experiments. A visual analytics interface for evolutionary data has been discussed in [13]. More traditional 2D scatterplots are used to inspect and categorization data. In contrast, *evoVision3D* expands the data presentation to 3D and across multiple evolutionary sessions.

2.2 VR Technology and 3D Game Engines

Following *evoVersion* [10] and Shepherd's genome browser [15], *evoVision3D* is built using the Unity 3D game engine [4]. With the recent increase in public availability for professional-quality game engines and their active developer communities, these engines have proven to be an extremely valuable asset in the development of visualization systems. A notable example of this is Unity's built-in support for virtual reality systems such as the Oculus Rift [3], allowing for easy integration of these systems into immersive data display solutions.

2.3 *evoVersion*

evoVersion is a system designed to collect, store and visualize interactive evolutionary data. *evoVersion* utilizes the iterative storage methodology of software version control systems such as Git [2] and Subversion [1] and applies it to evolutionary computation in order to record, organize and analyze the resulting data. It consists of three primary components: interactive selection, data storage and basic 3D visualization. The interactive selection component handles user-driven evaluation and evolution of the phenotype population. The data storage component records all iterations of the population on a remote SQL server. The visualization component takes the data stored in the database and visualizes histories of evolutionary designs in a column-based format (Fig. 2). *evoVision3D* builds upon this system and focuses on improving the functionality and performance of the visualization component with regard to the existing data collection and storage mechanics while using the data sets produced by *evoVersion* as the primary data source.

3 The *evoVision3D* System

evoVision3D seeks to build upon the visualization scheme seen in *evoVersion* and provide the user with intuitive and efficient means of visualizing complex evolutionary data sets from various evolutionary experiments. To achieve this *evoVision3D* combines *evoVersion*'s data arrangement with an additional set of features in order to further assist the user as they examine the data visualization space. These features include: (1) a spatial arrangement of the visual data representations, (2) a multi-level abstraction of data, (3) genealogy tracing for specific elements, (4) similarity filtering, and (5) a set of dynamic interface panels summarizing key details and statistics of a given element.

3.1 Data Arrangement

In *evoVision3D*, each of a user's evolutionary experiments are treated as a distinct event with a series of discrete populations arranged in ascending historical order. Each evolutionary experiment, or session, is represented by a single vertical column (Fig. 1). The height of this column reflects the number of generations during that experiment, allowing the viewer to easily determine which sessions were the most active. Session columns are arranged in a spiral pattern, growing from previous sessions towards the center to the newest around the outer edge.

3.2 Multiscale Abstraction

A common problem encountered when visualizing these kinds of data sets in 3D is the limits of computer memory and rendering capabilities, making it impractical to fully render each individual phenotype at once when dealing with larger data sets. To address this challenge, *evoVision3D* utilizes multiscale abstraction of the data sets to reduce the computational overhead incurred during the visualization, improving both load times and frame rate significantly. Similar to [5], *evoVision3D* uses multiple levels of visualization. Each level differs in terms of breadth and detail of the data portrayed in order to collect both general and specific details with regards to evolutionary design histories.

Based on a hierarchal storage structure, our system currently operates on multiple levels of detail: sessional, generational, and individual (Fig. 2). The sessional level data is represented as a series of stacked disks arranged to form a column. Each of these disks represents a single generation of that session. Disks are arranged in ascending order of creation, placing the first generation at the bottom and the most recent generation at the top. The color of the disk denotes the average fitness of the entire population at that point in time during the experiment. Each color lies along a linear gradient between red and green, where

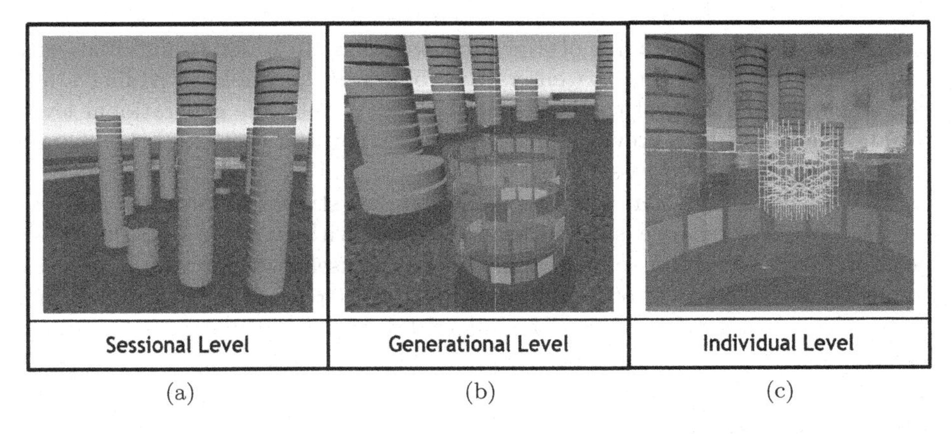

Fig. 2. The three levels of scale used in *evoVision3D* in descending order of detail. Automatic scale transition is triggered by the user approaching a specific object. (Color figure online)

red denotes a fitness of 0 (worst rating) and green represents a fitness of 100 (best rating). Alternatively, an object colored gray either has yet to be evaluated or has had its coloring toggled off by the user. By observing the color of each disk within the column the user can get a general feel for the quality of a session (Fig. 1).

At the generational level (Fig. 2b), each disk allows the user to view a series of nodes arranged in a ring within the disk. Each node, depicted as a distinct 3D object, represents a single member of a population. The color of each node represents the specific fitness of its corresponding element. This gives a more detailed breakdown of a generation's population without the need to render each individual phenotype, while also providing a quick visual summary of their fitness ratings.

At the individual level (Fig. 2c) the system renders the individual phenotypes of the elements within a generation. The nodes from the generational level lose their transparency and a representation of that element's phenotype is rendered inside the node. This presumes that a visual representation is available for each element. This allows the user to see a depiction of the element in combination with its fitness, represented by the color hue.

Each of these levels are rendered dynamically on demand. This keeps the memory overhead for the system minimal, while also reducing the amount of content loaded when the visualization engine initializes. This allows the system to maintain a high degree of efficiency even when rendering large data sets. The transition between each level of detail can be both manually and automatically triggered as needed. Automatic transitions are triggered based on the user's position relative to the session columns in the scene. A generation disk enters the generational level of detail when the distance between their position and the center point of a given disk is less than a user-defined value (Fig. 3). When this distance once again becomes greater than this user-set value the disk will return to its previous sessional level of detail (an opaque colored disk). The individual level of detail is triggered when the distance between the user and a disk is less than the radius of a generation's disk. In order to automatically trigger this transition the user enters the column in question, providing a 360 degree panoramic view of the local population. As with the generational level, the disk returns to its previous level of detail when the user exits the column space of that particular session.

Manual transitions can be invoked through key strokes at any time and will set the entire scene to a specific level of detail without regard to the user's position in the scene. These manual modes can be useful when trying to identify trends at a certain level of detail, allowing the user to navigate through and inspect the visualization space. This allows the system to maintain a minimal amount of wait time to load each scene.

3.3 Genealogy Tracing

One of the key aspects to evolutionary systems is their application of iterative development. New elements are derived from pre-existing elements through a

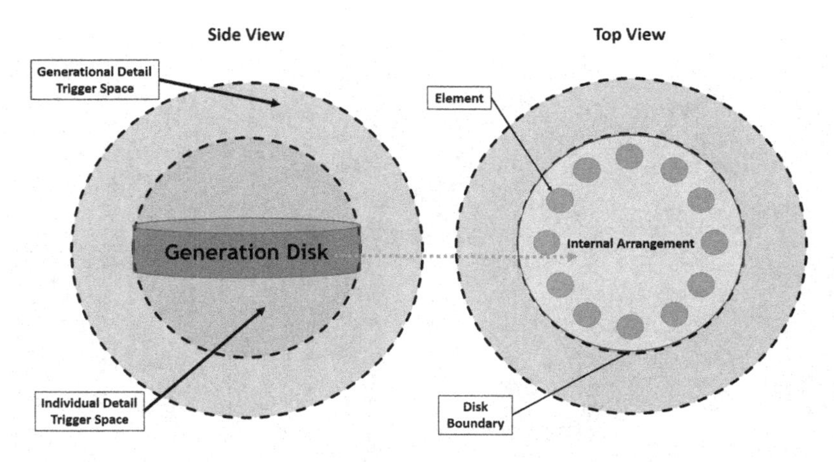

Fig. 3. An illustration of the conceptual boundaries used to trigger scale transitions in the visualization space. As the user approaches a generation disk its visual models become more and more detailed.

combination of crossover and mutation operations. Mutation is the process of applying a random modification to an existing genotype, while crossover is the act of combining two or more genotypes to produce a new child that shares a part of each parent's genotype combined together [10]. It is therefore quite valuable to maintain an understanding of an element's ancestry in order to help identify trends and patterns created through inheritance. In *evoVision3D* one can trace through a targeted element's genealogical history in order to identify and visualize the relationship it shares with its ancestors (Fig. 4). Serving as a filter, the genealogy trace removes all objects not related to the selected element from the scene and sets all remaining nodes to the individual detail level. The system then procedurally generates a series of line segments, where each line represents the relationship between a parent and its children. These connections produce a 3-dimensional family tree for the selected element, allowing the user to observe the genetic changes that culminated in the production of the target element. This operation uses a breadth-first expansion down the generations, allowing the user to examine the connections from more recent generations to older generations.

3.4 Similarity Filtering

Similarity filtering allows the user to identify what sections of a user's experiment set occupy the same genotype neighborhood. It allows the user to select an element and calculate its similarity to all other elements present in the scene. All elements whose similarity falls below a user-set threshold are filtered out, leaving only those individuals that have significant similarities to the selected element (Fig. 5). The colors of these individual nodes and the encompassing generational disks are changed from visualizing fitness to instead reflect the

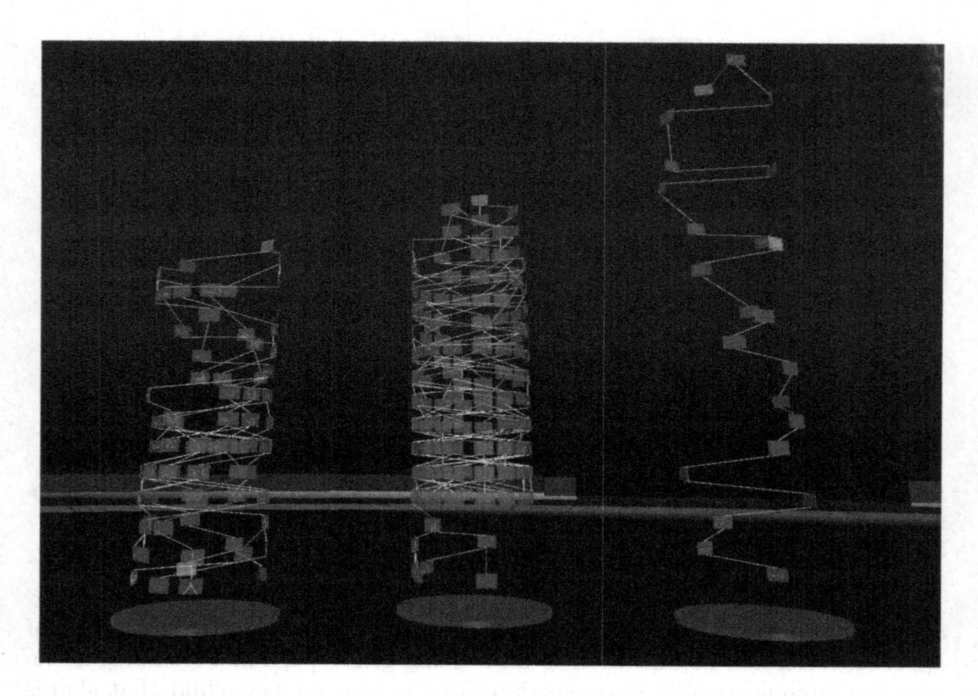

Fig. 4. Visualizing an element's genealogy: line segments illustrate the relationship between parent and child (from top to bottom in each column). Cyan lines indicate a mutation while yellow lines denote a crossover relationship, allowing for quick identification of development patterns within a session. The left column shows a mix of mutation and crossover, the middle session evolved mostly by crossovers, whereas only mutations created the individuals in the session on the right. (Color figure online)

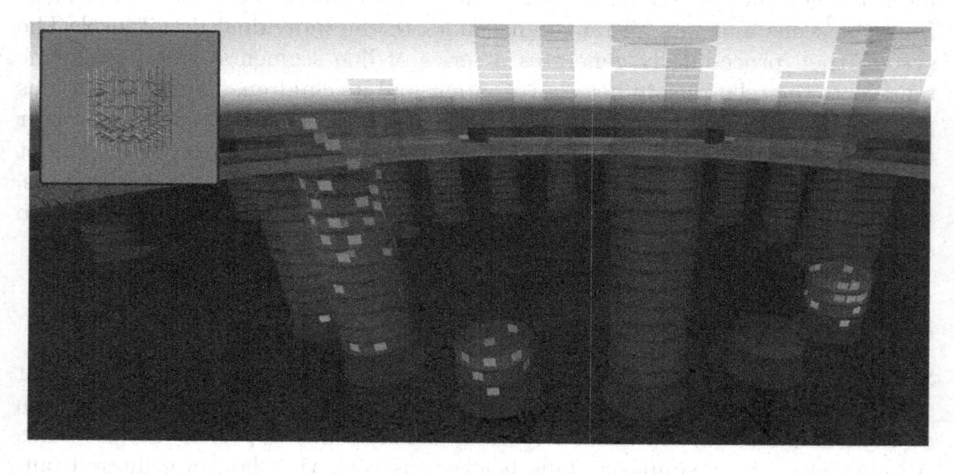

Fig. 5. Filtering the scene based on genetic similarity. The chosen phenotype is presented in the top left of the screen and only the nodes that have reached a user-defined degree of similarity are rendered in the scene. (Color figure online)

degree of similarity to the chosen element. Green denotes high similarity, while red denotes low similarity.

3.5 Dynamic Summary Panel

The multiscale representation and color encoding provide an effective summary of an element's phenotype and fitness. *evoVision3D* supplements this by providing three dynamic interface panels with a summary of an object's data and statistics. This function dynamically loads the data of any selected object in the scene, producing a summary of the underlying data, a close up view of its phenotype (if applicable) and a comparison of its fitness compared to all other objects in the scene (Fig. 6).

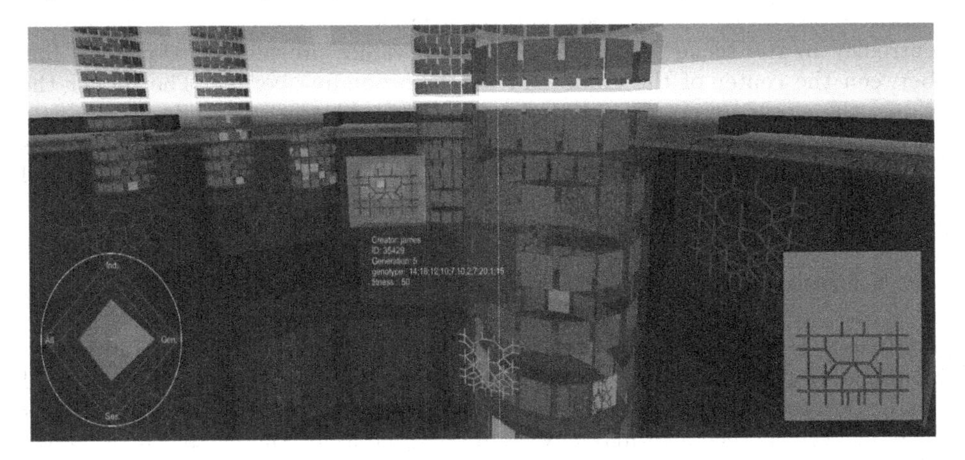

Fig. 6. Panel overlay. By selecting an object in the visualization space a summary view of key information is displayed, including summary data (center), phenotype viewer (to the right) and fitness statistics (to the left).

Data Panel. The data panel serves as a summary of an item's data as stored in the SQL database. Appearing next to the user's cursor, it consists of a translucent back panel and a textual output of key information. The information changes depending on the detail level. At the sessional level, a generation's disk displays high level information such as the population size at that generation and the ID number of the associated session. Alternatively, at the individual and generational levels the panel instead displays information for the now visible individual.

Phenotype Viewer. Displayed in the bottom right corner of the screen (Fig. 6), this panel allows the user to quickly check an element's phenotype without having to trigger the individual level of detail. This is useful for inspecting individual elements while operating on the generational level of detail, without the need to reposition the camera while still allowing the user to rotate and magnify the viewer space. This can assist in the identification of patterns shared by elements such as, for example, a certain fitness range via manual observation and selection.

Radial Glyph. Expanding on *evoVersion*'s radial fitness rings [10], we added a glyph display contained inside a circular boundary, depicted in the bottom left corner (Fig. 6) This panel consists of two primary components: a measurement guide and a data map. The measurement guide is a translucent blue diamond that serves as a form of relative measurement for the given data. Each of the four points on this diamond represents a specific attribute. Starting from the topmost point and moving clockwise these points measure the fitness of the current target (Ind), the average fitness of the current target's generation (Gen), the average fitness of the target session (Ses) and the average fitness for every single session currently in the scene (All). The respective values are plotted along the line between the center of the guide and its corresponding corner. The higher the value, the closer its point is plotted to the outer edge of the guide. The points are connected to form an irregular n-sided shape that, for example, represents the target's fitness rating compared to other elements in the scene. This model of representation can easily be expanded to include more than four values.

4 Conclusion and Future Work

evoVision3D is a promising avenue for visualizing and exploring evolutionary histories. Implemented in the Unity game engine [4], it provides a robust suite of navigation and filtering tools for evolutionary data inspection and analysis. This level of performance also makes VR support viable for the system as a whole, allowing for a new avenue of immersive visualization to be explored. At this point our future work is threefold: the expansion of existing features, integration of gesture control to support immersive VR interaction and extending use of the system to more complex evolutionary systems in order to test its viability in the context of dense evolutionary data visualization [11] and collaborative coevolution [16]. We plan to expand upon the current color encoding and radial glyph graphs used to analyze and compare the fitnesses of individual elements and generations. We are also working on a flexible search tool capable of visualizing the similarity between population elements in terms of both genotype and phenotype features to assist in identifying commonalities and trends within the data set. The addition of gesture control to supplement peripheral systems such as the Oculus Rift Virtual Reality Headset [3] makes navigation through the virtual space more natural and intuitive, providing a logical alternative to the traditional mouse and keyboard interaction. Finally, in order to further reduce the computational overhead and allow for rendering of even larger scenes we plan to apply a dynamic octree implementation similar to that seen in Shepherd's genome exploration system [15] to further improve performance.

References

1. Apache subversion; enterprise-class centralized version control for the masses. https://subversion.apache.org/
2. git -local-branching-on-the-cheap. https://git-scm.com/
3. Oculus. https://www.oculus.com/en-us/
4. Unity 3d game engine. https://unity3d.com/
5. Stolte, C., Tang, D., Hanrahan, P.: Multiscale visualization using data cubes. IEEE Trans. Vis. Comput. Graph. **9**(2), 176–187 (2003)
6. Coello, C.A.C., Van Veldhuizen, D.A., Lamont, G.B.: Evolutionary algorithms for solving multi-objective problems, vol. 242. Springer, New York (2002)
7. Daida, J.M., Hilss, A.M., Ward, D.J., Long, S.L.: Visualizing tree structures in genetic programming. Genet. Programm. Evolvable Mach. **6**(1), 79–110 (2005)
8. Davison, T., von Mammen, S., Jacob, C.: *EvoShelf*: a system for managing and exploring evolutionary data. In: Schaefer, R., Cotta, C., Kołodziej, J., Rudolph, G. (eds.) PPSN XI. LNCS, vol. 6239, pp. 310–319. Springer, Heidelberg (2010)
9. Etemad, K., Carpendale, S., Samavati, F.: Node-ring graph visualization clears edge congestion. In: Proceedings of the IEEE VIS Arts Program (VISAP), pp. 67–74
10. Kelly, J., Jacob, C.: evoVersion: visualizing evolutionary histories. In: IEEE Congress on Evolutionary Computation, CEC 2016. IEEE (2016) (in print)
11. Koçer, B., Arslan, A.: Transfer Learning in Genetic Algorithms (2012)
12. Koza, J.R.: Genetic Programming: On The Programming of Computers by Means of Natural Selection, vol. 1. MIT Press, Cambridge (1992)
13. Lutton, E., Fekete, J.D.: Visual analytics of EA data. In: Proceedings of the 13th Annual Conference on Genetic And Evolutionary Computation - GECCO 2011, pp. 145–146 (2011)
14. Miller, R., Mozhayskiy, V., Tagkopoulos, L., Ma, K.L.: EVEVis: a multi-scale visualization system for dense evolutionary data. In: 2011 IEEE Symposium on Biological Data Visualization (BioVis), pp. 143–150 (2011)
15. Shepherd, J.J., Zhou, L., Zhang, Y., Zheng, J., Tang, J.: Exploring genomes with a game engine. In: Proceedings - 2013 IEEE International Conference on Bioinformatics and Biomedicine, IEEE BIBM 2013, vol. 169(207890), pp. 26–30 (2013)
16. Yang, Z., Tang, K., Yao, X.: Large scale evolutionary optimization using cooperative coevolution. Inf. Sci. **178**(15), 2985–2999 (2008)

Landscape Features for Computationally Expensive Evaluation Functions: Revisiting the Problem of Noise

Eric O. Scott(✉) and Kenneth A. De Jong

Department of Computer Science, George Mason University, Fairfax, VA, USA
{escott8,kdejong}@gmu.edu

Abstract. When combined with machine learning, the black-box analysis of fitness landscapes promises to provide us with easy-to-compute features that can be used to select and configure an algorithm that is well-suited to the task at hand. As applications that involve computationally expensive, stochastic simulations become increasingly relevant in practice, however, there is a need for landscape features that are both (A) possible to estimate with a very limited budget of fitness evaluations, and (B) accurate in the presence of small to moderate amounts of noise. We show via a small set of relatively inexpensive landscape features based on hill-climbing methods that these two goals are in tension with each other: cheap features are sometimes extremely sensitive to even very small amounts of noise. We propose that features whose values are calculated using population-based search methods may provide a path forward in developing landscape analysis tools that are both inexpensive and robust to noise.

Keywords: Parameter tuning · Landscape analysis · Meta-learning · Noisy evaluation

1 Introduction

Tuning the parameters of large, stochastic simulations in science and engineering is becoming an increasingly important and popular application domain for evolutionary algorithms (EAs) and metaheuristics (ex. [6,16,17]). These applications tend to involve fitness functions that are very expensive to compute—each evaluation taking on the order of seconds, minutes, or even hours to complete. To approach problems of this kind effectively, the algorithm designer must have some means of quickly and efficiently gathering information about the problem that can help reduce the number of generations that are necessary for a search method to reach a satisfactory solution. In applications where a thorough analytical understanding of the problem is not available, this information-gathering process is often restricted to learning about the problem by directly sampling the evaluation function, which is treated as a *black box*.

J. Handl et al. (Eds.): PPSN XIV 2016, LNCS 9921, pp. 952–961, 2016.
DOI: 10.1007/978-3-319-45823-6_89

Finding ways of characterizing salient properties of fitness landscapes via empirical data—and, especially, of predicting what kinds of algorithms are likely to perform well on them—has been a fundamental goal of metaheuristics and evolutionary computation research since the early days of the field. Researchers have leveraged a number of different mathematical ideas over the years (such as epistasis, correlational properties, and information theory) to produce several families of black-box landscape features [11,14,18,19]. Because these statistical methods are based solely on queries made to the objective function, they can be used even on poorly-understood problems, where little or nothing is known *a priori* about the relationships among variables.

In order for black-box landscape analysis to be useful in practice, however, the information it provides about how to solve a given problem must outweigh the cost of calculating the statistical features. The 'budget' of computational effort that can be spared for up-front analysis is especially small in applications whose evaluation functions involve expensive scientific simulations. A number of landscape features have been proposed that can be computed effectively with especially few queries to the evaluation function, at least on deterministic (noiseless) test functions [1]. Real-world fitness landscapes, and stochastic simulations in particular, often display some degree of noise, however.

In this paper, we are concerned about the intersection of noisy fitness landscapes and the calculation of informative landscape features for computationally intensive applications. In some circumstances, noise may interfere, not only with the progress of a search algorithm as it seeks a global optimum, but also with the attempts of a landscape analysis tool to accurately estimate properties of the task. The problem of noisy fitness functions was heavily studied in the 1990's and early 2000's, and a variety of well-understood approaches are available for configuring evolutionary algorithms to cope with noise [3,10]. Coping with noise does not come for free, however—it often requires extra fitness evaluations which we may not be able to afford when the evaluation function is computationally intensive.

We find it necessary, then, to revisit the well-studied question of noise, now in the context of a pressing need for effective landscape analysis tools that make as few queries as possible to the evaluation function. In this study, we examine several cheap-to-evaluate landscape features and show that a subset of them are extremely sensitive to even very small amounts of noise. Furthermore, we find that the error that this noise introduces into feature estimation can be difficult to correct for in an efficient way. As an alternative, we propose features that use population-based methods as a means of gathering information about the landscape in a way that is both inexpensive and robust to noise.

1.1 Research Questions

Intuitively, it's clear that qualitative features of an objective function such as multimodality, deceptiveness, or the correlation of traits among parents and offspring [2,13] convey a great deal of information about whether a given search strategy is well-suited to particular task. Early work on landscape analysis sought to identify ways in which a problem might be "easy" or "hard" for a particular algorithm of choice (namely the genetic algorithm, ex. [8,9]). But as the philosophy of the research community moves toward "solving the problem at hand in the best way possible, rather than promoting a certain metaheuristic" [5], the primary purpose of landscape analysis has shifted to serving as a predictive aid in the design or selection of a custom algorithm that is well-suited to the given task [15]. Landscape features can be used as input data for machine learning algorithms, which are increasingly being used to learn predictive models for use in algorithm selection and configuration (ex. [4]). Even if a particular statistical feature is difficult for an engineer to interpret in terms of intuitive concepts like multimodality, the feature may be useful if it provides salient or complementary information to a machine learner in conjunction with other features.

If there is a great deal of error or bias in an estimate of a feature, however, its usefulness as a basis for learning may in some cases be greatly diminished. There is a practical need, then, for landscape features that are both (A) inexpensive to estimate, and (B) accurate in the presence of small to moderate amounts of noise. Table 1 details a number of features, taken from Abell et al., that can typically be computed in on the order of a few hundred or a few thousand fitness evaluations, but which are still sufficiently informative to enable a portfolio method to perform well on a suite of noiseless benchmark functions [1]. These satisfy our criterion of inexpensiveness (A), but how do they fair with noise (B)?

Research Question 1: How sensitive to noise are the 8 landscape features identified in Table 1?

Next we begin an investigation into how error in the estimation of features can be corrected for. A straightforward way to do this is to seek to approximate the features of the *expected fitness landscape* $\hat{F}(\boldsymbol{x})$ by taking several fitness samples each time the landscape is queried and returning their 'explicit average' [10].

Research Question 2: Is using explicit averaging an effective means of correcting for noise when measuring these features?

Finally, the features in Table 1 rely heavily on the results of a number of runs of a hill-climbing method as a means of exploring the structure of the landscape. Trajectory methods such as this are notorious for their sensitivity to noise. We consider the possibility that a population-based method may be more effective at identifying informative local optima in the presence of noise:

Research Question 3: Can population-based algorithms serve as a useful alternative to hill-climbers for quickly gathering information about noisy fitness landscapes?

Table 1. Landscape features used this study.

	Feature	Description
1	`MeanPairwiseLocalOptDist`	Mean pairwise distance between optima found by a number of hill climbers.
2	`StdPairwiseLocalOptDist`	Standard deviation of (1).
3	`MeanLocalToBestDist`	Mean distance between the best known optimum and the optima found by a number of hill climbers.
4	`StdLocalToBestDist`	Standard deviation of (3).
5	`FractionBest`	The ratio of local optima found by the hill-climbers that have fitness equal to the best known optimum.
6	`MeanRandomToLocalDist`	Mean distance from a number of random points to the nearest optimum found with a hill climber.
7	`StdRandomToLocalDist`	Standard deviation of (6).
8	`FDC`	Local fitness distance correlation, based on the best result of the hill-climbers

2 Methodology

2.1 Test Functions

Our experiments are conducted on 10-dimensional instances from the suite of 24 test functions that are implemented in version v15.03 of the COmparing Continuous Optimisers (COCO) platform, a framework that has been used for a number of years in the Black-Box-Optimization-Benchmarking (BBOB) workshops held at GECCO and CEC. This test suite includes many well-known unimodal and multimodal real-valued functions, such as the sphere, Rastrigin, and Rosenbrock functions, along with rotated variants, etc., all of which are defined on a range of $[-5, 5]$ in each dimension. The COCO source code is available from http://coco.gforge.inria.fr/.

In addressing **RQ1**, our independent variable will be the amount of noise on the landscape. We opt to use a multiplicative noise model of the form

$$F(\boldsymbol{x}) = f(\boldsymbol{x}) + p \cdot |f(\boldsymbol{x}) - f(\boldsymbol{x}^*)| \cdot \epsilon, \quad (1)$$

where $f(\boldsymbol{x})$ is the original (noiseless) test function, $f(\boldsymbol{x}^*)$ is the fitness of the global optimum, and $\epsilon \sim \mathcal{N}(0,1)$ is a standard Gaussian random variable. The constant p controls the strength of the noise. In this model, the amount of noise that is added to the landscape at the point $\boldsymbol{x}$ is proportional to the difference between its fitness and the global best fitness—so, the poorer a solution is, the nosier it is. This qualitative rule holds in many applications, where poor solutions often correspond to solutions that have especially unstable behavior.

2.2 Features

All of the features in Table 1 make some use of the result of a number of independent runs of a hill-climbing algorithm. We implement the hill climber as a (1+1)-style evolutionary algorithm, and we run this method 100 times to gather a set of representative local optima from which features may be computed. Each individual in the EA is represented as a point $\boldsymbol{x} \in \mathbb{R}^L$, with $L = 10$, and we apply a 1-dimensional Gaussian mutation operator to each element of the offspring with probability $1/L$. We let each hill climber run for 3,000 steps, so as to get a stable estimate of the features. It is worth noting, however, that features based on hill climbing can be informative even if they are run only for a very small number of steps [1].

Features 1–4 are computed directly from the best individuals found by the 100 runs. Feature 5 (`FractionBest`) denotes the fraction of the 100 hill-climbing runs whose best individual has fitness equal to the overall best individual found in all 100 runs. The intent of this feature is to measure the frequency with which a greedy search method converges on a local optimum. There is always some variation, however, in just how closely a given climber will converge to the true local optimum. For the purposes of calculating this feature, then, we consider two individuals to have 'equal' fitness if and only if the difference between their fitnesses is less then an arbitrary threshold value of 0.01.

We compute features 6 and 7 using 1,000 random points. For feature 8, we use a local variant of Jones' well-known fitness distance correlation (FDC) [11]. Classical fitness distance correlation requires knowledge of the global optimum to be computed. Since we are using synthetic test functions, we do have knowledge of the global optimum. The purpose of this study, however, is to examine the behavior of landscape features as exploratory, black-box analysis tools. We follow Kallel and Scipemaier in defining a local FDC simply by substituting the best *known* optimum for the global optimum [12]. In our case, the "best known optimum" refers to the best optimum found by the 100 hill-climbing routines.

2.3 Coping with Noise

A test of **RQ2** involves performing the feature measurements as described above, but we now replace the fitness function $F(\boldsymbol{x})$, which is a random variable, with a constant estimate $\hat{F}(\boldsymbol{x})$ of the expected fitness landscape like so:

$$\hat{F}(\boldsymbol{x}) = \frac{1}{N}\sum_{i=1}^{N} F(\boldsymbol{x}). \tag{2}$$

We will test this method's effectiveness by empirically examining the relationship between the observed error in feature estimates and the number of samples N.

To test **RQ3**, we replace the (1+1)-style EA used in the feature calculations with a $(\mu + \lambda)$-style EA. We vary the value of μ and keep $\lambda = \mu$.

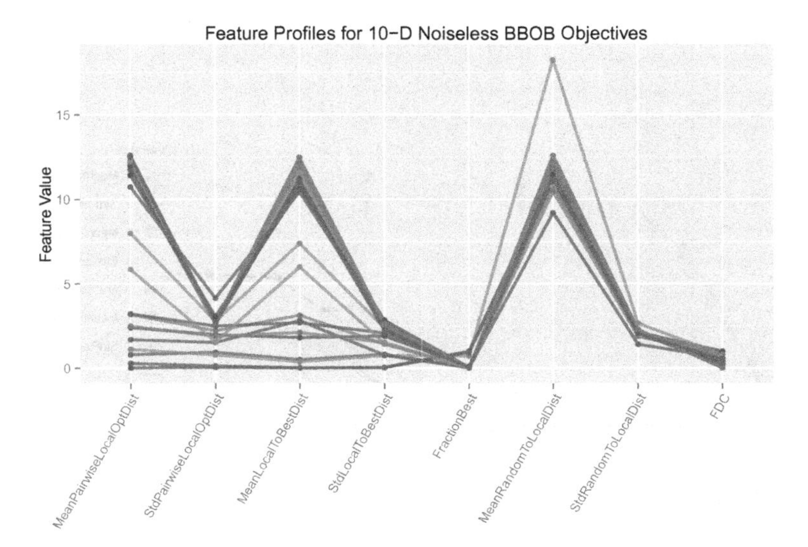

Fig. 1. Mean feature profiles, averaged across 50 ten-dimensional instances of each of the 24 noiseless test functions in the BBOB suite.

3 Results

3.1 Sensitivity Analysis

The features we have implemented provide us with an eight-dimensional characteristic profile of each test function. The parallel plot in Fig. 1 visualizes these profiles for all 24 of the noiseless test functions. We consider the profile calculated from each noiseless landscape to be the 'true' feature values. The question is how our estimate of those feature values incurs error as noise increases (**RQ1**).

Figure 2 shows the value of each feature estimate, averaged over all 24 test functions, as we increase the value of p (see Eq. 1). While there is a great deal of variance in behavior across the 24 functions (not shown), in general we find that features 1–4 are extremely sensitive to noise: the estimate becomes inaccurate as soon as p reaches a value of about 10^{-3}. The remaining feature estimators (6–8) appear to be reasonably robust to small amounts of noise—but they suddenly become inaccurate when p reaches a threshold of about 0.25.

This answers **RQ1**: The features under study are highly sensitive to noise in the fitness landscape.

3.2 Explicit Averaging

We've shown that we can make the error in feature estimation explode by adding small amounts of artificial noise. Now we turn to the question of whether we can attenuate this error through explicit averaging of more than one fitness sample. We implemented explicit averaging for fitness evaluation during the hill-climber

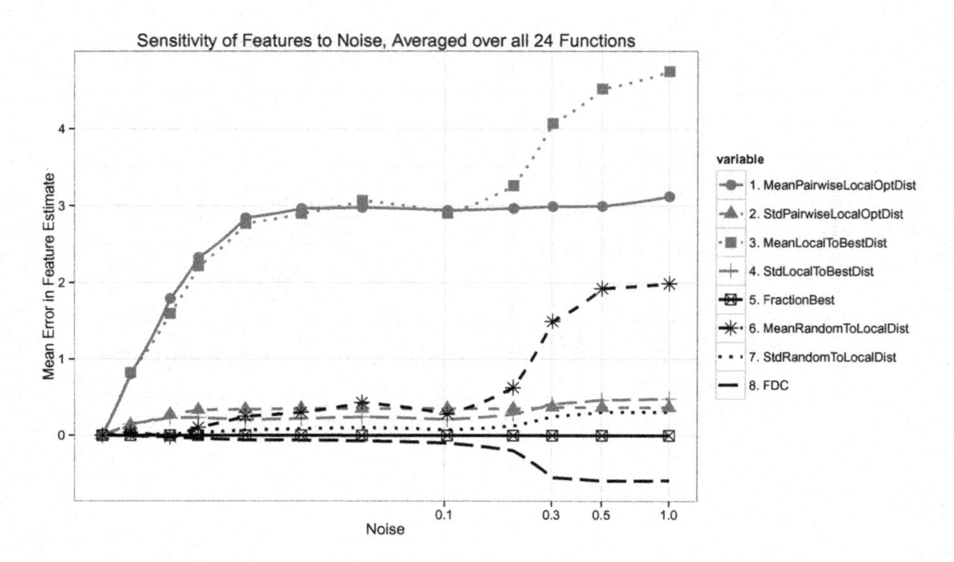

Fig. 2. Deviation between the estimated and true feature values as noise increases.

runs—that is, in this section we use the same (1+1)-style EA to compute feature estimates, but now every time an individual has its fitness evaluated, some $N \geq 1$ fitness samples are taken and averaged according to Eq. 2.

Let the magnitude of the noise be fixed at the small value of $p = 5 \cdot 10^{-4}$. Figure 3 shows the result of feature estimation averaged over all 24 instances while allowing the number of samples N to vary. We find that using explicit averaging of fitness has very little discernible impact on the accuracy of fitness measurements. Even at $N = 15$, a great deal of error remains.

Our answer to **RQ2** is thus negative: it seems that explicit averaging is not an effective means of correcting for noise.

3.3 Population-Based Search

It is well known that population-based search methods can perform a kind of 'implicit averaging' that makes their performance robust to noise. This is borne out in our experiments with the $(\mu + \lambda)$-EA, shown in Fig. 4. We see a sharp reduction in error when we increase μ from 1 to 2. As μ grows, however, we see stark, systematic deviations from the true feature values. This may be because, while the population-based EA is not significantly affected by small amounts of noise, it also has a tendency to converge to high-quality or *global* optima instead of the *local* optima that the features based on the (1+1)-EA are designed to seek out.

Our answer to **RQ3** is mixed, then: Replacing the hill-climbers in these features with a population-based algorithm does overcome noise, but it changes the kind of information that the features gather from the landscape. Whether this information is useful for prediction or not is a question that is beyond the scope of this study.

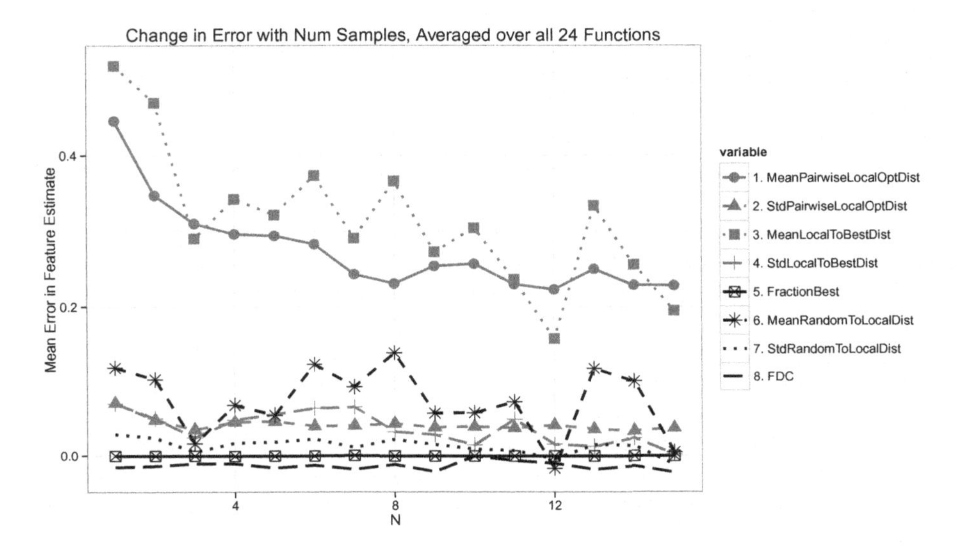

Fig. 3. Deviation between the estimated and true feature values as the number of explicit fitness samples increases.

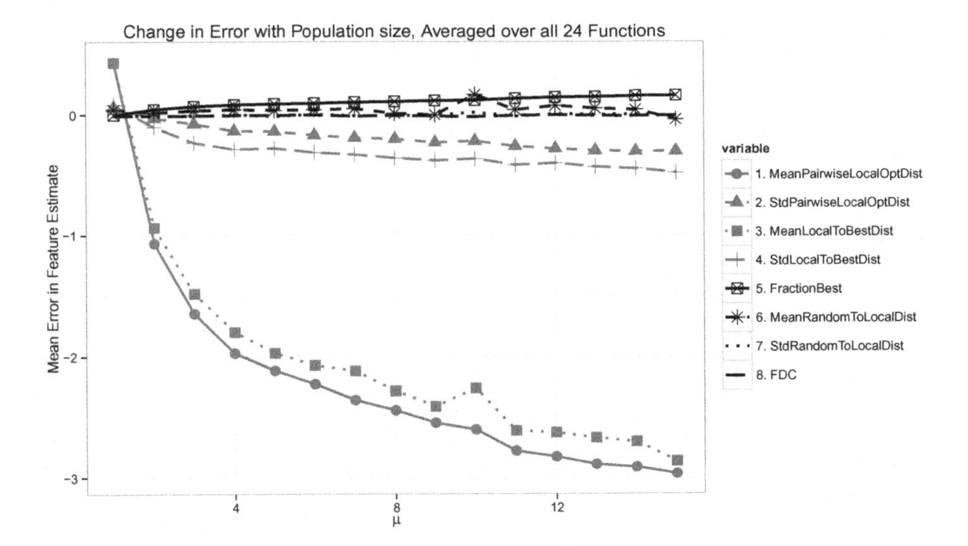

Fig. 4. Deviation between the estimated and true feature values as the population size increases.

4 Conclusion

Noise poses a particularly difficult challenge to the solution of computationally expensive problems. We find that the features under study, which are based on identifying local optima with a greedy search method, are sensitive to noise, yielding a positive answer to **RQ1**. Even very, very small quantities of noise are sufficient to entirely frustrate efforts to accurately measure features of a fitness landscape with these methods. Furthermore, we found that explicit averaging of many fitness samples is not sufficient to substantially attenuate the error caused by noise. A large number of fitness samples may be necessary to fully counteract even the impact that a very minuscule quantity of noise has on feature measurements—answering **RQ2** in the negative. Consequently, the features we have studied here, while they are initially appealing for computationally intensive applications because of their low cost, become computationally infeasible in the presence of noise.

We have shown that modifying these features to use a population-based algorithm in place of the hill-climbers is a promising approach, allowing us to overcome the issue of noise (**RQ3**). Because these algorithms are less greedy than a hill-climber, however, they gather different information about the landscape, and are less effective at collecting a representative sample of diverse local optima.

Our findings suggest that landscape analysis researchers should look toward the design of features that use population-based algorithms to gather information about the landscape. Future work might, for instance, explore replacing hill-climbers with state-of-the-art multimodal optimization methods. These may be able to overcome the noise problem while still gathering a representative sample of local optima [7]. Such an approach may be able to maintain some of the computational efficiency of the hill-climbing approach while also attaining some robustness to moderate amounts of noise.

Acknowledgments. This work was funded by U.S. National Science Foundation Award IIS/RI-1302256.

References

1. Abell, T., Malitsky, Y., Tierney, K.: Features for exploiting black-box optimization problem structure. In: Nicosia, G., Pardalos, P. (eds.) LION 7. LNCS, vol. 7997, pp. 30–36. Springer, Heidelberg (2013)
2. Bassett, J.K.: Methods for improving the design and performance of evolutionary algorithms. Ph.D. thesis, George Mason University, Fairfax, VA (2012)
3. Beyer, H.-G.: Evolutionary algorithms in noisy environments: theoretical issues and guidelines for practice. Comput. Methods Appl. Mech. Eng. **186**(2), 239–267 (2000)
4. Bischl, B., Mersmann, O., Trautmann, H., Preuß, M.: Algorithm selection based on exploratory landscape analysis and cost-sensitive learning. In: Proceedings of the 14th Annual Conference on Genetic and Evolutionary Computation, pp. 313–320. ACM (2012)

5. Blum, C., Puchinger, J., Raidl, G.R., Roli, A.: Hybrid metaheuristics in combinatorial optimization: a survey. Appl. Soft Comput. **11**(6), 4135–4151 (2011)
6. Carlson, K.D., Nageswaran, J.M., Dutt, N., Krichmar, J.L.: An efficient automated parameter tuning framework for spiking neural networks. Front. Neurosci. **8**(10), 168 (2014)
7. Das, S., Maity, S., Bo-Yang, Q., Suganthan, P.N.: Real-parameter evolutionary multimodal optimization–a survey of the state-of-the-art. Swarm Evol. Comput. **1**(2), 71–88 (2011)
8. Deb, K., Goldberg, D.E.: Sufficient conditions for deceptive and easy binary functions. Ann. Math. Artif. Intell. **10**(4), 385–408 (1994)
9. Forrest, S., Mitchell, M.: What makes a problem hard for a genetic algorithm? some anomalous results and their explanation. Mach. Learn. **13**(2–3), 285–319 (1993)
10. Jin, Y., Branke, J.: Evolutionary optimization in uncertain environments-a survey. IEEE Trans. Evol. Comput. **9**(3), 303–317 (2005)
11. Jones, T., Forrest, S.: Fitness distance correlation as a measure of problem difficulty for genetic algorithms. In: Sixth International Conference on Genetic Algorithms (ICGA 1995), vol. 95, pp. 184–192 (1995)
12. Kallel, L., Schoenauer, M.: Alternative random initialization in genetic algorithms. In: Seventh International Conference on Genetic Algorithms (ICGA 1997), pp. 268–275 (1997)
13. Manderick, B., de Weger, M., Spiessens, P.: The genetic algorithm and the structure of the fitness landscape. In: Proceedings of the 4th International Conference on Genetic Algorithms, pp. 143–150. Morgan Kaufmann, San Mateo, CA (1991)
14. Naudts, B., Kallel, L.: A comparison of predictive measures of problem difficulty in evolutionary algorithms. IEEE Trans. Evol. Comput. **4**(1), 1–15 (2000)
15. Smith-Miles, K.A.: Cross-disciplinary perspectives on meta-learning for algorithm selection. ACM Comput. Surv. (CSUR) **41**(1), 6 (2008)
16. Tayarani, M.-H., Yao, X., Hao, X.: Meta-heuristic algorithms in car engine design: a literature survey. IEEE Trans. Evol. Comput. **19**(5), 609–629 (2015)
17. Van Geit, W., De Schutter, D., Achard, P.: Automated neuron model optimization techniques: a review. Biol. Cybern. **99**(4–5), 241–251 (2008)
18. Vassilev, V.K., Fogarty, T.C., Miller, J.F.: Information characteristics and the structure of landscapes. Evol. Comput. **8**(1), 31–60 (2000)
19. Watson, J.P.: An introduction to fitness landscape analysis and cost models for local search. In: Gendreau, M., Potvin, J.Y. (eds.) Handbook of Metaheuristics. International Series in Operations Research & Management Science, vol. 146, pp. 599–623. Springer, Heidelberg (2010)

Towards Analyzing Multimodality of Continuous Multiobjective Landscapes

Pascal Kerschke[1(✉)], Hao Wang[2], Mike Preuss[1], Christian Grimme[1], André Deutz[2], Heike Trautmann[1], and Michael Emmerich[2]

[1] Information Systems and Statistics Group, University of Münster, Münster, Germany
{pascal.kerschke,mike.preuss,christian.grimme, heike.trautmann}@wi.uni-muenster.de
[2] LIACS, Leiden University, Leiden, The Netherlands
{h.wang,a.h.deutz,m.t.m.emmerich}@liacs.leidenuniv.nl

Abstract. This paper formally defines multimodality in multiobjective optimization (MO). We introduce a test-bed in which multimodal MO problems with known properties can be constructed as well as numerical characteristics of the resulting landscape. Gradient- and local search based strategies are compared on exemplary problems together with specific performance indicators in the multimodal MO setting. By this means the foundation for Exploratory Landscape Analysis in MO is provided.

Keywords: Multiobjective optimization · Multimodality · Landscape analysis · Hypervolume gradient ascent · Set based optimization

1 Introduction

In multiobjective optimization, respective algorithms are particularly challenged by multimodality of the underlying landscape caused by the interaction of objective functions. Thus, sophisticated Exploratory Landscape Analysis (ELA, [7]) features which are able to assess the level and type of multimodality based on an initial problem sample have huge potential for understanding algorithm behaviour, automated algorithm selection and algorithm design. Despite the success of ELA in single-objective continuous black-box optimization (e.g. [1,4]) multiobjective optimization has not been appropriately addressed apart from limited approaches in the combinatorial context (e.g. [6,9]) or expert-based characteristics such as the Pareto front shape, the dimensionality and some intuitions on multimodality. We here lay the groundwork for constructing such experimental features systematically by providing formal definitions of multimodality in terms of distinguishing between local and global efficient sets. A versatile problem generator is introduced for designing multimodal mixed sphere problems with predefined characteristics. Bringing together theoretical analysis and experiments, and contrasting gradient and local search based methods, highly increases understanding of the problem domain multimodality in multiobjective optimization as well as the explorative algorithm.

J. Handl et al. (Eds.): PPSN XIV 2016, LNCS 9921, pp. 962–972, 2016.
DOI: 10.1007/978-3-319-45823-6_90

Section 2 introduces topological definitions, Sect. 3 details the problem generator and theoretically analyzes the resp. multimodal structures, Sect. 4 discusses the exploration algorithms, Sect. 5 presents algorithm and problem characteristics, and Sect. 6 provides experimental results. Conclusions are drawn in Sect. 7.

2 Multimodality

We present an approach of defining multimodality in that we distinguish between global and local efficient sets in $\mathbb{R}^d$ (the decision space). We are aware that most parts can be generalized to other spaces. We first recall some topological notions:

1. Let $A \subseteq \mathbb{R}^n$. The set A is called *connected* if and only if there do not exist two open, *disjoint* subsets U_1 and U_2 of $\mathbb{R}^n$ such that $A \subseteq U_1 \cup U_2$, $U_1 \cap A \neq \emptyset$, and $U_2 \cap A \neq \emptyset$.
2. Let $B \subseteq \mathbb{R}^n$. A subset $C \subseteq B$ is a *connected component* of B iff C is connected, and any subset of B which is a strict superset of C is not connected, and C is non-empty.

Pareto concepts are given next: Let $\mathbf{f} : \mathcal{X} \to \mathbb{R}^m$ be a multiobjective function where $\mathcal{X} \subseteq \mathbb{R}^d$ is the decision space. We will denote the component functions of $\mathbf{f}$ by $f_i : \mathcal{X} \to \mathbb{R}, i = 1, \ldots, m$. Given a totally ordered set $(T, \leq)$ where $\leq$ denotes the total order, we can define as usual the Pareto order, denoted by $\prec$, on T^k for any $k \in \mathbb{N}$ as follows. Let $\mathbf{t}^{(1)} = (t_1^{(1)}, \ldots, t_k^{(1)}), \mathbf{t}^{(2)} = (t_1^{(2)}, \ldots, t_k^{(2)})$ be elements of T^k. We say $\mathbf{t}^{(1)} \prec \mathbf{t}^{(2)}$ iff $t_i^{(1)} \leq t_i^{(2)}, i = 1, \ldots, k$ and $\mathbf{t}^{(1)} \neq \mathbf{t}^{(2)}$. Specializing this to the reals with their natural, total order we obtain the Pareto order on $\mathbb{R}^m$. A point $\mathbf{x} \in \mathcal{X}$ is called *Pareto efficient* or *global efficient* or for short *efficient* iff there does not exist $\tilde{\mathbf{x}} \in \mathcal{X}$ such that $\mathbf{f}(\tilde{\mathbf{x}}) \prec \mathbf{f}(\mathbf{x})$. The subset of $\mathcal{X}$ consisting of all the efficient points of $\mathcal{X}$ is denoted by $\mathcal{X}_E$ and is called the *efficient subset* of $\mathcal{X}$ (or the *efficient* set of $\mathbf{f}$). The image of $\mathcal{X}_E$ under $\mathbf{f}$ is called the Pareto front of $\mathbf{f}$. To define local efficient points in $\mathcal{X}$ and local efficient *sets* in the multiobjective case, we propose the following definitions:

Definition 1 (Efficiency of Points/Sets). *A point $\mathbf{x} \in \mathcal{X}$ is called a* locally efficient point *of $\mathcal{X}$ (or of $\mathbf{f}$) if there is an open set $U \subseteq \mathbb{R}^d$ with $\mathbf{x} \in U$ such that there is no point $\tilde{\mathbf{x}} \in U \cap \mathcal{X}$ such that $\mathbf{f}(\tilde{\mathbf{x}}) \prec \mathbf{f}(\mathbf{x})$. The subset of all the local efficient points of $\mathcal{X}$ is denoted by $\mathcal{X}_{LE}$.*

A point $\mathbf{x} \in \mathcal{X}$ is called a global efficient point *of $\mathcal{X}$ (or of $\mathbf{f}$) if there is no point $\tilde{\mathbf{x}} \in \mathbb{R}^d \cap \mathcal{X}$ such that $\mathbf{f}(\tilde{\mathbf{x}}) \prec \mathbf{f}(\mathbf{x})$. The subset of all the global efficient points of $\mathcal{X}$ is termed* efficient set *of $\mathbf{f}$ and denoted by $\mathcal{X}_E$.*

A subset $A \subseteq \mathcal{X}$ is a local efficient set *of $\mathbf{f}$ if A is a connected component of $\mathcal{X}_{LE}$ (= the subset of $\mathcal{X}$ which consists of the local efficient points of $\mathcal{X}$).*

Definition 2 (Local Pareto Front). *A subset P of the image of $\mathbf{f}$ is a local Pareto front of $\mathbf{f}$, if there exists a local efficient set E such that $P = \mathbf{f}(E)$.*

The (global) Pareto front (PF) of $\mathbf{f}$ is obtained by taking the image under $\mathbf{f}$ of the union of the connected components of the set of global efficient points of $\mathcal{X}$. If $\mathcal{X}_E$ is connected, then the (global) Pareto front of $\mathbf{f}$ is also connected, provided $\mathbf{f}$ is continuous on $\mathcal{X}_E$. One might also consider definitions related to connected components in the objective space [8]. However, we will omit this for brevity.

3 Analytics on Simple Mixed Sphere Problems

We use a sophisticated problem generator based on the Multiple Peaks Model 2 (MPM2, [10]) to illustrate the proposed topological definitions and further analyse the behavior of explorative algorithms.

$$f(\mathbf{x}) = 1 - \max_{1 \le i \le N} \{g_i(\mathbf{x})\}, \quad \mathbf{x} \in \mathbb{R}^d \tag{1}$$

$$g_i(\mathbf{x}) = H_i \left(1 + \left(\sqrt{(\mathbf{x}-\mathbf{c}_i)^T \mathbf{D} (\mathbf{x}-\mathbf{c}_i)}\right)^{s_i} / R_i\right)^{-1}, \quad i = 1, \ldots, N \tag{2}$$

Note that the g_i functions define peaks with center $\mathbf{c}_i$, depth H_i, radius R_i and shape s_i. $\mathbf{D}$ is the inverse of the covariance matrix while we concentrate on spherical peaks with isotropic level curves (*mixed spheres*), i.e. $\mathbf{D} = c\mathbf{I}, c \in \mathbb{R}_{\neq 0}$.

A bi-objective optimization problem $(f_1(\mathbf{x}|g_i), f_2(\mathbf{x}|g_i')) \rightarrow \min$ results in choosing two different parameter sets – parameters of f_2 are labeled by the prime symbol. Exemplary problems are illustrated in Figs. 1, 2 and 3.

In order to evaluate and compare the Pareto fronts obtained by the optimization algorithms used in this paper, the analytical Pareto front (and efficient set) is derived in the following. First, we focus on the simplest case where each objective function consists of only one peak. In this case the Pareto efficient set PE is the line segment connecting $\mathbf{c}$ to $\mathbf{c}'$: $PE : \{\alpha\mathbf{c} + (1-\alpha)\mathbf{c}' \mid 0 \le \alpha \le 1\}$.

Then the parametric form of the Pareto front can be derived by mapping an arbitrary point in the efficient set $\hat{\mathbf{x}} = \alpha\mathbf{c} + (1-\alpha)\mathbf{c}'$ through the objective functions, which – using the Mahalanobis distance $d(\mathbf{c}, \mathbf{c}'; \mathbf{D}) = \sqrt{(\mathbf{c}'-\mathbf{c})^T \mathbf{D} (\mathbf{c}'-\mathbf{c})}$ – finally results after algebraic transformation in f_2 as a function of f_1, for $\mathbf{c} \neq \mathbf{c}'$:

$$f_2 = 1 - H' \left(1 + \left(d(\mathbf{c}, \mathbf{c}'; \mathbf{D}') \left(1 - \frac{R^{1/s}}{d(\mathbf{c}, \mathbf{c}'; \mathbf{D})} \left(\frac{H}{1-f_1} - 1\right)^{1/s}\right)\right)^{s'} / R'\right)^{-1}$$

The range of f_1 is $[\min\{f_1(\mathbf{c}), f_1(\mathbf{c}')\}, \max\{f_1(\mathbf{c}), f_1(\mathbf{c}')\}]$.

Using the expression above, we could calculate the red part of the global Pareto front in Fig. 1. For multiple peaks the (local) efficient sets still settle on line segments connecting each pair of peaks. It is difficult to derive the analytical expression because the *effective* peak might change when traversing along the line segment connecting peaks and multiple local efficient sets could exit on the same line segment (check Fig. 3 for example). However, it is possible to approximate the local efficient sets numerically by uniformly sampling on the line segments and taking the maximal non-dominated subset of the samples.

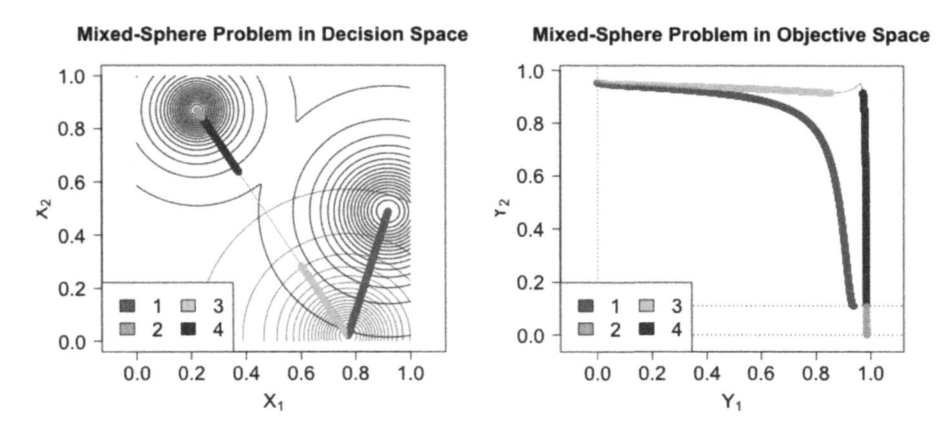

Fig. 1. Example of a simple mixed sphere problem in the decision (left figure) and objective space (right). Objectives are visualized in the decision space by pink (objective 1) and blue (objective 2) contour lines. The connections between peaks from the two objectives are shown as grey lines and the corresponding local efficient sets (or fronts) are colored. Here, the red and green parts form the disconnected global PF, whereas the cyan and purple parts show the remaining disconnected local PFs. The given scenario represents three disconnected local efficient sets (green/purple, cyan, red), two domination layers (red/green vs. cyan/purple) and four local Pareto fronts. (Color figure online)

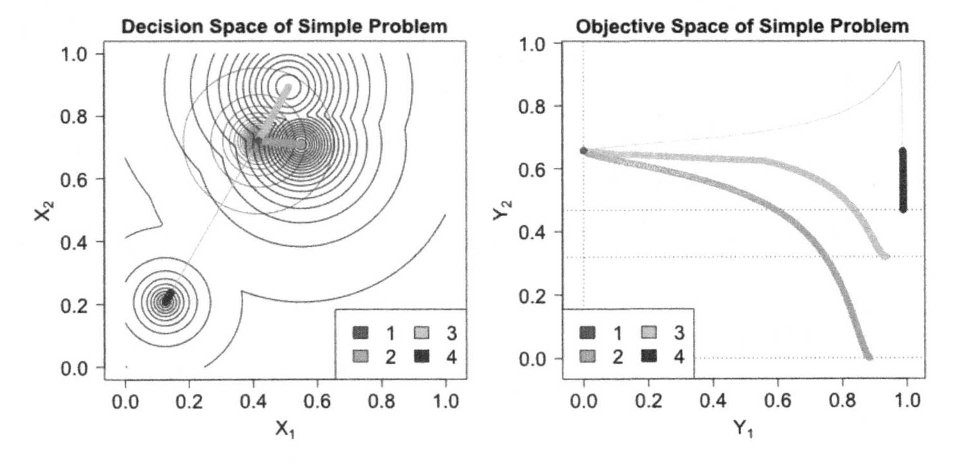

Fig. 2. Local Pareto fronts in the decision (left) and objective space (right) for a rather simple mixed sphere problem consisting of one peak in the first objective (pink contour lines) and three peaks in the second objective (blue contour lines). The red area is caused by the fact that it belongs to the same local efficient set as the cyan area, and at the same time to the global dominance layer. (Color figure online)

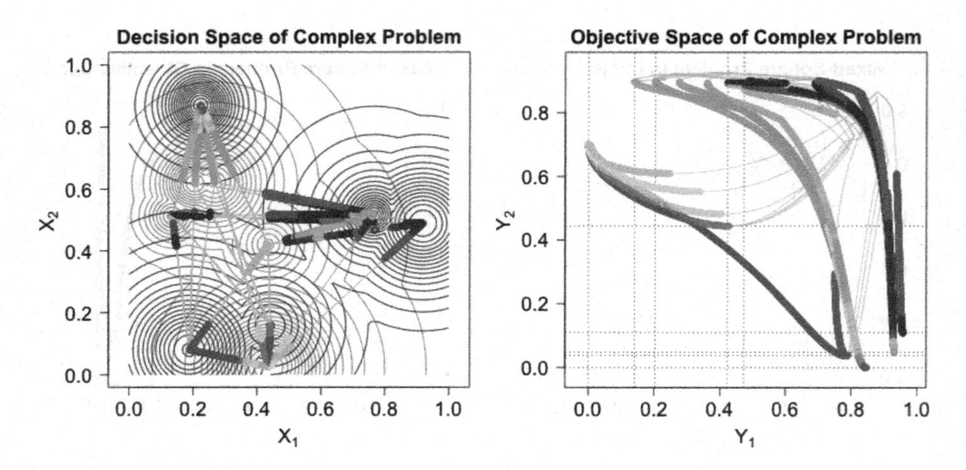

Fig. 3. Local Pareto fronts for a complex mixed sphere problem consisting of five peaks per objective, resulting in a total of 30 disconnected local efficient sets, 19 domination layers and 167 local Pareto fronts.

4 Explorative Algorithms

Hypervolume Indicator Gradient Ascent (HIGA-MO). Taking the advantage of the analytically computable gradient information of the mixed sphere problems, we choose the *Hypervolume Indicator Gradient Ascent* [2,3], because it is capable of generating well-distributed PF approximations and is almost free of control parameters. The corresponding pseudo-code is given in Algorithm 1. The basic idea is to maximize the Hypervolume indicator $\mathcal{H}$ of an approximation set of the true PF by using the gradient of $\mathcal{H}$. We denote the set of search points as $\{\mathbf{x}_1, \ldots, \mathbf{x}_\mu\}, \mathbf{x}_i \in \mathbb{R}^d$ and $\mathbf{X} = [\mathbf{x}_1^T, \ldots, \mathbf{x}_\mu^T]^T \in \mathbb{R}^{\mu d}$. As $\mathcal{H}$ can also be expressed as a function of the input vectors, one can calculate the Hypervolume indicator gradient $\partial\mathcal{H}(\mathbf{X})/\partial\mathbf{X}$. By applying the chain rule the so-called *subgradients* $\partial\mathcal{H}(\mathbf{X})/\partial\mathbf{x}_i = [\partial\mathcal{H}(\mathbf{X})/\partial\mathbf{y}_i] \cdot [\partial\mathbf{y}_i/\partial\mathbf{x}_i]$ [2] can be computed for $i = 1, \ldots, \mu$. In practice, a step-size control is used to adapt the step-size for each decision vector.[1]

HIGA-MO performs a fast local search and some individuals might get stuck in a local efficient set. However, in mixed sphere problems local efficient sets might be connected to the global one and the Hypervolume indicator gradient will steer the local efficient points towards the global one.

Stochastic Local Search (SLS). A simple local search strategy based on parallel perturbation and elitist selection is implemented. Essentially, each individual candidate solution of the current solution set is perturbed once per round. According to a simple (1+1)-selection scheme, for each pair of original and related perturbed solution the original solution is replaced when dominated by

[1] The HIGA-MO source code is available on moda.liacs.nl/index.php?page=code.

Algorithm 1. Hypervolume Indicator Gradient Ascent

1 initialize the search points $\mathbf{X}$ uniformly in the search space
2 **while** the termination criteria are not satisfied **do**
3 $\quad \mathbf{Y} \leftarrow$ evaluation of search points $\mathbf{X}$
4 $\quad$ layers $\{L_i\}_{i=1}^{q} \leftarrow$ non-dominated-sorting of population $\mathbf{R}$
5 $\quad$ **for** $i = 1$ to q **do**
6 $\quad\quad$ **for** every element $\mathbf{s}_j$ in layer L_i **do**
7 $\quad\quad\quad$ compute the subgradient $\partial\mathcal{H}(\mathbf{X})/\partial\mathbf{x}_j$
8 $\quad\quad\quad \mathbf{x}_j \leftarrow \mathbf{x}_j + \sigma \cdot [\partial\mathcal{H}(\mathbf{X})/\partial\mathbf{x}_j]$
9 $\quad\quad$ **end for**
10 $\quad$ **end for**
11 **end while**
12 **return** $\{L_i\}_{i=1}^{q}, \mathbf{X}, \mathbf{Y}$ with $\mathbf{y}_i = [f_1(\mathbf{x}_i),\ f_2(\mathbf{x}_i)]^T,\ i = 1, \ldots, \mu$

the perturbed one. Initially, μ independently random solutions are generated using a Latin hypercube design. In every iteration, each solution is modified by an upper bounded normal distributed perturbation with maximum step size of σ. Here the step-size is fixed. After the elitist and parallel selection process based on domination, μ solutions are available for the next round until the maximum number of iterations is reached.

The rational of using this simple approach is to contrast the HIGA-MO search approach with a local search representative that is unable to traverse along local Pareto fronts. We expect this approach to get stuck in local efficient solutions.

5 Problem and Algorithm Characteristics

Problem Characteristics. In contrast to sophisticated ELA features [4,7], we know the underlying objective functions and solely intend to quantify some obvious differences in landscapes. The *count ratio* describes the problems by ratios related to the number of all local fronts or sets: `count_ratio.global` computes the percentage of fronts that are global PFs, `count_ratio.conn_ps` the percentage of sets connected to any of the global efficient sets, while `count_ratio.conn_pf` denotes the analogous percentage for PFs. The *length ratio* characteristics compute ratios of the lengths of the fronts and sets: `length_ratio.global_ps` computes the ratio of the lengths of all global Pareto sets and all local sets, whereas `length_ratio.global_pf` denotes the analogous ratio of global and local PFs. While `length_ratio.conn_ps` captures the ratio of the total length of all sets connected to any of the global efficient sets and the length of all local sets, `length_ratio.conn_pf` measures the analogous ratio in objective space.

Algorithm Characteristics. We propose characteristics in order to capture differences in local search behavior of the considered explorative algorithms.

The *population* characteristics describe the distribution of the final set of individuals of an algorithm run. They measure the percentage of individuals that are located in the ε-environment of any of the global PFs (`pop.global_front`), a front that is connected to any of the global PFs (`pop.conn_global_front`), and any local front in general (`pop.local_front`). The *coverage* characteristics addresses the percentage of fronts that are reached by the final "population", i.e. at least one individual of that population is located in the ε-environment of the respective front.

We use the coverage of global fronts (`cover.global_front`), fronts connected to any of the global fronts (`cover.conn_global_front`), connected local fronts (`cover.conn_local_front`) and local fronts in general (`cover.local_front`). As the number of fronts might be larger than the population size (i.e. the number of considered individuals), we standardize each characteristic by its maximum.

6 Experiments

Experimental Setup. Two exemplary instances with different levels of multimodality (low, very high) were generated using the MPM-2 generator [5,10], which is e.g. available in the R-package `smoof`. For our experiments, we used the settings shown in Table 1. Our explorative algorithms (cf. Sect. 4), were run with a population size $\mu = 50$ and an initial step size of 0.01 (SLS) and 0.001 (HIGA-MO). Note that the step-size is adaptive in HIGA-MO and will increase largely during the optimization while it remains unchanged in SLS.

Experimental Results. As stated in the previous sections, we applied different algorithms (HIGA-MO and SLS) on two opposing multimodal, multiobjective problems. The analyzed problems can in fact be divided into a simple (cf. Fig. 2) and a complex scenario (cf. Fig. 3) as the corresponding problem characteristics show. The red line (representing the simple scenario) within the parallel coordinate plot (cf. Fig. 4) is always above the blue line of the complex scenario, which means that a higher ratio of the local fronts (or sets) are part of the global non-dominated front (set). These findings are supported by some count characteristics, which are listed in Table 2. As each of the peaks of one objective is connected to each peak of the other objective (and each of those connections can contain multiple connected components), there exist 30 connected components within the complex scenario. Given the fact that the points of a connected component often belong to multiple domination layers, the components can be

Table 1. Parameter configuration for the setup of the MPM2-generator.

Name of parameter		Simple scenario		Complex scenario	
in general	in R	Obj. 1	Obj. 2	Obj. 1	Obj. 2
Number of peaks	`n.peaks`	1	3	5	5
Dimensions	`dimensions`	2	2	2	2
Topology	`topology`	"random"	"random"	"random"	"random"
Seed	`seed`	1	3	2	5
Rotated peaks	`rotated`	FALSE	FALSE	FALSE	FALSE
Shape of peaks	`peak.shape`	"sphere"	"sphere"	"sphere"	"sphere"

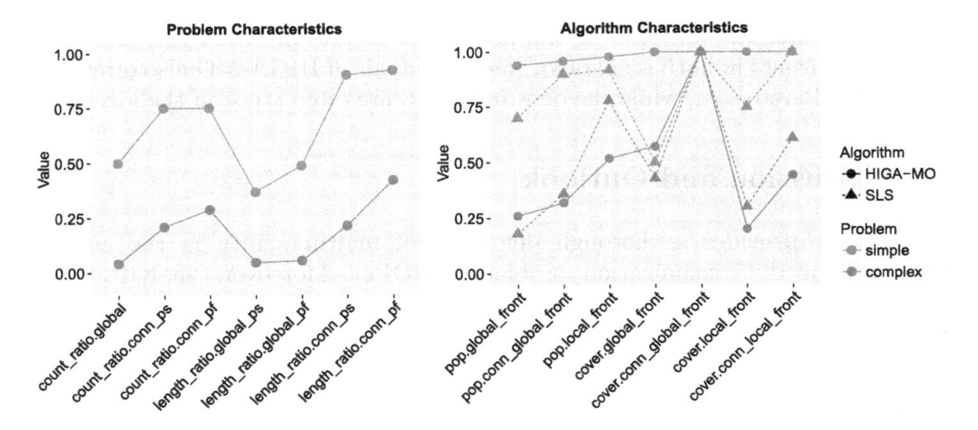

Fig. 4. The parallel coordinate plot on the left side distinguishes the two analyzed multiobjective, multimodal optimization problems from each other by means of rather simple problem characteristics, whereas the right figure visualizes the differences between HIGA-MO and SLS (Section 4) on these two problem instances. (Color figure online)

Table 2. Overview of the count chararacteristics of the problems.

Count characteristic	Scenario	
	Simple	Complex
Optima (obj. 1 vs. obj. 2)	1 vs. 3	5 vs. 5
Domination layers	3	19
Connected components	3	30
Sets connected to global efficient set	2	66
Fronts connected to global efficient front	2	12
Local (global) efficient sets	4 (2)	167 (7)

split into numerous local efficient sets, resulting in 167 local sets – seven of them being non-dominated.

In addition, the mixed-sphere problems come with a nice property: all local efficient sets are located on connections of two peaks and thus many of them are connected. For instance, almost 40 % of all local efficient sets (66 out of 167) in the complex scenario are connected to (at least) one of the seven non-dominated sets. In consequence, smarter optimization algorithms only need to find one of those sets and can then "travel" along the connected sets until they converge in one of the non-dominated sets. As shown in Fig. 5, HIGA-MO is able to exploit that property. At the beginning, it performs similar to SLS and tries to find any of the aforementioned local efficient sets. Once it finds one of them, it travels along the connections – the so-called *channels* – and afterwards often converges in one of the non-dominated sets. The channels are visible (as strong black lines) in Fig. 5. In contrast to HIGA-MO, a "regular" algorithm (such as SLS) likely stops, once it hits one of the local efficient sets. These findings can also be detected by our measures as the right plot of Fig. 4 shows. While both algorithms find the majority of the local efficient sets in either scenario (SLS finds more fronts in both scenarios), the individuals of HIGA-MO also often find the global Pareto sets, while the ones from SLS are often stuck in the local sets.

7 Discussion and Outlook

This paper provides a thorough definition of multimodality in the context of multiobjective optimization problems (MOPs). Moreover, analytical and

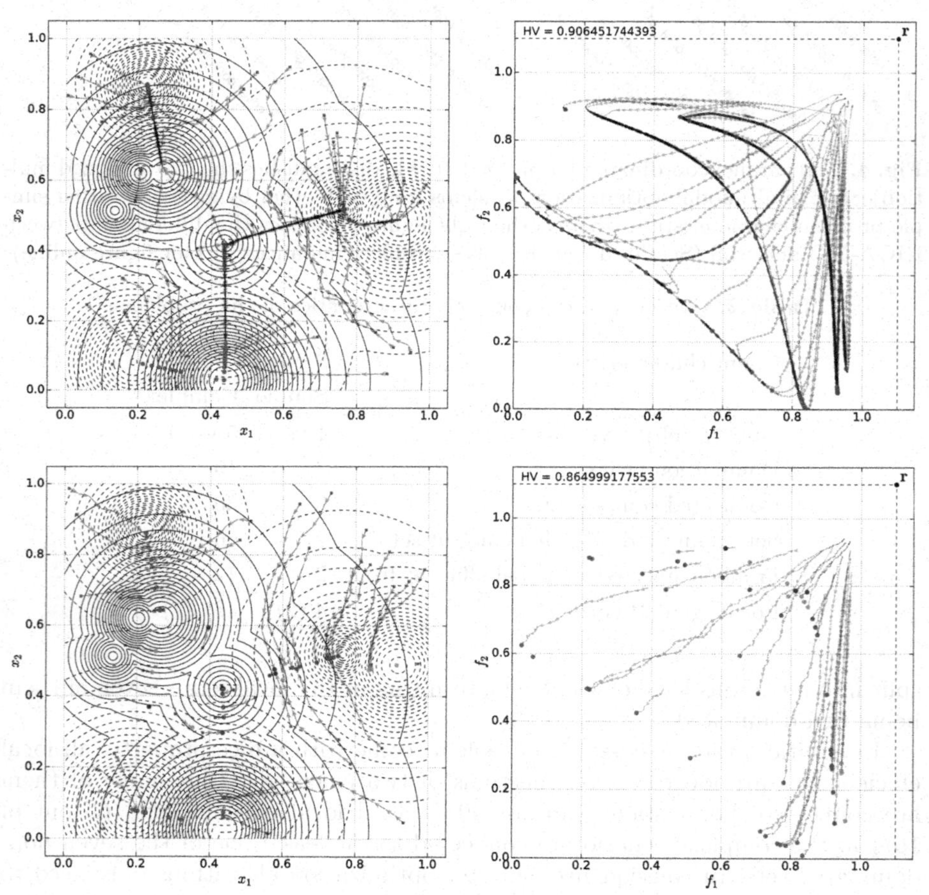

Fig. 5. Population of HIGA-MO (top) and SLS (bottom), respectively. The grey arrows visualize the trace of each individual and the colored points represent the elements from the final population. The different colors indicate the different domination layers – red points belong to the global PF, green points to the second layer, etc. (Color figure online)

experimental approaches are presented which derive or approximate the global and local Pareto fronts of a MOP. Specifically, mixed sphere test problems of different dimensionality are designed and the behavior of a sophisticated Hypervolume gradient ascent approach and a stochastic local search variant are contrasted on problems consisting of different levels of multimodality. It is reflected that multimodality is a crucial factor determining the difficulty of a problem, especially in case the optimization algorithm relies on local search techniques.

Moreover, indicators are derived which allow to assess algorithm behavior w.r.t. the detection of global and local Pareto fronts which can further be used for performance assessment. In combination with specific indicators for problem characteristics, the basis for systematically constructing respective Exploratory Landscape Features is formed which has huge potential w.r.t. algorithm benchmarking, selection and design, also for higher dimensional problems.

Acknowledgments. Heike Trautmann, Pascal Kerschke, Mike Preuss and Christian Grimme acknowledge support by the European Research Center for Information Systems (ERCIS).

References

1. Bischl, B., Mersmann, O., Trautmann, H., Preuss, M.: Algorithm selection based on exploratory landscape analysis and cost-sensitive learning. In: Proceedings of the 14th Annual Conference on Genetic and Evolutionary Computation, GECCO 2012, New York, NY, USA, pp. 313–320. ACM (2012)
2. Emmerich, M., Deutz, A.: Time complexity and zeros of the hypervolume indicator gradient field. In: Schuetze, O., Coello, C.A., Tantar, A.-A., Tantar, E., Bouvry, P., Moral, P.D., Legrand, P. (eds.) EVOLVE - A Bridge between Probability, Set Oriented Numerics, and Evolutionary Computation III. SCI, vol. 500, pp. 169–193. Springer, Heidelberg (2014)
3. Emmerich, M.T.M., Deutz, A.H., Beume, N.: Gradient-based/evolutionary relay hybrid for computing pareto front approximations maximizing the s-metric. In: Bartz-Beielstein, T., Blesa Aguilera, M.J., Blum, C., Naujoks, B., Roli, A., Rudolph, G., Sampels, M. (eds.) HCI/ICCV 2007. LNCS, vol. 4771, pp. 140–156. Springer, Heidelberg (2007)
4. Kerschke, P., Preuss, M., Wessing, S., Trautmann, H.: Detecting funnel structures by means of exploratory landscape analysis. In: Proceedings of the 17th Annual Conference on Genetic and Evolutionary Computation, pp. 265–272. ACM (2015)
5. Kerschke, P., Preuss, M., Wessing, S., Trautmann, H.: Low-budget exploratory landscape analysis on multiple peaks models. In: Proceedings of the 18th Annual Conference on Genetic and Evolutionary Computation. ACM (2016, accepted)
6. Liefooghe, A., Verel, S., Daolio, F., Aguirre, H., Tanaka, K.: A feature-based performance analysis in evolutionary multiobjective optimization. In: Gaspar-Cunha, A., Henggeler Antunes, C., Coello, C.C. (eds.) EMO 2015. LNCS, vol. 9019, pp. 95–109. Springer, Heidelberg (2015)
7. Mersmann, O., Bischl, B., Trautmann, H., Preuss, M., Weihs, C., Rudolph, G.: Exploratory landscape analysis. In: Proceedings of the 13th Annual Conference on Genetic and Evolutionary Computation, GECCO 2011, New York, NY, USA, pp. 829–836. ACM (2011)

8. Rudolph, G., Naujoks, B., Preuss, M.: Capabilities of EMOA to detect and preserve equivalent pareto subsets. In: Obayashi, S., Deb, K., Poloni, C., Hiroyasu, T., Murata, T. (eds.) EMO 2007. LNCS, vol. 4403, pp. 36–50. Springer, Heidelberg (2007)
9. Verel, S., Liefooghe, A., Jourdan, L., Dhaenens, C.: On the structure of multiobjective combinatorial search space: MNK-landscapes with correlated objectives. Eur. J. Oper. Res. **227**(2), 331–342 (2013)
10. Wessing, S.: Two-stage methods for multimodal optimization. Ph.D. thesis, Technische Universität Dortmund (2015). http://hdl.handle.net/2003/34148

Population Diversity Measures Based on Variable-Order Markov Models for the Traveling Salesman Problem

Yuichi Nagata(✉)

Graduate School of Science and Technology, Tokushima University, Tokushima, Japan
nagata@is.tokushima-u.ac.jp

Abstract. This paper presents entropy-based population diversity measures that take into account dependencies between the variables in order to maintain genetic diversity in a GA for the traveling salesman problem. The first one is formulated as the entropy rate of a variable-order Markov process, where the probability of occurrence of each vertex is assumed to be dependent on the preceding vertices of variable length in the population. Compared to the use of a fixed-order Markov model, the variable-order model has the advantage of avoiding the lack of sufficient statistics for the estimation of the exponentially increasing number of conditional probability components as the order of the Markov process increases. Moreover, we develop a more elaborate population diversity measure by further reducing the problem of the lack of statistics

1 Introduction

Maintaining the genetic diversity in the population is one of the most important factors for bringing out the potential of genetic algorithms (GAs). One of the approaches to maintain population diversity is to design an appropriate measure of population diversity, which are used as a trigger to activate diversification procedures [8,9] and a part of the fitness function to maintain population diversity in a positive manner [3,5,11].

As is well known in information theory, entropy is a measure of the uncertainty of a probability distribution and it has been used to design population diversity measures. Most of the entropy-based population diversity measures are defined as the sum of the entropies of the univariate marginal distributions of all variables in the form of $-\sum_{i=1}^{n}\sum_{a\in A} P(X_i = a)\log P(X_i = a)$. This type of population diversity measure is widely used in GAs for the knapsack problem [4], binary quadratic programming problem [10], traveling salesman problem [3,5,7,8], and others [11]. This entropy measure, however, does not have an ability to capture dependencies between the variables.

In our previous work [6], we proposed an entropy-based diversity measure that takes into account dependencies between the variables, and this measure was used to maintain population diversity in a GA for the traveling salesman

J. Handl et al. (Eds.): PPSN XIV 2016, LNCS 9921, pp. 973–983, 2016.
DOI: 10.1007/978-3-319-45823-6_91

problem (TSP). This diversity measure, denoted as H_m, was formulated as the entropy rate of a Markov process of order m, where the probability of occurrence of each vertex at a certain position was assumed to be dependent on the m preceding vertices in the population (tours). The use of the diversity measure H_m with an appropriate value of m $(= 4)$ improved the performance of the GA.

In the practical use of a fixed-order Markov model, there is seldom sufficient data to accurately estimate the exponentially increasing number of conditional probability components as the order of the Markov model increases. A variable-order Markov model is useful to reduce this problem, where the probability of occurrence of each symbol is assumed to be dependent on the preceding symbols of variable length, which varies depending on the available statistics. Variable-order Markov models have been successfully applied to areas such as machine learning [1] and bioinformatics [2]. In this paper, we develop an population diversity measure based on a variable-order Markov model, which models the probability distribution of individuals in the population. Moreover, we improve this diversity measure by further reducing the problem of the lack of data.

The remainder of this paper is organized as follows. In Sect. 2, we first describe the diversity measure H_m and its variant proposed in the previous work. Then, we propose two entropy-based diversity measures derived from variable-order Markov models. The GA framework used to evaluate the proposed diversity measures is described in Sect. 4. Computational results are presented in Sect. 5 and conclusion is given in Sect. 6.

2 Previous Work

In [6], we proposed an entropy-based population diversity measure that takes into account dependencies in sequences of vertices included in the population of the GA for the TSP. This section outlines this work.

Let S_i $(i = 1, \ldots, n)$ be a random variable representing the i-th vertex in the tours of the population, where n is the number of the vertices (cities). The probability of occurrence of each vertex at a certain position is modeled as a Markov process of order m, where it is assumed to be dependent on the m preceding vertices in the tours of the population. Given that each tour has a cyclic structure, the joint probability distribution $P(S_1 = s_1, S_2 = s_2, \ldots, S_n = s_n)$, which is denoted as $P(s_1, s_2, \ldots, s_n)$ for simplicity, is represented by the following formula, where index $i + n$ $(1 \leq i \leq m)$ corresponds to i.

$$P(s_1, s_2, \ldots, s_n) = \prod_{i=1}^{n} P(s_{i+m} \mid s_i, \ldots, s_{i+m-1}) \tag{1}$$

Given that each tour can start from an arbitrary vertex, the joint probability distribution of any subset of the sequence of random variables should be invariant with respect to shifts in the index. Therefore, the entropy H of this joint probability distribution is equivalent to nH_m (Eq. 2), where H_m (Eq. 3) is the entropy rate of the Markov process of order m that models the probability of occurrence of each vertex in the population. For a more detailed explanation of Eq. 2, see the previous work. Equation 3 can be easily transformed into Eq. 5.

In information theory, the entropy rate of a data source is the average number of bits per symbol needed to encode it. Therefore, the existence of the same sequence consisting of up to $m+1$ vertices in the population will decrease the value of H_m.

$$H = -\sum_{s_1} \cdots \sum_{s_n} P(s_1, \ldots, s_n) \log P(s_1, \ldots, s_n) = nH_m \tag{2}$$

$$H_m = -\sum_{s_1} \cdots \sum_{s_{m+1}} P(s_1, \ldots, s_{m+1}) \log P(s_{m+1} \mid s_1, \ldots, s_m) \tag{3}$$

$$= -\sum_{s_1} \cdots \sum_{s_{m+1}} P(s_1, \ldots, s_{m+1}) \log \frac{P(s_1, \ldots, s_{m+1})}{P(s_1, \ldots, s_m)} \tag{4}$$

$$= \overline{H_{m+1}} - \overline{H_m}, \tag{5}$$

where

$$\overline{H_k} = -\sum_{s_1} \cdots \sum_{s_k} P(s_1, \ldots, s_k) \log P(s_1, \ldots, s_k). \tag{6}$$

To compute $\overline{H_k}$ in the asymmetric TSP, all sequences of length k are sampled in the population, and $P(s_1, \ldots, s_k)$ is estimated by $\frac{N(s_1,\ldots,s_k)}{nN_{pop}}$, where $N(s_1, \ldots, s_k)$ is the number of a sequence of vertices $\{s_1, \ldots, s_k\}$ in the population consisting of N_{pop} tours. In the symmetric TSP, the sampling is conducted in both travel directions and $P(s_1, \ldots, s_k)$ is estimated by $\frac{N(s_1,\ldots,s_k)}{2nN_{pop}}$.

Another diversity measure, denoted as H'_m, was also proposed. This measure is defined as the sum of the diversity measures H_k $(k = 1, \ldots, m)$, which can be simplified as Eq. 7. This diversity measure was designed in an ad hoc way to reduce the problem of the lack of statistics for the accurate estimation of H_m.

$$H'_m = H_1 + H_2 + \cdots + H_m = \overline{H_{m+1}} - \overline{H_1} \tag{7}$$

3 Population Diversity Measures Based on Variable-Order Markov Models

3.1 Motivation

The population diversity measure proposed in this paper is also defined as the entropy rate of a Markov process. We denote a set of the symbols generated from an information source as L. In what follows, we use random variables S_i $(i = \ldots, -2, -1, 0)$ to represent a Markov process, where S_0 represents the symbol to be observed next and S_{-i} $(i > 0)$ represents the i-th preceding symbol. The expression of H_m is therefore given by the following formula.

$$H_m = -\sum_{s_{-m}} \cdots \sum_{s_{-1}} \sum_{s_0} P(s_{-m}, \ldots, s_{-1}, s_0) \log P(s_0 \mid s_{-m}, \ldots, s_{-1}) \tag{8}$$

In theory, the value of H_m gives the entropy rate of an Markov process of order k (*i.e.*, $H_m = H_k$) as long as $k \leq m$ because $P(s_0 \mid s_{-m}, \ldots, s_{-k}, \ldots, s_{-1}) = P(s_0 \mid s_{-k}, \ldots, s_{-1})$ in this case. Therefore, m should be set to a greater value so that the entropy rate H_m has an ability to capture higher-order dependencies in a sequence of symbols generated from an information source. If m is too large, however, H_m would not be a meaningful population diversity measure because there is seldom sufficient samples of sequences in the population to accurately estimate the conditional probability distributions $P(s_0 \mid s_{-m}, \ldots, s_{-1})$, $s_{-m}, \ldots, s_0 \in L$, which are estimated as $\frac{N(s_{-m},\ldots,s_{-1},s_0)}{N(s_{-m},\ldots,s_{-1})}$. Therefore, there is a tradeoff between the potential ability to capture higher-order dependencies and the estimate accuracy of the conditional probability distributions. The population diversity measures proposed in this paper aim to capture higher-order dependencies in sequences of vertices in the population while reducing the problem of the lack of data.

3.2 A Population Diversity Measure H_m^{tr1}

We model the probability of occurrence of a symbol (vertex) appearing in sequences of symbols (sequences of vertices in the population) as a variable-order Markov process. In a variable-order Markov process, the probability distribution of the next symbol s_0 depends on the preceding symbols of variable length k. The basic idea is to determine the value of k adaptively so that the number of samples $N(s_{-k}, ..., s_{-1})$ is a sufficient statistic for estimating the conditional probability distribution $P(s_0 \mid s_{-k}, \ldots, s_{-1})$. For example, if a specific sequence of symbols $\{\ldots, s'_{-3}, s'_{-2}, s'_{-1}\}$ is observed at a certain point, the conditional probability distribution of occurrence of the next symbol s_0 is modeled as $P(s_0 | s'_{-k}, ..., s'_{-1})$ such that the number of samples $N(s'_{-k}, ..., s'_{-1})$ is greater than a predefined minimum number of samples.

A variable-order Markov process is characterized by a set of the conditional probability distributions: $P(s_0|\boldsymbol{s}_c)$, $\boldsymbol{s}_c \in \mathcal{S}$, where $\mathcal{S}$ is a set of sequences of symbols for the conditioning variables and each element $\boldsymbol{s}_c$ represents a specific sequence of symbols of any length that is less than or equal to m. Here, we put the upper limit on the length of sequences for the conditioning variables because it is impractical to store all conditional probability components if m is too large (e.g. $m > 10$). For any sequence of symbols $\{\ldots, s_{-2}, s_{-1}\}$ at a certain point, the length of the sequence assigned to the conditioning variables must be uniquely determined. To represent set $\mathcal{S}$ that satisfies this requirement, a so-called *context tree* is useful. Let $\tilde{\boldsymbol{s}_c}$ be the reverse sequence of $\boldsymbol{s}_c$ and $\tilde{\mathcal{S}} = \{\tilde{\boldsymbol{s}_c} | \boldsymbol{s}_c \in \mathcal{S}\}$. The elements of $\tilde{\mathcal{S}}$ are represented as the leaf nodes of a context tree as illustrated in Fig. 1 (Left), where every node has either 0 or $|L|$ children.

The entropy rate of the variable-order Markov process, which we denote as H_m^{tr1}, is then defined by the following formula.

$$H_m^{tr1} = -\sum_{\boldsymbol{s}_c \in \mathcal{S}} \sum_{s_0 \in L} P(\boldsymbol{s}_c, s_0) \log P(s_0 \mid \boldsymbol{s}_c) = -\sum_{\boldsymbol{s}_c \in \mathcal{S}} \sum_{s_0 \in L} P(\boldsymbol{s}_c, s_0) \log \frac{P(\boldsymbol{s}_c, s_0)}{P(\boldsymbol{s}_c)}$$
$$= -\sum_{\boldsymbol{s}_c \in \mathcal{S}} \sum_{s_0 \in L} P(\boldsymbol{s}_c, s_0) \log P(\boldsymbol{s}_c, s_0) + \sum_{\boldsymbol{s}_c \in \mathcal{S}} P(\boldsymbol{s}_c) \log P(\boldsymbol{s}_c) \quad (9)$$

The entropy rate H_m^{tr1} is closely related to H_m. If a context tree $\tilde{\mathcal{S}}$ is represented as a perfect tree with depth m, H_m^{tr1} is equivalent to H_m, meaning that H_m^{tr1} is a generalization of H_m. In addition, H_m^{tr1} can be viewed as an approximation of H_m. In fact, H_m^{tr1} is obtained from H_m though the approximation of $P(s_0 \mid s_{-m}, \ldots, s_{-k}, \ldots, s_{-1}) = P(s_0 \mid s_{-k}, \ldots, s_{-1})$ for all $\{s_{-k}, \ldots, s_{-1}\} \in \mathcal{S}$.

Next, we describe how to determine set $\tilde{\mathcal{S}}$ (and equivalently $\mathcal{S}$). The corresponding context tree $\tilde{\mathcal{S}}$ is updated at fixed intervals (see Sect. 4) by the following procedure, where *ratio* is a parameter taking a value between 0 and 1.

1. $\tilde{\mathcal{S}}$ is initialized as the perfect tree of depth one, *i.e.*, $\tilde{\mathcal{S}} = \{s_{-1} | s_{-1} \in L\}$.
2. For each of the leaf nodes $\{s_{-1}, ..., s_{-k}\} \in \tilde{\mathcal{S}}$, if there exists at least one value $s'_{-(k+1)} \in L$ such that $N(s'_{-(k+1)}, s_{-k}, \ldots, s_{-1}) \geq N_{pop} * ratio$, this node is expanded to generate the new leaf nodes $\{s_{-1}, ..., s_{-k}, s_{-(k+1)}\}$, $s_{-(k+1)} \in L$.
3. Expansions of the leaf nodes are iterated until no expansion is possible or the depth of each leaf node reaches the predefined maximum number m. The resulting tree $\tilde{\mathcal{S}}$ is returned.

The aim behind the expansion of a leaf node $\{s_{-1}, ..., s_{-k}\} \in \tilde{\mathcal{S}}$ is to capture higher-order dependency expressed as the conditional probability distribution $P(s_0 | s'_{-(k+1)}, s_{-k}, \ldots, s_{-1})$ only when it is judged to be reliable. The parameter *ratio* balances the tradeoff between the potential ability to capture higher-order

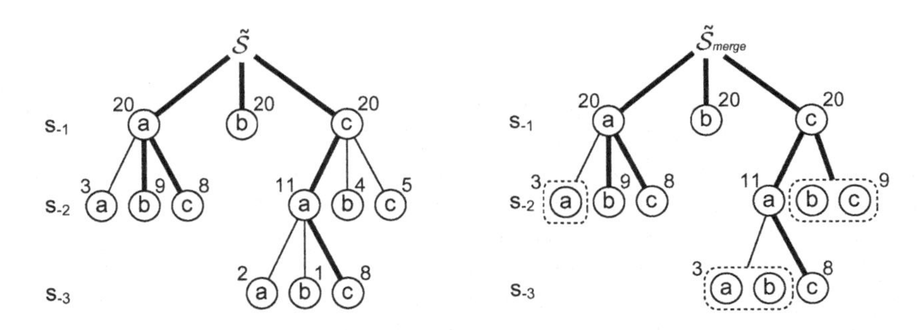

Fig. 1. (Left) A context tree representation of $\tilde{\mathcal{S}}$, where $L = \{a, b, c\}$ and the threshold is 8. Each node is connected by a thick link if the number of the corresponding sequence in the population (indicated beside each node) is greater than or equal to the threshold. The corresponding Markov process is defined as follows: $P(s_0|a, a)$, $P(s_0|b, a)$, $P(s_0|c, a)$, $P(s_0|b)$, $P(s_0|a, a, c)$, $P(s_0|b, a, c)$, $P(s_0|c, a, c)$, $P(s_0|b, c)$, and $P(s_0|c, c)$. (Right) A context tree representation of $\tilde{\mathcal{S}}_{merge}$ obtained from $\tilde{\mathcal{S}}$. Nodes in each dotted frame are merged. The corresponding Markov process is defined as follows: $P(s_0|a, a)$, $P(s_0|b, a)$, $P(s_0|c, a)$, $P(s_0|b)$, $P(s_0|a \vee b, a, c)$, $P(s_0|c, a, c)$ and $P(s_0|b \vee c, c)$.

dependencies and the estimate accuracy of the conditional probability distributions. However, this expansion collaterally generates unreliable conditional probability distributions $P(s_0|s''_{-(k+1)}, s_{-k}, \ldots, s_{-1})$ for $s''_{-(k+1)} \in L \setminus \{s'_{-(k+1)}\}$ because the values of $N(s''_{-(k+1)}, s_{-k}, \ldots, s_{-1})$ are less than the predefined threshold. Note that if expansion of a leaf node is allowed only if the number of samples is greater than the threshold for all child nodes, no expansion is likely to occur.

3.3 A Population Diversity Measure H_m^{tr2}

As suggested in the previous subsection, unreliable conditional probability distributions included in the formulation of H_m^{tr1} have a potentially harmful effect on the evaluation of population diversity. To reduce this problem, we modify the variable-order Markov model used to derive H_m^{tr1}.

The basic idea is to merge unreliable conditional probability distributions into a single one in order to increase the number of samples for the conditioning variables. One simple method is to merge the unreliable conditional parts $\{s''_{-(k+1)}, s_{-k}, \ldots, s_{-1}\}$, $s''_{-(k+1)} \in L \setminus \{s'_{-(k+1)}\}$ into a single one. Figure 1 illustrates an example where the unreliable conditional parts (nodes connected by thin links) in $\tilde{\mathcal{S}}$ are merged accordingly. We denote the resulting set of sequences of symbols for the conditioning variables and corresponding context tree as $\tilde{\mathcal{S}}_{merge}$. For example, a merged conditional probability distribution $P(s_0|b \vee c, c)$ is estimated by $\frac{N(b,c,s_0)+N(c,c,s_0)}{N(b,c)+N(c,c)}$. Although the number of samples for a merged conditional part may be still less than the predefined threshold, the problem of the lack of sufficient data will be alleviated. We denote the entropy rate of the variable-order Markov process defined by $\tilde{\mathcal{S}}_{merge}$ as H_m^{tr2}.

4 GA Framework

To evaluate the ability of the proposed population diversity measures H_m^{tr1} and H_m^{tr2}, we perform the GA proposed in [5] as in the case of the previous work [6]. Algorithm 1 gives the GA framework where brief comments are written directly in the algorithm. For more details, see the previous work [6].

An important point is that each of the population diversity measures is incorporated into the evaluation function used for selecting individuals to survive (line 8). Let L be the average tour length of the population and H the population diversity measure (H_m^{tr1} or H_m^{tr2}). For each individual $y \in \{c_1, \ldots, c_{N_{ch}}, p_A\}$, it is evaluated by the following evaluation function (Eq. 10), and the one with the smallest value is selected to replace the population member selected as p_A. Here, $\Delta L(y)$ and $\Delta H(y)$ denote the differences in L and H, respectively, when $x_{r(i)}$ $(= p_A)$ is replaced with an offspring solution y. This evaluation function is motivated to minimize $L - TH$ after the replacement, where T is a coefficient that takes a balance between the influences from L and H and it is adaptively updated (basically decreased) during the course of the search. Note that offspring

Algorithm 1. Procedure GA

1: Generate initial population $\{x_1, \ldots, x_{N_{pop}}\}$; // a simple local search is used;
2: **repeat**
3: Update $\tilde{S}$ or $\tilde{S}_{merge}$ based on the procedure described in Section 3 ;
4: Let $r(\cdot)$ be a random permutation of $1, \ldots, N_{pop}$;
5: **for** $i := 1$ to N_{pop} **do**
6: $p_A := x_{r(i)}$, $p_B := x_{r(i+1)}$; // set a pair of parents
7: $\{c_1, \ldots, c_{N_{ch}}\}$:= Crossover(p_A, p_B); // generate N_{ch} offspring solutions using edge assembly crossover
8: $x_{r(i)}$:= Select_Best($c_1, \ldots, c_{N_{ch}}, p_A$); // select the best individual to replaces the population member selected as p_A
9: **end for**
10: **until** a termination condition is satisfied
11: **return** the best individual in the population;

solutions that increase L are never selected in order to prevent the population from not converging, *i.e.*, no replacement occurs when p_A itself is selected.

$$Eval(y) = \begin{cases} \Delta L(y) - T\Delta H(y) \ (\Delta L(y) \leq 0) \\ \infty \ (\Delta L(y) > 0) \end{cases} \tag{10}$$

For every offspring solution y, $\Delta H(y)$ can be computed in $O(km)$ time, where k is the number of edges of an offspring solution y that do not exist in the parent p_A (k is usually much smaller than n). Each time p_A is replaced with the selected offspring solution, the values of $N(\cdot)$, which are stored in the form of a tree, can be updated in $O(km)$ time.

5 Experimental Results

5.1 Experimental Settings

To investigate the ability of the proposed population diversity measures H_m^{tr1} and H_m^{tr2}, we performed the GA described in the previous section by using each of the population diversity measures in the evaluation function (Eq. 10). The parameters for the GA were set as follows: $N_{pop} = 300$ and $N_{ch} = 30$. Note that the same settings were used in the previous work [6] for evaluating the population diversity measures H_m and H'_m. We tested the proposed population diversity measures with the following parameter settings.

- H_m^{tr1} ($m = 6$, $ratio = 0.05, \ 0.1, \ 0.2, \ 0.3$)
- H_m^{tr2} ($m = 6$, $ratio = 0.05, \ 0.1, \ 0.2, \ 0.3$)
- H_m^{tr2} ($m = 8$, $ratio = 0.1$)

For each setting, we performed the GA 30 times on 21 instances with sizes ranging from 10,000 to 25,000 in the well-known benchmark sets: TSPLIB (http://comopt.ifi.uni-heidelberg.de/software/TSPLIB95/), National TSPs (http://www.math.uwaterloo.ca/tsp/data/index.html), and VLSI TSPs.

5.2 Results

Table 1 shows the solution quality of the GA using the proposed population diversity measures H_m^{tr1} and H_m^{tr2} in the following format: the instance name (instance) together with the optimal or best known solution (Opt. or UB), the number of runs that succeeded in finding the optimal or best-known solution (#S), and the average percentage excess over the optimal or best-known solutions (A-Err). The result of the GA using the diversity measure H_4, which achieved the best solution quality among H_m $(m = 1, 2, 3, 4, 6)$ in the previous work, are also presented for a baseline comparison. We performed the one-sided Wilcoxon rank sum test for the null hypothesis that the median of the distribution of tour length obtained by the GA using each of H_m^{tr1} and H_m^{tr2} is greater than that of GA using H_4. If the null hypothesis is rejected as a significant level of 0.05, results in the table are indicated by the asterisk. In addition, results are also indicated by the dagger if the opposite null hypothesis is rejected.

Let us first summarize the results of the GA using the population diversity measures H_m and H'_m $(m = 1, 2, 3, 4, 6, 8)$ proposed in the previous work. Table 2 shows only averaged results taken from [6] (results of $m = 8$ are newly added). As indicated in Table 2, the diversity measure H_m improves the ability in evaluating population diversity with increasing the value of m up to 4, but the greater values of m deteriorates the ability due to the lack of available samples necessary to estimate the conditional probability distributions. The diversity measure H'_m also improves the ability in evaluating population diversity with increasing the value of m up to 6. Moreover, the result of H_6' is better than that of H_4. Considering the definition of H'_m, this result suggests that the diversity measure H'_6 achieves a better balance between the ability to capture higher-order dependencies and the estimate accuracy of the conditional probability distributions, although the definition of H'_m is somewhat ad-hoc.

Next, we focus on the results of the diversity measure H_6^{tr1}. Table 1 shows that the GA using H_6^{tr1} achieves the best solution quality when the parameter *ratio* is set to 0.2 or 0.3. For a smaller value of *ratio*, the solution quality is deteriorated. This is a predictable consequence because if the value of *ratio* is too small, it would not be likely to obtain a sufficient statistics from the population necessary for the accurate estimation of the conditional probability distributions. The use of H_6^{tr1} with the best parameter value for *ratio* (= 0.2 or 0.3), however, shows only a slight improvement over the use of H_4.

Next, we focus on the results of the diversity measure H_6^{tr2}. The GA using H_6^{tr2} achieves the best solution quality when the parameter *ratio* is set to 0.1, which is less than the best parameter value for H_6^{tr1}. Moreover, the best result of H_6^{tr2} is better than that of H_6^{tr1}. These observations indicate that the use of H_m^{tr2} succeeds in capturing higher-order dependencies while reducing the problem of the lack of sufficient samples for the accurate estimation of the conditional probability distributions. Compared to H_4, the use of H_6^{tr2} with the best parameter value for *ratio* (= 0.1) significantly improves the solution quality in four instances. However, this result is almost same as that of H'_6.

Table 1. Solution quality of the GA using the diversity measures H_m^{tr1} and H_m^{tr2}

		H_4		H_6^{tr1}							
				$ratio = 0.05$		$ratio = 0.1$		$ratio = 0.2$		$ratio = 0.3$	
Instance	Opt.(UB)	#S	A-Err	#S	A-Err	#S	A-Err	#S	A-Err	#S	A-Err
xmc10150	(28387)	24	0.00070	26	0.00047	29	0.00012*	29	0.00012*	28	0.00035
fi10639	520527	24	0.00011	21	0.00021	23	0.00016	23	0.00018	24	0.00010
rl11849	923288	25	0.00014	18	0.00037†	19	0.00030†	21	0.00026	27	0.00013
usa13509	19982859	22	0.00010	17	0.00017	24	0.00008	21	0.00010	19	0.00014
xvb13584	(37083)	29	0.00009	23	0.00081†	26	0.00036	27	0.00027	29	0.00009
brd14051	469385	23	0.00017	19	0.00026	26	0.00008	27	0.00005	26	0.00011
mo14185	(427377)	19	0.00014	20	0.00014	19	0.00018	18	0.00015	19	0.00016
xrb14233	(45462)	10	0.00279	9	0.00286	11	0.00279	10	0.00301	12	0.00271
d15112	1573084	16	0.00014	17	0.00008	16	0.00007	15	0.00005	17	0.00003
it16862	557315	6	0.00023	4	0.00044†	2	0.00040†	2	0.00039†	6	0.00030
xia16928	(52850)	24	0.00076	23	0.00050	18	0.00101	19	0.00095	16	0.00164†
pjh17845	(48092)	13	0.00132	17	0.00097	15	0.00125	15	0.00104	13	0.00118
d18512	645238	21	0.00009	20	0.00009	22	0.00008	23	0.00009	25	0.00007
frh19289	(55798)	30	0.00000	26	0.00030†	26	0.00024†	28	0.00012	30	0.00000
fnc19402	(59287)	19	0.00067	16	0.00079	18	0.00067	17	0.00079	19	0.00062
ido21215	(63517)	23	0.00058	18	0.00105	22	0.00058	27	0.00016	17	0.00110†
fma21553	(66527)	15	0.00090	10	0.00120	8	0.00120	16	0.00070	21	0.00050
vm22775	569288	0	0.00140	1	0.00141	0	0.00131	0	0.00121	1	0.00119
lsb22777	(60977)	21	0.00055	19	0.00060	22	0.00044	28	0.00011*	24	0.00033
xrh24104	(69294)	29	0.00005	28	0.00010	26	0.00019	28	0.00010	29	0.00005
sw24978	855597	9	0.00039	11	0.00047	14	0.00037	12	0.00031	11	0.00024
Average		19.1	0.00054	17.3	0.00063	18.4	0.00057	19.3	0.00048	19.7	0.00053

		H_6^{tr2}								H_8^{tr2}	
		$ratio = 0.05$		$ratio = 0.1$		$ratio = 0.2$		$ratio = 0.3$		$ratio = 0.1$	
Instance	Opt.(UB)	#S	A-Err	#S	A-Err	#S	A-Err	#S	A-Err	#S	A-Err
xmc10150	(28387)	26	0.00059	25	0.00070	28	0.00023	26	0.00047	28	0.00023
fi10639	520527	20	0.00013	25	0.00008	24	0.00010	28	0.00004	27	0.00005
rl11849	923288	23	0.00019	29	0.00004*	28	0.00006	28	0.00006	25	0.00013
usa13509	19982859	21	0.00011	23	0.00016	24	0.00009	25	0.00007	23	0.00010
xvb13584	(37083)	27	0.00036	29	0.00009	25	0.00045†	25	0.00045†	25	0.00045†
brd14051	469385	25	0.00015	26	0.00009	27	0.00007	29	0.00003*	27	0.00006
mo14185	(427377)	25	0.00005*	23	0.00009	22	0.00013	17	0.00020	24	0.00012
xrb14233	(45462)	9	0.00308	8	0.00330	3	0.00396†	6	0.00374†	12	0.00249
d15112	1573084	15	0.00007	18	0.00005	20	0.00004	18	0.00004	18	0.00003
it16862	557315	6	0.00033	5	0.00027	3	0.00037†	5	0.00032	2	0.00035
xia16928	(52850)	22	0.00082	22	0.00069	23	0.00088	9	0.00227†	25	0.00063
pjh17845	(48092)	11	0.00132	19	0.00083	19	0.00083	19	0.00076*	22	0.00062*
d18512	645238	19	0.00010	21	0.00007	19	0.00014	23	0.00006	24	0.00005
frh19289	(55798)	26	0.00042†	30	0.00000	29	0.00006	27	0.00024†	27	0.00018†
fnc19402	(59287)	22	0.00051	19	0.00062	20	0.00056	14	0.00118	20	0.00056
ido21215	(63517)	19	0.00079	25	0.00026	27	0.00021	23	0.00058	24	0.00031
fma21553	(66527)	23	0.00035*	22	0.00040*	19	0.00065	20	0.00055	26	0.00020*
vm22775	569288	1	0.00107*	2	0.00091*	3	0.00094*	1	0.00119	2	0.00097*
lsb22777	(60977)	23	0.00038	27	0.00016*	26	0.00022	26	0.00022	28	0.00011*
xrh24104	(69294)	28	0.00010	28	0.00010	28	0.00010	29	0.00005	29	0.00005
sw24978	855597	20	0.00020*	14	0.00033	17	0.00020*	12	0.00032	18	0.00023*
Average		19.6	0.00053	21.0	0.00044	20.7	0.00049	19.5	0.00061	21.7	0.00038

Table 2. Solution quality of the GA using the diversity measures H_m and H'_m

	$m = 1$		$m = 2$		$m = 3$		$m = 4$		$m = 6$		$m = 8$	
Div.	#S	A-Err	#S	A-Err	#S	A-Err	#S	A-Err	#S	A-Err	#S	A-Err
H_m	16.7	0.00085	18.1	0.00066	18.2	0.00065	19.1	0.00054	16.1	0.00063	11.8	0.00094
H'_m	16.7	0.00085	19.0	0.00069	19.8	0.00061	20.5	0.00053	21.5	0.00046	20.8	0.00047

Note: When no diversity measure is incorporated, #S and A-Err are 1.2 and 0.00544, respectively.

Next, we focus on the results of H_8^{tr2} with $ratio = 0.1$. We can see that the solution quality of H_8^{tr2} is better than that of H_6^{tr2}. This result also indicates that the core idea of H_m^{tr2} make it possible to successfully capture higher-order dependencies while reducing the problem of the lack of sufficient statistics. Moreover, the use of H_8^{tr2} achieves the best solution quality among all population diversity measures including H'_6.

6 Conclusion

The proposed population diversity measure H_m^{tr1} is defined as the entropy rate of the variable-order Markov process with the aim of capturing higher-order dependencies in sequences of vertices in the population while reducing the problem of the lack of sufficient statistics. The use of this diversity measure, however, has shown only a slight improvement in evaluating population diversity over the previously proposed entropy-based diversity measure H_m, which is based on the fixed-order Markov model. On the other hand, another variant of the proposed population diversity measure H_m^{tr2} has succeeded in improving the abilities of H_m and H_m^{tr1} by further reducing the problem of the lack of sufficient statistics. This research has shown a potential of entropy-based population diversity measures that take into account dependencies between the variables, and the efficacy of the proposed population diversity measures should be investigated on other permutation problems in the future work.

References

1. Begleiter, R., El-Yaniv, R., Yona, G.: On prediction using variable order markov models. J. Artif. Intell. Res. **22**, 385–421 (2004)
2. Colinge, J., Bennett, K.L.: Introduction to computational proteomics. PLoS Comput. Biol. **3**(7), e114 (2007)
3. Maekawa, K., Mori, N., Tamaki, H., Kita, H., Nishikawa, Y.: A genetic solution for the traveling salesman problem by means of a thermodynamical selection rule. In: Proceedings of 3rd IEEE Conference on Evolutionary Computation, pp. 529–534 (1996)
4. Mori, N., Kita, H., Nishikawa, Y.: Adaptation to a changing environment by means of the thermodynamical genetic algorithm. In: Ebeling, W., Rechenberg, I., Voigt, H.-M., Schwefel, H.-P. (eds.) PPSN 1996. LNCS, vol. 1141, pp. 513–522. Springer, Heidelberg (1996)

5. Nagata, Y., Kobayashi, S.: A powerful genetic algorithm using edge assembly crossover for the traveling salesman problem. Informs J. Comput. **25**(2), 346–363 (2013)
6. Nagata, Y., Ono, I.: High-order sequence entropies for measuring population diversity in the traveling salesman problem. In: Middendorf, M., Blum, C. (eds.) EvoCOP 2013. LNCS, vol. 7832, pp. 179–190. Springer, Heidelberg (2013)
7. Tsai, H., Yang, J., Tsai, Y., Kao, C.: An evolutionary algorithm for large traveling salesman problems. IEEE Trans. Syst. Man Cybern. B Cybern. **34**(4), 1718–1729 (2004)
8. Tsujimura, Y., Gen, M.: Entropy-based genetic algorithm for solving tsp. In: Proceedings of 2nd International Conference on Knowledge-Based Intelligent Electronic Systems, pp. 285–290. IEEE (1998)
9. Vallada, E., Ruiz, R.: Genetic algorithms with path relinking for the minimum tardiness permutation flowshop problem. Omega **38**(1), 57–67 (2010)
10. Wang, Y., Lü, Z., Hao, J.-K.: A study of multi-parent crossover operators in a memetic algorithm. In: Schaefer, R., Cotta, C., Kołodziej, J., Rudolph, G. (eds.) PPSN XI. LNCS, vol. 6238, pp. 556–565. Springer, Heidelberg (2010)
11. Zhang, C., Su, S., Chen, J.: Efficient population diversity handling genetic algorithm for QoS-aware web services selection. In: Alexandrov, V.N., Albada, G.D., Sloot, P.M.A., Dongarra, J. (eds.) ICCS 2006. LNCS, vol. 3994, pp. 104–111. Springer, Heidelberg (2006)

Convergence Versus Diversity in Multiobjective Optimization

Shouyong Jiang(✉) and Shengxiang Yang

Centre for Computational Intelligence (CCI),
School of Computer Science and Informatics, De Montfort University,
The Gateway, Leicester LE1 9BH, UK
shouyong.jiang@email.dmu.ac.uk, syang@dmu.ac.uk

Abstract. Convergence and diversity are two main goals in multiobjective optimization. In literature, most existing multiobjective optimization evolutionary algorithms (MOEAs) adopt a convergence-first-and-diversity-second environmental selection which prefers nondominated solutions to dominated ones, as is the case with the popular nondominated sorting based selection method. While convergence-first sorting has continuously shown effectiveness for handling a variety of problems, it faces challenges to maintain well population diversity due to the overemphasis of convergence. In this paper, we propose a general diversity-first sorting method for multiobjective optimization. Based on the method, a new MOEA, called DBEA, is then introduced. DBEA is compared with the recently-developed nondominated sorting genetic algorithm III (NSGA-III) on different problems. Experimental studies show that the diversity-first method has great potential for diversity maintenance and is very competitive for many-objective optimization.

1 Introduction

Multiobjective optimization problems (MOPs) widely exist in real-world applications, such as scheduling [11] and design [12]. MOPs often have several conflicting objectives for which any improvement in one objective inevitably aggravates another. Due to multiobjectivity, there is no single optimal solution. Instead, the optima of MOPs is a set of trade-off solutions, known as Pareto-optimal set (POS). Correspondingly, the image of the POS in the objective space is called Pareto-optimal front (POF).

Multiobjective optimization evolutionary algorithms (MOEAs) are a class of important methods for solving MOPs. MOEAs employ a population of candidate individuals and optimize them in an evolutionary manner. As a result, a set of solutions can be obtained in a single run. Besides, MOEAs do not necessarily require any knowledge and information of the MOPs to be optimized, i.e., continuousness or differentiability. All these features make MOEAs very suitable for solving MOPs. So far, a large number of MOEAs [2,9,10] have been proposed in the evolutionary computation community.

In the design of MOEAs, two goals should be considered: (1) minimizing the gap between candidate solutions and the true POS (convergence) and (2)

J. Handl et al. (Eds.): PPSN XIV 2016, LNCS 9921, pp. 984–993, 2016.
DOI: 10.1007/978-3-319-45823-6_92

maximizing the distribution of candidate solutions (diversity). However, these two goals are generally assumed to be conflicting [13]. In practice, most existing MOEAs achieve convergence by prior Pareto-based sorting of the evolving population and diversity by the additional calculation of individuals' density information. The well-known nondominated sorting genetic algorithm II (NSGA-II) [2] and strength pareto evolutionary algorithm 2 (SPEA2) [10] are representative examples of this method. Such a method actually performs environmental selection in a convergence-first-and-diversity-second manner. That is, nondominated individuals [2] are preferable to dominated ones although dominated individuals may contribute considerably to population diversity. While this method works well in two- and three-objective optimization problems, it has encountered great difficulties in many-objective optimization where problems have four or more objectives [6]. This is mainly because a large portion of the population becomes nondominated as the number of objectives increases. In this case, the convergence-first selection will consider only nondominated individuals and leave little room for diversity selection. If all nondominated individuals are themselves not diversified, it would lead to a detrimental diversity loss due to the convergence-first selection.

Inspired by the assumption that dominated individuals can contribute to population diversity, this paper proposes a new diversity-first sorting approach with the aid of a set of diverse reference directions. The approach sorts the popualtion into different fronts, each front representing a level of diversity and convergence. Then, a diversity-first sorting based evolutionary algorithm (DBEA) is introduced. Empirical studies and algorithm comparisons demonstrate the promise of DBEA for multi- and many-objective optimization.

The remainder of the paper is organized as follows. Section 2 reviews the classic nondominated sorting method. Section 3 presents a new diversity-first sorting method, followed by our detailed implementation of DBEA in Sect. 4. Experimental design and comparison results are presented in Sect. 5. Finally, Sect. 6 concludes the paper.

2 Classic Sorting Methods

Most existing MOEAs are convergence-first based methods, such as NSGA-II [2] and SPEA2 [10]. Convergence-first based methods prefer convergence to diversity. They sort population depending mainly on individuals' convergence[1]. One of the most important sorting methods is the nondominated sorting used in NSGA-II. In the following, we will briefly describe how nondominated sorting works, followed by some discussions on its advantages and disadvantages.

2.1 Nondominated Sorting

In every generation, when the parent population (P) and offspring population (Q) are combined to form a union population (R) of size $2N$, environmental

[1] Note that, although some algorithms like SPEA2 sort individuals by exploiting both convergence and diversity, convergence is priorly considered and emphasised.

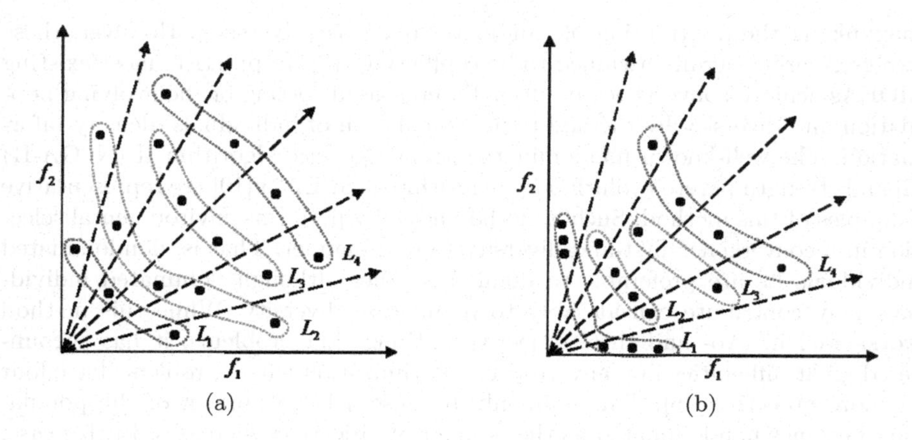

Fig. 1. Nondominated sorting.

selection should be carried out on R to construct a new parent population of size N for the next generation. The nondominated sorting strategy can be used for selection and works as follows. First, each individual is compared with all other individuals in R, and all nondominated solutions of R are identified and assigned to front L_1. Then, individuals in L_1 are removed from R, and the remaining individuals in R are compared with each other to determine the nondominated set, which are assigned to front L_2. The procedure is repeated until no individuals are left in R, i.e., all individuals have been assigned to a front.

Figure 1(a) gives a graphical illustration of the nondominated sorting. The main idea behind the nondominated sorting is to classify the entire combined population into different nondominated fronts according to individuals' convergence. After the non-dominated sorting, the new population can be constructed by selecting solutions of different non-dominated fronts, one at a time. The selection starts with individuals of the first front L_1 and continues with those of the second front L_2, followed by the rest of the fronts and so on. Since only N slots are allowable in the new population, not all fronts can be considered. When the last allowed front (e.g., L_l) is being considered, there may exist more individuals in L_l than the remaining slots in the new population. In this situation, niche-preservation strategies, such as crowding distance [2], the-farthest-the-first method [1], and nearest neighbour technique [10], are desirable for selecting the remaining number of individuals from front L_l in order to maintain diversity.

2.2 Advantages and Disadvantages

Advantages. The nondominated-sorting based selection favours convergence so that individuals in better fronts will be priorly preserved. The selection is helpful for a fast convergence speed.

Disadvantages. The nondominated-sorting based selection may undermine population diversity if well-converged individuals are not diversified. Figure 1(b) presents an example where diversity loss occurs if only six nondominated individuals are allowed to be preserved. The loss of diversity in the example could further incur evolutionary stagnation where overcrowded boundary regions are overexploited and intermediate regions are left unexplored.

3 Proposed Sorting Method

On the basis of discussions on convergence-first MOEAs, we propose and analyse a diversity-first sorting method in the following subsections.

3.1 Diversity-First Sorting

The proposed diversity-first sorting method works as follows. First, the objective space is partitioned into a number of subspaces with the aid of a reference direction set W. Reference directions in W are required to be uniformly distributed. Then, each individual (whose objective values need to be normalized beforehand) in the combined population is associated with a subspace. This can be done by identifying the nearest reference direction to the considered individual. In each subspace, individuals are assigned a fitness value that can reflect its convergence level. Potential fitness assignment approaches for this purpose can be scalarizing functions used in MOEA/D [9], strength fitness in SPEA2 [10], or nondominated ranks in NSGA-II [2], whichever is the easiest for users to implement. An individual with the best fitness from each subspace is assigned to front L_1. After that, the individual in L_1 are removed from the subspaces, and another with the best fitness from each subspace is assigned to front L_2. If multiple solutions have the same fitness, a random one is considered. This procedure continues until each individual in each subspace has been assigned to a front. Note that, in case that a subspace is empty, this subspace is skipped.

Figure 2(a) illustrates the outcome of diversity-first sorting, where population distribution is identical to that of Fig. 1(a). After the sorting, the new population can be constructed by selecting solutions of different fronts, one at a time. Similar to the nondominated sorting, not all fronts can be considered due to the limited number of slots in the new population. If the last allowed front (e.g., L_l) has more individuals than the remaining slots, random selection on L_l can be performed to fill up the new population. Note that, it is advisable to use techniques that are helpful for convergence to select individuals from L_l. For example, fitness assignment can be performed on L_l, and individuals with relatively good convergence are priorly selected.

3.2 Advantages and Disadvantages

Advantages. As can be seen from Fig. 2(b) (where the population distribution is the same as that of Fig. 1(b)), the diversity-first sorting enhances local diversity

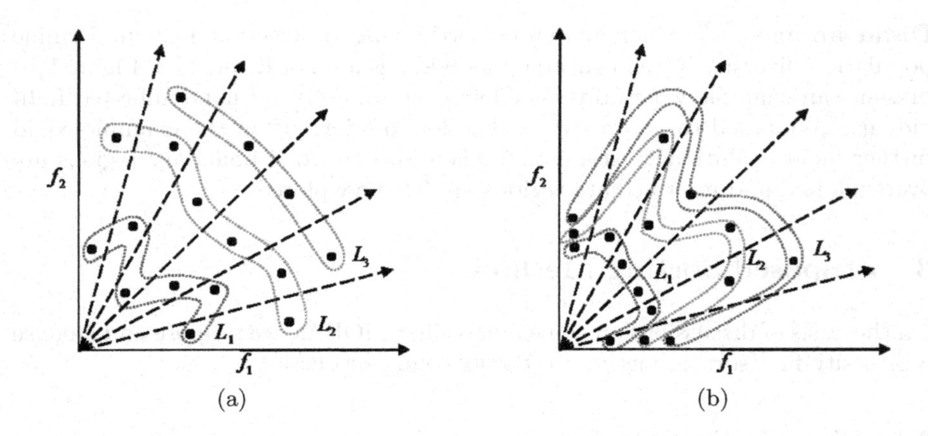

Fig. 2. Diversity-first sorting.

in each subspace. As a result, population diversity can be well maintained during the evolution. Besides, population convergence is also properly considered in the course of sorting. The sorting method can provide a good coverage and spread of approximation.

Disadvantages. Since the diversity-first sorting employs a reference direction set for population partition, the resulting population distribution depends largely on the uniformity of the reference direction set. The potential drawbacks remain unknown, and a future work will be devoted to these aspects.

4 Diversity-First Based Evolutionary Algorithm (DBEA)

In this section, we present an MOEA based on the proposed diversity-first sorting, called DBEA for short. The framework of DBEA is described in Algorithm 1. Several key components of DBEA are explained as follows.

Reference direction set W: W can be constructed on a unit simplex using Das and Dennis's systematic approach [4] if the number of objectives is small. Otherwise, W is constructed by two-layered approach mentioned in [6].

Objective normalization: similar to NSGA-III [6], DBEA identifies all extreme points and then use them to construct a hyperplane. The intercepts of objective axes and the hyperplane can be computed. DBEA uses these intercepts and the utopia point to normalize the objective values of individuals.

Population partition: for each population individual, DBEA computes the acute angle between its normalized objective vector and each reference direction. The reference direction having the minimum acute angle is considered the right subspace that the individual should reside in.

Algorithm 1. Framework of DBEA

1: **Input**: N (population size)
2: **Output**: approximated Pareto-optimal set
3: Generate a diverse reference direction set W:
4: Create an initial parent population P;
5: **while** *stopping criterion not met* **do**
6: Apply genetic operators on P to generate offspring population $\overline{P}$;
7: $Q := P \cup \overline{P}$; /*parent and offspring are combined*/
8: Normalize objectives of members in Q and partition Q into different subspaces;
9: $(L_1, L_2, \ldots)$:=diversity-first-sort(Q); /*diversity-first sorting is triggered*/
10: $(S_1, S_2, \ldots)$:=nondominated-sort(L_l); /*L_l is the last front to be included*/
11: Select continuously individuals from $(S_1, S_2, \ldots)$ until P is filled up;
12: **end while**

Diversity-first Sorting: to facilitate sorting, individuals should be distinguishable in terms of convergence. In this paper, DBEA simply applies nondominated sorting in each subspace, leading to each individual having a rank. Individuals with better rank values are priorly selected. In case there is a tie between individuals, those having the smallest perpendicular distance to the associated reference direction are preferred.

Selection on the last front to be included: DBEA performs the nondominated sorting on the last allowed front L_l, resulting in a series of subfronts $\{S_1, S_2, \ldots\}$. Then, DBEA selects individuals on these subfronts, starting from the first subfront S_1. If the last subfront to be included has more individuals than the remaining slots, individuals are randomly copied to the new population.

5 Experimental Study

To make a proper and fair comparison, algorithms to be considered should have similar framework other than different methodologies. In our experiments, we would like to compare DBEA with the recently-developed NSGA-III algorithm [6]. DBEA and NSGA-III have very similar framework but differ mainly in distinct sorting methods. For both algorithms, the simulated binary crossover (SBX) [3] and polynomial mutation [5] are used as variation operators. As suggested in [6], the crossover probability is $p_c = 1.0$ and its distribution index is $\eta_c = 30$. The mutation probability is $p_m = 1/n$ and its distribution $\eta_m = 20$. In the following subsections, DBEA and NSGA-III will be first tested on a hard problem that challenges algorithms' diversity performance. After that, these two algorithms will be compared on many-objective optimization.

5.1 Results on a Hard Three-Objective Problem

Liu *et al.* [8] introduced several hard-to-converge problems with considerably deceptive properties and strong variable linkages. As a testing example, we choose the three-objective MOP6 to distinguish the difference between diversity-first sorting and convergence-first sorting. MOP6 places deceptive attractors on

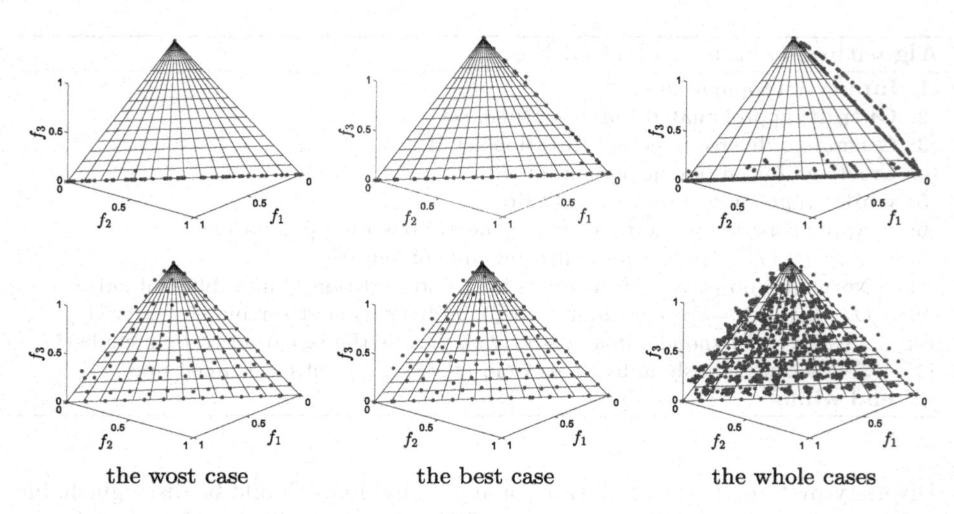

Fig. 3. Approximated POFs for MOP6 over 20 runs. Top: NSGA-III; bottom: DBEA.

the boundary of the objective space. If population diversity is not properly maintained, the search will get trapped into local optima. As a consequence, not all the POF regions can be found. Both algorithms use the systematic design method [4] to generate 91 reference directions or points. Correspondingly, the population size in DBEA and NSGA-III was set to 92, which, as suggested in NSGA-III, is the smallest multiple of four higher than reference points. The maximum number of generations was set to 5000, which is much higher than normal settings due to the hardness of this problem.

Figure 3 shows the worst-case, best-case, and whole MOP6 approximations of NSGA-III and DBEA over 20 independent runs. It can be clearly observed from the figure that, in all runs, NSGA-III prefers some boundary solutions and misses a large part of the POF of MOP6. In contrast, DBEA is always capable of obtaining a set of diversified solutions, although a few boundary solutions do not converge perfectly. The poor performance of NSGA-III is mainly caused by its convergence-first based selection. In NSGA-III, environmental selection is based on individuals' convergence level (nondominated sorting). That is, individuals on better sorting fronts have priority to be selected first. If the selected individuals all reside in a local search space (the boundary region in the case of MOP6), the evolution will experience a dramatic diversity loss, resulting in NSGA-III not being able to diversify the solution set any more. Thus, NSGA-III fails in this situation. On the contrary, the diversity-first based selection seems to be a wise option, as it can maintain population diversity at a high level. Therefore, DBEA shows better performance than NSGA-III on the considered test problem.

5.2 Results on WFG Problems

The previous subsection has demonstrated the superiority of DBEA over NSGA-III in the three-objective case in terms of diversity. One may wonder

Table 1. Best, mean, and worst IGD values of compared algorithms over 20 runs

	M=3			M=5			M=8		
Problem	NSGA-III	DBEA		NSGA-III	DBEA		NSGA-III	DBEA	
WFG4	**0.00946**	0.01068		**0.28054**	0.28103		0.71357	**0.71239**	
	0.01300	0.01313	–	0.28338	**0.28310**	≈	**0.72304**	0.72473	≈
	0.01714	0.01661		0.28749	**0.28557**		**0.73107**	0.74052	
WFG5	0.09445	**0.09386**		**0.34631**	0.34813		0.75591	**0.75563**	
	0.09683	**0.09483**	†	0.35063	**0.34879**	†	0.75870	**0.75820**	†
	0.09917	**0.09559**		0.35283	**0.35015**		0.76268	**0.76214**	
WFG6	**0.07036**	0.07828		0.34445	**0.33282**		**0.74635**	0.74721	
	0.12819	0.13388	–	**0.37310**	0.37950	≈	0.79313	**0.78546**	≈
	0.16738	0.19958		**0.40561**	0.41077		0.83568	**0.82690**	
WFG7	0.00993	**0.00920**		**0.27209**	0.27646		0.70625	**0.70584**	
	0.01232	0.01322	–	0.27733	**0.27720**	≈	0.72276	**0.72168**	†
	0.01553	0.01681		0.27974	**0.27943**		0.74034	**0.73439**	
WFG8	0.23728	**0.23639**		0.58692	**0.57160**		1.45400	**1.27610**	
	0.25205	**0.24869**	†	0.61987	**0.60271**	†	1.57690	**1.57910**	†
	0.26632	**0.26161**		0.62657	**0.61874**		1.73940	**1.72546**	
WFG9	0.07266	**0.05796**		0.39788	**0.38946**		**0.91319**	0.92767	
	0.11684	**0.11420**	†	0.42125	**0.41719**	†	1.00700	**0.99933**	†
	0.36338	**0.36184**		0.48428	**0.45785**		1.19820	**1.12270**	

whether DBEA can perform well in higher-dimensional cases. To this end, we test DBEA and NSGA-III on several WFG [7] test problems having three to eight objectives. Both algorithms use the same population size by setting identical reference directions or points with the two-layered method [6]. That is, 92, 210, and 156 for 3, 5, and 8 objectives, respectively. The maximum number of generations was 500, 1000, and 1500 for 3, 5, and 8 objectives, respectively. Each algorithm was executed 20 independent runs.

In order to quantify the performance of algorithms, we employ the reference point based inverted generational distance (IGD) suggested by [6] and hypervolume (HV) [14] as our performance metrics. The reference vector for the computation of HV was set as the nadir point of the true POF plus one. All reported HV values come from the normalization of originally computed HV values.

Tables 1 and 2 present the IGD and HV values of two compared algorithms, respectively, where the best results are highlighted in bold face. The Wilcoxon rank-sum test at a 0.05 significance level was employed to compare the statistical significance of difference between two algorithms. "†", "≈", and "–" in the tables denote DBEA is better than, equivalent to, and worse than NSGA-III, respectively. It is easy to see that, the performance of DBEA improves as the number of objectives increases. For three objectives, NSGA-III performs better than DBEA on WFG4, WFG6 and WFG7 whereas DBEA wins on the other

Table 2. Best, mean, and worst HV values of compared algorithms over 20 runs

Problem	M=3			M=5			M=8		
	NSGA-III	DBEA		NSGA-III	DBEA		NSGA-III	DBEA	
WFG4	**0.85281**	0.85230		**0.94849**	0.94844		0.98060	**0.98074**	
	0.85191	0.85191	−	0.94808	**0.94809**	≈	0.97992	**0.97995**	†
	0.85145	0.85151		0.94758	**0.94764**		**0.97908**	0.97836	
WFG5	0.82652	**0.82655**		0.91367	**0.91393**		0.93682	**0.93701**	
	0.82595	**0.82639**	†	0.91307	**0.91379**	†	0.93637	**0.93682**	†
	0.82541	**0.82614**		0.91244	**0.91321**		0.93527	**0.93642**	
WFG6	**0.83362**	0.83117		0.91607	**0.92151**		0.94384	**0.94624**	
	0.81705	0.81540	−	**0.90336**	0.90120	≈	0.92088	**0.92489**	†
	0.80628	0.79718		**0.89009**	0.88789		0.89975	**0.90262**	
WFG7	**0.85218**	0.85217		**0.94875**	0.94869		0.98069	**0.98071**	
	0.85180	0.85169	−	0.94838	**0.94841**	†	0.98030	**0.98042**	†
	0.85120	0.85120		0.94799	**0.94801**		0.97975	**0.97988**	
WFG8	0.80278	**0.80351**		0.88415	**0.88548**		0.91022	**0.92898**	
	0.80088	**0.80167**	†	0.88249	**0.88358**	†	0.89509	**0.90005**	†
	0.79854	**0.79838**		0.87989	**0.88171**		0.88229	**0.88473**	
WFG9	0.83951	**0.84112**		0.92595	**0.92771**		**0.94890**	0.94712	
	0.83030	**0.83188**	†	0.92242	**0.92445**	†	**0.93507**	0.93103	≈
	0.74526	**0.74771**		0.91616	**0.92146**		0.81770	**0.81948**	

problems. For five and eight objectives, DBEA generally obtains better results than NSGA-III in terms of IGD and HV. Since NSGA-III is a leading method for many-objective optimization, such observation implies DBEA can perform well in the case of many objectives. Thus, diversity-first based selection is effective and applicable to many-objective optimization.

6 Conclusions

While convergence-first based MOEAs have been increasingly reported to be effective in solving a variety of MOPs, they may come across difficulties in maintaining population diversity, resulting in a poor approximation of the POF. For this reason, this paper has suggested a new diversity-first sorting method to overcome the difficulty of convergence-first sorting. The advantages and disadvantages of convergence-first and diversity-first sorting methods have been briefly discussed. Afterwards, a new algorithm based on the proposed sorting method, i.e., DBEA, has been suggested.

The proposed DBEA has been examined and compared with the recently-developed NSGA-III algorithm on several test problems with different optimization difficulties. Experimental results have shown that DBEA has great advantages in maintaining diversity for problems where NSGA-III fails. Furthermore, DBEA has also great potential for many-objective optimization, as indicated by its outperformance over NSGA-III in many cases.

Inspired by these encouraging performance, we would like to extend the current work to other classes of MOEAs, such as indicator-based selection methods and decomposition-based MOEAs in the future. Also, the convergence part of DBEA needs to be investigated. Different fitness assignment techniques will be integrated into the diversity-first sorting, and their suitability and effectiveness will be investigated.

Acknowledgments. This work was funded by the Engineering and Physical Sciences Research Council (EPSRC) of U.K. under Grant EP/K001310/1.

References

1. Chen, B., Zeng, W., Lin, Y., Zhang, D.: A new local search-based multiobjective optimization algorithm. IEEE Trans. Evol. Comput. **19**(1), 50–73 (2015)
2. Deb, K., Agrawwal, S., Pratap, A., Meyarivan, T.: A fast and elitist multiobjective genetic algorithm: NSGA-II. IEEE Trans. Evol. Comput. **6**(2), 182–197 (2002)
3. Deb, K., Agrawal, R.B.: Simulated binary crossover for continuous search space. Complex Syst. **9**(4), 115–148 (1995)
4. Das, I., Dennis, J.: Normal-boundary intersection: a new method for generating the Pareto surface in nonlinear multicriteria optimization problems. SIAM J. Optim. **8**(3), 631–657 (1998)
5. Deb, K., Goyal, M.: A combined genetic adaptive search (GeneAS) for engineering design. Comput. Sci. Inf. **26**(4), 30–45 (1996)
6. Deb, K., Jain, H.: An evolutionary many-objective optimization algorithm using reference-point based non-dominated sorting approach, part I: solving problems with box constraints. IEEE Trans. Evol. Comput. **18**(4), 577–601 (2014)
7. Huband, S., Hingston, P., Barone, L., While, L.: A review of multiobjective test problems and a scalable test problem toolkit. IEEE Trans. Evol. Comput. **10**(2), 477–506 (2006)
8. Liu, H., Gu, F., Zhang, Q.: Decomposition of a multiobjective optimization problem into a number of simple multiobjective subproblems. IEEE Trans. Evol. Comput. **18**(3), 450–455 (2014)
9. Zhang, Q., Li, H.: MOEA/D: a multiobjective evolutionary algorithm based on decomposition. IEEE Trans. Evol. Comput. **11**(6), 712–731 (2007)
10. Zitzler, E., Laumanns, M., Thiele, L.: SPEA2: improving the strength pareto evolutionary algorithm. Technical report, ETH Zürich, Zürich, Switzerland (2001)
11. Salinas, S., Li, M., Li, P.: Multi-objective optimal energy consumption scheduling in smart grids. IEEE Trans. Smart Grid **4**(1), 341–348 (2013)
12. Fesanghary, M., Asadi, S., Geem, Z.W.: Design of low-emission and energy-efficient residential building using a multi-objective optimization algorithm. Build. Environ. **49**, 245–250 (2012)
13. Zitzler, E., Künzli, S.: Indicator-based selection in multiobjective search. In: Yao, X., et al. (eds.) PPSN 2004. LNCS, vol. 3242, pp. 832–842. Springer, Heidelberg (2004)
14. Zitzler, E., Thiele, L.: Multiobjective evolutionary algorithms: a comparative case study and the strength pareto approach. IEEE Trans. Evol. Comput. **3**(4), 257–271 (1999)

Tunnelling Crossover Networks for the Asymmetric TSP

Nadarajen Veerapen[1(✉)], Gabriela Ochoa[1], Renato Tinós[2], and Darrell Whitley[3]

[1] Division of Computing Science and Mathematics, University of Stirling, Stirling, UK
nve@cs.stir.ac.uk

[2] Department of Computing and Mathematics, University of São Paulo, São Paulo, Brazil

[3] Department of Computer Science, Colorado State University, Fort Collins, USA

Abstract. Local optima networks are a compact representation of fitness landscapes that can be used for analysis and visualisation. This paper provides the first analysis of the Asymmetric Travelling Salesman Problem using local optima networks. These are generated by sampling the search space by recording the progress of an existing evolutionary algorithm based on the Generalised Asymmetric Partition Crossover. They are compared to networks sampled through the Chained Lin-Kernighan heuristic across 25 instances. Structural differences and similarities are identified, as well as examples where crossover smooths the landscape.

1 Introduction

The global structure of fitness landscapes in combinatorial optimisation is far from being well-understood, and yet crucially impacts the dynamic of search heuristics. The operators within such algorithms usually restrict the search space in some way, potentially over-exploring or missing key parts of the actual landscape. Tools to better understand and visualise fitness landscapes are therefore needed. The symmetric Travelling Salesman Problem (TSP) has been widely studied. Its more general formulation, the Asymmetric TSP (ATSP) has received less attention but is useful to model real-world situations where symmetry is often a luxury. In this paper, we attempt to provide some insights into its landscape structure by studying local optima networks.

Local optima networks (LON) are graph-based models of combinatorial fitness landscapes, originally inspired by work on energy landscapes in computational chemistry [4]. A fitness landscape is compressed into a graph where nodes are local optima and edges possible search transitions among them [10,17]. The first model considered binary search spaces and the NK family of landscapes; nodes were local optima according to a best-improvement local search with bit-flip moves, and edges account for transition probabilities among basins of attraction [10]. This model required a full enumeration of local optima and basins, and was therefore impossible to scale to realistically sized landscapes. An alternative

J. Handl et al. (Eds.): PPSN XIV 2016, LNCS 9921, pp. 994–1003, 2016.
DOI: 10.1007/978-3-319-45823-6_93

definition of edges was later proposed to account for *escape* probabilities among optima, that is, probabilities to hop from a local optimum to another after a perturbation (large mutation) followed by local search [18]. Recently, sampling approaches have been developed using escape edges in order to model landscapes of realistic size [6,11–13]. In particular, work on the symmetric Travelling Salesman Problem has revealed intriguing landscape visualisations, providing compelling evidence of the existence of multiple valleys or clusters of local optima (also called funnels) on the studied instances [12,13]. Most local optima network models so far consider transitions based on perturbation operators. Ochoa et al. [9] proposed a model where transitions are based on recombination. Specifically, the deterministic Partition (Tunnelling) Crossover by Tinós et al. was considered [16], together with efficient procedures for extracting all the local optima of NK landscapes of string length up to 30, based on exploiting the structure of pseudo-Boolean problems with bounded epistasis [3].

The main goal of this article is to model tunnelling crossover networks for asymmetric Travelling Salesman instances of realistic size. More specifically, the contributions are:

1. First study of local optima networks for the asymmetric TSP.
2. An extension of the local optima network model to capture evolutionary algorithms. This is achieved by incorporating two types of edges, one based on mutation and another based on recombination.
3. A network sampling mechanism based on instrumenting an existing evolutionary algorithm.
4. Comparing the local optima network structure emerging from an evolutionary algorithm against a single-point heuristic (Iterated Local Search).

Following this introduction, the paper presents the crossover operator in Sect. 2. Section 3 provides key definitions for local optima networks and describes how the network data are gathered. Section 4 presents the instances, which are analysed in Sect. 5. The conclusion is found in Sect. 6.

2 Generalised Asymmetric Partition Crossover

Our study considers the Generalised Asymmetric Partition Crossover (GAPX), a deterministic recombination operator proposed by Tinós et al. [15] for the Asymmetric Travelling Salesman Problem. GAPX is based on the Generalised Partition Crossover (GPX), developed by Whitley et al. [19] for the symmetric TSP. GAPX and GPX recombine partial solutions that are not shared in common between two parent solutions. First, a union graph $G_u = G_1 \cup G_2$ is created from graphs G_1 and G_2 representing the parent solutions. Then, common edges are removed from G_u and connected components are identified. Some of the connected components are the *recombining components*, i.e., connected subgraphs that can be deterministically recombined. GAPX and GPX find the best recombinations among the recombining components in order to generate the offspring. If the number of recombining components is q, then the best of 2^q

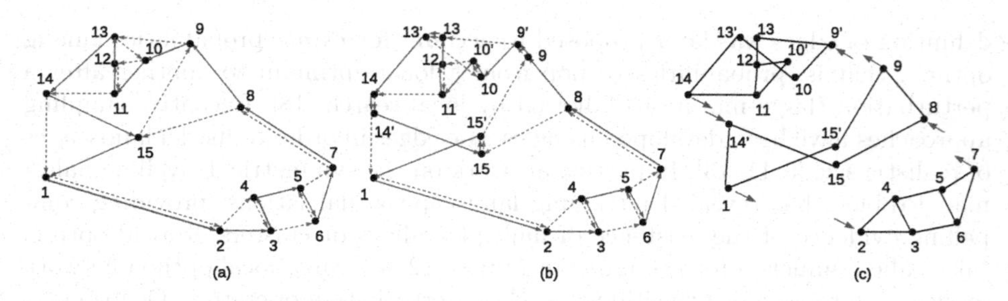

Fig. 1. Recombining two parent solutions using GAPX. (a) Parent solutions are shown by solid (blue) and dashed (red) lines. (b) In a first step, ghost vertices are inserted after vertices of degree 4. (c) Common edges are removed, allowing to identify 3 partitions. (Color figure online)

offspring is found at computational cost $O(n)$. This is possible because the partial evaluations of each one of the q recombining components are independently computed.

In GPX, the recombining components are the connected components separated from the rest of the graph by exactly two common edges. The remainder of the graph is also a recombining component. The Lin-Kernighan-Helsgaun (LKH) algorithm [5] includes a recombination operator, called Iterative Partial Transcription (IPT), which is similar in effect to GPX.

The GAPX includes enhancements to GPX that allow it to find many more recombining partitions than GPX and IPT. As a consequence, an exponentially higher number of offspring is explored. For example, when recombining the two parents shown in Fig. 1a, GPX (adapted to the asymmetric TSP) finds $q = 2$ partitions, while GAPX finds $q = 3$ partitions. Thus, while GPX finds the best of $2^2 = 4$ offspring in this example, GAPX finds the best of $2^3 = 8$ offspring.

One enhancement to GPX is that GAPX exploits cuts that break nodes of degree 4 of G_u as a site for crossover. This is possible by splitting every vertex of degree 4 in order to create "ghost" vertices (Fig. 1b). According to the the direction of flow given by the solutions, common edges between the original vertices and their respective ghost vertices can be created. Such common edges are candidate sites for crossover when the connected components of the new union graph are identified.

3 Local Optima Networks for TSP

Nodes and edges make up the networks. They are defined by the methodology for extracting the network data which is described in the next subsection. A full enumeration of the local optima for ATSP instances of non-trivial size is clearly unmanageable. Therefore, the networks are based on a sample of high-quality local optima in the search space. We first provide some basic definitions, below, before describing the sampling algorithm.

3.1 Definitions

Definition 1. A tour is a *local optimum* if none of its neighbours is shorter than it. The set of local optima is denoted by LO.

The neighbourhood is imposed by k-opt local search. The later is applied after a crossover or mutation operation, in the case of an evolutionary algorithm, and after a perturbation, in the case of an Iterated Local Search. A k-opt local search considers all the possibilities of exchanging k edges in a tour and picks the best. The local optimality criterion is, therefore, rather stringent since only a small number of tours are k-optimal.

Definition 2. *Edges* are directed and of different types based on crossover, mutation, or perturbation. There is an edge from local optimum LO_i to local optimum LO_j, if LO_j can be obtained after any of those operations to LO_i followed by k-opt search. The set of edges is denoted by E.

In the case of the crossover, a pair of edges is created: one starting from each parent and targeting the offspring as has been done in [9]. However, the local optima networks in the latter did not include a second type of edge based on a mutation operator, which we include here.

Definition 3. The *local optima network, LON*, is the graph $LON = (LO, E)$ where nodes are the local optima LO, and set E are the edges.

3.2 Gathering Network Data

The GAPX network data is generated by instrumenting and adapting the genetic algorithm from Tinós et al. [15] (see Algorithm 1). After each crossover or mutation operation, the solution obtained is transformed using 3-opt and each unique local optimum obtained is stored in LO. We also store, in E, an edge between the starting and end optima after one of these two operations. If no improving solution is found during 20 consecutive generations, all the solutions in the population, except the ones with best fitness, are replaced by random solutions improved by 3-opt.

Contrary to the GA from [15], a full 3-opt is performed, not a greedy version. This is done after all crossover and mutation operations. The algorithm can therefore be described as a fully hybrid algorithm, combining a GA and local search. The mutation operator consists of a sequence of up to 5 double-bridge moves, i.e., exchanges of 4 edges in a specific pattern. The algorithm is run 100 times with a population of 100 individuals, until a global optimum is found or 500 generations have elapsed. These parameters also depart from the original ones (300 individuals and 1500 generations), otherwise success rates of 100 % in finding a global optimum are observed on most instances. Discriminating between easier and harder instances would therefore be more difficult.

To provide a basis for the comparison of the GAPX network data, we use network data from an Iterated Local Search based on the well-known Lin-Kernighan

```
P ← popInit()
while termination condition is not satisfied do
    Q(1) ← bestSolution(P)
    for i ← 2 to maxPop do
        (p1, p2) ← selection(P)
        Q(i) ← 3opt(crossover(p1, p2))
        LO ← LO ∪ {Q(i)}; E ← E ∪ {(p1, Q(i)), (p2, Q(i))}
        if crossover did not improve the solutions then
            best ← chooseBest(p1, p2)
            Q(i) ← 3opt(doubleBridgeMutation(best))
            LO ← LO ∪ {Q(i)}; E ← E ∪ {(best, Q(i))}
        end
    end
    if best sol. did not improve in last 20 gen. then  Q ← immigration(P)
    P ← Q
end
```

Algorithm 1. Local optima network sampling in evolutionary algorithm.

(LK) heuristic [8]. We instrumented [13] the Chained Lin-Kernighan (Chained-LK) implementation by Applegate et al. [2] provided in the Concorde TSP solver [1]. LK applies 2, 3 and higher-order k-opt moves, with k chosen adaptively. The perturbation operator in Chained-LK is a double-bridge. Let us note that LK is designed for the symmetric TSP. A conversion step is required to handle ATSP instances. The process is described in the next section.

Ensuring fairness between two different algorithms when gathering local optima is not obvious. We chose to first run the hybrid algorithm and record the total number of edges that had been travelled across the 100 runs for each instance. Chained-LK was then executed such that it performed enough runs to have travelled across as many edges. Each of these runs ends when a global optimum has been found or when $20n$ perturbations have been performed, with n being the number of vertices in the original ATSP instance. On the instances where this method does not lead to at least 100 Chained-LK runs, additional runs are executed to reach 100.

4 Selected ATSP Instances

Our study considers the ATSP instances from TSPLIB [14] belonging to different types, as well as instances generated using the DIMACS symmetric TSP generator code. We use two types of generated instances: uniformly distributed cities (prefixed by E) and clustered cities (prefixed by C). The symmetric instances are transformed into asymmetric instances by inserting random Gaussian deviations (with standard deviation equal to $0.2d_{i,j}$) to each distance $d_{i,j}$, where i and j are any two cities. By considering this variety of instances, our aim is to discover structural differences distinguishing the hard from the easy to solve instances. The instances from TSPLIB are mostly real-life instances, originating from sequencing, scheduling, vehicle routing and stacker crane problems. Instances *ry48p* and *kro124p* are synthetic.

The Concorde exact solver was used to compute the minimal fitness for the generated instances. Since Concorde can only handle symmetric TSP instances, the ATSP instances were transformed into symmetric instances by doubling the number of vertices [7]. Given a set V of n vertices and the distance $d_{i,j}, \forall i, j \in V$, a new city $n+i$ is created $\forall i \in V$. The cost of edge $(i, n+i)$ is set to $0, \forall i \in V$, the cost of $(n+i, j)$ is set to $d_{i,j} + M, \forall i, j \in V$, where M is a sufficiently large number, and the cost of the remaining edges is set to ∞. The value nM is subtracted from the fitness. The same transformation was used to convert the ATSP instances into instances that are suitable for Concorde's Chained-LK heuristic.

5 Network Analysis

The execution and network data generated are summarised in Table 1. *GAPX* indicates results for the hybrid algorithm based on the GAPX. *CLK* indicates results based on Chained-LK. The column titled *Runs* indicates the number of Chained-LK runs required to traverse at least as many edges as where traversed by 100 runs of the GAPX-based algorithm. *Success* represents the proportion of runs that find a global optimum. *Unique Opt.* refers to the number of unique global optima. *Conn. Comp* refers to the number of connected components. *Edge Opt.* and *Mut.* show the proportion of GAPX edges where the end node is already 3-opt before local search and the proportion of edges that are mutation edges, respectively.

The first observation is that at least one global optimum has been found on most instances for each solving method. This indicates that, although the algorithms and their parameters may not be perfect, the best solutions are reachable using these two sampling approaches. We may therefore interpret the results with a minimal level of confidence that they represent a non-trivial part of the landscape. Furthermore, with the chosen parameters, Chained-LK sometimes has similar success rates to the GAPX-based algorithm on several instances, which would tend to show that the *edge budget* allocated is sufficient to fairly compare the two types of networks.

The smallest and easiest instance, br17, exhibits a smooth landscape under the different operators used. Global optima are found in the first generation or iteration, which is highlighted by the high number of connected components. This is an artefact of the sampling algorithms which terminates a run as soon as a global optimum is found. Thus plateaus of global optima are not fully explored.

At the opposite end, the largest instances (rbg) are very easy for the GAPX-based algorithm. Chained-LK, on the other hand, struggles and its 100 separate runs end up in 100 different funnels (connected components). However, the mean fitness of the Chained-LK nodes is lower than that of GAPX nodes. This showcases GAPX's ability to *tunnel through* what is a totally different landscape for

Table 1. Network data. *Success* – proportion of runs that find a global optimum. *Edge Opt.* and *Mut.* – proportion of GAPX edges where the end node is already 3-opt before local search and the proportion of edges that are mutation edges, respectively.

	Inst.	Size	Runs	Success		Unique Opt.		Conn. Comp.		Nodes		Edges		Edge Opt.	Edge Mut.
			CLK	GAPX	CLK	GAPX	CLK	GAPX	CLK	GAPX	CLK	GAPX	CLK	GAPX	
1	br17	17	28915	1.00	1.00	4843	23669	1966	10726	20332	25722	24821	17001	0.98	0.38
2	ftv33	34	1463	1.00	0.99	2	2	110	1	46854	2285	90360	6202	0.70	0.25
3	ftv35	36	3242	1.00	0.57	1	1	72	1	287402	6384	850330	13984	0.67	0.31
4	ftv38	39	3592	0.98	0.43	2	2	16	1	685874	9971	2125869	20918	0.67	0.31
5	p43	43	7766	0.87	0.47	98	3685	247	3439	2221484	988626	4464259	1340720	0.65	0.27
6	ftv44	45	891	1.00	0.64	16	16	64	1	131789	3959	298787	7905	0.68	0.28
7	ftv47	48	1615	1.00	0.55	16	16	10	1	287511	8853	701515	17052	0.68	0.28
8	ry48p	48	3035	1.00	0.33	1	1	24	1	644854	16212	1669819	28629	0.66	0.30
9	C50.0	50	9604	1.00	0.95	2	2	21	1	505885	28429	1314569	53336	0.65	0.32
10	E50.0	50	10069	1.00	1.00	2	2	50	1	121457	21012	255856	40388	0.70	0.29
11	ft53	53	932	1.00	0.27	1	1	4	1	214726	6611	555520	9784	0.64	0.32
12	ftv55	56	4622	1.00	1.00	85	128	121	1	208918	30718	443094	65671	0.70	0.27
13	ftv64	65	1683	1.00	0.71	95	132	31	4	341157	22646	748388	47802	0.70	0.26
14	ft70	70	7461	0.58	0.23	1	1	32	1	2496591	86694	6648585	132447	0.56	0.30
15	ftv70	71	4897	1.00	0.96	82	128	20	1	527355	61919	1185003	139180	0.71	0.26
16	kro124p	100	3109	1.00	0.86	1	1	34	1	597158	33167	1370484	52971	0.62	0.31
17	C100.0	100	4391	0.62	0.12	2	2	105	26	2509747	62616	5394010	96362	0.62	0.31
18	E100.0	100	614	1.00	0.38	2	2	48	2	470359	7602	921047	11930	0.64	0.32
19	ftv170	171	3357	0.55	0.36	79	1206	131	599	3310144	779809	6835402	1516150	0.68	0.26
20	C200.0	200	3044	0.06	0.02	1	2	166	877	4127102	117804	8171590	183914	0.62	0.28
21	E200.0	200	3070	0.02	0.01	1	2	135	894	4140274	97048	8085149	152355	0.61	0.30
22	rbg323	323	100	1.00	0.00	589	0	1868	100	20773	252775	22840	258353	0.20	0.11
23	rbg358	358	100	1.00	0.00	470	0	1939	100	20807	311891	22862	317813	0.17	0.10
24	rbg403	403	100	1.00	0.04	1418	4	2073	100	18241	534931	18904	540314	0.19	0.13
25	rbg443	443	100	1.00	0.22	1909	22	2091	100	18498	513508	18905	518744	0.12	0.09

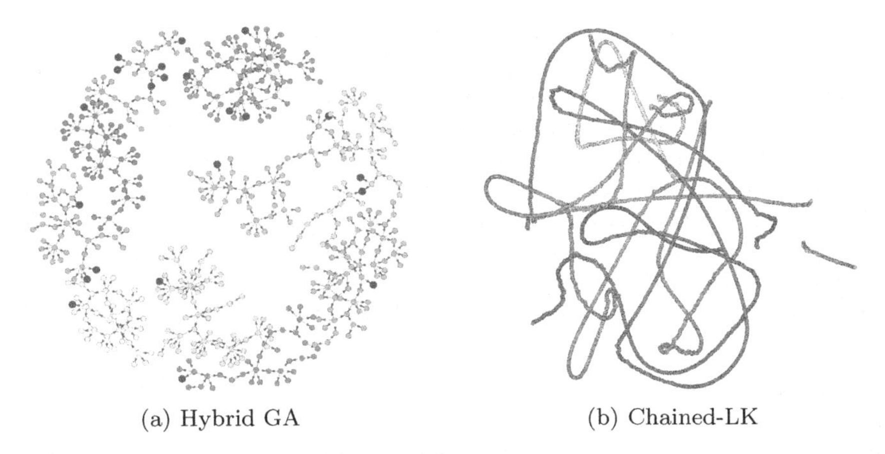

Fig. 2. Subsets of local optima networks for rbg323 under a fitness threshold of 1331. The hybrid GA network is further simplified by selecting only the 10 largest components (513 nodes). The global optima, with fitness 1326, are painted red. The Chained-LK network contains 3883 nodes in 7 components indicated by different colours, the smaller one on a plateau of fitness 1329, the others on a plateau of fitness 1331. (Color figure online)

Chained-LK. It is interesting to note that, as opposed to other instances, the majority of nodes generated through crossover are not 3-optimal and that there is a low proportion of mutation edges as well. GAPX combined with 3-opt are thus able to drive the search through the landscape largely without mutation. Figure 2 shows a subset of the local optima networks for rbg323 very close to the global optima. The structure difference is striking, with the Chained-LK network stuck on two plateaus.

Connected-component-wise, Chained-LK has a tendency to generate landscapes with fewer components while finding more unique global optima in general. This may seem surprising given that Chained-LK recorded fewer unique nodes and edges. These numbers are smaller since local optima that do not improve the current solution in a run are not recorded due to memory constraints. Otherwise these worsening solutions would make up the majority of nodes.

Pearson's correlation coefficients were computed for pairwise comparisons of several execution and landscape features. One of them is the mean normalised fitness of nodes, not displayed in Table 1, which is always under 0.3 units. It is negatively correlated to the number of edges (−0.54) and nodes (−0.60) for Chained-LK but there is no correlation for GAPX. It is the opposite for the number of connected components, which is not correlated to the mean normalised fitness for Chained-LK but shows a negative correlation (−0.65) for GAPX.

The number of connected components is strongly correlated (1.0) to the number of unique local optima for Chained-LK and to a lesser degree (0.72) for GAPX. In the context of crossover and mutation networks, the proportion of crossover edges that end in a local optima before the application of 3-opt is strongly correlated (0.92) to the proportion of mutation edges. This indicates that mutation is usually required to increase diversity.

This work is a first attempt at extracting crossover networks of moderately large permutation problem instances. As such it has some limitations. For example, the nodes' mean in-degree across instances ranges between 1 and 3. This is different from the results obtained by [9], where the mean in-degree ranged from 0.8 to 245 for exhaustively sampled NK landscapes. This could simply be due to the nature of ATSP landscapes, the use of mutation edges or be the result of some bias in the sampling. Further work will investigate such issues.

6 Conclusion

Local optima networks help to better understand the global structure of combinatorial landscapes by providing a relatively compact representation. Nevertheless, sampling is required to study instances on non-trivial sizes. We have done this here for the Asymmetric TSP with networks generated from an evolutionary and an iterated local search algorithm. We have presented evidence of their differences and similarities. On some larger real-life instances the crossover-based algorithm produced networks that were drastically different from the other approach, effectively demonstrating the tunnelling behaviour of carefully designed crossover operators.

This work only scratches the surface of the use of local optima networks for understanding evolutionary algorithms and the structure of non-trivial combinatorial problem instances. Further work will look at improving the sampling, both in the methodology and the quantity of data points gathered. We also intend to carry out systematic investigations of a wide range of instances with different characteristics.

Acknowledgement. N. Veerapen and G. Ochoa are supported by the Leverhulme Trust (award number RPG-2015-395) and by the UK's Engineering and Physical Sciences Research Council (grant number EP/J017515/1). R. Tinós is supported by FAPESP (grant 2015/06462-1) and CNPq. All data generated during this research are openly available from the Stirling Online Repository for Research Data (http://hdl.handle.net/11667/75). Results were obtained using the EPSRC funded ARCHIE-WeSt High Performance Computer (www.archie-west.ac.uk, EPSRC grant EP/K000586/1).

References

1. Applegate, D., Bixby, R., Chvátal, V., Cook, W.: Concorde TSP solver (2003). http://www.math.uwaterloo.ca/tsp/concorde.html
2. Applegate, D., Cook, W., Rohe, A.: Chained Lin-Kernighan for large traveling salesman problems. INFORMS J. Comput. **15**, 82–92 (2003)
3. Chicano, F., Whitley, D., Sutton, A.M.: Efficient identification of improving moves in a ball for pseudo-boolean problems. In: Proceedings of the Genetic and Evolutionary Computation Conference (GECCO 2014), pp. 437–444. ACM (2014)

4. Doye, J.P.K.: The network topology of a potential energy landscape: a static scale-free network. Phys. Rev. Lett. **88**, 238701 (2002)
5. Helsgaun, K.: An effective implementation of the Lin-Kernighan traveling salesman heuristic. Eur. J. Oper. Res. **126**(1), 106–130 (2000)
6. Iclanzan, D., Daolio, F., Tomassini, M.: Data-driven local optima network characterization of QAPLIB instances. In: Proceedings of the Genetic and Evolutionary Computation Conference (GECCO 2014), pp. 453–460. ACM (2014)
7. Jonker, R., Volgenant, T.: Transforming asymmetric into symmetric traveling salesman problems. Oper. Res. Lett. **2**(4), 161–163 (1983)
8. Lin, S., Kernighan, B.W.: An effective heuristic algorithm for the traveling-salesman problem. Oper. Res. **21**, 498–516 (1973)
9. Ochoa, G., Chicano, F., Tinós, R., Whitley, D.: Tunnelling crossover networks. In: Proceedings of the Genetic and Evolutionary Computation Conference (GECCO 2015), pp. 449–456. ACM (2015)
10. Ochoa, G., Tomassini, M., Verel, S., Darabos, C.: A study of NK landscapes' basins and local optima networks. In: Proceedings of the Genetic and Evolutionary Computation Conference (GECCO 2008), pp. 555–562. ACM (2008)
11. Ochoa, G., Veerapen, N.: Additional dimensions to the study of funnels in combinatorial landscapes. In: Proceedings of the Genetic and Evolutionary Computation Conference (GECCO 2016). ACM (2016, to appear)
12. Ochoa, G., Veerapen, N., Whitley, D., Burke, E.K.: The multi-funnel structure of TSP fitness landscapes: a visual exploration. In: Bonnevay, S., Legrand, P., Monmarché, N., Lutton, E., Schoenauer, M. (eds.) EA 2015. LNCS, vol. 9554, pp. 1–13. Springer, Heidelberg (2016). doi:10.1007/978-3-319-31471-6_1
13. Ochoa, G., Veerapen, N.: Deconstructing the big valley search space hypothesis. In: Chicano, F., Hu, B., García-Sánchez, P. (eds.) EvoCOP 2016. LNCS, vol. 9595, pp. 58–73. Springer, Heidelberg (2016). doi:10.1007/978-3-319-30698-8_5
14. Reinelt, G.: TSPLIB - a traveling salesman problem library. ORSA J. Comput. **3**(4), 376–384 (1991)
15. Tinós, R., Whitley, D., Ochoa, G.: Generalized asymmetric partition crossover (GAPX) for the asymmetric TSP. In: Proceedings of the Genetic and Evolutionary Computation Conference (GECCO 2014), pp. 501–508. ACM (2014)
16. Tinós, R., Whitley, L.D., Chicano, F.: Partition crossover for pseudo-boolean optimization. In: Proceedings of the 2015 ACM Conference on Foundations of Genetic Algorithms XIII, Aberystwyth, United Kingdom, 17–20 January 2015, pp. 137–149 (2015)
17. Verel, S., Ochoa, G., Tomassini, M.: Local optima networks of NK landscapes with neutrality. IEEE Trans. Evol. Comput. **15**(6), 783–797 (2011)
18. Vérel, S., Daolio, F., Ochoa, G., Tomassini, M.: Local optima networks with escape edges. In: Hao, J.-K., Legrand, P., Collet, P., Monmarché, N., Lutton, E., Schoenauer, M. (eds.) EA 2011. LNCS, vol. 7401, pp. 49–60. Springer, Heidelberg (2012)
19. Whitley, D., Hains, D., Howe, A.: Tunneling between optima: partition crossover for the traveling salesman problem. In: Proceedings of the Genetic and Evolutionary Computation Conference (GECCO 2009), pp. 915–922. ACM (2009)

Workshops and Tutorials at PPSN 2016

The Workshops at PPSN 2016

Christian Blum[1,2(✉)] and Christine Zarges[3]

[1] Department of Computer Science and Artificial Intelligence, University of the Basque Country UPV/EHU, San Sebastian, Spain
[2] IKERBASQUE, Basque Foundation for Science, Bilbao, Spain
christian.c.blum@gmail.com
[3] School of Computer Science, The University of Birmingham, Birmingham, UK
c.zarges@cs.bham.ac.uk

Abstract. Workshops have a longstanding tradition at PPSN. This document provides a short description of the workshops that were held at the 2016 edition of the conference. For each workshop we provide a general description of the aim and scope, in addition to the list of accepted papers.

1 Introduction

Workshops, in addition to the main program, have a longstanding tradition at the conference series on Parallel Problem Solving From Nature (PPSN). Consequently, the PPSN 2016 Organizing Committee invited proposals for workshops to be held in conjunction with the main PPSN conference. The workshops at PPSN are intended to be a forum for presenting and discussing, for example, new emerging approaches or critical reflections within a subfield. They often provide an excellent opportunity to meet people with similar interests, to be exposed to cutting-edge research and to exchange ideas in an informal setting. The responsibility of the workshops is solely in the hands of the organizers, who care for their coordination, publicity—that is, for sending out call for papers/abstracts—, collecting and reviewing the papers/abstracts, and maintaining a webpages providing the lists of accepted talks. As in previous years, workshop organizers were also in 2016 able to decide between half and full day workshops. The workshop format was up to the organizers. However, as workshop co-chairs we encouraged to facilitate interactive sessions and suggested to solicit concept papers or abstracts of not more than a few pages length, instead of full papers. All workshops were held during the first two days of PPSN, which are traditionally dedicated to workshops and tutorials.

2 The Four Workshops

At PPSN 2016, four workshops in very different areas of the field took place–some nicely complementing accepting tutorials presented earlier at the conference. In the following we provide a short description of the aim and scope of these workshops, together with their list of accepted papers.

J. Handl et al. (Eds.): PPSN XIV 2016, LNCS 9921, pp. 1007–1011, 2016.
DOI: 10.1007/978-3-319-45823-6_94

Workshop Title: 2nd International Workshop on Advances in Multi-modal Optimization

Organizers: Mike Preuss, Michael G. Epitropakis and Xiaodong Li
URL: http://www.epitropakis.co.uk/ppsn2016-niching/

Aim and Scope: This workshop aimed to bring together researchers from evolutionary computation and related areas who are interested in Multi-modal Optimization. This is a currently forming field, and the organizers aimed for a highly interactive and productive meeting that would make a step forward towards defining it. The workshop provided a unique opportunity to review the advances in the current state of the art in the field of Niching methods. Further discussion dealt with several experimental/theoretical scenarios, performance measures, real-world and benchmark problem sets and outline the possible future developments in this area. Positional statements, suggestions, and comments were very much welcomed by the organizers.

List of Accepted Abstracts/Papers

Authors	Title
C. Zarges	Towards theoretical analysis in Multi-modal Optimization
S. Wessing, G. Rudolph and M. Preuss	Assessing Basin Identification Methods for Locating Multiple Optima
S. Nallaperuma, K. Gao and F. Neumann	Feature based analysis on problem hardness
J. K. Pugh, L. B. Soros and K. O. Stanley	Quality Diversity: A New Kind of Multimodal Search
P. Kerschke and C. Grimme	Multi-modality in Continuous Multi-Objective Optimization
A. Moshaiov	Multi-concept Optimization vs. Multi-modal Optimization
K. Bibiks and J.-P. Li	Discrete species conserving cuckoo search for resource-constrained project scheduling problems

Workshop Title: Landscape-Aware Heuristic Search

Organizers: Nadarajen Veerapen and Gabriela Ochoa
URL: http://www.cs.stir.ac.uk/events/ppsn2016-landscape/

Aim and Scope: Fitness landscape analysis and visualization can provide significant insights into problem instances and algorithm behavior. The aim of this workshop was to encourage and promote the use of landscape analysis to improve search algorithms and their understanding. Examples include landscape analysis as a tool to inform the design of algorithms, landscape metrics for online adaptation of search strategies, mining landscape information to predict instance hardness and algorithm runtime. The workshop sought to bring together researchers interested in landscape analysis and in exploiting problem structure to develop informed search strategies. The workshop provided a unique opportunity to present existing work, propose new ideas or put forward position statements.

List of Accepted Abstracts/Papers

Authors	**Title**
V. Santucci and A. Milani	A Triple Interpretation of Combinatorial Search Spaces
S. Tari, M. Basseur, and A. Goëffon	Climbing Fitness Landscapes with the Maximum Expansion Pivoting Rule
P. A. Consoli, Y. Mei, L. L. Minku and X. Yao	Dynamic Selection of Evolutionary Operators Based on Online Learning and Fitness Landscape Analysis
P. Kerschke and H. Trautmann	Exploratory Landscape Analysis By Using the R-Package flacco
K. Alyahya and J. E. Rowe	Fitness Landscape Analysis of a Class of NP-Complete Binary Packing Problems
F. Daolio, A. Liefooghe, S. Verel, H. Aguirre and K. Tanaka	Fitness Landscape Analysis, Problems Features and Performance Prediction for Multi-objective Optimization
W. B. Langdon and M. Harman	Fitness Landscape of the Triangle Program
D. Whitley, F. Chicano and B. Goldman	Mk Landscapes Problem Structure
S. Verel, F. Daolio, G. Ochoa, and M. Tomassini	Toward Algorithm Portfolio Based on Local Optima Network Features

Workshop Title: Intelligent Transportation Workshop

Organizer: Neil Urquhart
URL: http://www.soc.napier.ac.uk/~40000408/ppsn/

Aim and Scope: This workshop aimed to bring together researchers using nature inspired computing to support intelligent transportation, allowing them to present and discuss ideas and concepts with their peers. Potential participants were asked to submit a one page abstract according to the instructions outlined on the workshop website. Abstracts were peer reviewed and the authors of successful abstracts were invited to give a presentation of 25 min of their work at the workshop. Solicited topics included the optimization of goods deliveries, the optimization of mobile workforce, the use of nature inspired computing techniques with real world transport related data and APIs, and traffic and transport management.

List of Accepted Abstracts/Papers

Authors	Title
J. Ouenniche	Invited Talk
A. Fernández-Ares, M. García-Arenas, P. García-Sánchez, V. Rivas-Santos and J. J. Merelo	Nowcasting traffic
A. Ekart, E. Ilie-Zudor and C. Buckingham	Combining Human Expertise and Machine Learning for Intelligent Transportation Resource Management
N. Urquhart and A. Fonzone	Using multiple real world objectives in mobile workforce problems
M. Adham and P. Bentley	Evaluating Fitness Functions within the Artificial Ecosystem Algorithm and their Application to Bicycle Redistribution
K. Sim and E. Hart	A Combined Generative and Selective Hyper-heuristic for the Vehicle Routing Problem

Workshop Title: Natural Computing in Scheduling and Timetabling

Organizers: Ahmed Kheiri, Rhyd Lewis and Ender Özcan
URL: http://ahmedkheiri.bitballoon.com/ppsn2016workshop/

Aim and Scope: The aim of this workshop was to bring together researchers and practitioners to share their experiences and report on emerging approaches in solving real-world scheduling problems. A particular interest was on approaches that give a deeper insight into scheduling problem classes, and that enable the exploitation of structural information during the automated search for a solution to a given problem. General purpose approaches used for the automated generation of heuristics for solving single and multi-objective scheduling problems and issues related to development of such approaches were also of particular interest.

List of Accepted Abstracts/Papers

Authors	Title
J. Branke	Invited Talk: Evolutionary Design of Production Scheduling Heuristics
J. Gasior and F. Seredynski	Multi-objective Scheduling in Unreliable Distributed Computing Environment
N. Pillay and E. Özcan	The Role of Generation Constructive Hyper-Heuristics in Educational Timetabling
A. Kheiri, R. Lewis, J. Thompson and P. Harper	Heuristic-based Method for Scheduling Surgical Procedures

3 Final Words

It is worth noting that the workshop on *Landscape-Aware Heuristic Search* that was originally planned for half a day, was extended to a 3/4-day workshop due to an unexpected large number of submissions. Summarizing we can say that all workshops, with their combined contribution of 26 presentations to the conference program, enjoy great popularity and made a significant contribution to PPSN 2016.

Acknowledgements. The writing of this document was supported by project TIN2012-37930-C02-02 (Spanish Ministry for Economy and Competitiveness, FEDER funds from the European Union). Additionally, we acknowledge support from IKERBASQUE.

Tutorials at PPSN 2016

Carola Doerr[1(✉)], Nicolas Bredeche[2], Enrique Alba[3], Thomas Bartz-Beielstein[4], Dimo Brockhoff[5], Benjamin Doerr[6], Gusz Eiben[7], Michael G. Epitropakis[8], Carlos M. Fonseca[9], Andreia Guerreiro[9], Evert Haasdijk[7], Jacqueline Heinerman[7], Julien Hubert[7], Per Kristian Lehre[10], Luigi Malagò[11], J.J. Merelo[12], Julian Miller[13], Boris Naujoks[4], Pietro Oliveto[14], Stjepan Picek[15,16], Nelishia Pillay[17], Mike Preuss[18], Patricia Ryser-Welch[13], Giovanni Squillero[19], Jörg Stork[4], Dirk Sudholt[14], Alberto Tonda[20], Darrell Whitley[21], and Martin Zaefferer[4]

[1] CNRS and Sorbonne Universités, Paris, France
Carola.Doerr@mpi-inf.mpg.de
[2] Sorbonne Universités, Paris, France
[3] University of Málaga, Málaga, Spain
[4] TH Köln, Cologne, Germany
[5] Inria Lille - Nord Europe, Villeneuve-d'ascq, France
[6] École Polytechnique, Palaiseau, France
[7] VU Amsterdam, Amsterdam, Netherlands
[8] Lancaster University, Bailrigg, UK
[9] University of Coimbra, Coimbra, Portugal
[10] University of Nottingham, Nottingham, UK
[11] Shinshu University, Matsumoto, Japan
[12] University of Granada, Granada, Spain
[13] University of York, York, UK
[14] University of Sheffield, Sheffield, UK
[15] KU Leuven, Leuven, Belgium
[16] University of Zagreb, Zagreb, Croatia
[17] University of KwaZulu-Natal, Durban, South Africa
[18] University of Dortmund, Dortmund, Germany
[19] Politecnico di Torino, Turin, Italy
[20] INRA, Paris, France
[21] Colorado State University, Fort Collins, USA

Abstract. PPSN 2016 hosts a total number of 16 tutorials covering a broad range of current research in evolutionary computation. The tutorials range from introductory to advanced and specialized but can all be attended without prior requirements. All PPSN attendees are cordially invited to take this opportunity to learn about ongoing research activities in our field!

1 Chairs' Welcome

Tutorials offer an efficient and interactive way of learning about ongoing research activities. While introductory tutorials are particularly targeted at researchers

J. Handl et al. (Eds.): PPSN XIV 2016, LNCS 9921, pp. 1012–1022, 2016.
DOI: 10.1007/978-3-319-45823-6_95

who have just recently entered (or are about to enter) the multifaceted research field of evolutionary computation, more specialized tutorials address both junior and senior researchers intending to intensify or refresh their knowledge about various topics of interest.

In response to our call for tutorials we have received a large number of high-quality tutorial proposals out of which 16 have been selected for presentation at the conference. These 16 tutorials will be presented in two days at PPSN 2016, on September 17 and 18, which are exclusively reserved for tutorial and workshop presentations.

The topics of the tutorials cover introductions to evolutionary computation in cryptography, multi-modal optimization, and hyper-heuristics.

More specialized tutorials discuss gray-box optimization, graph-based and cartesian genetic programming, intelligent systems for smart cities, and the importance of diversity in evolutionary optimization.

A classic in the tutorial landscape is the introduction to evolutionary multi-objective optimization (EMO), a topic also addressed in the tutorials on using the attainment function as a tool for the performance evaluation of EMO algorithms.

Those researchers wishing to learn more on the role of theory in our field should not miss the tutorial on theory of evolutionary computation. This tutorial is followed up by a basic introduction to runtime analysis of evolutionary algorithms (EAs) and one on the theory of parallel EAs. A forth theory-flavored tutorial aims at bridging the gap between the optimization over manifolds and evolutionary computation.

In addition, a hands-on guide to experiment with real hardware is proposed for evolutionary robotics, it is discussed how to efficiently implement EAs in the cloud, and how to save time and cost through meta-model assisted optimization. We invite all PPSN participants to explore the wide range of topics discussed in the selected tutorials and wish you an enjoyable conference!
Nicolas Bredeche and Carola Doerr
PPSN 2016 Tutorial Chairs

2 Abstracts of the Tutorials

2.1 A Bridge Between Optimization over Manifolds and Evolutionary Computation

Tutorial Speaker: Luigi Malagò, *Shinshu University (Japan)*

Tutorial Abstract: The aim of this tutorial is to explore the promising connection between the well-consolidated field of optimization over manifolds and evolutionary computation. In mathematics, optimization over manifolds deals with the design and analysis of algorithms for the optimization over search spaces with admit a non-Euclidean geometry. One of the simplest examples is probably the sphere, where the shortest path between two points is given by a curve, and not a straight line. Manifolds may appear in evolutionary computation in at least two contexts. The simplest one is the case when an evolutionary algorithm is employed to optimize a fitness function defined over a manifold, such as in the case of the sphere, the cone of positive-definite matrices, the set of rotation matrices, and many others. The second one is more subtle, and is related to the stochastic relaxation of a fitness function. A common approach in model-based evolutionary computation is to search for the optimum of a function by sampling populations from a sequence of probability distributions. For instance, this is the case of evolutionary strategies, probabilistic model-building genetic algorithms, estimation of distribution algorithms and similar techniques, both in the continuous and in the discrete domain. A strictly related paradigm which can be used to describe the behavior of model-based search algorithms is that of stochastic relaxation, i.e., the optimization of the expected value of the original fitness function with respect to a probability distribution in a statistical model. From this perspective a model-based algorithm is solving a problem which is strictly related to the optimization of the stochastic relaxation over a statistical model. Notably, statistical models are well-known examples of manifolds, where the Fisher information plays the role of metric tensor. For this reason, it becomes of great interest to compare the standard techniques in the field of optimization over manifolds, with the mechanisms implemented by model-based algorithm in evolutionary computation. The tutorial will consist of two parts. In the first one, a unifying framework for the description of model-based algorithms will be introduced and some standard well-known algorithms will be presented from the perspective of the optimization over manifold. Particular attention will be devoted to first-order methods based on the Riemannian gradient over a manifold, which in the case of a statistical model is known as the natural gradient. In the second part, we will discuss how evolutionary algorithms can be adapted to solve optimization problems defined over manifold, which constitutes a novel and promising area of research in evolutionary computation.

2.2 Advances on Multi-modal Optimization

Tutorial Speaker: Mike Preuss, *University of Dortmund (Germany),* and Michael G. Epitropakis, *Lancaster University (UK)*

Tutorial Abstract: Multimodal optimization is currently getting established as a research direction that collects approaches from various domains of operational research and evolutionary computation that strive for delivering multiple very good solutions at once. We discuss several scenarios and list currently employed and potentially available performance measures. Furthermore, many state-of-the-art as well as older methods are compared and put into a rough taxonomy. We also discuss recent relevant competitions and their results and outline the possible future developments in this area.

2.3 The Attainment Function Approach to Performance Evaluation in Evolutionary Multiobjective Optimization

Tutorial Speaker: Carlos M. Fonseca and Andreia P. Guerreiro, *University of Coimbra (Portugal)*

Tutorial Abstract: The development of improved optimization algorithms and their adoption by end users depend on the ability to evaluate their performance on the problem classes of interest. In the absence of theoretical guarantees, performance must be evaluated experimentally while taking into account both the experimental conditions and the nature of the data collected.

Evolutionary approaches to multiobjective optimization typically produce discrete Pareto-optimal front approximations in the form of sets of mutually non-dominated points in objective space. Since evolutionary algorithms are stochastic, such non-dominated point sets are random, and vary according to some probability distribution.

In contrast to quality indicators, which map non-dominated point sets to real values, and side-step the set nature of the data, the attainment-function approach addresses the non-dominated point set distribution directly. Distributional aspects such as location, variability, and dependence, can be estimated from the raw non-dominated point set data.

This tutorial will focus on the attainment function as a tool for the evaluation of the performance of evolutionary multiobjective optimization (EMO) algorithms. In addition to the theoretical foundations of the methodology, computational and visualization issues will be discussed. The application of the methodology will be demonstrated by interactively exploring example data sets with freely available software tools. To conclude, a selection of open questions and directions for further work will be identified.

2.4 Evolutionary Algorithms and Hyper-heuristics

Tutorial Speaker: Nelishia Pillay, *University of KwaZulu-Natal (South Africa)*

Tutorial Abstract: Hyper-heuristics is a rapidly developing domain which has proven to be effective at providing generalized solutions to problems and across problem domains. Evolutionary algorithms have played a pivotal role in the advancement of hyper-heuristics, especially generation hyper-heuristics. Evolutionary algorithm hyper-heuristics have been successful applied to solving problems in various domains including packing problems, educational timetabling, vehicle routing, permutation flowshop and financial forecasting amongst others. The aim of the tutorial is to firstly provide an introduction to evolutionary algorithm hyper-heuristics for researchers interested in working in this domain. An overview of hyper-heuristics will be provided. The tutorial will examine each of the four categories of hyper-heuristics, namely, selection constructive, selection perturbative, generation constructive and generation perturbative, showing how evolutionary algorithms can be used for each type of hyper-heuristic. A case study will be presented for each type of hyper-heuristic to provide researchers with a foundation to start their own research in this area. Challenges in the implementation of evolutionary algorithm hyper-heuristics will be highlighted. An emerging research direction is using hyper-heuristics for the automated design of computational intelligence techniques. The tutorial will look at the synergistic relationship between evolutionary algorithms and hyper-heuristics in this area. The use of hyper-heuristics for the automated design of evolutionary algorithms will be examined as well as the application of evolutionary algorithm hyper-heuristics for the design of computational intelligence techniques. The tutorial will end with a discussion session on future directions in evolutionary algorithms and hyper-heuristics.

2.5 Evolutionary Computation in Cryptography

Tutorial Speaker: Stjepan Picek, *KU Leuven (Belgium) and University of Zagreb (Croatia)*

Tutorial Abstract: Evolutionary Computation (EC) has been used with great success on various real-world problems. One domain abundant with numerous difficult problems is cryptology. Cryptology can be divided into cryptography and cryptanalysis where although not always in an obvious way, EC can be applied to problems from both domains. This tutorial will first give a brief introduction to cryptology intended for general audience. Afterwards, we concentrate on several topics from cryptography that are successfully tackled up to now with EC and discuss why those topics are suitable to apply EC. However, care must be taken since there exists a number of problems that seem to be impossible to solve with EC and one needs to realize the limitations of the heuristics. We will discuss the choice of appropriate EC techniques (GA, GP, CGP, ES, multi-objective optimization) for various problems and evaluate on the importance of that choice. Furthermore, we will discuss the gap between the cryptographic community and EC community and what does that mean for the results. By doing that, we give a special emphasis on the perspective that cryptography presents a source of benchmark problems for the EC community.

This tutorial will also present some live demos of EC in action when dealing with cryptographic problems.

2.6 Evolutionary Multiobjective Optimization

Tutorial Speaker: Dimo Brockhoff, *Inria Lille - Nord Europe (France)*

Tutorial Abstract: Many optimization problems are multiobjective, i.e., multiple, conflicting criteria need to be considered simultaneously. Due to conflicts between the objectives, usually no single optimum solution exists. Instead, a set of so-called Pareto-optimal solutions, for which no other solution has better function values in all objectives, does emerge.

In practice, Evolutionary Multiobjective Optimization (EMO) algorithms are widely used for solving multiobjective optimization problems. As stochastic blackbox optimizers, EMO approaches cope with nonlinear, nondifferentiable, or noisy objective functions. By inherently working on sets of solutions, they allow the Pareto-optimal set to be approximated in one algorithm run—opposed to classical techniques for multicriteria decision making (MCDM), which aim for single solutions.

Defining problems in a multiobjective way has two further advantages:

- The set of Pareto-optimal solutions may reveal shared design principles (innovization)
- Singleobjective problems may become easier to solve if auxiliary objectives are added (multiobjectivization).

Within this tutorial, we comprehensively introduce the field of EMO and present selected research results in more detail. More specifically, we

- explain the basic principles of EMO algorithms in comparison to classical approaches,
- show a few practical examples motivating the use of EMO, and
- present a general overview of state-of-the-art algorithms and selected recent research results.

2.7 Evolutionary Robotics—A Practical Guide to Experiment with Real Hardware

Tutorial Speaker: Jacqueline Heinerman and Gusz Eiben and Evert Haasdijk and Julien Hubert, *VU Amsterdam (Netherlands)*

Tutorial Abstract: Evolutionary robotics aims to evolve the controllers, the morphologies, or both, for real and/or simulated autonomous robots. Most research in evolutionary robotics is partly or completely carried in simulation. Although simulation has advantages, e.g., it is cheaper and it can be faster, it suffers from the notorious reality gap. Recently, affordable and reliable robots became commercially available. Hence, setting up a population of real robots is within reach for a large group of research groups today. This tutorial focuses on the know-how required to utilise such a population for running evolutionary experiments. To this end we use Thymio II robots with Raspberry Pi extensions (including a camera). The tutorial explains and demonstrates the work-flow from beginning to end, by going through a case study of a group of Thymio II robots evolving their neural network controllers to learn collecting objects on-the-fly. Besides the methodology and lessons learned, we spend time on how to code.

2.8 Graph-Based and Cartesian Genetic Programming

Tutorial Speaker: Julian Miller and Patricia Ryser-Welch, *University of York (UK)*

Tutorial Abstract: Genetic Programming is often associated with a tree representation for encoding expressions and algorithms. However, graphs are also very useful and flexible program representations which can be applied to many domains (e.g. electronic circuits, neural networks, algorithms).

Over the years a variety of representations of graphs have been explored such as: Parallel Distributed Genetic Programming (PDGP) , Linear-Graph Genetic Programming, Enzyme Genetic Programming, Graph Structured Program Evolution (GRAPE) and Cartesian Genetic Programming (CGP).

Cartesian Genetic Programming (CGP) is probably the best known form of graph-based Genetic Programming. It was developed by Julian Miller in 1999–2000. In its classic form, it uses a very simple integer address-based genetic representation of a program in the form of a directed graph. CGP has been adopted by a large number of researchers in many domains.

In a number of studies, CGP has been shown to be comparatively efficient to other GP techniques. It is also very simple to program. Since its original formulation, the classical form of CGP has also undergone a number of developments which have made it more useful, efficient and flexible in various ways. These include the addition of automatically defined functions (modular CGP), self-modification operators (self-modifying CGP), the encoding of artificial neural networks (GCPANNs) and evolving iterative programs (iterative CGP).

2.9 Gray Box Optimization in Theory and Practice

Tutorial Speaker: Darrell Whitley, *Colorado State University (USA)*

Tutorial Abstract: This tutorial will cover Gray Box Complexity and Gray Box Optimization for k-bounded pseudo-Boolean optimization. These problems can also be referred to at Mk Landscapes, and included problems such as MAX-kSAT, spin glass problems and NK Landscapes. Mk Landscape problems are a linear combination of M subfunctions, where each subfunction accepts at most k variables. Under Gray Box optimization, the optimizer is given access to the set of M subfunctions. If the set of subfunctions is k-bounded and separable, the Gray Box optimizer is guaranteed to return the global optimum with 1 evaluation. If a problem is not deceptive, the Gray Box optimizer also returns the global optimum after 1 evaluation. This means that simple test problems from ONEMAX to "Trap Functions" are solved in 1 evaluation in $O(n)$ time under Gray Box Optimization. If a tree decomposition exists with a fixed bounded tree width, then the problem can be solved using dynamic programming in $O(n)$ time. If the tree decomposition is bounded by $\lg(n)$, then the problem can be solved by dynamic programming in $O(n^2)$ time. Even for those problems that are not trivially solved, Gray Box optimization also makes it possible to exactly compute Hamming distance 1 improving moves in constant time. Thus, neither mutation nor enumeration of the Hamming neighborhood are necessary. Under many conditions it is possible to calculate the location of improving moves in a Hamming distance radius r neighborhood, thus selecting improving moves several moves ahead. This also can be done in constant time. There also exists deterministic forms of recombination that provably return the best possible offspring from a reachable set of offspring. Partition Crossover relies on localized problem decomposition, and is invariant to the order of the bits in the representation. The methods identify partitions of nonlinear interaction between variables. Variables within a partition must be inherited together. However, bits in different partitions can be linearly recombined. Given p partitions, recombination can be done in $O(n)$ time such that crossover returns the best solutions out of 2^p offspring. The offspring can also be proven to be locally optimal in the largest hyperplane subspace in which the two parents reside. Thus, Partition Crossover is capable of directly moving from known local optima to new, high quality local optima in $O(n)$ time. These innovations will fundamentally change both Local Search and Evolutionary Algorithms. Empirical results show that combining smart local search with Partition Crossover results in search algorithms that are capable of finding globally optimal solutions for nonlinear problems with a million variables in less than 1 min.

2.10 Implementing Evolutionary Algorithms in the Cloud

Tutorial Speaker: JJ Merelo, *University of Granada (Spain)*

Tutorial Abstract: Creating experiments that can be easily reproduced and converted in a straightforward way into a report involves knowing a series of techniques that are of widespread use in the open source and commercial software communities. This tutorial will introduce this techniques, including an introduction to cloud computing and DevOps for evolutionary algorithm practitioners, with reference to the tools and platforms that can make development of new algorithms and problem solutions fast and reproducible.

2.11 Intelligent Systems for Smart Cities

Tutorial Speaker: Enrique Alba, *University of Málaga (Spain)*

Tutorial Abstract: The concept of Smart Cities can be understood as a holistic approach to improve the level of development and management of the city in a broad range of services by using information and communication technologies.

It is common to recognize six axes of work in them: (i) Smart Economy, (ii) Smart People, (iii) Smart Governance, (iv) Smart Mobility, (v) Smart Environment, and (vi) Smart Living. In this tutorial we first focus on a capital issue: smart mobility. European citizens and economic actors need a transport system which provides them with seamless, high-quality door-to-door mobility. At the same time, the adverse effects of transport on the climate, the environment and human health need to be reduced. We will show many new systems based in the use of bio-inspired techniques to ease the road traffic flow in the city, as well as allowing a customized smooth experience for travelers (private and public transport).

This tutorial will then discuss on potential applications of intelligent systems for energy (like adaptive lighting in streets), environmental applications (like mobile sensors for air pollution), smart building (intelligent design), and several other applications linked to smart living, tourism, and smart municipal governance.

2.12 Meta-Model Assisted (Evolutionary) Optimization

Tutorial Speaker: Boris Naujoks and Jörg Stork and Martin Zaefferer and Thomas Bartz-Beielstein, *TH Köln (Germany)*

Tutorial Abstract: Meta-model assisted optimization is a well-recognized research area. When the evaluation of an objective function is expensive, meta-model assisted optimization yields huge improvements in optimization time or cost in a large number of different scenarios. Hence, it is extremely useful for numerous real-world applications. These include, but are not limited to, the optimization of designs like airfoils or ship propulsion systems, chemical processes, biogas plants, composite structures, and electromagnetic circuit design.

This tutorial is largely focused on evolutionary optimization assisted by meta-models, and has the following aims: Firstly, we will provide a detailed understanding of the established concepts and distinguished methods in meta-model assisted optimization. Therefore, we will present an overview of current research and open issues in this field. Moreover, we aim for a practical approach. The tutorial should enable the participants to apply up-to-date meta-modelling approaches to actual problems at hand. Afterwards, we will discuss typical problems and their solutions with the participants. Finally, the tutorial offers new perspectives by taking a look into areas where links to meta-modelling concepts have been established more recently, e.g., the application of meta-models in multi-objective optimization or in combinatorial search spaces.

2.13 Promoting Diversity in Evolutionary Optimization: Why and How

Tutorial Speaker: Giovanni Squillero, *Politecnico di Torino (Italy)*, and Alberto Tonda, *INRA (France)*

Tutorial Abstract: Divergence of character is a cornerstone of natural evolution. On the contrary, evolutionary optimization processes are plagued by an endemic lack of diversity: all candidate solutions eventually crowd the very same areas in the search space. Such a "lack of speciation" has been pointed out in the seminal work of Holland in 1975, and nowadays is well known among scholars. It has different effects on the different search algorithms, but almost all are quite deleterious. The problem is usually labeled with the oxymoron "premature convergence", that is, the tendency of an algorithm to convergence toward a point where it was not supposed to converge to in the first place. Scientific literature contains several efficient diversity-preservation methodologies that ranged from general techniques to problem-dependent heuristics. However, the fragmentation of the field and the difference in terminology led to a general dispersion of this important corpus of knowledge in many small, hard-to-track research lines.

Upon completion of this tutorial, attendees will understand the root causes and dangers of "premature convergence". They will know the main research lines in the area of "diversity promotion". They will be able to choose an effective solution from the literature, or design a new one more tailored to their specific needs.

2.14 Runtime Analysis of Evolutionary Algorithms: Basic Introduction

Tutorial Speaker: Per Kristian Lehre, *University of Nottingham (UK)*, and Pietro S. Oliveto, *University of Sheffield (UK)*

Tutorial Abstract: Evolutionary algorithm theory has studied the time complexity of evolutionary algorithms for more than 20 years. This tutorial presents the foundations of this field. We introduce the most important notions and definitions used in the field and consider different evolutionary algorithms on a number of well-known and important example problems. Through a careful and thorough introduction of important analytical tools and methods, including fitness- and level-based analysis, typical events and runs, and drift analysis. By the end of the tutorial the attendees will be able to apply these techniques to derive relevant runtime results for non-trivial evolutionary algorithms.

In addition to custom-tailored methods for the analysis of evolutionary algorithms we also introduce the relevant tools and notions from probability theory in an accessible form. This makes the tutorial appropriate for everyone with an interest in the theory of evolutionary algorithms without the need to have prior knowledge of probability theory and analysis of randomised algorithms.

Variants of this tutorial have been presented at GECCO 2013–2015, attracting well over 50 participants each time. The tutorial will be based on the 'Theoretical analysis of stochastic search heuristics' chapter of the forthcoming Springer Handbook of Heuristics.

2.15 Theory of Evolutionary Computation

Tutorial Speaker: Benjamin Doerr, *École Polytechnique (France)*

Tutorial Abstract: Theoretical research has always accompanied the development and analysis of evolutionary algorithms, both by explaining observed phenomena in a very rigorous manner and by creating new ideas. Since the methodology of theory research is very different from experimental or applied research, non-theory researcher occasionally find it hard to understand and profit from theoretical research. Overcoming this gap in our research field is the target of this tutorial. Independent of particular theoretical subdisciplines or methods like runtime analysis or landscape theory, we aim at making theory accessible to researchers having little exposure to theory research previously. In particular,

- we describe what theory research in EC is, what it aims at, and showcase some of key findings of the last 15 years,
- we discuss the particular strengths and limitations of theory research,
- we show how to read, understand, interpret, and profit from theory results.

2.16 Theory of Parallel Evolutionary Algorithms

Tutorial Speaker: Dirk Sudholt, *University of Sheffield (UK)*

Tutorial Abstract: Evolutionary algorithms (EAs) have given rise to many parallel variants, fuelled by the rapidly increasing number of CPU cores and the ready availability of computation power through GPUs and cloud computing. A very popular approach is to parallelize evolution in island models, or coarse-grained EAs, by evolving different populations on different processors. These populations run independently most of the time, but they periodically communicate genetic information to coordinate search. Many applications have shown that island models can speed up computation time significantly, and that parallel populations can further increase solution diversity. However, there is little understanding of when and why island models perform well, and what impact fundamental parameters have on performance.

This tutorial will give an overview of recent theoretical results on the runtime of parallel evolutionary algorithms. These results give insight into the fundamental working principles of parallel EAs, assess the impact of parameters and design choices on performance, and contribute to the design of more effective parallel EAs.

Author Index

Printed in the United States
By Bookmasters